HANDBOOK OF EPOXY RESINS

OTHER McGRAW-HILL HANDBOOKS OF INTEREST

BEEMAN · Industrial Power Systems Handbook
BELL · Petroleum Transportation Handbook
BLATZ · Radiation Hygiene Handbook
CARROLL · Industrial Instrument Servicing Handbook
COCKRELL · Industrial Electronics Handbook
CONSIDINE · Process Instruments and Controls Handbook
CROCKER · Piping Handbook
ETHERINGTON · Nuclear Engineering Handbook
FACTORY MUTUAL ENGINEERING DIVISION · Handbook of Industrial Loss Prevention
FRICK · Petroleum Production Handbook
GRANT · Hackh's Chemical Dictionary
GUTHRIE · Petroleum Products Handbook
HEYEL · The Foreman's Handbook
HUSKEY AND KORN · Computer Handbook
JURAN · Quality Control Handbook
KALLEN · Handbook of Instrumentation and Controls
KATZ · Handbook of Natural Gas Engineering
KING AND BRATER · Handbook of Hydraulics
KNOWLTON · Standard Handbook for Electrical Engineers
KORN AND KORN · Mathematical Handbook for Scientists and Engineers
LANGE · Handbook of Chemistry
LASSER · Business Management Handbook
MAGILL, HOLDEN, AND ACKLEY · Air Pollution Handbook
MANAS · National Plumbing Code Handbook
MANTELL · Engineering Materials Handbook
MARKS AND BAUMEISTER · Mechanical Engineers' Handbook
MAYNARD · Industrial Engineering Handbook
MAYNARD · Top Management Handbook
MEITES · Handbook of Analytical Chemistry
MOODY · Petroleum Exploration Handbook
MORROW · Maintenance Engineering Handbook
PERRY · Chemical Engineers' Handbook
PERRY · Engineering Manual
RICHEY, HALL, AND JACOBSEN · Agricultural Engineers Handbook
SHAND · Glass Engineering Handbook
STANIAR · Plant Engineering Handbook
STREETER · Handbook of Fluid Dynamics
TOULOUKIAN · Retrieval Guide to Thermophysical Properties Research Literature
TRUXAL · Control Engineers' Handbook

HANDBOOK OF EPOXY RESINS

HENRY LEE

Technical Director, The Epoxylite Corporation
South El Monte, California

KRIS NEVILLE

Project Engineer, The Epoxylite Corporation
South El Monte, California

New York
San Francisco
Toronto
McGRAW-HILL BOOK COMPANY
London
Sydney

HANDBOOK OF EPOXY RESINS

 Library of Congress Catalog Card Number 65-26165

36997

1234567890MP72106987

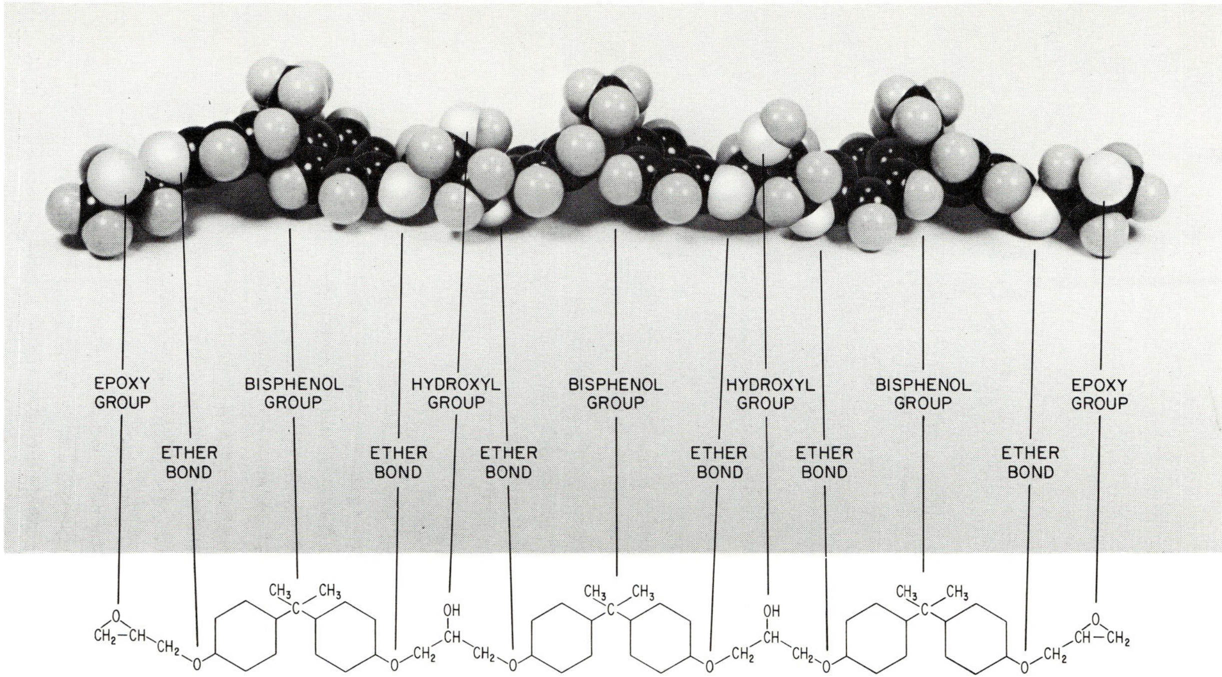

Atom model of widely used solid epoxy resin, diglycidyl ether of bisphenol A, degree of polymerization of 2 (DGEBA, $n = 2$). Approximate magnification: 100 million times. (*The Epoxylite Corporation.*)

PREFACE

The widespread acceptance by industrial readers in the United States and Europe of our first book on the epoxy resins has prompted the preparation of this second, expanded work. Since publication of the first book in 1957, the epoxy technology has advanced rapidly in terms of raw materials, applications, and basic chemistry. The number of distinct types of epoxy resins commercially available has increased from three to some twenty-five. Many new curing agents have been developed. Many reactions which were then poorly understood have since been described with reasonable accuracy. Applications then in their infancy, such as encapsulation, filament winding, fluidized bed coatings, concrete topping and repair, have, in the intervening years, attained stature and importance. In 1957, no epoxy transfer molding compounds were available commercially. By 1963, they were seriously challenging the alkyds and diallyl phthalate as electronic packaging materials. In 1957, fewer than 150 references seemed worthy of inclusion in the book; eight years later, over 2,000 deserved mention.

It seemed appropriate, then, to present this new technology in a completely revised and expanded format—once again to summarize the published papers, many of them generally inaccessible or ephemeral, and to organize the information in terms of our understanding of the technology.

The questions we set out to answer in the first volume are once again answered here—in greater detail and variety and informed by almost ten additional years of experience in the formulation of the epoxy resin compounds. How do the epoxy resins work, what are their strong points and their weak points and, most important, how can they be used intelligently for specific applications?

This book is intended to provide theoretical and practical information on these plastics and to answer many of the questions which will occur to the user. The book discusses the technical background and general chemistry of the resins and indicates the specific formulations that can be developed and the application techniques that have proved successful.

The chemist will find a review of the synthesis, curing mechanisms, and curing agents. The electronics engineer will find design data for the potting of electronic components and their performance in extreme environments; the power engineer will learn how epoxy resins have solved many of the application problems of electrical apparatus in corrosive atmospheres; the paint and varnish formulator will find data on the formulation of both solvent and solvent-free coatings of unusually durable properties; the tool and die maker will find a detailed discussion of the epoxies as they apply specifically to his field of interest; the aircraft and missile engineer will find suggestions for lightweight sandwiches, high-strength laminates, high-temperature adhesives, and chemically resistant sealants; the chemical engineer will find details on the use of epoxies for ducts and piping to handle corrosive fluids; the automotive engineer will meet a versatile metal adhesive and body patching compound; and the civil engineer will encounter a wide range of structural adhesives, floor coatings, and high-strength reinforcing materials.

The authors wish to acknowledge the helpful assistance of a number of people who reviewed individual chapters and the completed manuscript. Their number precludes a listing, and we can here but thank the individual companies, Ciba Products Co., Dow Chemical Co., Shell Chemical Corp., and Union Carbide Plastics Co. We wish, especially, however, to thank Ben Carroll of Western Electric, Reg Talley of UCLA, Tom F. Mika of Shell Development Co., C. F. Pitt of Union Carbide Plastics Co., Jerry Wolfe of Techform Laboratories, Inc., Don Loundy of the Mica Corporation, Donald F. Smith of Hughes Aircraft, John L. Hull of Hull Corp., and Dr. Harold H. Chen of Aerospace Corporation. We particularly are grateful for the assistance of Dr. Donald G. Stoffey, Dr. Roy G. Neville, Ieuan T. Smith, and Marion Stocker of the Epoxylite Corporation staff.

Finally, we wish especially to acknowledge the support of The Epoxylite Corporation—not as a purely legal person, but as a composite of flesh-and-blood individuals. During the more than six years required for the preparation of this volume, these individuals were a source of constant encouragement. Without this, the book could not have been prepared at all. Behind the Corporation, we wish also to acknowledge our indebtedness to the many, many customers who, in specifying Epoxylite formulations have, by their loyalty, directly contributed to the preparation of the present volume.

Henry Lee
Kris Neville

CONTENTS

Chapter 1

AN INTRODUCTION TO EPOXY RESINS

In a broad sense, the term *epoxy* refers to a chemical group consisting of an oxygen atom bonded with two carbon atoms already united in some other way. The simplest epoxy is a three-membered ring, to which the term α-epoxy or 1,2-epoxy is applied. Ethylene oxide (I) is an example of this type. The terms 1,3- and 1,4-epoxy are applied to trimethylene oxide (II) and tetrahydrofuran (III).

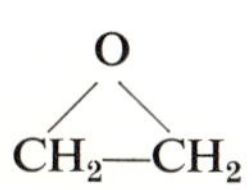

I
Ethylene oxide

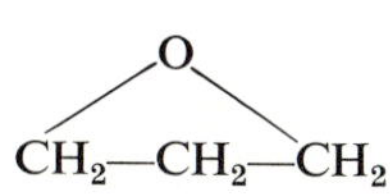

II
Trimethylene oxide
(oxetane)

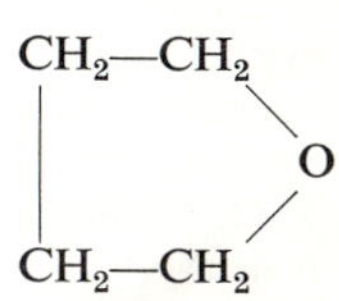

III
Tetrahydrofuran

In this book, we are concerned only with resins containing the three-membered rings (e.g., ethylene oxide derivatives).

There is no universal agreement on the nomenclature of the three-membered epoxy ring. There is division even on the term epoxy itself—the Europeans generally preferring the term *epoxide*, which is doubtless more correct than the American *epoxy*. The epoxies may be designated oxides, as in the case of ethylene oxide (epoxyethane) or cyclohexene oxide (1,2-epoxy-, or 1,2-oxidocyclohexane) (IV). The term *oxirane*, a trivial name for ethylene oxide, is also used in referring to the epoxy group. Several of the more common monoepoxies have trivial names, such as epichlorohydrin (V), glycidic acid (VI), and glycidol (VII). Glycidyl (VIII)

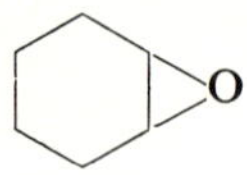

IV
Cyclohexene oxide

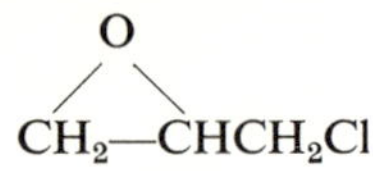

V
Epichlorohydrin

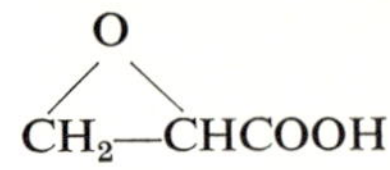

VI
Glycidic acid

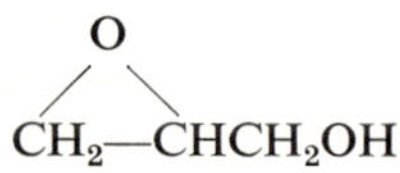

VII
Glycidol

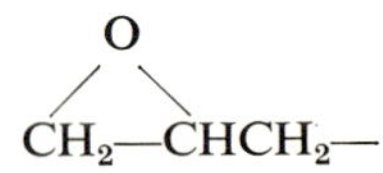

VIII
Glycidyl group

is used to refer to the terminal epoxy group, the name being modified by ether, ester, amine, etc., according to the nature of the group attached to the third carbon.

We prefer to follow the usage generally accepted in the United States. Trivial names are identified by structural formula to avoid confusion. The useful term *glycidyl* is used throughout the text, with epoxy and oxide following the current usage in the United States technology.

DEFINITION OF EPOXY RESIN

For the purpose of this book, an *epoxy resin* is defined as any molecule containing more than one α-epoxy group (whether situated internally, terminally, or on cyclic structures) capable of being converted to a useful thermoset form. The term is used to indicate the resins in both the thermoplastic (uncured) and thermoset (cured) state.

HISTORY

Epoxy resins are prepared commercially (1) by dehydrohalogenation of the chlorohydrin prepared by the reaction of epichlorohydrin with a suitable di- or polyhydroxyl material or other active-hydrogen-containing molecule; (2) by the reaction of olefins with oxygen-containing compounds such as peroxides or peracids, and (3) by dehydrohalogenation of chlorohydrins prepared by routes other than route 1.

Schrade [21] cites the first commercial attempt to prepare resins from epichlorohydrin as occurring in 1927 in the United States. However, credit for the synthesis of the materials first designated as epoxy resins—those derived specifically from epichlorohydrin and bisphenol A—is shared by Dr. Pierre Castan of Switzerland and Dr. S. O. Greenlee of the United States. These resins are still by far the most important in the technology.

In 1936 Dr. Castan produced a low-melting, amber-colored resin which was then reacted with phthalic anhydride to produce a thermoset compound. Dr. Castan, working for De Trey Frères of Switzerland, envisioned the use of such liquid resins in the manufacture of dentures and cast articles [13,14,16,17,21]. His developments were subsequently licensed to Ciba, Ltd.

In the United States, Dr. Greenlee, early in 1939, working for Devoe-Raynolds, explored the epichlorohydrin–bisphenol A synthesis route for the production of new resins for coatings, these resins not containing caustic-sensitive ester linkages [18,20].

The epichlorohydrin-bisphenol resins were the outgrowth of several decades of research by many workers on the broad front of substituted ethylene oxide derivatives. Without detracting from the credit due Drs. Castan and Greenlee, reference should be made to earlier discoveries which set the stage for the epoxy resins and provided considerable reaction background information which made possible the rapid commercial utilization of the diglycidyl ether of bisphenol A resins after World War II.

Briefly, reference should be made to:

1. McIntosh and Wolford [2,3], who in 1920 made plastics for application as coatings, molding powders, and waterproof impregnating materials by reacting phenol or cresol with epichlorohydrin in the presence of catalysts
2. Blumer [5], who in 1930 described a composition for the manufacture of coatings based on phenol-aldehyde compounds reacted with epichlorohydrin
3. Fonrobert and Lemmer [7], who in 1930 reported etherification of phenolic hydroxyls by dichlorohydrin in a caustic medium
4. Eisleb [4], who noted the reaction of epichlorohydrin with secondary amines and subsequent dehydrohalogenation with caustic in 1926, and Stallman [6], who reported production of diglycidyl amines from ammonia and epichlorohydrin in 1933
5. Van Peski and Hoeffelman, who produced butadiene dioxide as early as 1935 by dehydrohalogenation [12]
6. Groll and Hearne, who obtained a variety of diepoxies from chlorinated alcohols via dehydrohalogenation in 1934 [8]
7. Schlack [10,11], who in 1933 described a low-molecular-weight diglycidyl ether produced from 1 mole of bisphenol A and 10 moles of epichlorohydrin, reacted in a caustic medium
8. Block and Tischbein [15], who carried the research of Schlack somewhat further, but worked mostly with the low-molecular-weight diepoxies such as diglycidyl ether and butadiene dioxide

By the end of the 1930s, there was considerable work in several countries pointing toward the synthesis of epoxy resins and exploring chemical reactions that would later assume importance.

Reactions of epoxy resins with various active-hydrogen-containing compounds was known to the early technology. In 1934 Schlack [9] reported the reaction of diepoxy compounds with organic and inorganic acids, amines, and mercaptans. Cure of the epoxy resins by anhydrides was noted by Castan in his initial experiments. These reactions had all been earlier explored in the literature on monoepoxies.

A more detailed discussion of the historical and chemical background to the discovery of epoxy resins from the epichlorohydrin synthesis route is contained in [23,26].

From a base of zero, for all practical purposes, in 1948, production of the diglycidyl ether of bisphenol A resins (in the United States) increased to about 20 million pounds annually by 1954 and to about 30 million pounds in 1957. From 1957 to 1965 production increased to over 110 million pounds annually. In 1963 the United States Tariff Commission broke down the market for the diglycidyl ether of bisphenol A resins as follows:

Uses	*Pounds*
Bonding and adhesives	11,640,000
Protective coatings	39,398,000
Reinforced plastics	10,464,000
All other	19,497,000
Export	7,836,000
Total	88,835,000

By 1965, approximately 110 million pounds of epoxy resins were sold, and the growth rate appeared to stabilize at about a 10 per cent per year increase.

In the early years, the majority of these resins were of the solid type. The production of the liquid diglycidyl ether of bisphenol A drew abreast in the 1960s [28].

In the late 1950s and early 1960s, a number of new glycidyl ether resins were introduced commercially, these being synthesized from epichlorohydrin and polyhydric materials other than bisphenol A.

Simultaneously, a new route of synthesis for the epoxy resins was being explored. As early as 1909 it was known that olefins could be converted to epoxy groups through reaction with peracids [1] and with hypochlorous acid followed by dehydrohalogenation. Research work during 1944 to 1948 by Dr. Daniel Swern of the United States focused attention on this conversion, and by 1950 the reaction achieved commercial importance in the production of epoxidized soybean oil

Nighttime view of epoxy-resin production plant at Freeport, Tex., capacity 35 million lb. (*Dow Chemical Co.*)

plasticizers for vinyl resins [22,29,30]. Such epoxidized materials, however, were only sluggishly reactive with active-hydrogen-containing curing agents and were not particularly suitable for conversion to thermosetting plastics.

In the late 1950s, Dr. Swern's work was expanded to the synthesis of new epoxy resins having satisfactory reactivities with active-hydrogen-containing materials. This work was pioneered by several workers in the United States, among them being Dr. Benjamin Phillips of Union Carbide Chemicals and Dr. Frank Greenspan of Food Machinery Corporation. In Europe, pioneer work has been accomplished by Drs. Batzer, Ernst, Fisch, and Porret of Ciba, Ltd.

By 1963, these newer, so-called "peracetic acid resins" (excluding epoxidized oils) were being produced at the rate of about 5 million pounds a year, mostly for use in vinyl resin formulations.

By the mid-60s the diglycidyl ether resins and some of the newer types were being manufactured in the United States, Canada, England, France, Switzerland, West Germany, Poland, Czechoslovakia, and the U.S.S.R.

In addition to Lee and Neville [24], Paquin [26], Schrade [25], and Skeist [27] have presented booklength treatments of epoxy resins.

BASIC CHARACTERISTICS OF EPOXY RESINS

A number of properties have led to rapid growth of epoxy resins and their use in a wide range of industries.

Considering the liquid resins: These are low-viscosity liquids which readily convert to the thermoset phase upon the admixture of the proper curing agent. There are other liquid resins—phenolics, polyesters, acrylics, etc.—which cure in a similar fashion, but the epoxy resins possess a rather unique combination of properties:

1. Low viscosity: The liquid resins and their curing agents form low-viscosity, easy-to-process (or modify) systems.
2. Easy cure: Epoxy resins cure quickly and easily at practically any temperature from 5 to 150°C, depending on the selection of the curing agent.
3. Low shrinkage: One of the most advantageous properties of the epoxy resins is their low shrinkage during cure. Phenolic casting resins which evolve water reveal high shrinkage, as do the acrylic and polyester resins which must rearrange and reorient considerably in the liquid and semigelled phase. Epoxy resins react with very little rearrangement and with no volatile by-products being evolved.
4. High adhesive strengths: Because of the chemical makeup, chiefly the presence of polar hydroxyl and ether groups, the epoxy resins are excellent adhesives. The resins cure with low shrinkage, so that the various surface contacts set up between the liquid epoxy-resin formulation and the adherends are not disturbed during cure. Adhesive strengths, without the need for either open time or high pressures, are perhaps the best obtainable in the contemporary plastics technology.
5. High mechanical properties: The strength of properly formulated epoxy resins usually surpasses that of other types of casting resins. This is probably in part a result of their low shrinkage, which minimizes stresses that otherwise would weaken the mechanical structure.
6. High electrical insulation: Epoxy resins are excellent electrical insulators.
7. Good chemical resistance: The chemical resistance of the cured epoxy resin depends considerably on the curing agent used. Selectively outstanding chemical resistance can be obtained by specification of the proper material. Overall, most epoxy resins possess extremely high resistance to caustics and good to excellent resistance to acids.
8. Versatility: The epoxy resins are probably the most versatile of the contemporary plastics. The basic properties may be modified in many ways: by blending of resin types, by selection of curing agents, and by use of modifiers and fillers.

Considering the solid epoxy resins: The chief use of the solid diglycidyl ether of bisphenol A resins is in solution coatings. The higher-molecular-weight materials are cooked with conventional drying oils or reacted with methylol-containing resins. The epoxy-resin backbone imparts toughness, scuff resistance, and chemical resistance. Solid resins of intermediate molecular weight may be cross-linked by homo- or heteropolymerization through the epoxy groups. By this means room-temperature-curing films may be developed which provide properties equal to or exceeding those of many baking-type finishes.

The excellent adhesion of epoxy resins, ease of cure, mechanical strength, and high chemical resistance are as important advantages of the solid resins as they are of the liquids.

APPLICATIONS FOR EPOXY RESINS

Because of their versatility, the epoxy resins are used in thousands of industrial applications. Indicative of the range of applications are the following:

1. As adhesives for aircraft honeycomb structures, for paintbrush bristles and for concrete topping compounds
2. As body solders and caulking compounds for the repair of plastic and metal boats, automobiles, etc.

3. As casting compounds for the fabrication of short-run and prototype molds, stamping dies, patterns, and tooling
4. As caulking and sealant compounds in building and highway construction applications and in applications where high orders of chemical resistance are required
5. As potting and encapsulation compounds and impregnating resins and varnishes for electric and electronic equipment

Epoxy-resin production plant in Houston, Tex. The building and other installations shown here are protected with epoxy-based paints. (*Shell Chemical Co.*)

6. As laminating resins for airframe and missile applications, for filament-wound structures, and for tooling fixtures

Epoxy-based solution coatings are used as maintenance and product finishes, marine finishes, masonry finishes, structural steel coatings, tank coatings, aircraft finishes, appliance primers, automotive primers, can and drum linings, furniture finishes, and collapsible-tube coatings. They are used for concrete-floor paints, gym and floor varnishes, spar varnishes, etc.

Epoxy resins are also used in decorative floor applications; as chemically resistant mortars and floor topping compounds; in printing inks; in fabric-treating applications; in dental, surgical, and prosthetic applications; for breaking petroleum emulsions; and for lightweight, chemically resistant foams.

The epoxy resins are used as additives for a variety of other plastic materials, such as vinyl and acrylic resins and natural and synthetic rubbers.

Primarily industrial and manufacturing materials, epoxy resins were introduced as specialty-packaged adhesives into the consumer market in the early 1960s, and some epoxy-based paints have since gained limited acceptance on that level.

SUMMARY

The epoxy-resin technology had its genesis in research conducted in the United States and Europe just before World War II. The first resins—the reaction products of epichlorohydrin and bisphenol A—were produced commercially in 1947. In 10 years they achieved a production volume of some 30 million pounds, this being tripled 6 years later.

In the late 1950s, a number of new epoxy resins, different in kind from the earlier diglycidyl ethers, were introduced, and by the end of 1960 at least 25 distinct types of resins were commercially available. At this point the term *epoxy resin* became generic. It is now applied to a wide family of materials.

Epoxy resins belong to the thermosetting class of plastics and bear similarities to materials such as the phenolics and polyesters. The epoxy resins offer great versatility, low shrinkage, good chemical resistance, and outstanding adhesion.

REFERENCES

1911

1. Prileshajew, *Chem. Zentr.*, vol. II, p. 1279.

1927

2. McIntosh, Synthetic Resin and Method of Making Same, U.S. 1,642,079.
3. McIntosh and Wolford, Synthetic Resin and Method of Making Same, U.S. 1,642,978.

1931

4. Eisleb, Substituted 1,3-Di-amino-2-propanols, U.S. 1,790,042.

1933

5. Blumer, Process for the Manufacture of Resinous Products from Phenol Condensation Products and Polyfunctional Alcohols, Ger. 576,177.

1934

6. Stallmann, Vat Dyestuff Pastes, U.S. 1,977,250.

1935

7. Fonrobert and Lemmer, Phenolic Condensation Products, U.S. 1,990,985.

1936

8. Groll and Hearne, Conversion of Polyhalogenated Alcohols, U.S. 2,061,377.

1938

9. Schlack, Manufacture of Organic Compounds Containing Basic Substituents, U.S. 2,131,120.
10. Schlack, Manufacture of Amines of High Molecular Weight Which Are Rich in Nitrogen, U.S. 2,136,928.

1939

11. Schlack, Process for the Manufacture of High Molecular Weight Polyamines, Ger. 676,117.
12. Van Peski and Hoeffelman, Insecticide and Fumigant, U.S. 2,152,003.

1940

13. Castan, Process for Manufacture of a Hardenable Synthetic Resin, Swiss 211,116.
14. Castan, A Process for the Manufacture of Thermosetting Synthetic Resins by the Condensation of Alkylene Oxides with Anhydrides of Polybasic Acids, Brit. 518,057.

1943

15. Bock and Tischbein, Process for the Manufacture of Soluble Resins Convertible to the Insoluble State by Reaction with Nitrogenous Condensation Products, Ger. 731,030.
16. Castan, Process of Preparing Synthetic Resins, U.S. 2,324,483.

1948

17. Castan, Process for the Manufacture of Thermosetting Synthetic Resins by the Polymerization of Alkylene Oxide Derivatives, U.S. 2,444,333.

1950

18. Greenlee, Tall Oil Esters, U.S. 2,493,486.
19. Greenlee, Compositions of Resinous Epoxides and Aromatic Sulfonamide-Aldehyde Condensates, U.S. 2,494,295.
20. Greenlee, Phenol-Aldehyde and Epoxide Resin Compositions, U.S. 2,521,911.

1953

21. Schrade, Resins from Epichlorohydrin, *Kunststoffe*, vol. 43, no. 7.

1955

22. Epoxy Resins, *Chem. Process.*, August.
23. Wegler, Chemie der Polyepoxyde, *Angew. Chem.* **67** (19/20): 582–592.

1957

24. Lee and Neville, "Epoxy Resins," McGraw-Hill Book Company.
25. Schrade, "The Epoxy Resins," Dunod, Paris.

1958

26. Paquin, "Epoxydverbindungen und Epoxyharz," Springer-Verlag OHG, Berlin.
27. Skeist, "Epoxy Resins," Reinhold Publishing Corporation.

1959

28. Epoxy Resins, Market Survey and User's Reference, Graduate School of Business Administration, Harvard University.

1960

29. Lee and Neville, New Developments in Epoxy Resins, part 1, *Insulation*, December.
30. Wallace, Epoxidation, "Encyclopedia of Chemical Technology," second supplement, September.

Chapter 2

SYNTHESIS OF GLYCIDYL-TYPE EPOXY RESINS

The first commercial epoxy resins were the reaction products of epichlorohydrin and bisphenol A, the reaction giving the diglycidyl ether of bisphenol A and higher-molecular-weight species. The structure of the resin, ideally, is

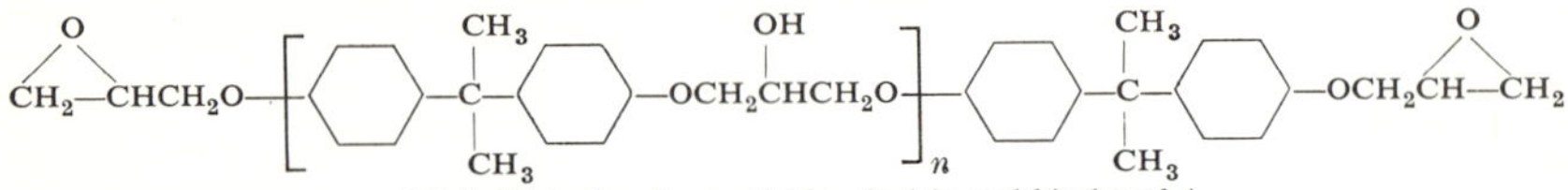

Diglycidyl ether from epichlorohydrin and bisphenol A

The low-molecular-weight liquid products have an n value about 1 or below. Above $n = 1$, the resins are brittle solids.

This chapter outlines the synthesis of these resins. It also describes the use of other hydroxyl-containing molecules to produce different glycidyl ethers. In addition, the synthesis of glycidyl resins from organic and inorganic acids, from nitrogen-containing compounds, and from silicones, via the epichlorohydrin route is described, as well as the synthesis of other glycidyl resins from monoepoxies by reaction through other functional groups present in the monoepoxy molecule.

General view of plant for production of epichlorohydrin at Toms River, N.J. (*Ciba Products Co.*)

In the synthesis of commercial epoxy resins, epichlorohydrin is the principal epoxidizing reactant. Other epihalohydrins or dihalohydrins may be used, but are not economically attractive. Methyl and ethyl epichlorohydrin and mixed isomers of more highly substituted epichlorohydrins have been suggested [143], but their use does not appear to provide commercial advantages.

DIGLYCIDYL ETHER OF BISPHENOL A (DGEBA)

The raw materials for the commercial synthesis of the DGEBA are epichlorohydrin and bisphenol A. Epichlorohydrin is synthesized from propylene and chlorine with dichlorohydrin as an intermediate. Bisphenol A is produced from phenol and acetone as follows:

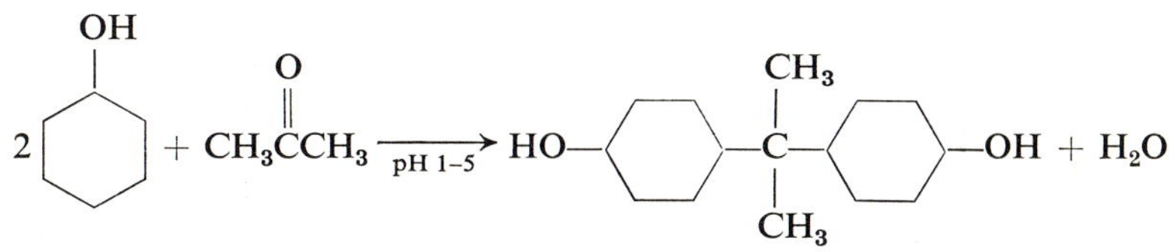

Phenol Acetone Bisphenol A Water

Because of the ready availability of the raw materials and the ease of manufacture, bisphenol A is the chief dihydric phenol used in epoxy-resin synthesis.

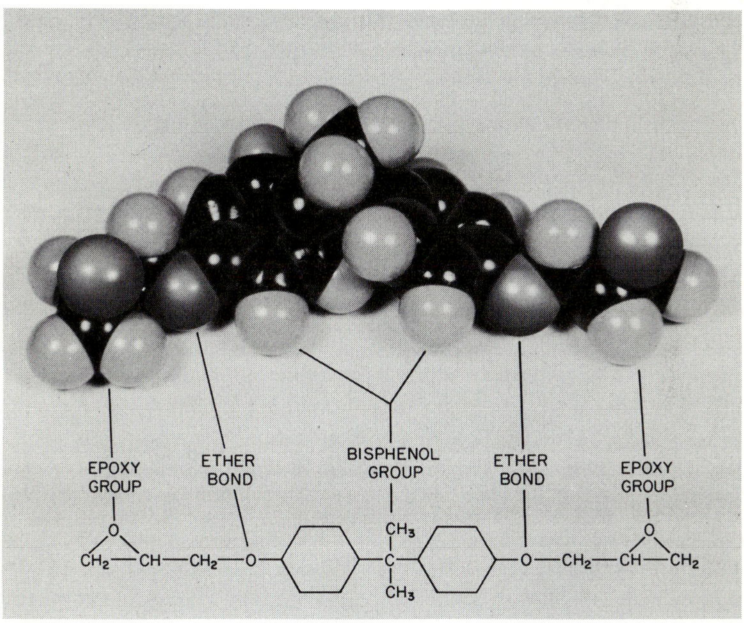

Atom model of the most widely used liquid epoxy resin, diglycidyl ether of bisphenol A (DGEBA, $n = 0$). Approximate magnification: 100 million times. (*The Epoxylite Corporation.*)

Synthesis of Monomeric DGEBA

DGEBA is obtained by reacting epichlorohydrin with bisphenol A in the presence of sodium hydroxide [3]. The reaction occurs in two steps: (1) the formation of a chlorohydrin intermediate and (2) the dehydrohalogenation of the intermediate to the glycidyl ether.

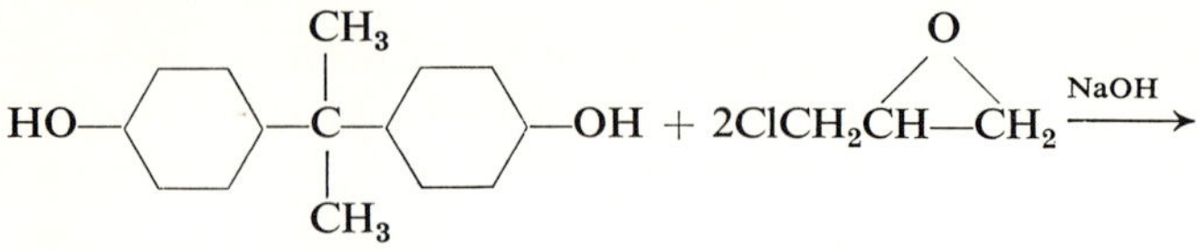

Bisphenol A Epichlorohydrin

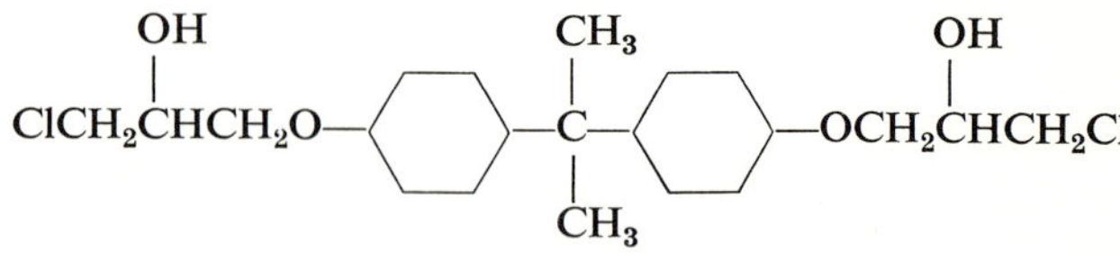

Chlorohydrin intermediate

Chlorohydrin intermediate + NaOH→DGEBA + salt and water

The caustic catalyzes the reaction to produce the chlorohydrin intermediate, acts as the dehydrohalogenating agent, and neutralizes the formed hydrochloric acid.

In order to obtain the monomeric DGEBA, 2 moles of epichlorohydrin is theoretically required for each mole of bisphenol A. However, when this ratio is employed, the monomeric yield is less than 10 per cent, with the remaining material being higher-molecular-weight polymerization products. In order to obtain high yields of the monomeric product, excess epichlorohydrin is employed, the stoichiometric amount being doubled or tripled. Yields of 90 per cent or higher are then possible. Amounts ten times in excess have also been suggested, the "purity" of the diglycidyl ether being dependent on the excess of epichlorohydrin.

Synthesis of the DGEBA (mol. wt. ~380)

The DGEBA having an average molecular weight of approximately 380 is the basic liquid epoxy resin in the commercial technology. The synthesis may be conducted as follows:

The apparatus employed was a heated reaction vessel with a thermometer, mechanical stirrer, dropping device for admitting aqueous caustic, and a vapor take-off to which was fitted a water-cooled condenser with a thermometer and a condensate collection and separating chamber provided with a return to convey the lower epichlorohydrin layer to the reaction vessel. The separating chamber also had a draw-off and collector for the upper water layer.

The reaction vessel was charged with 1,388 parts (15 moles) of epichlorohydrin and 342 parts (1.5 moles) of bisphenol. The stirrer was started and the solution of bisphenol in epichlorohydrin was heated to 119°C, whereupon the epichlorohydrin began refluxing at a temperature of 112°C. During a period of 3½ hours, 304 parts of 40 per cent aqueous sodium hydroxide (3.0 moles) were added to the boiling reaction mixture . . . . Heating was continued an additional 15 minutes after all the caustic was added . . . . The reaction mixture was then subjected to distillation for removal of unreacted epichlorohydrin . . . .

In order to separate salt from the crude product, approximately 550 parts of toluene was added with stirring and the mixture was filtered, the salt cake being washed with additional toluene and the washing combined with the filtrate. The mixture was then subjected to distillation for removal of toluene, the mixture being heated to a temperature of 170°C at 2 mm pressure. The resulting viscous product, obtained in amount of 509 parts, had a Durrans' Mercury Method softening point of 6°C, had an epoxy value of 0.519 epoxy equivalents per 100 grams, and contained 0.52 per cent chlorine [48].

Various aspects of the synthesis of the lower-molecular-weight species (mol. wt. 380) are covered in the literature. Part of the epichlorohydrin may be reacted, additional added, and the low-molecular-weight epoxy isolated by distilling out the epichlorohydrin azeotropically [104]. Improved yields are claimed by using caustic in solution with ethanol rather than water [86]. Sodium hydrosulfite may be added to prevent emulsification of resin during the washing operation [89]. An aliphatic secondary alcohol is suggested as an improved solvent to assist in salt removal [59]. Dilute acid is suggested for washing the resin [95]. Chlorine content can be reduced by using only 98 per cent stoichiometric amounts of NaOH

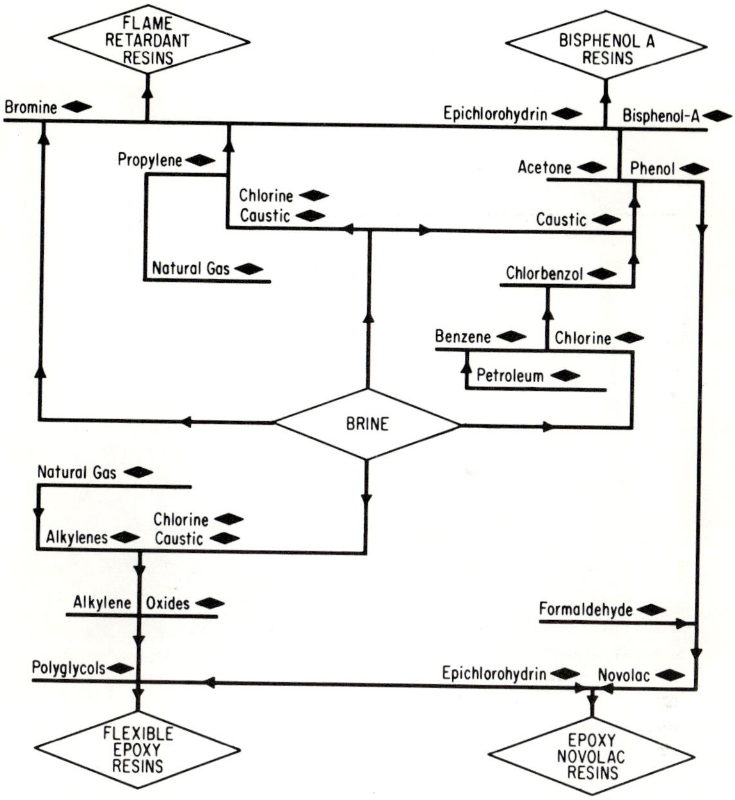

Flow chart showing derivation of commercial epoxy resins via epichlorohydrin route. (*Dow Chemical Co.*)

and removing excess epichlorohydrin before dehydrohalogenation [63]. Total reflux may be used during the addition of one-half the caustic, with the water formed during the reaction being removed during addition of the last half of the caustic [122]. Monohydric phenols [69] or polypropylene glycol [191] may be added during synthesis in order to reduce viscosity of the resin product. A continuous process for the production of DGEBA (mol. wt. 380) giving a yield of 50 to 80 per cent is described in [63].

An alternative two-step process reacts the bisphenols and epichlorohydrin in an anhydrous medium in the presence of a quaternary ammonium salt [108] or lithium hydroxide monohydrate [99]. The excess epichlorohydrin is then removed before dehydrohalogenation. Interfacial polycondensation techniques have also been used [211].

The intentional production of higher-chlorine-content resins is described in [32]. The anions generated by inorganic bases during the synthesis (see Chap. 5) usually add to the least substituted carbon atom of an unsymmetrical epoxide, but a small but measurable concentration of products arise from reaction at the most substituted carbon atom. These abnormal additions result in production of 1,3-chlorohydrins which are not subject to ring closure. Temperature has been shown to influence the extent to which abnormal addition occurs, being minimized when the initial reaction is accomplished at low temperatures (30 vs. 85°C) and subsequently increased [180]. When ethanol, rather than water, is used with the

Major structures in epoxy-resin plant are framed in arch of pipe at petrochemical complex in Freeport, Tex. (*Dow Chemical Co.*)

caustic, the primary alcoholic hydroxyls will not react until about 90 per cent of the phenolic hydroxyls have been consumed (i.e., at the point when the alkoxide begins to replace the phenoxide). Reaction of the primary alcohol in preference to reaction of the secondary alcohol along the polymer chain leads to some random chain terminations but eliminates chain branching from the main polymer.

Free-flowing crystals of pure DGEBA are obtained by addition of 10 per cent butyl glycidyl ether to the relatively pure product, cooling to 10°C, and seeding with 0.3 per cent of crystalline DGEBA. After 3 days at low temperature, a 66 per cent yield of pure crystalline material results from treatment of the standard commercial DGEBA resin having a molecular weight of approximately 380 [161].

Synthesis of Higher-molecular-weight DGEBA

As may be seen from the preceding discussion, it is possible to synthesize higher-molecular-weight epoxy resins by reacting epichlorohydrin and bisphenol A in the presence of excess caustic. The general formula for such a resin has been given previously. It should be noted that each such molecule contains as many alcoholic hydroxyls as there are repeating units, provided that the molecule is linear with no side branches and contains no more than two epoxy groups.

Synthesis of higher-weight molecules requires not only the consumption of the epoxy groups contained in the epichlorohydrin, but also, in order to link successive bisphenol A groups, the consumption of some of the epoxy groups formed by dehydrohalogenation.

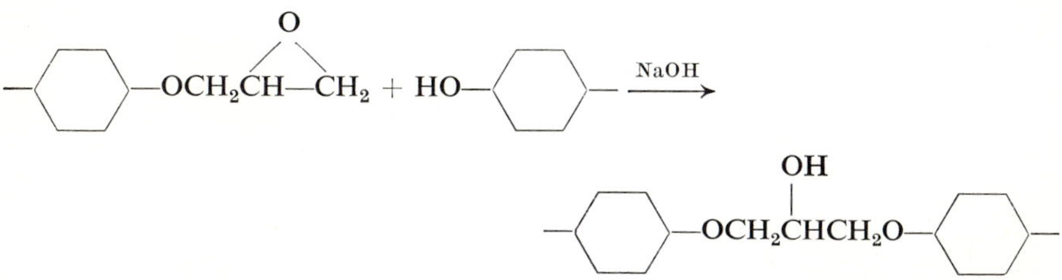

It should, therefore, be possible to govern the degree of polymerization by regulating the ratio of epichlorohydrin/bisphenol A when employing caustic in excess of the amount required to carry out the reaction [26]. Table 2-1 presents

Table 2-1. Effect of Varying Reactant Ratios on Molecular Weight of Epoxy Resins [10]

Mole ratio epichlorohydrin/ bisphenol A	Mole ratio NaOH/ epichlorohydrin	Softening point, °C	Molecular weight	Epoxide equivalent	Epoxy groups per molecule
2.0	1.1	43	451	314	1.39
1.4	1.3	84	791	592	1.34
1.33	1.3	90	802	730	1.10
1.25	1.3	100	1,133	862	1.32
1.2	1.3	112	1,420	1,176	1.21

data for various ratios in terms of resultant molecular weight. In actual practice, it is reported that a slight excess of epichlorohydrin (over the apparently stoichiometric ratio for a given molecular weight) is employed [43]. The need for some excess epichlorohydrin is accounted for by the reaction of some per cent of the epichlorohydrin with sodium hydroxide or with aliphatic hydroxyl groups present in impurities to form a variety of by-products, mostly glycerol derivatives. The reaction proceeds in a caustic medium slower than the phenol-epoxy reaction. A study of the kinetics of the overall reaction is given in [136].

A description from the literature illustrates the synthesis of higher-molecular-weight resins, using 1.12 moles epichlorohydrin/1 mole bisphenol A and employing heat and pressure to regulate the reaction.

The apparatus used was a 7½ gallon steam-jacketed pilot-plant kettle, capable of operating under pressure and equipped with a heavy-duty anchor-type agitator, a thermometer, a hot-water inlet, an adjustable level water take-off line, a gas pressure line with gage and pressure regulator, a vapor outlet line and a loading port and a sight glass in the top of the kettle.

The kettle was loaded with a mixture of 14,950 parts of water, 1,483 parts of commercial caustic soda (98.5 per cent NaOH), 17 parts of sodium orthosilicate, and 6,835 parts of bisphenol. After agitation for several minutes at 50°C, 3,000 parts of epichlorohydrin were added to the reaction mixture and the loading port was closed immediately, agitation being continued. With the exothermic heat of reaction and the external heat, the reaction temperature was brought to 115°C over a period of 30 minutes. Compressed air was introduced and the pressure inside the kettle was adjusted to 25 psi, and held during a 30 minute reaction period at 115°C. The pressure was then increased to 30 to 35 psi to cause the taffylike resinous product to settle to the bottom of the kettle, and as much as possible of the upper water layer was removed with the adjustable-level water take-off line. The taffylike resin was then washed free from the salt and alkali by introducing preheated hot water to the kettle and simultaneously allowing water to flow from the kettle. During the washing period, the resin was continuously agitated. The pressure in the kettle was

General view of supply lines, kettles, and controls for the production of epoxy resins. (*Ciba Products Co.*)

held at 30 to 35 psi, and the temperature was gradually increased from 120°C to 135°C. After 2½ hours of continuous washing, the water coming from the kettle was neutral. It is important that all of the unreacted alkali, when present, be removed to prevent further polymerization of the resin. After washing was completed, as much water as possible was removed from the kettle through the water take-off line. The air-pressure line was closed and the steam pressure in the jacket increased to raise the temperature of the resin to 137 to 138°C. The steam formed on heating the taffylike resin was slowly released, and, after the pressure had been completely released, the loading port was opened and the last traces of water were removed by heating the resin to 150°C. The resin was continuously agitated during the drying process. The hot resin was then poured from the kettle into a receptacle and allowed to cool [25].

In order to facilitate the washing operation, it has been suggested that an organic solvent, such as high-flash naphtha, be employed during resin synthesis [39]. Removal of the last traces of water may be accomplished by use of a stripping solvent, such as methyl isobutyl ketone [58].

A second process is available for the synthesis of higher-weight resins to eliminate the difficulties involved in washing the contaminants from the compounds melting above 80°C [16,17]. This process requires two steps: (1) the synthesis of a suitable low-molecular-weight resin, say predominantly $n = 2$, and (2) the reaction of this material with a calculated charge of bisphenol A in the presence of a trace amount of caustic, which further reacts the terminal epoxy groups with the terminal phenolic hydroxyls of the bisphenol A. The polymerization takes place without the evolution of by-products. In this case the chlorine content should be less than 0.3 per cent for the complete reaction with the charge of bisphenol A [63]. If the labile chlorine content is too high, it will consume the caustic, and if excess caustic is used to overcome this difficulty, erratic polymerization occurs. It has been suggested that lithium hydroxide [127] or lithium salts such as lithium chloride [126] may be employed at about 0.006 per cent (based on the weight of the bisphenol) in place of the caustic, to permit the reaction to be conducted at higher chlorine contents.

A study of the reactions in the synthesis of these higher-molecular-weight species indicates that with low chlorine contents, undesirable side reactions can be suppressed by using a reaction temperature no higher than about 170°C, to give, in the presence of caustic, under optimum conditions, about 93 per cent of the total reaction between the epoxy groups and the phenolic hydroxyls, and to achieve an even distribution of molecular weight [78].

Higher-molecular-weight resins with a narrower molecular-weight-range distribution have been reported when excess bisphenol A is reacted with epichlorohydrin to form a prepolymer and additional epichlorohydrin is subsequently added before dehydrohalogenation [164,210].

A small amount of divalent tin (e.g., stannous chloride [41]) or ferrous sulfate [209] may be added during synthesis to improve color of the higher-molecular-weight resins. Phosphoric acid may be added at the end of the synthesis to increase reactivity with urea or melamine resins [30].

Synthesis of Polyhydric Phenol Polyether Alcohols

It should be possible to create molecules of immense size from epichlorohydrin and bisphenol A. If an equimolar ratio were employed, the molecule formed would theoretically be equal to the combined weight of all the component adducts, with an aromatic hydroxyl at one end of the chain and an epoxy group at the other. In practice, of course, the reaction starts at many reactive points simultaneously, and as the molecular weight increases, the resin mobility decreases and the chain building terminates randomly. Nonetheless, when a 1/1 ratio is employed, very high molecular weight thermoplastic materials are developed [13].

For ester coatings, in which the hydroxyls on the chain are of greater interest than the epoxy groups, it is desirable to regulate the length of this polymeric chain, and monofunctional reactants, such as phenol or ethylene oxide, can be employed in calculated amounts to end-stop polymerization and yield polyhydric phenol polyether alcohols of almost any given molecular weight [9]. These higher-weight molecules [37] will contain a number of alcoholic hydroxyls, advantageously 20 or more [4], capable of direct esterification. Under suitable reaction conditions (i.e., ester interchange) epoxy groups may be preserved during the esterification operation [23].

A two-step process for the manufacture of the higher-molecular-weight species has been developed which involves addition of a calculated charge of bisphenol A, together with a calculated charge of a monofunctional drying acid, to the low-molecular-weight resin [113]. A base catalyst is used to ensure, first, the acid-epoxy reaction to regulate the molecular weight, and, second, the phenol-epoxy reaction to obtain the desired chain length. After completion of the epoxy-phenol reaction, the remaining drying acid can be added to the same vessel and reacted by conventional esterification procedures to produce the varnish.

As used in ester coatings, polyhydroxy polyethers derived from epichlorohydrin and bisphenol A, although containing few if any epoxy groups, are by custom referred to in the industry as *epoxy resins*, since they are synthesized from the basic reactants for diglycidyl ethers.

The very high molecular weight (e.g., mol. wt. 30,000 and higher) thermoplastic resins are designated *phenoxy resins* and are employed as coatings, adhesives, and molding compounds.

Properties of the phenoxy resins are varied by varying the molecular weight of the polymer. Additionally, variations may be obtained by using more than one type of bisphenol to obtain thermoplastic random and alternating copolymers [174]. Like the lower-molecular-weight species, the phenoxy resins may also be esterified to produce coatings resins [173].

The phenoxy resins based on epichlorohydrin and bisphenol A appear to be unbranched and highly linear below a molecular weight of about 70,000. Above that weight they are abruptly and increasingly branched to a very high degree [166].

Aside from a brief discussion on their use in coating formulations in Chap. 24, the polyhydric phenol polyether alcohols are not otherwise considered in this book.

GLYCIDYL ETHERS OF NOVOLAC RESINS

Novolac resins [70] are produced by the reaction of phenol and formaldehyde in acid solution. Such resins have the following idealized structure:

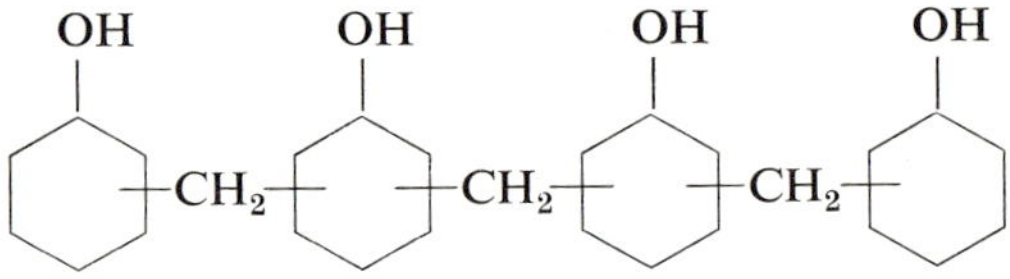

Novolac resin

The simplest novolac could be considered to be bisphenol F with the novolacs

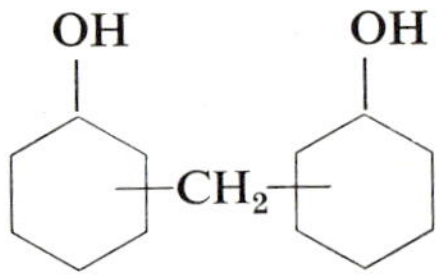

Bisphenol F

ranging upward in molecular weight from this to compounds containing 10 to 12 phenolic hydroxyls. Such resins are suitable intermediates for the synthesis of polyglycidyl ether resins.

The novolac-based epoxy resins are synthesized by reaction with epichlorohydrin in the same fashion as are the bisphenol A resins. The number of glycidyl groups per molecule in the resin is dependent upon the number of phenolic hydroxyls in the starting novolac, the extent to which they are reacted, and the extent to which the lowest-molecular-weight species are polymerized during synthesis. Theoretically all the phenolic hydroxyls may be reacted, but in practice selective epoxidation offers an advantage when the novolac contains more than three phenolic hydroxyls. Should complete epoxidation be accomplished, steric factors will prevent the reaction of some of the epoxy groups during cure [77].

Synthesis of a low-molecular-weight novolac, as a starting point for the production of an epoxy resin, may be conducted as follows:

94 parts of U.S.P. phenol (1 mole) and 94 of water were charged into a reactor. 2.9 parts of concentrated sulfuric acid (96.6 per cent) were added, to give a pH of 0.5. The acidified solution was heated to 80°C. Then 34.1 parts of 44 per cent aqueous solution of formaldehyde (0.5 mole) were added to the stirred acidified solution, over a period of 4 hours, while the temperature was maintained at 80°C.

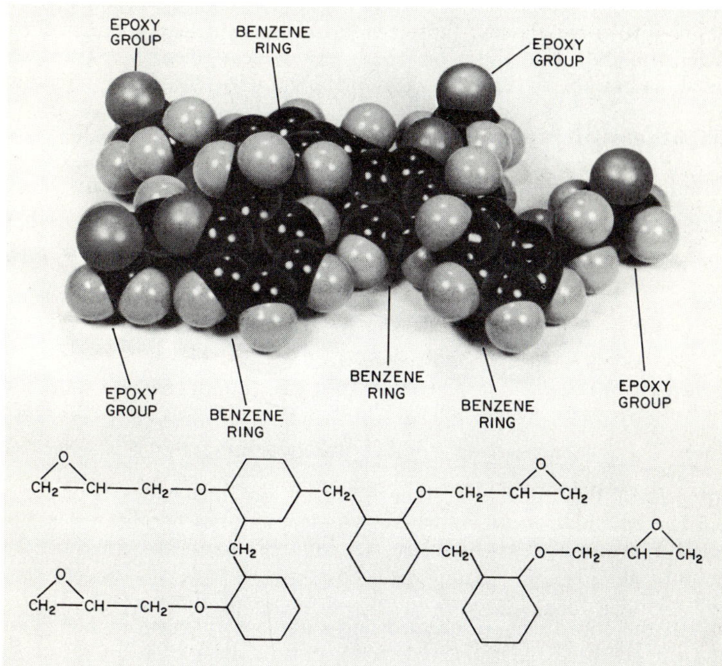

Atom model of epoxylated phenol-formaldehyde novolac resin showing massive, oxidation-resistant aromatic structure. Magnification approximately 100 million times. (*The Epoxylite Corporation.*)

Upon completion of the formaldehyde addition, the temperature was maintained at 80°C for an additional period of a half hour, after which 4 parts of sodium carbonate were added to neutralize the acid. The system was then placed under vacuum at a pressure of 30 mm mercury and water distilled therefrom by heating until the temperature reached 80°C. At this point, additional water was then added slowly to the system at a rate such that the distillation temperature was maintained at approximately 80°C. The water addition and distillation was continued until the total distillate collected amounted to 315 parts [62].

The resultant novolac resin is a solid at room temperature.

In addition to phenol, alkyl-substituted phenols such as *o*-cresol may be used as a starting point for the production of novolac. If at least four carbon atoms are present in the alkyl group, these novolacs can be epoxidized to yield resins soluble in paraffin hydrocarbons [31]. The novolac may be based on chlorinated phenols to impart flame resistance to the cured compositions [19].

The novolac resins may be partially esterified before the epoxidation reaction [75] or esterified afterward for use in ester coatings [24].

The glycidyl ethers of novolacs are of increasing commercial importance because of their higher functionality.

GLYCIDYL ETHERS OF OTHER AROMATIC COMPOUNDS

For convenience, we can divide the remaining aromatic-based glycidyl ether epoxy resins into three classes: those based on mononuclear phenols, those based on binuclear phenols, and those based on polynuclear phenols. These resins are synthesized, as before, by reaction with epichlorohydrin in a caustic medium.

Glycidyl Ethers of Mononuclear Di- and Trihydric Phenols

Representative of mononuclear di- and trihydric phenols and substituted phenols are:

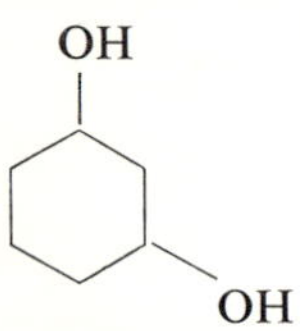

Resorcinol

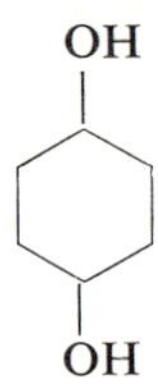

Hydroquinone

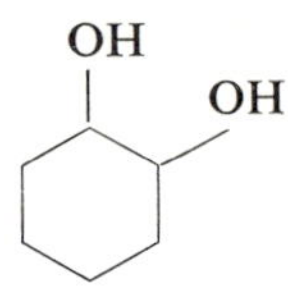

Catechol

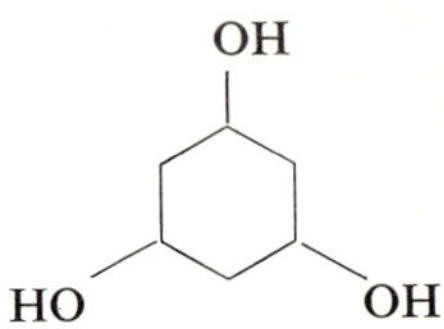

Phloroglucinol

The synthesis of the diglycidyl ether of resorcinol has been reported in detail [49,85]. The epoxidation of chlorinated hydroquinones has been described [22]. Substituted hydroquinones which are sterically hindered are epoxidized through the disodium salts [34].

Difficulties have been reported in the synthesis of the triepoxy from phloroglucinol. If the enol form is assumed and sufficient epichlorohydrin is employed to react with all three phenolic hydroxyls, only a diepoxy resin results, and the remaining phenolic hydroxyl reacts with the newly formed resin to produce a polymerized material during dehydrohalogenation. It is possible that changed reaction conditions might overcome the difficulties [134].

A description of the two-step epoxidation of saligenin, containing both an aliphatic and an aromatic hydroxyl, is given in [142], and epoxidation of vanillyl alcohol is described in [171].

Glycidyl Ethers of Bisphenols

Dihydric diphenols with a single carbon between the rings are most popular commercially, insofar as they are available from any aldehyde and many ketones and from any phenol having an open ortho or para position. Alternatives to bisphenol A were explored early in the technology, and resins from, for example, 4,4′-dihydroxybenzophenone were reported as early as 1942 [1]. Eighty-four representative bisphenols are given in [57].

With the bisphenols containing a single carbon bridge, a variety of substitutions on the benzhydryl carbon are possible; epoxidation of unsymmetrical compounds is described in [177]; the effect on epoxidation of chlorine substituents are reported [155,156], and the epoxidation of rosolic acid has been described [176]. With epoxidized aldehydes, an epoxy group may be attached to the benzhydryl carbon, as for example, when glycidaldehyde is used [120]:

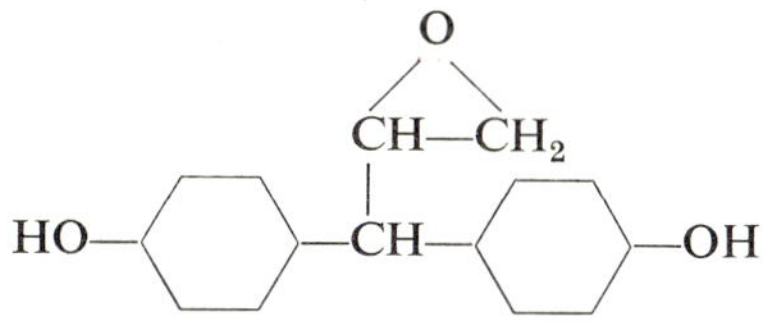

Bisphenol from glycidaldehyde

Bisphenols with more than one carbon atom between rings, although not as common, may be synthesized. The synthesis and epoxidation of a bisphenol from unsaturated polymers of butadiene and phenol is described in [135]. Long, aromatic-containing chains may be introduced by reacting *p*-xylene dichloride and bisphenol A [115], and may be obtained from substituted chlorinated xylenes [116].

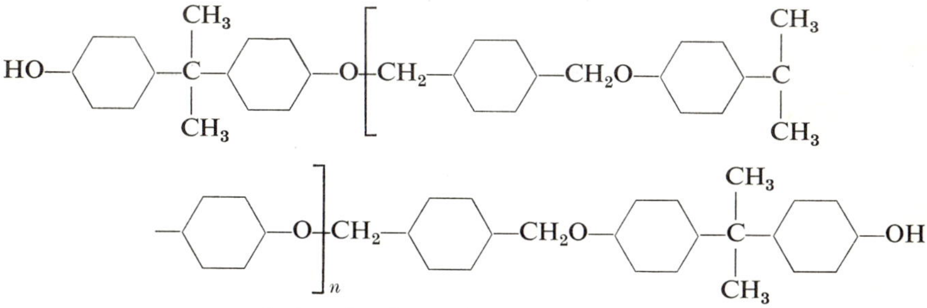

Product of *p*-xylene dichloride and bisphenol A

Phenol may be reacted with divinylbenzene to produce a three-ring structure, and hydroxyl-terminated polycarbonates have been epoxidized [160].

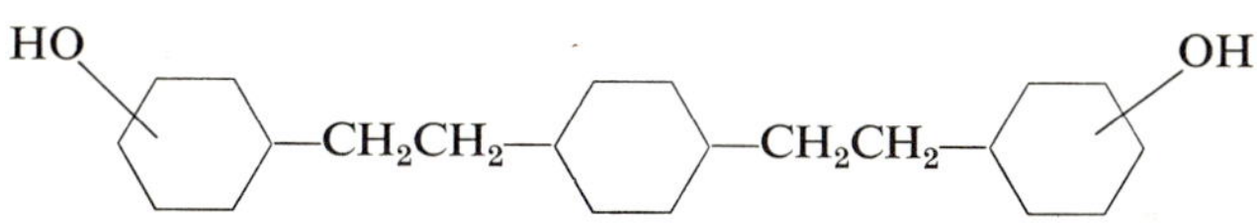

Product of phenol and divinylbenzene

Alternatively, the phenols may be directly connected, as is the case with bisphenol (4,4′-dihydroxybiphenyl) and resorcinol still residues which when epoxidized give 2,3′,4-tri(epoxypropoxy)diphenyl [194]. Or dihydroxynaphthalene may be used [83], as may dihydroxyfluoranthene [101], dihydroxydinapthylmethane [154,165, 207], and the yellow form of fluorescein [168].

The bisphenols may contain alkyl or aryl substituents [20,81], and the substituents may contain hydroxyl groups [29,205]. Hydroxyl-containing bisphenols such as 3,3-bis(4-hydroxyphenyl)-1,2-dihydroxypropane [82] may be used, as may mixed phenols from coal tars [72,157]. Phenol may be condensed with 3,9-divinyl-spirobi(*m*-dioxane) to give a diphenol containing a cycloaliphatic backbone that is similar to that of the cycloaliphatics from pentaerythritol described in Chap. 3 [188].

Of the substituents, perhaps the most interesting are the halogenated bisphenols. Both tetrabromo- and tetrachlorobisphenol A have been investigated. The synthesis of the tetrachloro species has been reported in detail [71,73], although the tetrabromo compound has proved superior and considerable work has been devoted to its synthesis. The diglycidyl ether of tetrabromobisphenol A may be crystallized from cyclohexane at purities of better than 99 per cent [199]. It has been reported that bromine may be added during synthesis of DGEBA to produce a brominated species whose flame resistance is further enhanced by the addition of antimony *sec*-butoxide at 110°C to the final resin [196]. Tetrachloro and bromo versions of the bisphenol from phenol and hexafluoroacetone have been epoxidized [213].

Suitable bisphenols, then, may range from the simplest structures, such as condensation products of phenol with acetylene [130], 2-butanone, and acetone with *o*-cresol; through materials of increased functionality, such as 2,4,4′-trihydroxydiphenyldimethylmethane; to compounds with fairly complex side chains, such as

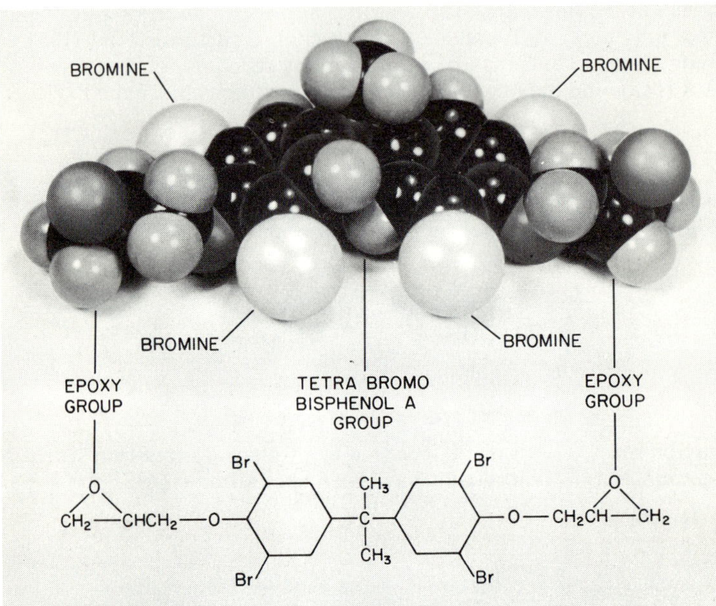

Atom model of diglycidyl ether of tetrabromobisphenol A. The marked change in physical properties often encountered when using chlorinated and brominated resins is clarified by examination of the significantly larger size of the chlorine or bromine atoms compared to the carbon, hydrogen, and oxygen atoms. Approximately 100 million times actual size. (*The Epoxylite Corporation.*)

1,8-bis(4-hydroxyphenyl)pentadecane from cashew nutshell liquid and phenol [183] or the similar compound based on substituted catechol (Bhilawanol) and phenol [192], and diphenols condensed with dihaloalkyl ethers (e.g., 2,2′-dichlorodiethyl ether) [118]; to compounds containing other atoms, such as, for example, sulfur in place of the benzhydryl carbon as with dihydroxydiphenylsulfone [33]. A review of sulfur-containing epoxy resins is contained in [131].

Carboxyl-containing epoxy resins, such as the epoxy resin from diphenolic acid,

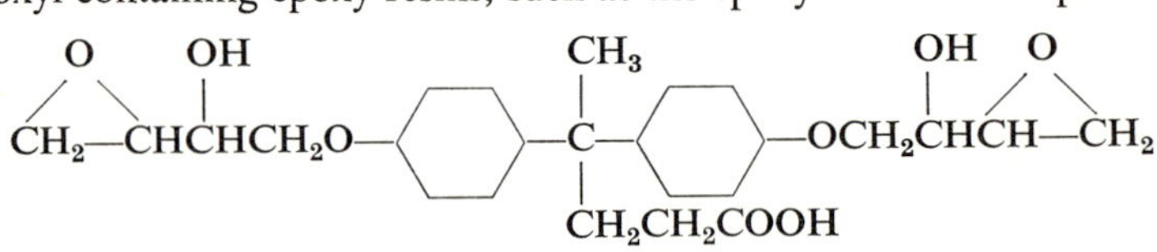

Epoxy resin from diphenolic acid

have been synthesized in a three-step process [93,100]. In the first step, the carboxyl group is converted to the carboxylate salt (COONa). In the second step, a diepoxy, such as butadiene dioxide is reacted with the phenolic hydroxyls in a strong solvent to prevent loss of epoxies. The carboxylate salt is then converted back to the carboxyl group by treatment with a strong acid. It is interesting to note that use of the diepoxy rather than the epichlorohydrin results in the production of additional hydroxyl groups, as indicated in the structure. In addition, the reaction when the diepoxy is employed, does not involve the evolution of by-products, and chlorine, even in trace amounts, will be absent from the final product since none need be introduced during synthesis. Diphenolic acid has been reacted with epichlorohydrin to produce a triepoxy resin, the third epoxy group being a glycidyl ester.

Glycidyl Ethers of Polynuclear Phenols

Tri- and tetranuclear phenols have been employed as coreactants with epichlorohydrin for the production of glycidyl ethers [52,56]. The synthesis of the tetraglycidyl ether from tetra(*p*-hydroxyphenyl)ethane is described in [65].

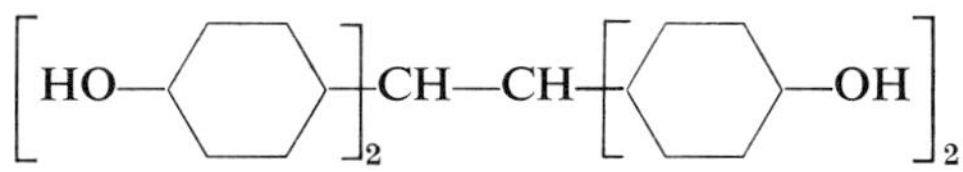

Tetra-(*p*-hydroxyphenyl)ethane

The synthesis of tri- and tetraglycidyl ethers from the reaction product of phenol and acrolein is described in [46]. The trisphenol from phosphoric acid and bisphenol A has been epoxidized [178]. Epoxidized polyphenols have been synthesized from phenols and chloroacetones [111], from phenols and polycarbonyl-substituted benzenes [175], from phenols and vanillyl alcohol [170], from phenol-xylene-formaldehyde [162], and from formaldehyde and 2,7-dihydroxynaphthalene [172]. The synthesis of polyglycidyl ethers from the reaction product of resorcinol and acetone is described in [112].

Many of these polyphenols are closely related to the novolac resins as previously discussed, in the main being derived from the condensation products of phenols and polyphenols with aldehydes, ketones, etc.

A variety of complex polyphenols, substantially different from aldehyde-phenol reaction products, may be developed from diphenolic acid.

Two moles of diphenolic acid may be reacted through the carboxyl groups with suitable difunctional molecules to produce tetraphenols of the following generalized structure:

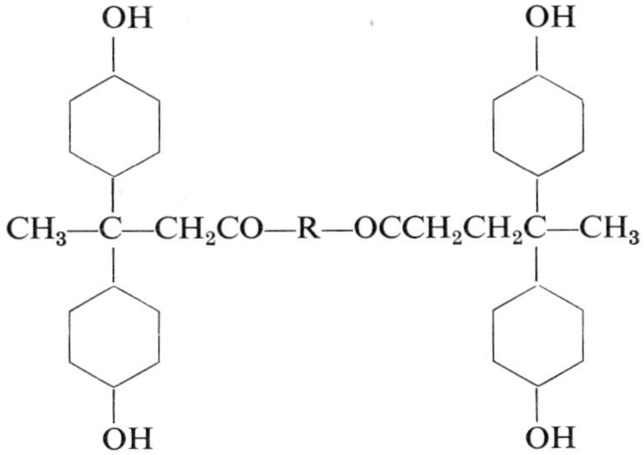

Tetraphenol from diphenolic acid

where R may be either the residue of an aliphatic alcohol such as diethylene glycol or a polyester (e.g., from phthalic anhydride or adipic acid and diethylene glycol). In this fashion, extremely complex structures may be conveniently synthesized for subsequent epoxidation through the phenolic hydroxyls.

GLYCIDYL ETHERS OF ALIPHATIC POLYOLS

Of the polyols available commercially, glycerol has, perhaps, been most used for the production of epoxy resins. Longer-chain polyols synthesized by the reaction of ethylene oxide with a diol or a triol are also used [72]. See Appendix to Chap. 4 for typical structures.

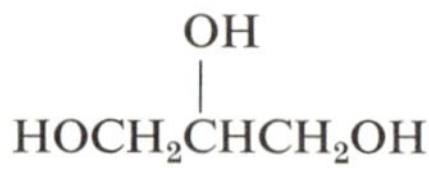

Glycerol

Suitable aliphatic polyols may be very simple, such as butanediol and hydrogenated bisphenol A, or may be quite complex, containing substitutions. For example, the following diol may be employed to synthesize an epoxy resin containing siloxane linkages:

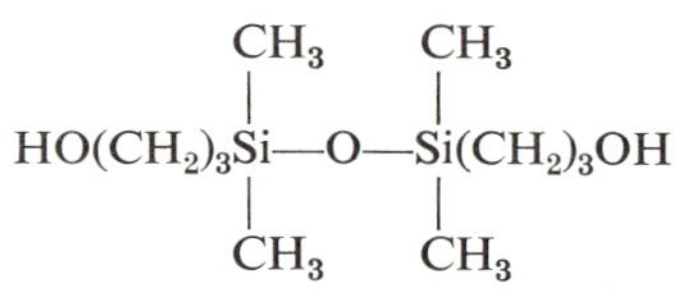

1,3-Bis(3-hydroxypropyl)tetramethyldisiloxane

Polyols containing the furan ring, such as

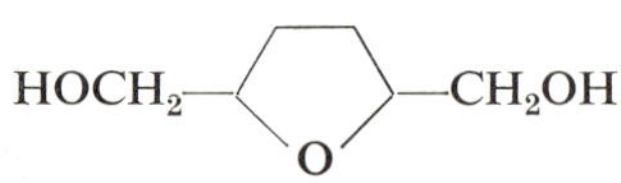

2,5-Bis(hydroxymethy)tetrahydrofuran

are suitable [137]. Epoxy resins have been derived from cyclic ether anhydro hexitols (i.e., 1,4:3,6-dianhydro-D-sorbitol) [146], from bis(hydroxymethyl)xylene and its polymers [145], from cyclohexene-containing polyols such as dihydroxymethylcyclohexene [138] which may be further reacted by Diels-Alder syntheses with unsaturated chlorinated dienes to obtain flame-retardant resins [150], from chlorinated butanediol [97], from acetylenic alcohols (e.g., 2-butynediol) [53], and from acetone-formaldehyde-glycol reaction products [182].

When epoxy resins are synthesized from aliphatic polyols, the reaction is customarily conducted in two steps [12]. Initially, the epichlorohydrin and polyols are reacted in the presence of a Lewis acid, such as BF_3, to produce the chlorohydrin intermediate. This is then dehydrohalogenated by sodium aluminate or the like. Dehydrohalogenation may be carried out in the presence of a water-immiscible solvent, which extracts the polyglycidyl ether as formed and minimizes the undesirable side reactions [158]. A one-step procedure, using caustic as a catalyst, is described in [91].

The synthesis of aliphatic epoxy resins is complicated by the fact that the hydroxyl group in the chlorohydrin intermediate, being also aliphatic, reacts at about the same rate as the initial polyol. Thus, the bischlorohydrin intermediate may react to give a higher-weight species:

Bischlorohydrin intermediate

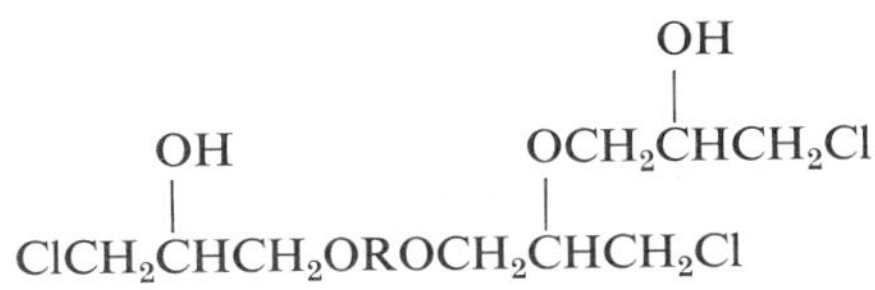

Chlorine-containing reaction product

When this product is subjected to dehydrohalogenation, a chlorine-containing diepoxy is the result:

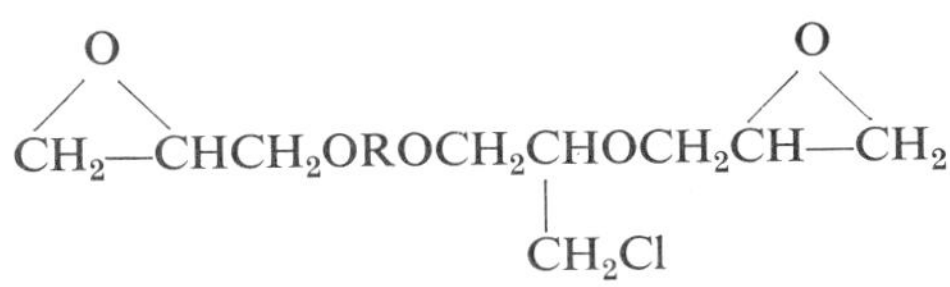

Chlorine-containing aliphatic diepoxy

The presence of chlorine is detrimental to color, reactivity, and electrical properties.

For applications where relatively high chlorine content is not disadvantageous, use can be made of the tendency of the epichlorohydrin to react with formed hydroxyls to produce polyepichlorohydrin [18,149]. In this instance, epichlorohydrin is reacted with a hydroxyl-bearing compound such as water to give the chlorohydrin intermediate. Additional epichlorohydrin is then added and reacted through the formed hydroxyl groups to extend the chain. At the appropriate

A section of resin kettles in epoxy-resin production plant. (*Ciba Products Co.*)

chain length, dehydrohalogenation is accomplished to produce polyepichlorohydrin:

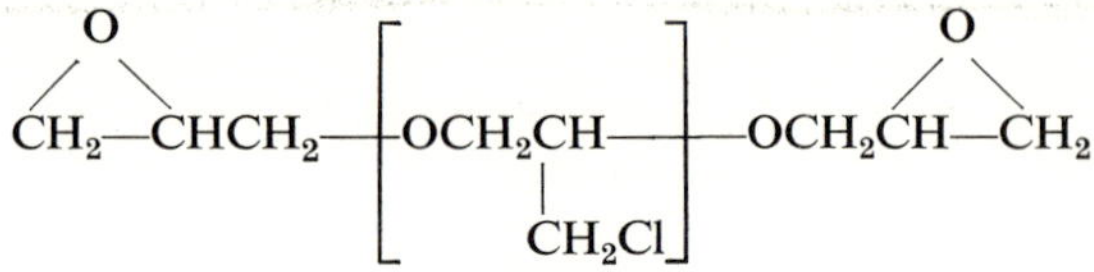

Idealized polyepichlorohydrin

The properties of the polyepichlorohydrin will be dependent on the type of initiator as well as the molecular weight, as indicated in Table 2-2.

Table 2-2. Properties of Polyglycidyl Polyepichlorohydrin Resins [181]

Initiator	Average mol. wt.	Weight per epoxy	Epoxide functionality	Total chlorine, %	Viscosity at 25°C, centipoises	Color, Gardner
Water	405	235	1.7	27.0	190	1
	608	402	1.5	31.1	2,010	2–3
	702	383	1.8	31.9	2,900	5–6
	723	484	1.5	31.8	3,700	5–6
	736	432	1.7	32.6	4,900	1–2
Bisphenol A...	725	457	1.6	23.0	30,300	6
Glycerin	615	259	2.4	24.0	1,200	5–6
	740	392	1.9	26.0	3,800	5–6

The aliphatic epoxy resins generally have low viscosities and good wetting action, and some, with numerous atoms between reactive sites, provide flexibility when cured with conventional curing agents. A number of them are water-soluble. Higher-molecular-weight polyhydroxyl resins from polyols and epichlorohydrin have been esterified for coatings applications [11].

OTHER GLYCIDYL EPOXY RESINS

Any molecule possessing an active hydrogen is suitable for reaction with epichlorohydrin. The resultant structure, after dehydrohalogenation, will then contain a glycidyl group, but the chemical properties may differ from those of the glycidyl ether group which we have considered in detail.

The nonether glycidyl resins produced by the epichlorohydrin route are not of the same order of commercial interest as are the glycidyl ethers, but they are of sufficient interest to warrant a brief examination.

Several types have been suggested in the literature: those resins derived from acid groups (glycidyl esters); from amide groups (glycidyl amides), from amine groups (glycidyl amines); from silanols (silicone glycidyl resins), from mercaptans (thioglycidyl resins), and from poly(difluoromethylene) chains.

Glycidyl Esters

A number of methods have been suggested for the preparation of glycidyl ester resins. Generally, they involve reaction of epichlorohydrin with organic acid or the sodium or potassium salt of the acid. An alternative procedure is to react glycidol with acid chlorides [50]. Most commonly the carboxyl group is reacted directly [102] in the presence of some catalyst other than sodium hydroxide, dehydrohalogenation being accomplished after completion of the initial reaction. When caustic is used as a catalyst, best results are reported when it is added in advance to form the disodium salt [140].

Quaternary ammonium hydroxide has been reported as a satisfactory catalyst for the initial reaction, as have ion-exchange resins, salts of acid-containing cation resins, and compounds with acid amide groups [129]. Use of a tertiary amine [88] or a quaternary salt such as benzyltrimethylammonium chloride is claimed to provide good yields [40]. Glycidyl esters of higher fatty acids may be produced in commercial yields by use of a quaternary salt with an alkali metal carbonate used as the hydrogen ion acceptor [125]. Epoxidation of chlorendic acid is accomplished by use of 10 to 20 times excess epichlorohydrin in the presence of a tertiary amine or a quaternary salt and then dehydrohalogenating with sodium hydroxide [167].

In some cases, the glycidyl ester epoxy resins have been reported as being somewhat unstable, reacting without catalyst at room temperature in 6 to 8 weeks [8]. In other cases, adequate stability is obtained.

Synthesis of diglycidyl phthalate acid [54] and tere- and isophthalic acid [195] have been reported, and polyglycidyl esters of long-chain fatty acids have been obtained in the two-step process involving the salt [105]. Synthesis of members of the series

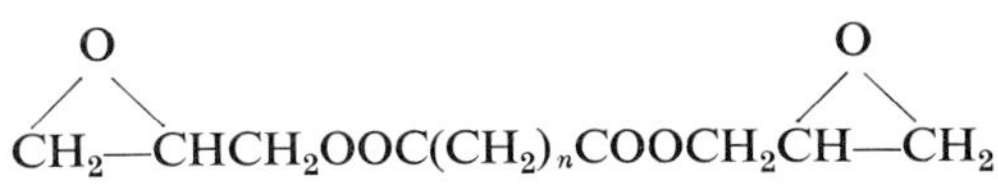

Aliphatic diacid glycidyl esters

(where n is 0,2,4,5,7, and 8) by the epichlorohydrin route has been described [163]. Synthesis of the epoxy from tetramethyl adipic acid also has been reported [204].

Epoxy resins from phenolphthalein are described in [102]. In this instance, both glycidyl ethers and glycidyl esters may be present, and different glycidyl products will result from the various forms of phenolphthalein. Two glycidyl ether groups may be present and the compound may be a triglycidyl material containing one glycidyl ester, or a material of the following structure:

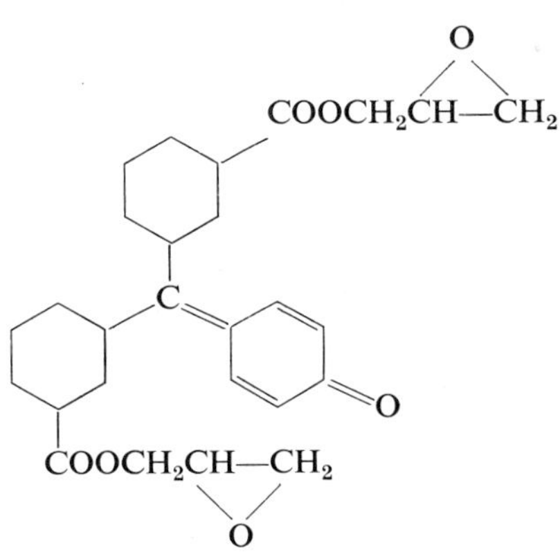

Epoxidized phenolphthalein

Epoxy resins from mixtures of bisphenol A and phenolphthalein have also been described [144].

Other epoxies containing both glycidyl ether and glycidyl ester groups have been suggested: for example, the epoxy resin from 2,4,6-tribromo-3-hydroxybenzoic acid [179].

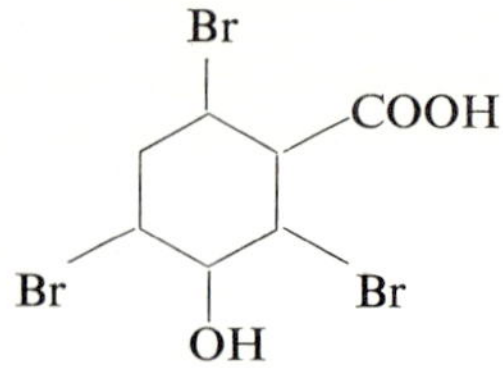

2,4,6-Tribromo-3-hydroxybenzoic acid

Epoxidation of other hydroxyl-containing acids is described [110].

The use of organophosphoric acid as a coreactant with epichlorohydrin has been

$$(HOCH_2O)_3PO$$

An organophosphoric acid

reported as producing a triglycidyl ester [106]. Epoxy esters have been synthesized from *a,a'*-(9,10-dihydro-9,10-anthrylene)disuccinic acid [96], mercaptocarboxylic acids [141], and shellac [197].

Glycidyl Epoxies Containing Nitrogen

Epoxy resins containing nitrogen may be developed from ammonia, from primary and secondary amides, from cyanuric acid, from melamines, and from primary and secondary amines. Resins from ammonia are described in [7].

Glycidyl Amides and Amide-containing Epoxies. Amides, generally, are rather sluggish in their reaction with an epoxy group, although reaction can be made to progress under proper conditions. Sulfonamides, because of their acidity, react readily with epichlorohydrin [43].

Amide-containing epoxy resins have been produced from linear polymeric amides terminated with phenolic hydroxyls [184] or primary amines [186] and from the reaction product of diphenolic acid and amines [76].

Cyanuric Acid. Cyanuric acid exists either as the acid or the amide form.

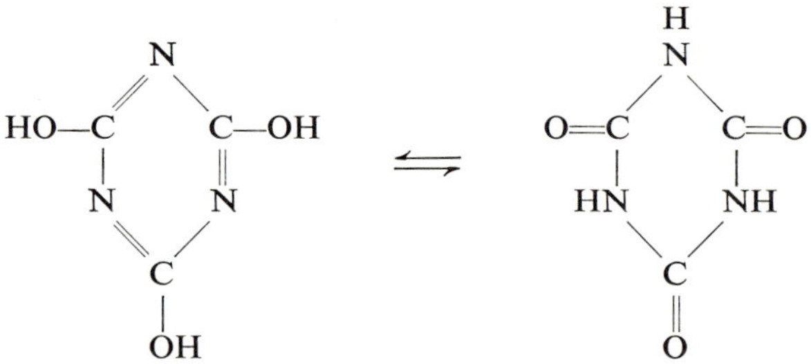

Acid form　　　　Amide form

The compound can react with epichlorohydrin to produce a glycidyl derivative.

The synthesis of resins from cyanuric acid and epichlorohydrin is reported in [45]. The synthesis of triglycidyl cyanurate simultaneously with the synthesis of the diglycidyl ether of bisphenol A has likewise been accomplished [55], and the production of higher-molecular-weight polyglycidyl ethers has been reported by the reaction of cyanuric acid with diepoxies [94,98].

Lower-functionality epoxies from cyanuric acid may be prepared by using phenol or a monoepoxy (other than an epichlorohydrin) to block some portion of the reactive sites [123]. Resins produced from cyanuric acid and epichlorohydrin are reported to be stable against self-polymerization.

Glycidyl Resins from Melamines. Glycidyl resins synthesized from *N*-methylolamino-1,3,5-triazines have been reported [132], these being produced conventionally from epichlorohydrin in a caustic medium. Alternatively, the epoxy resin may be produced by reaction with an epoxy alkanol or glycerol monochlorohydrin by transetherification [200].

Glycidyl Amines. Glycidyl amines may be synthesized from aliphatic or aromatic primary or secondary amines [90,92], and specifically the use of *m*-phenylenediamine and 4,4′-methylenedianiline in the synthesis of epoxy resins has been described [87], as has the use of these resins in combination with the glycidyl ether resins.

Dodecahydro-1,4,7,9*b*-tetraazaphenalene [153] and the alkylated species [187]

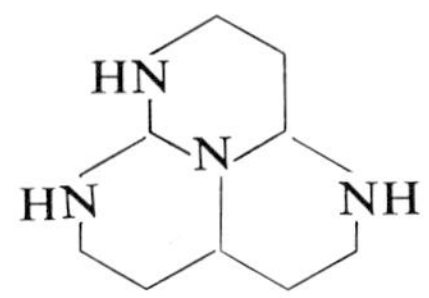

Dodecahydro-1,4,7,9*b*-tetraazaphenalene

have been reacted with epichlorohydrin to produce triepoxy resins. The triglycidyl ether amine of *p*-aminophenol has been synthesized [109], as has the diglycidyl ether

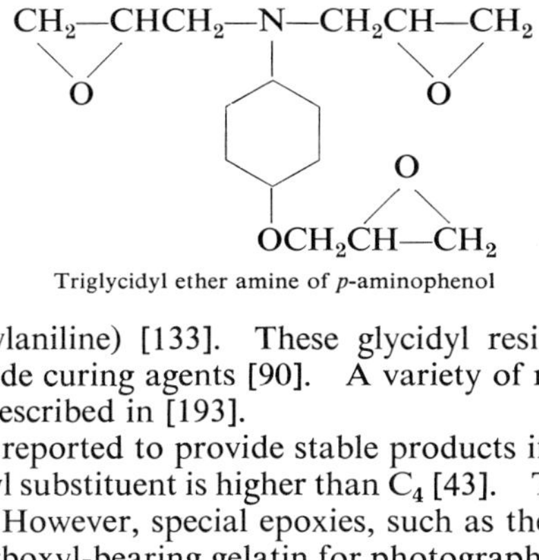

Triglycidyl ether amine of *p*-aminophenol

of methylenebis(methylaniline) [133]. These glycidyl resins provide heightened reactivity with anhydride curing agents [90]. A variety of resins from amines and epichlorohydrins are described in [193].

Glycidyl amines are reported to provide stable products if the amine is aromatic or, if aliphatic, the alkyl substituent is higher than C_4 [43]. The lower alkyl glycidyl amines are unstable. However, special epoxies, such as the following, have been used for hardening carboxyl-bearing gelatin for photographic layers (202).

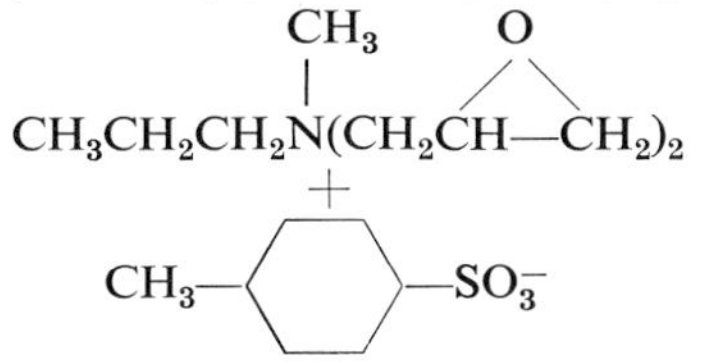

Bis(2,3-epoxypropyl)methylpropylammonium *p*-toluenesulfonate

A higher-molecular-weight water-soluble epoxy resin can be obtained by reacting epichlorohydrin with aliphatic amines such as hexamethylenediamine. In the synthesis of such resins, epichlorohydrin is used at 0.7 to 1.3 mole per equivalent of nitrogen to obtain the desirably high viscosity [60]. Use of larger amounts of epichlorohydrin produces lower-molecular-weight species, as would be expected from the previous discussion. It has also been reported that tetraethylenepentamine may be reacted with epichlorohydrin to produce an intermediate with three chlorohydrin groups and two secondary amines. This is then dissolved in water with a trace of NaOH and polymer formation is accomplished. The polymeric product is used to improve the wet strength of paper [14].

Glycidyl Triazines. Epoxidation of triazines containing C—Cl groups may be accomplished by the use of the chlorohydrin of glycerol as follows [47]:

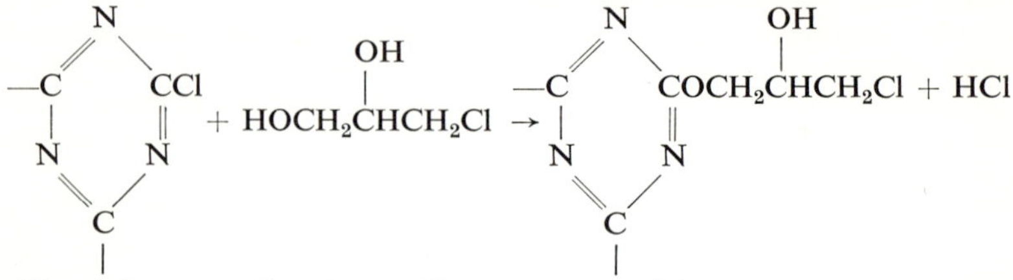

Chlorotriazine — Glycerol monochlorohydrin — Chlorohydrin ether — Hydrogen chloride

followed by dehydrohalogenation with caustic. These compounds are reported to be unstable at elevated temperatures [117]. Epoxy alcohols (such as 2,3-epoxypropanol) will react with cyanuric chloride or bromide in the presence of sodium hydroxide to produce epoxy resins [36]. Methoxycyanuric chloride was reacted with glycidol to obtain a diepoxy resin that self-polymerized explosively at about 110°C [139], and a diepoxy has been reported from ethoxycyanuric acid chloride by treating the cyanuric chloride with sodium ethoxide, followed by reaction with glycidol in the presence of sodium [198].

A review of glycidyl triazine resins is given in [201].

Thioglycidyl Resins

The epoxy group of the epichlorohydrin reacts with mercaptans (SH), and as a consequence epoxy resins can be obtained from such compounds. Synthesis of resins having the following generalized structure has been accomplished using epichlorohydrin [35].

Epoxidized bissulfide

The synthesis of epithio ether polymers has been reported from the reaction of the epoxy resin with ammonium thiocyanate to replace the epoxy oxygen with sulfur [114]. Such resins are reported to have heightened reactivity to nucleophilic curing agents.

Silicon–Glycidyl Resins

The synthesis of a silicon-substituted mononuclear glycidyl epoxy has been described in the literature [84].

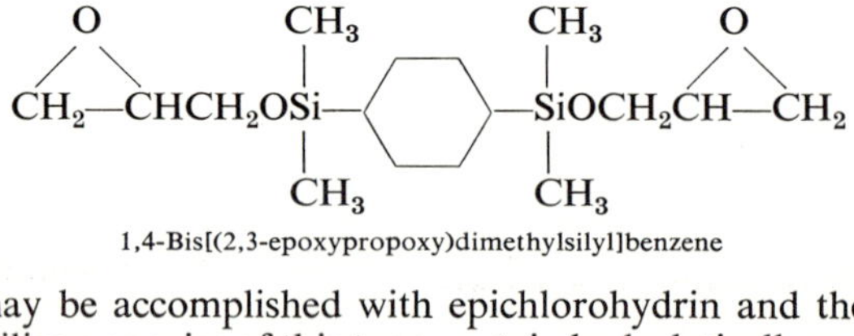

1,4-Bis[(2,3-epoxypropoxy)dimethylsilyl]benzene

The synthesis may be accomplished with epichlorohydrin and the sodium salt of the disilanol. Silicon epoxies of this type contain hydrolytically unstable Si—O—C links.

Fluorine–Glycidyl Resins

Fluorine-glycidyl resins have been obtained by a solvolysis reaction, with ring closure, according to the following general equation [208]:

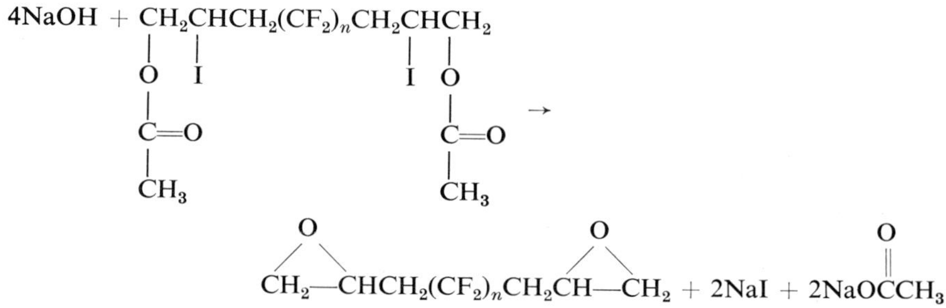

SYNTHESIS OF EPOXY RESINS FROM MONOEPOXIES OTHER THAN EPIHALOHYDRINS

Epoxy Resins from Unsaturated Monoepoxies

Unsaturated monoepoxies may be homopolymerized through the unsaturation or copolymerized with a variety of other unsaturated molecules. If the unsaturated monoepoxy is a glycidyl ether, the resulting epoxy resin will be a glycidyl ether. Polyallyl glycidyl ether [15] is an example:

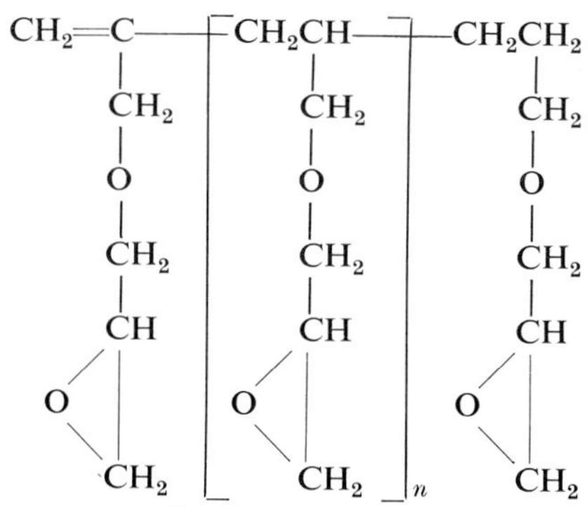

Polyallyl glycidyl ether

Glycidyl ethers derived from copolymers are discussed in [124]. Glycidyl ether copolymers from cyclic conjugated dienes [79] and compounds containing vinylidene groups [152] have been described. Copolymers of 4-vinylcyclohexene monoepoxide with other unsaturated molecules have been produced [28]. Allyl glycidyl ether has been copolymerized with vinyl acetate [51], vinylchloride [203], allyl chloride [212], and acrylonitrile [21]. Divinylbenzene monoepoxide has been polymerized to the polyepoxy compound [44].

Glycidyl methacrylate has been used to introduce epoxy groups into vinyl and acrylic resins (see Chap. 15). Glycidyl allyl phthalate has been reacted through the unsaturation to produce epoxy resins [6]. Glycidyl sorbate has been dimerized to give a typical structure as follows [2]:

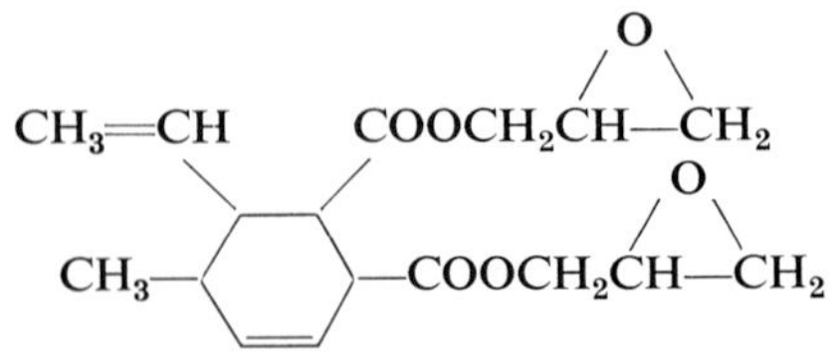

Idealized glycidyl sorbate dimer

Epoxy Resins from Monoepoxy Alcohols

A polyfunctional phosphorus-containing acid may be esterified by a hydroxyl-containing monoepoxy to yield phosphorus-containing polyglycidyl esters [67]. A variety of phosphorus-containing epoxy resins are obtained by reacting phosphoric acid monoesters with epoxidized alcohols or with unsaturated alcohols, which are subsequently epoxidized by a peracid (see Chap. 3). Alternatively, phosphorus oxychloride may be reacted with the epoxidized alcohol, to produce, with the liberation of HCl, the triepoxy ester. Instead of phosphoric acid, sulfur, selenium, or tellurium may replace either the double-bonded oxygen or one of the single-bonded oxygens [61]. Esterification of acid chlorides with hydroxyl-containing monoepoxies is reported as producing the diglycidyl terephthalate [64].

Epoxy Resins from Monoepoxies by Ester Interchange

Glycidyl esters have been produced through ester interchange between an epoxyalkyl acetate and a lower alkyl ester. A procedure using diethyl chlorendate (from chlorendic anhydride and ethanol) is as follows:

> In a 300 ml flask fitted with a thermometer, magnetic stirrer, and a distilling head, were placed 44.4 g of the diethyl ester of chlorendic acid, along with 116 g of glycidyl acetate and 1.6 g of potassium tertiary butoxide catalyst. The mixture was stirred at 110°C for four hours . . . . After standing overnight, the flask was heated to 140°C . . . . Distillation was continued at 140 to 155°C until 12 ml of ethyl acetate was collected. Then the mixture was distilled under vacuum while holding the temperature below 80°C . . . . The resulting crude product in the flask . . . was cooled, dissolved in toluene, stirred with charcoal, and then filtered. (After washing and final distillation), 480 g of the diglycidyl ester of chlorendic acid product was obtained [167].

Epoxy Resins from Glycidaldehyde

Polyepoxy compounds may be produced by reacting glycidaldehyde, an alcohol, and a ketone. Production of 1-epoxyethyl-2-oxabutyl acetate and 1,3-bis(epoxyethyl)-2,4-dioxahexyl acetate are given in [169]. Glycidaldehyde has been condensed with itself to give glycidyl glycidate [190].

MONOGLYCIDYL ETHERS CONTAINING UNSATURATION

There is an extremely wide variety of monoepoxy compounds available commercially, but few are produced using epichlorohydrin, since suitable monoepoxies can be produced more economically by epoxidation of unsaturation.

Insofar as the monoepoxies concern the epoxy-resin technology—either as materials for synthesis of polyepoxies or as reactive diluents for them—they are of interest within the scope of this work; however, since their main uses lie outside of the resin technology, their synthesis will not be emphasized.

One class of monoepoxies is of theoretical interest to the epoxy-resin technology: those containing both an epoxy group and unsaturation. An example would be [5]

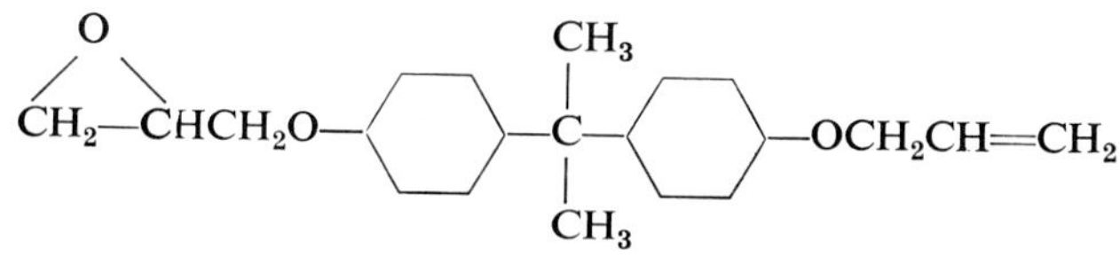

Mixed glycidyl/allyl ether of bisphenol A

A similar material may be produced by reacting the diglycidyl ether of bisphenol A with allyl alcohol in the presence of a Lewis acid [66]. The presence of the unsaturation indicates that the molecule may be polymerized in two cooperating steps: the first step consisting of reaction through the epoxy group, and the second consisting of reaction, under different catalytic conditions, through the unsaturation.

The unsaturation, itself, may be directly epoxidized such that an epoxy resin may be obtained in two steps from the compounds containing both hydroxyls and unsaturation. The synthesis of an epoxy from one such material, allylphenol, is

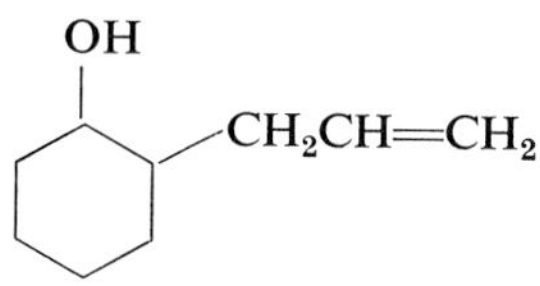

Allylphenol

presented in [68]. In [189] is a description of the synthesis of *o*- and *p*-(epoxycyclopentenyl) phenyl glycidyl ether and epoxydicyclopentenyl phenyl glycidyl ether by the two-step procedure.

In Chap. 3, the method of direct epoxidation is presented in some detail.

POLYGLYCIDYL COMPOUNDS CONTAINING UNSATURATION

Unsaturated groups may be added to the polyglycidyl ethers to provide additional reactive sites or may be present in the starting material. A description of such materials, including the allyl-substituted diglycidyl ether of bisphenol A is pre-

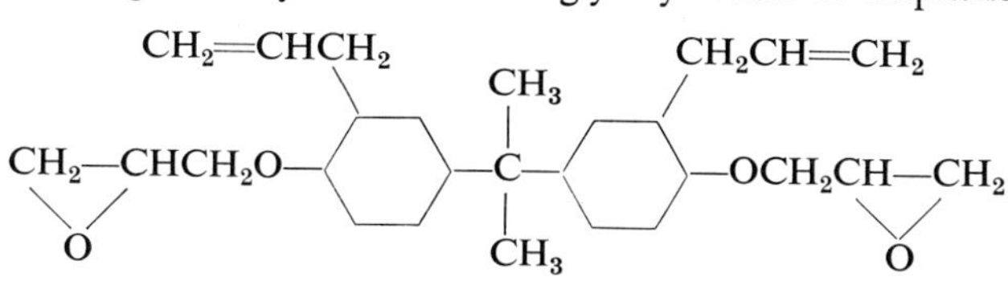

Allyl-substituted diglycidyl ether of bisphenol A

sented in [119,121]. Such materials as epoxidized polyester from tetrahydrophthalic anhydride and diethylene glycol may contain a plurality of epoxy and unsaturated groups. These compounds may subsequently be converted in coating formulations by esterification with organic acids and polymerization through the unsaturation [38]. Unsaturations, however, are more commonly present in compounds synthesized by the procedures described in Chap. 3.

EPOXY-RESIN ADDUCTS

From the foregoing discussion, and from a general consideration of the reactivity of the epoxy groups, it is evident that the resin itself may be reacted with di- or polyfunctional reactants to produce adducts containing more than one epoxy group. The number of such adducts is, potentially, quite large. Typical are phosphorus-containing epoxy resins prepared by reaction of diepoxies with acid dichlorides of phosphoric, phosphorous, or phosphinic acids at an acid/diepoxy ratio of 1/2 in the presence of $FeCl_3$ catalyst [185].

When the diepoxy contains additional reactive sites, such as unsaturation, adducts may be developed without the reaction of any of the epoxy groups, as from halogenated cyclopentadienes and unsaturated diglycidyl ether resins [128]. Adducts may likewise be prepared by the reaction of chain hydroxyls with phosphoryl isocyanate [206].

COMMERCIAL GLYCIDYL EPOXY RESINS

The commercial glycidyl epoxy resins are, to some extent at least, mixtures of high- and low-molecular-weight resins, the distribution of the molecular weights varying with the conditions of synthesis. This variation was indicated in the discussion of the synthesis of the diglycidyl ether of bisphenol A. Trace impurities will be present. For example: The pure diglycidyl ether of bisphenol A is a solid material melting at about 45°C. It is the impurities present in the commercial epichlorohydrin–bisphenol A resins which transform them to "supercooled" liquids [159]. In addition to trace amounts of impurities, commercial resins often contain a small amount of solvent, such as toluene, from the synthesis.

For a given type of resin, more than one range of molecular weights may be offered. Some may contain modifiers or diluents. Others may be composed of mixtures of two or more resin types. Typical of the commercial blends are such mixtures as combinations of brominated and nonbrominated DGEBA [147], blends of the pure DGEBA and a few per cent of an epoxidized novolac resin with a functionality of about 2 to prevent crystallization [151], blends of various molecular-weight DGEBA resins [42], blends of the DGEBA and the diglycidyl ester of dimerized fatty acid [148], and blends of the DGEBA with bis(2,3-epoxycyclopentyl) ether [107], or with epoxidized fatty-acid esters [27].

Properties of the commercial resins are given in Chap. 4, together with their idealized structures and IR curves.

Throughout the remainder of this book, the abbreviation DGEBA will be employed for the liquid commercial resin from epichlorohydrin and bisphenol A having an average molecular weight of approximately 380. Higher-molecular-weight species will be identified as such in parenthesis after the abbreviated designation.

SUMMARY

The diglycidyl ether of bisphenol A is synthesized by the reaction of epichlorohydrin and bisphenol A in a caustic liquid. The reaction proceeds in two steps: (1) The initial epoxy groups react with the phenolic hydroxyls, and (2) chlorine and hydrogen are then stripped out to produce new epoxy groups. Molecular weight of the resulting resin will depend upon the ratio of epichlorohydrin to bisphenol A employed. The greater the excess of epichlorohydrin, the lower the molecular weight of the resulting resin. Small excesses of epichlorohydrin result in the production of high-molecular-weight polyols. Although these contain few if any epoxy groups, they are, by custom, classified as epoxy resins when used in ester coatings; when molecular weights on the order of 30,000 are produced, they are called phenoxy resins.

Glycidyl ethers of other polyhydroxy compounds are synthesized in like fashion.

The polyhydroxy compounds may be aromatic or aliphatic, with the number of glycidyl groups in the resultant resin increasing with increasing hydroxyl content of the starting material.

Compounds containing active hydrogens are not restricted to hydroxyl-bearing ones. Other suitable reactants are acids, amines, amides, mercaptans, and silanols. Glycidyl-type epoxy resins have been synthesized from each of these classes.

Glycidyl ether and other glycidyl-type epoxy resins may be synthesized from monoepoxies containing unsaturation or hydroxyl groups.

The commercial glycidyl epoxy resins, predominantly the diglycidyl ethers of bisphenol A, are available in several molecular-weight ranges, and some may be composed of a mixture of more than one type of epoxy resin, and may contain, in addition, modifiers and diluents.

REFERENCES

1945

1. Britton and Slagh, Polyfunctional Ethers and Method of Making Same, U.S. 2,371,500.

1948

2. Boese, Lower Polymer of Glycidyl Sorbate and Method of Making It, U.S. 2,457,300.
3. Castan, Process for the Manufacture of Thermosetting Synthetic Resins by the Polymerization of Alkylene Oxide Derivatives, U.S. 2,444,333.
4. Greenlee, Synthetic Drying Compositions, U.S. 2,456,408.

1949

5. Shokal and Whitehill, Allyl Glycidyl Mixed Diether of Bis(4-hydroxyphenyl)-2,2-propane and Polymers Thereof, U.S. 2,464,753.
6. Shokal, et al., Glycidyl Allyl Phthalate and Related Mixed Esters Together with Polymers Thereof, U.S. 2,476,922.

1951

7. Daniel, et al., Process for Treating Nonfibrous Cellulose Film and Resulting Product, U.S. 2,573,957.
8. Erickson, Method of Polymerizing Glycidyl Compounds, U.S. 2,556,075.
9. Greenlee, Polymeric Polyether Polyhydric Alcohols, U.S. 2,558,949.
10. Newey and Shokal, Glycidyl Ether Compositions and Method of Using Same, U.S. 2,575,558.
11. Wittcoff and Roach, Resin Acid Esters, U.S. 2,572,086.
12. Zech, Epoxide Preparations, U.S. 2,538,072.

1952

13. Carpenter, et al., Production of Thermoplastic Resins and the Production from Such Resins of Threads, Fibres, Filaments, and the Like, U.S. 2,602,075.
14. Daniel and Landes, Wet Strength Paper and Process for the Production Thereof, U.S. 2,595,935.
15. Evans and Shokal, Carbo-polymerization Homopolymers of Unsaturated Alcohol Ethers of Glycidols and Their Homologs, U.S. 2,599,817.
16. Greenlee, Epoxide Resins, U.S. 2,615,007.
17. Greenlee, Epoxide Resins Compositions, U.S. 2,615,008.
18. Zech, Epoxide Compositions, U.S. 2,581,464.

1953

19. D'Alelio, Polymeric Compositions from a Chlorophenol, an Aldehyde, and a Chloroepoxyalkane, U.S. 2,658,884.

20. D'Alelio, Polymeric Compositions from a Hydrocarbon Saturated Phenol, an Aldehyde, and a Chloroepoxyalkane, U.S. 2,658,885.
21. Ham, Acrylonitrile, Copolymers, U.S. 2,650,151.

1954

22. Clemens, et al., Bis-glycidyl Ethers of Chlorohydroquinones and Process of Preparing Them, U.S. 2,682,547.
23. Crecelius, Epoxide Resins, U.S. 2,698,308.
24. D'Alelio, Polymeric Compositions, U.S. 2,695,894.
25. Greenlee, Manufacture of Epoxide Resins, U.S. 2,694,694.
26. Greenlee, Epoxide Resins, U.S. 2,698,315.
27. Newey, Compositions Containing Glycidyl Polyether of a Polyhydric Phenol and an Epoxidized Ester, U.S. 2,682,514.
28. Robertson, Epoxy-containing Copolymers of 4-Vinyl-cyclo-hexene Monoepoxide and a Polymerizable Ethylenically Unsaturated Compound, U.S. 2,687,404.
29. Wasserman, Glycidyl Ethers of Condensation Products of Certain Aromatic Hydroxyl Compounds with Phenols Having Unsaturated Hydrocarbon Side Chains, U.S. 2,665,266.
30. Whitehall and Taylor, Acid-modified Ether Resins, U.S. 2,686,771.
31. Bradley and Newey, Epoxy Resins from Alkyl Phenol Novolac Resins, U.S. 2,716,099.
32. Zech, Epoxide Compositions Containing Halogen, U.S. 2,712,000.

1956

33. Beavers, Preparation of Bis(4-glycidyloxyphenyl)-sulfone, U.S. 2,765,322.
34. Bell and McConnell, Bis-glycidyl Ethers of Hindered Dihydroxybenzenes and Process for Their Manufacture, U.S. 2,736,160.
35. Bender, et al., Diglycidyl Thio Ethers of Dithiols Containing Oxygen, U.S. 2,731,437.
36. Bradley and Mueller, Triglycidyl Cyanurate and Polymers Thereof, U.S. 2,741,607.
37. Earhart and Montague, Calculations of Compositions of Epoxy Resin Esters, ACS Symposium, Atlantic City, September.
38. Greenlee, et al., Film Curing by Simultaneous Esterification and Olefin Polymerization, ACS Symposium, Atlantic City, September.
39. Masters, Manufacture of Epoxide Resins, U.S. 2,767,157.
40. Mueller, Process for Preparing Epoxy Esters, U.S. 2,772,296.
41. Taylor and Whitehall, Phenolic Condensation Products Made with Divalent Tin Compound to Reduce Coloration, U.S. 2,738,339.
42. Wiles and Elam, Compositions Containing a Mixture of Glycidyl Polyethers of Dihydric Phenols, U.S. 2,735,829.
43. Wynstra, The Chemistry of Epoxy Resins, North New Jersey Section, ACS, March.

1957

44. Burk and Inskeep, Polymers of Divinyl Benzene Monoepoxide, U.S. 2,807,599.
45. Cooke, Process for Making Polyglycidyl Cyanurates, U.S. 2,809,942.
46. Farnham, Glycidyl Polyether of a Polyphenol, U.S. 2,801,989.
47. Frazier and Cadwell, Oxirane Triazine Resinous Compositions and Processes of Preparing the Same, U.S. 2,810,706.
48. Goppel, Process for the Manufacture of Glycidyl Ethers of Polyhydric Phenols, U.S. 2,801,227.
49. Moult and St. Clair, Preparation and Properties of Resorcinol Epoxides, ACS Symposium, New York City, September.
50. Mueller and Shokal, Epoxy Esters of Carbonic Acids and Their Polymers, U.S. 2,795,572.
51. Rothrock and Wilkinson, Film-forming Solid Polyepoxy Copolymers of Allyl Glycidyl Ether and Vinyl Acetate, U.S. 2,788,339.
52. Schwarzer, Polyglycidyl Ethers of Tetraphenols, U.S. 2,806,016.
53. Shokal and Wallingford, Epoxy Ethers of Acetylenic Alcohols and Their Polymers, U.S. 2,792,381.
54. Suen and Hicks, Diepoxy Resinous Compositions, U.S. 2,801,323.

1958

55. Cooke, Epoxide Resins, U.S. 2,864,805.
56. Davis and Hunter, Improvements in or Relating to Epoxide Resins, Brit. 789,392.
57. Erich, Literature Survey on the Synthesis of Epoxy Resins from Bisphenols, Government Report OTA Dept of Commerce, PB 151730.
58. Freeman and Gottschalk, Preparation of Epoxide Resins, U.S. 2,824,855.
59. Griffin and Long, Process for the Manufacturing of Epoxy Resins, U.S. 2,848,435.
60. Lehmann and Bayer, Stabilization of Basic Condensation Products of Epichlorohydrin, U.S. 2,849,411.
61. Mueller, et al., Epoxy Esters of Phosphorus Containing Acids and Their Use in Treating Textiles, U.S. 2,826,592.
62. Naprovnik and Lieb, Epoxide Resins, U.S. 2,829,124.
63. Pezzaglia, Process for Manufacture of Glycidyl Ethers of Polyhydric Phenols, U.S. 2,840,541.
64. Raecke, et al., Epoxy Alkyl Esters of Polybasic Aromatic Carboxylic Acids and Process of Making Same, U.S. 2,865,897.
65. Shepherd and Dearborn, Epoxy Resins, U.S. 2,857,362.
66. Shokal and Tess, Hydroxy-substituted Allylic Ethers and Derivatives, U.S. 2,864,804.
67. Smith, et al., Epoxy-substituted Esters of Phosphorus-containing Acid and Their Polymers, U.S. 2,856,369.

1959

68. Aelony, Some Polyglycidyl and Poly 2,3-Epoxybutyl Derivatives of Phenols, ACS Symposium, Atlantic City, September.
69. Anderson and Melstrom, Process for the Manufacture of Epoxy Resins, U.S. 2,901,462.
70. Applegath, et al., Epoxy Novolacs, *SPE J.*, January.
71. Cass and Reaville, Performance Characteristics of Epoxy Resins of Tetrachloro Bisphenol A, ACS Symposium, Atlantic City, September.
72. Clark and Winslow, Production of Polymeric Materials from Polyoxyalkylene Polyols and Butadiene Diepoxide, U.S. 2,897,163.
73. Cooke, A New Chlorinated Epoxy Resin, 15th ANTEC of SPE, New York City, January.
74. Dunbar and Lalonde, Epoxy Resins from Lignite Coal Tars, *Proc. N. Dakota Acad. Sc.*, vol. 13.
75. Forester, Process for the Preparation of Condensation Products, U.S. 2,868,768.
76. Greenlee, Compositions of 4,4-Bis (hydroxyaryl)penta-*n*-amide-ether Acids and Polyepoxides, U.S. 2,893,974.
77. Grover, New Developments in 100% Solids Epoxy Adhesives, 15th ANTEC of SPE.
78. Lidark, Manufacture of High-molecular Weight Resins: Reaction of 2,2′-Bis(*p*-hydroxyphenyl) propane, *Kunststoffe Plastics* **6** (1): 16–19.
79. Luvisi and Schmerling, Preparation of Halogen Substituted Polycyclic Epoxides, U.S. 2,882,279.
80. McKenzie, Glycidyl Ethers of Reaction Products of Certain Phenols with Polymerizable Aromatic Substances Containing at least One Vinyl Group, U.S. 2,865,887.
81. Martin, Epoxy Ethers of Polyhydroxy–Polyhydrocarbyl-substituted Benzenes and Their Polymers, U.S. 2,914,507.
82. Polyepoxy Ethers of Epoxy Substituted Polyhydric Phenols (to Shell Internationale Research Maatschappij N.V.), Brit. 873,016.
83. Process for the Preparation of Epoxy Resins (to N.V. de Bataafsche Petroleum Maatschappij), Brit. 814,695.
84. Rogers, et al., Epoxy Substituted Organosilicon Compounds and Method for Preparing the Same, U.S. 2,883,395.
85. St. Clair, Process for Preparing Epoxy Alkyl Aryl Ethers, U.S. 2,842,849.
86. St. Clair, Epoxy Resin, U.S. 2,892,809.
87. Shechter, et al., Hardener for Epoxy Resin Composition, U.S. 2,897,179.
88. Shokal and Mueller, Polymerization of Di(2,3-epoxypropyl Phthalate), U.S. 2,895,947.

89. Tess, Process for Manufacture of Glycidyl Ethers of Polyhydric Phenols, U.S. 2,879,259.
90. Wegler and Frank, Production of Cured Epoxide Resins, U.S. 2,884,406.
91. Zuppinger, et al., Process for the Manufacture of Reaction Products of Epichlorohydrin and Polyhydric Alcohols, U.S. 2,898,349.

1960

92. Andres, et al., Polyepoxide-Amine Compositions, U.S. 2,921,037.
93. Bader, Resinous Material, U.S. 2,933,472.
94. Belanger, Epoxide Resins, U.S. 2,947,726.
95. Bruin, et al., Process of Producing Improved Epoxy Resin Compositions, U.S. 2,921,923.
96. Budnowski, et al., Adhesive Prepared by Curing Polyepoxy Compounds, Ger. 1,103,579.
97. Chiddix and Wynn, Halogenated Epoxy Resins, U.S. 2,951,829.
98. Cooke, Polyepoxides, etc., U.S. 2,947,725.
99. Farnham, et al., Process for Preparing Glycidyl Polyethers of Polyhydric Phenols, U.S. 2,943,095.
100. Greenlee, Epoxide Acids, U.S. 2,933,471.
101. Kolbel, et al., Epoxy Resins, Ger. 1,091,328.
102. Lo, An Epoxy Resin from Phenolphthalien, *Ind. Eng. Chem.*, April.
103. Maerker, et al., Glycidyl Esters: I, Method of Preparation and Study of Some Reaction Variables, 34th Fall Meeting of American Oil Chemists' Society.
104. Moroson, Epoxy Resins and Process of Making and Curing the Same, U.S. 2,921,049.
105. Newey, Epoxy Esters of Long Chain Acids, U.S. 2,940,986.
106. P. May Help Epoxies, *Chem. Eng. News*, February.
107. Phillips, et al., Epoxide Compositions and Resins Therefrom, U.S. 2,935,488.
108. Reinking, Preparation of Monomeric Glycidyl Polyethers of Polyhydric Phenols, U.S. 2,943,096.
109. Reinking, et al., Glycidyl Derivatives of Amino Phenols, U.S. 2,951,825.
110. Schroeder, Epoxy Ether Esters, U.S. 2,925,426.
111. Schwarzer, Polyphenol Glycidyl Ethers, U.S. 2,965,611.
112. Segal and Rusta, Epoxy Resins from Resorcinol Acetone Condensation Products, *Ind. Eng. Chem.*, April.
113. Wynstra, et al., Glycidyl Ether Reactives in Varnishes from Liquid Epoxy Resins, *Ind. Eng. Chem.*, April.

1961

114. De Acetis, Epithio-ether Polymers, Belg. 609,375.
115. Bergmann, et al., Process for Manufacture of Epoxy Resins, Ger. 1,097,677.
116. Christenson, et al., Epoxy Resins with High Heat Distortions, U.S. 3,001,972.
117. Dickstein and Halpern, High Temperature Resistant Adhesives, AD 266 452.
118. Frank, et al., Glycidyl Polyesters, Ger. 1,104,696.
119. Hart, Interpolymer of Allylepoxy Resin and Polyamide Resin, U.S. 2,967,161.
120. Improvements in Polyepoxy Ethers of Epoxy Substituted Polyhydric Phenols (to Shell Internationale Research Maatschappij, N.V.), Brit. 873,017.
121. Kindall and Susman, High Temperature Resin Developments for Filament Windings, 17th ANTEC SPE.
122. Larkin and St. Clair, Method for Preparation of Glycidyl Ether of Polyhydric Phenols, U.S. 2,995,583.
123. Masters and Cooke, Epoxide Resins, U.S. 2,971,942.
124. Murdock and Schnieder, New Polymers of Unsaturated Glycidyl Ethers and Methods of Producing Them, U.S. 3,000,690.
125. Nevin and Fletcher, Preparing Glycidyl Esters of Higher Fatty Acids, U.S. 2,992,239.
126. Nikles, Epoxide Resin Process and Composition, U.S. 3,006,891.
127. Nikles, Epoxide Resin Compositions, U.S. 3,006,892.
128. Porret, et al., New Epoxy Resins, U.S. 2,992,193.
129. Process for the Preparation of Epoxide Resins, (to Henkel & Cie., G.m.B.H.), Brit. 884,033.

130. Vayser, Synthesis of Polycarbonates and Epoxy Resins, on the Basis of 1,1-(4,4-Dioxy)-diphenyl Ethane, *Himischeskaya Promyshlennost* (*USSR*) **7:** 24–25.
131. Wandelar, Concerning Sulfur-containing and Other Aromatic Substituted Mono and Diepoxies, Ph.D. thesis, Technische Hochschule, Zurich.
132. Zuppinger and Fisch, Glycidyl Ethers of *N*-Methylolamino-1:3:5-triazines, U.S. 2,980,676.

1962

133. Batzer and Nikles, Synthesis of New Epoxy Compounds, *Chimia* **16**:57–66.
134. Chen, et al., Development of High Temperature Laminating Resins, Progress Report 1, Contract AF 33 (657)-9063, Project 920K, June–August.
135. Epoxy Resins, (to Ciba, Ltd.), Belg. 612,682.
136. Fisch, Reaction of Epichlorohydrin with Bisphenols (Kinetics of Resin Manufacture), *Chimia* **16**:66–71.
137. Garber, et al., 2,5-Bis(2,3-epoxypropoxy-methyl)tetrahydrofuran, U.S. 3,025,307.
138. Glycidyl Ethers (to Ciba, Ltd.), Brit. 905,045.
139. Greenfield, Development of High Temperature Resistant Polymers, AD 285 370.
140. Hoppe, Process for the Production of Synthetic Resins, U.S. 3,067,171.
141. Improvements in Epoxy Resins (to Inventa A.G. für Forschung and Patentverwertung), Brit. 913,794.
142. Improvements in High Molecular Weight Resins (to Devoe & Raynolds, Inc.), Brit. 887,609.
143. Isagulyants and Balashova, Epoxy Resins Based on Substituted Epichlorohydrins, *Z. Prikl. Khim.*, November.
144. Lo, Epoxy Resin Compositions, U.S. 3,015,648.
145. Low Viscosity Diglycidyl Ether Polymers (to Pittsburgh Plate Glass Co.), Brit. 901,208.
146. Morrison, Polyglycidyl Ethers of Ether Anhydro Hexitols, Method of Production, and Aqueous Solutions Thereof, U.S. 3,041,300.
147. Nametz, Plastic Product Comprising Cured Mixture of a Brominated Polyepoxide and a Non-halogenated Polyepoxide, U.S. 3,058,946.
148. Newey, Flexible Epoxy Resin Compositions and Cured Products, U.S. 3,057,809.
149. Pannell, Epoxy Ethers, Their Preparation and Cure, U.S. 3,058,921.
150. Parret and Leuman, New Flame Resistant Epoxy Resins, *Chimia*, March.
151. Partansky, Process for Inhibiting Crystallization of the Diglycidyl Ether of Bisphenol A Comprising Incorporating Therein an Epoxylated Novolac Resin and Composition Obtained Thereby, U.S. 3,051,681.
152. Shokal, et al., Allyl Glycidyl Ether and Vinyl Acetate Copolymers, U.S. 3,040,010.

1963

153. Amine RC-94, Shell Chemical Co. Report PD-147, April.
154. Belaya, et al., Epoxide Resins Based on Dinaphthylolmethane and Its Mixtures with Diphenylolpropane, *Sintetich, Polikondensatsionnye Smoly., Urk. Nauchn-Issled. Inst. Plast. Mass. Sb. Statei*, pp. 84–94.
155. Bremmer, Diglycidyl Ethers of 2,2,2,2,-Trichloroethylidene Bisphenols, U.S. 3,102,874.
156. Brzozowski and Porejko, Self Extinguishing Epoxy Resins, Pol. 47,344.
157. Coating and Waterproofing Compositions (Th. Goldschmidt A.G.), Belg. 632,533.
158. Cohen and Haas, Synthesis of Polyglycidyl Ethers, *J. Am. Chem. Soc.*, April.
159. DeHoff, Effect of Various Curing Agents on the Corrosion Resistance of Epoxy Resins, Plastics Symposium, 16th Annual NACE Conference, Dallas.
160. Epoxidized Polycarbonates (to General Electric Co.), Brit. 942,538.
161. Fourie, et al., Production of High Purity Diepoxide, U.S. 3,093,661.
162. Kakiuchi and Tanaka, *Kobunshi Kagaku* **20**(222):619–628.
163. Maerker, et al., Reaction of Diglycidyl Esters and Cyclic Carboxylic Anhydrides, *J. App. Polymer Sci.* **7**:301–307.
164. Masters, Polymers of Glycidyl Polyethers, Belg. 633,173.
165. Moshchinskaya and Vishnevetskii, Diglycidyl Ester of 2,2′-Dihydroxyl-1,1′-dinaphthylmethane, U.S.S.R. 159,651.

166. Myers and Dagon, Distribution of Molecular Weight and Branching in High Molecular Weight Polymers from Bisphenol A and Epichlorohydrin, ACS Symposium, Los Angeles, April.
167. Newey and Holm, Epoxy Esters of Polyhalopolycyclic Acids, U.S. 3,106,536.
168. Novenko and Gavurina, Epoxy Resin, U.S.S.R. 159,649.
169. Payne, Epoxy Acetals, U.S. 3,107,253.
170. Process for Manufacture of Epoxy Resins (to Inventa A.G. für Forshung and Patentverwertung), Brit. 920,520.
171. Process for the Production of Hardenable Synthetic Epoxy Resins (to Inventa A.G. für Forshung and Patentverwertung), Brit. 923,772.
172. Reinforced Plastic Laminates for Long Time High Temperature Use, AD 451 221, April.
173. Reinking, et al., Polyhydroxyether: I, Effect of Structure on Properties of High Molecular Weight Copolymers from Dihydric Phenol Mixtures and Epichlorohydrin, ACS Symposium, Los Angeles, April.
174. Reinking, et al., Polyhydroxyether: II, Effect of Structure on Properties of High Molecular Weight Copolymers from Dihydric Phenol Mixtures and Epichlorohydrin, ACS Symposium, Los Angeles, April.
175. Schwarzer, Polyepoxy Ethers of Polyhydric Phenols and Cured Products Obtained Therefrom, U.S. 3,092,610.
176. Solov'eva, et al., Epoxy Resins with High Heat Stability, *Tr. Mosk. Khim-Tekhnol. Inst.*, vol. 42.

1964

177. Bisphenols in the Formulation of Epoxy Resin Adhesives (to Armour & Co.), Brit. 959,286.
178. Boozalis and McKeever, Tri-bisphenol-A Ester of Phosphoric Acid, U.S. 3,146,254.
179. Bremmer, Heat Stability of Brominated Epoxy Resins, *Ind. Eng. Chem. Prod. R & D*, March.
180. Brode and Wynstra, A Product Analysis of the Base Catalyzed Phenol-Epichlorohydrin Condensation, ACS Symposium, Chicago, September.
181. Burge and Hallstrom, Polyglycidyl Polyepichlorohydrin Resins for Casting and Laminating: an Evaluation, *SPE J.*, January.
182. Harvey, Compositions Comprising a Glycidyl Polyether and the Reaction Product of an Acetone-Formaldehyde Condensate and a Polyalkylene Glycol, U.S. 3,121,703.
183. Kamath, et al., Epoxy Resins Based on Cashew Nut Shell Liquid, *Paintindia* **14**(4):21–27.
184. Kornev and Litvinenko, Epoxy Monomers from Bis(methylolamides) of Dicarboxylic Acids and Napthols and Phenols, *Khim. Prom. Inform. Nauk-Tekhn. Zb.*, no. 3, pp. 36–37.
185. Kuznetsov, et al., Phosphorus-containing Epoxide Resins, U.S.S.R. 163,351.
186. Loshaek and Blommers, Epoxidized Polyamides, U.S. 3,125,552.
187. McClure, Derivatives of Dodecahydro-1,4,7,9*b*-tetraazaphenalene, U.S. 3,134,777.
188. Madden, et al., Designing Epoxy Resins Specifically for Filament Wound Deep Submergence Applications, 19th NANTEC (RIP Division) of SPI.
189. Norris, et al., Research on Improved Epoxy Resins, USNRL Report, Contract Nonr-4172(00)(x), Feb. 1.
190. Payne and Williams, Glycidyl Glycidate and the Preparation to Epoxyalkyl Esters of Epoxyalkanoic Acids from Epoxyalkanols, U.S. 3,136,788.
191. Perfetti, Epoxy Compositions, U.S. 3,145,191.
192. Potnis and Venugopal, Bhilawan Nutshell Liquid, part III, *Paintindia* **14**(4):29–36.
193. Ross, et al., Some Reactions of Epichlorohydrin and Amines, *J. Org. Chem.* **29**:824.
194. St. Clair and Waldron, Curable Tar-Resorcinol Still Residue Epoxy Ether Compositions, U.S. 3,133,033.
195. Schade, Glycidyl Esters of Terephthalic and Isophthalic Acids, Ger. 1,176,122.
196. Sidlow and Williams, Self-extinguishing Epoxide Polymers, Brit. 953,206.
197. Srinivasan and Kapur, Epichlorohydrin Modification of Shellac, *Indian J. Technol.* **2**(2):71–72.

198. Wende and Priebe, 2-Ethoxy-4,6-bis(β-glycidoxyethoxy)-1,3,5-triazine, Ger. 1,166,785.
199. Wiesner, et al., Isolation of Pure Diglycidyl Ethers of Tetrachlorodian and Tetrabromodian, Czech. 110,873.
200. Wohnsiedler, Process for Preparing Epoxy Alkyloxymethylamino-*s*-triazines, U.S. 3,145,207.

1965

201. Budnowski, Epoxy Resins from Triglycidyl Isocyanurate, *Kunststoffe* vol. 55, no. 8.
202. Burness, Photographic Composition Containing Epoxy Hardening Agents, U.S. 3,189,459.
203. Epoxy Adhesive and Coating Compositions for Chlorine-containing Thermoplastics (to Dynamit-Nobel A-G), Brit. 988,770.
204. Epoxy Alkyl Esters of Dicarboxylic Acids Containing Tertiary or Quaternary Carbon Atoms Adjacent to the Carboxylic Groups (to Shell Internationale Research), Neth. Appl. 286,209.
205. Epoxy Ethers (to Shell Internationale Research), Neth. Appl. 6,413,493.
206. Fielding and Ray, Phosphorus-containing Epoxy Resins, Brit. 982,933.
207. Jedlinski and Tokarzewska, Glycidyl Ether Reactions: II, Synthesis of Thermostable Epoxy Resins from Polynaphthols, *Makromol. Chem.* **83**:226–233.
208. Khrlakyan, et al., Fluorinated Mono- and Diepoxy Compounds, *Izv. Akad. Nauk, SSSR, Ser. Khim.* **1965**(1):72–75.
209. Kriz and Wiesner, Light-colored Epoxy Resins, Czech. 111,662.
210. Lewis, Low-melt-viscosity Epoxide Resins, Brit. 980,389.
211. Morgan, Condensation Polymers: by Interfacial and Solution Methods, p. 421, Interscience.
212. Nakano, et al., Epoxy Resins, Japan 7063(65).
213. Valles, et al., Halogenated Epoxy Type Resins: I, Preparation and Determination of Structure of Halogenated Bisphenol Type Compounds, ACS Symposium, Atlantic City, September.

Chapter 3

SYNTHESIS OF EPOXY RESINS FROM OLEFINS AND CHLOROACETYLS

The previous chapter discussed the synthesis of epoxy resins: (1) from active-hydrogen-bearing materials, wherein the epoxy content of the molecule was established by an epoxidation mechanism involving epichlorohydrin, and (2) from monoepoxies other than epichlorohydrin by synthesis routes which did not involve the consumption and regeneration of the epoxy groups.

This chapter discusses epoxidation of olefins and synthesis of resins from chloroacetyls.

Various olefin groups can be converted to epoxy groups, as noted below:

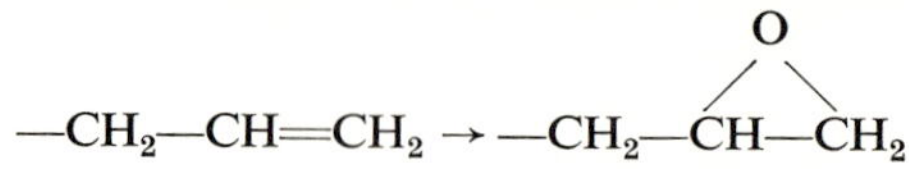

Allyl group — Glycidyl-type epoxy (a terminal epoxy)

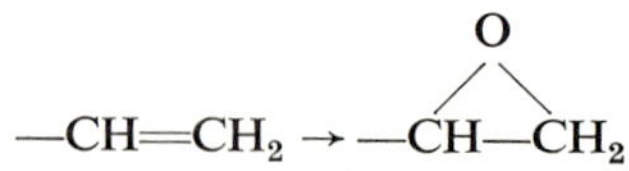

Vinyl group — 1,2-Epoxy (a terminal epoxy)

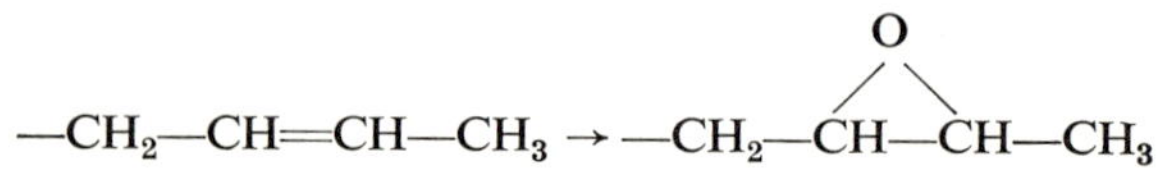

Olefin — 2,3-Epoxy, etc. (an internal epoxy)

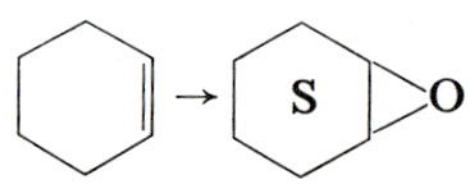

Cyclohexene — 1,2-Epoxycyclohexane (ring-situated epoxy)

In its simplest form, epoxidation of an olefin is represented as follows:

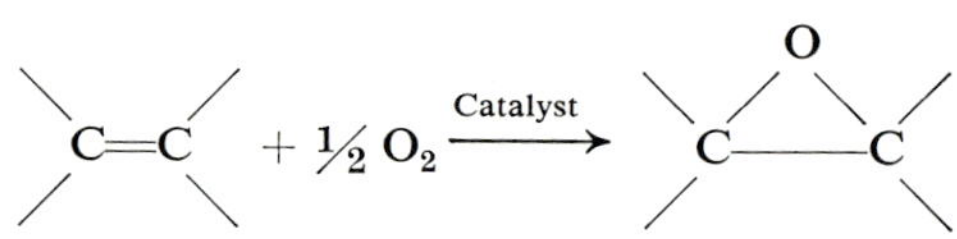

Olefin — Oxygen — Epoxy

This procedure is effective when converting ethylene ($CH_2{=}CH_2$) to ethylene oxide, in which case the reaction can be carried to about 60 per cent completion at temperatures from 90 to 150°C, in the presence of a heavy metal hydride catalyst [10]. Higher olefins are less efficiently epoxidized by this process.

Less direct methods are used in the synthesis of epoxy resins: the first involving the conversion of the olefin to a chlorohydrin group, followed by dehydrohalogenation; and the second, involving the direct addition of oxygen supplied from a source compound, most frequently a peracid, either preformed or formed in situ. The various methods by which olefins are epoxidized are indicated in Table 3-1.

Table 3-1. Methods for the Conversion of Olefins into 1,2-Epoxides [108]

1. Epoxidation by organic peracids and their derivatives:
   a. With preformed peracids
   b. In situ methods
2. Epoxidation by inorganic peracids
3. Epoxidation via halohydrins and their derivatives:
   a. With molecular halogen
   b. With preformed hypohalogenic acids
   c. With esters of hypochloric acid
   d. With other reagents which are able to donate positive halogen
   e. With chromyl chloride
4. Epoxidation by alkaline hydrogen peroxide
5. Epoxidation by various inorganic reagents:
   a. Oxygen
   b. Ozone
   c. Chromic acid, permanganate
6. Biological epoxidations

With regard to biological epoxidation, as indicated in the table, diepoxy species have been reported in nature, *trans-β*-carotene diepoxy having been identified in the extracts of the fresh blossoms of *C. moschata var.* Yellow Sultan [124]. A diepoxy has been isolated as substance D from the fruit of *Hyenanche globosa* [133], and epoxidized linalool has been found in the essential oils from freshly picked tea leaves and black tea [128]. Except for the biological production of monoepoxy compounds from some of the chlorinated insecticides (e.g., aldrin, isodrin, and heptachlor), there were no reports of simple foreign compounds being epoxidized in animals until 1965, when it was shown that rat-liver microsomal systems could produce a water-soluble epoxy from 1,4-epoxy-1,4-dihydronaphthalene [136].

In 1966 it was announced that a natural source of epoxy fatty acids had been discovered in seeds of *Euphorbia lagascae*, a plant that grows wild in southern Spain. A second wild plant, *Veronia anthelmintica*, also has been found to produce natural epoxidized fatty acids. The *Euphorbia* is twice as rich in oil as the *Veronia* and about 60 per cent of the oil is epoxy fatty acids. In the crude state, the oils may be used in low concentrations as stabilizers for plastics. Further laboratory epoxidation is required to obtain more highly epoxidized molecules normally required, but this can be done quickly and at less cost than with the completely nonepoxidized oil starting materials based on commercial sources [138].

In this chapter, the discussion is confined to the diepoxy compounds which have been suggested for their potential usefulness in the epoxy-resin technology. A variety of diepoxy resins, not referenced here, may be found in the periodic volumes issued by the Cancer Chemotherapy National Service Center of the National Institutes of Health under the report title, A Survey of Alkylating Agents.

EPOXIDATION BY HYPOCHLOROUS ACID

Olefins may be converted to chlorohydrins by treatment with hypochlorous acid. The chlorohydrin is then treated with base (e.g., NaOH) to strip off the hydrogen and the chlorine:

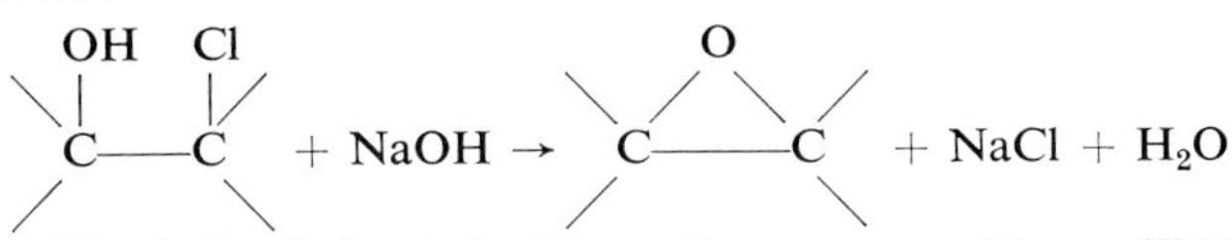

Chlorohydrin Sodium hydroxide Epoxy Salt Water

This procedure, dehydrochlorination, with the attendant difficulties involved in controlling the reaction and in removing the by-products from the resin, was discussed in Chap. 2.

In addition to the production of monoepoxies such as epichlorohydrin, the reaction is of commercial importance in the synthesis of butadiene dioxide from butadiene:

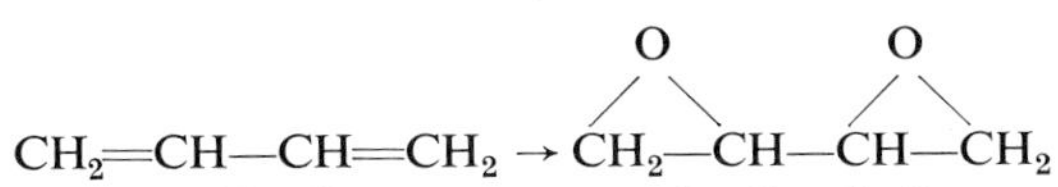

Butadiene Butadiene dioxide

Hypochlorous acid is used in cases of conjugated unsaturation, as above, because the second unsaturation is resistant to direct addition of oxygen via the peracid route [5]. However, when the polyunsaturation is not conjugated, the synthesis via hypochlorous acid is generally not so attractive as it is with the peracids. There have been numerous descriptions of such use, however, for example in the epoxidation of vinylcyclohexene [6,7,21] and aliphatic polyethers [9]. In place of hypochlorous acid, *tert*-butyl hypochlorite produces stereospecific monoepoxy isomers, useful as epoxy resin diluents [131].

EPOXY RESINS FROM ACETYL COMPOUNDS VIA CHLOROHYDRIN INTERMEDIATE

By reducing chloroacetyl compounds to chlorohydrin compounds it is possible to obtain materials suitable for dehydrohalogenation. A typical route is as follows:

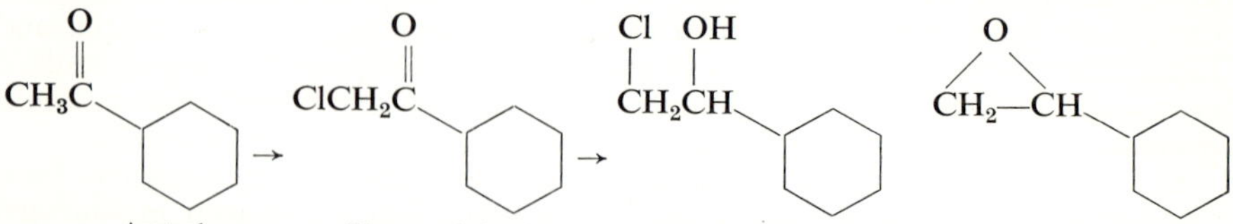

Acetophenone — Phenacetyl chloride — 2-Chloro-1-phenylethanol — Styrene oxide

By this route, a number of epoxy resins, as shown in Table 3-2, have been synthesized.

Table 3-2. Epoxy Resins Derived from Chloroacetyls [50]

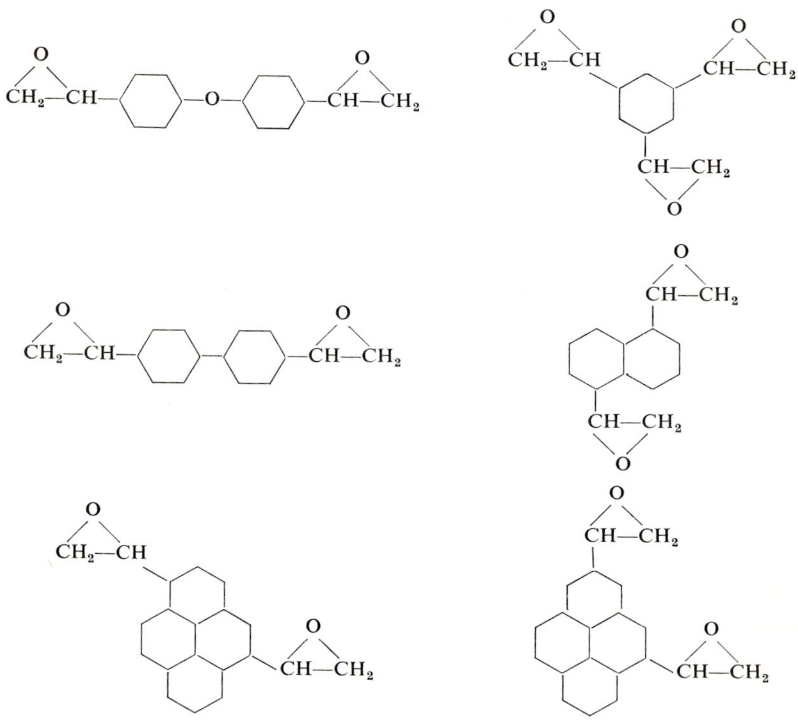

Similarly, epoxy resins can be produced by the Darzens condensation. Under basic conditions, haloesters and ketones or aldehydes give glycidic esters (with a chlorohydrin postulated as the intermediate):

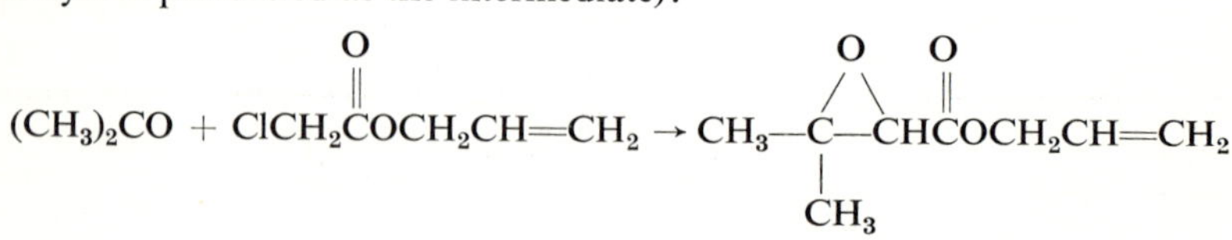

These can then be polymerized through the unsaturation to give the polyepoxy compound [137].

EPOXIDATION BY ACETALDEHYDE MONOPERACETATE

Acetaldehyde monoperacetate (AMP) can be produced as an intermediate in the

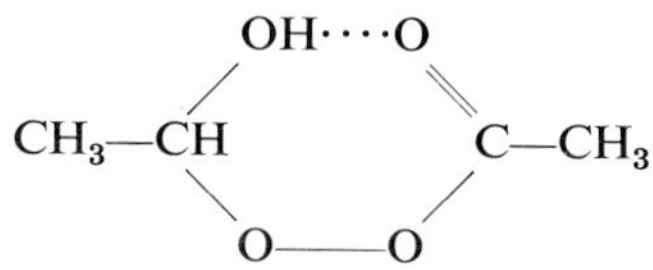

AMP

synthesis of acetic acid from acetaldehyde and molecular oxygen if the reaction is conducted at low temperature (−10°C) and catalyzed by ultraviolet light:

$$2CH_3CHO + O_2 \rightarrow AMP$$

Acetaldehyde oxygen

AMP is a crystalline solid, impact-sensitive, and explosive, and decomposes spontaneously and exothermically to acetic acid at room temperature.

In order to use AMP as the oxidizing agent, therefore, it is dissolved in an inert solvent to facilitate handling and is reacted with the unsaturated molecule at a temperature well below the decomposition point of the AMP [17,30].

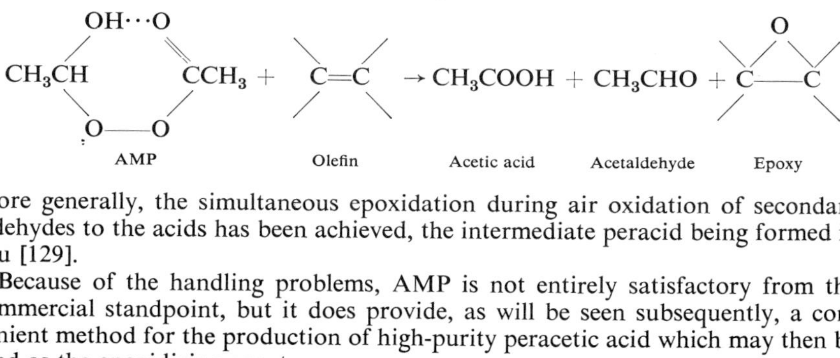

AMP Olefin Acetic acid Acetaldehyde Epoxy

More generally, the simultaneous epoxidation during air oxidation of secondary aldehydes to the acids has been achieved, the intermediate peracid being formed in situ [129].

Because of the handling problems, AMP is not entirely satisfactory from the commercial standpoint, but it does provide, as will be seen subsequently, a convenient method for the production of high-purity peracetic acid which may then be used as the epoxidizing agent.

EPOXIDATION BY PREFORMED PERACIDS

Shortly after the turn of the twentieth century it was found that perbenzoic acid could be made to react with olefins to give virtually quantitative yields of epoxy groups and benzoic acid.

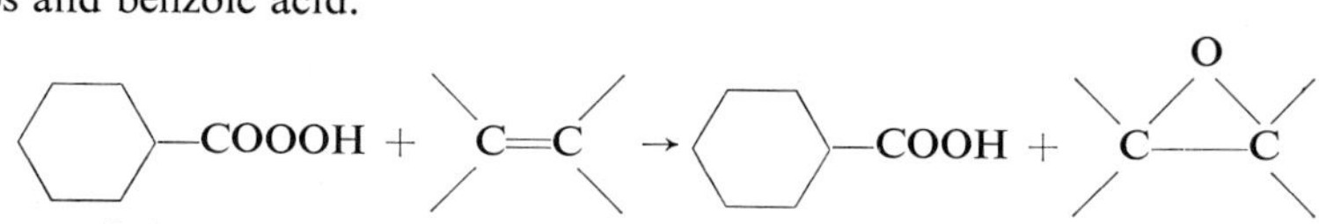

Perbenzoic acid Olefin Benzoic acid Epoxy

Although the reaction is specific and efficient, it is not commercially attractive because of the cost of perbenzoic acid.

Other peracids have subsequently been investigated [63], and the epoxidation reaction has been observed to occur, although with less efficiency. Of the organic peracids the most attractive from the commercial standpoint is peracetic acid. In addition to the AMP route, there are a number of preparations of organic peracids, as indicated in Table 3-3.

Table 3-3. Methods for the Preparation of Organic Peracids [108]

1. Carboxylic acids + hydrogen peroxide:

$$R{-}COOH + H_2O_2 \rightleftharpoons R{-}COOOH + H_2O$$

   a. With or without acid catalyst: equilibrium peracids
   b. Equilibrium displaced in favor of the peracid by removing water:
      (i) By addition of carboxylic anhydride
      (ii) With concentrated sulfuric acid
      (iii) By azeotropic distillation
   c. Equilibrium continuously displaced by consumption of the peracid: in situ method
2. Carboxylic anhydride + hydrogen peroxide (or inorganic peroxides):

$$R{-}\overset{\overset{\displaystyle O}{\|}}{C}{-}O{-}\overset{\overset{\displaystyle O}{\|}}{C}{-}R + H_2O_2 \rightarrow RCOOOH + RCOOH$$

3. Ester of carboxylic acid + hydrogen peroxide:

$$R{-}COOR' + H_2O_2 \rightleftharpoons R{-}COOOH + R'OH$$

4. Acid chloride + hydrogen peroxide (or inorganic peroxides):

$$R{-}COCl + H_2O_2 \rightarrow RCOOOH + HCl$$

5. Ketene + hydrogen peroxide:

$$CH_2{=}C{=}O + H_2O_2 \rightarrow CH_3COOOH$$

6. From diacyl peroxide:

   a. $$R{-}CO{-}OO{-}CO{-}R + H_2O \rightarrow R{-}COOOH + RCOOH$$

   b. $$R{-}CO{-}OO{-}CO{-}R + H_2O_2 \rightarrow 2R{-}COOOH$$

Synthesis of Peracetic Acid

Peracetic acid is synthesized commercially from acetic acid and hydrogen peroxide, or it may be generated from AMP.

Peracetic Acid from Acetic Acid. Peracetic acid is produced from acetic acid and hydrogen peroxide in the presence of an acid catalyst. The reaction is shown in Table 3-3.

The resultant peracetic acid is diluted with water and with the ingredients of the synthesis. The formed peracetic acid is too unstable to permit efficient removal of these impurities.

Peracetic Acid from Acetaldehyde Monoperacetate. If decomposition of AMP is conducted under reduced pressure and in the presence of an acid catalyst [31], acetaldehyde and peracetic acid are produced as intermediate products. If the acetaldehyde is removed as quickly as it is formed, peracetic acid is isolated [1]. Removal of acetaldehyde can be accomplished by conducting the reaction in the vapor phase, and only small amounts of acetic acid, produced from the reaction of peracetic acid and acetaldehyde, will be present.

$$\underset{\text{Peracetic acid}}{CH_3COOOH} + \underset{\text{Acetaldehyde}}{CH_3CHO} \rightarrow \underset{\text{Acetic acid}}{2CH_3COOH}$$

Effect of Impurities and Formed Acetic Acid

The impurities in the peracetic acid derived from acetic acid and hydrogen peroxide react with the epoxy groups as follows:

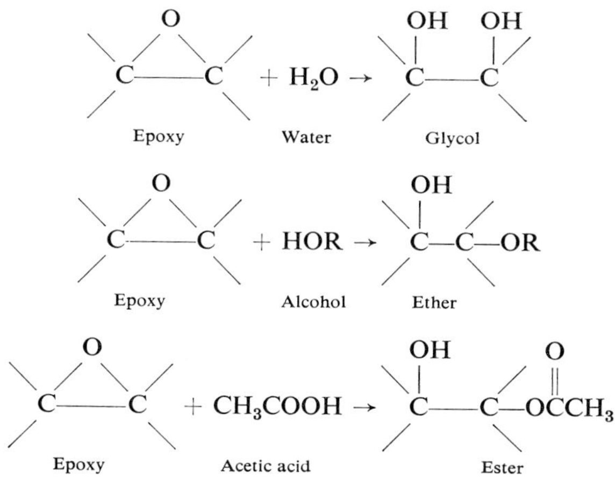

These reactions can be minimized by using a commercial grade of peracetic acid in which the sulfuric acid catalyst has been neutralized. The neutralized product is suitable for the epoxidation of natural oils by using low temperatures and short reaction times [11]; however, it has not successfully produced the more reactive epoxy groups.

If the peracetic acid from AMP is employed, side reactions are less of a problem. However, because acetic acid is formed as a by-product, they are not entirely eliminated. If the peracetic acid is supplied in an inert solvent (e.g., diethyl ether), the formation of glycol monoacetate, the main, initial side reaction will be suppressed [28]. To prevent subsequent reactions, all residual acetic acid should be removed from the resin as quickly as possible. One method suggested involves washing the resin in an inert solvent containing a quaternary ammonium ion-exchange resin [87].

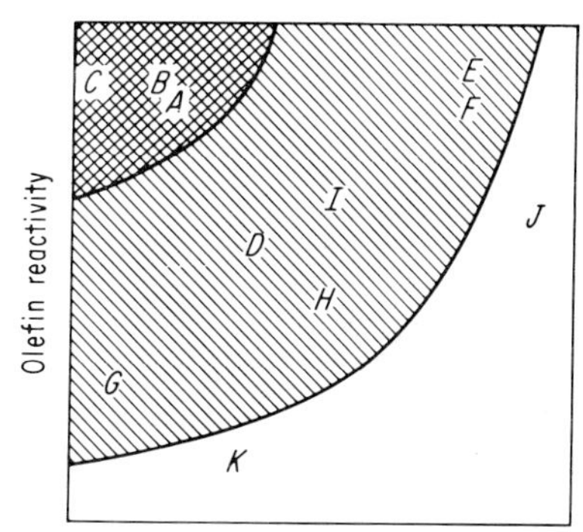

FIG. 3-1. Relative reactivities of various olefin groups with peracetic acid [28]. *A*, Soybean oil epoxide; *B*, alkyl epoxystearates; *C*, dicyclopentadiene dioxide; *D*, cyclohexene oxide; *E*, styrene oxide; *F*, diisobutylene oxide; *G*, 1,4-dichloro-2,3-epoxybutane; *H*, vinylcyclohexene dioxide; *I*, 3,4-epoxy-6-methylcyclohexylmethyl-3-4-epoxy-6-methylcyclohexane carboxylate; *J*, vinyl ether epoxides; *K*, ethylene oxide.

The relatively pure peracetic acid from AMP is a considerably more effective agent for the epoxidation of olefins than is the impure peracetic acid from acetic acid. Even so, when epoxidizing less reactive olefins, for example a second group of conjugated unsaturation, the reaction conditions required may be sufficiently severe to render the synthesis impractical, the epoxy groups being opened as quickly as formed [53]. At the other extreme, it is not possible to synthesize extremely reactive epoxy groups (for example vinyl ethers), since the normal conditions of reaction will promote their destruction [26]. A reactivity curve is presented in Fig. 3-1. Generally, alkyl substituents on the olefinic carbons and high acid concentrations favor isomerization of the epoxy groups to carbonyl groups.

The by-product of epoxidation with peracetic acid is equal molar quantities of acetic acid. When AMP is used, the peracetic acid is thus obtained virtually without raw-material cost as an intermediate in the synthesis of acetic acid from acetaldehyde and molecular oxygen. When acetic acid is used as the precursor of peracetic acid, the acetic acid may be recovered for reuse.

EPOXIDATION BY PERACIDS DERIVED IN SITU

Epoxidation by peracid derived in situ proceeds as follows:

$$H_2O_2 + RCOOH \rightarrow RCOOOH + H_2O$$

Hydrogen peroxide Organic acid Organic peracid Water

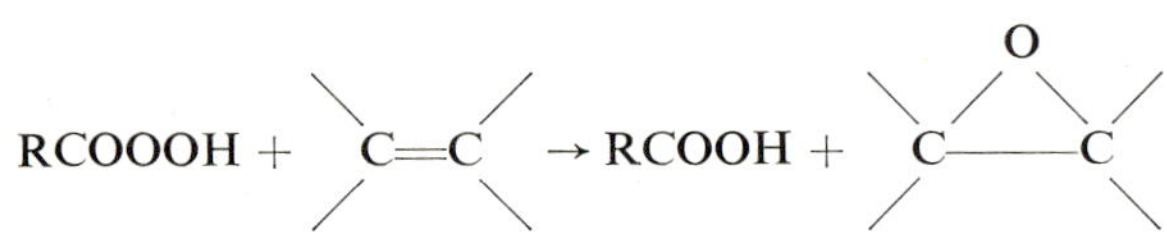

Organic peracid Olefin Organic acid Epoxy

Various acids have been suggested for use with the in situ technique, acetic acid being preferred from the commercial standpoint. The use of acetic acid, however, is somewhat hazardous [71], and *p*-nitroperbenzoic acid has been suggested as more suitable on the laboratory scale. Sulfuric acid is a suitable catalyst [25, 96], as are alkaline metal salts [40].

It is evident that the objections to the use of preformed peracetic acid are applicable to the use of the peracetic acid formed in situ, since the same by-products will be present.

The side reaction of these by-products with epoxy groups may be minimized by conducting the synthesis in such a way as to maintain phase separation between the epoxy compound and the peracetic acid reactants. This can be accomplished by the use of a large excess of water when the material being epoxidized is substantially water-insoluble, as with natural oils [4]; or, alternatively, anion [48] or cation [34] exchange resins may be used. If sulfonated polystyrene is used, for example, the bulk of the ion-exchange resin is such that it contains the majority of acetic acid, hydrogen peroxide, and sulfonic acid catalyst while excluding the higher-weight molecules being epoxidized. The ion-exchange catalyst is pretreated to rid it of metal, and a metal sequestering agent is used during the reactions. The ion-exchange resin may be recovered at the conclusion of synthesis.

In synthesizing water-sensitive epoxy molecules, it is reported advantageous to use the dehydrated form of the cation-exchange resin, the exchange resin additionally preventing emulsification by acting as a desiccant [74].

It is claimed that the in situ process may be improved by the addition of a small amount of acid anhydride to the acetic acid/hydrogen peroxide mixture [114]. Improved yields are obtained from the epoxidation reaction if the initial reaction is conducted at 20 to 25°C for a few hours, followed by ½ to 2 hours at 50 to 60°C. Eighty per cent of the epoxidation is accomplished at the lower temperature, and the higher temperature is required only for the last 10 to 15 per cent of the reaction [12].

EPOXIDATION BY INORGANIC PERACID

Use of a tungstic acid catalyst deposited on an inert solid support has been reported as permitting the direct epoxidation of olefins by hydrogen peroxide [45]. Pertungstic acid is the intermediate. Epoxidation of polyunsaturated materials and the very reactive vinyl ethers can be accomplished by this process.

SYNTHESIS OF HYDROXYLATED RESINS

Like the higher-molecular-weight epoxy resins prepared via the epichlorohydrin route, some of the resins from the peracid route are of primary interest for the numerous hydroxyls rather than for the epoxy groups per se. In these instances the reaction conditions may be made more severe, to promote ring opening by water to give hydroxylated products. The monoacetate resulting from reaction of acetic acid with the epoxy may be treated with caustic or water to obtain a hydroxyl group at the same position:

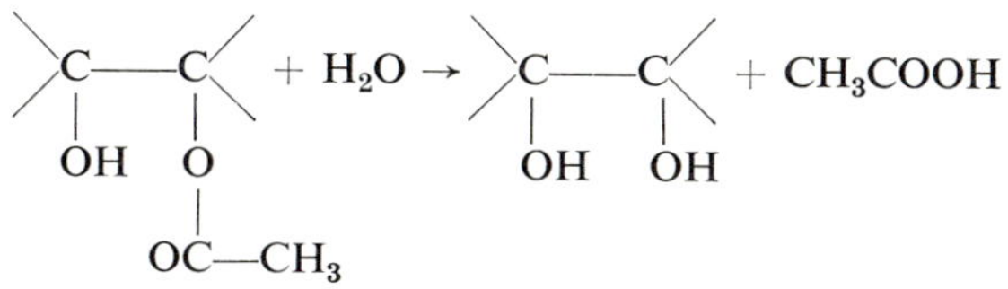

Glycol monoacetate   Water   Glycol   Acetic acid

Such hydroxylated products are of value as vinyl plasticizers.

POLYOLEFINIC RAW MATERIALS FOR SYNTHESIS OF EPOXY RESINS

Unsaturated raw materials for the production of epoxy resins are obtained from a variety of sources. The chief requirements are that they contain two or more olefin groups through which the oxidation can be conducted, and that the formed epoxy groups are stable to the conditions of the synthesis.

As a special case, epoxy resins can be derived from molecules containing only one carbon-to-carbon unsaturation, provided that epoxy groups or reactive groups suitable for epoxidation are also present in the molecules. Examples of these are given in Chap. 2.

The peracids are considered to be electrophilic, and consequently the more electronegative the olefin, the more readily it may be epoxidized. Thus, if substituents have electron-donating properties (e.g., alkyl groups) the epoxidation is facilitated; whereas if the substituents have electron-withdrawing properties (e.g., carboxyl groups), the epoxidation is retarded. Allyl ethers, because of the electron-attracting properties of the ethereal oxygen atom, are not readily epoxidized by peracetic acid; hence glycidyl ethers are not often produced by this route [108].

A wide variety of unsaturated materials are available for epoxidation and are discussed by types in the following sections.

Animal or Vegetable Oil

In natural oils (e.g., triesters of glycerol and long-chain unsaturated acids, usually C_{16}, and C_{18}, and C_{20} types) each acid chain usually contains one or more unsaturated linkages.

Partial view of panel and reactor used for production of commercial epoxidized oils. (*Swift & Co.*)

The composition of a typical triglyceride, soybean oil, is shown in Table 3-4. It is evident that with such natural oils some percentage of the composition will be nonreactive with the peracid used for synthesis, insofar as this percentage contains no unsaturation. It is likewise evident that of the unsaturation, some percentage will remain unreacted during the conditions of synthesis, and some percentage of the epoxy groups formed will undergo ring opening; thus the resulting resin will consist of a variety of species, differing in chemical reactivity.

The epoxidation of natural oils is usually conducted with the commercial peracetic acid available in mixture with acetic acid [8] or by the in situ process [33], although peracetic acid in solvent may be used [29]. The oils may be heat-treated in advance of epoxidation [130].

An example of the laboratory preparation of epoxidized lanolin using the in situ process is given in detail by way of illustration. Anhydrous lanolin of the following

Table 3-4. Chemical Composition of a Typical Triglyceride, Soybean Oil

Acidic component	Formula	Type	Weight %
Palmitic	$C_{16}H_{32}O_2$	Saturated	8.25
Stearic	$C_{18}H_{36}O_2$	Saturated	5.35
Arachidic	$C_{20}H_{40}O_2$	Saturated	0.90
Oleic	$C_{18}H_{34}O_2$	Unsaturated	24.90
Linoleic	$C_{18}H_{32}O_2$	Unsaturated	56.65
Linolenic	$C_{18}H_{30}O_2$	Unsaturated	7.90
Arachidonic	$C_{20}H_{32}O_2$	Unsaturated	0.05

characteristics was employed:

Iodine number (Hanus)	31.7 per cent
Acid number	2.5
Hydroxyl value	21
Hydroxyl content	0.68

To a 2-liter flask equipped with agitator, dropping funnel, reflux condenser and thermometer, were added 876 g lanolin, 220 g *n*-hexane, and 0.2–2 g 90 per cent formic acid (equivalent to 0.3 mole ratio formic acid/H_2O_2). The flask contents were warmed to obtain complete solution and then cooled to 25°C.

From a dropping funnel was added 89.4 g of 50 per cent H_2O_2 (1.32 mole) over a 1-hour period with agitation and cooling to maintain a temperature of 25°C. This temperature

Table 3-5. Epoxidation of Soybean Oil [53]

Iodine value	Extent of saturation, %	Theoretical oxirane value at point of saturation	Oxirane value found	Efficiency of epoxidation, %
130	0	0	0	
110	15.4	1.17	1.00	85.5
89.7	31.0	2.35	2.33	99.2
61.5	52.7	3.99	3.68	92.2
55.8	57.1	4.32	4.15	96.2
49.4	62.0	4.69	4.43	94.4
43.6	66.5	5.03	4.72	94.0
38.1	70.7	5.35	4.95	92.5
34.2	73.7	5.58	5.25	94.0
31.1	76.1	5.76	5.50	95.5
27.6	78.8	5.96	5.71	95.8
23.9	81.6	6.18	5.81	94.1
20.0	84.6	6.40	5.95	93.0
15.9	87.8	6.64	6.00	90.4
13.6	89.5	6.77	6.17	91.1
11.1	91.5	6.92	6.37	92.0
8.8	93.2	7.05	6.56	93.1
7.0	94.6	7.16	6.55	91.4
5.9	95.5	7.23	6.57	90.7
4.7	96.4	7.30	6.49	89.0
4.0	96.9	7.33	6.50	88.7
3.5	97.3	7.36	6.44	87.5
3.1	97.6	7.39	6.54	88.5
3.0	97.7	7.40	6.56	88.7
0	100	7.57		

was maintained with stirring for 6 hours, after which stirring was continued at room temperature (30–35°C) for an additional 16 hours. Titration of a sample of the reaction mixture showed that 65 per cent of the peroxide had been reacted.

The reaction mixture was diluted with 2 liters of benzene and the formic acid removed by extracting twice with 500-ml portions of water. The benzene solution was dried with 40 g anhydrous $MgSO_4$ and the $MgSO_4$ removed by filtration. The last traces of acid were then neutralized by stirring with 17 g finely divided $Ca(OH)_2$ for one half hour at 60°C, and the $Ca(OH)_2$ removed by filtration.

The major portion of benzene was then distilled off under water pump vacuum, using a maximum product temperature of 65°C. Benzene removal was completed by holding the product for 3 hours at 60–65°C and 1–2 mm Hg (mechanical vacuum pump).

Product analysis showed the following characteristics:

Oxirane (epoxide) oxygen	0.79 per cent
Iodine Number	27
Acid Number	1.2

A 79 per cent yield of oxirane oxygen was calculated based on an estimated unsaturated sterol content of 15 per cent cholesterol, 7.5 per cent lanosterol and 2.5 per cent agnosterol in the starting lanolin [56].

The epoxidation reaction is generally run under mild conditions to avoid ring opening of the formed epoxy groups. As the extent of epoxidation increases, the efficiency of the reaction is, to some extent, reduced. This is indicated in Table 3-5.

The effects of operating variables on the in situ epoxidation of anchovy oil have been investigated. An epoxidized oil with 7.8 per cent epoxy content and an I_2 value of 14.1 can be prepared when operating at 53 to 73°C by using a 33 weight per cent hydrogen peroxide solution in a proportion of 1.4 moles per mole of ethylenic unsaturation, a 15 per cent concentration of styrenesulfonic resin as a catalyst, and acetic acid in the proportion of 0.5 mole per mole ethylenic unsaturation [127].

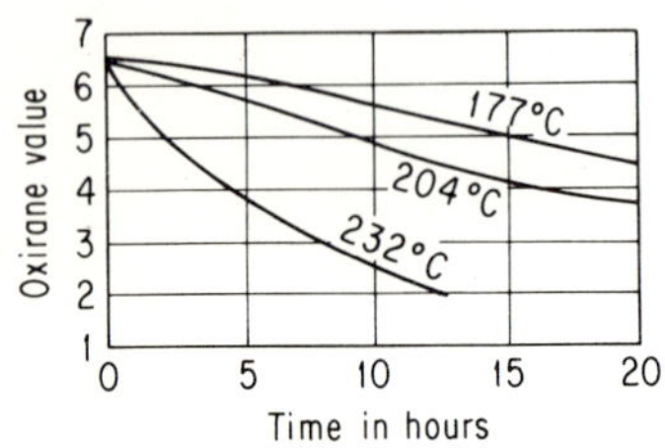

FIG. 3-2. Effect of temperature on destruction of epoxy groups in epoxidized soybean oil [53].

The separation of epoxidized triglyceride oils by use of solvents such as petroleum ether has been described [82]. About 1 to 5 parts of a lightweight petroleum solvent when added to epoxidized vegetable oils results in two fractions having different epoxy contents. The insoluble (clear) phase will contain the fraction with the higher epoxy content [117].

Epoxidized natural oils appear to be somewhat less stable than the epoxy resins described in Chap. 2. If, when uncatalyzed, they are exposed to elevated temperatures, there is progressive loss of epoxy groups (Fig. 3-2) and a corresponding increase in viscosity (Fig. 3-3).

The epoxidized natural oils find primary use as plasticizers/stabilizers for vinyl formulations. Approximately 45 million pounds was produced in 1960, over half being epoxidized soybean oil [73].

Epoxidized natural oils also may be cured with conventional epoxy-resin curing agents of the anhydride type. However, it is customary first to modify the epoxidized natural oils. The modification generally consists of introducing hydroxyl groups into the chain to shorten the distance between potential cross-linking points [64] and improve reactivity [57]. The introduction of the hydroxyl groups is usually achieved by reacting some portion of the epoxy groups present with either a phenol or a polyphenol. The resultant modified product possesses increased toughness and rigidity over the unmodified product when cured with an acid or anhydride. With amine curing agents, amide formation may occur, resulting in cured products of poor quality.

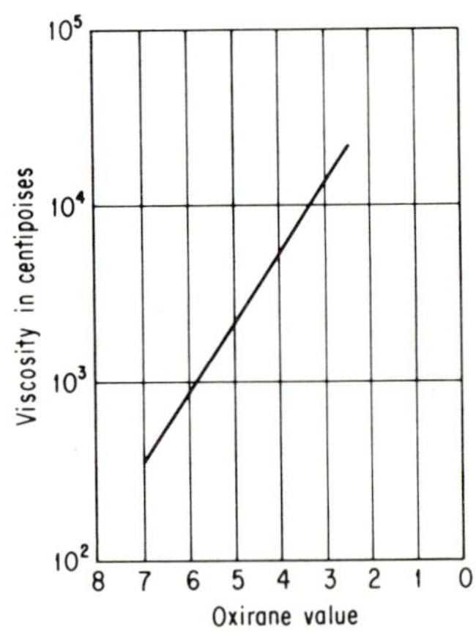

FIG. 3-3. Effect of destruction of epoxy groups on viscosity of epoxidized soybean oil [53].

Synthetic Polyesters

Suitable synthetic polyesters are derived from the reaction of organic polybasic acids (or anhydrides) with polyols:

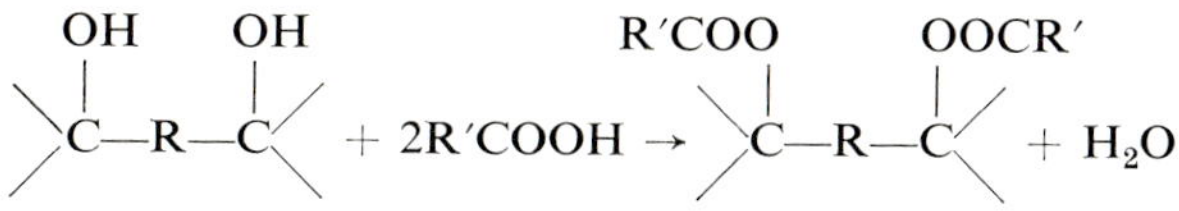

Diol Acid Diester Water

Either the acid or the polyol or both may contain the requisite unsaturation [16,52]. Long-chain polybasic acids may be reacted with glycerol or other alcohols to produce compounds closely related to the natural oils described above [58]. Short-chain polybasic acids, such as adipic, may be reacted with long-chain unsaturated alcohols to produce similar materials [20]. Both components may be relatively short chain, as is the case with 4,5-epoxypentyl-4,5-epoxypentanoate [100]. Di- and polyacids may be used [27]. Unconjugated, unsaturated diacids prepared by reacting ω-free radical acids with butadiene or cyclopentadiene, are esterified and epoxidized [14]. The ω-free radical acids are formed by the catalytic decomposition of cyclic peroxides (e.g., 1,1′-dihydroxydicyclohexyl peroxide). 1,2,4-Butanetricarboxylic acid has been used with aliphatic [116] and alicyclic [86] alcohols. Mono- and diepoxy resins have been produced from monocarboxylic acid esters of cresols and xylenols [132].

Di- and polyepoxy resins based on unsaturated polyesters from oleic and linoleic acids and ethylene glycol, glycerol, pentaerythritol, and dipentaerythritol have been produced with peracetic acid [54]. Resins based on unsaturated polyesters from linoleic and oleic acids, *cis*-4-cyclohexene-1,2-dicarboxylic acid and hexahydrophthalic anhydride and polyols from cyclohexene and cyclohexane as well as butane and butene diols, oleyl and linoleyl alcohols are described in [103]. Resins from unsaturated esters from oleoyl chloride or linoleoyl chloride and catechol, resorcinol, and hydroquinone are described in [102]. The epoxidation of α,β-olefinic esters is described in [76].

Tetrahydrophthalic anhydride (THPA) may be reacted with a glycol, and the unsaturation of the THPA then epoxidized [47]; the resultant structure may be modified with an unsaturated monocarboxylic acid [90]. Tetrahydrophthalyl esters of oleic and linoleic acid and oleyl and linoleyl esters of THPA have been epoxidized [122]. A more complex molecule may be developed involving the reaction of THPA, adipic acid, and a glycol [62]; or of bicyclo(2.2.1)hept-5-ene-2,3-dicarboxylic acid anhydride, propylene glycol, and oleic acid [75]. Unsaturated polyesters containing sulfur, phosphorus, silicones, etc., have also been epoxidized [32].

As with the naturally occurring oils, the principal potential use of these epoxidized polyesters is as vinyl plasticizers and stabilizers.

Unsaturated Polyethers

Polyethers may be obtained by a condensation reaction of an alcohol with a halomethyl group in the presence of base:

$$\underset{\text{Alkyl chloride}}{R'CH_2Cl} + \underset{\text{Alcohol}}{HOR} + \underset{\text{Sodium hydroxide}}{NaOH} \rightarrow \underset{\text{Ether}}{R'CH_2OR} + \underset{\text{Salt}}{NaCl} + \underset{\text{Water}}{H_2O}$$

The halomethyl group may be obtained from an alcohol by treating with a suitable halide; or it may be obtained by reacting an aromatic compound with formaldehyde in the presence of HCl (i.e., chloromethylation).

Alternatively, the polyether may be obtained by the addition reaction of an alcohol with an olefin [37] in the presence of an acid:

$$\underset{\text{Internal olefin}}{\text{—CH═CHCH}_3} + \underset{\text{Alcohol}}{\text{HOR}} \xrightarrow{\text{BF}_3} \underset{\text{Ether}}{\text{—CH}_2\text{—}\underset{\text{OR}}{\underset{|}{\text{CH}}}\text{—CH}_3}$$

An example of a third class of polyethers would be cyclic ethers from unsaturated aldehydes or ketones and polyols [44] such as pentaerythritol [43], to give compounds such as 3,9-divinylspirobi(*m*-dioxane) [85,95].

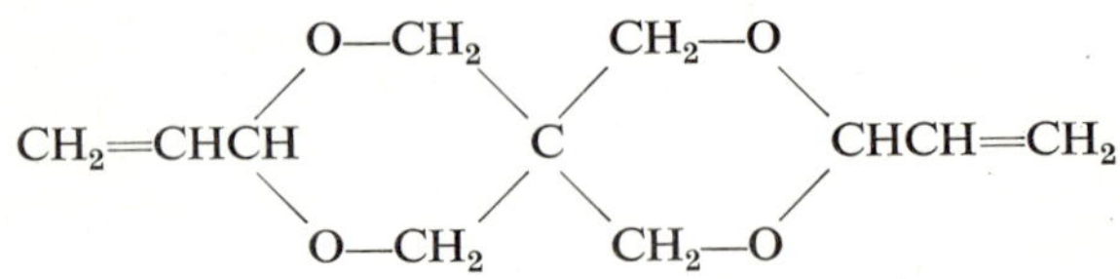

3,9-Divinylspirobi(*m*-dioxane)

An example of an ether obtained from an aromatic would be the material of the following structure [42]:

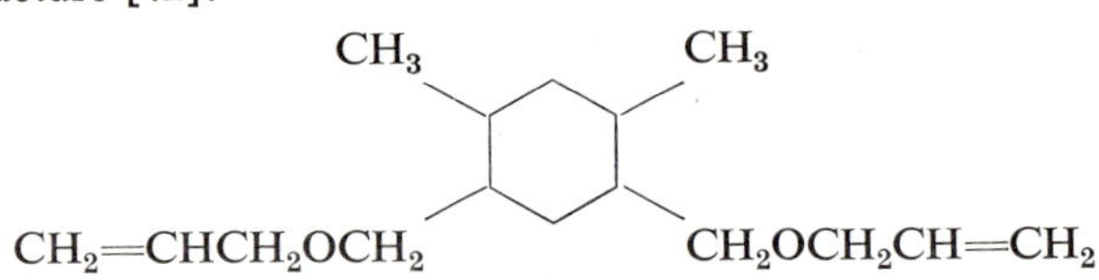

4,6-Bis(allyloxymethyl)-*m*-xylene

Other ethers may be developed from butadiene, as discussed subsequently, these having greater usefulness in the epoxy-resin technology.

In all cases, sufficient unsaturation must be available in the reaction product to permit epoxidation to the extent of more than one group per molecule. The unsaturation may be supplied by either or both of the reactants.

Resins from Butadiene

Butadiene (CH_2═CH—CH═CH_2) is obtained from cracked petroleum fractions, or may be synthesized from acetaldehyde, vinyl acetylene, etc. Most frequently, butenes are passed through superheated steam in the presence of a dehydrogenation catalyst to produce butadiene. The butadiene is then stabilized against polymerization by use of an inhibitor at 25 to 50 ppm. Butadiene is used chiefly for the production of synthetic rubbers such as Buna S (butadiene-styrene copolymer) and Buna N (butadiene-acrylonitrile copolymer).

Butadiene is a versatile intermediate in the epoxy-resin technology.

The epoxidation of butadiene has been indicated above. Butadiene may be combined with itself via Diels-Alder addition to produce vinylcyclohexene and this epoxidized to give the dioxide

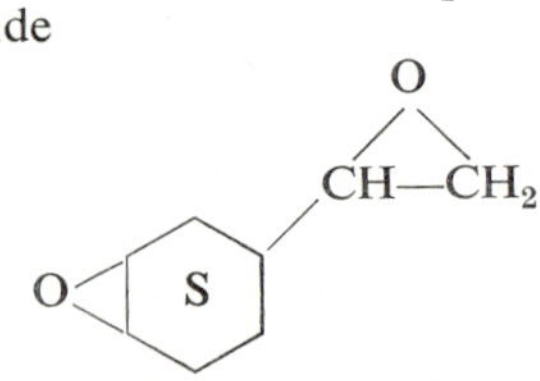

Vinylcyclohexene dioxide

Butadiene may be polymerized through the unsaturation to produce polybutadiene [41], and this epoxidized [55] to give

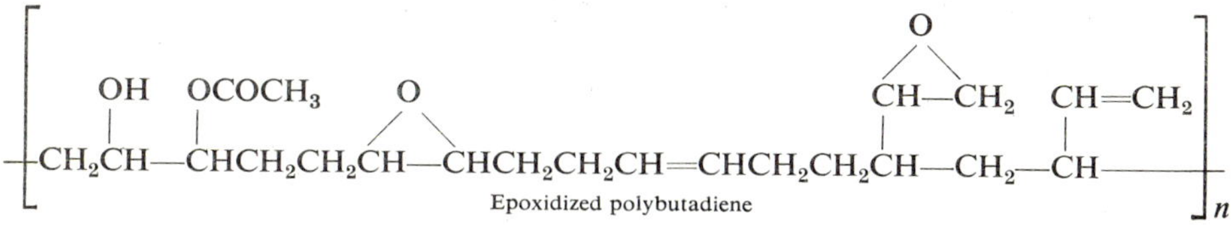

Epoxidized polybutadiene

The epoxidized polybutadiene contains residual unsaturation and may be stabilized with hydroquinone to increase shelf life [120]. Cyclic polybutadienes (e.g., 1,5-cycloöctadiene [84] and 1,5,9-cyclododecatriene [126]) have been epoxidized. Butadiene may be copolymerized with other unsaturated molecules (i.e., styrene [35], cyclopentadiene-styrene [99], and piperylene [107]). The epoxidation of high- and low-molecular-weight synthetic rubbers based on copolymers of butadiene has been reported [88]; and the efficiency of aqueous peracetic acid, anhydrous peracetic acid, and peracetic acid formed in situ has been compared in the epoxidation of these copolymers when using phosphoric acid and a cation-exchange resin [113]. Butadiene may be reacted with alcohols to produce ethers which are suitable intermediates in the synthesis of epoxy resins [2]. Resins of higher molecular weight have been reported by copolymerizing vinylcyclohexene dioxide with a monoepoxy such as propylene oxide [72].

The epoxidation of a liquid polybutadiene is given as an example:

100 g of polybutadiene, viscosity 1,500 cps at 100°F were dissolved in 100 ml of benzene. To this solution was added slowly, over a period of about one hour, 275 g of 40 per cent peracetic acid solution stabilized with dipicolinic acid and 10.6 g of sodium acetate to adjust the acidity of the reaction mixture. This amount of peracetic acid corresponds to the stoichiometric amount required to fully oxidize the polybutadiene. The temperature was maintained at about 20 to 25°C over a total reaction time of 3 hours 45 minutes. The product was then washed three times with a saturated solution of sodium chloride, twice with a saturated solution of sodium chloride containing potassium hydroxide to neutralize residual acid in the mixture and finally washed three times with distilled water. The mixture was then filtered through filter paper and excess solvent was removed at room temperature and a pressure of about one to two millimeters of mercury. Finally, the essentially solvent-free product layer was separated from the water layer and dried in a vacuum desiccator. The product was analyzed and found to contain 6.6 per cent oxirane oxygen [36].

Butadiene may be reacted with unsaturated aldehydes using Diels-Alder addition to produce useful intermediates in the production of epoxy resins. These are shown in Table 3-6. For example, crotonaldehyde may be used to produce the intermediate I. This may be trimerized to a substituted trioxan which contains three reactive unsaturations suitable for epoxidation [66]. The aldehyde group may be converted to a methylol (II) or carboxyl (III) group. A dibasic acid may then be used with 2 moles of II to produce IV [19], or a diol may be used with 2 moles of III to produce V [18]. Two moles of I may be condensed with pentaerythritol to give VI; one mole of III with II to give VII [13]. One mole of VIII is condensed with I to give IX or with crotonaldehyde to give X. Two moles of II may be condensed with one mole of aldehyde or ketone to give XI, which contains no ester link [80]. These and similar diolefins from butadiene and unsaturated aldehydes have been epoxidized [79]. For the epoxidation of V for example, see [38,39]. Other molecules may be developed from the intermediates shown in Table 3-6 as for example, the reaction of II with polyunsaturated fatty acids to produce a polyolefin suitable for epoxidation [24]. Nitrogen [65] and sulfur [68]

Table 3-6. Resin Intermediates from Butadiene and Aldehydes and Ketones

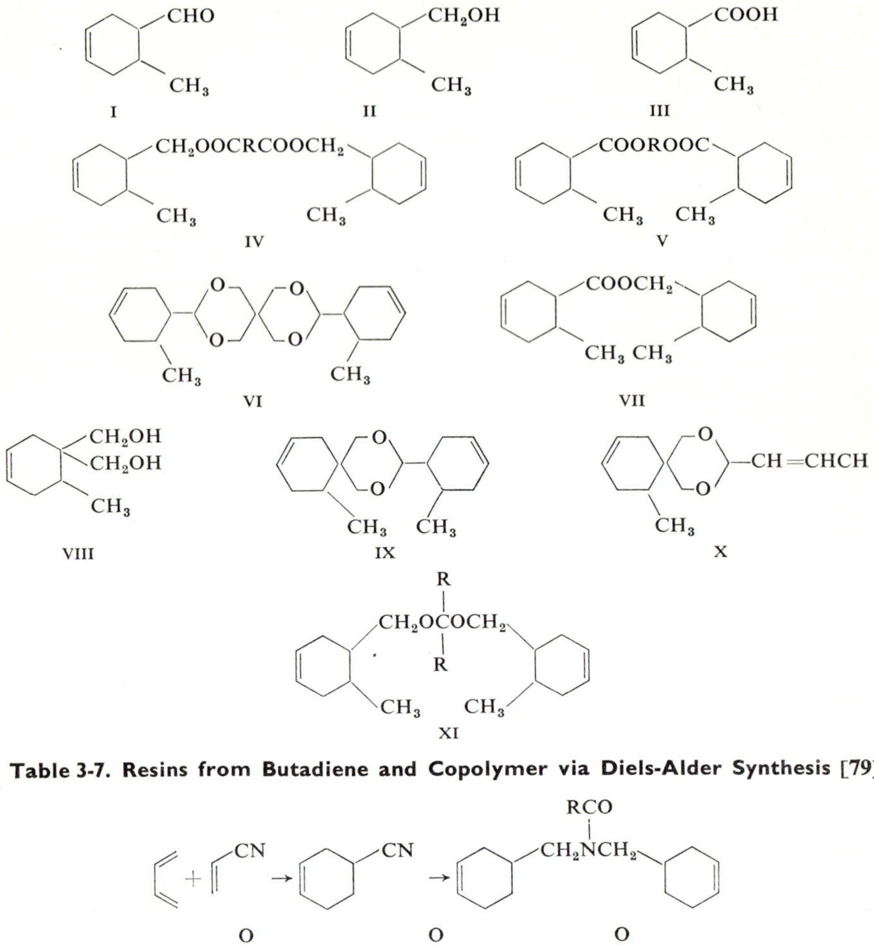

Table 3-7. Resins from Butadiene and Copolymer via Diels-Alder Synthesis [79]

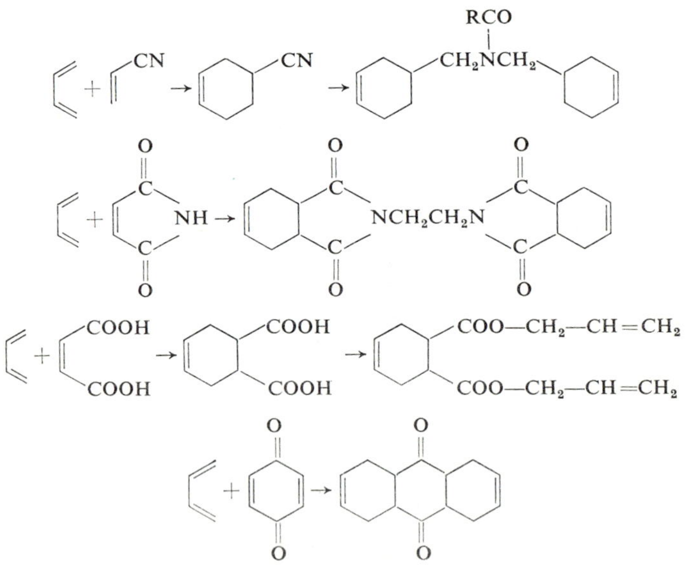

containing compounds similar to the above have been synthesized. Species without the 6-methyl group are common, these from acrolein, as for example, resins from unsubstituted structure V [98] and VI [94]; resins from butadiene and allyl alcohol [83] lack the 6-methyl group, and its replacement with a butyl ester group has been reported [115]. A number of epoxidized acetals from aldehydes and 1,1-dimethylol-3-cyclohexene have been described [101]. One from 3-chloropropylaldehyde is disclosed in [112]. Higher-molecular-weight polyether acetals are described in [125], and acetals from unsaturated aldehydes in [97]. Similarly, 1,1-dimethylol-3-cyclohexene may be esterified with long-chain monounsaturated acids [78], or monomethylol-3-cyclohexene may be used with the acids [23].

Butadiene may be reacted, via Diels-Alder addition, with a variety of other unsaturated compounds containing functional groups capable of further reaction and suitable as intermediates for resin formation. Typical are those given in Table 3-7.

Butadiene may be reacted with benzene to produce an aromatic of the following structure:

1-Phenyl-2-butene

By use of substituted benzene, a variety of diolefins may be developed.

Thus, the versatility of butadiene permits the production of a large family of linear, cyclic, and aromatic polyunsaturated compounds suitable for epoxidation by the peracid route.

Other Starting Materials

Similar to compounds from butadiene are those from cyclopentadiene. Cyclopentadiene may be reacted to produce a biscyclopentenyl ether [22].

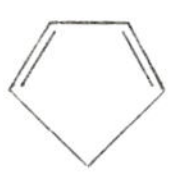

Cyclopentadiene

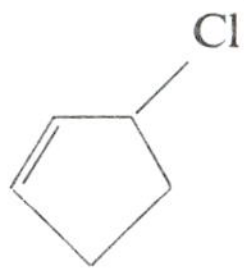

3-Chlorocyclopentene

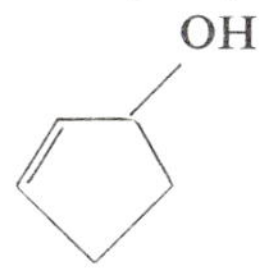

3-Hydroxycyclopentene

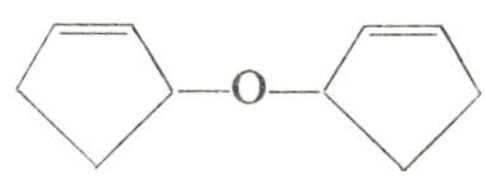

Biscyclopentenyl ether

Epoxidation of the molecule is described in [77].

The epoxy resin exists as two isomers, one solid,

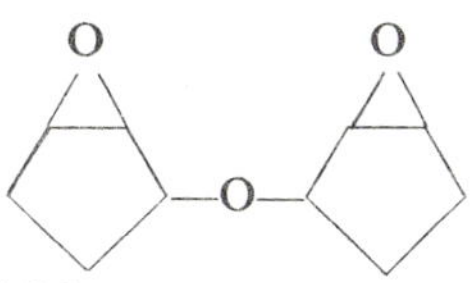

Bis(2,3-epoxycyclopentyl) ether

and one liquid,

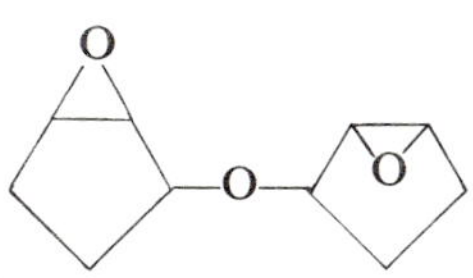

Bis(2,3-epoxycyclopentyl) ether

Cyclopentadiene may be reacted via Diels-Alder in a fashion similar to butadiene to produce hydroxyl-containing intermediates which may be reacted with di- or polybasic acids to produce esters suitable for epoxidation [93,118]. It may likewise be reacted with vinylcyclohexene to produce a diolefin [111].

Dicyclopentadiene gives resin intermediates, as shown in Table 3-8. It may be

Table 3-8. Resin Intermediates from Dicyclopentadiene

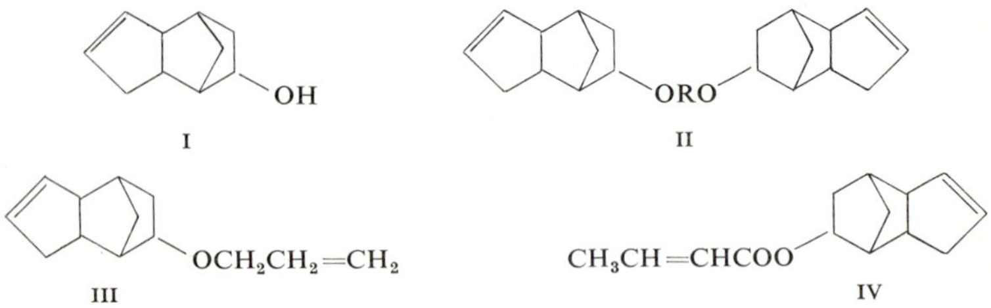

reacted with water to add a hydroxyl (I), and this reacted with a diol to produce II, with an unsaturated alcohol, such as allyl alcohol, to produce III, and with an unsaturated acid, such as crotonic acid, to produce IV. A variety of structure II type compounds have been disclosed in the patent literature [91,92,105,106]. Structure III type compounds are described in [110].

Unsaturated aliphatics from benzene may be produced as indicated in Table 3-9.

Table 3-9. Resin Intermediates from Benzene

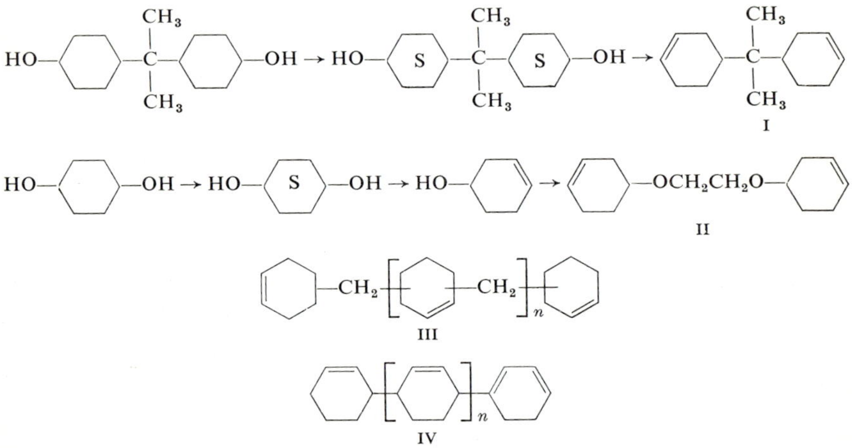

The epoxidation of dicyclohexenylpropane is described in [70]. Epoxidation of the hydrogenated novolac is reported in [51], and epoxidation of structure IV in [119]. *p*-Bis(1-cyclohexenyl)benzene [134] and poly 1,3-cyclohexadiene [135] have been epoxidized.

The synthesis of triepoxy acetals derived from the reaction of an epoxyalkoxyalkyl aldehyde and 2 moles of a mono-unsaturated alcohol has been reported [121].

Unsaturated polyamides, derived from the reaction of unsaturated fatty acids and amines [46] and unsaturated amines and acids [15] have been epoxidized, as have unsaturated polycarbamates derived from alcohols and isocyanates or from polyamines and chloroformates [32]. Water-soluble diepoxies have been produced by reacting triacrylylperhydrotriazine with an unsaturated alcohol (such as tetrahydrobenzyl alcohol) to give structures such as the following [81]:

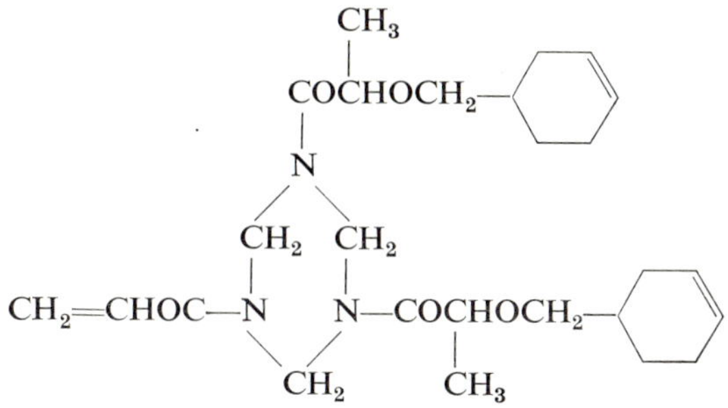

1,3-Bis-2-(cyclohex-3-enyl)methoxyproponyl-5-acrylylhexahydro-*s*-triazine

Epoxidation of triallyl isocyanurate has been reported [89].

Unsaturated side chains may be attached to aromatic compounds by methods described above and by other techniques of synthesis. For example [60], allyl-substituted phenols suitable for epoxidation are produced by Claisen rearrangement as follows:

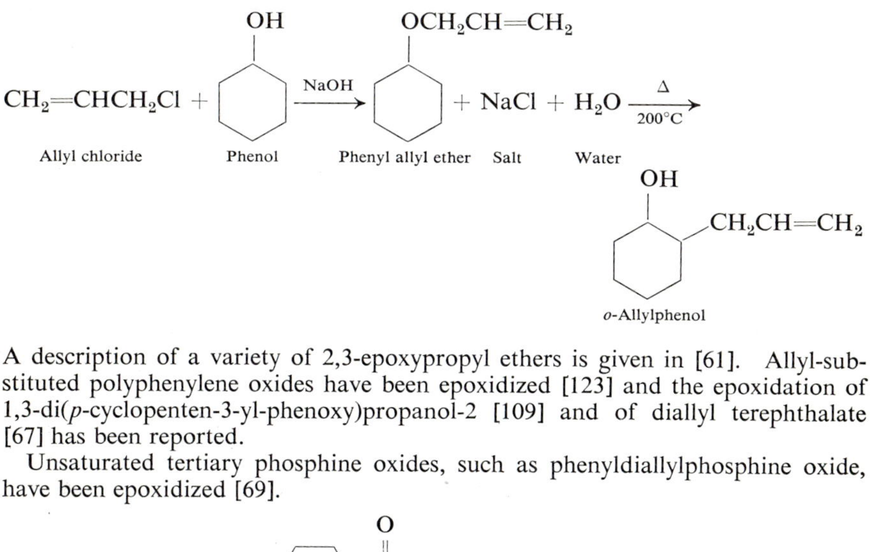

A description of a variety of 2,3-epoxypropyl ethers is given in [61]. Allyl-substituted polyphenylene oxides have been epoxidized [123] and the epoxidation of 1,3-di(*p*-cyclopenten-3-yl-phenoxy)propanol-2 [109] and of diallyl terephthalate [67] has been reported.

Unsaturated tertiary phosphine oxides, such as phenyldiallylphosphine oxide, have been epoxidized [69].

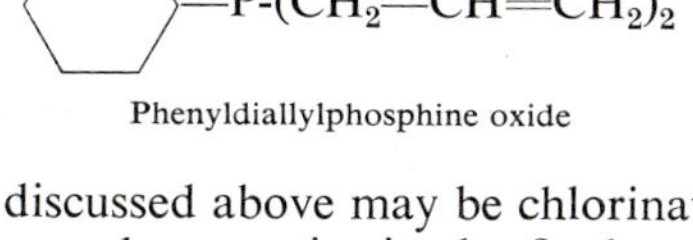

Phenyldiallylphosphine oxide

Many of the materials discussed above may be chlorinated [49,59], or otherwise substituted to obtain improved properties in the final epoxy resin. Unsaturated monoepoxies synthesized via the peracid route may be reacted as discussed in Chap. 2 through the unsaturation to produce polyepoxy resins. Production of a polyepoxy resin from 1-(1,2-epoxyethyl)-2-vinylcyclobutane is described in [104].

COMMERCIAL EPOXY RESINS FROM OLEFINS

The commercial epoxy resins derived from olefins may be divided in two general categories: those suitable primarily as vinyl stabilizers/plasticizers, and those of wider utility in the epoxy-resin technology.

The epoxy resins derived from animal or vegetable oils are primarily used as plasticizers, although by suitable modification they may be cured with the conventional epoxy-resin curing agents.

Chapter 4 presents the physical properties of the various commerical resins, classified as to type and supplier, together with their idealized chemical structures and IR curves.

The characterization of these resins and other epoxy resins is presented in greater detail in Chap. 4.

SUMMARY

Epoxy resins can be derived from all manner of polyolefinic compounds by the addition of oxygen to the unsaturation. Unsaturated animal and vegetable oils, synthetic polyesters, polyethers, and butadiene derivatives are the most common commercial raw materials employed. The epoxy group may be glycidyl, terminal, or pendant vinyl; or may be situated on cyclic aliphatic rings or internally along the chain.

Synthesis, ideally, would be by direct oxidation with molecular oxygen, but this process is practical only in the synthesis of ethylene oxide. For other materials the oxygen is supplied from a parent compound, either hypochlorous acid, a peracid, or hydrogen peroxide itself. In the first case chlorohydrin groups are produced which are converted to epoxies by dehydrohalogenation; in the second and third cases the peracid or hydrogen peroxide supplies the oxygen directly. The principal commercial processes make use of the peracid route, with peracetic acid being favored from the economic standpoint. The peracetic acid may be made from acetaldehyde monoperacetate as a step in the synthesis of acetic acid from acetaldehyde and molecular oxygen, with the desired acetic acid recovered following the epoxidation reaction.

REFERENCES

1916

1. Galitzenstein and Mugdan, Manufacture of Peracids from Aldehydes, U.S. 1,179,421.

1946

2. Bruson, Polydihydronordicyclopentadienyl Ethers, U.S. 2,393,610.

1949

3. Niederhauser and Koroly, Process for the Epoxidation of Esters of Oleic and Linoleic Acids, U.S. 2,485,160.
4. Terry and Wheeler, Process of Preparing Epoxy Derivatives from Unsaturated Aliphatic Compounds, U.S. 2,458,484.

1950

5. Elderfield (ed.), "Heterocyclic Compounds," vol. I, John Wiley & Sons, Inc.
6. Foster, Dicarboxylic Acid Anhydride-4-vinyl-cyclohexene Diepoxide Reaction Products, U.S. 2,527,806.

7. Polymeric 4-Vinylcyclohexene Dioxide, (to Canadian Industries, Ltd.), Brit. 642,800.

1951

8. Swern and Findley, Epoxidized Oils, U.S. 2,569,502.

1952

9. Improvements in or Relating to Epoxy Ethers (to Canadian Industries, Ltd.), Brit. 672,935.

1953

10. Glasson, et al., Process for the Manufacture of Epoxies, U.S. 2,650,927.
11. Greenspan and Gall, Epoxy Fatty Acid Ester Plasticizers, *Ind. Eng. Chem.*, December.

1954

12. Greenspan and Gall, Method of Epoxidation, U.S. 2,692,271.

1955

13. Frostick and Phillips, Diepoxides of Cycloaliphatic Esters, U.S. 2,716,123.

1956

14. Mika and Sullivan, Resinous Products Plasticized with Epoxidized Long Chain Polybasic Acid Esters, U.S. 2,764,497.
15. Payne and Smith, Amides of Epoxy Amines, U.S. 2,730,531.
16. Payne and Smith, Epoxy Resins of Polycarboxylic Acids, U.S. 2,761,870.
17. Phillips, Oxidation of Olefins, ACS Symposium, Dallas, April.
18. Phillips and Starcher, Diepoxides, U.S. 2,745,847.
19. Phillips and Starcher, Diepoxides, U.S. 2,750,395.
20. Ritter and Rowland, Epoxyoctadecyl Adipates, U.S. 2,771,472.
21. Strain, Polymers of Butadiene Dimer Epoxides, U.S. 2,765,296.

1957

22. Carlson, Bis-epoxycyclopentanyl Ether, U.S. 2,739,161.
23. Frostick and Phillips, Diepoxides of Cycloaliphatic Esters of Monoethylenically Unsaturated Fatty Acids, U.S. 2,786,066.
24. Frostick and Phillips, Triepoxides and Process for Making the Same, U.S. 2,786,067.
25. Greenspan and Gall, In Situ Epoxidation of Organic Esters with Sulfuric and Acetic Acid, U.S. 2,801,253.
26. Krummel, et al., New Epoxy Resins for Reinforced Plastics, *Western Plastics*, August.
27. Payne and Smith, Unsaturated Esters of Epoxy-substituted Polycarboxylic Acids and Polymers, U.S. 2,783,250.
28. Phillips, The Synthesis of Peracetic Acid and Epoxides, Peracetic Acid and Derivatives, Union Carbide Chemicals Co. Report.
29. Phillips and Frostick, Higher Fatty Ester Diepoxides and Process of Producing the Same, U.S. 2,779,771.
30. Phillips and Starcher, Process for Making Oxirane Compounds from Olefins and Aldehyde Monoperacylates, U.S. 2,785,185.
31. Phillips, et al., Production of Lower Aliphatic Peracids, U.S. 2,804,473.
32. Process for Preparing Epoxidized Condensation Polymers and the Resulting Products (to N.V. de Bataafsche Petroleum Maatschappij), Brit. 786,116.
33. Wahlroos, Method of Preparing Epoxidized Oils and the Like, U.S. 2,813,878.

1958

34. Greenlee, Compositions of Polyepoxide Polyesters and Aromatic Sulfonamide Aldehyde Condensates and Methods of Making Same, U.S. 2,847,394.

35. Greenspan and Light, Epoxy Resins, U.S. 2,829,130.
36. Greenspan and Light, Epoxidized Polybutadiene Resins, U.S. 2,829,135.
37. Newey, Polyepoxy Polyethers, Their Preparation and Polymers, U.S. 2,848,426.
38. Phillips and Starcher, 2-Methoxymethyl-2,4-dimethyl-1,5-pentanediol Bis(3,4-epoxycyclohexanecarboxylate), U.S. 2,853,498.
39. Phillips and Starcher, 3,4-Epoxycyclohexanecarboxylic Acid Esters of Di- and Triethylene Glycol, U.S. 2,853,499.
40. Rowland and White, In Situ Epoxidation Using Acetic Anhydride and a Basic Salt, U.S. 2,836,605.
41. Wheelock, Epoxidation of Liquid Polybutadiene, *Ind. Eng. Chem.*, March.

1959

42. Christenson and Jaruzelski, Epoxy Compounds, U.S. 2,874,150.
43. Epoxy Substituted Bicyclo Acetals (to Union Carbide Corp.), Brit. 887,593.
44. Fischer, Epoxidized Acetals and Polymers Thereof, U.S. 2,895,962.
45. Gable, Epoxidation Process, U.S. 2,870,171.
46. Gelb, et al., Epoxy Resins from Fats: IV, The Effect of the Structure of Epoxystearamides on the Physical Properties of Their Phthalic Anhydride Cured Resins, ACS Symposium, Atlantic City, September.
47. Greenlee, Esterified Epoxy Polyesters, U.S. 2,907,735.
48. Greenspan and Gall, Process of Epoxidation, U.S. 2,919,283.
49. Guest and Stansbury, Halogenated Diepoxide Compounds, U.S. 2,874,167.
50. Hopff and Keller, Concerning Aromatic Di- and Triepoxides, *Helv. Chim. Acta*, **42**(VII):264–265.
51. Kohler and Pietsch, Synthetic Resins and Process of Making Same, U.S. 2,915,484.
52. Payne and Smith, Esters of Epoxy Substituted Polycarboxylic Acids and Epoxy-substituted Alcohols, U.S. 2,870,170.

1960

53. Fath, Significance of Oxirane and Iodine Values in Epoxy Plasticizers, *Mod. Plastics*, April.
54. Gelb, et al., Effect of Structure of Epoxidized Esters on the Physical Properties of Their Phthalic Anhydride Cured Resins, *J. Chem. Eng. Data*, April.
55. Greenspan and Light, Resinous Products from Epoxidized Polybutadienes and Polymercaptans, U.S. 2,921,921.
56. Hydrogen Peroxide for Preparation of New Epoxidized and Hydroxylated Lanolin Derivatives, booklet by Solvay Process, Division of Allied Chemical & Dye Corp.
57. Ilardo and Schoepfel, Epoxy Resin Based on Epoxidized Soybean Oil and HET Anhydride, *Ind. Eng. Chem.*, April.
58. Newey, Epoxy Esters of Long Chain Acids, U.S. 2,940,986.
59. McGary and Patrick, Halogenated Diepoxide Compositions, U.S. 2,936,292.
60. Martin and Holm, Polyepoxy-substituted Aromatic Compounds, and Polymers, U.S. 2,938,875.
61. Martin and Pannel, Ethers of Epoxy-substituted Phenols and Their Polymers, U.S. 2,965,607.
62. Reiff, Epoxidized Resinous Polyester Containing Tetrahydrophthalic Anhydride and Compositions Containing Same, U.S. 2,928,805.
63. Smith, The Synthesis of Epoxides, *Paint Technol.*, August.
64. Wear, Aromatic Epoxidized Polyester and Method of Making, U.S. 2,944,035.

1961

65. Batzer and Nikles, New Nitrogenous Epoxy Compounds, U.S. 2,988,544.
66. Batzer and Nikles, New Epoxy Compounds, U.S. 2,991,293.
67. Diglycidyl Terephthalate (to Canadian Industries, Ltd.), Brit. 862,588.
68. Frostick and Phillips, Diepoxy Sulfone Compositions, U.S. 3,018,259.
69. Kindley, et al., Organophosphorus Epoxides, *Tappi* **44**:10.
70. Kohler and Pietsch, Hardened Epoxy Resins, Ger. 1,099,733.

71. Monger, Explosion Limits of Liquid Systems Containing Hydrogen, Peroxide, Water, and Oxygenated Organic Compounds, *J. Chem. Eng. Data*, January.
72. Morehouse and Horinibrook, Copolymers of 4-Vinylcyclohexene Dioxide, U.S. 2,555,500.
73. New Swift Process Turns out Epoxidized Oil, *Chem. Eng. News*, July 24.
74. Pearche, Process for Preparing Epoxide Compositions Using a Dehydrated Acid Cation Exchange Resin Catalyst, U.S. 2,976,265.
75. Phillips and Starcher, Epoxidized Polyesters from 3-Oxatricyclo(3.2.1.0^{2,4})-octane-6,7-dicarboxylic Acid, U.S. 2,969,336.
76. Phillips and Starcher, Epoxy Esters of Epoxy Carboxylic Acids, U.S. 2,969,377.
77. Phillips and Starcher, Bis(2,3-epoxycyclopentyl) Ether, U.S. 2,973,373.
78. Phillips and Frostick, Epoxides and Method of Preparing Same, U.S. 2,999,865.

1962

79. Batzer and Nikles, Synthesis of Newer Epoxy Resins, *Chimia*, March.
80. Batzer and Nikles, Epoxidized Hydroaromatic Compounds, U.S. 3,023,174.
81. Batzer, et al., Epoxy Derivatives of Triacrylyl Perhydro Triazine, U.S. 3,082,385.
82. Darrow, Separation of Epoxidized Triglyceride Oils, Fr. 1,370,461.
83. Epoxy Ethers (to Ciba, Ltd.), Belg. 619,149.
84. Frostick and Phillips, Resins from 1,2,5,6-Diepoxycyclooctane, U.S. 3,065,209.
85. Guest, et al., Epoxides of Bicyclo Acetals, U.S. 3,023,222.
86. Lynn, et al., Epoxy Esters of 1,2,4-Cyclopentane-tricarboxylic Acid, U.S. 3,057,880.
87. Pearce, Process for Purifying Acid Containing Compositions, U.S. 3,022,259.

1963

88. Belaya, et al., Preparation of Epoxy Resins by Epoxidation of Unsaturated Compounds *Sintetich, Polikondensatsionnye Smoly, Urk. Nauchn-Issled. Inst. Plast. Mass. Sb. Statei*, pp. 5–55.
89. Epoxy Isocyanurate Monomer and Polymer (to Spencer Chemical Co.), Brit. 941,507.
90. Greenlee, Drying Compositions Comprising Polyepoxide Polyesters of Tetrahydrophthalic Acid and Aliphatic Glycols Modified with Unsaturated Monocarboxylic Acid, U.S. 3,113,932.
91. Humphreys, et al., Esters Containing Epoxide Groups, Brit. 943,924.
92. Humphreys, et al., Resins Containing Epoxide Groups, Brit. 936,346.
93. Johnson, et al., Epoxy Esters, Brit. 937,412.
94. Lincoln and Chalmers, Diepoxy Acetals, Brit. 934,346.
95. McGary and Patrick, Curing Spirobi(metadioxane) Epoxides with Polycarboxylic Acid Anhydrides in Combination Modifiers Containing Active Hydrogen Atoms, U.S. 3,083,186.
96. Murata and Higo, Process for Manufacture of Epoxidized Fatty Acids, Alcohols, and Esters Thereof, U.S. 3,112,325.
97. New Epoxides and Process for Their Preparation (to Ciba, Ltd.), Brit. 932,642.
98. Newey, Polyepoxides from Epoxy-substituted Monocarboxylic Acids, Their Preparation and Polymers, U.S. 3,112,294.
99. Niki, et al., Epoxy Resin and Its Hardening, Japan 18,545.
100. Phillips and Frostick, 4,5-Epoxypentyl 4,5-Epoxypentonate and Process for Its Preparation, U.S. 3,074,973.
101. Porret, et al., Diepoxides of Ether Acetals of 1,1-Bis(hydroxymethyl)cyclohexene, U.S. 3,072,678.
102. Port and Komanowsky, Resins from Phenolic Esters of Epoxidized Fatty Acids, *J. Appl. Polymer Sci.* **7**:281,285.
103. Scholnick, et al., Epoxy Resins from Fatty Esters Derived from Cyclohexane and Epoxycyclohexane, *J. Am. Oil Chem. Soc.*, June.
104. Tinsley and MacPeek, 1-(1,2-Epoxyethyl)-2-vinylcyclobutane and Polymers Thereof, U.S. 3,096,315.
105. Tinsley, et al., Epoxy Ethers and Curable and Resinous Compositions Containing Them, Brit. 932,144.
106. Tinsley, et al., Improvements in and Relating to a Process for the Preparation of Diepoxides, Homopolymers, and Copolymers Thereof, Brit. 932,817.

1964

107. Andrei, Epoxy Resins with Piperylene, *Rev. Chim.* (*Bucharest*) **15**(5):274–275.
108. Batzer, Cycloaliphatic Epoxides, *Chem. Ind.* (*London*), Feb. 1.
109. Cutukovic, 1,3,Di(*p*-cyclopenten-2-yl-phenoxy)propanol-2, U.S. 3,118,953.
110. Epoxide Compounds (to Ciba, Ltd.), Brit. 950,285.
111. Epoxides of 6-(1-Cyclohexen-4-yl)bicyclo(2.2.1)-2-heptene (to Consortium für Elektrochemische Industrie, G.m.b.H.), Fr. 1,363,684.
112. Epoxy Acetals (to Ciba, Ltd.), Brit. 951,399.
113. Gorbunov, et al., Epoxidation of Butadiene Oligomers, *Plasticheskie Massy*, no. 8, pp. 7–10.
114. Greenspan and Linder, Improved Epoxidation Process with Liquid Aliphatic Peroxy Acids, U.S. 3,130,207.
115. Hosotani, et al., Epoxy Compounds, Japan 2,473.
116. Lynn, et al., Epoxy Esters of Polycarboxylic Acids, U.S. 3,122,568.
117. Manufacture of Drying Oils of High Epoxy-Oxygen Content (to Solvay & Cie.), Belg. 634,507.
118. Medved and Christie, Bicyclic Diepoxides, *J. Chem. Eng. Data* **9**(2):240.
119. Naumova, et al., Epoxide Resin, U.S.S.R. 159,978.
120. Nowlin and Reich, Stabilization of Epoxypolybutadienes with Hydroquinone, U.S. 3,133,092.
121. Phillips and Starcher, Tri-epoxy Acetals, U.S. 3,151,130.
122. Port and Scholnick, Epoxy Monomers and Resins Thereof, U.S. 3,147,236.
123. Tsou, et al., Epoxidation of Poly-2-methyl-6-allyl-1,4-phenylene Oxide and Copolymer of 2-Methyl-6-allyl-4-bromophenol and 2,6-Dimethyl-4-bromophenol, *J. Polymer Sci.* (A) **2**:4425–4431.
124. Tsukida and Yokota, Epoxy Carotenoids: VI, Carotenoids of *Centaurea Moschata var.* Yellow Sultan, *Bitamin* **30**(1):9–12.
125. Walk, High Molecular Weight Epoxidized Polyether Acetals, Ger. 1,180,530.
126. Willersinn, et al., Light Stabilization of Polyester Compositions, Ger. 1,177,336.
127. Wisniak, et al., Epoxidation of Anchovy Oils: a Study of Variables, *Ind. Eng. Chem. Prod. R. & D.* **3**(4):306–311.
128. Yamanishi, et al., Linalool Epoxides in the Essential Oils from Freshly Picked Tea Leaves and Black Tea, *Agr. Biol. Chem.* (*Tokyo*) **28**(9):653–655.

1965

129. Dickey, et al., Epoxidation of Olefins during Simultaneous Air Oxidation of Secondary Aldehydes, Fr. 1,400,305.
130. Duck and Simmons, Epoxy Graft Copolymers of Polybutadiene and Unsaturated Oils, Brit. 996,090.
131. Duke, et al., Epoxy Alcohols and Ethers Useful as Diluents for Epoxy Resins, Brit. 991,453.
132. Jaruzelski, et al., Polyvinyl Chloride Plasticized with Epoxy Containing Monocarboxylic Acid Esters of Cresols and Xylenols, U.S. 3,184,425.
133. Jommi, et al., Constituents of Hyenanche Globosa. Structure of Substance D, *Chim. Ind.* (*Milan*) **47**(8), 865–866.
134. Meiya and Nikolaev, Synthesis of 1,4-Dialkenylbenzenes and Their Oxides, *Zh. Organ. Khim.* **1**(2), 296–299.
135. Naumova and Kovaleva, Production and Some Properties of Epoxy Resin IFOX-1, *Dokl. Akad. Nauk Belorussk. SSR* **9**(5), 312–315.
136. Sims, The Metabolism of 1,4-Dihydronaphthalene in Rats, *Biochem. J.* **95**, 608–611.
137. Warson, et al., Unsaturated Epoxy Esters for Thermosetting Epoxy Resins, Brit. 995,726.

1966

138. Natural Source Discovered for Epoxy Fatty Acids, *Am. Paint J.*, vol. 50, no. 29, Jan. 10.

Chapter 4

CHARACTERIZATION OF UNCURED EPOXY RESINS

Proper formulation of epoxy-resin compounds requires knowledge of the molecular structure of the resins as well as of the chemical and physical properties of the commercial-production batches.

The properties of primary importance in commercial batches of resin are viscosity and epoxide equivalent. Properties which are of secondary importance, but which may be critical for some formulations, are softening point, hydroxyl equivalent weight, iodine number (when applicable), molecular weight and molecular-weight distribution, color, specific gravity, refractive index, chlorine content, and specialized properties such as odor and defoam time under high vacuum.

The structure of the resin determines the physical and chemical properties, particularly the reactivity of the epoxy and other groups present. The number and location of the reactive sites determines the functionality and the cross-linking density; those in turn, together with the curing agent, stoichiometry, and cure,

Table 4-1. Infrared Band Assignments for Epoxy Groups [21]*

Type of epoxy group	*Absorption wavelength, microns*
Monosubstituted (terminal epoxy) 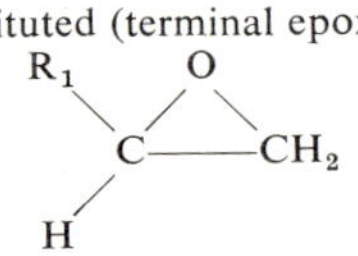	11.4–12.4
Disubstituted (internal epoxy) 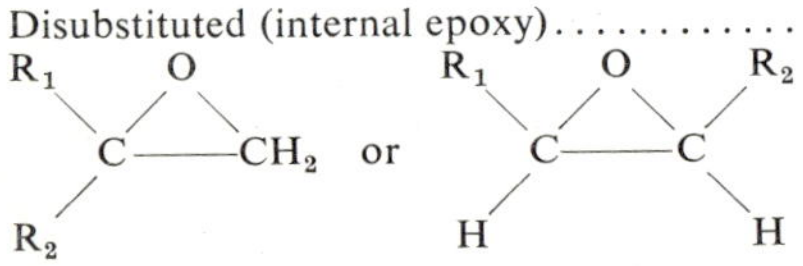	11.8–12.9
Trisubstituted 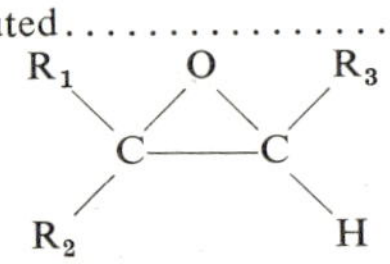	13.0–13.3

* Courtesy of *Analytical Chemistry*.

establish the rigidity, thermal stability, and solvent resistance, etc., to be expected in the cured system. Additionally, the structure will determine the viscosity of the resin.

The viscosity of the resin is particularly important. Low-viscosity liquids are useful as casting compounds and as 100 per cent solids adhesives, laminating, and coating resins. The high-viscosity and the solid materials are useful in hot-melt systems, in molding powders, and in thin-film applications, where they can be deposited from solutions or as fluidized bed coatings.

The epoxide equivalent permits calculations establishing the correct amounts of curing agents and modifiers that should be employed to obtain complete reaction. With some curing agents, and under some reaction conditions, it may be equally as important to know the hydroxyl equivalent weight and iodine number. Other properties such as color, although critical for some applications, are not generally of as great concern.

It is the purpose of this chapter to discuss the importance of the individual characteristics, as well as to indicate test methods normally employed for their determination.

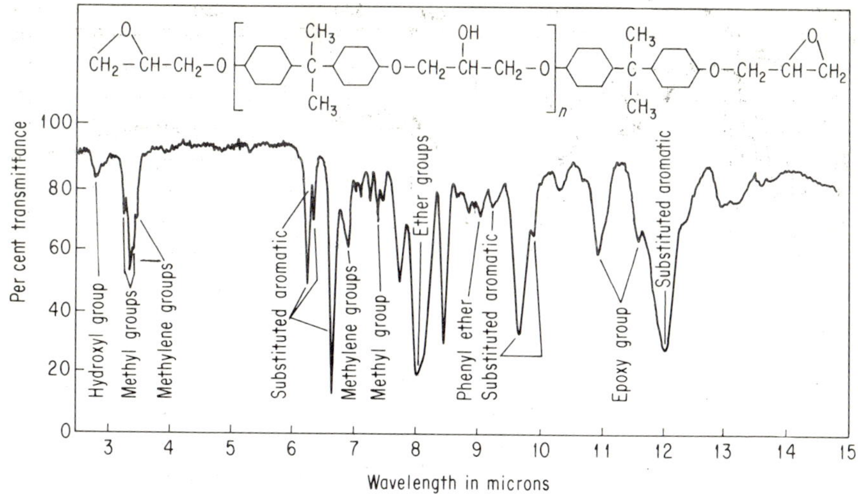

FIG. 4-1. Infrared band assignments for DGEBA resin [30].

DETERMINATION OF STRUCTURE

Determination of Structure by Infrared Spectroscopy

The structure of epoxy resins is generally determined (or verified) by means of infrared spectrophotometers. These instruments measure the absorption of infrared energy as a function of wavelength and plot an absorption or transmission spectrum for the resin. Because each functional group absorbs infrared energy at characteristic wavelengths which are dependent upon the molecular environment, spectroscopists can determine the structures accurately [9,10,11,14,16,19,21,25,30].

The infrared spectra of typical epoxy resins, together with their structural formulas, are presented in Appendix 4-1 to this chapter. Band assignments for various epoxy groups are shown in Table 4-1, and band assignments for DGEBA are shown in Fig. 4-1. The presence of diluents or additives is easily noted, as shown in Fig. 4-2. The use of infrared in quantitative determination of epoxide equivalent is discussed in a subsequent section.

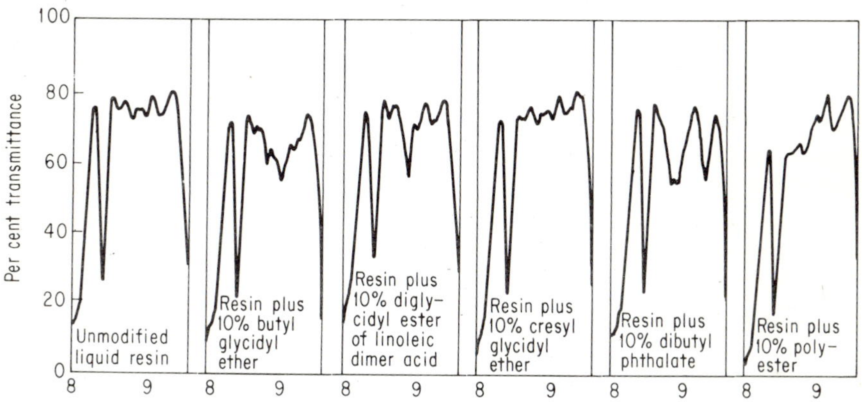

FIG. 4-2. Sensitivity of 9-micron region of the spectra of DGEBA to 5 to 10 per cent of additive [30].

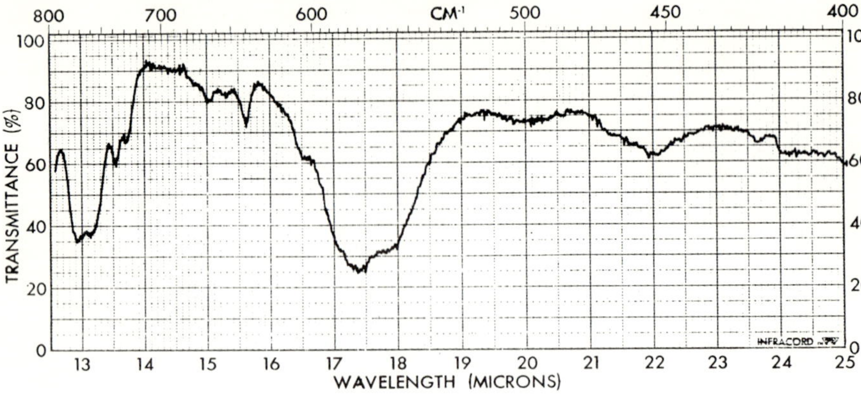

FIG. 4-3. Far-infrared spectra of DGEBA.

Determination of Structure by Far-infrared Spectroscopy

The far-infrared region of the spectrum (approximately 15 to 25 microns) is little used in epoxy-resin technology. Although absorption bands are observed (Figs. 4-3 and 4-4), they add very little information to that available in the conventional infrared region.

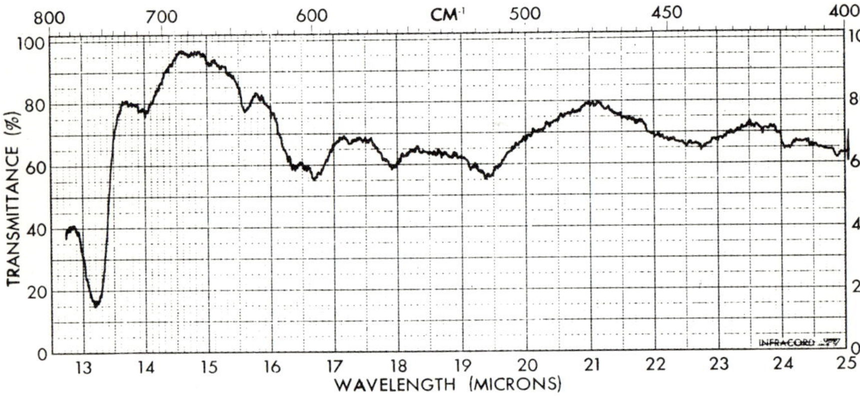

FIG. 4-4. Far-infrared spectra of liquid epoxidized novolac.

Determination of Structure by Nuclear Magnetic Resonance Spectroscopy

Nuclear magnetic resonance spectroscopy is a useful tool in studying the exact structure of a resin, since the atomic environment of each hydrogen atom (proton) may be detected as shown in Fig. 4-5.

Effect of Structure on Reactivity

The type of epoxy group and its location within the molecule influence reactivity. Some locations prefer to react with acid curing agents; some with bases; some are extremely reactive with certain catalysts, others virtually inert; and some possess good overall reactivity with most classes of curing agents.

Steric factors will influence reactivity, so that compact molecules containing a number of epoxy groups, (e.g., the more highly functional epoxidized novolac resins) may not under normal conditions be capable of complete reaction. Bulky side chains developed during the course of polymerization will also effectively block reactive sites.

Reactive groups other than the epoxy rings may be present in the molecule—hydroxyl and olefin groups most commonly—and their number and location will influence the reaction equilibrium with specific curing agents and may retard or accelerate the overall reaction rate.

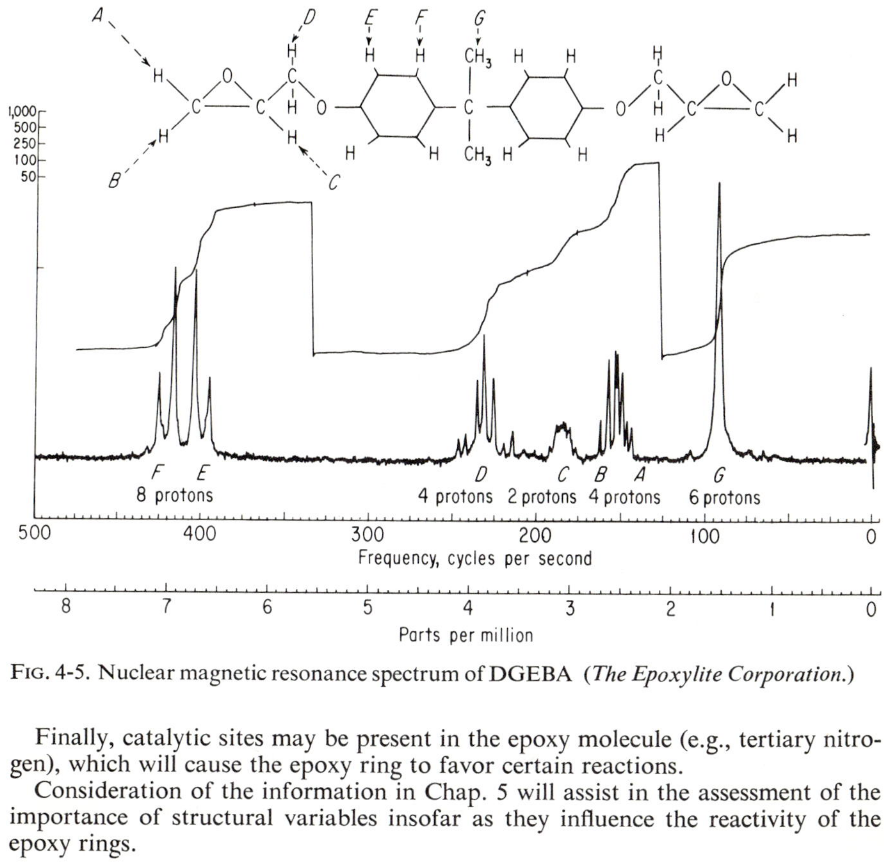

FIG. 4-5. Nuclear magnetic resonance spectrum of DGEBA (*The Epoxylite Corporation.*)

Finally, catalytic sites may be present in the epoxy molecule (e.g., tertiary nitrogen), which will cause the epoxy ring to favor certain reactions.

Consideration of the information in Chap. 5 will assist in the assessment of the importance of structural variables insofar as they influence the reactivity of the epoxy rings.

Effect of Structure on Functionality

In epoxy resins, each epoxy group may have a functionality of 1 or 2, depending upon the curing agent employed and the reaction conditions; the hydroxyl groups present may have a functionality of 0 or 1; and any reactive unsaturation will have a functionality of 0 to 2.

If the epoxy resin is reacted in such a fashion as to provide a *system* functionality of 2, only linear, thermoplastic molecules will result. As the system functionality is increased, cross-linked products are obtained. System functionality may be increased by increasing the functionality of the resin, of the curing agent, or of both (see Chap. 6).

Within a given class of epoxy resins, the more highly functional molecules will provide improved thermal stability and solvent resistance in the cured system. For example, substantial improvements in these properties may be obtained by increasing the number of epoxy groups in an epoxidized novolac resin from approximately two to approximately three [23].

However, just as there is a practical lower limit to the functionality, so, too, there is a practical upper limit to the functionality. For example, epoxidized novolacs having more than three epoxy groups do not, with the primary polyfunctional amine curing agents at least, provide substantially greater thermal stability than those with only three epoxy groups per molecule [23].

Commercial epoxy resins usually have lower than theoretical functionality. The standard commercial DGEBA, for example, instead of containing two epoxy groups per molecule, contains, on the average, about 1.85—the 0.15 loss in epoxy functionality being of significance in some high-temperature applications.

Effect of Cross-linking Density

Besides the functionality of the resin, it is necessary to know how far apart the functional groups are spaced, since this determines cross-link density. Cross-linking density may be defined as the number of *effective* cross-links per unit volume.

In the DGEBA ($n = 0$), the epoxy groups are separated by seven units (the aromatic ring being counted as one unit). In other resins, the groups may be separated by 30 or 40 units. In general, short spacings (3 to 10 units) give rigidity, medium spacings (10 to 30 units) give semirigid properties, and long spacing (30 or more units) gives flexibility or softness.

An additional factor arises when polar groups are found along the polymer chains. These increase the molecular interactions between adjacent chains and hence reduce the chain flexibility. Bulky side groups, such as aromatic rings, also hinder segmental motions and therefore decrease flexibility. Hence, to produce a given softness, or flexibility, at a given temperature, shorter chain lengths will be required for an aliphatic-based resin compared to an aromatic-based resin.

If resin properties are studied at elevated temperature, then the complications due to polar interactions between chains may be reduced or eliminated. In this case, considerations of chain flexibility revert to the simple concept based on the degree of cross-linking.

The effect of introducing substitution on the backbone is generally toward the reduction of cross-linking *density*. The effect of chain substituents on properties of practical importance can be seen in data on the water-vapor permeability and oxygen permeability of the thermoplastic phenoxy resins (Table 4-2). As the backbone becomes more highly substituted, the permeability increases, presumably because the chains are less closely spaced. If this concept is transferred to the cross-linked phenoxy resin (i.e., homopolymerized DGEBA), the substituents would, likewise, force the chain segments in the area further apart without, of course, changing the number of primary chain atoms between cross-linking points. The effect would be to decrease the number of cross-links in a unit volume of polymer.

Compact, ring-situated epoxies, in providing higher cross-link densities, can give higher deflection temperatures over less compact glycidyl ether resins based on the more stable aromatic rings. From model measurements, the distance between cross-linking points on the DGEBA ($n = 0$) is almost three times the distance on bis(2,3-epoxycyclopentyl) ether, and about 25 per cent greater than the distance

Table 4-2. Effect of Type of Dihydric Phenol Backbone on Oxygen Permeability and Moisture-vapor Transmission of Phenoxy Resin [41]

Backbone structure	Oxygen permeability, cc-mil/(100 sq in.) (24 hr)(atm)	Moisture-vapor transmission, g-mil/(100 sq in.) (24 hr)(38°C)(90% R.H.)*
	0.5	3
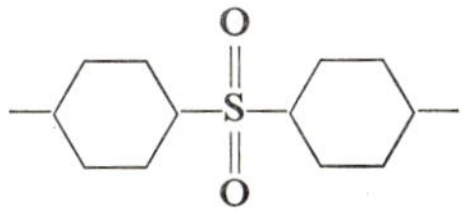	2	9
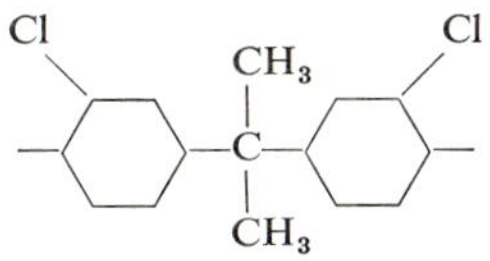	3	3
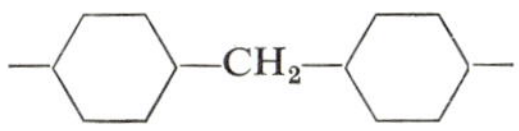	4	
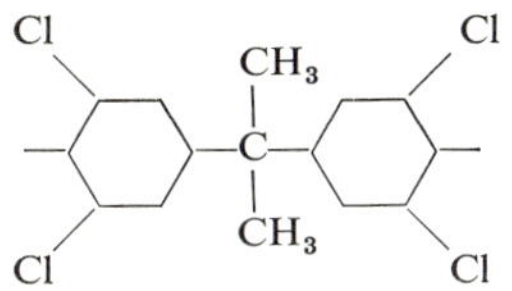	4	3
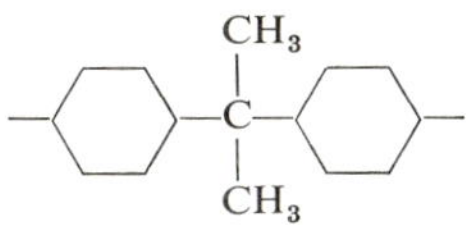	5	3
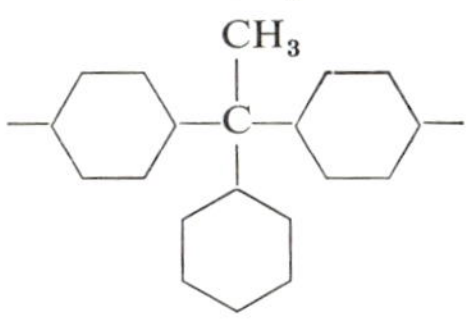	8	7
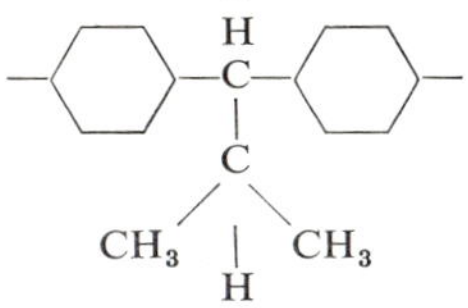	8	7
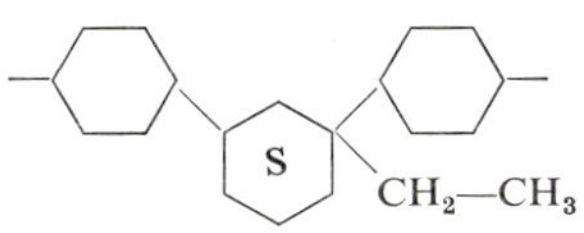	15	7
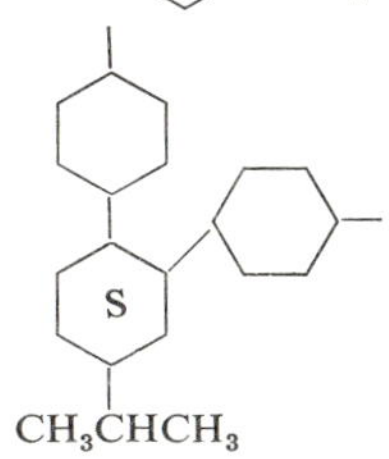	75	15

* R.H.: Relative humidity.

Table 4-3. Glass-transition Temperature vs. Type of Linkage between Aromatic Rings and Type of Dihydric Phenol of Phenoxy Resins [41]

Structure	*Tg, °C*
Connecting linkage:	
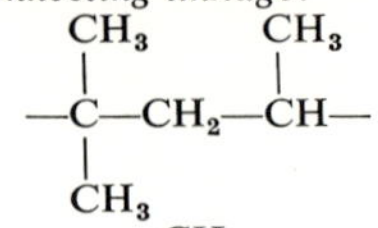	75
$-CH_2-$	80
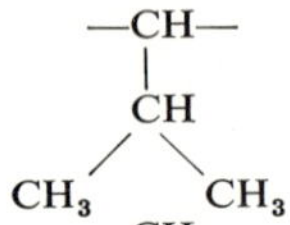	95
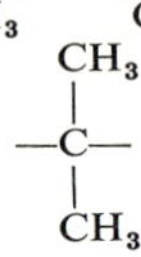	100
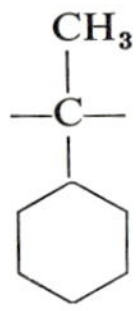	115
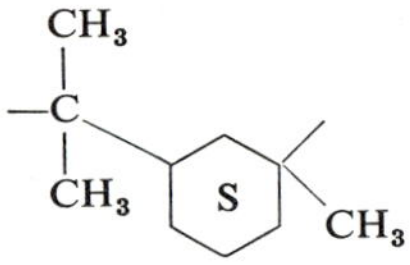	135
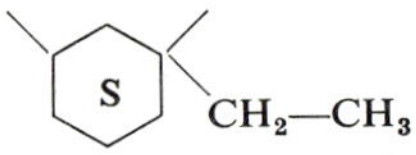	140
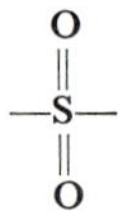	155
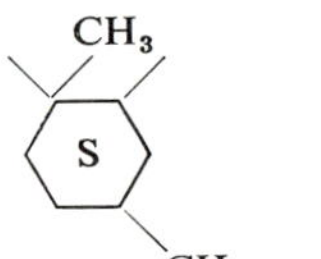	175
Dihydric phenol:	
	60
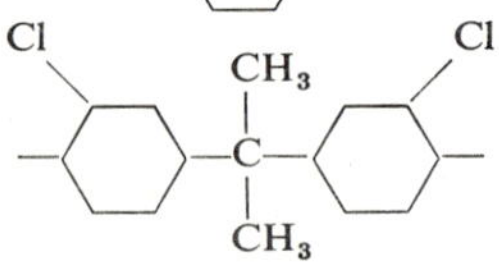	85

Table 4-3. (*Continued*)

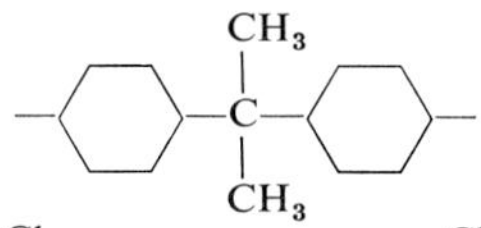	100
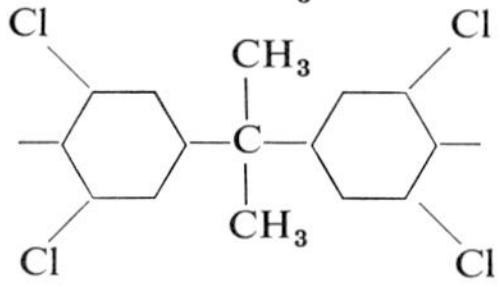	115
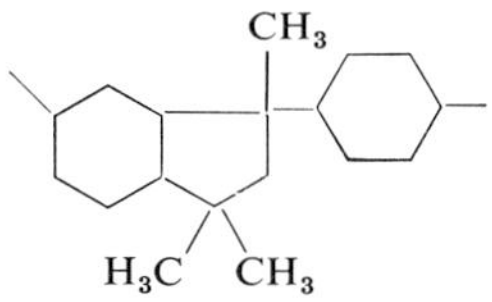	120

on 3,4-epoxy-6-methylcyclohexylmethyl 3,4-epoxy-6-methylcyclohexane carboxylate. The difference in cross-linking density, in practice, is probably substantially greater, since the commercial DGEBA will have an average n number somewhat more than zero.

Branch-chain and linear aliphatic epoxies may provide unexpectedly low cross-link densities in that chain flexibility permits the reaction of more than one epoxy group on a given molecule with more than one reactive site on an adjacent one, thus tending toward cyclic polymerization rather than cross-linking.

An extended study with phenoxy resins [41] indicates the general effect of the connecting links and the effect of various bisphenols on the glass-transition temperature of these high-molecular-weight thermoplastic molecules (Table 4-3).

Cyclization, together with randomly occurring areas of low bond density for other reasons (high chlorine content, presence of monofunctional reactants and other impurities), will lead to internal flaws within the polymer network to create notch-sensitive defects.

Effect of Structure on Thermal Stability

Thermal stability in the cured epoxy-resin system will depend in part on the chemical structure and bonds, as well as on the functionality and cross-linking density of the resin.

Commercial epoxy resins contain preponderantly carbon-to-carbon bonds. The resins synthesized via the epichlorohydrin route will contain other linkages introduced by the requirement of the synthesis—most commonly these are ether links.

Aromatic ether and aromatic ester links provide good thermal stability in epoxy formulations. Somewhat inferior are the aliphatic ethers and esters.

Pyrolysis studies of the uncured epoxy resins indicate that thermal decomposition progresses from etherification (homopolymerization) or rearrangement of the epoxy group to dehydration and then cleavage. Products for the DGEBA and an epoxidized novolac are indicated in Tables 4-4 to 4-7. With the epoxidized novolacs (at 475°C), the initial cleavage is believed to occur at methylene bridges, followed by cleavage at the aliphatic carbon of the aromatic ether to give phenol and variously substituted mono- and dinuclear phenols as the products. With DGEBA, the initial cleavage is believed to be at the aliphatic carbon of the aromatic ether, the parent bisphenol being recreated. The aliphatic portion of the

Table 4-4. Volatile Products from Pyrolysis of DGEBA Resins [37] (In mole per cent)

Compound	Mass/charge ratio m/e	Mol. wt. 380		Mol. wt. ~10,000	
		At 350°C	At 450°C	At 350°C	At 450°C
Methane	16	0.73	1.87	....	0.56
Water	18	34.10	11.10	40.70	20.50
Ethylene	28	....	1.09	0.88	1.41
Carbon monoxide	28	....	1.30	....	1.60
Ethane	30	0.49	1.50	0.44	0.54
—CH_2OH	31	1.21	1.24	0.88	0.91
Allene	40	....	0.15	0.93	0.34
Propylene	42	....	....	0.35	0.42
Acetaldehyde	44	....	....	4.34	12.90
Carbon dioxide	44	0.16	1.27	9.38	4.60
Methyl chloride	50, 52	....	0.60	2.93	14.80
Acrolein	56	0.07	0.57	....	0.09
—CH_3CH_2CO	57	3.89	3.75		
Acetone	58	....	....	0.77	1.32
Propionaldehyde	58	0.13	0.22		
Vinyl chloride	62, 64	0.92	0.51		
—COCl	63, 65	0.66	0.80		
Mass 70 (and 97)	70, 97	....	....	0.16	0.31
Cyclopentadiene	66	....	0.25		
Benzene	78	....	0.16		
Methylcyclopentadiene	80	0.38	16.50	0.22	0.20
Toluene	92	57.20	57.20	38.00	39.50

molecule is split out except for a minor amount of cyclization. The bisphenol then cleaves at the benzhydryl carbon to give phenolic fragments. To the extent that the system has undergone etherification, it is believed that dehydration may occur to form, in part, an allyl ether, which may then split at the aliphatic carbon of the aromatic ether or undergo cyclization with the adjacent benzene ring [46].

Pyrolysis studies of uncured DGEBA using tagged benzhydryl carbon at 300 to 400°C in vacuum show that most of the gaseous breakdown products are from the aliphatic portion of the molecule. The main products, carbon monoxide and methane, probably are formed from the epoxy groups by a free radical mechanism. At 800°C, it is suggested that cleavage at the benzhydryl carbon and subsequent decomposition proceeds as shown in Table 4-8. The tagged atoms in the solid residue have been shown by electron paramagnetic resonance spectra to participate in the formation of systems with conjugated unsaturation [35].

Table 4-5. High-boiling Products from Pyrolysis of DGEBA [37] (475°C under vacuum)

Compound	Mass/charge ratio m/e	Mol. wt. ~10,000	Mol. wt. 380
Phenol	94	Largest	Largest
Cresols	108	Major	Major
C_2H_5 phenols	122	Major	Major
Isopropylphenol	136	Major	Major
Isopropenylphenol	134	Major	Major
Bisphenol A	228	Minor	Minor
Unidentified masses	...	Small	Small

Table 4-6. Volatile Products from Pyrolysis of Epoxidized Novolac Resin [37] (In mole per cent)

Compound	Mass/charge ratio *m/e*	At 350°C	At 450°C
Hydrogen	2	0.39	1.49
Methane	16	1.05	
Water	18	25.80	7.42
Acetylene	26	....	1.47
Ethane	30	0.40	13.40
Formaldehyde	30	.....	8.52
$—CH_2OH$	31	0.22	2.52
Allene	40	.....	0.55
Acetaldehyde	44	.....	5.32
Propane	44	.....	2.42
Carbon dioxide	44	0.35	
Methyl chloride	50, 52	.....	1.08
Acrolein	56	0.11	4.24
$—CH_3CH_2CO$	57	1.12	6.75
Acetone	58	.....	0.49
Propionaldehyde	58	.....	1.41
Vinyl chloride	62, 64	0.95	
$—COCl$	63, 65	1.12	
$—(CH_2)_3Cl$	76, 78	.....	0.94
Allyl chloride	76, 78	.....	9.19
Methylcyclopentadiene	80	0.26	0.38
Toluene	92	68.30	32.40

Table 4-7. High-boiling Products from Pyrolysis of Epoxidized Novolac Resin [37] (475°C under vacuum)

Compound	Mass/charge ratio *m/e*	Amount
Cresols	108	Largest
Phenol	94	Major
$(HOC_6H_4)_2CH_2$	200	Major
C_2H_5 phenols	122	Minor
$(HOC_6H_4)_2(CH_2)_2$	214	Minor
Unidentified masses	...	Small

Table 4-8. Probable Decomposition Path of DGEBA Pyrolyzed at 800°C in Vacuum [35]

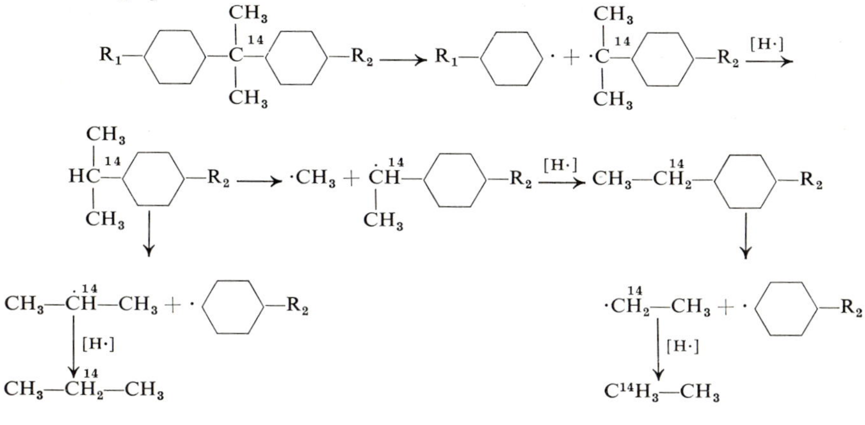

At lower temperatures, approximately 200°C, it is probable that the method of chain scission is governed by oxidative degradation, in which case, for example with DGEBA, scission occurs in the polymer chain close to a benzene ring to give a radical [39] such as:

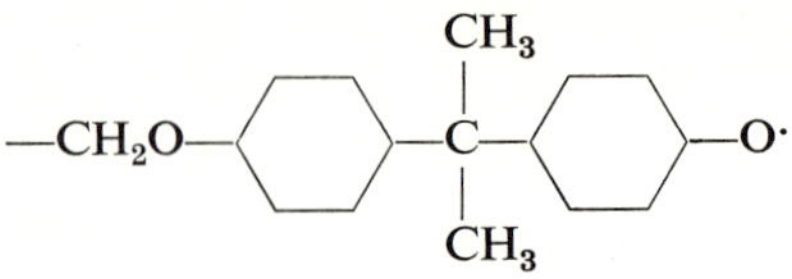

Substituents have been investigated in order to improve thermal stability, and in certain cases where they do not adversely effect cross-linking density, have led to improvements.

In bisphenols coupled in the para position,

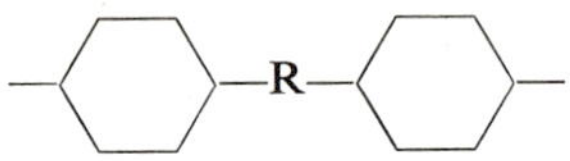

the nature of the link has been shown [27] to influence thermal stability in the following order;

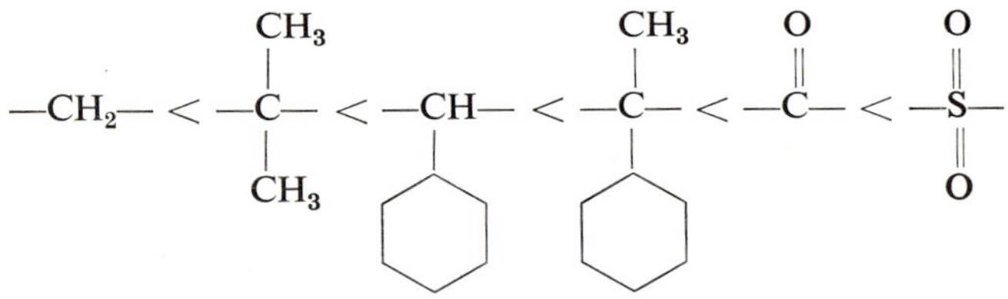

In this sequence, however, the degree of improvement in thermal stability from the DGEBA to the DGE 2,2-bis(*p*-hydroxyphenyl)-2-phenylethane is not dramatic [38].

The para-position-coupled bisphenols appear to provide less thermal stability than the ortho-position-coupled bisphenols [15]. Thus a methylene link in the ortho position

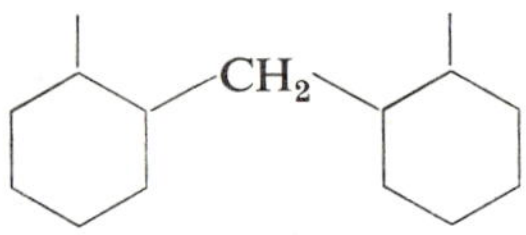

is superior, in terms of thermal resistance, to at least some of the substituted methyl links in the para position.

When chlorine is introduced as a substituent to impart flame resistance, the effect is to lower the thermal stability, particularly in the case of amine cures, where the chlorine substituent is brought in close proximity with a hydroxyl during the curing process, since heat aging may strip out hydrochloric acid. With brominated resins, lower bromine content gives somewhat greater stability at 200°C than higher bromine content, and above 200°C the brominated resins are relatively unstable. With these brominated resins, increasing the chlorine content also provides decreasing stability [36].

Other reactive sites present in the epoxy-resin molecule may produce thermal instability to a greater degree than would be anticipated from an inspection of bond energies and substituents. The hydroxyl group for example is vulnerable to oxidation, being converted to an aldehyde. Since some resins also contain catalytic centers, such as tertiary amines, the problem is further complicated.

Effect of Structure on Viscosity

Although viscosity cannot be determined from an examination of chemical structure, and generalizations are hazardous at best, the following relationships between structure and viscosity warrant mention:

1. Aromatic-based epoxies with three or more epoxy groups per molecule are semisolid or solid at room temperature. Blends of aromatic-based epoxies containing on the average more than two but fewer than three epoxy groups may yield room-temperature liquids, depending upon the molecular weight of the species involved.

2. Epoxies based on saturated-ring structures (i.e., cycloaliphatics) give lower viscosities than comparable resins based on aromatic rings.

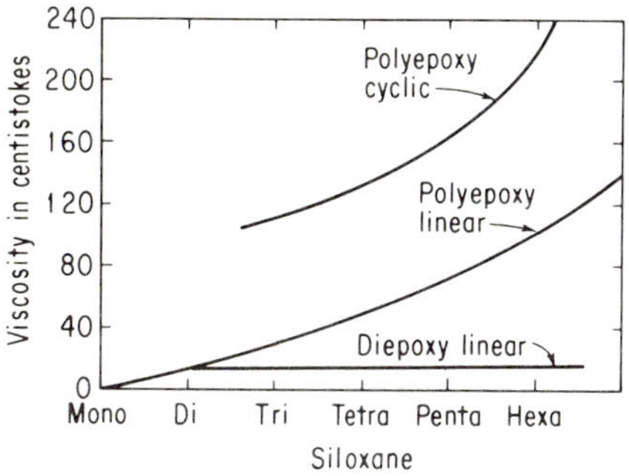

FIG. 4-6. Effect of structure on viscosity of siloxane-based epoxy resin [26].

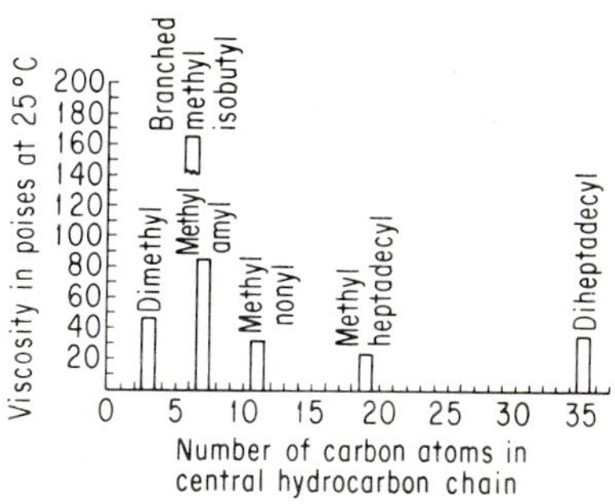

FIG. 4-7. Viscosity vs. length of alkyl groups attached to benzhydryl carbon of DGE of bisphenols [29].

3. Linear aliphatic epoxy resins give lower viscosities than do comparable cycloaliphatic epoxy resins.

4. The effect of chain length on the viscosity of linear aliphatic *di*epoxies is, within limits, of little significance; with linear aliphatic *poly*epoxies, however, increasing chain length results in a corresponding increase in viscosity. This relationship for various aliphatic siloxane epoxy resins is indicated in Fig. 4-6.

5. In general, the effect of substitution is to decrease the viscosity compared with the unsubstituted product, but the viscosity does not necessarily decrease in direct proportion to the length of the substituted chain (Fig. 4-7), nor do all substituents decrease viscosity. For example, the diglycidyl ether of tetrabromobisphenol A has a viscosity of about 98,000 centipoises at 60°C, against a comparable unsubstituted viscosity (for the same *n*-number resin) of about 8,700 centipoises at that temperature; and the diglycidyl ether of tetrachlorobisphenol A softens at 35°C against a comparable unsubstituted viscosity of about 40,000 centipoises at room temperature.

6. For a given resin type, viscosity increases in proportion to the higher-molecular-weight species present.

7. With some cyclic aliphatic epoxy resins, viscosity differences will be found among the isomers.

VISCOSITY

Viscosities of liquid resins are usually measured with a rotating spindle instrument, such as the Brookfield viscometer. A typical procedure is indicated in Table 4-9. Viscosity vs. temperature for various of the liquid epoxy resins is indicated in Fig. 4-8. The sharp dependence of viscosity on temperature should be noted.

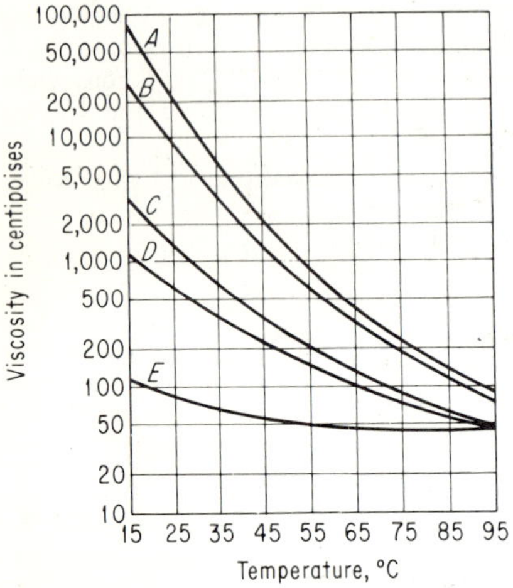

FIG. 4-8. Viscosity vs. temperature of five types of epoxy resins. *A*, Epoxidized phenol-formaldehyde novolac (*f* 2.3); *B*, DGEBA; *C*, 3,4-epoxy-6-methylcyclohexylmethyl-3,4-epoxy-6-methylcyclohexane carboxylate; *D*, diglycidyl ester of linoleic acid; *E*, DGE polypropylene glycol.

Solid resins are usually dissolved to a 40 per cent solution in butyl Carbitol. Solution viscosity is measured by spindle viscometers, by bubble viscometers such as the Gardner-Holdt tubes, or by the capillary viscometers, such as the Ostwald. Viscosity data for various solid resins in solution are given in Chap. 24.

EPOXIDE EQUIVALENT

The term *epoxide equivalent* (or *epoxy equivalent weight*) is defined as the weight of resin in grams which contains one gram equivalent of epoxy. If the resin chains are assumed to be linear with no side branching and it is further assumed that an epoxy group terminates each end, then the epoxide equivalent is one-half the average molecular weight of a diepoxy resin, one-third the average molecular weight of a triepoxy, etc. The term *epoxy value* is also employed and represents the fractional number of epoxy groups contained in 100 grams of resin. The two terms are convertible. Dividing the epoxy value into 100 gives the epoxide equivalent.

Determination of Epoxide Equivalent by Wet Analysis

The most frequently reported and, in general, the most specific and practical methods for determination of epoxide equivalents are based on the addition of a hydrogen halide to the epoxy group.

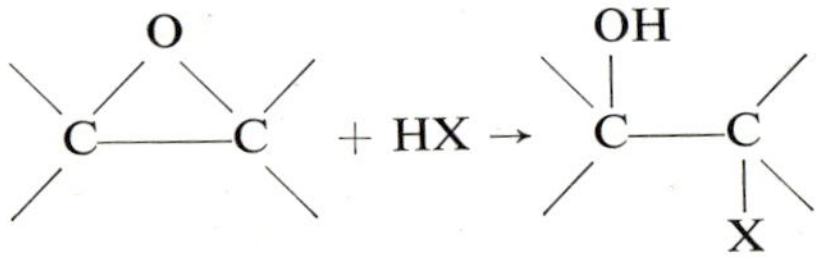

The difference between the amount of acid added and the amount unconsumed, determined by titration with standard base, is a measure of the *epoxy content*.

In practice, hydrogen chloride, hydrogen bromide, or hydrogen iodide is used (Table 4-10). One procedure involves the use of hydrogen bromide generated in situ from perchloric acid and a quaternary ammonium bromide and is reported to provide faster titrations and sharper end points than with HBr–acetic acid [45].

The four most common methods are the pyridinium chloride–pyridine method, the hydrogen bromide–acetic acid method, the hydrochloric acid–potassium iodide

Table 4-9. Test Method for Viscosity of Epoxy Compounds [32]

Apparatus:

Brookfield viscometer, model RVF
Thermometer, centigrade with 0.1° divisions
600-ml glass beaker or quart open-top paint can

Procedure:

1. Place 500 ml of the sample to be tested in the clean beaker or can. Insert the selected spindle and guard into the sample, being careful to avoid the trapping of air under the spindle plate.
2. Bring the sample to 25 ± 0.1°C.

Testing:

1. Attach the spindle and guard to the Brookfield viscometer, and adjust the spindle depth to the notch on the spindle shank.
2. Start the viscometer. Allow the spindle to rotate for ½ min. Stop the indicator by use of the clutch. Then stop the instrument, and read the dial. Allow the spindle to rotate an additional 3 to 4 times, and take a second reading. If the original reading agrees with the second reading, record this figure. If, however, the readings differ, continue spindle rotation for a reasonable period of time until a constant value is obtained.
3. Convert the reading obtained from the dial to viscosity in centipoises in accordance with the conversion table which accompanies the Brookfield viscometer.
4. Record the following information for each determination:
   a. Viscosity in centipoises
   b. Spindle
   c. Speed
   d. Temperature of sample to the nearest 0.1°C
   e. Ambient room temperature to the nearest 0.5°C

method, and potentiometric titration with hydrochloric acid. Procedures for conducting these methods are detailed in Tables 4-11 to 4-14.

The applicability of each method depends somewhat on the resin type being analyzed, as well as on other factors. First, the reagents should provide good solvent properties for the sample, particularly if the sample is a solid. Thus, the aqueous magnesium chloride–hydrochlorination method has a limited scope because only such epoxies as glycidol are readily soluble in the aqueous salt solution, although less soluble epoxies may dissolve slowly and consequently may react at a somewhat slower rate. The alcoholic magnesium chloride–hydrochlorination method is more general, in that many higher epoxies, such as liquid aromatic glycidyl ethers, are soluble in the reagent. However, high-molecular-weight epoxies such as the more highly polymerized aromatic glycidyl ether epoxy resins are not soluble in the alcohol medium. The hydrochloric acid–ethyl ether reagent is a good solvent for many monoepoxies but is not satisfactory for epoxy resins. The hydrochloric acid–Cellosolve reagent is a fairly good solvent for epoxy resins, but most samples go into solution rather slowly, requiring heating and mixing over a considerable length of time. Pyridine, chloroform, and dioxane are all good solvents for epoxies, including aromatic glycidyl ether resins.

Another factor in the applicability of a method depends on the extent to which undesirable side reactions occur. Any reaction of the epoxy is undesirable which does not result in the net consumption of one mole of acid per equivalent of epoxy group. For example, under certain conditions, some epoxies tend to isomerize to the corresponding carbonyl compounds and some of the epoxies or their chlorohydrins are susceptible to hydrolysis or alcoholysis. These reactions offer competition for the desired hydrochlorination reaction.

Various other organic radicals present in commercial resins may interfere with the desired reaction. For example, phenols interfere in the pyridinium chloride–chloroform method by causing poor end points; organic acids lower precision and accuracy, and in some methods various esters partially hydrolyze. A detailed study of such interfering factors is found in [12].

Table 4-10. Features of Eleven Epoxy Assay Methods [12, 17, 28, 42, 44]

Method	Reagent	Apparatus required	Reaction temperature	Reaction time	Titrant	Indicator
Aq. magnesium chloride–hydrochlorination	Satd. aq. magnesium chloride containing 0.1 *N* HCl	Capped bottle	Room temperature	15–30 min	Aq. 0.1 *N* NaOH	Methyl orange
Alc. magnesium chloride–hydrochlorination	0.5 *N* HCl–magnesium chloride in ethanol	Glass-stoppered flask	Room temperature	30 min	Aq. 0.5 *N* NaOH	Bromocresol green
Hydrochloric acid–Cellosolve	0.2 *N* HCl in Cellosolve	Capped bottle and oven	65°C	4 hr	Aq. 0.1 *N* NaOH	Bromothymol blue
Hydrochloric acid–ethyl ether	0.2 *N* dry HCl in ethyl ether	Glass-stoppered flask	Room temperature	3 hr	Aq. 0.1 *N* NaOH	Phenolphthalein
Hydrochloric acid–dioxane	0.2 *N* HCl in dioxane	Glass-stoppered flask	Room temperature	15 min	Methanolic 0.1 *N* NaOH	Cresol red
Pyridinium chloride–pyridine	0.2 and 1 *N* HCl in pyridine	Flask and reflux apparatus	Reflux, ca. 115°C	20 min	Aq. 0.5 *N* or methanolic 0.1 *N* NaOH	Phenolphthalein
Pyridinium chloride–chloroform	0.2 and 1 *N* dry HCl in pyridine and chloroform	Flask and reflux apparatus	Reflux, ca. 60°C	30 min (2 hr for resins)	Methanolic 0.1 or 0.5 *N* NaOH	Phenolphthalein
Hydrogen bromide–acetic acid	0.1 *N* hydrogen bromide–acetic acid in chlorobenzene	Flask or buret	Room temperature	15–30 min	....	Crystal violet
Hydrogen iodide	Potassium iodide–hydrochloric acid in *n*-propanol	Flask and reflux apparatus	Reflux, approx. 60°C	30 min	1 *N* HCl	Bromophenol blue
Hydrochloric acid–methyl ethyl ketone	Dil. HCl in MEK	Flask	Room temperature	1 hr	Alc. KOH	Cresol red
Potentiometric titration procedure	1 *N* HCl in distilled H_2O and 4 *N* HNO_3 in distilled H_2O–acetone	Flask, potentiometer	Room temperature	20 min	1 *N* $AgNO_3$	Electrode system

Table 4-11. Procedure for Determination of Epoxide Equivalent by the Pyridinium Chloride Method [22]

Reagents:

1. 0.2 *N* pyridinium chloride in pyridine (16 ml conc. HCl per liter of pyridine)
2. Standardized 0.5 *N* methanolic NaOH (20 g NaOH per liter)
3. Phenolphthalein indicator (0.1 g phenolphthalein per 100 ml methyl alcohol)
4. Methyl alcohol, reagent grade

Procedure:

Place a weighed sample of epoxy compound (2 to 4 milliequiv.) into a 200-ml round-bottom flask. Pipet 25 ml of pyridinium chloride–pyridine solution into another 200-ml flask. This will be used as a blank throughout the procedure. Swirl the solution until all the sample has been dissolved, heating gently if necessary. Use a heating mantle plus a magnetic stirrer. After the sample has dissolved, add a reflux condenser, and reflux and stir the solution for 25 min. After refluxing, cool the solution with the condenser in place. Add 50 ml of methyl alcohol through the reflux condenser and let drain. Remove the reflux condenser. Add 15 drops of phenolphthalein indicator. Titrate with 0.5 *N* methanolic NaOH from a 10-ml buret to a pink end point.

Calculations:

$$\text{Epoxide equivalent} = \frac{(16)(\text{sample weight, g})}{(\text{g of oxirane oxygen in sample})}$$

$$\text{Grams of oxirane oxygen in sample} = (A - B)(N)(0.016)$$

A = milliliters of NaOH for blank
B = milliliters of NaOH for sample
N = normality of NaOH
0.016 = milliequivalent weight of oxygen in grams

Table 4-12. Procedure for Determination of Epoxide Equivalent by the Hydrogen Bromide Method [42]

Reagents:

1. Acetic acid, glacial, reagent grade.
2. Hydrobromic acid (30 to 32% in acetic acid).
3. Crystal violet indicator solution. Prepare a 0.1% solution in glacial acetic acid.
4. Sodium carbonate, primary standard grade, dried to constant weight at 120 to 140°C.
5. Chlorobenzene, reagent grade.
6. Hydrogen bromide in acetic acid, 0.1 *N* maximum. Normality of HBr shall not exceed 0.1. Weigh 52 g hydrobromic acid (30 to 32% in acetic acid) into a 2-liter container fitted with a glass stopper. Fill to 2 liters with glacial acetic acid and mix thoroughly. Standardize the HBr-HAc solution daily. Standardize against 0.1 g anhydrous Na_2CO_3 dissolved in 10.0 ml glacial HAc (accurately measured by pipet) and titrate to the blue-green end point of the crystal violet indicator.

Procedure:

Use that quantity of sample which contains 0.001 to 0.002 equiv. of epoxide (approximately 0.25 to 0.50 g). Weigh the amount of sample desired (to the nearest milligram) into a 125-ml flask. Add 10.0 ml chlorobenzene to the sample, and use a magnetic stirrer until sample is dissolved. Add 4 to 6 drops crystal violet indicator solution, place the rubber stopper in position, and lower the buret tip to a point just above the solution. Titrate the solution with 0.1 *N* HBr in HAc to the blue-green end point of the crystal violet indicator. The titration rate shall be slowed near the end point to allow sample time for the reaction to go to completion. An effort should be made to obtain the same color at the end point as that obtained during standardization of the reagent.

Determine the reagent blank, using the same reagents and procedure, but omitting the sample. Run samples in duplicate and report average value.

Calculations:

Standardization of HBr in HAc:

$$N \text{ of HBr} = \frac{(\text{wt. of } Na_2CO_3)(1{,}000)}{(53)(\text{ml of HBr})}$$

Epoxide equivalent:

$$\text{Weight per epoxide equivalent} = \frac{(w)(1{,}000)}{(S - B)(N)}$$

w = weight of sample
S = milliliters of HBr used in titrating sample
B = milliliters of HBr used in titrating blank
N = normality of HBr

Table 4-13. Procedure for Determination of Epoxide Equivalent by the Potassium Iodide Method [44]

Reagents:

1. Normal hydrochloric acid (1 *N* HCl).
2. Potassium iodide solution (3.0 g KI in 5.0 g distilled water). KI should be low in iodates (0.0003% or less).
3. Indicator solution (0.1 g bromophenol blue in 99.9 g *n*-propanol).
4. *n*-Propanol

Procedure:

Weigh 1.0 to 1.5 g of the resin accurately to 0.001 g into a tared 125-ml Erlenmeyer flask. To the resin is added 25 ml of *n*-propanol. The Erlenmeyer flask is assembled on a gyrotherm and the resin *n*-propanol mixture brought to reflux while stirring. When the mixture is refluxing and the resin has dissolved, raise the buret with attached adapter, and quickly add 20 drops of the indicator solution and the freshly prepared KI solution. The resulting mixture is brought to reflux and titrated with 1 *N* HCl until a yellow end point is attained (i.e., when one drop of the HCl changes the color from blue to yellow and yellow persists for 1 min.)

During the titration, as the HCl is added, the blue solution immediately turns yellow, indicating an excess of HCl. This excess reacts with the KI and forms HI, which in a few seconds adds to the epoxy group. When the HI has completely reacted, the solution immediately turns blue. The speed of the HI addition depends upon the temperature. The titration is carried out under total reflux.

The resin solutions may become turbid during the titration. If the solution becomes turbid, add an additional few milliliters of *n*-propanol, bring to reflux, and continue titration.

Calculation:

$$\text{Epoxy equivalent, moles epoxy/kg} = \frac{\text{ml HCl used} \times N\ \text{HCl}}{\text{sample wt., g}}$$

Resultant values will be 1 to 3% less than theoretical.

Table 4-14. Procedure for Determination of Epoxide Equivalent by Potentiometric Titration [42]

Reagents:

1. Acetone, reagent grade
2. Dioxane, reagent grade
3. Hydrochloric acid—approximately 1 *N*, 85 ml conc. HCl diluted to 1,000 ml with distilled water. NOTE: It is convenient to adjust the normality of the HCl to just below 1.0 *N* in order to obtain blank readings less than 10.0 ml with the microburet.
4. Nitric acid—approximately 4.0 *N*, 250 ml conc. HNO_3 diluted to 1,000 ml with distilled water.
5. Silver nitrate—1.0 *N*, 84.5 g $AgNO_3$ dissolved in distilled water to a volume of 500 ml. Standardize against sodium chloride. CAUTION: Silver nitrate solutions are corrosive and stain very badly. Use utmost care when handling such solutions.

Procedure:

Use that quantity of sample which contains 0.002 to 0.004 equiv. of epoxide (approximately 0.4 to 0.8 g).

Weigh the amount of sample desired (to the nearest milligram) into a 250-ml beaker. Add 25 ml dioxane to the sample, cover with a watch glass, and use a magnetic stirrer until sample is dissolved. Carefully add 10 ml of 1 *N* HCl. Continue stirring for 20 min. Add 50 ml acetone, 35 ml distilled H_2O, and 2 ml of 4 *N* HNO_3. Use a measuring pipet to add the HNO_3.

Insert metallic silver and silver–silver chloride electrode system on pH meter. Titrate with 1 *N* $AgNO_3$, while stirring rapidly with a magnetic stirrer, to the potentiometric end point of 300 mv.

Run three blanks with each series of samples that are run at one time, using the same reagents and procedure, but omitting the sample. Use the average of the three determinations for the blank value. Run samples in duplicate, and report average value.

Calculations:

$$\text{Weight per epoxide equivalent} = \frac{(w)(1{,}000)}{(B - S)(N)}$$

w = sample weight
B = milliliters of $AgNO_3$ used in titrating blank
S = milliliters of $AgNO_3$ used in titrating sample
N = normality of $AgNO_3$

Determination of Epoxide Equivalent by Infrared Spectroscopy

Infrared spectroscopy may be used to measure epoxide equivalent, although this procedure is not in as common use as wet analysis.

Table 4-15. Infrared Band Intensity vs. Molecular Weight of the DGEBA [13]

Wavelength, microns	Average molecular weight of resin						
	350–400	450	700	900–1,000	1,400	2,900	3,800
2.90	0.126	0.186	0.332	0.402	0.409	0.411	0.461
3.37	0.736	0.749	0.701	0.791	0.665	0.726	0.764
6.21	1.00	1.00	1.00	1.00	1.00	1.00	1.00
7.45	0.300	0.284	0.319	0.324	0.267	0.253	0.296
8.45	1.904	2.25	1.88	2.11	1.910	2.22	1.988
9.05	0.340	0.401	0.540	0.600	0.565	0.611	0.593
10.95	0.712	0.578	0.450	0.337	0.260	0.204	0.193
11.60	0.592	0.480	0.393	0.289	0.228	0.180	0.157
12.06	2.01	2.03	1.63	1.81	1.682	1.71	1.573

In going from lower- to higher-molecular-weight species of the same resin, bands in the infrared spectrum change in intensity as the percentage composition of the various groups change. With the DGEBA resin, these changes in intensity have been related to changes in molecular weight, using the absorption band at 6.21 microns as a reference (Table 4-15). When these relative absorbances are plotted as a function of either the epoxide equivalent or the chemically determined molecular weight, reasonably smooth curves are obtained. If the relative absorbance is corrected to take into account the residual absorption at 10.95 microns in the absence of epoxy groups, a corrected absorbance, $R_C = R - 0.173$, is obtained. A plot of log R_C vs. epoxide equivalent (Fig. 4-9) shows that a linear relationship results. Similar values have been obtained with the band at 11.60 microns [13].

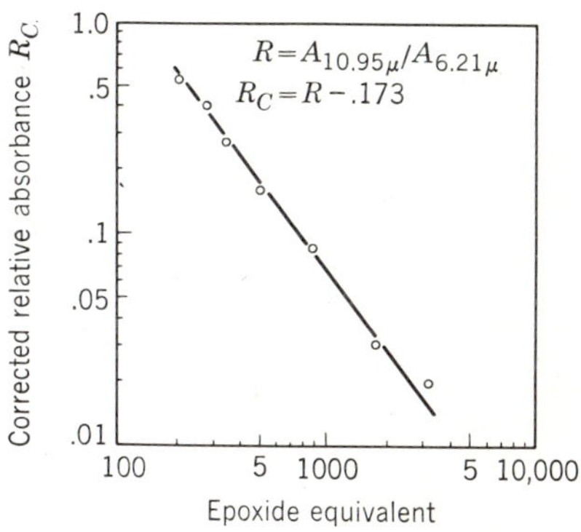

FIG. 4-9. Corrected infrared absorbance vs. epoxide equivalent of DGEBA resins [13].

The specific epoxy resin will determine which bands should be used for the calculation of epoxide equivalent. With the brominated DGEBA, for example, the combined reference bands at 6.21 and 13.2 microns are used rather than the 6.21- and 10.95-micron bands used for DGEBA [49].

Determination of Epoxide Equivalent by Near-infrared Spectroscopy

Near-infrared spectroscopy covers the spectral interval from about 0.6 to 2.5 microns in contrast to the conventional infrared region which extends from 2 to about 15 microns. The absorption bands seen in the near-infrared region are mostly first or second overtones of, or combinations of, the frequencies observed in the conventional region.

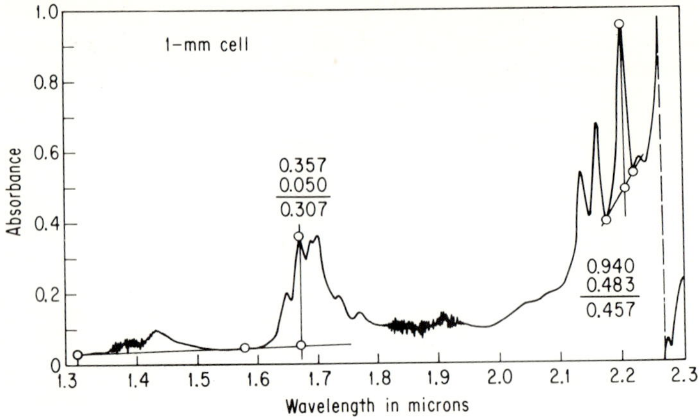

FIG. 4-10. Near-infrared spectrum of DGEBA showing evaluation of epoxy peak at 2.05 microns and reference peak at 1.668 microns [34].

Table 4-16. Near-infrared Absorbance for Epoxy Groups [34]

DGEBA	Density, d_4^{20}	Epoxy value by wet analysis		Absorbance			
		equiv./100 g	equiv./liter	2.205μ, 1-mm cell	1.668μ, 1-mm cell	$\frac{A_{2.205}}{A_{1.668}}$	1.159μ, 10-mm cell
Crystallized	1.162	0.585	6.79	0.707	0.311	2.27	0.164
Distilled	1.165	0.573	6.68	0.685	0.311	2.20	0.161
Mol. wt. 380	1.165	0.522	6.08	0.620	0.309	2.015	0.140
Mol. wt. 450	1.163	0.395	4.60	0.457	0.307	1.490	0.090
Mol. wt. 950–1,000	1.188	0.197	2.34	0.208	0.293	0.710	0.015
Mol. wt. 1,400	1.177	0.099	1.164	0.221 (3 mm)	0.937 (3 mm)	0.184	

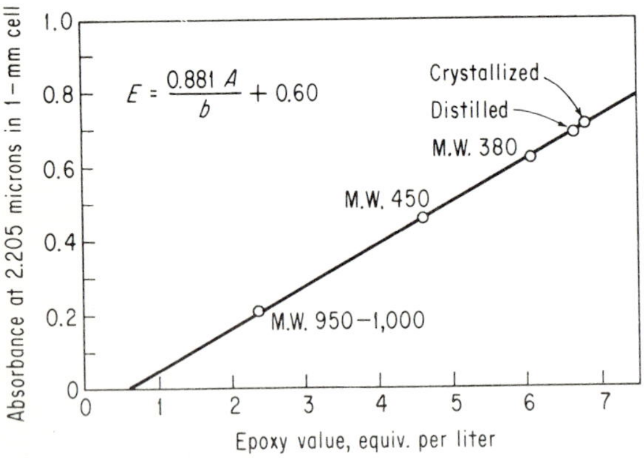

FIG. 4-11. Determination of epoxy values (direct method) vs. molecular weight of DGEBA [34].

A well-defined peak for α-epoxies is that at 2.205 microns, which probably represents a combination of a C—H stretching fundamental ($\nu = 3{,}050\ cm^{-1}$) with a CH deformation ($\nu = 1{,}460\ cm^{-1}$) [34]. A typical near-infrared absorption spectrum for the DGEBA is shown in Fig. 4-10. A calibration curve, comparing

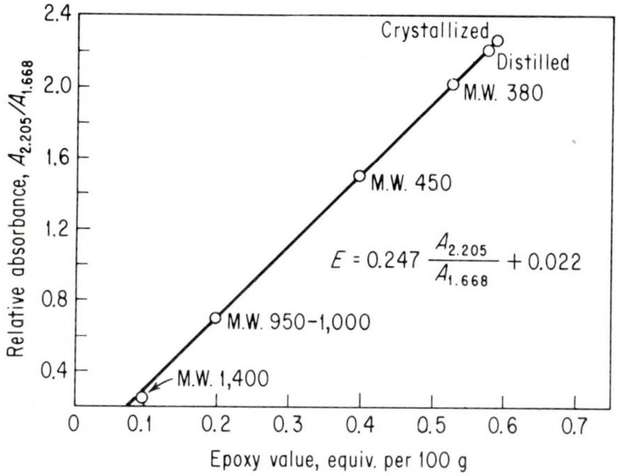

FIG. 4-12. Relative absorbance in near infrared vs. molecular weight of DGEBA [34].

epoxy value determined by wet analysis with absorbance at 2.205 microns, for five epichlorohydrin–bisphenol A resins is presented in Fig. 4-11. The data are tabulated in Table 4-16.

Indirect methods have also been used for determining epoxide equivalent. The variable absorbance of the epoxy group can be compared against a relatively nonvariable absorbance, such as the first overtone of the aromatic carbon–hydrogen stretching mode at 1.668 microns for the DGEBA resins [34]. Data for such a correlation are given in Table 4-16 and Fig. 4-12.

The absorption band at 1.159 microns has also been employed to determine epoxide equivalent (Fig. 4-13).

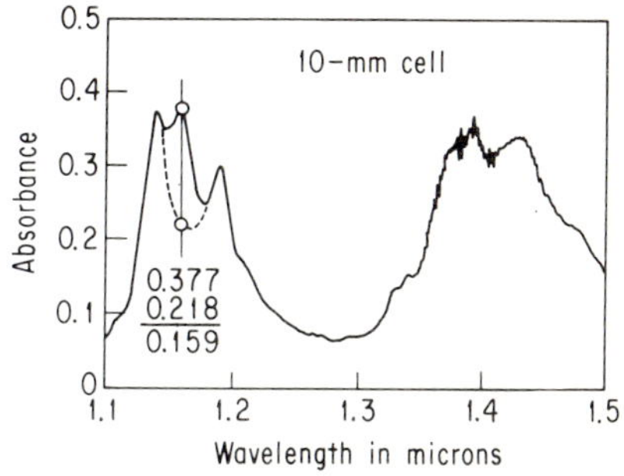

FIG. 4-13. Infrared spectrum of DGEBA showing evaluation of epoxy peak at 1.159 microns [34].

HYDROXYL EQUIVALENT

Hydroxyl equivalent is the weight of the resin containing one equivalent weight of hydroxyl group. It may also be expressed as equivalents per 100 grams. Hydroxyl equivalents may be determined by several methods, the most common being (1) esterification with acids, (2) reaction with acetyl chloride, (3) reaction with lithium aluminum hydride, and (4) reaction with phenyl isocyanate. Near-infrared spectroscopy may also be used.

Wet analysis, such as the benzyltrimethylammonium periodate method, is used for determining α-glycol content of the epoxy resins.

The hydroxyl equivalent of the resin is of importance in estimating average molecular weights and, since the hydroxyl groups are potentially reactive sites, in determining system functionality. In addition, with certain curing agents, the hydroxyl groups may activate the curing reaction or accelerate the cure.

Esterification Method

The esterification method involves esterifying the resin with about twice the theoretical amount of linseed oil acids necessary to react with all the hydroxyl groups and epoxy groups at a temperature of 225°C, until a constant acid value is obtained, and then back-titrating the unreacted linseed acids and calculating the hydroxyl groups plus epoxy content, one epoxy group being taken as equivalent to two hydroxyl groups.

Acetyl Chloride Method

The acetyl chloride method for determining hydroxyl equivalents is perhaps the most common, and the procedure is given in detail in Table 4-17.

Table 4-17. Acetyl Chloride Method for Determining Hydroxyl Equivalents

Reagents:

1. Acetone, reagent grade
2. Dioxane, reagent grade
3. 1.5 *N* acetyl chloride in toluene
4. Pyridine, C.P.
5. Toluene
6. 0.5 *N* alcoholic sodium hydroxide
7. Indicator solution (cresol red 0.1% solution in 50% ethanol)

Procedure:

Dissolve 1 or 2 g epoxy in 10 ml dioxane heated to 60°C. After solution has cooled to room temperature, add 10 ml acetyl chloride solution from a pipet. Swirl gently, and chill to 0°C. Add 2 ml pyridine by pipet. Stopper flask, shake thoroughly, and place in water bath at 60 ± 1°C. The stopper is loosened momentarily to release pressure. The mixture is held at temperature for 1 hr and shaken vigorously every 10 min. Cool flask in ice-water bath, and add 25 ml chilled distilled water. Shake well and allow to stand in the ice bath 30 min, with occasional shaking to decompose the excess reagent. Add 25 ml chilled acetone to prevent emulsification. Add a few drops indicator solution, and titrate with sodium hydroxide solution to first definite violet color. Any slow fading of the indicator color at the end point is disregarded.

Duplicate blank determinations are made under identical conditions but omitting the resin sample.

Calculation:

The difference between the average blank and the sample titration is a measure of the hydroxyl content.

Lithium Aluminum Hydride Method

In the lithium aluminum hydride method, the epoxy resin is dissolved in a measured quantity of tetrahydrofuran and allowed to react with an excess of lithium aluminum hydride at 0°C, and the evolved hydrogen is measured gasometrically. A blank determination is made using an identical volume of tetrahydrofuran. After suitable correction for water, acid, or other active hydrogen in the sample, the net volume of hydrogen evolved during the reaction is taken as a measure of the hydroxyl content of the sample [24].

Phenyl Isocyanate Method

The phenyl isocyanate method for determining hydroxyl equivalent weights is based on the reaction of the isocyanate group with hydroxyls. Details are given in [40].

Measuring Hydroxyl Equivalent by Near-infrared Spectroscopy

Measurement of hydroxyl equivalent by infrared or near-infrared spectroscopy is not as straightforward as the measure of epoxy equivalent weight. The hydroxyl groups do not exist as one species, giving precise absorption bands, but rather the hydroxyl groups are present in a variety of hydrogen-bonded forms. That is, the hydrogen atom of the hydroxyl group, besides being bonded to the oxygen atom of the hydroxyl group, is also partially bonded to the oxygen atom of adjacent oxygen or other polar groups. Hydrogen bonding may occur intra- or intermolecularly. The bands due to free and hydrogen-bonded hydroxyl groups vary in intensity and wavelength.

Studies in which the equilibrium between the various hydrogen bonds were shifted by elevated temperatures and then allowed to re-form during cooling have revealed the existence of a point where absorption was independent of the concentration for the various intra- and intermolecularly bonded forms [34]. A curve showing this isosbestic point at 1.474 microns for the diglyceryl ether of bisphenol A is presented in Fig. 4-14. A similar isosbestic point was located for two other model compounds (Figs. 4-15 and 4-16 and Tables 4-18 and 4-19). At these points the absorption was a function only of *total* hydroxyl concentration, regardless of its bonded type.

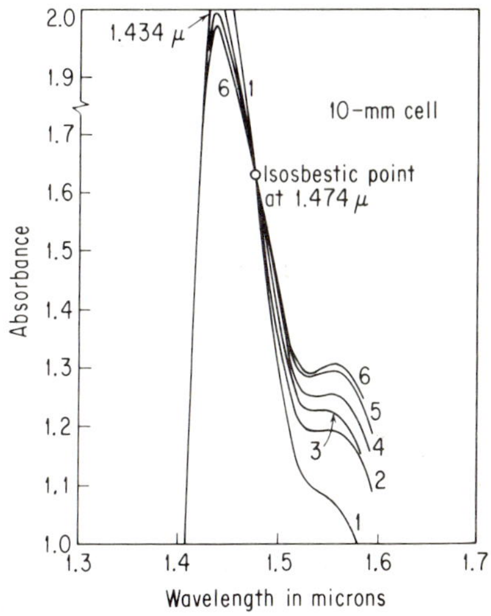

FIG. 4-14. Spectra of diglyceryl ether of bisphenol A cooling from 100°C [34]. Numbers on curves indicate sequence of spectra taken during cooling.

Use of the isosbestic point results in an analysis of hydroxyl equivalent weight having good agreement with wet methods (Table 4-20 and Fig. 4-17).

On samples having unknown path lengths, internal calibration can be used. Table 4-20 and Fig. 4-18 present a determination of hydroxyl equivalents when using the 1.188 micron peak (the second harmonic of a carbon-hydrogen stretching frequency of the methyl group.)

SOFTENING POINT

Uncured resins may be qualitatively graded by the ring-and-ball softening-point method or Durran's mercury method. These measure the temperature at which the resin reaches an arbitrary softness. Resins melting above room temperature are often characterized by this method in preference to solution viscosities or viscosity at elevated temperature. Typical softening-point values for the DGEBA resins are presented in Fig. 4-19.

In Durran's mercury method, a known weight of the resin is melted in a standard test tube and cooled, and a known quantity of clean mercury is placed on top of the solidified sample. The tube and contents are again heated, and the temperature at which the molten resin rises to the top of the mercury is recorded as the melting point. Since resins are not single crystalline chemical entities, it is not possible to

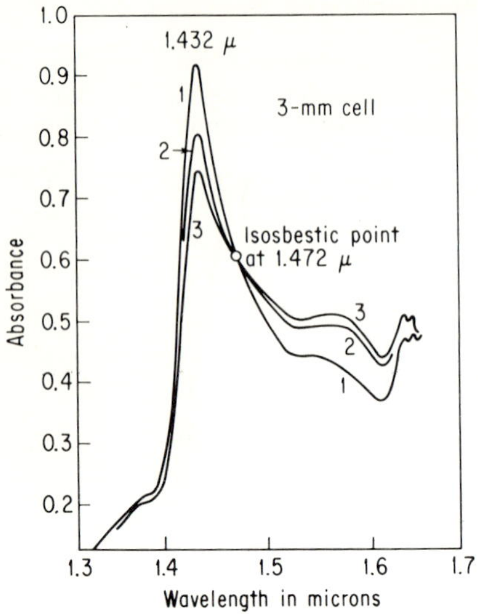

FIG. 4-15. Spectra of glyceryl *a*-phenyl ether cooling from 115°C [34]. Numbers on curves indicate sequence of spectra taken during cooling.

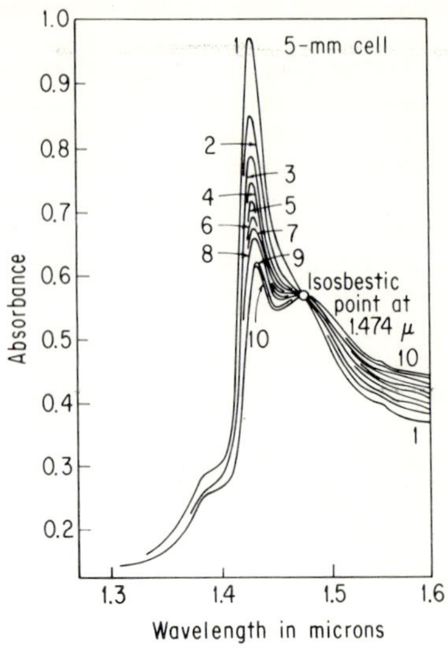

FIG. 4-16. Spectra of adduct from 4 moles phenyl glycidyl ether and 1 mole ethylenediamine cooling from 122°C [34]. Numbers on curves indicate sequence of spectra taken during cooling.

Table 4-18. Near-infrared Absorbance of an Adduct from 4 Moles Phenyl Glycidyl Ether and 1 Mole Ethylenediamine, Showing Variability of OH Band with Temperature [34]

Cooling time, minutes*	Absorbance values			Apparent relative hydroxyl value†
	Base line at 1.08μ	OH peak at 1.428μ	Reference peak at 1.722μ	
3	0.126	0.973	0.973	1.044
11	0.109	0.840	0.911	0.911
17	0.101	0.770	0.906	0.831
23	0.101	0.726	0.905	0.777
31	0.101	0.695	0.915	0.730
37	0.102	0.674	0.917	0.702
44	0.103	0.655	0.922	0.674
52	0.102	0.639	0.922	0.655
92	0.105	0.606	0.928	0.609
120	0.105	0.599	0.932	0.597
235‡	0.105	0.597	0.931	0.596

* Sample was heated to 122°C and allowed to cool in the spectrophotometer while spectra were recorded at intervals.

† Calculated as $(A_{1.4281} - A_{1.08})/(A_{1.722} - A_{1.08})$ to compensate for change of density.

‡ Sample solidified to a glass.

Table 4-19. Near-infrared Spectroscopic Isosbestic Data for Model Compounds and Epoxy Resins [34]

Compound	Path length b, cm	Position of OH peak, microns	Position of Isosbestic point, microns	Absorbance at isosbestic point, $\frac{A_{isos} - A_{backgr}}{b}$	OH value, equiv./100 g	Absorptivity, at isosbestic point, g/cm-equiv.
Glyceryl α-phenyl ether	0.3	1.432	1.472	1.625	1.077*	151
Diglyceryl ether of bisphenol A	1.0	1.434	1.474	1.336	1.12*	119
Adduct of 4 moles phenyl glycidyl ether + 1 mole ethylene-diamine	0.5	1.428	1.474	0.884	0.605†	146
DGEBA, mol. wt.:						
380	0.5	1.429	1.456			
450	0.5	1.427				
950–1,000	0.5	1.427	1.456			
1,400	0.5	1.428				
2,900	0.5	1.427				

* By chemical analysis.
† Theoretical value.

Table 4-20. Determination of Hydroxyl Values of Uncured DGEBA by Near-Infrared Spectroscopy [34]

Values	DGEBA, molecular weight 350	380	450	950–1,000	1,400
Analysis (chemical):					
Epoxy value, equiv./100 g	0.573	0.522	0.395	0.198	0.112
Hydroxyl value, equiv./100 g	0.016	0.056	0.129	0.273	0.306
Density, d_4^{20}	1.1653	1.165	1.163	1.188	1.177
Hydroxyl value, equiv./liter	0.186	0.653	1.500	3.24	3.60
Spectroscopic conditions:					
Path length b, cm	1.0	1.0	0.5	0.5	0.5
Temperature	Room	Room	Room	Room	Hot
Absorbance values:					
$A_{1.080}$(background)	0.017	0.040	0.038	0.048	0.134
$A_{1.188}$(methyl)	0.226	0.285	0.171	0.190	0.276
$A_{1.456}$(OH, isosbestic)	0.184	0.305	0.239	0.376	0.479
Evaluation:					
$\frac{A_{1.456} - A_{1.080}}{b}$	0.201	0.265	0.402	0.656	0.690
$\frac{A_{1.456} - A_{1.080}}{A_{1.188} - A_{1.080}}$	0.827	1.081	1.512	2.31	2.43
Resin n number	0	0.2	0.6	1.6	4
CH_3, %	8.83	9.05	9.40	9.81	10.15
Correction factor F	1	1.025	1.055	1.11	1.15
$F \times \frac{A_{1.456} - A_{1.080}}{A_{1.188} - A_{1.080}}$	0.827	1.11	1.61	2.56	2.80

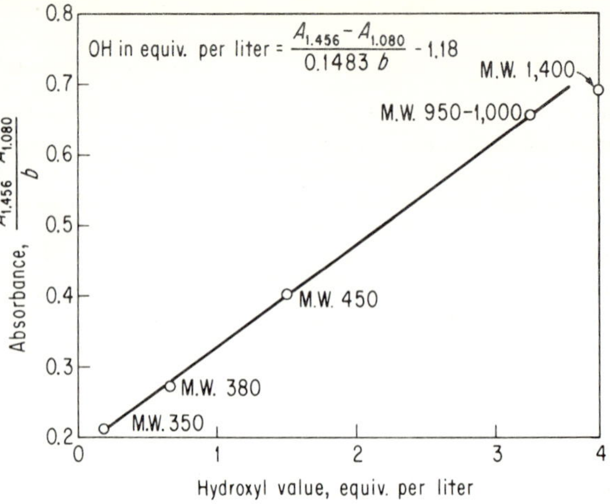

FIG. 4-17. Determination of hydroxyl values in DGEBA resins by near-infrared spectroscopy (direct method) [34].

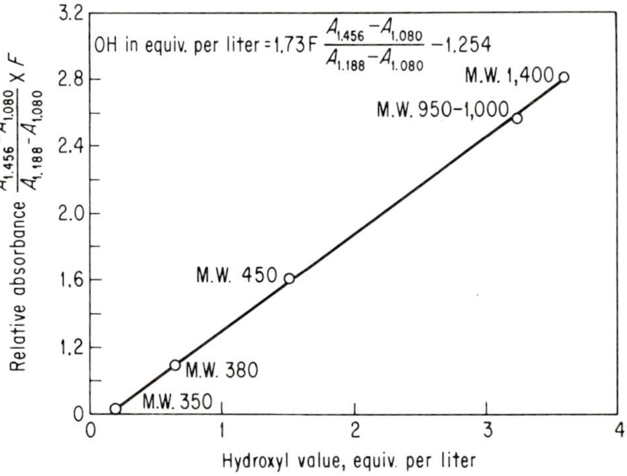

FIG. 4-18. Determination of hydroxyl values in DGEBA resins by near-infrared spectroscopy (relative method) [34].

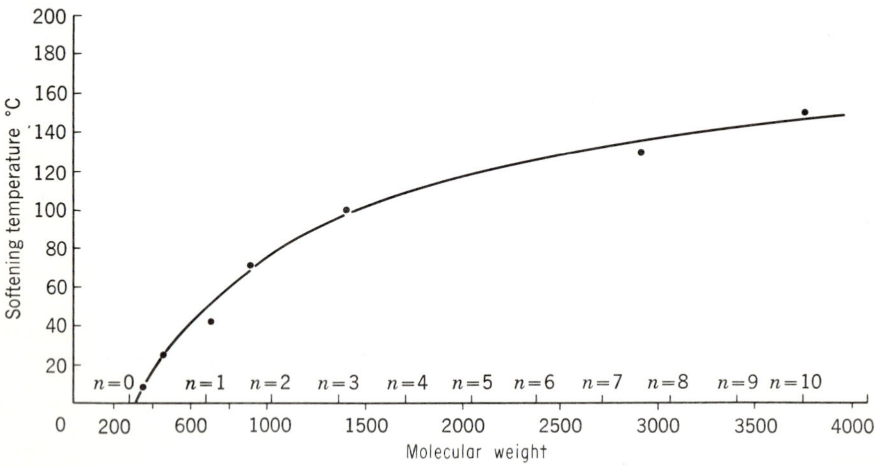

FIG. 4-19. Typical softening-point values for uncured DGEBA vs. molecular weight [6].

obtain sharp melting points. Any value for the melting point of the resin is empirical and is determined partly by the conditions under which the test is conducted. Durran's mercury method for determining melting points appears to stabilize these melting-point conditions and to eliminate to a considerable degree the possible personal bias of the observer.

An alternative method is to employ a dilatometer to obtain second-order transition temperatures, i.e., points at which a sharp break in thermal-expansion characteristics (or other physical properties) occurs, indicating an internal spatial rearrangement caused by expansion of the resin from the glassy state to the rubbery state.

IODINE NUMBER

The iodine-number—the number of milligrams of iodine absorbed by one gram of the compound—is given as a measure of the unsaturation present in the epoxy molecule, a low iodine value implying a low degree of unsaturation. Iodine number may be determined for example, by the Hanus method [8].

The iodine number used in conjunction with the per cent of oxirane oxygen can be used to determine the efficiency of the peracetic acid type synthesis.

MOLECULAR-WEIGHT DISTRIBUTION

In commercial resins, particularly of the DGEBA and epoxidized novolac types, the structure of the resinous mass may vary considerably because of a tendency of the epoxy groups to coreact during synthesis and produce a number of repeated units per molecule.

Conventional fractional precipitation from a solvent by a nonsolvent has been used to obtain DGEBA fractions varying in molecular weight [43]. Newer fractionation techniques which have been used on epoxy resins are gel-permeation chromatography [47] and preparative thin-layer chromatography [48].

Studies have shown, for example, that the highly purified DGEBA is over 99 per cent DGEBA of $n = 0$, whereas the commercial resins of 190 epoxy equivalent weight are approximately 88 per cent of $n = 0$, 10 per cent of $n = 1$, and 2 per cent of $n = 2$. Similarly, a high-molecular-weight solid DGEBA resin such as used in solution coatings and having an epoxy equivalent weight of 540, which is representative of a polymer of $n = 2$, contains, actually, over 50 per cent of $n = 3$, 4, and 5; about 15 per cent of $n = 2$; 15 per cent of $n = 1$; and 20 per cent of $n = 0$.

The ratio of molecular-weight species present, obviously, can be varied by manufacturing conditions, so that resins which are similar on an equivalent weight basis can differ widely in their viscosity, melting point, reactivity, film-forming characteristics, tensile strength, impact resistance, and other properties.

Another species which molecular analysis shows can be present is the 1,3-chlorohydrin isomer which is formed when bisphenol reacts with epichlorohydrin:

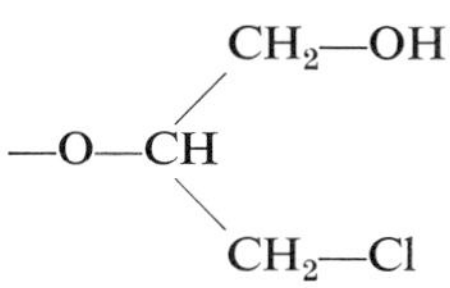

This species is especially undesirable because it does not dehydrohalogenate to form an epoxy group, and thus it reduces the functionality of the final resin to the extent of its presence.

COLOR

Color of the uncured resins is usually reported in terms of the Gardner liquid color standards or the Hellige color comparator or by means of APHA numbers.

Resins free of phenolic nuclei give better color in the uncured state, and in the cured state have less tendency to discolor on aging. With the aromatic-based materials, generally, the lower the molecular weight of the resin, the better will be the color.

SPECIFIC GRAVITY

The specific gravity of a resin is the weight of a given volume of resin compared with the weight of an equal volume of water.

Table 4-21. Test for Specific Gravity of Epoxy Compounds [31]

Apparatus:

200-ml cylinder

Procedure:

Method *A*: For low-viscosity unfilled liquids, ASTM D 901–56 hydrometer
Method *B*: For high-viscosity liquids or pastes

Bring a sample of at least 250 ml of ingredients to exact temperature by use of the water bath (25° ± 1°C), unless otherwise specified. Tare a clean, empty 200-ml cylinder on the triple-beam balance, and then obtain the average of 10 weighings of the cylinder filled with distilled water. Fill the cylinder by adding successive small portions of the ingredient, tapping to remove trapped air. Fill above the top, and then scrape off flush with top by use of a sharp, straight edge. Weigh the filled cylinder.

Calculation

$$\text{Specific gravity} = \frac{\text{wt. of cylinder charge, g} - \text{tare wt. of cylinder, g}}{(\text{wt. of cylinder} + \text{water}) - \text{tare wt. of cylinder, g}}$$

Average three determinations, each a separate filling of the cylinder.

A variety of methods may be used for the determination of specific gravity. Common instruments used with the liquid resins are the hydrometer and pycnometer. See Table 4-21 for methods.

REFRACTIVE INDEX

Refractive index may be defined as the ratio between the speed of light at a given wavelength in vacuum to the speed of light at the same wavelength in the substance measured.

The refractive index of epoxy resins is seldom reported. The commercial materials are not pure species of controlled molecular weight, and batch-to-batch variation in refractive index will be experienced. In view of the general unavailability of such data, values for a number of commercial resins and diluents, together with blends of resins and diluents, are given in Table 4-22.

CHLORINE CONTENT

Epoxy resins synthesized via the epichlorohydrin route will contain at least some hydrolyzable chlorine (i.e., present as 1,2-chlorohydrin as the result of incomplete dehydrohalogenation). This chlorine is considered "active."

Less active chlorine may also be present as 1,3-chlorohydrin (see above), as chloromethyl groups along the polymer chain, etc. Hydrolyzable chlorine may be determined by the procedure indicated in Table 4-23. Total chlorine content is measured by Parr bomb analysis.

Table 4-22. Refractive Index of Various Epoxy Resins and Diluents at 25°C

Material	*Index of refraction* n_D^{25}
Butyl glycidyl ether	1.4170
Diglycidyl ether of propylene glycol	1.4556
Diglycidyl ether of butanediol	1.4589
Limonene dioxide	1.4673
Vinylcyclohexene dioxide	1.4758
Diglycidyl ester of linoleic dimer acid	1.4846
Mixed isomers of bis(2,3-epoxycyclopentyl) ether	1.4877
Bis(3,4-epoxy-6-methylcyclohexylmethyl) adipate	1.4877
Bis(2,3-epoxycyclopentyl) ether (liquid isomer)	1.4895
3,4-epoxy-6-methylcyclohexylmethyl 3,4-epoxy-6-methylcyclohexane carboxylate	1.4921
p-Butylphenyl glycidyl ether	1.5148
Dicyclopentadiene dioxide (65% exo, 35% endo isomer)	1.5175
Diglycidyl ether of bisphenol-hexafluoroacetone	1.5186
Cycloaliphatic acetal epoxy (Ciba CY 175)	1.5209
Phenyl glycidyl ether	1.5291
50% bis(2,3-epoxycyclopentyl) ether in DGEBA	1.5296
Adduct of 2½ moles of DGEBA with 1 mole of hydrogenated castor oil	1.5325
DGEBA (mol. wt. 1,000) 60% solids in xylene and butyl Cellosolve (1:1)	1.5350
DGEBA (mol. wt. 4,000) 60% solids in xylene and Cellosolve acetate (1:1)	1.5375
DGEBA diluted with a glycidyl ether of a coke ester (C-8 or 9)	1.5400
DGEBA diluted with 37.5% bis(2,3-epoxycyclopentyl) ether	1.5414
DGEBA, 38%, DGE resorcinol, 38%, glycidyl ether of nonylphenol, 24%	0.5478
DGEBA diluted with 13% butyl glycidyl ether	1.5487
DGEBA diluted with 20% bis(2,3-epoxycyclopentyl) ether	1.5540
Triglycidyl *p*-aminophenol	1.5580
DGEBA diluted with dibutyl phthalate	1.5580
Triglycidyl *p*-aminophenol (molecularly distilled)	1.5660
Blend of DGEBA, 88%, DGEBA (mol. wt. 1,000), 12%, and xylene, 10%	1.5681
DGEBA (molecularly distilled)	1.5683
DGEBA (mol. wt. 360)	1.5689
DGEBA	1.5701
Epoxidized phenol-formaldehyde novolac ($n = 0.3$)	1.5804
Epoxidized phenol-formaldehyde novolac ($n < 1.6$)	1.5866
Triglycidyl ether of trihydroxydiphenyl	1.5895
Epoxidized phenol-formaldehyde novolac ($n = 1.6$)	1.5963

Table 4-23. Determination of Hydrolyzable Chlorine Content of Epoxy Resins [33]

Reagents:

1. 0.1 *N* alcoholic potassium hydroxide solution. Dissolve 5.61 g of reagent-grade KOH in 1 liter of methyl alcohol. No standardization is necessary.
2. 0.1 *N* hydrochloric acid solution—standardized.
3. Toluene, reagent grade.
4. Methyl alcohol, reagent grade.

Procedure:

Introduce 6 to 8 g of sample to be tested into each of two Erlenmeyer flasks. Pipet 50 ml of the 0.1 alcoholic KOH into each sample and into two blanks. Add 15 ml toluene from a graduated cylinder into each flask. Stopper and swirl until resin is in solution. Add 3 or 4 glass beads. Connect flasks to reflux condensers, and place on hot plate. Reflux gently for 15 min. Cool to room temperature with condensers in place. (When cool, wash condenser down with 5 ml methyl alcohol). Add 3 drops phenolphthalein indicator. Titrate with 0.1 *N* hydrochloric acid to the first disappearance of the pink color.

Calculations:

$$\frac{(B - A) \times N \times 3.55}{\text{wt.}} = \%\ \text{hydrolyzable chlorine}$$

A = milliliters of acid required to titrate sample
B = milliliters of acid required to titrate blank
N = normality of hydrochloric acid solution
wt. = weight of sample used

The active chlorine is capable, under some conditions, of blocking reactions with Lewis base catalysts, and when the amount of catalyst is fairly critical, variations in hydrolyzable chlorine can result in significant variations in cured properties [18]. Additionally of course, an inorganic base may react with the chlorine to form the neutral salt. However, the principal effect of increasing the chlorine content in the system, irrespective of the curing agent, is to decrease the system functionality and cross-linking density [20].

MISCELLANEOUS PROPERTIES

Perhaps characteristic of the unusual properties that may be required of the base epoxy resin are odor and defoam time under high vacuum.

Defoam Time under High Vacuum

In some critical potting applications, it is essential, in order to assure that viscosity during processing remains within acceptable limits, that the vacuum operation be completed within a minimum time period. This requires a high vacuum and requires that the bubbles produced by deaeration break with extreme rapidity. Since surface tension of the resin may vary considerably between commercial production batches, adequate test procedures must be established to permit selection of appropriate materials. A defoam under high-vacuum specification procedure is given in Table 4-24.

Table 4-24. Specification for Determining Defoam Time of Epoxy Resin under High Vacuum [42]

Procedure:

Thoroughly stir resin and curing agent separately. Weigh into 1,000-ml beaker sufficient curing agent to give 220 g of ultimate mixture. Weigh to ±0.1 g. Weigh out about 10% excess resin into a small glass beaker. Preheat the components to 54 ± 1°C.

Place the separate containers in a vacuum chamber, and evacuate the material at pressure of 2.5 ± 0.5 mm Hg abs. for 5 min ± 30 sec.

Reheat the components to 54 ± 1°C. Add exact amount of resin required (to an accuracy of ±0.1 g) to the preweighed curing agent. Thoroughly mix two components by hand for 1 min ± 5 sec. Immediately place the mixture in a vacuum chamber and reduce the pressure to 2.5 ± 0.5 mm Hg abs. in 30 ± 5 sec.

Record the time in which the foam collapses to a volume of 400 ml. This time shall be measured from the start of evacuation. The time requirement shall be waived if the maximum volume of foam rise is 400 ml or less.

Record the maximum volume of foam rise.

Odor

Odor is usually of minor concern with epoxy resins, since a number of the common curing agents possess strong or characteristic odors which mask any odor of the resin. However, in some applications female workers using epoxy resins at assembly benches have complained of the odor of one resin as compared with another.

Odor in epoxy resins can arise from a number of sources. The two chief sources are impurities such as sulfur in the bisphenol used to make the resin or from traces of solvent remaining in the resin after stripping it of expended caustic during manufacture.

A practical test is to put 100 grams of resin in a pint can, place the lid on the can loosely, and put the can in an oven at 125°C. After half an hour, the can is removed and smelled carefully. By doing this with several resins from different manufacturers, comparative odors may be determined.

COMMERCIALLY AVAILABLE EPOXY RESINS

A large number of epoxy resins are available commercially, and new varieties are, from time to time, being added.

The DGEBA resins account for the majority of the production, and are available from a number of United States suppliers. The epoxidized olefin resins, discussed in Chap. 3, are available in greater variety from a smaller number of producers. The listing in Appendix 4-2 to this chapter presents typical data on epoxy resins commercially available in the United States and Europe. Basic suppliers offer extensive literature on the resins [1 to 7].

SUMMARY

The properties of the uncured epoxy resin which are of most interest are the epoxide equivalent and viscosity. The epoxide equivalent indicates the number of grams containing 1 gram-mole of epoxy groups. Once the epoxide equivalent is known, it is then possible to calculate, in some cases, the optimum amount of curing agent required for the resin. The viscosity of the resin is a convenient index to its handling characteristics.

The equivalent weight of other reactive groups, such as hydroxyls and olefins may be required when the curing mechanisms are such that these groups will participate in the curing reaction. Other properties, of more limited interest, are the softening point, useful in characterizing various resins, the molecular weight and molecular-weight distribution, the color, specific gravity, refractive index, chlorine content, defoam time under high vacuum, and odor.

Much may be learned from an examination of the epoxy-resin structure. On the basis of the location of the epoxy rings within the molecule, rather reliable estimates may be made on the specific reactivity of the ring. The nature of the chemical backbone will indicate, in a general way, the flexibility, the thermal stability, and the solvent resistance which may be expected in the cured product, and to a less reliable extent, the viscosity that may be expected of the resin.

It should be stressed, however, that the epoxy resins are but building blocks in the development of an epoxy-resin formulation. The formulation will ultimately involve one or more epoxy resins together with curing agents; and modifiers such as plasticizers, flexibilizers, accelerators, stabilizers, flow control agents, and selected fillers. These aspects of formulation are presented in detail elsewhere in the book.

REFERENCES

1. Ciba Products Co., technical literature.
2. Dow Chemical Co., technical literature.
3. FMC Corporation, technical literature.
4. Jones-Dabney Co., technical literature.
5. Reichhold Chemicals, Inc., technical literature.
6. Shell Chemical Corp., technical literature.
7. Union Carbide Corp., technical literature.

1939

8. Furman (ed.), "Scott's Standard Methods of Chemical Analysis," D. Van Nostrand Company, Inc.

1950

9. Field, et al., Infrared Data on Epoxy Structures, *J. Chem. Phys.* **18**:1298.

1951

10. Shreve, et al., Infrared Absorption Spectra of Some Epoxy Compounds, *Anal. Chem.* **23**:277–282.

1953

11. Gunthard, et al., Infrared Spectra of Stereoisomeric Steroid Epoxides, *Helv. Chim. Acta* **36**:1900–1907.
12. Jungnickel, et al., Determination of the Alpha-epoxy Group, in Mitchell, et al. (eds.), "Organic Analysis," vol. 1, pp. 127–154, Interscience Publishers, Inc.

1954

13. Kagarise and Weinberger, Infrared Spectro-photography of Plastics and Resins, Government Report PB 111438, May.
14. Patterson, Infrared Absorption Bands Characteristic of the Oxirane Ring, *Anal. Chem.* **26**:823–835.

1956

15. Bassin, The Tolerance of Novolac Epoxy Resins to Converter Ratio, North New Jersey Section, ACS, March.
16. Dannenberg and Harp, Determination of Cure and Analysis of Cured Epoxy Resin, *Anal. Chem.* **28**:86–90.
17. Durbetaki, Direct Titration of Oxirane Oxygen with Hydrogen Bromide in Acetic Acid, *Anal. Chem.*, December.
18. Wynstra, The Chemistry of Epoxy Resins, North New Jersey Section, ACS, March.

1957

19. Henbest, et al., Detection of the Epoxide Group by Infrared Spectroscopy, *J. Chem. Soc.*, pp. 1459–1462.

1958

20. Belanger and Schulte, The Effect of Chlorine Contents on Electrical and Physical Properties of Cured Epoxy Resins, Epoxy Resin Symposium of SPE, Minneapolis, Minn., October.
21. Bronstein, Infrared Spectra of Oxirane Compounds, Correlation with Structure, *Anal. Chem.* **30**:544–546.
22. Knoll, et al., Determination of Epoxide Equivalents, ACS Symposium, Chicago, September.

1959

23. Wynstra, et al., Structure vs. Thermal Stability of High-temperature Performance Epoxy Resins, ACS Symposium, Atlantic City, September.

1960

24. Determination of Hydroxyl Group in Organic Compounds, Shell Method Series 237/55, Shell Chemical Corp.
25. Lee and Neville, New Developments in Epoxy Resins, *Insulation*, December.
26. Plueddemann, Epoxysiloxane Casting Resins, *J. Chem. Eng. Data*, January.
27. Wynstra, Some Correlation Between Mechanical Properties and Structure of Epoxy Resin Coatings, Union Carbide Plastics Co. Report.

1961

28. Jung and Kleeberg, Process for the Determination of Epoxy Group Concentration in Epoxy Resins, *Kunststoffe* **51**:714-715.

29. Kelley, et al., The Relationship of Molecular Structure and Resin Performance for a Series of Diglycidyl Ethers, ACS Symposium, Chicago.
30. Lee and Vincent, The Use of Infrared Spectroscopy in the Quality Control of Epoxy Resins, *Adhesives Age*, September.
31. Method of Test for Specific Gravity of Epoxy Compounds, SPI Method, August.
32. Method of Test for Viscosity of Epoxy Compounds and Related Compounds, SPI Method, August.

1962

33. Ciba Products Co. Technical Service Note 54.
34. Dannenberg, Determination of Functional Groups in Epoxy Resins by Near-infrared Spectroscopy, Shell Chemical Corp, Report P.1117.
35. Molseyev, et al., Thermal Destruction of Condensation Resins: Investigation of the Thermal Destruction of Epoxy Resins Using Tagged Atoms, *Plastmassy* **6**:11–15.

1963

36. Bremmer, Factors Influencing the Heat Stability of Fire Retardant Epoxy Resins, ACS Symposium, Los Angeles, April.
37. Lee, Thermal Stability and Degradation Schemes of Epoxy Resins, Battelle Symposium on Thermal Stability of Polymers, Columbus, Ohio, December.
38. May, The Influence of Bisphenol Structure on the Bond Strength of Epoxy Adhesives, TD Report ASD-TDR-396, April.
39. Ovenall, Electron Spin Resonance of Free Radicals in Epoxide Resins, *Polymer Letters* **1**.
40. Reed, et al., Phenyl Isocyanate Method for Determination of Hydroxyl Equivalent Weights of Polyalkylene Compounds, *Anal. Chem.*, April.
41. Reinking, et al., Polyhydroxyethers, part I, ACS Symposium, Los Angeles, April.
42. Quant, Carroll, and Moll, Materials Laboratory, Sandia Corp., private communication.

1964

43. Conley, Mechanisms of Thermal and Oxidative Degradation of Condensation Polymers, SPE Regional Technical Conference, Baltimore, June 4.
44. Dorman, Ciba Products Co., private communication.
45. Jay, Direct Titration of Epoxy Compounds and Aziridines, *Anal. Chem.*, March.
46. Lee, Mechanisms of Thermal Degradation of Phenolic Condensation Polymers, part III, ACS Symposium, Chicago, September.
47. Miles, Application of Gel Permeation Chromatography to the Characterization of Epoxy Resins, ASC Symposium, Chicago, September.
48. Spell and Eddy, The Characterization of Epoxy Resins with the Combined Technique of Thin Layer Chromatography and Infrared Spectroscopy, ACS Symposium, Chicago, September.

1965

49. MacKinney and Spalik, Infra-red Spectroscopy of Brominated Epoxy Resins, *SPE Trans.*, January.

Appendix 4-1

IR SPECTRA OF UNCURED EPOXY RESINS*

* From The Epoxylite Corp. Research Report 64–128, September, 1964 (revised December, 1965), by Art Cupples, Okko Koning, and Bette Wolfskill.

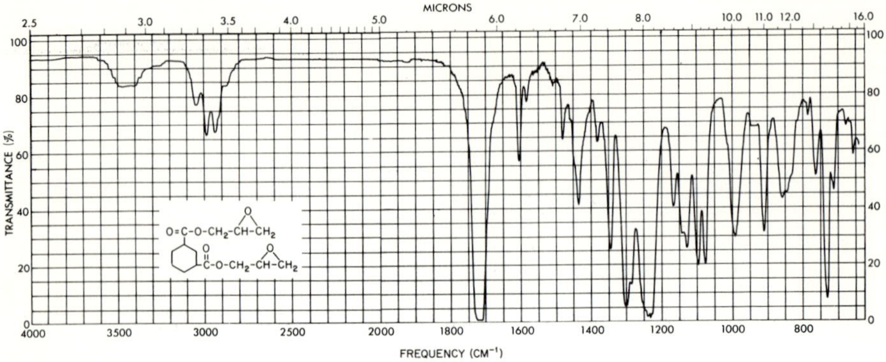

FIG. 4-1-1. Diglycidyl isophthalate (di-2,3-epoxypropyl isophthalate).

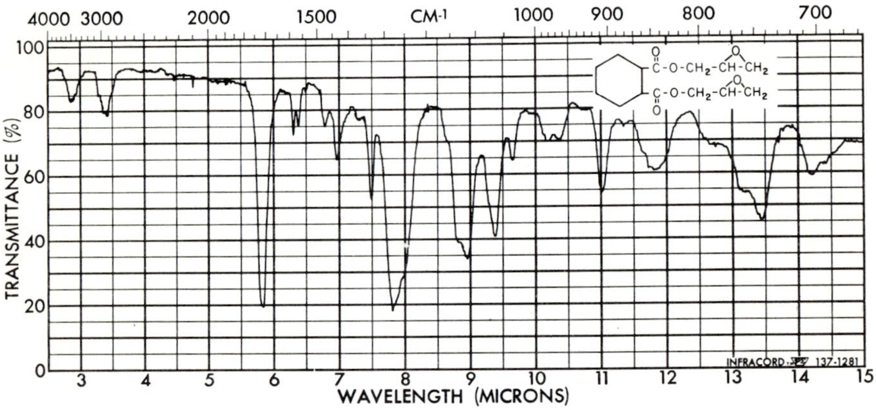

FIG. 4-1-2. Diglycidyl phthalate (di-2,3-epoxypropyl phthalate).

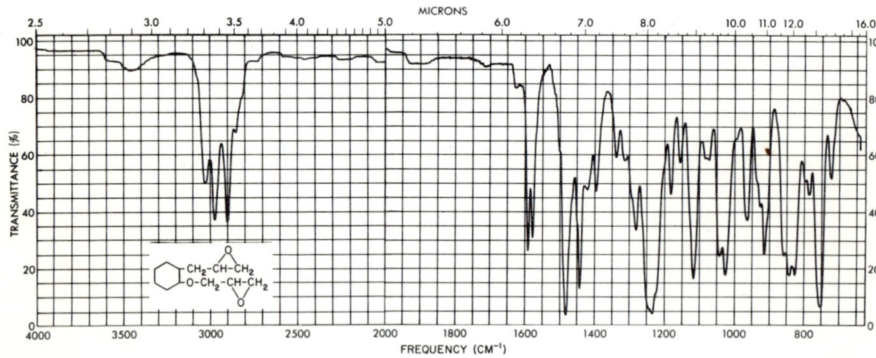

FIG. 4-1-3. *o*-Glycidyl phenyl glycidyl ether [*o*-(2,3-epoxypropyl)phenyl 2,3-epoxypropyl ether].

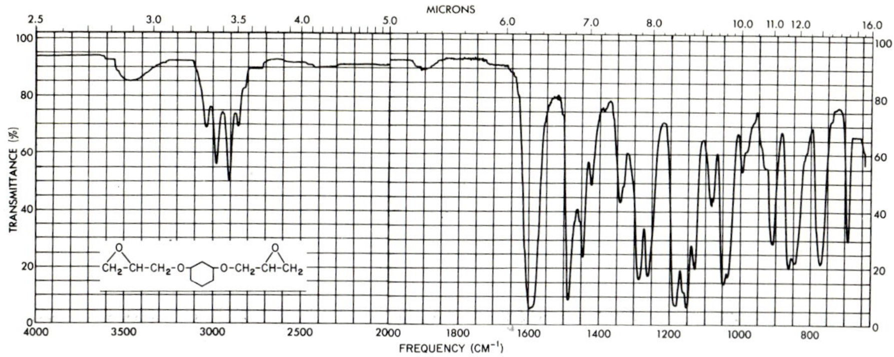

FIG. 4-1-4. Diglycidyl ether of resorcinol [1,3-bis-(2,3-epoxypropoxy)benzene].

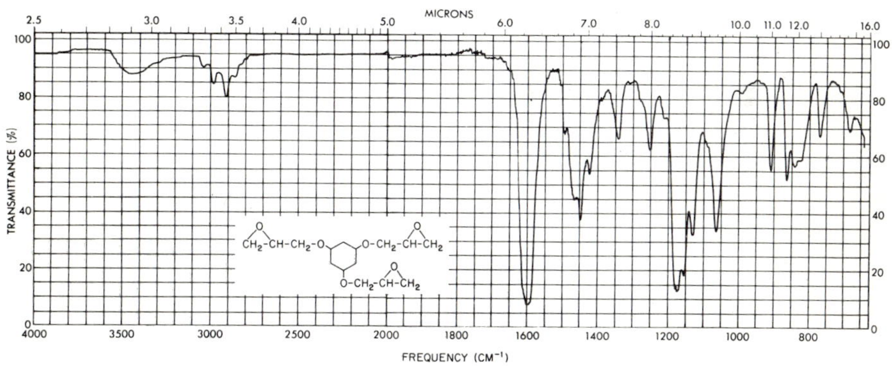

FIG. 4-1-5. Triglycidyl ether of phloroglucinol [tris-1,3,5-(2,3 epoxypropyl)phloroglucinol].

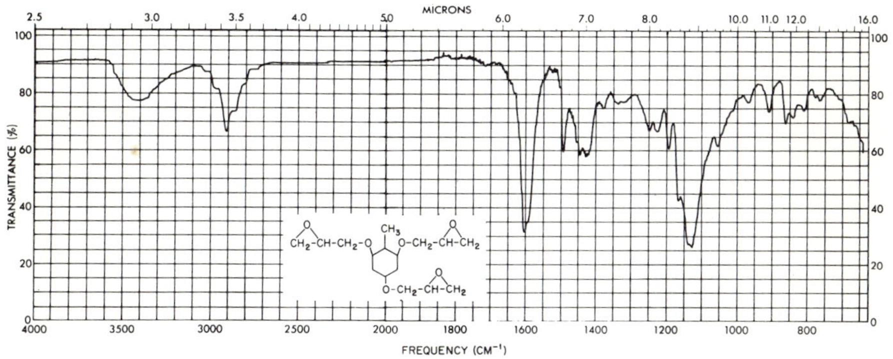

FIG. 4-1-6. Triglycidyl ether of methyl phloroglucinol [tris-2,4,6-(2,3-epoxypropyl)methyl-phloroglucinol].

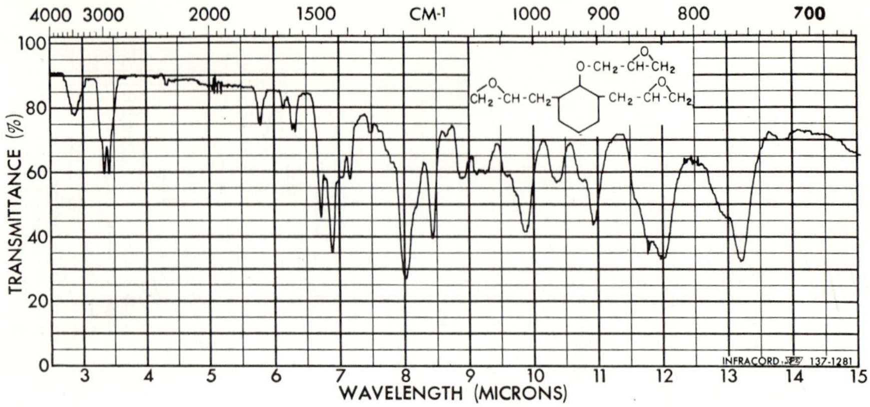

FIG. 4-1-7. 2,6-(2,3-Epoxypropyl)phenylglycidyl ether [2,6-(2,3-epoxypropyl)phenyl 2,3-epoxypropyl ether].

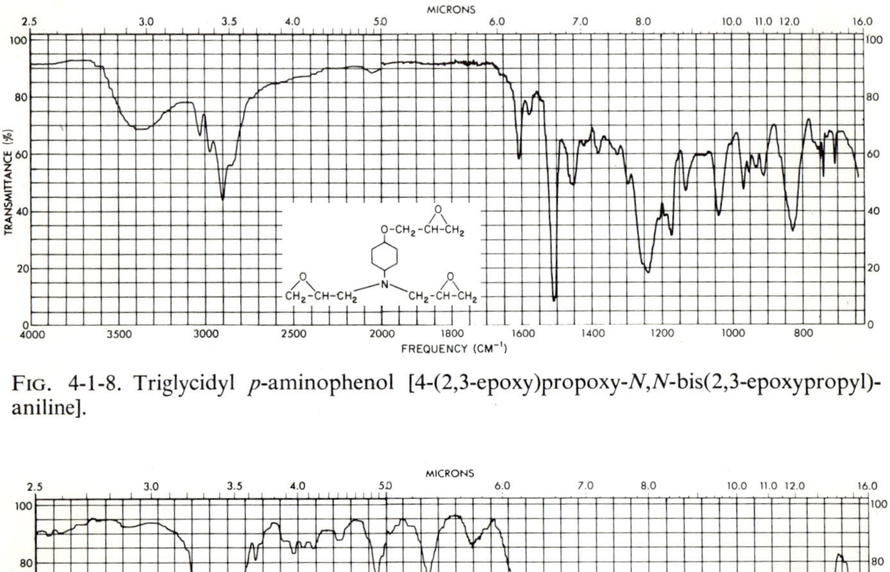

FIG. 4-1-8. Triglycidyl *p*-aminophenol [4-(2,3-epoxy)propoxy-*N*,*N*-bis(2,3-epoxypropyl)-aniline].

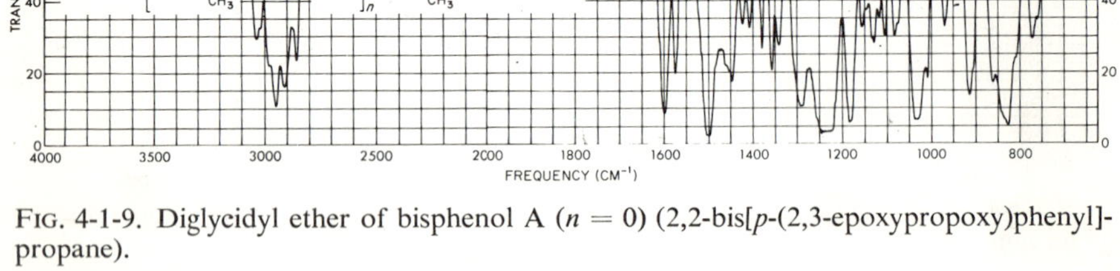

FIG. 4-1-9. Diglycidyl ether of bisphenol A ($n = 0$) (2,2-bis[*p*-(2,3-epoxypropoxy)phenyl]-propane).

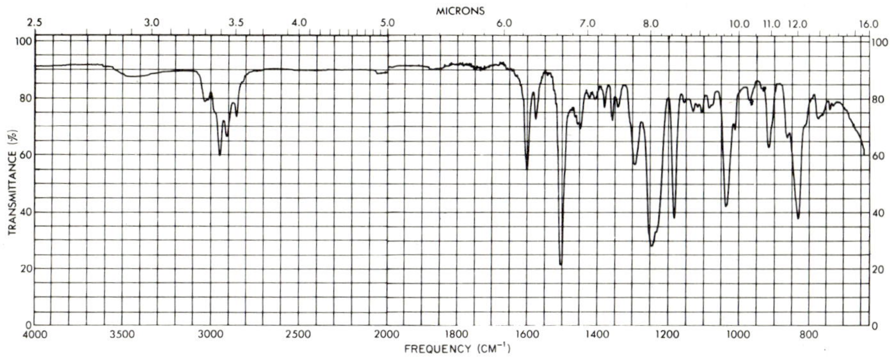

FIG. 4-1-10. Standard commercial resin (mol. wt. approx. 380). (See formula in Fig. 9.) Diglycidyl ether of bisphenol A ($n = 0.2$).

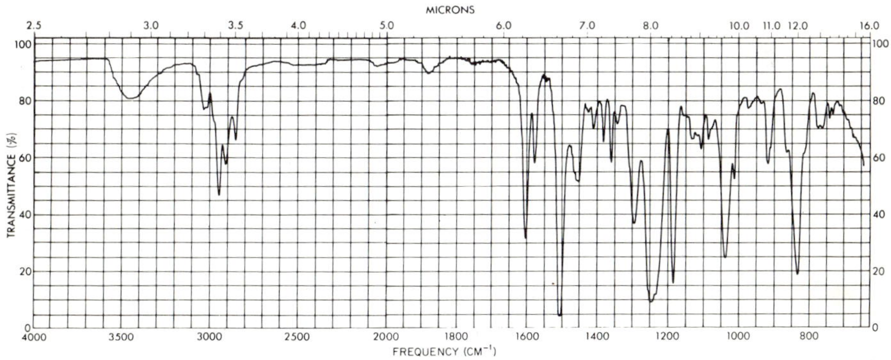

FIG. 4-1-11. Diglycidyl ether of bisphenol A ($n = 1$). (See formula in Fig. 9.)

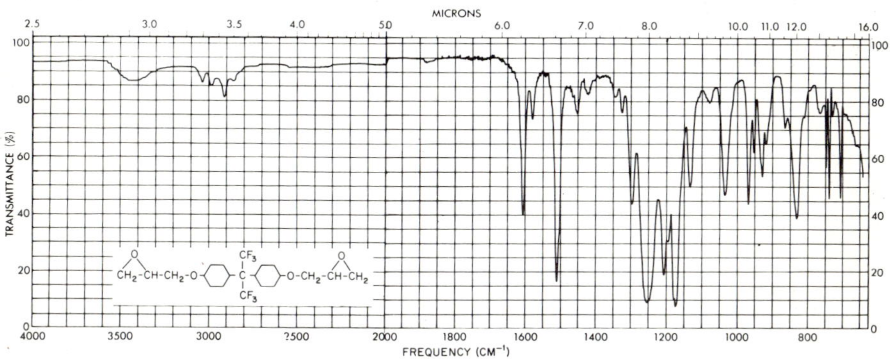

FIG. 4-1-12. Diglycidyl ether of bisphenol–hexafluoroacetone [2,2-bis(*p*-2,3-epoxypropoxyphenyl)-1,1,1,3,3,3-hexafluoropropane].

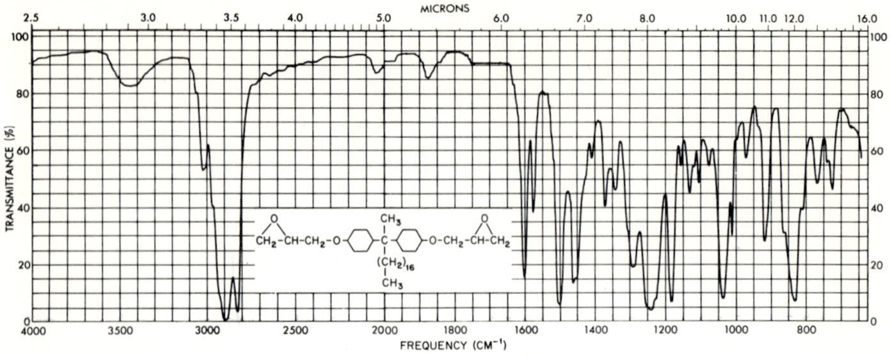

FIG. 4-1-13. Diglycidyl ether of 2,2-bis(4-hydroxyphenyl)nonadecane (2,2-bis[*p*-(2,3-epoxypropoxy)phenyl]nonadecane).

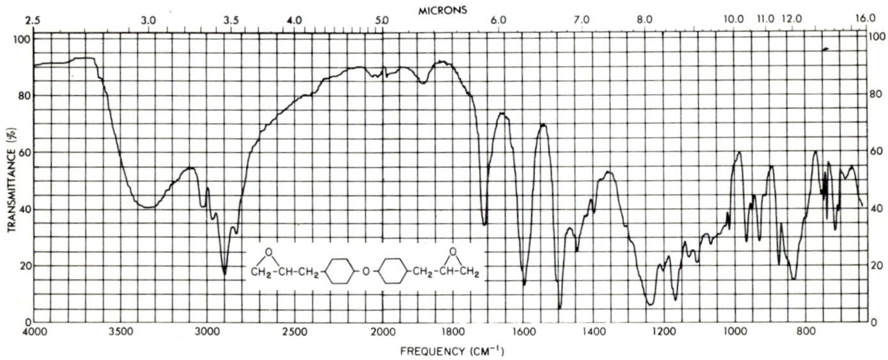

FIG. 4-1-14. Diglycidyl phenyl ether [4,4-bis(2,3-epoxypropyl)phenyl ether].

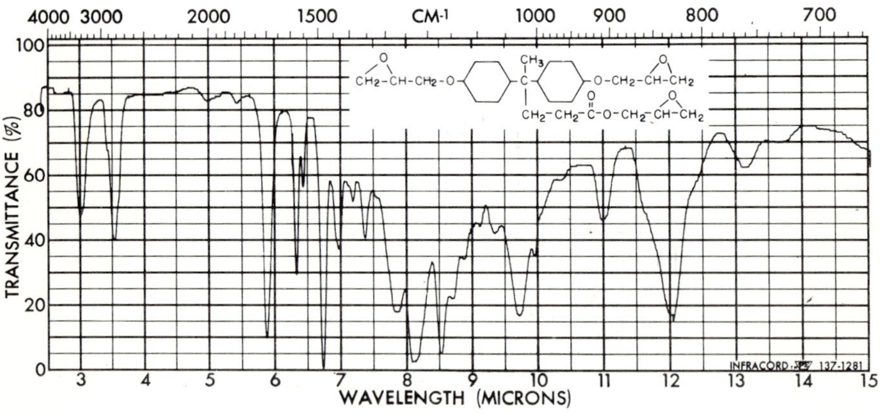

FIG. 4-1-15. Triglycidyl 4,4-bis(4-hydroxyphenyl)pentanoic acid (2,3-epoxypropyl 4,4-bis[*p*-(2,3-epoxypropoxy)phenyl]pentanoate).

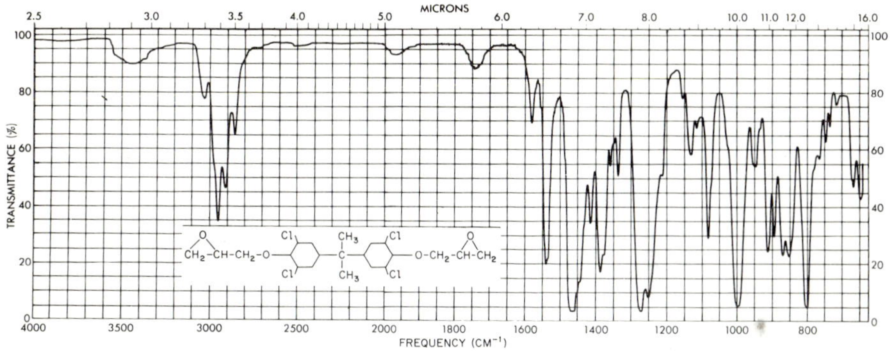

FIG. 4-1-16. Diglycidyl ether of tetrachlorobisphenol A (2,2-bis[4-(2,3-epoxypropoxy)-3,5-dichlorophenyl]propane).

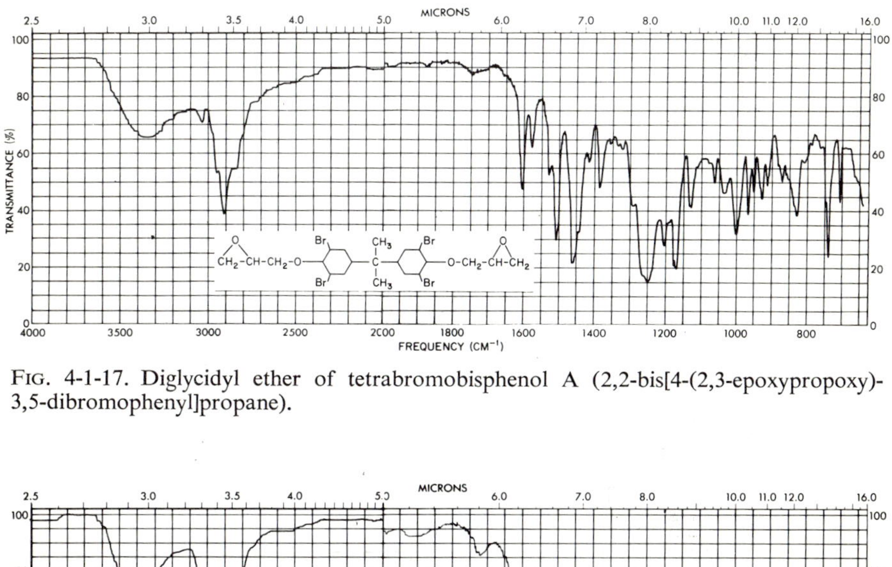

FIG. 4-1-17. Diglycidyl ether of tetrabromobisphenol A (2,2-bis[4-(2,3-epoxypropoxy)-3,5-dibromophenyl]propane).

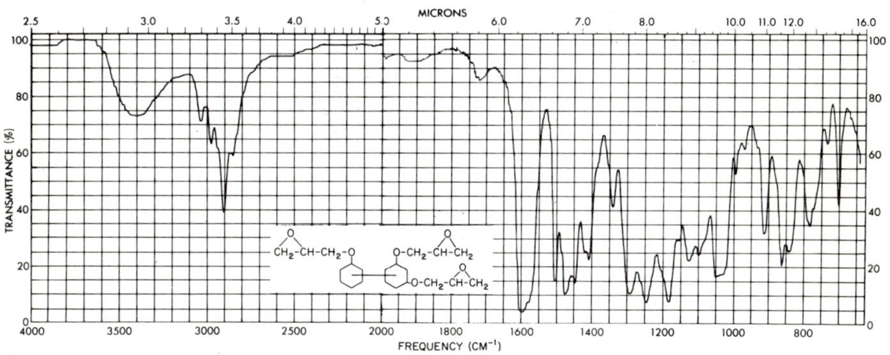

FIG. 4-1-18. Triglycidyl ether of trihydroxybiphenyl [2,4,4′-tris(2,3-epoxypropoxy)-biphenyl].

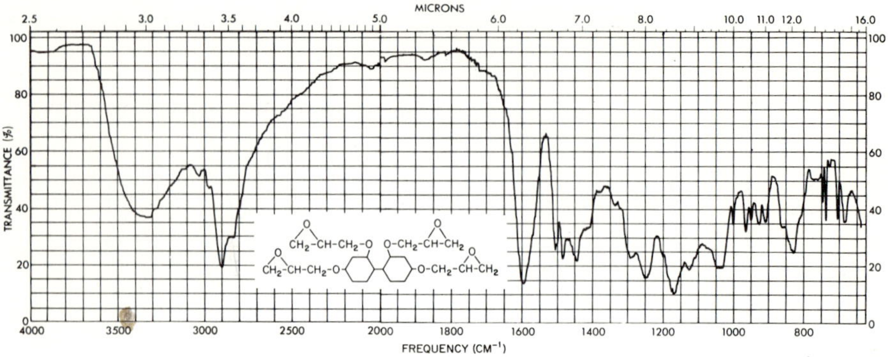

FIG. 4-1-19. Tetraglycidoxy biphenyl [2,2′,4,4′-tetrakis(2,3-epoxypropoxy)biphenyl].

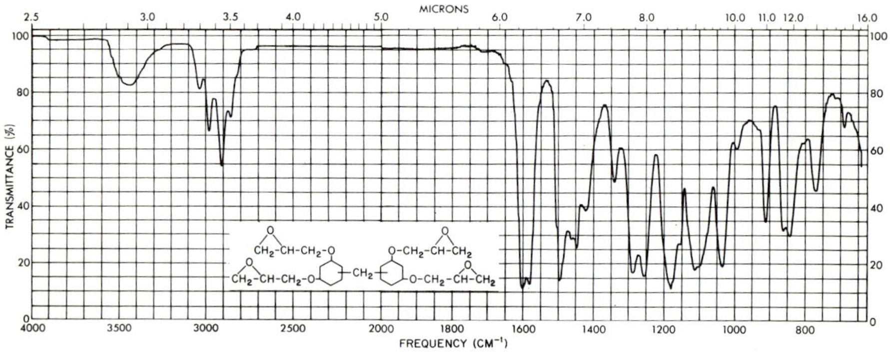

FIG. 4-1-20. Tetraglycidyl ether of bisresorcinol F [tetrakis(2,3-epoxypropoxy)diphenylmethane].

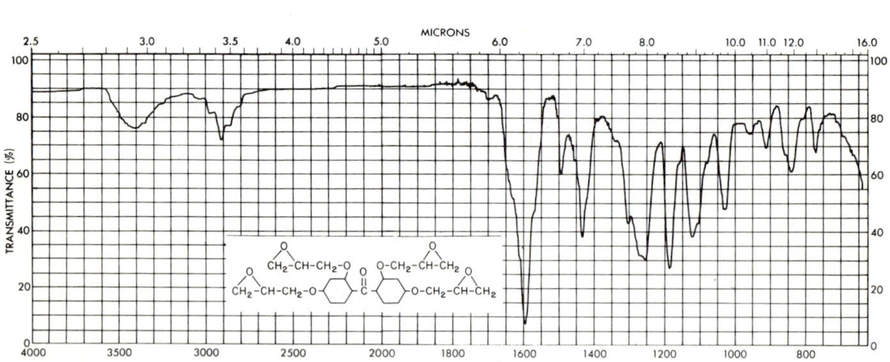

FIG. 4-1-21. Tetraglycidyl ether of resorcinol ketone [2,2′,4,4′-tetrakis(2,3-epoxypropoxy) benzophenone].

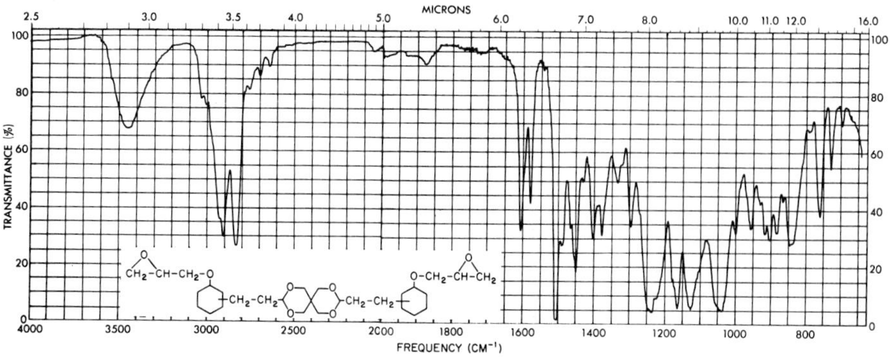

FIG. 4-1-22. Diglycidyl ether of bisphenol PA (3,9-bis[2-(2,3-epoxypropoxy)phenylethyl]-2,4,8,10-tetraoxaspiro [5,5]-undecane).

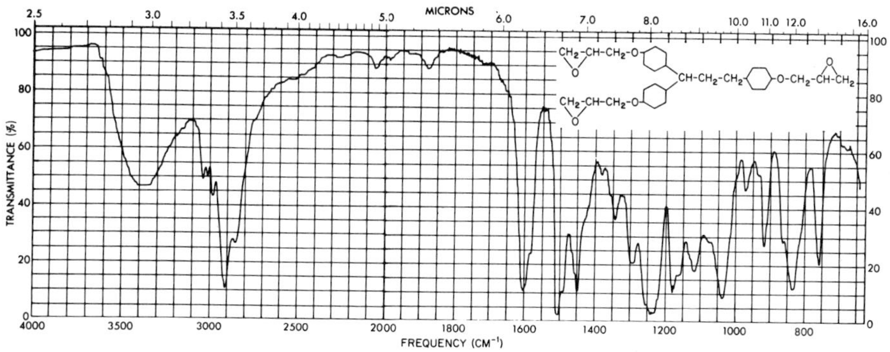

FIG. 4-1-23. Triglycidoxy-1,1,3-triphenylpropane (1,1,3-tris[p-(2,3-epoxypropoxy)phenyl]-propane).

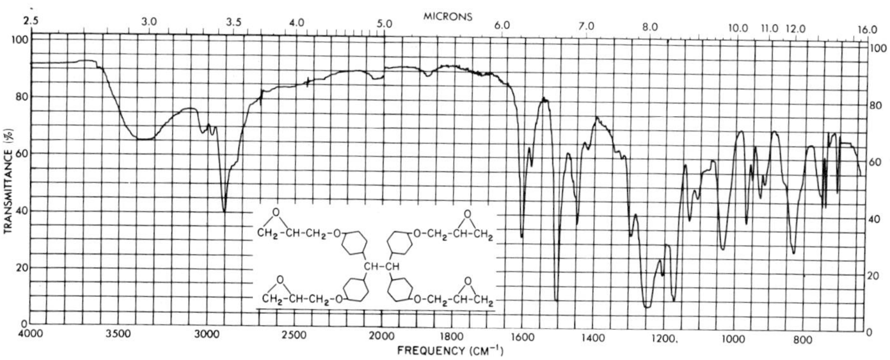

FIG. 4-1-24. Tetraglycidoxy tetraphenylethane (1,1,2,2-tetrakis[p-(2,3-epoxypropoxy)-phenyl]ethane).

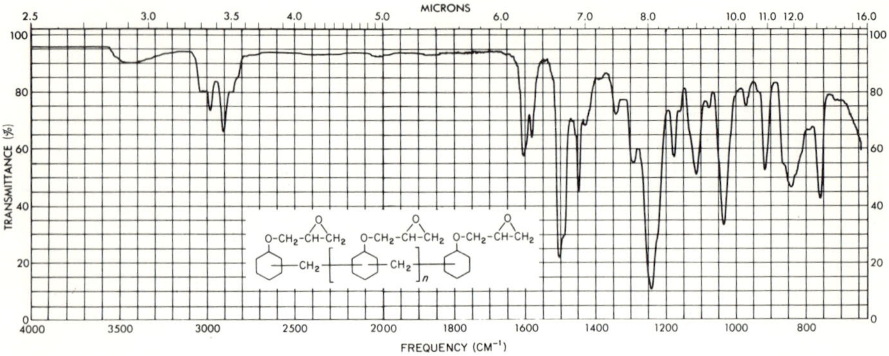

FIG. 4-1-25. Polyglycidyl ether of phenol-formaldehyde novolac [polyphenolformaldehyde poly(2,3-epoxypropyl)ether].

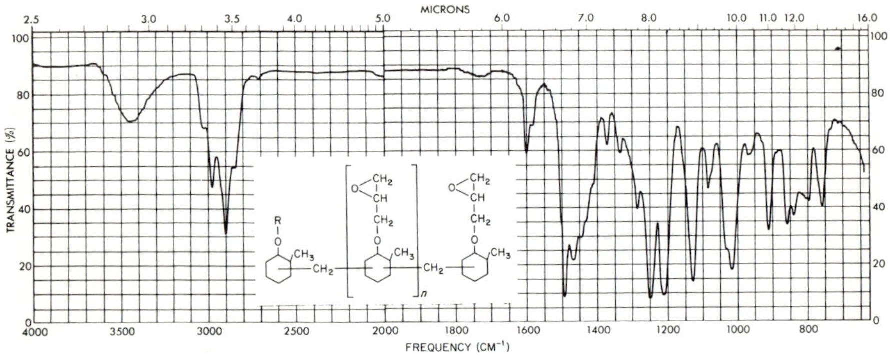

FIG. 4-1-26. Polyglycidyl ether of *o*-cresol-formaldehyde novolac [poly-*o*-cresolformaldehyde poly(2,3-epoxypropyl)ether].

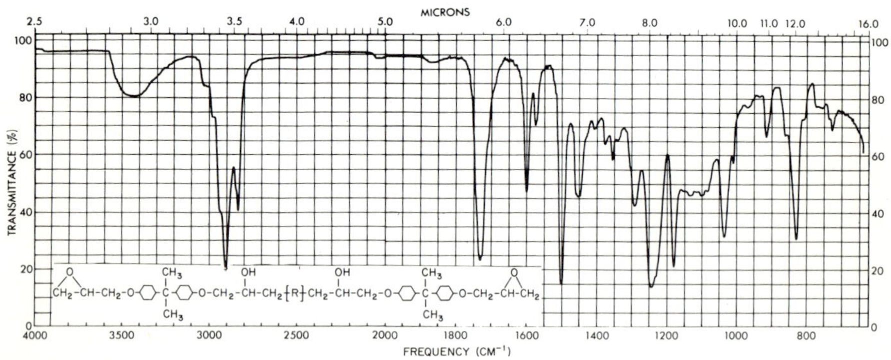

FIG. 4-1-27. Adduct of 2½ moles of the diglycidyl ether of bisphenol A with 1 mole of hydrogenated castor oil.

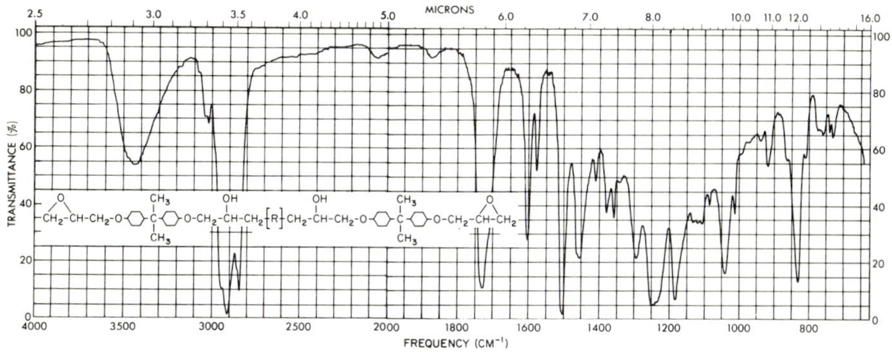

FIG. 4-1-28. Adduct of 2 moles of the diglycidyl ether of bisphenol A with 1 mole of linoleic dimer acid (*R* = linoleic dimer acid radical).

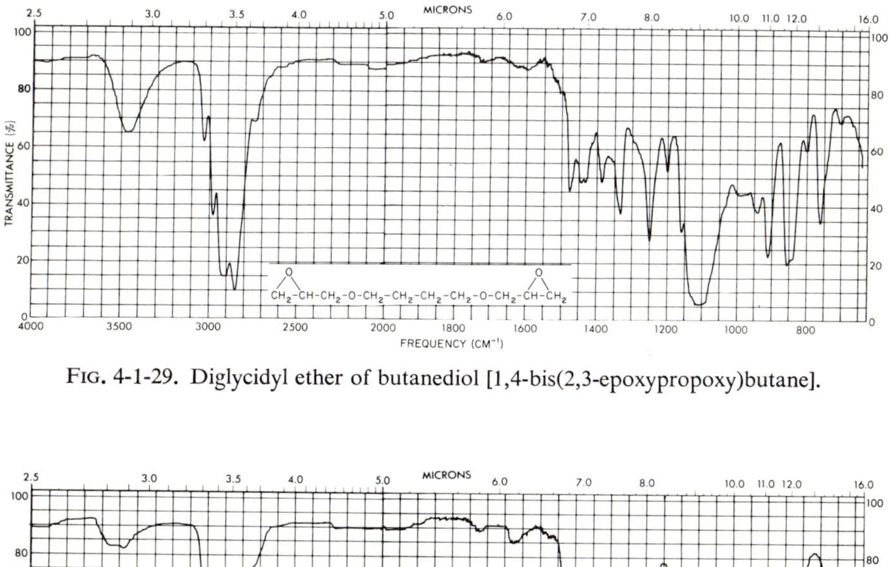

FIG. 4-1-29. Diglycidyl ether of butanediol [1,4-bis(2,3-epoxypropoxy)butane].

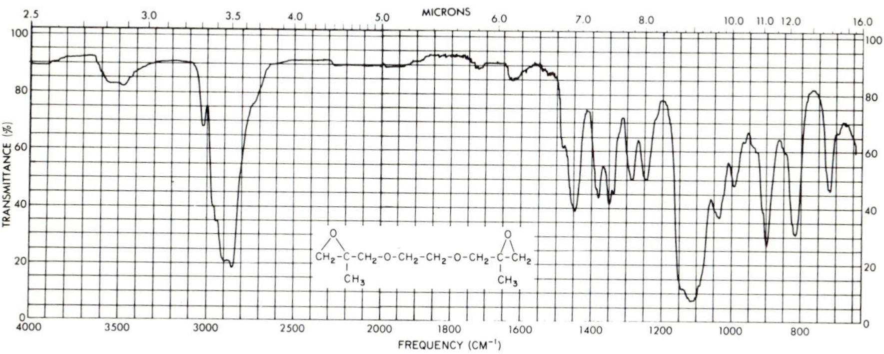

FIG. 4-1-30. Di(2-methyl)glycidyl ether of ethylene glycol [1,2-bis(2,3-epoxy-2-methylpropoxy)ethane].

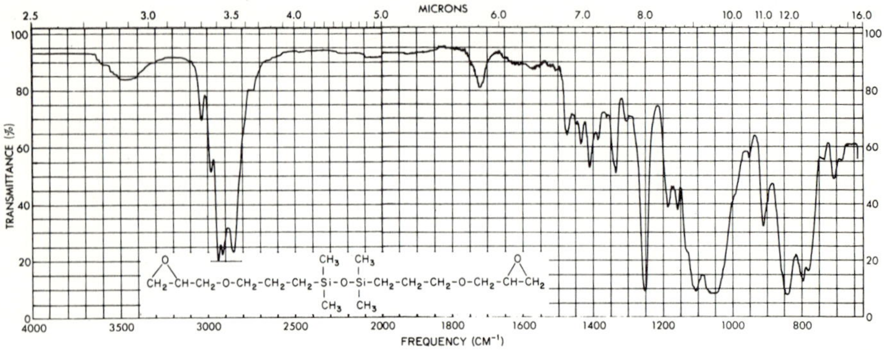

FIG. 4-1-31. 1,3-Bis[3-(2,3-epoxypropoxy)propyl]tetramethyldisiloxane.

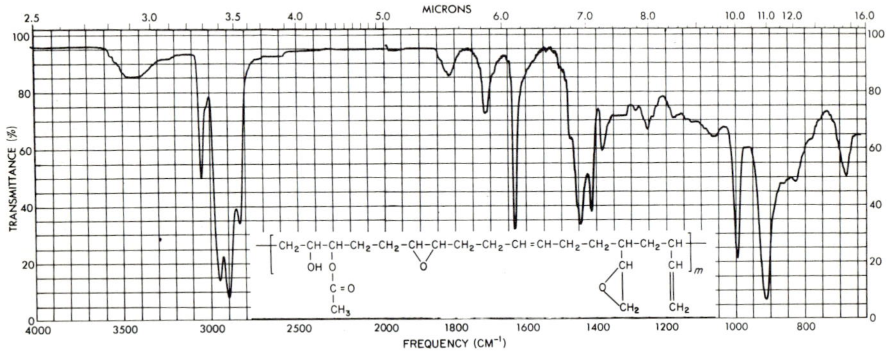

FIG. 4-1-32. Epoxidized polybutadiene.

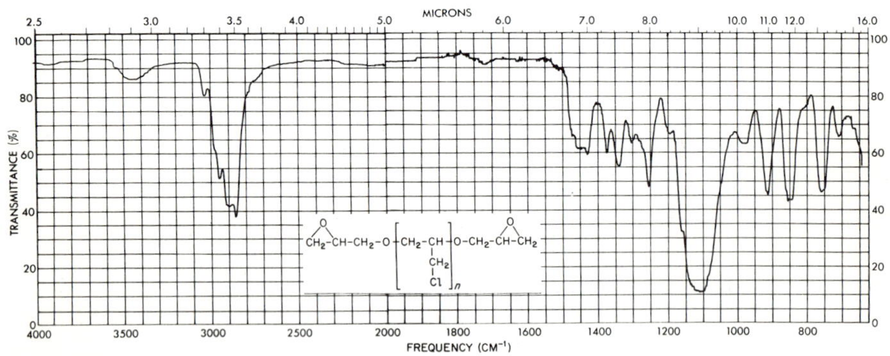

FIG. 4-1-33. Glycidyl ether of water-initiated polymerized epichlorohydrin [polyepichlorohydrin di(2,3-epoxypropyl)ether].

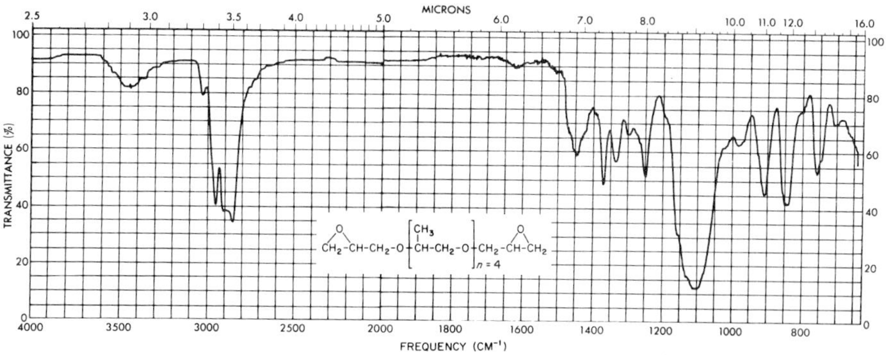

FIG. 4-1-34. Diglycidyl ether of polypropylene glycol [polypropylene glycol di(2,3-epoxypropyl)ether].

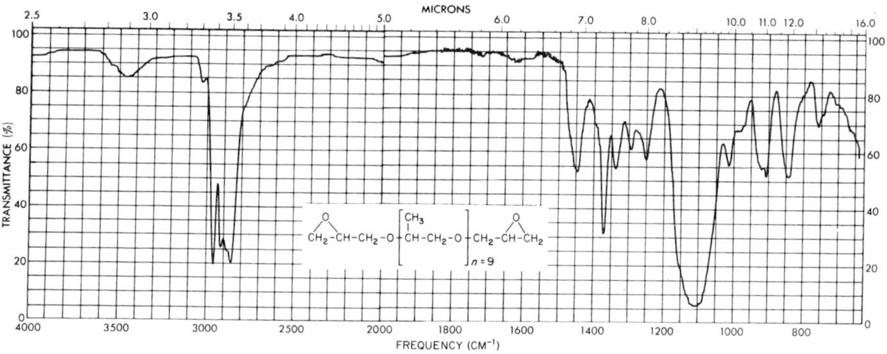

FIG. 4-1-35. Diglycidyl ether of polypropylene glycol [polypropylene glycol di(2,3-epoxypropyl)ether].

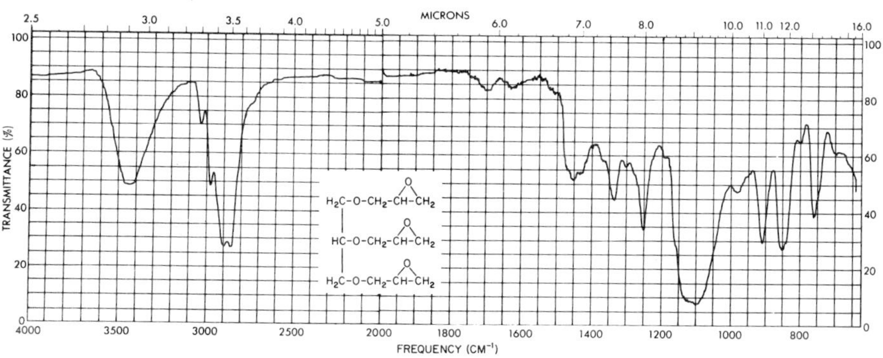

FIG. 4-1-36. Triglycidyl ether of glycerol [1,2,3-tris(2,3-epoxypropoxy)propane].

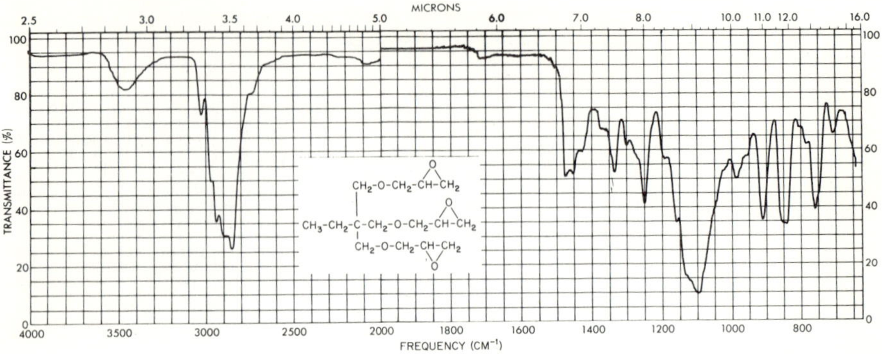

FIG. 4-1-37. Triglycidyl ether of trimethylolpropane [2,2-bis(2,3-epoxypropoxymethyl)-1-(2,3-epoxypropoxy)butane].

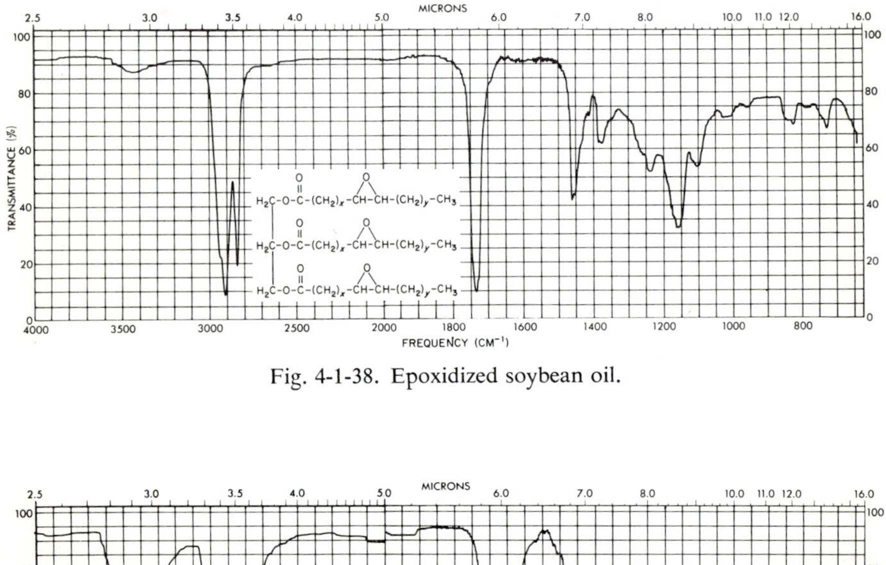

Fig. 4-1-38. Epoxidized soybean oil.

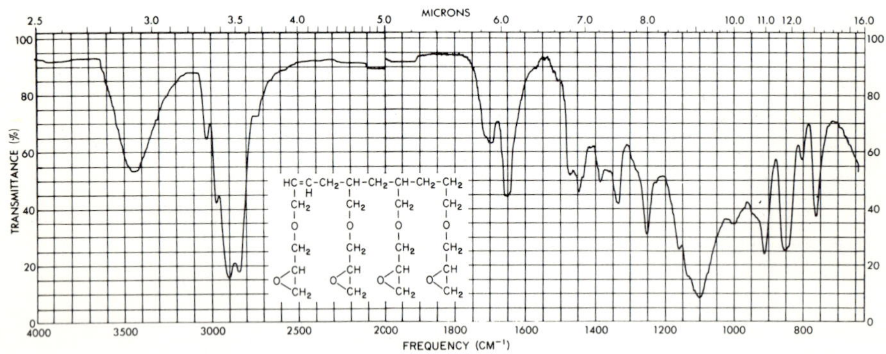

FIG. 4-1-39. Polyallyl glycidyl ether (2,3-epoxypropyl allyl ether telomer).

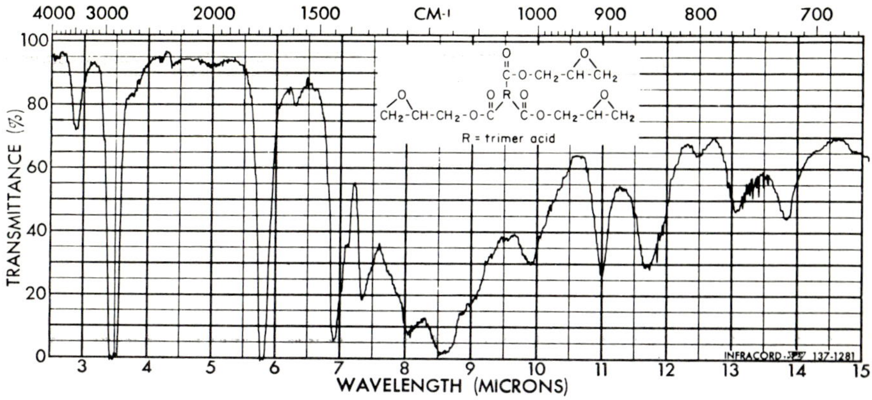

FIG. 4-1-40. Triglycidyl ester of linoleic trimer acid.

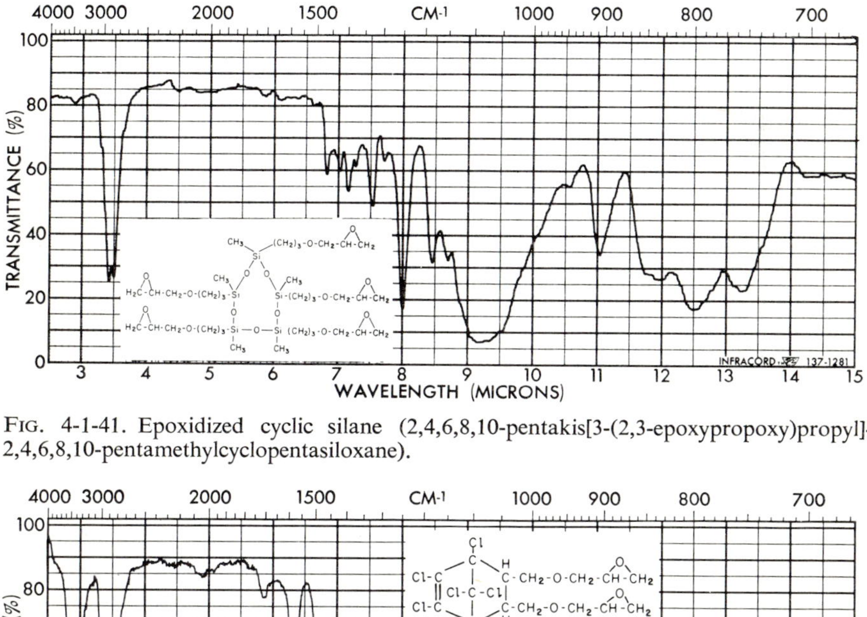

FIG. 4-1-41. Epoxidized cyclic silane (2,4,6,8,10-pentakis[3-(2,3-epoxypropoxy)propyl]-2,4,6,8,10-pentamethylcyclopentasiloxane).

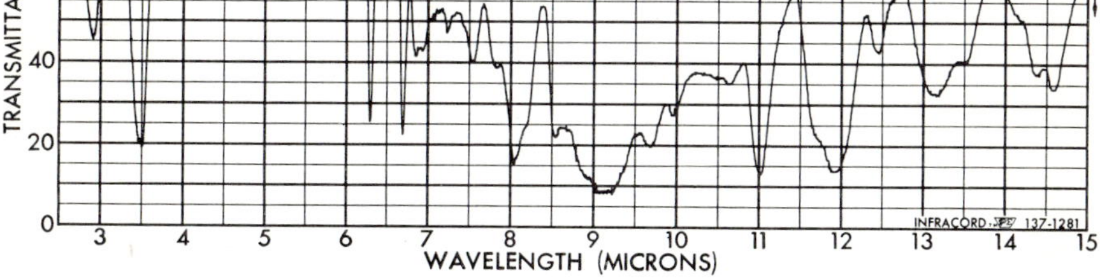

FIG. 4-1-42. Diglycidyl ether of chlorendic diol [5,6-bis(2,3-epoxypropoxymethyl)-1,2,3,4,7,-7-hexachloro-2-norbornene].

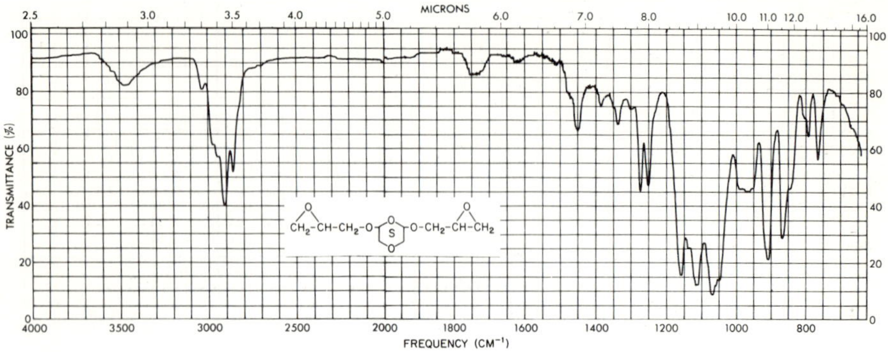

FIG. 4-1-43. Diglycidyl ether of dioxanediol [2,6-bis(2,3-epoxypropoxy)-*p*-dioxane].

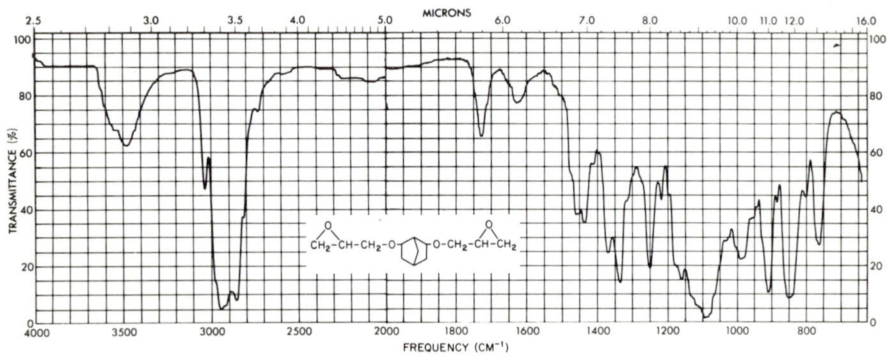

FIG. 4-1-44. Diglycidyl ether of endomethylene cyclohexanediol [2,6-bis(2,3-epoxypropoxy)-norbornene].

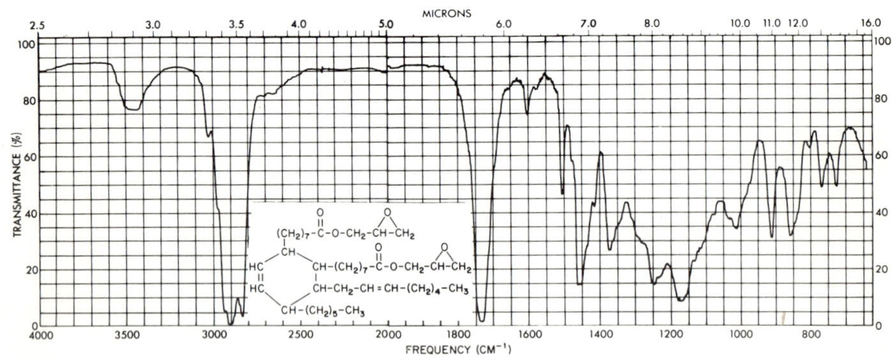

FIG. 4-1-45. Diglycidyl ester of linoleic dimer acid.

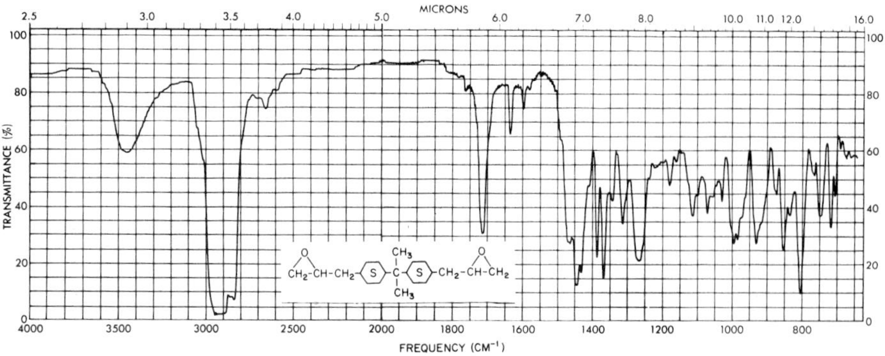

FIG. 4-1-46. 2,2-Bis[4-(2,3-epoxypropyl)cyclohexyl]propane.

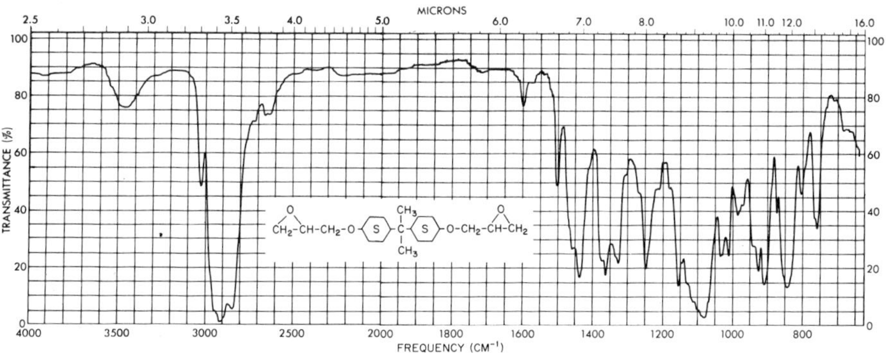

FIG. 4-1-47. Diglycidyl ether of hydrogenated bisphenol A (2,2-bis[4-(2,3-epoxypropoxy)-cyclohexyl]propane).

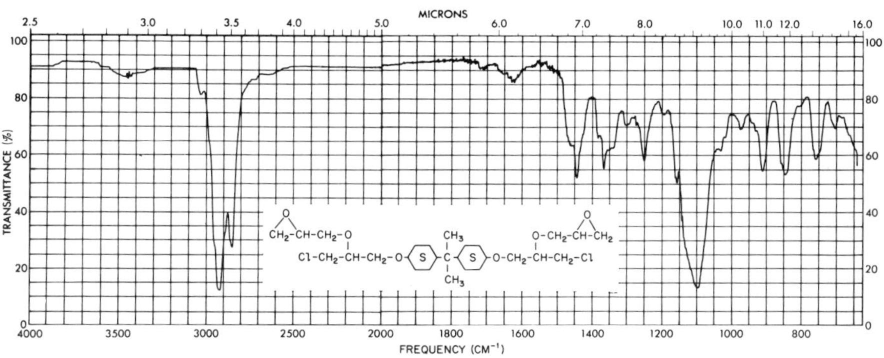

FIG. 4-1-48. 2,2-(4-[3-Chloro-2-(2,3-epoxypropoxy)propoxyl]cyclohexyl)propane.

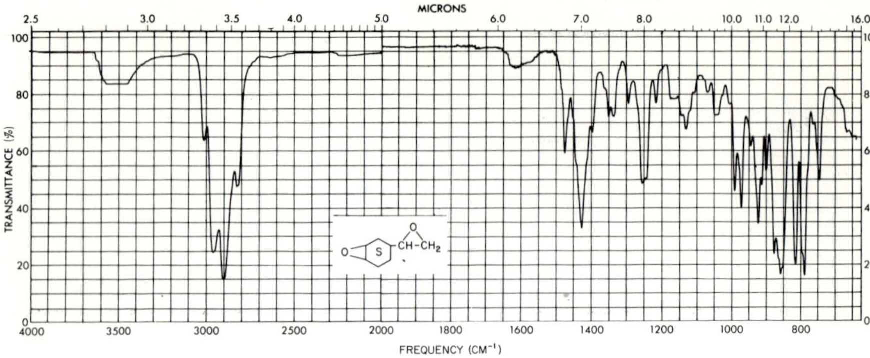

FIG. 4-1-49. Vinylcyclohexene dioxide [4-(1,2-epoxyethyl)-1,2-epoxycyclohexane].

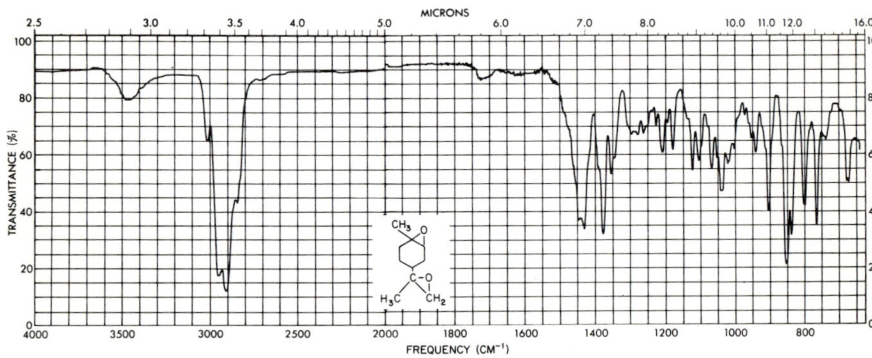

FIG. 4-1-50. Limonene dioxide (1,2-8,9-diepoxy-*p*-menthane).

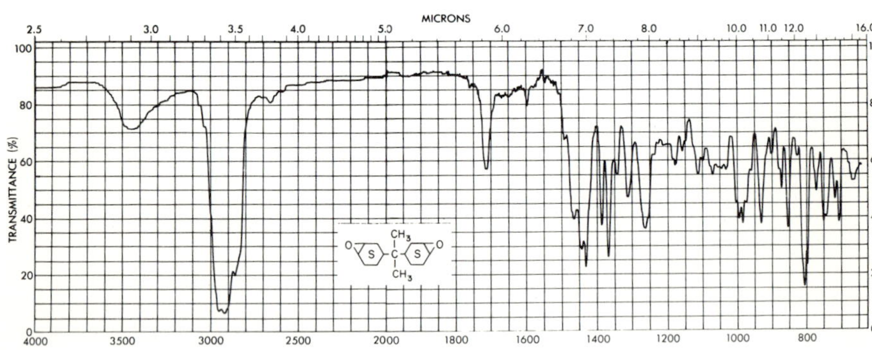

FIG. 4-1-51. 2,2-Bis(3,4-epoxycyclohexyl)propane.

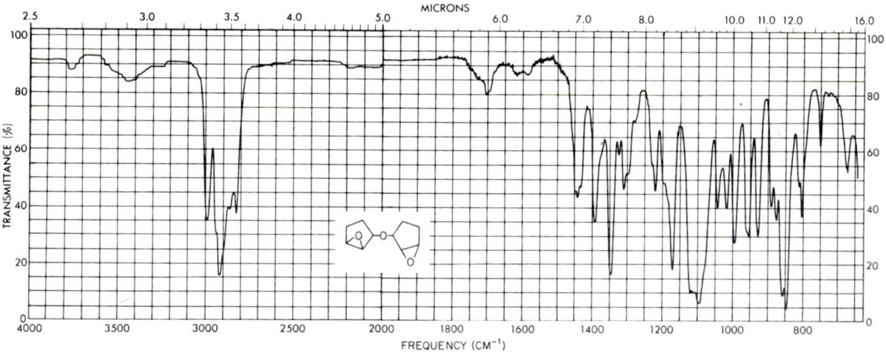

FIG. 4-1-52. Bis(2,3-epoxycyclopentyl)ether (liquid isomer).

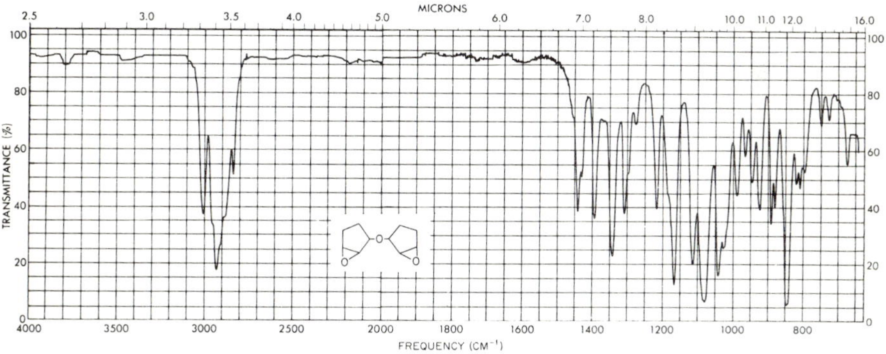

FIG. 4-1-53. Bis(2,3-epoxycyclopentyl)ether (solid isomer).

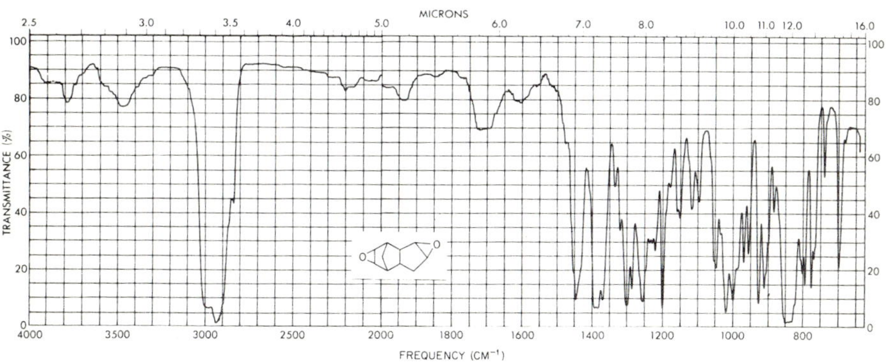

FIG. 4-1-54. Dicyclopentadiene dioxide (65% exo, 35% endo isomer) (1,2,-5,6-diepoxy-4,7-hexahydromethanoindane).

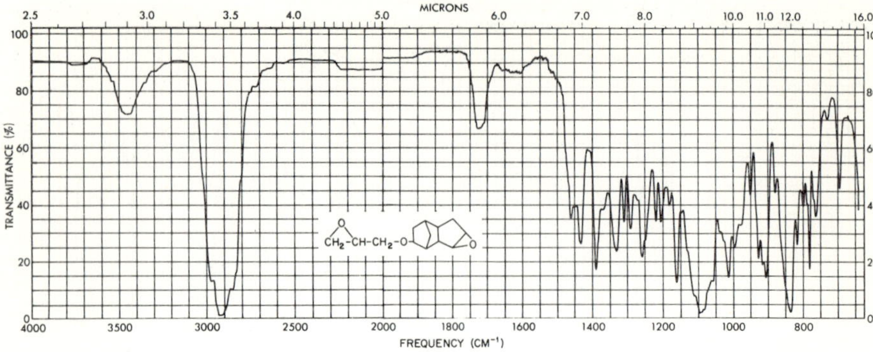

FIG. 4-1-55. 1,2-Epoxy-6-(2,3-epoxypropoxy)hexahydro-4,7-methanoindane.

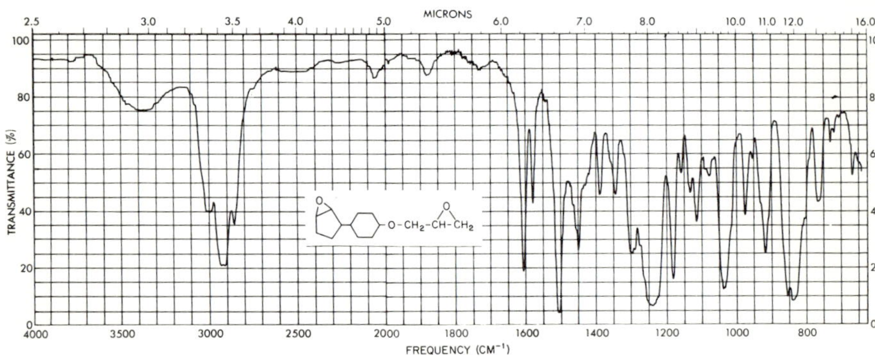

FIG. 4-1-56. *p*-Epoxycyclopentenylphenyl glycidyl ether [*p*-(2,3-epoxy)cyclopentylphenyl 2,3-epoxypropyl ether].

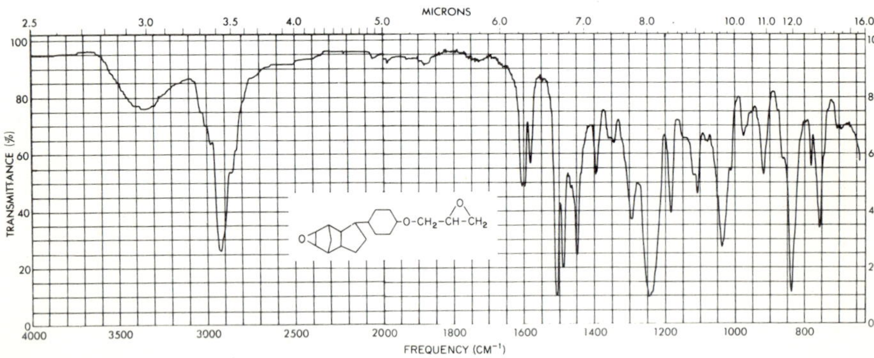

FIG. 4-1-57. Epoxydicyclopentenylphenyl glycidyl ether [1-(2,3-epoxypropoxy)phenyl-5,6-epoxyhexahydro-4,7-methanoindane].

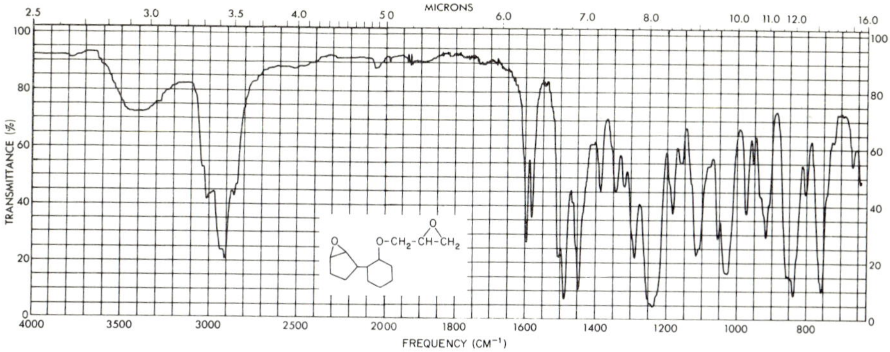

FIG. 4-1-58. *o*-Epoxycyclopentenylphenylglycidyl ether [*o*-(2,3-epoxy)cyclopentylphenyl-2,3-epoxypropyl ether].

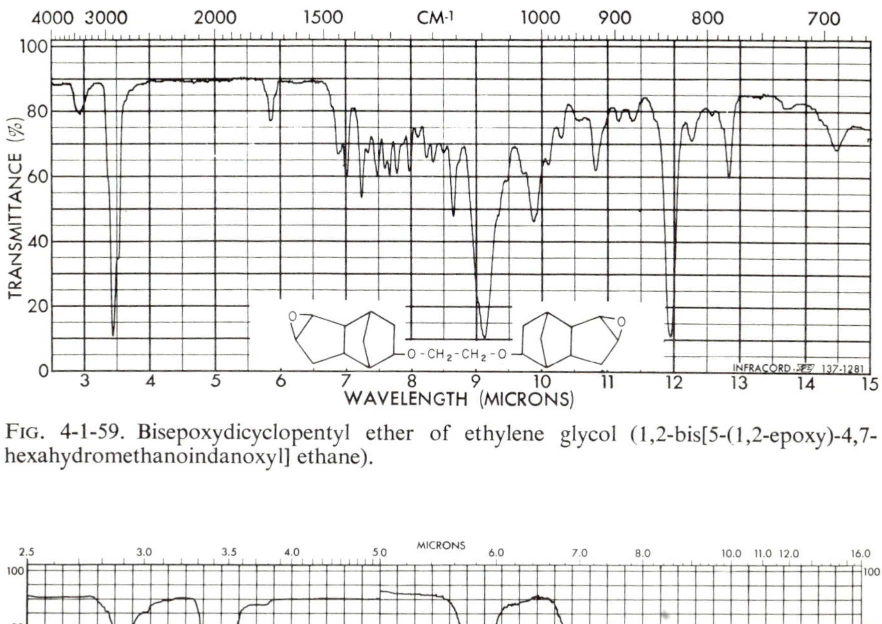

FIG. 4-1-59. Bisepoxydicyclopentyl ether of ethylene glycol (1,2-bis[5-(1,2-epoxy)-4,7-hexahydromethanoindanoxyl] ethane).

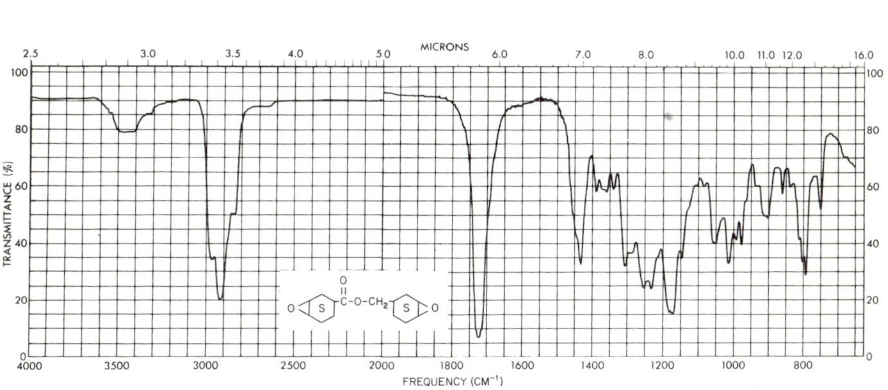

FIG. 4-1-60. 3,4-Epoxycyclohexylmethyl-(3,4-epoxy)cyclohexane carboxylate.

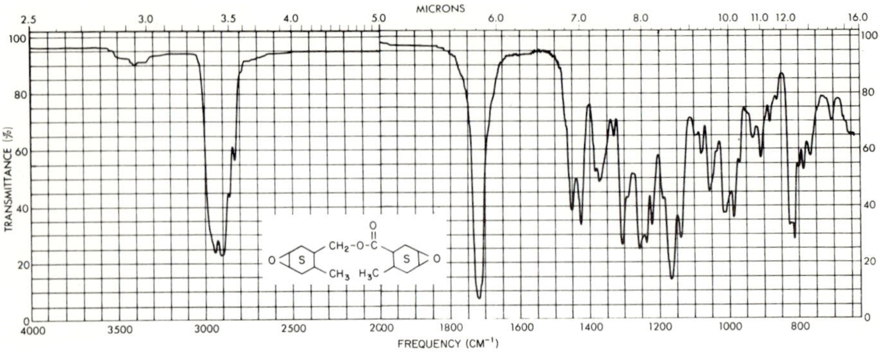

FIG. 4-1-61. 3,4-Epoxy-6-methylcyclohexylmethyl-4-epoxy-6-methylcyclohexane carboxylate.

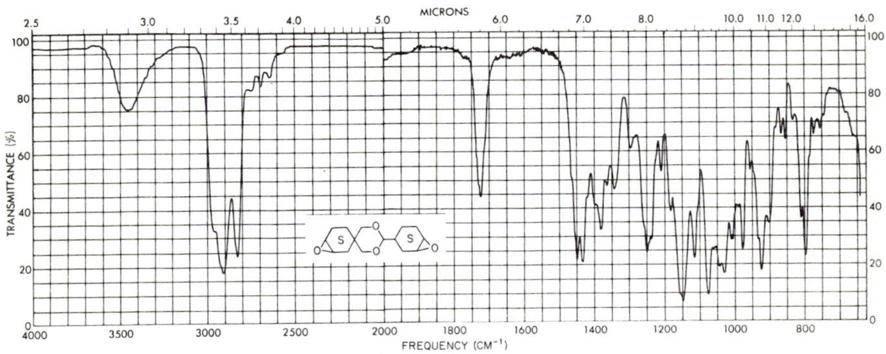

FIG. 4-1-62. Dicycloaliphatic diether diepoxy [2-(3,4-epoxy)cyclohexyl-5,5-spiro(3,4-epoxy)-cyclohexane-*m*-dioxane].

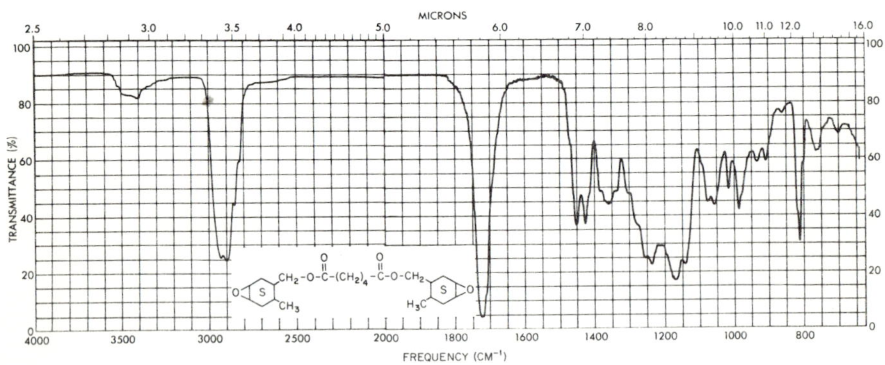

FIG. 4-1-63. Bis(3,4-epoxy-6-methylcyclohexylmethyl)adipate.

Appendix 4-2

PROPERTIES OF COMMERCIAL EPOXY RESINS AND DILUENTS

Ciba Products Co.—Trademark "Araldite"

Number	Epoxy equivalent weight	Viscosity, centipoises, at 25°C (Durran's m.p., °C)	Comments and representative IR spectra (Appendixes 4-1 and 13-1)
RD-1	143	2–3	Butyl glycidyl ether. IR no. 13-1-2
RD-2	136	19	1,4-Butanediol diglycidyl ether. IR no. 4-1-29
RD-4	76	20	Vinylcyclohexene dioxide. IR no. 4-1-49
Epoxide no. 7	227	4	Decyl glycidyl ether. IR no. 13-1-6
CY-175	157	150000	Cycloaliphatic-acetal epoxy resin. IR no. 4-1-62
CY-178	213	900	Bis(3,4-epoxy-6-methylcyclohexylmethyl) adipate. IR no. 4-1-63
CY-179	140	350	3,4-epoxycyclohexylmethyl-3,4-epoxycyclohexane carboxylate. IR no. 4-1-60
405 HXD-35	1180–1430	340	Special solid DGEBA resin supplied at 35% solids in cyclohexanone, xylol, and diacetone alcohol. For coatings
471 X-75	425–550*	4630–9850	Araldite 7071 supplied at 75% solids in xylol
485 E-50	2000–2500*	1290–2270	A proprietary solid (7000 series) DGEBA resin supplied at 50% solids in Cellosolve acetate. (Coatings resin)
488 E-32		1290–2700	High-molecular-weight copolymer of bisphenol A and epichlorohydrin supplied at 32% solids in Cellosolve acetate (Phenoxy resin)
488 N-40		627–1070	High-molecular-weight copolymer of bisphenol A and epichlorohydrin supplied at 40% solids in MEK (Phenoxy resin)
497 C-55	2000–5000*	4630–9850	Araldite 7097 supplied at 55% solids in Cellosolve
502	233–250	2100–3600	Araldite 6010 diluted with 17% dibutyl phthalate
506	172–185	500–700	Araldite 6010 diluted with 11% butyl glycidyl ether
507	185–192	500–700	Araldite 6010 diluted with cresyl glycidyl ether
508	400–455	2000–5000	Araldite 6010 diluted with an epoxidized polyol
509	189–200	500–700	Araldite 6010 diluted with Epoxide no. 7
540 X-90	232–279*	14800–59000	Araldite 6040 supplied at 90% solids in xylol
571 KX-75	425–550*	2700–6340	Araldite 6071 supplied at 75% solids in xylol (35%) and MIBK (65%)
571 T-75	425–550*	2700–6340	Araldite 6071 supplied at 75% solids in toluol
571 K-70	425–550*	1070–2270	Araldite 6071 supplied at 70% solids in MIBK
597 ET-55	2000–2500*	2700–6340	Araldite 6097 supplied at 50% solids in Cellosolve acetate and toluene (1:1)

*Based on solids.

Ciba Products Co.—Trademark "Araldite" (continued)

Number	Epoxy equivalent weight	Viscosity, centipoises, at 25°C (Durran's m.p., °C)	Comments and representative IR spectra (Appendixes 4-1 and 13-1)
EPN 1138	176–181	60000 at 52°C	Polyglycidyl ether of phenol-formaldehyde novolac resin. IR no. 4-1-25
EPN 1139	172–179	1700 at 52°C	Polyglycidyl ether of phenol-formaldehyde novolac resin. IR no. 4-1-25
ECN 1235	215	(35)	Polyglycidyl ether of ortho-cresol-formaldehyde novolac. IR no. 4-1-26
ECN 1273	225	(73)	Polyglycidyl ether of orthocresol-formaldehyde novolac. IR no. 4-1-26
ECN 1280	230	(78–81)	Polyglycidyl ether of orthocresol-formaldehyde novolac. IR no. 4-1-26
ECN 1299	235	(99)	Polyglycidyl ether of orthocresol-formaldehyde novolac. IR no. 4-1-26
ERE 1359	128	500	Resorcinol diglycidyl ether. IR no. 4-1-4
6004	185	5000–6000	Liquid DGEBA resin. Very low viscosity
6005	182–189	7000–10000	Liquid DGEBA resin. Low viscosity
6010	185–196	12000–16000	Liquid DGEBA resin. IR no. 4-1-10
6020	196–208	16000–20000	Liquid DGEBA resin. Medium viscosity
6030	196–222	25000–32000	Liquid DGEBA resin. High viscosity
6040	233–278	(20–28)	Semisolid DGEBA resin. IR no. 4-1-11
6060	385–500	(60–75)	Solid DGEBA resin
6071	425–550	(65–75)	Solid DGEBA resin
6075	560–770	(85–95)	Solid DGEBA resin
6084	825–1025	(95–105)	Solid DGEBA resin
6097	2000–2500	(125–135)	Solid DGEBA resin
6099	2500–4000	(145–155)	Solid DGEBA resin
7065	455–500	(68–78)	Solid DGEBA resin for laminating
7065 A-80	455–500*		Araldite 7065 supplied at 80% solids in acetone
7071	450–530	(67–75)	Solid DGEBA resin
7072	550–700	(75–85)	Solid DGEBA resin
7097	1650–2000	(113–123)	Solid DGEBA resin
8011	455–500	(70–80)	Solid epoxy flame-retardant resin for laminating, 21% Br
8011 A-75	455–500*		Araldite 8011 supplied at 75% solids in acetone
8011 K-75	455–500*		Araldite 8011 supplied at 75% solids in MIBK
9147	222	370 at 70°C	Flame-retardant resin, 20% Br

*Based on solids.

Developmental Products

Number	Epoxy equivalent weight	Viscosity, centipoises, at 25°C (Durran's m.p., °C)	Comments and representative IR spectra (Appendixes 4-1 and 13-1)
DP 452	333	(43)	Low melting, semisolid based on tetrabromo-bisphenol A, 40% Br

Dow Chemical Co.—Trademark "DER" and "DEN"

Number	Epoxy equivalent weight	Viscosity, centipoises, at 25°C (Durran's m.p., °C)	Comments and representative IR spectra (Appendixes 4-1 and 13-1)
DER 330	182–189	7000–10000	Low-viscosity DGEBA resin
DER 331	186–192	11000–14000	Liquid DGEBA resin. IR no. 4-1-10
DER 331VC	186–192	11000–14000	Liquid DGEBA resin. Low foam. IR no. 4-1-10
DER 332	172–176	4000–5500	Pure, liquid DGEBA resin
DER 332LC	170–175	3000–6400	Pure, liquid DGEBA resin. Low chlorine
DER 332VC	170–175	3000–6400	Pure, liquid DGEBA resin. Low foam
DER 334	178–186	500–700	DER 331 diluted with 12% butyl glycidyl ether
DER 335	170–180	150–210	DER 331 diluted with 20% butyl glycidyl ether
DER 336	181–192	4000–8000	DER 331 diluted with phenyl glycidyl ether
DER 337	230–250		Semisolid DGEBA resin for adhesives and coatings. IR no. 4-1-11
DER 337 X-90	230–250*	10000–20000	DER 337 supplied at 90% solids in xylene
DER 511	445–520	(68–80)	Brominated epoxy resin, 18–20% Br
DER 511 A-80	445–520*	1070–2270	DER 511 supplied at 80% solids in acetone
DER 511 EK-80	445–520*	1070–2270	DER 511 supplied at 80% solids in MEK
DER 542	330–380	(51–61)	Brominated epoxy resin based on tetrabromo-bisphenol A, 44–48% Br. IR no. 4-1-17
DER 557	1600–2000	(113–123)	Solid DGEBA resin
DER 660	425–475	(65–74)	Low-melting DGEBA resin
DER 661	475–575	(70–80)	Low-melting DGEBA resin
DER 661PR	475–525	(70–78)	DER 661 with narrower m.p. and WPE ranges
DER 661 A-80	475–575*	3500–8500	DER 661 supplied at 80% solids in acetone
DER 661 EK-75	475–575*	1000–4500	DER 661 supplied at 75% solids in MEK
DER 661PRA-80	475–550*	3500–8500	DER 661 PR supplied at 80% solids in acetone
DER 662	575–700	(80–90)	Solid DGEBA resin. Slightly higher m.p. than DER 661
DER 664	875–975	(95–105)	Solid DGEBA resin

*Based on solids.

Dow Chemical Co.—Trademark "DER" and "DEN" (continued)

Number	Epoxy equivalent weight	Viscosity, centipoises, at 25°C (Durran's m.p., °C)	Comments and representative IR spectra (Appendixes 4-1 and 13-1)
DER 667 EAX-55	1600–2000*	2500–10000	DER 667 supplied at 55% solids in xylene and Dowanol EE acetate (1:1)
DER 668	2000–3500	(120–140)	Solid DGEBA resin
DER 669	3500–5500	(135–155)	Solid DGEBA resin
DER 670 MK-75	425–475*	3000–8000	DER 660 supplied at 75% solids in MIBK
DER 670 T-75	425–475*	2000–10000	DER 660 supplied at 75% solids in toluene
DER 670 X-75	425–475*	4000–14000	DER 660 supplied at 75% solids in xylene
DER 671 MK-75	475–575*	4500–9500	DER 661 supplied at 75% solids in MIBK
DER 671 XM-75	475–575*	5500–12000	DER 661 supplied at 75% solids in MIBK and xylene (65:35)
DER 671 T-75	475–575*	7000–16000	DER 661 supplied at 75% solids in toluene
DER 732	305–335	55–100	Diglycidyl ether of propylene glycol ($n = 9$). IR no. 4-1-35
DER 736	175–205	30–60	Diglycidyl ether of propylene glycol ($n = 4$). IR no. 4-1-34
Dow BGE	143	3	Butyl glycidyl ether. IR no. 13-1-2

Dow Epoxy Novolacs

Number	Epoxy equivalent weight	Viscosity, centipoises, at 25°C (Durran's m.p., °C)	Comments and representative IR spectra (Appendixes 4-1 and 13-1)
DEN 431	172–179	1400–2000 at 52° C	Polyglycidyl ether of phenol-formaldehyde novolac. IR no. 4-1-25
DEN 438	176–181	35000–70000 at 52° C	Polyglycidyl ether of phenol-formaldehyde novolac. IR no. 4-1-25
DEN 438 A-85	176–181*	500–1200	DEN 438 supplied at 80% solids in acetone
DEN 438 EK-85	176–181*	600–1600	DEN 438 supplied at 85% solids in MEK

Dow Experimental Resins

Number	Epoxy equivalent weight	Viscosity, centipoises, at 25°C (Durran's m.p., °C)	Comments and representative IR spectra (Appendixes 4-1 and 13-1)
QX 2174	226–238	(55–60)	Low-melting, solid epoxy novolac resin. Similar to DEN 438 but higher functionality. IR no. 4-1-25
QX 2174 A-80	226–238*		QX 2174 supplied at 80% solids in acetone

*Based on solids.

Dow Experimental Resins (continued)

Number	Epoxy equivalent weight	Viscosity, centipoises, at 25°C (Durran's m.p., °C)	Comments and representative IR spectra (Appendixes 4-1 and 13-1)
QX 2174.1	230–245	(65–75)	Low-melting, solid epoxy novolac resin. Higher functionality. IR no. 4-1-25
QX 2221	182–188	500–700	DER 331 diluted with cresyl glycidyl ether
QX 2232	137–144	17000	Blend of 80 parts DEN 438 with 20 parts of vinylcyclohexene dioxide
QX 2242.1		6340–38800	High-molecular-weight copolymer of bisphenol A and epichlorohydrin supplied at 40% solids in MEK (Phenoxy)
QX 2242.2		627–1760	High-molecular-weight copolymer of bisphenol A and epichlorohydrin supplied at 40% solids in MEK (Phenoxy)
QX 2242.3		1290–2270	High-molecular-weight copolymer of bisphenol A and epichlorohydrin supplied at 32% solids in Dowanol EE Acetate (Phenoxy)
QX 2380	214–221	5000	Flame-retardant resin. DGEBA resin diluted with a Br containing diluent, 16–18% Br
QX 2633.12	375–415	(52–60)	Low-molecular-weight, solid DGEBA resin
QX 2633.14	186–192	11000–14000	Liquid DGEBA resin. Modified for faster gel with amines
QX 2633.17	193–203	16000–22000	High-viscosity DGEBA resin
QX 2633.18	195–205	18000–35000	High-viscosity DGEBA resin
QX 2633.19	198–220	23000–40000	High-viscosity DGEBA resin
QX 2638.6	194–210*	2200–5800	QX 2638.7 supplied at 85% solids in MEK
QX 2638.7	194–210	250000–550000 at 52°C	Polyglycidyl ether of phenol-formaldehyde novolac. Higher functionality than DEN 438. IR no. 4-1-25
QX 3441.4	450–550	(65–80)	Solid, flame-retardant epoxy resin. Contains 13.5–14.5% Br. For NEMA FR-4 laminates
QX 3441.4 A-80	450–550*		QX 3441.4 supplied at 80% solids in acetone
QX 3441.5	550–625	(80–90)	Higher Br (22–24%) than QX 3441.4. Also for NEMA FR-4 laminates
QX 3442.1	330–350	(51–61)	Semisolid epoxy resin containing 44–48% Br. Based on tetrabromobisphenol A. Higher purity than DER 542
QX 3450	225–300	13000–65000 at 38°C	Liquid epoxy resin containing 24–26% Br
QX 3476	135	7000 at 70°C	Tetraglycidyl ether of bisresorcinol F. Tetrafunctional epoxy resin. IR no. 4-1-20

*Based on solids.

Farbenfabriken Bayer A.G.—Germany—Trademark "Lekutherm"

Number	Epoxy equivalent weight	Viscosity, centipoises, at 25°C (Durran's m.p., °C)	Comments and representative IR spectra (Appendixes 4-1 and 13-1)
E-801	320	9000	Special bisphenol epoxy resin
K-2	190	30000	DGEBA resin
K-80	320	9000	Special bisphenol epoxy resin
X-20	190	28000	DGEBA resin
X-30	360	Solid	DGEBA resin
X-50	120	200	Aromatic amine-based epoxy resin
X-60	190	4000	Aromatic amine-based epoxy resin

FMC Corporation—Trademark "Oxiron"

Number	Epoxy equivalent weight	Viscosity, centipoises, at 25°C (Durran's m.p., °C)	Comments and representative IR spectra (Appendixes 4-1 and 13-1)
2000	177	180000	Epoxidized polybutadienes. IR no. 4-1-32.
2001	145	16000	
2002	232	1500	Contains styrene monomer

Jones-Dabney Company (Celanese Corp.)—Trademark "Epi-Rez"

Number	Epoxy equivalent weight	Viscosity, centipoises, at 25°C (Durran's m.p., °C)	Comments and representative IR spectra (Appendixes 4-1 and 13-1)
201	1600–2000*	435–550	Epi-Rez 540-C supplied at 45% solids in butyl alcohol, Cellosolve acetate, and xylene
202	1600–2000*	320–435	Epi-Rez 540-C supplied at 40% solids in Cellosolve acetate
242	235–275*	4630–9850	Epi-Rez 515-B supplied at 90% solids in xylene
285		1760–3620	Epi-Rez 530-C supplied at 60% solids in xylene and butyl Cellosolve (1:1)
501	132–138	2–3	Butyl glycidyl ether. IR no. 13-1-2
504	170–180	150–210	Epi-Rez 510 diluted with 20% butyl glycidyl ether
507	375–410	450–650	Glycol-modified glycidyl ether
508	171–177	3600–5500	Molecularly distilled DGEBA resin. IR no. 4-1-9
509	178–193	7000–10000	Liquid DGEBA resin
510	180–200	10000–16000	Liquid DGEBA resin. IR no. 4-1-10
515-B	240–285	(26–34)	Semisolid DGEBA resin. IR no. 4-1-11
520-C	450–525	(65–75)	Solid DGEBA resin, low melting
522-C	550–650	(75–85)	Solid DGEBA resin
530-C	860–1015	(95–105)	Solid DGEBA resin
540-C	1600–2000	(127–133)	Solid DGEBA resin
550	2400–4000	(145–155)	Solid DGEBA resin
560	4000–6000	(165–180)	Solid DGEBA resin

*Based on solids.

Jones-Dabney Company—Trademark "Epi-Rez" (Continued)

Number	Epoxy equivalent weight	Viscosity, centipoises, at 25°C (Durran's m.p., °C)	Comments and representative IR spectra (Appendixes 4-1 and 13-1)
2036	450–525*	2270–4630	Epi-Rez 520-C supplied at 75% solids in MIBK and xylene
2050	450–525*	2270–4630	Epi-Rez 520-C supplied at 75% solids in MIBK
2083	450–525*	3620–9850	Epi-Rez 520-C supplied at 75% solids in methyl Cellosolve
5011	170–190	5–50	Aromatic monoglycidyl ether diluent, cresyl glycidyl ether. IR no. 13-1-12
5071	180–195	500–900	Epi-Rez 510 diluted with 12% butyl glycidyl ether
5077	185–192	500–700	Epi-Rez 510 diluted with cresyl glycidyl ether
5091	185–200	4000–9000	Epi-Rez 510 diluted with 2–5% phenyl glycidyl ether
5108	175–185	10000–15000	High-purity DGEBA resin. Low-hydrolyzable chlorine
5132	400–450	40000–60000	Flexibilized DGEBA resin
5136	200–230	1700–3750	Modified Epi-Rez 510
5155	175–190	30000–90000 at 52°C	Polyglycidyl ether of phenol-formaldehyde novolac. IR no. 4-1-25
5156	185	100000 at 52°C	Polyglycidyl ether of phenol-formaldehyde novolac. IR no. 4-1-25
5161	308	(34.5)	Flame-retardant resin. 30% Cl. IR no. 4-1-16
5163	450	(60)	Flame-retardant resin. 50% Br. IR no. 4-1-17
50810	171–175	4000–5000	Low-viscosity, low-chlorine DGEBA resin. IR no. 4-1-9

Koppers Co., Inc.—Trademark "Kopox"

Number	Epoxy equivalent weight	Viscosity, centipoises, at 25°C (Durran's m.p., °C)	Comments and representative IR spectra (Appendixes 4-1 and 13-1)
170	147	88000 at 50°C	Polyglycidyl ether of polyhydroxy-polyphenyls
171	144	113000	Triglycidyl ether of trihydroxy-biphenyl. IR no. 4-1-18

Reichhold Chemicals, Inc.—Trademark "Epotuf"

Number	Epoxy equivalent weight	Viscosity, centipoises, at 25°C (Durran's m.p., °C)	Comments and representative IR spectra (Appendixes 4-1 and 13-1)
37–128	190–210	500–1000	Epotuf 37-140 diluted with *p*-tertiary butylphenyl glycidyl ether
37–130	185–200	500–900	Epotuf 37–140 diluted with 12% butyl glycidyl ether
37–134	240–260	2100–3100	Epotuf 37–140 with dibutyl phthalate cooked in for trans esterification

*Based on solids.

Reichhold Chemicals, Inc.—Trademark "Epotuf" (continued)

Number	Epoxy equivalent weight	Viscosity, centipoises, at 25°C (Durran's m.p., °C)	Comments and representative IR spectra (Appendixes 4-1 and 13-1)
37–135	180–200	4000–8500	Epotuf 37–140 diluted with 2–5% phenyl glycidyl ether
37–139	182–195	5000–10000	Low-viscosity liquid DGEBA resin
37–140	185–200	10000–19000	Low-viscosity liquid DGEBA resin. IR no. 4-1-10
37–141	195–205	20000–35000	High-viscosity liquid DGEBA resin
37–144	230–280	(20–25)	Semisolid DGEBA resin. IR no. 4-1-11
37–151	450–520	40000–50000	Adduct of 2½ moles DGEBA and 1 mole hydrogenated castor oil. IR no. 4-1-27
37–160	310–340	1500–2000	Glycol-modified glycidyl ether
37–170	175–185	(38)	Polyglycidyl ether of phenol-formaldehyde novolac. IR no. 4-1-25
37–200	245–265	650000–850000	Flame-retardant epoxy resin, 24% Br
37–300	400–450	(60–70)	Low-melting solid DGEBA resin
37–301	450–525	(60–68)	Low-melting solid DGEBA resin
37–302	575–725	(75–85)	Low-melting solid DGEBA resin
37–304	875–1000	(93–104)	Solid DGEBA resin
37–307	1550–2000	(114–124)	Solid DGEBA resin
37–309	2400–4000	(145–155)	Solid DGEBA resin
38–500	400–450*		Epotuf 37–300 supplied at 75% solids in xylol
38–501	450–525*		Epotuf 37–301 supplied at 75% solids in MIBK and xylol (66:33)
38–503	1550–2000*		Epotuf 37–307 supplied at 55% solids in Cellosolve acetate
38–504	1550–2000*		Epotuf 37–307 supplied at 55% solids in toluene and Cellosolve acetate (1:1)
38–505	450–525*		Epotuf 37–301 supplied at 75% solids in xylol
38–506	450–525*		Epotuf 37–301 supplied at 70% solids in MEK and toluene (1:1)
38–507	450–525*		Epotuf 37–301 supplied at 75% solids in toluene
38–509	2400–4000*		Epotuf 37–309 supplied at 45% solids in Solvesso 100 and diacetone alcohol (1:3)

Reichhold Experimental Resins

Number	Epoxy equivalent weight	Viscosity, centipoises, at 25°C (Durran's m.p., °C)	Comments and representative IR spectra (Appendixes 4-1 and 13-1)
AZ-3314		7200	Epotuf 37–128 further diluted with STF-6
STF-5	125–140		Diglycidyl aniline
STF-6	162	400	Triglycidyl ether of trimethylol-propane. IR no. 4-1-37
90–002			Epotuf 37–304 supplied at 60% solids in xylene and butyl Cellosolve (1:1)
90–027			Epotuf 37–301 supplied at 80% solids in acetone

*Based on solids.

Shell Chemical Company—Trademark "Epon" and "Eponol"

Number	Epoxy equivalent weight	Viscosity, centipoises, at 25°C (Durran's m.p., °C)	Comments and representative IR spectra (Appendixes 4-1 and 13-1)
152	172–179	1400–2000 at 52°C	Polyglycidyl ether of phenol-formaldehyde novolac ($n > 0$). IR no. 4-1-25
154	176–181	3500–7000 at 52°C	Polyglycidyl ether of phenol-formaldehyde novolac ($n > 1$). IR no. 4-1-25
812	140–160	100–170	Aliphatic resin, triglycidyl ether of glycerol. IR no. 4-1-36
815	175–195	500–700	Epon 828 diluted with 11% butyl glycidyl ether
820	180–195	4000–10000	Epon 828 diluted with 2–5% phenyl glycidyl ether
826	180–188	6500–9500	Liquid DGEBA resin
828	185–192	10000–16000	Liquid DGEBA resin. IR no. 4-1-10
830	190–210	15000–22500	Liquid DGEBA resin
834	230–280	(35–40)	Viscous DGEBA resin. IR no. 4-1-11
834-X-90	230–280*	6340–59000	Epon 834 supplied at 90% solids in xylene
836	290–335	(40–45)	Viscous DGEBA resin
836-C-75	290–335*	300–627	Epon 836 supplied at 75% solids in MIBK
840	330–380	(55–68)	Solid DGEBA resin
871	390–470	400–900	Aliphatic polyepoxide; diglycidyl ester of linoleic dimer acid. IR no. 4-1-45
872	650–750	Semisolid	Flexible semisolid; adduct of 2 moles Epon 828 with 1 mole linoleic dimer acid. IR no. 4-1-28
872-X-75	650–750*	1500–2500	Epon 872 supplied at 75% solids in xylene
1001	450–550	(65–75)	Solid DGEBA resin
1001-A-80	450–550*	1290–6340	Epon 1001 supplied at 80% solids in acetone
1001-BT-70	450–550*	275–884	Epon 1001 supplied at 70% solids in MEK: toluene (1:1)
1001-CX-75	450–550*	1760–6340	Epon 1001 supplied at 70% solids in MIBK: xylene (65:35)
1001-T-75	450–550*	2270–9850	Epon 1001 supplied at 75% solids in toluene
1001-X-75	450–550*	2270–14800	Epon 1001 supplied at 75% solids in xylene
1002	600–700	(75–85)	Solid DGEBA resin
1004	875–1025	(95–105)	Solid DGEBA resin
1007	2000–2500	(125–135)	Solid DGEBA resin
1007-CT-55	2000–2500*	1290–3620	Epon 1007 supplied at 55% solids in MIBK: toluene (1:1)
1009	2500–4000	(145–155)	Solid DGEBA resin
1010	4000–6000	(155–165)	Solid DGEBA resin
1028-B-70	235–280*	400	"B"-staged Epon resin supplied at 70% solids in MEK
1031	210–240	Solid	Tetrafunctional; polyglycidyl ether of tetraphenylene ethane. IR no. 4-1-24
1031-B-80	210–240*		Epon 1031 supplied at 80% solids in MEK

*Based on solids.

Shell Chemical Company—Trademarks "Epon" and "Eponol" (continued)

Number	Epoxy equivalent weight	Viscosity, centipoises, at 25°C (Durran's m.p., °C)	Comments and representative IR spectra (Appendixes 4-1 and 13-1)
53	80000 Molecular weight	Solid	Epichlorohydrin–bisphenol A copolymer. High-molecular-weight phenoxy
53-B-40	80000*	3000	Eponol 53 supplied at 40% solids in MEK
55	200000 Molecular weight	Solid	Epichlorohydrin–bisphenol A copolymer. High-molecular-weight phenoxy
55-B-40	200000*	2700	Eponol 55 supplied at 40% solids in MEK
55-C-32	200000*	1290–2700	Eponol 55 supplied at 32% solids in Cellosolve acetate

Other Epon Resins and Diluents

Number	Epoxy equivalent weight	Viscosity, centipoises, at 25°C (Durran's m.p., °C)	Comments and representative IR spectra (Appendixes 4-1 and 13-1)
Allyl glycidyl ether	114	1	1-Allyloxy-2,3-epoxypropane. IR no. 13-1-7
Phenyl glycidyl ether	150	6	1,2-Epoxy-3-phenoxypropane. IR no. 13-1-11
Poly(allyl glycidyl ether)	132	510	PAGE. IR no. 4-1-39
Cardura E ester	240–250	7–8	Glycidyl ester of Versatic 911 acid (Coke ester of C-8 or 9)
H-16	195–215	500–700	Epon 828 diluted with Cardura E Ester
X-22	170		Pure DGEBA resin ($n = 0$). IR no. 4-1-9
X-24	173		Pure DGEBA resin (with n almost equal to 0). IR no. 4-1-9
H-25	175–183		Pure DGEBA resin, slightly less pure than X-24. IR no. 4-1-9
H-45	430		Brominated resin, 18% Br
H-45-A-80	430*	4900	Epon H-45 supplied at 80% solids in acetone

Union Carbide Plastics Division—Trademarks "Bakelite" and "Unox"

Number	Epoxy equivalent weight	Viscosity, centipoises, at 25°C (Durran's m.p., °C)	Comments and representative IR spectra (Appendixes 4-1 and 13-1)
ERR-0100	190–220	(85–100)	Polyglycidyl ether of phenol-formaldehyde novolac ($n = 5$). IR no. 4-1-25
ERSA-0103	190–220*		ERR-0100 supplied at 80% solids in MEK
ERSB-0111	270–330	100–200	ERLA-0447 supplied at 65% solids in MEK

*Based on solids.

Union Carbide Plastics Division—Trademarks "Bakelite" and "Unox" (continued)

Number	Epoxy equivalent weight	Viscosity, centipoises, at 25°C (Durran's m.p., °C)	Comments and representative IR spectra (Appendixes 4-1 and 13-1)
ERSA-0112	289–325	125–150	Unknown composition. Used as laminating and filament-winding varnish
ERR-0153	208–240	(93)	Tetrafunctional epoxy resin. Polyglycidyl ether of tetraphenylene ethane. IR no. 4-1-24
ERNA-0246	156–158	22900†	Experimental epoxy resin. Epoxydicyclopentyl phenyl glycidyl ether. IR no. 4-1-57
ERNA-0256	128–129	129†	Experimental epoxy resin. *p*-(2,3-Epoxycyclopentyl) phenyl glycidyl ether. IR no. 4-1-56
ERR-0300	90–95	(60)	Bis(2,3-epoxycyclopentyl) ether (Solid isomer). IR no. 4-1-53
ERNA-0386	123	247†	Experimental epoxy resin. *o*-(2,3-Epoxycyclopentyl) phenyl glycidyl ether. IR no. 4-1-58
ERLA-0400	90–95	30–50	Bis(2,3-epoxycyclopentyl) ether (Liquid isomer). IR no. 4-1-52
ERLB-0447			Polyglycidyl ether of phenol-formaldehyde novolac ($n = 2+$). IR no. 4-1-25
ERLB-0448	200	88000 at 55°C	Polyglycidyl ether of phenol-formaldehyde novalac ($n = 1$). IR no. 4-1-25
ERL-0500	105–115	1500–5000	Triglycidyl paraaminophenol. IR no. 4-1-8
ERL-0510	97–101	500–700	Molecularly distilled ERL-0500. IR no. 4-1-8
ERLA-0520	105–115	1500–5000	Low-foam version of ERL-0500. IR no. 4-1-8
ERLA-0625			Based on tetrachloro-bisphenol A. IR no. 4-1-16
ELS-0703	550–700	20–100	Diglycidyl ether of tetrachloro-bisphenol A supplied at 60% solids in an unspecified solvent
ERL-0810	135–150	5–15†	Butyl glycidyl ether. IR no. 13-1-2
ERL-0833	225–250	15–30†	*p*-Butyl phenyl glycidyl ether. IR no. 13-1-14
ERSZ-0901	276–325	50–150	Triglycidyl ether of 1,1,3-triphenylpropane supplied at 63.5% solids in toluene
EXRL-1330	174–186	7000–19000	Low-foam version of ERL-3794
EXRL-1771		8100	Mixture of ERLB-0448 and ERL-4206 (100:40)
EKR-2002	450–525	(63–77)	Low-melting solid DGEBA resin
EKS-2002	450–525*	3500–5500	EKR-2002 supplied at 75% solids in toluene
EKR-2003	875–975	(93–104)	Solid DGEBA resin
ERR-2010	1600–1900	(110–119)	Solid DGEBA resin
ERR-2011	450–550	(63–71)	Solid DGEBA resin which contains some trisphenol for branching and toughness

*Based on solids.
†Viscosity in centistokes.

Union Carbide Plastics Division—Trademarks "Bakelite" and "Unox" (continued)

Number	Epoxy equivalent weight	Viscosity, centipoises, at 25°C (Durran's m.p., °C)	Comments and representative IR spectra (Appendixes 4-1 and 13-1)
ERR-2012	550–650	(71–82)	Solid DGEBA resin which contains some trisphenol for branching and toughness
EKS-2012	450–520	7000–16000†	EKR-2002 supplied at 75% solids in xylene
ERR-2013	900–1000	(91–99)	Solid DGEBA resin which contains some trisphenol for branching and toughness
EKRB-2014	1600–1900	(113–130)	Solid DGEBA resin
EKSB-2014	2600–3400	3500–7500†	EKRB-2014 supplied at 60% solids in MIBK
EXLR-2017			Low-foam and molecularly distilled ERL-3794
EKRA-2018	2400–4000	(130–146)	Solid DGEBA resin
EKSA-2018	4300–7500	3500–11000†	EKRA-2018 supplied at 55% solids in MIBK
EKRA-2053	875–975	(93–104)	Solid DGEBA resin
EKS-2102	60–700	7000–16000†	EKR-2002 supplied at 75% solids in xylene
ERL-2200	175–185	7000–9000	Low-foam liquid DGEBA resin
ERL-2255	150–170	1800–2300	ERL-2774 diluted with 20% ERLA-0400
ERL-2256	140–150	500–900	ERL-2774 diluted with 37.5% ERLA-0400
ERL-2258	120–140	100–500	ERL-2774 diluted with 50% ERLA-0400
ERL-2270	140–150	500–700	ERL-2774 diluted with ERL-4206
ERL-2271	140–150	175–275	ERL-2774 diluted with ERL-4206
ERL-2360	170–200	500–1200	Blend of 38 parts ERL-2774, 38 parts of resorcinol diglycidyl ether, and 24 parts of glycidyl ether of nonyl phenol
ERL-2400	180–195	11000–13500	Low-foam liquid DGEBA resin. IR no. 4-1-10
ERSB-2439	700	350	ERR-2011 supplied at 70% solids in toluene
ERSA-2440	700	4200	ERR-2011 supplied at 70% solids in MEK
ERSA-2450	1500–1700	200–300	ERR-2011 and diglycidyl ether of tetrabromo-bisphenol A supplied as solvent solution
ERLA-2500	179–194	500–700	ERL-2400 diluted with 13% butyl glycidyl ether
ERLB-2527	195–205	700–900	ERL-2774 diluted with an aromatic diluent
ERLA-2546	195–205	700–900	Low-foam version of ERLB-2527
ERLA-2600	200–205	16000–18000	Low-foam and high-viscosity DGEBA resin
ERL-2710	171–181	4000–5000	Molecularly distilled ERL-2774. IR no. 4-1-9
ERLA-2733	195–205	1800–2200	ERL-2774 diluted with aromatic diluent
ERL-2772	175–185	7000–9000	Low-viscosity DGEBA resin
ERL-2774	180–195	11000–13500	Liquid DGEBA resin. IR no. 4-1-10
EKS-2775	260–305	3500–4500†	ERL-2774 and EKR-2002 (88:12) supplied at 90% solids in xylol

†Viscosity in centistokes.

Union Carbide Plastics Division—Trademarks "Bakelite" and "Unox" (continued)

Number	Epoxy equivalent weight	Viscosity, centipoises, at 25°C (Durran's m.p., °C)	Comments and representative IR spectra (Appendixes 4-1 and 13-1)
ERL-2795	179–194	500–700	ERL-2774 diluted with 13% butyl glycidyl ether
ERL-2796	170–185	170–185	Unknown
ERL-2895	185–200	4000–5000	ERL-2774 diluted with butyl glycidyl ether (similar to ERL-2795)
ELS-3001	350–500	50–140	Epoxy-phenolic blend supplied at 70% solids in toluene
ERL-3794	174–186	7000–19000	Polyglycidyl ether of phenol-formaldehyde novolac blended with DGEBA resin
ERLA-4051	255–265	600	Unox Epoxide FE-51 diluted with ERL-4201
ERLA-4052		300	Unox Epoxide FE-51 diluted with ERL-4201
ERLA-4053			Unox Epoxide FE-51 catalyzed
ELS-4071			Unox Polyepoxide S-71
ERL-4201	145–156	1600–2000	3,4-Epoxy-6-methylcyclohexyl-methyl-3,4-epoxy-6-methyl-cyclohexane carboxylate. IR no. 4-1-61
ERL-4206	74–78	8	Vinylcyclohexene dioxide. IR no. 4-1-49
ERL-4221	133	350–450	3,4-Epoxycyclohexylmethyl-3,4-epoxycyclohexane carboxylate. IR no. 4-1-60
EP-207	82	(184)	Dicyclopentadiene dioxide. IR no. 4-1-54
ERL-4269	85	8	Dipentene dioxide
ERL-4289	220	900	Bis(3,4-epoxy-6-methyl-cyclo-hexylmethyl) adipate. IR no. 4-1-63

†Viscosity in centistokes.

Chapter 5

EPOXY-RESIN CURING MECHANISMS

The most valuable single property of the epoxy resins is their ability to transform readily from the liquid (or thermoplastic) state to tough, hard thermoset solids. The conversion normally occurs without the evolution of by-products.

The conversion is accomplished by the addition of a chemically active compound known as a curing agent (hardener, activator, or catalyst). Some curing agents promote curing by catalytic action; others participate directly in the reaction and are chemically bound into the resin chain. Depending on the particular agent, curing may be accomplished at room temperature with heat produced by exothermic reaction, or may require application of external heat.

The epoxy ring 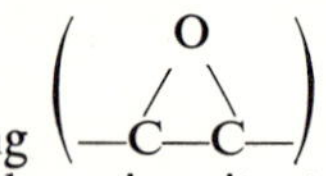 will differ in reactivity, depending on whether it is terminal, internal, or ring-situated. Suitably located, and under proper conditions, it has been found to react with well over 50 different chemical groupings, as summarized in Appendix 5-1.

Basically the cured structure may be a homopolymer (one composed essentially of epoxy-resin molecules linked together through their own reactive sites) or a heteropolymer (one composed essentially of epoxy-resin molecules linked together through the reactive sites of the curing agent) or a mixture of both types.

The epoxy group may react in one of two different ways: anionically and cationically. Both types of reaction are of importance in epoxy-resin technology. In the anionic mechanism, the epoxy group may be opened in various fashions to produce an anion:

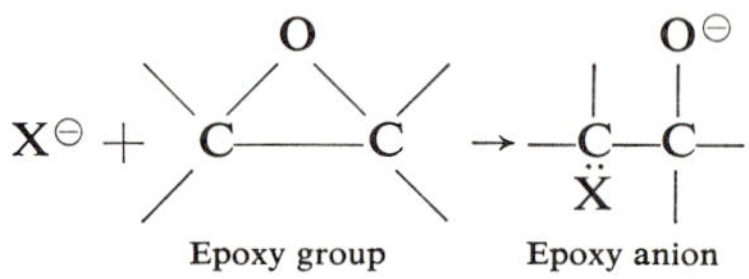

The anion is an activated species, capable of further reaction. In the cationic mechanism, the epoxy group may be opened by an active hydrogen (i.e., hydrogen atoms which are replaceable by a metal, such as metallic sodium) to produce a new chemical bond and a hydroxyl group. The cationic mechanism may proceed in a number of ways, such as the following:

(1) 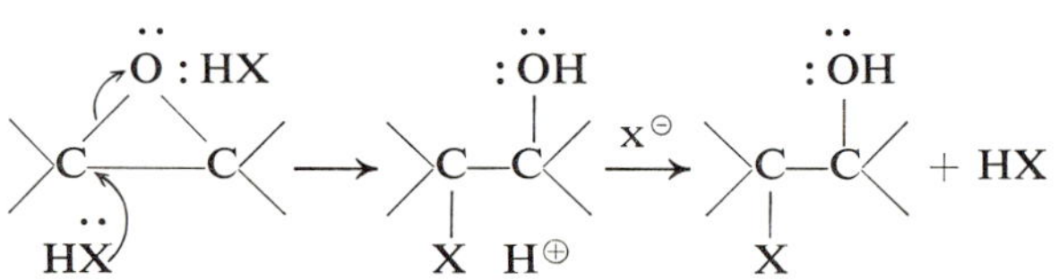

(2) 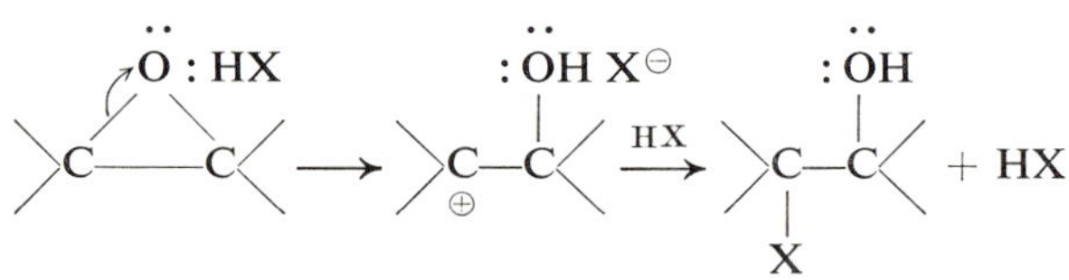

(3) 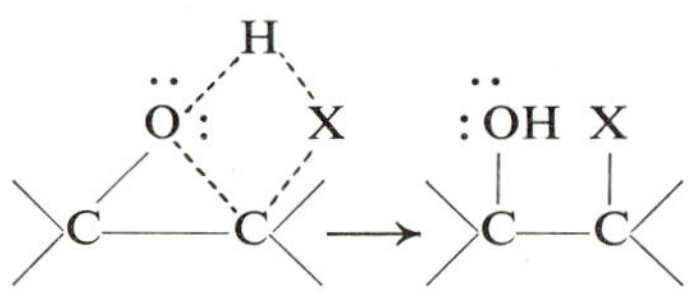

As will be indicated in this chapter, the mechanisms for the formation of epoxy anions and cations are rather more complex than indicated by the above simplified descriptions.

In order to present the information in a direct fashion, the initial discussion will concern itself with a more-or-less idealized epoxy group: that is to say, one situated in a molecule so as to undergo the reaction being described. Later, we will indicate in a more general way the reactivities to be expected of the various types of epoxy groups.

DETERMINATION OF REACTION MECHANISMS

Because the epoxy resins cure to hard, infusible materials, the reaction products are not readily amenable to chemical analysis. This fact, coupled with the complexity of the reactions that may occur, poses a number of problems for the investigator. These can, to an extent, be overcome by employing monofunctional reactants, selected for their chemical resemblance to the constituent parts of the actual resin, as models, and by controlling the conditions of reaction, isolating the competing mechanisms. Thus, phenyl glycidyl ether

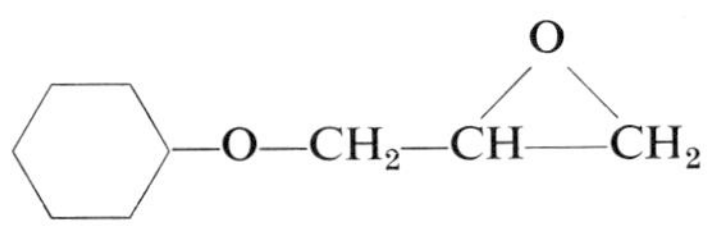

and tolyl glycidyl ether

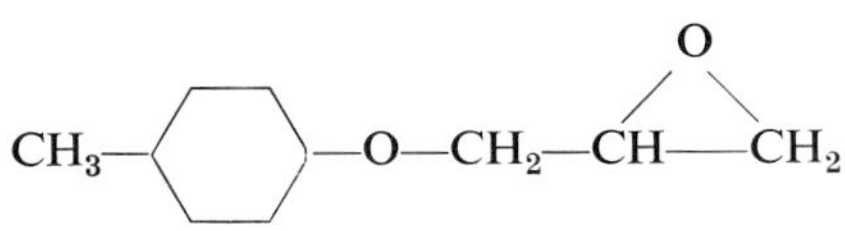

are representative of half of a DGEBA unit. Each "half" is free to react as though it were in the resin, but the un-cross-linked reaction products are soluble and amenable to analysis. Similarly, ditolyl glyceryl ether

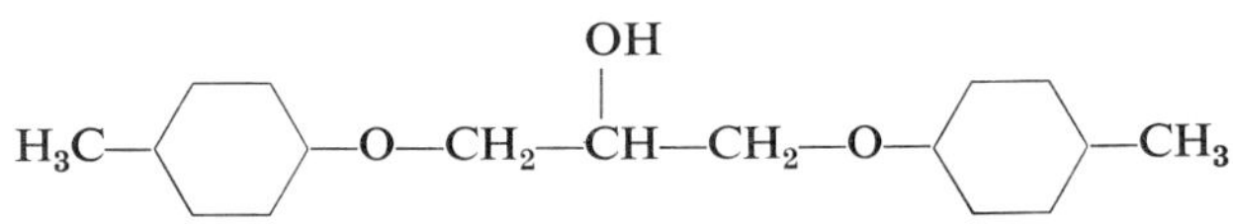

serves as a model of the middle section of the higher-molecular-weight DGEBA. By combining properly selected model compounds at the proper ratio, a close approximation of the actual resin under study may be obtained.

The general discussion presented here represents conclusions obtained from a number of experiments with model compounds and resins. The discussion is based chiefly on the work of Shechter, Wynstra, and Kurkjy [5,6,7]; Narracott [2]; Fisch, Hofman, and Koskikallio [3,20]; Newey [9]; Gough and Smith [16]; Smith [25]; and Belanger and Denham [11]. We are also deeply indebted to Mr. Thomas F. Mika of Shell Development Co., Mr. Ben Carroll of Western Electric Corp., and Dr. Donald G. Stoffey and Mr. I. T. Smith of The Epoxylite Corporation for criticisms in manuscript.

A large body of chemical literature dealing with the reactions of ethylene oxide and similar monoepoxies is not mentioned in this discussion. Excellent reviews are available [1,12,13]. The danger in interpolating from this literature (as, indeed, from any monoepoxy data) is that it is possible that the reaction rates and even directions of the reactions observed with these monofunctional compounds, usually measured in solutions, may not always indicate the conditions and results which obtain with polyfunctional resins in the liquid, nonsolvent phase.

BASIC CURING AGENTS

The basic curing agents employed in the epoxy-resin technology are Lewis bases, inorganic bases, primary and secondary amines, and amides.

Lewis Bases

Lewis bases are those compounds containing an atom with an unshared electron pair in its outer orbital. They are attracted to areas of reduced electron density in the molecules with which they react. The organic bases, such as tertiary amines (R_3N:), are representative of the more reactive-type Lewis bases suitable for curing epoxy resins.

The tertiary amine approaches one of the carbon atoms of the epoxy group and attaches itself, generally from the rear. In all probability, reaction then requires the presence of a hydroxyl group or the equivalent, which provides hydrogen bonding for the epoxy oxygen. The transition state is three-membered and favors the creation of an alkoxide ion from the original hydrogen donor. Once the alkoxide ion is available, it will then react directly with an available epoxy group to generate a new alkoxide ion. The process continues in a chain mechanism:

(1) 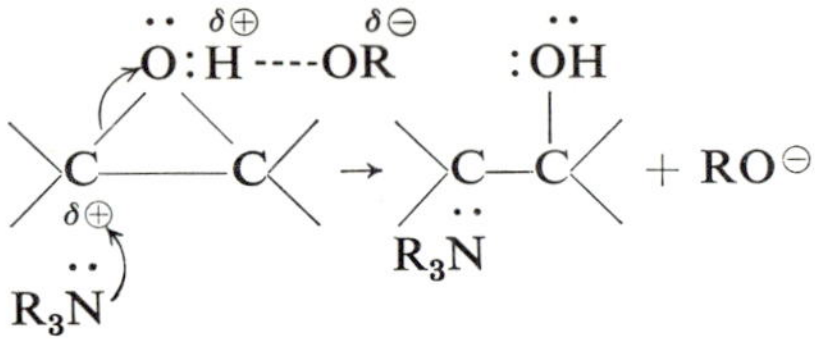

(2) 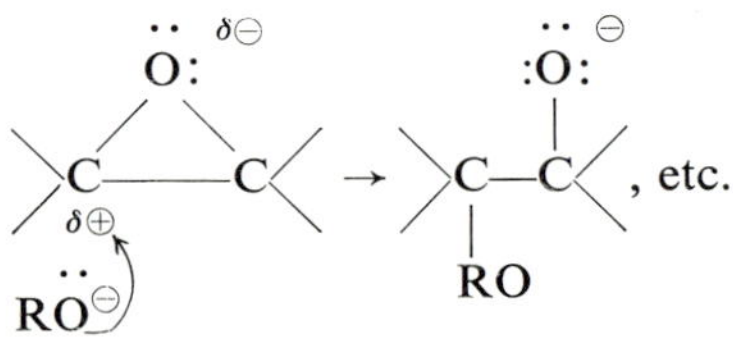

The presence of oxygen atoms with their attendant pairs of unshared electrons stabilizes the quaternary ammonium complex as the free hydroxyl group stabilizes the alkoxide ion by hydrogen bonding.

As an alternative to the above mechanism, it may be suggested that, in part, at least, the initial reaction does not form a separate alkoxide ion, but an activated complex as follows:

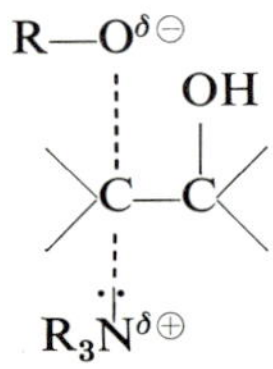

Additional epoxy groups could add to the complex without increasing the charge separation:

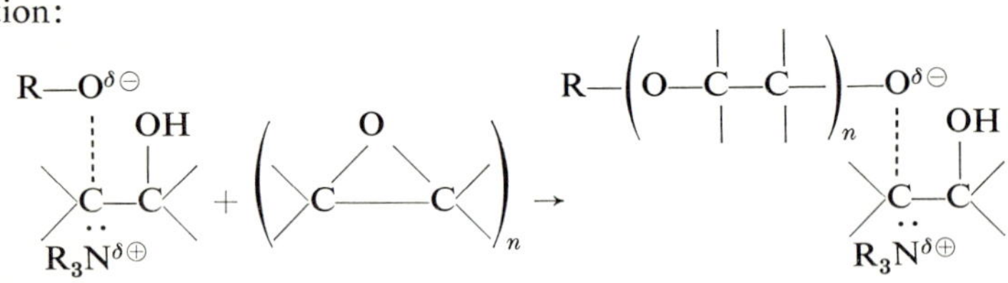

Termination could then occur as shown below with the generation of a new hydroxyl group and unsaturation, without the necessity for the development of a free alkoxide ion.

Table 5-1. Reactivity with DGEBA vs. Basicity of Tertiary Amine [9]

Amine	Ionization constant	Gel time, hours
Benzyldimethylamine	8.5×10^{-6}	5.3
Benzyldiethylamine	3.6×10^{-5}	>300
Triethylamine	6.4×10^{-4}	11.2
Tripropylamine	5.5×10^{-4}	29
Pyridine	2.3×10^{-9}	54

Tertiary amines, as sole curing agents, are normally required in fairly high concentrations, indicating that chain termination frequently occurs.

An analysis of the reaction products of phenyl glycidyl ether catalyzed by a tertiary amine indicated the presence of nitrogen-free as well as nitrogen-containing individual polymers and the presence of both new hydroxyl groups and unsaturation [9]. It was suggested that chain termination and the elimination of the charge separation may occur as follows:

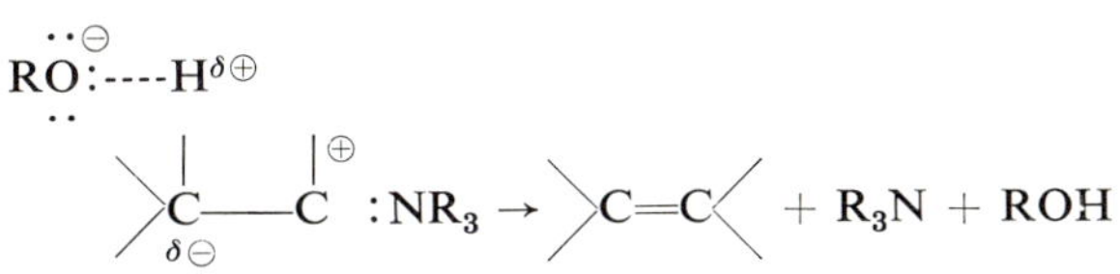

It is also probable that some of the quaternary amine complex may be dealkylated by an adjacent alkoxide ion:

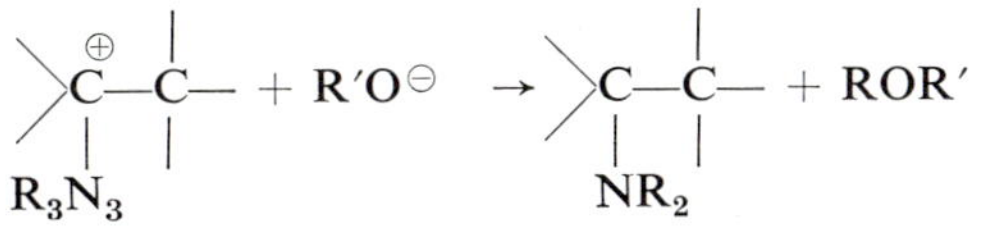

In the presence of an alcohol, a proton transfer may occur, with the formed alkoxide ion from the alcohol then directly replacing the tertiary amine.

In [9] it was shown that the activity of the tertiary amine as an epoxy-resin curing agent does not correlate with basicity (Table 5-1). Steric factors are more important. With two groups larger than ethyl, reactivity drops rapidly, as shown in Table 5-2. Efficiency of the tertiary amine as a curing agent is thus related to the

Table 5-2. Effect of Substituents on Tertiary Amine on Reactivity of DGEBA [9]

Amine	*Gel time, hours*
Benzyldimethylamine	5.3
Benzyldiethylamine	>300
Dimethylethanolamine	4.3
Diethylethanolamine	17.2
Triethylamine	11.2
Tripropylamine	29
Tributylamine	33
Pyridine	54
2-Picoline	112
4-Picoline	58
2,6-Lutidine	>500

ease with which the unshared pair of electrons can approach the epoxy group. It is probable that the tertiary amines are too bulky to approach any but the relatively unblocked carbons of non-ring-situated epoxy groups. A sidewise approach, such as would be required for the ring-situated epoxies, is apparently not practical for tertiary amines; hence their unsuitability as curing agents for these classes of resins.

Influence of Active Hydrogens on Lewis Base Cures. In [6] it has been shown that the anion generated initially from the tertiary amine–epoxy complex may react with an active hydrogen, present as water, alcohol, phenol, or carboxyl, to generate a new anion. This new ion is capable of opening a second epoxy group, and the reaction can thus proceed through first one group and then the other. The nascent hydroxyls formed during the initiating step may also participate in further reaction.

The type of alcoholic hydroxyl influences the direction of the reaction. In polymerization studies with phenyl glycidyl ether in the presence of various alcohols, it was found that all polymer chains contained an alcoholic residue when ethanol was used; only one-half of the polymer chains when isopropanol was used; and only one in seven, when *tert*-butanol was used [9]. The preference can then be written for Lewis base catalysis of epoxy-hydroxyl reactions as primary > secondary > tertiary alcohols. This order of reaction has also been observed in the uncatalyzed system (Fig. 5-1). The speed of the reaction is dependent on the amount of alcohol present, as indicated in Fig. 5-2.

When the active hydrogen is present as phenol or carboxyl, the reaction will be virtually exclusively through the corresponding phenolate or carboxylate ion, thus reacting those compounds into the cured network. The following direction of reactivity for the anions from the epoxy group has thus been established [6]:

Carboxyl > phenol > alcohol/epoxy

It has also been established that the speed of the reaction is in the reverse order [6]. This is to say that given the option, under Lewis base catalysis, the epoxy group

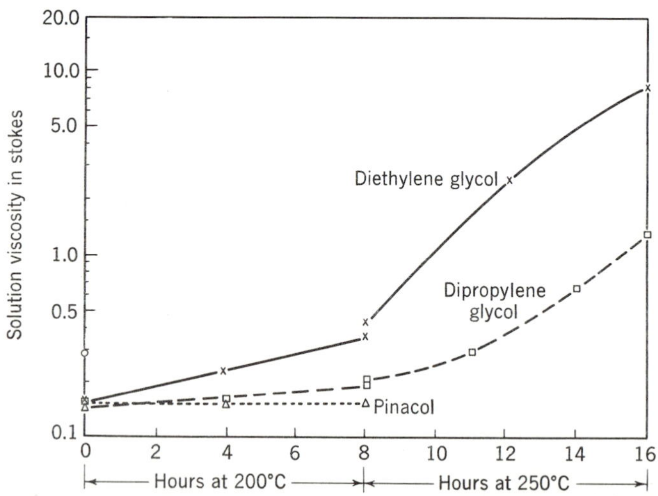

FIG. 5-1. The reaction of equivalent amounts of epoxy and hydroxyl held without catalyst indicates the order of reactivity is primary > secondary > tertiary alcohol [5].

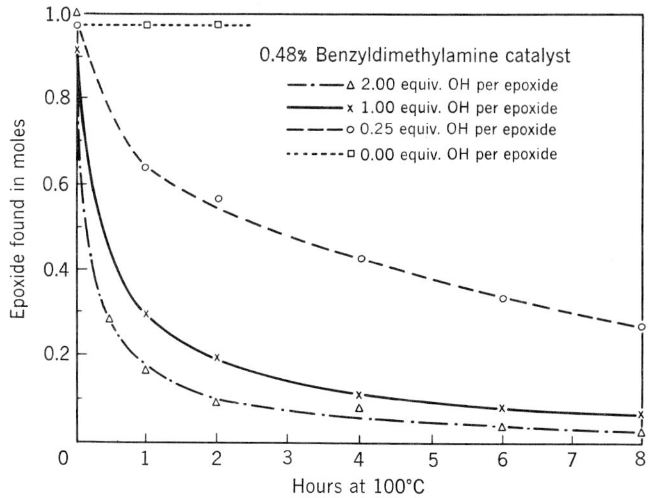

FIG. 5-2. The speed of Lewis-base-catalyzed epoxy-hydroxyl reaction is a function of the concentration of the alcohol [5].

will react until all carboxyl groups are exhausted; whereupon the faster phenol reaction will take over until all the phenols are exhausted; then the reaction will proceed even more rapidly as the alcohols present from the preceding reactions or present in the system in more reactive form will initiate the final polymerization. This situation is shown in Fig. 5-3, where the epoxy-carboxyl reaction continues until the acid is exhausted, whereupon the inherently faster alcohol-initiated polymerization takes over to react the remaining epoxy groups.

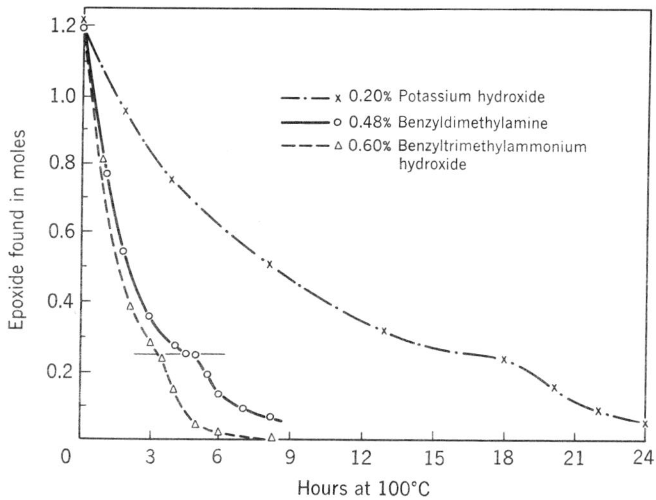

FIG. 5-3. In the presence of a Lewis base, reaction proceeds through epoxy and carboxyl groups until the acid is exhausted, at which time, the inherently faster alcohol-initiated polymerization takes over [5].

Inorganic Bases

Inorganic bases, such as alkali metal hydroxides, have been shown to react according to the ionic mechanism proposed above for tertiary amines [5].

The base generates an anion by reaction with the epoxy group:

This configuration, perhaps because of the nearness of the nascent hydroxyl, does not appear to require a hydrogen-bonded transition state for reaction with an epoxy to be initiated, the speed of the reaction being essentially independent of alcoholic concentration (Fig. 5-4).

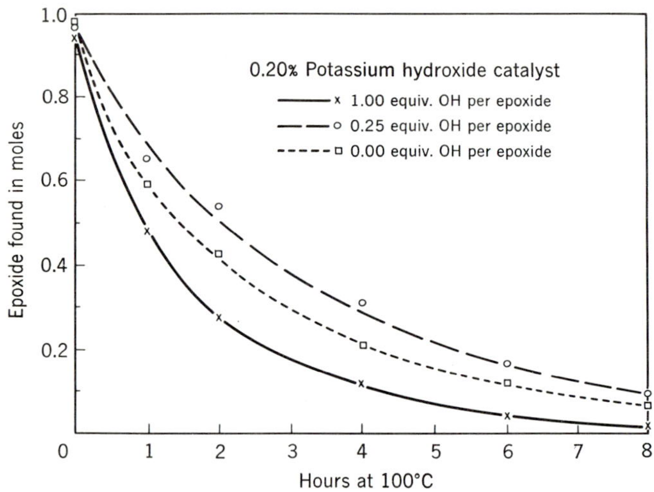

FIG. 5-4. Inorganic bases react with the epoxy group through the readily ionizable base hydroxyl. The speed of the reaction is little effected by the concentration of alcohol [5].

If phenol is present, the reaction proceeds through the phenolic hydroxyl (Fig. 5-5),

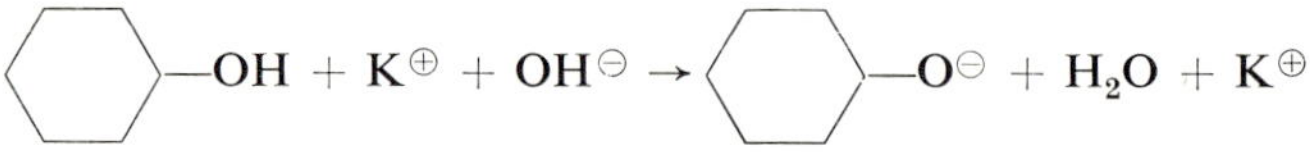

and if carboxyls are present, preferably through them (Fig. 5-6). Thus, as is the case with the tertiary amines, the reaction proceeds from the acid side, and to the extent that the base enters, the reaction can be placed:

$$\text{Carboxyl} > \text{phenol} > \text{base}$$

Because of this preference, and the tendency to suppress the epoxy-alcohol reaction, inorganic bases are preferred catalysts for the synthesis of resins from phenols and epichlorohydrin.

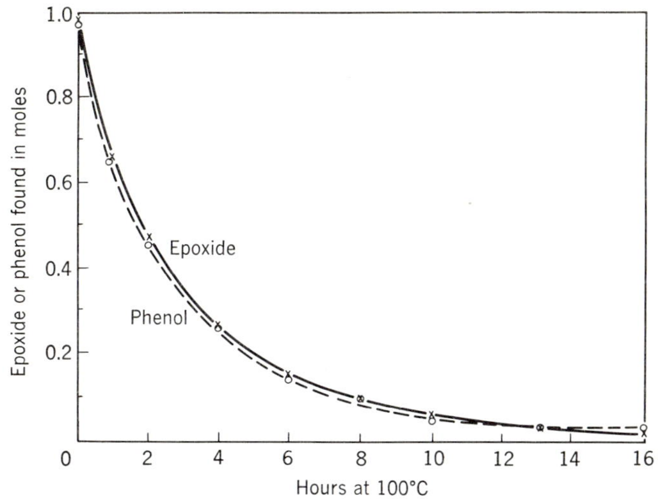

FIG. 5-5. The reaction between an inorganic base and an epoxy is suppressed in the presence of phenol, the phenol-epoxy reaction predominating [5].

Secondary Amines

Secondary amines have been shown to react with the epoxy group by the following simplified mechanism [6]:

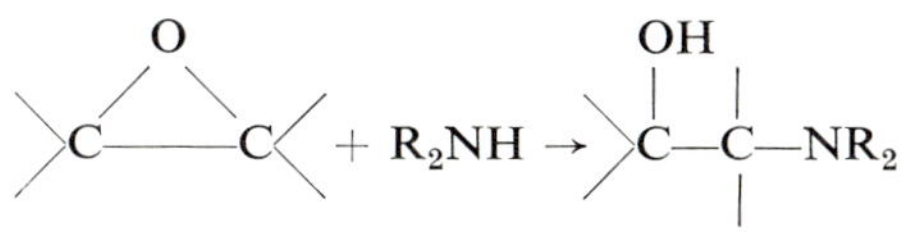

Amine addition reaction

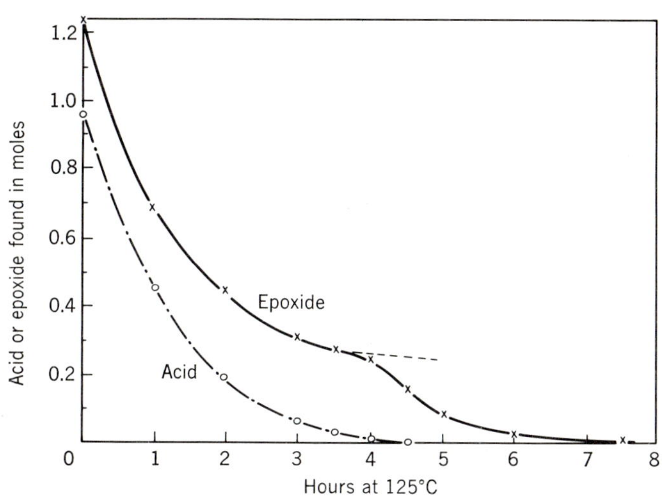

FIG. 5-6. The reaction between an inorganic base and epoxy is suppressed in the presence of carboxyl groups, the carboxyl-epoxy reaction predominating [5].

As a result of the reaction, a tertiary amine is formed. To the extent that the tertiary amine is not otherwise hindered, further reaction may occur by the mechanisms proposed for Lewis bases.

Progress of Amine Addition Reaction. That the amine addition reaction cannot progress in the absence of a hydroxyl group or other hydrogen donor has been indicated [16]. The specific donor will change the speed of the reaction but not the direction. The effect of several hydrogen donor compounds on the speed of the reaction is indicated in Fig. 5-7.

The hydrogen donor, itself, does not react in such a way as to be bound into the system and may be recovered, apparently unchanged.

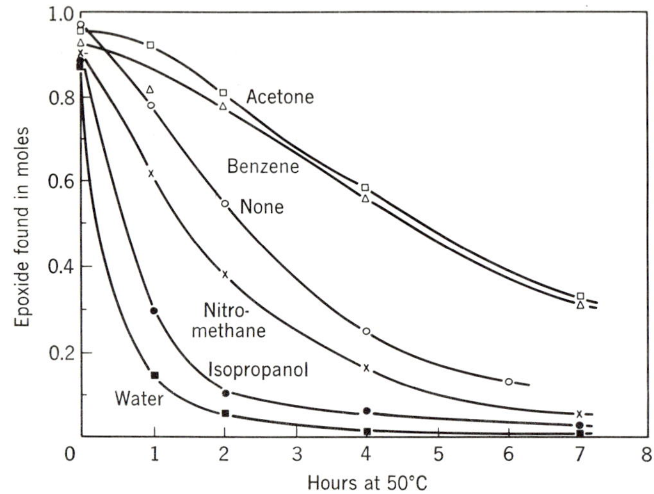

FIG. 5-7. Relative reactivity rates of various hydrogen donors as they influence the addition reaction of a secondary amine and an epoxy group [6].

In a series of studies [25], there was developed an explanation involving two sequential bimolecular reactions. The net effect is to transfer the proton from the donor to the nascent hydroxyl, and the proton from the amine to the donor. In simplified form, the reaction proceeds as follows:

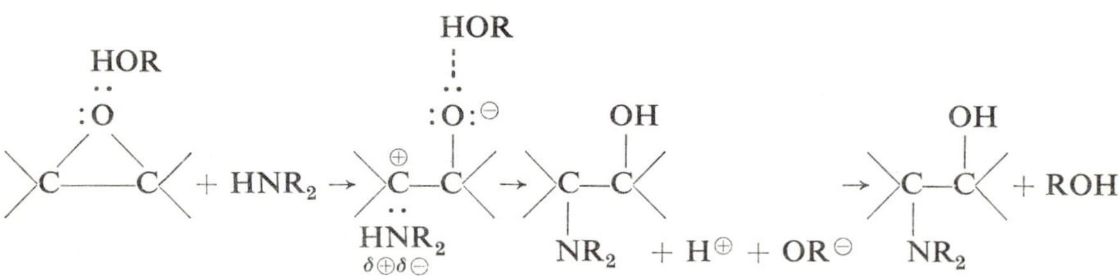

Because each amine addition reaction results in the generation of a nascent hydroxyl, the reaction might be expected to proceed at a satisfactory rate when only very few alcoholic hydroxyls are initially present, the supply increasing as the reaction progresses. This is indeed the case.

Phenol has been found [6] to be a more powerful donor compound for the reaction than alcohols. Organic acids might be expected to provide even more powerful catalytic activity than do the phenols, although the reaction is complicated by a number of factors [6,16], including the formation of salt by reaction of the acid and the secondary amine. Among the acids suggested as accelerators are salicylic acid (in high-boiling alcohol) and lactic acid [27].

As in the case of the tertiary amines, the approach of the secondary amine to the epoxy carbon atom is probably optimum from the rearward direction. Although the transition state is probably somewhat strained, some secondary amines are capable of sluggish reactivity with the ring-situated epoxy groups, whereas the tertiary amines are not. The more sterically hindered the secondary amine, the more sluggish is the amine addition reaction, as is shown in Fig. 5-8.

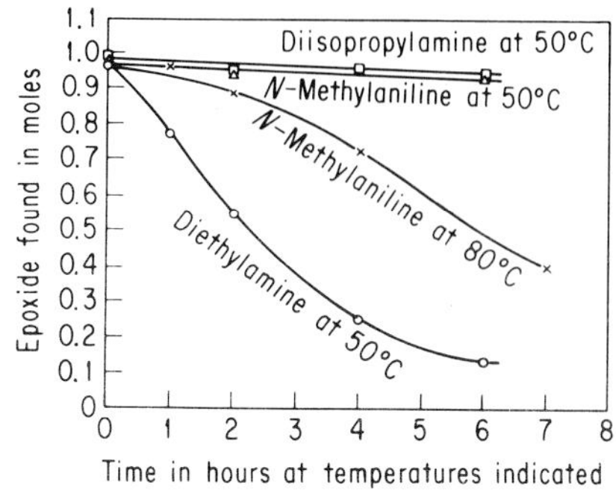

FIG. 5-8. Influence of steric factors on secondary amine–epoxy addition reaction [6].

Progress of Tertiary Amine Catalysis. The tertiary amine formed from the reaction of a secondary amine with an epoxy may or may not be sufficiently active to catalyze further reactions. The reactivity will depend on steric factors (the accessibility of the nitrogen) and on the electron density at the unshared electron pair.

Studies of secondary amines using a phenol accelerator show that with *N*-methylaniline the addition reaction goes well (Fig. 5-9) and that the formed tertiary amine is sufficiently active to introduce catalysis of the phenol-epoxy reaction. With diethylamine and with piperidine, although the addition reaction goes well, neither formed tertiary amine is strong enough to catalyze the phenol-epoxy reaction. Upon exhaustion of the secondary amine, a reaction takes over which is in the reverse direction of that expected under tertiary amine catalysis, favoring the epoxy-hydroxyl reaction over the epoxy-phenol reaction (Fig. 5-10 and 5-11). This is a phenol-catalyzed reaction, as discussed subsequently. However, in the absence of phenol, the tertiary amine from the diethylamine is an ineffective catalyst, although

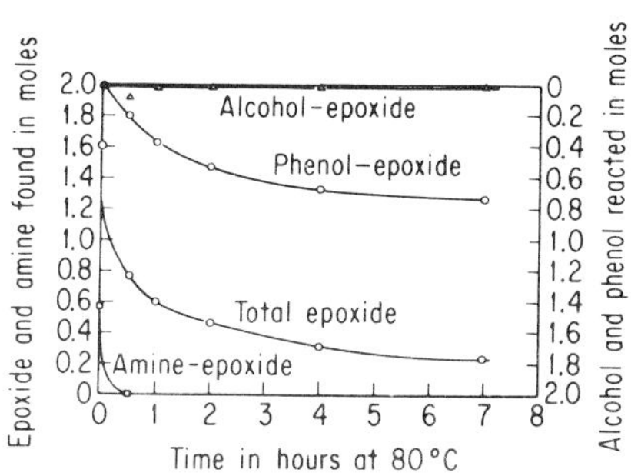

FIG. 5-9. The amine-addition reaction between *N*-methylaniline and epoxy is accelerated by phenol. The formed tertiary amine promotes the epoxy-phenol reaction [6].

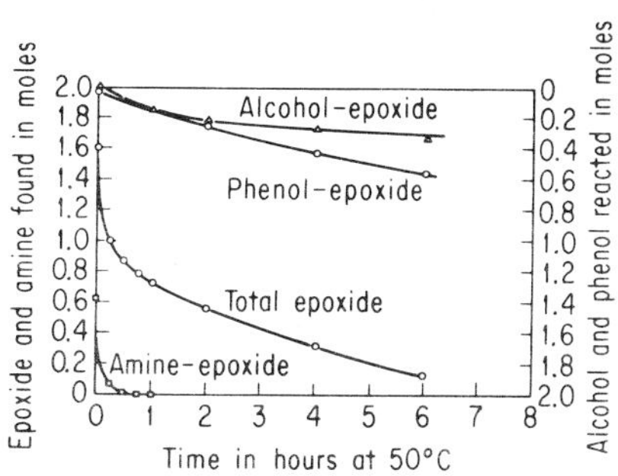

FIG. 5-10. The amine-addition reaction between diethylamine and epoxy is accelerated by phenol. The formed tertiary amine does not catalyze the epoxy-phenol reaction; rather the phenol then catalyzes an alcohol-epoxy reaction [6].

the tertiary amine from piperidine does give catalytic activity. Thus, insofar as tertiary amine activity is concerned, the cyclic amine, piperidine, stands somewhere between the aromatic-substituted *N*-methylaniline and the aliphatic diethylamine.

From the discussion, it can be concluded that in order for secondary amines to serve as effective curing agents, the amine molecule must contain more than two

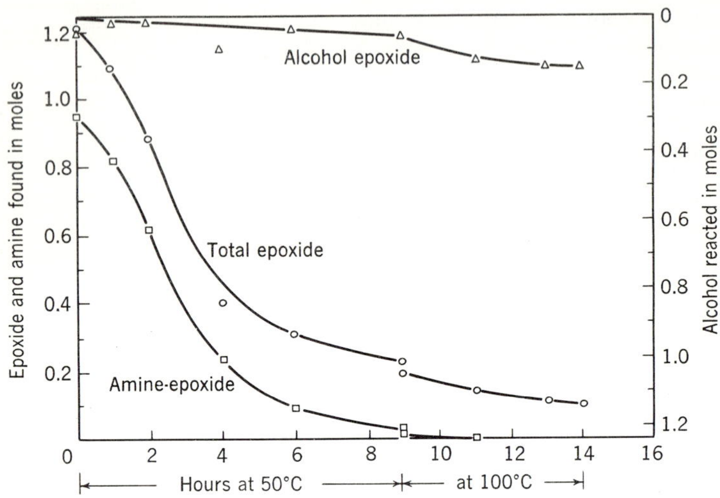

FIG. 5-11. The tertiary amine from diethylamine is too weakly catalytic to promote alcohol-initiated epoxy polymerization [6].

secondary amines (for a difunctional epoxy resin) or the formed tertiary amine must be sufficiently unhindered and electronegative to promote further reaction by the Lewis base mechanism.

Primary Amines

Primary amines react in a fashion similar to secondary amines. The amine addition reaction proceeds, in simplified form:

(1) 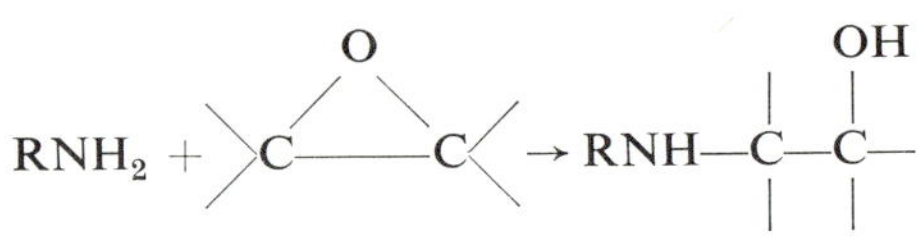

(2) 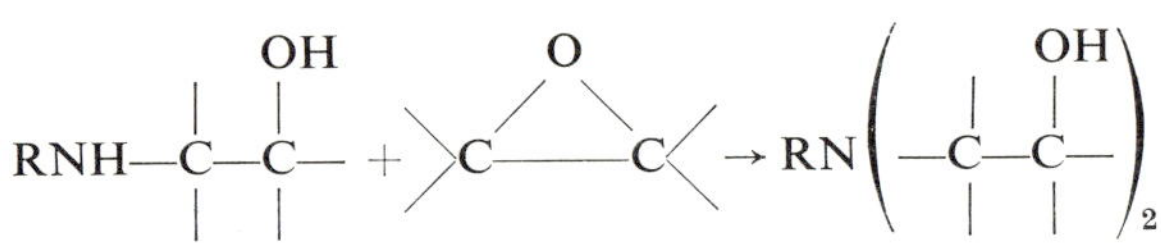

Although rate studies show that the primary amine reacts about twice as fast as the secondary amine [17,23], in epoxy-resin systems some tertiary amine is formed almost immediately (Fig. 5-12).

Generally, the tertiary amines formed by the reaction of two epoxy groups with a primary amine are ineffective catalysts. The primary amines, however, can be used effectively as epoxy curing agents, since the primary amine group is difunctional with the resin. If more than one primary amine is present per molecule, cross-linked structures can result through reaction with diepoxies.

Since primary amines have only one bulky substituent, they may approach epoxy carbon atoms from other than an absolutely linearly rearward direction, making them suitable for use with ring-situated epoxy groups; however, some reactivity is lost, and in these cases it is probable that at least some per cent of the formed secondary amine may be blocked.

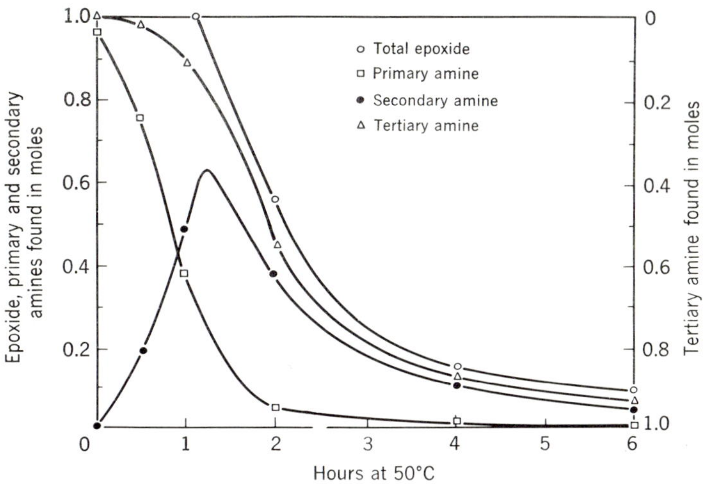

FIG. 5-12. When primary amines react with epoxies, the initial reaction product is an alcoholic hydroxyl and a secondary amine. The secondary amine then reacts, but the resultant tertiary amine is too weakly catalytic to promote additional polymerization [6].

Amides

Many so-called "amides" used in epoxy-resin formulations are actually amido-polyamines containing sufficient free-amine hydrogens to react by the procedures described above. The hydrogens of the amide group are much more sluggishly reactive, with elevated temperatures being required for a satisfactory rate [4]. The reaction mechanisms have not been developed in the literature on the epoxy resins, although presumably the mechanisms are similar to those encountered with the primary and secondary amines. The reaction generally occurs at temperatures of about 140°C with the glycidyl ether resins [26].

ACID CURING AGENTS

The acid curing agents normally used in the epoxy-resin technology are carboxylic acid anhydrides, dibasic organic acids, phenols, and Lewis acids.

Lewis Acids

Lewis acids are those compounds containing empty orbital positions in the outer shell of one of the atoms. They are attracted to areas of increased electron density.

Studies of the polymerization of ethylene oxide with boron trifluoride (BF_3) indicate that the reaction is quite complex and may involve production of inner oxonium salts from epoxy groups, anhydrous ethers, and the BF_3, these reacting with hydroxyl groups [18]. Other Lewis acids, such as aluminum chloride, are believed to react in a similar manner; whereas metal halides such as ferric chloride can initiate polymerization through the carbonium ion: $Cl_3Fe^{-}OCH_2CH_2^{+}$.

The Lewis acids most normally used in epoxy-resin technology are of the BF_3 type. Hydroxyls are usually present, and the reaction may proceed, in simplified form, as follows [31]:

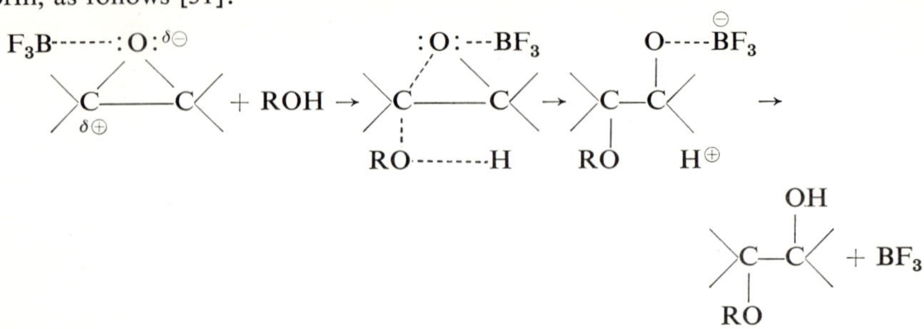

Alternatively, the reaction may proceed as follows, the total attack being a directly frontal one:

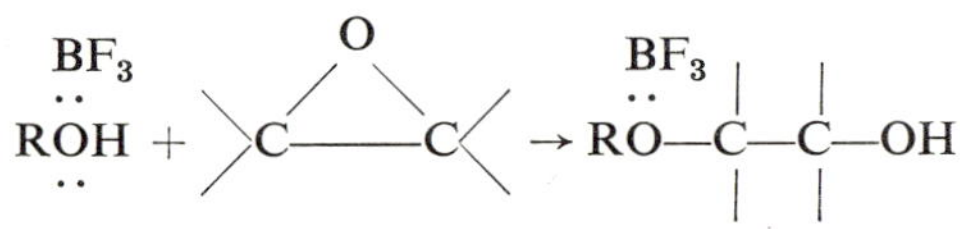

or

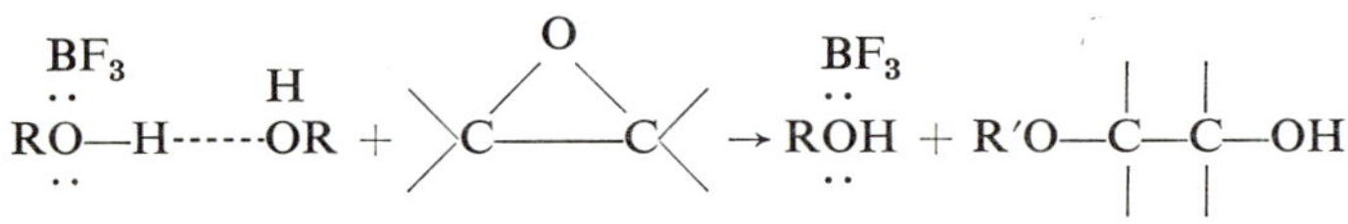

Or it may be that an oxonium salt is formed, together with a carbonium ion, as follows [29]:

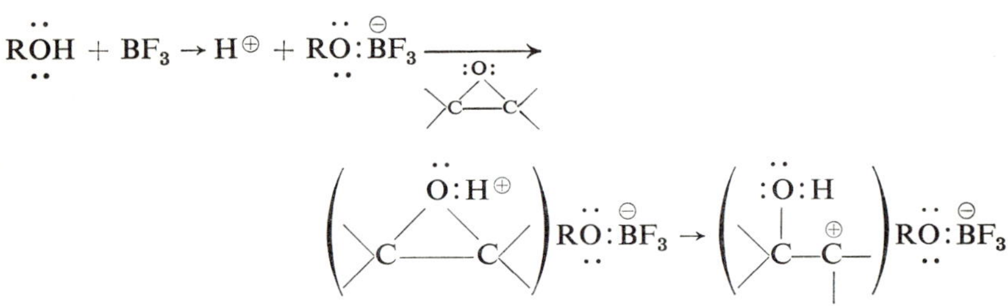

The carbonium ion may then proceed to provide for chain propagation.

In the absence of hydroxyls, the reaction may proceed as follows, with ultimate catalyst deactivation [31]:

(1)

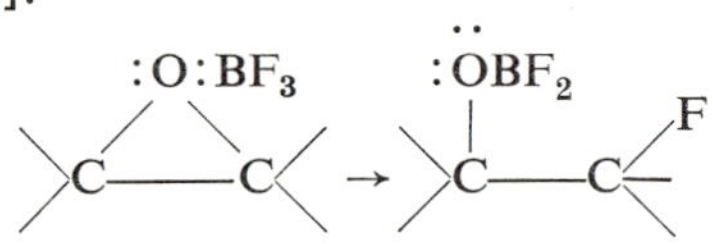

(2)

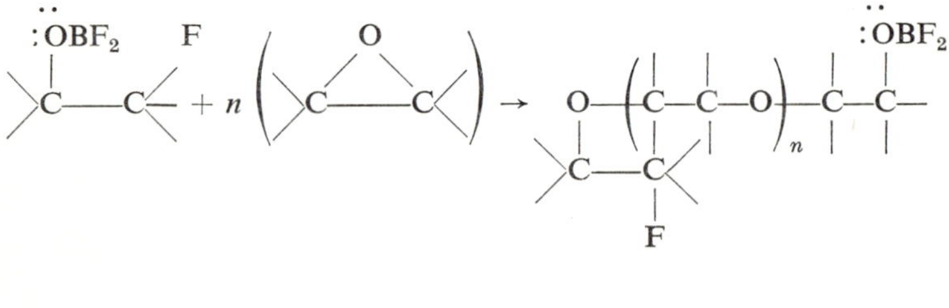

(3) 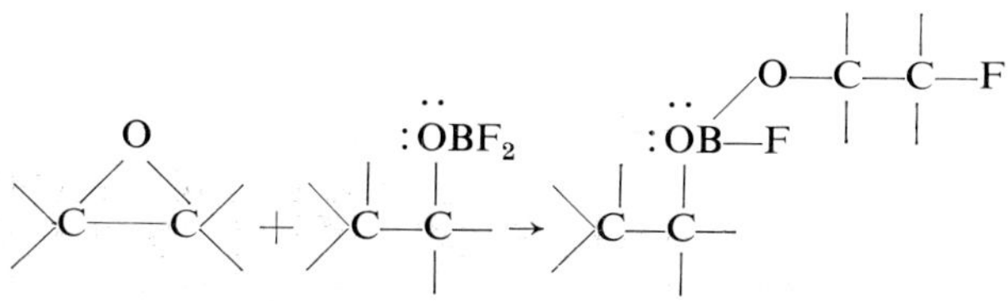

In the epoxy technology, when the hydroxyls are supplied separately from the resin, a heteropolymer will result. If only a few hydroxyls are originally present, the reaction will proceed through nascent hydroxyls to give, essentially, a homopolymer very similar to that obtained from tertiary amine cures. With the Lewis acids, however, functionality is not lost by nonreactive chain-termination mechanisms, and the cured polymers will consequently be more tightly cross-linked.

It might be reasoned that the Lewis acids would be very effective catalysts for the more basic active-hydrogen-containing compounds, such as the primary amines, since presumably the proton transfer from the amine would be favored over the proton transfer from the alcohol. Indeed, in some cases, BF_3 complexes may be used to accelerate the cure of sluggishly reactive primary amines. BF_3, itself, may be complexed with the amine, thereby coordinating the unshared electron pair of the nitrogen with the empty orbitals of the boron, leading to blocked reactivity until elevated temperatures are used to dissociate the complex. Such amine–Lewis acid complexes, as well as ether–Lewis acid complexes, are useful curing agents.

In the case of secondary amine–BF_3 complexes, the species may dissociate to

$$R_2N{:}BF_3^{\ominus} + H^{\oplus}$$

which then reacts with an epoxy to give

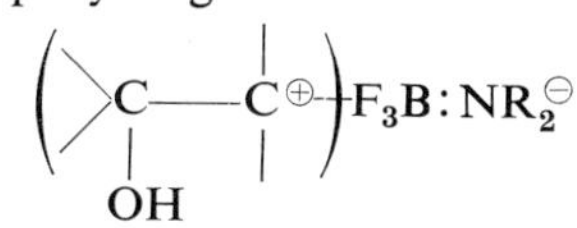

which is capable of chain propagation. Alternatively, dissociation may be $BF_3 + R_2NH$ to give independent cure by two mechanisms.

With anhydride cures, the intermediate

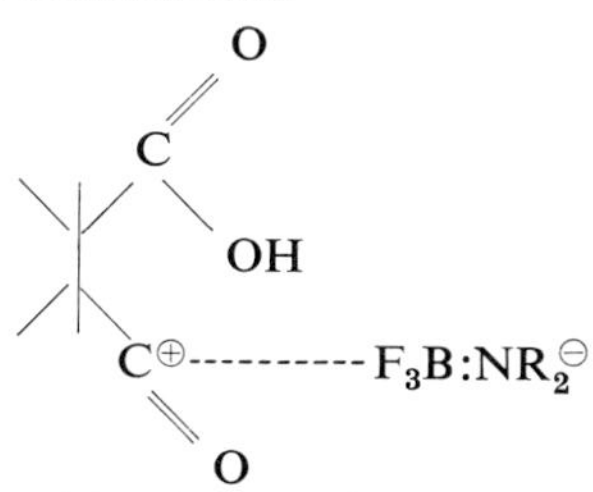

may be formed which reacts with a hydroxyl to give [29]:

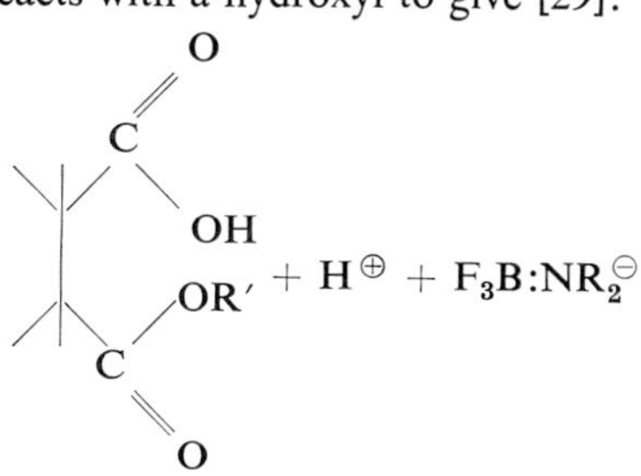

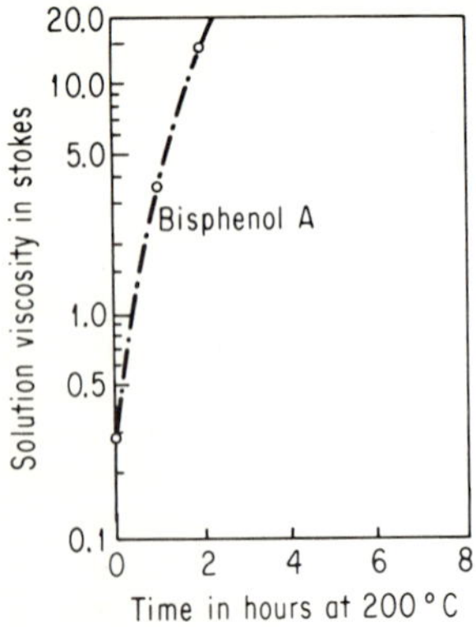

FIG. 5-13. Phenols react rapidly with epoxy resins at temperatures of 200°C [6].

The reactivity of BF_3-type curing agents varies sharply with the substituents on the epoxy group. Assigning a reactivity of 1 to phenyl glycidyl ether, epichlorohydrin has a reactivity of 0.7, and propylene oxide a reactivity of 3.5 [31]. BF_3 is extremely reactive with the glycidyl ether type of epoxy resins, the reaction going to substantial completion in just a few seconds at room temperature.

Phenols

Phenols have been shown to react with epoxy resins if elevated temperatures are employed (Fig. 5-13). During the course of the polymerization, the phenol will, in some fashion, bring about independent etherification through the reaction of available hydroxyl groups with epoxy groups (Fig. 5-14). Even in the absence of substantial initial hydroxyl groups, the reaction will proceed preferentially through nascent hydroxyls (Fig. 5-15). The reaction is thus in the reverse order of what would be expected with the mechanism proposed for tertiary amine cures.

The exact mechanisms have not been elucidated in the literature, but it is likely that the reaction proceeds, in part, through the formation of an intermediate oxonium ion:

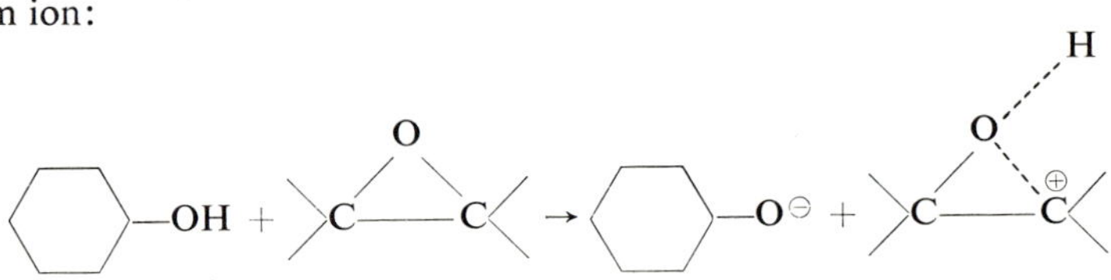

The resultant phenoxide ion may react as indicated previously. The carbonium ion from the oxonium complex may react with a hydroxyl group:

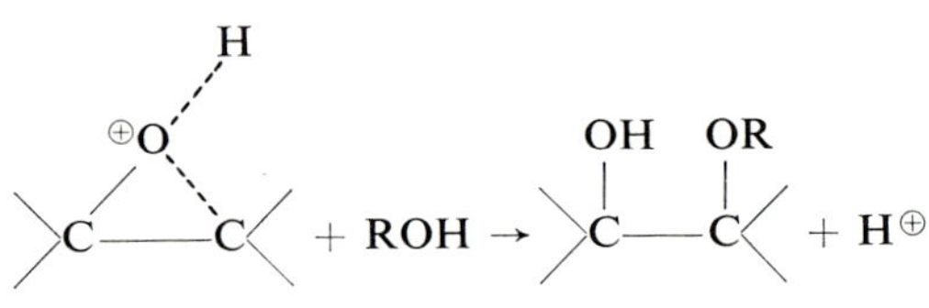

The free proton from this reaction may complex with an additional epoxy group or may react with an available phenolate or alkoxide ion. It might also be that a three-membered transition state is involved, with the phenolic hydroxyl group being regenerated in a fashion similar to that described for alcoholic hydroxyls during cure by secondary amines.

When phenol is reacted with an epoxy resin in the presence of an inorganic base or a tertiary amine, the reaction (as previously indicated) proceeds almost exclusively through the phenolic hydroxyls, the alcohol-epoxy reactions being suppressed.

In base-catalyzed reactions, the phenol may react with the base to produce the phenolate ion, which then reacts with an epoxy group. The new alkoxide ion then regenerates a phenolate ion to continue the reaction. With tertiary amines, it is probable that the tertiary amine is protonated to release the initiating phenolate

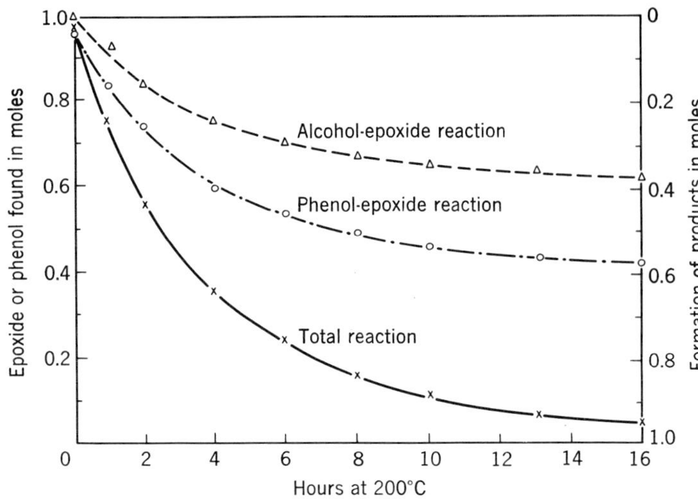

FIG. 5-14. When phenol and epoxy are reacted in the absence of a base catalyst, alcoholic hydroxyls are produced, which then react with the epoxy under the influence of the phenol [6].

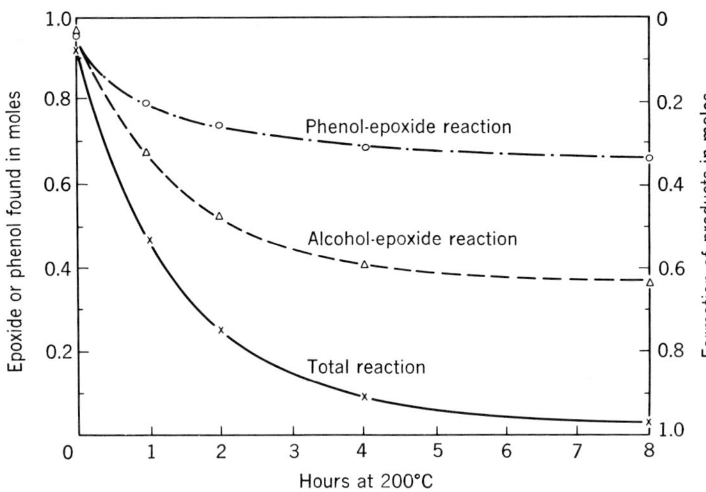

FIG. 5-15. When phenol and epoxy are reacted in the absence of a base catalyst and in the presence of excess alcohol, the epoxy preferentially reacts with the alcohol under the influence of the phenol [6].

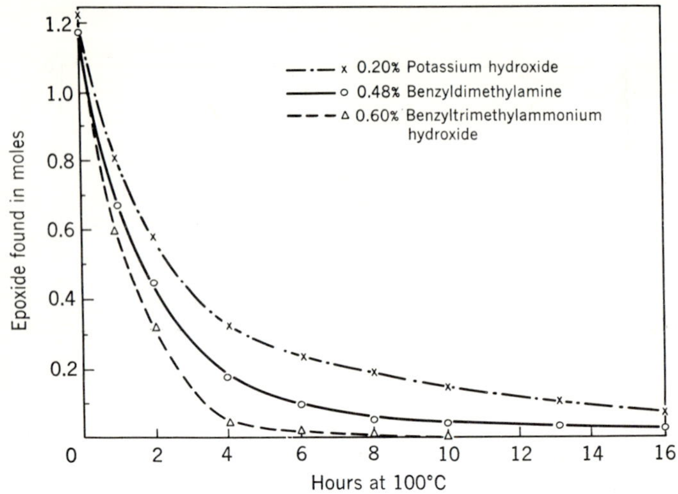

FIG. 5-16. When phenol is reacted with epoxy in the presence of a base catalyst, the phenol-epoxy reaction is preferred. The specific base affects the speed but not the direction of the reaction [6].

ion. The specific catalyst affects the rate but not the direction of the reaction (Fig. 5-16).

When phenol is present in an anhydride/epoxy system, it is probable that a termolecular transition state is involved, with the final product [33] being

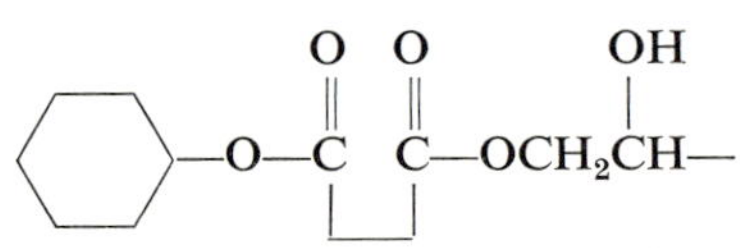

Organic Acids

Organic acids, i.e., those materials containing carboxyl $\left(-\overset{\overset{\large O}{\|}}{C}-OH\right)$ groups, would be expected to react in a fashion similar to the phenols.

The initial reaction would be *addition* esterification:

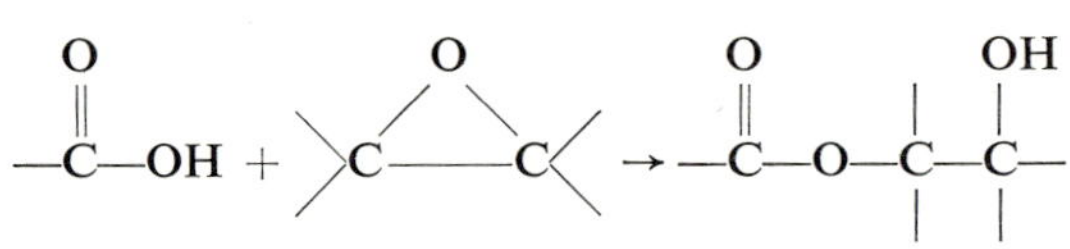

The formed alcoholic hydroxyl would then react preferentially by the mechanism suggested for phenol cures. However, a temperature-dependent competing reaction, condensation esterification, would necessarily be involved:

$$-\overset{\overset{\large O}{\|}}{C}-OH + ROH \rightleftharpoons -\overset{\overset{\large O}{\|}}{C}-O-R + H_2O$$

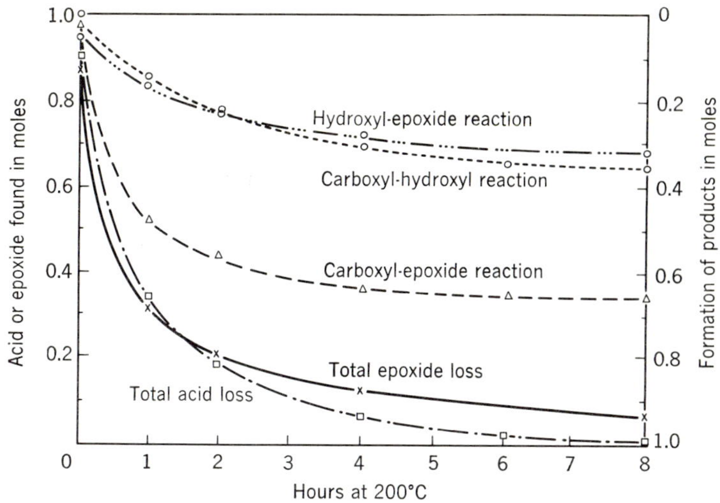

FIG. 5-17. When caprylic acid is reacted with DGEBA at 200°C with water of esterification being continuously removed, the rate of reaction is such that approximately two addition esterification reactions (carboxyl-epoxy) occur for each condensation esterification (carboxyl-hydroxyl) and etherification (epoxy-hydroxyl) reaction [6].

Experiments conducted at 200°C [5] indicate the relationship to be expected among the three reactions (Fig. 5-17). If the alcohol is present in excess, the epoxy-alcohol (etherification) and the carboxyl-alcohol (condensation esterification) reactions dominate (Fig. 5-18). Much of the excess alcohol is introduced into the resin system with the evolution of one molecule of water for each free hydroxyl group so reacted.

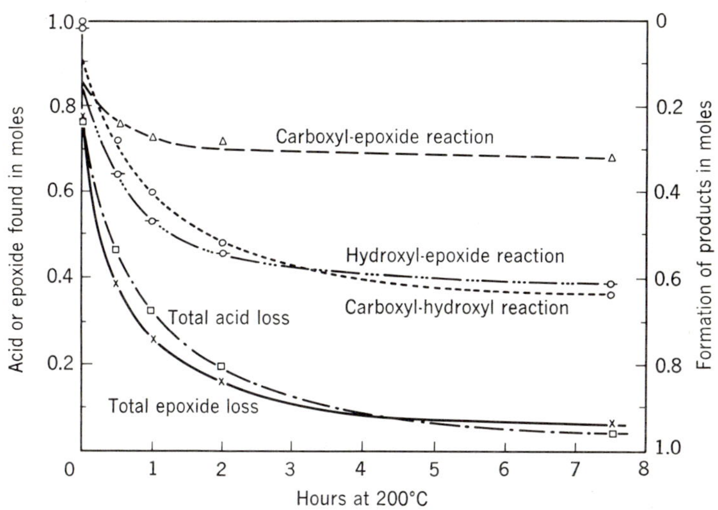

FIG. 5-18. When caprylic acid is reacted with DGEBA at 200°C with water of esterification being continuously removed and dipropylene glycol being present, the addition esterification is suppressed, and the etherification and condensation esterification reactions proceed at about an equal rate [6].

It should be noted in the above experiment that the water of condensation was removed upon formation. It might be expected, in actual practice, because of equilibrium conditions, that the condensation esterification would be virtually suppressed [20], although it is probable that some of the water might be removed through reaction with the epoxy groups to form alcohols or might evaporate during elevated-temperature cures.

Evidence developed from the study of the various organic acid/epoxy-resin systems indicates that the two reactions, addition esterification and etherification,

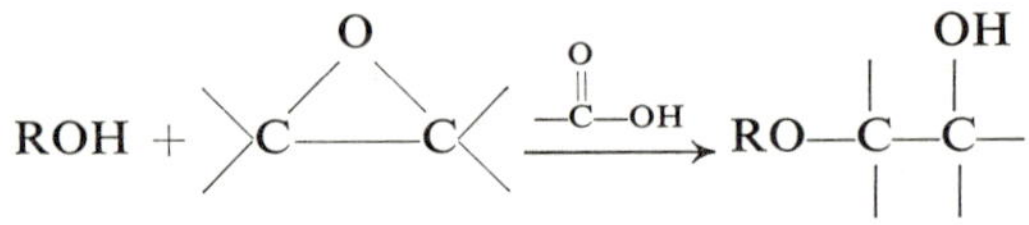

play the major role during cure and are temperature-dependent, with higher temperatures generally favoring increasing addition esterification and lower temperatures favoring increasing etherification. The extent to which etherification will occur is also dependent on the nature of the carboxyl group and the location of the epoxy group in the molecule [24].

As was the case with the phenols, when a base catalyst or a tertiary amine is employed, the reaction is specific. The epoxy-carboxyl reaction dominates, and only when the acid has been reacted does the inherently faster etherification reaction take over. This is in accordance with what is expected in base-catalyzed systems generally. The speed of the reaction is governed by the base selected, but the reaction direction is unaffected.

From the above we can conclude that mild reaction conditions, in the absence of a base catalyst, favor lower carboxyl-to-epoxy ratios and increased etherification; higher temperatures suppress the etherification reaction, thereby requiring higher carboxyl-to-epoxy ratios. If a base catalyst is present, the etherification reaction will be suppressed and the introduction of alcoholic residues into the cured system will not occur until the acid has been expended. If an acid catalyst is present, the effect will be further to promote the etherification reaction and suppress the esterification reaction, thus leading to even lower carboxyl requirements for complete reaction.

In order to permit the use of excess alcohol (as is often desired with acid cures), it is customary to employ the anhydride rather than the acid as a curing agent, since in both uncatalyzed and base-catalyzed systems, alcoholic residues may conveniently be introduced into the cured network.

Anhydrides

In epoxy chemistry only cyclic anhydrides are used. These anhydride groups

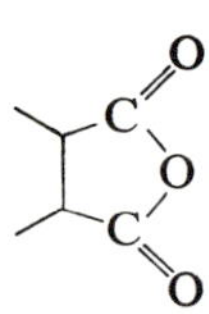

will not react directly with epoxy groups. In order for the reaction to occur, it is first necessary that the anhydride ring be opened. The ring may be opened either by active hydrogens present as water, or hydroxyls, or by a Lewis base. The reaction between the epoxy and the anhydride will be governed by the method of ring opening.

Ring Opening by Active Hydrogens. The anhydride ring may be opened by an active hydrogen to produce one or two carboxyl groups.

Ring opening by water:

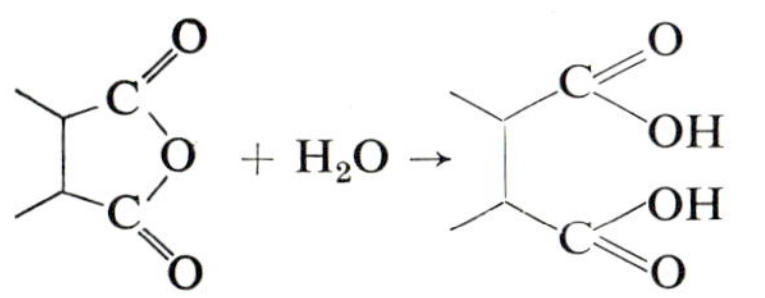

Ring opening by hydroxyl:

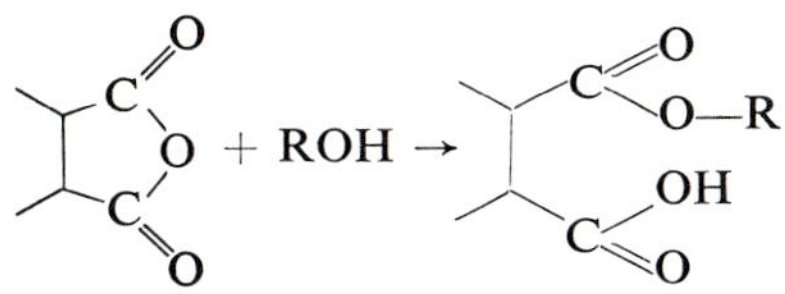

As a result of the ring opening, carboxyl groups are formed which will react with the epoxy in a fashion described previously for organic acids.

With the anhydrides, as with the organic acids, it has been concluded that the condensation esterification may virtually be ignored since, in the absence of removal of water, the reaction equilibrium will not favor it. However, a new, reversible reaction is introduced, when the monoester is formed preparatory to the carboxyl-epoxy reaction,

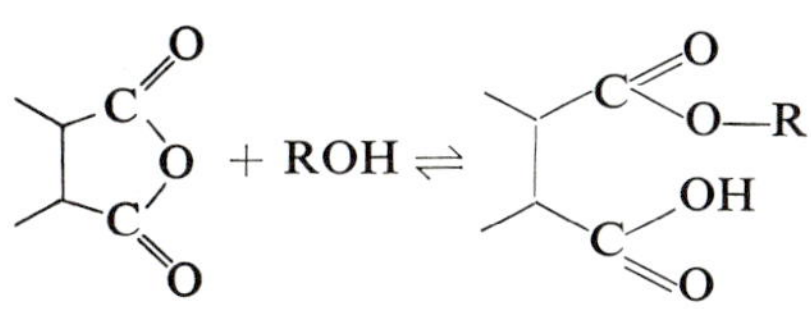

and volatile compounds which are part of the equilibrium may be lost by evaporation from the surface. Thus, for example, phthalic anhydride, when used to cure DGEBA (mol. wt. 450), will be present in reacted form at a little better than 50 per cent of the expected amount on the surface (0.01 to 0.02 mm) of the casting [20]. It is probable, too, that much of the weight loss experienced with anhydride systems at temperatures in the 150 to 200°C range may be attributed to the splitting out of the incompletely reacted anhydride initially present as the monoester (Fig. 5-19), and the subsequent volatilization of the anhydride. Increasing cure temperatures will effect improvements (Fig. 5-20).

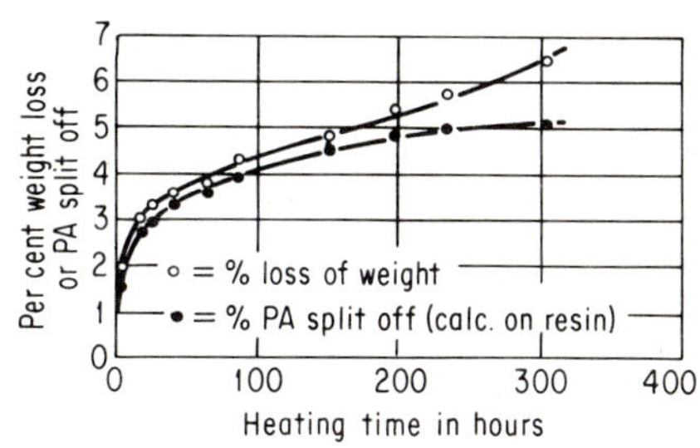

FIG. 5-19. Weight-loss studies of PA-cured DGEBA at 150°C under reduced pressure indicates that the main constituent of the volatile product is PA split from the resin [3].

As is true of acid cures generally, considerable etherification might be expected in anhydride-cured systems. That this is not necessarily so, and that the esterification reaction progresses rather strongly in some systems, is somewhat surprising.

However, if we make the simplifying assumption that the anhydride-hydroxyl reaction dominates the epoxy-hydroxyl reaction, we can conclude that substantial etherification need not, in fact, occur.

For example, if only one active hydrogen were initially present, the reaction would proceed (1) anhydride-hydroxyl,(2) carboxyl-epoxy, (3) anhydride-hydroxyl, (4) carboxyl-epoxy, (5) anhydride-hydroxyl, etc. Such a system would be expected to provide optimum properties at an A/E (anhydride/epoxy) ratio of 1/1, and no etherification would be possible.

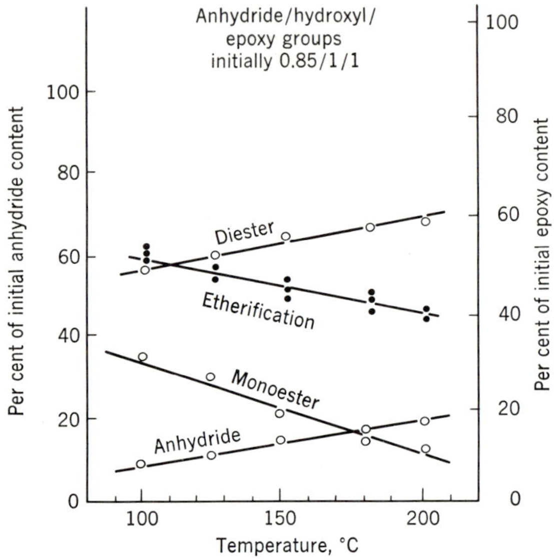

FIG. 5-20. Increasing cure temperature reduces the amount of monoester and increases the amount of the more stable diester in PA-cured DGEBA [3].

To the extent that hydroxyls are present, etherification would be anticipated, the extent to some degree governed by the hydroxyl concentration and type of hydroxyl. If sufficient hydroxyls are present to open all anhydride rings for example, essentially an acid-cured system would be involved, so that we might expect considerable etherification to be present in the reaction product, the exact amount determined by the nature of the carboxyl group and the reaction temperature. Intermediate concentrations of hydroxyls might be expected to favor ratios between an optimum A/E of about 0.4/1 (for a very strongly acid-catalyzed system with considerable etherification) and 1/1 (for a very weakly acid-catalyzed system with virtually no etherification).

This reasoning, of course, may represent an oversimplification. It is not altogether evident that in all cases the anhydride ring is first opened followed by the reaction, at a later time, of the formed carboxyl group with an available epoxy group. Reaction rate studies indicate that the opening of the anhydride ring may in some cases involve a termolecular transition state, producing two ester links simultaneously, somewhat as follows [28]:

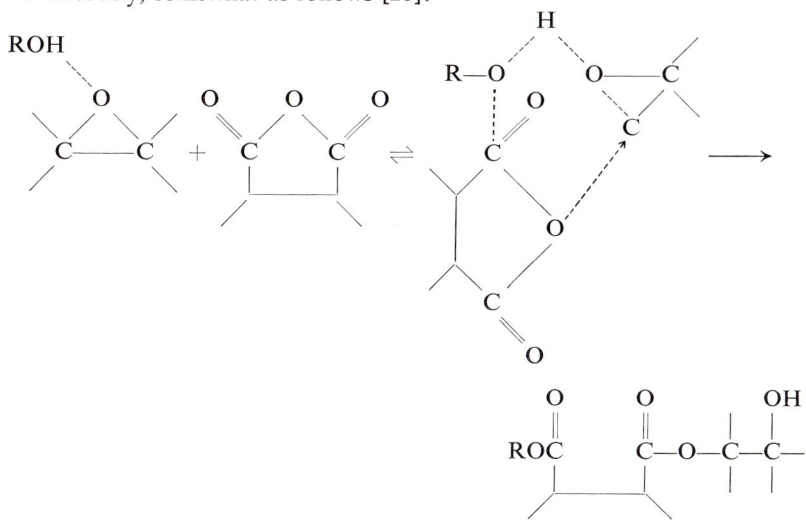

A product analysis of DGEBA (mol. wt. 380) cured with HHPA and NMA indicated that no etherification occurred during the course of the initial reaction to the gel point, lending support to this termolecular transition state [33]. To the extent that the termolecular reaction occurs, it seems evident that the extent to which etherification will progress will be a function of the amount of half-ester formed by direct hydroxyl-anhydride reaction. Since it has been shown that the etherification reaction does not progress until at least the gel point is reached, it should follow that higher temperatures would promote increased reaction prior to gelation, and it might thus be reasoned that higher temperatures of the initial cure would favor higher A/E ratios. This is in accordance with experimental observations.

It was found that when reacting *cis*-cyclohexanedicarboxylic acid anhydride with DGEBA (mol. wt. 450), the reaction during the first four hours at 155°C was primarily half-ester formation (55.2 per cent anhydride reacted vs. 13.5 per cent epoxy). In the middle stage (4 to 24 hours), diesters were formed: 87.7 per cent of the anhydride having reacted, but 43.5 per cent of the epoxy remaining intact. The last stage of the reaction was characterized by the rapid decrease of the epoxy groups, and at the end of 48 hours at temperature, both epoxy and anhydride groups had completely disappeared [32].

Obviously the reaction mechanism is quite complex. Depending on the nature of the formed carboxyl, some anhydrides will be more reactive with the epoxy group, and others will more readily promote etherification. The activity will be governed to some extent by reaction temperature (Fig. 5-21) and to some extent by concentration of hydroxyls (Figs. 5-22 to 5-24). Thus, no satisfactory general equations can be developed for the determination of optimum A/E ratios for all anhydride curing agents.

However, that optimum ratios should be developed empirically for each system is evident. As indicated previously, if anhydride is present in excess (for the reaction conditions involved), the reversible monoester reaction will result in the splitting off of the anhydride, and elevated-temperature weight losses will be high. Likewise, weight-loss studies indicate that when insufficient anhydride is present, the excess resin or resin fragments in the commercial materials will be driven off at

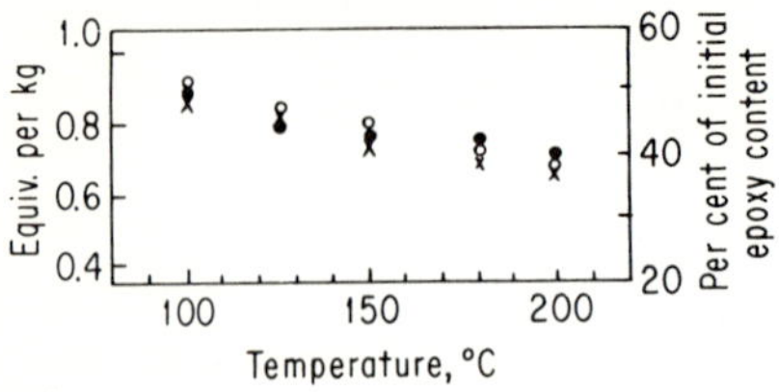

FIG. 5-21. The etherification reaction decreases from about 90 per cent of the total reaction to about 40 per cent of the total reaction in PA-cured DGEBA (A/E/hydroxyl = 0.85/1/1) as the temperature increases from 100 to 200°C [3].

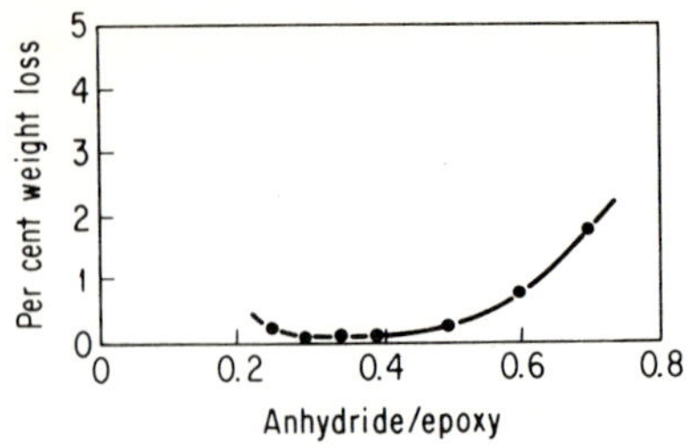

FIG. 5-22. When one hydroxyl is initially present for each anhydride group, the optimum A/E ratio is about 0.3 to 0.4 for 3,4-epoxy-6-methylcyclohexylmethyl-3,4-epoxy-6-methylcyclohexane carboxylate.

aging temperatures in the range of 240°C. The amount of weight loss by the uncatalyzed resin (Table 5-3) is sufficient to explain the weight loss encountered in the anhydride-poor systems at these temperatures.

From this it can be concluded that for optimum performance with an uncatalyzed epoxy/anhydride system, the hydroxyl content should be controlled, the cure temperatures should be reasonably high and long, and the A/E ratio should be adjusted to the specific cure temperature employed and the extent to which the specific carboxyl group promotes etherification at that temperature.

The specific reactivity of the anhydride will depend on two factors: (1) the speed with which it is converted to the half-ester, and (2) the speed with which the half-ester then reacts. This is indicated in a general way in Figs. 5-25 and 5-26. Electron-withdrawing substituents on the alcohols have been shown to increase reaction rates [33].

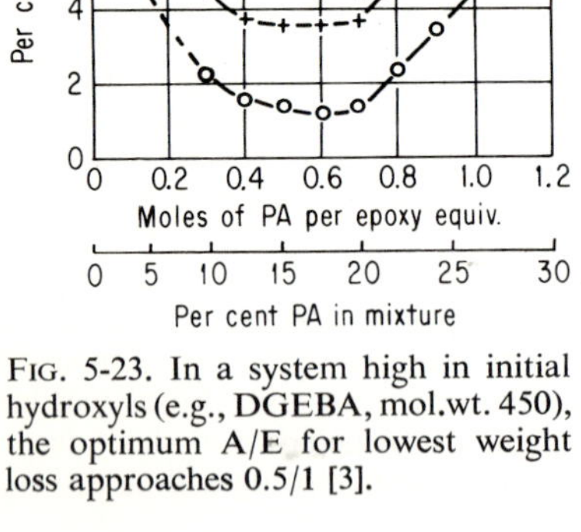

FIG. 5-23. In a system high in initial hydroxyls (e.g., DGEBA, mol.wt. 450), the optimum A/E for lowest weight loss approaches 0.5/1 [3].

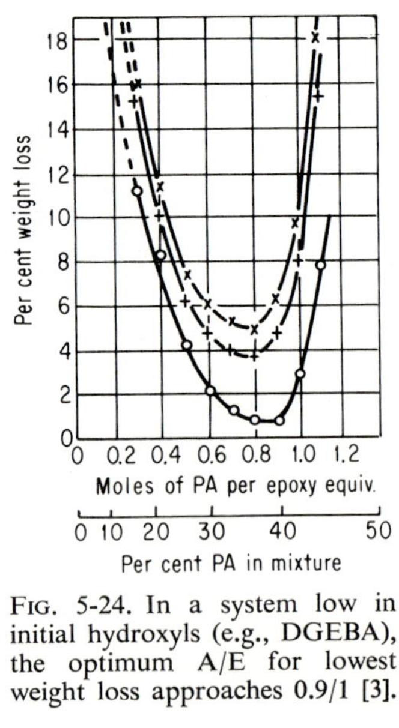

FIG. 5-24. In a system low in initial hydroxyls (e.g., DGEBA), the optimum A/E for lowest weight loss approaches 0.9/1 [3].

Table 5-3. Per Cent Loss of Weight of Pure Resins at 240°C [3]

Resin	% weight loss		
	6 hours	24 hours	48 hours
DGEBA (mol. wt. 450)	8	11	12
DGEBA (mol. wt. 350)	42	51	52

Ring Opening by Lewis Bases. The Lewis bases, as has been shown, can react with an epoxy group to release an anion capable of continuing the reaction

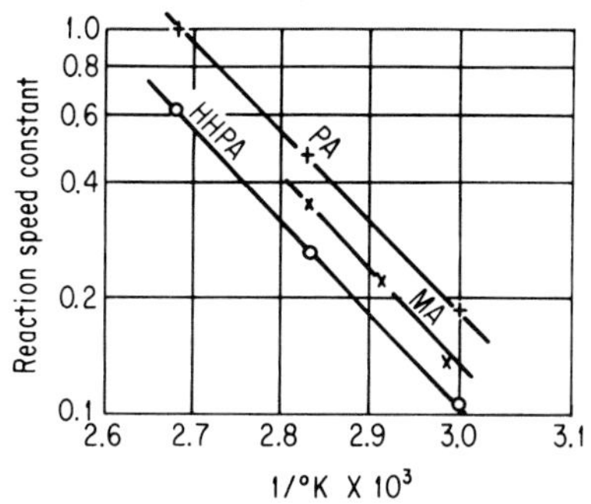

FIG. 5-25. Relative reaction rates of anhydrides with hydroxyls to form half-esters [8].

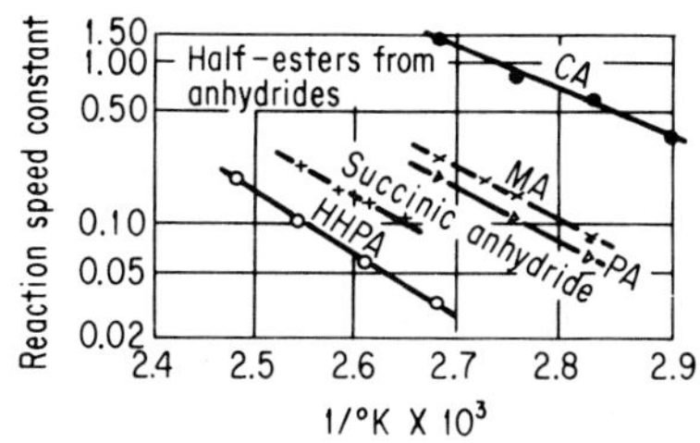

FIG. 5-26. Relative reaction rates of various half-esters with DGEBA [8].

through other chain elements. The Lewis base also can react, and reacts preferentially, with an anhydride to generate a carboxyl ion:

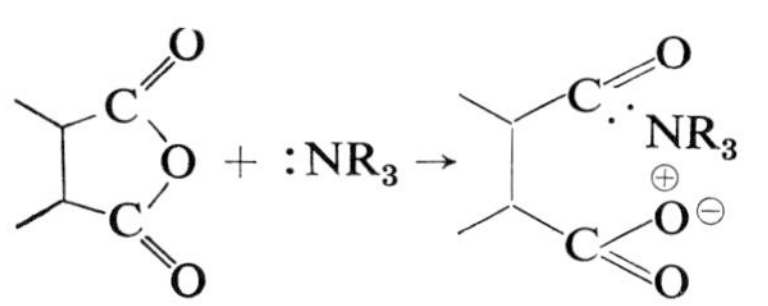

The carboxyl ion then reacts with an epoxy, and an ion is generated, which may then open a second anhydride ring. The absence of competing reactions is indicated in Fig. 5-27. The reaction rate is dependent on accelerator concentration, as shown in Fig. 5-28 and discussed in [10,15]. In [30] kinetic data are presented to indicate that the reaction is solely dependent on accelerator concentration.

When active hydrogens are present, the following mechanism has been proposed as the initiating step [33]:

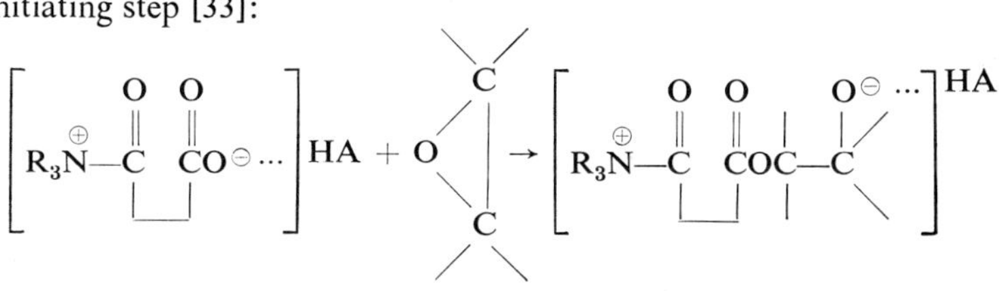

In support of the mechanism, it was found that cocatalysts of amine/alcohol or amine/phenol were more effective accelerators than either the amine or alcohol

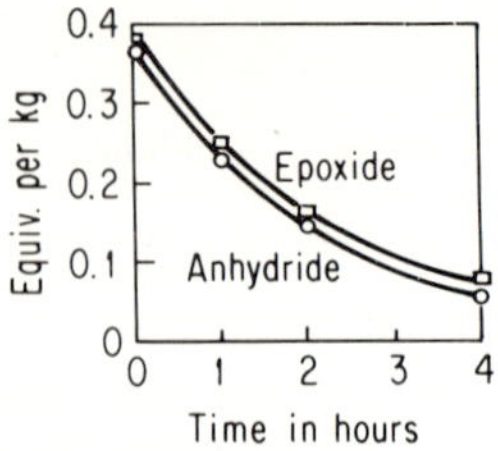

FIG. 5-27. Epoxy from allyl glycidyl ether and anhydride disappear at equal rates in presence of tertiary amine catalyst to form linear polyester [15].

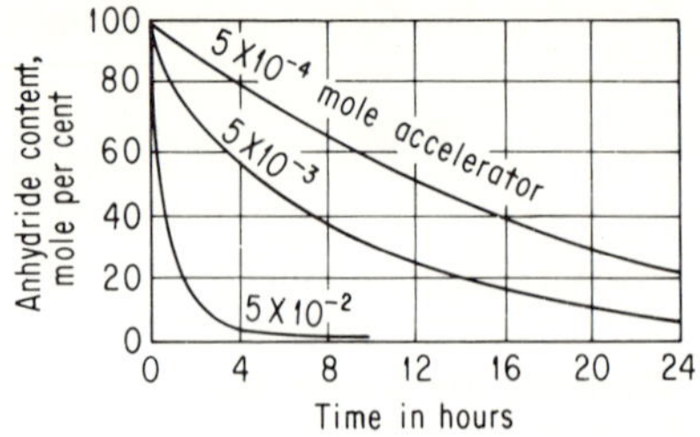

FIG. 5-28. Reaction speed as a function of accelerator concentration. Phenyl glycidyl ether and PA in presence of benzoic acid salt of dicyclohexylamine at 150°C [10].

alone. After initiation, the reaction continues in such a way as to introduce alcoholic residues into the system. It may also be that, to some extent, the polyol reacts with the anhydride (similar to the uncatalyzed state) to generate carboxyls [19], and the carboxyls, under Lewis base catalysis, then react as previously described. It has been suggested that the Lewis base reacts first with the epoxy group, and the resultant alkoxide ion reacts with the anhydride ring [22], but reaction rate calculations have been advanced to discount the suggestion [28].

POLYMERIZATION THROUGH OLEFINS

Because some epoxy resins contain olefins and suitable unsaturation may be present in the epoxy curing agent or may be added in other forms, the reaction between olefin groups is of importance to the epoxy-resin technology.

The reaction of one olefin with another may be initiated by a free radical, most commonly the decomposition product of benzoyl peroxide. Alternatively, polymerization may progress through carbonium ions which may be generated during the cure of epoxy resins with acid-type curing agents.

When polyunsaturation is present, cross-linked structures may occur, since the functionality will exceed 2. In addition, monoölefins may be added to the system to promote linear polymerization and regulate cross-linking density. The monoölefin, styrene

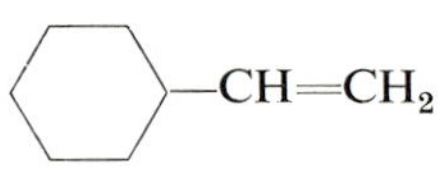

being the most popular, can be used for viscosity reduction, and accordingly may be used with epoxy resins containing unsaturation when it is possible to bind it into the system.

Curing agents containing unsaturation, for example maleic anhydride, may also be used to increase system functionality.

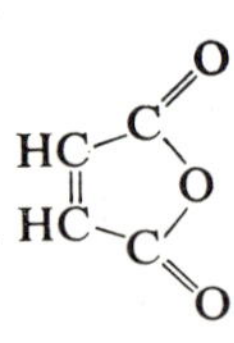

Maleic anhydride may react through the unsaturation with an available carboxylate ion to produce a trifunctional species capable of further reaction.

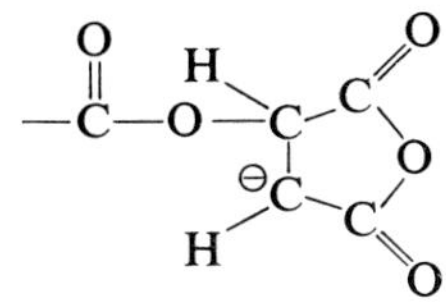

Maleic anhydride was found to provide satisfactory cures when used in up to five times excess of that required to react with the epoxy groups; and improved performance was obtained with maleic/epoxy systems by the addition of mono-ölefins in concentrations sufficient to degrade the properties if polymerization through unsaturation were not occurring [14].

RELATIVE REACTIVITIES OF EPOXY GROUPS

The previous general discussion has assumed that an ideal epoxy group is involved—one which will, in fact, undergo the reaction as described.

Not all epoxy groups, however, will undergo all reactions. The location of the epoxy group within the molecule, and the nature of the adjacent groups, will determine the reactivity.

Basically, the reactivity is determined by the accessibility of the epoxy group and the electronic nature of the epoxy oxygen In those reactions requiring rearward or sidewise approaches, steric factors are frequently very important. Given accessibility, then the nature of the substituents is the more important.

Electron-attracting substituent groups increase the rate with the nucleophilic curing agents such as the amines and the inorganic bases. Substituents such as methylene and vinyl groups improve reactivity to acidic-type, electrophilic curing agents. Propylene oxide, for example, is more acid reactive than a glycidyl ether or epichlorohydrin; whereas the reverse is true with base-type curing agents. The effect of methylene groups on amine reactivity is shown in Fig. 5-29.

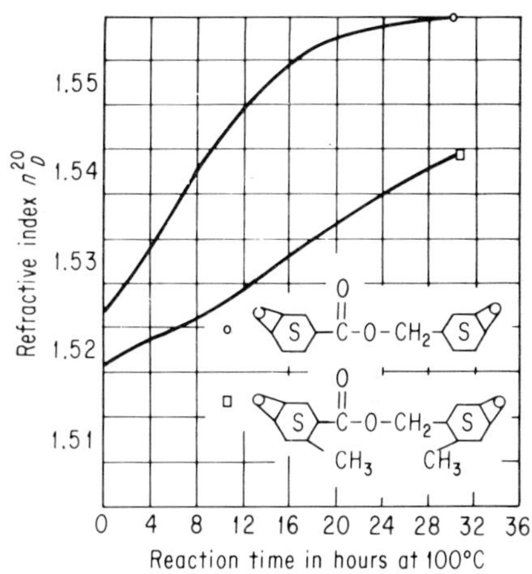

FIG. 5-29. The presence of methyl groups greatly reduces the rate of reaction of cyclohexyl epoxy resins with aniline [21].

In general, ring-situated epoxy groups favor reactions with acidic-type curing agents as do internal epoxy groups in aliphatic chains.

When organic acids and anhydrides are used as curing agents in the presence of a base catalyst, the nature of the epoxy group will influence the extent to which etherification progresses. Thus, with glycidyl ether resins, 99 per cent of the reaction will be through the anhydride and the epoxy; with epoxidized cyclohexene or vinylcyclohexene, 75 to 80 per cent; and with internal epoxies, such as alkylepoxy stearates, 60 to 65 per cent [15].

SUMMARY

Epoxy resins can be converted to thermosetting compounds by the action of basic or acidic curing agents. These may be of either the Lewis acid or base type or may contain active hydrogens. Both the Lewis acids and bases may be used to accelerate the reactions of active-hydrogen-bearing species.

With Lewis bases the generated alkoxide ion (via an intermediate hydroxyl ion) reacts preferentially as follows:

$$\text{Anhydride} > \text{carboxyl} > \text{phenol} > \text{alcohol-epoxy}$$

The same order is observed for organic acid and anhydrides catalyzed with the Lewis bases.

With the acid and anhydride curing agents uncatalyzed, the order is not so specific. Some amount of etherification will progress. With Lewis acid curing agents, the reaction will progress preferentially through hydroxyls present initially or through nascent hydroxyls formed during the course of the reaction.

Not all epoxy groups will react with all curing agents, and significant rate differences will be observed. In general, the attack by the basic type of curing agents is on the terminal carbon atom, but the acid-catalyzed reactions proceed through attack of the oxygen atom by the oxonium ion. Thus, substituents which are electron-attracting increase the positive charge on the terminal carbon and decrease the negative charge on the oxygen atom, favoring nucleophilic reaction; and, conversely, electron-repelling groups increase the negative charge on the oxygen atom, favoring formation of the oxonium ion. Additionally, the accessibility of the epoxy group to rearward approach will also be significant in determining reactivity.

REFERENCES

1950

1. Elderfield (ed.), "Heterocyclic Compounds," vol. 1, John Wiley & Sons, Inc.

1953

2. Narracott, The Curing of Epoxide Resins, *Brit. Plastics* **26**:120–123.

1956

3. Fisch, Hofmann, and Koskikallio, The Curing Mechanism of Epoxy Resins, *J. Appl. Chem.*, October.
4. Greenlee, Amine Amide–Epoxy Compositions, U.S. 2,760,944.
5. Shechter and Wynstra, Glycidyl Ether Reactions with Alcohols, Phenols, Carboxylic Acids, and Acid Anhydrides, *Ind. Eng. Chem.* **48**:86–93.
6. Shechter, Wynstra, Glycidyl Ether Reactions with Amines, *Ind. Eng. Chem.* **48**:94–97.
7. Shechter, Wynstra, and Kurkjy, The Chemistry of Styrene Oxide: a comparison with Phenyl Glycidyl Ether in Model Compound Reactions, ACS Symposium, Atlantic City, September.

1957

8. Kannebley, Kinetics of the Cure of Epoxy Resins, *Kunststoffe*, December.
9. Newey, The Cure of Epoxy Resins by Amines, Shell Chemical Corp. Report P-629.
10. Schwenke, The Synthesis of Polyesters from Epoxies and Dicarboxylic Acid Anhydrides, thesis, Technishe Hochschule, Stuttgart.

1958

11. Belanger and Denham, The Catalyzed Reactions of Various Glycidyl Ether Compounds with Amines, Acids, Phenols, and Alcohols, ACS Symposium, Chicago, September.

1959

12. Chapman et al., The Mechanism of Epoxide Reactions, part I, *J. Chem. Soc.*, June.
13. Parker and Isaacs, Mechanisms of Epoxide Reactions, *Chem. Rev.* **59**:737–799.

1960

14. Ehlers, Correlation between Structure and Thermal Stability of Epoxy Resins, *Polymer*, vol. 1, no. 3.
15. Fischer, Polyesters from Epoxides and Anhydrides, *Ind. Eng. Chem.*, April.
16. Gough and Smith, Accelerated Amine Curing of Epoxy Resins, *J. Oil Colour Chemists' Assoc.* **43**:408–418.
17. Isaacs and Parker, Mechanisms of Epoxide Reaction, part II, *J. Chem. Soc.*, September.
18. Merral et al., Cationic Polymerization of Ethylene Oxide: IV, The Propagation Reactions, *Can. J. Chem.* **38**:1967–1975.

1961

19. Budde and Matson, Resins and Methods of Making the Same, U.S. 2,993,920.
20. Fisch and Hofmann, Reaction Mechanisms, Chemical Structures, and Changes in Properties during the Curing of Epoxy Resins, *Plastics Technol.*, August.
21. May and Nixon, Reactive Diluents in Aromatic Amine Cured Epoxy Adhesives, *Chem. Eng. Data*, April.

1962

22. Arnold and Barshatky, Quantitive Differential Thermal Analysis of Epoxy Resins and Hardeners, 17th NANTEC (RIP Division) of SPI.
23. Harrod, Reactions of Primary Amines with Epoxides, *J. Appl. Polymer Sci.* **6**:24.
24. Hofmann and Fisch, Reaction between Carboxylic and Epoxy Groups and Their Influence on Film-forming Properties, *Fed. Assoc. Technicien Ind. Peintures, Vernis, Emaux Eneres Imprimerie Europe Continentale* (*FATIPEC*), book, May.
25. Smith, The Mechanism of the Crosslinking of Epoxide Resins by Amines, *Polymer* **2**:95–108.
26. Wood, Identification of the Urethane-Epoxy Reaction Product, *SPE Trans.*, July.

1963

27. Carr and Stolton, Improvements in or Relating to a Process for Curing Epoxy Resins, Brit. 929,373.
28. Tanaka and Kakiuchi, Study of Epoxy Compounds, part I, *J. Appl. Polymer Sci.* **7**: 1063–1081.

1964

29. Arnold, Acceleration of Glycidyl Epoxy Resin–Anhydride Reaction, *Mod. Plastics*, April.
30. Feltzin et al., Kinetics of Epoxy Polymerization, ACS Symposium, Chicago, September.
31. Landua, BF_3 Catalyzed Reactions of Epoxy Compounds, ACS Symposium, Chicago, September.
32. Shimazaki, Chemical Changes in an Epoxy Resin during Cure with an Organic Acid Anhydride Hardener, *Kogyo Kaguki Sasshi* **67**(8):1304–1309.
33. Tanaka and Kakiuchi, Study of Epoxy Compounds, part VI, *J. Polymer Sci.* (A) **2**:3405–3430.

Appendix 5-1

REACTION OF EPOXIDES*

* Prepared by Dr. D. L. MacPeak for "Peracetic Acid and Derivatives," 2d ed., Union Carbide Corp., 1963.

Epoxides react with a variety of reagents to produce chemicals of broad interest. To illustrate their potential as intermediates, a number of typical reactions are given below. Every epoxide, however, will not undergo all reactions shown. The course of any individual reaction depends upon the structure of the epoxide, the solvent and catalyst used, the nature of the other reactant, and the reaction conditions in general. The numbers after each reaction refer to pertinent references in the bibliography.

Acetals:

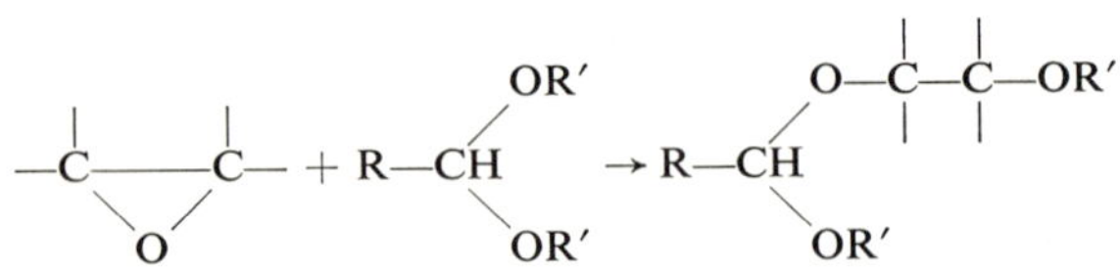

[1]

Acetoacetates:

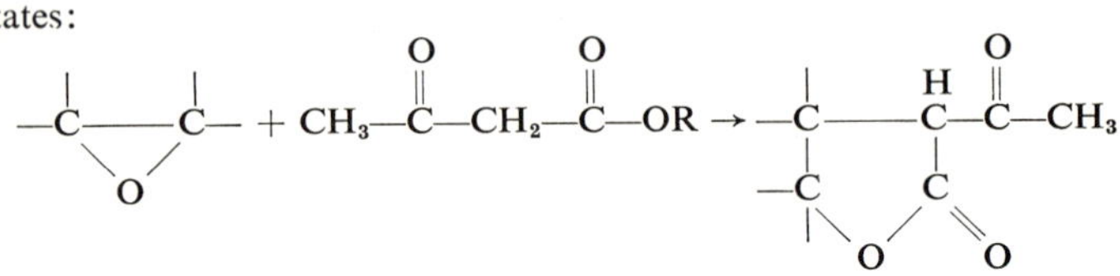

[2]

Acetonitriles:

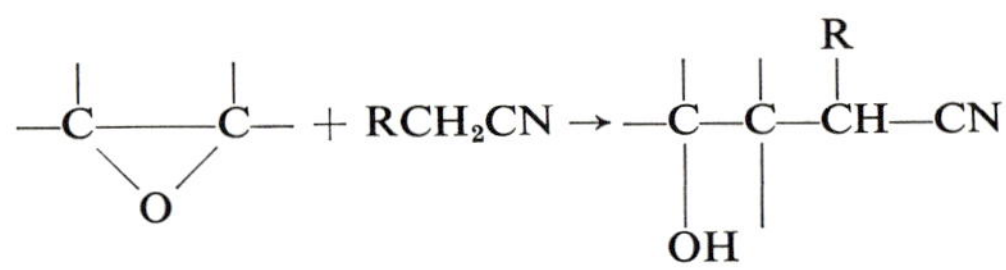

[3]

Acetylenes:

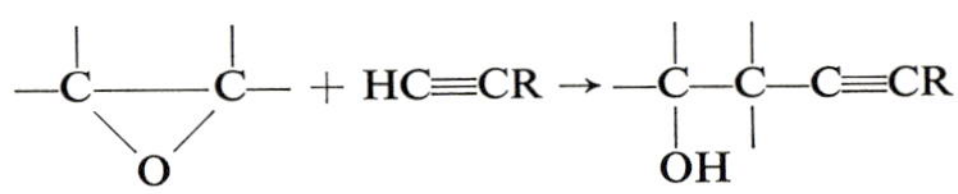

[4]

Acid anhydrides:

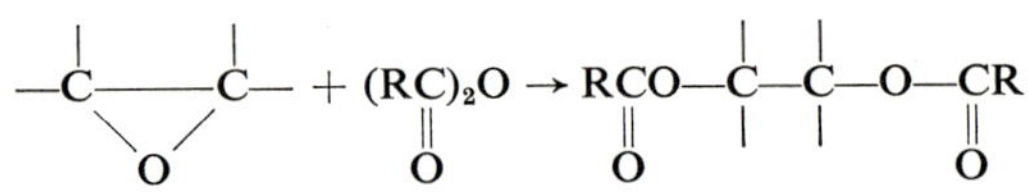

[5]

Acyl halides:

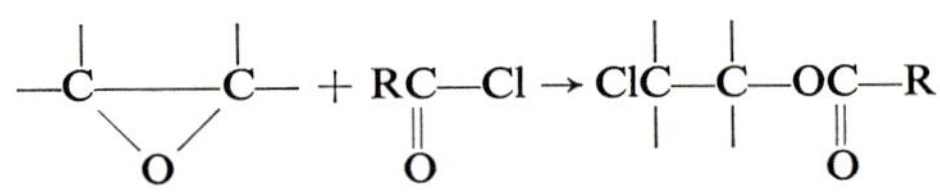

[6]

Alcohols:

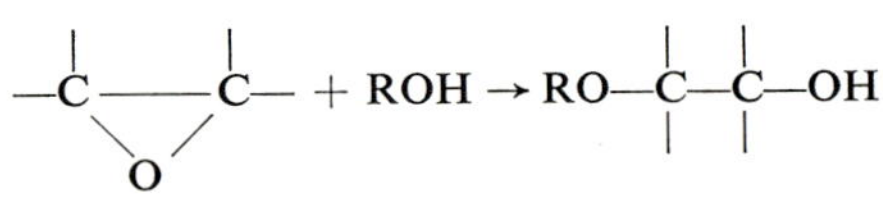

[7]

Aldehydes:

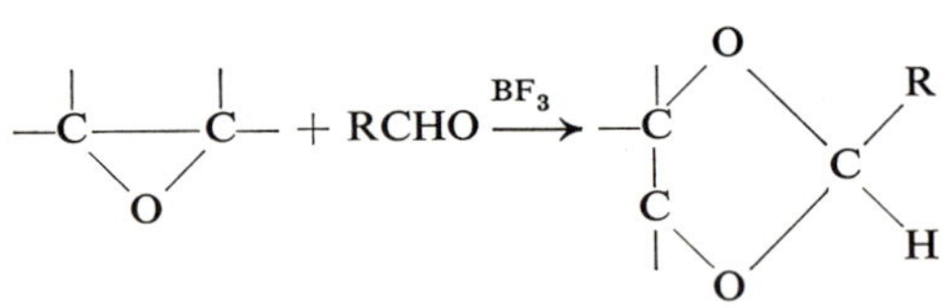

[8]

Alkyl halides:

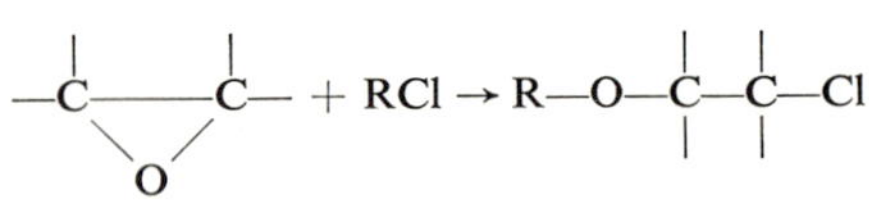

[6]

Alkyl hydroperoxides:

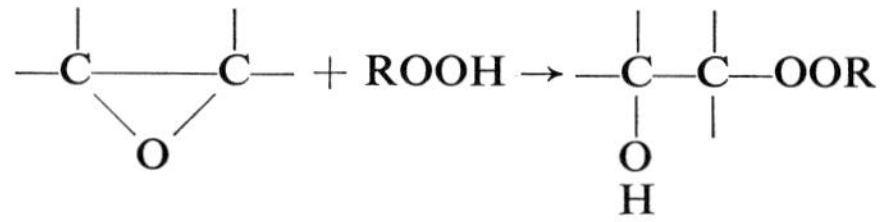

[9]

Amides:

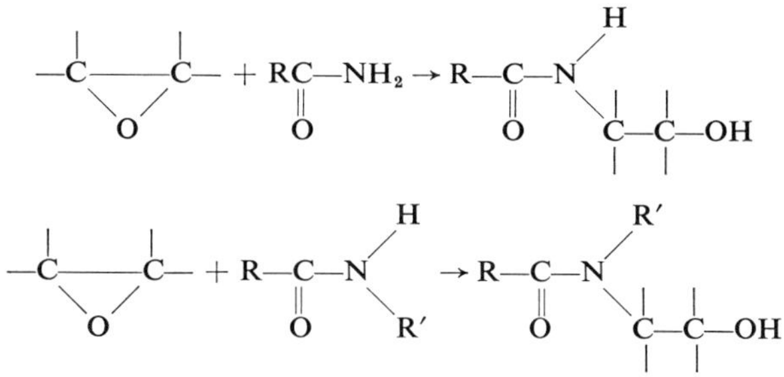

[10]

Amines—primary:

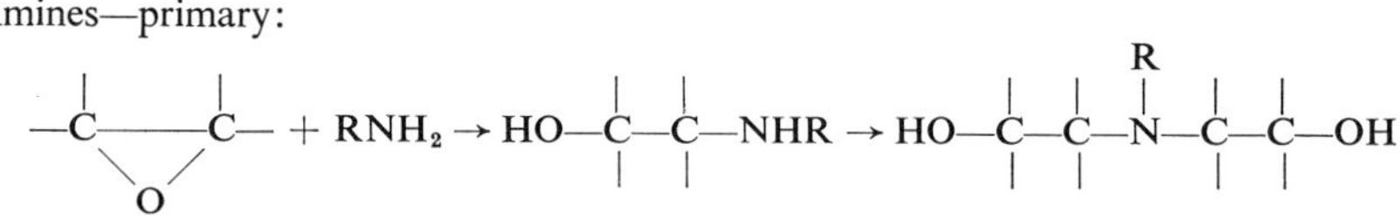

[11]

Amines—secondary:

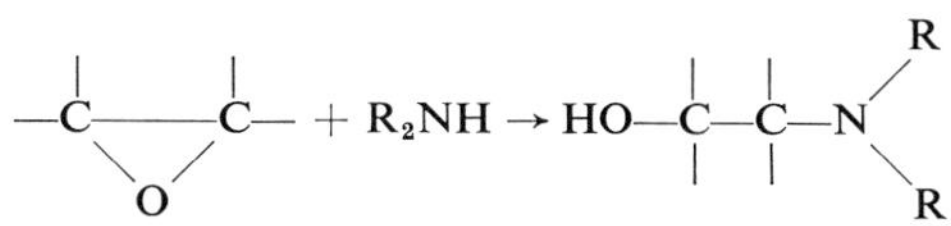

[12]

Amines—tertiary:

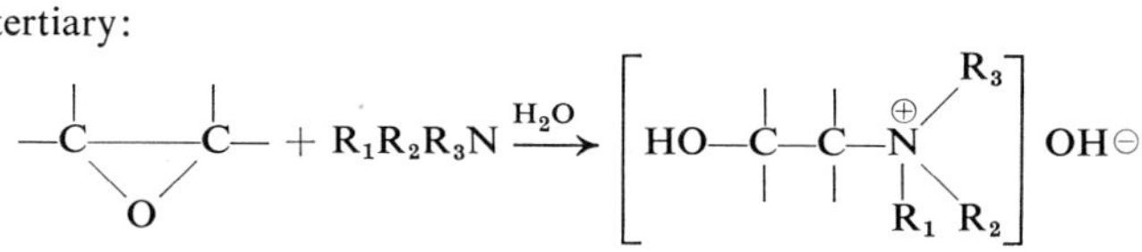

[13]

2-Aminothiols:

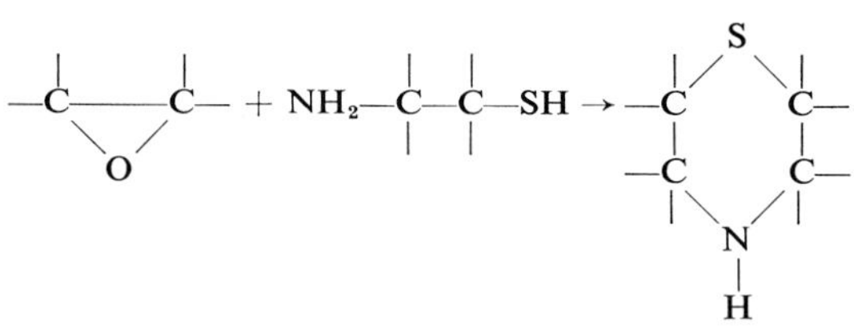

[14]

Ammonia:

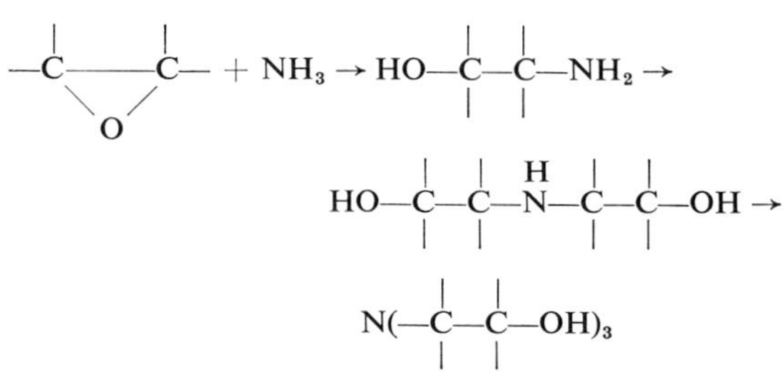

[15]

Antimony trihalides:

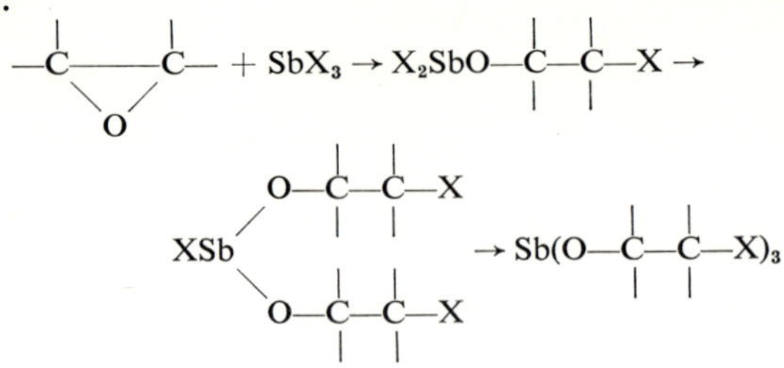 [16]

Arsenic trichloride:

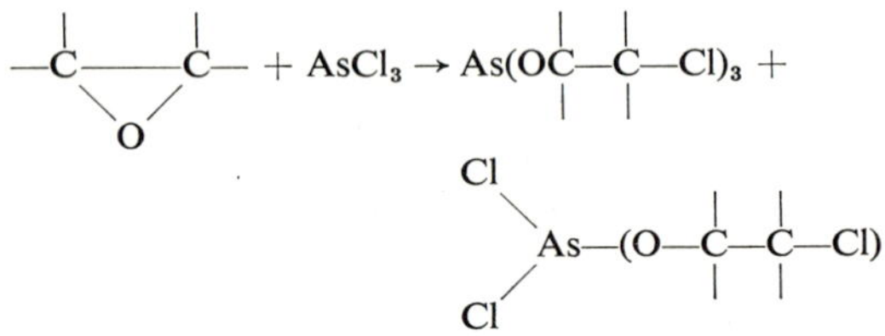 [17]

Aryl dichloroarsines:

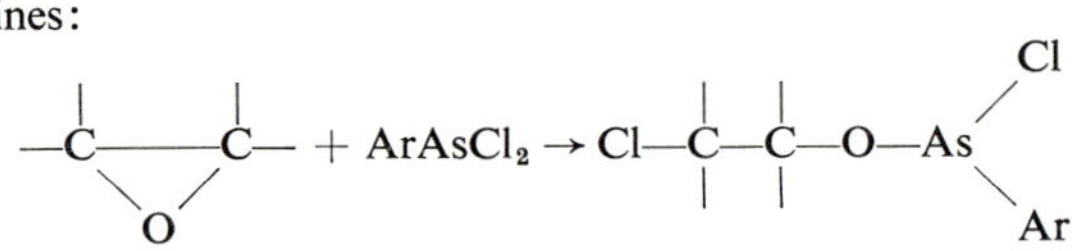 [17]

Aryllithiums:

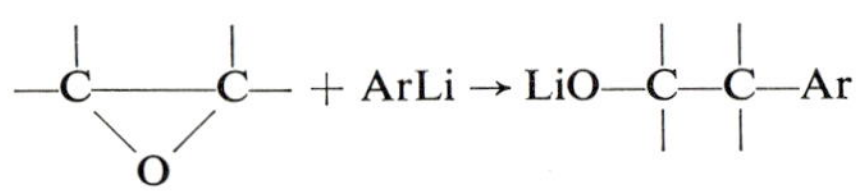 [18]

Azides:

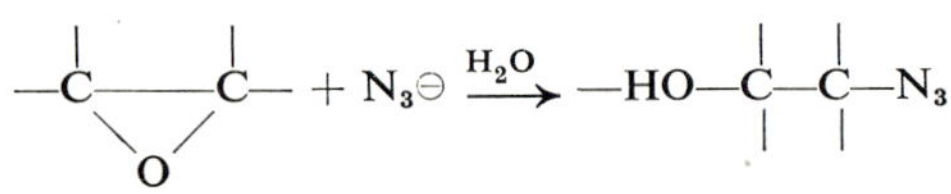 [19]

Bismuth halides:

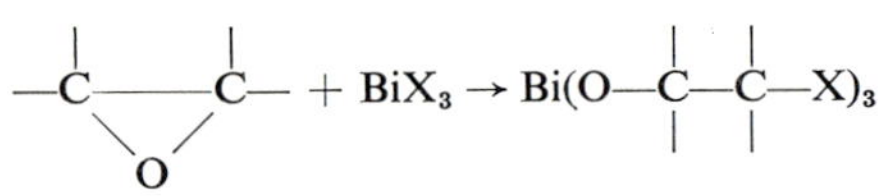 [16]

Carbamyl chloride:

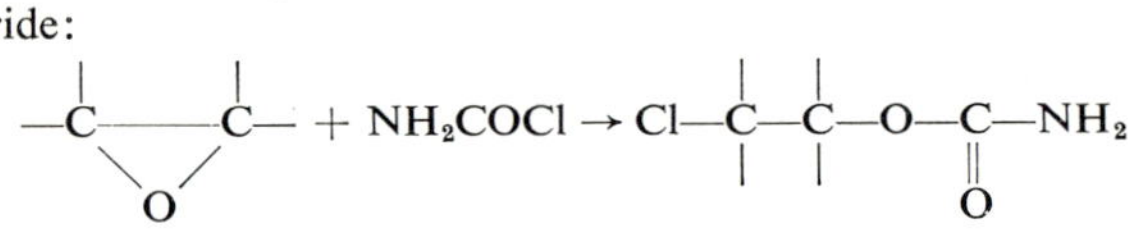 [20]

Carbon dioxide:

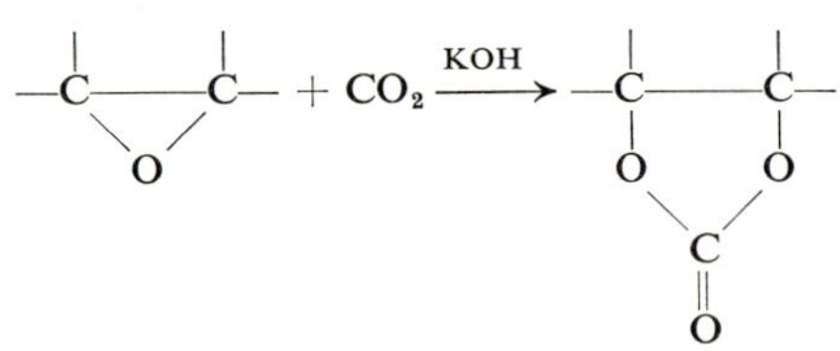 [21]

Carbon disulfide:

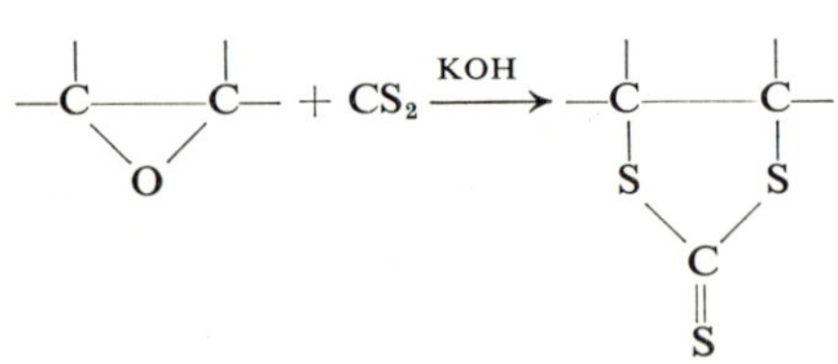 [22]

Carboxylic acids:

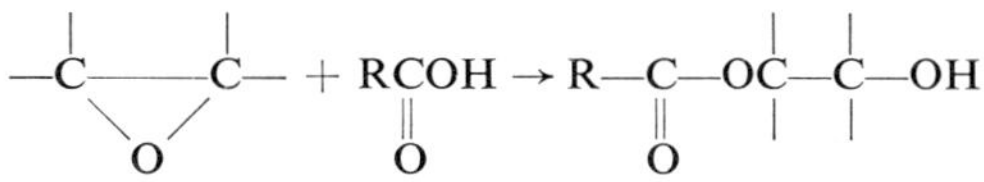

[23]

Cyanates:

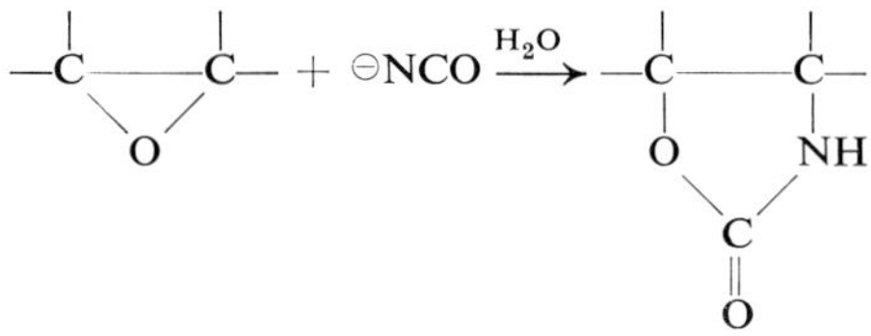

[24]

Cyanoacetates:

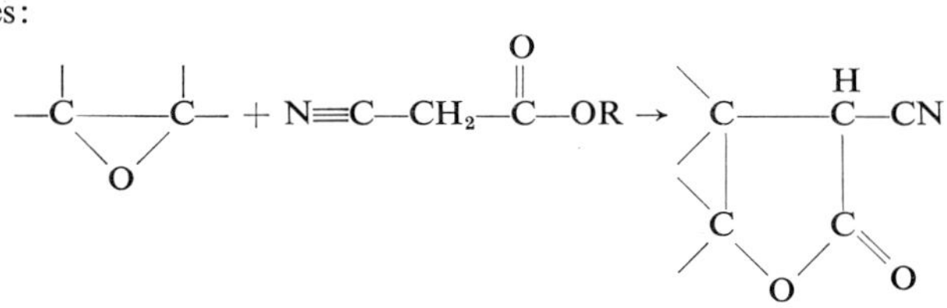

[25]

Diborane:

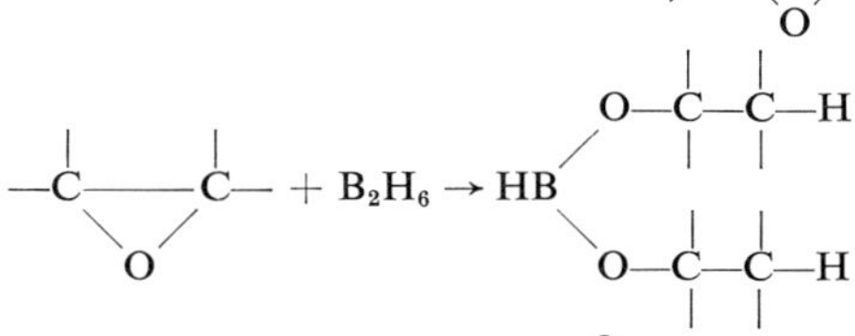

[26]

Dioxane formation:

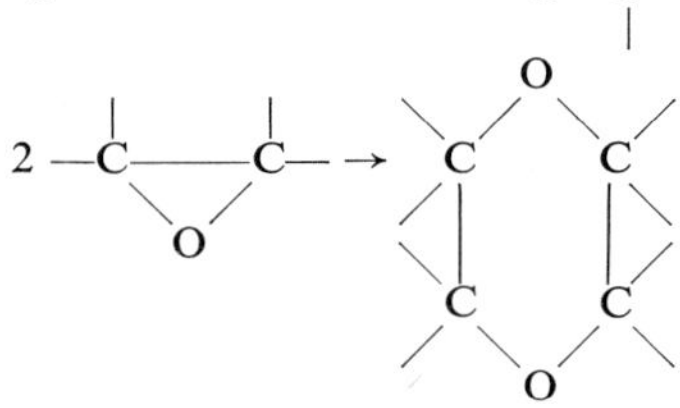

[27]

Ethylene imines:

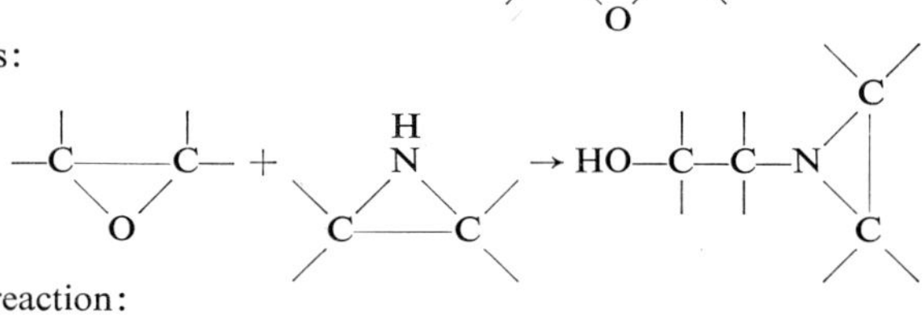

[28]

Friedel-Crafts reaction:

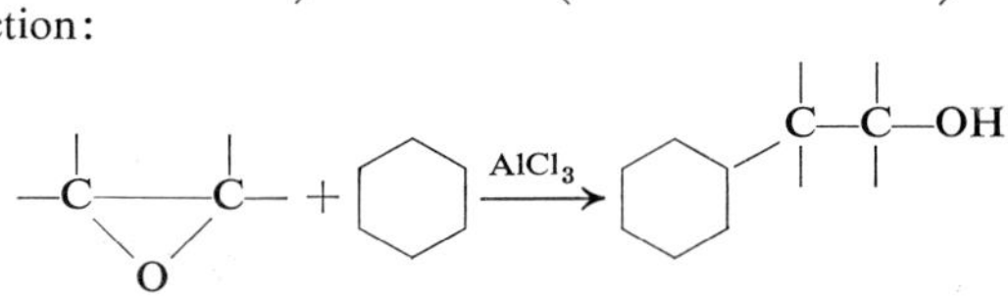

[29]

Grignard reagents:

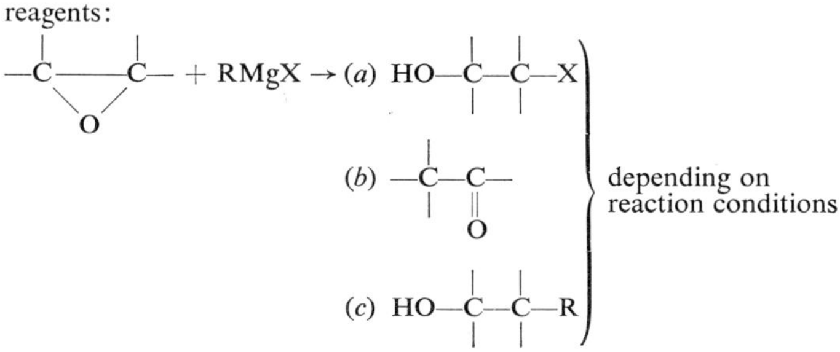

[30]

Halohydrins:

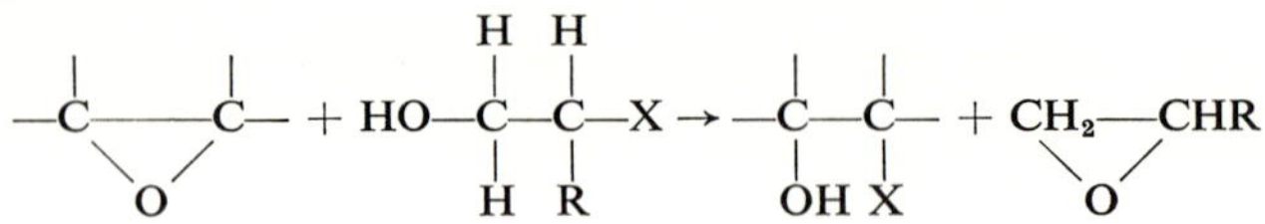

[31]

Hydrogen cyanide:

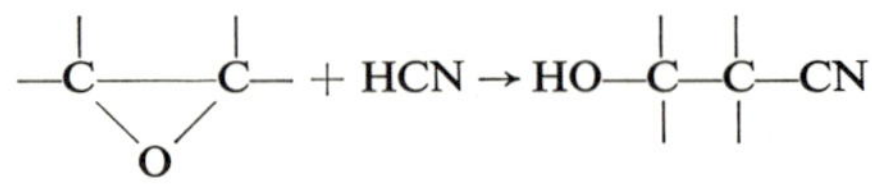

[32]

Hydrogen sulfide:

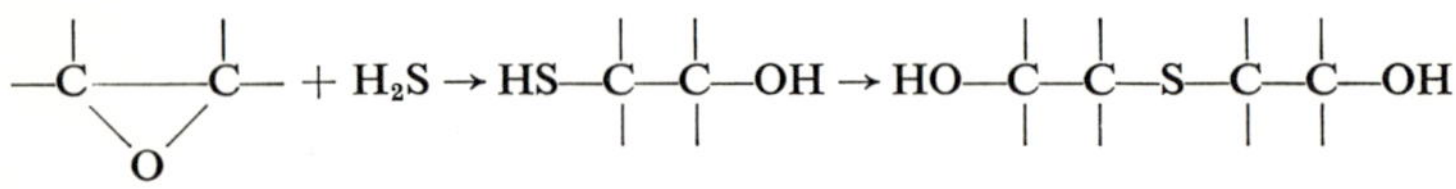

[33]

Ketones:

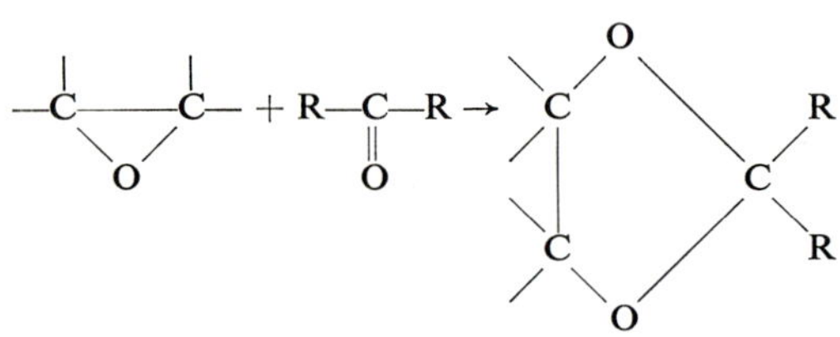

[34]

Malonates:

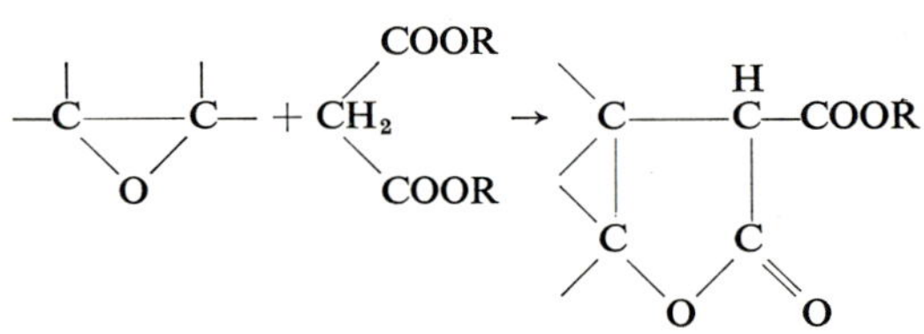

[35]

Nitrogen dioxide:

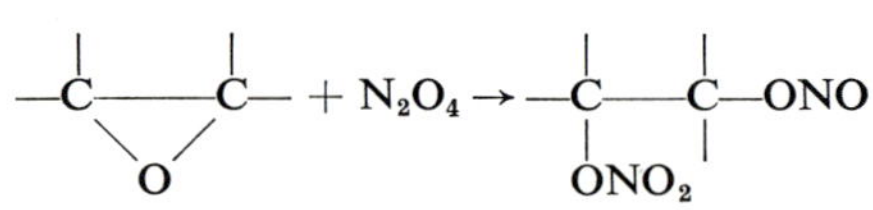

[36]

Nitrosyl chloride:

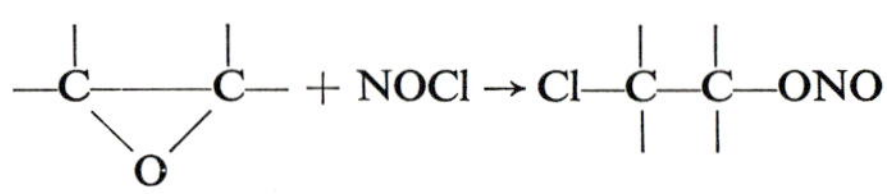

[37]

Oxidation:

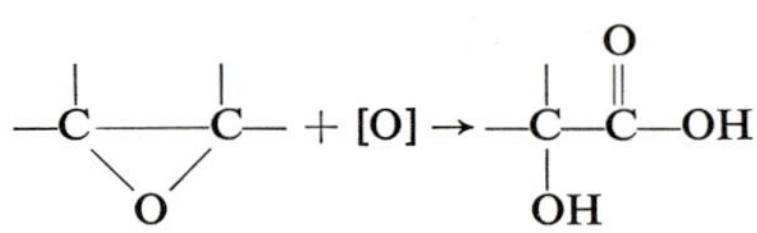

[38]

Phenols:

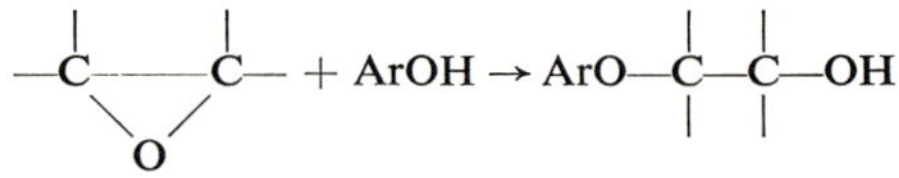

[39]

Phosgene:

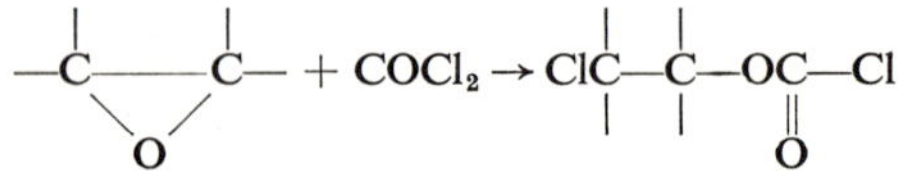

[40]

Phosphines:

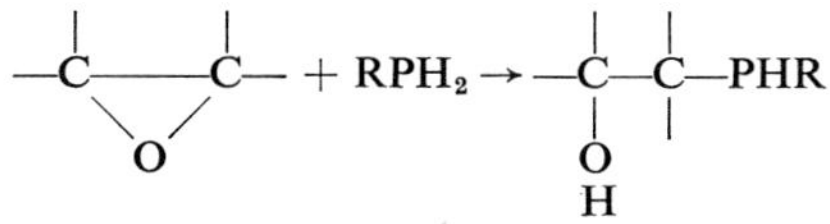

[41]

Phosphoric acid:

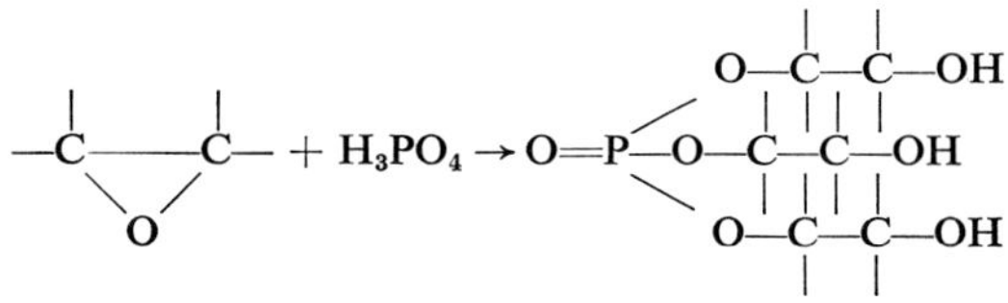

[42]

Phosphorous acid:

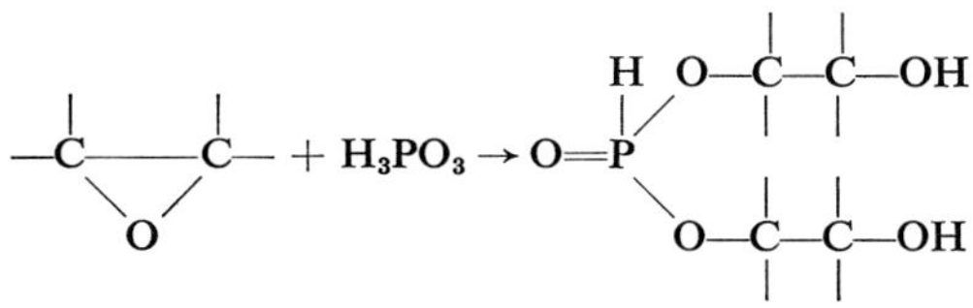

[42]

Phosphorus oxychloride:

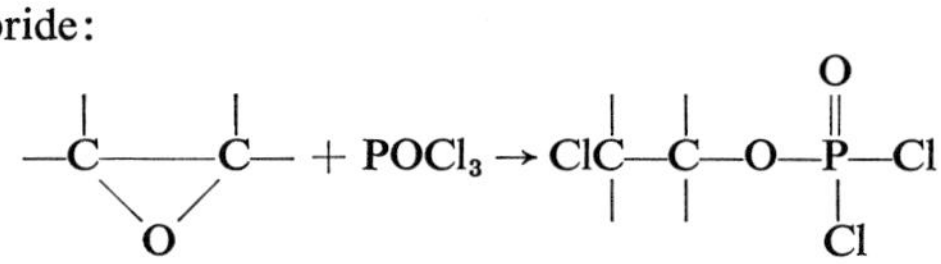

[43]

Phthalimide:

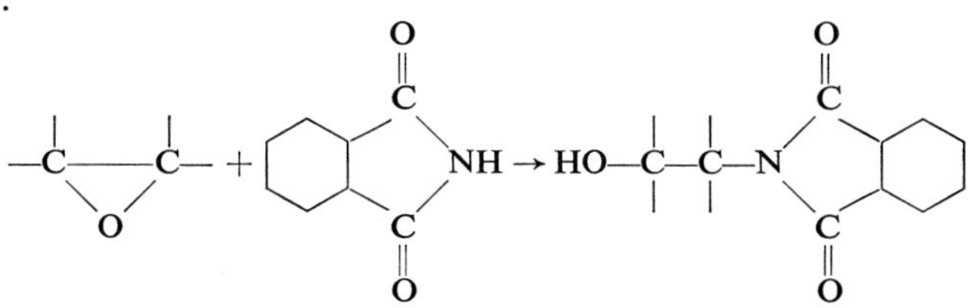

[44]

Polymerization:

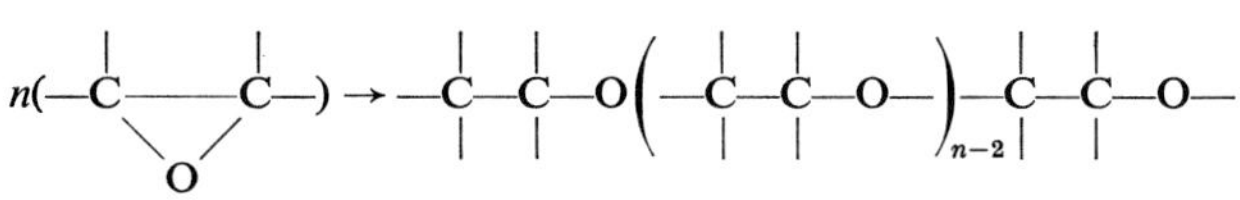

[45]

Rearrangement (catalytic or thermal):

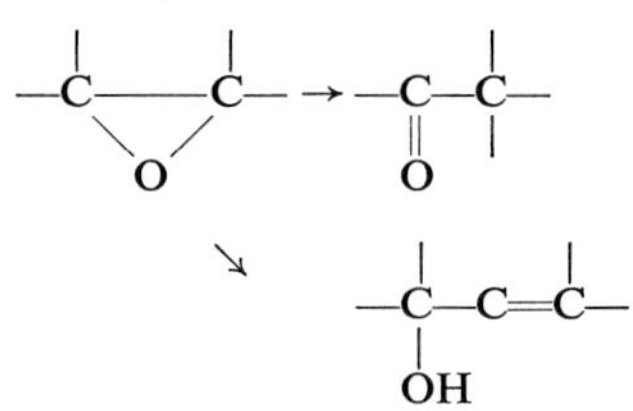

[46]

(depending on oxide structure and conditions employed)

Reduction:

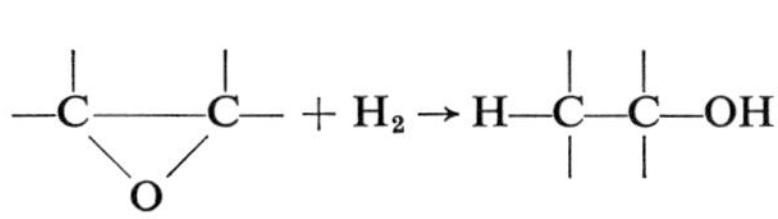

[47]

Silicon tetrahalides:

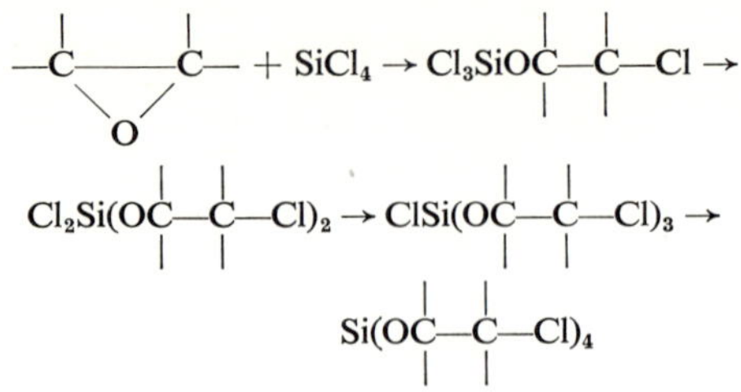

[48]

Sodium sulfite:

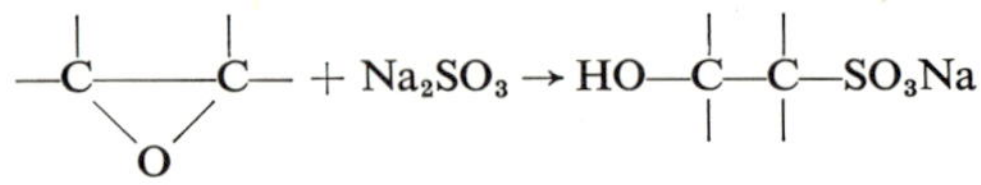

[49]

Sulfur dioxide:

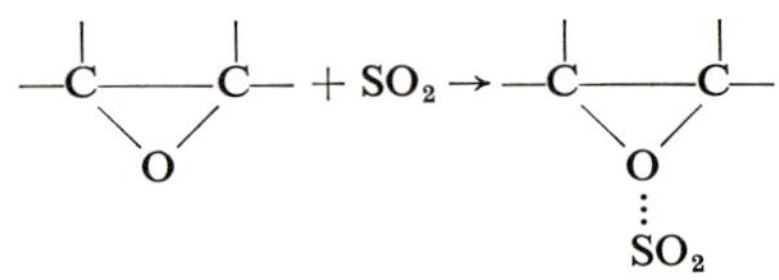

[50]

Sulfuryl chloride:

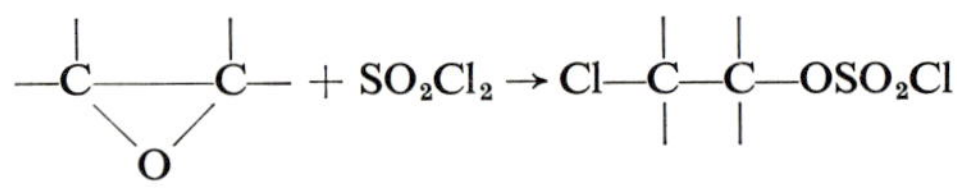

[51]

Thiocyanates:

[22]

Thiocyanic acid:

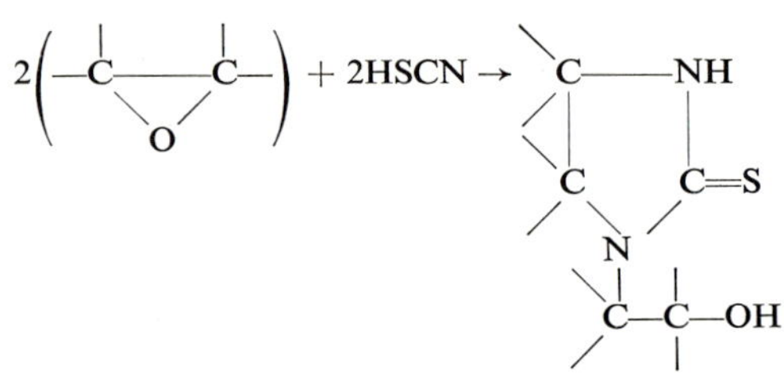

[52]

Thiolacids:

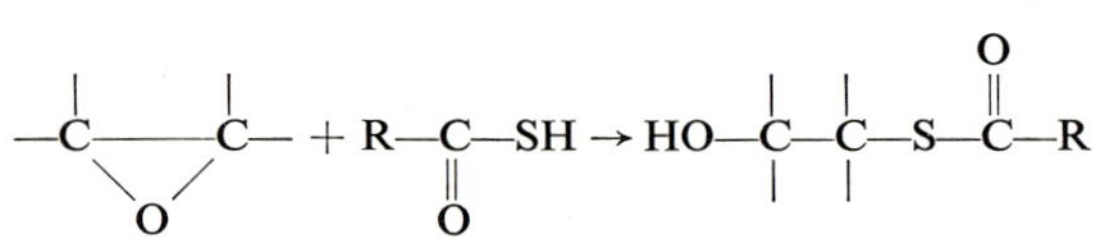

[53]

Thiols:

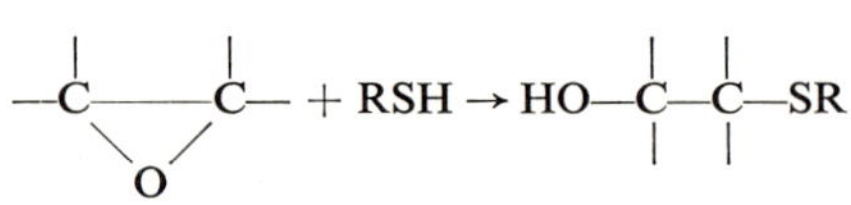

[54]

Thionyl chloride:

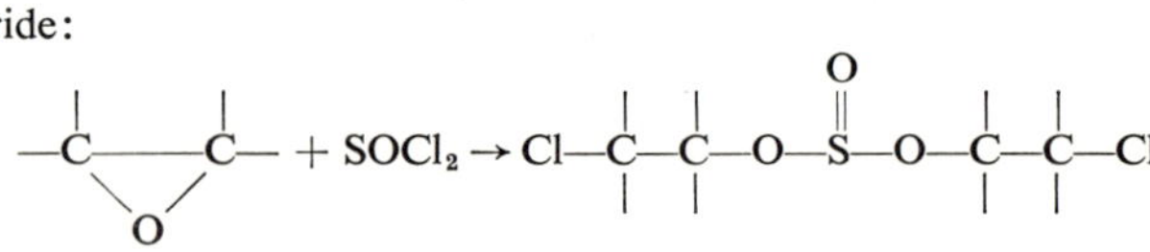

[55]

Thiosulfates:

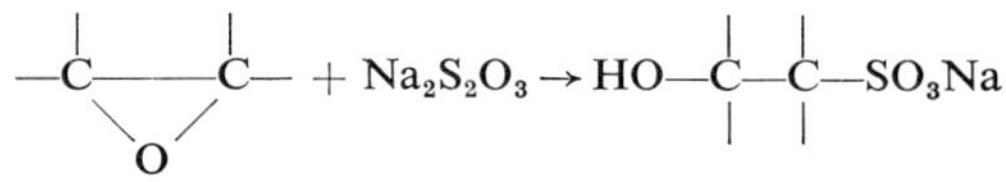 [33]

Water:

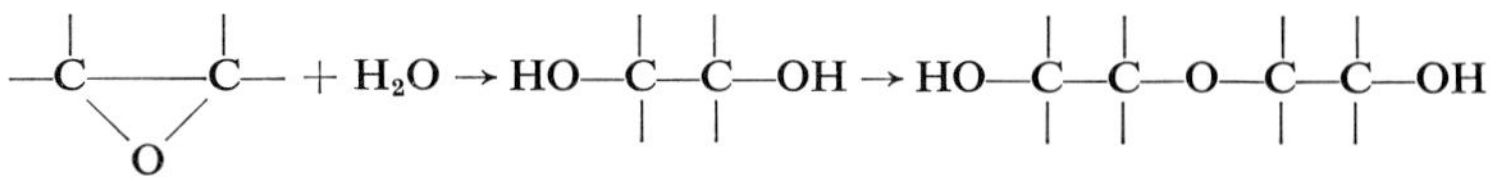 [56]

REFERENCES

1. Dermer and Durr, *J. Am. Chem. Soc.* **76**:912–913 (1954).
2. Haller and Blanc, *Compt. Rend.* **137**:1203 (1903).
3. Mousseron and Canet, *Bull. Soc. Chim. France* (5)**19**:190–197 (1952).
4. Raphael and Roxburgh, *J. Chem. Soc.* **1952**:3875.
5. Bedos, *Compt. Rend.* **183**:750–753 (1926).
6. Bedos, *Compt. Rend.* **183**:562–565 (1926).
7. Winstein and Henderson, *J. Am. Chem. Soc.* **65**:2196 (1943).
8. Bogert and Roblin, *J. Am. Chem. Soc.* **55**:3741 (1933).
9. Barusch and Payne, *J. Am. Chem. Soc.* **75**:1987 (1953).
10. Carnes, U.S. 2,520,381 (to American Cyanamid Co.).
11. Brunel, *Ann. Chim. et Phys.* (8)**6**:249–253 (1905).
12. Kotz and Merkel, *J. Prakt. Chem.* (2)**113**:63 (1926).
13. Schmidt, *Ann.* **337**:116–121 (1904).
14. King, Acheson, and Spensley, *J. Chem. Soc.* **1948**:1366.
15. Brunel, *Compt. Rend.* **137**:198–199 (1903).
16. Malinovski and Romantsevich, *Sb. Statei Obs. Khim.* **2**:1366–1369 (1953).
17. Malinovski, *J. Gen. Chem. USSR* **10**:1918–1929 (1940).
18. Cook, Hewett, and Laurence, *J. Chem. Soc.* **1936**:71.
19. Van der Werf, Heisler, and McEwan, *J. Am. Chem. Soc.* **76**:1231–1235 (1954).
20. Boberg and Schultze, *Ber.* **88**:275 (1955).
21. Cline, U.S. 2,667,497 (to Olin Mathieson Chemical Corp.).
22. Culvenor, Davies, and Pausacker, *J. Chem. Soc.* **1946**:1050.
23. Fraenkel-Conrat and Olcott, *J. Am. Chem. Soc.* **66**:1420 (1944).
24. Paterno et al., *Gazz. Chim. Ital.* (1)**38**:243–246 (1908).
25. Glickman and Cope, *J. Am. Chem. Soc.* **67**:1012 (1945).
26. Stone and Emeleus, *J. Chem. Soc.* **1950**:2755.
27. Favorsky, *J. Russ. Phys. Chem. Soc.* **38**:741 (1906).
28. Funke and Benoit, *Bull. Soc. Chim. France* **1953**:1021–1023.
29. Cologne and Rochas, *Compt. Rend.* **223**:403 (1946).
30. Gaylord and Becker, *Chem. Rev.* **49**:413 (1951).
31. Bradley, Forrest, and Stephenson, *J. Chem. Soc.* **1951**:1589.
32. Hormann, *Ber.* **12**:23 (1879).
33. Culvenor, Davies, and Heath, *J. Chem. Soc.* **1949**:278–282.
34. Willfang. *Ber.* **74**:145 (1941).
35. McRae et al., *Can. J. Res.* **21B**:186 (1943).
36. Pujo, Boileau, and Frejacques, *Bull. Soc. Chim. France* **1955**:974.
37. Malinovski and Medjanzewa, *J. Gen. Chem. USSR* **23**:84–86 (1953).
38. Wurtz, *Ann. Chim.* (3)**69**:317 (1863).
39. Mousseron, Jacquier, and Fontaine, *Bull. Soc. Chim. France* (5)**19**:767–776 (1952).
40. Malinovski and Medjanzewa, *J. Gen. Chem. USSR* **23**:221–223 (1953).
41. Knunyants and Sterlin, *Dokl. Akad. Nauk SSSR* **56**:49 (1947).
42. Adams, U.S. 2,372,244 (to Standard Oil Co., Indiana).
43. Tachimori and Matsushima, *J. Soc. Chem. Ind. Japan* **46**:1270 (1943).
44. Gabriel and Ohle, *Ber.* **50**:819 (1917).
45. Rothrock, U.S. 2,054,099 (to E. I. duPont de Nemours & Co.).
46. Rice and Stallbaumer, *J. Am. Chem. Soc.* **64**:1527–1530 (1942).

47. Brunel, *Compt. Rend.* **137**:62 (1903).
48. Patnode and Sauer, U.S. 2,381,137.
49. Suter and Milne, *J. Am. Chem. Soc.* **65**:582 (1943).
50. Campbell and Fernelius, U.S. 2,413,405.
51. Malinovski, *J. Gen. Chem. USSR* **17**:1559–1563 (1947).
52. Sergeev and Kolychev, *J. Gen. Chem. USSR* **7**:1390–1396 (1937).
53. Nylen and Olsen, *Svensk Kem. Tidskr.* **53**:274 (1941).
54. Mousseron and Jacquier, *Compt. Rend.* **229**:374–376 (1949).
55. Pechukas, U.S. 2,576,138 (to Pittsburgh Plate Glass Co.).
56. Boeseken and Van Giffen, *Rec. Trav. Chim.* **39**:184 (1920).

Chapter 6

CHARACTERIZATION OF EPOXY-RESIN CURING-AGENT SYSTEMS

Epoxy resins, as a class, are available in a very wide range of molecular structures, suitable for reaction with a large variety of different curing agents, for a multitude of end uses. Over half a hundred different resins are available, and when the hundreds of different curing agents are considered, it is easy to realize the diversity of properties available in cured epoxy resins.

The purpose of this chapter is to outline how cured epoxy resins are characterized—test methods, general range of properties to be expected, and the general effect of molecular structure upon the properties of the cured product. The properties of specific resins and curing agents are discussed in detail in subsequent chapters.

From a consideration of data in this chapter, one can come to a realistic appreciation of both the strengths and weaknesses of the epoxy resins.

THERMODYNAMICS OF CURE

Heat of Activation

Heat of activation, or activation energy, is that energy which must be supplied for the formation of the transition state, i.e., the difference between the initial energy and the maximum energy content during cure.

The apparent activation energy of DGEBA reacted with *n*-butylamine is 11.6 to 11.8 kcal per mole; with ethylenediamine, 13.2 kcal per mole [56]. Tertiary amines give somewhat higher values as a result of epoxy-epoxy polymerization [72]. Apparent activation energy of DGEBA and DGEBA (mol. wt. 950) cured with HHPA and triethanolamine is 13.7 to 14.1 with the first resin and 14.4 with the second [81]. DGEBA/NMA/triethanolamine with catalytic amounts of benzoic acid, benzyl alcohol, or phenol gives values of 14.7 [95].

Heat of Polymerization

The heat of polymerization of an epoxy resin indicates the amount of heat given off during the curing reaction and is expressed in kilocalories per mole of epoxy.

Table 6-1. Measured Values of the Heat of Polymerization of Phenyl Glycidyl Ether and DGEBA with Various Curing Agents [26]

Curing agent	Monomer	Moles curing agent per epoxy equivalent	Heat of polymerization, kcal/mole (mean value)
Primary amine curing agents:			
Benzylamine	PGE	0.48	27.62
	DGEBA	0.53	26.83
m-Phenylenediamine	PGE	0.52	26.26
	DGEBA	0.30	25.42
Tertiary amine curing agents:			
Pyridine	PGE	0.19	26.17
	DGEBA	0.37	26.98
Benzyldimethylamine	PGE	0.13	22.14
	DGEBA	0.19	22.44
2,4,6-Tris(dimethylaminomethyl)phenol	PGE	0.09	21.90
	DGEBA	0.10	22.31
2,4,6-Tris(dimethylaminomethyl)phenol/tri-2-ethylhexoic acid	PGE	0.022	20.47
	DGEBA	0.028	19.88
Boron trifluoride:			
BF_3–diethyl ether complex	PGE	0.34	22.15
Mixed-type amine curing agents:			
Dimethylaminopropylamine	PGE	0.15	22.66
	DGEBA	0.19	22.78
Piperidine	PGE	0.21	22.29
	DGEBA	0.29	21.43
BF_3–monoethanolamine complex	PGE	0.15	25.39
	DGEBA	0.17	24.98

Heats of polymerization have been determined for a variety of epoxy/curing-agent systems, primarily with the aim of elucidating the nature of the curing reactions. Results of one such study are presented in Table 6-1. With DGEBA resins, the teat of polymerization has been shown to be about 25 kcal per mole of epoxy when a primary amine curing agent is employed and about 22 kcal per mole when a hertiary amine curing agent is employed [26].

Exotherm. In the polymerization reaction, the heat, if liberated during a short course of time, will result in a substantial increase in the temperature of the reactive mixture. This temperature increase, which is readily measured by a thermocouple, is referred to as *exotherm.* A typical series of exotherm profiles are shown in Fig. 6-1. These illustrate the temperature dependence of the reaction. With DGEBA systems, peak exotherms may reach 260°C without external heating with

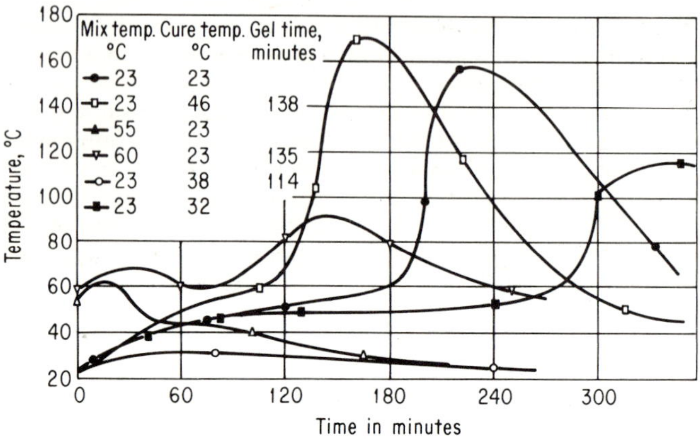

FIG. 6-1. Exotherm profile of DGEBA cured with tertiary amine salt at various temperatures [18]. All runs made with 260-g DGEBA and 34-g DMP-30 salt. No filler.

room-temperature curing systems, even in relatively small batches from 100 grams or so to one pound. With the heat curing systems, particularly the anhydrides, the exothermic reaction may not be more than 20 to 30°C over the cure temperature, even in 50-lb batches or more.

Exotherm considerations are discussed further in Chap. 17.

Pot Life

Pot life is defined as the time available for use of the epoxy system after the resin and curing agent are mixed. Some systems become progressively more viscous, and for these, in applications requiring low viscosity, pot life generally ends at a viscosity of about 5,000 centipoises at processing temperature. In applications where high viscosities are used, the end of the pot life is normally considered to be that point where application is no longer conveniently possible. For systems retaining low viscosity until a few minutes prior to gelation, the gel point can be regarded as the limit of the pot life. Pot life of epoxy-resin systems may be varied from a few seconds to several years.

Formulas have been developed for calculating relative gelation times with amine curing agents, and these have been shown to coincide with observed values [68], but values of a given system are generally measured directly in industrial practice.

MORPHOLOGY

Epoxy resins of the DGEBA type are rigid, amorphous, glasslike solids. Examination of their molecular structure suggests that each artifact cast from a mixture of resin and curing agent could be, in fact, one gigantic molecule. However, there is no real evidence of this, and instead modern theories and evidence from electron-microscope studies suggest that epoxy resins consist of relatively small macromolecules—three-dimensional branched networks or agglomerates of moderate molecular weight—embedded in a surrounding matrix of lower-molecular-weight material. The ultimate size which the macromolecules achieve during cure is limited by steric hindrance that minimizes the occurrence of intermolecular reactions. As the molecules polymerize, their rotational and translational freedom is reduced, which reduces the chances for primary bonds to be set up with adjacent molecules (see Micrographs 1 and 2). Additional evidence for low cross-linking density has been developed from dynamic mechanical properties, where the temperature dependence of Young's modulus and internal friction is characteristic of slightly cross-linked polymers.

This amorphous agglomerate concept is useful in explaining improvements in solvent resistance, DT, and hardness, which occur when resins are given postcures at elevated temperatures—the increased thermal agitation at elevated temperature permits an additional number of the sterically hindered epoxy groups to react, and thus the size of the agglomerates is increased at the expense of the surrounding lower-molecular-weight material, which has lower solvent resistance, low DT, and lower hardness than the highly cross-linked agglomerates. The postcuring also

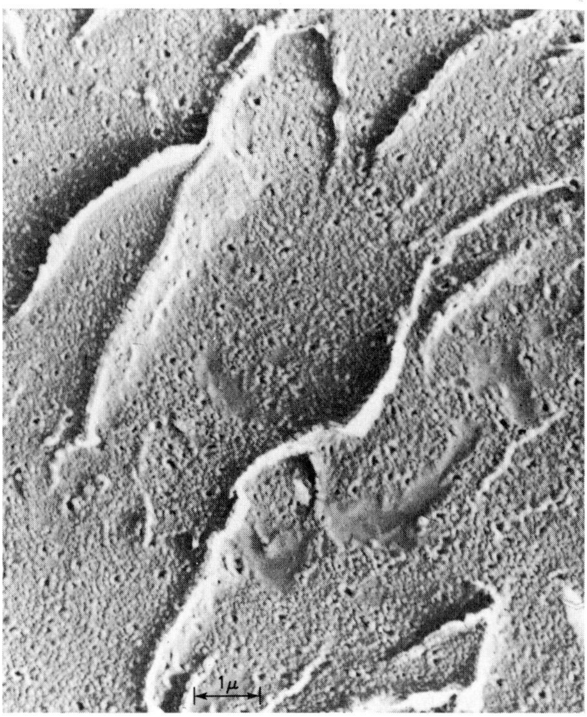

Micrograph 1.* Electron micrograph (3,000×) of preshadowed carbon replica of freshly fractured surface of a cured epoxy resin. The resin aggregates appear to be randomly distributed and discreet, about 0.85 micron in diameter, and there is no apparent spatial relationship between neighboring particles that could be invoked to support the idea that a large skeletal network exists throughout the material.

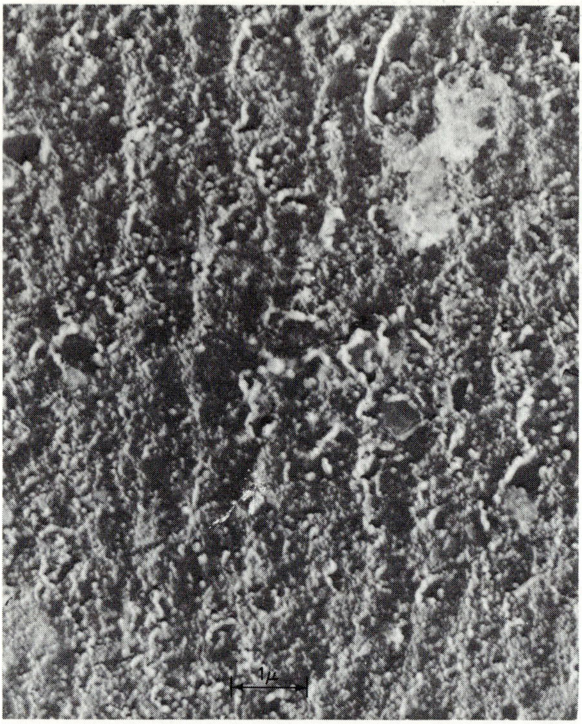

Micrograph 2.* Electron micrograph (3,000×) of preshadowed carbon replica of freshly fractured cured epoxy resin which has been leached with acetone. Comparison of this micrograph with micrograph 1 shows that although the solvent has leached large quantities of material from the surface, the many individual particles have not decreased in size, supporting the concept that the cured resin consists of macromolecular agglomerates or particles embedded in a substrate of lower-molecular-weight, less highly cross-linked (and therefore more soluble) matter.

* Micrographs 1 and 2 are from Morphology of Polymers, *Journal of Polymer Science:* part C, Polymer Symposia no. 3, Colloidal Particles in the Thermosetting Resins by Edward H. Erath and Morton Robinson, Interscience Publishers, New York 1963, reproduced by permission of the copyright owner.

probably serves to cross-link and raise the molecular weight of the low-molecular-weight matrix.

The agglomerate concept also explains other phenomena in a mechanistic manner.

Resins which have lower viscosity often develop higher properties than similar species of slightly higher viscosity; the lower viscosity enhances molecular mobility, and hence larger, better cross-linked agglomerates can be found. These effects are often obtained even by using 2 to 3 phr of a nonreactive diluent such as xylene, as shown in Chap. 13.

Similarly, the tensile and compressive strengths of epoxy resins are increased if the distance between cross-links is shortened. Shorter cross-link distances imply high volume concentration of epoxy groups and hence a higher probability of reaction of all the epoxy groups, so that the boundary or demarcation between the high-cross-linked network agglomerates and the surrounding lower-molecular-weight matrix may fade away or the differences in the properties of the two areas may be markedly reduced.

DEGREE OF CURE

During the early stages of cure of an epoxy resin, before the molecules are all cross-linked, the resin exists as a thermoplastic or B-stage resin. It is hard but frangible and soluble in solvents such as acetone. At this stage, the cure reaction has been initiated at a number of widely separated points in the reactive mass. With additional cure, cross-linking becomes general, and the compound assumes its thermoset cured conditions. Shear stress measurements, for example, taken during cure indicate this: The stress increases at first slowly (formation of branched and linear molecules) and then rapidly near the end of the reaction (formation of the final network) [106].

A theoretical method for calculating the optimum cure cycle for a given compound has not been advanced. The usual procedure is to determine degrees of cure empirically in terms of optimizing the desired end properties. Statistical evaluations have been employed to obtain optimum cured properties for several epoxy systems [21,34,109]. Results indicate that the cure cycle producing the optimum value for one physical property will not necessarily produce the optimum value for another. Results of one such study, in terms of optimizing the anhydride/epoxy ratio, catalyst concentration, gel time and temperature, and postcure time and temperature, are presented in Table 6-2.

Some epoxy/curing-agent systems favor initial low-temperature cures, followed by high-temperature postcures. In these, the low-temperature cures promote, presumably, linear polymerization, and the postcures provide for the necessary cross-linking. Such systems will give a different, and often less desirable, cured network when the initial cure temperature is increased; or again may provide improved properties in smaller, low-exotherm masses over larger, high-exotherm masses. On the other hand, some systems form molecules that are sterically hindered when a low-temperature cure is employed, and conversion cannot be completed even under the influence of long, high-temperature postcures [6]; these favor initial elevated-temperature cures. So-called "shock" cures, subjecting the slightly reacted mass to high temperature, can result in poorly cured, brittle masses. If a volatile curing agent is used, bubbling may occur and the reaction will be incomplete.

Anhydride curing agents catalyzed with tertiary amines appear to be particularly temperature-sensitive, and the initial cure temperatures will determine, to some extent, the type of bonds present in the cured system and hence its cured properties. Epoxidized novolacs appear, likewise perhaps because gelation occurs at a lower per cent of reaction, more sensitive to cure cycles than DGEBA [83].

In general, however, irrespective of the curing agent, postcuring at temperatures higher than the initial cure cycle will produce improvements, insofar as, toward the end of the reaction, more frequent collisions resulting from increased thermal energy are required to make the remaining reactive groups react.

In the epoxy-resin technology, the degree of cure has, by custom, come to refer to the extent to which the epoxy groups have been consumed, and, when present, the extent of consumption of the reactive groups in the curing agent. For practical purposes, a thoroughly cured system is considered one in which the degree of cross-linking is sufficient to provide optimum physical properties for a particular application.

Table 6-2. Optimization of Physical Properties by Statistical Analysis for DGEBA Cured with Nadic Methyl Anhydride [21]

LEVELS OF EXPERIMENTAL VARIABLES AND DT OBTAINED

Factor levels in natural units	Factor levels in units of the design		
	−1	0	+1
A. Gel temperature, °C	90	110	130
B. Gel time, hours	1	2	3
C. Anhydride concentration, phr*	80	85	90
D. Catalyst concentration, phr	2.5	3	3.5
E. Postcure temperature, °C	183	205	227

Factors					Flexural strength, psi	DT, °C	Tensile strength, psi
A	*B*	*C*	*D*	*E*			
−1	−1	−1	−1	−1	21,200	146.0	5,730
1	−1	−1	−1	1	20,600	158.0	5,970
−1	1	−1	−1	1	20,170	160.5	4,060
1	1	−1	−1	−1	19,640	147.0	4,200
−1	−1	1	−1	1	12,880	163.0	4,770
1	−1	1	−1	−1	20,500	138.0	10,390
−1	1	1	−1	−1	22,000	150.0	10,770
1	1	1	−1	1	17,290	165.0	7,740
−1	−1	−1	1	1	18,380	156.0	5,120
1	−1	−1	1	−1	20,480	144.0	9,890
−1	1	−1	1	−1	20,340	145.0	4,610
1	1	−1	1	1	17,520	156.0	8,270
−1	−1	1	1	−1	21,980	144.0	10,130
1	−1	1	1	1	15,950	163.0	7,020
−1	1	1	1	1	18,420	165.0	8,270
1	1	1	1	−1	22,920	144.0	11,560

ANALYSIS OF VARIANCE—DT

Effect	Mean effect	Degrees of freedom	Mean square†	Estimates of linear and interaction coefficients
A	−1.81	1	13.1406‡	$b_1 = 0.9062$
B	2.56	1	26.2656§	$b_2 = 1.2812$
C	2.44	1	23.7656§	$b_3 = 1.2188$
D	−1.31	1	6.8906	$b_4 = 0.6562$
E	16.06	1	1,032.0156§	$b_5 = 8.0312$
AB	−0.31	1	0.3906	$b_{12} = 0.1562$
AC	−1.19	1	5.6406	$b_{13} = 0.5937$
AD	1.06	1	4.5156	$b_{14} = 0.5312$
AE	1.19	1	5.6406	$b_{15} = 0.5312$
BC	1.44	1	8.2656	$b_{23} = 0.7187$
BD	−1.81	1	13.1406‡	$b_{24} = 0.9062$
BE	−0.94	1	3.5156	$b_{25} = 0.4687$
CD	−3.75	1	6.8906	$b_{34} = 0.6562$
CE	3.94	1	62.0156§	$b_{35} = 1.9687$
DE	−0.31	1	0.3906	$b_{45} = 0.1562$

* Parts per hundred of resin, by weight.

† To be significant at the 5% and 1% level of $F_{1,12}$ the mean square based on one degree of freedom must be as great as $2.25 \times 4.75 = 10.69$, and $2.25 \times 9.33 = 20.99$, respectively.

‡ Denotes significant, $F_{1,12}$ ratio $>$ 5% value but $<$ 1% value.

§ Denotes highly significant $F_{1,12}$ ratio $>$ 1% value.

Table 6-2 (*continued*)

OPTIMIZED VALUES

Variable	DT, °C	Flexural strength, psi	Tensile strength, psi
Gel temperature, °C	107	115	90
Gel time, hours	2.8	2.2	3.0
Anhydride concentration, phr	89	90	90
Benzyldimethylamine, phr	2.2	2.5	2.5
Postcure temperature, °C	256	188	161
Optimum values	218	22,980	13,070

During cure, two phenomena are involved: *conversion* (the disappearance of reactive groups) and, more important, *cross-linking* (the coupling of the molecules into three-dimensional networks through reactive sites to form the desired thermoset resins as opposed to mere formation of long, linear non-cross-linked resins).

Both are significant. To give a single example of the importance of cross-linking: If a tertiary amine curing agent is employed, the nature of the network and the extent of cross-linking will depend on the concentration of the curing agent as well as on the temperature of the cure. Analyses indicating the extent of conversion of the cured system (i.e., disappearance of the epoxy groups) will be equal under all conditions, whereas the properties of the cured system will vary quite widely.

Methods for determining both conversion and cross-linking are presented below.

Infrared Absorption for Determining Conversion

Infrared spectroscopy provides a fast and accurate means for not only identifying the curing agent used in curing an epoxy resin (Appendix 6-I) but also determining the extent of conversion.

Samples for recording spectra are prepared as films about 0.025 mm thick between rock salt plates. The curing then takes place in the cell, and the spectrum is recorded at various time intervals. Absorption at various times during cure are

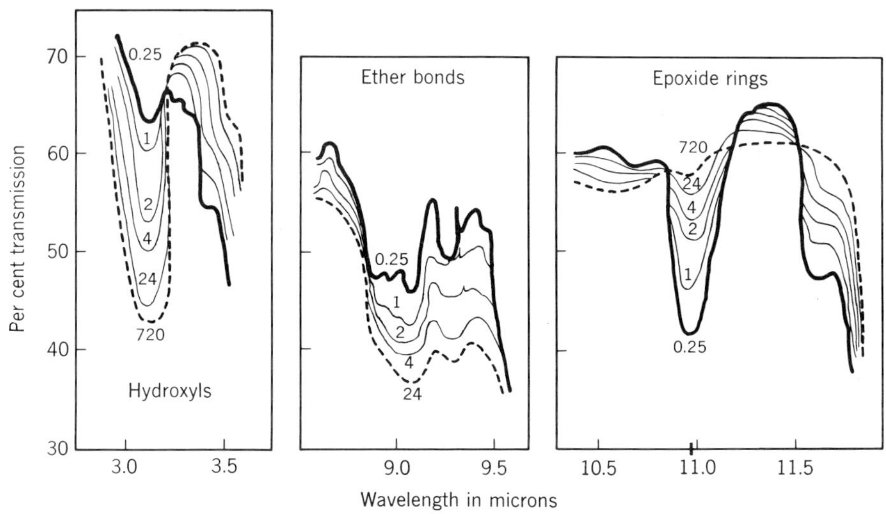

FIG. 6-2. Infrared adsorption of DGEBA during cure with primary aliphatic polyamine [3]. Numbers indicate cure time in hours.

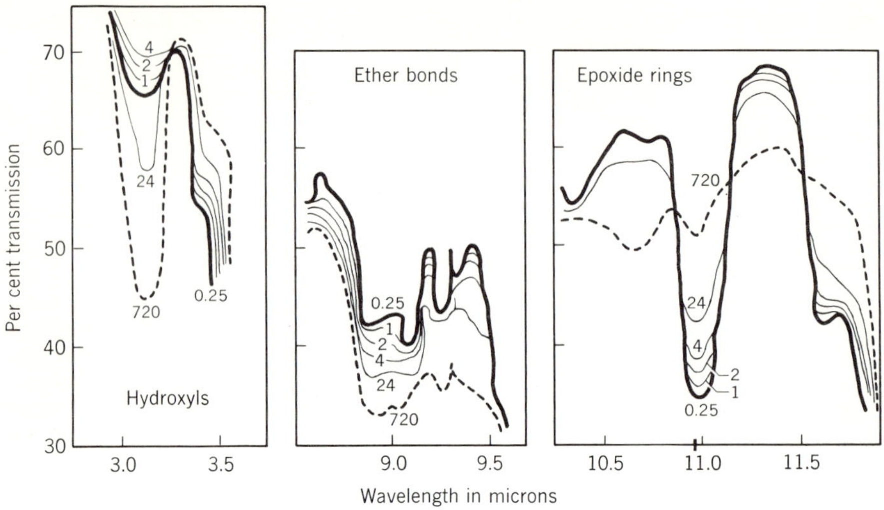

FIG. 6-3. Infrared adsorption of DGEBA during cure with tertiary amine [3]. Numbers indicate cure time in hours.

presented in Figs. 6-2 and 6-3. The epoxy groups at 10.96 and 11.6 microns disappear steadily as the reaction proceeds. Hydroxyl groups form at 3.1 microns. It is interesting to note that in Fig. 6-2, where the epoxy is reacting with a primary amine, hydroxyl concentration increases throughout the reaction, since hydroxyls are formed directly, whereas in Fig. 6-3, where the resin is being cured with a tertiary amine, the hydroxyl concentration initially decreases, as it participates in the reaction, but subsequently, after the hydroxyls have been almost entirely consumed, hydroxyl concentration increases markedly because of the elimination of charge separation (see Chap. 5).

Near-infrared spectroscopy may also be used to follow epoxy (Tables 6-3A and 6-3B) and hydroxyl concentration (Table 6-3B and Fig. 6-4). Besides identifying

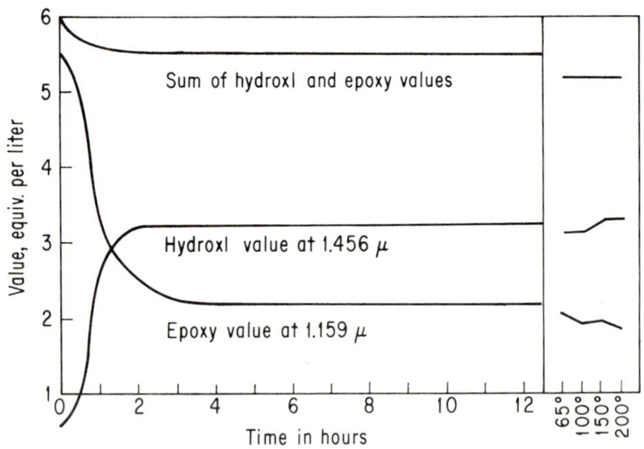

FIG. 6-4. Comparison of epoxy and hydroxyl values during cure of DGEBA with 5 phr ethylenediamine [53].

Table 6-3A. Spectroscopic Analysis of the Cure and Postcure of DGEBA + 5 phr Ethylenediamine [68]

Cure		Light path	Absorbance at		Ratio,	Epoxy value		
Time, hr:min	Temperature, °C	b, cm	2.205 μ	1.668 μ	$A_{2.205\mu}/A_{1.668}$	equiv./100 g[a]	equiv./100 g[b]	equiv./g[c]
0[d]	....	0.227	1.417	0.704	2.013	0.519	....	6.10
0:15	~25	0.228	1.190	0.653	1.822	0.472	....	5.20
0:31	~25	0.228	1.153	0.649	1.777	0.461	....	5.06
1:01	~25	0.228	1.075	0.642	1.674	0.436	....	4.75
1:35	~25	0.228	0.864	0.630	1.371	0.361	....	3.94
2:10	~25	0.228	0.737	0.613	1.202	0.319	....	3.65
2:53	~25	0.228	0.619	0.599	1.033	0.277	....	2.99
19:53	~25	0.228[e]	0.534	0.585	0.913	0.247	....	2.66
91:35	~25	0.228	0.508	0.579	0.877	0.239	0.224	2.56
+26:00	65[f]	0.228	0.380	0.578	0.657	0.184	....	2.07
+4:00	100[f]	0.228	0.323	0.579	0.558	0.160	....	1.85
+1:00	150[f]	0.228	0.287	0.581	0.494	0.144	....	1.71
+0:30	200[f]	0.228	0.243	0.582	0.418	0.125	....	1.54

[a] Using formula $0.247(A_{2.205}/A_{1.668}) + 0.022$ (relative method, see Chap. 4).
[b] Determined by chemical analysis.
[c] Using formula $(0.881 \times A_{2.205})/b + 0.60$ (direct method, see Chap. 4).
[d] No curing agent added.
[e] Length of light path uncertain owing to shrinkage: a value of 0.228 was assumed for calculation by formula c.
[f] In addition to the cure at the lower temperatures.

Table 6-3B. Change of Epoxy and Hydroxy Content of DGEBA during the Cure with 5 phr Ethylenediamine [68]

Cure time, hr:min	Cure temperature, °C	Epoxy value,* equiv./liter	Hydroxy value,† equiv./liter	Sum of epoxy + hydroxy values
0:08	~25	5.42	0.56	5.98
0:18	~25	5.27	0.56	5.83
0:28	~25	4.93	0.74	5.67
0:38	~25	4.79	0.95	5.74
0:48	‡	3.99	1.54	5.53
0:58	‡	2.97	2.62	5.59
1:08	‡	2.74	2.87	5.61
1:28	‡	2.57	3.08	5.65
2:08	~25	2.60	3.21	5.81
3:08	~25	2.20	3.29	5.49
4:08	~25	2.32	3.29	5.61
8:13	~25	2.20	3.38	5.53
12:13	~25	2.23	3.29	5.52
18:13	~25	2.17	3.29	5.46
24:13	~25	2.17	3.25	5.42
+26:00	65§	2.03	3.14	5.17
+4:00	100§	1.95	3.19	5.14
+1:00	150§	1.95	3.29	5.24
+0:30	200§	1.89	3.33	5.22

* Calculated as $(30A_{1.159})/1.055 + 1.89$.
† Calculated as $(A_{1.456}/A_{1.188})/0.52 - 0.73$.
‡ Temperature increasing through heat of reaction.
§ In addition to the cure at lower temperatures.

the curing agent used and determining the extent of conversion, infrared measurements are capable of giving additional information about the cured sample. For example, studies over the range of 30 to 200°C found a shift in the hydrogen-bonded OH band to lower frequencies with increasing temperature [71]. This was attributed to a shift from long-range to short-range bonds. Even at 200°C there were relatively few unbonded hydroxyl groups in the amine-cured resin.

Chemical Analysis for Determining Conversion

Chemical analysis for determining conversion is usually performed by the swollen-particle method. The resin sample is reduced to a very small particle size and

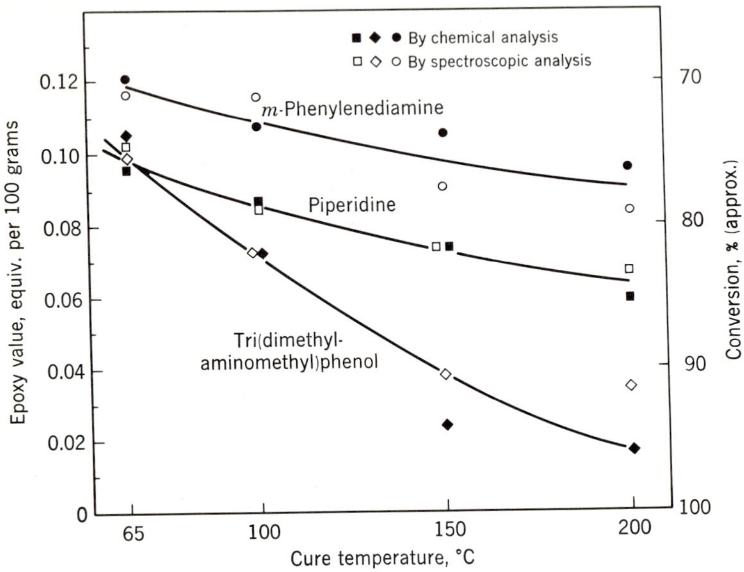

FIG. 6-5. Conversion of DGEBA as a function of curing agents and curing temperatures [2].

suspended in a liquid which acts as a swelling agent. The analytical reagent is then allowed to diffuse into the swollen particle and perform the analytical reaction, and the reaction products diffuse out of the particle. The values obtained by the chemical method are compared with the values obtained by infrared techniques in Fig. 6-5.

Epoxy-resin films have been exposed to chloroform for one day at room temperature to dissolve out the soluble, unreacted epoxy. The per cent insolubles may then be used to estimate extent of cure. Values so obtained correlate well with hot cut-through resistance (the temperature at which a wire under pressure makes electric contact with a coated bar) [111].

Vicat Softening Point for Estimating Extent of Cross-linking

The Vicat softening-point apparatus (ASTM 1525) may be modified [24] to permit its use in determining the softening point of higher-temperature-resistant thermosetting resins. The modified apparatus is shown in Fig. 6-6. Essentially, the device records the penetration of a Vicat needle vs. temperature. For epoxy-resin systems, a penetration of about 0.25 mm under a 5-kg load will occur very near the deflection temperature (see below). This relationship is shown in Fig. 6-7.

As the Vicat softening point increases to the maximum value, cure has correspondingly progressed: although with aging at temperature and with higher-temperature testing, the physical condition of the surface will more strongly affect the Vicat reading than the DT.

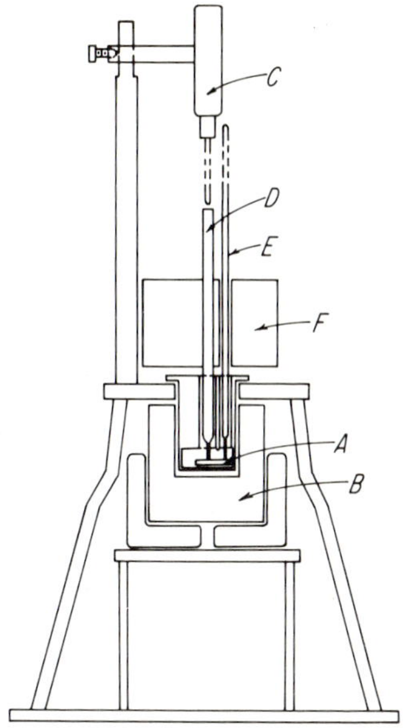

FIG. 6-6. Modified Vicat apparatus [24]. *A*, Sample; *B*, aluminum block; *C*, gage; *D*, Vicat needle; *E*, thermometer; *F*, lead-filled steel container.

Refractive Index for Estimating Extent of Cure

The refractive indices of various catalyzed epoxy-resin systems have been studied during cure [23], and it was found that the total increase from mixing to final cure is approximately 280 $\times 10^{-4}$ for tertiary and secondary amines and approximately 400 $\times 10^{-4}$ for primary amines. The change in refractive index as cure progresses with DGEBA, using piperidine as a curing agent, is shown in Fig. 6-8.

Electrical Resistivity Measurements for Estimating Extent of Cross-linking

Electrical measurements may be conducted under isothermal conditions to determine the progress of the polymerization continuously from the liquid through the solid state.

Volume-resistivity measurements indicate that the rate of change of resistivity levels off only after the reaction has advanced considerably beyond the gel point [13]. The progress of the reaction is indicated in Figs. 6-9 and 6-10. Volume-resistivity measurements on various epoxy systems during cure are described in [20]. Other electrical measurements have also been recorded during cure [15], as indicated in Figs. 6-11 and 6-12. The relationship between mechanical and electrical properties is discussed in [8]. In general, correlations have been obtained between 10^3 cps dissipation factor and tensile strengths for a variety of cure cycles and formulations.

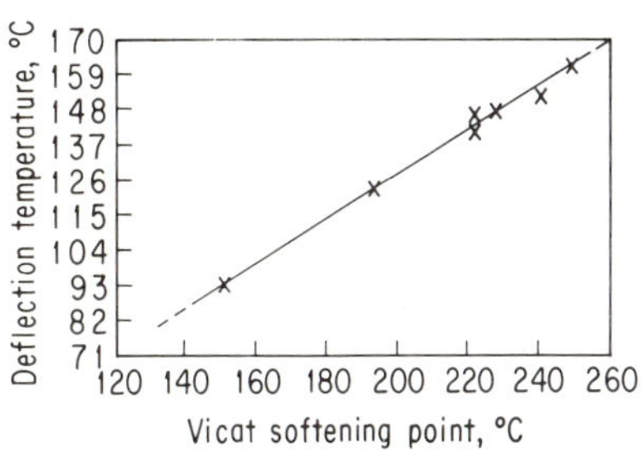

FIG. 6-7. Relationship between Vicat softening point and DT [24].

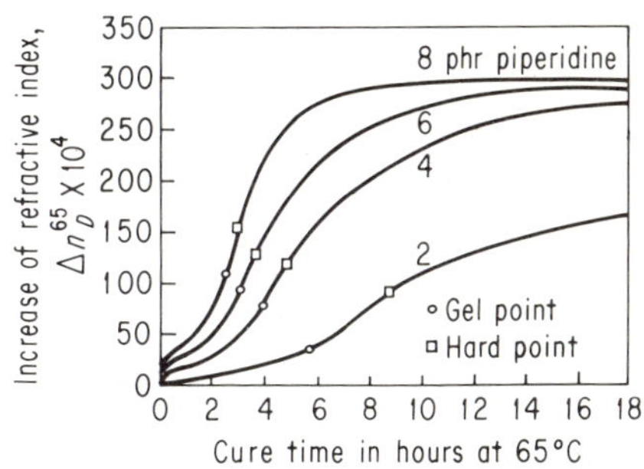

FIG. 6-8. Extent of cure vs. curing-agent concentration as measured by refractive index for DGEBA cured with piperidine [23].

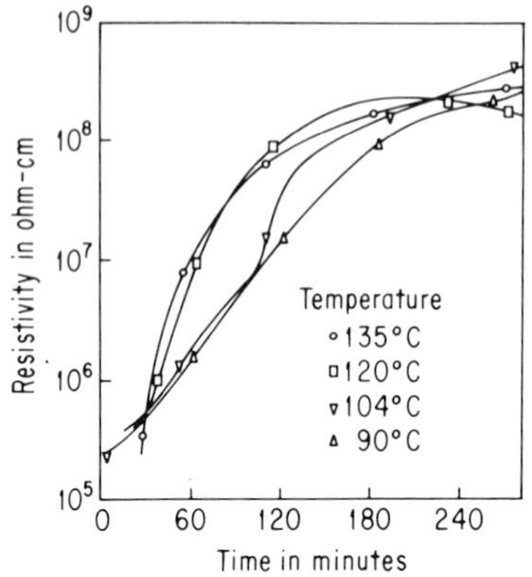

FIG. 6-9. Volume resistivity during isothermal polymerization of DGEBA with MA [13].

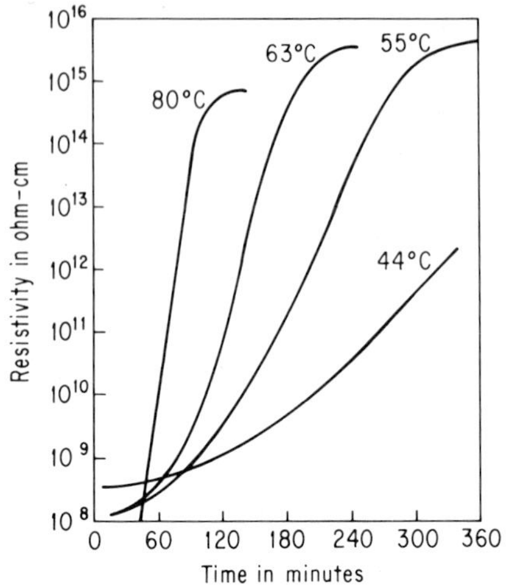

FIG. 6-10. Volume resistivity during isothermal polymerization of DGEBA with MPDA [13].

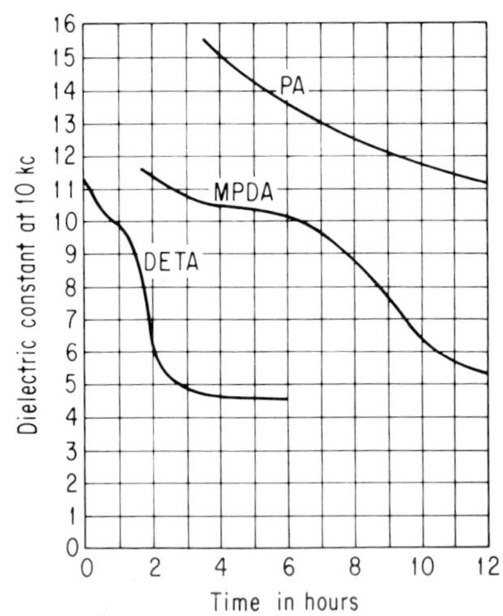

FIG. 6-11. Dielectric constant of DGEBA during polymerization with various curing agents [15].

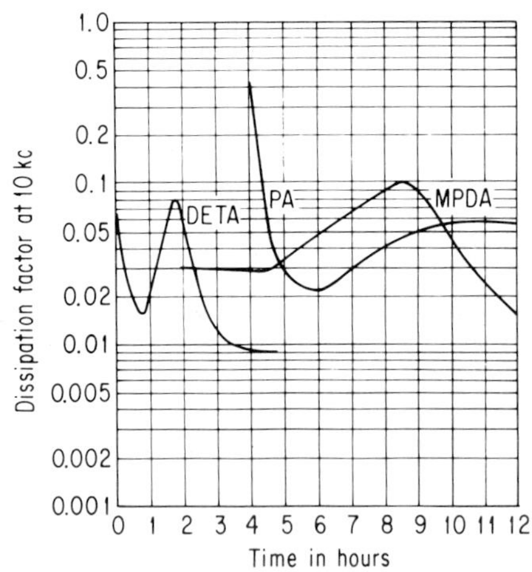

FIG. 6-12. Dissipation factor of DGEBA during polymerization with various curing agents [15].

Solvent Swell-volume Studies for Estimating Extent of Cross-linking

A correlation between swell volume and softness (measured by using a hot needle in a procedure somewhat like the Vicat test) has been found, as indicated in Fig. 6-13.

DGEBA cured with hexanediamine and with a fatty polyamide resin, when exposed to swelling in chloroform and in dioxane, gives values for the molecular weight per cross-linked unit which agree well with values estimated from chemical considerations [73].

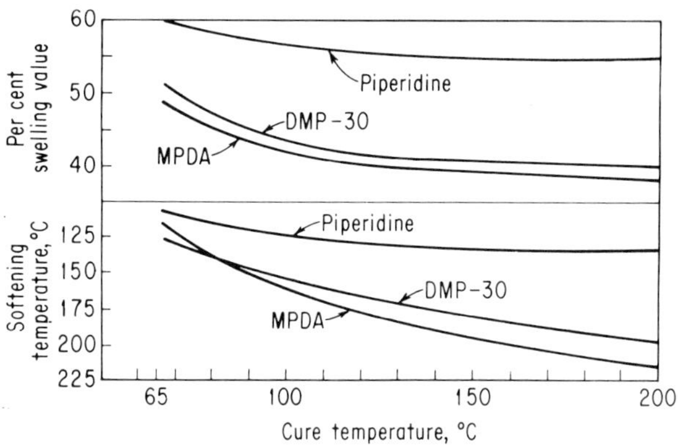

FIG. 6-13. Comparison of swell volume and softening temperature of cured DGEBA [2].

Deflection Temperature (DT) for Estimating Extent of Cross-linking

The deflection temperature is a useful guide to developing an optimum cure cycle and in general indicates the extent of cross-linking.

For ascertaining the effect of curing temperature, it is often desirable to plot cure time divided by DT against cure time for various temperatures. If the DT is independent of curing temperature within the range of temperatures under consideration, then all the lines so plotted will coincide. If the DT is dependent on the cure temperature, the lines will differ in slope [5].

The DT is determined by ASTM Procedure D 648–56. In this test, a standard bar of cured resin, supported at each end, is loaded in the middle with a weight designed to give a uniform fiber stress of 264 psi. The sample is then immersed in an oil bath with a temperature control and a deflection-measuring dial indicator. The dial indicator is set at zero and the bath warmed at a rate of 2°C per minute. When deflection of the bar reaches 0.010 in., the temperature of the bath is reported as the DT. The ASTM procedure specifies bar dimensions of $\frac{1}{8}$, $\frac{1}{4}$, or $\frac{1}{2}$ in. wide by $\frac{1}{2}$ in. high by 5 in. long. Since values are determined in part by heat transmission, the $\frac{1}{2}$- by $\frac{1}{2}$-in. bar gives higher and more consistent values. It apparently is more commonly used.

EFFECT OF SYSTEM FUNCTIONALITY ON CURED PROPERTIES

System functionality may be defined as the average functionality per molecule of the system components.

In order to compute system functionality, it is necessary to determine the potential functionality of the reactive sites within the molecules involved, to examine the type of reaction these sites undergo during cure, and to determine the amount of functionality lost because of the reaction conditions.

The potential functionality of a given reactive group can be determined from curing mechanisms (Chap 5). Thus a secondary amine, a carboxyl group, and a hydroxyl group each have a potential functionality of 1. A primary amine, an anhydride group, an epoxy group, and a reactive unsaturation each have a potential functionality of 2.

Functionality may be lost from the molecule in any one of several ways: because of the direction of the reaction; because of the absence of coreactant groups; because of steric hindrance; or through molecular immobility.

Loss of Functionality as Influenced by Reaction Conditions

The conditions of reaction may be such as to exclude certain potentially functional groups. For example, if a primary amine is reacted with an epoxy, the formed hydroxyls should not normally be considered functional, and the functionality of the epoxy group in this case is therefore 1.

Loss of Functionality as Influenced by Coreactant

In a copolymer system, in order for a reactive site to assume functionality, it is necessary that a coreactant be available. System functionality is therefore dependent upon the ratio of reactants, and may be reduced if one or the other is available in excess. System reactants, therefore, should be combined at the optimum ratio to avoid loss of effective functionality.

Loss of Functionality as Influenced by Steric Factors

Reactives sites normally active may be blocked from the reaction by steric factors. For example, in a primary amine cure, the formed tertiary amine may be too hindered to promote cure by the self-propagating ionic mechanism, such that any epoxy present in excess of stoichiometric will not be cured.

Loss of Functionality through Immobility

Increasing functionality results in gelation increasingly early in the reaction. Immobilized reactants generally require increasingly severe reaction conditions for cure to obtain substantial completion. Thus, it might be expected that increasingly severe postcures would be required with increasingly functional systems. Increasing the potential functionality beyond a point does not appear to offer any substantial advantage, since increasing immobility results in increasing loss of functionality.

Functionality Calculations

Primary Amine Cures. The functionality of a resin and a primary amine (at stoichiometric) is determined by assigning a functionality of 1 to each epoxy group and 1 to each active hydrogen of the amine.

Catalytic Cures. In catalytic cures, it might be assumed that each epoxy group has a functionality of 2 and the functionality might therefore be written as

$$f = 2\text{E}$$

where E is the number of epoxy groups per molecule. Thus, going from a di- to a triepoxy increases the system functionality from 4 to 6.

Although this is theoretically true, it must be modified by two considerations:

1. If the catalytic reaction proceeds in such a way as to block potentially reactive sites, as for example,

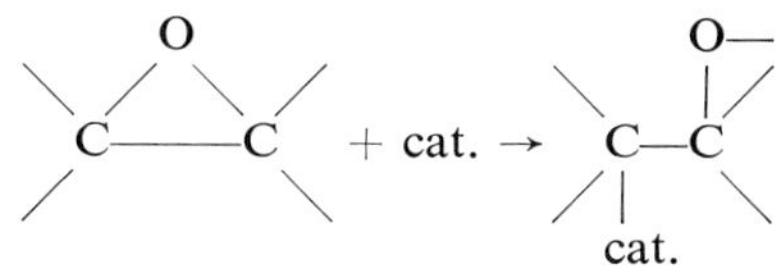

then system functionality will be reduced in proportion to the amount of catalyst employed.

2. The mechanism for elimination of the charge separation that occurs during catalytic cure may also result in the elimination of reactive sites in the system, thus reducing the system functionality. This is of increasing significance when larger amounts of tertiary amines are used as catalysts and the reaction is conducted at elevated temperatures.

Because of these two considerations, it is not possible to assign system functionality to a catalytically cured system, since, empirically, the functionality will vary from curing agent to curing agent, and from cure temperature to cure temperature.

Anhydride Cures. A consideration of the mechanisms involved in anhydride cures (Chap. 5) leads to the conclusion that the optimum ratio of anhydride groups to epoxy groups (A/E) may vary over the range from about 0.4/1 to about 1/1.

Functionality calculations for anhydride-cured systems are by no means straightforward. A variety of reactions are occurring during cure, and the extent to which hydroxyls are available in part determines the functionality. Gel-point calculations for anhydride-cured systems have been developed and verified experimentally [80].

All other things equal, the more highly functional systems provide for increased heat resistance, improved chemical resistance, and somewhat better electrical properties. These changes are attained, however, at the expense of increasing apparent brittleness in the cured product and lower elongation.

EFFECT OF CROSS-LINKING DENSITY ON CURED PROPERTIES

For systems having the same functionality, certain properties will vary with increasing or decreasing cross-linking density [40].

Lowering of cross-linking density, at the expense of functionality, can be accomplished by (1) employing monofunctional reactants which effectively eliminate cross-linking points, (2) by using nonreactive modifiers or plasticizers, which lubricate chain movement and physically separate cross-linking points, or (3) by using reactive modifiers containing long side chains, which also serve physically to separate the cross-linking points (internal plasticization). Change in cross-linking density without loss of functionality may be accomplished by the selection of ingredients having reactive sites suitably spaced on the molecules.

Increasing the cross-linking density—usually by selection of short-chain molecules—has the effect of improving the thermal and chemical resistance; whereas decreasing the cross-linking density—usually accomplished by the selection of long-chain molecules—has the effect of improving toughness, increasing elongation, and decreasing shrinkage.

Tests have been conducted to determine the effect of cross-linking density on a series of cycloaliphatic epoxy resins with MPDA. Increasing cross-linking density, that is, reducing distance between reactive sites, from 28 to 2 angstroms, resulted in increasing compressive modulus and yield, tensile modulus and yield, DT, and Izod impact. However, increasing the cross-linking density also resulted in

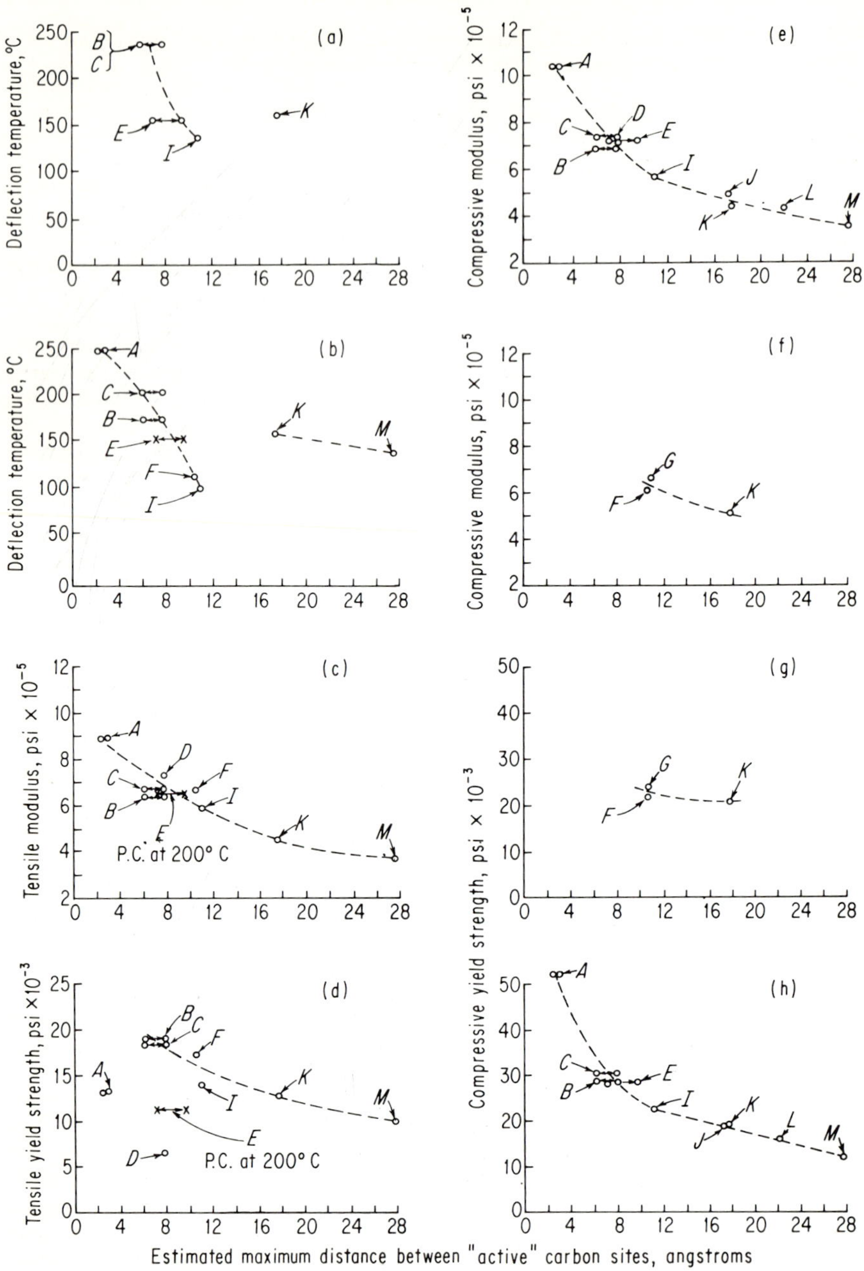

FIG. 6-14. Effect of cross-linking on properties of various epoxy resins [100]. Letters on curves are identified in Table 6-4. (*a*) Cured with MPDA and postcured at 200°C. (*b*) Cured with MPDA and postcured at 160°C. (*c*) Cured with MPDA and postcured at 160°C. (*d*) Cured with MPDA and postcured at 160°C. (*e*) Cured with MPDA and postcured at 160°C. (*f*) Cured with NMA and postcured at 160°C. (*g*) Cured with NMA and postcured at 160°C. (*h*) Cured with MPDA and postcured at 160°C.

decreasing resistance to crack propagation and increasing brittleness [100]. The considerable data generated on the influence of cross-linking distances during this research is presented in Fig. 6-14*a* to *h*. The maximum distances between cross-linking points for these resins are indicated in Table 6-4. In a separate series of tests with varying-molecular-weight species of DGEBA cured with various amine curing agents, the same general results were obtained. Tensile strength was found to correlate with reasonable accuracy to DT and Tg below an upper limit. However, some of the higher-temperature systems revealed low tensile strengths, which was attributed to increased flaw sensitivity with the more tightly cross-linked DGEBA systems. This was further demonstrated by negative correlation between Izod impact and Tg, indicating, further, an increased flaw sensitivity over that encountered with the cycloaliphatic resins [101].

Table 6-4. Maximum Molecular Distances between Active Sites on Epoxy Resins and Curing Agents (from model measurements) [100]

Resin or curing agent	*Maximum distance, angstroms*
A. 1,4-cyclohexadiene dioxide	
bis(2,3-epoxycyclopentyl)ether	$o = 2.50$, $p = 2.75$
B. Liquid isomer	6.05
C. Solid isomer	7.75
D. Glycidyl 2,3-epoxycyclopentyl ether	7.80
E. *o*-2,3-epoxycyclopentyl PGE (mixed isomers)	
Isomer 1	7.05
Isomer 2	9.50
F. DGE dioxane	10.50
G. DGE sulfolane	10.50
H. Poly-2,3-epoxycyclopentyl PGE many isomers (range)	7.05–11.00
I. *p*-2,3-Epoxycyclopentyl PGE	11.00
J. DGE bisphenol O	17.20
K. DGE bisphenol A	17.50
L. DGE bisphenol hydroquinone	22.00
M. DGE bisphenol pentaerythritol	27.60
MPDA	5.20
MDA	16.6
NMA	~6.40

When intentionally reducing cross-linking density by use of long-chain additives, physical effects will not necessarily be a function of additive concentration. For example, if the selected long-chain element is a DGE of an aliphatic diol and is used in small amounts, the effect may be to reduce the strength and lead to stress cracking, probably because the long chain serves to concentrate mechanical stresses rather than to relieve them. Increasing the amount of diol and further reducing the cross-linking density will overcome the difficulty [36].

Studies have indicated that fillers will significantly change the apparent cross-linking density of epoxy-based films. This appears to be accomplished by an alteration in segmental mobility of macromolecules, and the effect extends well into the bulk polymer, away from the interface of the polymer and filler particle. DGEBA (mol. wt. 1,000) cross-linked with 1,6-hexanediamine gives an average molecular weight between cross-links (M_c) of 240, calculated from solution data. When TiO_2 is added, up to 14.31 per cent by volume, the M_c is progressively reduced to 30 [108].

EFFECT OF STRUCTURE ON CURED PROPERTIES

The structure of a cured epoxy resin can be varied by changes in the resin type and the curing-agent type. The large variety of molecular compositions possible, together with variables of functionality and cross-link density, leads to an unusually large range of cured properties for a polymer system.

Thermal Stability

In general, the presence of aromatic groups in the cured resin network tends to increase thermal stability. In nonaromatic resins, the primary consideration is probably the cross-linking density of the resin, as was demonstrated with a series of epoxysiloxane resins [39], the cyclics being most stable, the linear next, and the branched least stable of all in terms of weight loss at 260°C (Fig. 6-15A). The same order prevails in terms of DT, as shown in Fig. 6-15B. Investigation to determine the effect of structure of bisphenols derived from various ketones on DT and other

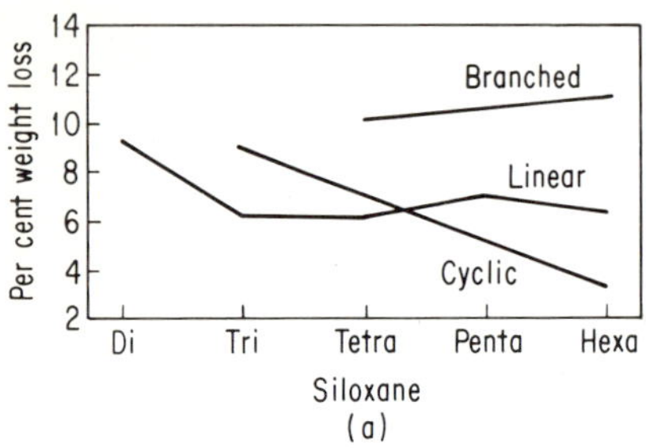

FIG. 6-15A. Effect of structure of poly-epoxy siloxanes on weight loss after 100 hours at 260°C [39].

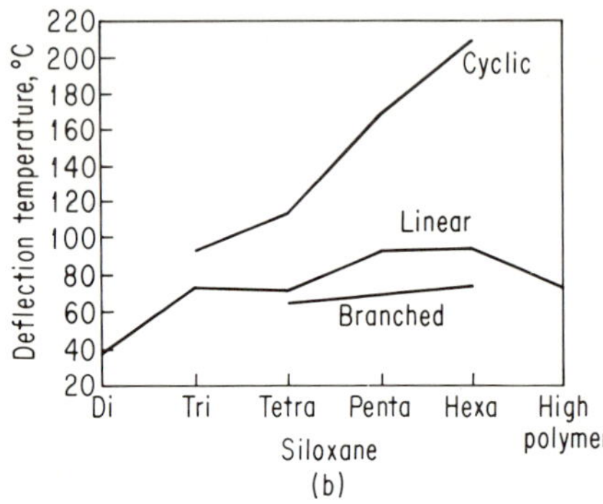

FIG. 6-15B. Effect of structure of poly-epoxy siloxanes on DT [39].

properties gave results as indicated in Table 6-5. Investigations of elevated-strength properties of adhesive formulations indicated improved elevated-temperature stability when symmetrical substituents were present on the benzhydryl carbon of the bisphenol; with unsymmetrical aliphatic substituents strength decreased with increasing chain length of the pendant carbon chain [58]. In a separate series of experiments improvements were shown when one of the methyl groups was replaced by phenyl [77]. With cyclic substituents, only the DGE of 1,1-bis(4-hydroxyphenyl)cyclohexane showed improvement over DGEBA [35].

Table 6-5. Performance Tests on Cured Diglycidyl Ethers of Phenols [58]

Substituents on benzhydryl carbon	DT, °C		Notched Izod impact resistance, ft-lb/in. of notch		Rockwell hardness, M Scale		% weight loss on aging *A*	
	A	*B*	*A*	*B*	*A*	*B*	24 hr at 260°C	7 days at 260°C
Methyl methyl	164	146	0.98	0.61	107	111	2	23
Methyl ethyl	139	129	0.93	0.73	110	109	2	23
Methyl isobutyl	154	135	0.89	0.65	105	104	2	29
Methyl amyl	137	125	0.79	0.63	97	101	2	24
Methyl nonyl	105	98	0.74	0.61	81	88	2	23
Methyl heptadecyl ...	76	72	0.72	0.52	28	37	2	18
Diheptadecyl	54	50	0.41	0.70	−98	18	3	17

A. MPDA/MDA/PGE curing agent.
B. Phthalic anhydride curing agent.

The ether links and the ester links are relatively good in terms of thermal stability. The amino groups provide for relatively poor thermal stability in terms of weight-loss studies. Hydroxyl groups present in the cured system appear to be particularly vulnerable points for degradation at elevated temperatures. High chlorine contents in glycidyl ether resins are detrimental to thermal stability, as shown in Fig. 6-16. Increased high-temperature instability has been noted for amine-cured chlorinated and brominated DGEBA resins over what is obtained with the non-halogenated resin [86]. Thermal stability in terms of weight loss and infrared

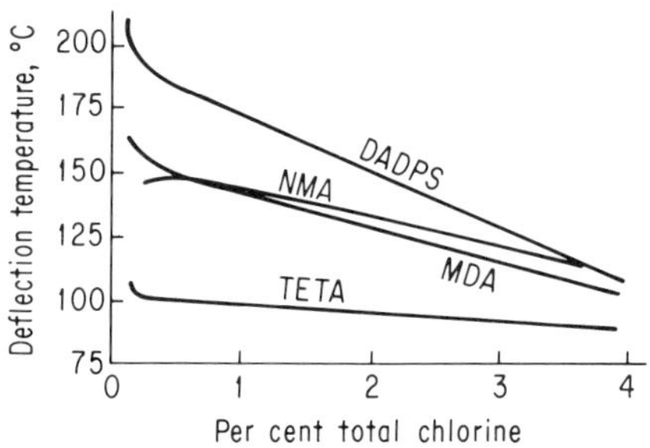

FIG. 6-16. Effect of chlorine content of DGEBA resin on thermal stability of cured system [14].

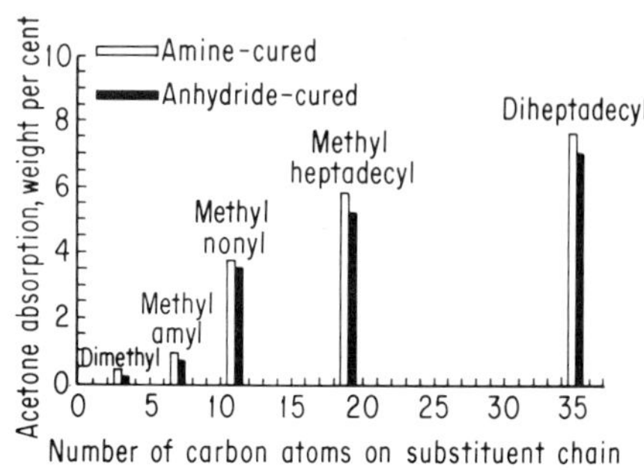

FIG. 6-17. Acetone absorption (24-hour immersion) vs. length of alkyl groups on benzhydryl carbon of DGEs of bisphenols [58].

measurements on narrow molecular-weight fractions of DGEBA showed that the systems from the lower-weight fractions degraded the fastest [88]. It was concluded that the weak segment in the resin was the amine-alcohol residue:

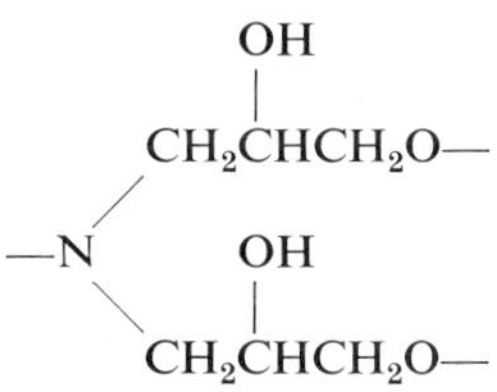

Chemical Resistance

In general, those factors tending to promote thermal stability also tend to improve chemical resistance. This is particularly true in regards to the molecular structure, less true in regards to the nature of the chemical linkages.

In terms of chemical linkages, the following generalizations are usually valid:

In cured systems, ether links are fairly stable against most inorganic and organic acids as well as against caustics.

The carbon-amine nitrogen link is fairly stable in the presence of most inorganic acids, poor in organic acids, and good in caustics. As would be expected from thermal stability, the aromatics outperform the aliphatics in terms of chemical resistance.

Ester links are good against organic acids, and fair to poor against inorganic acids and caustics. The ester link is more subject to hydrolysis than the carbon-nitrogen bond in amine-cured resins.

Again it should be borne in mind that the structure of the cured system will often provide results different from what would be expected in terms of the nature of the chemical linkages alone. The effects of substituents on the benzhydryl carbon on acetone resistance is indicated in Fig. 6-17.

Electrical Properties

Increasing the polarity of the molecules in the cured structure results in a reduction of electrical insulating properties. Polar ester links are less detrimental to electrical insulating properties than are the polar linkages introduced during primary and secondary amine cures. Tertiary amines appear to give somewhat better electrical properties than the primary amines. Lewis acids, such as BF_3MEA, provide more highly polar systems, with attendant reduction of electrical properties [55]. The effect of chlorine content is indicated in Figs. 6-18 and 6-19. A study

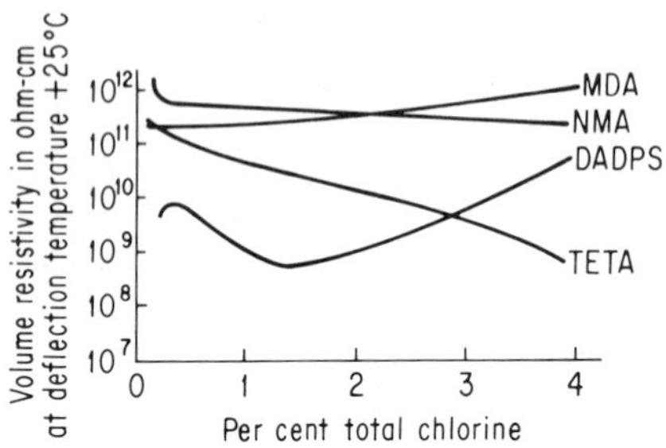

FIG. 6-18. Effect of chlorine content of DGEBA resin on volume resistivity of cured systems [14].

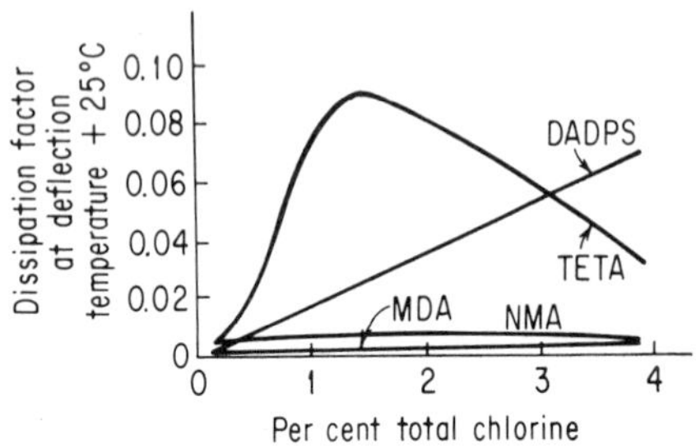

FIG. 6-19. Effect of chlorine content of DGEBA resin on dissipation factor of cured systems [14].

of DGEBA cured with TETA, MPDA, MDA, and DADPS indicates that the characteristic dielectric absorption could be ascribed to the orientation of the OH groups in the cured resin [98].

Such generalizations, without taking into account the effect of moisture and temperature, are of course simplifications, but they may be used with care as intelligent guides.

PHYSICAL PROPERTIES OF CURED RESINS

The physical properties of the cured epoxy resins, which are useful in design studies, are normally described according to ASTM test procedures or according to modified procedures designed especially for the epoxy resins. The Epoxy Resin Formulators of the SPI also periodically issue test procedures especially adapted for the epoxies. Suitable microminiature test procedures are described in [110].

The results of such tests are meaningful but they should be approached with reservations. Samples may have been prepared using unusual precautions in mixing and curing, such that they are not truly representative of the properties to be expected in production. The test procedures, in addition, often fail fully to characterize the property under consideration. For example, the Izod impact test employs a notched specimen, which often does not sufficiently discriminate, for practical purposes, differences between cured compounds. Further, the test results may be of little design value unless they are reported over a sufficiently wide range of temperatures to indicate performance in actual service.

And lastly, and by far the most important: The tests are conducted under laboratory conditions on samples of a particular molecular weight and structure. There is no assurance that performance will be identical if the configuration is changed in actual use. For example, even the best thermal shock-resistance tests may not be able to duplicate the worst conditions of field service that can be expected on unusual configurations.

In the measurements of physical properties of epoxy-resin castings, the method by which initial failure occurs and is propagated is of considerable interest. Tests conducted with diluted DGEBA cured with DETA have indicated that fracture is initiated at bubbles, inclusions, and surface scratches. The fracture spreads slowly outward in a radial direction until it reaches a critical size corresponding to the applied stress, at which point it becomes unstable and propagates rapidly. A "clean" specimen will give tensile strengths of 14,000 psi vs. 8,000 psi for a "dirty" one. Gross flaws will produce premature failure and rapid crack propagation, as indicated by failure on the positive slope of the stress-strain curve. Samples containing only minor flaws propagate fractures slowly, and failure will occur on the negative slope of the stress-strain curve. Low strain rates permit propagation by decohesion prior to the critical stress level, whereas at high strain rates the stresses are built up too fast for decohesion to be significant. Under creep conditions, surface appearance of fractured samples (either broken in the test or stressed and then broken) indicate (1) some slow fracture growth and (2) perhaps stress-induced orientation of the macromolecular chains along the axis of tension which occurs by breaking of cross-links. This latter however, might also be explained on the basis of cross-linking density and thoroughness of cure. When the system is flexibilized, the cross-link density apparently is reduced to the point where failure occurs strictly by slow decohesion rather than brittle failure [55].

The mechanism of initiation and propagation of failure in epoxy castings has been further investigated by introducing measurably large defects in tensile specimens. Energy expended in the fracture process was calculated and interpreted as representing the fracture surface work. Tests were conducted with several amine-cured DGEBA systems (as representative of highly cross-linked polymers) in an effort to substantiate the conclusion that previous experience with polystyrene (which gave values many times the theoretical) could be attributed to a flow mechanism in the immediate region of the fracture which oriented the molecules and absorbed large amounts of work. Contrary to expectations, the fracture surface work of the epoxies was in the same range as with polystyrene (4×10^5 ergs per sq cm) and the value was even improved somewhat with additional cross-linking (as indicated by increasing DT after postcuring). It seems unlikely that this can be explained by the concept of plastic flow, and the experimenters were able to advance no firm hypothesis: although it seemed to them most likely that as the materials become more brittle, there was a tendency, under load, for local shattering in a great number of areas, and so that the apparent fracture area (from the built-in flaw) was considerably less than the real area created by the process [76].

Tensile Strength

Tensile strength is a measure of the pull required to break a sample of material and is determined by dividing the force at failure by the area of the initial cross section, to give a value in pounds per square inch. Measurement procedures are recorded in ASTM 638–52TD. Tensile strengths of unmodified DGEBA cured with amine curing agents are about 9,000 psi.

Compressive Strength and Compressibility

The compressive strength of a material is the measure of its resistance to a crushing force, expressed in pounds per square inch at failure. The value is meaningful when it records the force at which a fracture occurs. If deformation continues without fracture, compressive yield, in pounds per square inch, provides a better criterion. Compressive yield is the point at which the deformation increases without additional load. Both are measured in accordance with ASTM D 695–54. A typical value is 15,000 psi yield, 30,000 psi ultimate.

Compressibility data are given in Table 6-6. If the compressibility is due to squeezing out of free volume, the limit is about 15 per cent. Samples of DGEBA cured with DMP-30, DDSA, and MPDA tested at up to 12,000 atmospheres returned to their original dimensions after testing, indicating no permanent change. Pressure-induced Tg's were observed with all systems. Phase diagrams for various temperatures permitted extrapolation to Tg at 1 atmosphere with values in good agreement with experimental data for DMP-30 and DDSA. With the more rigid curing agent, MPDA, the extrapolated Tg was 196°C as against measured Tg's of about 160°C [115].

Table 6-6. Compressibility Data for Cured DGEBA [97]

Curing agent	Pressure, atm	Temperature, °C	Compressibility, %
DDSA	12,000	23	14
DDSA	10,000	182	15
MPDA	12,000	23	13
MPDA	12,000	138	14
DMP-30 salt	12,000	23	13
DMP-30 salt	12,000	193	15

NOTE: with DMP-30 salt, Tg increases from 80°C at 2,000 atm to 200°C at 5,000 atm.

Flexural Strength

Flexural strength is the resistance of a material to bending stresses. A sample beam of test material is subjected to center loads, and the value at which permanent distortion or failure occurs is recorded in pounds per square inch. The test method is described in ASTM D 790–49T. A typical value for flexural strength is 18,000 psi.

Modulus of Elasticity

Young's modulus or stiffness, which is the ratio of stress to strain, may be determined for tension, compression, and flexure by the respective ASTM test procedures cited above. Typical values of all three are 400,000 to 600,000 psi.

Poisson's Ratio

Poisson's ratio is the ratio of the transverse unit deformations to the longitudinal unit deformations. A typical value for a cured rigid epoxy system is 0.34.

Impact Resistance

Impact resistance is a measure of the ability of a material to withstand sudden loading and is usually expressed in foot-pounds or foot-pounds per inch of notch width of specimen at failure. The usual test is the Izod type, in which the specimen is held as a cantilever beam and is broken by a blow delivered at a fixed distance from the edge of the specimen clamp. The specimen is notched to produce a standard degree of stress concentration. This test is described in ASTM D 256–54T.

A second method, the falling-ball test, records the failure of a test specimen subjected to a sudden impact delivered by standard balls falling a standard distance, the weight of the balls being increased throughout the test. This method is described in MIL-I-16923, Ships. A typical value by the Izod test is 0.2 ft-lb per in. of width; by the falling-ball test, 2 ft-lb.

For impact produced by high-*g* loadings (as for example in the case of gun-launched components), it has been found that rigid potting compounds are preferred, insofar as the flexible materials allow intrapackage component movement. For this application, best possible compressive strengths coupled with lowest practical densities are required [66].

Dynamic Mechanical Properties

The term *dynamic properties* refers to those properties obtained by a method in which the results depend upon the time scale of the technique used and the sample does not attain an equilibrium response to an applied force.

Dynamic mechanical properties of epoxy resins have been studied and when determined near or above the Tg, have been found to be descriptive of the cross-linked network [25,38].

One series of experiments used a torsion pendulum in the frequency range of one cycle per second and observed two transition peaks in the dynamic loss tangent (i.e., the ratio of the imaginary to the real part of the complex modulus) [61]. It was argued that the low-temperature transition was caused by rotation of the glycidyl ether portion of the molecule, and the high-temperature transition (corresponding to the DT) was governed by the substituents on the central carbon bridge of the bisphenol residue in the molecule. The mobility of the glycidyl ether residues, as reflected by the low-temperature transition, can be limited by using a curing agent which gives a higher cross-link density by causing the epoxy group to react with a functionality of 2 (e.g., etherification curing agents).

Second-order Transition Temperature

One of the most useful properties in characterizing a resin and determining its useful temperature range is the second-order transition temperature or glass temperature, Tg. A first-order transition temperature is one at which a definite change in volume occurs. A second-order transition is one at which a change occurs only in the rate of change of volume with temperature.

Below their Tg polymers are rigid, glasslike materials. Above the Tg, they are soft, elastomeric materials.

Second-order transitions are caused by thermal agitation freeing portions of the resin chain to rotate around its bonds.

Second-order transition temperatures may be measured by hardness tests, bouncing-ball tests, tests with a dilatometer, or DT tests. Because molecular segment rotations occur at definite rates, test methods which subject the polymer to a rate of loading faster than the polymer is capable of adapting to, will give a higher value of Tg than methods which load the polymer slowly, at a rate less than its relaxation time. The two chief slow methods are deflection temperature, DT, and thermal expansion rate.

Deflection Temperature. Deflection temperature, also referred to as the heat-distortion temperature, has been described previously. The DT of DGEBA with common curing agents is 90 to 180°C, but with highly functional resins and curing agents values as high as 300°C have been measured. Flexibilized aliphatic epoxy resins can have values below zero.

Extensions of the DT test procedure also have given increased understanding of the nature of the cured polymer network.

If the DT test is continued beyond the arbitrary 10-mil deflection, profiles such as are shown in Fig. 6-20 are developed. These indicate in a general way the performance to be expected under elevated-temperature loading.

In order to account for the observed results, it has been postulated [16] that the initial sharp break in the curve, in the region of the DT, may be accounted for by

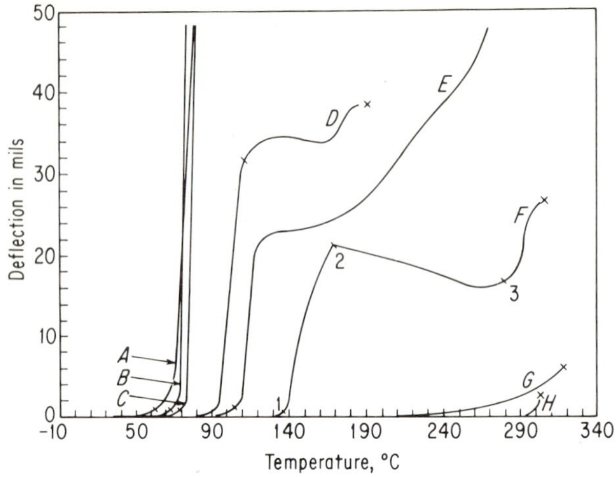

FIG. 6-20. Performance of various cured DGEBA systems under 264 psi fiber stress vs. temperature [16]. (1) Deflection temperature; (2) tolerance point; (3) slippage point.

the lessening of intermolecular Van der Waals forces and reduction in resistance of chain segments to rotation. Following this, the system may then deform in an almost linear fashion, the extent of deformation dependent upon the freedom of the chain elements to rotate, twist, lengthen, and compress. Chain stretching has been verified by an equation developed as a result of an investigation of the rubbery plateau regions of a series of amine-cured DGEBA systems [58]. Ultimately, depending upon the nature of the cross-linked network, a point is reached where little or no further rotation or stretching is possible, the bonds having been stretched to maximum. Following this, thermal expansion forces may actually provide for

Deflection temperature test setup. (*Ciba Products Co.*)

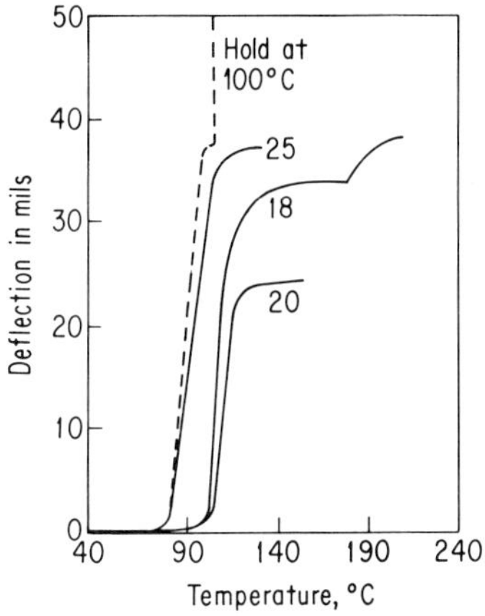

FIG. 6-21. Performance of cured DGEBA under 264 psi fiber stress at 100°C [16]. Numbers on curves indicate phr curing agent.

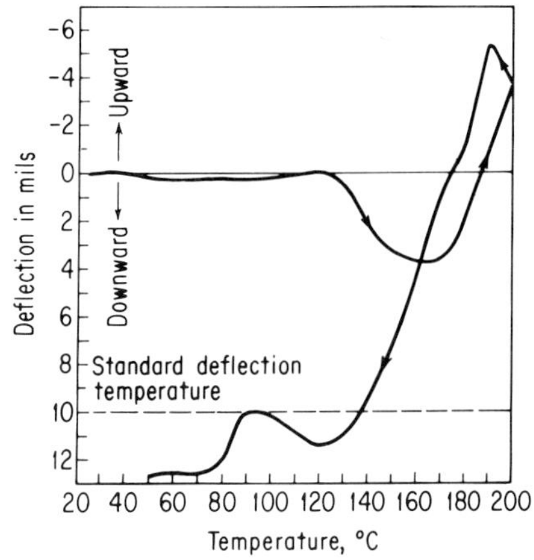

FIG. 6-22. Performance of cured DGEBA filled with chopped glass fibers under 264 psi fiber stress vs. temperature [37]. Net flexural deflection versus temperature. Data corrected for thermal expansion.

some recovery, up to the point where failure of primary valence bonds dictates increased deformation and sample failure.

It is interesting to note that in a system having low cross-link density, once the system has become unfrozen, deformation continues rapidly without additional increase of either load or temperature (Fig. 6-21).

In order to investigate the observed effect of chopped-glass fiber fillers in improving load-bearing performance vs. temperature, a series of experiments were conducted with the standard DT test setup [37]. It was observed that the initial deflection increased with temperature up to about the DT of the unfilled resin, after which increasing temperature resulted in recovery of the sample (even when corrected to eliminate the effect of thermal expansion). When the sample was cooled under load, the DT expected of the unfilled sample was obtained at about the proper point and the deflection increased somewhat with further cooling. This is shown in Fig. 6-22.

Tests such as these are probably more valuable for characterizing the cured resin than is the DT itself.

Thermal Expansion Rate. Thermal expansion tests are conducted with a dilatometer. The sample is placed in a fuzed quartz tube, and an indicator positioned atop the sample. The sample is then heated in an oil bath or in air, and

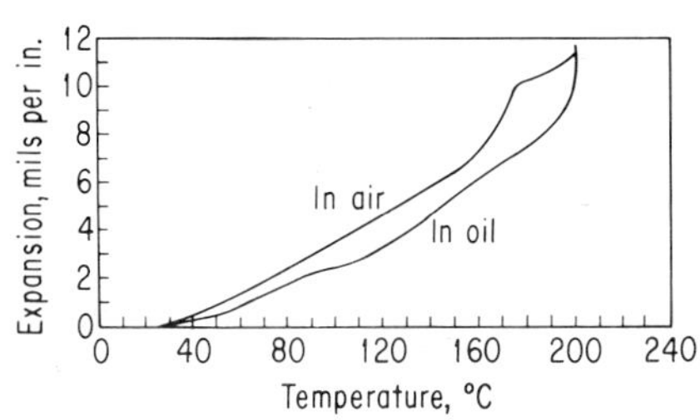

FIG. 6-23. Effect of heat-transfer media on thermal expansion data [7]. Rate of temperature rise, 2°C per minute.

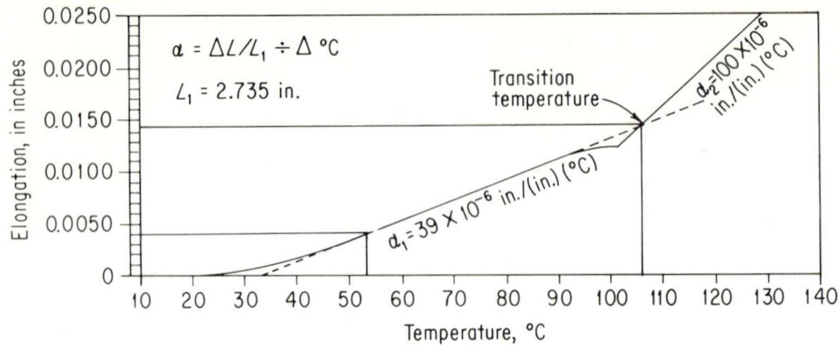

FIG. 6-24. Typical thermal expansion curve for cured DGEBA [16].

recordings are taken. The effect of the heat-transfer medium is indicated in Fig. 6-23.

A typical thermal expansion curve will show a sharp break in linearity in the region of the DT. This is the second-order transition temperature. Above the Tg, the thermal expansion rate increases more rapidly than below it, and, unlike high-temperature-deflection test data does not achieve a plateau. A typical thermal expansion curve indicating the Tg, is shown in Fig. 6-24.

A series of tests have been conducted with DGEBA cured with aliphatic- and aromatic-based anhydrides. It was possible to predict the Tg for the aliphatic-cured systems from the normal laws of polymer properties relating to Tg. However, with the aromatic-cured systems, because of the bulk of the aromatic rings, the Tg's were higher than the predictions from theory [78].

Thermal expansion and contraction rates can be correlated with physical performance (see Chap. 17). They may, likewise, be calculated from physical performance data. For example, foil strips, coated on one side with a cured epoxy film, will deflect during temperature cycling in accordance with thermal expansion rates, which may be calculated from the deflection [102].

Measurement of Tg by Electrical and Dynamic Mechanical Methods. Second-order transition temperatures of epoxy resins have also been determined by electrical resistivity measurements [28]. Some results are depicted in Fig. 6-25. It will be seen that a flexibilized resin exhibits no sharp break, the material being above the Tg at 25°C; likewise, the high heat-resistant epoxy system exhibits no sharp break, its Tg being above the range investigated [65].

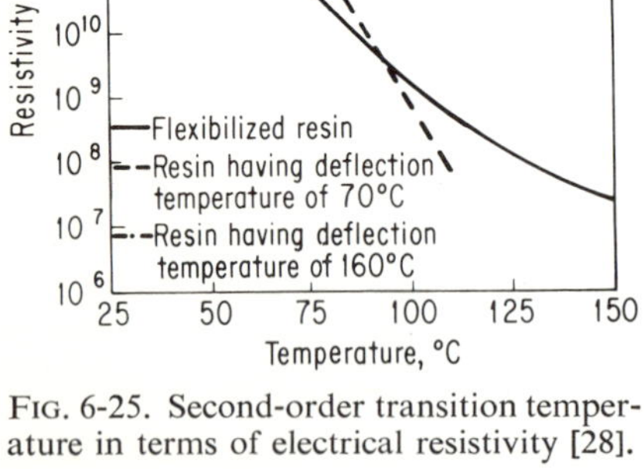

FIG. 6-25. Second-order transition temperature in terms of electrical resistivity [28].

With DGE's of bisphenols, the nature of the substitution on the benzhydryl carbon has been shown to influence the Tg, obtained by a dynamic mechanical method (Fig. 6-26) in the order given in Chap. 4. A second, low-temperature transition point was also detected with some systems, this being dependent on the glycidyl ether portion of the molecule, as indicated in Fig. 6-27.

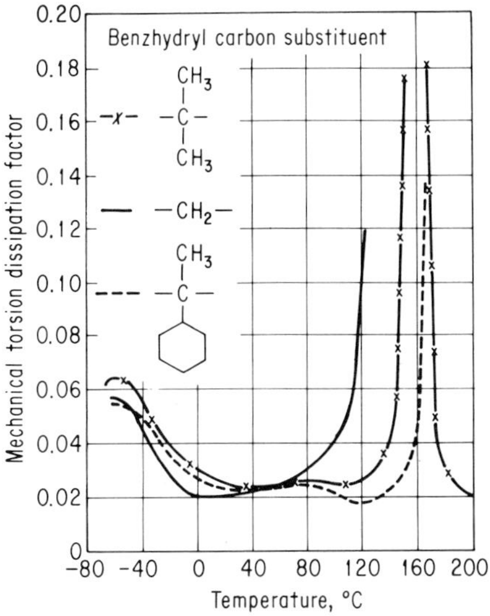

FIG. 6-26. Dynamic properties vs. temperature as a function of resin structure [61].

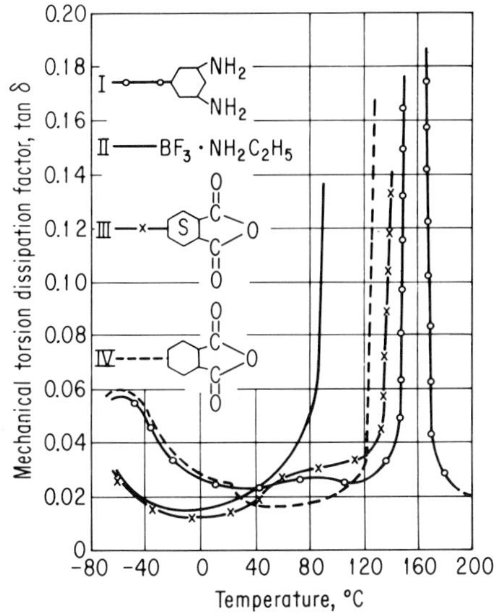

FIG. 6-27. Dynamic properties vs. temperature as a function of curing agent [61].

Shrinkage

Shrinkage is the dimensional change that occurs in a material over a stated period of time. Volume shrinkage is usually expressed as per cent change in volume. Linear shrinkage is expressed as the inch change in length, per inch.

Shrinkage occurs while the epoxy/curing-agent system is in a liquid state, while it is in a thermoplastic state, and after it has gelled. Design values are difficult to obtain from laboratory test data, insofar as the specific configuration of the structure, the mass of material involved, and the type of release agent employed all significantly influence the results of the test. In general, shrinkage occurs in the direction toward the bulk of the material.

Shrinkage is of two kinds: curing shrinkage caused by the reaction and rearrangement of the molecules into a more compact configuration; and thermal shrinkage, brought about by the cooling of the specimen from higher processing temperatures [44].

Volumetric shrinkage may be determined [58] by use of a closed mold to which is attached an oil-filled pipet. The oil level decreases as the resin volume decreases. The mass of resin is sufficiently small to minimize the effect of thermal expansion on the results. With unfilled room-temperature curing compounds, at least, volumetric shrinkage begins immediately after the system is mixed, and reaches values as high as 3½ per cent within a few hours, the ultimate value being as high as 5 per cent at the end of a week. The majority of shrinkage occurs before gelation, so that the effect can be largely offset when making open-mold castings by feeding in additional resin through a riser under gravity head. On closed-mold pressure castings the volumetric shrinkage assumes increasing importance in design.

Linear shrinkage is normally measured by casting the mixed compound into a machined trough coated with release agent. The standard trough is 36 in. long with a ½-in. radius. The casting is secured to one end of the mold. The small

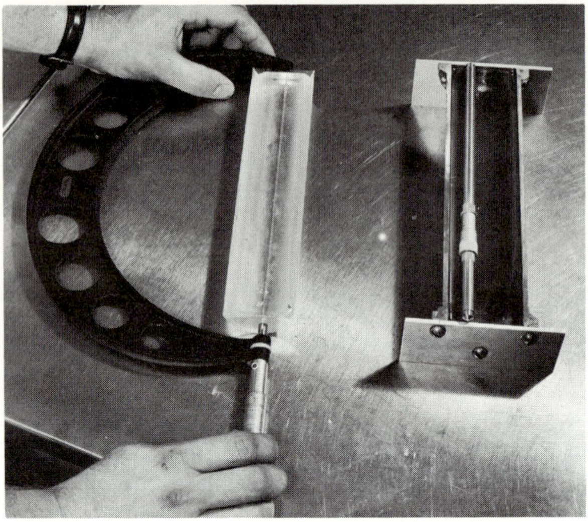

Epoxy-resin casting is measured after cure to determine shrinkage. (*Ciba Products Co.*)

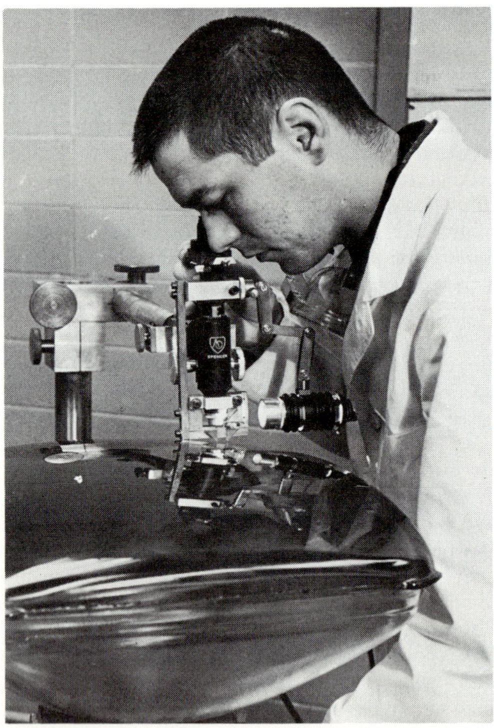

Notched polarimeter is used to measure stresses developed during cure of epoxy resin used for bonding glass safety panel to TV picture tube. (*Dow Chemical Co.*)

cross section reduces exotherm and the influence of thermal expansion during cure. With a satisfactory low-friction release system, linear shrinkage for a room-temperature curing system will not be detected until after gelation, and will thereafter increase with time to reach a value on the order of 0.001 in. per in. of length. With more highly functional epoxy systems, gelation occurs at a lower percentage of the total reaction, and shrinkage in the solid phase will be increased, some of the 4 to 5 per cent liquid-phase shrinkage now occurring in the solid phase.

During shrinkage, considerable pressures may be exerted on any components within the epoxy-resin compound. Readings from strain-gage-instrumented transducers indicated that a filled, amine-cured DGEBA system will apply pressures of 900 psi during the first 70 hours of room-temperature cure, this increasing to 2,300 psi at the end of 12 months [79].

Shrinkage and its effects are discussed more fully in Chap. 17.

Thermal Shock Resistance

Thermal shock resistance is the resistance of a cured resin to cracking or shattering when exposed to rapid changes in temperature. The tendency for resins to crack under temperature cycling is, of course, aggravated by metallic inserts having both sharp corners and significantly different coefficients of thermal expansion.

The basic epoxy resins, although possessing good thermal shock resistance compared to various other casting resins, usually are modified with flexibilizers when this parameter is critical. A discussion of test methods and typical results is presented in Chap. 17.

Properties at Cryogenic Temperatures

Recent aerospace research has heightened interest in the performance of plastics at cryogenic temperatures. Out of this work, to date, a few generalizations have developed. For the epoxies, the modulus of elasticity increases with decreasing temperature as do the tensile strengths. At cryogenic temperatures, up to 100 per cent improvement in tensile properties may be obtained. Brittleness increases as temperature decreases, and significant dimensional changes occur. Resins having the best toughness (defined as the area under the stress-strain curve) perform best at cryogenic temperatures. Thus, a number of rigid systems outperform flexibilized systems at $-196°C$.

Hardness

Hardness is a measure of the ability of a material to withstand localized indentation and may be expressed in a number of different test scales, depending on the degree of rigidity. Rockwell hardness is determined by procedures described in ASTM Method E–18, with the M scale being most common for epoxies. The plastics test procedure is explained in ASTM D 785–51. Durometers (usually Shore type) are used to measure indentation in softer materials (ASTM D 676–55T). A cured rigid epoxy resin gives Rockwell M readings in the neighborhood of 100. Barcol hardness testers are frequently used for materials in the Rockwell range. A typical value is 40 Barcol.

Special hot-hardness devices, involving a blunt-pointed heated stylus will provide results which correlate well with the degree of cure of the more rigid epoxy systems [91]. A special, 90° $\frac{1}{4}$-in.-wide needle has been shown useful for testing in the Shore hardness range [42].

Specific Heat

Specific heat of a material is the ratio of its heat capacity to that of water at 15°C. The heat capacity is the quantity of heat necessary to produce a unit change of

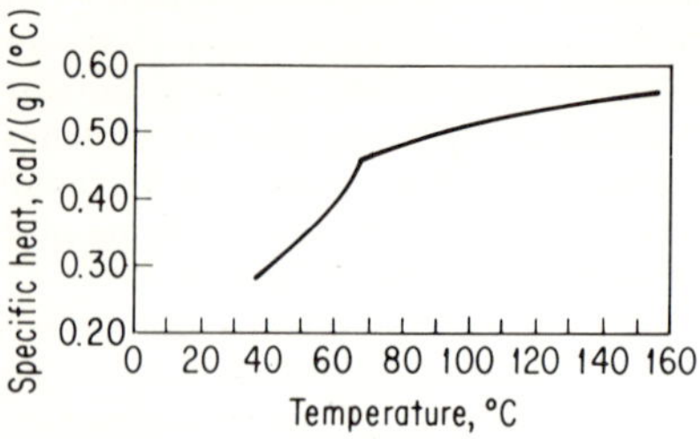

FIG. 6-28. Specific heat of DGEBA cured with salt of DMP-30 [19].

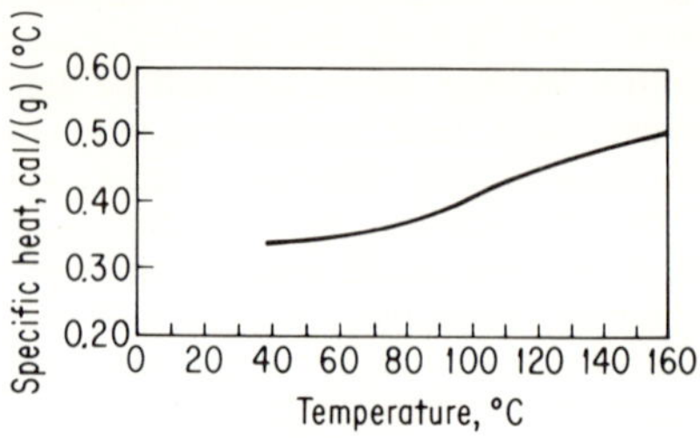

FIG. 6-29. Specific heat of DGEBA cured with MPDA [19].

temperature in a unit mass of the material. The average specific heat may be determined over a temperature range by means of a calorimeter [19].

Specific heat values for a number of cured epoxies have been determined [19,29]. Results of typical determinations are shown in Figs. 6-28 to 6-30.

Thermal Conductivity

Thermal conductivity is a measure of the ability of a material to conduct heat. A device for measuring thermal conductivity is described in MIL-I-16923. A typical setup consists of a heat source and a receiver and a thermally insulated block of copper of known thermal capacity. Thermocouples are embedded in the source and the receiver. The sample is then tightly clamped between the source and the receiver, and periodic readings taken for up to about 90 minutes. The readings are plotted on semilog paper. The thermal conductivity is then calculated from the thickness of the specimen, mass, area, specific heat of the copper, and slope of the semilog curve. A variety of other test methods have been suggested. One quick test is reported in [4] and defined in [45].

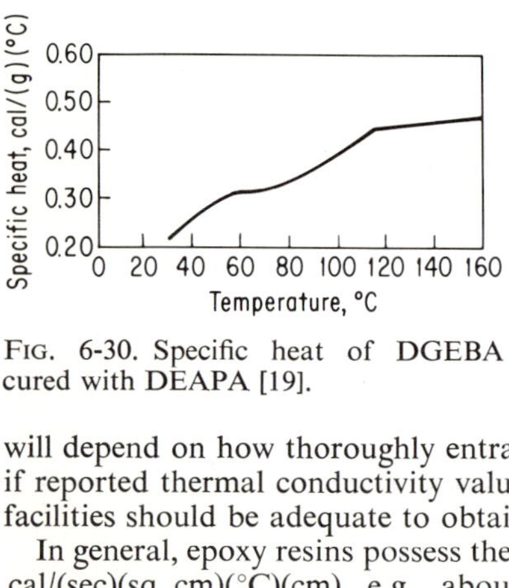

FIG. 6-30. Specific heat of DGEBA cured with DEAPA [19].

A number of practical difficulties stand in the way of reproducible thermal conductivity measurements. To mention only one: With filled epoxy-resin systems, results will depend on how thoroughly entrapped air is removed from the specimen—and if reported thermal conductivity values are to be used in design work, production facilities should be adequate to obtain the necessary deaeration.

In general, epoxy resins possess thermal conductivity values of about 500×10^{-6} cal/(sec)(sq cm)(°C)(cm), e.g., about 0.1 Btu/(hr)(sq ft)(°F)(ft). This is in the general range for other rigid plastics, and is characteristic of materials having high electrical resistance, since the properties go hand in hand. Fillers increase thermal conductivity somewhat (see Chap. 14), but it is not possible to match the values obtained by metals and other materials of high thermal conductivity.

Flammability

Flammability is a measure of the ability of a given material to support combustion. Most plastics, when exposed to a flame, will ignite and burn. The epoxies are no exception. The flammability of such materials is then determined in terms of the flame-out time, once the heat source is removed. If flame-out occurs within an arbitrary period of time, the materials are considered to be self-extinguishing. Otherwise, the burning rate in inches per minute is reported. The epoxies are generally classified as self-extinguishing or slow-burning by the test procedures of ASTM D 757–49. By this test, a sample 6 by $\frac{1}{2}$ by $\frac{1}{8}$ in. is prepared, and the end is placed in the hottest part of a Bunsen burner flame for 10 seconds. The flame is removed, and the time, in seconds, to flame-out (or the burning rate per minute) is recorded.

Flammability as determined from such a test procedure is not always indicative of the performance of the epoxy systems on a specific production item. Whether the material should be regarded as self-extinguishing or slow-burning will depend on the configuration of the unit and the ability of the unit to dissipate heat upon removal of the flame. For this reason, unsatisfactory flammability performance by laboratory test may often be disregarded in favor of actual performance on the production unit.

The method generally employed to convert normally slow-burning epoxy systems to self-extinguishing ones is through the incorporation of chlorine, bromine, or phosphorus into the cured structure. This can be accomplished by the use of chlorinated or brominated resins or curing agents, or by the incorporation of modifiers containing these atoms. Such cured systems, however, often exhibit poor thermal stability.

Specific Gravity

Specific gravity of the cured resin systems is determined by the water displacement method, as described in Federal Specification LP-4-6, Method 5011. Specific gravities for cured unfilled formulations generally fall in the range of 1.2 to 1.3.

Outgassing under Vacuum

Stability at high vacuums is a property of interest to the space technology. It is usually measured in terms of weight loss.

Low weight losses have been experienced with proprietary epoxy systems tested at 95 to 120°C and 5×10^{-6} to 5×10^{-7} torr for about 100 hours. Typical weight-loss values: anhydride curing agents 0.15 per cent, amine curing agents 2.2 per cent, fatty-polyamide curing agents 3.5 per cent [85].

In general, with epoxy-resin systems, stability in vacuum with simultaneous exposure to radiation and/or heat is greater than in air.

THERMAL AND RADIATION STABILITY

Differential Thermal Analysis (DTA)

Differential thermal analysis [33] can be used as a guide to thermal stability. In such analysis, the samples are heated by progressive advance of temperature, and the exothermic and endothermic reactions are recorded. In samples without curing agents, an exothermic reaction is detected in the range of 240 to 360°C. In freshly mixed samples, with curing agents, an exotherm at lower temperatures is encountered as cure progresses; this then levels off until about 300°C, at which time a second exothermic reaction occurs.

The exothermic reaction encountered in uncured systems appears to be somewhat independent of oxidation (Figs. 6-31 and 6-32), indicating that it may be accounted for at least partly by homopolymerization. The reduced exotherm in the cured systems (Figs. 6-33 and 6-34) perhaps indicates that some unreacted epoxy groups are present which are not eliminated, even with a thorough postcure (Fig. 6-35).

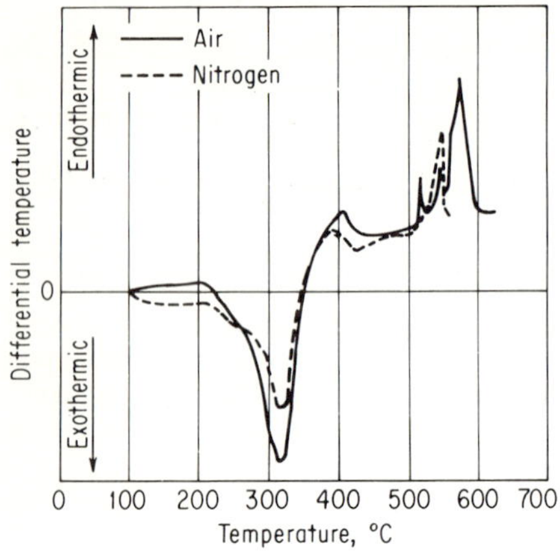

FIG. 6-31. Differential thermogram of DGEBA [75].

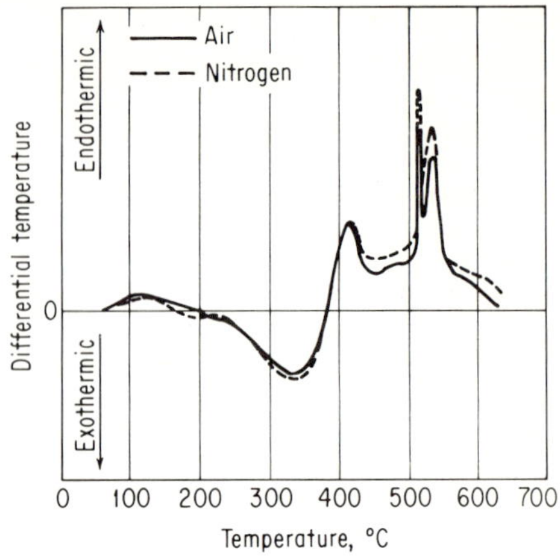

FIG. 6-32. Differential thermogram of DGEBA (mol. wt. 10,000) [75].

These exothermic peaks may be caused by homopolymerization or to some extent (in air) by isomerization of epoxy groups to aldehydes. The extent to which isomerization occurs, however, is a matter of some debate. It has been demonstrated that if it occurs at all, it does not occur by some form of free-radical attack, since DTA curves for systems containing a free-radical initiator, such as MEK peroxide, do not differ from those without the initiator [99]. An ionic mechanism remains a possibility.

Differential thermal analysis can also be used to determine the relative reactivity of various curing agents. For example, Fig. 6-36 indicates the comparative reactivity of maleic and the slower NMA. DTA appears to discriminate more between curing agents than resin types, and is less preferable than TGA for determining thermal stability.

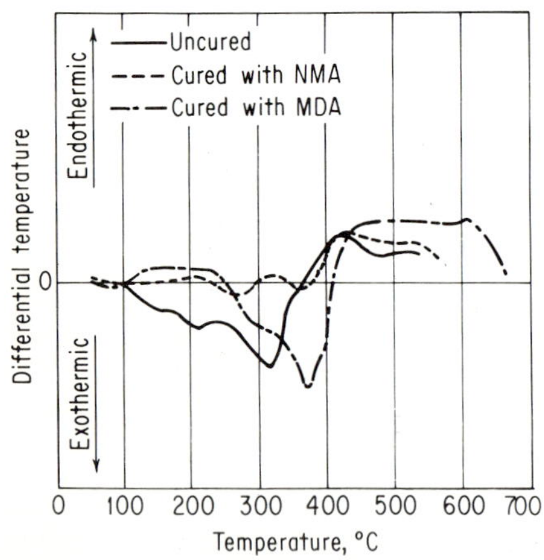

FIG. 6-33. Differential thermogram of cured and uncured epoxidized novolac resin [75].

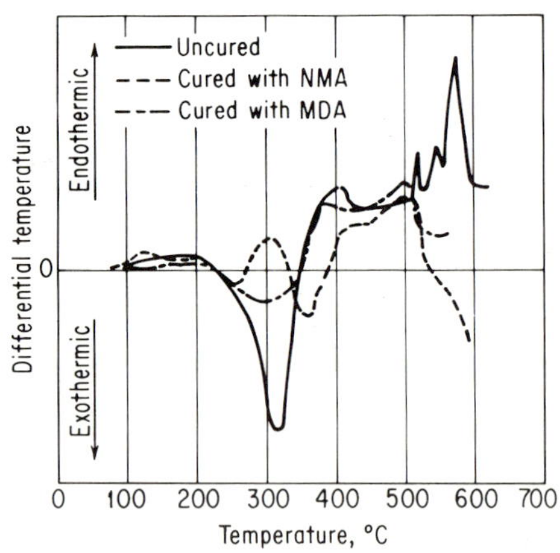

FIG. 6-34. Differential thermogram of cured and uncured DGEBA [75].

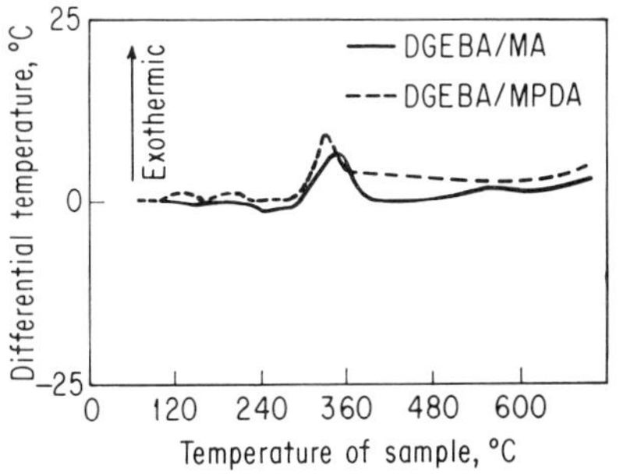

FIG. 6-35. Exothermic reaction at elevated temperature indicates incomplete cure, even with long cure time [33]. Cure cycle: 24 hours at 100°C + 4 hours at 150°C + 4 hours at 200°C. (*Courtesy of Analytical Chemistry.*)

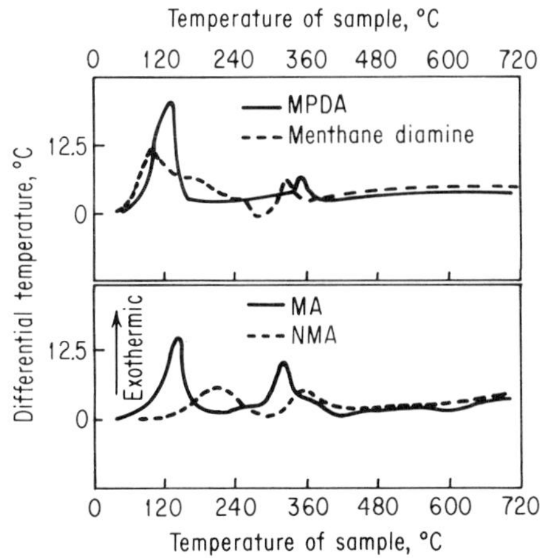

FIG. 6-36. Differential thermal analysis indicates relative reactivity of curing agents with DGEBA [33]. (*Courtesy of Analytical Chemistry.*)

Thermogravimetry

Perhaps the most convenient method for determining thermal stability is elevated-temperature weight loss.

Weight-loss values may be reported in terms of long-time aging at a constant temperature; or may be determined over a short period of time by means of a recorder that registers weight loss vs. temperature as the temperature is slowly

The effect of curing agent on heat stability of cured DGEBA is readily apparent from cured specimens exposed to 260°C for 2 hours. Curing agent from left to right was aliphatic polyamine, aromatic diamine, aromatic anhydride, and aromatic dianhydride. (*The Epoxylite Corporation.*)

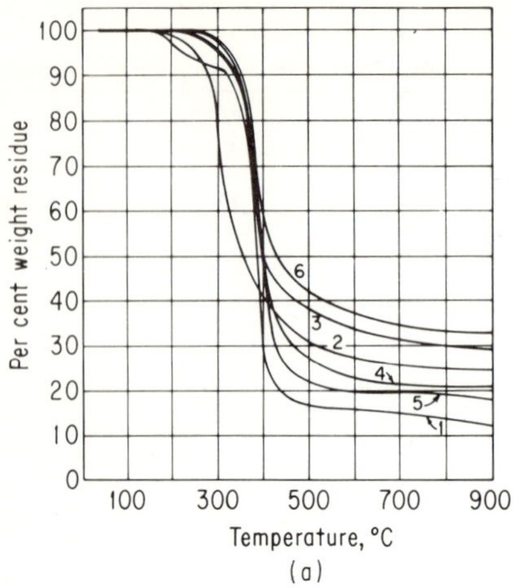

FIG. 6-37A. Weight residue of cured epoxy resin heated at a rate of 150°C per hour [53]. (1) DGEBA + PA (A/E = 1/1); (2) DGEBA + CA (A/E = 1/1); (3) tetraglycidyl ether of tetraphenylethane + 1 phr BF_3MEA; (4) DGEBA + PMDA (A/E = 0.5/1); (5) DGEBA + MA (A/E = 1/1); (6) tetraglycidyl ether of tetraphenylethane + MA (A/E = 1/1).

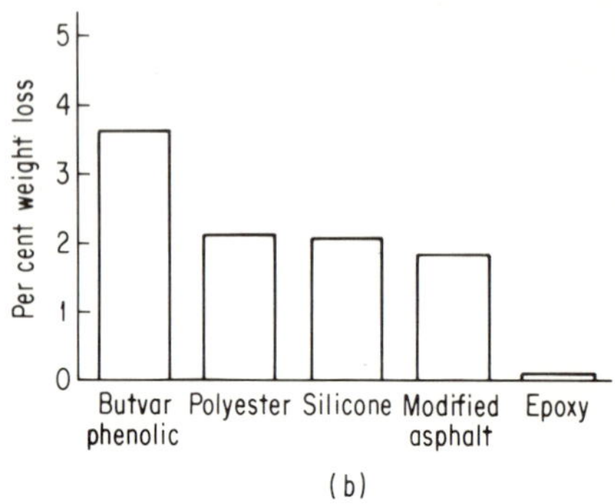

FIG. 6-37B. Comparison of cured DGEBA with various other polymers in terms of weight loss after 15 hours at 175°C [49].

increased. For example, a cured epoxy system which gives about 1.2 per cent weight loss after aging for 16 hours at 175°C, will give a total weight loss in air of less than 1 per cent at 300°C if the temperature is advanced at a constant rate of 150°C per hour [49]. Data obtained by the two methods are given in Figs. 6-37A and B. Elevated-temperature outgassing under vacuum has been investigated, and it was shown that, at least in the case investigated, this did not provide a useful guide to thermal stability, although it did develop useful design values, as shown in Fig. 6-38.

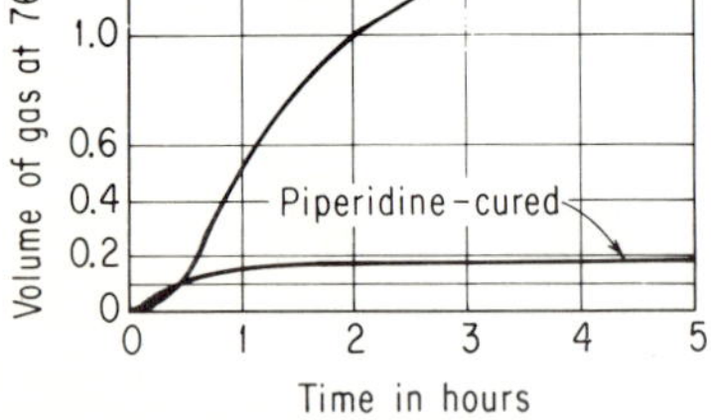

FIG. 6-38. Gas evolution at 150° C for two cured DGEBA specimens [49].

The efficiency of thermogravimetric analysis (TGA) has been compared with DTA for studying thermal stability and been found to provide more meaningful results. With DTA, endo- and exothermic reactions may be occurring simultaneously at any given temperature, and the results therefore cancel out on the recording. To indicate this, NMA (A/E:1/1) with 1 phr BDMA was used to cure dipentene dioxide and vinylcyclohexene dioxide, and TGA and DTA measurements were recorded at a heating rate of 5°C per minute. By DTA both systems showed equivalent thermal stability (peaks being at about 275°C), whereas by TGA vinylcyclohexene dioxide gave a maximum rate of weight loss at 375 vs. 265°C for dipentene dioxide [41].

It should be noted, however, that neither DTA nor TGA is satisfactory for determining the effect of heat on physical properties. In a series of tests in which DGE resorcinol was reacted with *o*-, *m*-, and *p*-phenylenediamine (tetrafunctional) and with aniline (difunctional), all polymers performed about the same on DTA, TGA, and pyrolysis studies, thus eliminating cross-linking as a factor in judging thermal stability by these methods [99]. This is consistent with data developed on new linear polymers, where it can be shown that the molecular weight of the polymer (above a lower limit) does not substantially alter the thermal stability observed by TGA [117].

Electrical Resistivity Tests

Electrical resistivity studies can be conducted at elevated temperature to determine relative thermal stability data [50]. The change in resistivity vs. temperature for various pyrolysis temperatures is indicated in Fig. 6-39 for a cured epoxidized novolac resin.

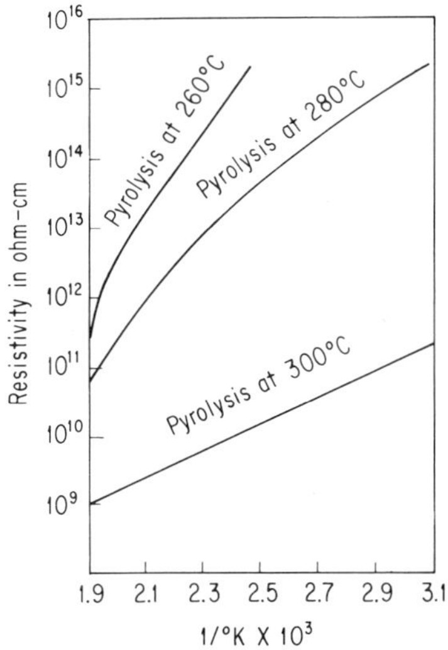

FIG. 6-39. Electrical resistivity of cured DGEBA after 15 hours' pyrolysis [50].

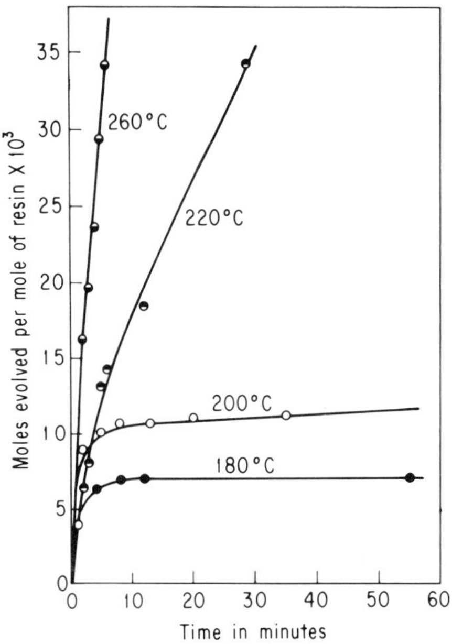

FIG. 6-40. Pyrolysis curves for 3,4-epoxy-6-methylcyclohexylmethyl-3,4-epoxy-6-methylcyclohexane carboxylate cured with MA [52].

Pyrolysis

A number of epoxy-resin systems have been investigated at pyrolysis temperatures and the degradation products studied. When the resin is cured with a primary amine, the unreacted epoxy is volatilized, and the higher-molecular-weight fractions are degraded to liberate H_2, CO, CH_4, C_2H_4, C_2H_6, C_3H_6, and C_3H_8, the latter four possibly from decomposition of polyamine radicals [62]. Typical of the decomposition products are those listed in Tables 6-7 to 6-11 for various temperatures and curing agents. Typical of gas evolution are data presented in Fig. 6-40.

Table 6-7. Analysis of Volatile Products from Pyrolysis of Commercial-grade 1,1,2,2-Tetrakis(2,3-Epoxypropoxyphenyl)-ethane (mol. wt. 700) cured with BF_3MEA [43]

Component	Weight, %		
	At 500°C	At 800°C	At 1200°C
Hydrogen	.....	0.8	2.1
Carbon monoxide	3.1	11.2	25.9
Carbon dioxide	6.0	3.7	1.8
Methane	0.8	1.6	4.3
Acetylene	.....	....	2.5
Ethylene	.....	3.0	3.0
Acetone	2.2		
Propane	1.1		
Propylene	2.3	2.2	
Ethane	.....	1.6	
Cyclopentadiene	.....	....	0.6
Pentane	0.5		
Benzene	1.3	2.8	8.1
Others (>0.5% each), total	0.7	0	0.8
Residue, nonvolatile at 23°C	82.0	73.1 (avg. mol. wt.) 352	50.9
Total	100.0	100.0	100.0
Volatilization, % of sample	75	86	87

Table 6-8. Volatile Products via Mass Spectra from Pyrolysis of Epoxy Resins Cured with Methylenedianiline [75]
(In mole per cent)

Component	Mass/charge ratio *m/e*	Epoxidized novolac		DGEBA	
		At 350°C	At 450°C	At 350°C	At 450°C
Hydrogen	2	.....	1.02	.....	1.12
Methane	16	.....	3.59	0.14	2.59
Water	18	97.70	17.00	96.70	15.40
Carbon monoxide	28	.....	25.40	.....	16.80
Ethylene	28	.....	0.77	0.13	1.58
Ethane (or HCHO)	30	.....	0.37		
—CH_2OH	31	.....	0.63	0.13	0.66
Allene (or methylacetylene)	40	.....	0.70	0.10	0.79
Propylene	42	.....	0.97	.....	0.76
Acetaldehyde	44	.....	4.85	.....	7.71
Carbon dioxide	44	1.70	13.90	0.79	9.95
Methyl chloride	50, 52	.....	6.19	.....	6.92
Acrolein	56	.....	0.11	.....	0.22
Acetone	58	.....	0.93	.....	0.81
Propionaldehyde	58	.....	0.41	.....	0.23
Vinyl chloride	62, 64	.....	0.21	.....	0.26
Ethyl chloride	64, 66	.....	0.21	.....	0.26
Cyclopentadiene	66	.....	0.79	0.03	1.26
Benzene	78	.....	0.33	.....	0.54
Methylcyclopentadiene	80	.....	21.00	1.44	31.50
Toluene	92	0.56	0.56	0.58	0.62

Table 6-9. Volatile Products via Mass Spectra from Pyrolysis of Epoxy Resins Cured with NMA [75]
(In mole per cent)

Component	Mass/charge ratio *m/e*	Epoxidized novolac		DGEBA	
		At 350°C	At 450°C	At 350°C	At 450°C
Hydrogen	2	6.06	1.30	4.21	0.55
Methane	16	0.59	1.11	1.12	1.63
Water	18	38.40	4.32	19.00	5.78
Carbon monoxide	28	8.89	7.48	5.16	6.38
Ethylene	28	0.85	1.22	1.11	1.24
Ethane (or HCHO)	30				
$—CH_2OH$	31	0.33	0.57	0.37	0.64
Allene (or methylacetylene)	40	0.91	0.52	0.79	0.69
Propylene	42	.....	0.22	.....	0.22
Acetaldehyde	44				
Carbon dioxide	44	21.90	40.80	23.40	30.50
Methyl chloride	50, 52	0.71	1.52	1.41	1.58
Acrolein	56	.....	0.09	.....	0.07
Acetone	58	.....	0.27	0.25	0.25
Propionaldehyde	58				
Vinyl chloride	62, 64	.....	0.19	0.26	0.33
Ethyl chloride	64, 66	.....	0.56	0.24	0.69
Cyclopentadiene	66	1.05	4.12	2.52	5.41
Benzene	78	0.23	0.49	0.50	0.54
Methylcyclopentadiene	80	19.10	35.00	38.90	43.20
Toluene	92	1.03	0.27	0.76	0.30

Table 6-10. High-boiling Products via Mass Spectra from Pyrolysis of Epoxidized Novolac at 475° under Vacuum [75]

Compound	Mass/charge ratio *m/e*	Cured with MDA	Cured with NMA
Cresols	108	Largest	Largest
Phenol	94	Major	Major
$(HOC_6H_4)_2CH_2$	200	Major	Major
C_2 phenols	122	Minor	Minor
$(HOC_6H_4)_2(CH_2)_2$	214	Minor	Minor
C_3H_5 phenol	134	Minor	Minor
C_3H_7 phenol	136	Minor	Minor
Water	18	Minor	Minor
C_3H_6O	58	Minor	Minor
Unidentified masses	...	Small amounts	Small amounts

Table 6-11. High-boiling Products via Mass Spectra from Pyrolysis of DGEBA at 475° under Vacuum [37]

Compound	Mass/charge ratio *m/e*	Cured with MDA	Cured with NMA
Phenol	94	Largest	Largest
Cresols	108	Major	Major
C_2 phenols	122	Major	Major
Isopropylphenol	136	Major	Major
Isopropenylphenol	134	Major	Major
Bisphenol A	228	Minor	Major
Water	18	Major	Minor
Unidentified masses	...	Small amounts	Small amounts

The presence of hydroxyl groups in amine-cured systems may be one of the causes for their relatively poor thermal stability, perhaps dehydration preceding cleavage. In addition, it is possible that amine hydrochloride is formed and decomposes to HCl which initiates cleavage of the bisphenol A structure. In

Table 6-12. Gases Evolved by Burning Cured Halogen-containing Epoxy Systems in Continuous Flame [82]

Product	Brominated DGEBA		Chlorinated DGEBA		DGEBA chlorendic anhydride
	NMA	MDA	NMA	MDA	
Per cent halogen	33.5	41.5	18	24	35.4
Acetylene	......	0.034	......	0.022	0.11
Ethylene	......	0.034	......	0.059	0.036
Ethane	0.14	0.124	0.21	0.11	0.016
Oxygen	......	0.17	......	0.092	0.29
Carbon dioxide	1.71	1.98	2.13	3.24	2.60
Propane	......	0.035	0.11		
Methyl chloride	......	0.007	......	0.008	
Butanes	......	......	0.020		
Acetone	0.13	0.33	......	0.018	
Cyclopentadiene	......	......	0.013		
Pentenes	......	......	0.013		
Pentanes	......	......	0.020		
Benzene	0.33	0.34	0.33		
Methylcyclopentadiene	0.04	......	0.12		
Hexanes	......	......	0.029		
Toluene	0.078	0.034	0.033		
Methyl bromide	......	0.047			
Heptanes	......	......	0.013		
Ethylbenzene	0.016	......	0.002		
Ethyl bromide	0.006	0.030			
Allyl bromide	0.086	0.020			
Methylene bromide	......	0.004			
Butyl bromide	0.016	0.11			
Hydrogen bromide	190.0	410.0			
Hydrogen chloride	......	......	190.0	500.0	590.0
Chlorine	......	......	3.1	3.8	6.5
Bromine	6.0	9.0			

Unit of quantity is 10^{-4} mole.

anhydride-cured systems, the diester can undergo beta elimination to release an olefin and carbon dioxide,

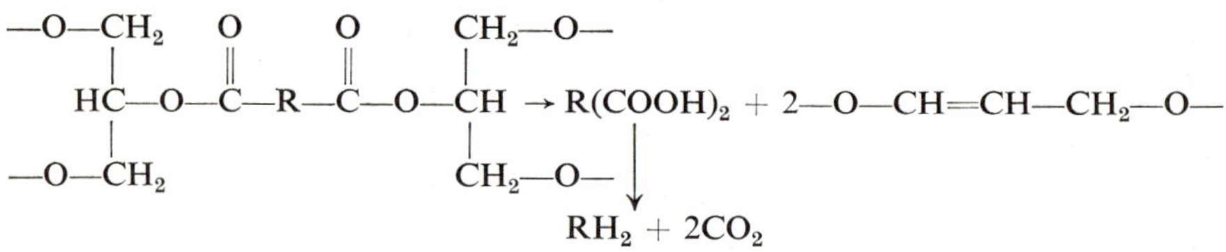

with other decomposition products preceding through cleavage at the ether links or following dehydration [75].

Pyrolysis studies of epoxy-resin systems indicate formation of enough gas to cause an explosive hazard if the gas is sufficiently confined. Thus, 6.3 grams of cured resin pyrolyzed in a 500 cu in. air volume would reach the lower explosive limit [116].

Of particular concern has been the nature of the pyrolysis products of cured epoxy resins containing halogens. Several systems were burned in a continuous flame and the gases evolved were analyzed (Table 6-12). It was found that the toxic fumes present were of the highly acid type, quite painful to the lungs, and

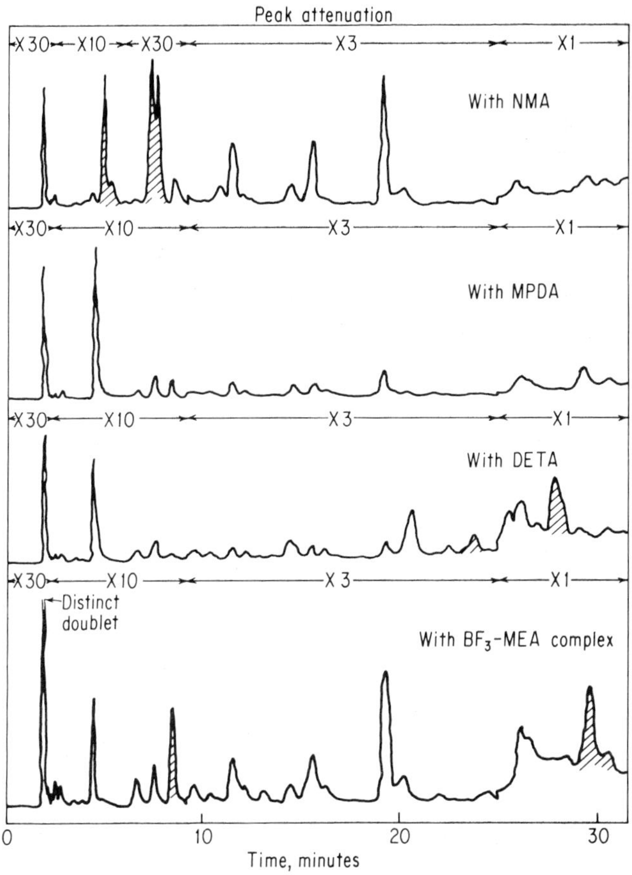

FIG. 6-41. Pyrolysis chromatograms of condensates of cured DGEBA [103]. Volatile products were swept with nitrogen carrier gas through a 15-ft separating column packed with 10 per cent polyethylene glycol supported on Chromosorb W. The column was program-heated at 5°C per minute from an initial temperature of 40°C. The components were monitored with a flame ionization detector.

therefore, providing adequate signals of danger. No insidious phosgene was developed [82].

Pyrolysis has been employed as a method for analyzing the composition of cured epoxy resins. The volatile gases are analyzed by means of gas chromatography to provide characteristic and distinctive curves, as shown in Fig. 6-41. The unique pyrolysis curves permit identification of curing agents through comparison with known curves. Further work to identify characteristic peaks could extend the usefulness of this method [103]. Preliminary studies which involve both amine- and anhydride-cured DGEBA resins indicate that the starting anhydride may be

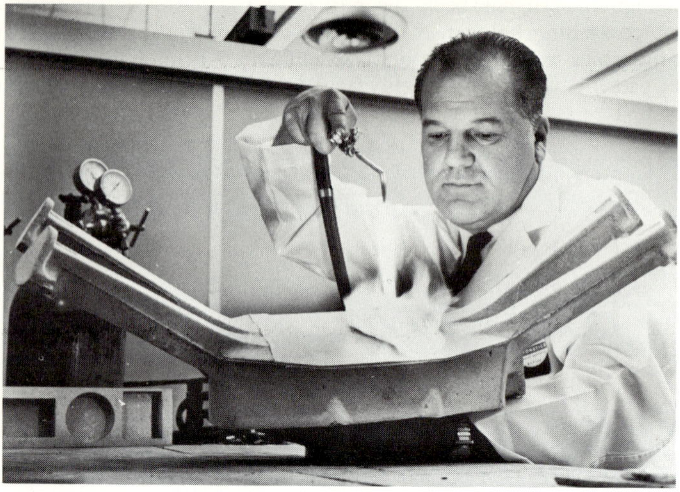

Testing of heat-shielding coating based on epoxy resins for electronic equipment exposed to rocket-engine exhaust gases. (*North American Aviation, Inc.*)

isolated and identified in the pyrolysis products. This appears to be true for the monoanhydrides, but with the dianhydride, PMDA, and with the amines, the curing agent could not be isolated [114].

Radiation Resistance

Radiation resistance is the measure of a material to withstand exposure to high-energy particles and is usually reported in terms of the per cent change in some physical property (such as flexural strength) after a given dosage of radiation.

Several types of radiation may be involved—fast neutrons from atomic piles; gamma radiation from spent fuel; and electron (beta) irradiation from linear accelerators. Additionally present in the piles and in the spent fuel are slow (or thermal) neutrons. Correlation of various radiation units is approximately as follows [59]:

$$\begin{aligned}1 \text{ roentgen} &= 10^{10}\text{–}10^{11} \text{ slow (or thermal) neutrons per sq cm}\\ &= 10^{8}\text{–}10^{9} \text{ fast neutrons per sq cm}\\ &= 2 \times 10^{9} \text{ 1-Mev photons per sq cm}\\ &= 5 \times 10^{7} \text{ 1-Mev electrons per sq cm}\end{aligned}$$

Tests conducted to determine the effects of the various types of radiation indicate [31]:

1. Thermal neutrons do not substantially contribute to degradation.
2. Fast neutrons, gamma rays, and accelerated electrons all provide comparable degradation, which is a function of the total dosage rather than dosage rate, when irradiation is conducted in a vacuum.
3. If the environment may react with the irradiated sample, the dosage rate becomes important. For example, electron radiation may produce free radicals at a greater rate than oxygen can diffuse into sample thereby causing cross-linking reactions to predominate. In the case of gamma and pile irradiations, free-radical production is sufficiently low to allow the diffusion of oxygen to influence the degradative process, causing chain shortening and inhibition of cross-linking.

Because of the interdependency of dosage rate and environment, aging studies in atmosphere under high dosage rates to obtain accelerated data may not adequately define the performance under actual service conditions.

During irradiation, both cross-linking and cleavage occur [32]. Epoxy systems containing carbon-nitrogen bonds appear to be particularly vulnerable to cleavage [10]. In general, the more heat-resistant the epoxy compound, the more radiation-resistant it will be [27]: with the most heat-resistant epoxies being among the more radiation-resistant of the plastics. Radiation-resistance data are presented in

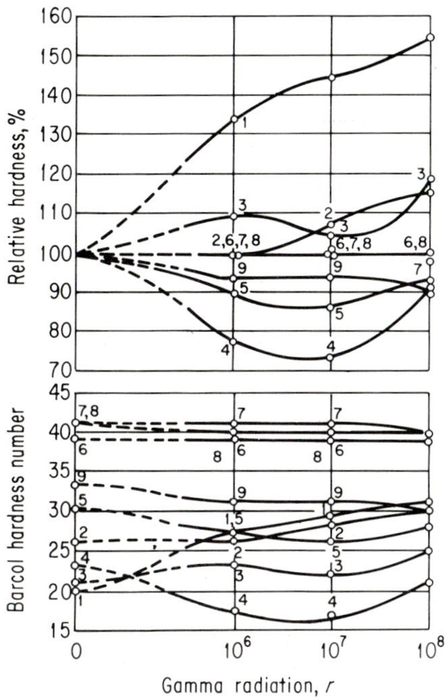

FIG. 6-42. Effects of gamma radiation on hardness of epoxy systems defined in Table 6-13 [7]. (*From Harper, "Electronic Packaging with Resins," McGraw-Hill,* 1961.)

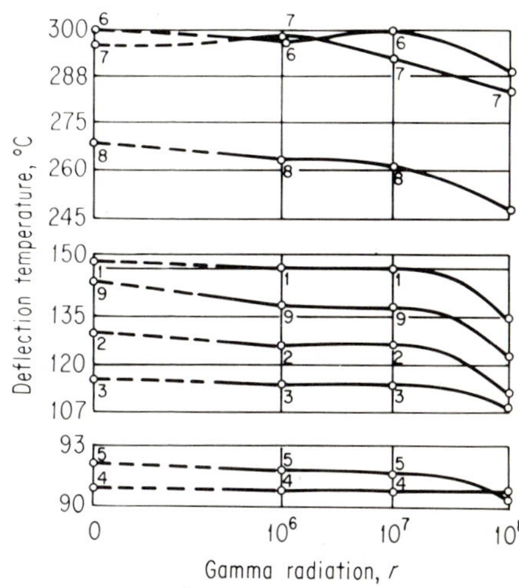

FIG. 6-43. Effects of gamma radiation on DT of epoxy systems defined in Table 6-13 [7]. (*From Harper, "Electronic Packaging with Resins," McGraw-Hill,* 1961.)

Figs. 6-42 to 6-44, with the resin systems employed being defined in Table 6-13. Relative radiation stability for various thermosetting plastics is shown in Fig. 6-45. The threshold degradation for a rigid, silica-filled epoxy system is at a total radiation dose of 5×10^{10} ergs per gram, with major deterioration occurring at 5×10^{11} ergs per gram. Epoxy groups in the uncured DGEBA have been shown to undergo polymerization when exposed to 775 megarads gamma radiation from a Co^{60} source [112].

Irradiated samples are degraded physically in a fashion similar to those exposed to high-temperature aging: surface discoloration, cracking, and blistering.

A rather unusual method for determining radiation resistance of the bulk material involves the use of microtransducers embedded in the bulk material. The transducers contain ferrite cores sensitive to hydrostatic pressures and stable to radiation up to 10^{16} roentgens. They may thus be used to follow static and dynamic changes in the bulk of the cured epoxy resin during irradiation [113].

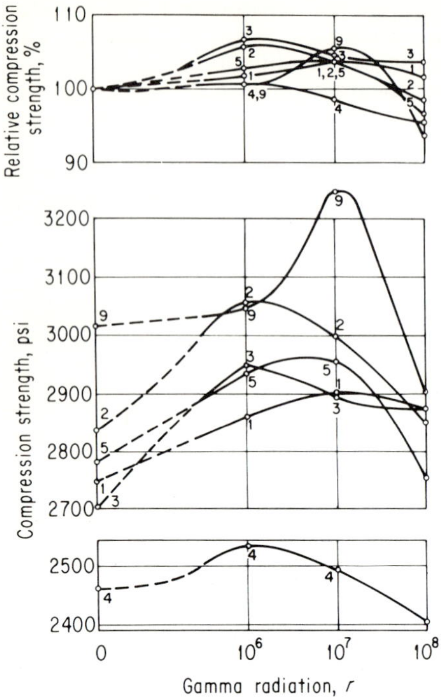

FIG. 6-44. Effects of gamma radiation on compressive strength of epoxy systems defined in Table 6-14 [7]. (*From Harper, "Electronic Packaging with Resins," McGraw-Hill*, 1961.)

With regard to chemical resistance after radiation, it has been found that the number of decontaminations an epoxy-resin coating can withstand without degradation is not affected by doses below 10^7 rads. Above this dosage, the useful life is reduced to two cycles [46].

Color Stability

Color stability is usually measured by exposure to sunlight, or exposure to artificial sunlight sources such as the Weatherometer or S-1 sunlamp (ASTM 620).

The color stability of the cured epoxy resins of the DGEBA type is poor. Clear, water-white systems have been developed from these, but, in the main, they discolor at about 100°C and reveal poor color stability when exposed to ultraviolet radiation. More recently several proprietary systems have been developed which reveal good UV resistance [90].

CHEMICAL PROPERTIES

The chemical resistance of the cured resins may be determined by exposing standard samples to various chemical atmospheres and recording changes in physical properties, such properties being determined before and after exposure by standard test methods. A method reporting the changes in weight, dimensions, and appearance after immersion in various chemical reagents is discussed in ASTM D 543–52T. Reference [69] discusses the use of electrical measurements to determine chemical resistance. Tests may be conducted to determine the resistance of the compound to accelerated weathering by exposing a sample to artificial

Table 6-13. Identification of DGEBA Formulations Employed in Radiation-resistance Studies [7]

Formulation no. and cure cycle, hr/°C	Diluent, phr		Curing agent, phr		
	Styrene oxide	AGE	MDA	PMDA	MPDA
1. 4.5/165	..	....	26.5		
2. 16/23 + 2/85 + 4.5/165	12	....	31.6		
3. 16/23 + 2/85 + 4.5/165	..	9.8	31.6		
4. 16/23 + 2/85 + 4.5/165	..	25.4	39.3		
5. 16/23 + 2/85 + 4.5/165	30	....	39.3		
6. 20/220	..	....	0.025	29.4	
7. 2/100 + 20/220	..	9.8	0.035	34.8	
8. 2/100 + 20/220	12	....	0.035	34.8	
9. 0.5/115 + 1/210	..	....	......	....	15

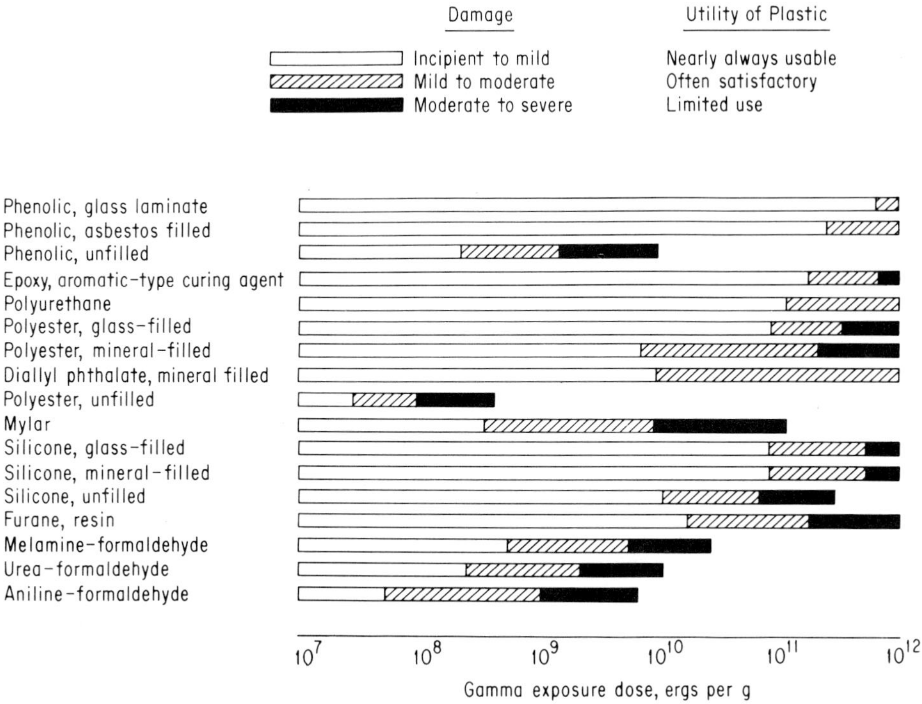

FIG. 6-45. Relative radiation resistance of thermosetting plastics [87].

sunlight and fog. Other tests, such as salt spray and oxidation, are based on the same principle.

Moisture Resistance

Most moisture-absorption data reported in the literature are based on 24-hour immersion at room temperature, per ASTM D-570. Under these test conditions, many epoxy-resin systems appear similar. Few data have been reported on such systems in accordance with the long-term immersion provisions of the specification at higher temperatures in order to establish differences of equilibrium values between the various systems. A summary of such long-time or elevated-temperature data as are available is presented in Figs. 6-46 to 6-57. A key to the formulations employed is given in Table 6-14. From these data it can be seen that there are considerable differences among curing agents with regard to the attainment of equilibrium conditions with cured DGEBA.

A study of the effect of moisture on physical properties indicates that the primary effect is as a lubricant. The changes are completely reversible on drying. The effect of moisture contents below 1 per cent on physical properties of cured epoxies is slight [105].

Resistance to Acids, Bases, and Solvents

The chemical resistance of cured epoxy resins varies markedly with the type of curing agent and the degree of cure. However, a well-cured epoxy resin is fairly resistant to most caustics, to most solvents, and to all but the strongest oxidizing acids. The influence of structure on chemical resistance was discussed previously, and typical chemical resistance data are given throughout the book.

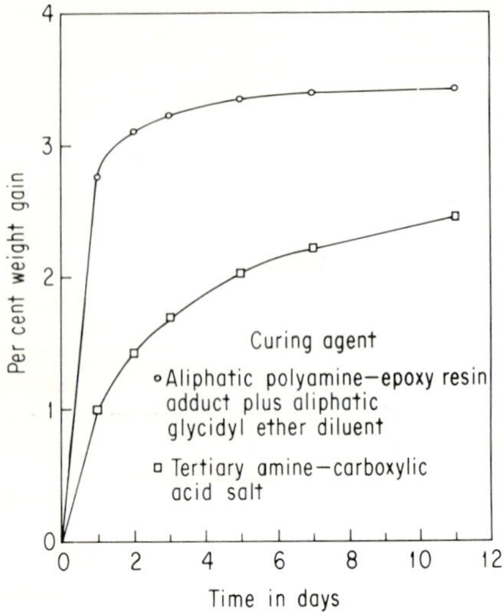

FIG. 6-46. Water absorption at 40°C of epoxy resin formulas *A* and *B* [1].

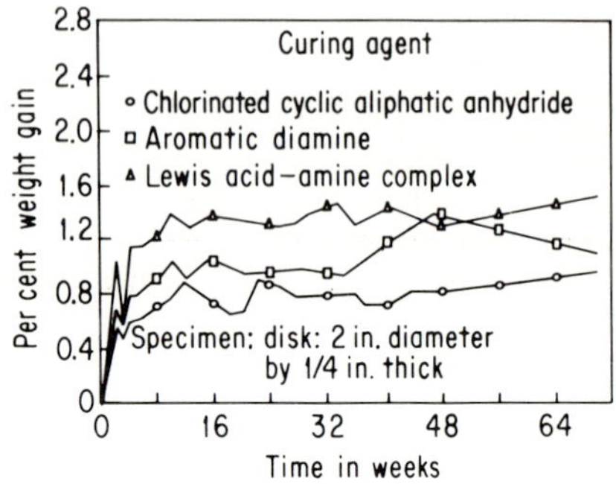

FIG. 6-47. Water absorption in distilled water at 72°C of epoxy-resin formulas *C*, *D*, and *E* [17].

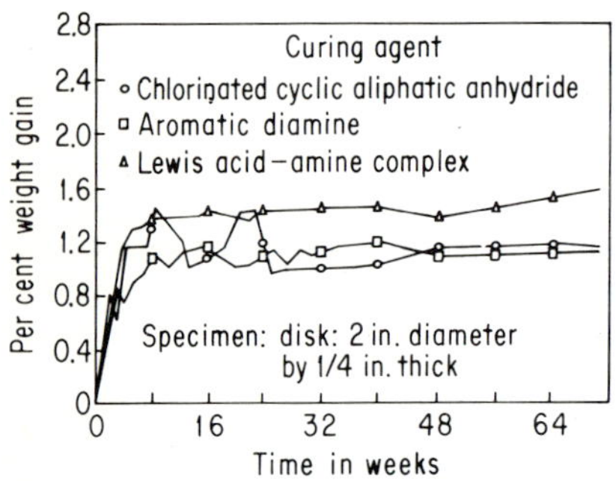

FIG. 6-48. Water absorption at pH 11 and 72°C of epoxy-resin formulas *C*, *D*, and *E* [17].

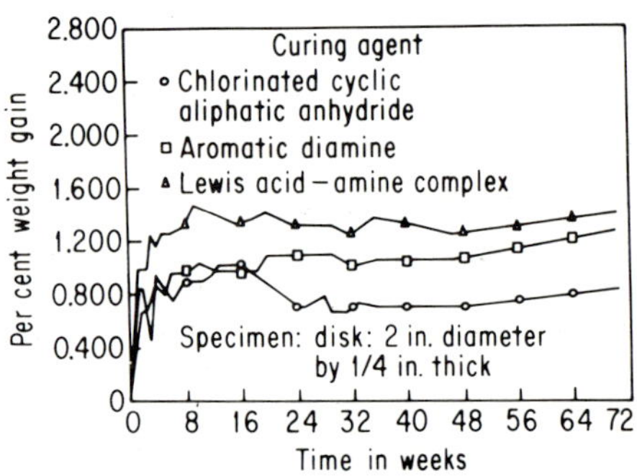

FIG. 6-49. Water absorption at pH 2 and 72°C of epoxy-resin formulas *C*, *D*, and *E* [17].

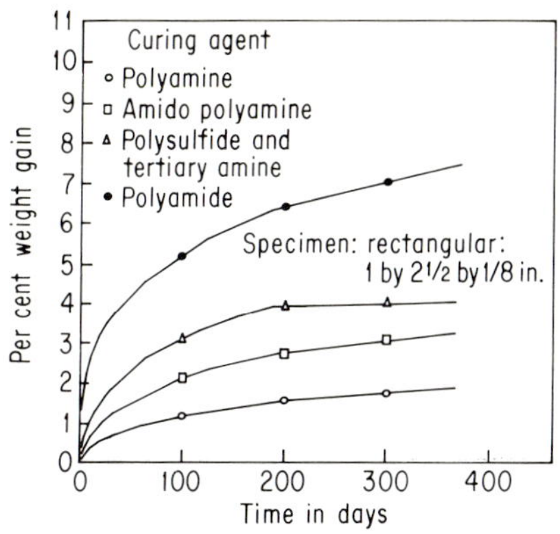

FIG. 6-50. Water absorption at 22°C of epoxy-resin formulas *G*, *H*, *I*, and *J* [84].

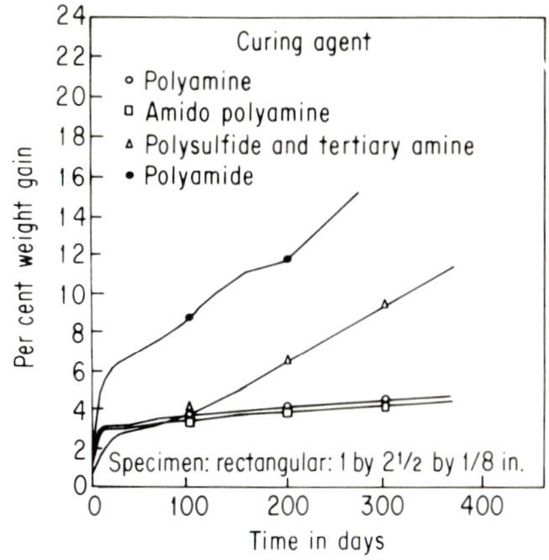

FIG. 6-51. Water absorption at 66°C of epoxy-resin formulas *G*, *H*, *I*, and *J* [84].

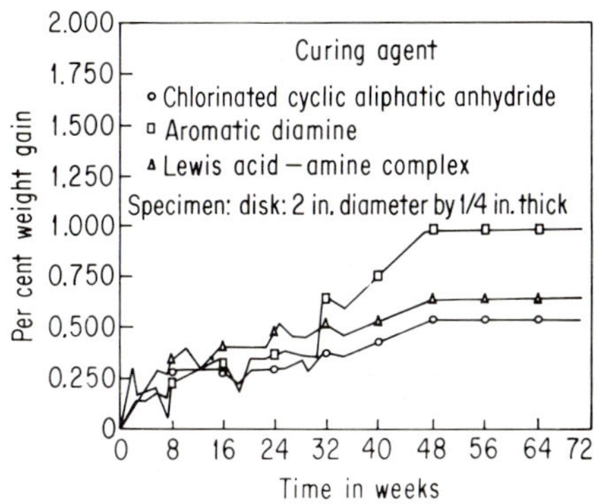

FIG. 6-52. Water absorption in moist air of epoxy-resin formulas *C*, *D*, and *E* [17].

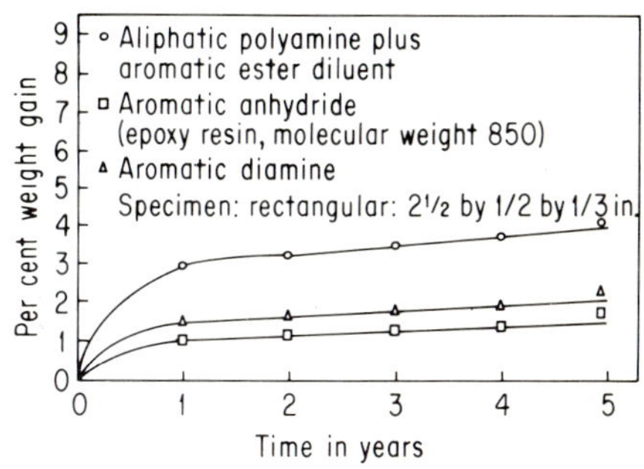

FIG. 6-53. Water absorption at 22°C of epoxy-resin formulas *K*, *L*, and *M* [89].

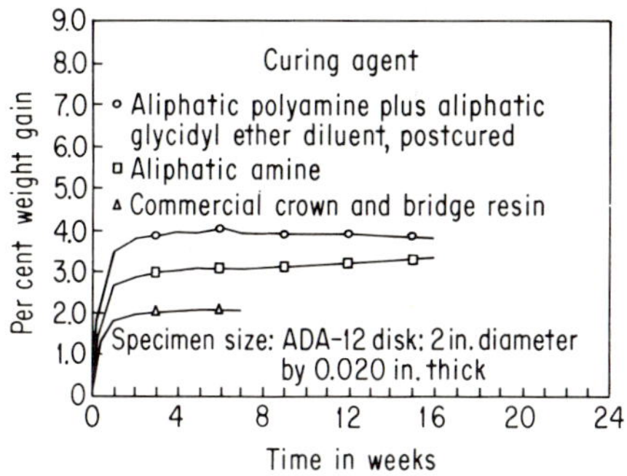

FIG. 6-54. Water absorption in distilled water at 37°C of epoxy-resin formulas *O*, *P*, and *Q* [74].

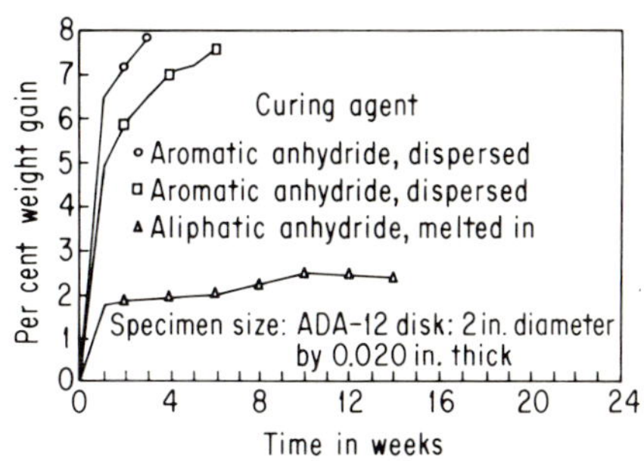

FIG. 6-55. Water absorption in distilled water at 37°C of epoxy-resin formulas *W*, *X*, and *Y* [74].

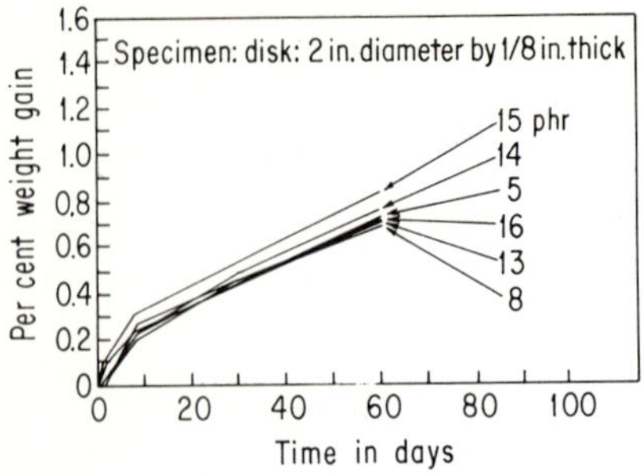

FIG. 6-56. Water absorption at 22°C of formula *N* at various curing-agent ratios [74].

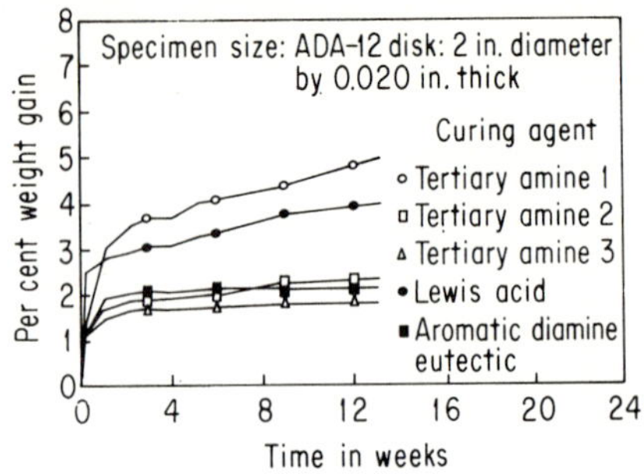

FIG. 6-57. Water absorption in distilled water at 37°C of epoxy-resin formulas *R*, *S*, *T*, *U*, and *V* [74].

Table 6-14. Test Conditions of Water-absorption Studies of Unfilled DGEBA Resins [74]

Reference	Specimen sizes	Temperature and time	Curing agent and additives
1	Rectangular: 1.18 × 0.78 × 1/16 in.	11 days at 40°C	*A*. Aliphatic glycidyl ether as diluent, aliphatic polyamine–epoxy-resin adduct as curing agent *B*. Tertiary amine–carboxylic acid salt as curing agent
17	Disk: 2 in. diam. × 1/4 in. thick	504 days (1⅓ years) at 72°C	*C*. Chlorinated cyclic aliphatic anhydride *D*. Aromatic diamine *E*. Lewis acid–amine complex
51	Dentures	540 days (18 months) at 22°C	*F*. Unsaturated aliphatic glycidyl ether as diluent, aliphatic polyamine as curing agent
84	Rectangular: 1 × 2½ × 1/8 in.	365 days (1 year) at 22 and 66°C	*G*. Polyamine *H*. Amidopolyamine *I*. Polysulfide and tertiary amine *J*. Fatty polyamide
89	Rectangular: 2½ × ½ × ⅓ in.	1,825 days (5 years) at 22°C	*K*. Aromatic ester as diluent, aliphatic polyamine as curing agent *L*. DGEBA (mol. wt. 850) aromatic anhydride *M*. Aromatic diamine
	Disk: 2 × 1/8 in.	60 days at 22°C	*N*. Aliphatic amine over range of stoichiometry
	Disk: 1.97 in. diam. × 0.020 in. thick	120 days at 37°C	*O*. Same as *F* but postcured *P*. Same as *N* at stoichiometric *Q*. Commercial crown and bridge resin *R*. Tertiary amine 1 *S*. Tertiary amine 2 *T*. Tertiary amine 3 *U*. Lewis acid *V*. Aromatic diamine eutectic *W*. Aromatic anhydride, dispersed *X*. Aromatic anhydride, dispersed *Y*. Aliphatic anhydride, melted in

Weathering Resistance

Weathering of epoxies is fairly good. Standard rigid cast formulations show about 10 per cent reduction in tensile and 20 per cent reduction in flexural strength after 3 years' exposure in various climates—surface erosion probably accounting for the majority of this reduction. Epoxy laminates are not greatly degraded mechanically by weathering. Color changes may range from slight to pronounced, and an orange-peel effect has been experienced with flexible systems [93].

Tests conducted to determine the effect of ultraviolet (in the range of 2,000 to 600 angstroms) at reduced pressures (10^{-6} torr) indicate only a very slight loss of strength on epoxy laminates after 500 hours' exposure, and weight losses of about 1 per cent [64]. Weathering resistance of several typical formulations are given in Figs. 6-58 to 6-60, with the formulations identified in Table 6-15.

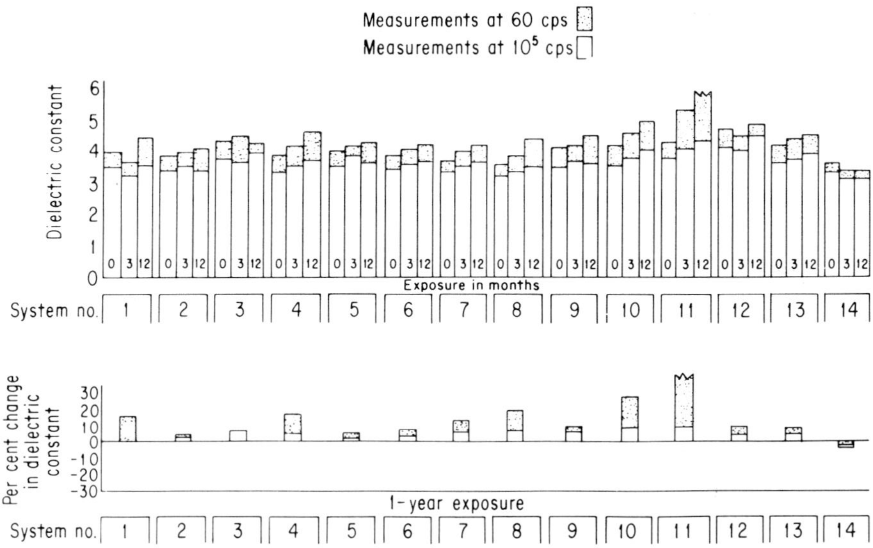

FIG. 6-58. Change in dielectric constant with weathering for epoxy-resin formulations given in Table 6-16 [47].

The minor degradation created by UV on castings and laminates is not generally of consequence. In thin films, however, more severe effects are detected. Weathering of epoxy paints, for example, is generally regarded only as fair, because of discoloration and chalking. Physical deterioration will occur which will result in reduced physical-strength properties of the films. The degradation products encountered with UV exposure of DGEBA (mol. wt. 1,000) films cured with a primary amine adduct and with a fatty polyamide are indicated in Table 6-16.

Permeability

The water-vapor permeability of cured epoxy-resin systems may be measured in accordance with the procedure in MIL-I-16923. Special permeability cups are employed. A typical value is < 0.01 g/(sq ft)(24 hr)(in.) thickness.

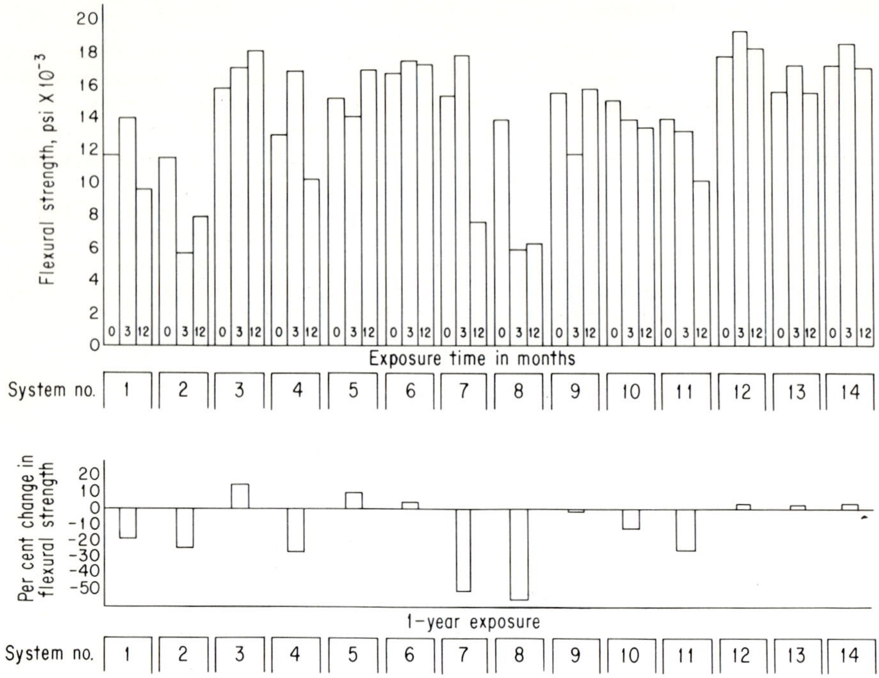

FIG. 6-59. Change in flexural strength with weathering for epoxy-resin formulations given in Table 6-16 [47].

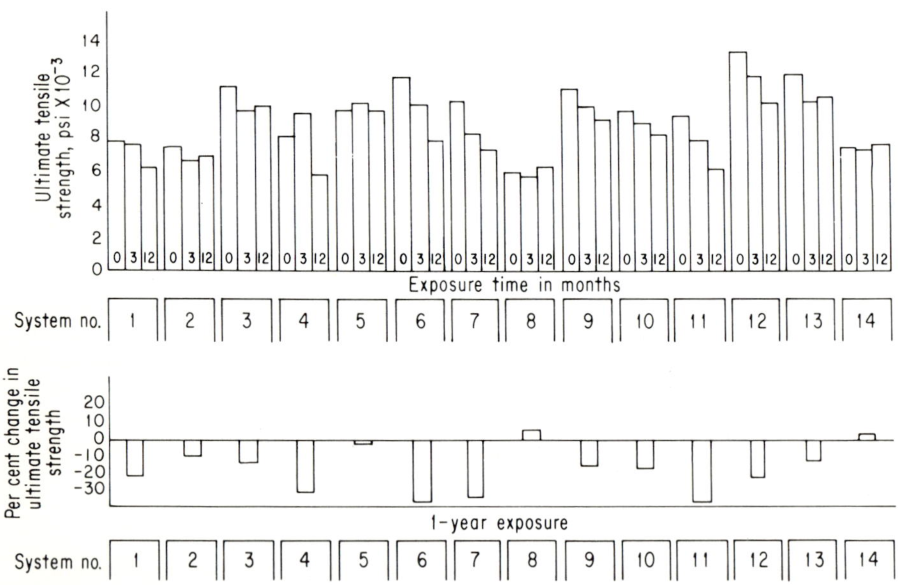

FIG. 6-60. Change in tensile strength with weathering for epoxy-resin formulations given in Table 6-16 [47].

Table 6-15. Formulations Used for Weathering Studies [47]

	DGEBA		Curing agent, phr of resin + additive—cure, hrs/°C					Additive, phr		
Formulation	Mol. wt. 350	Mol. wt. 380	TETA	Propylene oxide adduct of TETA	MPDA	MDA	NMA/DMP—30	Dibutyl phthalate	Butyl glycidyl ether	Butanediol diglycidyl ether
1	..	1	10—2/100	....	....	....	....	17		
2	..	1	10—160/23	....	....	....	....	17		
3	1	..	....	25—2/100						
4	1	..	....	25—168/23						
5	1	..	12.5—2/100							
6	1	..	13—8/100	....	....	....	....	..	7.2	
7	1	..	13.2—8/100	....	....	....	....	..	13.5	
8	1	..	13.4—8/100	....	....	....	....	..	21	
9	1	..	13.5—8/100	....	....	....	....	..	....	16
10	1	..	13.9—8/100	....	....	....	....	..	....	23
11	1	..	14.5—8/100	....	....	....	....	..	....	55
12	1	..	....	....	13.5—2/85 + 4/150					
13	1	..	....	....	....	26—2/85 + 4/150				
14	1	..	....	....	....	....	80/2—3/150 + 4/260			

Table 6-16. Volatile Products from Epoxy Films [107]

Absorption peaks, cm^{-1}	Probable assignments	% absorption after 6-hr irradiation	
		Epoxy-amine	Epoxy–fatty polyamide
721	Carbon dioxide	2.0	2.4
730	Acetylene	4.0	2.4
912	Vinyl group	13.2	3.4
992	Vinyl group	2.2	0.6
1,034	Methanol	5.2	3.6
1,140		2	4
1,217	Acetone	4.6	2.4
1,370	Carbonyl	3	2
1,740	Carbonyl	8	4
2,120	Carbon monoxide	6.2	6.0
2,160	Carbon monoxide	7.4	7.0
2,360	Carbon dioxide	51	54
2,960	C—H stretching	8	6

Little information regarding gas permeability for epoxy resins has been reported in the literature. A test method for determining this is described in [60], where it is indicated that permeability of epoxies is highly dependent on composition, values ranging from 0.0086 to 0.14 for carbon dioxide permeability, and from 0.0049 to 0.16 for oxygen permeability. Hydrogen diffuses slightly faster than its heavier isotopes (by a factor of from 1.2 to 1.7) and helium permeates more readily than hydrogen [70].

Corrosive Effects

The reaction of the cured epoxy system on metallic environments has been investigated [11]. Results varied with the epoxy system and the type of metal. The corrosivity was caused by direct chemical attack, which is accelerated by electric currents. Corrosion is retarded if insoluble products are formed on the surface, and, conversely, conditions tending toward removing such products accelerate the corrosion rate. The presence or absence of water is significant. Also of importance are the concentration and conductances of ions existing within the epoxy, permeability of the epoxy to water and electrolytes, and the type and concentrations of filler present. Specifically in relation to copper, epoxy systems range from nil to bad in terms of corrosivity [92]. Epoxy decomposition products have been shown to cause severe corrosion of copper—the corrosion more pronounced in the proximate rather than in the contact area [104].

Stability with Explosives

The reaction of the cured epoxies in terms of stability of explosives in contact with them has been investigated [12], the conclusion being that stability is dependent upon choice of curing agent.

Fungus Resistance

A variety of cured epoxy-resin compounds have been exposed to standard fungus-resistance tests, and all have been shown to be nonnutrient to fungus. Some modifiers, particularly if of a vegetable origin, could cause the resin to be nutrient.

Cured epoxy-resin compounds have been found not to be attacked by marine bacteria when incubated at either 5 or 20°C [48].

ELECTRICAL PROPERTIES

The chief electrical properties reported for the cured resin system are dielectric constant, power (or dissipation) factor, and loss factor (ASTM D 150), resistivity (ASTM 257–61), dielectric strength (ASTM D 149), and arc resistance (ASTM D 495–48T). A review of the various standard electrical tests is given in [63].

It has been suggested that a capacitor may conveniently be used as a test piece, the plates serving as electrodes. The capacitor is encapsulated, and dielectric strength and volume resistivity are calculated from capacitance and resistance readings. The effect of temperature and accelerated aging on the properties may conveniently be determined; and finally, since materials of differing thermal expansion rates are involved, the capacitor may be used as a thermal shock specimen [22].

Since electrical properties are sharply influenced by temperature, frequency, and relative humidity, values are often reported for several conditions. Typical values as given below indicate the values to be expected from the cured DGEBA tested at 10^3 cps and 25°C.

Dielectric Constant

Dielectric constant is the ratio of the capacitance of a given dielectric material to the capacitance of a vacuum in the same configuration. For comparison, the dielectric constant of air is taken as approximately 1. A typical value for a cured epoxy resin is 3.8.

Power Factor

Power factor is a measure of the electric power loss in a dielectric. It is, in insulations, close in value to the dissipation factor. (The relationship $\cos\theta \sim \sin\delta \sim \tan\delta \sim \delta$ is approximately true, where θ is the alternating-voltage phase angle and δ is its complement.) Typical values of power factor or dissipation factor are 0.003 to 0.030.

Loss Factor

Loss factor is a product of the power factor and the dielectric constant and represents a quantitative measure of power loss. Typical values are 0.01 and 0.08.

Dielectric Strength

Dielectric strength is a measure of the ability to withstand voltage. It is calculated from the voltage required to break down an insulation section of given thickness and is expressed in volts per mil. The value will be highly dependent on the thickness of the sample—the thinner the specimen, the higher the calculated dielectric strength per unit thickness. Typical values are 425 volts per mil (0.125 in. thickness), and 2,000 volts per mil (0.001 in. thickness).

Resistivity

Resistivity is the ability of a material to resist passage of an electric current. Resistivity may be recorded in terms of volume resistivity (ohm-centimeters) or in terms of surface resistivity (ohms). When reported as electrical resistance, it is a function of both values. A typical value of volume resistivity is 10^{13} ohm-cm.

Volume resistivity test (*Ciba Products Co.*)

Arc Resistance

Arc resistance is the ability of an insulating material to withstand bridging by high voltage and low current close to the surface. The breakdown value is reported in seconds and indicates that a path has been formed across the insulation surface. A typical value of an unmodified cured resin is 60 seconds.

Corona Resistance

Corona resistance (ionization resistance, brush or slot discharge resistance) is the time during which an insulation and electrode system will withstand the action of a specified level-field intensified ionization that does not result in the immediate complete breakdown of the insulation. Corona resistance is a complex phenomenon and is dependent on the total electric system. It is usually reported in terms of minutes or hours of resistance. Reported values for epoxies vary widely.

SUMMARY

This chapter has been designed to familiarize the reader with the general properties by which the resin may be characterized.

The epoxy resin cures, generally, with the liberation of heat. Systems which cure at room temperature liberate substantial quantities of heat within a short period of time, resulting in a rise in temperature of the reaction mass, thereby limiting its size. Systems which require a heat cure generally liberate heat over a substantially greater period of time, thereby permitting the casting of larger masses.

In general, the rigid epoxy resins—those with DTs higher than room temperature—offer very similar mechanical and electrical properties when tested below the DT. The significant difference is the heat resistance of the systems: many drop off sharply in properties at about 100°C; whereas others may provide acceptable properties to 150 or even 200°C.

Chemical resistance, however, is more dependent upon composition, with, in general, the more heat-resistant systems providing improved chemical resistance over the less heat-resistant systems.

REFERENCES

1955

1. Formo and Bolstad, Where and How to Use Epoxies, *Mod. Plastics*, July.
2. Polymer Progress no. 2, Shell Chemical Corp., October.

1956

3. Damusis, Amines as Curing Agents for Ether Resins, ACS Symposium, Atlantic City, September.
4. Lang, A Quick Thermal Conductivity Test on Insulating Materials, *ASTM Bull.*, September.
5. Partansky and Schrader, Some Thermal Properties of Epoxylated Novolac and Bisphenol A Type Casting Resins, ACS Symposium, Atlantic City, September.
6. Peerman et al., A Study of the Reaction of Polyamide Resins, ACS Symposium, Atlantic City, September.

1957

7. Colichman and Strong, Effect of Gamma Radiation on Epoxy Plastics, *Mod. Plastics* October.
8. Delmonte, Relationships between Electrical and Mechanical Properties of Epoxy Plastics, *ASTM Bull.*, September.
9. Lascoe, Plastic Tooling Shrinkage Test Method, ASTE Research Report 6, August.
10. Mixer, Nuclear Deterioration of Structural Adhesives, ACS Symposium, New York City, September.
11. Perry, et al., The Corrosivities of Plastics and Rubbers, 13th ANTEC of SPE.

1958

12. Anderson, The Effect of Different Amine Cured Epoxy Resins on the Stability of Explosives, ACS Symposium, Chicago, September.
13. Aukward, et al., Change in Electrical Resistivity of Some High Polymers during Isotherm Polymerization, *J. Polymer Sci.*, January.
14. Belanger and Schulte, The Effect of Chlorine Contents on Electrical and Physical Properties of Cured Epoxy Resins, Epoxy Resin Symposium of SPE, Minneapolis, Minn., October.
15. Delmonte, Electrical Properties of Epoxy Resins during Polymerization, Epoxy Resin Symposium of SPE, Minneapolis, Minn., October.
16. Klopfenstein and Lee, Some Thermal Aspects of Epoxy Resins, *Insulation*, February.
17. Murray and Stein, Evaluating Epoxy Resin Degradation Systems, *Plastics Technol.*, February.
18. Warfield, Test Methods for the Study of the Properties of Casting Resins, NAVORD Report 6029.
19. Warfield, et al., An Adiabatic Calorimeter for the Determination of the Specific Heats of Casting Resins, NAVORD Report 6094.
20. Warfield and Petree, The Temperature Dependence and Activation Energy of Electrical Conduction in High Polymers, NAVORD Report 6246.
21. Weiss, Optimum Conditions for Curing Methyl Nadic Anhydride–Epoxy Casting Resins, Epoxy Resin Symposium of the SPE, Minneapolis, Minn., October.

1959

22. Christensen, Evaluating Encapsulating Materials, *Elec. Equipment Eng.*, September.
23. Dannenberg, Refractive Index Method for Determining Cure Rates of Epoxy Resins, *SPE J.* **15**:875–880 (October).
24. Ehlers, A Modified Vicat Type Apparatus for Measuring the Softening Point of Thermosetting Plastics and Laminates, *ASTM Bull.*, February.
25. Kaelble, The Dynamic and Tensile Properties of Epoxy Resins, ACS Symposium, Atlantic City, September.

26. Klute and Viehamann, Heat of Polymerization of Phenyl Glycidyl Ether and of an Epoxy Resin, Diamond Ordnance Fuse Laboratories, DA 506-01-010.
27. Lyons and Leininger, Radiation Effects on Insulation, Wire and Cable, *Insulation*, May.
28. Warfield, A New Method for Determining the Second Transition in Polymers, *SPE J.*, August.
29. Warfield, et al., The Specific Heat of Thermosetting Polymers, NAVORD Report 6255.
30. Warfield, et al., The Specific Heat of High Polymers, *SPE J.*, December.

1960

31. Aitken, et al., Irradiation Studies on Anhydride Cured Epoxy Resins, AERE R-3339 (Engineering Division, Harwell, Berkshire, England).
32. Aitken and Ralph, Some Effects of Radiation in Cast Epoxide Resin Systems, AERE R-3085 (Engineering Division, Harwell, Berkshire, England).
33. Anderson, Differential Thermal Analysis of Polyepoxides, *Anal. Chem.* **32**:1592–1595.
34. Bolson, Optimum Curing Conditions for Solid Epoxy Resins Determined by Statistical Evaluation, ACS Symposium, New York City.
35. Erich and Bodnar, Effect of Molecular Structure on Mechanical Properties of Epoxy Resins, *J. Appl. Polymer Sci.* vol. 3, no. 9.
36. Helmreich and Harry, Two Flexible Resins, ACS Symposium, New York City.
37. Hollinger and Doyle, Stress Response of Glass-filled Resins above Their Glass Transition Temperatures, SPI Symposium, Chicago, February.
38. Kline, Dynamic Mechanical Properties of Polymerized Epoxy Resins, *J. Polymer Sci.* **47**:237.
39. Plueddemann, Epoxy Siloxane Casting Resins: a Study in Geometry and Functionality, *J. Chem. Eng. Data*, January.
40. Wynstra, Some Correlations between Mechanical Properties and Structure of Epoxy Resin Castings, Union Carbide Plastics Co. Report.

1961

41. Anderson, Differential Thermogravimetry vs. Differential Thermometry in Studying the Pyrolysis of Polymers, *Nature*, Sept. 9, pp. 1088–1089.
42. Hodge, Hardness of Epoxy Resins by the H-H Method, AD 259 420.
43. Madorsky and Straus, Stability of Thermoset Plastics at High Temperature, *Mod. Plastics*, February.
44. Nelson, et al., Casting to Close Tolerances with Unfilled Epoxy Resins, *SPE J.*, March.
45. Norris and Fitzroy, A Quick Thermal Conductivity Test on Insulating Materials, *Mater. Res. Std.*, September.
46. Otto, Decontamination Capacity of Epoxy and Vinyl Resin Coatings Contaminated by Radioactive Materials, *Kernenergie* **4**(10–11):801–807.
47. Pschorr, Weathering of Epoxy Resins Systems, 17th ANTEC of SPE.
48. Steinberg, The Resistance of Organic Materials to Attack by Marine Bacteria at Low Temperatures, *Bell System Tech. J.* **40**(5):1369–1395.
49. Torossion and Jones, A Study of Polymers Applicable to Insulation Systems, 17th ANTEC of SPE.
50. Warfield, Electro-thermal Analysis of Thermosetting Polymers, 17th ANTEC of SPE.
51. Woefel, et al., Changes in Dentures during Storage in Water and in Service, *J. Am. Dental Assoc.*, June.

1962

52. Barenholtz and Rutzler, Correlation of Some Thermal Properties of Epoxy Adhesives and Shear Strength of Steel Laminates, ACS Symposium, Washington, D.C., March.

53. Ehlers, Thermogravimetric Analysis of Polymers, TD Report ASD-TR 61-622.
54. Feltner, On the Mechanical Behavior and Fracture Morphology of Epoxy Resins, AD 287 159.
55. Feuchtbaum, Electronic Insulation: a Critical Factor in High Reliability Design, *Insulation*, September.
56. Kakurai and Noguchi, Studies on the Gelation of Epoxy Resins with Amines by Viscosity Determination, *J. High Polymer* (*Japan*) **19**:209.
57. Katz and Tabolsky, Rubber Elasticity in a Highly Crosslinked Epoxy System, AD 280 279.
58. Kelly, et al., The Relationship of Molecular Structure and Resin Performance for a Series of Diglycidyl Ethers, *J. Polymer Sci.*, vol. 6, no. 22.
59. Linnenbom, The Effects of Radiation on Materials, *Insulation*, January.
60. Major and Kammermeyer, Gas Permeability of Plastics, *Mod. Plastics*, July.
61. May and Weir, Dynamic Mechanical Properties of Epoxy Resins, *SPE Trans.* **2**:207–212, (July).
62. Neiman, et al., The Thermal Degradation of Some Epoxy Resins, *J. Polymer Sci.* **56**:383–389.
63. Scott, Standard Tests for Electrical Properties, *SPE J.*, September.
64. Wahl, Effects of Ultraviolet and Vacuum on Properties of Plastics, Cornell Aeronautical Laboratories, Inc., Buffalo 21, N.Y.
65. Ward and Bobalek, Some Correlations between Thermal, Electrical, and Mechanical Properties of Alkyd-Epoxy Copolymer Adhesives, ACS Symposium, Atlantic City, September.
66. Young, An Examination of Epoxy Systems Useful in Packaging "High G" Radio Telemeters, AD 273 681.

1963

67. Asanuma and Noguchi, Gelation in Epoxy Resin–Amine Curing Systems, *Kobunshi Kagaku* **20**(213):27–34.
68. Dannenberg, Determination of Functional Groups in Epoxy Resins by Near-infrared Spectroscopy, *SPE Trans.* **3**:78–88 (January).
69. Delmonte and O'Neal, Evaluating Chemical Resistance by Dielectric Property Changes, *Plastics World*, May.
70. Evans, et al., Permeation and Diffusion of Hydrogen Isotopes and Helium through Polymeric Materials, AWRE 0–45/63, (Aldermaston, Berkshire, England).
71. Harrod, Hydrogen Bonding in Amine Epoxide Adducts, *J. Polymer Sci.* (A) **1**:385.
72. Kakurai and Noguchi, Gelation in the Reaction of Epoxy Resins with Amines, *Kobunshi Kagaku* **20**(213):17–21.
73. Kwei, Swelling of Highly Crosslinked Network Structure, *J. Polymer Sci.* (A) **1**:2977.
74. Lee, Attaining Equilibrium in Water Absorption Tests of Epoxy Casting Resins, *Mater. Res. Std.*, vol. 3, no. 11.
75. Lee, Thermal Stability and Degradation Schemes of Epoxy Resins, Battelle Symposium on Thermal Stability of Polymers, Columbus, Ohio, December.
76. McGarry and Selfridge, Fracture Surface Work of Modified Polystyrenes and Certain Crosslinked Polymers, Department of Civil Engineering Research Report R-63-46, MIT, September.
77. May, The Influences of Bisphenol Structure on the Bond Strength of Epoxy Adhesives, TD Report ASD-TDR-63-396, April.
78. Shigayama and Kodama, Effects of Crosslink Density and Swelling upon the Properties of Epoxy Resins, *Kobunshi Kagaku* **20**(222):587–590.
79. Steele, Internal Stresses Developed in an Epoxy Resin Potting Compound during Long Term Storage, AD 412 514.
80. Tanaka and Kakiuchi, Gelation Point of Epoxy Resin Anhydride Systems, *J. Appl. Polymer Sci.* **7**:1951.
81. Tanaka and Kakiuchi, Study of Epoxy Compounds, part I, *J. Appl. Polymer Sci.* **7**:1063–1081.
82. Thrune, Gases Released When Fire Resistant Resins Are Burned, ACS Symposium, Los Angeles, April.

83. Ward and Bobalek, Some Correlations among Thermal, Electrical and Mechanical Properties of Alkyd Epoxy Copolymer Adhesives, *Ind. Eng. Chem. Prod. R & D*, June.
84. Wasser, Jones-Dabney Co., private communication.

1964

85. Anderson, Adhesives for the Installation of Solar Cells on Photovoltaic Concentrating Panels, SPE Conference, Garden City, N.Y., May.
86. Bremmer, Heat Stability of Brominated Epoxy Resins, *Ind. Eng. Chem. Prod. R&D.*, March.
87. Broadway and Palinchak, The Effect of Nuclear Radiation on Elastomeric and Plastic Components and Materials, AD 454 056.
88. Conley, Mechanism of Thermal and Oxidative Degradation of Condensation Polymers, SPE Regional Technical Conference, Baltimore, June 4.
89. Dorman, Ciba Products Co., private communication.
90. Lee and Watson, Ultraviolet Resistance of Epoxy Resins, 20th ANTEC of SPE.
91. McSheehy, Quality Control of the Cured State of Encapsulated Electronic Modules, *Plastics Technol.*, November.
92. Rider, Bonded Metal Assembly, *Prod. Eng.* May 25.
93. Rugger, Weathering Resistance of Plastics, *Mat. Design Eng.*, January.
94. Space Environmental Effects on Materials and Components, vol. 1, appendices A–F, RSIC-150, Redstone Arsenal.
95. Tanaka and Kakiuchi, Study of Epoxy Compounds, part VI, *J. Polymer Sci.* (A)**2**:3405–3440.
96. Tanner, The Feasibility of Using Chlorine-containing Epoxy Resins in Adhesive Formulations, TM 1396, Feltman Research Laboratory, Picatinny Arsenal.
97. Warfield, The Compressibility of Bulk Polymers, AD 445 119.
98. Yano, et al., Dielectric Properties of Epoxy Resins Cured by Amines, *Deni Kaguka* **32**(3):220–226.

1965

99. Anderson, Synthesis and Pyrolysis of Model Epoxide Compounds and Polymers, AD 610 390.
100. Burhans, et al., Designing Epoxy Resins Specifically for Filament-wound Deep Submergence Applications, RIP Division of SPI, Chicago, February.
101. Christie, The Effects of Resin Parameters on the Properties of Cured Epoxy Resins, RIP Division of SPI, Chicago, February.
102. Dannenberg, Determination of Stresses in Cured Epoxy Resins *SPE J.*, July.
103. Eggertsen, et al., Pyrolysis-Gas Chromatographic Analysis of Epoxy Resins, Shell Development Co., Emeryville, Calif.
104. Garland, Effect of Decomposition Products from Electrical Insulation on Metal and Metal Finishes, 6th Electrical Insulation Conference (IEEE and NEMA), September.
105. Glowacki and MacDonald, Protective Coatings for Filament-wound Deep Submergence Vessels, AD 460 364.
106. Gul and Maizel, Effect of the Special Structure of Electroconducting Materials on Their Electrical Conductivity, *Plasticheskie Massy* **1965**(5).
107. Hearst, Degradation of Organic Coatings by Irradiation with Light, AD 458 231.
108. Kumins, Long Range Effects of Polymer Pigment Interaction in the Solid State, *Offic. Dig.*, November.
109. Levitan, The Use of A Statistically Designed Experiment for Studying Resin Parameters, RIP Division of SPI, Chicago, February.
110. Sherr, Preliminary Evaluation of Polymer Properties, *SPE.*, January.
111. Smith, Measuring Cure of Thermoset Coatings, *Mod. Plastics*, January.
112. Sorokin, et al., The Influence of Gamma-irradiation on Some Epoxide Resins, *Lakokaraschnye Materialy i ikh Primenenie* **1965**(2).
113. Stucki, Radiation Effects in Epoxies Used as Encapsulants for Electronic Packages, *Advan. Electron. Circuit Packaging* **5**, 278–283.

114. Sugita and Ito, Thermosetting Resins, I: The Infrared Analysis of Cured Epoxy Resins, *Bull. Chem. Soc. Japan* **38**(10), 1620–1627.
115. Warfield, The Compressibility of Crosslinked Polymers, *NOLTR* 64–243.
116. Weigand and Hanna, Evaluation of the Explosive Hazard of Gases Generated by Transformer Potting Materials When Subjected to Electrical Overloads, 6th Electrical Insulation Conference (IEEE and NEMA), September.

1966

117. Lee, et al., "New Linear Polymers," McGraw-Hill Book Company.

Appendix 6-1

INFRARED SPECTRA OF DGEBA CURED WITH TYPICAL CURING AGENTS*

* Prepared by Art Cupples, Okko Koning, and Bette Wolfskill of The Epoxylite Corporation.

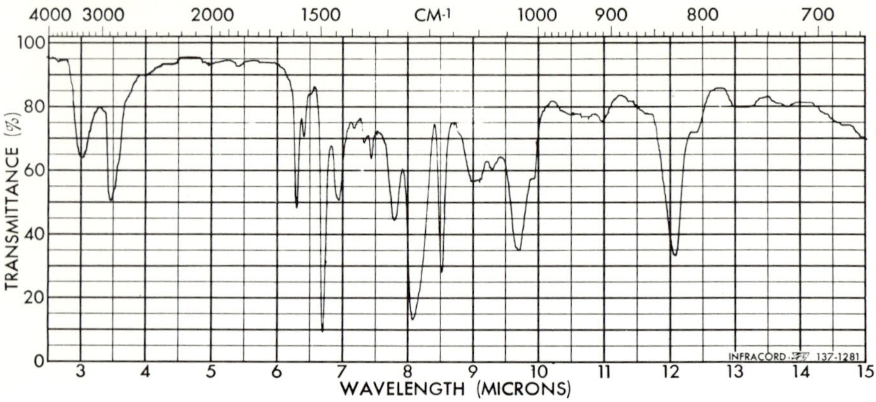

Fig. 6-1-1. DGEBA/DETA.

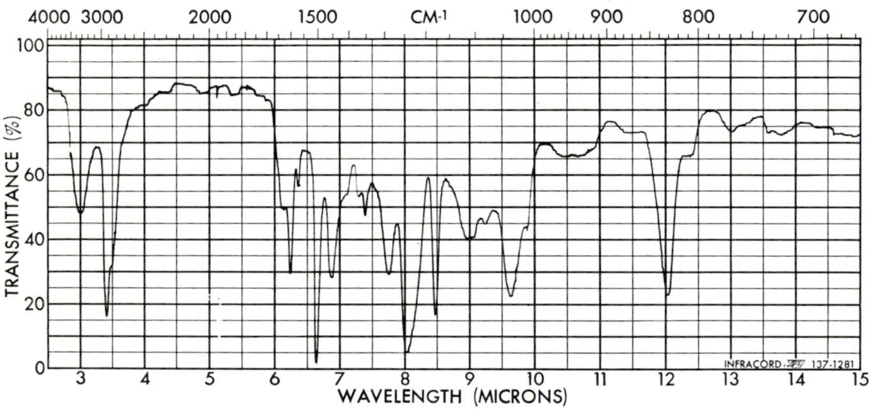

Fig. 6-1-2. DGEBA/monoamide of tall-oil acid–TETA.

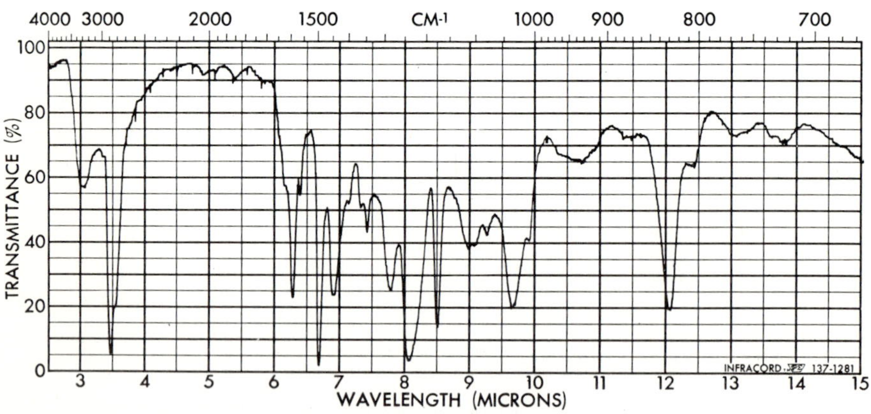

Fig. 6-1-3. DGEBA/fatty polyamide.

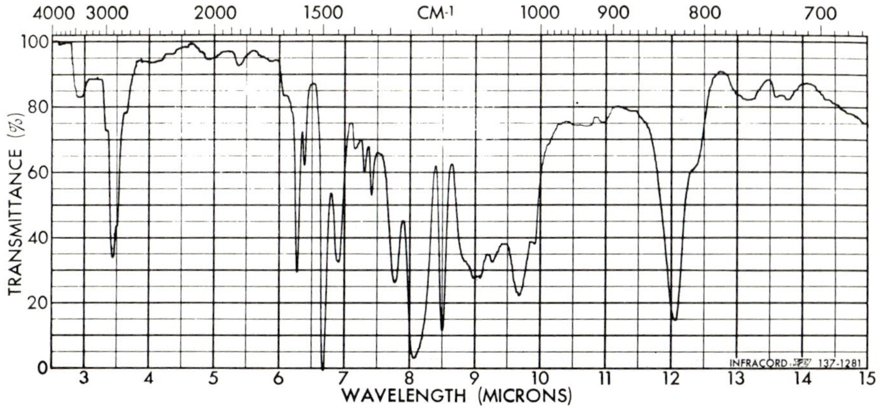

FIG. 6-1-4. DGEBA/DMP-30.

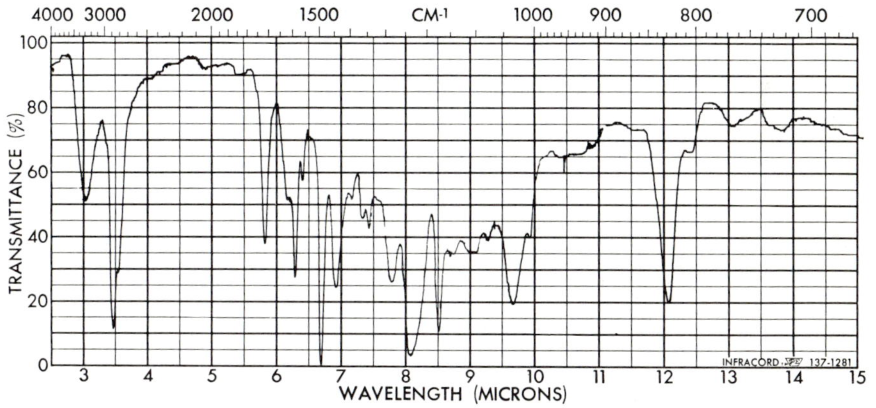

FIG. 6-1-5. DGEBA/DDSA/DMP-30.

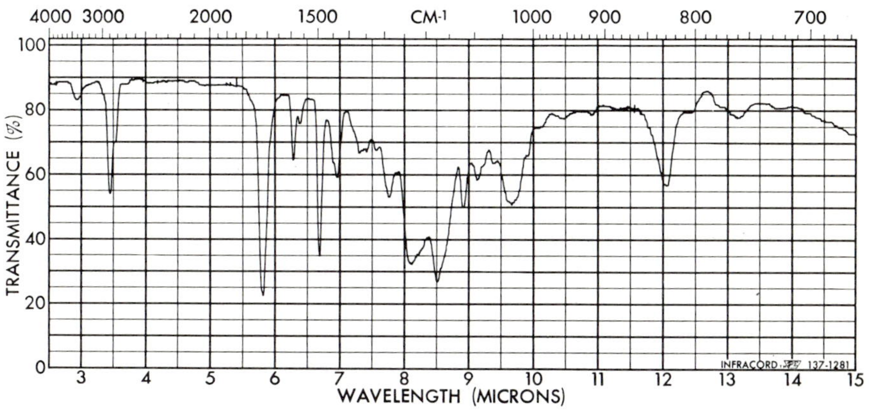

FIG. 6-1-6. DGEBA/HHPA/BDMA.

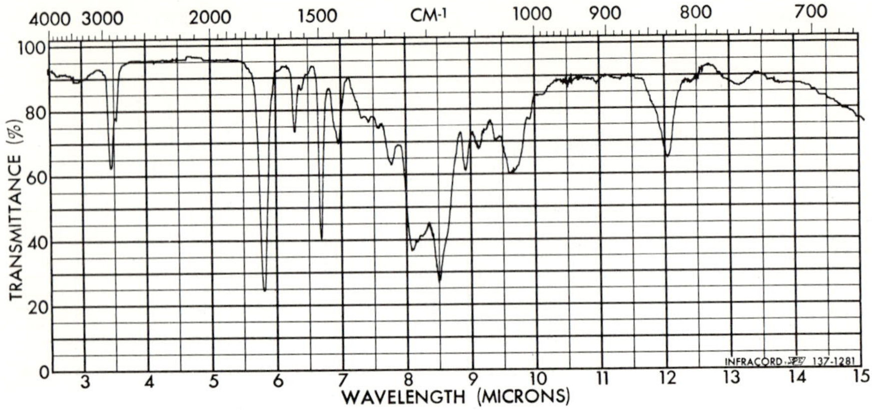

FIG. 6-1-7. DGEBA/CA–HHPA eutectic/DMP-30.

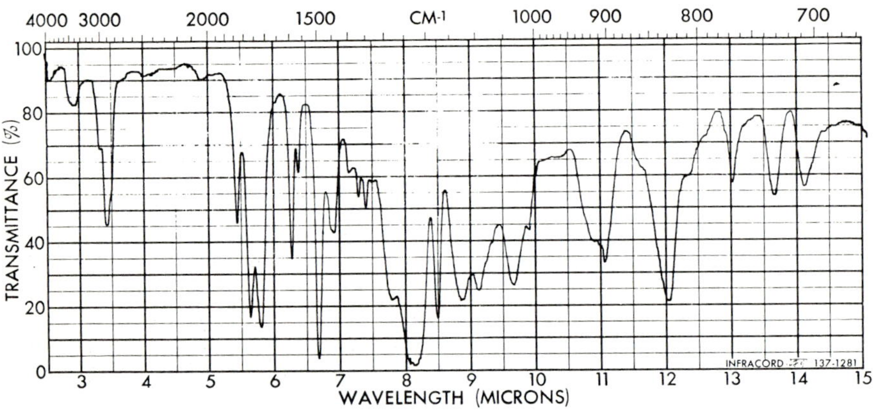

FIG. 6-1-8. DGEBA/PMDA.

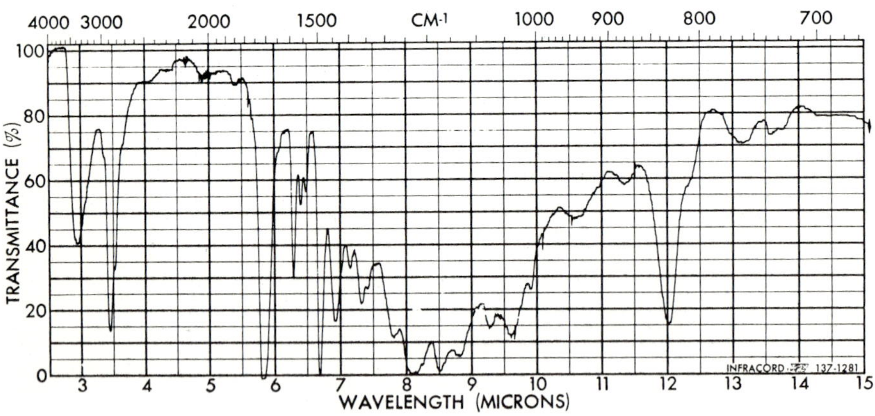

FIG. 6-1-9. DGEBA/hydroxyl-terminated polyester/DMP-30.

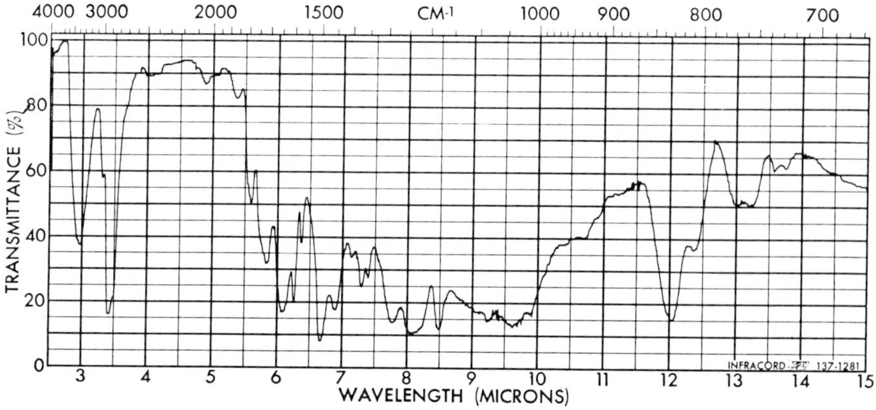

FIG. 6-1-10. DGEBA/urea-formaldehyde resin.

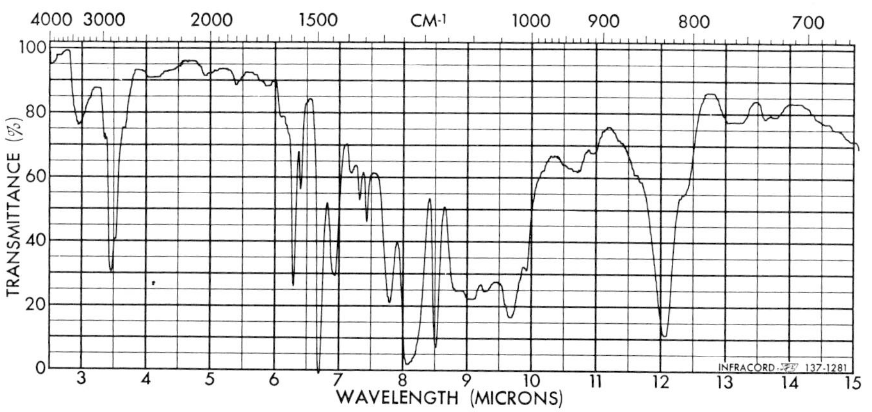

FIG. 6-1-11. DGEBA/trimethylolallyloxybenzene/phosphoric acid.

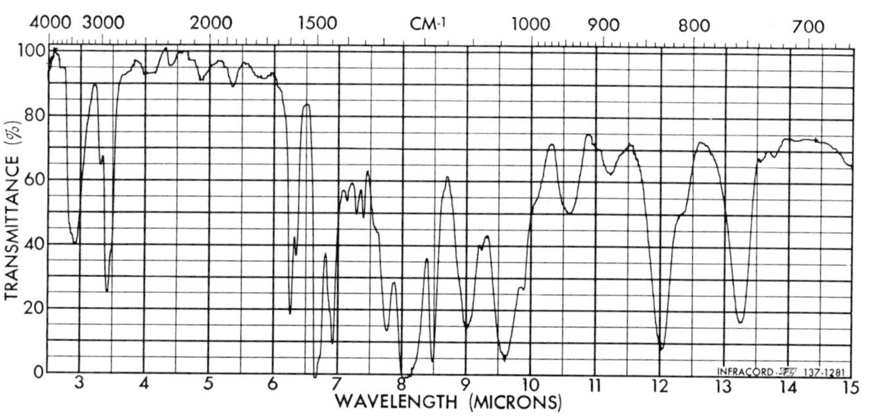

FIG. 6-1-12. DGEBA/phenolic resin/BDMA

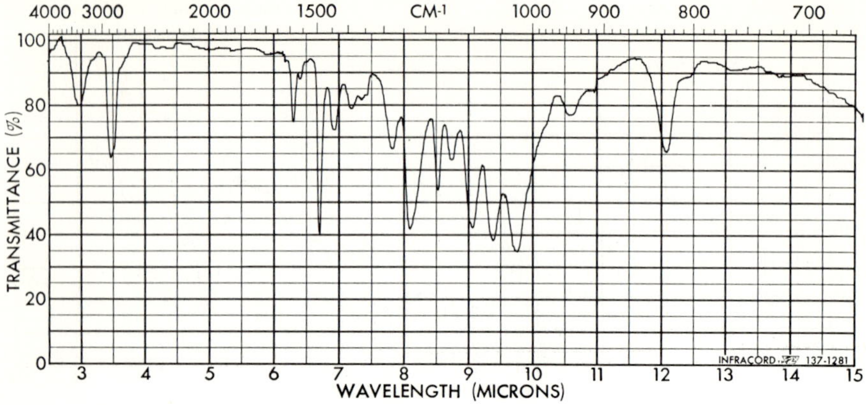

FIG. 6-1-13. DGEBA/polysulfide/DMP-30.

Chapter 7

ALIPHATIC PRIMARY AMINES AND THEIR MODIFICATIONS AS EPOXY-RESIN CURING AGENTS

Amines are derivatives of ammonia (NH_3). Depending on the number of hydrogen atoms replaced by hydrocarbon groups, the amines are defined as primary, secondary, or tertiary. Depending on the number of amine groups in the molecule, the amine is a mono-, di-, tri-, or polyamine. The amines are further classified as aromatic, alicyclic, or aliphatic. In this chapter, we are concerned only with the aliphatic and alicyclic di-, tri-, and polyamines and their derivatives.

The primary aliphatic amines were among the first materials to gain acceptance in the United States as curing agents for the epoxy resins [1,3,20]. With glycidyl ether resins, they give good cures at room temperature with excellent all-around properties and temperature resistance to 100°C, depending on the functionality and aromatic structure of the resin. With the nonglycidyl epoxy resins, the reaction is sluggish at room temperature [30,39,68], and heat and/or acidic accelerators are required to obtain satisfactory reaction rates. Because of sluggish reaction rates, aliphatic primary amines are little used with the commercial epoxidized olefin resins. Accordingly, most of this chapter will be confined to a discussion of the glycidyl ethers, with emphasis on the most useful class, the aromatic glycidyl ethers.

Although aliphatic primary amines are generally quite reactive with glycidyl ether epoxy resins, accelerators are sometimes used. Gamma rays from a cobalt-60 source have been found to accelerate the reaction but not to provide complete cures below room temperature [48]. Generally, active-hydrogen-containing compounds are used as accelerators. In a series of tests with phenol, *m*-nitrophenol, *p*-cresol, resorcinol, cyclohexanol, 1-hexanol, and 1-butanol, a linear relationship between the acidity of the accelerator and the reaction rate of the amine with DGEBA was established when the reaction was conducted at 20°C. At 5°C, phenol proved an effective accelerator, the other phenols being somewhat sluggish and the alcohols ineffective. With DGEBA (mol. wt. 950), phenol proves less effective than with the lower-molecular-weight resin [63].

When primary amines are used to cure DGEBA (mol. wt. 950) in coatings applications, accelerators are frequently employed. DMP-30 is generally satisfactory and superior to phenol. Without accelerators, and at lower temperatures, an oily surface may occur on the coating. This is ascribed to reaction of the amine primary hydrogen atoms with carbon dioxide and hydrogen [60]. The resultant amine carbonates are not reactive with epoxy resins at low temperatures. The amine carbonates, however, may be removed with heating, and when low ambient temperatures are encountered during the cure of small masses of DGEBA, heating is frequently recommended.

ALIPHATIC POLYAMINES AS CURING AGENTS

Typical aliphatic polyamines used as curing agents for epoxy resins are shown in Tables 7-1 and 7-2. When used in stoichiometric proportions (i.e., one active amine hydrogen for each epoxy group), these amines, as a class, provide similar cured properties below their deflection temperature (DT). The DT is about 120°C, maximum, for the aromatic diepoxies, and about 150°C, maximum, for the aromatic epoxies with functionalities of 2.5 or greater.

In practice, because of the critical effect of cure cycle, mixing, and weighing on this property, the DTs are generally considerably lower than maximum, falling in the 80 to 90°C range for the aromatic diepoxies and the 110 to 120°C range for the aromatic polyepoxies. The aliphatic-polyamine-cured systems are resistant to attack by alkalies and some inorganic acids; they are less resistant to most organic acids. Solvent resistance is high, and water resistance is good.

A nomograph for use in determining stoichiometric quantities of aliphatic polyamines for various epoxy resins is shown in Fig. 7-1.

Aliphatic Diamines (Table 7-1)

The straight-chain polymethylene diamines ($H_2NCH_2NH_2$), with *n* varying from 0 to 6, have been investigated as curing agents for DGEBA. Experiments to

Table 7-1. Structures of Typical Commercial Primary Amines Used as Epoxy-resin Curing Agents

ALIPHATIC DIAMINES

$H_2N(CH_2)_nNH_2$

Polymethylene diamines

$H_2N(CH_2)_nO(CH_2CH_2O)_m(CH_2)_nNH_2$

Polyether diamines

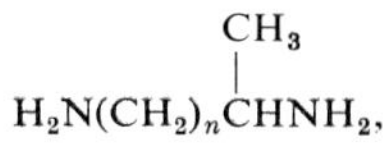

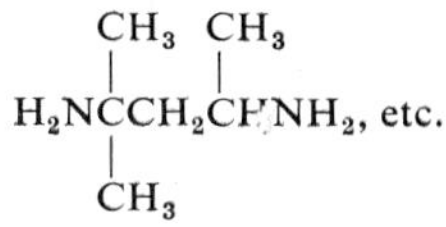

Branched-chain polymethylene diamines

LINEAR AND BRANCHED ALIPHATIC POLYAMINES

$H_2N(CH_2)_nNH(CH_2)_nNH_2$

Diethylenetriamine ($n = 2$)
Iminobispropylamine ($n = 3$)
Bis(hexamethylene)triamine ($n = 6$)

$H_2N[(CH_2)_2NH]_n(CH_2)_2NH_2$

Triethylenetetramine ($n = 2$)
Tetraethylenepentamine ($n = 3$)
Pentaethylenehexamine ($n = 4$)

$H_2N(CH_2)_2NH(CH_2)_2OH$

Aminoethylethanolamine

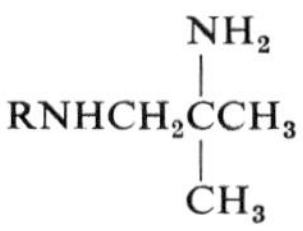

Substituted 1,2-diamines
(R = isopropyl, phenyl, etc).

$CH_3(CH_2)_n(CH_2CH_2NH)_2CH_2CH_2NH_2$

Substituted polyamines

$[CH_3(CH_2)_n]_2N(CH_2)_3NH_2$

Dimethylaminopropylamine ($n = 0$)
Diethylaminopropylamine ($n = 1$)

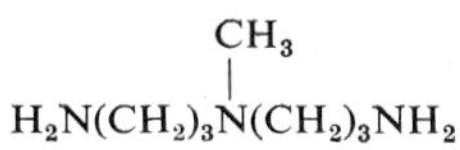

Methyliminobispropylamine

ALICYCLIC POLYAMINES

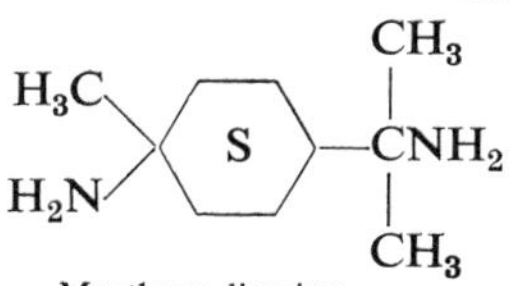

Menthane diamine

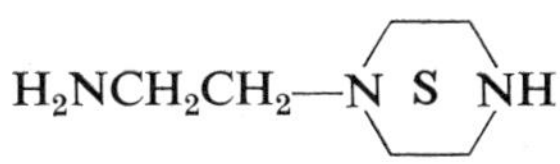

N-Aminoethylpiperazine

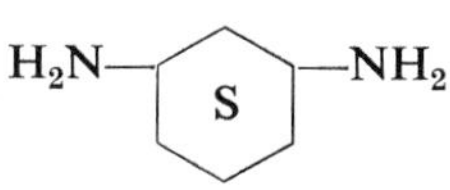

1,3-Diaminocyclohexane

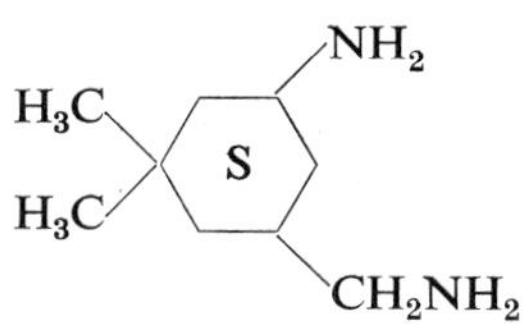

Isophoronediamine

ALIPHATIC AMINES CONTAINING AROMATIC GROUPS

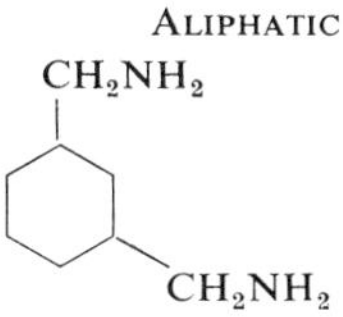

m-Xylylenediamine

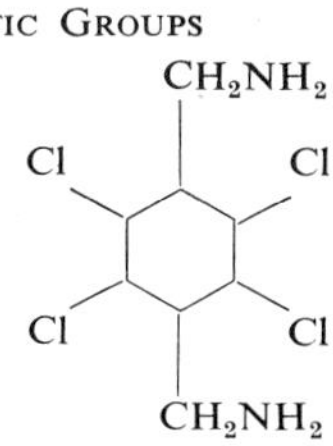

Tetrachloro-*p*-xylenediamine

Table 7-2. Properties of Commercial Aliphatic Polyamines and Reactivity with DGEBA

Amine	Mol. wt.	Active hydrogens	Calc. phr	State	Vapor pressure, mm Hg at 20°C	Sp. gr. at 23°C	Pot life at 23°C 50 g, minutes	Exotherm, 100 g, °C
Diethylene glycol bispropylamine . . .	210	4	28	Liquid	<0.001	1.01	60	
DETA	103	5	11	Liquid	0.2	0.95	25	235
Iminobispropylamine	131	5	14	Liquid	<0.01	0.93	35	220
Bis(hexamethylene)diamine	215	5	23	Solid	<0.01	0.88 at 35°C	120	
TETA.	146	6	13	Liquid	0.01	0.98	26	233
TEPA	189	7	14	Liquid	<0.01	0.99	27	228
DEAPA	130	2	7	Liquid	1.2	0.82	120 (1 lb)	170 (1 lb)
Menthane diamine . .	170	4	22	Liquid	. . . . . .	0.91	480 (2.5 lb)	93 (2.5 lb)
AEP	128	3	23	Liquid	. . . . . .	0.98	17	

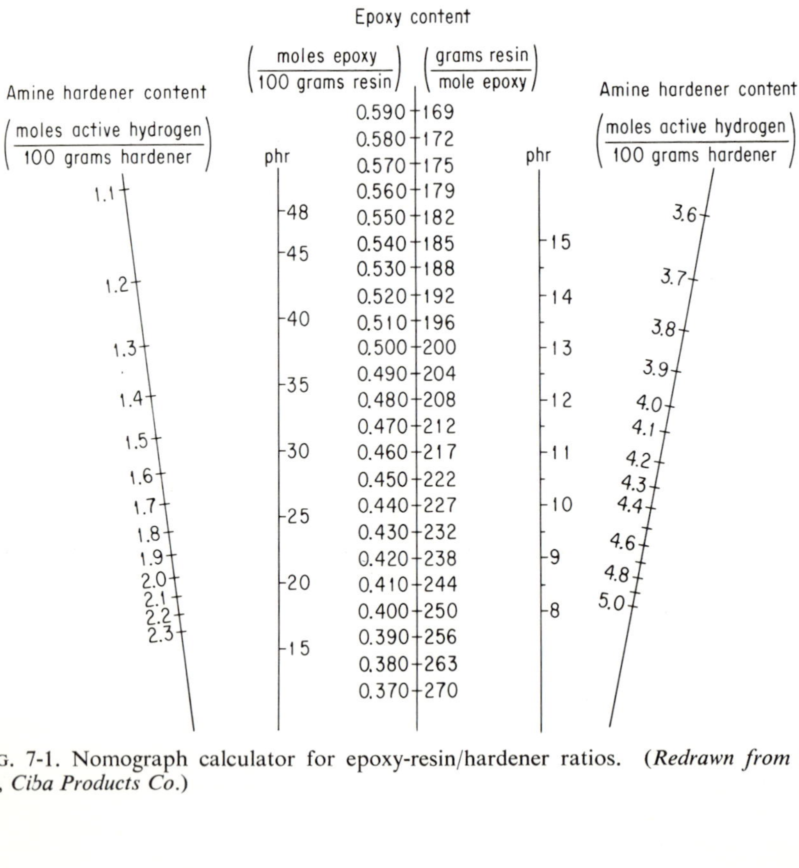

FIG. 7-1. Nomograph calculator for epoxy-resin/hardener ratios. (*Redrawn from TSN 99, Ciba Products Co.*)

determine the effect of increasing the chain length, starting from what may be considered the simplest diamine, hydrazine ($n = 0$), through hexamethylenediamine ($n = 6$), indicate that when the two primary amine groups are closely spaced, the initial reaction, involving one group, is quite rapid; however, the reaction leads to the second group being hindered sterically with further cure being achieved only by catalytic activity [8]. Ethylenediamine ($n = 2$) and 1,3-diaminopropane ($n = 3$) give short pot lives, but the reaction products are somewhat brittle. Butylene, amylene, and hexamethylene diamines all give cures without brittleness. The last has been suggested as a curing agent for epoxy coatings [19]. Experiments designed to establish comparative electrical properties indicate a slight improvement with increasing chain lengths [18]. Amines generally are hygroscopic and somewhat sensitive to moisture, as illustrated in Figs. 7-2 and 7-3 for an ethylenediamine-cured system.

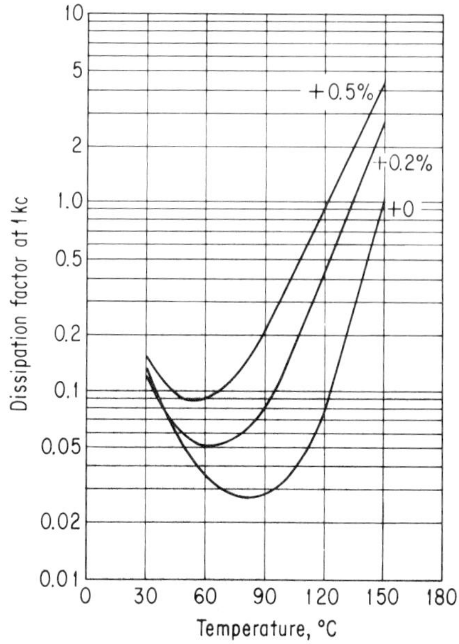

FIG. 7-2. Effect of water content on dissipation factor of DGEBA cured with ethylenediamine [18].

The polyether diamines produced by the reaction of ammonia with ethylene or propylene glycol or by other synthesis routes, and similar materials obtained by reacting halohydrin ethers (from epichlorohydrin and polyols) with amines [41] are used as epoxy-resin curing agents to provide some improvement in toughness and flexibility over the polymethylene diamines at the expense of more sluggish reaction rates [49].

A typical polyether diamine, diethylene glycol bispropylamine gives cured properties with DGEBA as shown in Table 7-3. Typical shrinkage values for a system cured with a longer-chain polyether diamine are shown in Table 7-4.

The more common branched-chain polymethylene diamines provide cured properties similar to those obtained with the straight-chain polymethylene diamines, but with less tendency for film discoloration on aging and somewhat better initial clarity [2]. Properties of DGEBA cured with a long, branched-chain polymethylenediamine are given in Table 7-5.

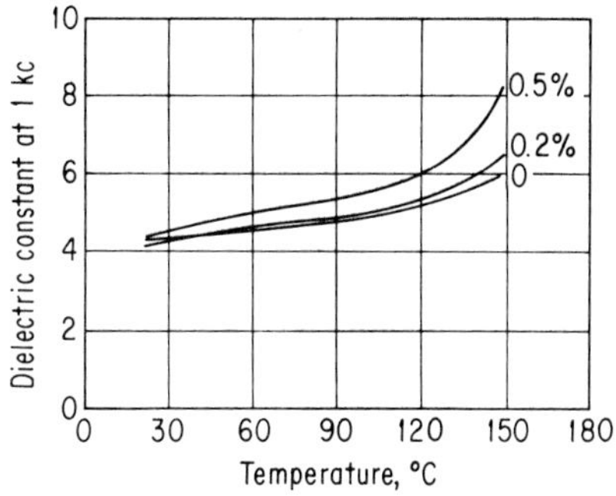

FIG. 7-3. Effect of water content on dielectric constant of DGEBA cured with ethylenediamine [18].

Linear and Branched Aliphatic Polyamines (Table 7-1)

The linear and branched aliphatic polyamines are the most widely used of the commercial aliphatic polyamines in the epoxy-resin technology. Representatives of these are discussed below.

Diethylenetriamine. Diethylenetriamine (DETA) is a low-viscosity liquid

Table 7-3. Physical and Electrical Properties of DGEBA Cured with a Polyether Diamine [4-7]*

PHYSICAL PROPERTIES (room-temperature cure)

Property	*Value*
DT, °C	66
Tensile strength, psi	8,880
Modulus, psi $\times 10^{-6}$	0.39
Flexural strength, psi	12,800
Modulus, psi $\times 10^{-6}$	0.40
Impact strength, Izod, ft-lb	1.53

ELECTRICAL PROPERTIES (postcured 3 hours at 120°C)

Property	*At 23°C*	*At 60°C*
Arc resistance, seconds	110	
Dielectric strength, volts/mil	496	517
Dielectric constant:		
At 60 cps	4.99	5.12
At 10^3 cps	4.94	5.08
At 10^6 cps	3.90	4.44
Dissipation factor:		
At 60 cps	0.0038	0.0079
At 10^3 cps	0.0011	0.0050
At 10^6 cps	0.547	0.0554
Volume resistivity, ohm-cm	6×10^{13}	2.6×10^{13}

* Double numbers in brackets are for references in a previous chapter.

Table 7-4. Shrinkage of DGEBA Cured with 40 phr of a Polyether Diamine [57]

Property	*Value*
Mixing temperature	23°C
Cure, hr/°C	15/23 + 5/46 + 15/74
Peak exotherm	76°C
Shrinkage, %;	
Total shrinkage	4.78
Pregelation shrinkage	3.18
Postgelation shrinkage	−1.21
Cooling shrinkage	2.80

Table 7-5. Physical Properties of DGEBA Cured with a Long-chain Polymethylenediamine [21]

(Cure cycle: 24 hr at 23°C + 16 hr at 120°C)

Property	*Value*
Tensile strength, psi	4,730
Modulus, psi $\times 10^{-6}$	0.29
Flexural strength, psi	9,110
Modulus, psi $\times 10^{-6}$	0.24
Compressive strength, psi	24,800
Modulus, psi $\times 10^{-6}$	0.24
Strain, %	57.3
Impact strength, Izod	0.23
Hardness, Rockwell L	87
Coefficient of thermal expansion, cm/(cm)(°C) $\times 10^5$:	
Below Tg	7.79
Above Tg	20.0
Glass transition Tg, °C	47
Modified Vicat penetration temperature, °C	51

which fumes in contact with air at room temperature. Properties and reactivity are indicated in Table 7-2.

With DGEBA, DETA is normally used at the stoichiometric concentration of 11 phr*, though it is also sometimes used at as low as 75 per cent of stoichiometric to reduce the exotherm and permit casting of larger masses. When used at less than stoichiometric concentration, the most significant effect is reduction of DT as illustrated in Fig. 7-4. Table 7-6 indicates the effect of varying amounts on tensile strength.

With the glycidyl ether resins, cure will occur at room temperature, even in thin films, although low ambients will produce some degree of brittleness. Cure is substantially completed in 24 hours at 25°C, but the ultimate value is not obtained until about 7 days at this temperature. Properties of the room-temperature-cured

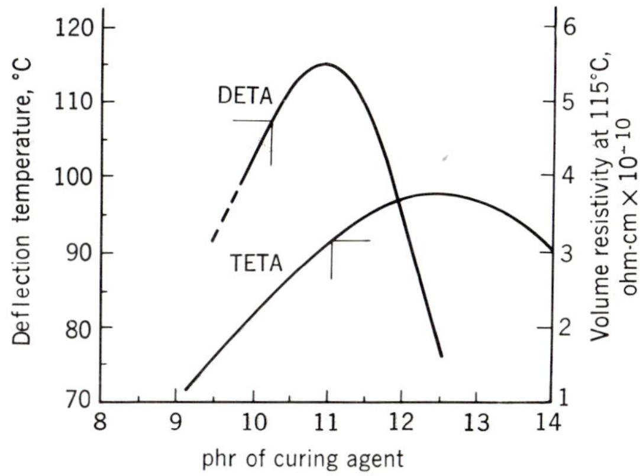

FIG. 7-4. Effect of concentration of DETA and TETA on DT of DGEBA [7].

system will be improved by a postcure of 1 to 2 hours at 65 to 95°C. With DGEBA ($n = 0$), the pot life will be extended. Addition of a small amount of polyol will accelerate the cure [6].

The physical properties of a DETA-cured system are summarized in Table 7-7. Representative of the electrical properties are the values given in Table 7-8. Volumetric shrinkage with DGEBA is about 4.5 per cent with the majority of the shrinkage (3.3 per cent) occurring while the material is in the liquid state.

DETA may be used with the epoxidized olefins to provide long pot-life systems. With divinylbenzene dioxide the pot life will be about 2 days at room temperature [31]. Stoichiometric considerations are not necessarily straightforward. With 3,4-epoxy-6-methylcyclohexylmethyl-3,4-epoxy-6-methylcyclohexane carboxylate, for example, DETA is employed at 1.3 to 1.6 amine hydrogens per epoxy group to provide a gel time of 40 minutes at 120°C and a DT of about 100°C. The increase over stoichiometric is presumably necessary because of the sluggish reactivity of at least some of the amine hydrogens toward ring-situated epoxy groups.

DETA has been used as a general-purpose curing agent with the aromatic glycidyl ether epoxy resins for small castings, rigid laminates, medium-strength adhesives, and baking-type solution coatings. It is a skin irritant and has been replaced in many cases by less irritating substances.

* The abbreviation phr as used here and subsequently throughout the book, stands for parts per hundred parts of resin, by weight. Throughout the epoxy technology, this is the preferred method for expressing the concentration of various ingredients.

Table 7-6. Tensile Strength of DGEBA Cured with DETA, as a Function of Temperature and Curing-agent Concentration [4–6]

Test temperature, °C	phr DETA	Ultimate, psi	Yield stress at 0.2% offset, psi	Modulus, psi	Elongation at yield, %	Elongation at ultimate, %
100	8	Too soft to test				
50	8	9,800	6,100	400,000	1.7	6.5
25	8	11,300	8,300	490,000	1.8	3.7
0	8	9,200	......	550,000	....	1.8
−25	8	11,700	11,400	550,000	2.4	2.4
100	10	2,700	1,200	175,000	0.80	24.0
50	10	8,600	6,600	430,000	1.7	3.6
25	10	11,400	7,100	435,000	1.8	5.7
0	10	11,300	6,900	530,000	1.5	3.2
−25	10	11,400	9,000	490,000	2.0	2.8
100	12	3,700	2,400	220,000	1.3	18.2
50	12	9,000	5,500	340,000	1.9	7.3
25	12	10,000	6,100	380,000	1.8	5.3
0	12	10,900	6,000	480,000	1.4	3.9
−25	12	13,800	8,400	420,000	2.2	4.9
100	14	3,300	1,600	150,000	1.2	27.7
50	14	8,700	5,400	330,000	1.8	7.3
25	14	9,900	5,900	370,000	1.7	5.8
0	14	10,700	6,200	450,000	1.5	4.0
−25	14	14,000	8,200	430,000	2.1	5.4

Table 7-7. Physical Properties of DGEBA Cured 2 Hours at 120°C with Stoichiometric Amount of DETA [21]

Property	*Value*
DT, °C	91
Tensile strength, psi:	
At 23°C	6,160
At 100°C	5,670
Modulus, psi × 10^{-6}:	
At 23°C	0.5070
At 100°C	0.844
Elongation, %:	
At 23°C	1.35
At 100°C	14.5
Flexural strength, psi:	
At 38°C	15,000
At 100°C	4,230
Modulus, psi × 10^{-6}:	
At 38°C	0.36
At 100°C	0.16
Compressive strength, psi:	32,600
Modulus, psi × 10^{-6}	0.271
Strain, %	50
Impact strength, Izod	0.22
Hardness, Rockwell L	118
Coefficient of thermal expansion, cm/(cm)(°C) × 10^5:	
Below Tg (122°C)	6.00
Above Tg	17.9
Modified Vicat penetration, °C	93

Table 7-8. Electrical Properties vs. Temperature for DGEBA Cured with DETA [11]

Property	Frequency, cps					
	10^2	10^3	10^4	10^5	10^6	10^7
Power factor:						
At −40°C	0.0220	0.0200	0.0195	0.0175	0.0165	0.0170
At −20°C	0.0300	0.0500	0.0500	0.0700	0.0900	0.1000
At 20°C	0.0090	0.0400	0.0480	0.0800	0.1000	0.1300
At 60°C	0.0045	0.0100	0.0600	0.0750	0.1100	0.1500
At 100°C	0.0175	0.0165	0.0300	0.0600	0.1200	0.2000
Dielectric constant:						
At −40°C	3.4	3.35	3.3	3.3	3.2	3.1
At −20°C	3.8	3.8	3.75	3.7	3.6	3.4
At 20°C	4.1	4.2	4.2	4.1	4.2	4.1
At 60°C	4.3	4.4	4.6	4.6	4.5	4.4
At 100°C	4.5	4.6	4.7	4.8	4.9	5.0
Volume resistivity, ohm-cm:						
At 25°C				2×10^{16}		
At 50°C				6×10^{14}		
At 75°C				5×10^{13}		
At 100°C				5×10^{12}		
At 125°C				5×10^{11}		

Iminobispropylamine. Iminobispropylamine (Table 7-2) like DETA is sometimes used at less than stoichiometric.

When used as 10 phr with DGEBA it gives cured properties as indicated in Table 7-9.

Table 7-9. Physical Properties of DGEBA Cured with Iminobispropylamine [4-4]

Property	*Value*
Tensile strength, psi	9,028
Modulus, psi $\times 10^{-6}$	0.178
Elongation, %	4.56
Flexural strength, psi	15,000
Modulus, psi $\times 10^{-6}$	0.550
Compressive strength, psi	17,250
Modulus, psi $\times 10^{-6}$	0.539
Impact strength, Izod	0.427

Bis(hexamethylene)triamine. Bis(hexamethylene)triamine (Table 7-2) is characterized by an unpleasant odor and must be dissolved in the resin at 50°C. The material is of interest insofar as it provides an extended pot life with DGEBA resins together with a room-temperature cure.

Triethylenetetramine. Triethylenetetramine (TETA) (Table 7-2), is a widely used general-purpose curing agent for diglycidyl ether resins. TETA is frequently used at about 70 per cent stoichiometric. Cured physical properties of such a system are presented in Table 7-10 and chemical resistance is given in Table 7-11. DGEBA cured with stoichiometric amounts of TETA shows virtually no change in tensile strength after exposure to weathering for one year; flexural strength increases slightly (12 per cent) and dielectric constant increases 5 per cent [45].

When TETA is used with epoxidized polybutadiene, the pot life at room temperature is quite long, and even with an acid accelerator cure times of 2 to 4 hours at 150°C are required to develop properties similar to those achieved at room temperature with DGEBA.

Table 7-10. Physical Properties of DGEBA Cured with 10 phr TETA [4–1]

Property	*Value*
DT, °C	66
Tensile strength, psi	9,000
Flexural strength, psi	18,000
Modulus, psi $\times 10^{-6}$	0.54
Compressive strength, psi	24,000
Impact strength, Izod	0.3–0.4
Hardness, Rockwell *M*	108
Thermal conductivity, cal/(sec)(sq cm)(°C)(cm)	4.8×10^{-4}
Flammability, in./min	0.34
Shrinkage during cure, in./in.	0.005
Water absorption after 24-hr immersion, %	0.12
Moisture-vapor transmission at 38°C, relative humidity 95%, g/(sq ft)(24 hr)(in. thickness)	<0.01

The TETA-cured systems are quite similar in overall performance to the systems cured with DETA. The lower vapor pressure generally favors the use of TETA.

Tetraethylenepentamine. Tetraethylenepentamine (TEPA) (Table 7-2) provides reactivities and cured properties quite similar to those obtained with TETA.

As with other aliphatic polyamines, the concentration of TEPA with the aromatic glycidyl ether resins becomes less critical as the functionality of the resin increases, and with the more highly functional resins when heat cures are employed, the preference is for a greater excess than when room-temperature cures are used, presumably because of steric considerations. This effect, as reflected in the DT of an epoxidized phenol-formaldehyde novolac, for both room-temperature and elevated-temperature cures, is indicated in Fig. 7-5.

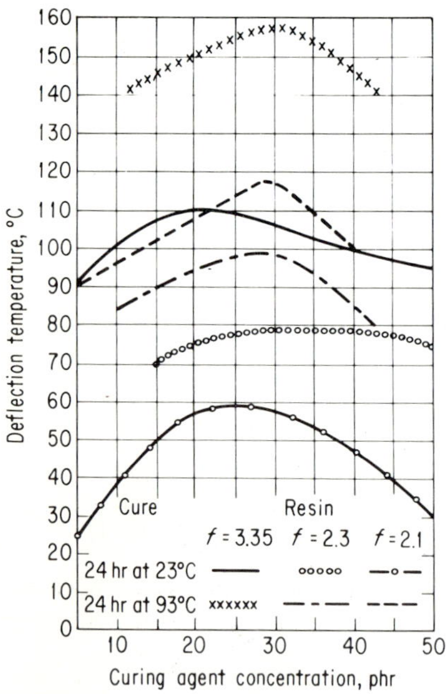

FIG. 7-5. Effect of concentration of TEPA on DT of various epoxidized novolacs [27].

TEPA is perhaps most widely used in the epoxy-resin technology as a curing agent for solution coatings based on DGEBA (mol. wt. 950).

Diethylaminopropylamine. Diethylaminopropylamine (DEAPA) (Table 7-2) is representative of the aliphatic polyamines containing tertiary amine groups. DEAPA possesses only two active hydrogens, which serve to tie the amine into the cured system, but cross-linking must be effected by catalytic activity. The tertiary

Table 7-11. Chemical Resistance of DGEBA Cured with 10 phr TETA [4-1]

3 months' immersion in listed fluids at 23°C	Concentration	Resistance	Remarks
Accumulator acid	32%	Resistant	
Acetaldehyde, conc.	99%	Relatively resistant	Impact flexural strength unchanged, flexural strength 45% decrease, increase in weight 10%
Acetone...............	100%	Nonresistant	Completely destroyed
Acetone, dil...........	50%	Nonresistant	Completely destroyed
Acetone, dil...........	10%	Relatively resistant	Impact flexural strength 30% decrease, flexural strength, 25% decrease, increase in weight 2%
Ammonia, dil..........	10%	Resistant	
Ammonia, dil..........	3%	Resistant	
Aniline	100%	Nonresistant	Completely destroyed
Beer, dark.............	100%	Relatively resistant	
Benzene...............	100%	Nonresistant	After 1 month resistant, after 3 months considerably distorted
Benzine	100%	Resistant	Impact flexural strength 50% decrease, flexural strength 10% decrease, increase in weight, 2%
Butanol	100%	Resistant	
Butyl acetate	100%	Relatively resistant	Impact flexural strength 30% decrease, flexural strength 60% decrease, increase in weight 4%
Carbon disulfide	100%	Nonresistant	Swollen test castings
Carbon tetrachloride ...	100%	Resistant	Impact flexural strength unchanged, flexural strength, 20% decrease, increase in weight 8%
Caustic soda solution ...	10 g/100	Resistant	
Caustic soda solution ...	50 g/100	Relatively resistant	Impact flexural strength 30% decrease, flexural strength 10% decrease, increase in weight 1%
Chlorobenzene	100%	Nonresistant	
Clophene	100%	Resistant	
Cooking salt solution ...	10 g/100	Relatively resistant	Impact flexural strength 50% decrease, flexural strength 20% decrease
Cooking salt solution ...	20 g/100	Relatively resistant	Impact flexural strength 30% decrease, flexural strength 1% decrease
Cooking salt solution	26 g/100	Resistant	
Cyclohexanol..........	100%	Resistant	
Cyclohexanone	100%	Nonresistant	Test castings destroyed
Dioxane	100%	Nonresistant	
Ethyl alcohol	96%	Nonresistant	Impact flexural strength unchanged, flexural strength 80% decrease, increase in weight 10%
Ethyl ether	100%	Resistant	Flexural strength 30% decrease
Ferrochloride solution ..	45 g/100	Resistant	
Fluoric acid	100%	Nonresistant	Completely destroyed
Glycerin	100%	Resistant	Decrease in weight 0.3%
Hydrochloric acid, conc.	37%	Relatively resistant	Impact flexural strength 25% decrease, flexural strength 20% decrease, increase in weight 7.5%
Hydrochloric acid, dil. ..	20%	Resistant	Impact flexural strength 20% decrease, flexural strength 8% decrease, increase in weight 1.5%
Hydrochloric acid	4%	Relatively resistant	Impact flexural strength 55% decrease, flexural strength 15% decrease, increase in weight 2%
Hydrochloric acid, gaseous	....	Resistant	Increase in weight 5%

Table 7-11. (contd.)

3 months' immersion in listed fluids at 23°C	Concentration:	Resistance	Remarks
Hydrogen peroxide, dil.	30%	Nonresistant	Impact flexural strength 90% decrease, flexural strength 70% decrease, increase in weight 13%
Hydrogen sulfide, dry	100%	Resistant	Impact flexural strength, 20% decrease
Hydrogen sulfide, dil.	3%	Resistant	
Javelle water	....	Resistant	14% active chloride
Lactic acid, conc.	91%	Relatively resistant	Impact flexural strength unchanged flexural strength 20% decrease, increase in weight 7%
Mineral oil	100%	Resistant	
Nibrene wax D 88	100%	Resistant	
Nitric acid, dil.	33%	Nonresistant	
Nitric acid, dil.	7%	Relatively resistant	Impact flexural strength 30% decrease, flexural strength 20% decrease, increase in weight 2.7%
Nitrobenzene	100%	Nonresistant	
Photographic developer	....	Resistant	Composition: 1,000 cc water, 1 g methanol, 25 g sodium sulfite, 4.5 g hydroquinone, 17 g soda, 0.5 g potassium bromide
Photographic fixing bath	....	Resistant	Composition: 1,000 cc water, 200 g sodium sulfite, 20 cc lye of sodium bisulfite
Relative humidity	100%	Resistant	
Saltwater spray	....	Resistant	Daily spraying with 2 liters salt solution
Sodium bichromate solution	66 g/100	Resistant	
Sodium bisulfite solution	20 g/100	Relatively resistant	Impact flexural strength 55% decrease, flexural strength 15% decrease, increase in weight 1.4%
Sodium chlorite solution	10 g/100	Resistant	
Soft soap, dil.	50%	Resistant	
Stearic acid	100%	Resistant	
Sulfuric acid, dil.	75%	Nonresistant	Swollen test castings
Sulfuric acid, dil.	50%	Relatively resistant	Impact flexural strength 60% decrease, flexural strength unchanged, increase in weight 2.5%
Sulfuric acid, dil.	10%	Relatively resistant	
Teepol	40%	Resistant	
Tetrachloroethane	100%	Nonresistant	Test castings destroyed
Tetrahydronaphthaline	100%	Resistant	Impact flexural strength 20% decrease, flexural strength 20% decrease, increase in weight 2.7%
Toluene	100%	Nonresistant	
Triethanolamine	100%	Resistant	
Vinegar	....	Relatively resistant	Impact flexural strength 60% decrease, flexural strength 20% decrease, increase in weight 3.6%
Water, distilled	....	Resistant	

amine possesses heightened reactivity by virtue of the electron-releasing effect of the substituted alkyl chains, such that if DEAPA is reacted with a monofunctional epoxy, such as butyl glycidyl ether, to the extent that amine hydrogens are eliminated, it may still be used as a catalytic-type curing agent.

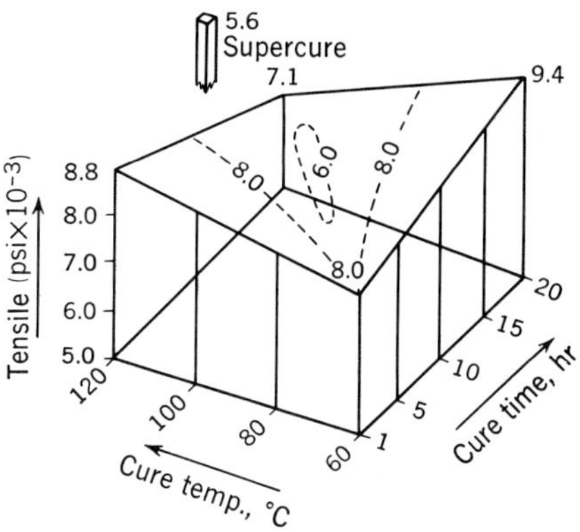

FIG. 7-6. Tensile strength of DGEBA cured with DEAPA [4–7].

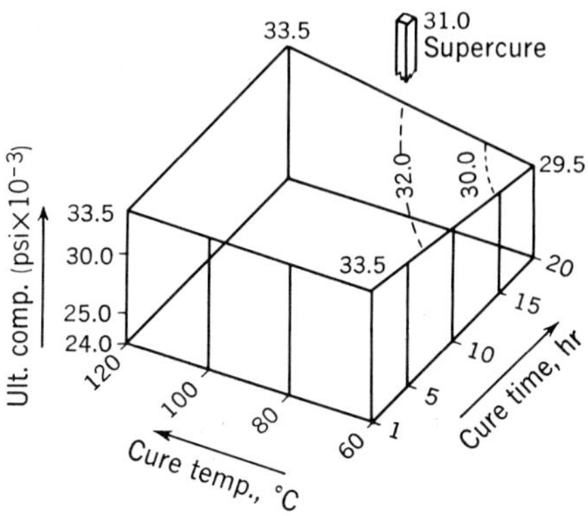

FIG. 7-7. Compressive strength of DGEBA cured with DEAPA [4–7].

With DGEBA, DEAPA is usually used at 4 to 8 phr [4]. At 8 phr, 95 per cent of the epoxy groups will be consumed on a cure cycle of 24 hours at 25°C plus 24 hours at 65°C [35]. With DGEBA (mol. wt. 450 to 580), room-temperature cure will be adequate, probably because of the accelerating effect of the hydroxyls present on the resin chain.

The effects of cure temperature and time on the various physical properties of the DEAPA-cured DGEBA are presented in Figs. 7-6 to 7-11. Shrinkage data are given in Table 7-12.

Alicyclic Polyamines (Table 7-1)

The commercial alicyclic polyamines may contain primary, secondary, or tertiary amine groups. Typical of these are menthane diamine [34], N-aminoethylpiperazine [38] 2,6-diaminocyclohexanol [62], and hydrogenated aromatic

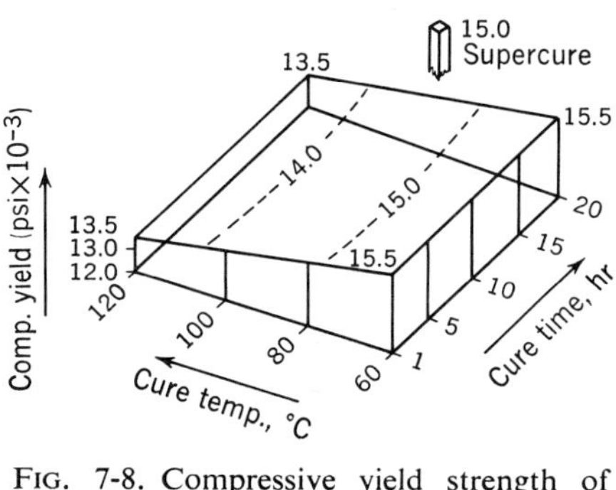

FIG. 7-8. Compressive yield strength of DGEBA cured with DEAPA [4–7].

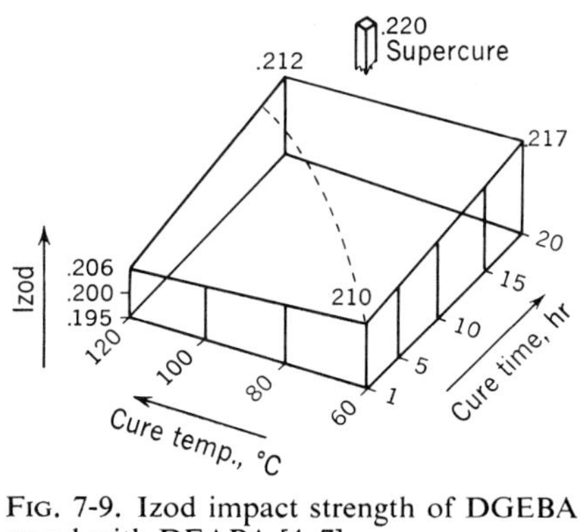

FIG. 7-9. Izod impact strength of DGEBA cured with DEAPA [4–7].

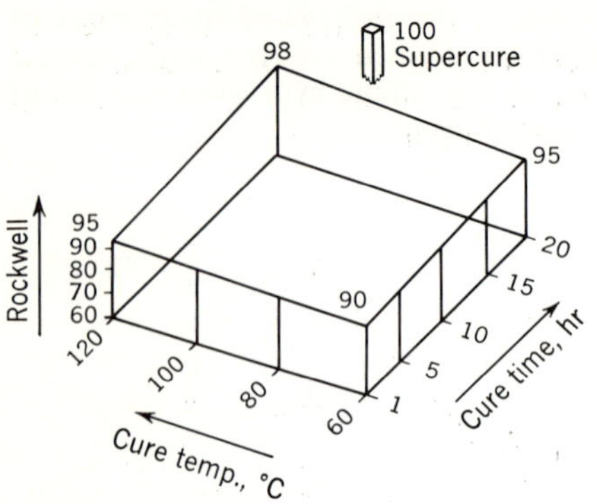

FIG. 7-10. Rockwell hardness M of DGEBA cured with DEAPA [4–7].

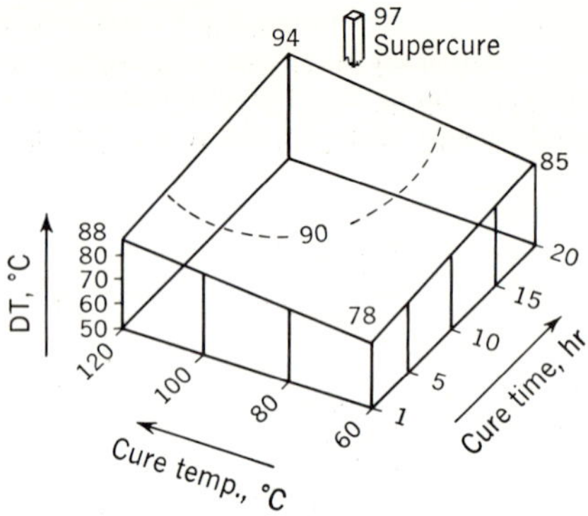

FIG. 7-11. DT of DGEBA cured with DEAPA [4–7].

diamines such as 1,3-diaminocyclohexane [13]. The synthesis of a number of other cycloaliphatic amines and their use as epoxy curing agents is described in [14 to 17].

Menthane Diamine. Menthane diamine (Table 7-2) is used principally with the aromatic glycidyl ether resins to provide long pot lives with fast, moderate cure cycles, from 2 hours at 80°C to 30 minutes at 130°C. When a postcure of 150°C for overnight is employed, the DT is about 160°C for DGEBA. Although menthane diamine provides DTs as high as are experienced with some aromatic amine

Table 7-12. Shrinkage of DGEBA Cured with a 6-phr Diethylaminopropylamine [57]

Property	*Value*
Mixing temperature	23°C
Cure, hr/°C	15/23 + 5/46 + 15/74
Peak exotherm	103°C
Shrinkage, %	
Total shrinkage............	3.75
Pregelation shrinkage	2.87
Postgelation shrinkage....	0.25
Cooling shrinkage......	1.12

curing agents, the thermal stability is more characteristic of the aliphatic amines, and substantial deterioration and outgassing will occur when the cured system is aged at 160 to 200°C. Other physical properties fall in the range provided by the linear aliphatic primary amines.

***N*-Aminoethylpiperazine.** *N*-Aminoethylpiperazine (AEP) (Table 7-2) contains a tertiary amine in the cyclic ring which promotes catalytic cure only sluggishly, if at all. Reacting the compound with excess butyl glycidyl ether results in an adduct not capable of effecting any substantial reaction in DGEBA even after 4 hours at 120°C. Increasing the temperature to 150°C gives a slow reaction. It is probable that the tertiary nitrogen is even more greatly hindered in the actual resin system, so that for practical purposes it can be disregarded as a catalytic site unless long-time, elevated-temperature postcures are employed. However, AEP appears to have increased reactivity such that it is perhaps more reactive than the aliphatic primary amines generally.

At stoichiometric concentration with DGEBA, AEP provides a DT of 110°C. Izod impact strengths are on the order of 1.3 ft-lb, a threefold increase over a comparable TETA-cured system. The high impact resistance, however, is only obtained when the system is cured for 2 or more hours at 100°C. Room-temperature cure, even for 30 days, provides Izod values in the 0.4 to 0.5-ft-lb range. Improved impact resistance is probably attributable to the reduction in system functionality. It may be that rapid immobilization occurs during cure, such that the postcure is required to react the somewhat hindered formed secondary amines.

As might be expected, AEP may be used at less than stoichiometric concentrations to increase pot life and lower exotherm. But even when used at 13 phr in DGEBA it is quite active, providing a pot life of 17 minutes in a one-pound mass with a peak exotherm of about 200°C. Such a system, postcured for 2 to 4 hours at 80 to 100°C gives properties falling in the general range of TETA, but with reduced chemical resistance.

Aliphatic Amines Containing Aromatic Groups (Table 7-1)

Representative of the aliphatic amines containing aromatic groups are *m*-xylylenediamine and tetrachloro-*p*-xylylenediamine.

Commercial *m*-xylylenediamine is a blend of the meta and para forms, with approximately 97 per cent of the former. It provides excellent heat cures with diglycidyl ether resins [51], but few data have been published on its use in this connection.

Tetrachloro-*p*-xylylenediamine imparts flame resistance to the cured epoxy-resin structure [44].

In addition to these compounds, di(*m*-xylylene)triamine [25], and the reaction products of phenol and formaldehyde [54,69,70], and various phenols [46] with aliphatic amines have been suggested as epoxy-resin curing agent. Ethylenediamine, likewise, has been reacted with various bromo- and chloromethyl phenyl ethers to produce curing agents [61].

MODIFIED AMINE CURING AGENTS

A number of modifications have been developed to improve performance of the commercial amines as epoxy-resin curing agents. Adducts have been made from alkyl mono- and diepoxies, from ethylene and propylene oxide, from acrylonitrile, and from aldehydes. Schiff's bases and amine-acid salt complexes have been introduced.

Amine Adducts with Mono- and Diepoxies

The aliphatic polyamines may be adducted with monofunctional glycidyl ethers [47], with pentachlorophenyl glycidyl ether [66], with specialized reaction products of epichlorohydrin and bisphenol A containing phenolic hydroxyls and less than one epoxy group per molecule [32], with epoxidized olefins from unsaturated fatty-acid glycerides [28], and with DGEBA and its higher-molecular-weight species. The net effect is to obtain a β-hydroxyalkyl-substituted amine. The general reaction is as follows:

$$\underset{\text{Amine}}{H_2NRNH_2} + \underset{\text{Alkyl epoxy}}{R'CH\overset{O}{\diagup\diagdown}CH_2} \rightarrow \underset{\beta\text{-Hydroxyalkylamine}}{H_2NR\overset{H}{N}CH_2\overset{OH}{\overset{|}{C}}HR'}$$

These adducts have lower volatility due to a higher molecular weight and usually somewhat heightened reactivity due to the accelerating effect of increased hydroxyl concentration [22]. Because of the higher molecular weight, the adducts are used at higher ratios. Amine-resin adducts are typical of this group of curing agents.

Amine-Resin Adducts. A typical amine-resin adduct is produced by reacting DETA with DGEBA in a jacketed kettle. DETA is heated to 75°C, and the resin is added slowly, with exothermic heat being limited to 100°C by cooling. At the conclusion of the reaction, the product is cooled to 55°C and 4 phr phenol is added to accelerate cure in thin films.

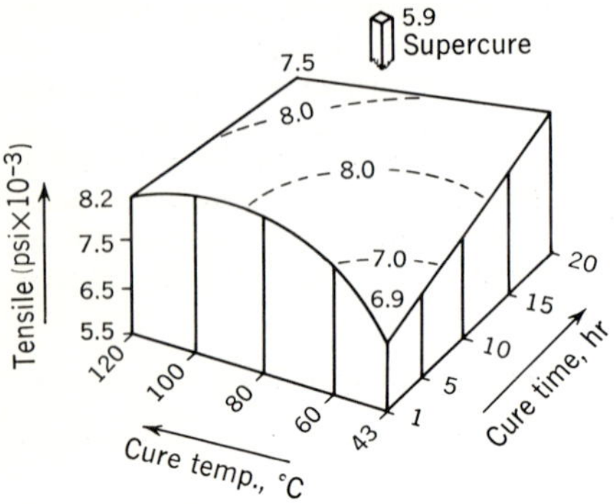

FIG. 7-12. Tensile strength of DGEBA cured with amine-resin adduct [4–7].

The adduct gives a viscosity of about 8,000 cps at room temperature and is used at about 25 phr in DGEBA. At this ratio, the exotherm will reach about 250°C in small masses, and the pot life will be about 20 minutes at room temperature in a 50-gram mass.

Solid amine adducts for solution-coating formulations are made by reaction of DETA with higher-molecular-weight species of DGEBA in solution, with a small amount of styrene oxide added. Again, accelerated cures are obtained, with the styrene oxide providing increased abrasion resistance and reduced yellowing [43].

Figures 7-12 to 7-26 present physical properties for three systems cured with a typical liquid amine-resin adduct, indicating the variations to be expected between the DGEBA, a fluidized version, and a liquid epoxidized novolac resin. Tables 7-13 and 7-14 compare electrical properties and chemical resistance for typical adduct curing agents.

Table 7-13. Electrical Properties of DGEBA Cured with Amine Adducts [4–7]

Adduct	Power factor			Dielectric constant		
	At 60 cps	At 10^3 cps	At 10^6 cps	At 60 cps	At 10^3 cps	At 10^6 cps
Amine-resin	0.009	0.019	0.032	4.4	4.2	3.7
Amine–ethylene oxide	0.015	0.027	0.047	4.7	4.5	3.6
Cyanoethylated amine	0.0128	0.0195	0.0354	4.22	4.11	3.55
Highly cyanoethylated amine	0.019	0.025	0.054	5.1	4.9	4.0

Table 7-14. Chemical Resistance of DGEBA Cured with Amine Adducts [4–7]

7 days immersion in listed fluid at 23°C	Resistance, % increase in weight			
	Amine-resin adduct	Ethylene oxide adduct	Cyanoethylated product	Highly cyanoethylated product
10% caustic	Negligible	0.40	Negligible	Negligible
Ethyl alcohol	Negligible	....	Negligible	Negligible
Kerosene	Negligible	....	Negligible	Negligible
Toluol	Negligible	....	Negligible	Negligible
Ethyl acetate	Negligible	....	Negligible	Negligible
30% sulfuric acid	0.88	5.10	2.05	4.0
Ethyl ether	0.55	1.05	1.32	0.68
Chloroform	1.81	Disintegrated	17.30	0.55
Acetic acid	2.88	Disintegrated	8.53	6.03
Acetone	0.83	Plasticized	5.91	0.03
Trichloroethylene	0.13	Plasticized	1.73	0.03

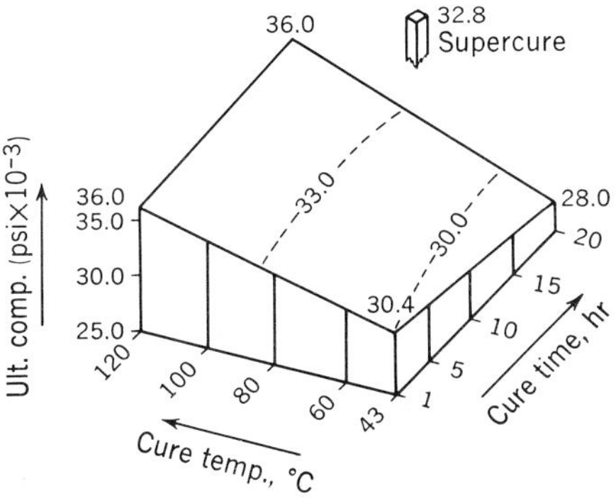

FIG. 7-13. Compressive strength of DGEBA cured with amine-resin adduct [4–7].

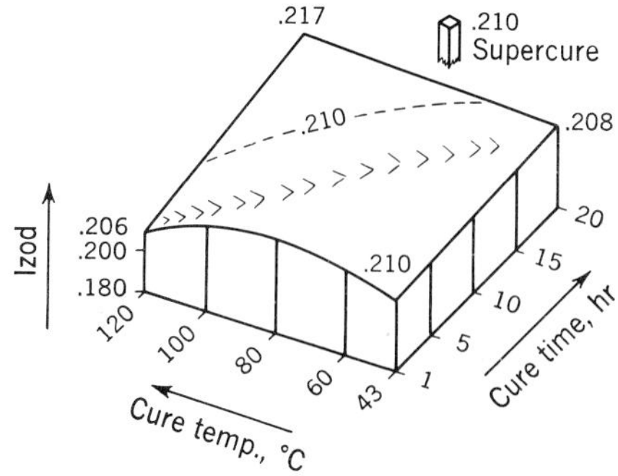

FIG. 7-14. Izod impact strength of DGEBA cured with amine-resin adduct [4–7].

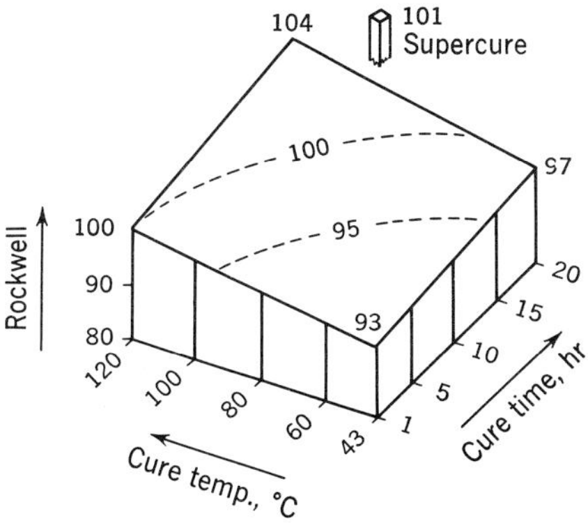

FIG. 7-15. Rockwell hardness M of DGEBA cured with amine-resin adduct [4–7].

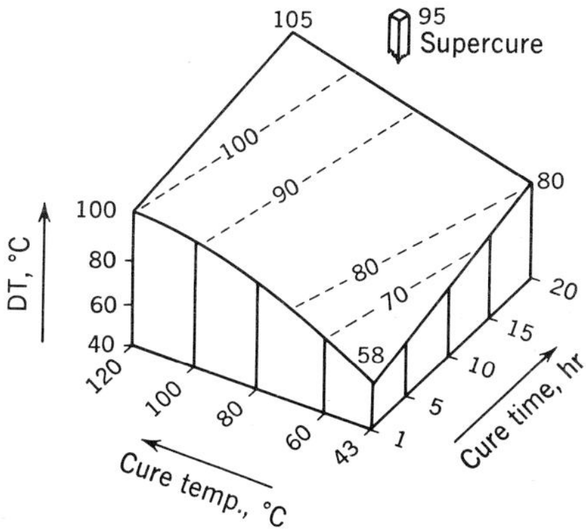

FIG. 7-16. DT of DGEBA cured with amine-resin adduct [4–7].

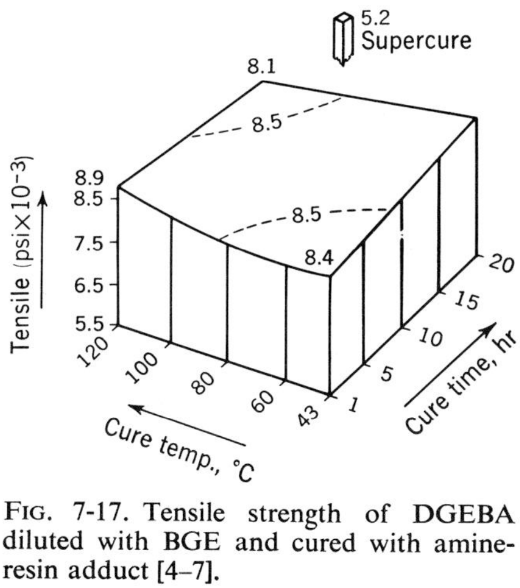

FIG. 7-17. Tensile strength of DGEBA diluted with BGE and cured with amine-resin adduct [4–7].

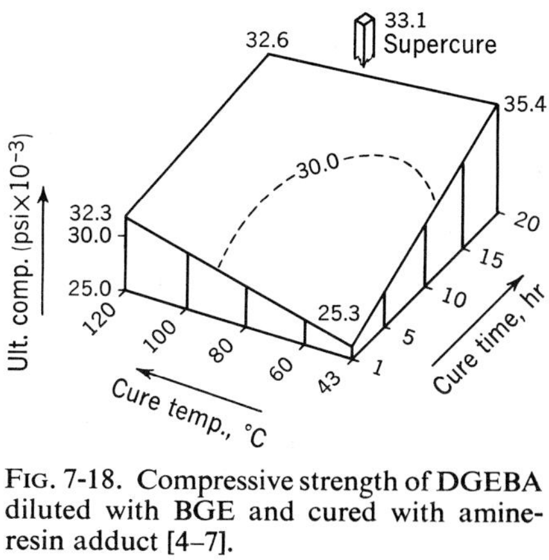

FIG. 7-18. Compressive strength of DGEBA diluted with BGE and cured with amine-resin adduct [4–7].

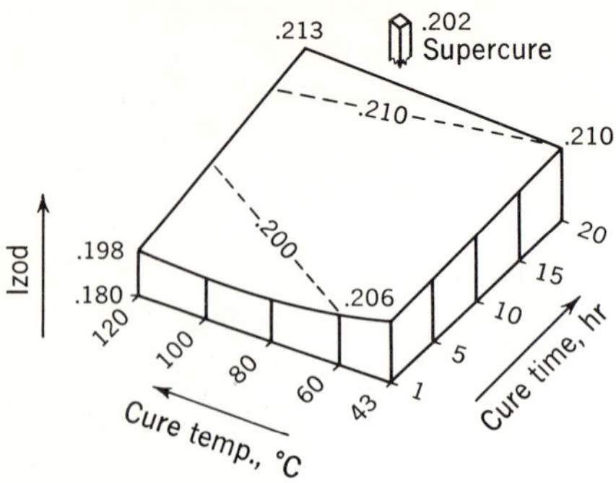

FIG. 7-19. Izod impact strength of DGEBA diluted with BGE and cured with amine-resin adduct [4–7].

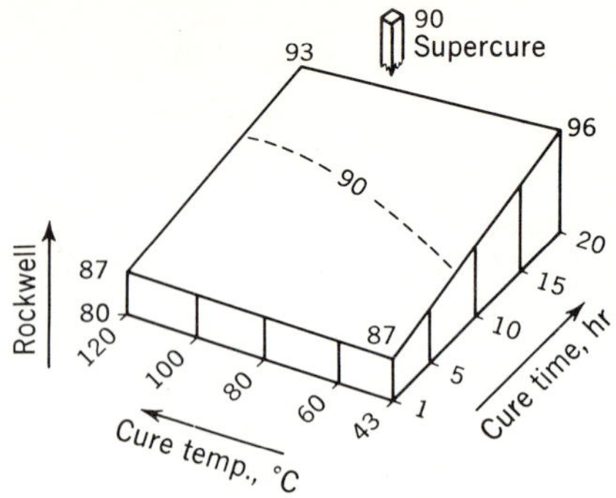

FIG. 7-20. Rockwell hardness M of DGEBA diluted with BGE and cured with amine-resin adduct [4–7].

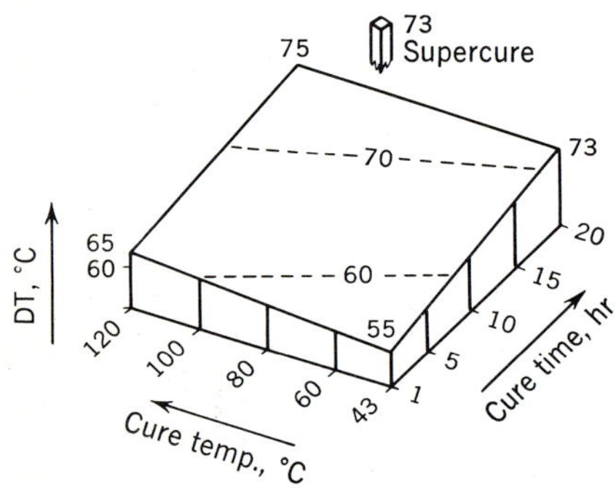

FIG. 7-21. DT of DGEBA diluted with BGE and cured with amine-resin adduct [4–7].

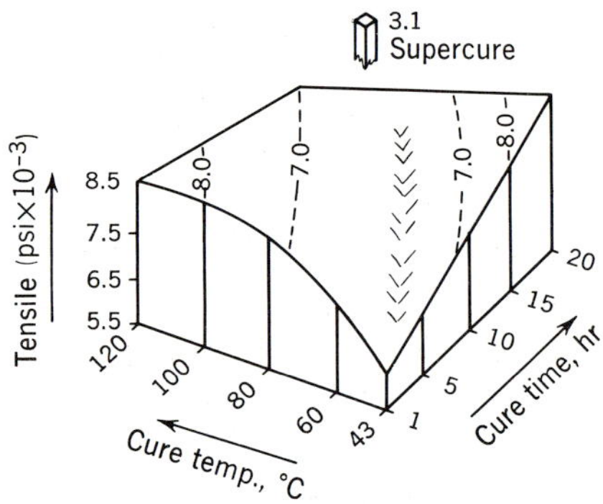

FIG. 7-22. Tensile strength of liquid novolac resin cured with amine-resin adduct [4–7].

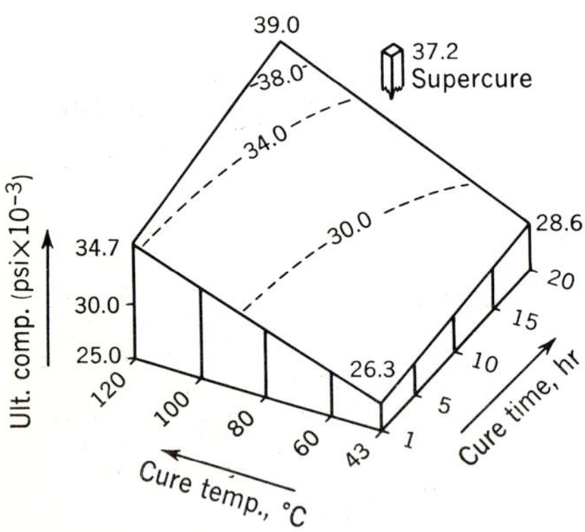

FIG. 7-23. Compressive strength of liquid novolac resin cured with amine-resin adduct [4–7].

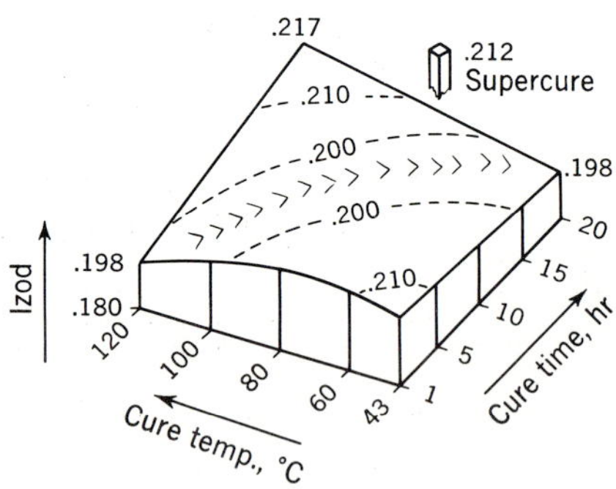

FIG. 7-24. Izod impact strength of liquid novolac resin cured with amine-resin adduct [4–7].

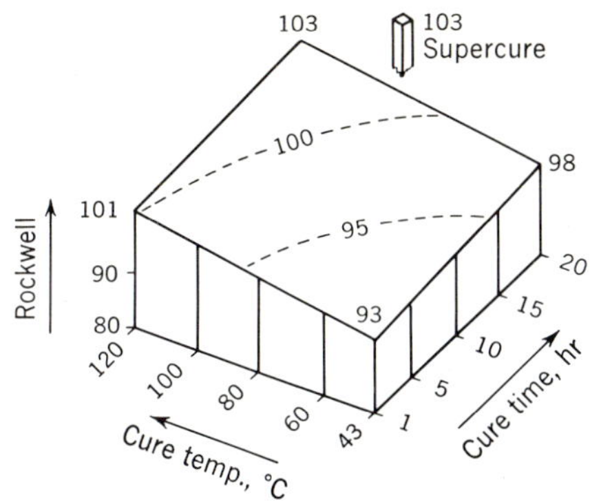

FIG. 7-25. Rockwell hardness M of liquid epoxidized novolac resin cured with amine-resin adduct [4–7].

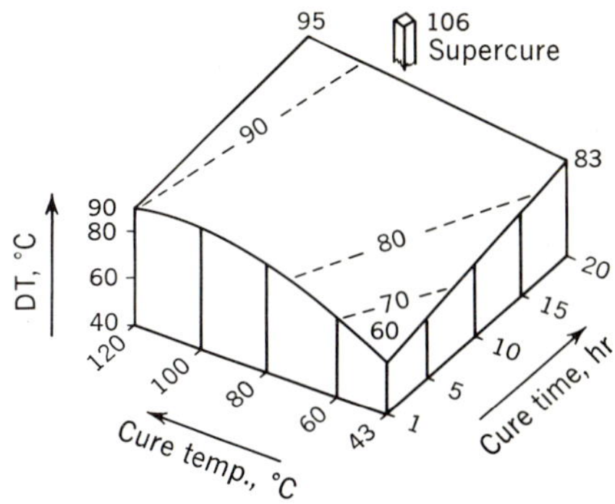

FIG. 7-26. DT of liquid epoxidized novolac resin cured with amine-resin adduct [4–7].

A wide variety of primary amines may be employed in the manufacture of resin-amine adducts. Hydrazine, for example, has been reported as satisfactory [53]. Likewise, resins other than DGEBA have been used. For example, a curing agent imparting flexibility has been developed by reacting the aliphatic polyamines with DGE of polyoxypropylene glycol [52].

The amine-resin adducts are used principally in adhesive and laminating formulations where thin sections must be cured rapidly. The higher system viscosity and the short pot life (in the glycidyl ether resins) militate against the use of the adducts in casting formulations. Despite reduced volatility, these adducts have a skin-irritation potential similar to the unadducted species.

Ethylene and Propylene Oxide Adducts. Ethylene or propylene oxide adducts of amines are obtained by reaction with the amine in the presence of water. Depending upon the ratio of reactants and the reaction conditions, the end product will consist predominantly of the mono- or bishydroxyalkyl derivatives. Typical are the following:

$$H_2N(CH_2CH_2\overset{H}{N})_2CH_2CH_2OH$$

N-Hydroxyethyldiethylenetriamine

$$HO(CH_2CH_2\overset{H}{N})_3CH_2CH_2OH$$

N,N′-Bis(hydroxyethyl)diethylenetriamine

$$H_2NCH_2CH_2\overset{H}{N}CH_2\overset{CH_3}{\overset{|}{C}}HOH$$

N-(2-Hydroxypropyl)ethylenediamine

When either ethylene or propylene oxide is employed in sufficient amounts, the reaction can be continued to produce a tetrahydroxyethyl- or propylamine [5] such as

$$(HO\overset{CH_3}{\overset{|}{C}}HCH_2)_2NCH_2CH_2N(CH_2\overset{CH_3}{\overset{|}{C}}HOH)_2$$

N,N,N′,N′-Tetrahydroxypropylethylenediamine

As the extent of the reaction progresses, the resulting compound contains fewer and fewer active hydrogens. With bis(hydroxyethyl)diethylenetriamine a polysecondary amine is formed; and with tetrahydroxypropylethylenediamine, a tertiary amine results. Polysecondary amines react similarly to primary amines, and provide cure by cross-linking with diepoxy resins when the number of active hydrogens exceeds two. In general, the tertiary amines obtained by this route are too sluggish to provide satisfactory cure, even with the glycidyl ether epoxy resins, and are normally used as accelerators in amine blends, as discussed subsequently.

Hydroxyethyl amines may also be obtained by heating the carbonate salts of mixtures of diethanolamine and monoethanolamine at elevated temperatures and superatmospheric pressures [24].

Introduction of the hydroxyethyl or propyl group on the nitrogen reduces the skin-irritation potential of the amine [22,26]. Although of reduced irritation potential, the hydroxylalkylated derivatives of ethylenediamine are, to some extent, skin-sensitizing, whereas the corresponding derivatives of DETA may be classified as MIP (minimum irritation potential) curing agents.

The hydroxyalkyl amines are hygroscopic. For example, the water pickup of *N*-(2-hydroxypropyl)ethylenediamine at 21°C and 60 per cent relative humidity in a film 4 mm thick is as follows [40]:

Time, hours	*Increase in weight, %*
5	14.5
10	25
20	36.8
30	46.5
40	56
50	59.5
100	79.5
150	90

Because of moisture pickup, curing agents of this type provide poor cures in thin films, particularly under humid conditions. This can be offset to a considerable extent by the addition of phenol, nonyl phenol, bisphenol or other acid-type accelerators.

As epoxy curing agents, the monohydroxyalkyl amines are more satisfactory than

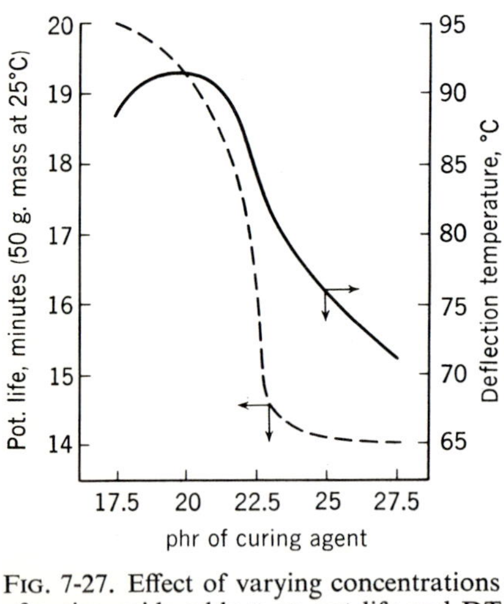

FIG. 7-27. Effect of varying concentrations of amine-oxide adduct on pot life and DT of DGEBA [4–7].

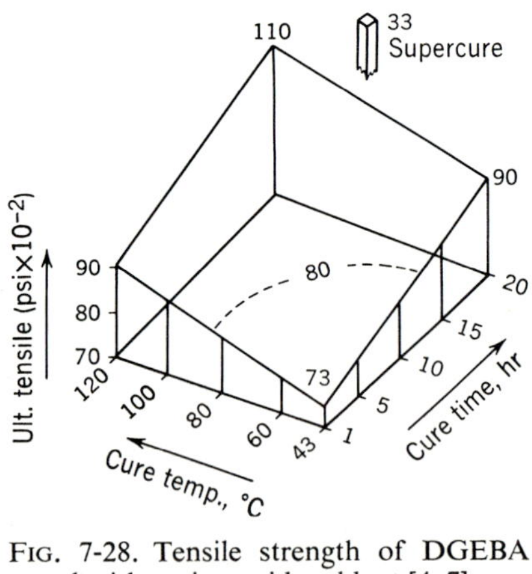

FIG. 7-28. Tensile strength of DGEBA cured with amine-oxide adduct [4–7].

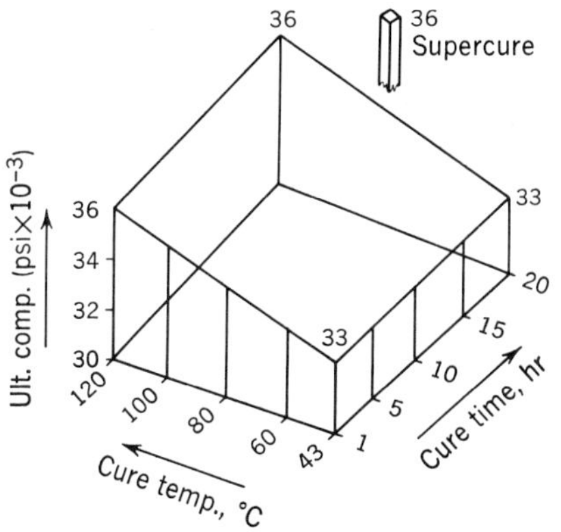

FIG. 7-29. Compressive strength of DGEBA cured with amine-oxide adduct [4–7].

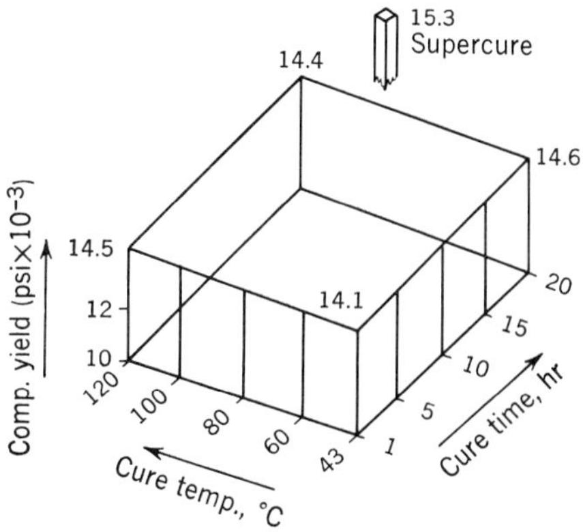

FIG. 7-30. Compressive yield strength of DGEBA cured with amine-oxide adduct [4–7].

the more highly substituted materials. In addition to reduced functionality, the di- and trisubstituted amines are only soluble with difficulty. Thus, bis(hydroxyethyl)diethylenetriamine cannot be completely dissolved into DGEBA until exothermic heat promotes solution, whereas hydroxyethyldiethylenetriamine dissolves in the resin at room temperature. Propylene oxide adducts are more satisfactory in this respect than ethylene oxide adducts, the terminal methyl groups improving solubility, even in compounds containing numerous hydroxyls [10].

Hydroxyethyldiethylenetriamine. Commercial hydroxyethyldiethylenetriamine (containing about 15 per cent bis- and 85 per cent monohydroxyethyldiethylenetriamine) has a viscosity of 200 to 500 centipoises at room temperature. When used at 20 phr in DGEBA, its pot life is 18 minutes and the peak exotherm is 250°C for a 100-gram mass. Figure 7-27 shows the effect of mixing ratio on pot life and DT. Typical physical properties for cured systems are given in Figs. 7-28 to 7-32. Tables 7-13 and 7-14 present electrical and chemical resistance data.

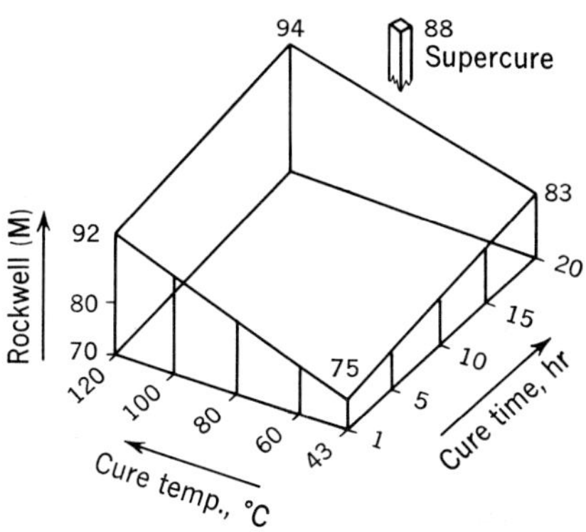

FIG. 7-31. Rockwell hardness M of DGEBA cured with amine-oxide adduct [4–7].

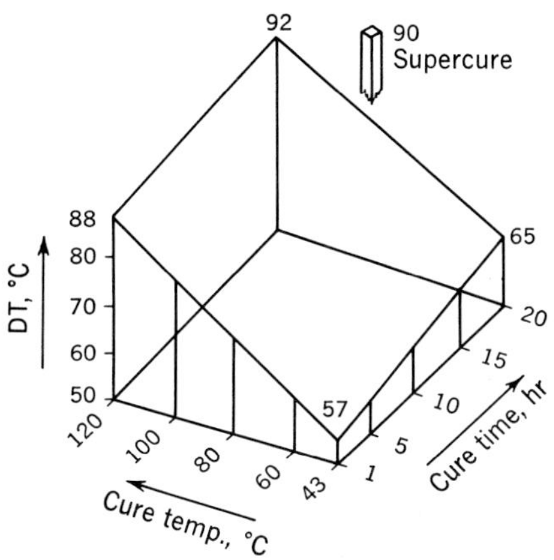

FIG. 7-32. DT of DGEBA cured with amine-oxide adduct [4–7].

Table 7-15. Pot Life of Various Amine–Ethylene Oxide Adducts When Combined at Stoichiometric with DGEBA [10]

Compound	*Pot life of 50 g, minutes*
N-(2-Phenyl-2-hydroxyethyl) DETA	18
N-(Hydroxyethyl) DETA	19
N,N′-Bis(hydroxyethyl) DETA (mixed isomers)	24
N,N′-Bis(hydroxyethyl)-3,3′-iminobis(propylamine)	25
N-(Hydroxypropyl) DETA	27
N,N′-Bis(hydroxyethyl) TETA	28
N,N′-Bis(hydroxypropyl) DETA	29
N,N′-Bis- and Tris(hydroxypropyl) DETA	34
N,N′-Bis- and Tris(hydroxyethyl) DETA	54
N-(2-Hydroxy-2,4,4-trimethylpentyl) DETA	57
N-(Hydroxypropyl)-1,2-diaminopropane	67
N-(Hydroxypropyl)-*m*-phenylenediamine	250

Pot lives of various other hydroxyalkyl amines are presented in Table 7-15. Table 7-16 gives the physical properties of resins cured with these compounds.

Because some ethylene and propylene oxide adducts are no more sensitizing than cocoa butter (used in cosmetic formulations), they are quite popular as general-purpose room-temperature curing agents for DGEBA resins.

Table 7-16. Physical Properties of DGEBA Cured with Various Amine–Ethylene Oxide Adducts [10]

Curing agents	DT, °C	Compressive properties, psi		Flexural properties, psi	
		Strength	Yield	Strength	Modulus
N,N′-Bis(hydroxyethyl) DETA	58	45,000	12,000	14,500	0.500
N,N′-Bis(hydroxypropyl) DETA	68	41,500	13,500	16,500	0.525
N,N′-Bis(hydroxyethyl) TETA	76	41,500	13,000	15,500	0.400
N-(Hydroxyethyl) DETA	96	41,000	15,000	14,000	0.410
N-(Hydroxyethyl) DETA, 85 parts, + 15 parts *N,N′*-bis(hydroxyethyl) DETA	92	43,000	15,000	15,500	0.420
N-(Hydroxyethyl) DETA, 85 parts, + 15 parts *N,N′*-bis(hydroxyethyl) DETA + 15 parts bisphenol A	84	35,000	14,600	16,900	0.475
N-(Hydroxypropyl) DETA	94	36,500	15,000	15,500	0.420
N-(Hydroxypropyl)-*m*-phenylenediamine	116	45,000	19,000	22,100	0.570
N-(2-hydroxy-2,4,4-trimethylpentyl) DETA	90	35,000	14,000	17,500	0.460
N-(2-phenyl,2-hydroxyethyl) DETA	93	39,000	15,000	12,500	0.525

Cyanoethylation Products

In addition to reaction with epoxy groups, amines may be modified by reaction with other organic compounds capable of addition to the amine groups through the active hydrogens. A large number of such compounds have been suggested, but few give any advantage over the unmodified amines. However, because accelerating hydroxyls are not developed, the reaction of amines with acrylonitrile (cyanoethylation) yields curing agents with reduced reactivity. Depending on the extent of the modification, the degree of reactivity can be regulated rather markedly [9].

Table 7-17. Effect of Varying Degrees of Cyanoethylation of DETA on Reactivity with DGEBA [7]

Property	DETA/acrylonitrile ratio			
	1/0	1/1	1/1.33	1/2
Viscosity at 25°C, centipoises	7	30	50	110
Mixture by weight, resin curing agent	9/1	5/1	4/1	3/1
Pot life at 25°C, minutes:				
50 g	40	80	150	240–300
1 lb	20	40	90	120
Peak, exotherm, °C				
1-in. cube	198	178	None	None
2-in. cube	...	223	200	48
3-in. cube	263	231	226	150

Cyanoethylation involves addition to the double bonds of acrylonitrile:

$$\underset{\text{Polyamine}}{H_2NRNH_2} + \underset{\text{Acrylonitrile}}{CH_2{=}CH{-}C{\equiv}N} \rightarrow \underset{\text{Monocyanoethylamine}}{H_2NR\overset{H}{N}CH_2CH_2C{\equiv}N} \xrightarrow[\text{Acrylonitrile}]{CH_2{=}CH{-}C{\equiv}N}$$

$$\underset{\text{Biscyanoethylamine}}{N{\equiv}CCH_2CH_2\overset{H}{N}R\overset{H}{N}CH_2CH_2C{\equiv}N}$$

Normally, DETA is used with acrylonitrile to produce epoxy-resin curing agents, although the cyanoethylated hexamethylenediamine has been reported as a flexibilizing co-curing agent [42].

Table 7-17 presents various parameters relative to the extent of cyanoethylation of DETA. Biscyanoethyl DETA represents about the limit for a useful product.

A typical cyanoethylation product from a near-equimolar ratio of acrylonitrile and DETA has a room-temperature viscosity of 30 to 40 centipoises. It is used stoichiometrically at 22 phr with DGEBA, reducing the viscosity of the mix to 100 to 800 centipoises. At the 22-phr ratio, it gives only a slight improvement in pot life over the unmodified amine, and the peak exotherm reaches 230°C for a 50-gram batch. The mechanical properties of the cured resin system are presented as a function of cure time in Figs. 7-33 to 7-35. Tables 7-13 and 7-14 present electrical properties and chemical resistance of the cured system.

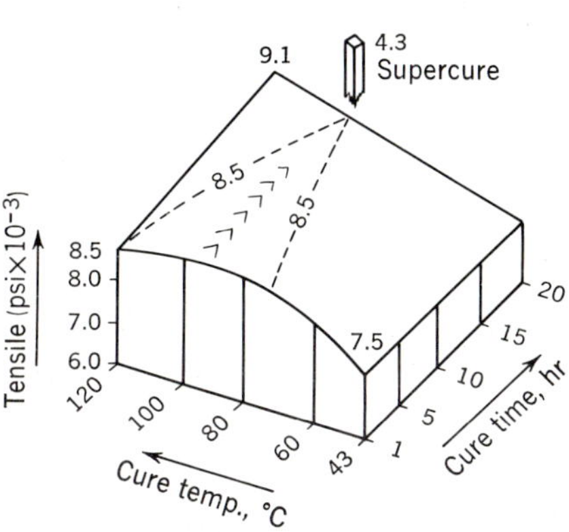

FIG. 7-33. Tensile strength of DGEBA cured with cyanoethylation product [4–7].

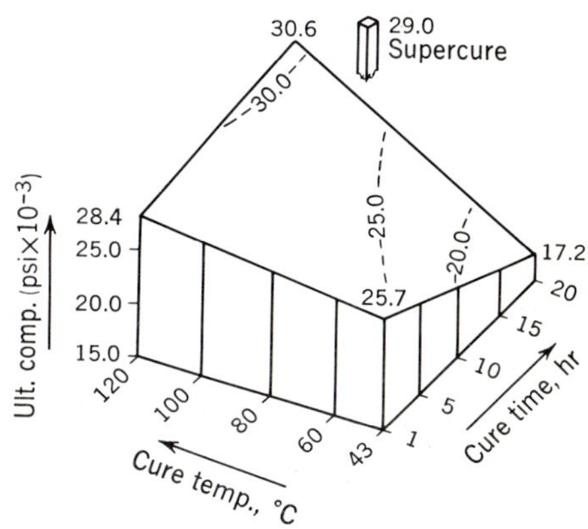

FIG. 7-34. Compressive strength of DGEBA cured with cyanoethylation product [4–7].

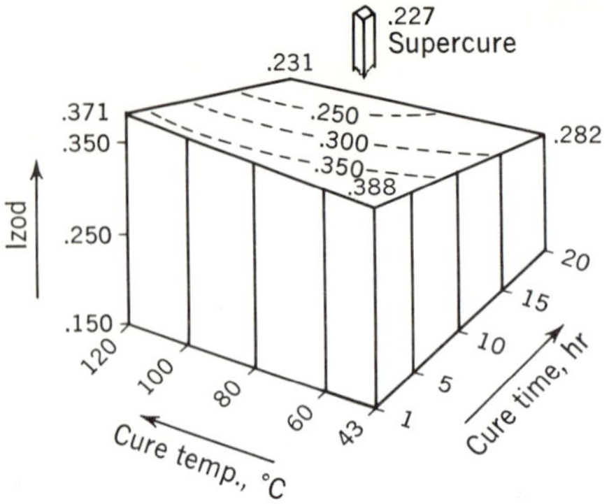

FIG. 7-35. Izod impact strength of DGEBA cured with cyanoethylation product [4–7].

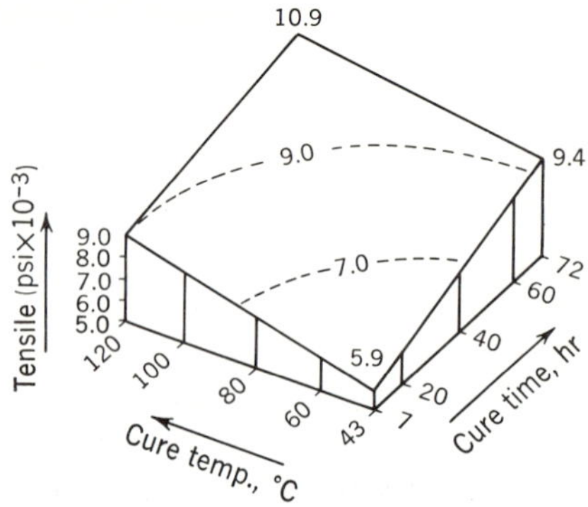

FIG. 7-36. Tensile strength of DGEBA cured with highly cyanoethylated product [4–7].

A more highly modified cyanoethylation product from DETA has a room-temperature viscosity in the 90- to 125-centipoise range. The stoichiometric ratio for DGEBA is about 38 phr, providing a mixed system with an initial viscosity of 800 to 1,600 centipoises. Pot life, in 50-gram batches, is in excess of 2.5 hours at room temperature; in 1-lb batches, 60 to 80 minutes. Exotherm is negligible in smaller amounts; in 1-lb batches, it reaches about 180°C. A recommended optimum cure cycle is 2 hours at 70°C, 3 hours at 100°C, and 1 hour at 110°C. Higher temperatures can be employed to provide reduced cure time at the expense of somewhat higher exotherms. Typical physical properties for the cured system are presented in Figs. 7-36 to 7-40. Electrical properties and chemical resistance data are presented in Tables 7-13 and 7-14.

The cyanoethylation products are useful in laminating and adhesive formulations, and as the extent of the modification increases, the wetting properties of the resin mix are improved. The reduced exotherms of the more highly modified curing agents make them useful for casting applications. The vapor pressure also is reduced as the extent of modification proceeds—from 0.2 mm Hg at 25°C for DETA to less than 0.1 mm Hg at 25°C for biscyanoethyl DETA, making the modifications of value for vacuum casting. Like the commercial polyamines, the cyanoethylation products are skin sensitizers.

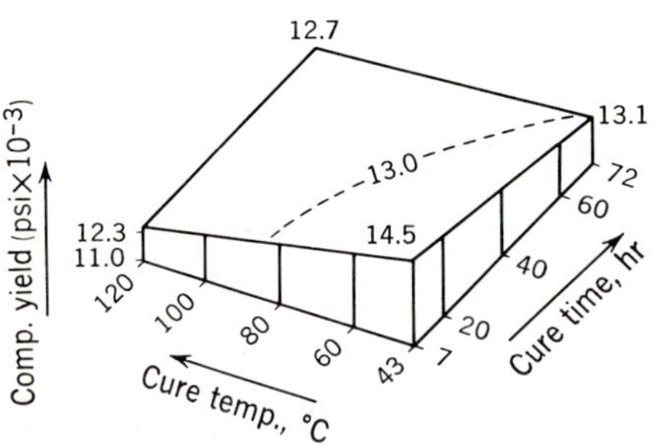

FIG. 7-37. Compressive yield strength of DGEBA cured with highly cyanoethylated product [4–7].

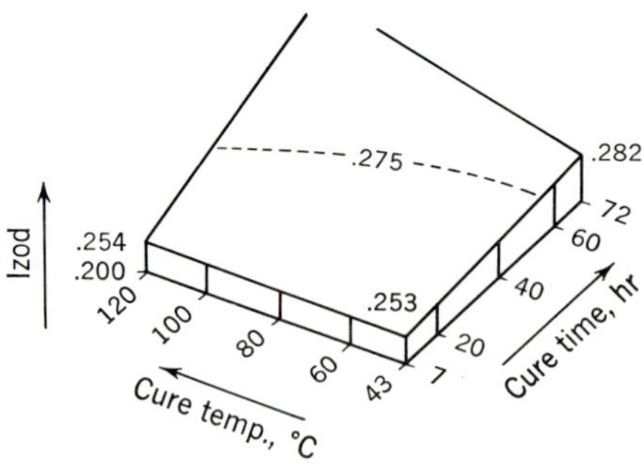

FIG. 7-38. Izod impact strength of DGEBA cured with highly cyanoethylated product [4–7].

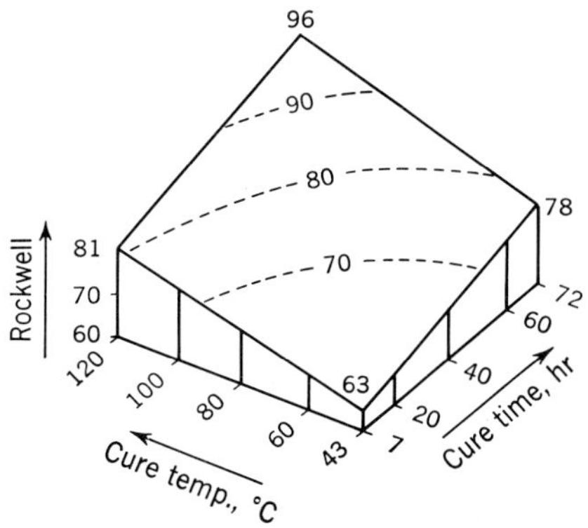

FIG. 7-39. Rockwell hardness M of DGEBA cured with highly cyanoethylated product [4–7].

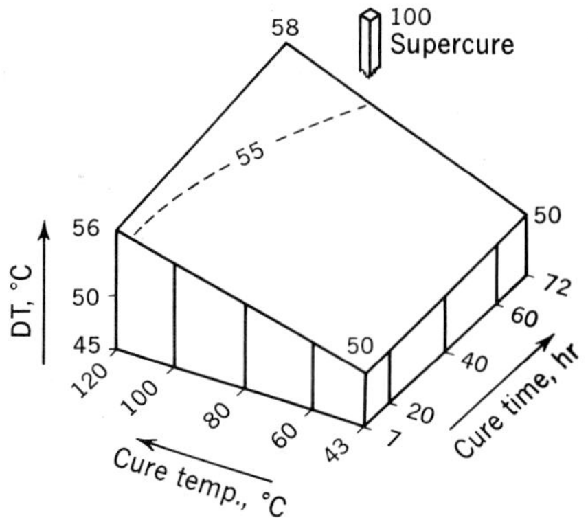

FIG. 7-40. DT of DGEBA cured with highly cyanoethylated product [4–7].

Schiff's Bases and Ketone-blocked Polyamines

Amines may be condensed with aldehydes to produce Schiff's bases and with ketones to produce alkylnitrilo (1-alkyl) alkylidyne compounds (sometimes incorrectly called "ketimines"), which are useful semilatent curing agents. For example, 2 moles of ketone reacted with diethylenetriamine

$$\underset{\text{Ketone}}{R_2C{=}O} + \underset{\text{Diethylenetriamine}}{H_2NCH_2CH_2\overset{H}{N}CH_2CH_2NH_2} \rightarrow$$

$$\underset{\text{Ketone-blocked Polyamine}}{R_2C{=}N{-}CH_2CH_2\overset{H}{N}CH_2CH_2{-}N{=}CR_2} + \underset{\text{Water}}{H_2O}$$

yield a compound in which both primary amines are blocked. Reaction of the imine with 1 mole of phenyl glycidyl ether eliminates the active hydrogen on the secondary amine:

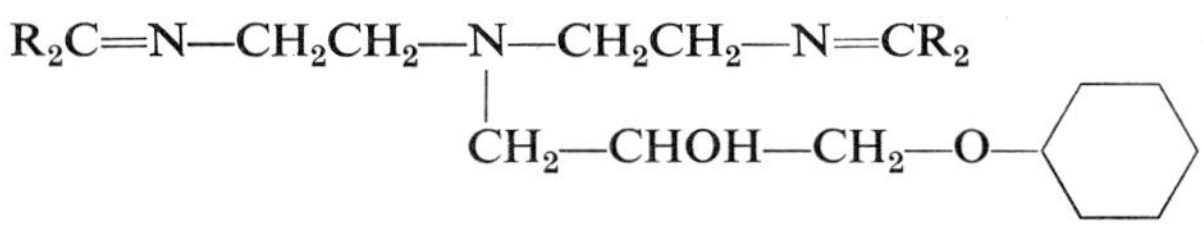

Such a compound may be incorporated into an epoxy resin, and will prove very nonreactive. However, because of hydrolytic instability of the double bond in these compounds, hydrolysis occurs with moisture from the air once the resin is applied, and the regenerated primary amines provide cure. The ketone which is regenerated diffuses to the surface and evaporates [55,56].

Such imines have gained considerable popularity in long-pot-life, room-temperature-curing 100 per cent solids-coating formulations. They are most frequently used with DGEBA at about 25 to 30 phr. Such mixtures provide fluid systems with pot lives up to 8 hours, and yet are capable of curing in a few hours at room temperature in films, the atmospheric moisture liberating the amine. Film thicknesses obtainable generally fall in the 4 to 8-mil range, considerably in excess of the thickness normally achieved by a one-coat solvent system.

In coating formulations, the use of simple mixtures of ketones and amines rather than the condensates will produce some of the advantages of the condensates without the need to remove the water of reaction. For example, DETA may simply be mixed with methylethyl ketone to give a low-viscosity complex which, when added to DGEBA provides a pot life of 6 hours or so in a mass, and yet gives fairly rapid cures at film thicknesses on the order of 5 mils as the solvent dissociates from the amine and evaporates. Figure 7-41 shows the effect of ketones on the pot life of an amine-catalyzed DGEBA.

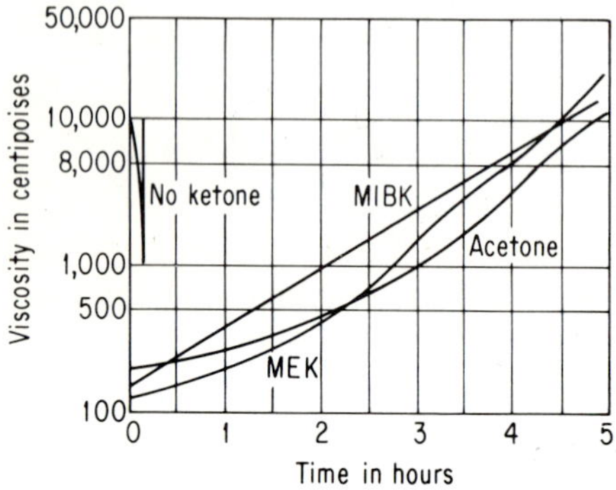

FIG. 7-41. Effect of ketones on working life of DGEBA reacted with modified aliphatic polyamine [4–7].

Acid Salts–Primary Amine Complexes

Amines may be complexed with the metallic salts of either organic [23] or inorganic [36] acids to obtain long-pot-life curing agents which break down to provide cure at about 120°C. The metal forms a valence bond with the active hydrogen to block reactivity at lower temperatures. It has also been reported that the inorganic acid salts of primary amines may be dissociated with water [58], and acid salts of amine-resin adducts have been used as curing agents in fabric-treating applications [33]. Diesters and hydrochloric acid salts of monoethanolamine have been reported as curing agents [59].

Although such complexes involving primary amines have been investigated, the acid salts of tertiary amines are the more commonly used curing agents, as discussed in Chap. 9. The acid salts of the primary amines would appear to offer fast cure in baking-type coatings, but their use in castings would be of questionable value. Zinc benzoate salts of DETA have been reported as accelerators for epoxy/phenol-formaldehyde blends [50].

AMINE BLENDS

It is often desirable to replace a portion of an aliphatic polyamine with a second amine in order to develop a curing agent offering properties somewhere between those provided by the two materials separately.

A primary amine, such as DETA, may be blended with a fatty monoamine of the general structure

$$CH_3(CH_2)_nNH_2$$

Fatty amine

(where n is from 8 to 22) to provide a curing agent imparting some degree of flexibility to the cured system. A primary amine may be blended with a secondary amine, such as ethylenediiminodi(*o*-cresol)

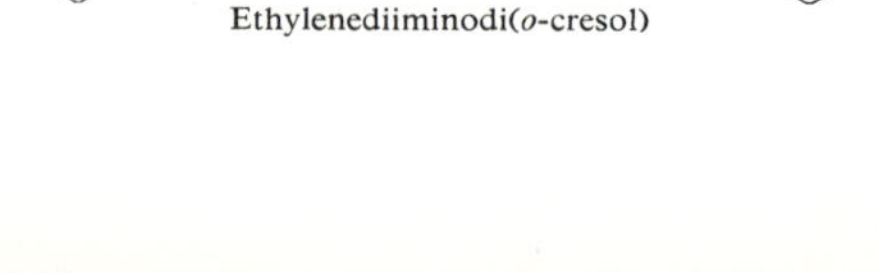

Ethylenediiminodi(*o*-cresol)

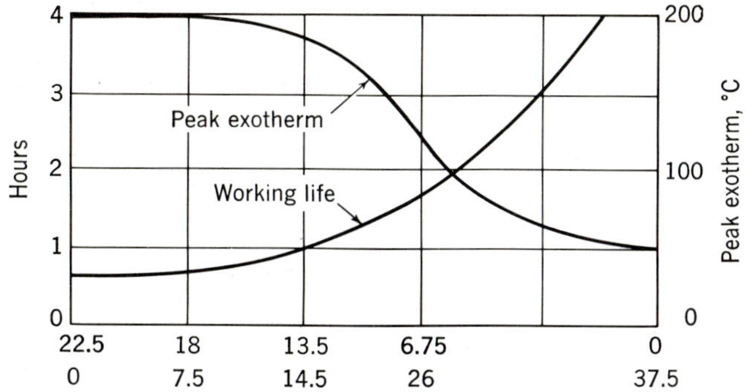

FIG. 7-42. Effect of blend of cyanoethylated products on pot life and exotherm of DGEBA [4–7].

to incorporate phenolic residues into the cured structure. Or, as is more often the case, a primary amine may be blended with a tertiary amine [53].

A blend of TETA and triethanolamine has been reported to provide a synergistic lengthening of pot life [12]. Other hydroxylated tertiary amines have also been employed with primary amines to produce long-pot-life, low-exotherm systems [29]. The reaction product of ammonia and ethylene glycol produces a blend of amines suitable for use as an epoxy curing agent, the blend containing a small amount of

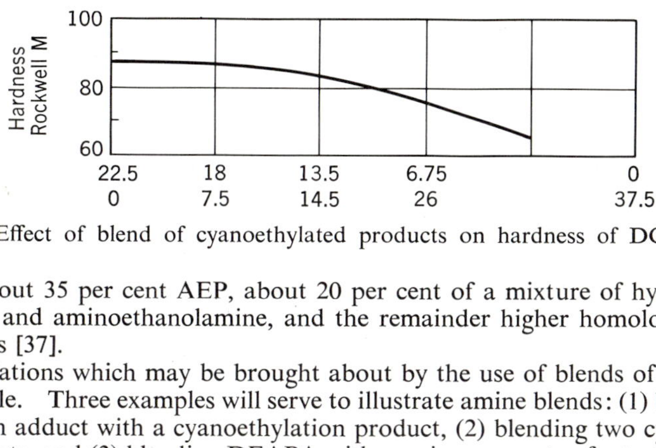

FIG. 7-43. Effect of blend of cyanoethylated products on hardness of DGEBA [4–7].

DETA, about 35 per cent AEP, about 20 per cent of a mixture of hydroxyethylpiperazine and aminoethanolamine, and the remainder higher homologs of these compounds [37].

The variations which may be brought about by the use of blends of amines are considerable. Three examples will serve to illustrate amine blends: (1) blending an amine-resin adduct with a cyanoethylation product, (2) blending two cyanoethylation products, and (3) blending DEAPA with varying amounts of a tertiary amine.

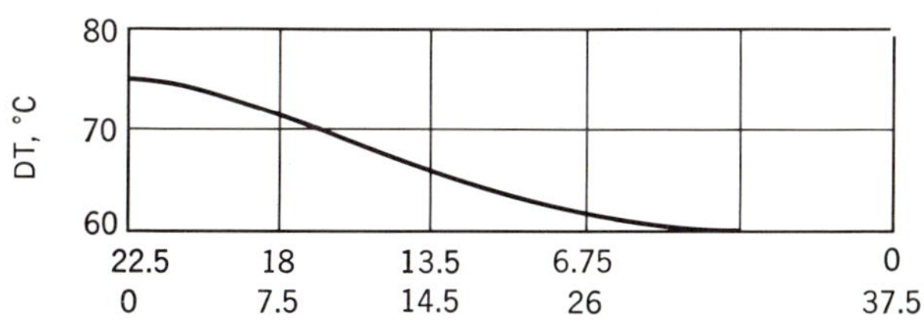

FIG. 7-44. Effect of blend of cyanoethylated products on DT of DGEBA [4–7].

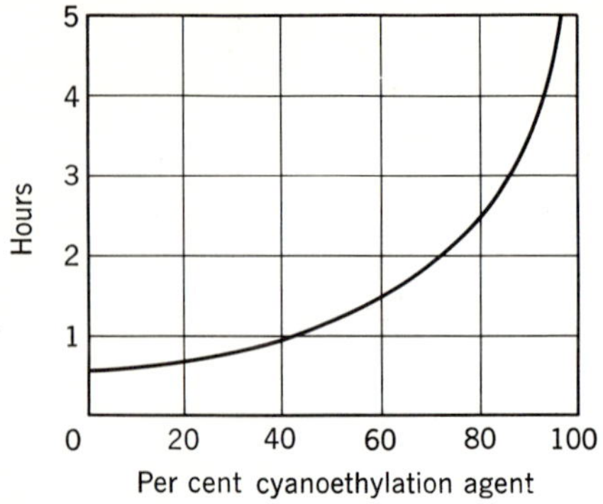

FIG. 7-45. Effect of blends of cyanoethylation product and amine-resin adduct on pot life of DGEBA [4–7].

When a cyanoethylation product synthesized from a 1/1 molar ratio of reactants is blended with a more highly modified cyanoethylation product, the effect is to increase the pot life of the mixture and decrease the exotherm over what would be expected of the less modified compound. These data are presented in Fig. 7-42. The effect on the hardness and DT is presented in Figs. 7-43 and 7-44.

Similarly, to improve the pot life of an amine-resin adduct, a highly modified cyanoethylation product can be employed. Figure 7-45 presents the effect on pot life realized with increasing percentages of the cyanoethylation product.

The tertiary amines are particularly effective co-curing agents for the primary amines. Such blends may be formulated to reduce viscosity and to regulate the

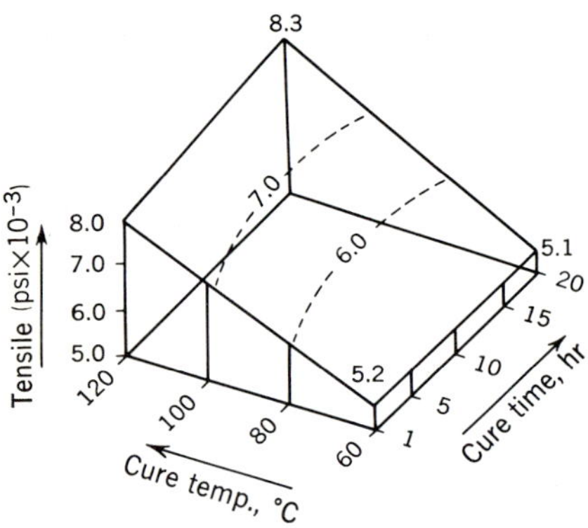

FIG. 7-46. Tensile strength of DGEBA cured with a blend of 5 phr primary and 2 phr tertiary amine [4–7].

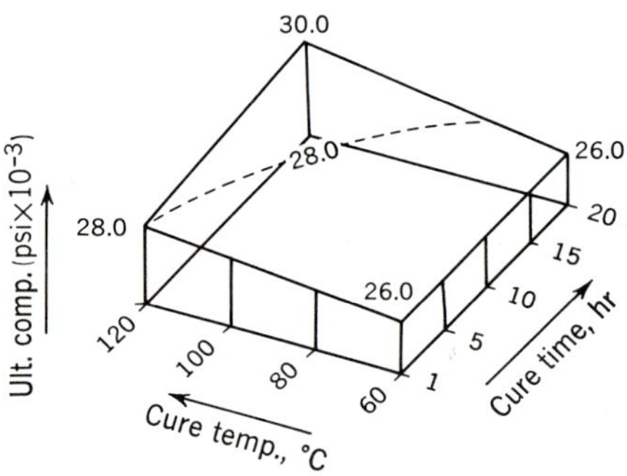

FIG. 7-47. Compressive strength of DGEBA cured with a blend of 5 phr primary and 2 phr tertiary amine [4–7].

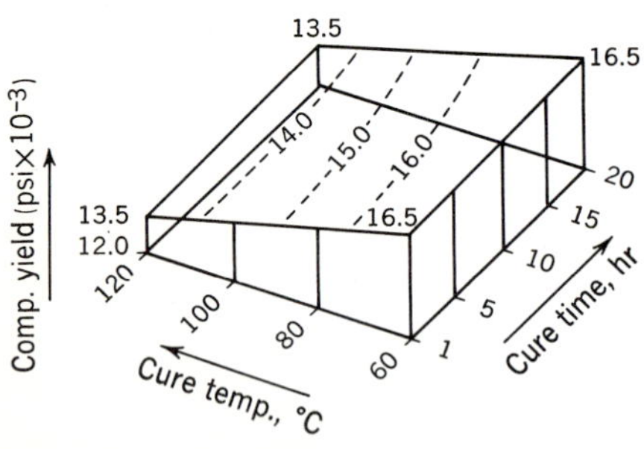

FIG. 7-48. Compressive yield strength of DGEBA cured with a blend of 5 phr primary and 2 phr tertiary amine [4–7].

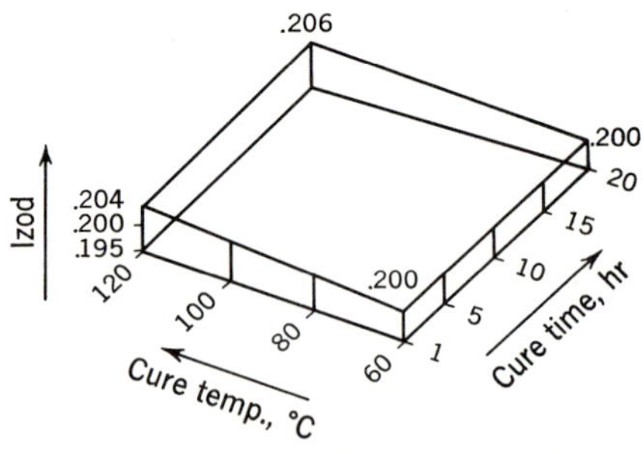

FIG. 7-49. Izod impact strength of DGEBA cured with a blend of 5 phr primary and 2 phr tertiary amine [4–7].

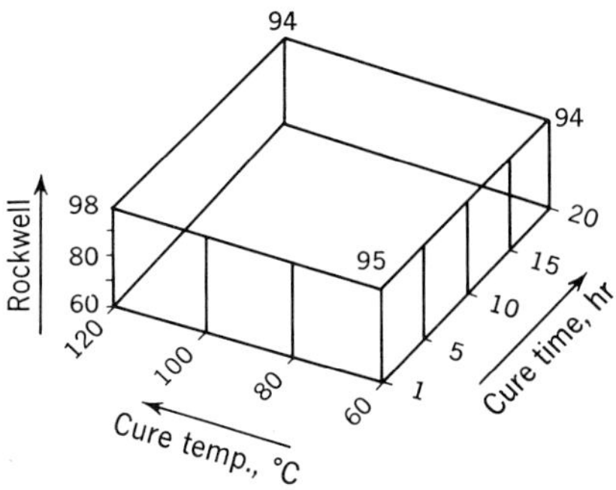

FIG. 7-50. Rockwell hardness M of DGEBA cured with a blend of 5 phr primary and 2 phr tertiary amine [4–7].

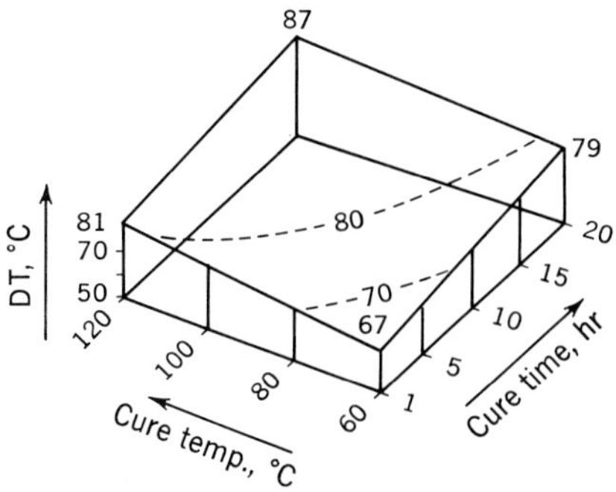

FIG. 7-51. DT of DGEBA cured with a blend of 5 phr primary and 2 phr tertiary amine [4–7].

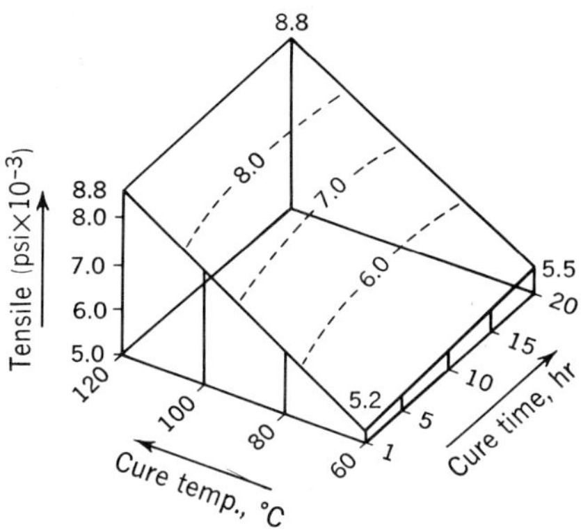

FIG. 7-52. Tensile strength of DGEBA cured with a blend of 3 phr primary and 4 phr tertiary amine [4–7].

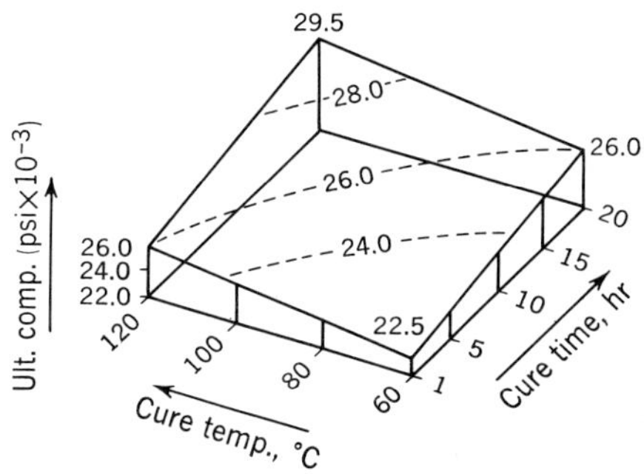

FIG. 7-53. Compressive strength of DGEBA cured with a blend of 3 phr primary and 4 phr tertiary amine [4–7].

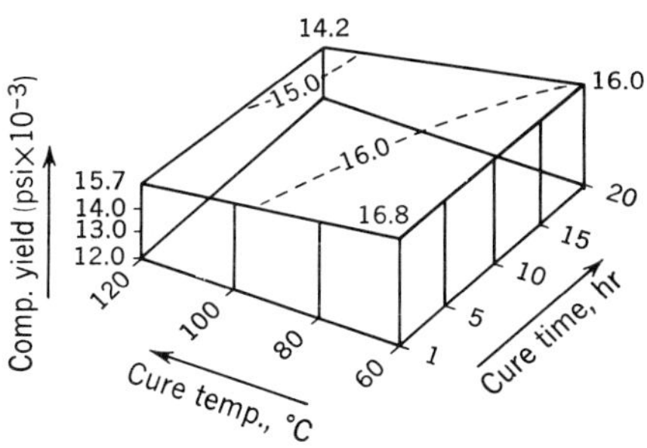

FIG. 7-54. Compressive yield strength of DGEBA cured with a blend of 3 phr primary and 4 phr tertiary amine [4–7].

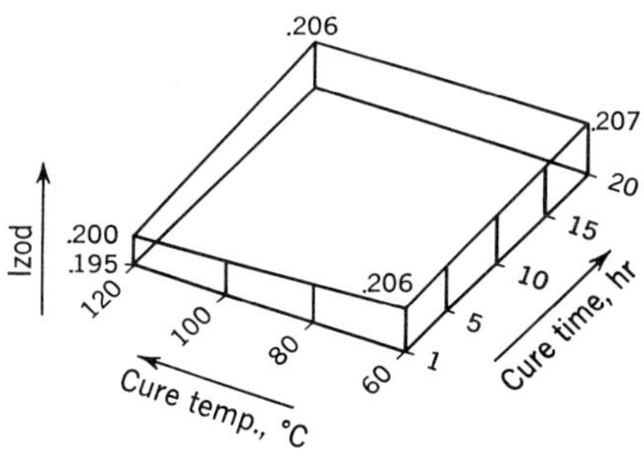

FIG. 7-55. Izod impact strength of DGEBA cured with a blend of 3 phr primary and 4 phr tertiary amine [4–7].

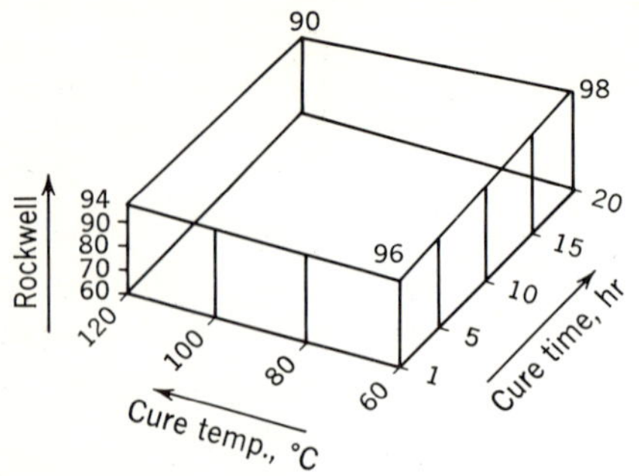

FIG. 7-56. Rockwell hardness M of DGEBA cured with a blend of 3 phr primary and 4 phr tertiary amine [4–7].

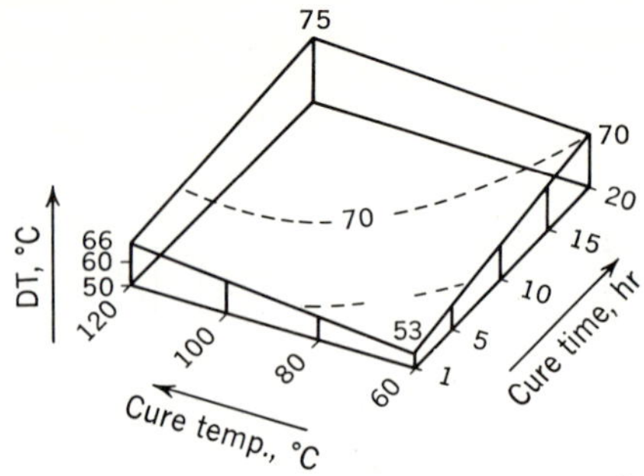

FIG. 7-57. DT of DGEBA cured with a blend of 3 phr primary and 4 phr tertiary amine [4–7].

speed of the reaction. Physical properties for DEAPA in combination with 1-hydroxyethyl-2-heptadecenylglyoxalidine at a 5 phr/2 phr ratio are presented in Figs. 7-46 to 7-51. The same compounds at 3 phr/4 phr ratio were used to cure the resin system presented in Figs. 7-52 to 7-57. These data indicate the range of systems that may be realized by the use of a single primary-tertiary amine blend.

SUMMARY

The amines were the first materials to gain general acceptance in the United States as curing agents for the epoxy resins, being used in the DGEBA resins since the beginning of the technology. Although suitable for use with the more recently introduced epoxidized olefins, the amines provide few if any advantages over the acid curing agents for this type of resin.

Of the amines, the most popular curing agents are the aliphatic primary polyamines and their modifications. The modifications consist of adducts derived from resins, from monoepoxies, or from acrylonitrile. The polyamines and their modifications may be used alone or in combination with other amines.

With the glycidyl ether resins, the aliphatic polyamines, as a class, provide for rapid, room-temperature reaction, with the attendant liberation of exothermic heat. Cure is usually obtained in masses of 20 grams or more within about one hour. In films, the cure requires overnight. A mild postcure can be used to improve system properties.

With the epoxidized olefin resins, heat and usually some form of acid accelerator are required to bring about a convenient reaction rate.

In general, when they do not contain tertiary nitrogen, the primary amines are used at or near stoichiometric (i.e., one amine hydrogen for each epoxy group). If an active tertiary nitrogen is present, the amount of amine employed may be diminished.

The primary amines generally provide well-cured resins, even with only room-temperature cure. Except for certain long-chain amines, amine-cured resins are rigid, adhesive, and chemically resistant.

Modifications to the aliphatic polyamines have been undertaken to bring about a selective improvement in their handling or other properties. The amine-resin adducts provide more convenient mixing ratios and accelerated room-temperature cures. They give properties similar to those of the unmodified amines. The adducts of polyamines and ethylene and propylene oxide provide curing agents with reduced irritation potential but somewhat reduced mechanical and chemical properties. The cyanoethylation products provide convenient mixing ratios and lower vapor pressures but reveal rather marked reduction in physical properties. Amine blends are used to permit the regulation of pot life and to modify selectively a generally suitable curing agent.

REFERENCES

1948

1. Castan, Process for the Manufacture of Thermosetting Synthetic Resins by the Polymerization of Alkylene Oxide Derivatives, U.S. 2,444,333.

1950

2. Bradley, Compositions of Matter Containing Epoxy Ethers and Diamines, U.S. 2,500,600.

1952

3. Greenlee, Amine-Epoxide Compositions, U.S. 2,585,115.

1953

4. Newey and Shokal, Curing Glycidyl Polyethers with *N,N*-Dialkyl-1,3-propanediamines, U.S. 2,642,412.

1954

5. Lundsted et al., Totally Hydroxypropylated Alkylene Diamines, U.S. 2,697,118.

1955

6. May and Shokal, Compositions Containing a Polyhydric Alcohol and Glycidyl Ethers, U.S. 2,728,744.

1956

7. Allen and Hunter, Some Characteristics of Epoxide Resin Systems, *J. Appl. Chem.*, October.
8. Damusis, Amines as Curing Agents of Ether Resins, ACS Symposium, Atlantic City, September.
9. Farnham, Epoxy Resins Cured with Amines Having Cyanoethyl Groups, U.S. 2,753,323.
10. Pitt and Paul, Low Toxicity Aliphatic Amines as Cross-linking Agents for Polyepoxy Resins, ACS Symposium, Atlantic City, September.
11. Tucker et al., Dielectric Properties of Casting Resins, *Electron. Equipment*, July.

1957

12. Homan, Method of Producing a Solidified Curing Resin, U.S. 2,783,214.
13. Shokal and Newey, Process of Curing Polyepoxides and Resulting Products, U.S. 2,817,644.
14. Smiley, Bis(1-nitrocycloalkylmethyl) Amines, U.S. 2,816,927.
15. Smiley, Bis(1-aminocycloalkyl methyl) Amines, U.S. 2,816,928.
16. Smiley, *N,N'*-Bis(1-nitrocycloalkyl)methylalkylene Diamines, U.S. 2,816,925.
17. Smiley, *N,N'*-Bis(1-aminocycloalkyl)methylalkylene Diamines, U.S. 2,816,926.

1958

18. Delmonte and Cressey, Influence of Curing Agents on Electrical Properties of Epoxy Plastics, SPE Conference, Stanford Research Institute, October.

19. Greenlee, Amine-Epoxide Compositions, U.S. 2,865,888.
20. Greenlee, Polyethylene Polyamine–Epoxide Compositions, U.S. 2,865,886.
21. Kaelble, The Dynamic Mechanical Properties of Epoxy Resins, Epoxy Resin Symposium of SPE, Minneapolis, Minn., October.
22. Newey, Process for Curing Polyepoxides by Amine Adducts and Resulting Products, U.S. 2,864,775.
23. Smith and Hall, Curing of Epoxide Resin with a Chelate, U.S. 2,819,233.
24. Steele, Di(hydroxyethyl)imidazolidone-2 and Production of *N,N'*Di(2-hydroxyethyl) Ethylene Diamine, U.S. 2,847,418.
25. Wilkes, Di(methoxylylene)triamine, U.S. 2,844,631.

1959

26. Auerbach and Ingberman, Curable Glycidyl Polyether-Polyamine Compositions, U.S. 2,901,461.
27. Bassin, The Tolerance of Novolac Epoxy Resins to Converter Ratio, ACS Symposium, Atlantic City, September.
28. Foerster, Hardening Mixtures of Epoxy Resins and Polyamide-like Condensation Products and Process for Producing Same, U.S. 2,890,184.
29. Lieb, Curing Epoxy Resins, U.S. 2,907,748.
30. Phillips et al., Resins from Bis(2,3-epoxy-cyclopentyl) Ether, U.S. 2,918,439.
31. Phillips et al., Polymers of Divinylbenzene Dioxide, U.S. 2,912,389.
32. Reese, Epoxy Resin Compositions of Improved Heat Hardening Characteristics and the Process for Their Production, U.S. 2,906,723.
33. Schroeder, Salts of Polyamine Polyepoxide Adducts and Their Use as Curing Agents for Polyepoxides, U.S. 2,909,448.
34. Shechter et al., Hardener for Epoxy Resin Compositions, U.S. 2,897,179.
35. Anderson, The Effect of an Amine Cured Epoxy Resin on the Stability of Trinitrotoluene, *SPE J.*, November.
36. Bishop, Thermosettable Epoxide Resin Compositions, Brit. 836,695.

1960

37. Harwell and Peppel, Resinous Compositions and Methods for Producing Same, U.S. 2,923,696.
38. Newey, Process for Curing Polyepoxides and Resulting Products, U.S. 2,965,609.
39. Phillips et al., Diepoxide Polymers, U.S. 2,932,626.
40. Wyandotte Chemicals Corp., technical literature.

1961

41. Belanger, Preparation and Use of Halohydrin Ether–Amine Condensates, U.S. 2,921,050.
42. George et al., Epoxy Resin Compositions, Brit. 868,733.
43. Hood, Catalyzed Epoxy Coatings and Articles Coated Therewith, U.S. 2,967,172.
44. Prescott and Bressler, Flame Resistant Epoxy Resins, U.S. 2,989,502.
45. Pschorr, Weathering of Epoxy Resins, 17th ANTEC of SPE.
46. Salensky, Coating Composition and Methods, U.S. 2,974,121.
47. Schmidle and Riley, Preparation of Hydroxyalkyl Amines, U.S. 2,989,263.
48. Shinohara and Kasamatus, Radiation-induced Crosslinking of Epoxy Resins, *Nippon Isotope Haigi Hobunshu* **4**:377–380.
49. Wegler et al., Hardening of Polyepoxy Resins with Polyamines, Ger. 1,102,391.
50. Wille and Jellinek, Accelerated Curing of Mixtures of Epoxy Resins and Phenol Formaldehyde Condensates, Ger. 1,097,132.
51. Zonsveld and Edwards, Hardening of Polyepoxides, Ger. 1,117,871.

1962

52. Bressler and Presley, Epoxy Curing Agent, U.S. 3,029,286.
53. Danusis, Method of Amine Curing Epoxy Resins and Composition Therefor, U.S. 3,023,190.

54. Halewood, Curing Agents for Epoxy Resins, Brit. 886,767.
55. Hirosawa and Delmonte, Curing Agent Comprising the Reaction Product of an Aliphatic Polyamine and an Aldehyde, and Epoxy Resin Cured Therewith, U.S. 3,026,285.
56. Improvements in or Relating to an Epoxy Resin Composition (to Shell Internationale Research Maatschappij N.V.), Brit. 905,725.
57. Spencer, Cure Shrinkage in Casting Resins, Sandia Corp. Report SCDS 2568.
58. Wagner and Weller, Dry Epoxy Resin Adhesive Composition and Process for Preparing Same, U.S. 3,050,493.

1963

59. Brotherton and Lynn, Production of Diamine and Diisocyanate Esters, Belg. 633,465.
60. Juneau, Polyamine Carbonates, 6th Symposium on Materials for Space Vehicle Use, Seattle.
61. Sonnabend, Polyamine Curing Agents for Epoxy Resins, Belg. 629,098.
62. Stansbury and Manning, Diamino Cyclohexanols, U.S. 3,112,345.

1964

63. de la Court, Catalysis of the Epoxide–Amine Reaction at 20 and 5°C, *FATIPEC Cong.* **7,** 154–159.
64. Curable Epoxy Resin Compositions (to Norton Co.), Brit. 974,139.
65. Epoxy Resin–Polyamine Compositions (to Rohm & Haas Co.), Brit. 972,988.
66. Halogen-containing Polyamines for Curing Epoxy Resins (to Ciba, Ltd.), Brit. 962,020.
67. Hydroxy-containing Imines for Use as Curing Agents and Binders for Laminates for Epoxide Coatings (to Shell Internationale Research Maatschappij N.V.), Brit. 960,236.
68. Phillips et al., Diepoxide Polymers, U.S. 3,153,018.
69. Pritchett and Lewis, Polyepoxide Compositions, Brit. 975,702.
70. Wille and Jellinek, Resinification of Compounds with More than One Epoxy Group per Molecule, Ger. 1,162,076.

1965

71. Isophoronediamine and Trimethylhexamethylenediamine, Technical Bulletin, Scholven-Chemie Aktiengesellschaft.

Chapter 8

AROMATIC PRIMARY AMINES AS CURING AGENTS

An aromatic amine is defined as one in which the amine nitrogen is attached directly to an aromatic ring. Three of the more common aromatic primary amines are:

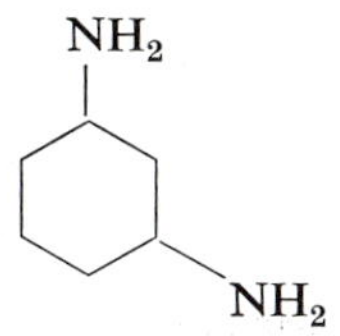

m-Phenylenediamine (MPDA)

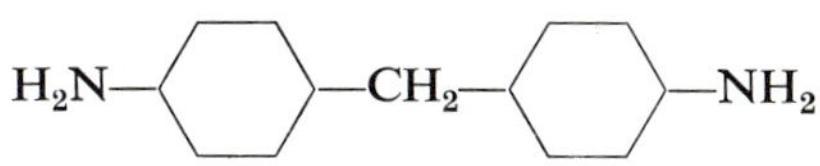

4,4′-Methylenedianiline (MDA)

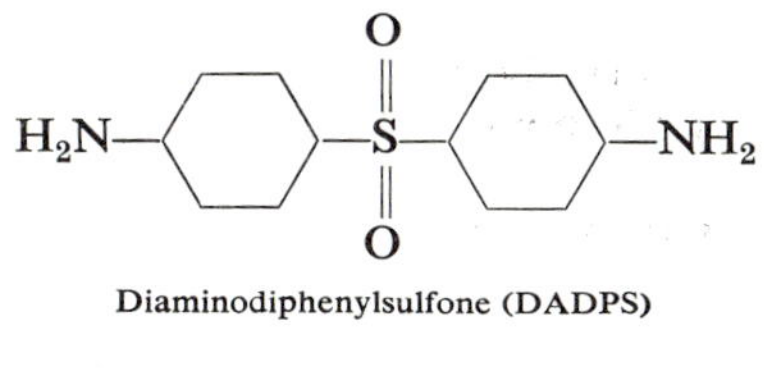

Diaminodiphenylsulfone (DADPS)

In the United States, the aromatic diamines were introduced into the epoxy technology to improve heat and chemical resistance over that attained with the aliphatic polyamines. Much of the initial work was conducted with MPDA and MDA as curing agents for DGEBA resins used in glass cloth laminates. These systems were thoroughly characterized at an early date.

The curing reaction of the aromatic amines is described in Chap. 5. The tertiary amines formed as the active hydrogens react are too hindered sterically to bring about catalytic activity. For this reason, the aromatic amines, like the aliphatic primary amines, are normally used with the glycidyl ether resins in amounts so as to provide one active hydrogen for each available epoxy group, although in a few cases a slight excess may be used.

The reaction of the epoxy groups with aromatic amines is slower than that of the aliphatic amines, owing to lower basicity, to steric factors of the aromatic ring, and to the formation of high-melt solids in the B-stage during cure. With DGEBA resins and most aromatic amines, the reaction rate and the exothermic heat at 85 to 100°C are about the same as those generated by the aliphatic primary amines when reacted at room temperature. With most epoxidized olefin resins, the reaction is unsatisfactorily slow, even at temperatures up to 150°C. With the glycidyl ether resins and sluggish aromatic diamines, such as DADPS, and with the epoxidized olefins and most if not all aromatic amines, it is customary to employ acidic accelerators. Phenols and boron trifluoride complexes are normally used. Chlorendic anhydride accelerator has been suggested as providing for a room-temperature cure with some systems [23], and maleic anhydride has been used to reduce gel times at 100°C to about one-fifth of the unaccelerated system.

Aromatic amines are generally solids at room temperature. They have been converted into liquid form by four techniques: (1) melting (and supercooling), (2) eutectic blending, (3) adduct formation, and (4) modifying with diluents.

For a given resin, the aromatic diamines provide similar cured properties. The deflection temperature (DT) will be about 150°C for most aromatic-based diepoxy resins, and up to 200°C for epoxidized novolac resins or aromatic glycidyl ether resin blends with an average of 2.5 or more epoxy groups per molecule. DADPS provides heat resistance somewhat superior to that obtained with the other commonly used aromatic diamines.

Aromatic-amine-cured systems are very resistant to chemical attack, being superior in this property to the aliphatic-amine-cured systems (Chap. 7) and superior to anhydride-cured systems (Chap. 12) against some chemicals, primarily alkalies and solvents. The heat resistance and long-time aging characteristics are superior to those obtained with the aliphatic polyamines but inferior to those obtained with some of the acid and anhydride curing agents. The aromatic amines, like the aliphatic amines, are primary skin irritants and skin-sensitizing agents.

m-PHENYLENEDIAMINE (MPDA)

MPDA is a colorless crystalline solid with a molecular weight of 108 and a melting point of about 62°C. If exposed to air, MPDA will darken noticeably because of oxidation.

MPDA possesses four active hydrogens and is used stoichiometrically with DGEBA at 14.5 phr.

Technical-grade MPDA may be heated to about 80°C and then supercooled slowly to room temperature to obtain a liquid which is stable for at least 9 months at 25°C [33].

If the solid material is to be used, two mixing procedures may be employed. The first involves melting the MPDA at about 65°C and warming the resin to the same temperature. The two components may then be blended. The other procedure involves heating the resin to about 80°C and then, while stirring continuously, dissolving the curing agent into the resin. In either case, a considerable amount of objectionable and irritating fumes will be liberated during the mixing operation.

After mixing, the system may be allowed to cool to room temperature in order to prolong the pot life. In ½-lb batches, the pot life will be about 6 hours at 25°C for DGEBA.

At elevated temperatures, or in larger batches, the MPDA/glycidyl ether resin systems generate substantial exotherm. For this reason, it is customary to gel the mixture at about 85°C for 2 hours and then postcure at 150°C.

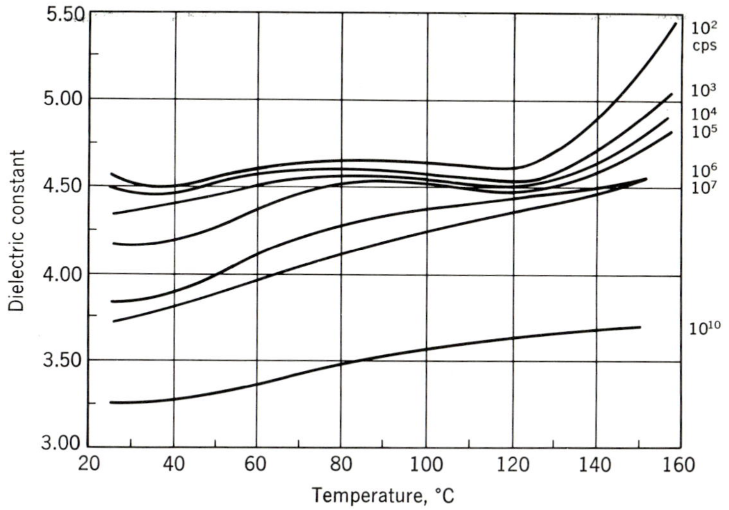

FIG. 8-1. Dielectric constant vs. temperature of DGEBA cured with MPDA [4–6].

The physical properties of DGEBA cured with MPDA are presented in Table 8-1; the chemical resistance in Table 8-2; and the electrical properties in Table 8-3 and Figs. 8-1 and 8-2. When MPDA is used to cure DGE of tetrachlorobisphenol A, properties are as indicated in Table 8-4.

With the epoxidized novolac resins, a resin functionality of something under 3 appears to offer optimum DTs. A resin with a functionality of 3.5 will give a DT of 200°C, but the functionality can be cut back by adding up to 33 phr DGEBA without substantial reduction of the DT. Even when 100 phr is used, the DT will only be reduced to 180°C. This performance of the more functional epoxidized novolacs appears to be characteristic, irrespective of the aromatic amine curing agent employed.

Epoxidized polybutadiene, cured with MPDA gives properties similar to those obtained by curing DGEBA with TETA. Phenol, or, preferably, a dihydroxy aromatic such as resorcinol as an accelerator provides for cures in 4 hours at 150°C. Table 8-5 gives the physical properties of one such system.

The ability of the aromatic amines to partially cure to dry, brittle compounds (B-stage) which soften and flow upon application of heat makes them useful in the preparation of dry lay-up laminates [5]. The life of the B-stage at room temperature is limited for MPDA and DGEBA to a few days.

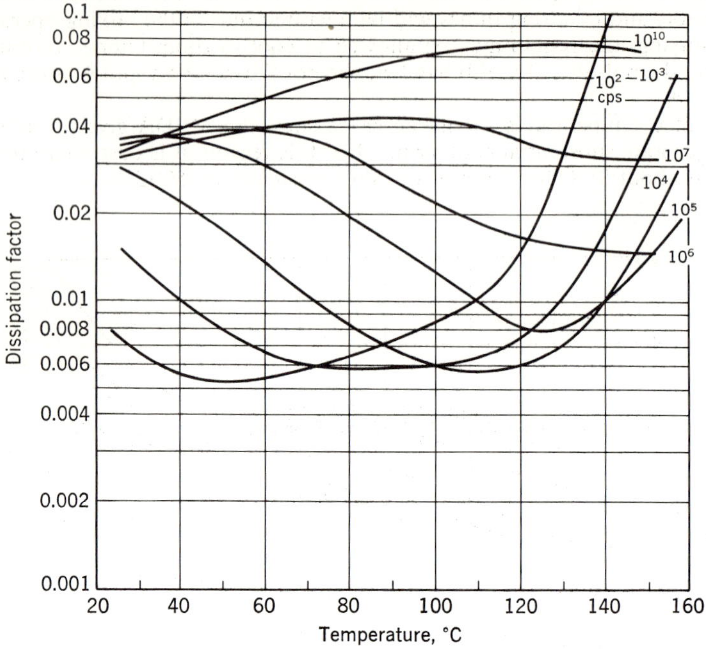

FIG. 8-2. Dissipation factor vs. temperature of DGEBA cured with MPDA [4–6].

Table 8-1. Physical Properties of DGEBA Cured with Aromatic Diamines

Property	MPDA	MDA	DADPS	Eutectic blend
DT, °C	155	155	175	150
Tensile strength, psi	8,000	9,500	9,000	9,500
Modulus, psi × 10^{-6}	0.458	0.36	0.46	0.43
Elongation, %	3.5	4.5	6	4
Flexural strength, psi	19,000	17,000	16,000	18,000
Modulus, psi × 10^{-6}	0.461	0.39	0.424	0.45
Compressive strength, psi	34,000	35,000	34,000	35,000
Modulus, psi × 10^{-6}	0.310	0.320	0.815	0.28
Impact strength, Izod, ft-lb	0.41	0.40		
Hardness, Rockwell M	112	110	116	114
Coefficient of thermal expansion, in./(in.)(°C)	0.00048	0.0005		

Table 8-2. Chemical Resistance of DGEBA Cured with MPDA and DADPS [4-6]

30 days' immersion in listed fluids	Weight change, %		Volume change, %	
	MPDA	DADPS acid-accelerated	MPDA	DADPS acid-accelerated
At 23°C:				
Sodium hydroxide, 50%	−0.2	−0.4	−0.7	−1.0
Sulfuric acid, 25%	0.7	0.7	−0.2	0.0
Hydrochloric acid, 25%	0.9	0.2	0.2	−0.2
Acetic acid, 25%	0.9	1.0	0.0	0.2
Trichloroethylene, 100%	0.3	0.2	−0.2	−0.2
Formaldehyde, 40%	0.6	0.7	0.0	0.2
Sodium hypochlorite, 6%	0.8	0.9	0.0	0.2
Chromic acid, 25%	0.1	−0.5	−0.7	−1.0
Distilled water	0.8	0.9	0.0	0.0
Sodium aluminum sulfate, 28.5%	0.7	1.0	0.2	0.7
At 54°C:				
Hydrochloric acid, 25%	2.0	0.6	1.0	−0.2
Acetic acid, 25%	1.7	1.8	1.3	1.0
Trichloroethylene, 100%	8.8	9.3	6.3	6.6
Sodium hypochlorite, 6%	0.5	0.6	−0.2	−0.2
Distilled water	1.6	1.6	0.7	1.2

Table 8-3. Electrical Properties of DGEBA Cured with Aromatic Diamines [4-7]

Property	MPDA	MDA	DADPS	Eutectic blend
Arc resistance, seconds	98	97	......	100
Dielectric strength, volts/mil	400	420	410	430
Volume resistivity, ohm-cm	1×10^{16}	0.18×10^{16}	2×10^{8}	$>2 \times 10^{8}$
Surface resistivity, ohms	$>4 \times 10^{10}$	$>4 \times 10^{10}$	4×10^{7}	$>4 \times 10^{7}$
Dielectric constant at 25°C:				
At 60 cps	4.42	4.35	4.22	4.20
At 10^3 cps	4.34	4.27	4.15	4.13
At 10^6 cps	3.80	3.72	3.94	3.58
Dielectric constant at 80°C:				
At 60 cps	4.74	4.61	4.71	4.47
At 10^3 cps	4.70	4.58	4.68	4.43
At 10^6 cps	4.33	4.22	4.35	3.86
Dielectric constant at 100°C:				
At 60 cps	4.70	4.58	4.71	4.72
At 10^3 cps	4.68	4.56	4.62	4.66
At 10^6 cps	4.42	4.35	4.38	4.12
Dissipation factor at 25°C:				
At 60 cps	0.0068	0.0066	0.0044	0.0055
At 10^3 cps	0.0183	0.0172	0.0128	0.0133
At 10^6 cps	0.0346	0.0352	0.0268	0.0318
Dissipation factor at 80°C:				
At 60 cps	0.0038	0.0029	0.0084	0.0335
At 10^3 cps	0.0035	0.0025	0.0059	0.0523
At 10^6 cps	0.0353	0.0358	0.0273	0.0352
Dissipation factor, 100°C				
At 60 cps	0.0030	0.0021	0.0177	0.0407
At 10^3 cps	0.0032	0.0022	0.0076	0.0627
At 10^6 cps	0.0268	0.0260	0.0214	0.0422

Table 8-4. Properties of DGE of Tetrachlorobisphenol A Cured with 8.8 phr MPDA [4-4]

Property	*Value*
Gel time at 120°C	20 min
DT, °C	122
Flexural strength, psi	16,750
Impact, ft-lb/in. of notch	0.30
Hardness, Rockwell M	111
Water absorption, %, in 24 hr	0.06
Weight loss, %, in 168 hr:	
At 150°C	1.01
At 177°C	5.76
Flame extinction, seconds:	
Horizontal	1
Vertical	1

Table 8-5. Physical Properties of Epoxidized Polybutadiene Cured with MPDA [4-3]

Property	*Value*
DT, °C	90
Tensile strength, psi	9,000
Modulus, psi $\times 10^{-6}$	0.145
Elongation, %	6.5
Flexural strength, psi	16,000
Modulus, psi $\times 10^{-6}$	0.489
Elongation, %	5.9
Hardness, Rockwell M	114

MPDA finds its chief use in the epoxy technology as a curing agent for DGEBA in glass cloth laminates to provide good performance in the 120 to 150°C range. MPDA is also used in some filament wet-winding applications, particularly as curing agent for blends of glycidyl ether and epoxidized olefin resins, where a relatively long pot life is offered. The B-stage systems are used in combination with reinforcing fillers as stick solders and molding compounds. Unreinforced potting and adhesive formulations are somewhat brittle and of limited usefulness.

MPDA fumes stain the skin upon contact. The stains resemble walnut shell stains and are very difficult to remove.

The para isomer is reported to be carcinogenic, but the meta isomer is free from this disadvantage.

4,4′-METHYLENEDIANILINE (MDA)

MDA is a light-brownish solid having a molecular weight of 198 and a melting point of 89°C. Several commercial grades are available.

MDA possesses four active hydrogens and is used stoichiometrically with DGEBA at about 26 phr.

With DGEBA [3], varying the amount of MDA indicates that stoichiometric quantities provide maximum DTs. At 22 phr, the DT will be 120°C; at 27 phr, 160°C; at 32 phr, 135°C. As with other aromatic amines, the use of DGEBA ($n = 0$) results in about a 10°C increase in DT.

Three mixing procedures may be employed. MDA may be dissolved in hot resin; it may be heated to 100°C and added to the resin heated to 90°C; or it may be heated to 100°C and mixed with equal parts of resin at 100°C and the mixture then added to the remainder of the resin at room temperature. The last procedure is preferred when larger batches are involved.

After mixing, it is customary to cool the mixture to prolong the pot life. The best working temperature is 50 to 60°C, pot life and viscosity permitting. The pot life of the MDA/DGEBA blend ranges from 20 hours in a 50-gram mass at 23°C to 30 minutes in a 1-lb mass at 100°C.

At elevated temperatures, or in large batches, the MDA/glycidyl ether resin blends generate substantial exotherms, reaching, for example, 260°C in 1-lb batches worked at 85°C. If larger batches are involved, it is advisable to allow the material to gel overnight at room temperature and then to postcure it for 4 hours at about 150°C. Gel temperatures below 25°C should be avoided, since the MDA may crystallize out and there is little likelihood of redissolving it before gelation.

Table 8-6. Chemical Resistance of Epoxidized Phenol-Form-aldehyde Novolac Resin Compared with DGEBA, Both Cured with MDA [10]

After 1 year's immersion at 23°C in listed fluids	% weight change	
	Novolac (f 3.5)	DGEBA
Acetone	1.9	12.4
Ethyl alcohol	1.0	1.5
Ethylene dichloride	2.6	6.5
Distilled water	1.6	1.5
Glacial acetic acid	0.3	1.0
Sulfuric acid, 30%	1.9	2.1
Sulfuric acid, 3%	1.6	1.2
Sodium hydroxide, 10%	1.4	1.2
Sodium hydroxide, 1%	1.6	1.3
Ammonium hydroxide, 10%	1.1	1.3

The physical properties of DGEBA cured with MDA are shown in Table 8-1, and the electrical properties in Table 8-3.

With the epoxidized novolac resins, it is customary to employ MDA in a 3 to 5 per cent excess over stoichiometric for optimum properties. The chemical resistances of epoxidized novolac and DGEBA/MDA cured systems on the basis of 1-year immersion tests are compared in Table 8-6.

With aromatic amines, as with the primary aliphatic amines, the epoxidized novolacs exhibit greater sensitivity to cure temperatures than does DGEBA. With MDA, for example, DGEBA gives progressively higher DTs when the cure temperature increases from 100°C (DT 125°C) to 150°C (DT 155°C). Equivalent results can be obtained by employing a stage cure. With the epoxidized novolac resins and MDA, higher cure temperatures are required and the stage cure is not so effective. This is illustrated by Figs. 8-3 and 8-4 showing the effects of cures between 140 and 180°C on the two systems.

With MDA employed to cure epoxidized olefin resins, properties are similar to those obtained with MPDA; like MPDA, MDA requires an acid accelerator.

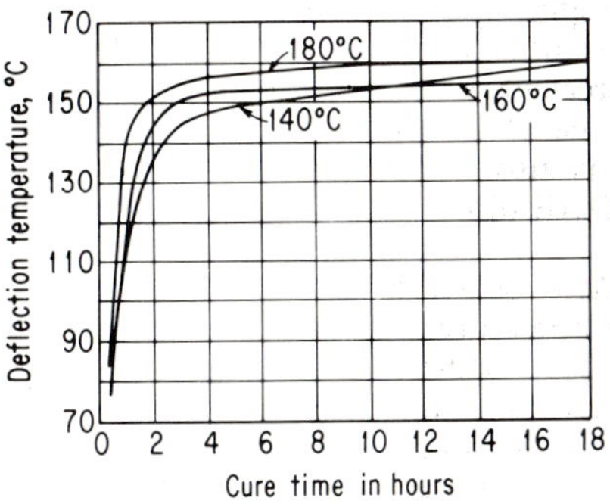

Fig. 8-3. DT vs. cure cycle for DGEBA cured with MDA [4].

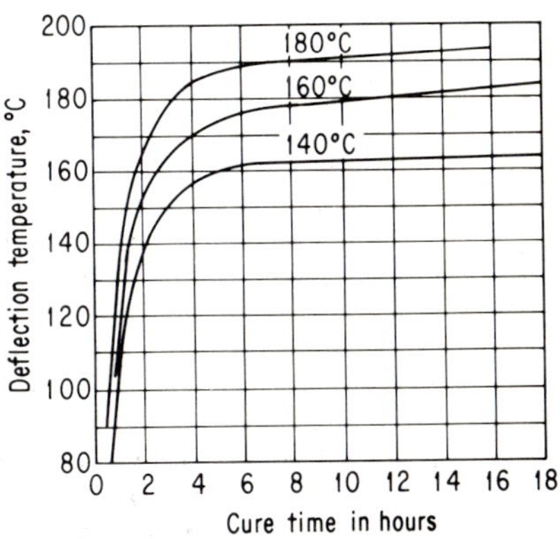

Fig. 8-4. DT vs. cure cycle for epoxidized novolac (f 3.5) cured with MDA [4].

MDA finds its chief use in the epoxy technology in laminating applications where good performance is required in the 120 to 150°C range. It is used to a lesser extent in stick solders and reinforced molding powders. Like MPDA, the unreinforced castings are somewhat brittle.

MDA does not stain as does MPDA. For this reason, and because of its lower cost, it is often preferred to MPDA in epoxy work.

DIAMINODIPHENYLSULFONE (DADPS)

DADPS is a free-flowing, reddish-to-yellow powder having a melting point of 175°C and a molecular weight of 256.

DADPS is employed stoichiometrically with DGEBA at 33.5 phr. Because of sluggish reactivity arising from low basicity, it is customary to employ the DADPS at about a 10 per cent excess to obtain best properties in unaccelerated systems. The effect of curing-agent concentration on the DT of the cured DGEBA is shown in Table 8-7. The physical and electrical properties of the DADPS-cured system

Table 8-7. Effect of DADPS Concentration on DT of DGEBA [4-1,4-6]

DADPS concentration, phr	DT, °C, no acceleration		DT, °C, with 1 phr BF_3MEA Cured 2 hr at 125°C +2 hr at 200°C
	Cured 24 hr at 120°C	+4 hr at 175°C	
10	...	...	157
14	...	...	160
18	...	...	176
20	...	...	175
22	...	...	173
24	...	...	172
26	117	138	175
28	...	...	175
30	131	162	175
32	...	...	172
33	138	169	
34	...	...	170
36	144	184	169
38	...	...	166

are presented in Tables 8-1 and 8-3. The diglycidyl ether resins may be used in connection with a more highly functional aromatic resin to improve performance at elevated temperature. The effect of increasing amounts of commercial-grade polyglycidyl ether of tetraphenylethane on the DT of a DGEBA/DADPS-cured system is indicated in Fig. 8-5. DADPS has also been used as curing agent for other temperature-resistant resins, such as DGE of 4,4′-dihydroxydiphenylsulfone [25].

When DADPS is employed with DGEBA, it provides a pot life of 3 hours at 100°C and requires a rather extended high-temperature postcure for the attainment of ultimate properties. To reduce the cure time, at the expense of cutting the pot life to about 1 hour at 100°C, an acid accelerator may be used at about 0.5 to 2 phr. When the accelerator is used, the optimum concentration of DADPS is shifted to something less than stoichiometric, as indicated in Table 8-7. The effect of postcure temperatures on the DT of the accelerated systems is indicated in Fig. 8-6.

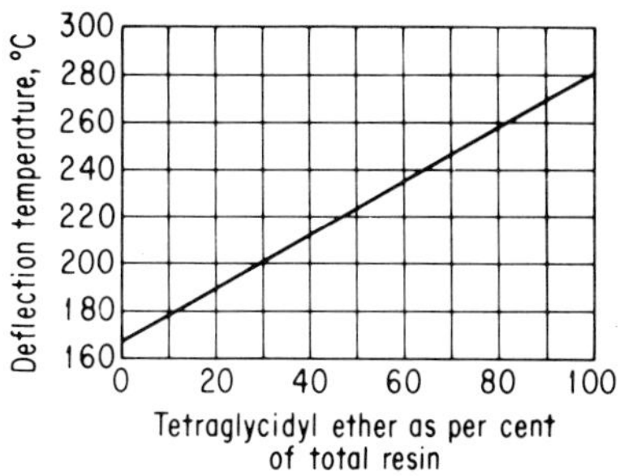

FIG. 8-5. Effect of increasing amount of polyglycidyl ether of tetraphenyl ethane on DT of DADPS-cured DGEBA [6].

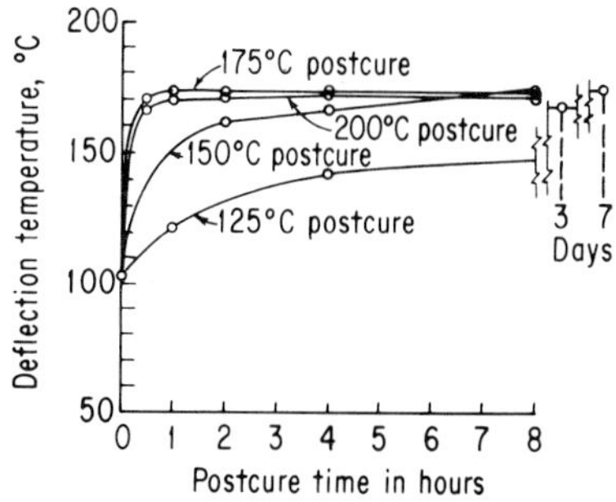

FIG. 8-6. DT vs. postcure for DGEBA/DADPS/BF_3MEA (100/30/1) [6].

The presence of Lewis-acid-type accelerators degrades the electrical properties, as indicated by Figs. 8-7 to 8-10. The chemical resistance is little influenced. The chemical resistance of the accelerated system is given in Table 8-3.

DADPS is normally employed in a hot-melt system, the room-temperature viscosity with aromatic epoxies being quite high. The usual procedure is to dissolve DADPS into DGEBA which has been heated to 135°C. After solution, the temperature is lowered to 100°C and acid catalyst is added. The mixture is gelled at 100 to 120°C to prevent the occurrence of excessive exotherm. After gelation, it is postcured for 2 or 3 hours at 175 to 200°C.

In the epoxy technology, DADPS is used primarily as a curing agent for wet and dry lay-up laminating

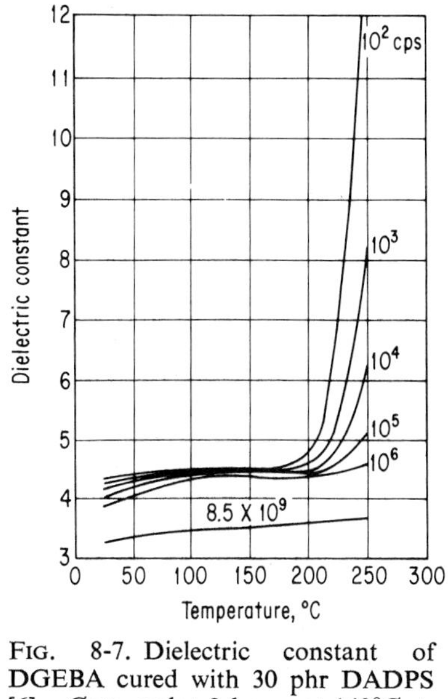

FIG. 8-7. Dielectric constant of DGEBA cured with 30 phr DADPS [6]. Cure cycle: 2 hours at 140°C + 32 hours at 200°C.

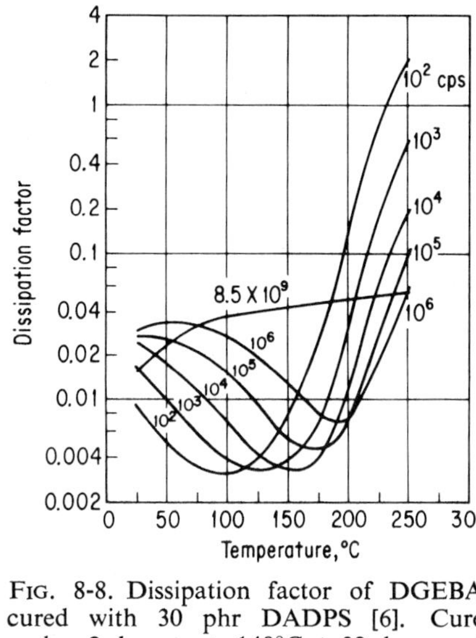

FIG. 8-8. Dissipation factor of DGEBA cured with 30 phr DADPS [6]. Cure cycle: 2 hours at 140°C + 32 hours at 200°C.

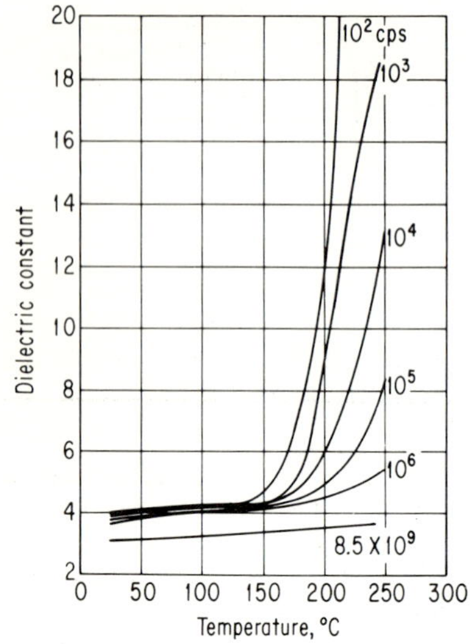

FIG. 8-9. Dielectric constant for DGEBA/DADPS/BF_3MEA (100/20/1) [6]. Cure cycle: 2 hours at 120°C + 2 hours at 200°C.

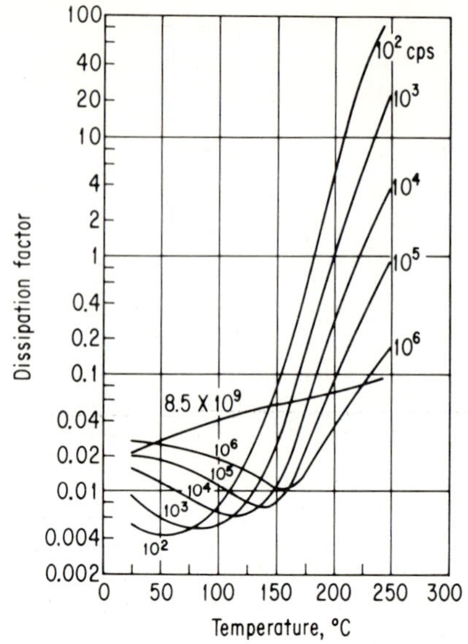

FIG. 8-10. Dissipation factor for DGEBA/DADPS/BF_3MEA (100/20/1) [6]. Cure cycle: 2 hours at 120°C + 2 hours at 200°C.

To some degree, DADPS may be a skin irritant, and contact with its fumes should be avoided. It has been administered orally for a number of years in the successful treatment of leprosy without exhibiting undesirable side effects and can therefore be classified as nontoxic, at least in the concentration which would normally be ingested by accident.

MISCELLANEOUS AROMATIC AMINES

In addition to MPDA, MDA, and DADPS, a number of other aromatic di- and polyamines have been suggested for use in the epoxy-resin technology, and cured properties have been developed for a few systems. In general, these aromatic amines have not been shown to possess advantages, and considerations of cost and toxicity have militated against their use.

The listing in Table 8-8 is representative of the aromatic amines which have been suggested.

Perhaps the most interesting of the compounds, benzidine exhibits quite good high-temperature properties and extremely good radiation resistance (Fig. 8-11); however, the material is a known carcinogen.

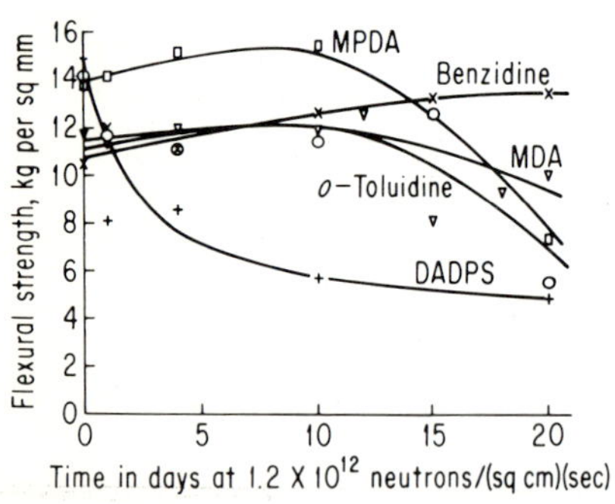

FIG. 8-11. Radiation resistance of DGEBA cured with aromatic diamines [15].

Diaminodiphenyl ether and diaminodiphenyl sulfide have been reported to be

Table 8-8. Formulas and Melting Points for Aromatic Amines Suggested as Epoxy-resin Curing Agents

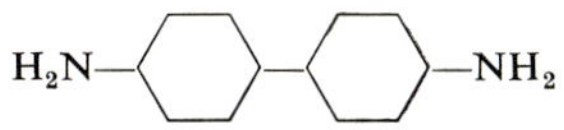

Benzidine
(m.p. 81°C)

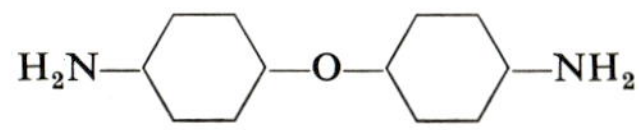

Diaminodiphenyl ether
(m.p. 186°C)

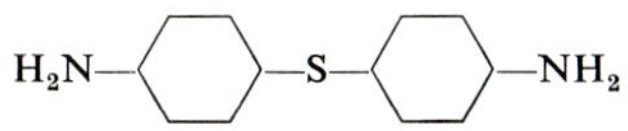

4,4′-Thiodianiline
(m.p. 108°C)

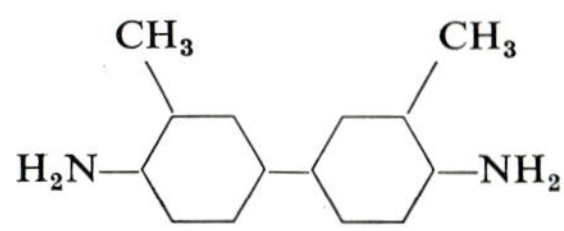

4,4′-Bis(*o*-toluidine)
(m.p. 130°C)

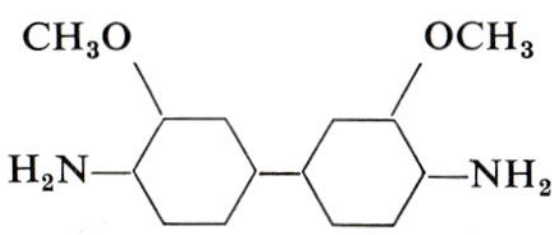

Dianisidine
(m.p. 138°C)

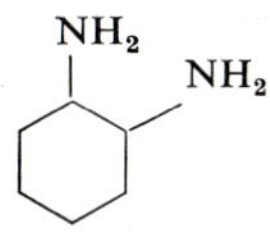

o-Phenylenediamine
(m.p. 102°C)

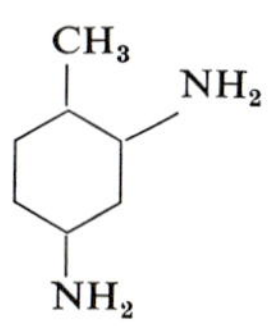

2,4-Toluenediamine
(m.p. 98°C)

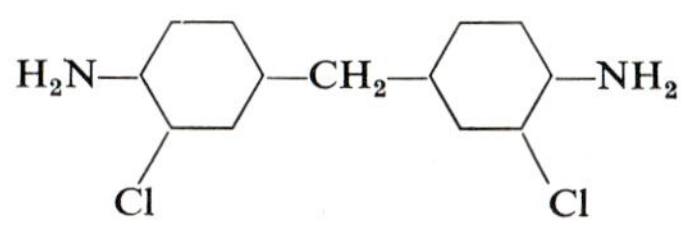

Methylenebis(*o*-chloroaniline)

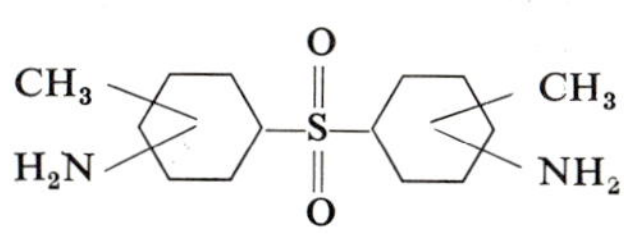

Diaminoditolylsulfone
(m.p. 70–105°C)

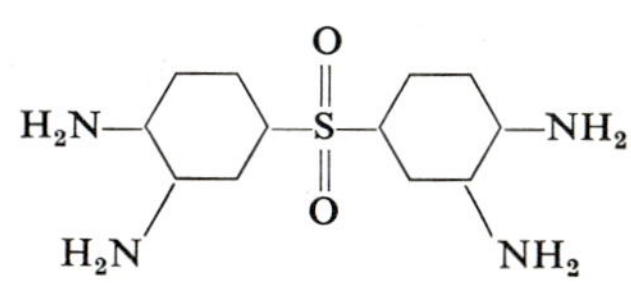

Bis(3,4-diaminophenyl)sulfone

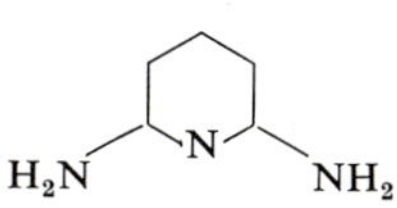

2,6-Diaminopyridine
(m.p. 121°C)

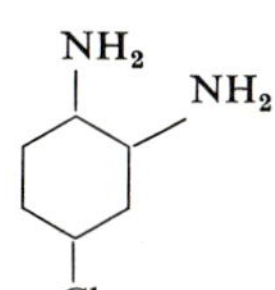

4-Chloro-*o*-phenylenediamine
(m.p. 72°C)

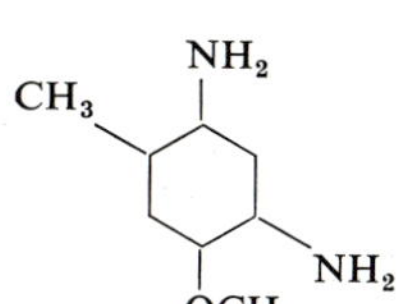

4-Methoxy-6-methyl-*m*-phenylenediamine
(m.p. 98°C)

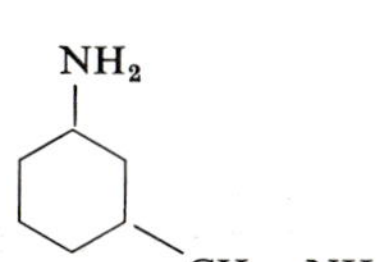

m-Aminobenzylamine
(m.p. 38°C)

Table 8-9A. Physical Properties of DGEBA Cured with Representative Diamines [4-4]

(Cure cycle: 2 hours at 120°C + 2 hours at 204°C)

Property	4-Chloro-*o*-phenylene-diamine, 18.8 phr	Benzidene, 24 phr	2,6-Diamino-pyridine, 14.3 phr	*m*-Amino-phenol, 18.2 phr
DT, °C	140	166	110	128
Tensile strength, psi	11,077	11,946	14,360	14,087
Modulus, psi $\times 10^{-6}$	0.230	0.216	0.240	0.227
Elongation, %	5.4	7.8	8.9	10.9
Flexural strength, psi	14,516	15,069	23,475	22,934
Modulus, psi $\times 10^{-6}$	0.464	0.414	0.538	0.529
Compressive strength, psi	17,129	16,631	18,227	19,126
Modulus, psi $\times 10^{-6}$	0.358	0.294	0.417	0.403
Impact, Izod, ft-lb	0.521	0.904	0.671	0.823
Hardness, Rockwell M	113	111	101	108

of reduced toxicity and increased solubility in the resins [11]. 2,4-Toluenediamine is reported to provide room-temperature cures when used with short-chain aliphatic glycidyl ether resins [19,20]. Bis(3,4-diaminophenyl)sulfone has been reported as providing DTs to 244°C with DGEBA [18]. Bis(aminoxylyl)sulfone has been used [31]. The polyamine-substituted pyridines are described as curing agents in [9] and the diaminodiphenyl ketones in [29]. Tris(4-aminophenyl)phosphate may be used as a latent curing agent for fluidized bed coatings [35].

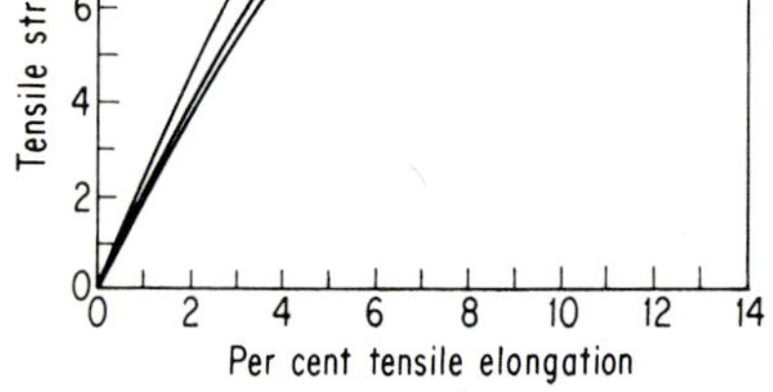

FIG. 8-12. Tensile strength vs. elongation for amine-cured DGEBA [30].

A list of the cured properties of several representative aromatic diamines is given in Table 8-9A. *o*-Phenylenediamine gives lower tensile strengths, elongations, and modulus than does the meta isomer [34].

m-Aminobenzylamine (MABA) is particularly interesting in that it contains both aliphatic and aromatic amine groups. Cured property data are given in Table 8-9B and Fig. 8-12. When used with cis-trans bis(2,3-epoxycyclopentyl) ether, MABA provides ultimate tensile strengths of 20,100 psi, with a modulus of 766,000 psi [34]. This represents highest tensile properties reported in the literature for epoxy resins. MABA is a waxy solid and requires refrigerated storage to prevent oxidation. It has been produced experimentally by the Naval Research Laboratory.

AROMATIC AMINE EUTECTICS

A number of liquid aromatic amine eutectics are available. They may be regarded as supercooled liquids and generally are characterized by crystallization when stored for extended periods of time. Minor amounts of other compounds are frequently added to prevent the crystallization. When crystallization occurs, the mixture may be returned to the liquid state by warming.

Table 8-9B. Tensile Properties of Amine-cured DGEBA [30]

Sample	MPDA			*m*-Xylylenediamine			*m*-Aminobenzylamine		
	Tensile strength, psi	Elongation, %	Elastic modulus, psi	Tensile strength, psi	Elongation, %	Elastic modulus, psi	Tensile strength, psi	Elongation, %	Elastic modulus, psi
1	12,900	7.5	427,000	10,800	7.4	410,000	12,300	11.0	426,000
2	12,600	6.8	451,000	10,800	7.8	427,000	12,200	10.3	403,000
3	12,400	6.3	488,000	10,000	4.9	433,000	12,200	10.0	402,000
4	13,000	8.7	451,000	10,800	7.6	445,000	12,100	11.5	410,000
5	12,900	8.1	496,000	10,400	5.6	464,000	12,100	9.5	406,000
Average	12,800	7.5	463,000	10,600	6.7	436,000	12,200	10.5	409,000
Standard deviation..	200		28,000	360		20,400	<100		9,700

Table 8-10. Liquid Mixtures of Phenylenediamine and Second Aromatic Amine [7]

Amines	*Weight of second amine,* %
m-Phenylenediamine and 2,4-toluenediamine	45–15
N-Methyl-*p*-phenylenediamine	95–15
N-Methyl-*o*-phenylenediamine	95–15
Mixed diaminonaphthalenes resulting from dinitration of naphthalene and subsequent reduction	95–20
Mixed toluenediamine resulting from dinitration of toluene and subsequent reduction	65–15
o-Phenylenediamine and *N*-methyl-*o*-phenylenediamine	95–50
p-Phenylenediamine and *N*-methyl-*p*-phenylenediamine	95–75

The use of eutectic blends of aromatic amines has the obvious advantage of extending the pot life of the mixture by permitting the mixing operation to be conducted at room temperature.

The most commonly used eutectic blends consist of 60 to 75 per cent MPDA and 40 to 25 per cent MDA [13]. These give properties for most practical purposes equal to those provided by either MPDA or MDA alone. When they are stabilized with phenols or with monoepoxies, their properties are somewhat reduced.

Table 8-10 furnishes data on other aromatic amine blends which give liquid eutectics. Three-component liquid eutectics of MPDA, orthophenylenediamine and 4-chloro-*o*-phenylenediamine are described in the literature [12]. The use of aniline and related substances as a third component for liquid blends has also been described [13]. The three-component aromatic diamine blends are reported not to require stabilization [6].

Properties of DGEBA cured with a typical MDA/MPDA eutectic (stabilized with bisphenol A) are shown in Tables 8-1 and 8-3. Shrinkage of a system containing phenyl glycidyl ether is indicated in Table 8-11, and electrical properties of this system are shown in Figs. 8-13A and B.

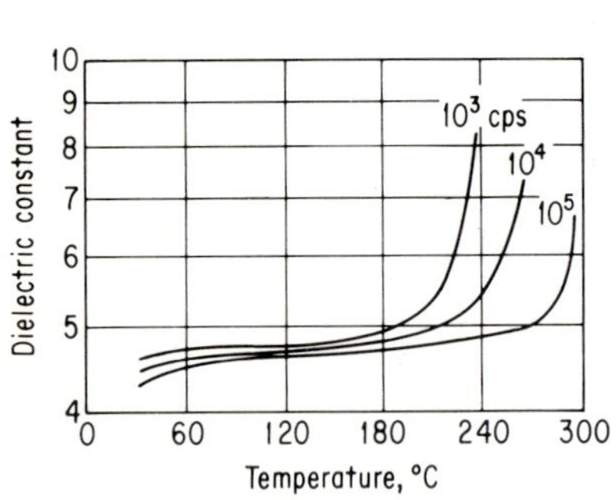

FIG. 8-13A. Dielectric constant of DGEBA cured with MPDA/MDA/PGE [26].

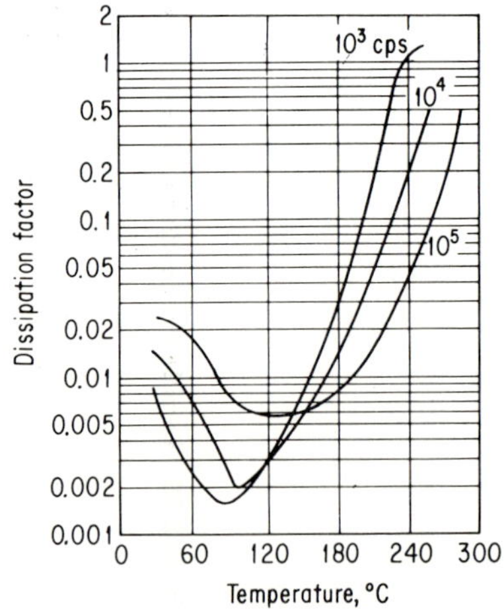

FIG. 8-13B. Dissipation factor of DGEBA cured with MPDA/MDA/PGE [26].

Table 8-11. Shrinkage of DGEBA Cured with MPDA/MDA/PGE [22]

Property	*Value*
Mixing temperature	23°C
Cure, hr/°C	8/23 + 15/49 + 8/100
Peak exotherm, °C	>121
Shrinkage, %:	
Total shrinkage	4.01
Pregelation shrinkage	2.43
Postgelation shrinkage	−1.01
Cooling shrinkage	2.59

Use of a stabilized MPDA/MDA eutectic blend for curing blends of DGEBA and bis(epoxycyclopentyl) ether provides low-viscosity systems having good pot lives and improved tensile elongation, as indicated in Fig. 8-14. A typical blend of the two resins and the curing agent at 27 phr gives strength properties as indicated in Table 8-12 when cured 4 hours at 150°C after an initial period of 2 hours at 85°C. Such systems are used in filament-winding applications.

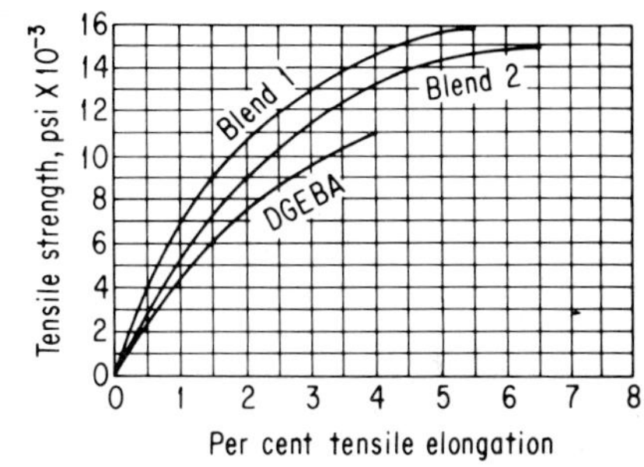

FIG. 8-14. Tensile strength vs. elongation for DGEBA and blends with bis-(epoxycyclopentyl) ether cured with stabilized MPDA/MDA eutectic [4-7].

AROMATIC AMINE ADDUCTS AND BLENDS

Like the aliphatic polyamines, the aromatic diamines may be adducted with a variety of compounds, such as a monoepoxy, i.e., styrene oxide [16,17] or PGE [32], a ketone [21], or a diepoxy. Typical of the last is a 10/1 blend of DADPS and DGEBA. The aromatic diamine may be reacted with DMP-30 at about 120°C to liberate dimethylamine and produce a liquid primary and tertiary amine containing a phenolic hydroxyl and capable of providing cured properties similar to those obtained with the aliphatic amines [8].

Amine-copper chelates, i.e., copper–mono-MPDA sulfate, has been reported as providing reduced weight loss when used as a curing agent for DGEBA over that experienced with MPDA after 200 hours at 260°C [27]. MPDA and MDA have been complexed with metal salts from copper, iron, cadmium, etc., to provide systems that are stable in DGEBA at room temperature but which break down to cure at elevated temperatures [28].

Table 8-12. Properties of a 50/50 Blend of DGEBA and Bis(epoxycyclopentyl) Ether Cured with Stabilized MPDA/MDA Eutectic [4-7]

Property	*Value*
DT, °C	145–150
Tensile strength, psi	14,500
Modulus × 10^{-6} psi	0.42
Flexural strength, psi	22,500
Modulus × 10^{-6} psi	0.527
Compressive strength, psi	27,000
Modulus × 10^{-6} psi	0.437

The aromatic diamines may be blended with other aromatic amines to produce desirable intermediate properties. For example, MPDA may be blended with 4-isopropyl MPDA to extend the pot life from 40 to 50 per cent at all temperatures over what is expected for MPDA alone.

The aromatic diamines may be liquefied with diluents, some 41 of which are given in [14], including dimethylacetamide and dimethylformamide.

The effect of these modifications is usually to reduce somewhat the cured properties of the system, with the diluted and some of the adducted versions reducing the DT to the range expected with the aliphatic polyamines. Properties of DGEBA cured with a proprietary liquefied aromatic diamine of this general type are given in Table 8-13.

Table 8-13. Physical Properties of DGEBA Cured 6 Hours at 120°C with Liquefied Aromatic Diamine [4-1]

Property	*Value*
DT, °C	115
Tensile strength, psi	12,000
Flexural strength, psi	20,000
Compressive strength, psi	16,000
Impact strength, Izod, ft-lb	0.90
Water absorption after 4 hr in boiling water, %	0.32
Weight loss after 188 hr at 240°C, %	4.34

A commercial aromatic amine blend composed of 70 per cent MDA and 30 per cent polyaniline high-molecular-weight tars preserves the elevated-temperature resistance of the aromatic amines. This mixture melts to a liquid stable at room temperature for about 2 days, after which it resolidifies.

AROMATIC POLYAMINES (ANILINE-FORMALDEHYDE RESINS)

Aniline-formaldehyde resins may be considered as aromatic polyamines, the analogs to the aliphatic polyamine curing agents [1].

Aniline-formaldehyde (A/F) resins may be of varying molecular weights, depending on the degree of polymerization. They are synthesized by reacting the two intermediates in a strongly acid medium to produce a polymer of the structure

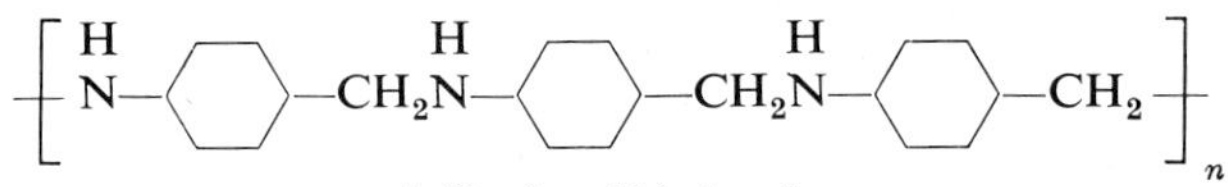

Aniline-formaldehyde resin

When an A/F ratio of 1/0.75 is employed, a low-melting-point (60°C) solid is formed; used in an equimolar ratio, a high-melting-point partially cross-linked solid results.

The lower-melting A/F resins have been investigated as curing agents for the liquid epoxy resins [2]. Use of the higher-melting-point A/F resins in solution coatings appears feasible.

The low-melting-point A/F resins can be incorporated into DGEBA at 100°C. At that temperature, the pot life will be about 30 minutes. The mix may be cooled to increase the pot life, but it will necessitate reheating to obtain sufficiently low viscosities. Exotherms are low, and a recommended cure cycle is 16 hours at 120°C.

When used at 35 phr with DGEBA, an A/F resin of 1/0.7 composition will provide DTs of about 155°C. Solvent and chemical resistance equal those provided by aromatic-diamine-cured systems. Special proprietary materials provide electrical properties which are virtually independent of temperature up to 200°C, the properties changing little after 1,000 hours' aging at that temperature [24]. These curing agents, in that respect, more closely resemble anhydrides such as HHPA than aromatic amines.

SUMMARY

The aromatic diamines were introduced at an early date into the epoxy-resin technology in order to provide improved heat and chemical resistance over that attained with the aliphatic amines. They have been used with success in laminating applications from the early days of the epoxy technology to the present, and find limited use in casting and adhesive applications, where their improved properties warrant the handling difficulties involved. Aromatic polyamines from aniline-formaldehyde resins have been investigated and provide properties generally falling in the upper range of the aromatic diamines.

The aromatic amines generally react sluggishly with both glycidyl ether and epoxidized olefin resins, and usually require a heat cure. Cure is usually conducted in two steps, one at lower temperature to minimize exotherm and one at higher temperature to attain maximum properties.

The commercial aromatic amines are solid at room temperature and must be melted or dissolved into the resin. Liquid versions are available under proprietary designations, these consisting of eutectic blends, adducts, or diluent-containing materials.

In general, the aromatic diamines are used at or near stoichiometric, although when employed with acid accelerators, something less than the stoichiometric amount may be optimum. Certain properties may be optimized by use of ratios above or below stoichiometric.

The aromatic amines, generally, are skin-sensitizing agents, and the fumes attendant upon their application and cure are objectionable.

REFERENCES

1950

1. Greenlee, Compositions of Resinous Epoxides and Aromatic Amine Aldehyde Condensates, U.S. 2,511,913.

1956

2. Bishop, The Use of Aniline-Formaldehyde Resins as Curing Agents for Epoxide Resins, British Society of Chemical Industry Symposium, London, April.
3. Formo et al., Epoxy Curing Compositions Containing 4,4′MDA, U.S. 2,773,048.
4. Partansky and Schrader, Some Thermal Properties of Epoxylated Novolac and Bisphenol A Type Casting Resins, ACS Symposium, Atlantic City, September.

1957

5. DeHoff and Parry, Curing Glycidyl Polyethers, U.S. 2,801,229.
6. Philipson, Liquid Amine Hardeners for High Temperature Epoxy Resins, *Western Plastics*, February.

1958

7. Bloom and Welch, Liquid Aromatic Diamines as Curing Agents for Epoxy Ether Resins, U.S. 2,853,467.

8. Delmonte, Amine Curing Agent for Epoxy Resins, U.S. 2,837,497.
9. Newey, Process for Curing Polyepoxides and Resulting Products, U.S. 2,824,082.

1959

10. Applegath, et al., Epoxy Novolacs, *SPE J.*, January.
11. Cyba, Curing of Epoxy Resins, U.S. 2,899,407.
12. Philipson, Method and Composition for Curing Epoxy Resins, U.S. 2,891,927.
13. Steckler et al., Curing Agents for Epoxy Resins, U.S. 2,904,530.
14. Steckler et al., Liquefied *m*-Phenylene Diamine Compositions for Curing Epoxy Resins, U.S. 2,893,873.

1960

15. Aitken and Ralph, Some Effects of Radiation in Cast Epoxide Resins Systems, AERE R-3085, (Engineering Division, Harwell, Berkshire, England).
16. DeHoff and Parry, Adducts of Aromatic Polyamines and Monoepoxides, U.S. 2,938,004.
17. Mayurnik, Epoxy Curing Agent, U.S. 2,926,156.

1961

18. Duckworth et al., New Resin Compositions and Curing Agent Therefor, U.S. 3,006,890.
19. St. Clair and Moult, Curable Aliphatic Polyepoxide Compositions, U.S. 2,981,701.
20. St. Clair and Moult, Curable Epoxy Phenol Aldehyde Aliphatic Polyepoxide Compositions, U.S. 2,981,702.

1962

21. Smith and Graham, Condensation Product of *m*-Phenylene Diamine and a Dialkyl Ketone Epoxy Resin Composition Containing Same and Process of Curing, U.S. 3,032,526.
22. Spencer, Cure Shrinkage in Casting Resins, Sandia Corp. Report SCDS 2568.

1963

23. Landolt, Resinous Compositions, U.S. 3,094,498.
24. Lemon, Effect of Heating at 200°C on the Properties of Some Epoxide Resins, *Brit. Plastics*, June.
25. Manufacture of Hardened Products from Epoxides (to Schering A.G.), Brit. 915,106.
26. Quant, Sandia Corp., private communication.
27. Reinforced Plastic Laminates for Long Time High Temperature Use, AD 451 215.

1964

28. Curing Catalysts for Epoxy Resins (to Leicester, Lovell & Co., Ltd.), Fr. 1,351,709.
29. Garnish and Rainton, Hardening of Epoxide Resins, Brit. 961,987.
30. Griffith and Whisenhut, Filament-winding Plastics, part 1: Molecular Structure and Tensile Properties, NRL Report 6047.
31. Lehto, Hardening of Epoxide Resins, Fr. 1,350,348.
32. Tanaka, Polyamine Ethers Suitable for Hardening Epoxy Resins, Japan 1,841.
33. Warfield et al., Supercooled MPDA as a Hardener for Epoxide Resins, *Polymer Letters* **2**: 173–174.

1965

34. Griffith, Filament Winding Plastics, part 3: Some Effects of Curing Agent Structure on Epoxide Plastic Physical Properties, NLR 6313.
35. Heat Hardenable Epoxy Resin Precondensates (to Ciba, Ltd.), Neth. Appl. 6,414,446.

Chapter 9

TERTIARY AND SECONDARY AMINES AS EPOXY-RESIN CURING AGENTS

TERTIARY AMINES AS CURING AGENTS

Tertiary amines are formed when the three hydrogen atoms of the ammonia molecule are substituted by hydrocarbon groups. Tertiary amines serve three roles in the cure of epoxy resins: (1) as sole curing agents; (2) as co-curing agents for primary amines (Chaps. 7 and 8), fatty amides (Chap. 10), and related compounds; and (3) as accelerators for dicarboxylic acid and anhydride cures (Chaps. 11 and 12).

In all three cases, reactivity results from the unshared electron pair of the nitrogen. The structure of the tertiary amine (i.e., the size, shape, and composition of the various groups) determines how available or reactive the unshared electron pair will be. Reactivities may vary considerably. Figure 9-1 shows the effect of chain length on reactivity for a series of tertiary amines when used as catalysts for the reaction of phenyl glycidyl ether in the presence of dipropylene glycol as initiator.

Similar effects are obtained with glycidyl ether resins. Substituents which favor increasing the electron density on the nitrogen increase reactivity; those which reduce electron density on the nitrogen reduce reactivity. Bulky side chains may contribute steric hindrances which reduce reactivity over what would be indicated by electronic considerations.

Tertiary amine curing agents are satisfactorily reactive with the glycidyl ether epoxy resins but only very sluggishly reactive or completely nonreactive when used as sole curing agents with the epoxidized olefin resins.

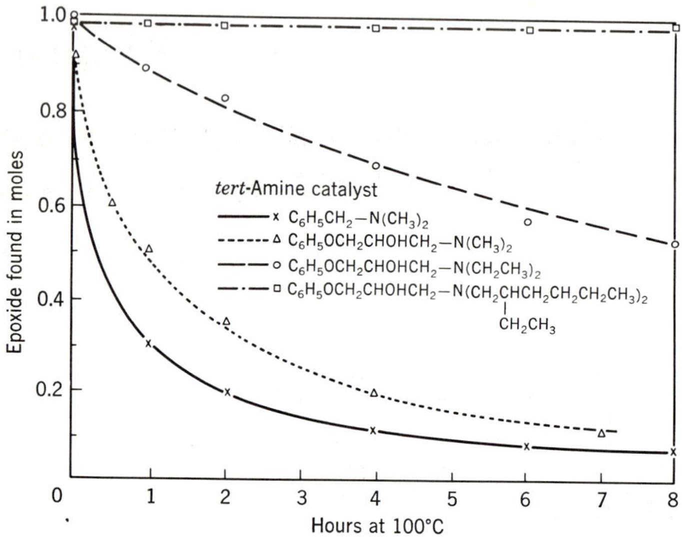

FIG. 9-1. Effect of structure of tertiary amine upon reactivity of epoxy group [6].

Tertiary amines are little used as sole curing agents for epoxy casting formulations. Cured properties are, to some extent, temperature-dependent and vary with the mass of the casting—exothermic temperatures being considerably higher at the center of the casting than at the periphery. In addition, the reduction in functionality (particularly with higher concentrations of tertiary amine) leads to an overall reduction in cured properties.

Tertiary amines, as sole curing agents, have found some favor in coatings and adhesives formulations. In these cases exotherm is negligible and reproducibility is thereby improved. More important, however, tertiary amines provide improved adhesion, in part because of good surface-wetting properties, in part because of decreased strains arising from end stopping of polymerization (Chap. 5), and in part because of the presence of polar hydroxyl groups in the cured system.

When used as a sole curing agent, tertiary amines provide systems containing predominantly ether linkages. A number of hydroxyl and vinyl groups are present from the chain terminations. Deflection temperatures (DTs) are usually 80 to 100°C, the values being in this low range because tertiary amine curing generally results in low-molecular-weight polymers due to a limited propagation step. Physical properties below the DT are in the range expected of amine-cured systems generally. Radiation resistance is rather poor, although elevated-temperature weight loss may be quite low.

Most tertiary amines require at least a mild heat cure when reacted with the liquid glycidyl ether resins. With higher-molecular-weight resins, containing a considerable number of hydroxyl groups, satisfactory cures may be obtained at room temperature.

Linear and Branched Aliphatic Tertiary Amines (Table 9-1)

The linear and branched aliphatic tertiary amines containing at least two tertiary amine groups and no other reactive sites have been reported to provide cures of epoxy coating systems at 80 to 100 per cent relative humidity without clouding [8].

The hydroxylated versions and adducts of these with acids and anhydrides [9] have been suggested for use with casting formulations. The hydroxyl groups present in these amines provide for accelerated cure, but even with the more reactive dimethylaminoethanol, a mild heat cure is normally required when the amine is used as the sole curing agent for the liquid glycidyl ether resins.

Alicyclic Tertiary Amines (Table 9-1)

Alicyclic tertiary amines have been employed in special-purpose applications. Triethylenediamine, i.e., 1,4-diazabicyclo(2.2.2)octane, provides extremely fast reaction rates with glycidyl ether resins. Hexamethylenetetramine, which reacts with phenolic resins, and particularly with the novolac type, has been used in adhesive formulations as a curing agent for epoxy/phenolic blends.

Unsaturated-ring Heterocyclic Tertiary Amines (Table 9-1)

The unsaturated-ring tertiary amines are little used in the epoxy technology. The imidazolines without primary or secondary amines are sometimes reported as co-curing agents for primary amines and accelerators for anhydrides.

Vinyl pyridines have been suggested as giving long pot lives (approximately 3 to 4 days at room temperature) with DGEBA [11]. Heavy pyridine bases, containing quinoline in combination with related materials available as by-products of coke production, offer low cost and give pot lives of 48 hours with glycidyl ether resins [19]. These systems require heat cures of approximately 80°C or above for several hours.

Aromatic-ring-containing Aliphatic Tertiary Amines (Table 9-1)

The aromatic-ring-containing aliphatic tertiary amines are more widely used in the technology than are the materials previously discussed.

Benzyldimethylamine and α-methylbenzyldimethylamine used with DGEBA at 6 to 10 phr cure at room temperature in about 6 days [1]. The pot life is generally 1 to 4 hours, depending on the mass and the concentration of amine. α-Methylbenzyldimethylamine is useful in catalyzing heat-curing epoxy/phenolic blends, particularly in casting and laminating formulations.

Preferred as sole curing agents for casting formulations are the tertiary amines derived from phenol (Mannich bases). The phenolic hydroxyl accelerates the curing rate markedly. Dimethylaminomethylphenol (DMP-10) and tris(dimethylaminomethyl)phenol (DMP-30) are used in the 5- to 10-phr range with DGEBA to give pot lives of approximately 30 minutes to an hour, with correspondingly high exotherms. Cure occurs overnight at room temperature even in thin films. If heat curing is employed, the amount of amine may be reduced to the range of 2 to 3 phr, and cure will occur in about 4 hours at 120°C.

DMP-30 is often used as the curing agent for epoxy-polysulfide blends (Chap. 16) in which it is normally used at about 10 phr for room-temperature cures, 6 phr when heat cures are used.

Table 9-1. Typical Commercial Tertiary Amines

LINEAR AND BRANCHED ALIPHATIC TERTIARY AMINES

$(CH_3)_2N(CH_2)_nN(CH_3)_2$

Linear diamines

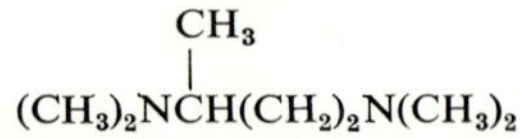

N,N,N′,N′-Tetramethyl-1,3-butanediamine

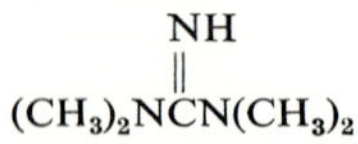

Tetramethylguanidine

$N[(CH_2)_nCH_3]_3$

Alkyl tertiary monoamines

$N(CH_2CH_2OH)_3$

Triethanolamine

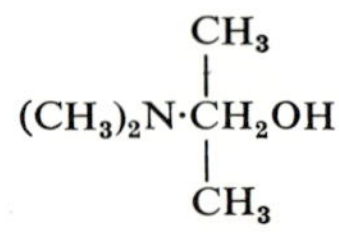

2-Dimethylamino-2-hydroxypropane

$[CH_3(CH_2)_n]_2N(CH_2)_2OH$

Dialkyl aminoethanols

ALICYCLIC TERTIARY AMINES

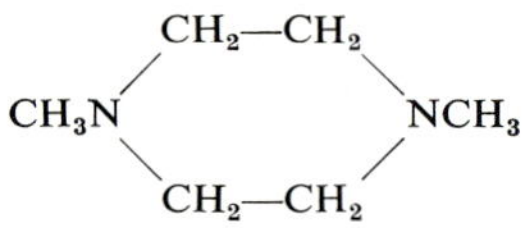

N,N′-Dimethylpiperazine

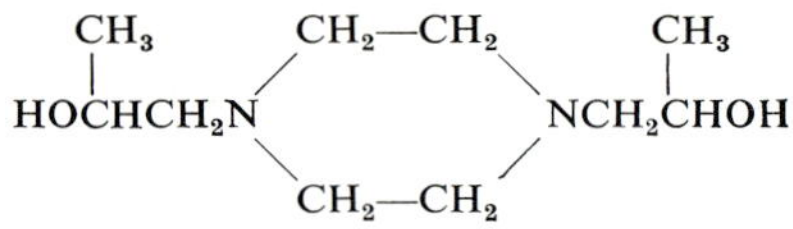

N,N-Bis[(2-hydroxy)propyl]piperazine

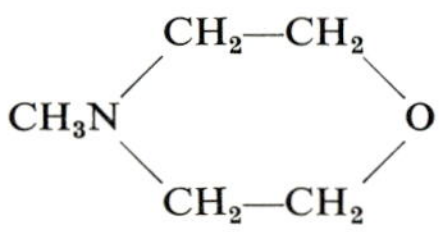

N-Methylmorpholine

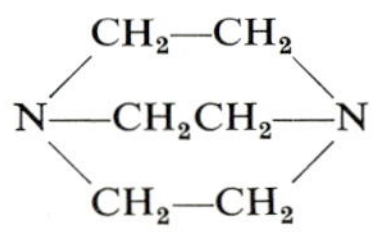

1,4-Diazabicyclo(2.2.2)octane (triethylenediamine)

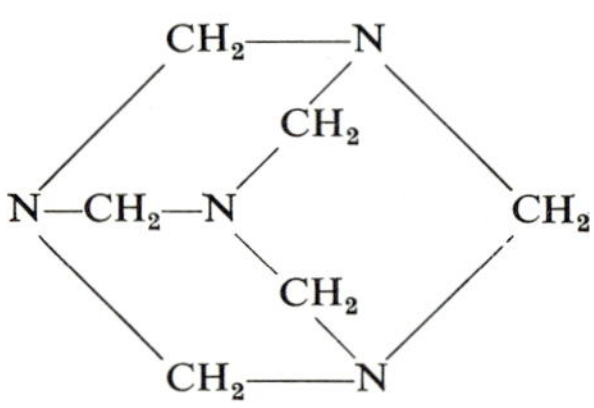

Hexamethylenetetramine

UNSATURATED-RING TERTIARY AMINES

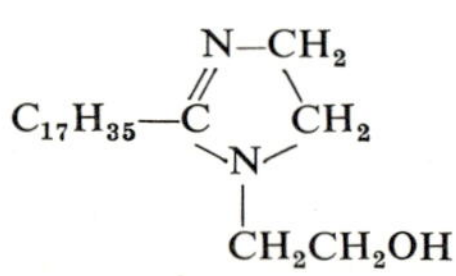

1-Hydroxyethyl-2-heptadecylgloxalidine

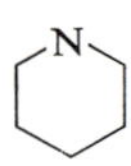

Pyridine

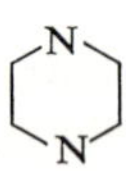

Pyrazine

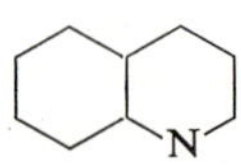

Quinoline

Table 9-1 (contd.)

AROMATIC-RING-CONTAINING ALIPHATIC TERTIARY AMINES

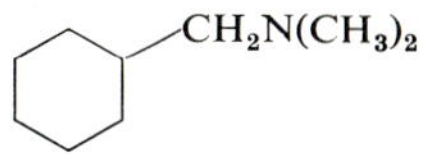

Benzyldimethylamine

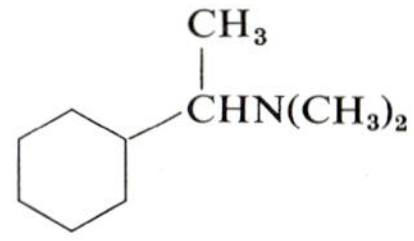

α-Methylbenzyldimethylamine

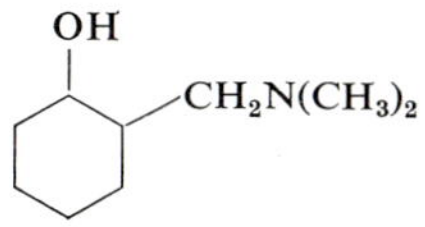

2-(Dimethylaminomethyl)phenol (DMP-10)

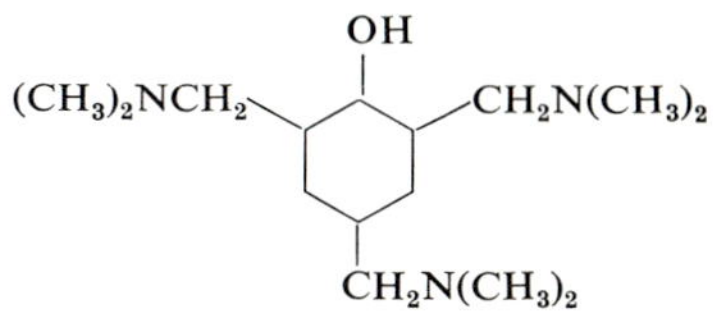

2,4,6-Tris(dimethylaminomethyl)phenol (DMP-30)

TERTIARY AMINE SALTS AND QUATERNARY BASES AS CURING AGENTS

Tertiary amine salts obtained from organic and inorganic acids may provide latency as epoxy curing agents. The amine and acid are generally dissociated with heat to provide rapid cures. Quaternary bases are likewise dissociated by heat to produce tertiary amines.

Tertiary Amine Salts

Organic fatty-acid salts of tertiary amines, neutral and free of other chemical groups reactive with the epoxy group, provide latency in epoxy formulations [5].

The tertiary amine may be one of those previously discussed or may be obtained by the reaction of a secondary amine with an epoxy resin [2]. The tri-2-ethylhexoate salt of DMP-30 is the salt most commonly employed commercially.

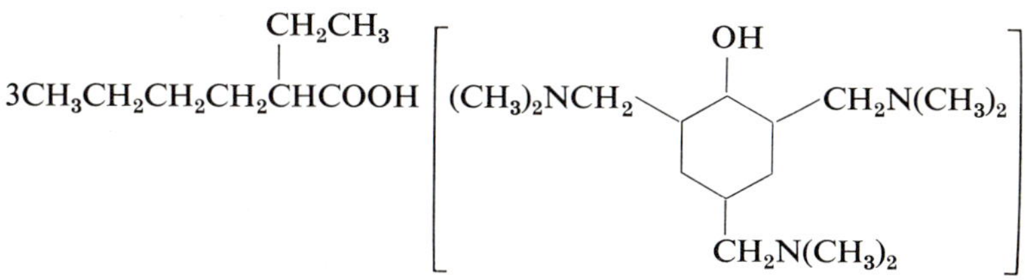

Tri-2-ethylhexoate salt of DMP-30

Because of the presence of the phenolic hydroxyl, the salt does not provide true latency, but it does offer extended pot life (4 to 6 hours) over what would be expected of the amine alone. The salt is normally used at 10 to 14 phr with DGEBA.

It has been suggested [16] that the curing mechanism involves two steps, the first and rate-determining one being the removal of the fatty acids by esterification with some of the epoxy groups, the tertiary amines thereafter initiating polymerization through the mechanism proposed in Chap. 5.

Because it provides extended pot life at room temperature with a mild cure requirement, the hexoate salt of DMP-30 has found use in small casting formulations. Figure 9-2 presents an exotherm profile for various batch sizes of DGEBA

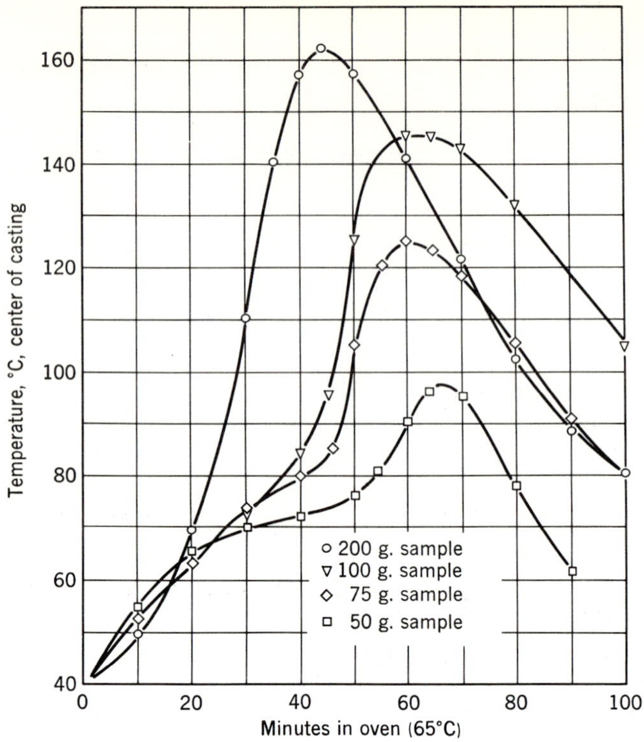

FIG. 9-2. Exotherm profile of DGEBA cured with DMP-30 salt [4–6].

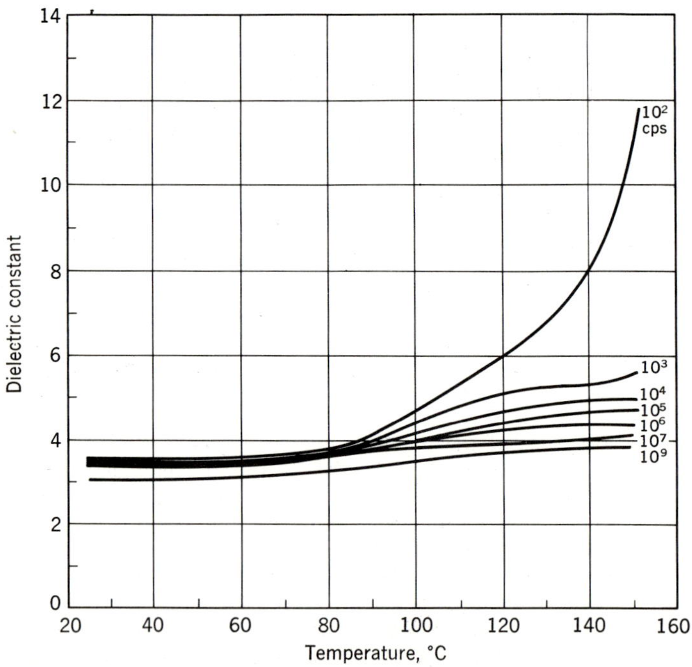

FIG. 9-3. Dielectric constant vs. temperature for DGEBA cured with DMP-30 salt [4–6].

cured with 10.5 phr of the material. Table 9-2 presents physical properties and chemical resistance. Dielectric constant and dissipation factor are shown in Figs. 9-3 and 9-4.

Table 9-2. Physical Properties and Chemical Resistance of DGEBA Cured with 10.5 phr 2-Ethylhexoate Salt of Tris-(dimethylaminomethyl)phenol [4-6]

PHYSICAL PROPERTIES

Property	*Value*
Tensile strength at 23°C, psi	8,000
Modulus, psi × 10^{-6}	0.66
Elongation, %	1.6
Tensile strength at 54°C, psi	8,000
Modulus, psi × 10^{-6}	0.53
Elongation, %	1.5
Tensile strength at −57°C, psi	9,000
Modulus, psi × 10^{-6}	0.65
Elongation, %	1.7
Flexural strength at 23°C, psi	21,000
Modulus, psi × 10^{-6}	0.54
Compressive strength at 23°C, psi	18,400
Compression, ultimate, %	5.5
Impact strength, Izod	0.36
Hardness, Barcol	0.36

CHEMICAL RESISTANCE

System tested in	*Weight gain, %*
Ethyl alcohol for 1 month	0.75
Benzene for 1 month	0.26
Water absorption:	
In 24 hr	0.07
In 1 week	0.20
In 1 month	0.47
Water-vapor transmission, g/(sq m)(24 hr)	1.0

A variety of other amine salts have been suggested from time to time. Morpholinium *p*-toluenesulfonate, a crystalline solid melting at 126°C, employed in a solvent, provides a curing agent stable at room temperatures but giving fast cures in the 100 to 150°C range [4]. The salt is formed by neutralizing *p*-toluenesulfonic acid with morpholine. DMP-30 hydrochlorides have been reported as curing agents for epoxidized novolac molding compounds.

Triethylenediamine, blocked with hydroquinone to get a complex of the following structure,

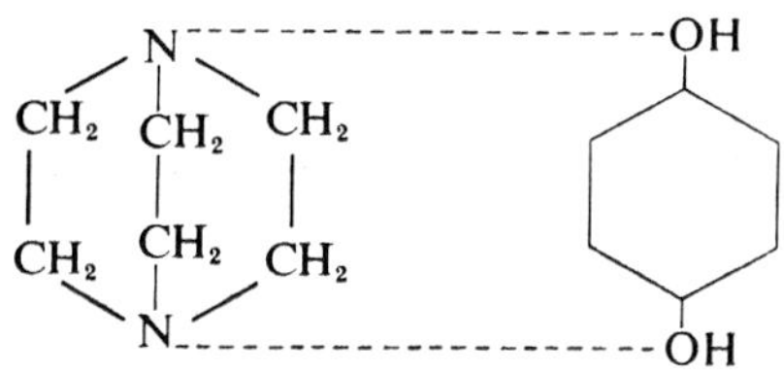

Hydroquinone-blocked triethylenediamine

has also been reported as an epoxy curing agent [18].

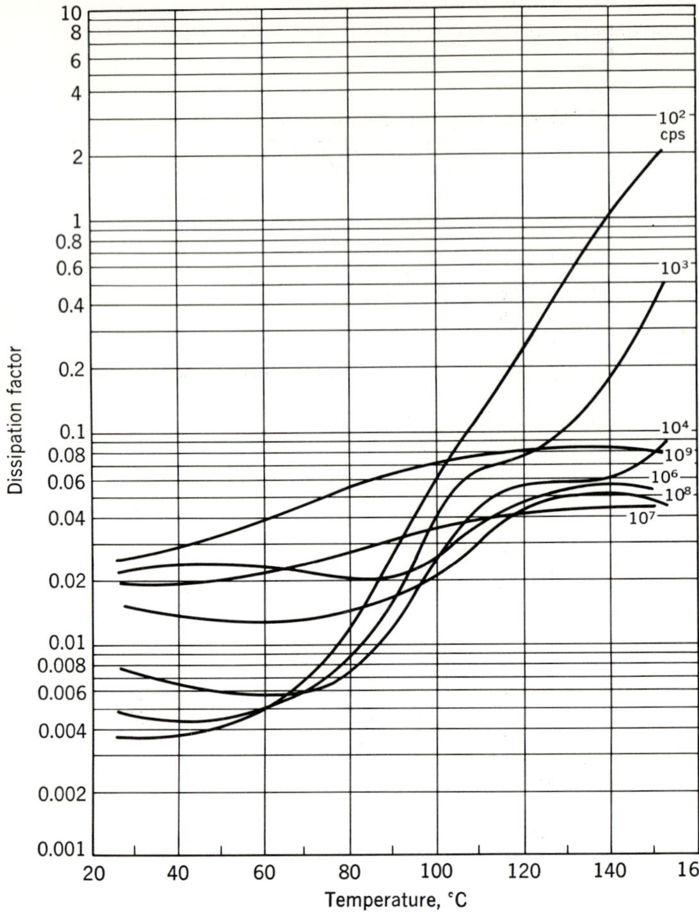

FIG. 9-4. Dissipation factors vs. temperature for DGEBA cured with DMP-30 salt [4–6].

Quaternary Bases

Typical of the quaternary bases is benzyltrimethylammonium chloride (BTMAC).

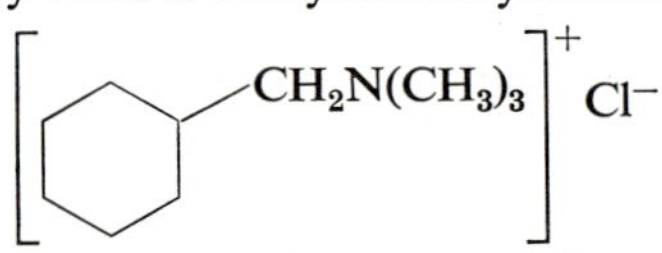

Benzyltrimethylammonium chloride

BTMAC decomposes to benzyl chloride and trimethylamine at 135°C and may be used to provide room-temperature stability with fairly fast gel times at 150°C [13]. BTMAC is also reported as an accelerator for dicyandiamide [14], for primary and secondary amines [15], and for phenol and anhydride [17] curing agents, as well as curing agent for blends of epoxy and polyols [12].

Quaternary ammonium salts of tertiary-amine-containing diepoxy resins are used as curing agents for gelatin and other carboxyl-containing polymers used in photographic layers [22].

Betaine has been reported to provide 3-month pot life with DGEBA when used at 1 phr [28].

$$(CH_3)_3\overset{\oplus}{N}CH_2COO^{\ominus}$$

Betaine

Oxides of tertiary amines, especially of 1-methylpyrrolidine, tribenzylamine, and *N,N*-dimethyl-*p*-nitrobenzylamine, will gel DGEBA in about 1 hour at 100°C and provide full properties with a postcure of 2 hours at 200°C [23].

SECONDARY AMINES AS CURING AGENTS

Secondary amines (R_2NH) have been shown (Chap. 5) to react in a fashion similar to the primary amines.

Mono- and disecondary amines used as curing agents are incapable through reaction of active hydrogens alone of promoting complete cross-linking with diepoxy resins. Additional functionality is obtained by the reaction of excess epoxy groups through the catalytic influence of the formed tertiary amines or of tertiary amines already present in the system. It will thus be seen that the secondary amine cures can be considered as a special case of tertiary amine cures. For example, a tertiary amine curing agent can be provided by reacting excess secondary amine with an epoxy to form a tertiary adduct [3].

Secondary amine cures provide results superior to tertiary amine cures in coating formulations insofar as the tertiary amines are formed in situ, thereby being tied tightly into the cured system, and are not carried to the film surface by escaping solvents. When disecondary amines are employed, linear polymerization may be promoted to high molecular weights.

A number of aliphatic secondary amines may be employed, such as diethanolamine, ethyl- and methylethanolamine, dimethylamine, diethylamine, methylethylamine and methyl-*n*-propylamine. Specialized secondary amines from the

$$HN(CH_2CH_2OH)_2$$

Diethanolamine

hydrogenated reaction products of acetoacetic esters [7] or acrylic acid esters [26], and aliphatic and aromatic primary amines have been suggested. *N′*, N^3-Di-*sec*-butyldiethylenetriamine has been used alone or in blends with DETA to give cured DGEBA systems of improved flexibility [25].

Representative of the alicyclic secondary amines commonly used are those shown in Table 9-3. Although piperazine itself does not provide for cure by tertiary amine catalysis, the substituted piperazines do. Though containing only one nitrogen, both piperidine and pyrrolidine are excellent catalysts, with the latter giving very fast cures under mild heat. Anabasine has been reported as a satisfactory epoxy resin catalyst, presumably as a result of the tertiary activity of the alicyclic nitrogen [24,29,30]. 2,3-Bipiperidine is reported as a nondermatitic curing agent providing cures with DGEBA in 2 hours at 80°C [31].

Piperidine and diethanolamine can be considered as representative of the secondary amines.

Piperidine

Piperidine was one of the first amines to be used with the epoxy resins on a commercially significant scale. It is a clear, colorless liquid with a characteristic amine odor.

With DGEBA it is customarily used at 5 to 7 phr. If an appreciable bulk of resin is involved, upon mixing at 25°C there is a prompt rise in temperature to about

Table 9-3. Typical Alicyclic Secondary Amines Used as Epoxy Curing Agents

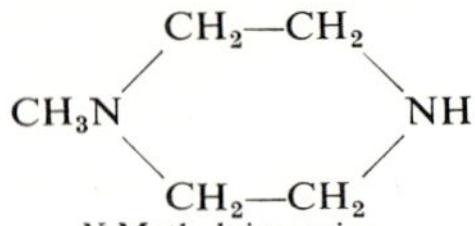

N-Methylpiperazine

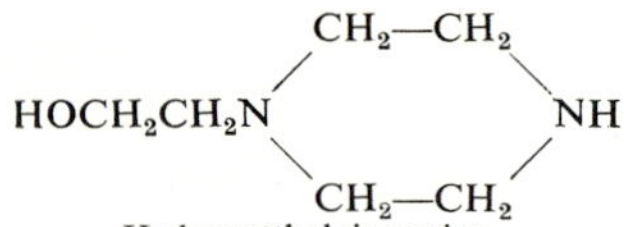

Hydroxyethylpiperazine

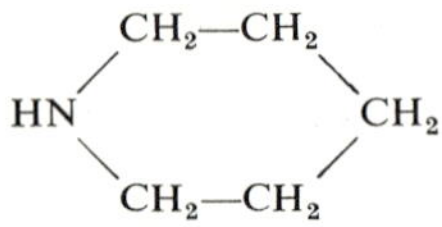

Piperidine

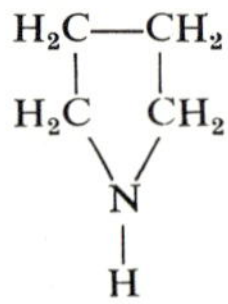

Pyrrolidine

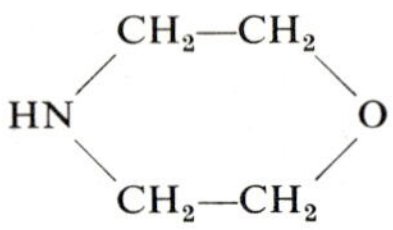

Morpholine

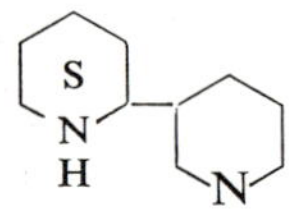

Anabasine

50°C, caused by the reaction of the active hydrogen. The temperature then returns to ambient to provide a pot life of about 2 to 6 hours. If it is allowed to gel at room temperature, there is no appreciable subsequent exotherm, and the mixture sets to a brittle solid in 1 or 2 days. The properties of the system do not thereafter improve greatly with room-temperature aging, a postcure being required. Cure cycles range from overnight at 60°C to about 3 hours at 100°C. If the system is cured at elevated temperatures before gelation, an exothermic rise will occur. Figure 9-5 presents an exotherm profile for a 100°C cure.

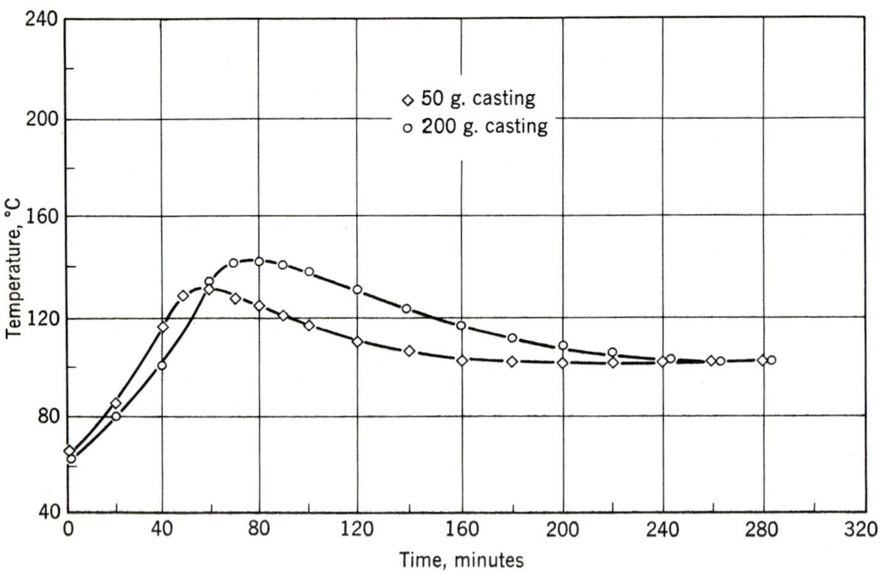

FIG. 9-5. Exotherm profile for DGEBA cured with 5 phr piperidine at 100°C [4–6].

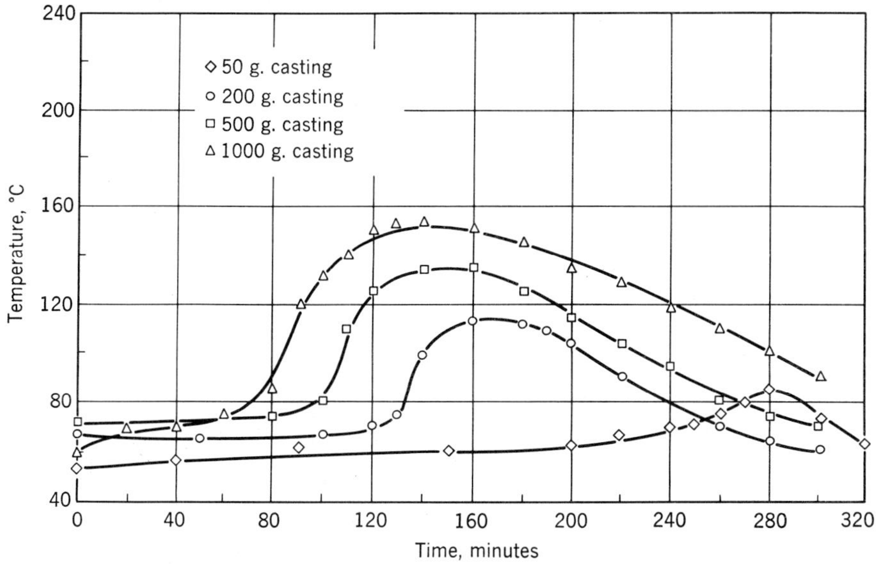

FIG. 9-6. Exotherm profile for DGEBA cured with 5 phr piperidine at 60°C [4–6].

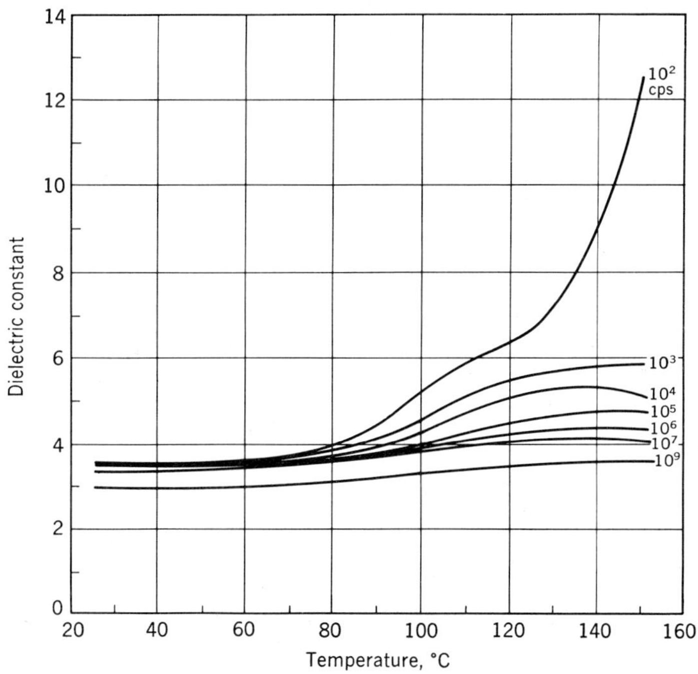

FIG. 9-7. Dielectric constant vs. temperature for DGEBA cured with 5 phr piperidine [4–6].

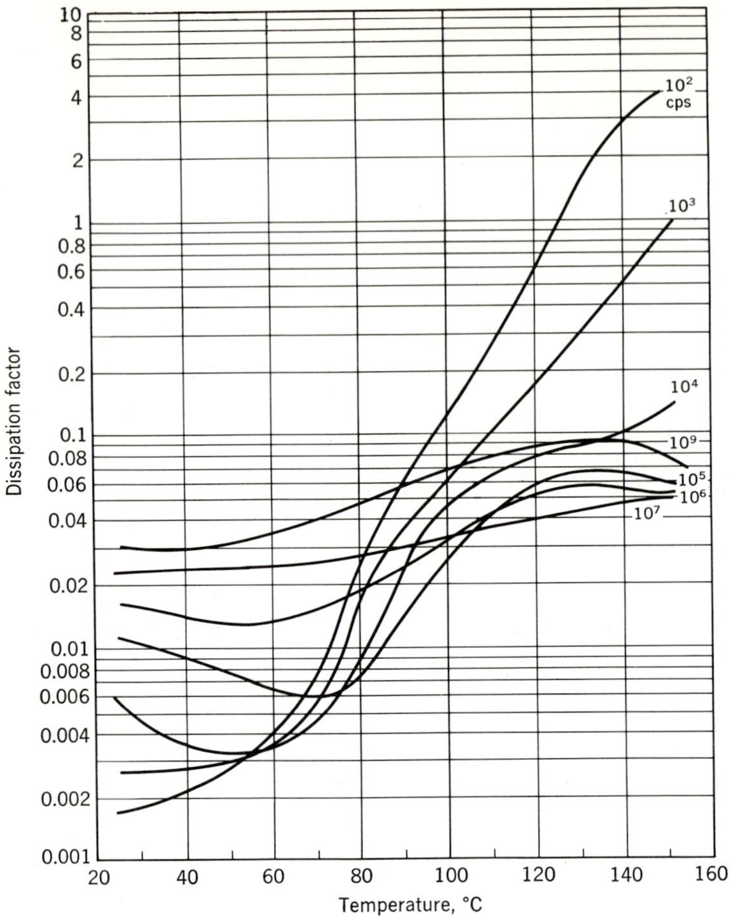

FIG. 9-8. Dissipation factor vs. temperature for DGEBA cured with 5 phr piperidine [4–6].

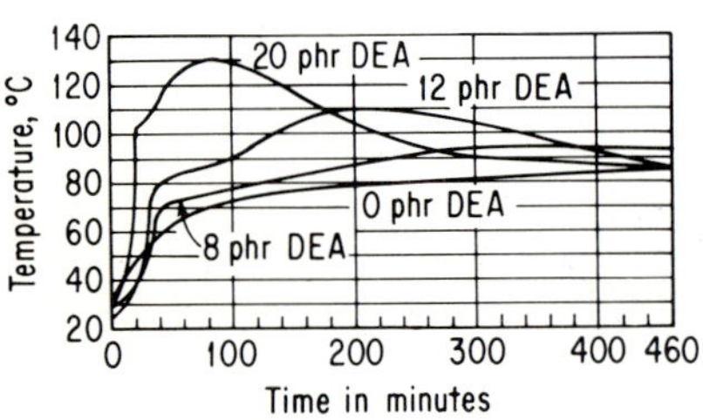

FIG. 9-9. Exotherm of DGEBA filled with 66 phr mica and cured with DEA [20].

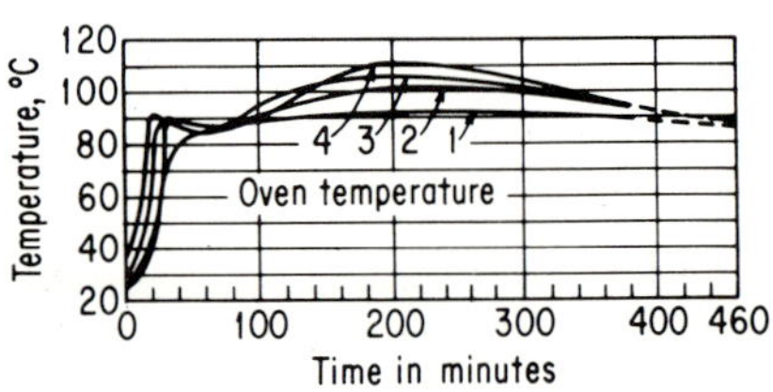

FIG. 9-10. Mass vs. exotherm of DGEBA filled with 66 phr mica and cured with 12 phr DEA [20]. (1) 230 grams, 2.40 in. diameter by 2.15 in. high; (2) 460 grams, 3.35 in. diameter by 2.15 in. high; (3) 690 grams, 3.35 in. diameter by 3.75 in. high; (4) 920 grams, 4.15 in. diameter by 2.8 in. high.

Table 9-4. Shrinkage of DGEBA Cured with 6 phr Piperidine [21]

Property	*Value*
Mixing temperature	23°C
Cure. hr/°C	15/23 + 5/46 + 15/74
Peak exotherm	88°C
Shrinkage, %:	
Total shrinkage..........	3.82
Pregelation shrinkage...	1.20
Postgelation shrinkage..	1.39
Cooling shrinkage	1.24

Since exotherm control presents a problem with amine curing agents generally, it is frequently customary to gel the resin mix at lower temperatures before the final high-temperature cure. In the case of piperidine, if the initial temperature is moderate (Fig. 9-6), the exotherm, likewise, will be moderate. Once it has gelled, the temperature can then be elevated for postcure without significant exotherm being encountered.

The physical properties of a piperidine-cured system, below the DT, fall into the same general range as primary-aliphatic-polyamine-cured systems. The DT will be about 75°C for a moderately cured system. Electrical properties are likewise typical, as indicated in Figs. 9-7 and 9-8. Shrinkage is indicated in Table 9-4. Adducts of piperidine and DGEBA (mol. wt. 450) give improved DTs and higher-temperature electricals.

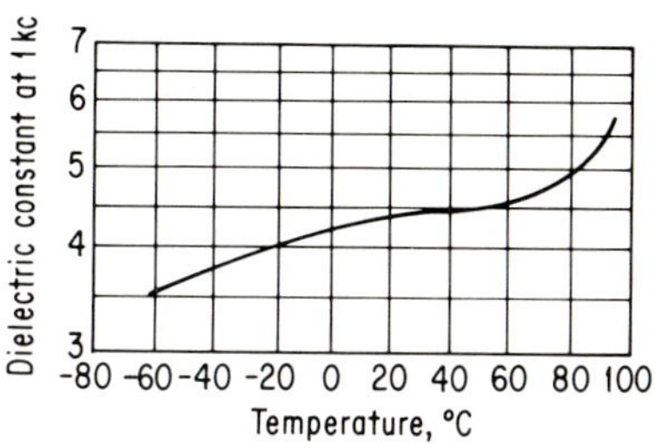

FIG. 9-11. Dielectric constant vs. temperature for DGEBA cured with 12 phr DEA [20].

Diethanolamine (DEA)

DEA is a solid melting at about 28°C. When added to warm resin, it is initially insoluble, but as the active hydrogen reacts, an in situ adduct is formed and the system clears.

With the DGEBA, DEA is usually employed in the 12-phr range. Varying the DEA concentration affects the exotherm of 2-lb castings of DGEBA containing mica filler, as indicated in Fig. 9-9, at a cure temperature of 85°C. The effect of mass on exotherm for the system cured with 12 phr DEA is indicated in Fig. 9-10. Dielectric constant data are shown in Fig. 9-11.

SUMMARY

Tertiary amines may serve as sole curing agents, as co-curing agents, and as accelerators for epoxy-resin formulations. When used as sole curing agents, they provide satisfactory reactivity with the glycidyl ether resins, but with the epoxidized olefins reactivity is sluggish.

The electronic nature of the tertiary amine will, in part, determine its efficiency as a curing agent. The curing reaction proceeds by an ionic mechanism which is sensitive to both temperature and the presence of active-hydrogen impurities. Because of this sensitivity, the tertiary amines are little used as curing agents for epoxy casting formulations, but because they do provide improved adhesion, are sometimes used as sole curing agents for adhesive and coating formulations.

Tertiary amine salts have been employed to extend the pot life over that obtained with the tertiary amines, and these generally provide satisfactory cure only at elevated temperature.

Secondary amines, as curing agents, may be considered a special class of tertiary amines. The secondary amine reacts in a fashion identical with a primary amine. After this reaction, subsequent cross-linking occurs through the catalytic influence of the formed tertiary amines.

Tertiary- and secondary-amine-cured systems, generally, have limited temperature resistance and provide somewhat poorer chemical resistance than is obtained with the primary aliphatic polyamines. The physical and electrical properties, below the DT, fall in the range expected of amine-cured systems generally.

REFERENCES

1951

1. Newey and Shokal, Glycidyl Ether Compositions, U.S. 2,553,718.

1953

2. Parry and Wiles, Fatty Acid Salts of Nitrogen Containing Polyethers, U.S. 2,640,037.
3. Shokal and Newey, Nitrogen Containing Polyethers and Process for Curing Glycidyl Polyethers to Resinous Products, U.S. 2,643,239.

1954

4. Dannenberg, Resin-forming Compositions Containing Amine Salts of Sulfonic Acids, U.S. 2,687,397.
5. Wiles and Elam, Glycidyl–Polyether Resin Forming Compositions Containing Amine Salts of Fatty Acids, U.S. 2,681,901.

1956

6. Shechter and Wynstra, Glycidyl Ether Reactions with Alcohols, Phenols, Carboxylic Acids, and Acid Anhydrides, *Ind. Eng. Chem.* **48**:86–93.

1958

7. New Curing Agents for Epoxy Resins (to Henkel & Cie., G.m.b.H.), Ger. 1,025,884.
8. Ott et al., Compositions of Matter Comprising an Ethoxyline Resin and an Alkylene Polyamine Containing at Least Two Tertiary Amine Groups, U.S. 2,839,480.

1959

9. Cyba, Curing of Epoxy Resins, U.S. 2,868,767.
10. Dannenberg, Refractive Index Method for Determining Cure Rate of Epoxy Resins, *SPE J.*, October.

1960

11. Barb, Improvements in Epoxide Resins, Brit. 846,706.
12. Belanger, Aliphatic Alcohol–Epoxide Resin Compositions, U.S. 2,928,810.
13. Belanger, Curing of Polyepoxides, U.S. 2,928,808.
14. Belanger, Epoxide Resin–Amide Compositions, U.S. 2,928,811.
15. Belanger and Cooke, Curing of Polyepoxides, U.S. 2,928,794.
16. Bondi and Parry, A Mechanism for the Curing of Epoxide Resins by Amine Salts, Shell Chemical Corp. Report.
17. Hicks, Epoxide Resin–Quaternary Ammonium Salt Compositions, U.S. 2,982,809.

1962

18. Erner et al., Addition Compounds of Diazabicyclooctane, U.S. 3,037,026.
19. Fadeyev et al., Hardening of Epoxy Resins with Quinoline Derivatives, *Plastmassy* **6**:59–60.
20. Quant, Sandia Corp., private communication.
21. Spencer, Cure Shrinkage in Casting Resins, Sandia Corp. Report SCDS 2568.

1963

22. Burness, Hardening of Photographic Layers, U.S. 3,091,537.
23. Searles, Tertiary Amine Oxide Hardeners for Epoxy Resins, Belg. 623,038.

1964

24. Forostyan et al., Hot Hardening of Epoxy Resins, U.S.S.R. 160,307.
25. Hayes and Larson, Curing Epoxide Resins, Belg. 635,338.
26. Humphreys et al., (3-Dialkylamino-propylamino)propionic Acid Esters, Brit. 960,622.
27. Kaplan et al., Cold Curing of Epoxide Compositions, U.S.S.R. 165,303.
28. Smith and Thompson, Stable Epoxide Resin Compositions, Brit. 961,023.

1965

29. Forostyan et al., Hardening of Epoxy Resins, U.S.S.R. 167, 301.
30. Forostyan et al., Anabasine: Hardener for Epoxy Resins, *Plasticheskie Massy* 1965(3), 60–62.
31. Forostyan et al., Hardening of Epoxy Resins by α,β-Dipiperidyl, *Plasticheskie Massy*, 1965(1), 16–17.

Chapter 10

AMIDES AND MISCELLANEOUS NITROGEN COMPOUNDS AS EPOXY-RESIN CURING AGENTS

In addition to primary, secondary, and tertiary amines, a large number of other nitrogen-containing compounds may be used to cure epoxy resins. Nitrogen compounds which react with hydroxyl groups of high-molecular-weight epoxy resins are also described in this chapter.

AMIDES

Amides may be considered as carboxylic acids whose hydroxyl groups have been replaced by a basic group derived from ammonia. As is the case with amines, the amide may be primary, secondary, tertiary, or polymeric:

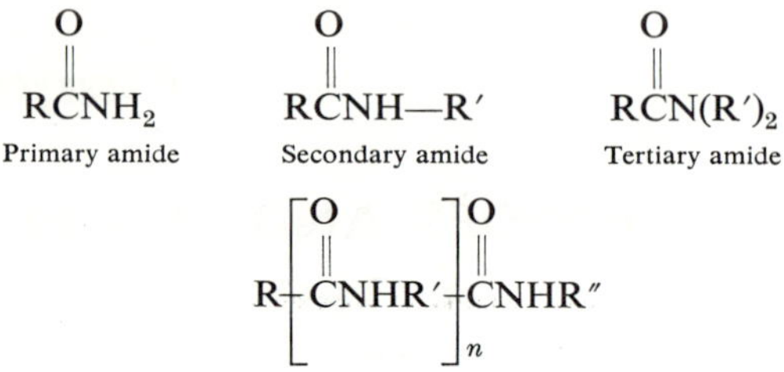

Primary amide Secondary amide Tertiary amide

Polymeric amide

The hydrogen of primary and secondary amides is weakly reactive with epoxy groups, and such simple amides as acetamide, benzamide, and adipamide have been used to cure epoxy resins. However, they are low in functionality and low in solubility, and cure with such materials must usually be conducted at high temperatures (150°C) in the presence of catalysts such as sodium or potassium hydroxide, sodium phenoxide, or bisphenol [6,11].

Whereas such simple amides are little used commercially as curing agents per se, they are used to some extent as accelerators in promoting cure with acid anhydrides [35]. In this case, the amides probably serve as acidic accelerators. The unshared electron pair of the nitrogen is no longer available for acceleration by the Lewis base mechanism:

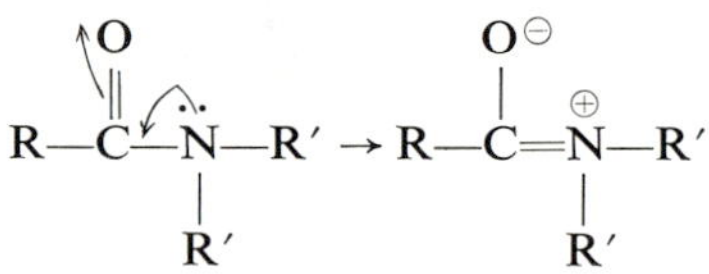

The tertiary amide, dimethylformamide, although not having been reported as a sole curing agent, has been described as a co-curing agent (and solvent) for dicyandiamide and metaphenylenediamine [59]. Polyphosphorylamide has been used as a fire-retardant in epoxy urea- or melamine-formaldehyde coatings [13]. *N,N*-Bis-1,2-ethylene sebacamide has been used as a co-curing agent for amines and anhydrides [66].

Condensation products of biurea [79] and biurets [78] with aliphatic polyamines, these containing primary and secondary amides and secondary amines, have been used as sole epoxy-resin curing agents. Polymeric amides, of which the copolymer of *N*-diethylaminopropylacrylic acid amide, tertiary butylacrylic acid amide, and ethyl acrylate (20/50/30) is typical, have been reported as curing agents for glycidyl ether resins [60].

Amidopolyamines

Although simple amides are not much used as curing agents, amine-containing materials referred to in the trade as "amides" are used. With these substances, cure is achieved not by the amide groups but by the amine groups present. Thus, one class of amide-type curing agents for epoxy resins is that known as the amidopolyamines.

Amidopolyamines are derived from a monobasic carboxylic acid and an aliphatic polyamine:

$$\underset{\text{Carboxylic acid}}{CH_3(CH_2)_nCOOH} + \underset{\text{Polyamine}}{H_2NCH_2CH_2NHCH_2CH_2NH_2} \rightarrow$$

$$\underset{\text{Amidopolyamine}}{CH_3(CH_2)_n\overset{\overset{\displaystyle O}{\|}}{C}NHCH_2CH_2NHCH_2CH_2NH_2}$$

The acids employed in such syntheses are usually of the C_{16}, C_{18}, and C_{19} types derived from fats and oils, particularly soya, tall oil, and ricinoleic acids. A number of special acids have been used, such as, for example, the copolymer of castor oil fatty acid and styrene or vinyltoluene [75], and rice oil, which contains free fatty acid and is otherwise a complex blend of glycerides [70]. The polyamines used range from DETA, TETA, and TEPA (see Chap. 7) to piperazine. Amidopolyamines have been produced by ester interchanges of aliphatic acids and aliphatic polyamines [83], typical being the interchange between ethyl ester of tall-oil fatty acid and TETA [82].

Table 10-1. General Specifications of Amidopolyamine from Tall-oil Fatty Acid and TETA [2]

Property	*Value*
Color, Hellige	8–9
Viscosity at 27°C, centipoises	200–300
Specific gravity, 25°C	0.95
Weight per gallon, lb	7.91
Moisture content, %	0.2–0.3
pH	11.8
Refractive index, n_D^{25}	1.4998

Amidopolyamines useful as curing agents are low-viscosity, amber-colored fluids. Table 10-1 lists the properties of a typical amidopolyamine based on tall-oil fatty acid and TETA. Figure 10-1 shows viscosity vs. temperature for DGEBA, an amidopolyamine, and a blend of the two.

Although such products have an active-hydrogen equivalent weight, they are often used over a range of mixture ratios to achieve desired handling properties and controlled flexibility, usually to the detriment of the other properties, such as water or solvent resistance or age hardening. Table 10-2 presents a summary of properties of a typical amidopolyamine (tall-oil fatty acid and TETA) used over a range of concentrations.

The advantages of amidopolyamines over the straight polyamines are (1) reduced volatility, (2) reduced—although not entirely eliminated—skin-irritation potential, (3) more convenient mixing ratios, and (4) increased flexibility and impact strength in the cured products. The disadvantages are (1) poor hot strength, leading to cracking in castings at peak exotherms, and (2) some incompatibility with the epoxy resin—the extent of incompatibility depending markedly upon the acid chains used, there being substantial differences between various commercial products. The amidopolyamines require very thorough stirring, followed by an induction period before use to allow the material to react partially and become compatible.

Because of the above advantages and disadvantages, the chief uses for the amidopolyamine curing agents are in construction applications as curing agents for

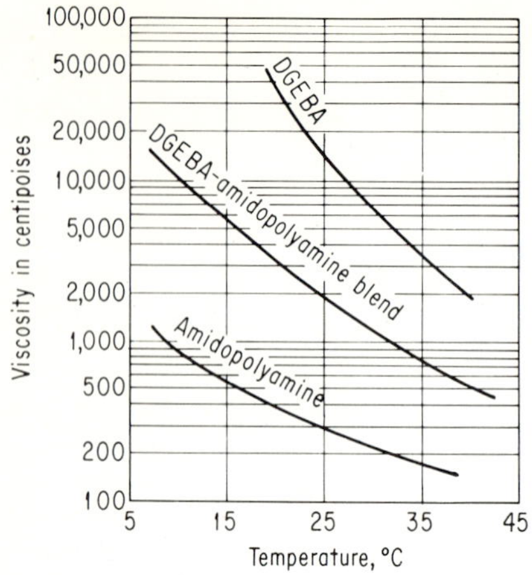

FIG. 10-1. Viscosity vs. temperature for DGEBA, amidopolyamine, and blend of the two [4–4].

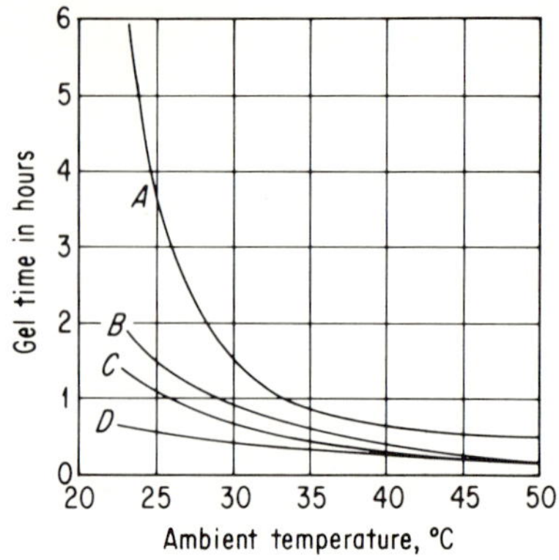

FIG. 10-2. Effect of accelerators on gel time of thin films of DGEBA cured with typical amidopolyamine [4–4]. *A*, no accelerator; *B*, triphenyl phosphite, 25 phr; *C*, nonyl phenol, 25 phr; *D*, resorcinol, 5 phr.

concrete bonding and troweling compounds. Special amidopolyamines are used as curing agents for DGEBA formulations to provide tile-setting adhesives which cure in the presence of water. A typical formulation of this type involves the use of 14.6 parts adipic acid and 40 parts *N*-octadecenetrimethylenediamine [87]. It has also been reported that amidopolyamine curing agents may be modified by addition of 5 to 50 per cent water or glycerol to produce stable gels which impart thixotropic properties to DGEBA formulations without the use of fillers [99].

When amidopolyamides are applied as thin films, it is necessary to use accelerators to promote cure within a reasonable period of time. Typical accelerators are

Table 10-2. Properties of DGEBA Cured with an Amidopolyamine from Tall-oil Fatty Acid and TETA [2]

Property	Amidopolyamine, phr			
	50	75	100	150
Pot life, minutes	75	63	57	60
Gel time, minutes	78	66	61	135
Exotherm, °C	209	190	169	133
Tensile strength, psi, cured 7 days at 25°C	8,400	6,000	1,780	200
Elongation, %	9	12	70	93
Tensile strength, psi, cured 7 days at 25°C + 70 hr at 100°C	8,620	6,160	2,670	620
Elongation, %	8	10	55	130
Hardness, Shore D, after 24-hr cure	80	74	56	10
Ultimate hardness/time elapsed, days	80/1	76/2	60/2	15/2
Volume resistivity, ohm-cm	1.2×10^{14}	3.4×10^{13}	4.5×10^{12}	1.1×10^{11}
Surface resistivity, ohms	1.4×10^{11}	4.7×10^{9}	1.2×10^{9}	1.9×10^{9}
Dielectric constant	3.4	3.5	4.4	7.2
Dissipation factor	0.039	0.062	0.126	0.238

aliphatic polyamines or tertiary amines (which increase the hazard of skin irritation), phenols such as bisphenol A, nonyl phenol, resorcinol, or semi-inorganic-derived accelerators such as triphenyl phosphite and toluene-*p*-sulfonic acid. Figure 10-2 shows the effect of accelerators on gel times.

Imidazoline Polyamines

Another class of amide-type curing agents is that of the fatty imidazoline polyamines. These substances are produced in a manner similar to the amidopolyamines previously discussed by using higher temperatures during the reaction to cause cyclization as shown below:

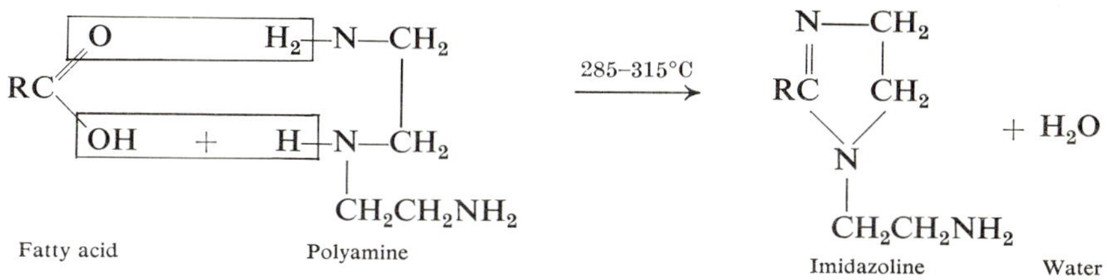

Imidazoline compounds may be used as curing agents in very much the same manner as the amidopolyamines; however, since the imidazoline compounds have fewer active hydrogens, larger percentages must be employed [43,44,51]. However, the functionality is often such that complete cure cannot be achieved by these substances alone, and they must be used in blends with other curing agents. In this role, their chief function is often that of a surfactant—they improve wetting of the epoxy resin in relation to some surfaces and thereby improve the adhesion.

Diimidazolines from both long [38] and short [32] chain aliphatic and from aromatic [39] dibasic acids have also been used to cure epoxy resins. The diimidazolines, having increased functionality, provide somewhat better thermal stability than the imidazolines.

Fatty Polyamides

Higher-melting linear polyamides of the nylon type, containing carboxyl and amine end groups and amide groups along the chain, have been used at high loading volumes in epoxy-resin adhesive formulations to improve flexibility (see Chap. 21), but their use appears to be restricted to such formulations. Substituted nylons (e.g., methoxymethylpolycaprolactam or methylpolycaprolactam) are used as epoxy-resin curing agents [69,74] as are aliphatic/aromatic nylons [poly(hexamethylene isophthalamide)] [92].

Typically, use is made of methanol-soluble nylons (such as the interpolyamide from hexamethylenediammonium adipate, hexamethylenediammonium sebacate, and caprolactam). Dissolved in a methanol/water mixture, these nylons are used, along with an epoxy-resin curing agent, to provide flexible films and adhesives [89].

However, similar to the diimidazolines are a large class of resinous amidopolyamines produced from dibasic fatty acids, and these are referred to in the trade as *fatty polyamides.*

The first patent covering the production of the fatty polyamides from dimer- and trimerized fatty acids and primary amines was issued in 1945 [3]. The tendency of the unsaturated fatty acids to decarboxylate during thermal processing forced all the early workers to polymerize their esters and either use such as reactants or to saponify them for recovery of the dimer and trimer acids [63]. The first good commercial process for direct polymerization of polyene fatty acids was disclosed in 1949 [4].

The fatty polyamides used as epoxy curing agents are generally derived from dimerized linoleic acid and ethylenediamine or diethylenetriamine:

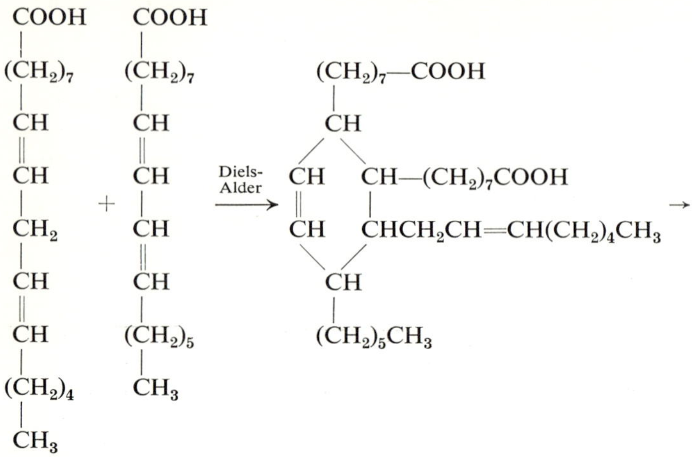

9,12-Linoleic acid 9,11-Linoleic acid Linoleic acid dimer

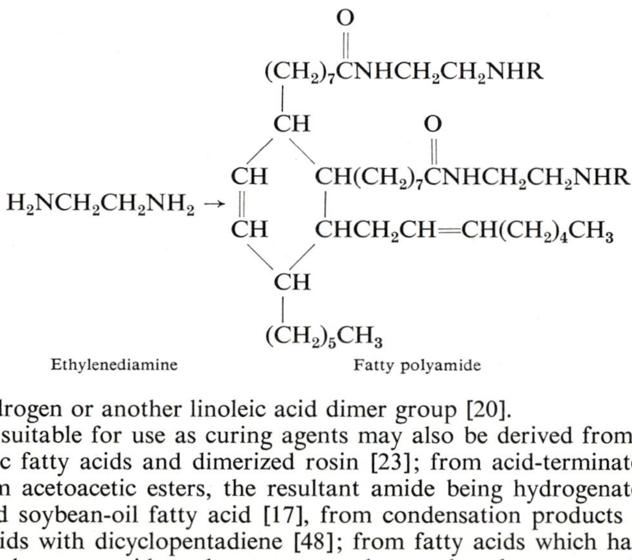

Ethylenediamine Fatty polyamide

where R may be hydrogen or another linoleic acid dimer group [20].

Fatty polyamides suitable for use as curing agents may also be derived from a mixture of polymeric fatty acids and dimerized rosin [23]; from acid-terminated polyesters [14]; from acetoacetic esters, the resultant amide being hydrogenated [29]; from dimerized soybean-oil fatty acid [17], from condensation products of unsaturated fatty acids with dicyclopentadiene [48]; from fatty acids which have been reacted with carbon monoxide and water to produce carboxyl groups at the double bonds [61]; from castor oil–DGEBA–styrene oxide adduct and TETA [62]; from ethylene glycol, adipic acid, and ethanolamine [81]; and from the reaction product of maleic anhydride or acid with an olefinic compound and DETA [85].

It has been reported that the fatty polyamides obtained by using an admixture of polyamides provide decreased viscosity and increased reactivity, believed to be caused by chain lengths being interrupted at irregular intervals, thus reducing hydrogen bonding between the molecules of the fatty polyamide [53]. Reduced viscosity and increased reactivity can also be obtained by reacting the fatty polyamide with an additional charge of polyamine. Thus, a fatty polyamide with an amine number of about 225 may be reacted to effect a partial amide interchange

Table 10-3. Typical Properties of Commercial Fatty Polyamides [1]

Property	Values		
Amine value...........	83–93 (85–95)*	210–230 (230–246)*	290–320 (330–360)*
Melting point, °C	43–53	Fluid	Fluid
Viscosity, centipoises:			
At 25°C.............	....	....	40,000–60,000
At 40°C.............	....	80,000–120,000	
At 150°C............	10,000–15,000		

* Values based on test procedure for amine nitrogen as adopted by General Mills in December, 1962. Throughout this book, values based on the earlier test procedures are employed to maintain consistency with the bulk of the published literature.

with a primary amine to produce a fatty polyamide with an amine number of approximately 300 [31].

Commercial fatty polyamides are amber-colored thermoplastic compounds with molecular weights up to 10,000 and melting points ranging up to 190°C. The higher-melting compounds, however, are only compatible in low concentrations with the epoxies and are of little interest to the epoxy technology, although epoxy resins may be used at less than 10 per cent as curing agents for the high-molecular-weight fatty polyamides used in adhesive formulations [54].

Properties of three commercial grades of lower-molecular-weight fatty polyamides which are used as epoxy curing agents are presented in Table 10-3. The amine number is a guide to the degree of reactivity expected and may be used to calculate the stoichiometric amounts to employ. The amine value is expressed as the number of milligrams of KOH which is equivalent to the base content of 1 gram of fatty polyamide as determined by titration with HCl.

Since the polyamide is the curing agent, the extent to which the mixture ratios can be varied without loss of properties is limited by chemical as well as physical considerations. Calculations based on deflection temperature (DT) indicate that, for the commercial fatty polyamides (amine value approximately 225), 80 phr provides stoichiometric amounts for DGEBA. For fatty polyamides having amine values in the 300 range, approximately 45 phr provides stoichiometric amounts. These ratios can be increased to achieve more flexible materials. Within acceptable limits, the higher the ratio of fatty polyamide to epoxy, the greater the flexibility. At lower concentrations, cure can be accomplished by the addition of a second curing agent to provide rigid systems with some degree of impact resistance.

The range of hardness obtained with a fluidized DGEBA and a fatty polyamide (amine value approximately 300) is presented in Table 10-4. In stoichiometric

Table 10-4. Hardness of Cured DGEBA as a Function of Varying Ratios of Fatty Polyamide [1]

Fatty polyamide (amine value 290–320), phr	Hardness	
	Barcol*	Shore A
66	60–65	
100	20–25	
150	.....	90
185	.....	85
234	.....	50
300	.....	30
400	.....	5

* Impressor model GYZJ-935.

Table 10-5. Typical Physical Properties of DGEBA Cured with Fatty Polyamides

Property	*Value*
DT, °C	50–70
Tensile strength, psi	4,500–7,000
Elongation, %	9–11
Flexural strength, psi	7,000–9,000
Modulus, psi × 10^{-6}	0.2–0.3
Compressive strength, psi	7,000–12,000
Impact, Izod	0.2–0.3
Impact, falling ball, ft-lb	9–10
Shrinkage, linear, in./in.	0.008–0.009
Coefficient of thermal expansion, in./(in.)(°C) × 10^{-6}	90–100

amounts, the DT will be higher for the fatty polyamides with higher amine values than for those with lower values. If higher DTs are desirable, a portion of the fatty polyamide may be replaced by an aromatic amine. For instance, with DGEBA, if 50 phr of a fatty polyamide (amine value approximately 225) is employed in connection with 3.8 phr metaphenylenediamine, the DT of the cured system is about 104°C as compared with 80°C for a system cured by stoichiometric amounts of the fatty polyamide.

Physical properties of DGEBA cured with typical fatty polyamides at slightly greater than stoichiometric amounts to provide increased impact resistance are shown in Table 10-5. Electrical properties are given in Table 10-6; chemical resistance in Table 10-7.

Chemical and solvent resistance of fatty-polyamide-cured epoxies is lower than that obtained with aliphatic-polyamine-cured systems [56]. Some improvement, however, may be obtained in water resistance. Water-absorption data for one year's immersion are presented in Table 10-8. The improved moisture resistance has been attributed to the fatty groups shielding the amide and the hydroxyl groups so that they are not so readily attacked by water as would otherwise be the case [88].

Table 10-6. Electrical Properties of DGEBA Cured with Fatty Polyamides [4-9]

Property	Frequency, cps		
	10	10^3	10^6
Fatty polyamide (amine value 210–230), 100 phr:			
Power factor	0.0090	0.0108	0.0170
Dielectric constant	3.20	3.14	3.01
Loss factor	0.0357	0.0339	0.0572
Dielectric strength, volts/mil		470	
Volume resistivity, ohm-cm		1.5×10^{14}	
Arc resistance, seconds		76	
Fatty polyamide (amine value 290–320), 66 phr:			
Power factor	0.0085	0.0108	0.0213
Dielectric constant	3.37	3.32	3.08
Loss factor	0.0285	0.0359	0.0656
Dielectric strength, volts/mil		430	
Volume resistivity, ohm-cm		1.1×10^{14}	
Arc resistance, seconds		82	

Table 10-7. Chemical Resistance of DGEBA Cured with Fatty Polyamide Amine No. 290–320 [4-1]

Test media	Weight change, %		
	In 24 hr	In 7 days	In 30 days
Water	0.2	0.4	1.2
Boiling water	1.8		
Acetone...........	2.6	5.8	14.0
NaOH, 20%.......	0.2	0.4	0.6
HNO_3, 5%........	0.3	0.7	1.3
Acetic acid, 10%...	1.1	3.3	5.7
HCl, 30%.........	0.6	1.8	3.1
Toluene...........	1.3	6.7	12.0

To the extent that imidazolines are present in the fatty polyamides, as is usually the case with commercial products, they improve the solvent and acid resistance and increase the heat resistance of the films. Presence of imidazolines, however, will somewhat increase the curing time. Commercial blends of fatty polyamides and imidazolines have been developed with suitable synthesis conditions [80].

Oxygen-permeability data on various cured coating systems have been developed. Values in cubic centimeters at 23°C and 50 per cent room humidity/(100 sq in.)(24 hr)(atm)(mil) are as follows: phenolic, 302; alkyd, 74; polyamide/epoxy, 49; vinyl, 33; and aliphatic polyamine/epoxy, 8 [88].

The room-temperature viscosity of the commercial fatty polyamides is high. For this reason, they are often used in connection with fluidized resins or are worked at elevated temperatures. Figure 10-3 presents viscosity as a function of temperature for two commercial fatty polyamides.

At room temperature, it is possible to obtain pot lives of 1 to 2 hours with production quantities of fatty polyamides and DGEBA blends. In 9-lb masses, the pot life will be 90 minutes and the peak exotherm about 150°C. Pot lives at higher

Table 10-8. Per Cent Water Absorption vs. Time for DGEBA Cured with Fatty Polyamide Amine No. 290–320 [4-4]

Immersion time, days	Fatty polyamide, phr					
	50		67			
			Mixed at 23°C		Mixed at 93°C	
	At 25°C	At 66°C	At 25°C	At 66°C	At 25°C	At 66°C
1	0.29	1.17	1.34	2.06	0.16	0.74
2	0.41	1.74	1.70	2.89	0.24	1.27
5	0.64	2.66	2.26	4.21	0.42	2.10
6	0.70	2.85	2.37	4.51	0.46	2.25
7	0.75	3.00	2.48	4.72	0.51	2.39
8	0.80	3.15	2.57	4.92	0.55	2.52
9	0.84	3.26	2.64	5.12	0.59	2.63
12	0.96	3.53	2.85	5.59	0.70	2.88
19	1.22	3.88	3.22	6.15	0.92	3.15
26	1.43	4.05	3.54	6.49	1.10	3.24
33	1.64	4.18	3.79	6.75	1.29	3.29
62	2.17	4.40	4.58	7.53	1.79	3.36
96	2.63	4.59	5.20	8.77	2.24	3.47
124	2.93	4.75	5.67	10.10	2.52	3.56
155	3.13	4.89	6.02	11.21	2.74	3.61
190	3.30	5.02	6.42	11.62	3.00	3.75
365	3.90	6.38	7.44	20.88	3.64	4.37

temperatures are correspondingly shorter and fall in the same general range as that expected of the aliphatic polyamines reacted at room temperature. Typical cure cycles for thin sections vary from 20 minutes at 150°C to 3 hours at 65°C. Larger masses require no external heat for cure.

The fatty-polyamide curing agents are most frequently used with glycidyl ether epoxy resins, although they have been reported as providing satisfactory cures in thin films at temperatures of 175°C when used with epoxidized polyester resins [25].

The principal uses for the fatty-polyamide curing agents are in construction applications, as general-purpose adhesives, for bonding can seams [52], and as

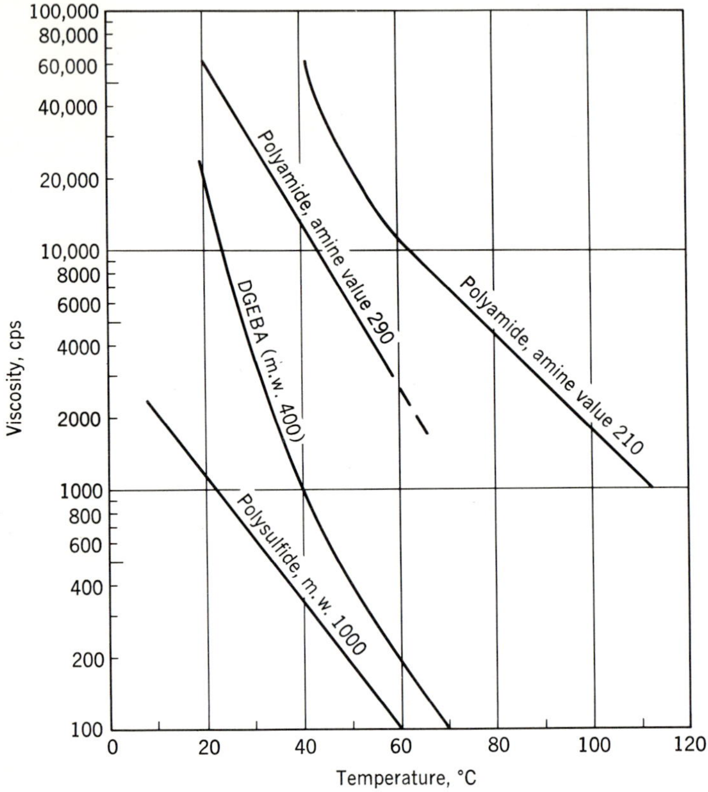

FIG. 10-3. Viscosity vs. temperature for commercial fatty polyamides [1].

curing agents for coating formulations. In coating formulations, the fatty polyamides having lower amine values are preferred. With amine values of about 300, properties in films are comparatively unsatisfactory [21].

Unlike the unmodified polyamines and the amidopolyamines, the fatty polyamides from unsaturated dimer acids and polyamines are not primary skin irritants or skin-sensitizing agents.

Dilution of Amides

Because of the high viscosities of the fatty polyamides and some of the imidazolines, it is desirable to lower their viscosities to permit more convenient processing.

Table 10-9. Accelerating Effect of Various Additives on Cure of DGEBA/Imidazoline Blend [37]

Additive (at 5.6% unless noted)	Peak exotherm, °C	Gel time, minutes	
		With additive	Without additive
Acids:			
Formic	130	88	243
Acetic	127	111	243
Butyric	124	112	243
Capric	128	134	237
Lactic (crude)	127	104	237
Oxalic, dihydrate	131	90	191
Maleic	109	130	191
Benzoic	128	101	268
Phthalic	119	179	191
Salicylic	134	71	191
p-Toluenesulfonic, monohydrate	140	59	243
Alcohols and phenols:			
Water, 5%	159	47	192
Methanol	127	112	212
Propane-1,2-diol	130	123	242
Polyethylene glycol, mol. wt. 300	118	180	237
Glycerol	125	111	237
Phenol	130	98	212
Resorcinol	135	87	268
Amides and sulfonamides:			
Acetamide	129	100	196
Dimethylformamide	91	285	218
	106	233	196
Urea, 2%	133	195	300
Fatty polyamide, 5%	137	115	300
Salicylamide	135	85	191
Benzenesulfonamide	149	95	266
N-Ethyl-*o*- and *p*-toluenesulfonamides (crude mixture)	132	128	191
Hydroxyamines:			
Diethanolamine	135	118	237
Triethanolamine	134	97	243
Tri(hydroxymethyl)aminomethane	128	117	191
o-Aminophenol	133	112	268
p-Aminophenol	128	110	268
m-Diethylaminophenol	125	143	237
Tetraethanolammonium hydroxide solution	135	111	237
Miscellaneous compounds:			
Toluene	83	315	260
Xylene	87	323	250
Dioxane	84	320	260
Salicylaldehyde	117	207	237
Methylethyl ketone	62	390	260
Ethyl acetate	74	365	260
Ethylene sulfite	141	160	250
Acetonitrile	104	280	250
Nitromethane	149	235	260
Nitrobenzene	95	265	260

NOTE: Maleic acid, phthalic acid, acetamide, and urea were incompletely dissolved. The following additives may have suffered partial decomposition or reaction with curing agent: tri(hydroxymethyl)aminomethane, oxalic acid dihydrate, maleic acid, phthalic acid, ethylene sulfite, and nitromethane.

This is generally accomplished by the use of a few parts of a nonreactive diluent, such as dibutyl phthalate [46], or of a tertiary amine. It has been reported that the viscosity can be somewhat lowered, while improving the heat resistance, by reacting the fatty polyamide with an aromatic amine at about 150°C [41]. Viscosity reduction can sometimes be obtained by reacting a portion of the active hydrogens with a monoepoxy [49,76] or with acrylonitrile via the cyanoethylation route discussed in Chap. 7 [71]. Formaldehyde has been reported as a diluent which extends the pot life of epoxy/fatty-polyamide coating systems while decreasing the drying time of films [19].

Acceleration of Amides

The fatty polyamides and imidazolines provide rather sluggish cures in smaller masses, and it is often desirable to accelerate the reaction rate. This may be done by the addition of small amounts of primary aliphatic polyamines or tertiary amines; by the addition of phenols; or by the addition of other accelerators. Phenol-containing tertiary amines are particularly preferred [55]. The accelerating effect of a variety of these chemicals is indicated in Table 10-9.

Use of Amides in Amine/Amide Blends

The fatty polyamides and the imidazolines are frequently used in blends with tertiary amines, the tertiary amines being present in substantial quantity. Typical is a blend of dimethylaminoethanol [44] with a fatty polyamide (amine value approximately 225) at a ratio of 30/70. The blend, at 10 phr, is then used to cure

Table 10-10. Chemical Resistance of DGEBA Cured with 30/70 Blend of Dimethylaminoethanol/Fatty Polyamide [4-1] (Amine no. approx. 225)

Test media	Weight change, %		
	In 24 hr	In 7 days	In 30 days
Water	0.2	0.4	1.3
Boiling water	1.4		
Acetone	0.4	1.4	4.0
NaOH, 20%	0.2	0.5	0.7
HNO_3, 5%	0.3	0.6	1.2
Acetic acid, 10%	0.2	0.6	1.2
HCl, 30%	0.1	0.5	0.9
Toluene	...	0.08	0.2

DGEBA. The blend provides a room-temperature viscosity of about 800 centipoises and reduces the mixed viscosity to about 6,000 centipoises. The pot life is somewhat shortened and the exotherm increased. Physical properties of the cured system fall in the same general range as those of the fatty polyamide, although there is some improvement in heat resistance, with the DT being about 100°C. Chemical resistance for this system is shown in Table 10-10. Like the fatty polyamides, the dimethylaminoethanol blends are reported to possess low skin-irritation properties.

A variety of other blends is, of course, possible. To cite only one example: the use of a fatty amine having aliphatic hydrocarbon radicals from 8 to 22 carbon atoms in blend with the fatty polyamides [47].

UREAS, SUBSTITUTED UREAS, AND UREA-FORMALDEHYDE CURING AGENTS

Urea

$$H_2N\overset{O}{\overset{\|}{C}}NH_2$$

Urea

Urea, or carbamide, may be made to react with the epoxy resins at elevated temperatures and may be used at stoichiometric or less than stoichiometric amounts to produce cured products [16]. More commonly, however, the urea is reacted with formaldehyde, and these reaction products are used as curing agents.

Urea-Formaldehyde

Urea-formaldehyde compounds are produced from the reaction of urea and formaldehyde to give amide-like products of the following structures:

$$\underset{\text{Urea}}{H_2N\overset{O}{\overset{\|}{C}}NH_2} + \underset{\text{Formaldehyde}}{CH_2O} \rightarrow$$

$$H_2N\overset{O}{\overset{\|}{C}}NHCH_2OH$$

Monomethylol urea

$$HOCH_2NH\overset{O}{\overset{\|}{C}}NHCH_2OH$$

Dimethylol urea

$$HOCH_2NH\overset{O}{\overset{\|}{C}}\left[NHCH_2NH\overset{O}{\overset{\|}{C}}\right]_n NHCH_2OH$$

Urea-formaldehyde resin

The methylol groups in these compounds are capable of condensing with hydroxyl groups when heated, and thus have proved of particular value as curing agents for the high-molecular-weight DGEBA resins [8,9,24]. The urea-formaldehyde reaction products may be used as accelerators of anhydride cures and, in connection with acid-terminated short-chain polyesters, as co-curing agents for epoxidized novolac baking coatings [64].

Substituted Ureas

Substituted ureas of the generalized structure

$$RHN\overset{O}{\overset{\|}{C}}NHR$$

Substituted ureas

where R = 1 to 12 carbon atoms have been reacted with an organic acid in proportions such that some secondary amide groups are preserved. These urea amides have then been used as curing agents for DGEBA to provide a pot life of 8 days at 66°C and cured products having improved thermal shock resistance [28].

MELAMINES AND OTHER *s*-TRIAZINE-TYPE CURING AGENTS

Melamines

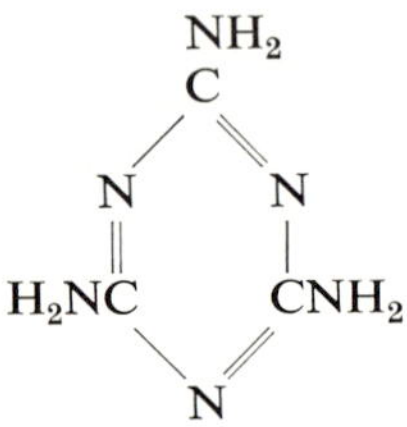

Melamine

The amide of cyanuric acid, melamine, has been used as an epoxy-resin curing agent [10,36], and may be accelerated with sodium hydroxide [27]

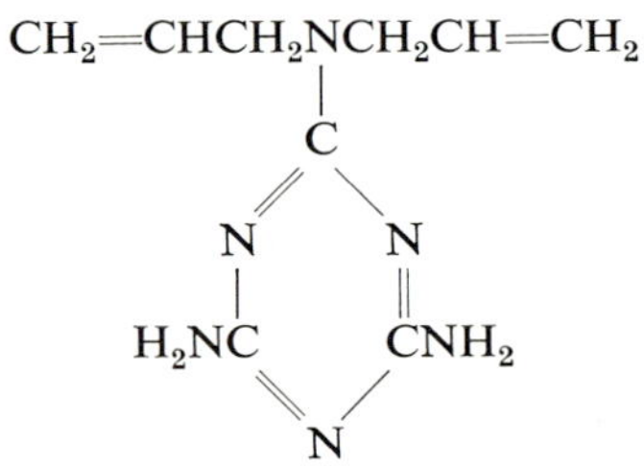

N,N-Diallylmelamine

N,N-Diallylmelamine, however, is generally preferred to melamine. *N,N*-Diallylmelamine provides a long pot life and gives fast cures at 130°C with DGEBA as indicated in Fig. 10-4. Physical properties of the cured systems are indicated in Table 10-11. B-staged fluidized coating systems with DGEBA have been developed [65].

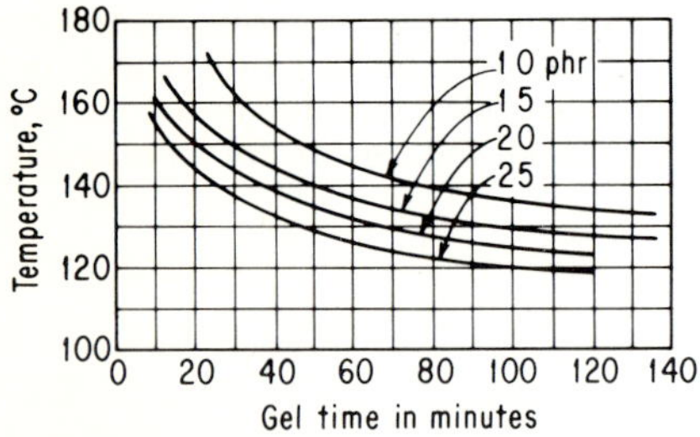

FIG. 10-4. Gel time vs. curing agent concentration for DGEBA cured with *N,N*-diallylmelamine [73].

Melamine-Formaldehyde Resins

Melamine may be condensed with formaldehyde in a fashion similar to urea, and the resulting melamine-formaldehyde resin used as a curing agent in epoxy coating formulations [8,90].

Hexamethoxymethylmelamine

Condensation products of melamine with excess formaldehyde may subsequently be reacted with excess methyl alcohol to produce hexamethoxymethylmelamine, which may be used, in the

Table 10-11. Physical Properties of DGEBA as a Function of Concentration of *N,N*-Diallylmelamine [73]

Curing agent, phr	Cure cycle, °C for 4 hr	DT, °C	Flexural strength, psi	Hardness, Rockwell
10	130	73	21,500	98
10	145	94	20,400	100
15	130	104	21,700	100
15	145	111	22,300	103
20	130	108	20,900	101
20	145	107	22,600	103

presence of an acid catalyst, as a curing agent for the hydroxyl-containing, higher-molecular-weight DGEBA resins, the reaction occurring by condensation with hydroxyl groups. It has been reported as compatible with DGEBA and DGEBA

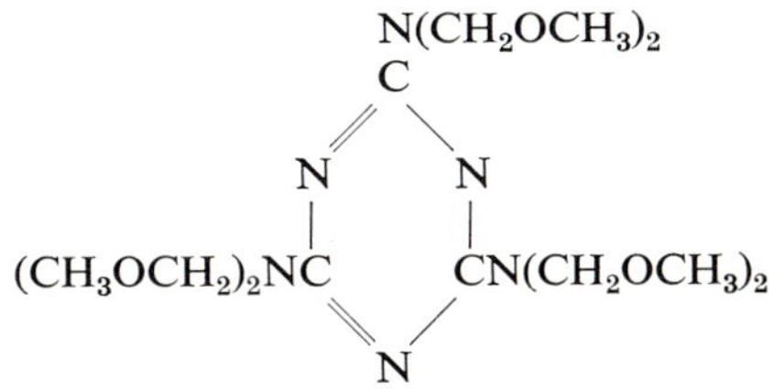

Hexamethoxymethylmelamine

esters and as being superior to urea-formaldehyde resins as a curing agent for baked finishes [68].

Triallyl Cyanurate

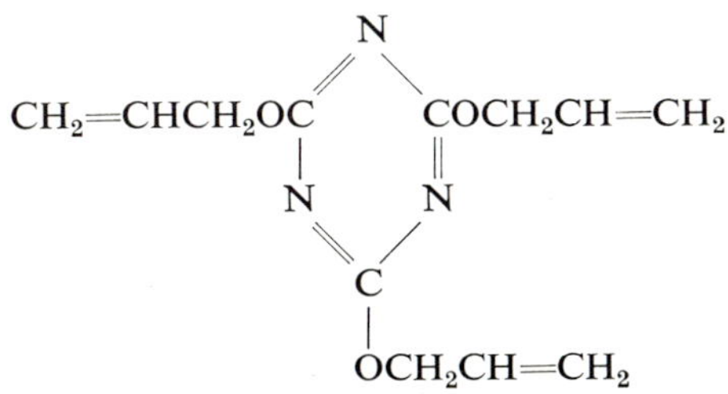

Triallyl cyanurate

Triallyl cyanurate has been employed as an epoxy-resin curing agent providing long pot life coupled with improved elevated-temperature properties in the cured formulation [18]. The allyl unsaturation may be reacted by using vinyl curing agents, thus adding to cross-linking density and rigidity.

Cyanuric Chloride

Cyanuric chloride has been reported as providing a latent curing agent for DGEBA [57].

Guanamines

Various guanamines other than *N,N*-diallylmelamine, have been reported as epoxy-resin curing agents [96]. Typical of the structures are those containing aliphatic chains with or without secondary amines [72] and more highly functional species of the following generalized structure,

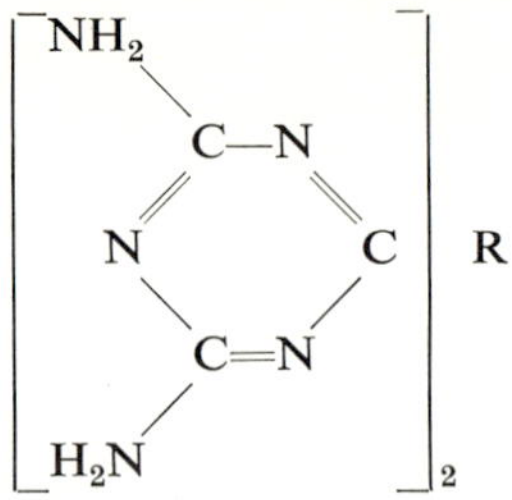

where R is from a dimerized unsaturated fatty acid [67]. In addition, cocoguanamines (C_{16} to C_{18} chains) have been specifically claimed as epoxy curing agents [86,91]. 2-Phenyl-4,6-diamino-*s*-triazine (benzoguanamine) has been reported as a curing agent [100]. It gives latency and can be B-staged at 177°C in 30 minutes, with cure occurring at 205°C [98]. Also reported as a curing agent is 2,4-diamino-6-(2,5-*endo*-methylene-3-cyclohexenyl)-*s*-triazine [94]. More generally it has been reported that 5 to 20 per cent by weight of guanamines cure DGEBA resins at temperatures from 80 to 120°C in times from 3 to 10 hours [93].

DICYANDIAMIDE

Of the more complex, nonpolymeric amides, as distinct from ureas and melamines, the most commonly used in the epoxy technology is dicyandiamide.

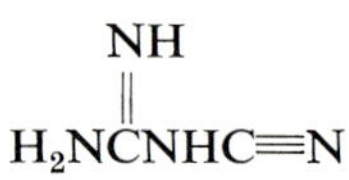

Dicyandiamide

Dicyandiamide cures through all four nitrogren-containing functional groups, and consumes both epoxy and hydroxyl groups in the resin, the final crosslinking reaction yielding guanylurea groups —C═O—NH—R [84,101].

Dicyandiamide is used with solid DGEBA epoxy resins for laminating applications, since it provides latency at room temperature coupled with rapid cures at 145 to 165°C. With these solid resins, it is normally used in solution, being dissolved in a solvent such as acetone-water or dimethylformamide. Hydroxy-containing ethers may also be used [77]. It may be dissolved in amines such as aminoethylpiperazine for use with liquid resins.

When dicyandiamide is stirred into DGEBA, it will provide a usable life of about 24 hours, after which time it separates out. If milled into the system with sufficient thoroughness, it will provide a one-container system stable for at least 6 months. Such one-container systems have been used as adhesive formulations, but if any substantial mass is heated to curing temperature, the exothermic heat is virtually uncontrollable. Such cured systems provide DTs of about 125°C.

Dicyandiamide has been used as an accelerator for epoxy/anhydride [12] and epoxy/phenol-formaldehyde [5] blends.

ACRYLAMIDES

Polymerizable amides, such as acrylamide, have been reported as curing agents for DGEBA [34]. These polymerizable amides have been reacted with unsaturated monomeric compounds [40], and the resultant product reacted with formaldehyde for use in coating formulations.

IMIDAZOLES

Substituted imidazoles have been reported as curing agents for the epoxy resins; these provide pot lives with DGEBA from about 2 to 20 days, depending on the type of substitution on the imidazole, and cure in about 2 hours at 100°C. The substituted imidazoles are also useful accelerators for anhydride cures.

Table 10-12. Properties of DGEBA Cured with 2-Ethyl-4-methylimidazole [4-6]

Property	Curing agent, phr, and cure, hr/°C		
	10 phr, 8/55	4 phr, 8/55	4 phr, 4/55 + 2/120
DT, °C	110	65	130
Tensile strength, psi:			
At 23°C	9,700	8,500	5,700
At 100°C	5,100	4,200	
Modulus, psi:	510,000	540,000	540,000
At 23°C			
At 100°C	150,000	130,000	
Elongation at break, %:			
At 23°C	2.3	1.8	1.8
At 100°C	10.3	7.2	
Flexural strength, psi	21,000	19,900	
Modulus, psi	540,000	563,000	

Properties of DGEBA cured with 2-ethyl-4-methylimidazole are shown in Table 10-12.

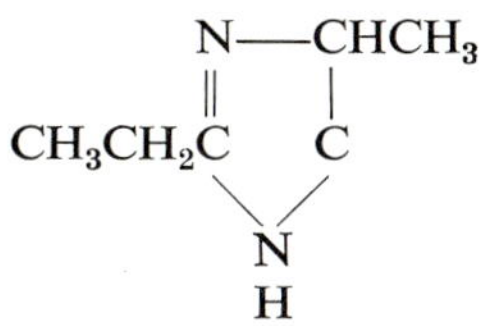

2-Ethyl-4-methylimidazole

The imidazoles generally offer, when used at low phr's, DT's to 150°C with DGEBA and other elevated-temperature properties in the range expected of aromatic amine cures. This improved heat resistance is obtained with a supercooled liquid having a viscosity of 4,000 to 8,000 centipoises and providing a long pot life [95].

HYDRAZIDES

Hydrazides of the formula

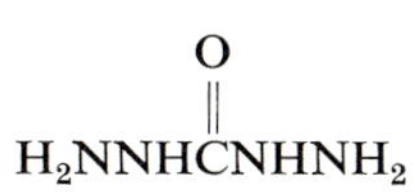

Carbohydrazide

have been used at about 12 phr to cure DGEBA for adhesive formulations. Cure occurs in about 1 hour at 150°C. The cured product provides good bond strengths and retains properties to 150°C [50].

Longer-chain dihydrazides of the following structure,

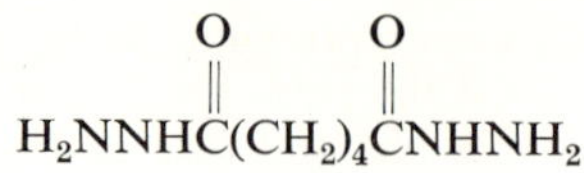

Adipic acid dihydrazide

have been reported as providing pot lives with DGEBA of about 4 months or more at room temperature [30]. These systems may be accelerated without substantial reduction in pot life by the use of equal parts of lead octoate and the product obtained by reacting equal molar quantities of zinc octoate and diethylaminopropylamine.

GUANIDINES

Compounds such as tetramethylguanidine and heptamethylisobiguanidine, which contain the guanidine structure

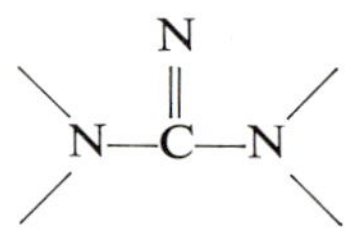

Guanidine structure

have been found to be effective heat-curing catalysts for DGEBA. They are used at 0.5 to 2 phr and give pot lives at room temperature in excess of 4 hours with cures occurring in about 1 hour at 160°C. DT's are 80 to 90°C [97].

NITROSAMINES

Ethylenically unsaturated α,β-nitrosamines (i.e., containing the following group)

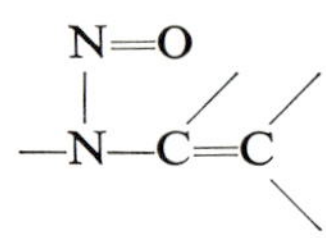

Nitrosamine group

are reported to be curing agents which react with epoxy resins in the presence of light [58].

ETHYLENE IMINES

Polyalkylene imines have been used with epoxy resins for fabric treatment [22],

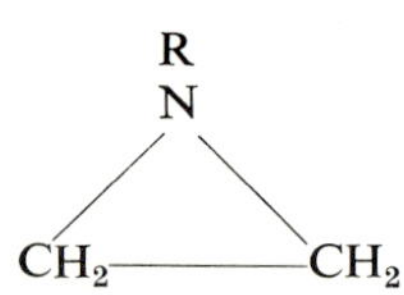

Ethyleneimine group

and esters containing ethyleneimine groups have been reported as curing agents [33].

THIOUREAS

Polythioureas (from the reaction of an aliphatic diamine and carbon disulfide) having the generalized formula

$$\overset{\displaystyle S}{\overset{\|}{}}\qquad\\ \text{-(NHCNHR)-}_n$$

Generalized polythiourea

have been found to react with DGEBA. The more reactive polythioureas cure at room temperature [42].

SULFONAMIDES

Sulfonamides of the following generalized structure,

$$\begin{array}{c} O \\ \uparrow \\ RSNH_2 \\ \downarrow \\ O \end{array}$$

Generalized sulfonamide

may be used as curing agents for DGEBA resins [15], or the sulfonamide may be reacted with formaldehyde [7,26] and the product, like the urea-melamine- and phenol-formaldehyde compounds, may be employed as a curing agent for the polyhydric polyether alcohol epoxy resins in coating formulations.

SUMMARY

Amides, as a class, provide sluggish reactivity, even with the glycidyl ether epoxy resins. They are normally used in heat-cured systems and give some degree of room-temperature latency. The monomeric amides find use in adhesive and laminating formulations. Urea-formaldehyde or melamine-formaldehyde resins are widely used in coating formulations.

Fatty amides, obtained generally from fatty acids and primary aliphatic polyamines, are used as room-temperature-curing agents. These compounds contain free primary and secondary amine groups which are the reactive sites. The fatty amides vary from materials with very little modification (usually low-viscosity liquids) to materials with substantial amidation (viscous liquids). As the extent of amidation increases, the reactivity is reduced, and the impact strength and flexibility of the resultant cured epoxy is increased. The more highly modified materials do not cause skin irritation or skin sensitization.

The amides are generally employed with the glycidyl ether epoxy resins, and are less frequently used, if at all, with the epoxidized olefin type.

REFERENCES

1. General Mills, Inc., technical literature.
2. Thiokol Chemical Corp., technical literature.

1945

3. Bradley, Amides of High Molecular Weight Carboxylic Acids, U.S. 2,379,413.

1949

4. Goebel, Polymerization of Unsaturated Fatty Acids, U.S. 2,482,761.
5. Ott and Kraus, Phenol Oxide Resin Lacquers and Their Manufacture, U.S. 2,458,796.

1950

6. Greenlee, Amide-Epoxy-Phenol Compositions, U.S. 2,510,886.
7. Greenlee, Compositions of Resinous Epoxides and Aromatic Sulfonamide-Aldehyde Condensates, U.S. 2,494,295.
8. Greenlee, Epoxide Resin Composition, U.S. 2,528,360.
9. Greenlee, Polyepoxide-containing Compositions and Reaction Products, U.S. 2,528,359.

1951

10. Gams et al., Polyglycidyl Ether of a Polyhydric Phenol Hot Setting Adhesive and Method of Uniting Articles Therewith, U.S. 2,786,794.

1952

11. Greenlee, Amide-Epoxide Compositions, etc. U.S. 2,589,245.

1953

12. Ott, Epoxy-Resin-Polycarboxylic Acid Dicyandiamide Reaction Products, U.S. 2,637,715.

1954

13. Marotta, Fire Retardant Coating Composition Containing a Reaction Product of Phosphoryl Chloride and Anhydrous Ammonia and Articles Coated Therewith, U.S. 2,676,162.

1955

14. Fisch, Elastic Infusible Products from Epoxy Compounds and Crosslinking Agents, U.S. 2,712,535.
15. Greenlee, Sulfonamide–Epoxide Compositions, U.S. 2,712,001.
16. Greenlee, Urea–Epoxy Compositions, U.S. 2,713,569.
17. Renfrew and Wittcoff, Thermosetting Resinous Compositions from Epoxy Resins and Polyamides Derived from Polymeric Fat Acids, U.S. 2,705,223.
18. Skiff and Finholt, Ethoxyline–Triallyl Cyanurate Resinous Compositions, U.S. 2,707,177.
19. Wittcoff, Stabilized Solutions of Mixtures of Polyamide Resins and Epoxy Resins, U.S. 2,707,708.

1956

20. Floyd et al., Characteristics of the Polyamide–Epoxy Resin System, British Society of Chemical Industry Symposium, London, April.

1957

21. North, General Resistance Properties of Epoxy–Polyamide Systems, *FATIPEC*, book.
22. Pardo, Shrinkproofing of Wool with Epoxy Resins and Polyalkyleneimines, U.S. 2,817,602.

1958

23. Floyd, Polyamide–Epoxy Resin Reaction Product, U.S. 2,823,189.
24. Greenlee, Compositions Containing Polyepoxide Polyesters and Amine–Aldehyde Condensates and Methods of Making Same, U.S. 2,850,475.

25. Greenlee, Compositions of Polyamides and Polyepoxy Polyesters, U.S. 2,852,477.
26. Greenlee, Compositions of Polyepoxide Polyesters and Aromatic Sulfonamide–Aldehyde Condensates and Method of Making Same, U.S. 2,847,394.
27. Greenlee, Polyethylene Polyamine–Epoxides, U.S. 2,865,886.
28. Jenkins et al., Ureamide–Epoxide Compositions, U.S. 2,855,372.
29. New Curing Agents for Epoxy Resins (to Henkel & Cie., G.m.b.H.) Ger. Appl. H 25 016, March.
30. Wear, Stable Heat-curing Epoxy Resin Compositions, U.S. 2,847,395.

1959

31. Peerman and Floyd, Polyamide Resin Process and Product, U.S. 2,881,194.
32. Peterson, Epoxy Resins including Imidazoline Curing Agent, U.S. 2,878,234.
33. Starck et al., Process for Preparing High Molecular Weight Compounds, U.S. 2,901,443.
34. Vogel, Coating Composition Comprising an Epoxy Resin and an Aldehyde-reacted Polymer of Acrylamide and Metallic Surface Coated Therewith, U.S. 2,870,117.

1960

35. Bruin and Zonsveld, Curing Polyepoxides with an Acid Anhydride and an Acid Amide, U.S. 2,955,101.
36. Donimerski, Epoxy Resin Adhesives and Joints Therefrom, *Biul. Wojskowgi Akad. Tech.* **9**(98):74–92.
37. Gough and Smith, Accelerated Amine Curing of Epoxy Resins, *J. Oil Colour Chemists Assoc.* **43**:409–418.
38. Harrison, Curing of Epoxy Resins, Ger. 1,074,856.
39. Harrison, Hardeners for Epoxy Resins, Ger. 1,079,320.
40. Maeder et al., Curing of Epoxy Resins, Ger. 1,085,667.
41. Renfrew and Peerman, Novel Compositions and Method of Curing Epoxy Resins, U.S. 2,930,773.
42. Wesp, Polythiourea and Polyepoxide Co-cured Resins, U.S. 2,957,844.

1961

43. Arnold, Metallic Containers, U.S. 2,994,455.
44. Capron and Lederman, Synthetic Resins, U.S. 2,994,673.
45. Delmonte and Dewar, Epoxy Resin Compositions and Methods, U.S. 2,994,685.
46. Floyd and Peerman, Epoxy–Polyamide–Ester Reaction Product, U.S. 2,999,825.
47. Glaser, Composition Comprising an Epoxy Resin, an Amino Polyamide and a Polyamine, U.S. 2,990,383.
48. Hovey, Liquid Polyamide Composition and Process for Preparing Same, U.S. 3,001,960.
49. Katz et al., Compositions of a Monoepoxide, Modified Polyamide and an Epoxy Resin, U.S. 2,970,971.
50. Levine, Carbohydrazide Curing Agent for Epoxide Type Resins, U.S. 3,014,009.
51. Lescisin, Process for the Production of Substituted Imidazolines, U.S. 2,992,230.
52. Peerman, Metallic Containers, U.S. 2,994,456.
53. Peerman and Floyd, Reaction Product of Epoxy Resins and Polyamides from Blends of Aliphatic Polyamines and Aromatic Polyamines, U.S. 2,999,826.
54. Schniepp et al., Polyamide–Epoxy Adhesives, U.S. 2,986,539.
55. Zumstein, Rapidly Hardenable Mixtures of Epoxy Resins and Polyamide Resins, U.S. 2,977,332.

1962

56. Burge and Brown, Epoxy Formulations in Warm Corrosives, *Mater. Protection*, June.
57. Faerber, Epoxy Resin Hardener, Ger. 1,128,650.
58. Improvements in or Relating to Photo-sensitive Epoxy Resin Compositions (to Minnesota Mining & Manufacturing Co.), Brit. 912,022.

59. Katz et al., Catalyst Composition, U.S. 3,028,342.
60. A Process for the Manufacture of Hardened Artificial Resin Materials and Compositions Therefor, Brit. 897,726.
61. Rogier, Amino–polyamide Resins and Reaction Product Thereof with an Epoxy Resin, U.S. 3,062,773.

1963

62. Amides Containing Amine Groups (to Ciba, Ltd.), Belg. 630,621.
63. Bradley, T. F., private communication, July.
64. Drubel and Buege, Coating Compositions of Epoxy Novolac Resins, Amino Resins and Acid Esters of Diethylene Glycol and Dibasic Carboxylic Acids, U.S. 3,097,183.
65. Floyd, Epoxy Resins, Belg. 635,726.
66. Fram and Charbonneau, Epoxy Resin Copolymers, U.S. 3,079,367.
67. Fulmer and Peerman, Partially Crosslinked Epoxy Resins, Belg. 630,605.
68. Hensley, HMMM with Epoxy Resins, *Official Digest*, January.
69. Kujawa-Penczek and Skora, Adhesives Based on Epoxy Resins and Poly(*N*-methylolamides), *Polimery* **8**(6):241–244.
70. Meigs, Epoxy Resin–Amine Rice Oil Compositions and Method of Making Same, U.S. 3,099,634.
71. Mika, Cyano-substituted Polyamides and Their Use as Curing Agents for Polyepoxides, U.S. 3,091,595.
72. Peerman, Epoxide–Guanamine Compositions, Belg. 630,100.
73. Resinous Compositions (to American Cyanamid Co.), Brit. 928,284.
74. Skora et al., Polyamide–Epoxy Adhesives, Pol. 48,112.
75. Swann and Evans, Synthetic Resin Compositions, Brit. 919,880.

1964

76. Battles et al., Composition and Method for Coating Leather, U.S. 3,130,069.
77. Buser and Helbig, Laminating Epoxy Resins, Ger. 1,162,017.
78. Diamido Polyamine Compounds for Hardening Epoxy Resins (to Inventa A.G. für Forshung und Patentverwertung), Neth. Appl. 300,217.
79. Epoxy Resin Hardeners (Inventa A.G. für Forschung und Patentverwertung), Neth. Appl. 300,178.
80. French, Epoxy Resins, Brit. 952,322.
81. Gazzard, Polymerizable Epoxide Compositions for Coating of Metals, Brit. 956,268.
82. Humphreys and Hinton, Curing Agents for Epoxide Resins, Brit. 973,002.
83. Humphreys and Ratcliffe, Hardening Agents for Epoxide Resins Prepared by Reaction of Polyamines with Amino Esters, Brit. 962,242.
84. Levine, Reaction of Dicyandiamide and Epoxy Resins, ACS Symposium, Chicago, September.
85. *N*-Aminoalkylamides and *N*-Aminoalkylimidazoline–Resin Curing Agents (to Schering A.G.), Brit. 956,709.
86. Peerman, Epoxy Resin Formulations Stable at Ambient Temperature, Brit. 969,257.
87. Wagner, Adhesive Compositions, U.S. 3,140,566.
88. Wittcoff, Polyamide–Epoxy Blends: a Corrosion Inhibiting Vehicle for Paints, *Oil & Colour Chemists Assoc.*, **47**:4.

1965

89. Adhesive Compositions (to 3M Co.), Brit. 986,190.
90. Bloomfield, Hardening Epoxide Resins, Brit. 998,801.
91. Epoxide Resins (to General Mills, Inc.), Fr. Addn. 84,401 (addition to Fr. 1,359,377).
92. Gorton, Polyamide-polyepoxide Adhesive Composition, Fr. 1,386,572.
93. Hardening of Polyurethanes and Epoxy Resins (to Olin Mathieson Chemical Corp.), Neth. Appl. 6,411,810.
94. Mori et al., Curing Epoxy Resins, Japan 1660 (1965).
95. New Cure for Epoxies, *Chem. Week*, July 31.

96. Pazenko et al., Hardening of Epoxy Resins with Aminotriazines, U.S.S.R. 167,300.
97. Sherr and Krupnik, Base Catalysts for the Cure of Epoxy Resins, *J. Appl. Polymer Sci.*, **9**:2707–2714.
98. Thermosetting Plastic Compositions Partially Hardened and Stable at Room Temperature (to General Mills, Inc.), Neth. Appl. 284,328.
99. Thixotropic Epoxy Resin Composition as Adhesive (Sels et Produits Chimiques, S.A.), Neth. Appl. 6,411,560.
100. Vertnik, Epoxy Resin Compositions for Impregnating Fibrous Materials, Fr. 1,390,235.

1966

101. Saunders et al., A Mechanistic Study of the Dicyandiamide–Epoxy Curing Process, ACS Symposium Pittsburgh, March.

Chapter 11

ACID CURING AGENTS FOR EPOXY RESINS

A particularly large and diverse class of curing agents for epoxy resins is that designated as acid curing agents. This class contains the Lewis acids (or Friedel-Crafts) catalysts, the phenols, the inorganic (or mineral) acids, and the carboxylic acids. Under the broad classification "Lewis acids and related compounds" are included a variety of boron compounds, metal-organics, and amphoteric oxides.

LEWIS ACIDS AND RELATED COMPOUNDS

Lewis Acids

As used in the epoxy technology, Lewis acids consist of such compounds as aluminum chloride, aluminum bromide, zinc chloride, boron trifluoride, silicon tetrachloride, stannic chloride, ferric chloride, and titanium tetrachloride.

It is probable that the Lewis acid catalyst coordinates with the epoxy oxygen and/or the oxygen of an adjacent hydroxyl to assist in the transfer of a proton and the formation of an ether link. The Lewis acids differ from the Lewis bases (e.g., tertiary amines) in that they are effective catalysts for glycidyl ether resins and both linear [109] and cyclic [8,79] epoxidized olefin resins. Because Lewis acids serve merely to assist hydrogen transfer, they do not, as do the Lewis bases, reduce the system functionality and in general can be used at lower concentrations than can the Lewis bases.

The more reactive Lewis acids provide highly exothermic reactions with both glycidyl ether resins and epoxidized olefins at room temperature. For this reason, they are useful curing agents only for small masses, when pot lives of about 30 seconds or so can be tolerated. In larger masses the exotherm is sufficiently high to preclude their use. However, with higher-molecular-weight resins containing numerous hydroxyls, the resin may assume the deactivating characteristic of a polyhydric alcohol [59], and elevated temperatures may be required to achieve a satisfactory reaction rate. With systems having high hydroxyl contents, it is also probable that water content becomes increasingly critical, and the actual effect of elevated temperature is to remove water from the system. From the authors' research, it appears probable that excess water will cause complexing with the Lewis acid and will promote, preferentially, the hydrolysis of epoxy groups, which, under these conditions, appears to occur at a very slow rate. The net effect, until the water is removed, is to consume the epoxy groups slowly in side reactions which do not contribute to cross-linking.

The Lewis acids are usually employed as curing agents in combination with some carrier solvent. Stannic chloride, for example, may be complexed with polyols or monoethers to produce gel times at 37°C from about 1 to 5 minutes, depending on the concentration of polyol and catalyst [113]. Antimony pentachloride is reported as effective when carried in triethylene glycol [149]. Adducts of mono- and diepoxies with BF_3 and hydrofluoroboric acid may be dissolved in solvents [141]. Antimony trichloride, one of the more sluggish Lewis acids, is reported as providing, in blend with MEK-peroxide, gel times with DGEBA of about 4 hours at room temperature [116].

The Lewis acids have generated considerable interest as curing agents for epoxy resins in aqueous media for fabric-treating applications, where it is desired to bring about reaction of epoxy groups with hydroxyls in the fibers [39]. With DGEBA (mol. wt. 950), the best of the catalysts for this purpose were found to be zinc, magnesium, and aluminum fluoroborate. Zinc fluoroborate is also suitable for use with bis(2,3-epoxycyclopentyl) ether in fabric-treating applications [170]. Elevated-temperature cures are employed to drive off the water.

Because of the speed with which the most suitable Lewis acid compounds react with the epoxy resins, it is desirable to block the reactivity in some fashion, in those cases where water cannot be used for that purpose.

Blocked Lewis Acids. A variety of techniques may be employed to block the reactivity of the Lewis acids. Normally, such blocked compounds do not react, or react but sluggishly, at room temperature. In special cases, temporary blocking can be employed, this having the effect of extending the pot life to permit more mixing time but still providing for room-temperature cure. One method by which this may be accomplished is to absorb boron trifluoride diethyl etherate $BF_3.O(C_2H_5)_2$ in a particle carrier, such as pumice, and coat the carrier with paraffin. When this is incorporated into the resin, the Lewis acid leaks from the capsule and initiates exotherm to melt the paraffin, thus in turn releasing more curing agent [42]. The technique, of course, is also suitable for use with other curing agents [72,128,129]. Alternatively, room-temperature cure may be obtained by blocking BF_3 with tetrahydropyran, the pyran ring being opened and the $—(CH_2)_5—O—$ group acting as a chain extender [57].

A more generally useful method for blocking reactivity involves complexing the Lewis acid with amines. Such blocked materials provide for latency at room temperature but give rapid cures in both glycidyl ether and epoxidized olefins [100] when the temperature is elevated to the point at which the complex dissociates. These complexes may also be used as accelerators for amines [43], anhydride [73], and dicyandiamide [151] curing agents. It is probable that cure and acceleration is obtained not only by the Lewis acid but also by the influence of the tertiary nitrogen and such primary and secondary amines as are present.

Commonly, the Lewis acid complexed with amines is the gas BF_3 [22]. A variety of such complexes has been suggested, ranging in dissociation temperature from the monoethylamine complex dissociating at about 90°C to the pyridine complex (dissociating above 300°C). Various blends of the BF_3-amine complexes may be used, such as a blend of the monoethylamine complex with the piperidine complex [85].

BF_3-amine adducts are prepared by treating BF_3 etherate with amines in solvents under normal conditions or by introducing BF_3 into solution of the amine under nitrogen at −20 to 50°C. One series of tests with a variety of BF_3-amine adducts noted large differences in reactivity toward DGEBA at temperatures of 120 to 150°C, with the adduct of benzylamine being nearest to latent [168]. Although, in general BF_3 adducts with aniline and its derivatives are reactive with epoxy resins at room temperature [168], monohydrofluoroborate is reported as providing stable adducts with aromatic amines such as MPDA [159]. The adducts of the above may be stabilized by the addition of excess amine, the stabilizing effect also reducing the elevated-temperature reactivity, which with some systems, may be quite rapid.

Phosphorus-group pentahalides such as PF_5, AsF_5, and SbF_5 give satisfactory latency and cure of epoxy resins when complexed with amines, but they require higher cures than do the BF_3 complexes [175].

Representative of the Lewis acid complex curing agents are BF_3 monoethylamine and BF_3 piperidine.

BF_3 *Monoethylamine* (BF_3MEA). Boron trifluoride monoethylamine is a yellowish-white crystalline solid which melts close to its dissociation temperature

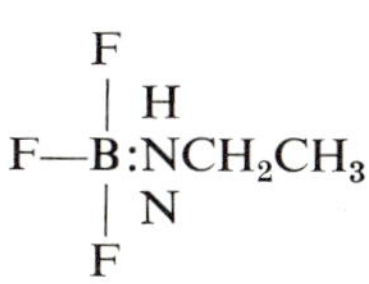

BF_3MEA

(about 90°C). Upon exposure to moist air, it hydrolyzes to a viscous liquid unsuitable as a curing agent.

BF_3MEA is normally dissolved in the epoxy resin. For smaller batches, the procedure is to heat the resin to about 85°C and stir in the curing agent. After the catalyst is thoroughly stirred in, the mixture is cooled to room temperature. For larger batches, about 3 parts by weight of the catalyst is stirred into about 5 parts

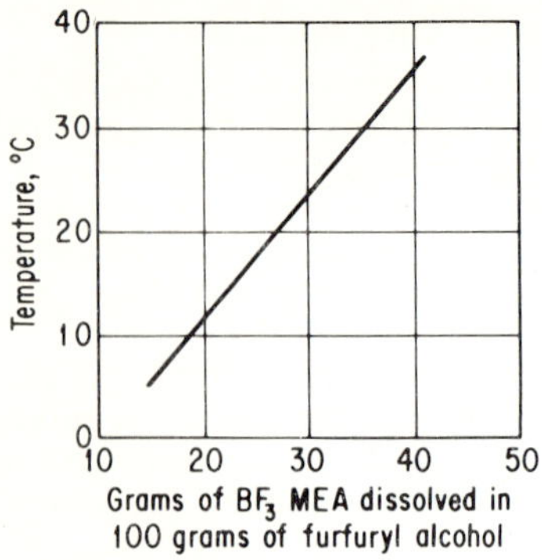

FIG. 11-1. Solubility of BF_3MEA in furfuryl alcohol [103].

by weight of the resin preheated to about 50°C; this produces a smooth paste which may then be added to the remainder of the resin heated to 85°C. Alternatively, the BF_3MEA may be dissolved in an hydroxyl-rich carrier, such as a glycol [51], furfuryl alcohol, or bis(tetrahydrofurfuryl) phthalate [64], and this solution is used as a curing agent. A typical solubility curve is given in Fig. 11-1. The carrier generally reduces viscosity and, when present at 10 phr or so, will impart some flexibility to the cured system, although, since it is a coreactant through hydroxyl groups with the epoxy resin, the reduction in cured properties is less than that obtained with a nonreactive diluent. Typical of viscosity reduction is that obtained with furfuryl alcohol in DGEBA, as indicated in Fig. 11-2. The effect of increasing amounts of the alcohol on the flexural strength of the cured system is indicated in Fig. 11-3. The increased strength may arise from polymerization through the double bonds of the alcohol.

BF_3MEA, when used with DGEBA, can be considered a semilatent catalyst. Optimum properties (except for thermal stability) are obtained when it is used at about 3 phr. At this concentration, there will be a substantial increase in viscosity at the end of 6 months' storage at room temperature. Viscosity vs. concentration of catalyst for top room-temperature storage conditions is shown in Fig. 11-4. Some fillers will, in contact with the catalyst, neutralize it over a period of time, at least to some extent, and fillers such as mica, asbestos, and alkaline materials such as calcium carbonate should be avoided in the formulation. The tendency of BF_3MEA to advance DGEBA has been used to advantage in laminating applications. Higher-molecular-weight resins have been prepared by cooking DGEBA with BF_3MEA to a suitable viscosity and cooling to obtain a room-temperature dry B-stage material [126].

With the glycidyl ether resins, the curing reaction is rapidly triggered and highly temperature-dependent. For example, with 3 phr BF_3MEA in DGEBA, a mass as

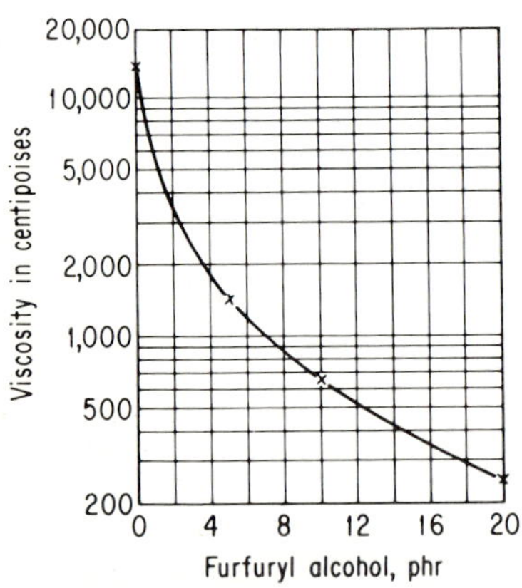

FIG. 11-2. Effect of furfuryl alcohol on viscosity of DGEBA [103].

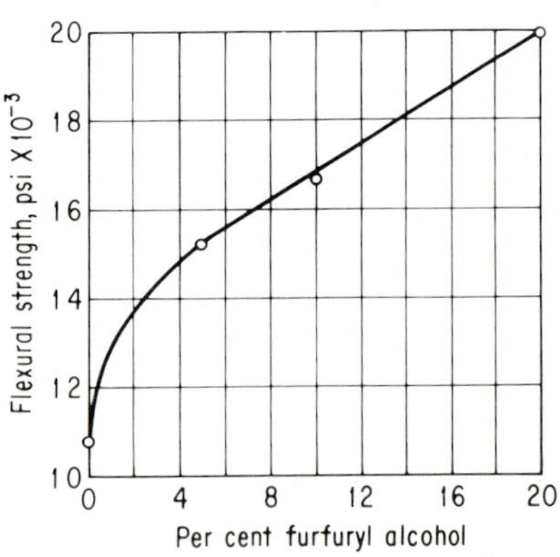

FIG. 11-3. Effect of furfuryl alcohol on flexural strength of DGEBA cured with BF_3MEA [103].

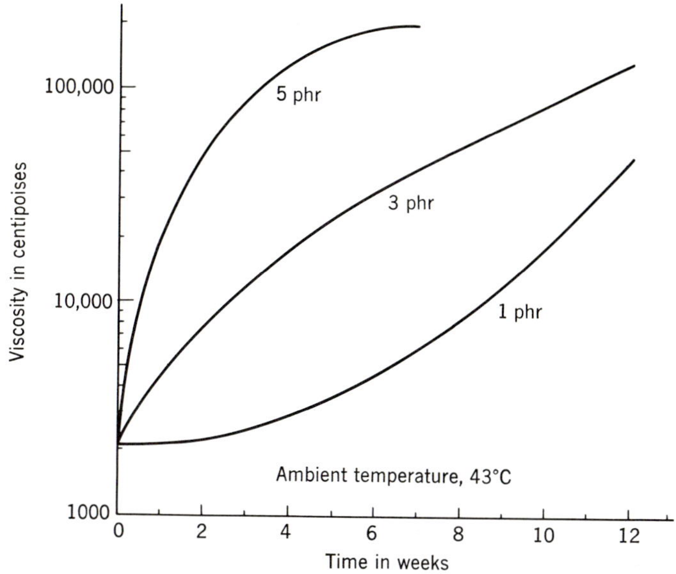

FIG. 11-4. Viscosity vs. time for liquid epoxidized novolac catalyzed with BF_3MEA [4–7].

small as 3 grams will cure in 2 minutes at 230°C. In a 25-gram mass, increasing the curing temperature from 100 to 107°C reduces the gel time from 3 hours to 30 minutes and the cure time proportionally. The reaction is somewhat difficult to control, and reproducibility will depend upon accuracy in measuring the ingredients and upon the elimination of moisture pickup in the catalyst prior to incorporation into the resin. When significant masses are involved, use of high initial-cure temperatures results in exotherms sufficiently high to degrade physical properties; the larger the mass, the more serious the consequences.

Because of exotherm, it is customary to gel the BF_3MEA/DGEBA systems at about 120°C, following this with a postcure. Generally the higher the postcure temperature, within practical limits, the higher the deflection temperature (DT) (Table 11-1). With DGEBA and a 4-hour postcure, the DT will be about 25°C below the selected postcure temperature over the 125 to 200°C range. It has been reported [110] that DTs may be improved by the addition of about 4 phr triaryl phosphite.

In spite of the relatively high DTs, thermal stability is rather poor, and for maximum thermal stability, lower concentrations of curing agent are preferred [83], thus necessitating relatively long high-temperature cure cycles. During such cures, it is probable that some per cent of the BF_3 gas is driven out of the system [130], this being reflected in reduction in dielectric constant and dissipation factor of the cured system. Radiation resistance is poorer than would be expected from the relatively high DTs (Fig. 11-5).

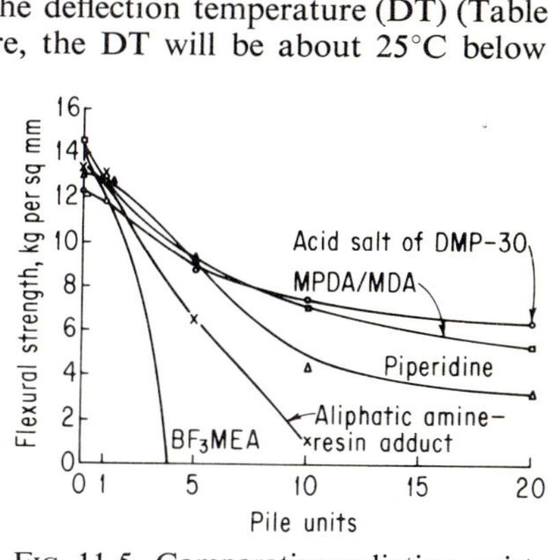

FIG. 11-5. Comparative radiation resistance of BF_3MEA-cured DGEBA and other relatively poor radiation-resistant DGEBA systems [84].

Table 11-1. DT of DGEBA Cured with BF_3MEA [4-4]

Aging conditions	DT, °C		
	1 phr BF_3MEA	3 phr BF_3MEA	5 phr BF_3MEA
Hours at 150°C:			
4	32	131	175
24	...	152	189
48	51	170	190
72	58	177	191
96	68	194	192
120	...	195	193
144	88	200	187
168	96	205	186
Hours at 204°C:			
2	73	186	170
4	98	176	162
6	110	176	163
8	126	169	159
16	163	160	153
24	165	158	148
48	156	148	136

Table 11-2. Physical Properties of DGEBA Cured with BF_3MEA [4-4]

Property	BF_3MEA, phr, and cure, hr/°C		
	1 phr, 24/105 + 7/150	3 phr, 16/93 + 2/150	5 phr, 16/80 + 2/150
DT, °C	65	128	142
After 168 hr at 150°C	81	210	161
Tensile strength, psi	8,400	8,500	8,400
Elongation, %	1.8	2.3	2.4
Flexural strength, psi	16,400	16,300	20,000
Izod impact	0.30	0.27	0.36
Hardness, Rockwell M	93	108	112
Weight loss after 168 hr at 150°C, %	0.71	0.24	0.38

Table 11-3. Chemical Resistance of DGEBA Cured with 4 phr BF_3MEA [4-1]

After 6 months' immersion in fluid	*Change in flexural strength, %*
Salt resistance:	
10% sodium chloride	+8
25% sodium chloride	−3
10% sodium bisulfite	+8
Acid resistance:	
3% sulfuric	−8
10% sulfuric	−33
30% sulfuric	+2
10% hydrochloric	−4
10% nitric	−12
10% glacial acetic	+20
Base resistance:	
10% caustic soda	+3
25% caustic soda	−3
25% sodium carbonate	−1
25% ammonia solution	−16
Solvent resistance:	
Gasoline	−2
Benzene	+1
Ethyl alcohol	+20
Ethylene glycol	+18
Water resistance:	
At 25° C	−22
Boiling	−39

Physical properties of the cured DGEBA using 1, 3, and 5 phr BF_3MEA are shown in Table 11-2. Chemical resistance for the system cured with 4 phr catalyst is given in Table 11-3. Further chemical resistance data are given in Fig. 11-6. The effect of high-temperature aging and the effect of high relative humidities on dielectric constant and dissipation factor are indicated in Figs. 11-7 and 11-8, from which it will be seen that humidity resistance is relatively low.

With the epoxidized olefin resins, in general, the BF_3MEA complex is more reactive than with the glycidyl ether resins. Cure occurs at lower temperatures, and mixed storage life is reduced.

BF_3 *Piperidine.* With 3,4-epoxy-6-methylcyclohexylmethyl 3,4-epoxy-6-methylcyclohexane carboxylate, and with other epoxidized olefin resins, it is desirable to use a less reactive BF_3 complex, such as BF_3 piperidine. This combination will provide only semilatency at room temperature, viscosities increasing from 1,500 to 10,000 centipoises in about 10 days at room temperature. However, the addition of a liquid polyol, in concentrations of 30 per cent or more, will provide systems of reduced initial viscosity with storage lives at room temperature up to 4 months or so. With such a system,

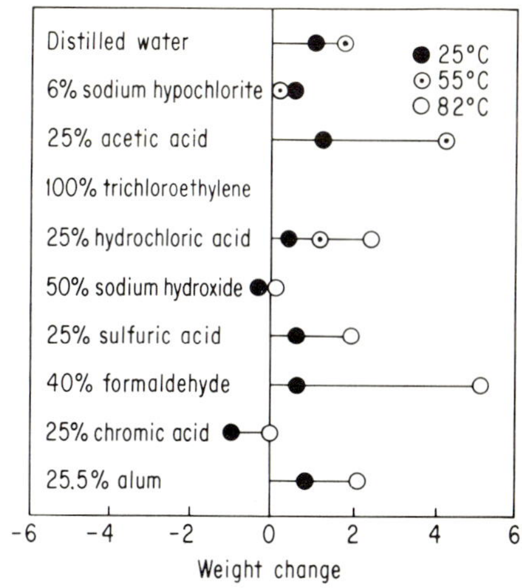

FIG. 11-6. Weight change (180 days' immersion) vs. chemicals for BF_3MEA-cured DGEBA [6].

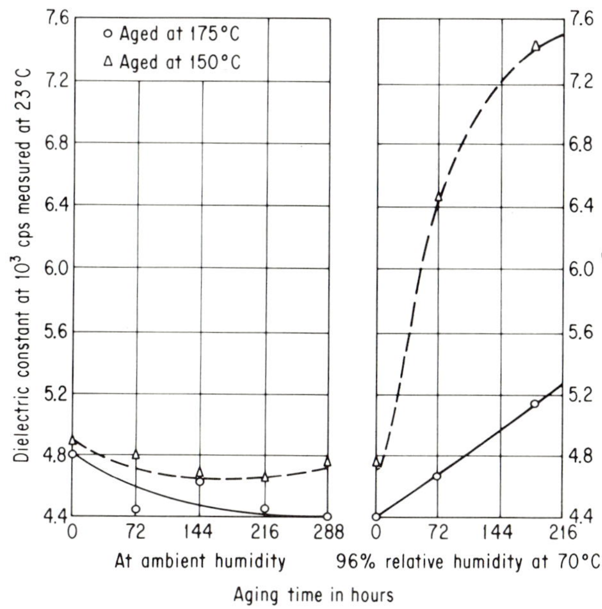

FIG. 11-7. Dielectric constant at 25°C vs. aging for BF_3MEA-cured DGEBA [130].

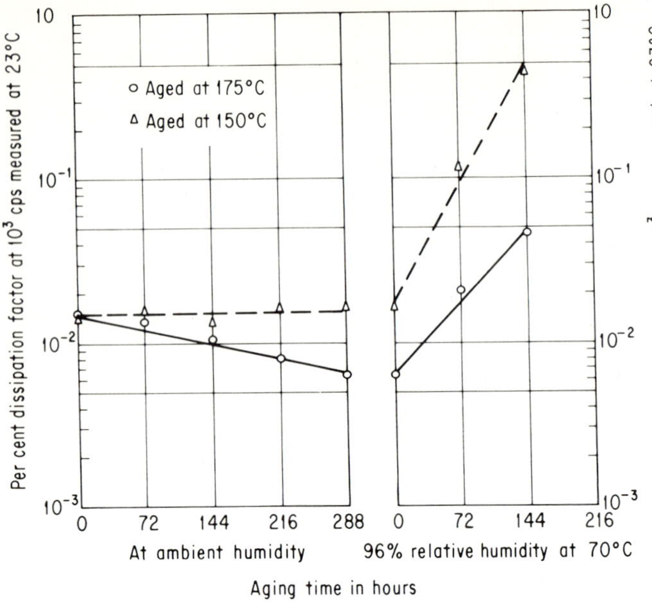

FIG. 11-8. Dissipation factor at 25°C vs. aging for BF_3MEA-cured DGEBA [130].

cure can be accomplished in 1 to 4 hours at 160 to 200°C. By varying the concentration and type of polyol, cured products having a wide range of flexibility, coupled with good thermal shock resistance, may be produced [111].

Boron-containing Catalysts Other than BF_3

Boron-Nitrogen Complexes. Many borane-amine complexes have been investigated. Di- and trimethylamine-boranes [162], pyridine-borane, morpholine-borane, and trisoctylamine-borane were found [97] to provide elevated-temperature cures for DGEBA with fairly good storage stability, although they appeared to offer no advantages over the BF_3 complexes. These compounds were satisfactory accelerators for anhydride cures, but appeared to react only sluggishly with the epoxidized olefin resins investigated, perhaps indicating that the majority of the catalytic activity could be attributed to the nitrogen rather than the boron constitutent. B-Trichloroborazine and B-trichloro-*N*-trimethylborazine provide limited solubility in DGEBA but are capable of effecting cure [135,178].

Boric acids and boronic acids may be reacted with ethanolamines to produce amine borates and boronates which provide latency with elevated-temperature cures in glycidyl ether resins. Typical of these materials are triethanolamine

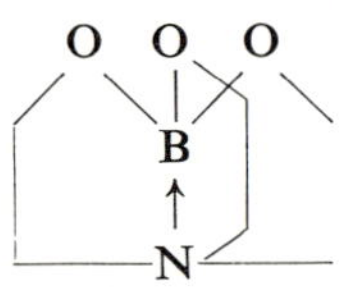

Triethanolamine borate

borate [30,152], diethanolamine borate [106], *N*-ethyldiethanolamine phenylborate [62], and other borate esters [167].

In these cases, it is probable that a spatial rearrangement occurs within the molecule upon heating to dissociate the coordinating N—B bond, thus releasing the unshared electron pair of the nitrogen and the unfilled orbitals of the boron to provide, essentially, cure by the Lewis acid and base mechanisms. The reaction, however, is rather sluggish, even at elevated temperatures, and a metal chelate catalyst [71] to coordinate with the unshared electron pairs of the nitrogen, thus freeing the boron orbitals, may be employed to accelerate the reaction. One such catalyst is chromium acetylacetonate.

Boroxine-amine complexes, such as diethyl- or dibutylamine-boroxine have been found to provide pot lives of about 6 months at room temperature with DGEBA but require cures for 15 to 20 hours at 175 to 200°C [135]. In situ complexes of trimethoxyboroxine (TMOBX) and primary amines have been reported to produce foamed products, the TMOBX, reacting with the amine to form B—N bonds with the liberation of methanol [127]. Foams from aromatic primary amines and TMOBX are self-extinguishing and stable to 300°C. These boroxine-amine reaction products provide catalytic cure by a push-pull mechanism between the Lewis acid and base [143].

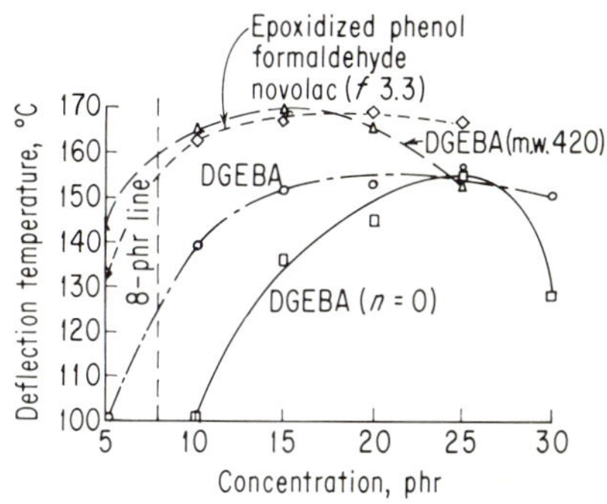

FIG. 11-9. Effect of concentration of 2(3-dimethylaminoethoxy)-4-methyl-1,3,2-dioxaborinane on DTs of several glycidyl ether resins [122]. Cured 16 hours at 120°C.

Boroester-nitrogen complexes have also been employed as latent curing agents for DGEBA, the nitrogen being supplied by an amine [88], or di- or trialkanolamine titanate esters [36,61], or an amine-silicone adduct [90]. Compounds such as *n*-propyl phenylboronate and trihexylene glycol diborate [144] have proved satisfactory in this connection. In these cases, it is probable that cure at elevated temperature occurs by a cooperating reaction upon the epoxy group by the relatively hindered amine and the relatively hindered boron complex. Accelerators, such as BF_3 piperidine, may be employed to speed the reaction [36].

2-(2-Dimethylaminoethoxy)-4-methyl-1,3-dioxa-2-borinane has been investigated

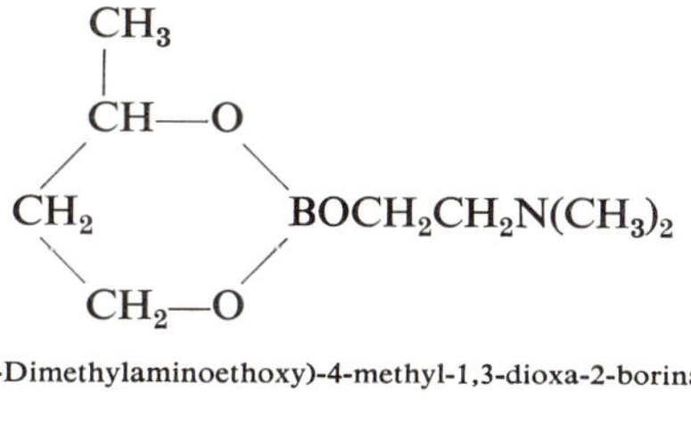

2(2-Dimethylaminoethoxy)-4-methyl-1,3-dioxa-2-borinane

in some detail. A mobile liquid, it provides pot lives of 1 to 2 hours with DGEBA and cures in about 2 to 3 hours at 120°C [122]. DTs range from 100°C with DGEBA to 165°C for an epoxidized phenol-formaldehyde novolac (*f* 3.3). The influence of curing-agent concentration on DT is shown in Fig. 11-9.

Aminodiboron curing agents have been used with all types of commercial epoxy resins. Typical is tetrakis(dimethylamino)diboron. Heat cures are required with these compounds [156].

Hindered phenyl borates, such as

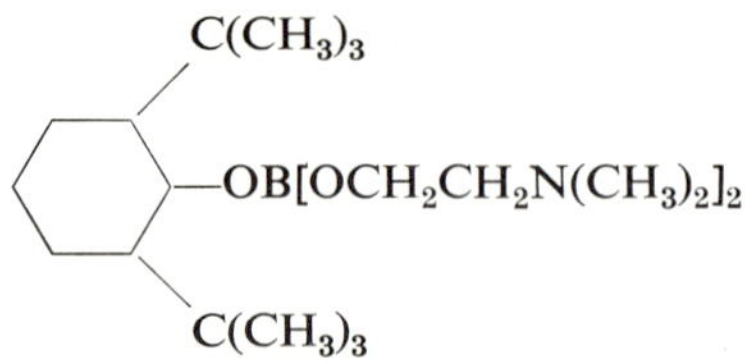

Bis(β-dimethylaminoethyl)2,6-di-*tert*-butylphenyl borate

have been reported as curing agents [157] as have aminoboryl benzenes [169].

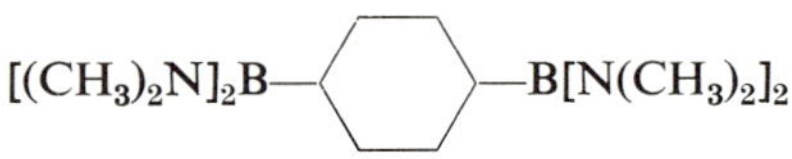

1,4-Bis[bis(dimethylamino)boryl]benzene

Non-nitrogen-containing Boron Compounds. Among the non-nitrogen-containing boron compounds which have been investigated, boric oxide is reported to give pot lives of several days, rapid cures at elevated temperature, and DTs from 160 to 200°C with DGEBA [153]. Boric acid has been reported as a curing agent for epoxidized fatty-acid esters [123] and other epoxy resins [147]; and metaboric acid has been found to provide cures [124]. Benzene boronic acid cures DGEBA at 150°C, foam being produced at this temperature [135]. Sodium and potassium borohydrides, while insoluble, provide satisfactory reactivity when carried into DGEBA in a solvent [97]. Alkyl boric acids are soluble in the epoxy resins but do not cure them. Methyl borate is likewise nonreactive, and sodium tetramethoxyborate is insoluble [97]. The methylpolyborates have been found to provide pot lives of about 30 minutes with DGEBA, but relatively long, high-temperature postcures are required to develop ultimate properties [135]. Some species react even more quickly, giving pot lives of 1 minute with DGEBA at room temperature [179]. Tricresyl borates [164] react vigorously with the epoxidized olefin resins, providing short pot lives and DTs of 150°C with 3,4-epoxy-6-methylcyclohexyl-methyl 3,4-epoxy-6-methylcyclohexane carboxylate after postcure at 120°C (Fig. 11-10). Tricyclohexyl borate, triallyl borate, trifurfuryl borate, etc., cure glycidyl ether resins at temperatures of 200°C and epoxidized olefin resins at temperatures of 100°C [134,180]. Thioborates have been used [173]. Triphenylborane, sodium tetraphenylboronate, etc., have been reported to provide somewhat flexible heat-cured products with DGEBA [161].

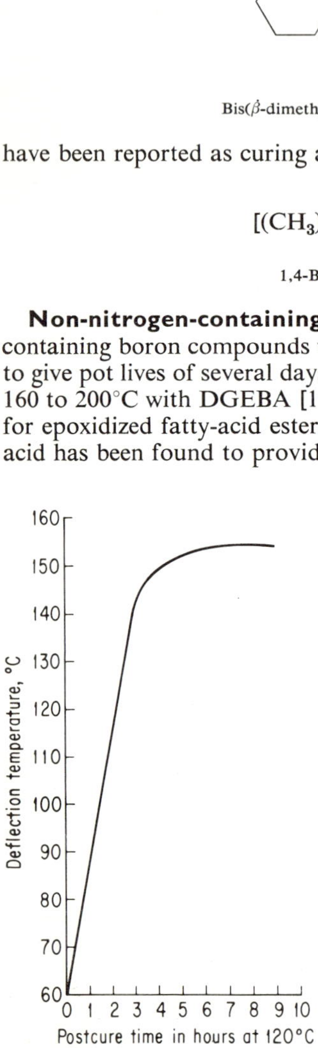

FIG. 11-10. Effect of postcure on DT of 3,4-epoxy-6-methylcyclohexylmethyl 3,4-epoxy-6-methylcyclohexane carboxylate cured with tricresyl borate [12].

Among the boroxines (or metaborates) of general structure

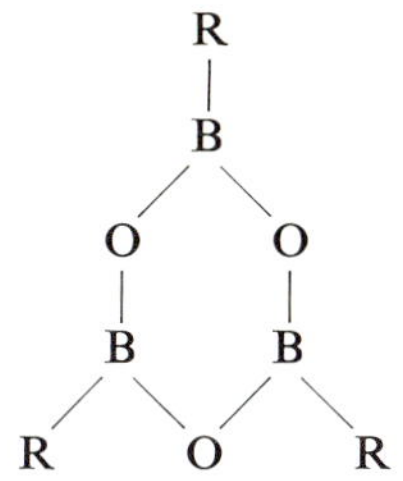

Idealized boroxine

both trimethoxyboroxine (TMOBX) [137] and tributoxyboroxine [158] provide cures with DGEBA, with the former providing increased reactivity [135].

When considered as an anhydride and employed at about 25 phr in DGEBA, TMOBX provides room-temperature cures and gel times of about 20 minutes. With 3,4-epoxy-6-methylcyclohexylmethyl 3,4-epoxy-6-methylcyclohexane carboxylate, reaction at room temperature is even more rapid, rigid foams being produced [97]. Because of its high reactivity, best results are obtained when TMOBX is used at 6 to 7 phr with DGEBA, considerable etherification occurring. TMOBX used with the distilled resin at 6.5 phr produces water-white castings. Ultraviolet stability is exceptionally good for an epoxy system, parallel-light transmission decreasing from 89 to 80 per cent during 520 hours' alternate exposure to black light and sun lamps [104].

Chelate compounds based on boron have been reported as curing agents for the epoxies. These decompose at 80 to 200°C to provide cure. Typical of the compounds is boron fluoride acetylacetonate [142]:

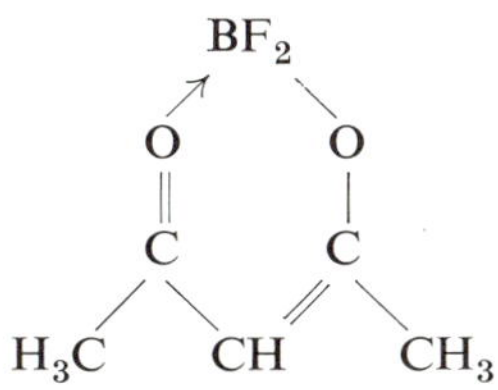

Boron fluoride acetylacetonate

Although some of the boron-containing compounds provide latency, they generally seem to give rather high dielectric constants and dissipation factors with poor overall humidity resistance.

Metal Salts and Oxides

Organic salts of polyvalent metals (dibutyltin dilaurate [35], aluminum stearate [50], lead naphthenate [38], zinc stearate [91]), metallic phenates [53,165,174], metallic alkoxides [32,139], esters [96], and triethanolamine aluminate [163] have been evaluated as epoxy-resin curing agents. These metallic reaction products are primarily suggested as curing agents in epoxy coating formulations [40], but their use is limited because of their low resistance to hydrolysis. The phenates from phenol-formaldehyde are reported to be more resistant to hydrolysis than either the phenates from phenol or the alkoxides [52]. The metallic compounds may be used in connection with a second curing agent. For example, the use of a titanate ester in conjunction with piperidine has been reported, the system giving pot lives of one month at room temperature with DGEBA but fairly good reaction at elevated temperature, the titanate and the amine presumably complexing at room temperature.

Of the metallic salts, perhaps the most useful is stannous octoate, $Sn(O_2CC_7H_{15})_2$. This is an effective catalyst for promoting the cure of resins containing the cyclohexene oxide or cyclopentene oxide groups. Good working life with room-temperature cures without rapid gelation or uncontrollable exotherms may be obtained [172]. Stannous octoate may also be used as an accelerator for these resins cured with polyfunctional amines, polycarboxylic acids and anhydrides, polyhydric phenols, polythiols, and polyisocyanates. Stannous octoate is reported to be sufficiently efficient for use with blends of epoxidized linseed oil and phthalic anhydride [181]. Stannous octoate (or similar stannous carboxylates, alcoholates, or phenolates) may be blended with alkali-metal alcoholates to give accelerators for ring-situated epoxies cured with anhydrides. These accelerators are reported to provide long pot lives [177]. A typical composition would be a blend of 4 parts of an alcoholate (from 8.2 grams sodium and 1,000 grams 3-hydroxymethyl-2,4-pentanediol) and 1 part Sn(II)2-ethylcaproate. The blend is used at about 15 phr.

The exact curing mechanisms have not been elucidated in the literature for these metal-organic compounds. The mechanisms are probably complex, with the reaction going in part through formed carboxyl or alkoxide ions. For example, it is probable that the metal alkoxides react as follows:

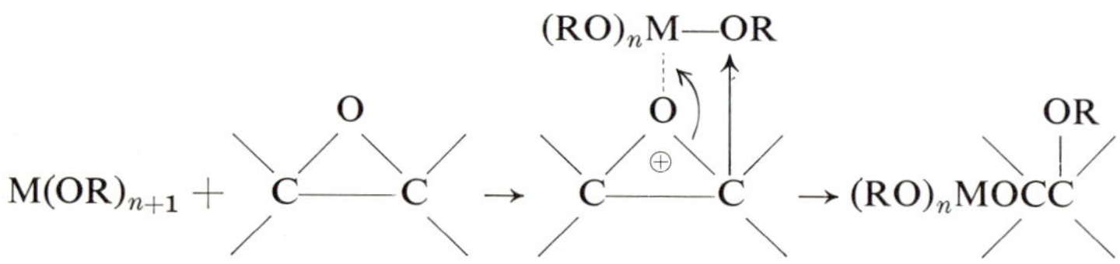

Butyl titanate has been reported as a curing agent, although it is probable that the Lewis acid activity is quite sluggish at best. In high-molecular-weight DGEBA resins, butyl titanate gives rapid gelation by esterification with hydroxyls. Alkyl titanates have been claimed as curing agents for the partially esterified higher-molecular-weight DGEBA resins [27], and as co-curing agents with primary amines for these species [26]. Tertiary amine titanates have been shown to provide good pot lives with DGEBA resins, cure occurring in 16 hours at 120°C [182].

Butyl titanate may be reacted with epoxies containing no more than one hydroxyl group per molecule to form a partial ester. Such partial esters are stable for several months at room temperature, and may be cured through the epoxy groups with conventional curing agents, such as anhydrides, to provide improved heat resistance [70].

In addition to chelates from boron, as mentioned previously, aluminum chelates have been used as curing agents [142], as have other metal chelates, such as from Cr(III) tris(acetylacetonate) [146]. Triethanolamine aluminate was synthesized on the theory that it would provide latency as an epoxy-resin curing agent and yet would react more rapidly than triethanolamine borate, due to the lower coordination strength of the aluminum as compared to that of boron. The material did provide latency and gelled in one-half to one-third the time required for the triethanolamine borate [176].

Amphoteric Oxides

Amphoteric oxides—such as those derived from phosphorus, arsenic, antimony, bismuth, tin, lead, and germanium—are compounds capable of acting as both an acid and a base. Such materials, used in small amounts, are stabilizers for elevated-temperature-resistant epoxy formulations; used in larger amounts, they are co-curing agents for anhydrides or other compounds; and when used in still larger amounts, may serve as the sole curing agent [116].

The most commonly used amphoteric oxide is arsenic pentoxide [136]. The exact mechanism by which cure is effected has not been developed, although etherification has been shown to occur. Arsenic trisulfide has been found to react in such a way that sulfur is chemically bound into the reaction product. An intermediate step is suggested where the epoxy oxygen is replaced by sulfur to form an unstable thiirane group ($-\overset{\displaystyle S}{C\text{———}C}-$) which immediately reacts with like groups within the system [148]. See also [183].

PHENOLIC CURING AGENTS

Diphenols and triphenols are reported as curing agents for the glycidyl ether resins [2,3,58,80], for epoxidized polybutadienes [47], and for other epoxidized olefins [75].

In addition, phenols may be used to accelerate other curing reactions. With anhydrides, monohydroxyl phenols have been reported as imparting flexibility and reducing the amount of anhydride required [56]. The monohydroxyl phenols markedly accelerate the curing rates of both aliphatic and aromatic polyamines; and the use of diphenols [5] and 1,3-bis(*p*-hydroxycumyl)benzene as coreactants for amine/epoxy blends has been reported.

Diphenols such as bisphenol A and diphenolic condensation products of diphenols and polycarbonates [101] are used, to some extent, as curing agents in the present technology, most usually with the epoxidized olefin resins. More commonly, however, phenol-formaldehyde resins are employed. These resins may be of either the novolac or the resole type.

Phenol Curing Agents

Diphenols should be reacted with diepoxy resins in such a fashion as to provide some etherification through alcoholic hydroxyls, since the functionality of the curing agent is no more than 2. When diphenols are employed at a ratio of 1 phenolic hydroxyl to 1 epoxy group, fairly low cross-linked structures are developed, as indicated in Fig. 11-11. However, with the more highly functional epoxy resins, such as the epoxidized phenol-formaldehyde compounds with 3 to 5 epoxy groups, the diphenol can be used satisfactorily at the 1 to 1 ratio.

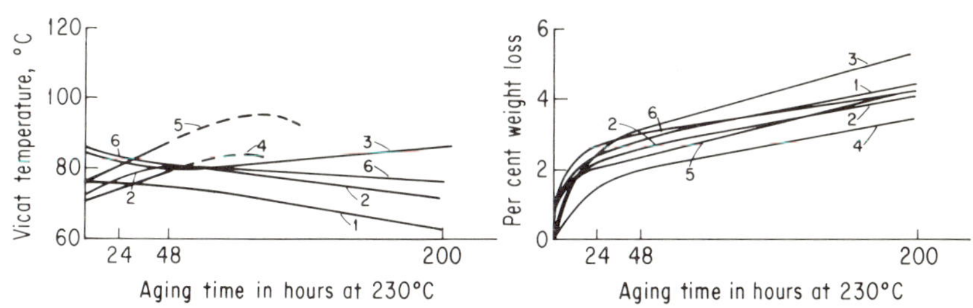

FIG. 11-11. Effect of phenol-type curing agents on high-temperature stability of DGEBA [87]. 1, hydroquinone; 2, resorcinol; 3, phloroglucinol; 4, 1,5-naphthalenediol; 5, 1,6-naphthalenediol; 6, 4,4′-dihydroxydiphenylsulfone.

Diphenols are generally used at a mole ratio of about 0.6/1 with diepoxies, and additional etherification is obtained, as described in Chap. 5. The reactions are sluggish, even with epoxidized olefin resins. Catalysts may be used which are capable of promoting etherification after the preferential exhaustion of the phenolic hydroxyls; benzyldimethylamine is usually preferred, although quaternary ammonium salts of strong acids [86] and inorganic bases may also be used. With the epoxidized olefins, acidic catalysts, such as sulfuric acid and stannic chloride may be employed to accelerate the reaction, the tertiary amine catalysts proving generally ineffective with these resins. With acidic catalysts, considerable etherification should be allowed for in calculating the amount of curing agent to use [112].

Other more highly functional phenols derived from the esterification of alcoholic diphenols have also been suggested [131]. Phenol addition products from unsaturated oils or butadiene have been considered [132], as well as specialized molecules such as the polymer from phosphonitrilic chloride [107]:

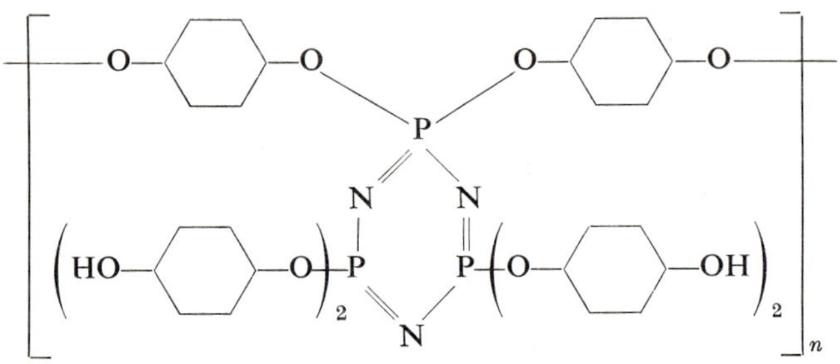

Polymeric phosphonitrilic derivative

Phenol-Formaldehyde Curing Agents

Phenolic resins from phenols or alkyl phenols [46] and formaldehyde may be of two types, the novolac resins and the resole resins. The novolac resins are represented by the formula

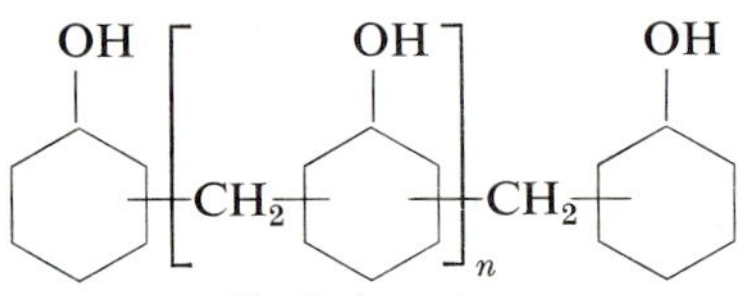

Idealized novolac

whereas, the resoles are represented by the generalized formula

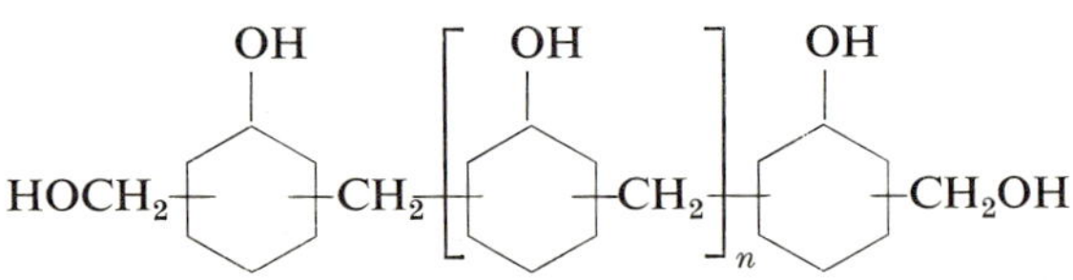

Idealized resole

or by partially condensed products from the reaction of a phenol with excess formaldehyde, the monomeric compound containing more than two hydroxymethyl groups.

Novolac Resins. When a suitable catalyst is employed to give direction to the reaction mechanism and suppress the alcohol-epoxy reaction, the novolac resins react with the epoxy groups through existing phenolic hydroxyls.

Table 11-4. Mechanical Properties of DGEBA/Phenolic Blend Catalyzed with 0.5 Per Cent α-Methyl BDMA [4-7]

Property	Curing cycle	
	40 min at 150°C	9 hr at 160°C
DT, °C	68.3	79.5
Tensile strength, psi	11,100	11,000
Flexural strength, psi	18,700	19,000
Modulus, psi × 10^{-6}	0.50	0.48
Yield, psi	16,500	15,500
Compressive strength, psi	46,000	40,600
Yield, psi	16,100	14,500
Impact strength, Izod	0.268	0.334
Hardness, Rockwell M	82	84
After 200 hrs at 200°C		
DT, °C	84	
Impact strength, Izod	0.22	
Hardness, Rockwell M	91	

The high-molecular-weight novolac resins are solid materials and are incorporated into the epoxy, usually the higher-melting-point DGEBA resins, in a solvent solution, where they are catalyzed by about 0.5 per cent benzyldimethylamine. They are used in laminating and adhesive applications to provide improved high-temperature strengths.

Lower-molecular-weight novolac resins, highly viscous at room temperature, may be melted and combined with DGEBA, and the mixture processed at 65 to 70°C, at which temperature the initial mixed viscosity is below 1,000 centipoises. Storage life of such uncatalyzed blends is several months, and cure cycles, with little exotherm, are 3 to 4 hours at about 150°C. Typical cured properties for a blend of DGEBA and a lower-molecular-weight novolac resin are given in Tables 11-4 and 11-5.

Table 11-5. Chemical Resistance of DGEBA/Phenolic Blend Catalyzed with 0.5 Per Cent α-Methyl BDMA [4-7]

System tested 7 days in	*Weight gain, %*
Distilled water	0.33
Soap solution, 5%	0.37
H_2SO_4, 30%	0.22
NaOH, 10%	0.27
NH_4OH, 10%	0.30
Ethanol	0.30
Glacial acetic acid	0.67
Acetone	9.1
Chloroform	11.9
o-Dichlorobenzene	0.08
Ethyl ether	0.35
Ethyl acetate	0.95
Ethylene dichloride	4.90
Toluene	0.01
Trichloroethylene	0.43

Resole Resins. Resole resins from phenol-formaldehyde, partially condensed monomers of the structure [37]

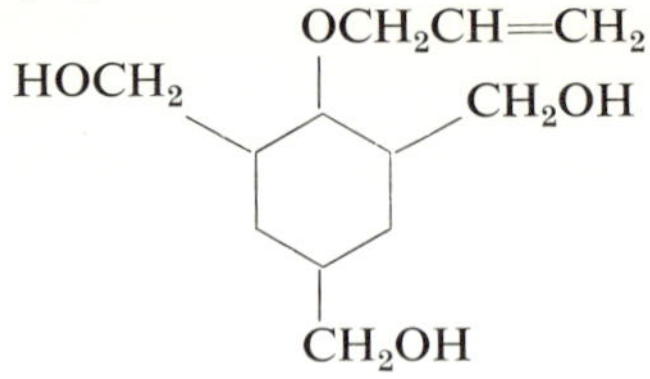

Trihydroxymethylallyloxybenzene

and other hydroxymethyl-containing benzene derivatives [23] are employed in baking coating formulations, usually with the higher-molecular-weight DGEBA resins [6,7]. They have also been reported as being used with the epoxidized polyesters [48] and with epoxidized polybutadiene [102]. Specialized resoles have been employed in high-temperature adhesive applications.

Proper selection of the resole resin is critical [28], since in addition to reaction with the epoxy group, through available phenolic hydroxyls, the resoles will condense with themselves through the hydroxymethyl groups and with the epoxy resins through available alcoholic hydroxyls. Generally, the more highly hydroxymethylated compounds prefer to condense with themselves, leaving the epoxy resin substantially unreacted. A blend of resole resins or special resoles low in hydroxymethyl groups are preferred as providing a balance between flexibility and solvent resistance as required for coating formulations.

NON-CARBOXYLIC ACID CURING AGENTS

Inorganic acids react quite rapidly with the epoxy resins at room temperature. The reaction of sulfuric acid with epoxidized olefin resins has been reported [78], as has the use of aromatic polysulfonic acids at about 0.5 phr for "instant-set" curing agents for the DGEBA resins [14] and blends of epoxy with urea-formaldehyde resins [15].

Phosphoric is the most useful of the acids reported in the literature. It is employed as a curing agent for baking films, where it provides long pot life in the presence of ketone solvents [82]. The use of phosphoric acid as a curing agent for partial esters of higher-molecular-weight DGEBA resins has been described, the reaction being through available hydroxyls [24]. Orthophosphoric acid has been described as a curing agent [9], and has been reacted with an epoxy to form a

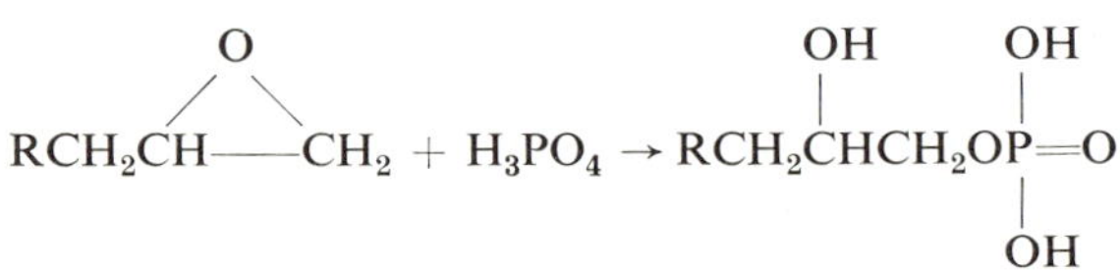

Orthophosphate ester

phosphoric acid ester [19] which may be employed as a curing agent [45]. Diepoxy adducts have also been used [16].

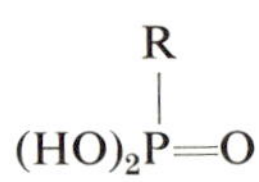

Phosphonic acid

Phosphonic acids have been reacted with epoxy resins to produce adducts capable of being thermoset by other curing agents [33].

Perchloric acid has been reported as a curing agent for water-soluble aliphatic epoxy resins used for textile treatment. The perchloric acid is supplied in dehydrated form (such as magnesium perchlorate) which upon addition of water during processing yields the acid [54]. Perchloric acid has also been reported as a curing agent at about 0.02 phr for 4-vinylcyclohexene dioxide [1].

Liquid alkyl phosphates of the following generalized structure [114],

$$(HO)_2POOROOP(OH)_2$$

Idealized alkyl phosphates

derived by reacting ethylene oxide with phosphoric acid—the R indicating that some homopolymerization of the ethylene oxide occurs under acid catalysis—provide room-temperature cures with both glycidyl ether and epoxidized olefin resins. The presence of the long carbon chain provides for internal flexibility and thereby reduces the brittleness of the cured product. With DGEBA, pot lives of 200-gram masses will be approximately 40 minutes with exotherms only reaching about 100°C. Tensile strengths will be about 2,300 psi, with an ultimate elongation of 45 per cent. With epoxidized polybutadiene and the alicyclic diepoxies, pot lives are approximately 5 minutes, with low tensile strengths being obtained in the cured system. With epoxidized novolac resins, because of high functionality and rapid reaction, gelation occurs within a minute or two.

The liquid alkyl phosphates are reported to be relatively free of irritation potential, but are hygroscopic [154]. The cured systems are stable in contact with explosives.

Trialkyl phosphites, P—$(OR)_3$, where R is a hydrocarbon radical containing 15 carbon atoms, have been suggested as epoxy-resin curing agents providing for latency with DGEBA and yet providing room-temperature cures when exposed in thin films to relative humidities in excess of 25 per cent [105].

ORGANIC ACID CURING AGENTS

As indicated in Chap. 5, when an organic acid is employed to cure an epoxy resin in an hydroxyl-free medium, the initial reaction is with the carboxyl group, followed by the reaction of the epoxy with the formed hydroxyl. The reaction is temperature-dependent, since at elevated temperatures addition esterification competes with the epoxy-hydroxyl reaction.

In order to eliminate complex side reactions it is customary, with organic acid cures of the glycidyl ether epoxy resins, to employ a base catalyst (usually a tertiary amine) which serves to bring about the carboxyl-epoxy reaction to the virtual exclusion of the epoxy-hydroxyl reaction and at the same time to accelerate the reaction rate. Use of polyhydric phenol accelerators blended with organic acids has been reported as providing satisfactory performance with alicyclic diepoxy resins [74]; and special metallic accelerators, such as stannous octoate, have been employed with these resins [98]. Strong inorganic and organic bases (e.g., potassium hydroxide and trimethylbenzylammonium hydroxide) are used as accelerators to promote the epoxy carboxyl reaction with 4-vinylcyclohexene dioxide [11]. With diepoxy resins, mixture ratios and reaction conditions should be adjusted to provide for cross-linking in addition to linear polymerization if the functionality of the selected acid is not greater than 2.

In general, the reaction of organic acids with the glycidyl ether resins is sluggish, requiring temperatures of 120°C or higher for a satisfactory rate of cure. With the epoxidized olefin resins, cures can be conducted at somewhat lower temperatures and somewhat shorter times. Exothermic reactions are low with the organic acid cures, although as the amount of accelerator is increased, the exotherm is also increased, so that, with some systems, in larger masses, it may become a problem.

The organic acids which have been suggested as curing agents in the epoxy technology can be classed as saturated acids, unsaturated acids and their adducts, carboxyl-terminated polyesters, aromatic-containing aliphatic acids, substituted phthalocyanines, and polymeric dimer and trimer fatty acids. In addition to being used as sole curing agents, the organic acids may be employed in conjunction with anhydride curing agents.

Saturated Acids

In order to promote etherification, dibasic saturated acids are employed at less than a 1:1 mole ratio with diepoxy resins; or alternatively, they may be used in substantial excess, in the presence of a suitable catalyst, cross-linking being accomplished through condensation esterification with the consequent liberation of water.

Among the materials which have been investigated are compounds represented by the formula

$$HOOC(CH_2)_nCOOH$$

Generalized dibasic acid

such as succinic ($n = 2$), glutaric ($n = 3$), adipic ($n = 4$), pimelic ($n = 5$), suberic ($n = 6$), azelaic ($n = 7$), and sebacic ($n = 8$) [44]. With these acids, using 3,4-epoxy-6-methylcyclohexylmethyl 3,4-epoxy-6-methylcyclohexane carboxylate, it was found that a mole ratio of about 0.6/1 provided optimum properties in terms of DTs when using a cure cycle of 160°C for 6 hours. The general effect, as expected, of increasing the chain length of the acid was to reduce the DT. Glutaric gave a DT of 93°C as compared with 67°C for sebacic [112]. Gelation occurs in 5 to 30 minutes at 120°C with the epoxidized olefin resins, whereas about 9 hours at that temperature is required for the diglycidyl ethers.

Other acids, such as the more reactive oxalic acid [4], phthalic acid, and *p*-aminobenzoic acid have also been suggested. *p*-Aminobenzoic acid has been reported to

$$HOOCCOOH$$

Oxalic Acid

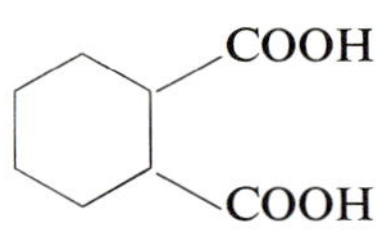

Phthalic Acid

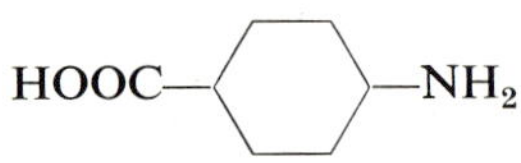

p-Aminobenzoic acid

give dry, stable preimpregnated laminates with epoxidized polybutadiene resins. Oxalic acid provides a pot life of several days when employed with the DGEBA and is cured at temperatures of 125 to 175°C.

Adducts of these and similar acids with epoxy resins have also been suggested as curing agents [17]. The monobasic salicylic acid is reported as being an effective accelerator for urea-formaldehyde epoxy-resin coatings [25].

Organic diacid chlorides have been reacted with DGEBA resins to produce fibers and films [171].

Unsaturated Acids and Their Adducts

Monobasic unsaturated acids have been reported as curing agents for glycidyl ether epoxy resins, among them being acrylic acid [49] and shellac (predominantly abietic acid) [29,65].

Use of the monobasic *b*-monoallyl itaconate has been reported, when used at

b-Monoallyl itaconate

30 phr with DGEBA, to give elevated-temperature stability to 200°C [115]. Other monobasic unsaturated acids have been used with unsaturated epoxy resins [133]. In these cases, substantial etherification must occur, and/or additional reaction must take place during cure through the unsaturation.

Unsaturated diacids, such as maleic [10],

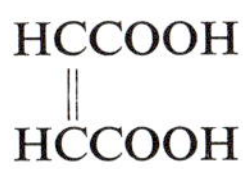

Maleic acid

and unsaturated triacids, such as aconitic [12],

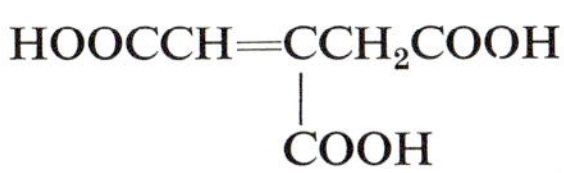

Aconitic acid

have been used as curing agents. Maleic acid cures DGEBA at 150°C and higher, and provides properties similar to those found with the saturated dibasic acids; aconitic acid provides foamed compositions with DGEBA and epoxidized olefins such as vinylcyclohexene dioxide.

Chlorendic acid has been used to cure epoxidized vegetable oils in coating formulations [125].

The unsaturated acids may be reacted with other unsaturated compounds (e.g., methyl methacrylate) and the resultant copolymer used as curing agent or, preferably, the copolymer may be partially esterified with a polyol and this employed [94]. In a similar fashion, maleic acid may be reacted with a tribasic unsaturated vegetable oil, and the resultant product partially esterified [60,108]; similarly, styrene may be reacted with itaconic acid [118]. Low-cost fatty acids may be reacted with piperazine to produce a carboxyl-terminated curing agent having tertiary amides in the structure [117]. By techniques like this, it is possible to build up, step by step, specialized organic acid curing agents, and quite a number of such curing agents have been developed [34,92,93].

Carboxyl-terminated Polyesters

Carboxyl-terminated polyesters have been employed as curing agents for both glycidyl ether and epoxidized olefin resins. A discussion of the use of a special polyester, made from adipic acid, diethylene glycol, and glycerol, as a curing agent for DGEBA is presented in [119]. A variety of suitable polyesters from anhydrides and alcohols [18,81,99,138] and polyhydric phenols [120] have been suggested, and suitable half-esters have been obtained from hydroxyl-bearing silicones [31]. A number of special polyfunctional carboxylic acids containing three or more carboxyl groups, produced by reacting pentaerythritol with a dibasic acid or anhydride, are reported as curing agents in [13]. Typical is the acid from 1 mole of pentaerythritol, 3 moles of phthalic anhydride, and 1 mole of maleic anhydride. The esterification of hydroxyl-containing long-chain fatty acid with diacids or anhydrides has been reported to give long-chain acid-curing agents of increased functionality [20]. In general, if the polyester contains about 14 to 24 carbon atoms between carboxyl groups, the compound will impart some measure of flexibility to the cured product [63].

Typical of suitable polyesters is one formed by reacting 5 moles of adipic acid with 4 moles of glycerol. This reaction product will have an average molecular

Table 11-6. DT of 3,4-Epoxy-6-methylcyclohexylmethyl 3,4-Epoxy-6-methylcyclohexane Carboxylate Cured with Carboxyl-terminated Polyesters [112]

Derivation of curing agent		DT, °C
Anhydride	Polyol	
Succinic	Ethylene glycol	65
Succinic	1,1,1-Trimethylolpropane	100
Succinic	1,2,6-Hexanetriol	100
Succinic	Glycerol	110
Succinic	Pentaerythritol	123
Methyltetra-hydrophthalic	Glycerol	135

weight of about 800. It can be reacted in stoichiometric concentrations, in the presence of a tertiary amine catalyst, with DGEBA at 160°C to yield an elastomeric cured product [21].

Examples of the carboxyl-terminated polyesters investigated as curing agents for 3,4-epoxy-6-methylcyclohexylmethyl 3,4-epoxy-6-methylcyclohexane carboxylate are those shown in Table 11-6. These polyesters were prepared by the direct esterification of one equivalent of hydroxyl groups from the polyol with one mole of anhydride.

An alkyd composition from

Phthalic anhydride	0.8
Trimellitic anhydride	0.2
Glycerol	1.0
Linoleic acid	0.4

may be used as a hot-melt curing agent for both glycidyl and nonglycidyl epoxy resins, and provides properties as given in [155].

A linear carboxyl-terminated polyester, providing room-temperature cure with epoxidized unsaturated long-oil-alkyds coating systems, may be prepared from chlorendic anhydride.

Polyesters containing quaternized amines have been reacted with DGEBA to produce antistatic coatings for films and fibers.

Aromatic-containing Aliphatic Acids

Tricarboxylic acids from diphenolic acid are produced by condensation with a halogen-containing aliphatic acid:

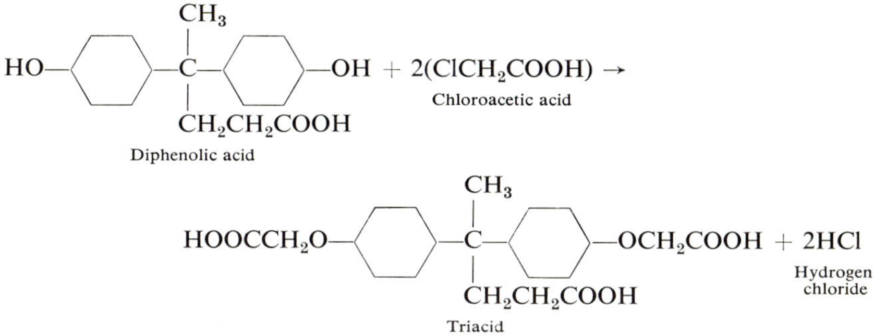

These have been used as curing agents in blends of DGEBA-, phenol-, or amine-aldehyde copolymers [69]. Alternatively, they may be partially amidified [68] to give substituted diacids or tetrabasic acids, and these may be used as curing agents in epoxy/phenol-aldehyde [66] or in amine/aldehyde blends [67].

Substituted Phthalocyanines

Phthalocyanines are produced by the reaction of aromatic anhydrides with urea. Species containing carboxyl groups may be obtained when the anhydride contains more than one anhydride ring or when additional carboxyl groups are present. Such substituted phthalocyanines, with and without metals as a central bridge, have been produced from both pyromellitic dianhydride and trimellitic anhydride. Typical is that from the former:

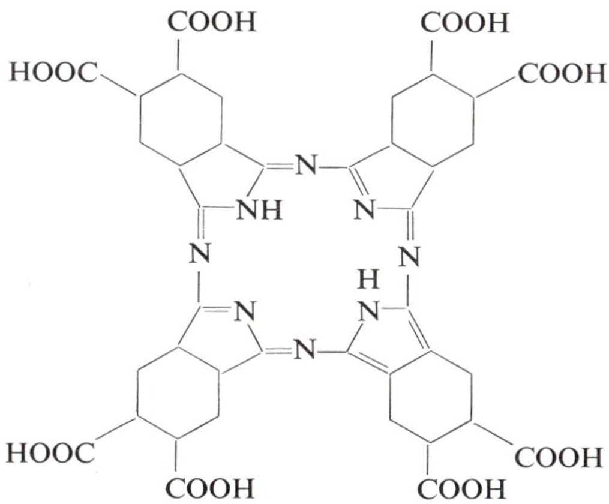

Octacarboxyl phthalocyanine

These compounds may be used as curing agents to produce compositions with extremely good heat resistance, capable of long-time exposures to 260°C [150].

Polymeric Dimer and Trimer Fatty Acids

Dimerized and trimerized fatty acids are formed by the polymerization of unsaturated fatty acids. Thus, a commercial C_{36} dibasic acid may be formed by addition at mid-molecule of two molecules of a C_{18} unsaturated fatty acid, and the C_{54} trimer by addition of three such fatty-acid molecules. Either the dimer or trimer may be used as curing agent for epoxy resins, or the dimer and trimer may be blended to obtain intermediate properties. A combination with a functionality greater than 2 is preferred. A commercial blend of 10 per cent C_{36} dimer with 90 per cent of the C_{54} trimer will be taken as typical.

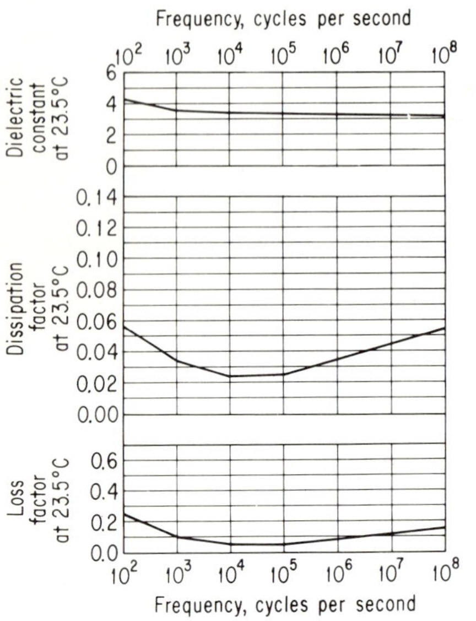

FIG. 11-12. Effect of frequency on electrical properties of DGEBA cured with trimer acid [145].

This mixture, dark brown in color, has a viscosity of 65,000 centipoises at room temperature and a specific gravity of 0.959. It is used at about 160 phr with the DGEBA. The system, accelerated with 1 per cent benzyldimethylamine and cured 3 hours at 105°C will provide a tensile strength of 2,500 psi and an ultimate elongation of 250 per cent. Used with 3,4-epoxy-6-methylcyclohexylmethyl 3,4-epoxy-6-methylcyclohexane carboxylate, and without catalyst, it cures in 3 hours at 175°C to provide a tensile strength of 2,500 psi and an ultimate elongation of 160 per cent. Typical room-temperature electrical properties vs. frequency, for the DGEBA cured with 160 phr of the acid and accelerated with 1 per cent benzyldimethylamine are shown in Fig. 11-12. The dimer/trimer blend is compatible with at least some of the epoxidized olefin resins at room temperature, but with DGEBA compatibility is only attained at a temperature of 100 to 110°C.

Use of Organic Acids in Anhydride/Epoxy Blends

Anhydrides (Chap. 12) tend, generally, to provide somewhat brittle cured products and sluggish reaction times. To overcome these disadvantages, it is frequently desirable to employ, as a co-curing agent, an organic acid. Most frequently, the acid employed is one of the long-chain acids, such as a carboxyl-terminated polyester [55,76,77], or a dimer- or trimerized fatty acid. For example, the use of a diacid from a polyalkylene glycol and an anhydride as a co-curing agent for tetrapropenylsuccinic anhydride is described in [89]. Adducts of epoxies and dibasic acids have been reported as co-curing agents [41], and adducts of monocarboxylic fatty acids and epoxies have been suggested as flexible additives for anhydride-cured system [95]. The general effect of increasing the acid co-curing agent and/or decreasing the functionality of the acid is to lower the DT, improve the toughness, and decrease the anhydride requirement.

Use of the dibasic acid, adipic, blended with phthalic anhydride as curing agent for 3,4-epoxy-6-methylcyclohexylmethyl 3,4-epoxy-6-methylcyclohexane carboxylate is illustrated in Table 11-7.

With the longer-chain acids, a variety of acid/anhydride curing-agent blends can be developed, the ultimate properties depending upon the proportions of the ingredients. With the commercial dimer/trimer acid discussed above, for example, and NMA 100:25, DGEBA can be cured to give a product with a tensile strength

Table 11-7. Effect of Organic Acid Additive on DT and Impact Resistance of 3,4-Epoxy-6-methylcyclohexylmethyl 3,4-Epoxy-6-methylcyclohexane Carboxylate Cured with Phthalic Anhydride [112]

Molar proportions			DT, °C	Impact, Izod
Epoxy	Anhydride	Acid		
1.0	1.00	0.00	200	0.2
1.0	0.50	0.25	170	0.4
1.0	0.25	0.50	117	0.4
1.0	0.00	0.60	89	0.5

of about 9,000 psi and an ultimate elongation of 40 per cent. Different blends and different anhydrides can be selected to vary cured properties widely.

SUMMARY

The acid curing agents employed in the epoxy technology can be divided into Lewis acids, phenols, inorganic acids, and organic acids.

The Lewis acids promote homopolymerization and provide extremely exothermic reactions with both glycidyl ether and epoxidized olefin resins at room temperature. Reactivity may be diminished by reaction of the Lewis acid (most commonly BF_3) with amines to permit use of the complexes in latent, one-component systems. Cure will occur rapidly at the dissociation temperature of the complex.

The phenolic curing agents are sluggishly reactive with both the epoxidized olefin and the glycidyl ether resins, and elevated temperature is required for cure. With the former, cure may be accelerated by the use of acid accelerators; with the latter, amine accelerators are normally employed. With diepoxies, dihydroxy phenolic curing agents are used at less than one phenolic hydroxyl per epoxy group in order to provide for cross-linked products. Unaccelerated, or acid-accelerated, the phenolic curing agents, in addition to reacting with the epoxy resin directly, promote the etherification of epoxy groups with alcoholic hydroxyls. Resinous phenolic compounds, of either the novolac or resole type, may be employed as curing agents and as coreactants. With the novolacs, the reaction is similar to that experienced with the monomeric phenols; with the resoles, autocondensation through methylol groups and with available alcoholic hydroxyls complicates the reaction mechanism.

The inorganic acid curing agents generally provide extremely fast cure for both epoxidized olefins and glycidyl ether type resins. Most commonly, phosphoric acid alone or as partial esters with an epoxy is employed.

Organic acids are employed as sole curing agents or as co-curing agents with anhydrides, and are reactive with both epoxidized olefins and glycidyl ether resins, although the reactions are more sluggish with the latter. With dibasic organic acids, etherification will be required to provide for cross-linked products when reacted with diepoxies; with more highly functional acids, etherification or condensation esterification is not required for cross-linked products.

REFERENCES

1949

1. Segall, Process for the Polymerization of 4-Vinylcyclohexene Dioxide, U.S. 2,460,195.

1950

2. Bender and Farnham, Thermosetting Resins from a Diphenol and a Diglycidyl Ether of a Diphenol, U.S. 2,506,486.

3. Bixler, Integral Brush Assembly, U.S. 2,512,997.
4. Bradley, Compositions of Matter Containing Glycidyl Ether and Oxalic Acid, U.S. 2,500,449.
5. Greenlee, Amine–Epoxy–Phenol Compositions, U.S. 2,510,885.
6. Greenlee, Phenol–Aldehyde and Epoxide Resin Compositions, U.S. 2,521,911.
7. Greenlee, Phenol–Aldehyde Polyepoxide Compositions and Reaction Products, U.S. 2,521,912.
8. Polymeric 4-Vinylcyclohexene Dioxide (to Canadian Industries, Ltd.), Brit. 642,800.

1951

9. Bradley, Compositions of Matter Containing Epoxy Ethers and Phosphorous-containing Compounds, U.S. 2,541,027.
10. Buck and Canlon, Coating Compositions and Processes of Applying Same to Glass, U.S. 2,569,920.
11. Segall and Allenby, Reaction Products of 4-Vinylcyclohexene Dioxide, U.S. 2,541,670.

1952

12. Koroly, Expanded Cellular Resinous Products Containing Polyepoxides and Aconitic Acid, U.S. 2,623,023.
13. Koroly, Preparation of Thermoset Resins, U.S. 2,609,357.

1953

14. Dannenberg, Curing Glycidyl Polyethers of Dihydric Phenols with Sulfonic Acids, U.S. 2,643,243.
15. Dannenberg, Curing Resin-forming Compositions with Sulfonic Acids, U.S. 2,631,138.
16. A Process for Curing Glycidyl Ethers and Resinous Products Formed therefrom (to N.V. de Bataafsche Petroleum Maatschappij), Brit. 691,543.
17. Shokal et al., Process for Forming Cured Glycidyl Ether Resinous Bonds between Two Solid Surfaces, U.S. 2,651,589.

1954

18. Cass, Acidic Polyester–Ethoxyline Resinous Compositions, U.S. 2,683,131.
19. Cupery, Phosphates of Styrene–Epoxy Compound Copolymers, U.S. 2,692,876.
20. Doyle, Oil-modified Acidic Polyester–Ethoxyline Resinous Compositions, U.S. 2,691,004.

1955

21. Fisch, Elastic Infusible Products from Epoxy Compounds and Crosslinking Agents, U.S. 2,712,535.
22. Greenlee, Compositions, U.S. 2,717,885.
23. Howard et al., Compositions Containing Glycidyl Polyethers of a Polyhydric Phenol, a Methylol Substituted Alkenyloxybenzene and a Polyvinyl Acetal, U.S. 2,713,565.
24. Narracott, Epoxy Resins Esterified with Drying Oil Fatty Acids, and Phosphoric Acid, U.S. 2,709,690.
25. Osdal, Salicylic Acid Cured Coating Compositions Comprising an Epoxy Polyhydroxy Polyether Resin and an Amine–Aldehyde Resin, U.S. 2,703,765.

1956

26. Beacham, Epoxy Resin Compositions Containing Titanate, U.S. 2,742,448.
27. Beacham, Fatty Acid Esters of Epoxy Resins and Alkyl Titanates, U.S. 2,733,222.
28. Bruin, Some Aspects of the Chemistry of Epoxide Resins in Relation to Their Application, British Society of Chemical Industry Symposium, London, April.
29. Flowers and Homberg, Shellac–Ethoxyline Resin Insulating Compositions, U.S. 2,769,739.

30. Langer and Elbling, Organic Reaction Mechanisms as a Basis for Selection of an Epoxy Resin Catalyst–Triethanolamine Borate, ACS Symposium, Atlantic City, September.
31. Miller et al., Silane Derivative–Epoxide Resin Composition and Coatings Therefrom, U.S. 2,768,150.
32. Schlenker and Starck, Epoxide Resins Cured by Alcoholates of Polyvalent Metals, U.S. 2,767,158.
33. Skokal, Polyepoxides Reacted with a Phosphonic Acid, U.S. 2,732,367.

1957

34. Foster, Synthetic Copolymer Resinous Insulating Compositions, U.S. 2,785,383.
35. Improvements in or Relating to the Hardening of Epoxide Resins (to Chemische Werke Albert), Brit. 783,764.
36. Langer, Process for Curing Glycidyl Polyethers and Products Produced Thereby, U.S. 2,809,184.
37. Nowacki, Coating Compositions Containing an Epoxy Resin, a Methylol Substituted 2-Alkenyloxybenzene and a Butylated Urea–Formaldehyde, U.S. 2,816,084.
38. Schlenker, Ethoxylin Resin Compositions and Process of Hardening Such Resins, U.S. 2,803,609.
39. Schroeder and Condo, Epoxy Resins in the Crease Proofing of Cotton, *Textile Res. J.*, February.
40. Starck and Schlenker, Epoxy Resin Compositions Cured with Metallic Salts of Enol-Keto Tautomeric Compounds, U.S. 2,801,228.
41. Suen and Hicks, Diepoxy Resinous Compositions, U.S. 2,801,232.
42. Wagner, Controlled Boron Trifluoride Catalysis of Epoxy Polymerizations, *J. Polymer Sci.* **26**:329–332.

1958

43. Carey, Process for Resinifying Polyepoxides and Resulting Products, U.S. 2,839,495.
44. Fang, Process of Making Glycidyl Methacrylate Polymer, Composition Containing Same, and Product Coated Therewith, U.S. 2,857,354.
45. Fang, Thermosetting Epoxide Polymer Mixtures, Method of Making Same, and Coatings Containing Same, U.S. 2,849,418.
46. Glazer and Vogel, Blends of Alkenylphenol–Aldehyde Resins with Polyepoxide Resins and Metallic Surfaces Coated Therewith, U.S. 2,856,094.
47. Greenspan, Polybutadiene Polyhydric Phenol Resins, U.S. 2,851,441.
48. Greenlee, Polyepoxide Polyester and Phenol–Aldehyde Compositions and Process of Making Same, U.S. 2,847,393.
49. Hall, Acrylic Acid Derivatives of Epoxide Resins and Method of Curing Same, U.S. 2,824,851.
50. Hardenable Compositions Comprising Epoxy Resins and a Process for Curing These Compositions (to N.V. de Bataafsche Petroleum Maatschappij), Brit. 806,188.
51. Parry and Hubbard, Compositions Containing BF_3 Complexes and Their Use as Curing Agents for Polyepoxides, U.S. 2,824,083.
52. Schlenker, Epoxy Resin Compositions Containing Polyvalent Metal Phenolates or Phenolate–Aldehyde Resins and the Process of Their Production, U.S. 2,837,493.
53. Schlenker, Hardenable Epoxide Resin Compositions, Can. 568,236.
54. Thomas, Process for the Treatment of Textile Materials with Epoxy Resins and for Curing Same, U.S. 2,829,072.

1959

55. Belanger, Polyepoxide Polyalkylene Glycol–Anhydride Compositions and Processes for Their Preparation, U.S. 2,908,660.
56. Belanger et al., Modified Epoxide Resins, U.S. 2,908,664.
57. Brueschweiler and Zuppinger, Resins Based on Polyepoxides, Ger. 1,099,162.
58. Bruin and Oosterhof, Process for Curing Polyepoxides and Resulting Products, U.S. 2,904,531.

59. Brunner, Curing Liquid Polyepoxide Compositions, Can. 575,667.
60. Cooke and Masters, Reaction Products of Polyepoxides and Monomeric Partial Esters of Polyhydric Alcohol with Maleic Anhydride–unsaturated Organic Acid Adducts and Methods and Compositions for Producing Same, U.S. 2,909,496.
61. Finestone, Catalysts for Glycidyl Polyethers and Products Produced Thereby, U.S. 2,871,222.
62. Finestone, Improvements in or Relating to Glycidyl Polyethers, Brit. 833,748.
63. Fisch, Composition Comprising a Reaction Product of a Polyester and a Polyepoxide, and Process for Preparation, U.S. 2,891,034.
64. Fisch et al., Polyepoxides Plastics, Ger. 1,113,566.
65. Flowers, Ethoxyline Resin-Shellac-Dicyandiamide Insulation Compositions, U.S. 2,899,399.
66. Greenlee, Compositions Containing Tricarboxylic Oxy-acids, U.S. 2,893,935.
67. Greenlee, Compositions of Polycarboxylic Acid Amides, Polyepoxides and Ammonia Derivative–Aldehyde Condensates, U.S. 2,893,968.
68. Greenlee, Compositions of 4,4 Bis(hydroxyaryl) Penta-*n*-amide Ether Acids, and Polyepoxides, U.S. 2,893,974.
69. Greenlee, Resin Compositions Containing Tricarboxylic Oxy-acids, U.S. 2,893,966.
70. Improvements in or Relating to the Production of Hardenable Epoxy Resin Compositions (to Maschinenfabrik Oerlikon), Brit. 819,191.
71. Langer, Glycidyl Polyether–Triethylamine Borate Composition and Product Treated Therewith, U.S. 2,871,454.
72. Meier and Kuhn, Curing Agents for Polyepoxides, U.S. 3,018,258.
73. Parry and Carey, Process for Resinifying Polyepoxides and Resulting Products, U.S. 2,909,494.
74. Phillips et al., Compositions Comprising a Diepoxide, a Polycarboxylic Acid, and a Polyhydric Compound, U.S. 2,890,197.
75. Phillips et al., Compositions Comprising a Diepoxide and a Polyhydric Compound, U.S. 2,890,195.
76. Phillips et al., Compositions Comprising Epoxides and Acid Anhydrides, U.S. 2,890,210.
77. Phillips et al., Compositions of Epoxides and Polycarboxylic Acid Compounds, U.S. 2,890,194.
78. Phillips et al., Diepoxide Polymers, U.S. 2,917,491.
79. Phillips et al., Polymers of 3,4-Epoxycyclohexylmethyl 3,4-Epoxy-cyclohexanecarboxylates, U.S. 2,890,209.
80. Process for Curing Epoxy Resins (to N.V. de Bataafsche Petroleum Maatschappij), Brit. 814,694.
81. Rosenberg, Carboxyl Rich Alkyd Resin–Ethoxyline Resin Compositions and Process for Their Preparation, U.S. 2,909,495.
82. Simms, Epoxide-substituted Vinyl and Acrylate Co-polymers, ACS Symposium, Atlantic City, September.
83. Sonneborn et al., Control of Variables in Heat Resistant Reinforced Plastics, AMC Technical Report 59-7-589.

1960

84. Aitken and Ralph, Some Effects of Radiation in Cast Epoxide Resin Systems, AERE R-3085, (Engineering Division, Harwell, Berkshire, England).
85. Bailey, Cured Epoxy Resin Compositions Containing Boron Trifluoride Complex of a Monoethylamine and Piperidine, Method of Curing Said Compositions, and Glass Fibers Impregnated Therewith, U.S. 2,938,880.
86. Belanger and Cooke, Curing of Polyepoxides, U.S. 2,928,803.
87. Ehlers, Correlation between Structure and Thermal Stability of Epoxy Resins, *Polymer*, vol. 1, no. 3.
88. Elbling and Thomas, Curing Agents Comprising Amine-Polyborate Esters for Epoxy Resins, U.S. 2,941,981.
89. Epoxy Resin (to Minnesota Mining & Manufacturing Co.), Brit. 857,073.

90. Finestone, Process for Curing Glycidyl Polyethers and Products Produced Thereby, U.S. 2,953,545.
91. Hecker et al., Process for Curing Liquid Epoxy Resins, Belg. 588,383.
92. Hicks, Carboxy–Copolymer Epoxide Compositions and Their Preparation, U.S. 2,934,516.
93. Hicks, Carboxy–Copolymer Epoxide Compositions and Their Preparation, U.S. 2,965,602.
94. Hicks, Film-forming Compositions Comprising Carboxy-Copolymers and Polyepoxides and Method of Making Same, U.S. 2,949,438.
95. Hicks et al., Modified Epoxide Resins, U.S. 2,934,506.
96. Kahn and Horner, Ethoxyline Resins, U.S. 2,962,410.
97. Lee and Neville, New Curing Agents for Epoxy Resins, *SPE J.*, March.
98. McGary and Patrick, Resins from Cyclohexene Oxide Derivatives and Fatty Acids, Union Carbide Chemicals Co. Research Report.
99. Phillips et al., Polymerizable Composition Containing Dicyclopentadiene Dioxide and Polycarboxylic Acid and Resins Prepared Therefrom, U.S. 2,962,469.
100. Phillips et al., Polymerizable Compositions and Resins Made Therefrom, U.S. 2,962,453.
101. Schnell and Krimm, Plastics Based on Ester-modified Resins, Ger. 1,113,085.
102. Wheelock, Resins from Epoxidized Liquid Diene Polymers and Phenol Aldehyde Resins and Method of Making Same, U.S. 2,927,094.

1961

103. Brown and Stiffer, Furfuryl Alcohol as a Resin Modifier, *Mod. Plastics*, December.
104. Christie and Medved, High Temperature Resistant Transparent Plastics, Midwest Research Institute Final Report, October.
105. Curing a Glycidyl Polyether (to Shell Internationale Research Maatschappij N.V.), Brit. 869,969.
106. Finestone, Catalysts for Glycidyl Polyethers, Products Produced Thereby and Method of Producing Said Catalysts, U.S. 2,970,130.
107. General Dynamics, technical literature: Resin XHU.
108. Greenfield, Composition and Method for Curing Epoxy Resins, U.S. 2,967,837.
109. Johnston and Greenspan, Oxiron Resins: a Series of New Epoxy Resins, *Mod. Plastics*, May.
110. McWhorter et al., Thermosetting Epoxy Resins, Ger. 1,106,493.
111. Patrick, Flexibilized Resins from Unox 201 and Polyols, Union Carbide Chemicals Co. Report.
112. Patrick et al., Resin Forming Reactions, Union Carbide Chemicals Co. Technical Report.
113. Rapidly Hardenable Compositions of Epoxide Components and Complexes of Friedel-Crafts Catalysts (to Ciba, Ltd.), Brit. 882,360.
114. St. Cyr, Hydroxy Alkyl Phosphate Esters as Curing Agents for Epoxy Resins, *SPE Trans.*, January.
115. Skiff, Epoxy-Resin B-Monoallyl Itaconate Containing Compositions, Their Preparation and Articles Containing Same, U.S. 2,975,148.
116. Susman and Janis, Curing Epoxy Resins with Inorganic Amphoteric Oxides, U.S. 3,014,893.
117. Wear et al., Resinous Compositions, U.S. 2,970,972.
118. Wille and Jellinek, Hardening of Diglycidyl Ethers or Glycidyl Polyethers, Ger. 1,099,730.
119. Wynstra, Flexible Polyester–Liquid Epoxy Resin Condensates, *SPE J.*, May.

1962

120. Alsthom, Hardening of Polyepoxides for Electrical Insulating Compositions, Belg. 613,247.
121. Bengelsdorf, Boroester 8, (Tri-*m,p*-cresyl Borate) a Unique Boron Containing Epoxy Resin Curing Agent, U.S. Borax & Chemical Corp. Report BR-270-13.

122. Bengelsdorf, USB 110, An Epoxy Resin Curing Agent with a Novel Boric Acid Ester Structure, U.S. Borax & Chemical Corp. Report BR-270-12.
123. Bradley and Wolff, Polymers from Polyepoxy Acid Esters and Boric Acid, U.S. 3,030,392.
124. Brunner, Polyepoxide Compositions, Brit. 912,115.
125. Budde, Protective Film Forming Compositions and Resulting Films, U.S. 3,050,480.
126. Carey, Process for Preparing High Molecular Weight Epoxy-containing Products from Lower Molecular Weight Polyepoxides, U.S. 3,067,170.
127. Chen, Foamed Resin Obtained by the Reaction of a Polyepoxide, a Trialkylboroxine and an Amine, U.S. 3,025,249.
128. Colclough, Polyepoxide Compositions, U.S. 3,018,264.
129. Colclough, Polyepoxide Compositions, U.S. 3,063,965.
130. Feuchtbaum, Electronic Insulation: a Critical Factor in High Reliability Design, *Insulation*, September.
131. Greenlee, Curable Epoxide Compositions Containing Polyhydric Phenol and Cured Product Thereof, U.S. 3,032,527.
132. Greenlee, Resinous Compositions, U.S. 3,057,614.
133. Greenlee and Pearce, Epoxide Conversion of Unsaturated Acids, U.S. 3,023,178.
134. Hardening of Polyepoxide Resins (to U.S. Borax & Chemical Corp.), Belg. 609,522.
135. Haworth and Pollnow, Boron Curing Agents for Epoxy Resins, *Ind. Eng. Chem. Prod. R&D*, September.
136. Levine, Arsenic Pentoxide Cure of an Organic Adhesive, *Ind. Eng. Chem. Prod. R&D*, June.
137. Polyepoxide Compositions (to Imperial Chemical Industries, Ltd.), Brit. 910,899.
138. Radlove, Method of Curing Epoxy Components and Heat Stable Epoxy Compositions, U.S. 3,031,434.
139. Schwartz et al., Reaction of Titanium Tetrachloride with Alcohols, *Proc. N. Dakota Acad. Sci.* **16**:43–44.

1963

140. Bartan and Schwartz, Modified Epoxy Resin, Belg. 635,199.
141. Buck et al., Composition of Matter, Brit. 923,078.
142. Chelate Hardeners for Epoxy Resins (to Ciba, Ltd.), Belg. 623,245.
143. Chen and Nixon, Heat-resistant Epoxy-Boroxine Foams, ACS Symposium, Los Angeles, April.
144. Composition of Matter Suitable as a Curing Catalyst for Epoxy Resins (to Westinghouse Electric Corp.), Can. 928,835.
145. Emery Industries, Inc., technical literature.
146. Epoxy Compound Polymers (to Mitsui Chemical Industry Co., Ltd.), Belg. 633,621.
147. Improvements in Resins Derived from Epoxy Compounds (to Devoe & Raynolds Co., Inc.), Brit. 931,000.
148. Levine, Curable Resin System, Cured Epoxy Resin and Process of Forming the Same, U.S. 3,092,604.
149. Lidarik et al., Cationic Polymerization of Glycidyl Ethers, *Vysokomolekul. Soedin.* **5**(11):1738–1747.
150. Parry, Heat Resistant Adhesives Based on Phthalocyanine Copolymers, TD Report ASD-TDR-63-396, April.
151. Partansky, Flameproof Epoxy Resin Articles, Belg. 631,287.
152. Scheibli, Acetone-soluble Epoxide Resins for Coatings and Articles, Belg. 628,249.
153. Skiff, Resin Compositions and Their Preparation, U.S. 3,073,799.
154. Thiokol Chemical Corp. Bulletin PE 13.
155. Ward and Bobalek, Some Correlations among Thermal, Electrical, and Mechanical Properties of Alkyl-Epoxy Copolymer Adhesives, *Ind. Eng. Chem. Prod. R&D*, June.

1964

156. Amino-diboron Curing Agents for Epoxide Resins (to U.S. Borax & Chemical Corp.), Brit. 955,748.
157. Bengelsdorf and Woods, Hindered Phenyl Borates, U.S. 3,135,782.

158. Brunner and Waghorn, Polyepoxide Compositions, U.S. 3,134,754.
159. Buck et al., Curing Epoxy Resins, Brit. 963,058.
160. Caldwell, Resinous Compositions Comprising Mixtures of Polyesters Containing Quaternized Amine Groups and Epoxy Resins, U.S. 3,128,265.
161. Curing Agents for Polyepoxide Resins (to U.S. Borax & Chemical Corp.), Brit. 955,873.
162. Curing Agents for Polyepoxide Resins (to U.S. Borax & Chemical Corp.), Brit. 956,044.
163. Elbling and Finestone, Insulation for Electric Machinery, Building Elements, or Apparatus with Hardened Epoxy Resin, Ger. 1,162,439.
164. Foamed Epoxy-resin Compositions (to U.S. Borax & Chemical Corp.), Brit. 972,952.
165. Heat-resistant Epoxy Resins (to High Polymer Chemical Industry, Inc.), Fr. 1,351,508.
166. Kudisch and Leebrick, Polyepoxides, Fr. 1,362,703.
167. Monoborate Esters for Use as Epoxy-resin Curing Agents, (to U.S. Borax & Chemical Corp.), Brit. 960,301.
168. Nowak and Saure, BF_3-amine Adducts as Latent Curing Hardeners for Epoxy Resins, *Kunststoffe* **54**(9):557–564.
169. Organoboron Compounds (to U.S. Borax & Chemical Corp.), Brit. 950,172.
170. Phillips and Starcher, Finishing of Cellulose Textiles, Ger. 1,176,603.
171. Porck, Synthetic Thermoplastic Linear Polyesters and Process, U.S. 3,129,202.
172. Proops and Fowler, Curable Mixtures Comprising Epoxide Compositions and Divalent Tin Salts, U.S. 3,117,099.
173. Thioborates as Curing Agents for Polyepoxide Resins (to U.S. Borax & Chemical Corp.), Brit. 955,301.
174. Tin Compounds in Hardening Polyepoxide (to Consortium für Elektrochemische Industrie, G.m.b.H.), Brit. 971,525.

1965

175. Case and White, Novel Lewis Acid Complexes as Latent Curing Agents for Epoxy Resins, ACS Symposium, Atlantic City, September.
176. Chen, Triethanolamine Aluminate as Latent Epoxy Curing Agent, ACS Western Regional Meeting, Los Angeles, November.
177. Curable Mixtures of Cycloaliphatic Polyepoxides, Hardeners, and Accelerators (to Ciba, Ltd.), Neth. Appl. 6,411,657.
178. Curing Epoxy Resins (to Allis-Chambers Mfgr. Co.), Brit. 998,267.
179. Curing of Polyepoxide Resins (to U.S. Borax and Chemical Corp.), Brit. 987,584.
180. Curing of Polyepoxide Resins (to U.S. Borax and Chemical Co.), Brit. 988,375.
181. Epoxy Resins for Moldings, Foams, and Varnishes (to Unilever-Emery N.V.), Fr. 1,386,502.
182. Hardeners for Epoxy Resins (to Dr. Beck and Co., G.m.b.H.), Brit. 994,717.

1966

183. Sander, Thiiranes, *Chem. Rev.*, vol. 66, no. 3, June.

Chapter 12

ACID-ANHYDRIDE CURING AGENTS FOR EPOXY RESINS

Acid anhydride curing agents are generally derived by the elimination of water from diacids in a reversible reaction:

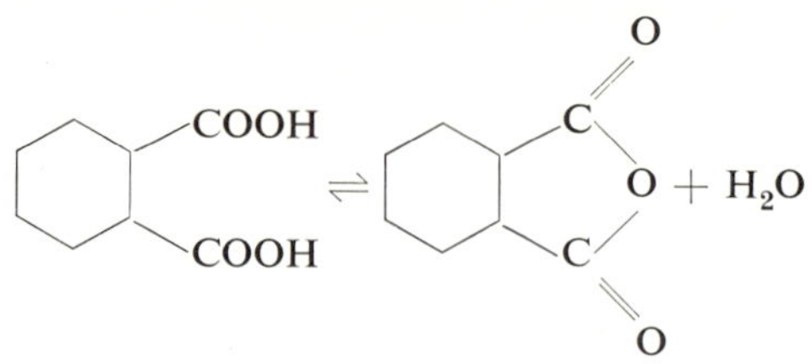

Phthalic acid Phthalic anhydride Water

Anhydrides found acceptance in Europe from the start of the technology but did not achieve importance as curing agents in the United States until the mid-1950s because of patent considerations [10]. By the mid-1960s, nearly 30 anhydrides were in use, perhaps a dozen of them being of major commercial importance.

With the glycidyl ether resins, anhydrides generally are selected to provide low viscosity and long pot life. Compared with amines, they have only moderate skin-irritation potential.

Anhydride-cured epoxy resins generally exhibit improved high-temperature stability over the amines and better physical and electrical properties above their deflection temperature (DT). As a class, they give improved radiation resistance.

Most anhydride formulations require elevated-temperature cures, with ultimate properties depending on postcures at temperatures of 150°C or higher.

In addition to the use of anhydrides with the glycidyl ether resins, anhydrides (together with carboxylic acids) are the chief type of curing agent used with the epoxidized olefin resins [33,34,53,54,55]. The Plastics Division of Allied Chemicals, a chief commercial supplier of anhydrides, has issued a detailed booklet covering the use of their materials in the epoxy-resin technology [107].

Reactions of the anhydride with the resin are complex and dependent on a number of factors: gel time and temperature, postcure time and temperature, presence or absence of accelerator, type of accelerator, amount of hydroxyl present, ratio of anhydride to epoxy (A/E), and amount of free acid.

In the absence of accelerator, the anhydride will not react directly with the epoxy group. If hydroxyl groups are present on the resin chain, the anhydride will react with them to produce carboxylic acids, which may then react with the epoxy groups to generate additional hydroxyls capable of reacting with additional anhydride groups. Hydroxyl groups may be provided separately, if required. The use of low-molecular-weight alcohols has been reported to reduce the cure times as much as 60 per cent when used in concentrations on the order of 5 phr with DGEBA [21]. The use of ethylene glycol with anhydride cures of dicyclodiepoxy carboxylate resins [52] and epoxidized polybutadiene [76] has been described. To the extent that water is present to convert the anhydride to the acid, cure will be accelerated, but with loss of system functionality, insofar as the functionality of the epoxy group is effectively reduced to 1 [102].

Polyols may be used in higher concentrations, such that it is probable that some polyester formation is occurring when the anhydride is used in excess. For example, excess anhydride and polyol have been used with bis(2,3-epoxycyclopentyl) ether to produce cured systems [70]. When the polyol is present in higher concentrations, it will serve to impart flexibility to the cured system.

ANHYDRIDE ACCELERATORS

In almost all commercial epoxy-resin/anhydride formulations, some form of accelerator is employed. Acid accelerators favor etherification, and basic accelerators favor esterification. The optimum A/E ratio and thus the cured properties are determined to a large degree by the accelerator used.

The variety of accelerators suggested for anhydride cures is considerable. Tertiary amines have been widely used with glycidyl ether resins [80], ring-situated epoxy resins, resins with internal epoxy groups [50], and epoxidized oils [72]. Use of sulfur-containing rubber-vulcanizing agents (e.g., tetraethyl thiuram disulfide) [22] and of benzyltrimethylammonium chloride [40] have been suggested as providing room-temperature stability. Thiodiglycols have also been used [43,58], as have a variety of mercaptan-containing compounds [93]. Phenols have been reported as accelerators [48], as have the reaction products of phenol with a tertiary amine or a quaternary ammonium hydroxide [106]. Metal chelates containing epoxy groups are reported as effective accelerators [91]. Tetra-*sec*-butyl titanate has been reported as providing cure for the sluggish epoxidized linseed oil and HHPA in 2 hours at 150°C [92]. Organophosphines (e.g., tricyclohexylphosphine), arsines, stibines, and bismuthines have been reported as providing room-temperature cures for blends of aliphatic- and aromatic-based glycidyl ether resins [16]. Chlorine-containing amides [65] and sodium salts of carboxylic acids have been reported as accelerators [18]. Stannous octoate has been suggested [115] and its use has found some commercial acceptance.

Tertiary amines are more conventionally used accelerators. Tertiary amines were first observed to catalyze the reaction of anhydrides with epoxies (i.e., epichlorohydrin) in 1905—with pyridine, or better still, dimethylaniline being employed [9]. Preferred tertiary amines for catalyzing the epoxy-resin–anhydride reaction are benzyldimethylamine (BDMA) and tris(dimethylaminomethyl)phenol (DMP-30), although some tertiary amines having low vapor pressure find increasing use in vacuum-processed resin formulations. Optimum amounts of tertiary amine accelerators in terms of cured properties fall in the 1- to 5-phr range with the exact value depending on the specific system. Thermal stability will be at an optimum when measured in terms of both elevated-temperature weight loss and DT [23].

LINEAR ALIPHATIC ANHYDRIDES

Linear anhydrides may be either monomeric or polymeric. The monomeric materials are unsuitable as curing agents, since the anhydride molecule splits to give monofunctional carboxylic acid molecules.

The linear polymeric anhydrides, derived from the *inter*-molecular condensation of organic diacids, are carboxyl-terminated polymers containing internal anhydrides. These provide difunctionality upon reaction and are therefore suitable as curing agents.

Representative are materials of the following generalized structure:

$$HO\text{+}OC(CH_2)_xCOO\text{+}_nH$$

Generalized linear polyanhydride

Polyanhydrides from sebacic and azelaic acid can be taken as typical.

Polysebacic Polyanhydride

Polysebacic polyanhydride (PSPA) has a melting point of 80°C and gives a viscosity of 160 centipoises at 90°C when used at 85 phr with DGEBA. At 90°C,

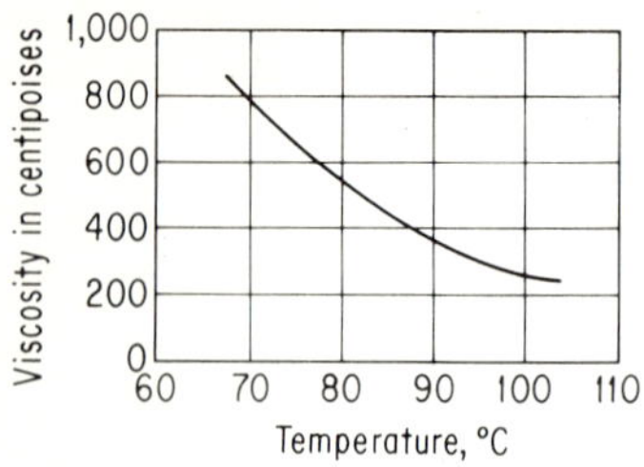

FIG. 12-1. Viscosity vs. temperature of PAPA [99].

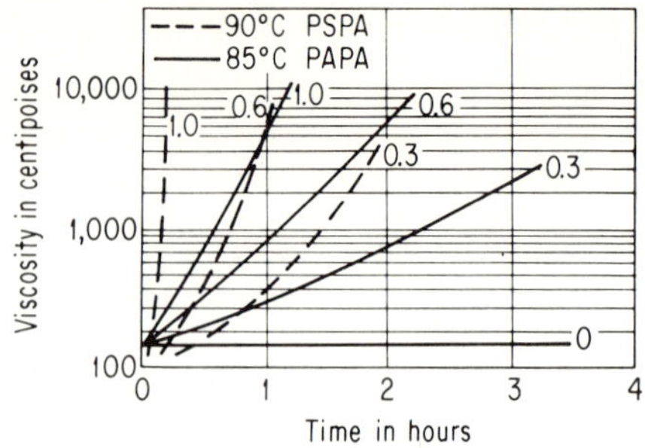

FIG. 12-2. Pot life of PAPA and PSPA with DGEBA [99]. Numbers on curves indicate phr BDMA.

the system with 2 phr DMP-30 gives a pot life of about 1 hour. Cured 2 hours at 120°C, it provides tensile strengths of 3,000 psi with up to 80 per cent elongation [4-1].

Polyazelaic Polyanhydride

Polyazelaic polyanhydride (PAPA) is available commercially at a molecular weight of about 2,300. Viscosity vs. temperature for the material is indicated in Fig. 12-1. Pot lives of PAPA and PSPA are compared in Fig. 12-2. DGEBA cured with 70 phr PAPA (and with DMP-30 as accelerator) give tensile strengths of about 5,000 psi with 100 per cent elongation. Weight loss at 150°C after 8 weeks is less than 1 per cent [99]. The effect of mixing ratios on hardness and thermal shock resistance is shown in Fig. 12-3.

The use of linear polymeric anhydrides blended with monomeric anhydrides has been reported as providing complete reactions where steric factors influencing one anhydride do not necessarily influence the other [47]. A melting-point diagram for anhydrides in blend with PAPA is shown in Fig. 12-4.

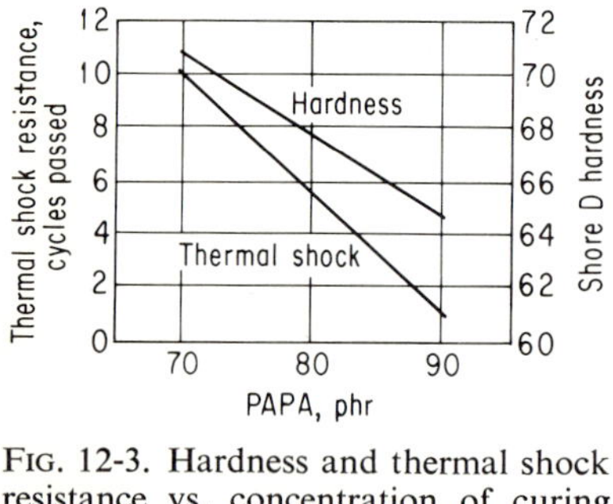

FIG. 12-3. Hardness and thermal shock resistance vs. concentration of curing agent for DGEBA cured with PAPA and 0.3 phr BDMA [99].

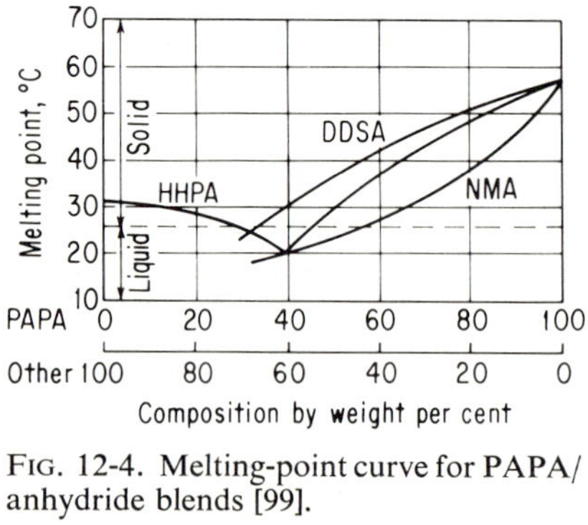

FIG. 12-4. Melting-point curve for PAPA/anhydride blends [99].

ALICYCLIC ANHYDRIDES

The alicyclic anhydrides are of two types: the simple alicyclic anhydrides, where the only ring is the cyclic anhydride group, and the multiring alicyclic anhydrides, where the anhydride group is attached to another cyclic structure.

Simple Alicyclic Anhydrides

Representative of the simple alicyclic anhydrides are succinic and substituted succinic anhydrides, anhydrides from citric acid, maleic anhydride, and special adducts of maleic anhydrides. These anhydrides are displayed in Table 12-1.

Succinic Anhydride and Its Modifications. Succinic anhydride (m.p. 120°C) has been evaluated as a curing agent for both glycidyl ether and epoxidized olefin resins. When used at 50 phr to cure DGEBA with 0.5 phr BDMA as accelerator, it provides tensile strengths of 9,000 psi with 6 per cent elongation and DTs of about 100°C [59]. With 3,4-epoxy-6-methylcyclohexylmethyl 3,4-epoxy-6-methylcyclohexane carboxylate, the DT will be approximately 100°C higher.

Methylsuccinic anhydride (m.p. 37°C) has been reported to provide somewhat improved properties and better clarity over succinic anhydride [39]. Cumylsuccinic anhydride has been used [114].

Citraconic and itaconic anhydride both cure the glycidyl ether resins, but during cure in the presence of a tertiary amine carbon dioxide is liberated [32].

Alkenyl anhydrides, essentially substituted succinic anhydrides of the generalized structure shown in Table 12-1, have been used with epoxy resins. When n is 9 to 11, the linear-chain materials are viscous liquids; from 11 to 19, light pastes or waxy solids. Anhydrides with branched chains are preferred, those having 8 to 11 carbon atoms in the linear portion being room-temperature liquids. Typical materials are nonenyl- and dodecenylsuccinic anhydride. Both give substantially the same properties when used as epoxy curing agents, and dodecenylsuccinic anhydride (or the isomeric form, tetrapropenylsuccinic) may be taken as typical.

Dodecenylsuccinic anhydride (DDSA) is a light-yellow, low-viscosity liquid (300 centipoises), having a molecular weight of 266. It may be manufactured by the reaction of maleic anhydride with tetrapropylene [61]. It is chiefly interesting for its long aliphatic chain, which reduces brittleness of the cured system.

When used with a glycerol-based glycidyl ether resin (mol. wt. 300) at 186 phr without accelerator, the pot life is on the order of 8 days and the initial viscosity is below 1,000 centipoises. DTs are about 40°C. It is probable that the chlorine normally present in aliphatic-based epoxies synthesized from the epichlorohydrin route accounts for the heightened reactivity of these resins with anhydride curing agents.

When used with the commercial polyglycidyl ether of tetraphenylethane and 1 phr BDMA, DDSA at an A/E from 0.2 to 0.6 gives maximum DTs of about 110°C, compared to 95°C at an A/E of 1/1 [4-6]. This indicates the flexibilizing effect of the side chain of DDSA and perhaps the effect of steric hindrances from the increased functionality. It is probable that at lower concentrations, a substantial amount of the curing reaction is brought about by etherification under tertiary amine catalysis or that the cure does not, in fact, progress to substantial completion.

When used to cure DGEBA, DDSA provides only slightly poorer electrical properties than when the more rigid anhydrides are used, as shown in Fig. 12-5. Additional flexibility may be obtained by the use of a polyol in blend with DDSA and the resin [110].

Table 12-1. Typical Alicyclic Anhydride/Epoxy-resin Curing Agents

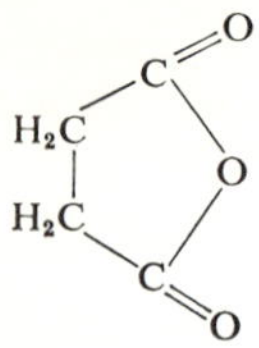

Succinic anhydride

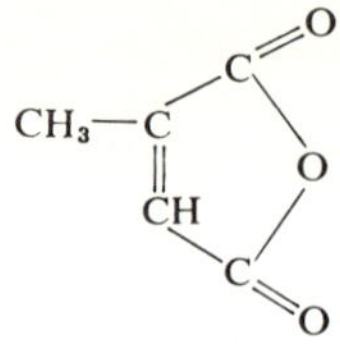

Citraconic anhydride

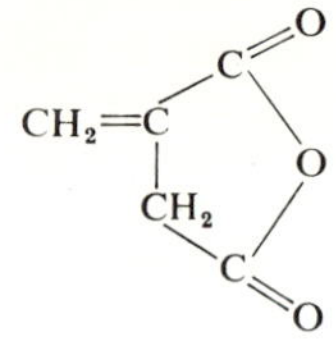

Itaconic anhydride

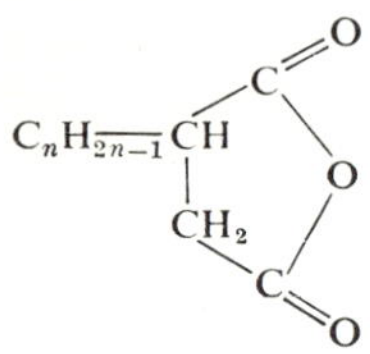

Alkenyl anhydrides

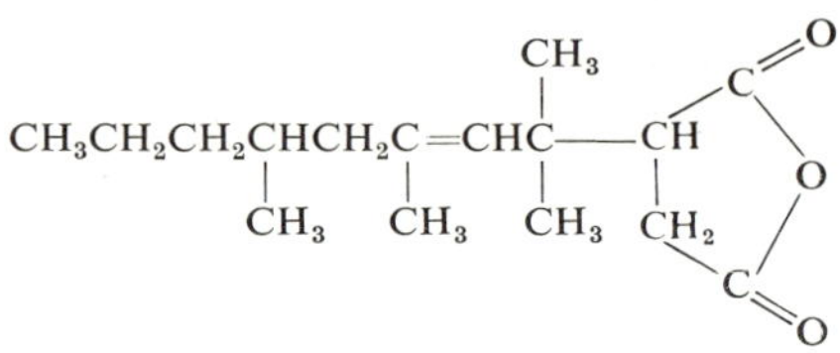

Dodecenylsuccinic anhydride

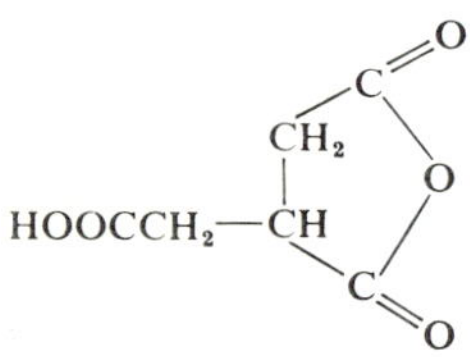

Tricarballylic anhydride

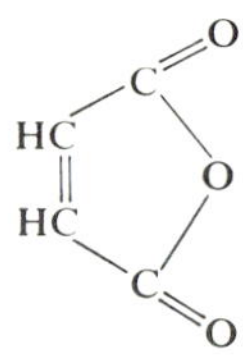

Maleic anhydride

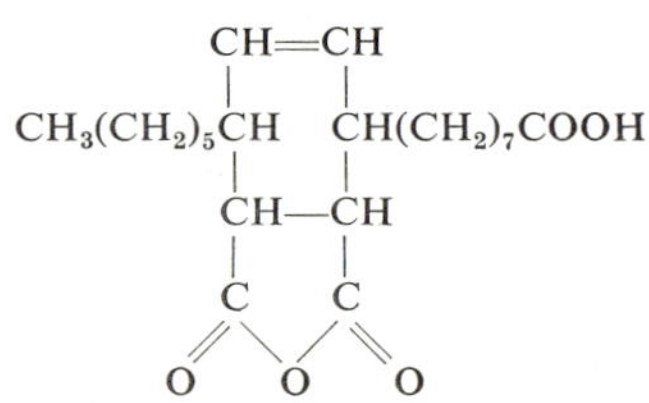

Linoleic acid adduct of maleic anhydride

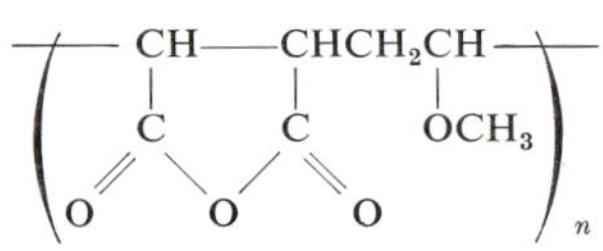

Copolymer of maleic anhydride and vinyl ether

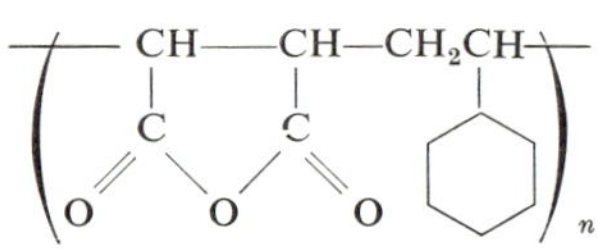

Copolymer of maleic anhydride and styrene

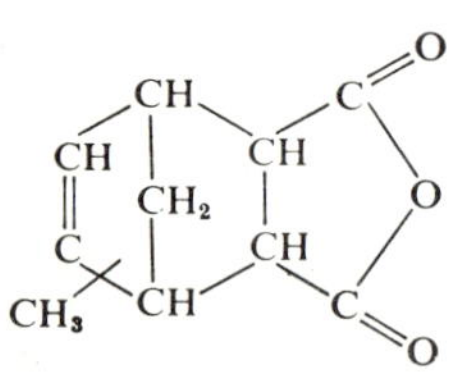

Maleic anhydride adduct of methylcyclopentadiene

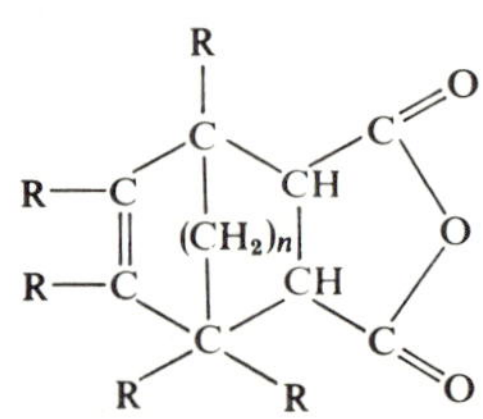

Alkylated endoalkylenetetrahydrophthalic anhydride

Table 12-1. (contd.)

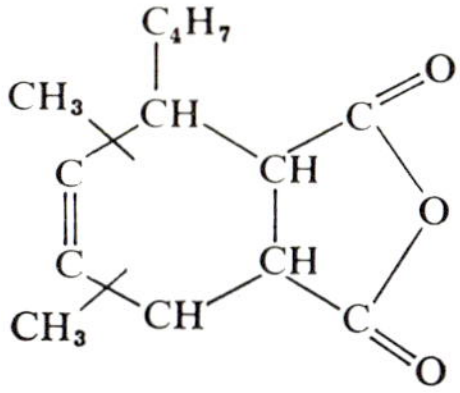

Dimethyl-substituted butenyltetrahydrophthalic anhydride

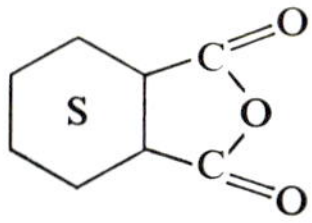

Hexahydrophthalic anhydride

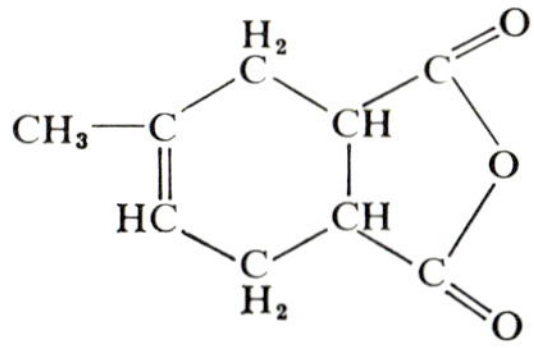

Methyltetrahydrophthalic anhydride

Anhydrides from Citric Acid. Several anhydrides from citric acid, including acetylcitric anhydride, aconitic anhydride, and tricarballylic anhydride have been investigated as epoxy-resin curing agents [32]. Of these only tricarballylic anhydride did not foam the resin. Because of the presence of the carboxyl group in the structure, which serves to promote etherification [62], the material is used at an A/E of 0.4/1 with DGEBA to provide a maximum DT of 147°C against 108°C for 0.5/1 and 65°C for 1/1 [32].

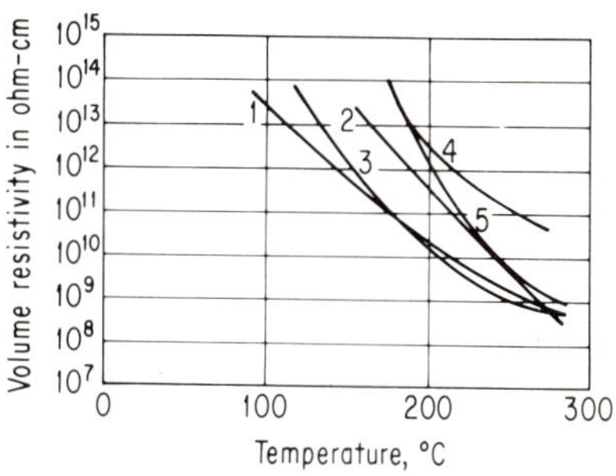

FIG. 12-5. Volume resistivity of DGEBA cured with anhydrides [4–6]. (1) 134 phr DDSA + 1 phr BDMA; (2) 90 phr NMA + 1 phr BDMA; (3) 78 phr HHPA + 1 phr BDMA; (4) 75 phr PA + 0.1 phr BDMA; (5) 110 phr chlorendic anhydride.

Maleic Anhydride. Maleic anhydride (MA) is a white crystalline solid with a molecular weight of 98 and a melting point of 53°C. It liberates irritating fumes at the temperatures normally required for incorporation into the resin. MA produces brittle castings with the more rigid epoxy resins and is normally used with a flexibilizer or in blend with other anhydrides.

MA has been investigated in combination with polyols and 3,4-epoxy-6-methylcyclohexylmethyl 3,4-epoxy-6-methylcyclohexane carboxylate [78] and epoxidized liquid polybutadienes [31]. The former is representative and illustrates the use of polyol modifiers in the anhydride epoxy technology.

Generally if the polyol is used at fairly low concentrations, it will have little effect upon the DT; but as the amount of polyol increases or as the functionality decreases and the molecular weight increases, the DT decreases. For example, the DTs increase as follows when the polyol is present at an A/E/H ratio of 0.5/1/0.33: polyethylene glycol (mol. wt. 200), 102°C; ethylene glycol, 170°C; 1,2,6-hexanetriol, 180°C; and glycerol, 214°C [78]. With dicyclopentadiene dioxide, the effect of polyols on the DT is indicated in Table 12-2. The polyol shifts the optimum A/E ratio toward lower values. This is illustrated with an uncatalyzed system in Table 12-3. Even at these low A/E ratios, however, when polyol is present, improved results can be obtained by use of BDMA at 1 phr, for example, in one case increasing the DT from 140 to 170°C.

Because of the reactivity of the double bond, other unsaturated compounds may be added to the MA/epoxy blend. Styrene, for example, may be used, and the addition reaction through the unsaturation may be obtained in situ [19,20,95].

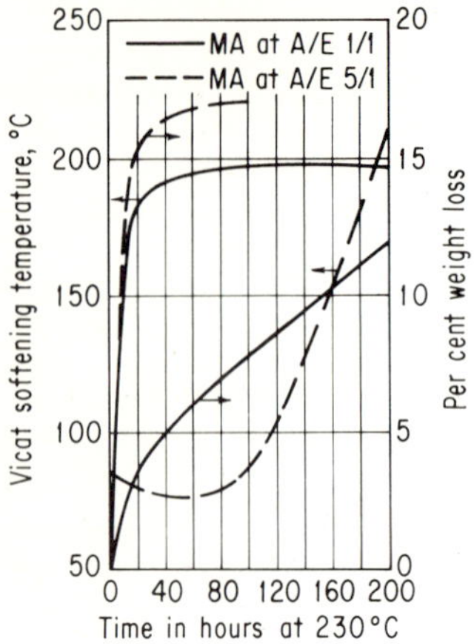

FIG. 12-6. Vicat softening temperature and weight loss vs. curing-agent concentration for DGEBA (mol. wt. 950) cured with MA [42].

MA itself may be employed in excess with addition occurring through the unsaturation at elevated temperatures [42]. Use of excess MA is indicated in Fig. 12-6, which shows an increase in Vicat softening temperature at 230°C upon extended aging, and a weight loss below what would be expected if no reaction involving MA were occurring. If MA is used with an unsaturated epoxy, such as epoxidized liquid polybutadiene, a peroxide catalyst may be employed to increase the DT from about 120 to 200°C [44] and, again, other unsaturated compounds may be added [81].

Infrared-spectroscopy studies on the reaction of maleic anhydride with DGEBA indicates a temperature-sensitive cis-trans isomerization. Long, low cures at 60°C followed by a postcure at 150°C yield very little trans form; whereas initial cures at 150 to 180°C cause formation of a considerable amount of the trans form in the cured resin [83]. At cure temperatures in the 180°C range, it is suggested that the double bonds in the maleate structure may undergo addition polymerization and thermal oxidation simultaneously, resulting in the formation of additional cross-links and new carbonyl and hydroxyl groups [83].

Table 12-2. Effect of Polyols on DT and Weight Loss of Dicyclopentadiene Dioxide Cured with 48 phr MA [89]

Polyol ($H/E = 0.2/1$)	DT, °C Gel at 80°C + 2 hr at 120°C + 2 hr			Weight loss, 1,000 hr at 260°C, %
	At 160°C	At 200°C	At 260°C	
Glycerol	202	257	294	21.3
Bisphenol A	187	234	277	18.2
Trimethylolpropane	183	240	294	19.6
1,2,6-Hexanetriol	173	239	292	22.0
Ethylene glycol	159	218	293	20.0
Neopentyl glycol	152	209	293	18.6
1,4-Butanediol	149	202	295	21.7
1,5-Pentanediol	134	184	294	21.8

Table 12-3. DT of 3,4-Epoxy-6-methylcyclohexylmethyl 3,4-Epoxy-6-methylcyclohexane Carboxylate Cured with MA as Function of A/E and H/E Ratios [78]

Anhydride/epoxy	Hydroxyl/epoxy	DT, °C
0.83/1	0.67/1	79
0.66/1	0.67/1	88
0.5/1	0.67/1	90
0.83/1	0.33/1	88
0.66/1	0.33/1	120
0.5/1	0.33/1	140
0.83/1	0.17/1	175
0.66/1	0.17/1	147
0.5/1	0’17/1	168

MA is reported as a particularly useful curing agent for higher-viscosity epoxidized novolac resins. For example, when mixed at 36 phr with an epoxidized novolac having a functionality of about 3.3, the mixture cooled to room temperature will provide a viscosity of 5,000 centipoises as against a mixed viscosity with the liquid NMA (used at 75 phr) of 20,000 centipoises at room temperature.

Because of the presence of unsaturation, MA may be reacted by Diels-Alder and other routes to obtain a variety of adducts.

Vinyl and styrene copolymers are available (Table 12-1). Use of heteropolymers from MA and butadiene-styrene copolymers as curing agents for epoxidized polybutadiene has been reported [36]. MA may be reacted with tetrapropylene (as discussed above) to produce DDSA or with an organic acid to produce 4-carboxy-3-methyl-1,2-butanecarboxylic anhydride [100]. It may be reacted with methylcyclopentadiene [63] or with other substituted cyclopentadienes [94]; with tetrahydrofurans to give materials such as diethyl (2-tetrahydrofuryl) succinate [85]; with butadiene to produce tetrahydrophthalic anhydride; with unsaturated fatty acids to produce compounds represented by the linoleic adduct shown in Table 12-1 [41]; with tung oil (a combination of palmitic, stearic, oleic, and eleosteric fatty acids) [96]; and with a variety of long-chain polyunsaturated diesters, diketones, diacids, and diamides to produce di- and trianhydrides suitable as curing agents [25].

Multiring Alicyclic Anhydrides

A number of multiring alicyclic anhydrides from Diels-Alder condensation of MA with compounds having conjugated unsaturation are liquids, and hence are preferable as epoxy-resin curing agents to the higher-melting compounds.

Hydrogenated phthalic anhydrides, such as tetra- and hexahydrophthalic anhydride, are sometimes used in preference to the aromatic-based material. The commercial tetrahydrophthalic anhydride is less preferable than the lower-melting hexahydrophthalic, although mixed isomers of the former as liquids would appear to offer advantages [86]. Hydrogenated pyromellitic dianhydride has been suggested as a co-curing agent for epoxy/monoanhydride blends [51].

Maleic Anhydride Adduct of Methylcyclopentadiene. The maleic anhydride adduct of methylcyclopentadiene (NMA), normally a solid, is made into a light-yellow liquid having a melting point below 12°C by the addition of about 0.1 per cent phosphoric acid. NMA (mol. wt. 178) employed with DGEBA is usually used at 80 to 90 phr with a tertiary amine accelerator, although good elevated-temperature stability has been reported [73] when it is used in concentrations as low as 21 phr with DMP-30 at 2 phr. Since DMP-30, alone, at 2 phr provides quite good cures with this resin, it is likely that when small amounts of NMA are used, the DMP-30 accounts for a high percentage of the cure.

DMP-30 is a comparatively active accelerator which causes the resin to thicken fairly rapidly at room temperature, giving a pot life on the order of 4 hours when used in the 2- to 3-phr range. BDMA is considerably slower, giving a pot life of 3 to 4 days.

The usual cure schedules for the aromatic glycidyl ether resins, with NMA and accelerator, call for a gel at temperatures on the order of 120°C with a postcure from 150 to 180°C. However, a statistical analysis of DGEBA cured with NMA at an optimized 89 phr and BDMA at an optimized 2.16 phr requires a gel temperature of 107°C and a postcure of 256°C [26]. A second analysis predicted a DT of 300°C using 0.5 phr BDMA with a cure of 2 hours at 120°C + six hours at 200°C + 24 hours at 260°C [4-6].

Physical properties of two NMA/DGEBA systems, one accelerated with DMP-30 and one with BDMA, are given in Table 12-4. Electrical resistivity vs. temperature

Table 12-4. Physical Properties of DGEBA Cured with Alicyclic Anhydrides

Property	NMA, 84 phr, +2.2 phr BDMA, cured 4 hr at 80°C + 15 hr at 150°C	NMA, 85 phr, +2.5 phr DMP-30, cured 4 hr at 93°C + 3 hr at 204°C	DBTA, 119 phr, +1 phr BDMA, cured 16 hr at 120°C + 1 hr at 180°C	HHPA, 85 phr, +0.93 phr BDMA, cured 2 hr at 90°C + 18 hr at 120°C
DT, °C	153	132	127	135
Tensile strength, psi	6,400	11,000		8,700
Elongation, %	2.38	2.7		
Flexural strength, psi	16,800	20,700	12,000	21,500
Modulus, psi × 10^{-6}	0.53	0.57	0.46	
Compressive strength, psi	16,900	19,300	15,700	15,900
Modulus, psi × 10^{-6}	0.106	0.40	0.38	

are shown for a 90-phr NMA system accelerated with 1 phr BDMA in Fig. 12-5. Radiation resistance is given in Fig. 12-7.

NMA-cured systems are also found to have good resistance to irradiation from the electron beam of electron microscopes [69]. Chemical resistance is given in Table 12-5.

Increasing the functionality of the glycidyl ether resin has the effect of increasing the DTs of the cured system. The relationship, for example, of the functionality f of an epoxidized *o*-cresol novolac to DT is as follows: $f = 2.7$, DT at 280°C; $f = 4.8$, 6-mil deflection, at 300°C, $f = 5.4$, 2.5-mil deflection, at 300°C [7].

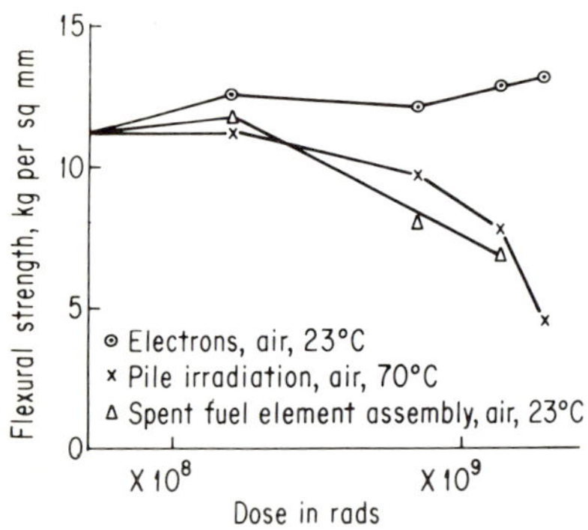

FIG. 12-7. Radiation resistance of DGEBA cured with NMA [38].

Studies with NMA and an epoxidized novolac (f 3.3) over an A/E of 0.85/1 to 1/1 with DMP-30 from 0.5 to 1.5 phr and with varying cure cycles showed that optimum tensile strengths of the casting at room temperature were independent of the variables. However, when the casting was tested at 175°C, the preference was for the higher accelerator concentration. The importance of establishing a sound network system during cure (before exposure to high-temperature postcure) was demonstrated in terms of increasing tensile strength and decreasing weight loss during postcure [105].

In general, weight loss of NMA-cured systems is low and strength retention at elevated temperatures is good when the curing conditions and accelerator concentrations are at optimum. However, there is some question as to its suitability for service at temperatures above about 250°C, since in that range there is a tendency for decarboxylation [74].

NMA-cured systems, in general, are somewhat too brittle for casting applications unless flexibilizers are present. With castor oil additives, NMA-cured systems may be flexibilized to virtually any degree, with systems having elongations up to 380 per cent being reported [98], but of course at the expense of heat resistance. The initial low viscosity of NMA coupled with the good elevated-temperature strength retention makes it a curing agent well suited to laminating applications.

Alkylated Endoalkylenetetrahydrophthalic Anhydride. Very similar in structure to NMA is the alkylated endoalkylenetetrahydrophthalic anhydride shown in Table 12-2, where R is hydrogen or lower alkyl and n is a small integer. The material is available commercially as a mixture of isomers consisting of a short homologous series [1]. It is a clear, reddish liquid with an average molecular weight of 275. The viscosity at room temperature is 2,800 centipoises. It generally requires use of an accelerator; when uncatalyzed, it is only sluggishly reactive, at best, with glycidyl ether resins. With DGEBA, this mixed anhydride is used at 95 to 110 phr with DMP-30 at 2 to 3 phr and cured at 140°C for best DTs (140°C). With an epoxidized novolac (*f* 3.3), the DT is some 20°C higher if the cure temperature is 160°C, but lower and higher temperatures reduce the value to about that obtained with DGEBA [1].

Table 12-5. Chemical Resistance of DGEBA Cured with NMA and PA [4-1]

Chemical environment, 6 months' immersion	Change in flexural strength, %	
	NMA, 80 phr, +2 phr DMP-30, cured 12 hr at 120°C + 12 hr at 180°C + 12 hr at 250°C	PA, 65 phr, cured 24 hr at 140°C
Salt resistance:		
10% sodium chloride	−12	−5
25% sodium chloride	+17	−14
10% sodium bisulfite	+15	+9
Acid resistance:		
3% sulfuric	−18	−16
10% sulfuric	+20	+13
30% sulfuric	+2	+12
10% hydrochloric	−5	+12
10% nitric	+13	−26
10% glacial acetic	+25	+19
Base resistance:		
10% caustic soda	−46	−80
25% caustic soda	−24	−78
25% sodium carbonate	+3	+8
25% ammonia solution	−23	Destroyed
Solvent resistance:		
Gasoline	+15	−2
Benzene	+20	−7
Ethyl alcohol	+20	+20
Ethylene glycol	+20	+20
Water resistance:		
At 25°C	−10	+8
Boiling	−38	Destroyed

Dimethyl-substituted Butenyltetrahydrophthalic Anhydride. The dimethyl-substituted butenyltetrahydrophthalic anhydride (Table 12-2) is a low-viscosity yellow liquid with a melting point of −10°C and a molecular weight of 234 [5]. With DGEBA and accelerator, it gives a pot life of 4 to 5 days and gives cured properties as indicated in Table 12-4, where the anhydride is identified as DBTA.

Methyltetrahydrophthalic Anhydride. MTHPA (Table 12-1) is a low-viscosity liquid (60 centipoises at 25°C) with a solidification point of 4°C. MTHPA has a molecular weight of 166 and a density of 1.195 at 25°C. When used with DGEBA at a ratio of 84 phr and 2 phr of ethylmethylimidazoline as accelerator, it reveals a gel time of 143 minutes at 70°C in comparison to 162 minutes for NMA. Deflection temperature is 155°C in comparison to 154°C for NMA. Electrical properties and chemical resistance also closely parallel those obtained when curing with NMA.

Although early researchers recognized that THPA possessed properties similar to those of NMA, and in some regards those of HHPA, MTHPA has been slower to gain widespread acceptance because it was not available commercially until 1965, some 10 years after the others.

Hexahydrophthalic Anhydride. Hexahydrophthalic anhydride (HHPA) is a waxy solid having a molecular weight of 154 and a melting point of about 35°C. When HHPA is mixed with DGEBA, a viscosity of about 200 centipoises at 40°C is obtained. Working life at 40°C is about one day. Cured properties are as shown in Table 12-4. Volume resistivity for an accelerated system is shown in Fig. 12-5.

The effect of a variety of accelerators on thermal stability in terms of elevated-temperature weight loss is indicated in Table 12-6. Optimization of the HHPA system by statistical analysis is discussed in [37].

Table 12-6. Effect of Amine Accelerators on Weight Loss of Anhydride-cured DGEBA [28]

Accelerators at 1.6 phr	Weight loss at 205°C, %					
	In 1 day		In 3 days		In 7 days	
	PA, 60 phr	HHPA, 60 phr	PA, 60 phr	HHPA, 60 phr	PA, 60 phr	HHPA, 60 phr
None	1.86	10.3	2.44	12.2	3.46	14.7
DMAPA	....	0.72	....	1.90	.....	3.39
DEAPA	0.33	0.98	1.68	1.91	5.28	3.06
Triethylamine	0.43	0.91	1.34	1.79	5.47	3.27
Diethylamine	1.03	3.11	2.42	4.07	7.08	5.58
Benzyldimethylamine	1.79	0.98	4.29	2.04	7.60	3.42
α-Methylbenzyldimethylamine	2.17	1.36	5.39	2.51	7.90	3.80
DMP-10	1.81	1.13	4.97	2.14	7.32	3.29
DMP-30	1.44	0.95	4.54	2.06	6.76	2.28
Diisopropylamine	1.21	2.75	3.18	3.84	5.99	5.04
Triethanolamine	1.30	0.90	4.15	1.91	6.61	3.32
Diethanolamine	2.85	3.04	5.51	4.00	7.75	5.31
Methylethanolamine	0.41	2.36	2.25	3.26	6.33	4.73
α-Methylbenzyldiethanolamine	2.17	6.61	4.28	7.96	6.83	9.33
Piperidine	1.54	3.10	2.43	3.96	6.31	5.44
Monoethanolamine	3.59	6.61	4.85	7.28	7.95	9.70
Nonylamine	6.11	14.20	7.49	16.21	10.05	17.40
N-Aminopropylmorpholine	0.65	1.64	1.53	2.48	5.12	3.70
Monoisopropanolamine	4.00	8.89	5.41	10.77	8.57	12.25
Methyldiethanolamine	2.50	1.58	4.34	2.43	8.71	3.62
Mercaptoethanol	3.97	2.24	5.24	2.91	6.03	3.57
Mercaptopropionic acid	3.65	2.38	4.98	3.21	5.80	3.87

When used with 3,4-epoxy-6-methylcyclohexylmethyl 3,4-epoxymethylcyclohexane carboxylate, DTs on the order of 184°C can be obtained under conditions of acidic catalysis (i.e., stannic chloride), with, as expected, the optimum A/E ratio being reduced to about 0.5/1. At an A/E of 0.75/1 with this system, the DT is some 40°C lower [78].

HHPA may be used with polyallyl glycidyl ether (PAGE) to provide for room-temperature cure when co-cured with 6 phr pyridine. Properties of this formulation are given in Table 12-7. Also shown are the properties of PAGE and polyglycidyl ether of glycerol accelerated with 1 phr BDMA and cured at 100°C.

When methyl-HHPA is employed as a curing agent, it is reported that a diisopropylcarbodiimide is used at about 3 per cent on the anhydride to stabilize it against conversion to the acid on atmospheric storage [109].

Multiring Alicyclic Dianhydrides

In addition to the anhydride copolymers shown in Table 12-1, a number of other di- and polyanhydride compounds may be produced for use as epoxy-resin curing

Table 12-7. Properties of Aliphatic-based Epoxy Resins Cured with HHPA [4-6]

Property	PAGE		Polyglycidyl ether of glycerol
	HHPA, 102 phr, +6 phr pyridine, cured 168 hr at 23°C	HHPA, 102 phr, +1 phr BDMA, cured 4 hr at 100°C	HHPA, 107 phr, +2 phr BDMA, cured 4 hr at 100°C
DT, °C	53	89	57
Tensile strength, psi	7,830	13,300	
Flexural strength, psi	13,400	18,300	17,500
Modulus, psi × 10^{-6}	0.52	0.47	0.55
Compressive strength, psi	16,400	16,200	15,000
Modulus, psi × 10^{-6}	0.57	0.74	0.66
Impact strength, Izod	0.33	0.39	
Hardness, Barcol	36	38	

agents. Cyclohexenone, for example, may be reacted with anhydrides containing unsaturated side chains such as DDSA to produce telomers suitable as curing agents [97].

A number of anhydrides based on biscyclopentadienyl alkanes have been reported [84], and the synthesis has been described in detail [108]. Of these, the commercially available bis Nadic anhydrylbutene is typical.

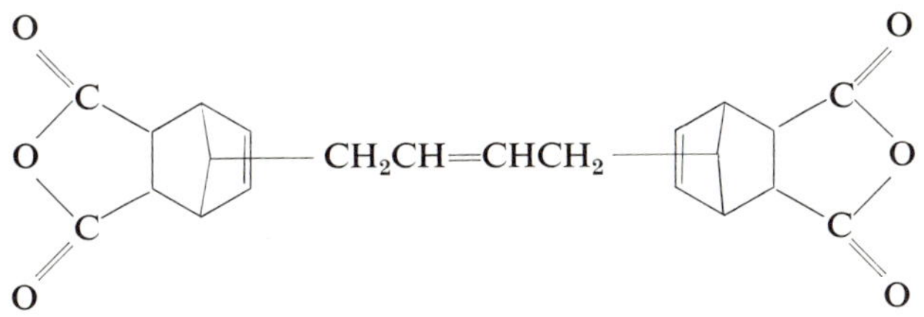

Bis Nadic anhydrylbutene

This compound has been investigated as a high-temperature curing agent for epoxy-resin filament-winding formulations. The material has been found to be reactive with itself at temperatures above 220°C through the unsaturation to provide additional cross-linking [112].

Additionally, multiring alicyclic dianhydrides useful as epoxy-resin curing agents have been obtained by Diels-Alder reaction of *trans*-cyclohexadiene-2,5-dicarboxylic acid derivatives with dienophiles [108].

AROMATIC ANHYDRIDES

Both aromatic mono- and dianhydrides are available to the epoxy technology. Phthalic anhydride is an example of the former; trimellitic represents the monoanhydrides containing additional functionality in the form of a carboxyl group; and the latter are represented by the compounds displayed in Table 12-8, of which pyromellitic dianhydride is a representative.

Phthalic anhydride, which did not gain the popularity in the United States that it enjoyed in Europe, has been almost completely replaced by the liquid materials in spite of their generally higher costs. Trimellitic anhydride is quite reactive with epoxy resins, and its use is therefore limited. Pyromellitic dianhydride and other

Table 12-8. Dianhydride Epoxy-resin Curing Agents [103]

Name	Pyromellitic dianhydride	Cyclopentanetetracarboxylic dianhydride	Benzophenonetetracarboxylic dianhydride
Structure	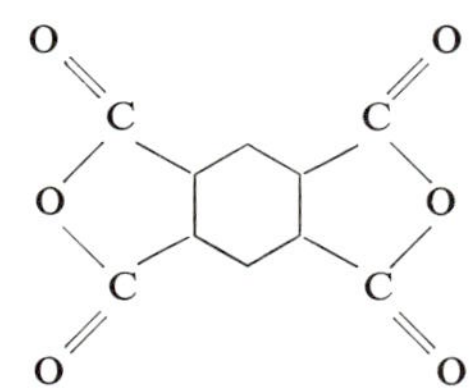	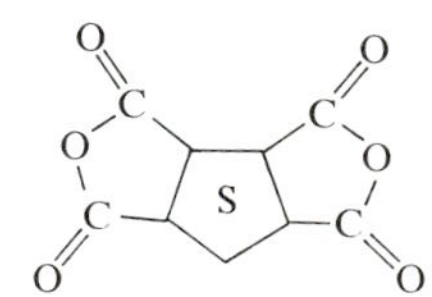	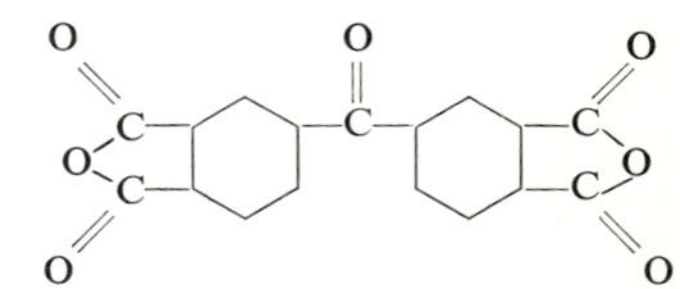
Molecular weight	218.12	210.2	322
Anhydride equivalent weight (A/E = 1/1)	109	105.1	161
Carboxyl equivalent weight	54.5	52.5	80.5
Physical appearance	White powder	White powder	White powder
Melting point	286°C	220–221°C	236°C
Solubility	Poor, 6–13% at reflux temperature, depending on solvent	Moderate in polar solvents: dimethylformamide, acetonitrile, butyrolactone, dimethyl sulfoxide	Poor, similar to PMDA

Table 12-8. (Cont'd.)

Name	Ethylene glycol bistrimellitate	Glycerol tristrimellitate
Structure	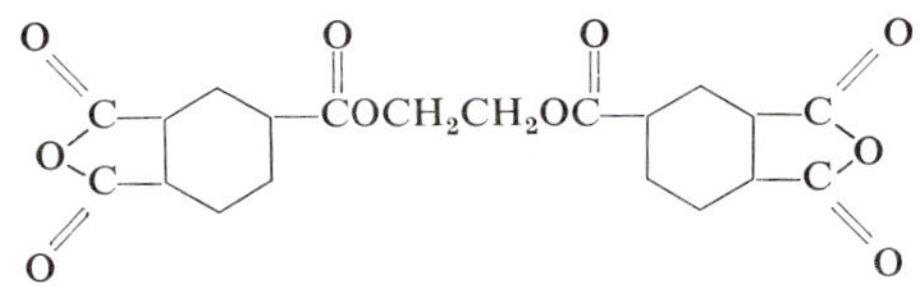	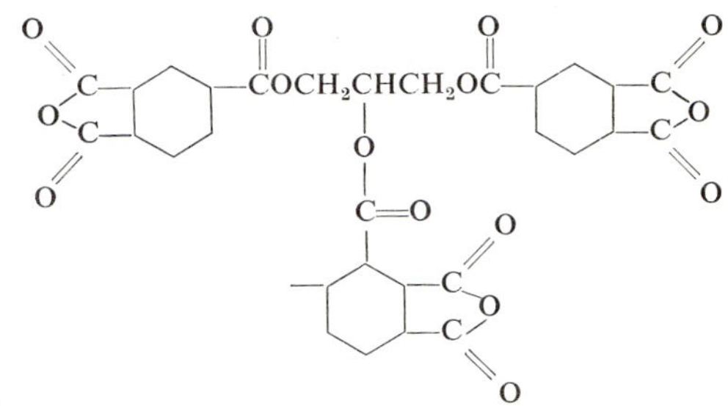
Molecular weight	410	613
Anhydride equivalent weight (A/E = 1/1)	205	204.3
Carboxyl equivalent weight	102.5	102.2
Physical appearance	White powder	White powder
Melting point		
Solubility	Forms 19% solids solution in acetone	Forms 80% solids solution in MEK

dianhydrides used alone or with monoanhydrides [24] are more or less specialty compounds offering exceptionally good high-temperature performance.

Phthalic Anhydride

Phthalic anhydride (PA) is a white powder having a molecular weight of 148 and a melting point of 128°C.

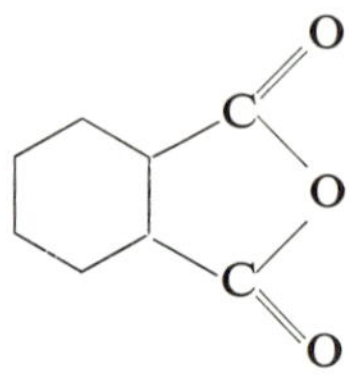

Phthalic anhydride

When it is used with DGEBA, the usual procedure is to heat the resin to 120 to 140°C before adding PA. The mixture is then stirred until homogeneous, a constant temperature being maintained during the mixing. The anhydride will partially sublime during this operation. Once the anhydride is thoroughly dissolved, the resin mix is cooled to 60°C to prolong the pot life. At that time, the accelerator, if employed, is added. At temperatures below 60°C, the PA will precipitate, but, pot life permitting, may be returned to solution upon reheating. The higher-molecular-weight species of DGEBA are usually processed at correspondingly higher temperatures, mixing being accomplished at 150 to 170°C. The initial cure cycles are generally about 120 to 130°C. Higher temperatures should be avoided, particularly in larger castings, to prevent uncontrollable exotherm.

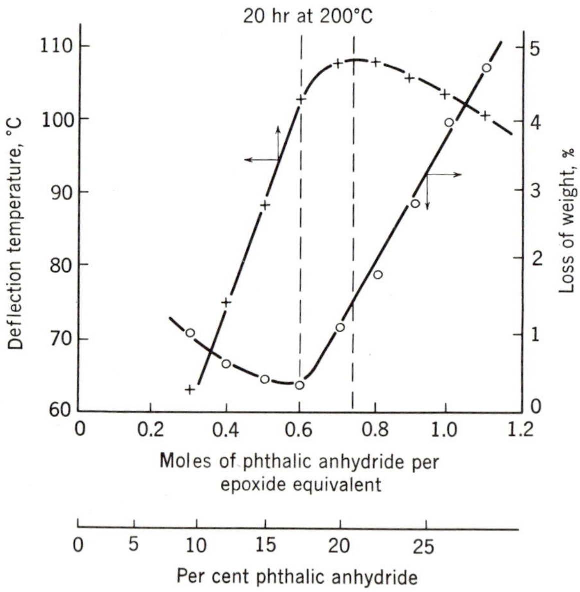

FIG. 12-8. Weight loss and DT vs. curing-agent concentration for DGEBA (mol. wt. 400) cured with PA [13].

In uncatalyzed systems, DGEBA (mol. wt. 800) is usually preferred. These resins, viscous liquids at room temperature, are sufficiently fluid at processing temperatures to permit use in casting applications. They provide improved toughness and thermal shock resistance over the lower-molecular-weight species and cure somewhat faster.

Because of the presence of hydroxyls on the higher-molecular-weight DGEBA, the PA is used at an A/E of less than unity, and the greater the amount of hydroxyls present in the chain, the more nearly the ratio comes to 0.5/1. The effect of the PA concentration on DT and weight loss for the 800-mol.-wt. species is shown in Fig. 12-8. A characteristic of anhydride cures generally is that the maximum weight loss and DT values are obtained at different A/E ratios.

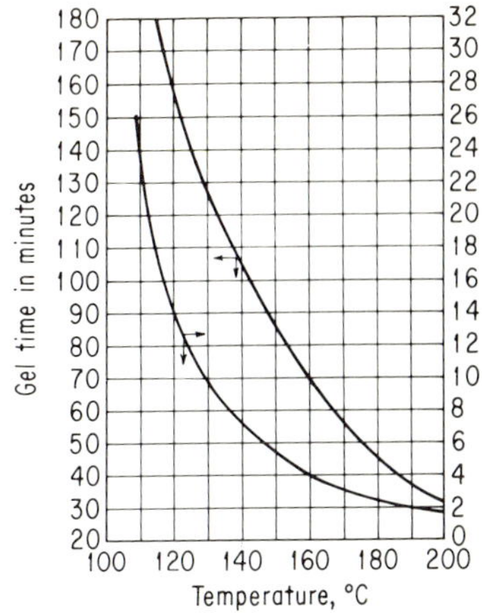

FIG. 12-9. Gel and cure time vs. temperature for DGEBA (mol. wt. 800) cured with PA [4–1].

The properties of DGEBA (mol. wt. 800) cured with 30 phr PA, without accelerator, are shown in Table 12-9. Gel and cure times vs. temperature are shown in Fig. 12-9. Power factor and dielectric constants vs. temperature are shown in Fig. 12-10, and dielectric strength is given in Fig. 12-11. Linear shrinkage vs. temperature is shown in Fig. 12-12. Chemical resistance is shown in generalized terms, for 6 months' immersion, in Table 12-10.

When PA is used to cure DGEBA with accelerator, it is used at about 75 phr. DTs are about 150°C, with the physical properties below the DT generally in the range given in the previous paragraph for the higher-molecular-weight species. Volume resistivity vs. temperature are shown in Fig. 12-5. Chemical resistance is given in Table 12-11. When PA is used to cure DGEBA in the absence of an accelerator, it is usually used at an A/E of about 0.85/1 or approximately 65 phr. Chemical resistance of such a system is indicated in Table 12-5.

When PA is replaced in part by phthalic acid, a less tightly cross-linked system is

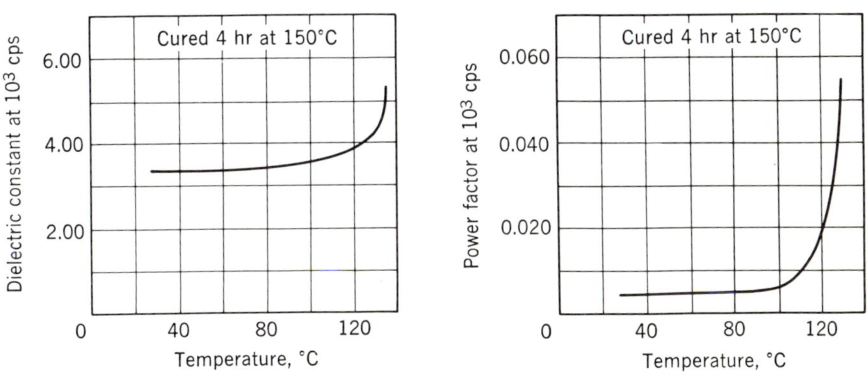

FIG. 12-10. Dielectric constant and power factor vs. temperature for DGEBA (mol. wt. 800) cured with PA [4–1].

Table 12-9. Properties of DGEBA (mol. wt. 800) Cured with PA [4-1]

Property	*Value*
DT, °C	109
Tensile strength, psi	12,000
Modulus, psi × 10^{-6}	0.45
Flexural strength, psi	19,000
Impact strength, Izod	0.70
Hardness, Rockwell	M100
Coefficient of thermal expansion, in./(in.)(°C) × 10^5	4.8
Thermal conductivity, cal/(sec)(sq cm)(°C)(cm) × 10^4	4.0
Thermal shock, MIL-I-16923	Type B
Flammability, in./min	0.32
Water absorption, 24 immersion, %	0.08
Moisture-vapor transmission at 93°C, relative humidity 95%, g/(sq ft)(24 hr)(in. thickness)	<0.01

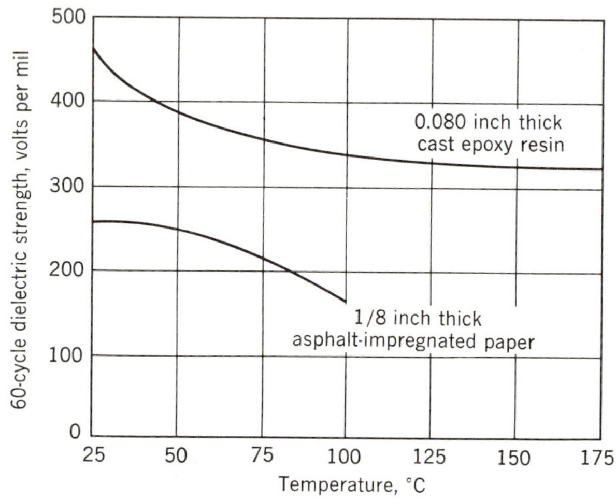

FIG. 12-11. Dielectric strength vs. temperature for DGEBA (mol. wt. 800) cured with PA [4–1].

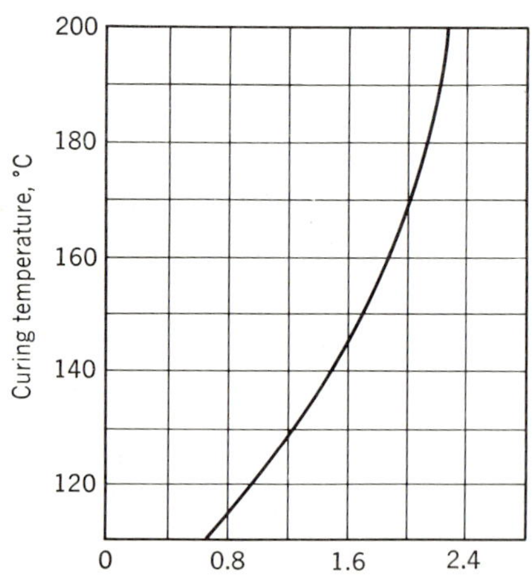

FIG. 12-12. Linear shrinkage vs. temperature for DGEBA (mol. wt. 800) cured with PA [12].

Table 12-10. Chemical Resistance of DGEBA (mol. wt. 800) Cured with PA [11]

6 months' immersion in fluid	Concentration	Temperature of test, °C	Resistance
Accumulator acid	32%	20–40	Resistant
Acetaldehyde	Conc.	20	Limitedly resistant
	40%	20	Resistant
	40%	70	Nonresistant
Acetic acid, glacial	100%	20	Resistant
	100%	60–100	Nonresistant
Acetone	100%	20	Nonresistant
	50%	20	Limitedly resistant
	1–10%	20	Resistant
Ammonia, aq.	30%	20	Limitedly resistant
	10–20%	20	Resistant
Aniline	100%	20	Nonresistant
Benzene	100%	20	Resistant
	100%	60–80	Nonresistant
Benzine	100%	20–80	Resistant
Butanol	100%	20	Resistant
Butyl acetate	100%	20	Nonresistant
Butyric acid	100%	20	Resistant
Carbon disulfide	100%	20	Limitedly resistant
	100%	45	Nonresistant
Carbon tetrachloride	100%	20	Resistant
Caustic soda, aq.	10 g/100	20	Resistant
	10 g/100	60–100	Limitedly resistant
	25–50 g/100	20	Limitedly resistant
Chlorosulfonic acid	100%	20	Nonresistant
Chromium bath liquid	....	40	Resistant
Clophen	100%	20–95	Resistant
Cooking salt, aq.	10–20 g/100	20–70	Resistant
	10–20 g/100	100	Limitedly resistant
Cresol, tech.	....	20	Nonresistant
Crotonaldehyde	100%	20	Nonresistant
Cyclohexanol	100%	20	Resistant
Cyclohexanone	100%	20	Limitedly resistant
Developer, photographic	....	20	Resistant
Dioxane	100%	20	Nonresistant
Ethyl alcohol	100%	20–40	Resistant
	100%	60–80	Nonresistant
Ethyl ester	100%	20	Resistant
	100%	35	Limitedly resistant
Ferrous chloride	45 g/100	20	Resistant
Fixing bath, photographic	....	20	Resistant
Formic acid	Conc.	20	Resistant
	Conc.	60	Limitedly resistant
	Conc.	100	Nonresistant
Glycerin	100%	20–100	Resistant
Hydrochloric acid	10%	20–100	Resistant
	50%	20–60	Resistant
	50%	100	Limitedly resistant
	100%	20	Resistant
	100%	60–100	Limitedly resistant
Hydrofluoric acid	Conc.	20	Resistant
Hydrogen peroxide, aq.	30%	20	Resistant
Javelle water (14% active chlorine)	....	20–100	Resistant
Liquid air	100%	−180	Resistant (10-min test)
Methyl chloride	100%	20	Nonresistant
Mineral oil	100%	20–100	Resistant

Table 12-10. (cont'd.)

Nitric acid	Conc. 50%	20	Nonresistant
	10%	20	Resistant
Nitrobenzene	100%	20	Nonresistant
Potassium cyanide bath liquid	....	80	Resistant
Potassium permanganate	6 g/100	20–100	Resistant
Sodium bisulfite, aq.	20 g/100	20	Resistant
Sodium chlorite, aq.	10 g/100	20–100	Resistant
Sodium dichromate, aq.	60 g/100	20	Resistant
Soft soap, aq.	50%	20–100	Resistant
Stearic acid	100%	20–60	Resistant
Sulfuric acid	10%	20–100	Resistant
	50%	20–60	Resistant
	50%	100	Nonresistant
	75%	20	Resistant
	Conc.	20	Nonresistant
Thionyl chloride	100%	20	Nonresistant
Toluene	100%	20	Resistant
Trichloroethylene	100%	20	Nonresistant
Triethanolamine	100%	20	Resistant
Water, distilled	....	20–40	Resistant
		70	Limitedly resistant
		100	Nonresistant

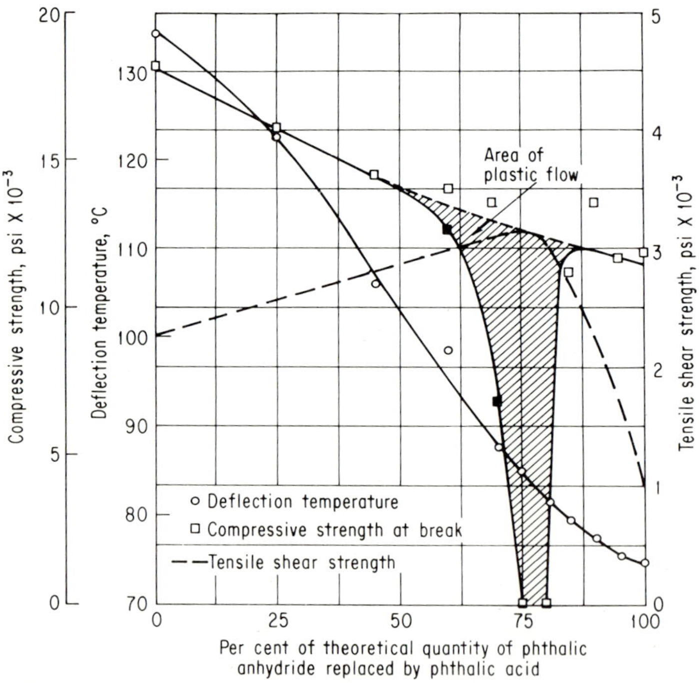

FIG. 12-13. Effect of phthalic acid concentration on DT and compressive and tensile shear strengths of DGEBA cured with PA [90].

Table 12-11. Chemical Resistance of DGEBA Cured with 75 phr PA + 0.25 phr BDMA and with 110 phr Chlorendic Anhydride [59]

Chemical	Volume change % at 23°C		Weight change % at 23°C		Volume change % at 55°C		Weight change % at 55°C	
	PA	CA	PA	CA	PA	CA	PA	CA
Sodium hydroxide, 50%:								
1 day	0	−0.3	0	0				
7 days	−0.3	−0.8	−0.2	−0.1				
30 days	−0.5	−0.8	−0.3	−0.2				
Sulfuric acid, 25%:								
1 day	−0.2	−0.5	0	0				
7 days	−1.0	0	0.2	0.2				
30 days	−0.2	0.5	0.4	0.4				
Hydrochloric acid, 25%:								
1 day	−0.2	−0.2	0	0	0.2	0	0	0
7 days	−0.7	−0.8	0	0.1	−0.2	−0.5	0.1	0
30 days	−0.2	0	0.1	0.3	0.2	0	0.2	0
Acetic acid, 25%:								
1 day	−0.5	0	0	0	−0.2	−0.3	0.2	0.2
7 days	−0.7	−0.4	0.2	0.2	−0.2	0.3	0.5	0.5
30 days	0.2	0.7	0.5	0.5	0.7	1.3	0.7	0.7
Trichloroethylene, 100%:								
1 day	−0.2	0.2	0	0	0	0.5	0.5	1.8
7 days	−1.1	0.2	0	0.2	2.6	8.6	4.3	11.3
30 days	0.2	7.0	0.9	0.5	13.1	25.6	17.7	13.7
Formaldehyde, 40%:								
1 day	−0.2	0.2	0	0				
7 days	−0.6	−0.5	0.2	0.2				
30 days	0	0.5	0.3	0.4				
Sodium hypochlorite, 6%:								
1 day	0	0	0	0	1.6	−1.2	0.2	0.2
7 days	−0.7	−0.7	0.2	0.1	0.2	0.2	0.5	0.4
30 days	0.2	0.7	0.5	0.5	0.2	0.7	0.6	0.5
Chromic acid, 25%:								
1 day	−1.0	−0.5	0	0				
7 days	−0.2	0.2	0.2	0.1				
30 days	0	0.5	0.4	0.5				
Distilled water:								
1 day	−0.2	−0.9	0	0	−0.5	−0.2	0.2	0.2
7 days	−0.4	−1.1	0.3	0.2	−1.4	0.7	0.5	0.5
30 days	0	−0.4	0.4	0.6	0.5	1.0	0.7	0.8
Sodium aluminum sulfate, 28.5%:								
1 day	−0.2	0	0	0				
7 days	−0.5	−0.2	0.2	0.2				
30 days	0.5	0.7	0.4	0.4				
24-hr boiling water, %			0.5	0.58				
3-hr boiling acetone, %			3.0	1.2				

developed because of loss of functionality. This is illustrated in Fig. 12-13 for DGEBA accelerated with a tertiary amine.

When PA is used to cure 3,4-epoxy-6-methylcyclohexylmethyl 3,4-epoxymethylcyclohexane carboxylate, DTs in the range of 200°C may be obtained with either a caustic soda or a tertiary amine. This is indicated in Table 12-12, which also shows the catalyst concentration to be critical, with increasing catalyst favoring the higher A/E ratios.

Table 12-12. Effect of Concentration and Type of Base Catalyst on DT of 3,4-Epoxy-6-methylcyclohexylmethyl 3,4-Epoxy-6-methylcyclohexane Carboxylate Cured with PA [78]

Catalyst	Catalyst, %	DT, °C	
		A/E 0.5/1	A/E 0.75/1
BDMA	0.01	212	186
BDMA	0.05	198	197
BDMA	0.1	174	193
BDMA	0.5	130	213
KOH	0.004	216	175
KOH	0.02	197	194
KOH	0.04	140	220

When PA is used to cure the DGE of dihydroxyphenylsulfone, no acceleration is required, DTs are sharply dependent on anhydride concentrations (Fig. 12-14), and cure occurs rapidly at 150°C (Fig. 12-15).

When PA is used to cure the chlorinated DGEBA, properties are, as indicated in Fig. 12-16, again sharply dependent on PA concentration.

PA, because of the processing temperatures required, has found little use in laminating applications, and is more suitable for castings, where it is normally employed with fairly high filler loading volumes to overcome brittleness.

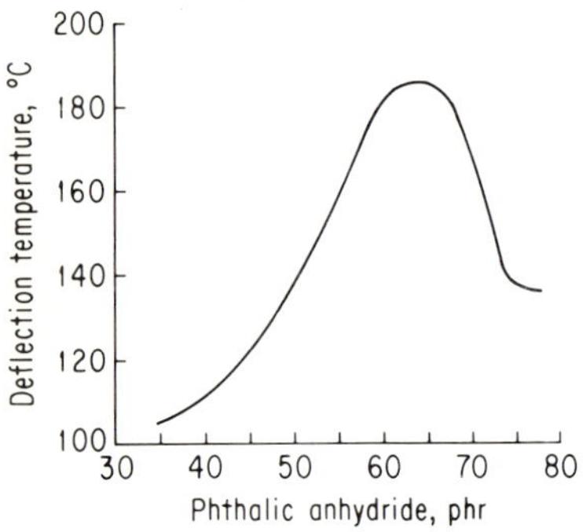

FIG. 12-14. Effect of PA concentration on DT of DGE diphenylsulfone ($n = 0.053$ to 0.450) [79]. Cured 4 hours at 150°C.

Trimellitic Anhydride

Trimellitic anhydride (TMA) is a white crystalline powder with a molecular weight of 192 and a melting point of 168°C.

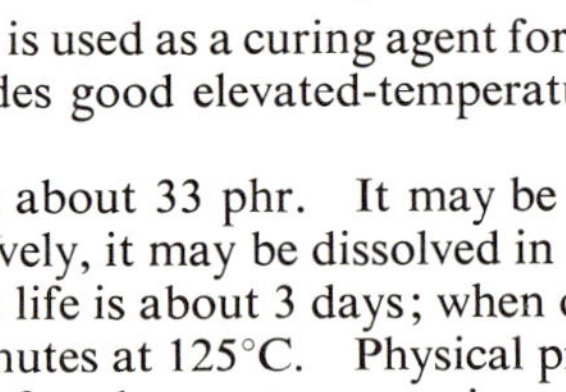

Trimellitic anhydride

TMA is used as a curing agent for DGEBA resins and for epoxidized novolacs. It provides good elevated-temperature properties and high DTs.

When used with DGEBA, it is employed at about 33 phr. It may be dispersed into the resin by thorough stirring; or alternatively, it may be dissolved in the resins at 125°C. With the dispersed system, the pot life is about 3 days; when dissolved, TMA provides a pot life on the order of 10 minutes at 125°C. Physical properties, chemical resistance, and electrical properties of such a system are given in Tables 12-13 and 12-14.

TMA may be dissolved in acetone and added to the epoxy resin in the solvent. Such solution systems have been evaluated in laminating applications.

When accelerators are practical with this curing agent, aromatic amines have been found to provide for satisfactory performance [66].

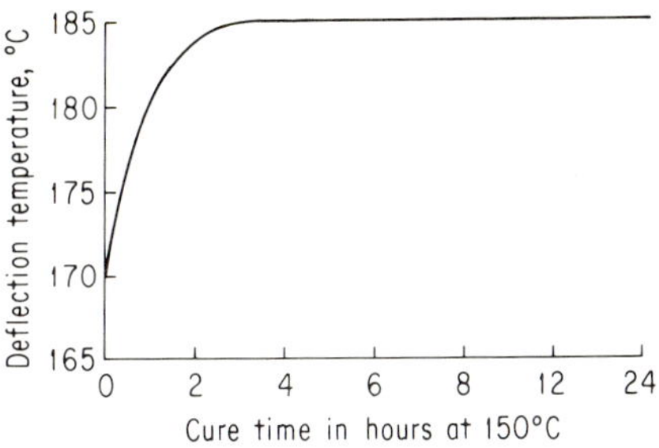

FIG. 12-15. Effect of cure time on DT of DGE diphenylsulfone ($n = 0.053$ to 0.450) cured with 64 phr PA [79].

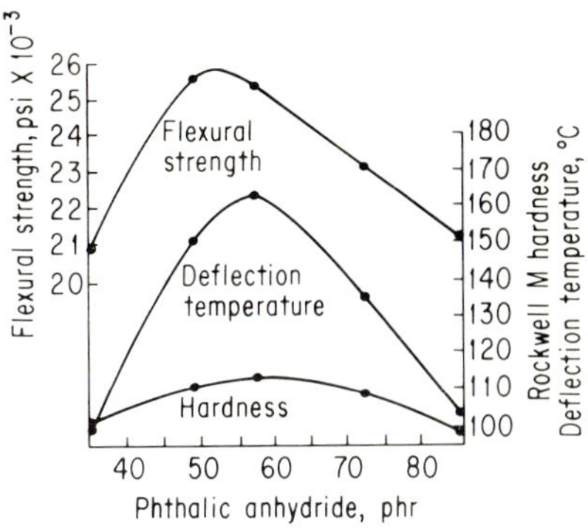

FIG. 12-16. Flexural strength, DT, and hardness vs. curing-agent concentration of DGE tetrachlorobisphenol A cured with PA and 0.1 per cent DMP-30 [27].

Trimellitic anhydride may be reacted with a diol to provide a dianhydride curing agent, such as shown in Table 12-8 [104,111]. Tetraanhydride curing agents may be produced from TMA and pentaerythritol [88]. TMA may be reacted with acetic anhydride to produce an internal anhydride group which will open during cure [74]. The carboxyl group of TMA may react with an ester or an amine, e.g., acetanilide, to produce curing agents [87].

Pyromellitic Dianhydride

Pyromellitic dianhydride (PMDA) (Table 12-8) is insoluble in DGEBA at room temperature but quite soluble at elevated temperatures.

Several systems have been developed to permit the use of PMDA curing agent for DGEBA. The first involves dissolving the PMDA into the resin at elevated temperatures, using a monoanhydride to reduce reactivity. The second involves dissolving the PMDA in acetone and using PMDA-resin solution. The third involves reacting the PMDA with a glycol to produce a material soluble in the more powerful solvents, also for use in solutions. The fourth involves suspending the PMDA in the resin at room temperature, subsequent cure temperatures being sufficient to promote substantially complete reaction.

PMDA in Blends with Monoanhydrides. As few as 5 phr of a dianhydride in a monoanhydride system has been found to increase heat resistance substantially

Table 12-13. Physical Properties of DGEBA Cured 6 Hours at 250°C with 33 phr TMA [2]

Property	*Value*
DT, °C	189
Tensile strength, psi	3020
Modulus, psi $\times 10^{-6}$	0.371
Elongation, %	0.9
Flexural strength, psi	3160
Modulus, psi $\times 10^{-6}$	0.341
Impact strength, Izod	0.24
Hardness, Rockwell, M	116
Density, g/ml	1.2404

Table 12-14. Chemical Resistance and Electrical Properties of DGEBA Cured 6 Hours at 250°C with 33 phr TMA [2]

Chemical resistance	Dimensional increase, %			
	Weight	Length	Width	Thickness
Distilled water	0.7349	0.0166	0.0498	0.4173
NaOH, 10%	0.3742	0.0166	0.1498	0.2038
Acetic acid, 10%	0.5651	0.0332	0.0998	0.0000
Sulfuric acid, 10% ...	0.4222	0.0665	0.0000	0.2052

Electrical properties	At 23°C	At 100°C	At 175°C
Dielectric strength, volts/mil	396		
Arc resistance, seconds..............	124		
Volume resistivity, ohm-cm $\times 10^{-13}$...	5.96	5.96	
Dielectric constant:			
At 60 cps	4.03	4.06	4.06
At 10^3 cps......................	4.00	4.03	4.01
At 10^6 cps......................	3.52	3.82	3.78
Dissipation factor:			
At 60 cps	0.0069	0.0037	0.0061
At 10^3 cps......................	0.0072	0.0054	0.0058
At 10^6 cps......................	0.0232	0.0185	0.0142

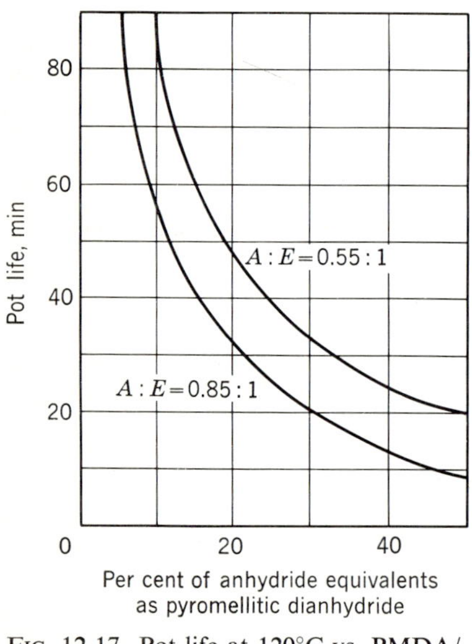

FIG. 12-17. Pot life at 120°C vs. PMDA/PA ratio [17].

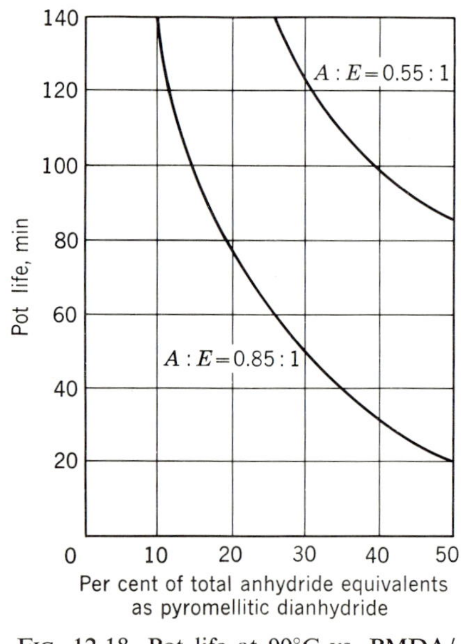

FIG. 12-18. Pot life at 90°C vs. PMDA/PA ratio [17].

[35]. With this aim, PMDA was introduced to the technology in blends with monoanhydrides such as MA [56].

When PMDA is used in blends with a monoanhydride, the total anhydride is maintained at an A/E of about 0.85/1 with the unaccelerated DGEBA, by allowing for the differences in the equivalent weight of the anhydrides involved. The higher the concentration of the PMDA in the curing agent, the more difficult the mixing and the shorter the pot life. With DGEBA, the anhydride mixture is added separately or as a solid mix to the resin which is preheated at 70°C. Heating is continued to 120°C with agitation. Once the anhydrides are completely dissolved, temperature may be returned to 90°C to prolong the pot life. By this procedure, anhydride blends containing up to 65 per cent PMDA may be employed, but at higher ratios gelation is too rapid at mixing temperatures [17]. Pot-life data on PMDA/PA and PMDA/MA blends are presented in Figs. 12-17 to 12-19.

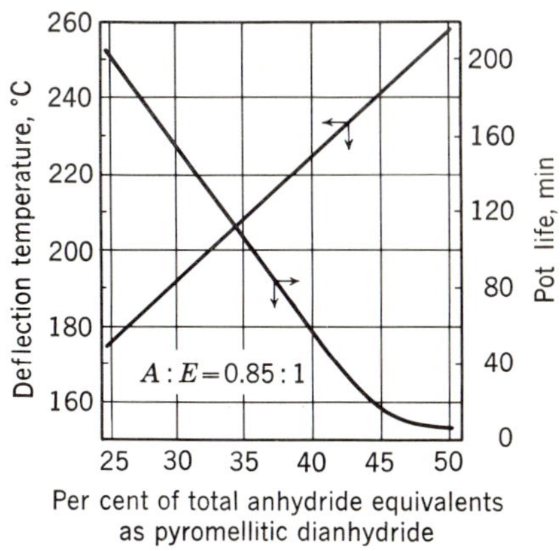

FIG. 12-19. Pot life at 90°C and DT vs. PMDA/MA ratio [4].

With the PMDA/monoanhydride blends, the DT is a function of the cure cycle (Figs. 12-20 and 12-21) and the modifying anhydride. In terms of heat resistance, MA is superior to PA as a modifying anhydride. The superiority is evidenced when considering weight loss at 200°C over a period of 500 hours. A PMDA/PA system at 20/28 phr gives a value of 23.7 per cent; a PMDA/MA system at 21/29 phr gives a value of 2.3 per cent [4]. Other properties of the two cured systems are compared in Tables 12-15 and 12-16.

PMDA in Solvents. PMDA may be dissolved in resin-acetone solutions at reflux temperatures, with the mixture providing for stability for about 1 week. Such solvents as dimethylformamide, dimethylsulfoxide, and dimethylacetamide, may also be employed to obtain PMDA solutions. The solution systems are principally useful in laminating formulations, and details on the properties of PMDA laminates are presented in Chap. 22.

PMDA-Glycol Adducts. PMDA may be reacted with glycols to produce adducts of the following general structure [75]:

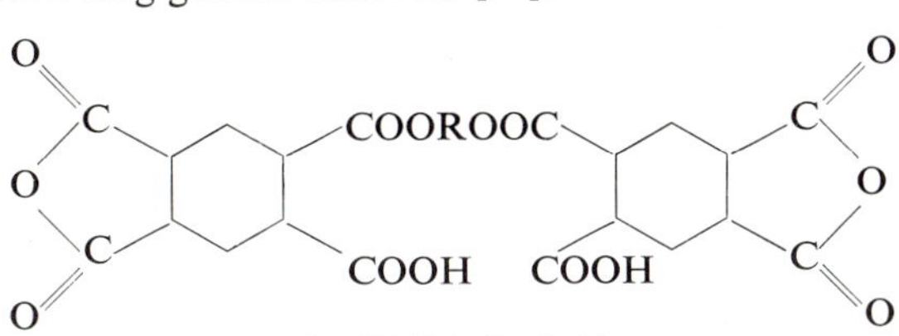

Idealized PMDA-glycol adduct

The adducts are formed in the presence of solvent, under dry nitrogen, the reaction being continued until a clear solution is obtained [4]. Table 12-17 indicates the solubility characteristics of a number of such adducts.

The adducts of PMDA are employed in coatings and high-temperature adhesive formulations.

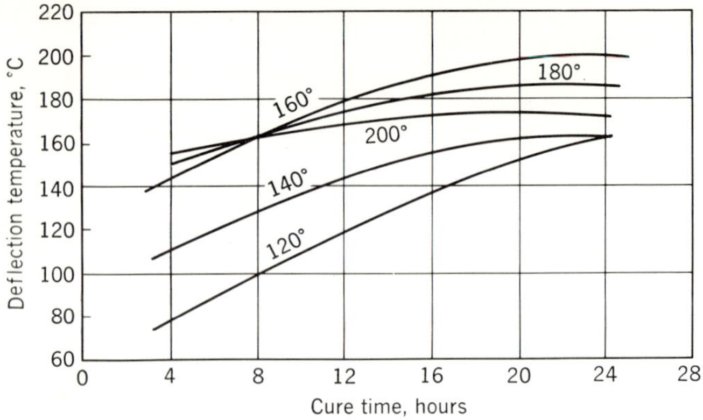

FIG. 12-20. Cure time and temperature vs. DT of DGEBA cured with PMDA/PA [4].

PMDA Dispersions. Finely powdered PMDA may be incorporated into liquid epoxy resins at room or only slightly elevated temperature by stirring. No noticeable settling will occur in resins having initial viscosities over 5,000 centipoises, even after centrifuging the dispersion at 2,000 rpm for 30 minutes at 38°C [4].

Because of its functionality, PMDA can be used as a curing agent for monoepoxies, such systems providing DTs on the order of 150°C [46]. The use of PMDA with *N*-glycidylphthalimide produces castings with DTs of about 280°C [68].

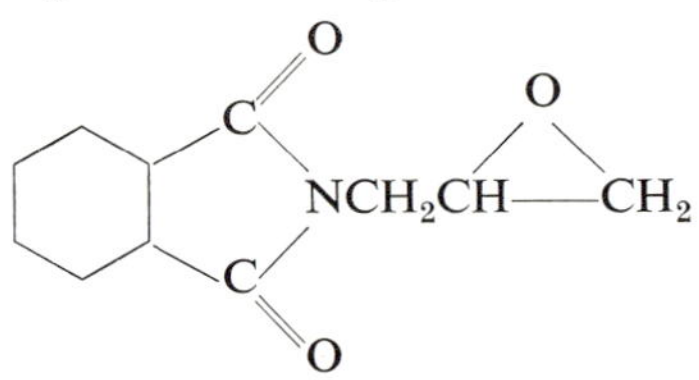

N-Glycidylphthalimide

When dispersed in DGEBA, PMDA is normally employed at an A/E of 0.4–0.5/1, the system being considered as acid-catalyzed [77]. Cure times are relatively long, as indicated in Fig. 12-22. Cure may be accelerated by the addition of glycols or

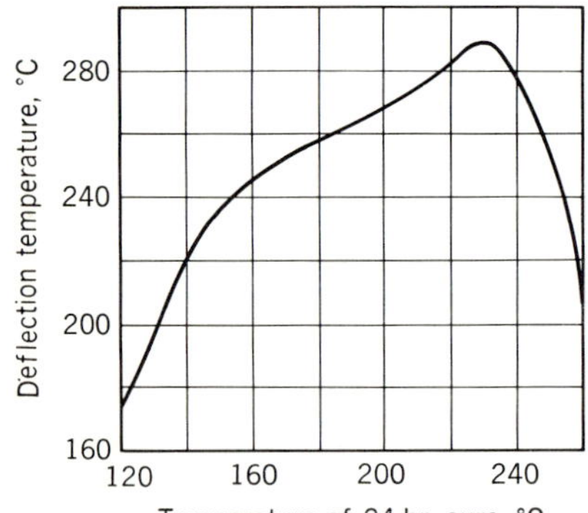

FIG. 12-21. Cure temperature vs. DT of DGEBA cured with PMDA/MA [4].

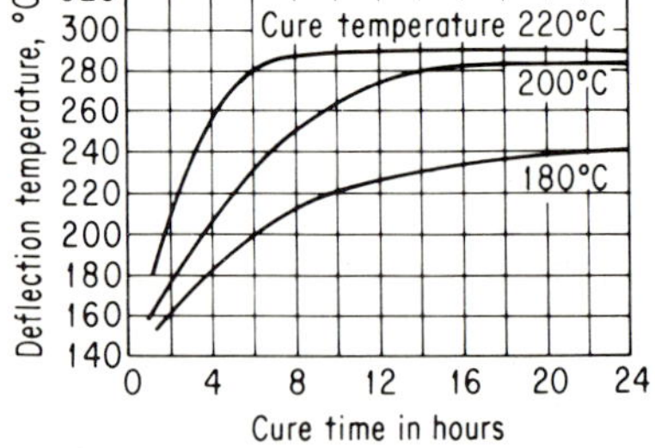

FIG. 12-22. Cure temperature vs. DT of DGEBA cured with PMDA [4].

Table 12-15. Physical and Electrical Properties of DGEBA Cured with PMDA/MA and PMDA/PA Blends [4]

Formulation and properties	MA		PA	
	A	*B*	*C*	*D*
Formulation				
A/E	0.85/1	0.85/1	0.85/1	0.55/1
Per cent total anhydride as PMDA	50	40	50	50
PMDA, phr	21	17	20	14
Second anhydride, phr	19	23	28	20
Pot life at 90°C, minutes	10	56	20	90
Physical properties				
DT, °C	290	255	207	150
Tensile strength, psi	3,830	3,670	4,500	6,000
Modulus, psi $\times 10^{-6}$	0.42	0.40	0.54	0.54
Flexural strength, psi:	11,100	10,940	13,000	13,000
At 150°C	4,600	5,100	3,660	
At 200°C	3,900	3,800		
Modulus, psi $\times 10^{-6}$	0.40	0.42	0.53	0.56
At 150°C	0.24	0.18	0.13	
At 200°C	0.18	0.13		
Compressive strength, psi	41,600	46,000	43,000	
Impact strength, Izod	0.30	0.34	0.30	0.25
Hardness, Rockwell	M109	M109	E84	M119
Coefficient of thermal expansion, in./(in.)(°C)	5.2×10^{-5}	5.37×10^{-5}	4.5×10^{-5}	2.46×10^{-5}
Weight loss at 230°C %:				
In 7 days	1.7			
In 15 days	2.4			
In 22 days	3.4			
Electrical properties				
Power factor:				
At 60 cps	0.0129	0.00706	0.0078	
At 10^3 cps	0.0144	0.01030	0.0077	0.0025
At 100°C	0.0057			
At 150°C	0.0052			
At 200°C	0.0057			
At 10^6 cps	0.0295	0.02690	0.0325	
Dielectric constant:				
At 60 cps	4.06	3.73	3.75	
At 10^3 cps	3.98	3.67	3.70	4.12
At 100°C	3.83	3.90		
At 150°C	3.86	3.86		
At 200°C	3.88	3.88		
At 10^6 cps	3.56	3.34	3.39	
Dielectric strength, volts/mil	324	349	398	400
Volume resistivity, ohm-cm	1.59×10^{15}	8.10×10^{15}	1.27×10^{16}	1.8×10^{14}
Surface resistivity, ohm	5.66×10^{15}	5.66×10^{15}	2.51×10^{15}	2×10^{12}

acidic accelerators, but usually at the expense of elevated-temperature performance. Physical properties of the unaccelerated systems are given in Table 12-18 and electrical properties in Table 12-19.

CHLORINATED AND BROMINATED ANHYDRIDES

A number of chlorinated and brominated anhydrides have been suggested for use in the epoxy technology, among them being tetrabromo- and tetrachlorophthalic, monochlorophthalic, dichloromaleic, dichlorocitraconic, and chlorendic anhydride.

Chlorinated anhydrides, used to impart flame resistance to the cured epoxy resins, are more reactive than the nonchlorinated species, owing to being more acidic

Table 12-16. Chemical Resistance of DGEBA Cured with PMDA/PA and PMDA/MA Blends [4]

System tested in	Weight gain (or loss), %			
	PMDA/PA			PMDA/MA
	After 7 days' immersion	After 30 days' immersion	After 90 days' immersion	After 90 days' immersion
Sulfuric acid:				
95%	......	......	−51.5	−37.8
3%	0.18	0.35	0.53	1.42
70%	−0.06	−0.09	−0.11	−0.05
Hydrochloric acid:				
35%	0.12	0.20	0.38	0.49
10%	0.12	0.12	0.47	1.24
Nitric acid:				
50%	0.20	0.15	1.23	2.20
30%	0.10	0.24	0.40	1.10
Water (distilled)	0.16	0.33	0.40	1.51
Sodium hydroxide				
1%	0.18	0.37	0.56	1.45
50%	0.04	−0.049	−0.04	0.01
Sodium sulfate, 30%	0.18	0.34	0.47	1.35
Sodium chloride, 20%	−0.06	0.08	0.21	1.19
Phosphoric acid:				
88%	0.07	0.09	−0.09	0.08
10%	0.18	0.38	0.54	1.68
Acetic acid:				
Glacial	−0.01	0.06	0.21	0.32
10%	0.20	0.70	5.44	1.45
Oxalic acid, sat.	0.17	0.37	0.56	1.44
Acetone	0.77	1.82	4.04	0.98
Ethyl acetate	0.16	0.54	1.54	
Ethylene glycol	−0.045	−0.07	−0.05	−0.04
Ethylene dichloride	0.70	2.51	1.73	0.61
Butyl Cellosolve	0	0.08	0.20	0.47
Hydrogen peroxide, 30%	0.10	0.34	0.61	1.67

Table 12-17. Solubility of PMDA Adducts [4]

Glycol	g/100 g of PMDA	Reactants, wt. %	MEK, reflux		MIBK/MEK (9/1), reflux		Cellosolve acetate, 120° C	
			Time, hr	Result	Time, hr	Result	Time, hr	Result
Ethylene	15.5	28	...	...	...	...	3	Sol.
Butanediol-1,4	22.5	29	...	...	3	Ppt.	4	Ppt.
2-Butenediol-1,4	22.0	29	...	...	...	...	2	Sol.
Pentanediol-1,5	26.0	30	...	...	...	...	3	Sol.
Neopentyl	26.0	30	...	...	9	Ppt.	2	Sol.
Diethylene	26.5	30	...	...	8	Ppt.	2	Sol.
Triethylene	37.5	32	...	...	9	Ppt.	2	Sol.
Polyethylene (mol. wt. 300)	75.0	36	2	Sol.	2	Sol.		
Dipropylene	33.5	31	...	...	8	Ppt.	4	Sol.
Tripropylene	48.0	33	3	Sol.	8	Ppt.		

Table 12-18. Physical Properties of DGEBA Cured with PMDA [4]

Property	*Value*
DT, °C	
Initial	280–290
After 200 hr:	
At 205°C	270–280
At 260°C	230–240
Hardness, Rockwell M	110
Shrinkage during cure, linear, %	2.0
Coefficient of thermal expansion, in./(in.)(°C)	5.0×10^{-5}
Impact strength, Izod	0.1
Taber abrasion, weight loss, mg (1,000-g load, 1,000 cycles, CS-17 wheel)	11.6
Weight loss, %, after 200 hr:	
At 205°C	0.3
At 230°C	5.4
At 260°C	19.0
Tensile strength, psi:	
At 23°C	3,200
At 150°C	2,000
At 205°C	800
At 260°C	400
Modulus, psi $\times 10^{-6}$	0.39
Flexural strength, psi:	
At 23°C	8,600
At 150°C	5,000
At 205°C	3,000
At 260°C	2,000
Modulus, psi $\times 10^{-6}$:	
At 23°C	0.38
At 150°C	0.20
At 205°C	0.15
At 260°C	0.12
Compressive strength, psi	37,000
Modulus, psi $\times 10^{-6}$	0.35

because of the electron-withdrawing or inductive effect of halogen atoms [49]. Of the chlorinated anhydrides, chlorendic anhydride is the one most widely used, although dichloromaleic will also be discussed as representative of the others.

Dichloromaleic Anhydride

Dichloromaleic anhydride is a white solid having a molecular weight of 167. It may be incorporated into liquid epoxy resins either by dissolving it at elevated

Table 12-19. Electrical Properties of DGEBA Cured with PMDA [4]

Temperature, °C	Dielectric constant and dissipation factor									
	At 60 cps		At 10^3 cps		At 10^6 cps		At 10^7 cps		At 8.5×10^9 cps	
	K	tan	*K*	tan	*K*	tan	*K*	tan	*K*	tan
23	3.57	0.0067	3.52	0.0079	3.34	0.0223	3.26	0.0208	2.89	0.0130
93	3.70	0.0050	3.65	0.0044	3.52	0.0189	3.48	0.0260	2.92	0.0193
150	3.73	0.0084	3.70	0.0043	3.61	0.0118	3.58	0.0230	2.95	0.0265
205	3.76	0.0062	3.71	0.0053	3.64	0.0078	3.59	0.0159	2.98	0.0343
260	3.84	0.0131	3.78	0.0086	3.69	0.0071	3.60	0.0109	2.98	0.0464
300	4.14	0.0800	3.98	0.0200	3.79	0.0145	3.60	0.0115	2.98	0.0570

Surface resistivity, ohms			
23°C $> 1 \times 10^{13}$	100°C $> 1 \times 10^{13}$	150°C $> 1 \times 10^{13}$	200°C, 3×10^{12}

temperatures or by dispersing it into the resin at room temperature. If it is dissolved in DGEBA at 70 to 80°C, the pot life is quite short, but with room-temperature mixing procedures, involving grinding the anhydride into the resin with a ball-mill or heavy-duty agitator, pot lives on the order of 8 hours may be obtained.

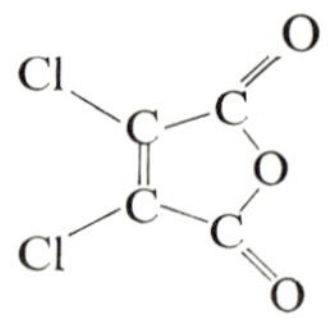

Dichloromaleic anhydride

For DGEBA, cure temperatures, even without accelerators, are reasonably low and short with 4 hours at 85°C being adequate. Longer and higher-temperature cures, however, are necessary to obtain the best DTs, with values of 100°C being obtained after a one-hour cure at 120°C [8]. Flame resistance is improved over that of the nonchlorinated anhydrides.

Chlorendic Anhydride

Chlorendic anhydride is prepared by Diels-Alder addition of maleic anhydride

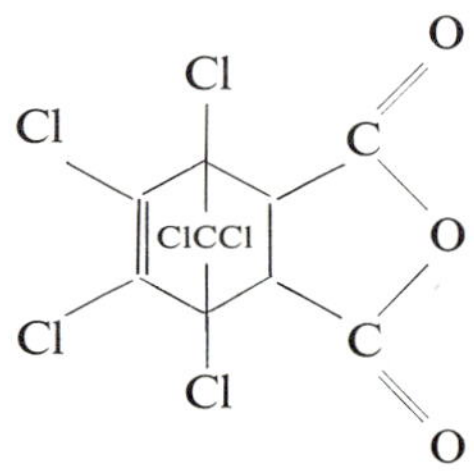

Chlorendic anhydride

and hexachlorocyclopentadiene [15]. It is a white powder having a molecular weight of 370 and a melting point of 239°C.

When used as a curing agent for the epoxies, it imparts flame resistance, but has limited thermal stability. Chlorendic anhydride breaks down with temperature as follows [30]:

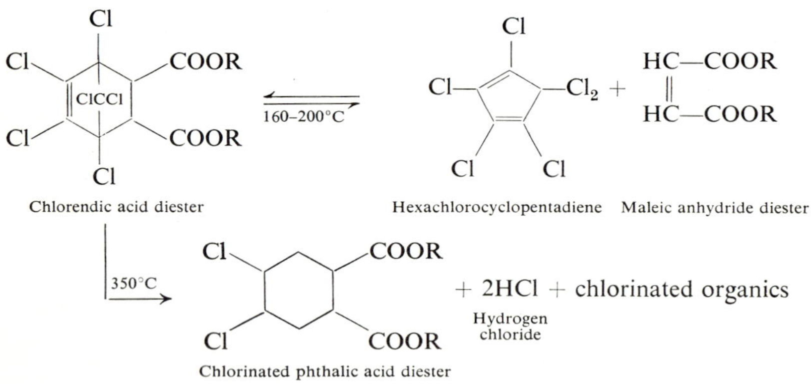

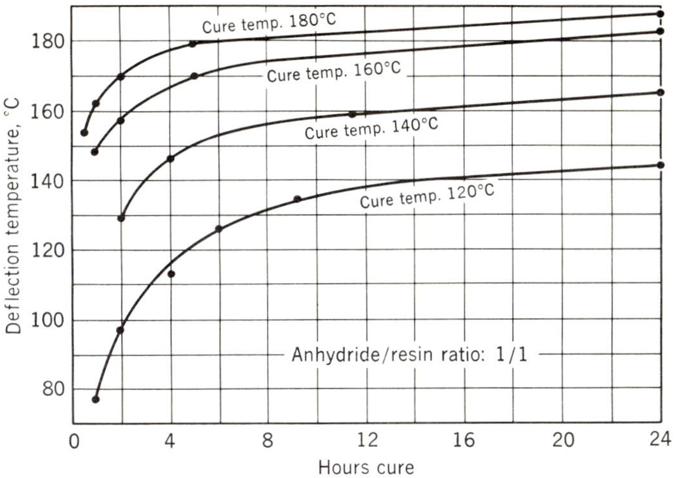

FIG. 12-23. Cure time and temperature vs. DT of DGEBA cured with CA [6].

At processing temperatures (120°C) the pot life is quite short, usually under 30 minutes even with glycidyl ether resins, and for this reason, it is normally employed in blends with other anhydrides, as discussed under the following section. Because of the increased reactivity, chlorendic anhydride can be considered to contain a built-in acid accelerator. With DGEBA, the A/E ratio is between 0.5 and 0.6/1, (i.e., 110 phr). The effect of cure time and temperature on the DT of the cured system is shown in Fig. 12-23. Figure 12-24 presents the effect of anhydride concentration on the DT and Fig. 12-25 on the weight loss. Physical properties of the cured system are given in Table 12-20, chemical resistance in Table 12-11. Volume resistivity vs. temperature are shown in Fig. 12-5.

Chlorendic anhydride, in common with other anhydrides, tends to pick up moisture upon exposure to the atmosphere. The usual effect of moisture pickup is to decrease the solubility of the anhydride and to accelerate the cure rate. The

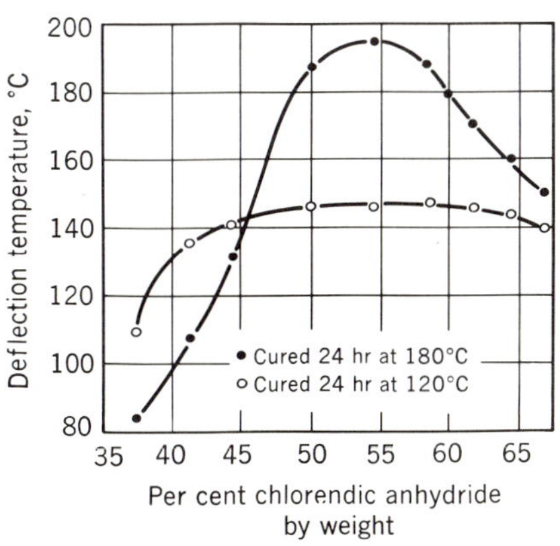

FIG. 12-24. Curing-agent concentration vs. DT of DGEBA cured with CA [6].

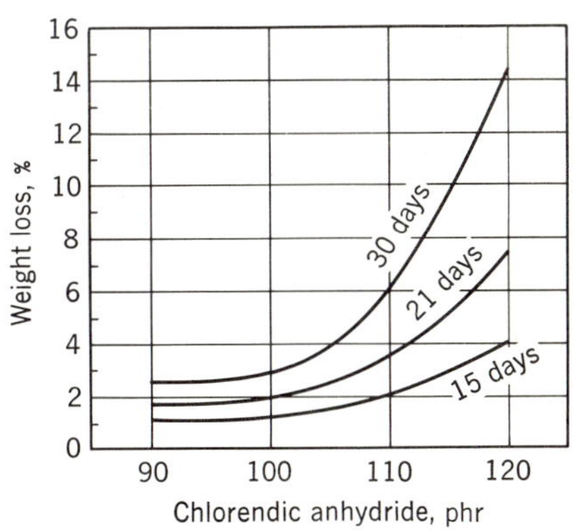

FIG. 12-25. Curing-agent concentration vs. weight loss of DGEBA cured with CA [6].

Table 12-20. Physical Properties of DGEBA Cured with Chlorendic Anhydride and Blends

Property	Cured with		
	CA	NMA/CA	HHPA/CA
Tensile strength, psi	12,000	9,785	12,620
Modulus, psi × 10^{-6}	0.46	0.39	0.50
Flexural strength, psi	17,000	15,417	15,780
Modulus, psi × 10^{-6}	0.52	0.414	0.50
Compressive strength, psi	20,500	18,147	19,300

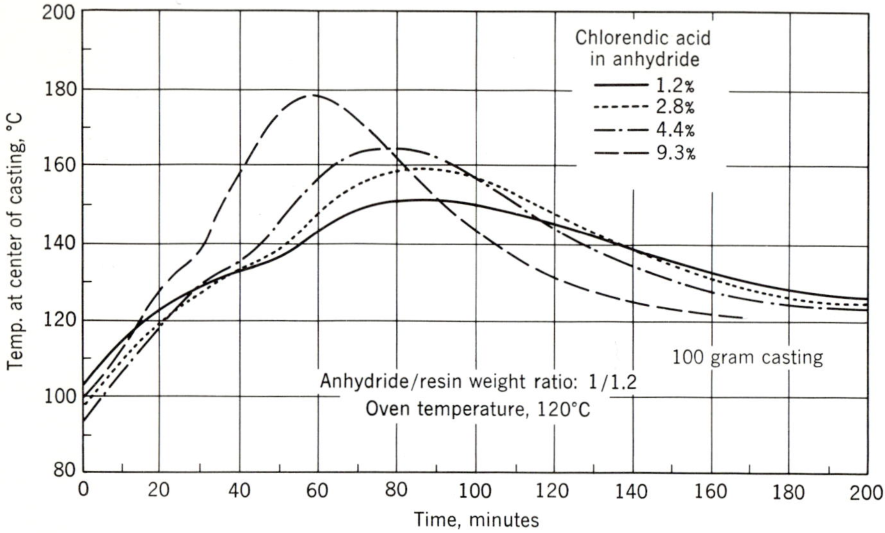

FIG. 12-26. Chlorendic acid concentration vs. exotherm of DGEBA cured with CA [6].

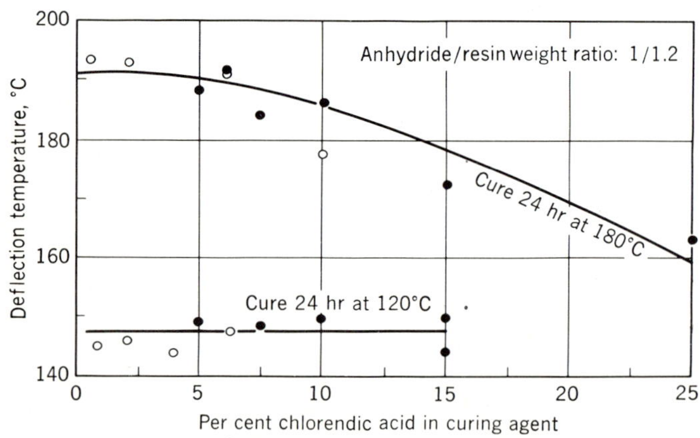

FIG. 12-27. Chlorendic acid concentration vs. DT of DGEBA cured with CA [6].

effect of chlorendic acid on the exotherm of the chlorendic anhydride–cured system is shown in Fig. 12-26. The effect of the acid on the DT of the system is shown in Fig. 12-27.

Chlorendic anhydride is particularly useful in applications where flame resistance is of significance. DGEBA cured with chlorendic anhydride has a burning rate of 0.15 in. per minute, a time to ignite of 65 seconds, and a time to flame out of 115 seconds [14]. Chlorendic anhydride is suitable for use with the relatively sluggish epoxidized vegetable oils [67].

LIQUID EUTECTICS AND ADDUCTS

As in the case of the aromatic amines, it is possible to develop liquid eutectics and adducts with anhydrides.

The adducts are obtained through the reaction of the anhydride, in excess, with an epoxy resin: for example the reaction of 80 parts HHPA with 20 parts of DGEBA. In this case, the excess HHPA remains in a eutectic blend with the adduct, but may crystallize out on standing. Liquid adducts of the resin with a blend of HHPA, PA, and tetrahydrophthalic anhydride; with a blend of NMA and HHPA; and with a blend of MA and PMDA have also been described [29]. In addition, acetone-soluble reaction products of DGEBA and substantially less than stoichiometric amounts of anhydride have been described as suitable for use in coating formulations, final cure being obtained with primary amines [71].

More commonly, however, a eutectic blend of anhydrides is employed. A variety of these have been reported, for example, eutectics of azelaic anhydride or PMDA with phthalic, succinic, and chlorendic anhydride [57] and 78 parts HHPA with 13 parts THPA [101], or HHPA (50), THPA (25) and PA (25). Representative of liquid eutectic blends, with the temperature of initial crystallization, are those given in Table 12-21.

The most widely used are blends with chlorendic anhydride or NMA.

Table 12-21. Typical Liquid Blends of Anhydrides

Ratio of anhydrides by weight	*Temperature of initial crystallization,* °C
50 NMA/50 HHPA	−40
80 NMA/20 HHPA	−30
20 NMA/80 HHPA	15
20 NMA/80 DDSA	−10
30 NMA/70 DDSA	−30
60 NMA/40 DDSA	−30
65 NMA/35 THPA	−20
80 NMA/20 PA	5
15 PA/85 HHPA	23
80 DDSA/20 THPA	20
90 DDSA/10 THPA	−20
50 DDSA/50 HHPA	10
70 DDSA/30 HHPA	−20
40 CA/60 HHPA	10

Eutectic Blends with Chlorendic Anhydride

Chlorendic anhydride may be blended with a second anhydride by melting the two compounds together at elevated temperature; upon cooling to room temperature, a liquid or a low-melting point solid will result depending upon the selection of the anhydrides. Several such mixtures, with their solidification temperatures,

Table 12-22. Solidification Temperatures of Typical Eutectic Blends with CA [45] (In degrees centigrade)

Coanhydride	CA/coanhydride ratio			
	80/20	60/40	40/60	20/80
HHPA	114–115	74–75	15	15
DDSA	104–105	66–67	10	10
NMA	115–116	89–90	15	10
PA	....	108		
MA	....	86		

are indicated in Table 12-22. A melting-point diagram of the chlorendic/HHPA blend is given in Fig. 12-28.

As the amount of chlorendic anhydride is reduced, the burning time is increased and the DT is lowered. For example, if a blend of 79/21 chlorendic anhydride/HHPA is used to cure DGEBA, the DT will be about 185°C with a burning time of about 1 second; whereas a blend of 58/42 will produce a DT of about 145°C and a burning time of 8 seconds.

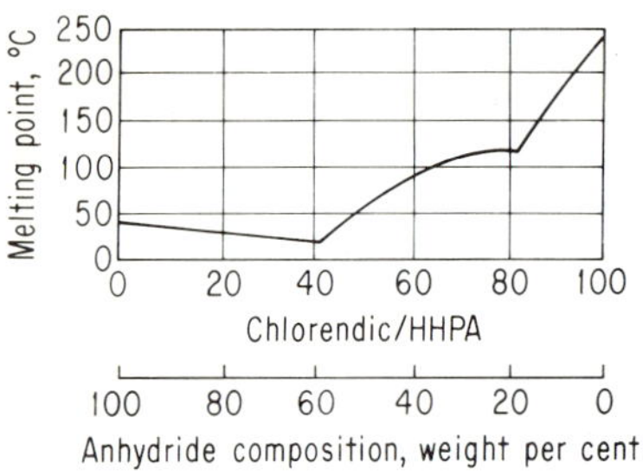

FIG. 12-28. Melting-point diagram of CA/HHPA blends [45].

As with anhydride curing agents generally, the DT is dependent upon catalyst concentration and cure time. This is indicated in Figs. 12-29 and 12-30 for DGEBA (mol. wt. 420) cured with a 60/40 blend of chlorendic and HHPA at an A/E ratio of 1/1, using α-methyl BDMA as catalyst. Typical physical properties of this system are given in Table 12-20. The effect of catalyst type on optimum concentration of curing agent is indicated in Fig. 12-31. The effect on DT of DDSA, HHPA, and NMA in increasing concentrations with chlorendic anhydride as curing agent for DGEBA (mol. wt. 420) at an A/E of 1/1 is shown in Fig. 12-32.

The eutectic blends of chlorendic anhydride may be processed at lower temperatures than chlorendic anhydride, with the consequent extension of the pot life.

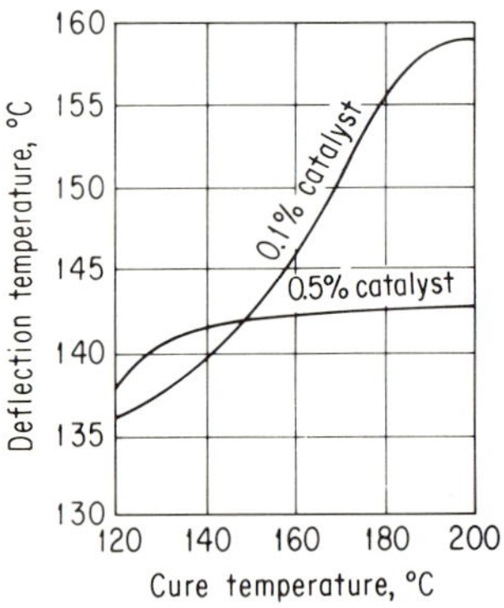

FIG. 12-29. Catalyst concentration vs. DT of DGEBA (mol. wt. 420) cured 20 hours with CA/HHPA blend [45].

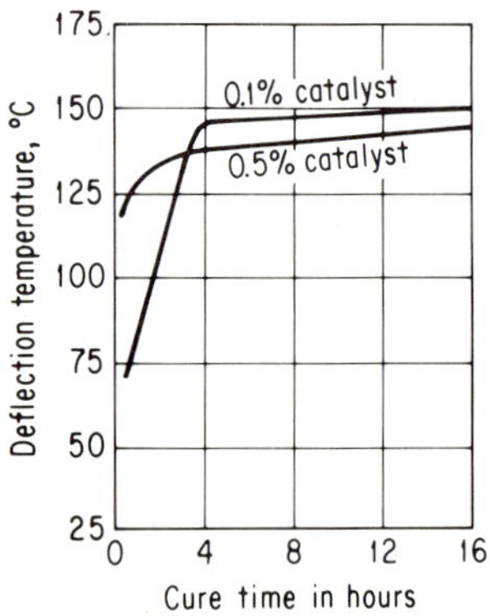

FIG. 12-30. Cure time vs. DT of DGEBA (mol. wt. 420) cured at 180°C with HHPA/CA blend [45].

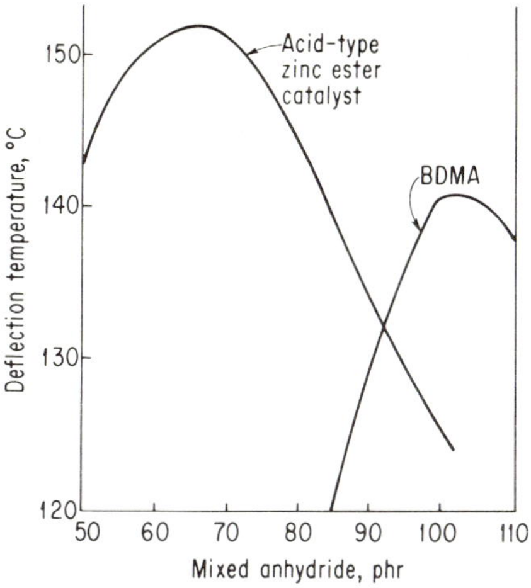

FIG. 12-31. Catalyst type and concentration vs. DT for DGEBA cured with HHPA/CA (21/32) [3].

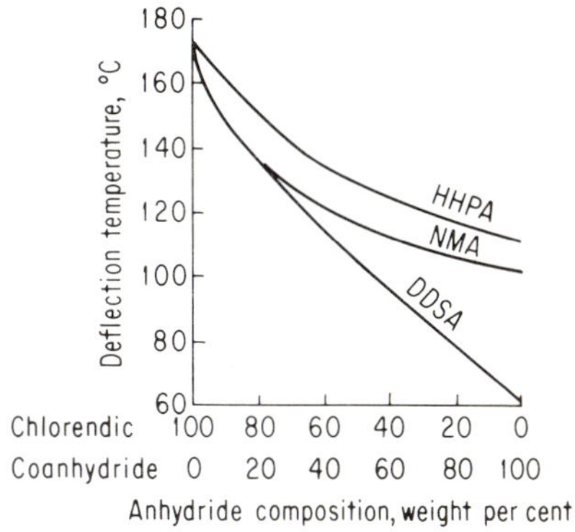

FIG. 12-32. Effect of coanhydride on DT of DGEBA (mol. wt. 420) cured with CA (A/E = 1/1) and 0.5 per cent α-methyl BDMA [45].

Liquid Blends with NMA

NMA, itself a liquid, may be blended with other anhydrides to produce suitable epoxy-resin curing agents.

A liquid blend of 60 parts NMA and 40 parts chlorendic anhydride, for example, will produce a mixture having a viscosity on the order of 3,000 centipoises. Used with DGEBA, it will provide a workable pot life on the order of 4 hours at room temperature with α-methyl BDMA as a catalyst and will gel in about 30 minutes at 100°C. DTs will be 140°C after 24 hours' cure at 150°C. Physical properties are given in Table 12-20.

It has been reported that chlorendic/NMA blends when used with an accelerator to cure diglycidyl ether resins are less exothermic than either is singly in the same system [64].

A liquid blend of HHPA and NMA has been used to cure an epoxylated novolac to provide a thermally stable casting resin. The compound, when cured 16 hours at 125°C and 5 hours at 150°C was then aged varying periods of time and the DT recorded. Data are given in Fig. 12-33.

In general, liquid blends with NMA are somewhat special-purpose curing agents, the NMA being used to upgrade the temperature resistance or to provide liquid blends which cannot otherwise be obtained.

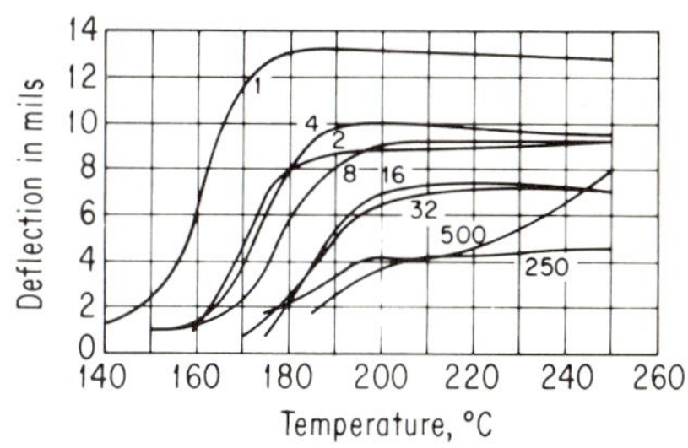

FIG. 12-33. Aging at 232°C vs. DT of novolac resin cured with NMA/HHPA blend [60]. Numbers on curves are the hours at 232°C before testing.

Miscellaneous Blends

In addition to obtaining liquid products, anhydrides may be blended to produce desired physical properties. For example, a blend of PA and HHPA will provide

Table 12-23. Tensile Properties of Systems Cured With Selected Anhydride Blends [116]

Anhydride blends, composition	Anhydrides, mole ratio			
	1:2:1 (lowest cost)	2:1:1 (most flexible)	1:1:2 (most heat-resistant)	1:1:2* (most heat-resistant)
	Tensile strength, psi			
PAPA/MA/CA	11,000	8,500	5,300†	6,300
PAPA/PA/NMA	12,600	Not tested	12,300	610
DDSA/PA/NMA	8,600	8,500	10,700	720
PAPA/PA/CA	12,600	10,600	7,700	7,000
DDSA/PA/CA	11,800	Not tested	Not tested	
	Tensile elongation, %			
PAPA/MA/CA	5.2	3.2	1.5†	6.1
PAPA/PA/NMA	5.4	3.4	5.1	30
DDSA/PA/NMA	2.1	Not tested	3.3	34
PAPA/PA/CA	7.1	6.4	2.2	11.8
DDSA/PA/CA	4.3	Not tested	Not tested	

* Values obtained at 100°C.
† Low value due to brittleness of system.

lower exotherm in DGEBA than will straight PA [113]. Tensile properties of typical blends are given in Table 12-23.

SUMMARY

Anhydride curing agents for the epoxy resins have gained in popularity since their widespread introduction into the United States technology in the middle of 1956. The first materials investigated were the solid anhydrides, but these have been progressively displaced by the liquid or low-melting-point solids, which greatly facilitate processing.

The reaction of the anhydride is complex; several reactions occur simultaneously. For this reason, the calculation of the proper anhydride-to-epoxy ratio is difficult. However, as a general rule, when the anhydride is employed without accelerator, the ratio is a function of the reactivity of the anhydride and the epoxy and the concentration of hydroxyls present. More active anhydrides, and more acid-reactive epoxy groups tend to shift the ratio to an A/E of less than 1. The greater the concentration of hydroxyls, other things equal, the more nearly the ratio approaches about 0.4/1. When a tertiary amine catalyst is present, the reaction mechanisms are substantially different, and an A/E of more nearly 1 is favored.

Anhydrides generally provide improved thermal stability over the amines. The majority of the anhydrides, however, with the exception of those containing flexibilizing chains, produce somewhat brittle castings and are usually modified with a flexibilizer for casting applications. Addition of the flexibilizer often has considerable effect on reducing the heat and the chemical resistance of the cured product.

Anhydride blends and adducts have been developed which produce liquid systems from solid anhydrides. Chlorinated anhydrides have been developed to impart flame resistance to cured products.

Nearly all the anhydrides are moisture-sensitive, the anhydrides being hydrolyzed to the corresponding diacids. Acids from the anhydrides generally have reduced solubility in the epoxy resins and are less desirable as curing agents.

Anhydrides do not possess the skin-sensitizing potential of the amine curing agents, although if elevated-temperature processing is required, some systems release irritating fumes. The organic acids from the anhydrides are capable of producing skin burns upon prolonged contact.

REFERENCES

1. Amchem Products, Inc., technical literature.
2. Amoco Chemical Corp., technical literature.
3. Argus Chemical Corp., technical literature.
4. E. I. duPont de Nemours & Co., Inc., technical literature.
5. Heyden Newport Chemical Corp., technical literature.
6. Hooker Electro-Chemical Co., technical literature.
7. Koppers Co., Inc., technical literature.
8. Westvaco, Chloro–Alkali Division, FMC Corp., technical literature.

1905

9. Weinschank, Condensation of Epichlorohydrin with Phthalic Anhydride under Influence of Tertiary Bases, *Chem. Ztg.*, p. 1311.

1943

10. Castan, Process of Preparing Synthetic Resins, U.S. 2,324,483.

1954

11. Meyerhans, Chemical Resistance of Ethoxyline Resins, *Kunststoffe*, vol. 44, no. 4.

1955

12. Farneth and Gallousis, Versatility Cast into New CT's, *Allis-Chalmers Elec. Rev.*, 3d quarter.

1956

13. Fisch et al., The Curing Mechanism of Epoxy Resins, British Chemical Industry Symposium, London, April.
14. Robitschek and Nelson, Flame and Heat Resistant Epoxy Resins, *Ind. Eng. Chem.*, October.
15. Rudoff, Chlorinated Maleic Adduct–Ethoxyline Resinous Compositions, U.S. 2,744,845.
16. Shokal, Process for Curing Polyepoxides and Resulting Products, U.S. 2,768,153.

1957

17. Field and Robinson, Pyromellitic Diahydride in Curing of Epoxy Resins, *Ind. Eng. Chem.*, March.
18. Schwenk, Synthesis of Polyesters through Reaction of Epoxies with Dicarboxylic Acid Anhydrides, thesis, Technische Hochschule, Stuttgart.
19. Updegraff, Polymerizable Compositions Containing Polyglycidyl Esters of Polycarboxylic Acids, U.S. 2,781,333.

1958

20. Eirich, Copolymerized and Crosslinked Epoxide Resins, U.S. 2,848,433.
21. Hardenable Compositions Containing Glycidyl Ethers (to N.V. de Bataafsche Petroleum, Maatschappij), Brit. 792,237.
22. Improved Resinous Compositions and Products (to Minnesota Mining & Manufacturing Co.), Brit. 792,995.
23. Pschorr, Epoxy Resin Compositions. U.S. 2,863,853.

24. Shepherd and Dearborn, Epoxy Resins Cured with Mixture of Poly-carboxylic Acid Anhydrides, U.S. 2,863,852.
25. Smith, Adducts of Long Chain Unsaturated Compounds, U.S. 2,858,323.
26. Weiss, Optimum Conditions for Curing Nadic Methyl Anhydride–Epoxy Coating Resins, Upper Midwest RETEC of SPE, Minneapolis, Minn.

1959

27. Cass and Reaville, Performance Characteristics of Epoxy Resins of Tetrachloro Bisphenol A, ACS Symposium, Atlantic City, September.
28. Cressy and Delmonte, Influence of Amine on Cure of Epoxy-Anhydride Systems, ACS Symposium, Atlantic City, September.
29. Delmonte, Epoxy Resin Compositions, U.S. 2,890,204.
30. Haroldson and Watrous, Thermal Stability of Glass Fabric Base Epoxy Resin Laminates, ACS Symposium, Atlantic City, September.
31. Heiberger et al., Epoxypolyolefins: Anhydride Polyol Cured Systems, ACS Symposium, Atlantic City, September.
32. Michelotti, Some Anhydride Curing Agents for Epoxy Resins, *J. Chem. Eng. Data*, January.
33. Phillips et al., Compositions of Polyepoxides and Polycarboxylic Acid Anhydrides, U.S. 2,884,408.
34. Phillips et al., Polyepoxide Compositions, U.S. 2,917,483.
35. Shepherd and Dearborn, Epoxy Resins Cured with Dianhydrides of Tetracarboxylic Acids, U.S. 2,871,221.
36. Snoddon, Low Temperature Curing 1,3-Diene Containing Elastomers and Method of Making Same, U.S. 2,915,494.
37. Weiss, Why Use Statistical Methods, *Dyestuffs*, December, 1958, and January.

1960

38. Aitken et al., Irradiation Studies on Anhydride Cured Epoxy Resins, AERE-R-3339 (Engineering Division, Harwell, Berkshire, England).
39. Bavley and Michelotti, Resin Compositions, U.S. 2,938,014.
40. Belanger, Curing of Polyepoxides, U.S. 2,928,807.
41. Cooke and Masters, Epoxide Compositions Containing Maleic Anhydride–unsaturated Organic Acid Adducts and Method of Forming Infusible Products Therefrom, U.S. 2,947,711.
42. Ehlers, Correlation between Structure and Thermal Stability of Epoxy Resins, *Polymer*, vol. 1, no. 3.
43. Hardenable Products (to Fabwerke Hoeschst A.G.), Brit. 855,871.
44. Heiberger et al., Epoxy Polyolefins: II, Anhydride-Polyol-Peroxide Cure Systems, ACS Symposium, Cleveland, April.
45. Ilardo and Schoepfle, Low Melting HET-Hexahydro Phthalic Anhydride Mixtures as Hardeners for Epoxy Resins, *SPE J.*, August.
46. Improvements in or Relating to Molding and Like Compositions (to E. I. duPont de Nemours & Co., Inc.), Brit. 842,073.
47. Mayurnik, Epoxy Resin Compositions, U.S. 2,934,520.
48. Masters et al., Epoxide Resin Compositions, U.S. 2,934,521.
49. Newey, Process for Curing Material Having Internal Epoxy Groups and Resulting Products, U.S. 2,935,492.
50. Newey, Process for Curing Materials Possessing Internal Epoxy Groups, U.S. 2,949,441.
51. Newey, Process for Curing Polyepoxides and Resulting Products, U.S. 2,965,610.
52. Phillips et al., Curable Compositions Comprising a Diepoxide, a Polycarboxylic Anhydride, and a Polyhydric Compound, U.S. 2,890,196.
53. Phillips et al., Polymerizable Compositions and Resins Made Therefrom, U.S. 2,934,508.
54. Phillips et al., Polyepoxide Compositions, U.S. 2,921,925.
55. Phillips et al., Polyepoxide Compositions, U.S. 2,921,929.
56. Robinson, Resinous Compositions, U.S. 2,948,705.

57. Rosenberg, Curing of Epoxy Resins, Ger. 1,091,326.
58. Schleede and Schultz, Acceleration of Resin Hardening, Ger. 1,070,364.
59. Teliha, Evaluation of Curing Agents: Acid Anhydrides, Shell Chemical Corp. Report PP-SD-1010-1.
60. Warburton and Norman, High Temperature Epoxide Resin Formulations, 3d National Conference on Application of Electrical Insulation, December.
61. Wear, Aromatic Epoxidized Polyester and Method of Making, U.S. 2,944,035.

1961

62. Bavley et al., Epoxy Resin Compositions, U.S. 2,989,500.
63. Delmonte and Hirosawa, Epoxy Resin Compositions Having High Heat Stability, U.S. 2,967,843.
64. Ernst, Heat Hardenable Compositions Comprising Epoxide Compounds and Dicarboxylic Anhydrides, U.S. 2,978,435.
65. Ernst et al., Procedures for the Acceleration of the Curing of Polyepoxy Resins by Acid Anhydrides, Ger. 1,120,133.
66. Garvey, Heat Curing Molding Materials from Epoxy Resins and Polycarboxylic Anhydrides, Ger. 1,112,831.
67. Ilardo et al., Polyepoxide–Dicarboxylic Anhydride Compositions, U.S. 2,992,196.
68. Ludwig, Reaction Product of *N*-Glycidyl-Phthalimide and a Dianhydride Compound, Process for Preparing Same, and Modification Thereof, U.S. 2,999,836.
69. Luft, Improvements in Epoxy Resin Embedding Methods, *J. Biophys. and Biochem. Cytol.* vol. 9, no. 2.
70. McGary and Patrick, Composition Comprising Bis(2,3-epoxycyclopentyl) Ether, Polycarboxylic Acid Anhydride and a Polyol, and Cured Product Thereof, U.S. 3,000,848.
71. Newey, Epoxy Containing Condensates of Polyepoxides and Acidic Materials, Their Preparation and Polymers, U.S. 2,970,983.
72. Port et al., Non-terminal Epoxy Resins, U.S. 2,975,149.

1962

73. Buchoff and Sherwin, Properties of High-temperature Epoxy Systems, 18th ANTEC of SPE.
74. Chen et al., Development of High Temperature Laminating Resins, Progress Reports 1 and 2, Contract AF 33(657)-9063, Project 920K, June–August.
75. Curing Agents for Epoxy Resins (to E. I. duPont de Nemours & Co., Inc.), Brit. 886,601.
76. Improvement in and Relating to Curing (to FMC Corp.), Brit. 888,133.
77. Lee and Neville, Epoxy Carboxylic Acid Dianhydride Compositions, U.S. 3,025,263.
78. Patrick, Resin Forming Reactions, Union Carbide Chemical Co. Report.
79. Singley and Christiansen, Cured Resins from Glycidyl Ethers of Dihydroxy-diphenylsulphone, U.S. 3,044,983.
80. Wear and Bowman, Heat Curing Epoxy Resin Compositions, U.S. 3,052,650.
81. Wheelock, Thermosetting Resins, U.S. 3,028,370.

1963

82. Bremmer, Factors Influencing the Heat Stability of Fire Retardant Epoxy Resins, Dow Chemical Co. Report.
83. Chang Te-Ho, A Study of the Infrared Spectra of Epoxy Resins during the Course of Cure, *Sci. Sinica*, vol. 12, no. 9.
84. Curing Agents for Epoxy Resins (to Ciba, Ltd.), Belg. 629,032.
85. Ecke and Jacobs, Monomeric Ether-Olefin Adducts and Their Preparation, U.S. 3,114,752.
86. Lincoln and Hawthorne, Epoxide Resins, Brit. 914,463.
87. Loncrini, Anhydride Hardening Agents for Epoxy Resins, Belg. 631,373.
88. Loncrini, Compositions of Heat-hardenable Epoxy Resins, Belg. 631,376.
89. McGary et al., Resins from *endo*-Dicyclopentadiene Dioxide, *J. Appl. Sci.* **7**: 1–14.

90. May, The Physical Significance of Acid-Anhydride Cures on Epoxy Adhesive Properties, *SPE Trans.* **3**:251–259 (October).
91. Metal Chelates Containing Epoxide Groups (to Ciba, Ltd.), Belg. 633,330.
92. Nishimura and Hata, Synthetic Resins from Epoxidized Oils, Japan 22,744.
93. Rumscheidt and Bruin, Process for Curing Polyepoxides with Anhydrides and Organic Sulfur Compounds, U.S. 3,114,731.
94. Renner, Bicyclo(2.2.1)hept-5-ene-2,3-dicarboxylic Acid Anhydrides, U.S. 3,105,839.
95. Schmitz-Johnson et al., Hardening of Epoxy Resins, Belg. 627,887.
96. Smith, An Epoxy Resin with Improved Electrical Properties, *Mod. Plastics*, July.
97. Telomers Containing Acid Anhydride Groups (to Ciba, Ltd.), Belg. 624,357.
98. Walton et al., Development of Flexible Polymers as Thermal Insulation in Solid-propellant Rocket Motors, AD 403 799.

1964

99. Black et al., PAPA Offers Many Advantages as a Converter for Epoxies, *Plastics Technol.*, March.
100. Curing Agents for Epoxy Resins (to Ciba, Ltd.), Brit. 970,444.
101. Fisch, Elastic Reaction Product of a Polyepoxide, a Glycol, and an Acid Anhydride, U.S. 3,142,657.
102. Kaplan, Effects of Moisture on Glycidyl Reactions with Carboxylic Acid Anhydrides, R&D Division, AVCO Corp. Report.
103. Lee and Neville, Use of Dianhydrides in High Temperature Epoxy Resin Formulations, SAMPE Symposium, Los Angeles, May.
104. McIntosh and Brooker, Epoxy Resin Coating Composition and Process, Brit. 963,557.
105. Mack, Formulation and Cure Cycle Study for an Anhydride-cured Epoxy-Novolac System, *Ind. Eng. Chem. Prod. R&D*, September.
106. Winfield and Ashcroft, Epoxide Resin Compositions, Brit. 959,028.

1965

107. Anhydride Hardeners for Epoxy Resins, Allied Chemical Plastics Division Booklet 81-112-5M.
108. Bicyclic Carboxylic Acid Derivatives (to Badische Anilin- & Soda-Fabrik A-G), Neth. Appl. 6,409,347.
109. Curing of Epoxide Resins (to Fabenfabriken Bayer A-G), Neth. Appl. 6,410,025.
110. Delmonte, Semi-rigid Epoxy Resin Compositions and Method, U.S. 3,188,362.
111. Hirsch and Van Strien, Esters of Trimellitic Anhydride, U.S. 3,183,248.
112. Jones et al., New Prepreg Epoxy Systems for Filaments, RIP Division of SPI, Chicago, February.
113. Kenny, Casting of a Low Exotherm Epoxy Resin, *J. Sci. Instr.* **42**(9):719–720.
114. Moldavskii et al., Epoxy Compositions, U.S.S.R. 170,654.
115. Proops and Fowler, Curable Mixtures Comprising Epoxide Compositions and Divalent Tin Salts, U.S. 3,201,360.

1966

116. Christie et al., Anhydride Blends as Curing Agents for Epoxy Resins, 22nd ANTEC of SPE, March.

Chapter 13

DILUENTS FOR EPOXY RESINS

Diluents are employed in the epoxy-resin technology primarily to reduce the viscosity of the mixed system.

In addition to providing for viscosity reduction, the diluent may be selected to provide modification of the cured properties. The diluent may be employed to increase the deflection temperature (DT) or to reduce it. It may be employed to improve, selectively, the chemical resistance; to change the electrical properties; to impart flame resistance, and within limits, to change impact resistance and physical properties.

The diluent may also be selected to reduce or extend pot life and to reduce or increase exotherm.

In general, with unfilled systems, the diluents find primary use in formulations involving room-temperature curing agents. When elevated-temperature curing is required, the cure temperature is usually adjusted to control viscosity more satisfactorily than through the use of diluents.

The diluents of interest to the epoxy-resin technology may be divided into three basic categories: (1) nonreactive, (2) reactive, epoxy-containing, and (3) reactive, containing functional groups other than epoxies.

NONREACTIVE DILUENTS

Nonreactive diluents of the solvent type (e.g., xylene) have been employed to lower the viscosity of the epoxy resins, principally in high-buildup coating formulations with nonvolatile contents on the order of 95 per cent or more. Here, it is probable that the diluent remains entrapped in the cured composition, but because it is present in amounts on the order of only 2 or 3 phr, the effect upon the cured properties is not pronounced. Use of nonreactive diluents may carry a system to a higher degree of cure than is obtained in the undiluted system. This is shown in terms of hot hardness and per cent extractables in Figs. 13-1 and 13-2.

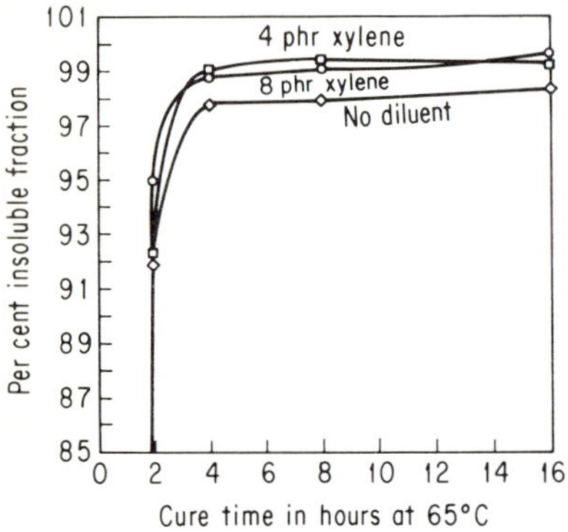

FIG. 13-1. Diluents vs. solubility of DGEBA cured with 4 phr DMP-30 at 65°C [4–6].

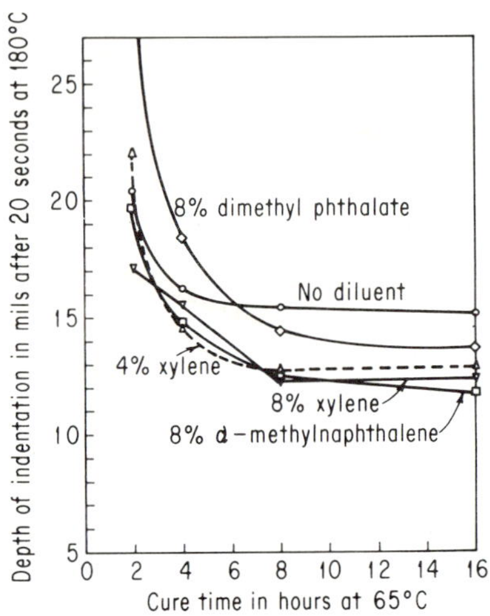

FIG. 13-2. Diluents vs. hot hardness at 180°C of DGEBA cured with 4 phr DMP-30 at 65°C [4–6].

Other materials which do not react with the epoxy resin, such as polymethoxy acetals [13], dibutyl phthalate [9], methyl methacrylate [8], glycol ether–esters, having molecular weights above 258, from ethylene glycol ethers and monobasic aliphatic acids [35], indene, pine oil, etc., have been suggested as diluents capable of being employed on the order of 10 to 20 phr without severe reduction of cured properties. Furfuryl alcohol has been reported as an accelerator for DGEBA cured with a blend of AEP/DMP-30 to give complete hardening in 10 minutes at room temperature [44].

Chlorinated diluents have been used with antimony oxide to impart flame resistance to the cured castings. A typical formulation of this type, based on DGEBA employs about 15 phr chlorinated phenol and 5 phr antimony oxide to produce self-extinguishing systems. With DETA cures the DT is reduced to about 90°C by addition of chlorinated phenol; with PA cures the DT is about 135°C. Some of the nonreactive diluents may be used to impart other special properties. For example, the polymethyl acetals may be used to improve adhesive strengths of room-temperature-curing formulations, and 4,4-dimethyl-5-hydroxymethylmetadioxane (prepared from condensing isobutylene with formaldehyde) when used at from 5 to 10 phr is reported as improving tensile strengths of DGEBA systems [28].

Halohydrins, such as dichlorohydrin, when used as nonreactive diluents, have been found to accelerate the curing rate of primary and secondary amines. The acceleration is more marked in the low-chlorine-content DGEBA than in the higher-chlorine-content DGEBA resins [19].

In general, with the nonreactive diluents, the effect upon physical properties, at low concentrations, is slight; the chemical resistance, and particularly the solvent resistance, is more strongly affected. As the concentration of nonreactive diluent increases, the properties of the cured system are correspondingly degraded. Some portion of the nonreactive diluent may be driven off during cure, consequently increasing the shrinkage and reducing the adhesion. Volatilization of the diluent during exothermic heating—or oven cure—prior to gelation can entrap diluent bubbles throughout the casting.

Dibutyl Phthalate

Of the various materials suggested, dibutyl phthalate has perhaps proved most successful.

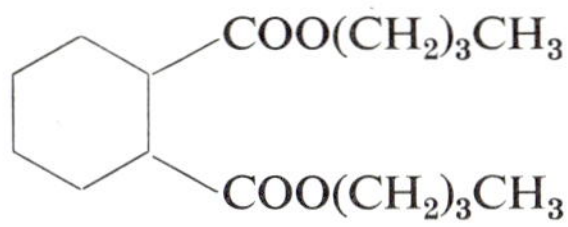

Dibutyl phthalate

It is employed with DGEBA, usually incorporated with heating, to reduce the viscosity from about 15,000 to about 4,000 centipoises when used at about 17 phr. In addition to reducing viscosity, dibutyl phthalate provides some plasticizing action by virtue of the side chains which somewhat decrease the cross-linking density. This serves to improve thermal shock resistance and adhesive strengths [4], although at the expense of elevated-temperature performance. In DGEBA, dibutyl phthalate does not exhibit migratory tendencies on aging. It can transamidify upon heat aging of amine-cured resins.

Physical properties of DGEBA containing about 17 phr dibutyl phthalate and cured with the stoichiometric amount of TETA, are given in Table 13-1. Chemical resistance is given in Table 13-2. Physical properties below the DT are little affected, whereas there is some reduction in chemical resistance as would be expected from the presence of the unreacted diluent.

The general effect of ester plasticizers on epoxy resins has been investigated. Within the series diethyl, dibutyl, etc., it was found that the decrease in Tg is

Table 13-1. Physical Properties of TETA-cured DGEBA Diluted with 17 phr Dibutyl Phthalate [4-1]

Property	*Value*
Tensile strength, psi	9,000
Modulus, psi $\times 10^{-6}$	0.45
Flexural strength, psi	15,000
Compressive strength, psi	18,000
Impact strength, Izod	0.39
Hardness, Rockwell M	83
Coefficient of thermal expansion, in./(in.)/(°C)	9.0×10^{-5}
Thermal conductivity, cal/(sec)(sq cm)(°C)(cm)	4.0×10^{-4}
Thermal shock, MIL-1-16923B	Type B
Flammability, in./min	0.34
Shrinkage during cure, linear, %	<2%
Water absorption in 24-hr immersion, %	0.12
Moisture-vapor transmission, at 93°C, relative humidity 95%, g/(sq ft)(24 hr)(in. thickness)	<0.01

Table 13-2. Chemical Resistance of TETA-cured DGEBA Diluted with 17 phr Dibutyl Phthalate [4-1]

3 months' immersion in listed fluids	*Weight change, %*
Acetaldehyde, conc.	10.28
Acetone, 10%	1.96
Acetone, 1%	1.39
Ammonia, 30%	1.22
Ammonia, 10%	1.14
Battery acid, 32%	1.04
Butyl acetate, 100%	3.86
Butyl alcohol, 100%	0.92
Carbon tetrachloride, 100%	7.93
Caustic soda, aq. 10/100	1.15
Clophen	1.03
Cyclohexanol, 100%	0.55
Distilled water	1.48
Ethyl alcohol	9.31
Ethyl ether	1.11
Ferrous chloride, 45 g/100 g	0.86
Gasoline	0.003
Glycerin, 100%	−0.32
Hydrochloric acid, conc.	7.48
Hydrochloric acid, 50%	1.53
Hydrochloric acid, 10%	2.18
Hydrochloric acid gas	5.22
Hydrogen peroxide, 30%	13.08
Hydrogen sulfide	0.79
Hydrogen sulfide, aq.	2.31
Javelle water	0.62
Lactic acid, conc.	7.09
Mineral oil	0.09
Nitric acid, 10%	2.76
Photographic developer	1.19
Photographic fixing solution	0.87
Relative humidity, 100%	0.94
Sodium chloride, aq. 10 g/100 g	1.18
Sodium chloride, aq. 20 g/100 g	0.99
Sodium chloride, aq. 26 g/100 g	0.85
Soldering fluid	0.70
Stearic acid, 100%	0.64
Sulfuric acid, 50%	2.61
Sulfuric acid, 10%	2.91
Tetrahydronaphthalene	2.75
Triethanolamine	−0.014
Vinegar	3.63

directly proportional to the molar and volume concentrations of the esters. Esters containing the largest number of CH_2 groups in the aliphatic portion of the ester molecule produced the maximum plasticizing effect. Esters with increased polarity have lesser plasticizing effects but are more compatible [30].

Styrene Diluents

Monomeric styrene may be used as a nonreactive diluent for the saturated epoxies. When it is used with DGEBA, typical properties are as given in Table 13-3. A formulation has been developed which employs DGEBA (mol. wt. 1,000) dissolved in 80 phr styrene and cured with 113 phr CA and 0.5 per cent by weight dimethylaniline and 1.5 per cent by weight *tert*-butyl peroxide [43].

Table 13-3. Effect of Concentration of Styrene on Properties of TETA-cured DGEBA [15]

Styrene, phr	Viscosity, centipoises	Pot life at 23°C, minutes	Exo-therm, °C	H_2O absorption, %		Flexural strength, psi	DT, °C	Weight loss 48 hr at 200°C	Hard-ness, Barcol	Dielectric constant		Dissipation factor	
				2-hr boil	30 days' immersion					at 60 cps	at 10^5 cps	at 60 cps	At 10^5 cps
None	....	40	200	0.73	0.79	20,300	120	1.89	87	4.17	3.67	0.0096	0.036
7.7	1500	71	156	0.87	1.08	16,220	93	3.84	86	3.66	3.39	0.0096	0.026
13.4	500	85	163	0.97	1.17	14,770	81	4.93	84	3.53	3.31	0.0075	0.022
20	200	106	144	0.79	1.26	11,297	65	Unstable	75	3.49	3.28	0.0087	0.020

Likewise, low-molecular-weight polystyrenes may be used as nonreactive diluents. These are compatible with DGEBA in proportions up to about 40 per cent concentration and serve to reduce exotherm and extend pot life at the expense of some reduction in impact strength, tensile strength, heat resistance, and chemical resistance. The low-molecular-weight polystyrenes are incompatible with most aliphatic primary amine curing agents such as DETA and TETA but are compatible with AEP, DEAPA, propylenediamine, and menthane diamine.

Phenolic Diluents

As was indicated in Chap. 11, bisphenols and phenolic resins may be employed, under suitable reaction conditions, as curing agents for both glycidyl ether and epoxidized olefin resins. Likewise, as acid-type catalysts, they may be employed to accelerate the cure of amines. With primary amine cures, the phenolic hydroxyls assist proton transfer but do not react into the system; with tertiary amine cures,

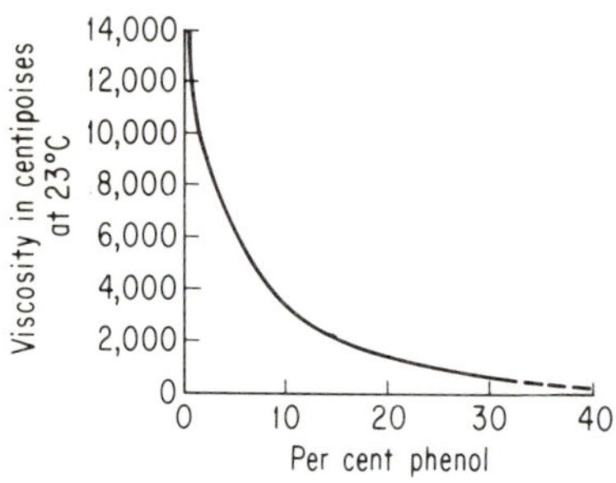

FIG. 13-3. Viscosity vs. 3-(pentadecyl)-phenol concentration in DGEBA [4–4].

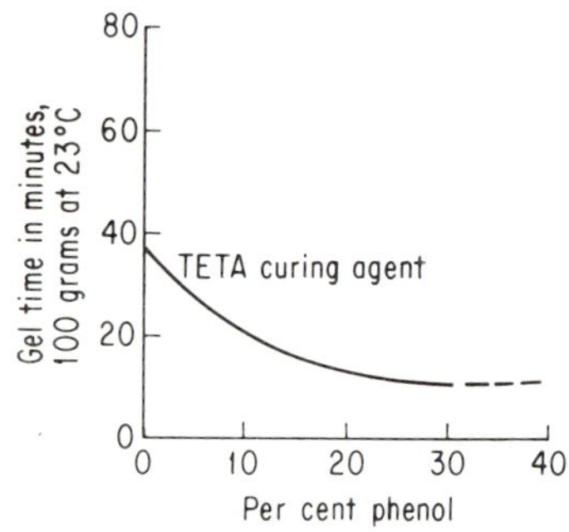

FIG. 13-4. Gel time vs. 3-(pentadecyl)phenol concentration in DGEBA [4–4].

the phenolic hydroxyls are reacted with the epoxy groups, and the phenol is therefore chemically bound into the cured structure.

Thus, when phenols are used as diluents, they may be either reactive (in the presence of a base catalyst or when elevated-temperature cures are employed) or nonreactive (when primary amine systems are involved).

Although phenol may be melted into DGEBA to produce a lower-viscosity mixture, the phenols usually employed as diluents are monohydroxyl long-chain phenols, typical being nonyl phenol. Most commonly they are used with aliphatic primary amines and fatty mono- and polyamide curing agents, where they serve both to reduce viscosity and accelerate the curing reaction. In these cases, it is probable that the phenols remain mechanically locked in the system and exert a plasticizing action similar to that obtained with dibutyl phthalate.

Typical of such phenols is commercial 3-(pentadecyl) phenol [17]. The material serves to reduce viscosity, as indicated in Fig. 13-3. The accelerating effect when TETA is used as a curing agent is indicated in Fig. 13-4. Physical and electrical properties for a TETA-cured system, as a function of the concentration of diluent, are indicated in Table 13-4.

When 3-(pentadecyl) phenol is epoxidized, a site reactive with a wider variety of curing agents is introduced, and, from the standpoint of ultimate cured properties, the epoxidized material is generally preferred.

Table 13-4. Physical and Electrical Properties of TETA-cured DGEBA Diluted with 3-(Pentadecyl)phenol [4-4]

(Cure cycle: 16 hours at 120°C + 2 hours at 93°C)

Properties	3-(Pentadecyl)phenol/TETA, phr		
	0/10	25/12.5	43/13
Gel time, 100 g at 23°C, minutes	35	10	10
Tensile strength, psi	11,814	8,745	4,922
Elongation, %	5.1	4.8	5.4
Modulus, psi $\times 10^{-6}$	0.23	0.19	0.11
Hardness, Shore D	85	85	80
Maximum exotherm, °C	161	185	194
Impact, Izod	0.36	0.36	0.34
DT, °C	87	67	54
Water absorption, %	0.14	0.15	0.26
Dissipation factor at 10^3 cps	0.022	0.014	0.017
Dielectric constant at 10^3 cps	4.178	3.716	3.614
Loss factor at 10^3 cps	0.093	0.052	0.061
Volume resistivity, ohm-cm:			
At 25°C	2.7×10^{15}	1.3×10^{15}	1.0×10^{15}
At 66°C	6.3×10^{14}	1.3×10^{14}	2.5×10^{11}
At 93°C	5.0×10^{11}	5.9×10^{10}	$<10^7$
At 130°C	1.6×10^{10}	$<10^7$	
At 150°C	$<10^7$		

EPOXY-CONTAINING REACTIVE DILUENTS

The epoxy-containing reactive diluents may be divided into two classes: (1) those containing a single epoxy group, and (2) the low-viscosity epoxy resins.

Monoepoxy Diluents

The monoepoxy-containing diluents may be regarded as "chain stoppers" since, unless they contain other reactive sites, they reduce system functionality and decrease system cross-linking density. The extent to which the cured properties are reduced is a function of the concentration of the diluent, with the greater amounts being preferred for viscosity reduction and the lower amounts being preferred for the preservation of properties.

The amount of curing agent used is adjusted to take into account the epoxy groups present in the diluent. This relationship is indicated in Fig. 13-5, where, to increase pot life, DETA is used at less than stoichiometric throughout the range.

The reactivity of the diluent with the curing agent will depend upon the location of the epoxy group in the molecule. As would be expected from Chap. 5, the diluents containing glycidyl ether groups are quite reactive with amines; whereas the diluents containing internal epoxies or ring-situated epoxies are more reactive toward the anhydrides and the acids. Thus, the diluent may be selected to modify the pot life of the formulation as well as to provide for a reduction in viscosity.

However, even with amine-reactive diluents (i.e., glycidyl ethers), amine-cured systems are somewhat slowed down, although the peak exotherms may be somewhat increased over those experienced with the undiluted resins.

The general effect of the monoepoxy diluents is to reduce the DT and the heat resistance, although with the longer-chain diluents, some improvement in flexural strength may be realized, together with a slight increase in impact resistance. Hardness is little affected unless relatively large amounts of diluent are used. The coefficient of thermal expansion of the cured resins will be increased. For example, a cured resin with no diluent will have a coefficient of thermal expansion of 50 to

55×10^{-6} in./(in.)(°C), which with 10 phr of a monofunctional diluent will increase to 60 to 65×10^{-6} in./(in.)(°C). Electrical properties may be somewhat improved below the DT. Aliphatic epoxy diluents, even though more efficient than aromatic-based epoxy diluents, more severely degrade the chemical and solvent resistance [31]; with the aromatic-based monoepoxies, resistance to acids and bases is little influenced although solvent resistance is somewhat lowered. Epoxidized 3-(pentadecyl)phenol has been reported as improving acetic acid resistance of primary aliphatic amine-cured systems [13]. Overall improved properties have been obtained

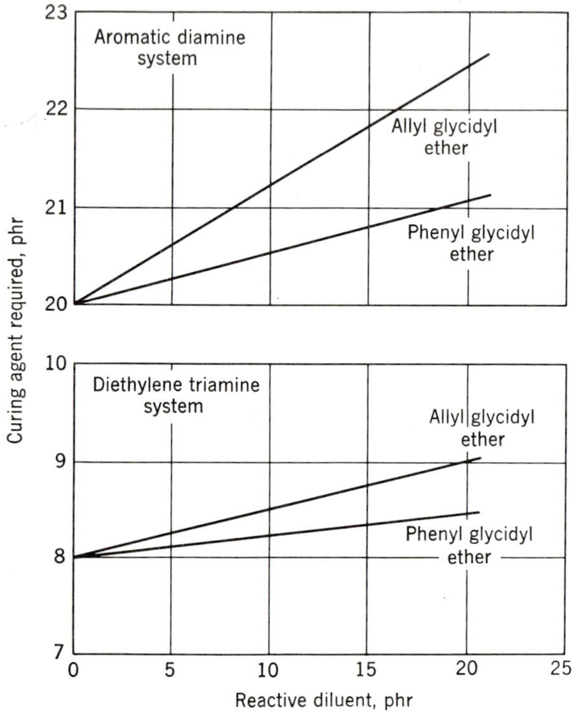

FIG. 13-5. Relationship between diluent concentration and curing-agent concentration for DGEBA [4–6].

with polyfunctional monoepoxies such as vinylcyclohexene oxide when unsaturation is present in the resin or curing agent and a peroxide catalyst is employed [37].

Monoepoxy diluents are employed to improve the wetting ability of adhesive and laminating formulations and to increase the filler loading volume of castings. Chlorinated monoepoxy diluents have been suggested to improve flame resistance [5,6].

In addition to improving the wetting action of adhesives (particularly of filled adhesive systems), the diluent also somewhat decreases the tendency of primary and secondary amine curing agents to volatilize during cure in thin films. In casting applications, when vacuum is employed, high-vapor-pressure diluents should be avoided to prevent volatilization during the vacuum cycle. A pressure-temperature curve for a typical reactive diluent is given in Fig. 13-6. Even at room temperature, volatility may offer problems. Butyl glycidyl ether, when used at 15 per cent in

DGEBA, gives a weight loss after 24 hours at atmospheric conditions of 8.3 per cent as against 0.70 per cent for phenyl glycidyl ether (used at 20 per cent) and 0.50 per cent for cresyl glycidyl ether (used at 24 per cent) [4-6].

Most monoepoxy diluents have some solvent power, and their compatibility with any foreign material in contact with the fluidized resin during cure should be established before use.

The monoepoxy diluents of interest to the epoxy technology are listed in Table 13-5. Infrared spectra are given in Appendix 13-1. Gas-liquid chromatography has been shown to be effective for the identification of monoepoxy diluents when they are removed from the resin (conveniently accomplished by dissolving the mixture in ethyl ether and freezing it). The supernatant liquid, containing the diluent, is then analyzed. Direct methods involving the injection of the resin and diluent blend into the instrument also provide satisfactory results, but the accumulation of carbonaceous material in the vaporization chamber is objectionable [41].

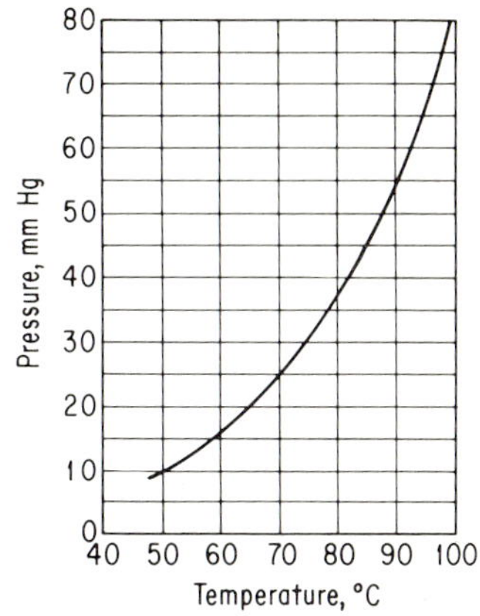

FIG. 13-6. Vapor-pressure curve for BGE [21].

Monoepoxy diluents are used predominantly with the glycidyl ether resins, and a considerable body of comparative data has been developed on their use with DGEBA. Typical data are given in Tables 13-6 to 13-9 and Figs. 13-7 to 13-12. Table 13-10 indicates the reactivity rates of the variously located epoxy groups—nonglycidyl terminal, glycidyl internal, and ring-situated.

In addition to the common monoepoxies, as listed in Table 13-5, other monoepoxies have been investigated. For example, hydroxyl-containing monoepoxies have been used, these based on spirocyclohexyl-*m*-dioxane with an alkyl substituent [32].

Some of the monoepoxy diluents, e.g., butyl glycidyl ether, are reported to be sensitive to minor amounts of impurities and the thixotropic properties of filled formulations varying with the pH of the diluent, with slightly basic conditions preferred for best thixotropy.

The monoepoxy diluents are generally more powerful skin-sensitizing agents than the resins themselves.

Epoxy-resin Reactive Diluents

In order to preserve physical properties at elevated temperature, di- or polyepoxy reactive diluents may be employed. When suitably reacted, these will not reduce system functionality and in some cases will result in an increase in system cross-linking density.

Typical of the epoxy resins which have been suggested as reactive diluents are the materials shown in Table 13-11. They may be used in higher concentrations than the monoepoxy diluents to obtain the desired reduction in viscosity. For example, when divinylbenzene dioxide is used at about 26 phr, the viscosity of DGEBA will be reduced to about 1,500 centipoises [24]; 40 phr DGE of diethylene glycol, 30 parts vinylcyclohexane dioxide, or 15 phr DGE of butanediol will be required to obtain similar reduction [16]. Use of epoxy diluents has been covered in the patent literature for a variety of compounds, representative being the use of bis(2,3 epoxycyclopentyl) ether to reduce the viscosity of DGEBA [20], the use of limonene

Table 13-5. Commercial Epoxy-containing Monofunctional Reactive Diluents

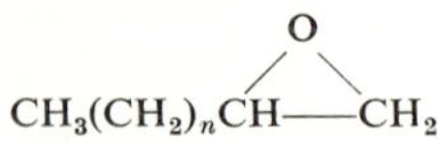
Olefin oxides, b.p. 168°C at 37 mm

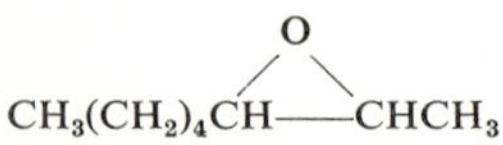
Octylene oxide (mol. wt. 128) 85%, 15% 1,2-isomer b.p. 77°C at 45 mm

Butyl glycidyl ether (mol. wt. 130)

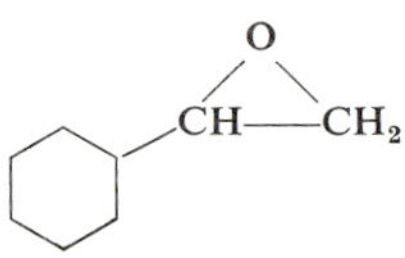
Styrene oxide (mol. wt. 120)

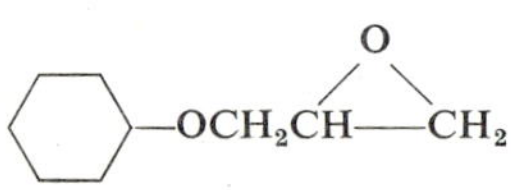
Phenyl glycidyl ether (mol. wt. 150)

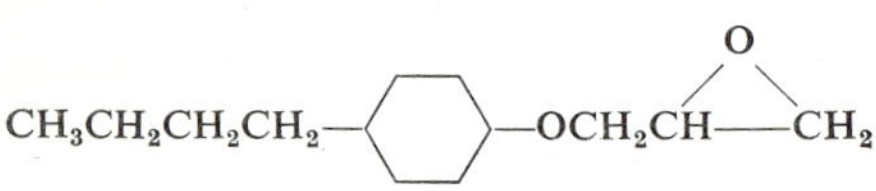
p-Butyl phenol glycidyl ether

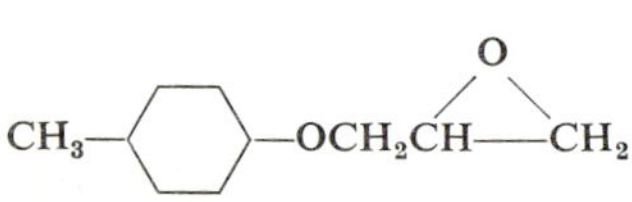
Cresyl glycidyl ether (mol. wt. 165)

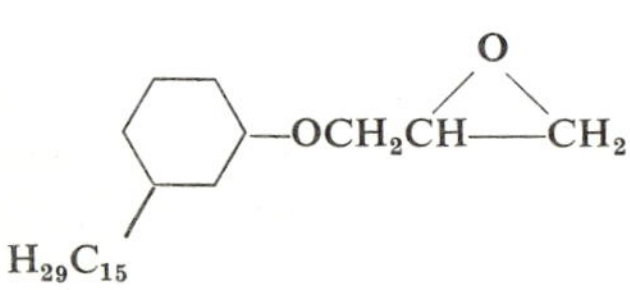
3 (Pentadecyl) phenol glycidyl ether

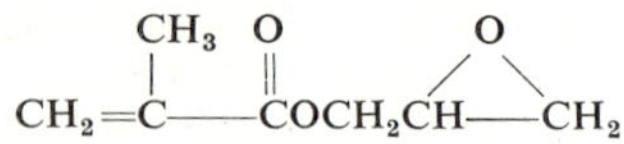
Glycidyl methacrylate (mol. wt. 142), b.p. 75°C at 10 mm

Allyl glycidyl ether (mol. wt. 114)

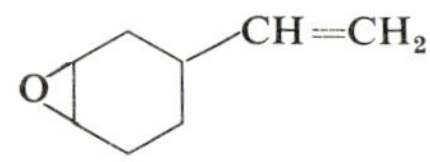
Cyclohexene vinyl monoxide (mol. wt. 124), b.p. 169°C at 760 mm

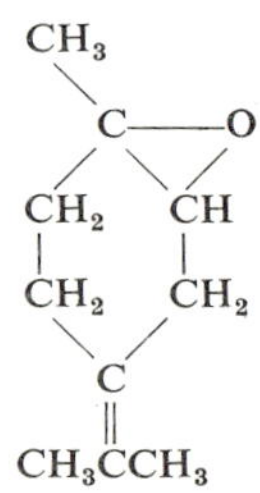
Dipentene monoxide (mol. wt. 152), b.p. 75°C at 10 mm

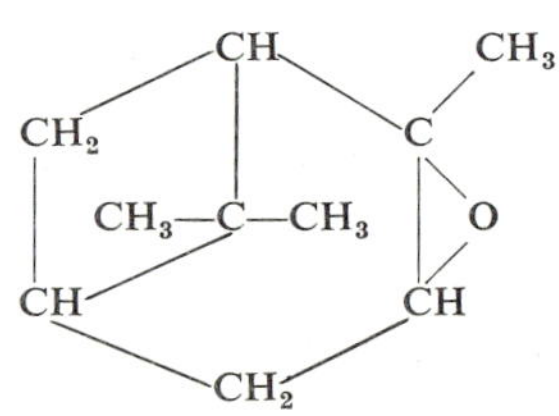
α-Pinene oxide (mol. wt. 152), b.p. 62°C at 10 mm

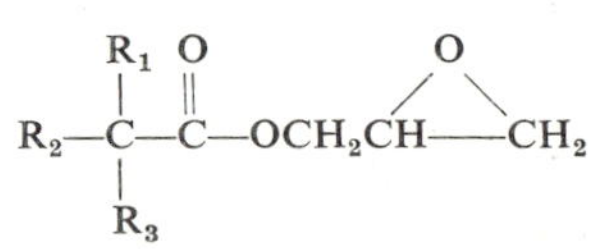
($C_{12-14}H_{22-26}O_3$) Glycidyl ester of *tert*-carboxylic acid (mol. wt. 240–250), b.p. 135°C at 760 mm

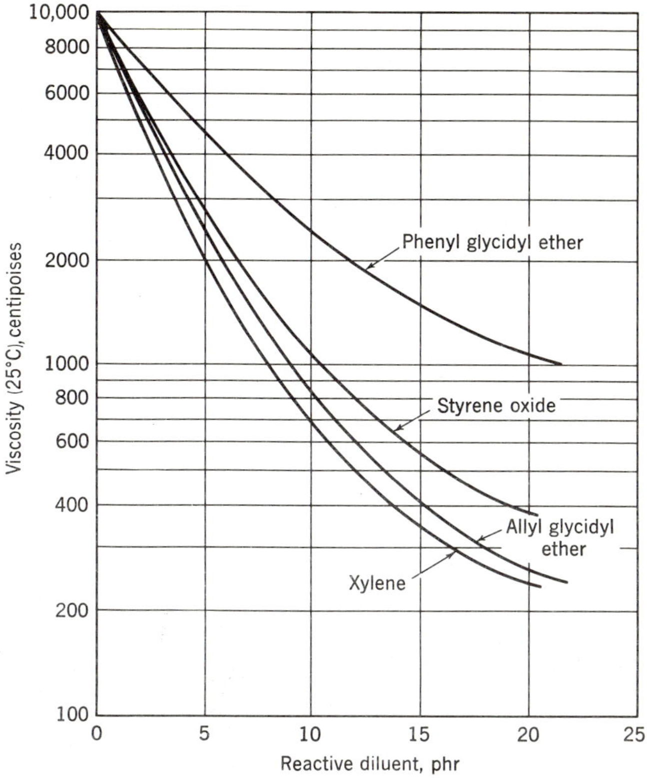

FIG. 13-7. Diluent concentration vs. viscosity for DGEBA [4–6].

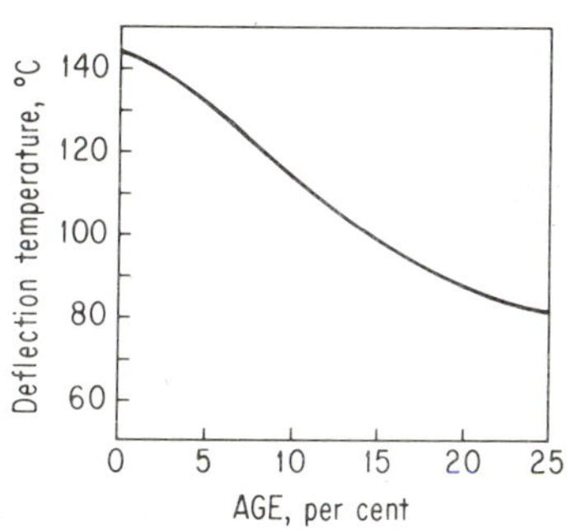

FIG. 13-8. AGE concentration vs. DT of DGEBA cured with MPDA [23].

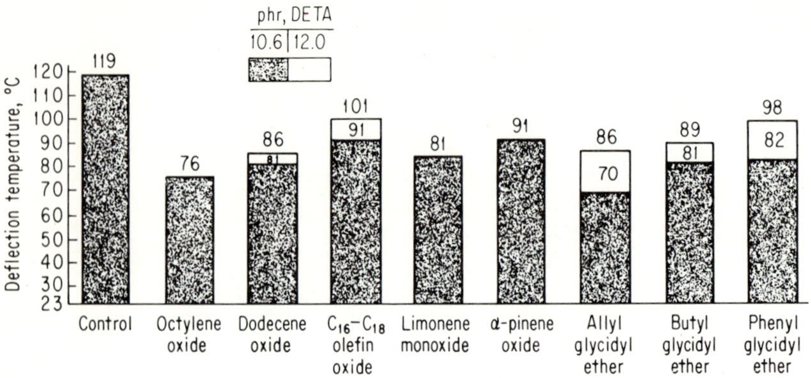

FIG. 13-9. Effect of 10 phr diluent on DT of DGEBA cured with DETA [7]. Initial cure: 20 hours at 23°C. Postcure: 2 hours at 100°C.

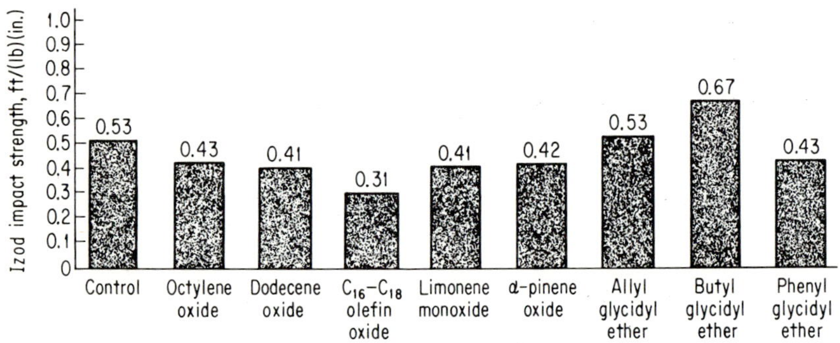

FIG. 13-10. Effect of 10 phr diluent on Izod impact strength of DGEBA cured with DETA [7]. Initial cure: 20 hours at 23°C. Postcure: 2 hours at 100°C.

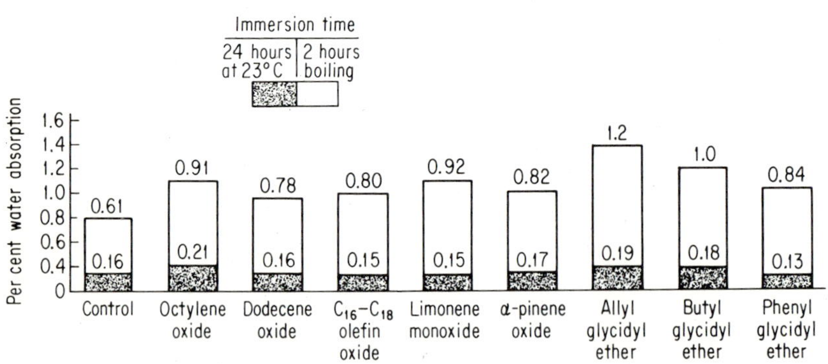

FIG. 13-11. Effect of 10 phr diluent on water absorption of DGEBA cured with DETA [7]. Initial cure: 20 hours at 23°C. Postcure: 2 hours at 100°C.

Table 13-6. Effect of Diluent on Properties of TETA-cured DGEBA [15]

Diluent and mixed viscosity, centipoises	Pot life at 23°C	Exo-therm, °C	Water absorp-tion, weight increase, %	Flexural strength, psi	DT, °C	Weight loss in 48 hr at 200°C, %
None	40	200	0.73	20,300	120	1.89
Octylene oxide:						
1,500	55	205	1.10	14,185	74	5.20
500	72	180	1.18	11,440	65	Unstable
200	100	164	1.04	7,797	58	Unstable
Styrene oxide:						
1,500	48	200	0.71	19,403	93	2.39
500	69	193	0.80	18,854	82	4.29
200	84	178	1.06	16,303	70	6.20
3(Pentadecyl) phenol glycidyl ether:						
1,500	54	164	1.19	14,010	76	2.84
500	68	148	1.48	6,443	55	4.15
Butyl glycidyl ether:						
1,500	55	204	1.01	18,433	101	2.12
500	63	207	1.32	17,237	85	2.72
200	76	193	1.63	16,400	75	3.77
Allyl glycidyl ether:						
1,500	48	204	1.11	18,817	95	1.86
500	59	194	1.51	18,047	76	2.37
200	66	189	1.72	15,827	74	3.14
Phenyl glycidyl ether:						
1,500	47	198	1.04	19.717	...	2.37
500	49	209	1.26	19,523	...	3.28
200	54	198	1.35	19,600	...	4.93
Cresyl glycidyl ether:						
1,500	46	202	0.98	22,067	92	2.98
500	46	204	1.24	20,000	78	4.40
200	51	201	1.35	15,230	63	6.58
Glycidyl methacrylate:						
1,500	41	208	1.30	18,507	105	2.93
500	43	224	1.64	17,510	95	3.78
200	46	224	2.02	18,400	87	4.43

Table 13-7. Viscosity Reduction Efficiency of Various Monofunctional Diluents Used with DGEBA [15]

Diluent	phr diluent required to obtain listed viscosity		
	1,500 centipoises	500 centipoises	200 centipoises
High-efficiency diluents:			
Allyl glycidyl ether	6.8	13.0	20.0
Octylene oxide	7.2	13.2	20.5
Butyl glycidyl ether	7.2	13.5	21.0
Styrene oxide	5.3	19.3	30.0
Medium-efficiency diluents:			
Glycidyl methacrylate monomer	10.0	19.0	30.0
Phenyl glycidyl ether	15.0	29.0	49.0
Low-efficiency diluent:			
Cresyl glycidyl ether	19.0	38.0	65.0
3-(Pentadecyl) phenol glycidyl ether	27.0	64.0	

Table 13-8. Effect of Diluent upon Electrical Properties of TETA-cured DGEBA [15]

Diluent and viscosity, centipoises	Dielectric constant		Dissipation factor	
	At 60 cps	At 10^5 cps	At 60 cps	At 10^5 cps
None:				
12,200	4.17	3.67	0.0096	0.036
Butyl glycidyl ether:				
1,500	3.94	3.57	0.009	0.031
500	3.72	3.45	0.009	0.029
200	3.70	3.50	0.009	0.027
Allyl glycidyl ether:				
1,500	4.14	3.68	0.012	0.038
500	4.72	4.01	0.037	0.048
200	4.03	3.68	0.012	0.032
Phenyl glycidyl ether:				
1,500	3.91	3.54	0.0078	0.032
500	3.70	3.46	0.0048	0.027
200	3.46	3.28	0.0052	0.020
3-(Pentadecyl) phenyl glycidyl ether:				
1,500	3.49	3.27	0.0078	0.019
500	3.55	3.45	0.021	0.021

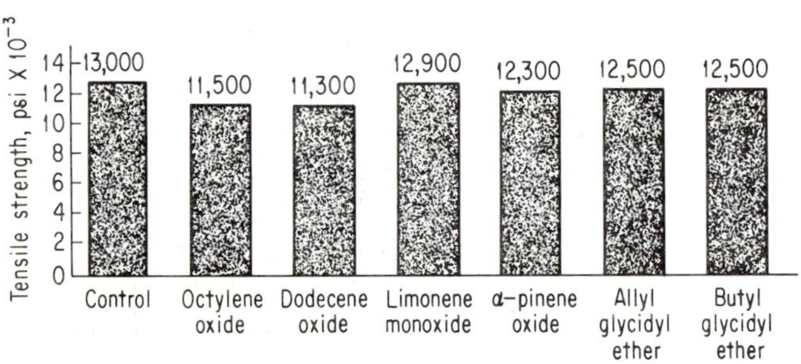

FIG. 13-12. Effect of 10 phr diluent on tensile strength of DGEBA cured with DETA [7]. Initial cure: 20 hours at 23°C. Postcure: 2 hours at 100°C.

Table 13-9. Effect of Diluent and Concentration on Viscosity of DGEBA [2]

Diluent	Viscosity, centipoises				
	Diluent concentration				
	0 phr	5 phr	10 phr	15 phr	20 phr
Octylene oxide	18,685	3,275	1,180	490	260
Dodecene oxide*	18,685	4,520	1,880	975	560
C_{16}–C_{18} olefin oxide†	18,685	4,540	2,480	1,775	
α-Pinene oxide	18,685	7,508	3,855	2,400	1,448
Dipentene monoxide	18,685	6,120	2,950	1,465	930

* Casting slightly turbid.
† Heat required to effect solution—settle out on cooling.

dioxide to reduce the viscosity of epoxidized polybutadiene [34], the use of allo-ocimene dioxide with DGEBA [36], and the use of aliphatic diepoxies with epoxidized halogenated aromatics [26].

Perhaps the most important use of the epoxy-resin reactive diluents is in combination with the more highly functional resins. Vinylcyclohexene dioxide, for example, may be used to convert the normally solid or semisolid epoxidized *o*-cresol novolacs to pourable liquids at room temperature with far less effect on elevated-temperature properties than would be obtained with a monofunctional diluent.

Table 13-10. The Effect of Catalysts on the Reaction of Monofunctional Diluents with Various Additives at 50°C [2]

Epoxy and additive	Time of ½ epoxy reaction, hours					
	No cat.	H_2SO_4	NaOH	NaOBu	NaOAc	$MeSO_3H$
1,2-Epoxyoctane:						
With *n*-butyl alcohol	. . . . .	18	2.5	2.75		
With acetic acid	15	15	. . . . .	. . . . . .	2.75	
With *n*-butylamine in chlorobenzene	100	. . . . .	. . . . .	. . . . . .	. . . .	30
2,3-Epoxyoctane:						
With *n*-butyl alcohol	. . . . .	6	100	100	. . . .	
With acetic acid	10	9.25	. . . . .	. . . . . .	4	
With *n*-butylamine in chlorobenzene	100	. . . .	. . . . .	. . . . . .	. . . .	30
Dipentene monoxide:						
With *n*-butyl alcohol	. . . . .	3				
With acetic acid	0.6					
With *n*-butylamine in chlorobenzene	100	. . . . .	. . . . .	. . . . . .	. . . .	30
Phenyl glycidyl ether:						
With *n*-butyl alcohol	. . . . .	2				
With acetic acid	24					
With *n*-butylamine in chlorobenzene	5	. . . . .	. . . . .	. . . . . .	. . . .	2.5

The shorter-chain materials and the cyclic structures of suitable reactivity have little effect on the DT of the cured systems. For example, with a system having a DT of 120°C without diluent, the addition of 15 per cent butadiene dioxide raises the DT to 137°C; of 3,4-epoxy-6-methylcyclohexylmethyl 3,4-epoxy-6-methylcyclohexane carboxylate, to 126°C.

With the linear aliphatic diepoxy reactive diluents, as the lengths of the chain between reactive sites is increased, the effect is to reduce the DT of the cured system. For example, while butadiene dioxide leads to an improvement in this value, DGE of butanediol, and the less efficient DGE of diethylene glycol, results in a reduction in the DT, the extent of the reduction being a function of the concentration of the diluent. As chain length is further increased, the diluents become increasingly less efficient in reducing viscosity and assume the character of flexibilizing modifiers.

When the diglycidyl ether diluents are used with the glycidyl ether resins, the extension of pot life obtained with the monoepoxy diluents is not realized. With non-glycidyl ether diluents are employed, the effect upon the pot life will depend on the type of curing agent employed, an increased reaction rate being obtained with acidic curing agents when the epoxy group is suitably situated, and a decreased reaction rate with amine curing agents. With bis(2,3-epoxycyclopentyl) ether, a general reduction in reactivity is obtained toward both the amine and anhydride curing agents.

Table 13-11. Typical Epoxy-resin Reactive Diluents

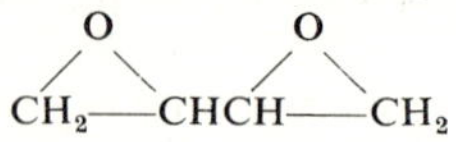

Butadiene dioxide (mol. wt. 86)

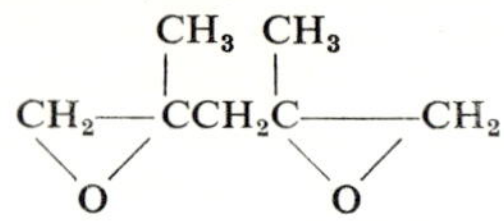

Dimethylpentane dioxide (mol. wt. 128)

Diglycidyl ether (mol. wt. 130)

Butanediol diglycidyl ether (mol. wt. 202)

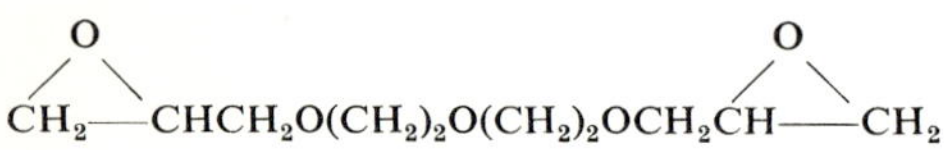

Diethylene glycol diglycidyl ether (mol. wt. 218)

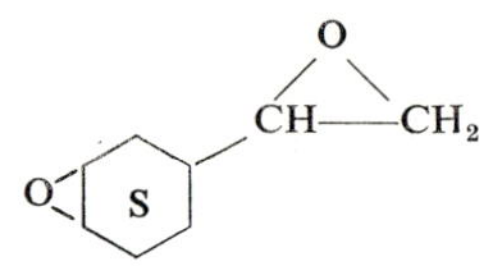

Vinylcyclohexene dioxide (mol. wt. 140)

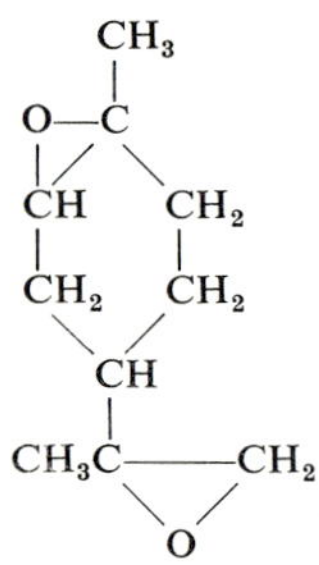

Limonene dioxide (mol. wt. 168)

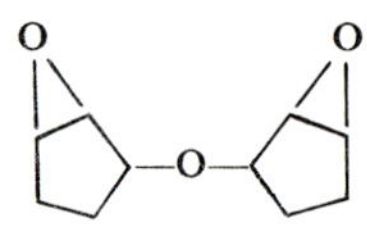

Bis(2,3-epoxycyclopentyl) Ether (mol. wt. 182)

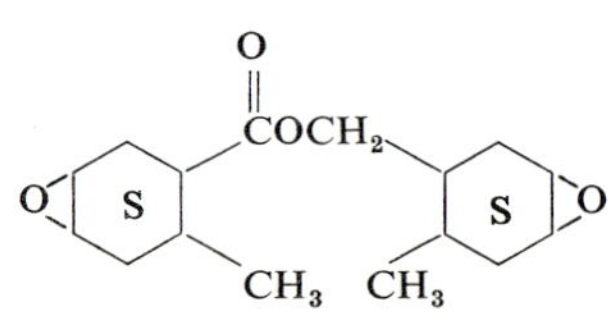

3,4-Epoxy-6-methylcyclohexylmethyl 3,4-epoxy-6-methylcyclohexane carboxylate (mol. wt. 280)

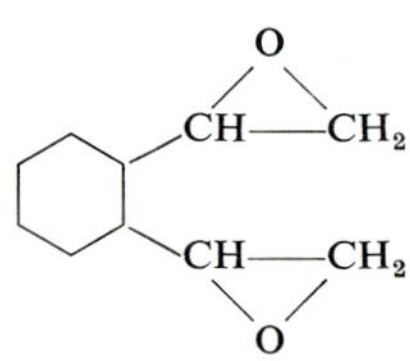

Divinylbenzene dioxide (mol. wt. 162)

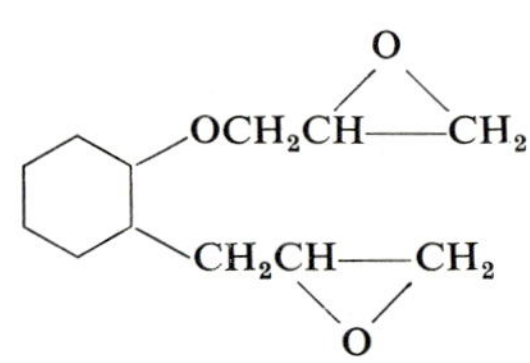

2-Glycidyl phenyl glycidyl ether (mol. wt. 206)

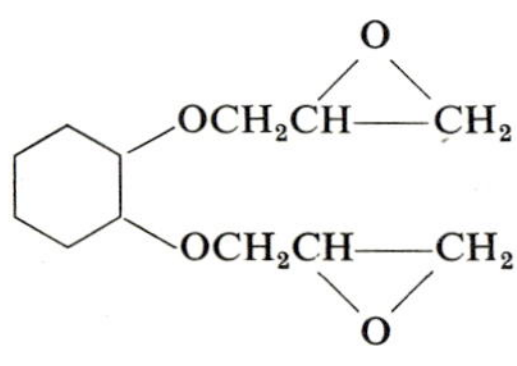

Diglycidyl ether of resorcinol (mol. wt. 222)

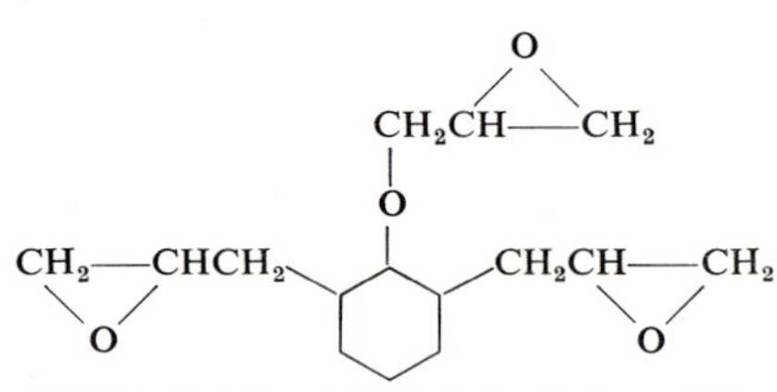

2,6-Diglycidyl phenyl glycidyl ether (mol. wt. 262)

Table 13-12. Effect of Concentration of DGE of Butanediol on Properties of TETA-cured DGEBA [4-1]

Property	DGE of butanediol, phr			
	0	16	33	55
Initial viscosity, centipoises	12,000	1,500	500	200
Pot life, minutes, at 23°C	40	41	42	43
Maximum exotherm, °C	200	217	230	225
DT, °C	120	102	87	80
Flexural strength, psi	20,300	14,700	15,100	13,100
Weight loss, 48 hr at 200°C	1.89	3.56	5.16	6.56
Hardness, Barcol	87	87	84	82
Water absorption, %:				
After 2-hr boil	0.72	1.13	1.75	2.37
After 30 days' immersion	0.79	1.30	2.04	3.86
Dielectric constant:				
At 60 cps	4.17	4.22	4.21	4.44
At 10^5 cps	3.67	3.73	3.76	3.96
Dissipation factor:				
At 60 cps	0.0096	0.012	0.012	0.015
At 10^5 cps	0.036	0.036	0.037	0.040

Vinylcyclohexene dioxide, when used at low phrs, will reduce the DT of primary-amine-cured DGEBA from 120 to about 109°C [16], and epoxidized unsaturated terpenes will reduce the DT to 80°C. In these cases, the reduction in DT probably indicates that these epoxy groups provide decreased reactivity with the curing agent such that full advantage is not taken of the functionality of the molecule.

The effect of typical monoepoxy diluents on the properties of DGEBA are presented in Tables 13-12 to 13-14.

Table 13-13. Physical Properties of DGEBA as a Function of Concentration of Bis(2,3-epoxycyclopentyl) Ether When Cured with Various Curing Agents [4-7]

Property	Bis(2,3-epoxycyclopentyl) ether, phr							
	43				100			
	A	*B*	*C*	*D*	*E*	*B*	*C*	*D*
DT, °C	45	140	161	129	63	165	161	123
Tensile strength, psi	7,500	11,400	3,570	......	......	9,000	3,400	2,270
Modulus, psi × 10^{-6}	0.462	0.470	0.372	......	......	0.486	0.415	0.447
Flexural strength, psi	11,000	20,100	9,970	12,800	19,700	16,000	11,000	8,820
Modulus, psi × 10^{-6}	0.335	0.504	0.439	0.410	0.455	0.551	0.388	0.427
Compressive strength, psi	15,200	31,000	33,000	20,200	......	36,000	31,400	19,700
Modulus, psi × 10^{-6}	0.521	0.526	......	......	......	0.529		
Hardness, Rockwell M	91	118	111	108	104	118	107	111

A. Ethylene oxide adduct of DETA/bisphenol A cured 13 days at 23°C.
B. MPDA/MDA eutectic/bisphenol A cured 2 hr at 80°C + 4 hr at 160°C.
C. 3 phr BF_3MEA gelled at 100°C and cured 9 hr stepwise to 200°C.
D. HHPA/0.2% BDMA gelled at 100°C and cured stepwise 7 hr to 180°C.
E. Curing agent *A* cured 2 hr at 88°C + 2 hr at 110°C.

Table 13-14. Electrical Properties of DGEBA Cured 2 Hours at 80°C + 4 Hours at 160°C with MPDA/MDA/Bisphenol A as a Function of Concentration of Bis(2,3-epoxycyclopentyl) Ether [4-7]

Property	Bis(2,3-epoxycyclopentyl) ether, phr					
	43			100		
	At 25°C	At 50°C	At 100°C	At 25°C	At 50°C	At 100°C
Arc resistance, seconds	....	....	....	82		
Dielectric strength, volts/mil	500			412		
Dielectric constant:						
At 60 cps	4.22	4.34	4.42	4.57	4.49	4.56
At 10^3 cps	4.03	4.31	4.41	4.48	4.45	4.51
At 10^6 cps	3.66	3.84	4.09	3.62	3.69	4.24
Power factor:						
At 60 cps	0.0155	0.00381	0.00637	0.0062	0.0038	0.0042
At 10^3 cps	0.0628	0.00688	0.00268	0.0121	0.0062	0.0028
At 10^6 cps	0.0412	0.0391	0.0267	0.0315	0.0365	0.0286
Loss factor:						
At 60 cps	0.0648	0.0166	0.0281	0.0281	0.159	0.0193
At 10^3 cps	0.107	0.0296	0.0118	0.0542	0.0274	0.0126
At 10^6 cps	0.151	0.150	0.109	0.114	0.135	0.143

NON-EPOXY-CONTAINING REACTIVE DILUENTS

The non-epoxy-containing reactive diluents consist of those materials which can properly be considered low-viscosity curing agents, those materials which may react with the curing agents rather than the epoxies, and those which, while not specifically curing agents, will, under suitable conditions, be caused to react with other functional groups present in the epoxy resin.

Low-viscosity curing agents, such as tertiary amines, may be blended with higher-viscosity curing agents, such as the fatty polyamides, in effect serving as reactive diluents for the epoxy system. A variety of such blends may be developed, and many are discussed in other parts of this book.

Unsaturated monomers, such as styrene and *d*-limone [40], may be employed with unsaturated epoxy resins, the reaction being brought about through unsaturation by peroxide curing agents or by temperatures of about 200°C. The effect on

Table 13-15. Effect of Acrylic Monomers on Properties of Epoxidized Polybutadiene [4-3]

Formulation and properties	Ingredient concentration, cure cycle, and property values							
	A	*B*	*C*	*D*	*E*	*F*	*G*	*H*
Epoxidized polybutadiene	100	100	100	100	100	100	100	100
Methyl methacrylate	20	30	30					
Butyl methacrylate	......	......	......	20				
Styrene	......	......	......	......	20	30	30	
Hexahydrophthalic anhydride	80	80	80	80	80	80	80	80
Propylene glycol	3	3	3	3	3	3	3	3
Lupersol 101	3	3	3	3	3	3	3	3
Cure cycle, hours at 155°C	24	4	24	24	24	9	24	24
Properties:								
DT, °C	122	......	137	120	110	......	......	168
Flexural strength, psi	13,700	13,200	14,400	10,600	13,100	12,500	11,700	13,100
Clarity	Yes	Yes	Yes	Yes	Cloudy	Cloudy	Cloudy	Yes

properties is indicated in Table 13-15. Unsaturated diluents may still be considered reactive when unsaturation is present in the curing agent or the resin; for example when maleic anhydride or an unsaturated acid is employed [39]. Conjugated unsaturated dienes have been reported as reactive diluents for epoxidized oils cured with unsaturated anhydrides, the reaction here being through unsaturation during cure [27].

Short-chain polyols may be employed, the reaction being either with the curing agent (i.e., anhydrides) or with the epoxy groups (i.e., Lewis-acid-catalyzed). Other compounds containing active hydrogens may also be considered as reactive diluents under suitable conditions, one example being the phenols [38] as previously discussed.

Internal amides or lactams such as ε-caprolactam [42] have been suggested as diluents when employed in conjunction with Lewis acids [12], amines [11], and titanium esters [10]. In the last case, pot lives of up to 4 months at room temperature have been obtained with DGEBA, and cures as rapid as 30 seconds at 150°C.

Reactive diluents may be employed to reduce functionality of the epoxy-resin molecule and to introduce side chains which have the effect of lowering the viscosity. In this connection, 3-ethylhexoic acid has been used to reduce the functionality and viscosity of epoxy resins derived from reacting excess diepoxy with cyanuric acid [18].

Reaction is also possible through other than unsaturation and epoxy groups. Triphenyl phosphite, for example, can be reacted into the system during cure by a trans-esterification involving the formed hydroxyl groups [14]. A hydroxyl on the resin chain reacts with the triphenyl phosphite as follows:

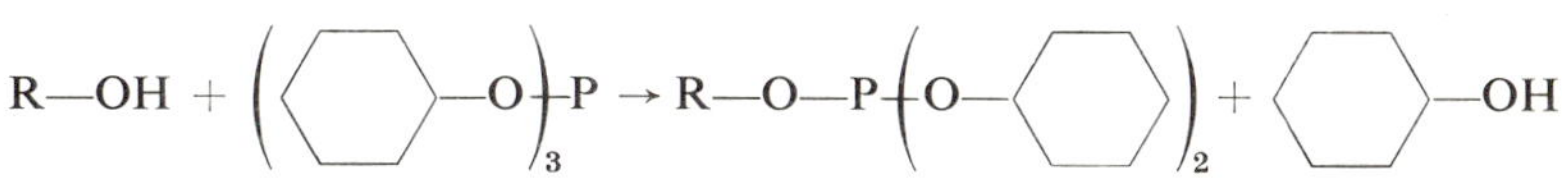

Hydroxyl Triphenyl phosphite Phosphorus ester Phenol

The reaction may continue to liberate three phenols, which in turn may react with the resin or accelerate the reaction [25].

Properties of DGEBA cured with TETA as a function of the concentration of triphenyl phosphite are indicated in Table 13-16. Triphenyl phosphite is somewhat moisture-sensitive, and is not stable in the epoxy resin.

Lactones may be employed as diluents [33], these reacting with amine curing agents, the reaction being

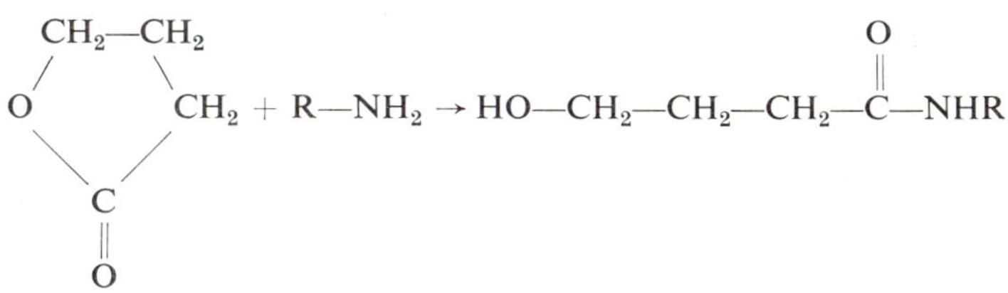

Butyrolactone Amine Hydroxybutyl amide

Table 13-16. Effect of Concentration of Triphenyl Phosphite on Properties of TETA-cured DGEBA [15]

TPP, phr	Viscosity, centipoises	Pot life at 23°C, minutes	Exo-therm, °C	H_2O absorption, 2-hr boil	H_2O absorption, After 30-days' immersion	DT, °C	Flexural strength, psi	Weight loss in 48 hr at 200°C	Hardness, Barcol
None....	......	40	200	0.73	0.79	120	20,300	1.89	87
27	1,500	16	>300	1.36	1.21	96	17,300	7.93	86
55	500	23	>300	1.11	1.81	79	12,033	Unstable	80
92	200	33	>300	–2.81	1.91	49	6,223	Unstable	68

Table 13-17. Effect of Butyrolactone on Properties of DETA-cured DGEBA [1]

Property	Butyrolactone, phr				
	0	5	10	15	20
Viscosity, centipoises	16,700	4,700	1,820	960	580
DT, °C	113	93	77	65	57
Weight loss, %:					
After 8 hr at 100°C	0.01	0.02	0.04	0.08	0.16
After 24 hr at 150°C	0.65	0.69	1.05	1.81	2.73
Weight change after 18 hr in boiling water, %	+1.39	.....	+4.88	....	+6.12
Hardness, Barcol:					
After 2 hr at 25°C	45	45	41	38	22
After 2 hr at 60°C	41	46	43	38	27

Butyrolactone, used with DGEBA gives properties as indicated in Table 13-17.

Acrylonitrile has been used as a diluent for amine cures [3], the reaction being as described in Chap. 7. It has also been described as a diluent for organic acid cures [22].

SUMMARY

Diluents for epoxy resins are used principally to lower the system viscosity, although they may also be employed to make minor changes in pot life, exotherm, and cured properties.

The diluents may be classified broadly as nonreactive, reactive epoxy-containing, and reactive non-epoxy-containing. The nonreactive diluents generally degrade cured properties, the extent of the reduction depending upon the concentration; except for special applications, they are not preferred.

The epoxy-containing diluents may be either monoepoxy molecules or epoxy resins. The former give somewhat reduced cured properties but generally provide significant viscosity reduction at relatively low concentrations. The latter may be used at rather high concentrations with little effect upon cured properties, sometimes even leading to selective improvements.

The reactive, non-epoxy-containing diluents consist of co-curing agents, unsaturated molecules, active-hydrogen-containing molecules, materials which are capable of promoting transesterification with available hydroxyl groups and materials capable of reacting with the curing agent. These will affect properties in general as they influence cross-linking density and system functionality.

Many of the diluents, particularly those containing epoxy groups, are more severe skin irritants than the resins themselves, and some may have undesirable systemic effects.

REFERENCES

1. Antara Chemicals Co., technical literature.
2. Becco Chemical Co., technical literature.

1951

3. Shokal and Mueller, Fluid Compositions Containing Glycidyl Polyethers, U.S. 2,548,447.

1954

4. Gams, et al., Adhesives and Articles United Therewith, U.S. 2,682,490.

1958

5. Bressler and Smith, Epoxide Resins and 1,4,5,6,7,7-Hexachloro 2-(2,3-Epoxypropoxymethyl)bicyclo(2,2,1) Hept-5-ene, U.S. 2,838,496.
6. Bressler and Smith, Hexachlorocyclopentadiene Adduct, U.S. 2,834,790.
7. Gall, et al., The Olefin and Terpene Oxides: Novel Reactive Diluents for Epoxy Resin Formulations, Epoxy Resin Symposium of SPE, Minneapolis, Minn., October.
8. Gollub and Wood, Primers for Vinyl Resin Coating, U.S. 2,842,459.
9. Heinrich, et al., Compositions of Matter Comprising an Ethoxyline Resin and an Alkylene Polyamine Containing at Least Two Tertiary Amino Groups, U.S. 2,839,480.
10. Kohn, Ethoxyline Resin Compositions and Their Preparation, U.S. 2,847,341.
11. Kohn, Ethoxyline Resin Compositions and Their Preparation, U.S. 2,847,342.
12. Kohn, Ethoxyline Resin Compositions and Their Preparation, U.S. 2,847,343.
13. Steckler and Werner, Epoxy Ether Resins: Polymethoxy Acetal Compositions, U.S. 2,853,468.

1959

14. Barnstorff, et al., Chemistry of Triphenyl Phosphite Modified Epoxy Resins, ACS Symposium, Atlantic City, September.
15. Pschorr and Dorman, Reactive Diluents in Epoxy Resin Systems, 15th ANTEC of SPE, New York City, January.
16. Thomas, Effect of Modifiers on Properties of Epoxy Resins, ACS Symposium, Atlantic City, September.
17. Wasserman, Compositions of Matter and Methods for Preparing Them, U.S. 2,900,364.

1960

18. Belanger, Epoxide Resins, U.S. 2,947,726.
19. Bruin and Sinnema, Process for Curing Polyepoxides and Resulting Products, U.S. 2,951,824.
20. Phillips, et al., Epoxide Compositions and Resins Therefrom, U.S. 2,935,488.
21. Rott, Vacuum Casting the 2012 Transformer, W.E.-BTL Casting Resin Conference, North Andover, Mass., May.

1961

22. Feuchtbaum, Coated Electrical Structure, U.S. 2,996,693.
23. May and Nixon, Reactive Diluents in Aromatic Amine-cured Epoxy Adhesives, *J. Chem. Eng. Data*, April. (The *Journal of Chemical and Engineering Data* is the original source of this material, and the American Chemical Society is the publisher and copyright owner of the journal.)
24. Phillips, et al., Polyepoxide-Divinylbenzene Dioxide Resinous Compositions, U.S. 2,982,752.

1962

25. Curing Epoxidized Resins and Curing Compositions Therefor (to Argus Chemical Corp.), Brit. 903,932.
26. Polyglycidyl Ether Resins (to Pittsburgh Plate Glass Co.), Brit. 897,290.
27. Radlove, Solidifiable Fluid Compositions Prepared from Enedioyl Acids, Epoxidized Hydrocarbon Drying Oils, and Conjugated Dienes, U.S. 3,050,478.
28. Russell, Epoxide Resin Compositions, U.S. 3,050,474.
29. Supplementary Bulletin, Cardolitc Epoxy Resin Flexibilizer and Reactive Diluent NC-513, Minnesota Mining & Manufacturing Co., January.

1963

30. Chernin, Thermomechanical Investigation of Compositions Based on Ester-plasticized Epoxy Resins, *Lakokrasochnye materialy i ikh Primenenie* **6**:16–18.
31. Cooke, et al., Chemical Resistance of Room Temperature Cured Epoxy Resin Compositions, ACS Symposium, New York City, September.

32. Epoxides and Epoxy Resins (to Ciba, Ltd.), Brit. 944,408.
33. Epoxy Resins (to Ciba, Ltd.), Belg. 624,986.
34. Greenspan, Composition Comprising Epoxy-Polybutadiene and Limonene Dioxide, U.S. 3,073,792.
35. Greenspan and Light, Thermoset Epoxy Resin Plasticized with a Glycol Ether-Ester, U.S. 3,073,793.
36. Kraft and Weisfield, Polyepoxide Resin Composition, U.S. 3,073,786.
37. Lincoln and Cooke, Improvement in and Relating to Epoxide Resin Compositions, Brit. 916,246.
38. Modified Ethoxylin Resins (to Allman Electriska Aktiebolaget), Brit. 927,527.
39. Nowlin, et al., Improvements in and Relating to Thermosetting Resins, Brit. 922,356.

1964

40. Pschorr and Weschler, Compositions Comprising a Liquid 1,2-Epoxy Resin and *d*-Limone, U.S. 3,132,115.

1965

41. Esposito, Identification and Determination of Reactive Epoxy Diluents by Gas-Liquid Chromatography, AD 615 955.
42. Gruenwald, Epoxy Resin Compositions, Fr. 1,384,243.
43. Modified Epoxy Resins (to Westinghouse Electric Corp.), Brit. 1,006,848.
44. Simm, Molded Articles from Epoxides, Ger. 1,190,664.

Appendix 13-1

INFRARED SPECTRA FOR TYPICAL EPOXY-RESIN DILUENTS*

* Prepared by Art Cupples, Okko Koning, and Bette Wolfskill of The Epoxylite Corporation.

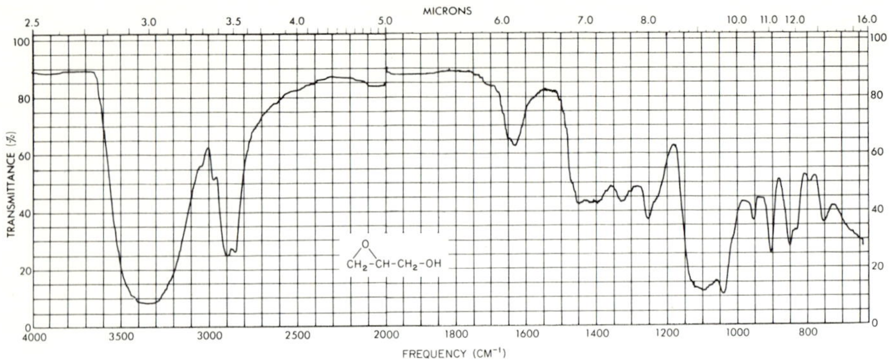

FIG. 13-1-1. Glycidol.

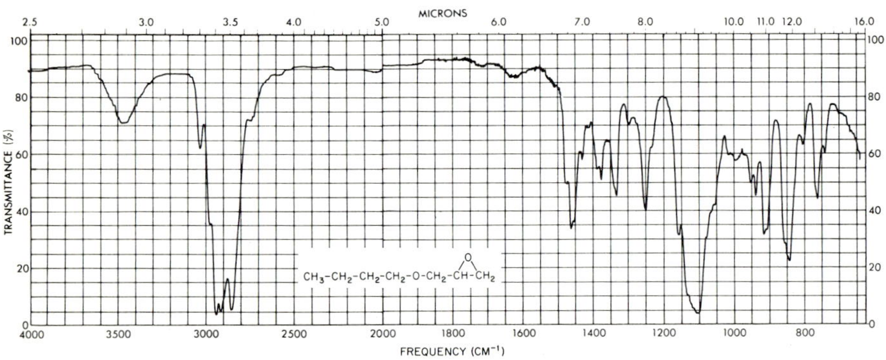

FIG. 13-1-2. Butyl glycidyl ether (butyl 2,3-epoxypropyl ether).

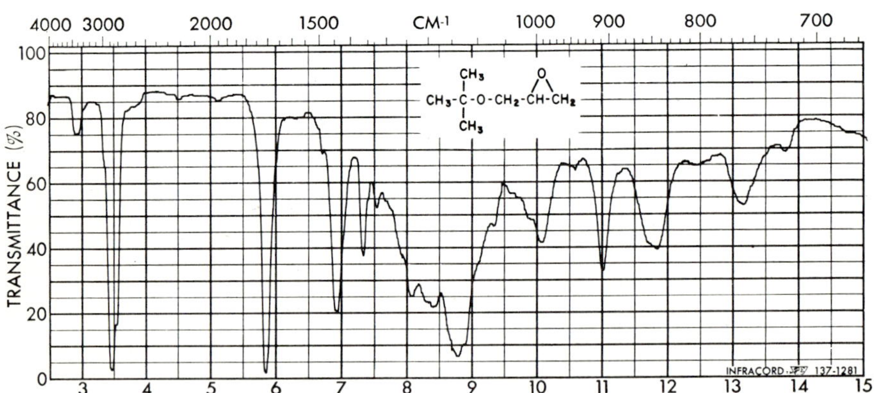

FIG. 13-1-3. Tertiary butyl glycidyl ether (*tert*-butyl 2,3-epoxypropyl ether).

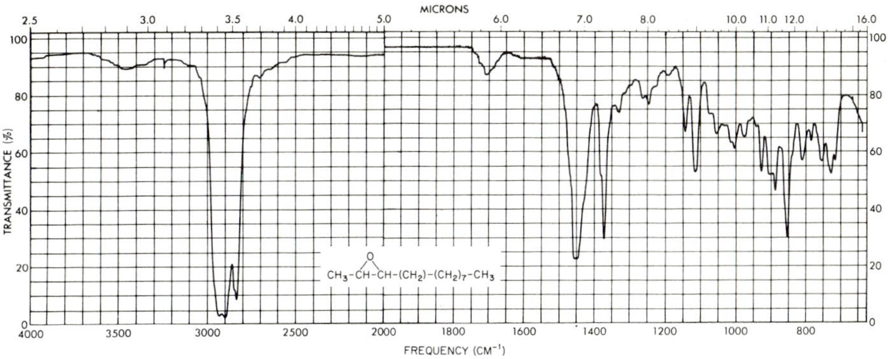

FIG. 13-1-4. Octylene oxide (2,3-epoxyoctane).

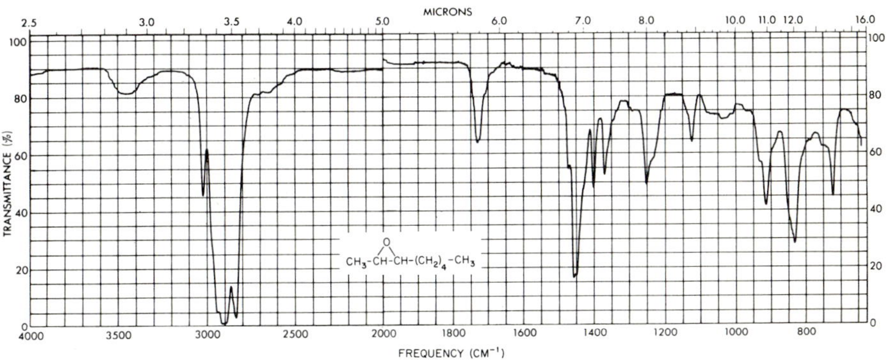

FIG. 13-1-5. Dodecene oxide (2,3-epoxydodecane).

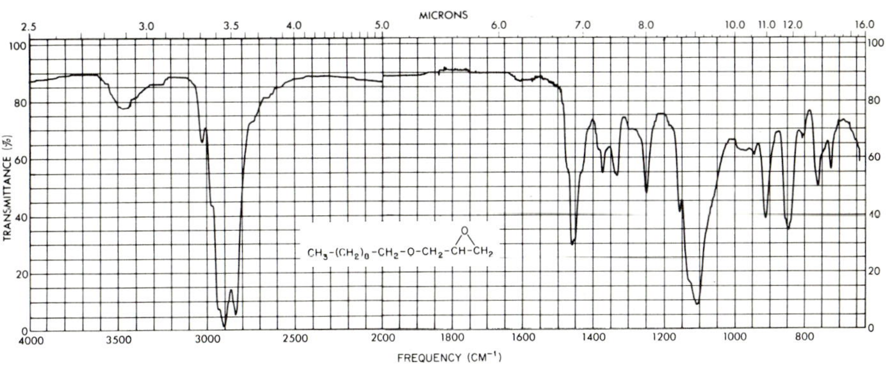

FIG. 13-1-6. Decyl glycidyl ether (decyl 2,3-epoxypropyl ether).

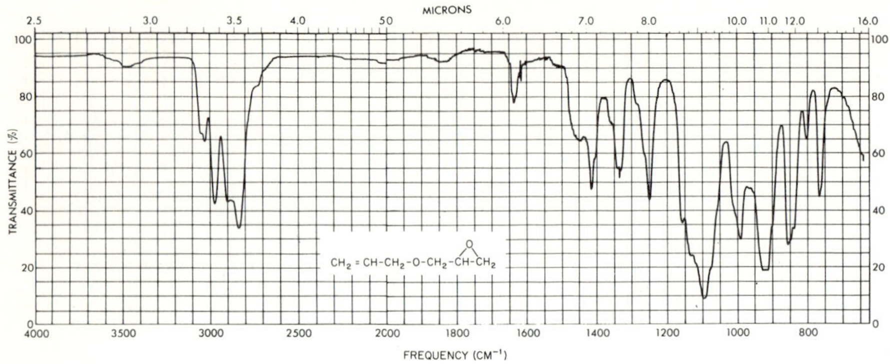

FIG. 13-1-7. Allyl glycidyl ether (allyl 2,3-epoxypropyl ether).

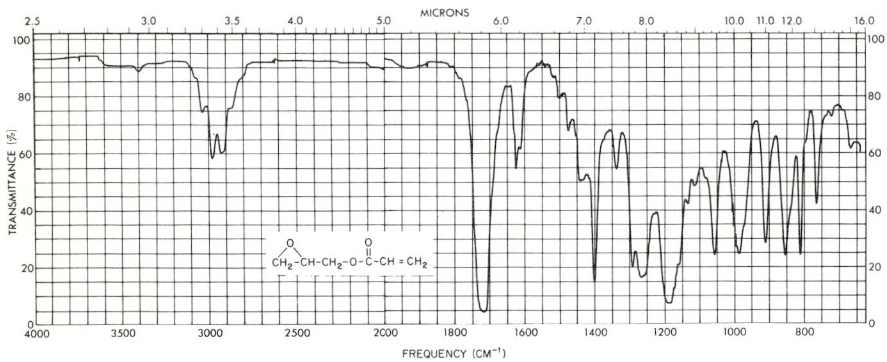

FIG. 13-1-8. Glycidyl acrylate (2,3-epoxypropyl acrylate).

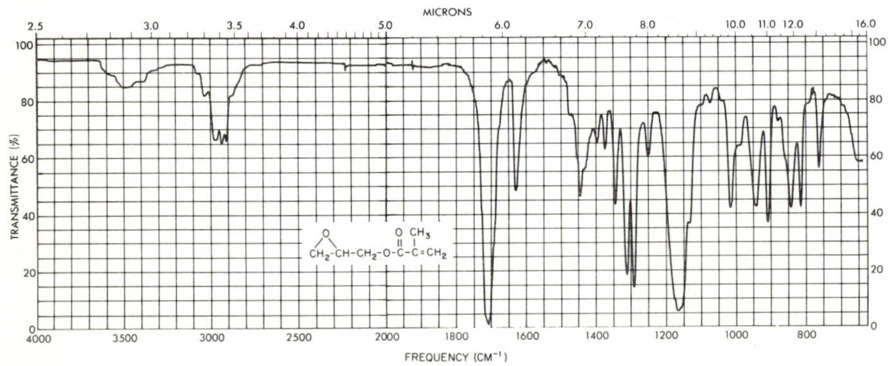

FIG. 13-1-9. Glycidyl methacrylate (2,3-epoxypropyl methacrylate).

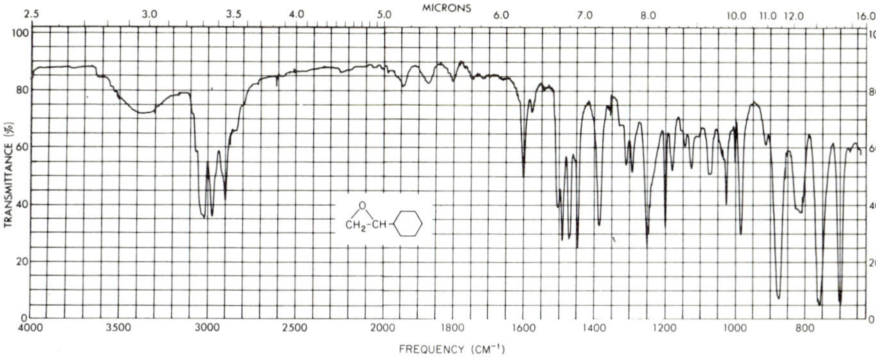

FIG. 13-1-10. Styrene oxide (1,2-epoxyethylbenzene).

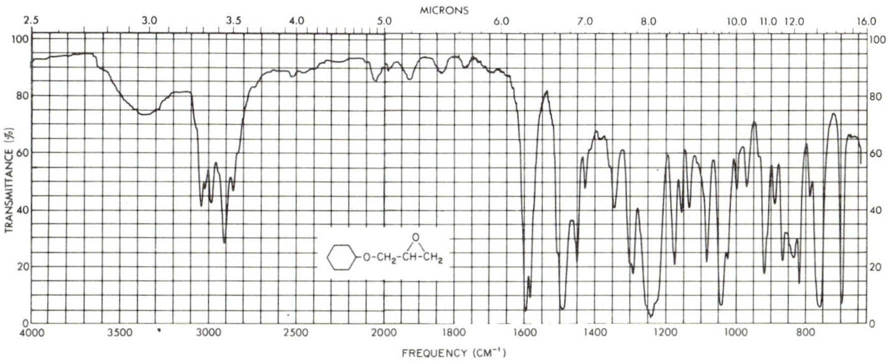

FIG. 13-1-11. Phenyl glycidyl ether (2,3-epoxypropyl phenyl ether).

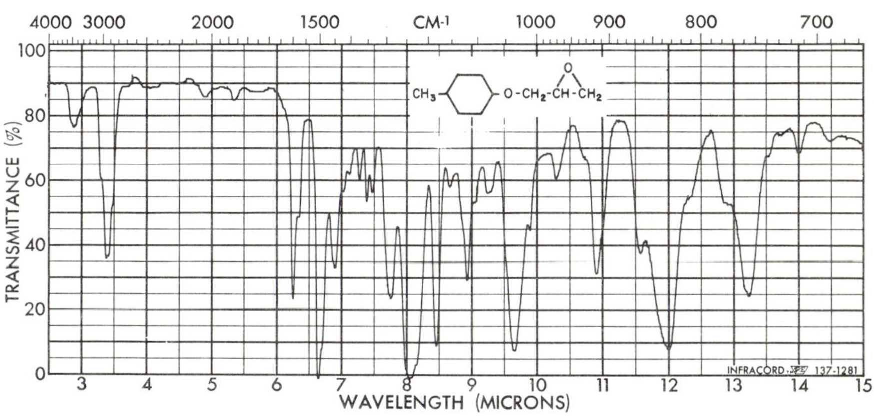

FIG. 13-1-12. Cresyl glycidyl ether (2,3-epoxypropyl *p*-tolyl ether).

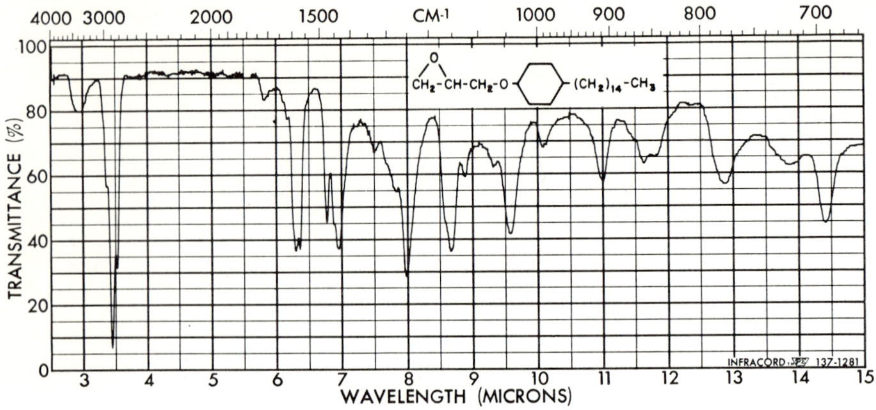

FIG. 13-1-13. 2,3-Epoxypropyl (*p*-pentadecylphenyl ether).

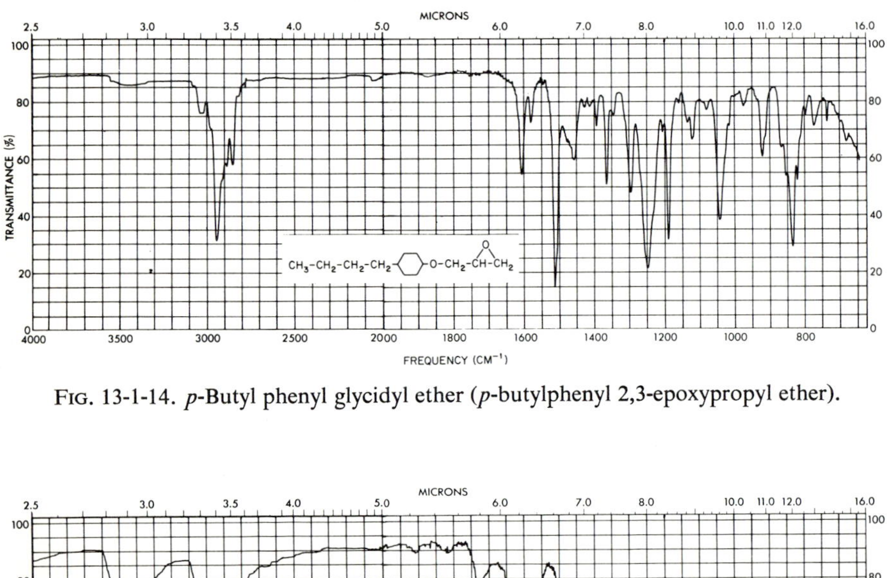

FIG. 13-1-14. *p*-Butyl phenyl glycidyl ether (*p*-butylphenyl 2,3-epoxypropyl ether).

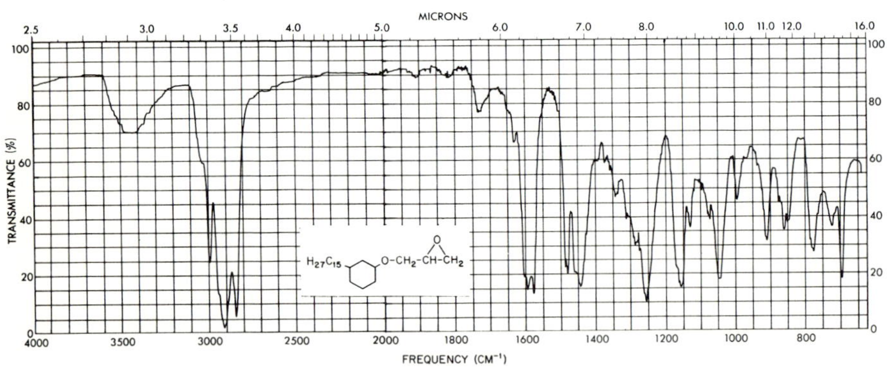

FIG. 13-1-15. 2,3-Epoxypropoxy-3-pentadecadienylbenzene.

Chapter 14

FILLERS FOR EPOXY RESINS

The purpose of this chapter is to discuss the effect of fillers on various important design properties of the epoxy resins. Fairly extensive tabular data are provided as a reference for the subsequent chapters on the applications of the epoxy resins in specific types of services. A special section contains information on molecular sieves.

COMMERCIAL FILLERS

There are available at least several hundred commercial fillers which are suitable for use in epoxy resins. Most range in particle size from about 200 mesh to about 0.015 micron. Typical commercial materials are shown in Table 14-1. Properties of various fillers are compared in Table 14-2.

Although the majority of the fillers used in the epoxy technology are finely divided, coarser materials (sand, gravel, etc.) may also be employed. These are used mainly in essentially 100 per cent solids epoxy floor-topping and gap-filling formulations and in some cast tooling applications. The principal function of the coarse fillers is bulk displacement of the more expensive resin.

Most commercially available inorganic fillers are suitable for use with most epoxy-resin systems. The filler should be neutral or only slightly basic, and nonreactive with the resin or the curing agent. This requirement precludes the use of some fillers with some classes of curing agents. Mica, when used at fairly high concentration, may cause undesirable reactions with DGEBA during elevated-temperature cures, particularly with some anhydride curing agents. Asbestos may react unfavorably in systems catalyzed by BF_3 complexes. Menthane diamine has been observed to react with zirconium silicate to produce bubbles [25]. Other fillers may vary in their effect, depending upon the water content and the tendency to pick up water during storage, some talcs being particularly bad in this respect. Mineralite 4X mica, mined from the deposit near Kershaw, S.C., appears to possess piezoelectric properties. A DGEBA system containing 4X mica, whose configuration is probably one of electrical asymmetric platelets, revealed a charge release during shock-wave transit time when the sample was exposed to explosive loading. This mica, therefore, should be avoided as an electronic-grade filler for encapsulation of units that will be subjected to stress or shock, since the charge release could result in the production of noise signals in sensitive circuits [73].

Organic fillers are comparatively little used in the epoxy technology, although several have been suggested for imparting specialized properties. Coarse Buna N improves thermal shock resistance and resiliency [19]. A cured, chlorinated polyester powder with antimony trioxide gives flame resistance [26]. Powdered polypropylene may be added to epoxy coatings to produce mat or semilustrous surfaces which resist dust, dirt, and scratches [63]. Organic surface-active agents have been used to treat fillers to increase or decrease viscosity at constant loading volumes [7]. Amine reaction products of special clays are used as antisettling agents for powdered curing agents such as dicyandiamide [13]. Low-density fillers, such as organic microballoons, have found use in special casting formulations, where they are employed to reduce weight. Copper chelating agents have been used as heat stabilizers in high-temperature adhesive formulations [28]. The most widespread use for the organic fillers, however, is as dyes or pigments.

Inorganic fillers, as well as organic ones, may be used to impart specialized properties. Graphite or molybdenum disulfide provides a surface-lubricating effect [30,67]. Zinc oxide with some systems is a cure accelerator. Special ceramic fillers improve radiation resistance [71].

The commercial fillers generally are composed of a variety of particle sizes as indicated, for example, in the particle-size distribution for commercial lithium aluminum silicate (Fig. 14-1).

In some cases, the filler may be treated either in advance or in situ to provide

Table 14-1. Typical Commercial Fillers for Epoxy Resins

Filler	Supplier	Chemical nature	Sieve analysis
Aluminum silicate	Edgar Brothers Co.	SiO_2 44.20% Al_2O_3 40.14% MgO 0.02% CaO 0.02% TiO_2 0.40% $Na_2O + K_2O$ 0.50% Fe_2O_3 trace	Through 400 mesh 100%
Calcium carbonate	Diamond Alkali Co.	$CaCO_3$ 98.43% $MgCO_3$ 0.37% SiO_2 0.01% Fe_2O_3 0.01% $CaSO_4$ 0.88% Al_2O_3 0.05%	Retained on 100 mesh 0.01% Retained on 200 mesh 0.05% Retained on 300 mesh 0.50% Particle size 1–5 microns 99.44%
Magnesium silicate	Whittaker, Clark & Daniels	SiO_2 57.06% MgO 28.68% CaO 3.07% Al_2O_3 0.81% Fe_2O_3 1.25%	Through 325 mesh 100%
Ceramic zircon	Orefraction Inc.	Maximum iron 0.10% TiO_2 0.25% SiO_2 0.50% Misc. 0.15% $ZrSiO_4$ 99.00%	Through 400 mesh 99%
Aluminum powder no. 44	Metals Disintegrating Co.	Aluminum 99.00% Other values not given	Through 40 mesh 100% Through 325 mesh 40%
Tabular alumina T-60	Aluminum Co. of America	Al_2O_3 89.37% L.O.I. 8.00% Na_2O 2.32% TiO_2 0.01% Fe_2O_3 0.21% SiO_2 0.09%	Through 325 mesh 100%
Kaolin no. 347	Whittaker, Clark & Daniels	SiO_2 46.25% Al_2O_3 39.15% Fe_2O_3 0.64% CaO 0.20% MgO trace L.O.I. 13.76% Lead 4.7 parts/10^6	Through 325 mesh 99.98%
Mica 4X	Mineralite Sales Corp.	SiO_2 58.38% Al_2O_3 29.63% K_2O 6.91% CaO 1.08% MgO 0.41%	1,000 mesh
Silica	United Clay Mines Corp.	SiO_2 99.92% Fe_2O_3 0.016% Al_2O_3 0.04% CaO trace	Through 325 mesh 93.5%
Zirconium silicate	National Lead Co.	ZrO_2 62.47% SiO_2 35.67% MgO 0.17% Al_2O_3 0.59%	Less than 1.5 microns
Hydrated alumina	Aluminum Co. of America	Al_2O_3 64.40% Combined H_2O 34.90% SiO_2 0.05% Fe_2O_3 0.02% NaO (total) 0.45%	0.08–0.12 micron

Table 14-1. (contd.)

Filler	Supplier	Chemical nature	Sieve analysis
Lithium aluminum silicate	Foote Mineral Co.	....	Through 200 mesh 0.5% Through 325 mesh 95% Average 10 microns
Colloidal silica	Godfrey L. Cabot Inc.	SiO_2 99.0–99.7% CaO, MgO 0.00% Fe_2O_3 0.004%	0.015–0.020 micron Surface area 175–200 sq m/g
Dimethyl dioctadecyl-ammonium bentonite	National Lead Co.	....	Platelets 0.05–1.0 by 0.002–0.004 micron thick Through 200 mesh less than 5%

improved properties. Volan A treated glass balloons are available. The treatment of silica with vinyl silanes has been described as improving adhesion between the filler and the resin [46]. Silane coupling agents have been added to the resin/silica combination for the same purpose; in this case, the silane migrates to the filler surface before cure [65]. Special molecules containing epoxy groups may also be used, for example the reaction product of silicon tetrachloride with a partially esterified epoxy resin [60]. Use of silane additives to metal-filled epoxies has been reported as improving flexural strengths as much as 94 per cent and compressive strengths as much as 75 per cent over those of comparable systems without the additives. In most cases, strengths were improved over original values after 72 hours' water boil [80].

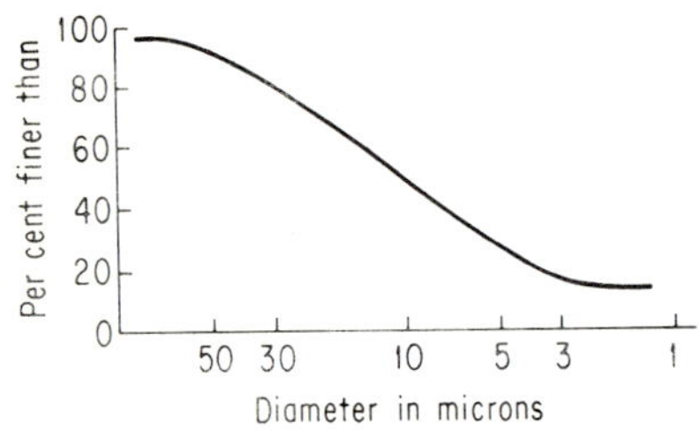

FIG. 14-1. Particle distribution of commercial lithium aluminum silicate [1].

EFFECT OF FILLER ON VISCOSITY

Fillers increase the viscosity of epoxy-resin systems: particulate fillers causing the least increase, fibrous fillers the most. Finer particle sizes, with attendant increase in wettable surface area, generally increase viscosity more than equal weights of larger particle sizes, although this is not always true. For example, with room-temperature curing systems, the epoxy resin may tolerate somewhat less silica than mica having a substantially larger surface area. Some fillers, such as clays, may be sensitive to pH changes such that deflocculation and viscosity increase will occur under alkaline conditions (i.e., with amine curing agents).

Most fillers when used in epoxy resins provide systems exhibiting viscosities relatively unaffected by shear rate. Some fillers, however, provide an apparent shear rate dependence: the viscosity being high at low shear rates, and low at high shear rates (thixotropy), or, more rarely, vice versa.

The thixotropic effect usually arises from flocculation of particles when they are thoroughly wetted. The particles are held together by Van der Waals forces at low shear rates, whereas at high shear rates the particles are forced apart to the extent that the forces are no longer of consequence. The effect is to produce systems into which the curing agent may be readily dispersed, but which, in the unagitated state, are nonsag in fairly substantial masses. Thixotropy may be obtained at fairly low loading volumes with colloidal silica, asbestos floats, methyldioctadecylammonium bentonite, metallic leafing powders, metallic soap powders, and special hydrated magnesium aluminum silicates.

Table 14-2. Comparison of Various Nonfibrous Fillers [58]

Filler	Specific gravity	Average particle size, microns	Particle shape	Bulk density, lb/ft	Thermal conductivity, cal/(sq cm)(sec)-(°C)(cm)	Coefficient of thermal expansion, in./(in.)(°C) × 10^6	Heat resistance, °C	Electrical resistivity, microhm-cm	Dielectric constant at 10^6 cps	Principal reasons for use
Metals:										
Aluminum	2.72	30–150	*a, b*	40–70	0.26–0.54	22–26	650^b	2.8–6.3	....	*c, d, e, f*
Copper	8.91	30–150	*b*	400–500	0.79–0.91	17	1080^g	1.7–2.0	....	*d, e*
Iron	7.8	30–150	*b*	350–450	0.08–0.15	8–14	1525^g	32–41	....	*d, e, h*
Silver	10.5	30–150	*a, b*	540–600	0.97	19.7	960^g	1.59	....	*d, e*
Steel	7.75	30–150	*b*	350–450	0.08–0.15	8–12	1600^g	19	....	*c, d, f*
Inorganics:										
Aluminum oxide	3.99	30–150	*b*	80–100	0.08	7.0	1425^i	$>10^{21}$	8.2–9.6	*f, h, j*
							2010^k			
Aluminum silicates (hydrated)	2.58	0.55–7.3	*a*	40–100	0.006	2–20	1870^k	$>10^{21}$	6.3	*c, f, l, m*
Antimony oxide	5.62	30–50	*b*	80–100	....	....	650^g	....	....	*n*
Barium titanate	5.52	30–50	*b*	80–100	....	16–19	1480^g	10^{18}	1,090–1,150	*h*
Colloidal silica	2.6	0.015–0.020	*b*	2.2–4	0.0032	0.15–1.3	1600^g	$>10^{21}$	3.8	*o*
Calcium carbonate	2.93	1–50	*b*	60–160	0.006	0.5–0.9	825^k	10^{15}–10^{17}	8.3	*c, f, m*
Calcium sulfate	2.96	20–50	*b*	40–60	0.0016	....	1450^g	....	....	*f, m*
Ferric oxide	5.2	20–50	*b*	250–300	....	....	1690^k	....	....	*c, h*
Fuller's earth (magnesium aluminum silicates)	2.3–2.5	0.1–18	*b*	15–35	0.009	....		....	....	*o*
Lithium aluminum silicate	2.38	20–30	*b*	40–59	0.004	0.6–0	1260^g	....	....	*f*
Mica	3.4	10–80	*a*	80–100	0.0016	18–27		10^{19}–10^{23}	5.4–8.7	*h, j, l*
Silica	2.6	80–100	*b*	80–120	0.0032	0.5–1.3	1600^g	$>10^{21}$	3.8	*e, f, j, m*
Silicone carbide	3.1	30–2500	*b*	100–140	0.04–0.11	....	2700^g	....		*f, h*
Talc	2.4	30–50	*b*	80–100	0.004	....		10^{14}–10^{15}	....	*m*
Titanium oxide	4.26	30–50	*b*	75–85	0.15	7.1–9.2	1480^i	10^{19}–10^{24}	95–105	*c, h, m*
							1900^g			
Quartz	2.6	20–300	*b*	80–120	0.0032	0.5	1920^i	$>10^{21}$	3.8	*e, f, j, m*
Zirconium oxide	5.5	30–50	*b*	75–85	0.01–0.015	....	2690^g	$>10^{20}$	8–12	*h*
Zirconium silicate	4.7	1–30	*b*	80–90	0.08	3.0	1200^i	$>10^{21}$	12.6	*f, h, j, l*
Organics:										
Carbon	1.8	5–45	*b*	20–40	0.12	1.3	3445^g	3,500–4,600	....	*c, d, e*
Graphite	2.26	5–45	*a*	40–70	0.28–0.48	0.6–4	3665^q	800–1,300	....	*c, d, e, r*

[a] Thin flat plate. [b] Random, irregular. [c] Pigmentation. [d] Thermal conductivity. [e] Electrical conductivity. [f] Dimensional stability.
[g] Melts. [h] Dielectric properties. [i] Softens. [j] Electrical resistivity. [k] Decomposes. [l] Chemical and moisture resistance.
[m] Extender. [n] Flame retardancy. [o] Thixotropic agent. [p] Needlelike. [q] Sublimes. [r] Abrasion resistance and lubricity.

Table 14-3. Effect of Various Fillers on Viscosity of DGEBA [50]

Filler	*Approximate phr at maximum pourable viscosity*
Black iron oxide	300
Tabular aluminum oxide	200
Atomized aluminum	150
Calcium carbonate	30
Silica	30
Graphite	50
Titanium dioxide	50
Asbestos	5

Colloidal silica, as representative of thixotroping agents, is usually incorporated into the resin at concentrations on the order of 10 phr. Triethanolamine acetate, water (0.5 phr), or a small amount of a polyhydric alcohol may be used to obtain improved thixotroping action. Systems based on the lower-viscosity resins generally tend to hold thixotroping action better at elevated temperatures than systems based on the higher-viscosity resins.

The reverse of thixotropy, high viscosity at high shear rates and low viscosity at low shear rates, is called *dilatance*. Some alumina-filled anhydride-cured systems

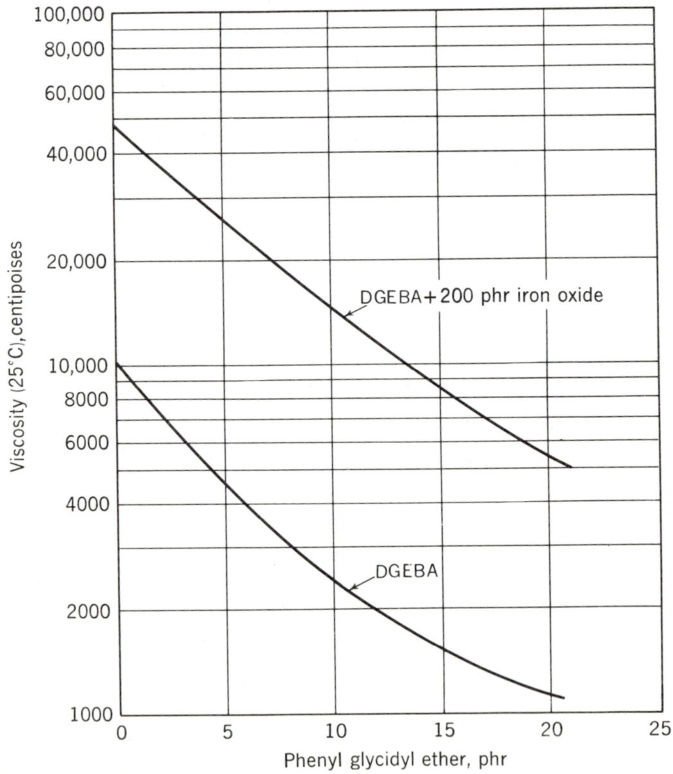

FIG. 14-2. Effect of reactive diluent concentration on viscosity of filled DGEBA [4–6].

have been found to be dilatant at processing temperatures on the order of 110°C [25].

In general, the filled epoxy systems become increasingly more pastelike as filler is added. In "gunk" molding compounds, and in molding powders, the upper limit imposed on filler loading volume is usually determined by wettability considerations, and rather high loading volumes are practical. When pourable viscosities are desired, the amount of filler is restricted. This limit, for several fillers, is indicated in Table 14-3 and Figs. 14-2 and 14-3. Figure 14-4 and Table 14-4 indicates the increase in loading that may be obtained upon addition of low-viscosity curing agents.

The upper limit for a casting formulation may be taken as 4,000 to 5,000 centipoises at working temperature. Loading volumes of several fillers giving this viscosity are shown in Table 14-5. If a diluted resin is employed in order to increase the loading volume of the filler, some increase may be obtained if the system is processed at or near room temperature (Figs. 14-5 and 14-6). At elevated temperatures, however, large percentage-changes in viscosity are of relatively little significance, use of the diluted resin does not offer any marked advantage (Fig. 14-7), and increasing the curing-agent requirement will in some systems result in higher viscosity than with the undiluted resin.

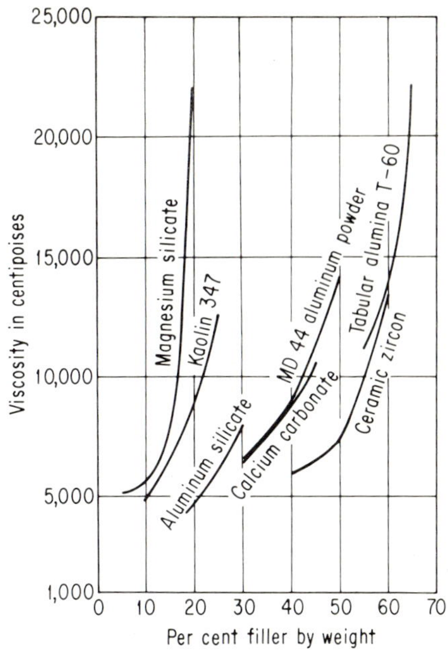

FIG. 14-3. Viscosity of DGEBA diluted with dibutyl phthalate as a function of filler and loading volume [4–1].

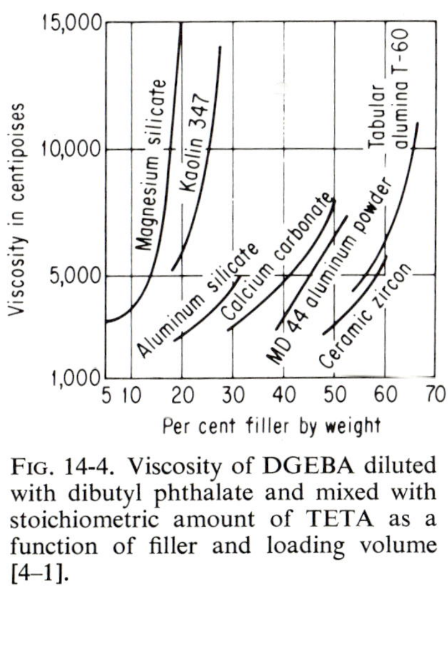

FIG. 14-4. Viscosity of DGEBA diluted with dibutyl phthalate and mixed with stoichiometric amount of TETA as a function of filler and loading volume [4–1].

SETTLING PROPERTIES OF FILLERS

Fillers generally tend to settle from the resin on storage. The degree of settling is dependent upon the specific gravity of the filler, the particle size of the filler, the specific resin, and the viscosity of the filled compound.

The more finely divided the filler, the less the settling tendency; conversely, the coarser the filler, the more rapid the settling. Fillers passing 325 mesh generally show little settling after 24-hour storage at room temperature in DGEBA. Passing 325 mesh, aluminum silicate (at 33 per cent), magnesium silicate (at 17 per cent), tabular alumina (at 64 per cent), and Kaolin 347 (at 23 per cent) show no tendency to settle, and ceramic zircon (at 50 per cent)

Table 14-4. Effect of Fillers on Viscosity, Exotherm, Modulus of Rupture, and Shrinkage of DGEBA Cured with Primary Amine Curing Agent [4–7]

Filler	Viscosity at 25°C, centipoises				Peak exothermic temperature rise, °C				Working life, minutes				Modulus of rupture, psi	Shrinkage, %
					% filler				% filler					
	20% filler	35% filler	50% filler	65% filler	20	35	50	65	20	35	50	65	20% filler	20% filler
Unfilled	1,100				223°C				48 min				18,174	0.91
Asbestine	3,850	23,650	100,000		158	122	55	..	53	60	73	...	10,920	0.48
Albalith	2,700	9,600	95,000		169	133	76	..	47	66	85	...	6,896	0.59
Silica	2,750	6,800	51,500		170	115	53	..	52	72	95	...	16,875	0.77
Mica	2,100	4,800	54,500		159	111	51	..	50	71	94	...	12,358	0.66
Surfex MM	2,200	3,500	22,600	100,000	186	125	55	33	49	57	70	80	11,970	0.64
Flint powder	2,200	4,300	39,500		178	113	53	..	52	75	96	...	17,566	0.41
Quartz	2,100	3,750	13,150	100,000	178	113	51	34	49	74	107	130	14,100	0.67
Cryolite	1,900	2,750	6,150	20,900	166	113	56	36	54	71	107	120	15,483	0.54
Portland cement	1,950	2,400	4,100	19,300	183	147	77	41	52	62	89	102	14,370	0.69
Limestone	1,700	3,250	10,000	31,700	179	128	59	37	49	71	90	95	12,433	0.47
Atomized aluminum 101	2,100	3,100	4,600	34,500	174	127	47	37	56	68	100	105	13,710	0.80
Barytes	1,700	2,450	4,200	17,200	179	152	83	38	49	60	84	116	17,417	0.71
Atomized aluminum 100	1,900	2,950	5,500	33,500	176	115	52	35	53	70	99	105		
Atomized aluminum 120	1,800	3,000	5,900	35,000	178	128	53	33	52	67	100	105	14,790	

Table 14-5. Per cent of Filler Required to Produce Initial Viscosity of 4,000 to 5,000 Centipoises for DGEBA with Listed Curing Agents [25]

Filler	DETA/ acrylonitrile (½) at 25°C	MDA at 50°C	HHPA + 0.1% BDMA at 110°C	Chlorendic anhydride at 110°C	Menthane diamine at 50°C
Mica, 1,000 mesh	35	67	62	47	48
Silica, 325 mesh	30	68	85	62	63
Zirconium silicate <1.5 microns	45	62	72	53	58
Hydrated alumina, 0.08–0.12 microns . . .	30	32	50	37	33

settles only slightly; whereas less finely divided aluminum powder settles rather quickly during the same period. To retard settling, it is customary to employ some per cent of a finely divided filler when coarse fillers are employed [62]. Aluminum stearate or other thixotroping agent may be used to reduce settling tendencies. The limestone fillers (calcite, dolomite) tend to settle rather badly, but they can be used in much higher loading volumes than can the talcs to obtain comparable viscosities [74].

A roughly qualitative test for determining the settling characteristics involves mixing the resin with 50 per cent 325-mesh silica powder, thoroughly deaerating the mixture, and observing it periodically during storage. The settling has been found to be of three types:

a. Homogeneous Soft Settle—A spatula penetrates to the bottom of the resin container (a glass jar) easily. This suspended filler will slowly settle out of the liquid leaving a clear layer above the filler. However, no stratification of the filler layer is observed and a spatula will easily penetrate even after extended periods of time.

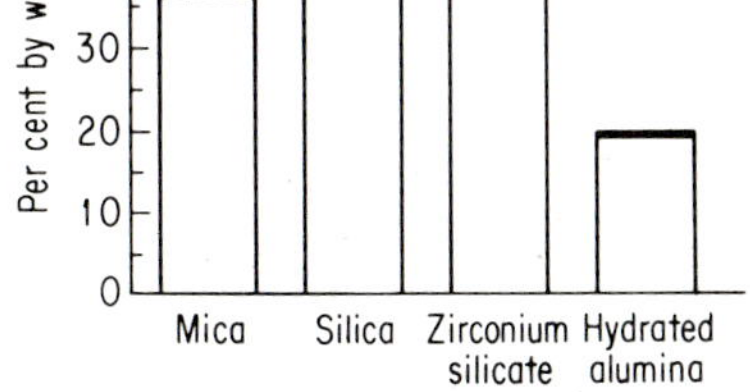

FIG. 14-5. Amount of filler producing initial viscosity of 4,000 to 5,000 centipoises at working temperature for DGEBA and DGEBA diluted with BGE when mixed stoichiometrically with DETA/acrylonitrile (1/1) at 25°C [25]. Shaded areas indicate diluted resin.

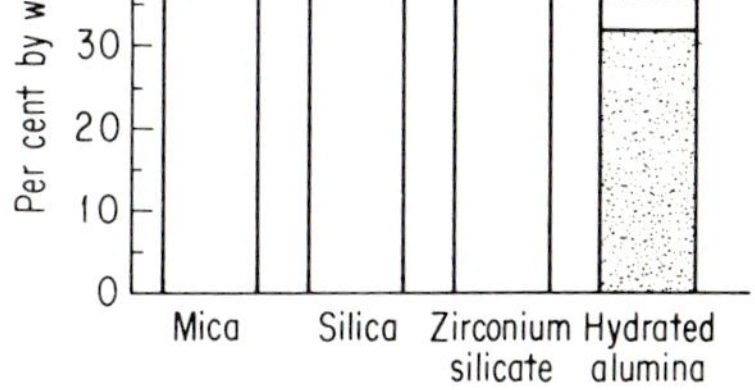

FIG. 14-6. Amount of filler producing initial viscosity of 4,000 to 5,000 centipoises at working temperature for DGEBA and DGEBA diluted with BGE when mixed stoichiometrically with MPDA/MDA/bisphenol A at 50°C [25]. Shaded areas indicate diluted resin.

b. Non-homogeneous Soft Settle—A spatula can penetrate this suspension with little difficulty; however, the filler layer exhibits visible stratification as evidenced by a deposit formed at the bottom of the jar. This deposit will offer some resistance when a spatula is inserted. The supernatant resin layer is generally clear.

c. Hard Settle—A heavy filler deposit forms on the bottom of the jar which is difficult to penetrate with a spatula and has sufficient tenacity to cause the sample jar to be lifted from the table when removal of the spatula is attempted. If this condition is developed in ten days, the sample is considered hard settle (17).

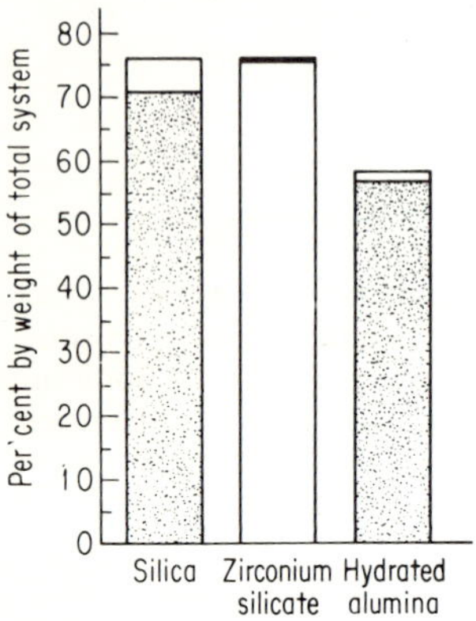

FIG. 14-7. Amount of filler producing initial viscosity of 4,000 to 5,000 centipoises at working temperature for DGEBA and DGEBA diluted with BGE when mixed stoichiometrically with PA and 0.1 per cent BDMA at 110°C [25]. Shaded areas indicate diluted resin.

The settling characteristics of fillers and resins are of considerable significance when long-time storage is involved or when elevated temperatures are employed during cure. The lower viscosity brought on by high-temperature cures often accelerates the settling tendency such that, upon gelation, undesirable stratification of fillers can be observed. The situation is particularly critical when centrifugal casting techniques are employed, such that use of fillers is usually impractical unless a resin-rich inner layer is desired, as for instance in centrifugally cast pipe.

EFFECT OF FILLERS ON POT LIFE, EXOTHERM, AND SHRINKAGE

The general effect of fillers is to change pot life and to reduce exotherm and shrinkage.

The extension of pot life is usually a function of the filler loading volume. At equal loading volumes, differences are observed between fillers due to differences in their specific heats and thermal conductivities (Table 14-4). If DGEBA is filled to provide a viscosity between 4,000 and 5,000 centipoises at working temperatures, substantial pot-life differences among fillers will be observed (Table 14-6). In not all cases do the fillers (particularly at lower concentrations) offer an extension of pot life. Hydrated alumina

Table 14-6. Pot Life and Exotherm of DGEBA vs. Filler, Loaded to Give an Initial Viscosity of 4,000 to 5,000 Centipoises at Working Temperature with Listed Curing Agents [25]

Filler	Pot life and exotherm in 454 g mass					
	DETA/acrylonitrile (½) at 25°C		MPDA/MDA/bisphenol A at 50°C		PA at 110°C	
	Pot life, min	Exotherm, °C	Pot life, min	Exotherm, °C	Pot life, min	Exotherm, °C
No filler	100	145	75	245	15	235
Mica, 1,000 mesh	150	87	60	165		
Silica, 325 mesh	175	95	100	117	87	115
Zirconium silicate, <1.5 microns	222	75	10	140	75	195
Hydrated alumina, 0.08–0.12 microns	140	80	85	175	10	148

may accelerate anhydride cures, for example, and mica and zirconium silicate accelerate cures with aromatic amines (Table 14-6).

In addition to bulk displacement of resin and absorption of exotherm heat, the efficiency with which the filler conducts heat from the casting will also contribute to a reduction in exotherm. Table 14-4 indicates the significant reduction which may be obtained when higher loading volumes of fillers are employed. In an investigation of a variety of fillers [24], it was determined that aluminum powder gives the lowest exotherm; but in some cases exotherm may be changed by a blend

Table 14-7. Per Cent Shrinkage of DGEBA Cured with 12 phr Diethanolamine [52]

(Mixed at 75°C and Cured 15 hr at 75°C)

Kind of shrinkage	Unfilled	Filled with 100 phr mica
Total	7.17	4.85
Pregelation	2.82	2.12
Postgelation ...	2.46	2.10
Cooling	1.90	0.63

of fillers. Barium sulfate and calcium carbonate together gave lower exotherms than either filler separately, and the reverse was true with calcium carbonate and aluminum powder.

The effect of fillers on shrinkage is similar to the case of exotherm. Shrinkage is reduced in proportion to the filler loading volume, although differences between fillers will be observed (Table 14-4). The reduction in shrinkage is brought about by bulk replacement of the resin with an inert compound which does not undergo permanent dimensional change during the cure cycle. A comparison of a filled and an unfilled system (Table 14-7) indicates the effect of filler on pregel, postgel, and cooling shrinkage.

EFFECT OF FILLERS ON CURED PHYSICAL PROPERTIES

The effect of two typical fillers on the physical properties of a high-temperature formulation is shown in Table 14-8. The effect of mica on physical properties of DGEBA cured with three different amine curing agents is given in Table 14-9. The effect of metallic fiber fillers is shown in Table 14-10.

The following discussion will indicate, in a general way, the effect of fillers upon specific properties. Although test results are given predominately for DGEBA, similar effects should be experienced with other resins, differing somewhat in degree.

Deflection Temperature

The general effect of particulate fillers on the DT of cured DGEBA is slight. Some fillers raise the DT slightly, as indicated in Fig. 14-8, the rise increasing with filler loading volume. In some cases, as with the acrylonitrile-DETA adduct as a curing agent, a lowering may be obtained (Table 14-11). Silica appears to work particularly well with HHPA whereas hydrated alumina gives better results with chlorendic anhydride and menthane diamine.

The increase in DT experienced with use of particulate fillers does not appear to be sufficiently great to overcome the reduction obtained from the use of a reactive diluent, as indicated in Figs. 14-9 to 14-11.

Tests on filled systems reinforced with ½-in. chopped glass fiber indicate a considerable increase in deformation resistance with fibrous fillers. Probably two separate phenomena account for the increased deformation resistance. As the glass-transition temperature is approached, the cast specimen becomes less rigid, and some of the load is transferred to the high-modulus reinforcement. Above the glass-transition temperature, rubbery retroactive forces resist deformation. As the testing bar is cooled slowly to room temperature, the cooling rate will be more uniform than the heating rate, and a transition state will be more evenly reached between the point at which the retroactive forces and Van der Waals forces contribute strength [31]. A 10-mil deflection will then occur at about the point expected of the DT of the unfilled casting (Figs. 14-12 and 14-13), and since this deflection is locked into position by further cooling, the total deflection may be considerably greater when room temperature is reached.

Table 14-8. Effect of Fillers on Properties of DGEBA Resins Cured with PMDA [4–1]

Property	DGEBA, 56 phr PMDA + 20 phr tetrahydrofuryl alcohol + 1% dicyandiamide, cured 30 min at 180°C		DGEBA (mol. wt. 420) + 27.6 phr PMDA, cured 16 hr at 220°C	
	Alumina T-60, 200 phr	Iron oxide (200 mesh), 150 phr	Alumina T-60, 150 phr	Iron oxide (200 mesh), 150 phr
DT, °C	228	227	300	229
After 200 hr at 204°C	267	264	299	288
After 200 hr at 260°C	300	275	300	263
Tensile strength, psi:				
At 23°C	5,900	5,400	5,700	3,800
At 150°C	4,200	4,200	3,600	2,200
At 260°C	760	880	1,400	660
Modulus, psi × 10^{-6}, at 23°C	0.98	0.65	0.90	0.71
Flexural strength, psi:				
At 23°C	10,200	10,200	9,600	9,700
At 150°C	6,900	7,700	5,800	6,100
At 260°C	910	640	2,200	1,600
Modulus, psi × 10^{-6}:				
At 23°C	1.14	0.66	1.06	0.76
At 150°C	0.55	0.56	0.36	0.44
At 260°C	0.06	0.03	0.28	0.17
Compressive strength, psi:				
At 23°C	34,900	36,800	38,000	38,200
At 150°C	19,200	19,600	21,400	19,500
At 260°C	7,500	6,300	11,200	6,700
Modulus, psi × 10^{-6}:				
At 23°C	0.69	0.47	0.66	0.56
At 150°C	0.14	0.11	0.13	0.10
At 260°C	0.04	0.03	0.08	0.06
Yield stress at 1% offset, psi:				
At 23°C	21,300	16,300	21,500	17,600
At 150°C	9,400	8,400	11,600	9,700
At 260°C	No offset	No offset	No offset	No offset
Hardness, Barcol	67	57	69	64
Linear shrinkage during cure, %	0.8	.....	1.0	
Weight loss after 200 hr, %:				
At 204°C	0.4	0.4	0.1	0.05
At 232°C	3.6	4.2	3.4	2.5
At 260°C	10.5	11.6	10.5	7.1

Table 14-9. Effect of Curing Agent on Properties of DGEBA Filled with Mica [3]

Property	Diethanolamine, 12 phr			Piperidine, 6 phr			*n*-Hydroxyethyl-piperazine, 6 phr		
Mica, phr	0	66.6	100	0	66.6	100	0	66.6	100
DT, °C	70	75	77	80	86	88	88		
Tensile strength, psi	11,000	8,400	7,700	10,700	7,700	6,700	12,500	7,880	7,110
Strain at maximum stress, %	4.430	1.468	0.906	2.84	1.105	0.683	4.5	1.27	0.93
Stress at 0.2% offset, psi	7,050	6,800	7,400	8,800	6,700	N/A	8,500	6,660	6,520
Strain at 0.2% offset, %	1.672	0.983	0.820	1.94	0.905	N/A	1.91	0.94	0.79
Modulus, psi × 10^{-6}	0.488	0.881	1.94	0.504	0.956	1.192	0.497	0.906	1.113
Flexural strength, psi	17,300	12,000	11,300	19,100	11,200	9,100	19,100	11,600	11,300
Strain at maximum stress, %	5.02	1.75	1.08	5.34	1.44	0.86	5.44	1.52	1.21
Stress at 5% fiber strain, psi	17,200	N/A	N/A	18,900	N/A	N/A	18,900	N/A	N/A
Stress at 0.2% offset, psi	12,900	10,900	N/A	14,300	11,200	N/A	13,800	N/A	N/A
Maximum fiber strain at 0.2% offset, %	3.01	1.52	N/A	3.06	1.44	N/A	2.90	N/A	N/A
Modulus, psi × 10^{-6}	0.469	0.827	1.215	0.501	0.899	1.169	0.503	0.878	1.097
Compressive strength, psi	25,300	23,000	17,970	23,300	20,790	18,500	57,500	18,700	18,700
Strain at maximum stress, %	44.4	36.0	18.45	38.9	25.78	14.03	54.10	8.02	5.66
Stress at yield, psi	15,000	15,500	17,900	17,800	18,080	18,200	17,100	18,200	18,500
Strain at yield, %	4.40	3.93	3.34	10.6	4.21	3.95	5.31	3.08	2.77
Stress at 0.2% offset, psi	10,700	12,100	14,100	8,080	13,900	14,200	13,400	12,100	12,600
Strain at 0.2% offset, %	2.12	2.06	1.59	2.0	2.11	1.94	2.81	1.69	1.50
Modulus, psi × 10^{-6}	0.560	0.740	1.025	0.457	0.822	0.943	0.571	0.812	0.972
Impact resistance, Izod	0.58	0.394	0.368	......	0.350	0.355	0.712		
Hardness									
Shore D	86	87	90	86	90	90	88	90	90
Barcol	16	34	48	30	40	45	32	42	47
Coefficient of thermal expansion from −55 to +75°C, in./(in.)(°C) × 10^6	59.5	38.6	36.8	.....	39.6	.....	55.6		

N/A. Material did not reach this value.

Table 14-10. Effect of Metal Fibers on Properties of Cured DGEBA [59]

EFFECTS OF INCREASING FIBER CROSS SECTION

Property	Cross section of aluminum fiber at 0.004 × 3/8 in.	at 0.007 × 3/8 in.	at 0.011 × 3/8 in.	Direction of improvement	Confidence limit of effect
Tensile strength, psi	4,780	3,890	3,680	Decreasing cross section	95
Compressive strength, psi	20,600	17,700	16,600	....	Negligible in relation to testing error
Modulus, psi × 10^{-6}	0.72	0.66	0.71	....	Negligible in relation to testing error
Thermal conductivity, Btu/(hr)(ft)(°F)	0.76	0.73	0.84	Increasing cross section	95

EFFECTS OF TYPE OF METAL

Property	Volume of metal, %	Aluminum, 0.007 × 3/8 in.	Copper, 0.006 × 1/4 in.	Direction of increase	Confidence limits of effect, %
Tensile strength, psi	10	1,810	7,740	Aluminum to copper	99
Compressive strength, psi	10	13,400	14,400	Aluminum to copper	95
Modulus, psi × 10^{-6}	10	0.54	0.48	Copper to aluminum	98
Thermal conductivity, Btu/(hr)(ft)(°F)	35	0.73	1.06	Aluminum to copper	92

PROPERTIES OF EPOXY RESIN REINFORCED WITH STEEL FIBER

Property	Steel, 1/4 × 0.004 in., by volume 5%	15%	25%
Tensile strength, psi	2,110	5,100	4,290
Compressive strength, psi	17,100	18,800	16,400
Modulus, psi × 10^{-6}	0.75	0.67	0.45
Thermal conductivity, Btu/(hr)(ft)(°F)	0.44	0.49	0.64

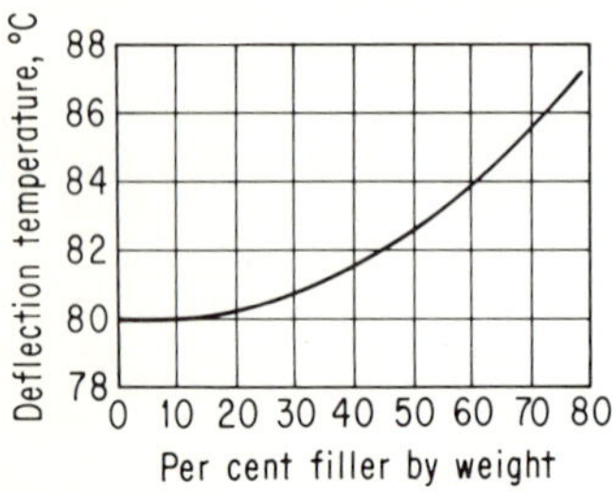

FIG. 14-8. Iron oxide concentration vs. DT of DGEBA diluted to 500 centipoises with BGE and cured with 10 phr TETA [1].

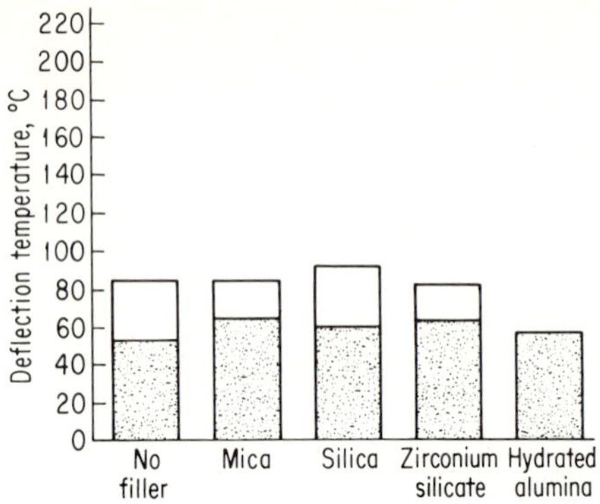

FIG. 14-9. DT of DGEBA diluted to 500 centipoises with BGE (shaded portion) vs. filler, loaded to provide an initial viscosity of 4,000 to 5,000 centipoises at working temperature and cured with stoichiometric amount of DETA/acrylonitrile (1/1) [25]. Shaded areas indicate diluted resin.

Similar retroactive forces are also encountered with systems containing particulate fillers (and with unfilled systems) above the DT, but, lacking reinforcement, they appear to be of lesser design consequence, although their significance would depend largely upon the stress levels encountered.

Thermal Stability

By bulk displacement, fillers reduce long-time shrinkage and elevated-temperature weight loss, as indicated in Fig. 14-14.

Tensile Strength

The general effect of particulate fillers is to reduce the room-temperature tensile strengths of the cured epoxies, although some improvement may be obtained at elevated temperature (Table 14-12).

Table 14-11. DT of DGEBA as a Function of Filler, Loaded to Provide an Initial Viscosity of 4,000 to 5,000 Centipoises at Working Temperature with Stoichiometric Amounts of Listed Curing Agents

(In degrees centigrade) [25]

Filler	DETA/ acrylonitrile (½)	MDA	HHPA + 0.1% BDMA	Chlorendic anhydride	Menthane diamine
No filler	78	168	120	164	136
Mica, 1,000 mesh	49	157	...	...	148
Silica, 325 mesh	40	158	132	180	139
Zirconium silicate, <1.5 microns	49	148	120	157	
Hydrated alumina, 0.08–0.12 microns	..	144	128	201	152

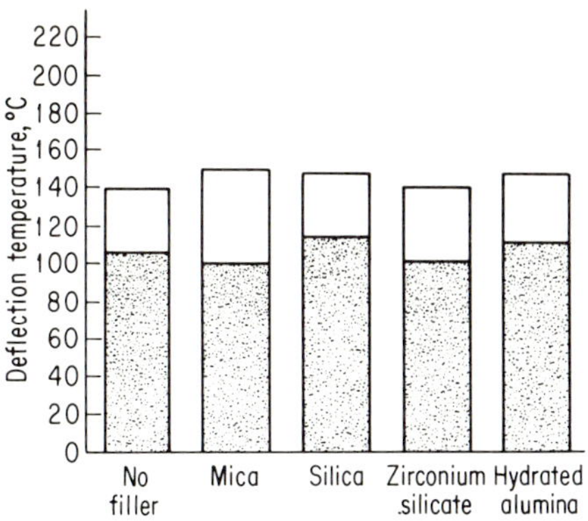

FIG. 14-10. DT of DGEBA and DGEBA diluted to 500 centipoises with BGE (shaded portion) vs. filler, loaded to provide an initial viscosity of 4,000 to 5,000 centipoises at working temperature and cured with stoichiometric amount of MPDA/MDA/bisphenol A [25]. Shaded areas indicate diluted resin.

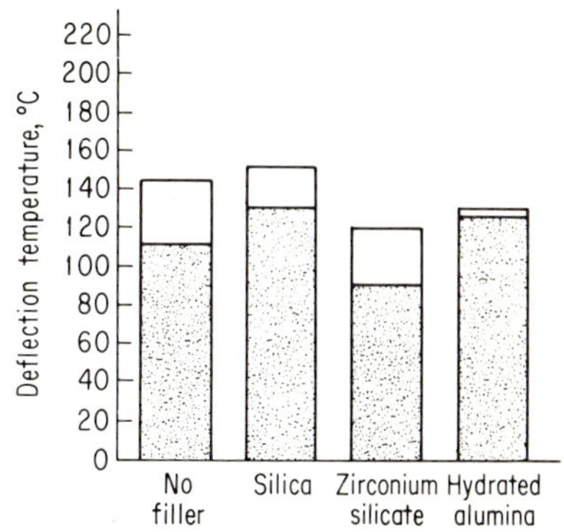

FIG. 14-11. DT of DGEBA and DGEBA diluted to 500 centipoises with BGE (shaded portion) vs. filler, loaded to provide an initial viscosity of 4,000 to 5,000 centipoises at working temperature and cured with stoichiometric amount of PA and 0.1 per cent BDMA [25]. Shaded areas indicate diluted resin.

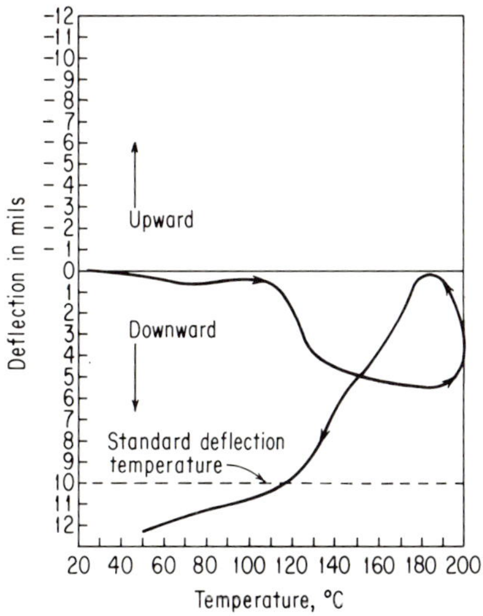

FIG. 14-12. Performance of filled DGEBA cured with stoichiometric amount of amine-resin adduct under 264 psi fiber stress upon heating to 200°C and cooling at 2°C per minute [31]. Data corrected for thermal expansion. Filler: 20 phr ½-in. chopped glass fiber, 180 phr alumina, 20 phr Santocel.

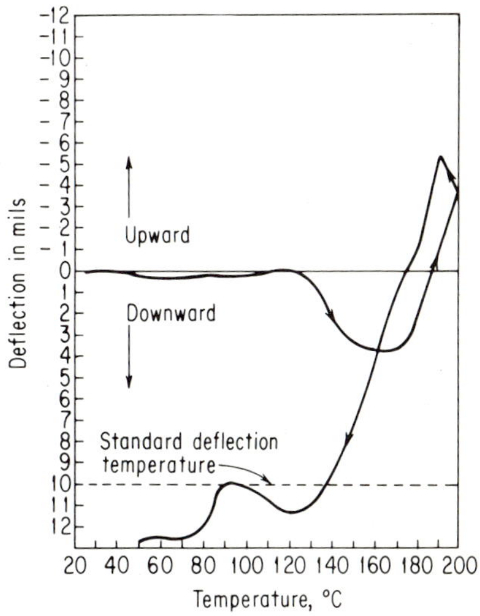

FIG. 14-13. Performance of filled DGEBA cured with stoichiometric amount of MDA under 264 psi fiber stress upon heating to 200°C and cooling at 2°C per minute [31]. Data corrected for thermal expansion. Filler: 50 phr ½-in. chopped glass fiber, 180 phr alumina, 20 phr Santocel.

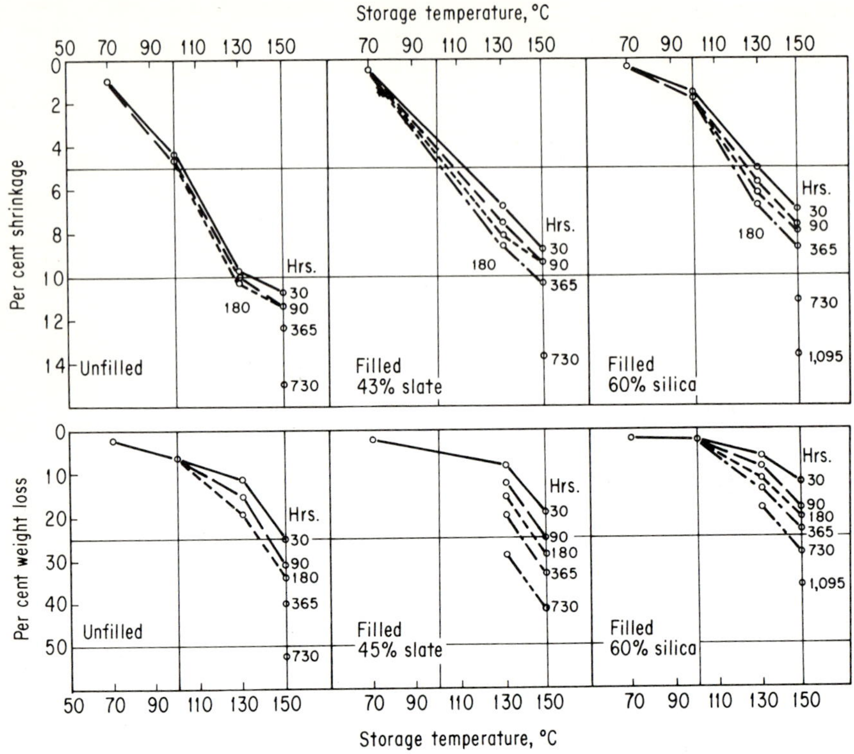

FIG. 14-14. Effect of fillers on shrinkage and weight loss of DGEBA (mol. wt. 400) cured with PA [41].

Table 14-12. Effect of Filler on Tensile Properties of Cured DGEBA [39]

Curing agent	Filler, phr*	At 25°C Ult., psi	At 25°C Ult., elong., %	At 150°C Ult., psi	At 150°C Ult., elong., %
MPDA	None	10,500	4	4,800	8
	Aluminum oxide, 300	10,700	1		
	Iron oxide, 400	4,500			
	Aluminum powder, 200	9,300	2		
	Hydrated aluminum silicate, 75	6,200	1		
	Mica, 75	6,300	2		
	Calcium carbonate, 175	5,100	1		
	Aluminum silicate, 75	5,000	1	4,000	8
	Silicate, 5 microns, 100	9,500	1	5,900	5
	Silica, 35 microns, 100	11,800	2	5,200	6
	30 microns, 175	9,500	1	5,600	4
	44 microns, 175	10,500	1	6,700	4
NMA	None	8,500	1	1,000	20
	Aluminum silicate, 100	5,100	1	2,800	12
	Silica, 5 microns, 150	11,500	1.5	3,100	18
	30 microns, 275	12,100	1	4,300	75
	44 microns, 350	7,300	0.4	4,100	7

* phr chosen to yield approximately comparable viscosities.

Table 14-13. Tensile Strength of Epoxy Resins Cured with NMA as a Function of Concentration and Type of Fibrous Metallic Fillers

(In pounds per square inch) [39]

Fiber volume, %	Dicyclopentadiene dioxide					DGEBA	
	Aluminum				Copper	Steel	Aluminum
	3/8 in.			3/4 in.	1/4 in.	1/4 in.	3/8 in.
	× 0.004 in.	× 0.007 in.	× 0.011 in.	× 0.004 in.	× 0.006 in.	× 0.004 in.	× 0.007 in.
5	....	....	....	....	....	2,110	
10	950	1,810	1,170				
15	....	....	....	....	....	5,100	
20	3,180	2,230	2,460	3,550	3,550	....	2,930
25	....	....	....	....	....	4,290	
35	4,780	3,890	3,680	3,680	....		

Fibrous fillers at low concentrations may also reduce tensile strength, but at higher concentrations will have the opposite effect and, in the case of laminates, will increase tensile strengths dramatically. Short fibers influence tensile strengths as indicated in Table 14-13, improvements generally occurring when the filler is present in concentrations greater than 10 per cent by volume.

With particulate fillers, the decrease in tensile strength appears more in evidence with the amine curing agents (Table 14-14) than with the anhydride curing agents (Table 14-15). With some anhydrides, tensile strengths may even be improved with fillers, although the improvement is marginal. Particle size of the filler has been shown to exert some effect on tensile strengths of the cured casting [45]. Silica seems generally to be a good choice of filler when least reduction of tensile strength is required, although it will prove more effective with some curing agents than with others.

Tensile modulus is generally increased and ultimate elongation reduced in proportion to the amount of filler when tested at room temperature; the reverse is usually true at elevated temperature.

Flexural Strength

Particulate fillers generally decrease the flexural strength and increase the flexural modulus, as indicated in Tables 14-16 and 14-17, although in the case of silica fillers, flexural strengths may be little influenced (Table 14-18). Powdered aluminum appears to bring about a more substantial reduction than most other fillers and

Table 14-14. Effect of Fillers on Strengths of DGEBA Cured with MPDA/MDA/PGE [50]

Filler, phr	Compressive strength, psi	Tensile strength, psi
None	28,000	12,500
Iron oxide, 400	18,000	4,500
Atomized aluminum, 200	15,000	6,500
Asbestos floats, 15	17,500	7,000
Water-ground mica, 15	20,000	7,500

Table 14-15. Tensile Properties of DGEBA Cured with Stoichiometric Amounts of Listed Anhydrides vs. Loading Volume of Silica [4-6]

Anhydride	Cure schedule, hr/°C	Filler concentration, parts/100 of mix	Tensile properties		
			Strength, psi	Modulus, psi × 10^{-6}	Elongation, %
DDSA + 1 phr BDMA ..	1/130 + 24/150	43	7,700	0.45	3.4
		85	8,000	0.65	2.9
		128	8,500	0.80	2.2
		213	8,800	1.2	1.1
NMA + 1 phr BDMA ..	1/140 + 24/150	52	9,700	0.8	1.3
		105	10,400	1.1	1.0
		157	10,000	1.5	0.8
		209	9,500	1.8	0.6
Chlorendic anhydride HHPA + 0.5 phr BDMA	1/120 + 24/150	50	9,400	0.8	1.3
		100	11,100	1.2	1.1
		150	7,800	1.5	0.6

Table 14-16. Effect of Fillers on Flexural Properties of DGEBA Diluted with 17 phr Dibutyl Phthalate and Cured with Stoichiometric Amount of TETA [4-7]

Filler, phr	Strength, psi	Modulus, psi × 10^{-6}
Aluminum silicate, 50	8,683	0.332
Calcium carbonate, 90	8,712	0.482
Magnesium silicate, 20	9,210	0.316
Ceramic zircon, 150	9,982	0.507
Aluminum powder, 100 ...	6,415	0.301
Alumina, 180	8,500	0.713
Kaolin no. 347, 30	8,050	0.367

Table 14-17. Flexural Properties of DGEBA Cured with Stoichiometric Amounts of Listed Anhydrides vs. Loading Volume of Silica [4-6]

Anhydride	Cure schedule, hr/°C	Filler concentration, parts/100 of mix	Flexural properties	
			Strength, psi	Modulus, psi × 10^{-6}
DDSA + 1 phr BDMA	1/130 + 24/150	43	12,800	0.4
		85	13,900	0.6
		128	15,100	0.7
		213	14,000	1.0
NMA + 1 phr BDMA	1/140 + 24/150	52	17,000	0.7
		105	19,100	1.0
		157	17,300	1.3
		209	17,400	1.6
Chlorendic anhydride/HHPA + 0.5 phr BDMA	1/120 + 24/150	50	16,800	0.7
		100	17,000	1.0
		150	17,300	1.2

Table 14-18. Effect of Filler on Flexural Properties of DGEBA Cured with MPDA [47]

Filler, phr	Strength, psi	Modulus, psi × 10^{-6}
None	19,300	0.400
Calcium carbonate, 175 ...	8,700	0.900
Silica, 5 microns, 100	20,100	0.700
44 microns, 175	18,400	1.000

causes substantial deformation during fatigue testing prior to failure, a condition not encountered with most other fillers. Studies investigating the influence of filler-resin bond on flexural strength indicate that improvement may be obtained by treating quartz with a silicone coupling agent, as indicated in Fig. 14-15.

Fibrous fillers, at higher loading volumes, increase flexural strength. Glass flakes, having a thickness on the order of 0.0002 in. and in random lengths, have been employed to make essentially glass laminates, and when the filler content is 75 per cent, flexural strengths of 45,000 psi have been obtained.

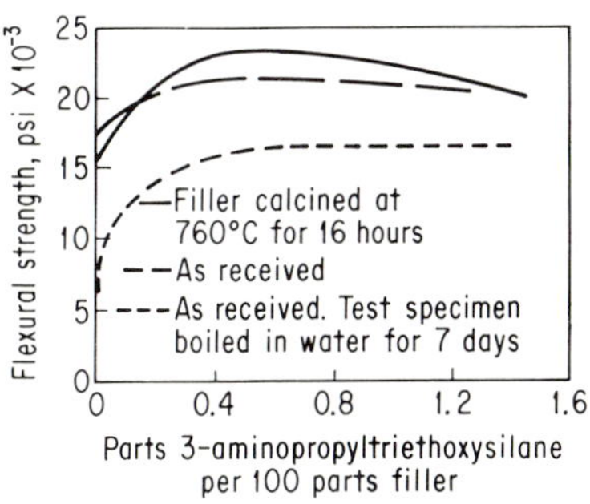

FIG. 14-15. Effect of surface treatment of quartz filler (50 per cent finer than 17 microns) on flexural strength of DGEBA cured with aromatic amine eutectic [53].

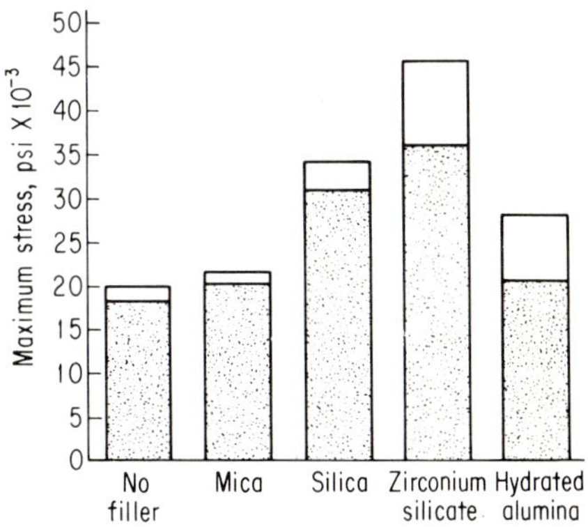

FIG. 14-16. Compressive strength of system shown in Fig. 14-10 [25].

Compressive Strength

Particulate fillers generally decrease compressive fatigue but increase ultimate compressive modulus and compressive yield strength because of stiffening effect. This is indicated in Tables 14-19 and 14-20, and in Figs. 14-16 to 14-18. Once again, aluminum brings about a lesser improvement than most other fillers.

Compressive strength as a function of temperature, is shown in the values for a DADPS-cured DGEBA/polyglycidyl ether of tetraphenylethane with 200 phr black iron oxide filler: 14,000 psi at 150°C; 10,000 psi at 204°C; and 4,000 psi at 260°C.

The effect of short-fiber metallic fillers is indicated in Tables 14-21 and 14-22. Ultimate compressive strengths and compressive modulus are not decreased as greatly as with particulate fillers.

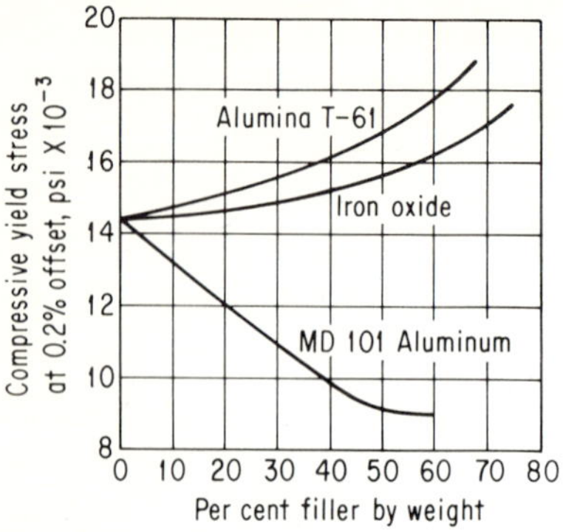

FIG. 14-17. Effect of filler concentration on compressive strength of DGEBA cured with 8 phr DETA [50].

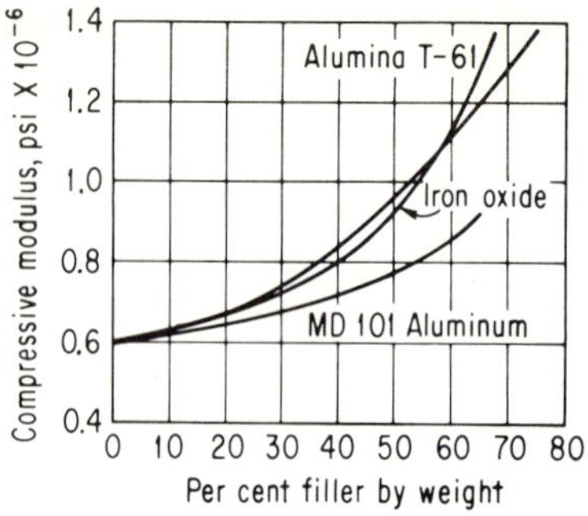

FIG. 14-18. Effect of filler and loading volume on compressive modulus of DGEBA cured with 8 phr DETA [50].

Table 14-19. Compressive Properties of DGEBA Cured with Stoichiometric Amounts of Listed Anhydrides vs. Loading Volumes of Silica [4–6]

Anhydride	Cure schedule, hr/°C	Filler, parts/100 of mix	Strength, psi	Modulus, psi × 10^{-6}	Yield at 0.2% offset, psi	Deformation at yield, %	Deformation at max., %
DDSA + 1 phr BDMA	1/130 + 24/150	43	11,200	0.4	9,000	2.8	5.2
		85	12,400	0.5	9,500	2.2	4.9
		128	14,300	0.6	9,900	1.8	6.0
		213	22,300	0.9	11,800	1.5	17.7
NMA + 1 phr BDMA	1/140 + 24/150	52	19,300	0.5	15,100	3.3	5.6
		105	23,400	0.7	18,200	2.9	6.1
		157	30,300	0.9	19,600	2.4	12.0
		209	34,800	1.1	22,900	2.3	9.8
Chlorendic anhydride/ HHPA + 0.5 phr BDMA ..	1/120 + 24/150	50	21,800	0.6	16,100	2.9	6.1
		100	26,800	0.9	18,800	2.3	7.8
		150	25,200	0.9	18,100	2.4	6.9

Table 14-20. Effect of Filler on Compressive Properties of DGEBA Cured with MPDA [47]

Filler, phr	Strength, psi	Modulus, psi × 10^{-6}	Deformation, ultimate, %
None	18,500	0.500	7
Aluminum oxide, 300	29,700	1.000	10
Aluminum powder, 200	18,300	0.900	11
Hydrated aluminum silicate, 75	20,200	0.600	11
Calcium carbonate, 175	23,900	0.900	9
Silica, 5 microns, 100	36,600	0.500	23
30 microns, 175	29,600	0.800	11
44 microns, 175	28,600	0.700	12

Table 14-21. Compressive Strength of Epoxy Resins Cured with NMA as a Function of Concentration and Type of Fibrous Metallic Fillers [39]

(In pounds per square inch)

Fiber volume, %	Dicyclopentadiene dioxide					DGEBA	
	Aluminum				Copper	Steel	Aluminum
	3/8 in.			3/4 in.	1/4 in.	1/4 in.	3/8 in.
	× 0.004 in.	× 0.007 in.	× 0.011 in.	× 0.004 in.	× 0.006 in.	× 0.004 in.	× 0.007 in.
5	.....	.....	.....	.....	.....	17,100	
10	13,000	13,400	16,500	.....	14,400		
15	.....	.....	.....	.....	.....	18,800	
20	16,500	13,000	16,800	22,600	17,500	.....	19,500
25	.....	.....	.....	.....	.....	16,400	
35	20,600	17,700	16,600	.....	16,500		

Table 14-22. Compressive Modulus of Epoxy Resins Cured with NMA as a Function of Concentration and Type of Fibrous Metallic Fillers [39]

(In psi $\times 10^{-6}$)

Fiber volume, %	Dicyclopentadiene dioxide					DGEBA	
	Aluminum				Copper	Steel	Aluminum
	3/8 in.			3/4 in.	1/4 in.	1/4 in.	3/8 in.
	× 0.004 in.	× 0.007 in.	× 0.011 in.	× 0.004 in.	× 0.006 in.	× 0.004 in.	× 0.007 in.
5	...	...	...	...	...	0.75	
10	0.42	0.49	0.48	...	0.53		
15	...	...	...	...	...	0.67	
20	0.56	0.54	0.55	0.54	0.48	...	0.42
25	...	...	...	...	...	0.45	
35	0.72	0.66	0.71	...	0.37		

Surface Hardness, Abrasion Resistance, and Machining Qualities

The general effect of fillers is to increase the surface hardness of the castings. The effect will vary from filler to filler. Alumina, for example, causes a substantial increase in surface hardness, whereas aluminum causes little increase (Rockwell M 90 as against M 62 for a typical system).

The effect of various fillers on abrasion resistance is indicated in Table 14-23. Silicon carbide has been reported as particularly effective in increasing abrasion resistance [11], presumably because of its high hardness, whereas aluminum powder decreases abrasion resistance.

The machinability of epoxies will vary with the filler. Molybdenum disulfide, calcium carbonate, magnesium silicate, aluminum, and talc provide easy machinability; aluminum silicate and ceramic zircon provides fair machinability; silica and alumina provide difficult machinability.

Table 14-23. Effect of Filler on Taber Abrasion Resistance of DGEBA Diluted with BGE to 500 Centipoises and Cured with 10 phr TETA [50]

Filler, weight %	*Loss, mg/1,000 cycles*
Graphite, 15	46
Titanium dioxide, 33	100
Nylon, 15	111
Graphite, 2.7	129
Molybdenum disulfide, 3.5	130
None	131
Atomized aluminum, 31	147
Iron oxide, 58	147

Thermal Conductivity

The general effect of particulate fillers is to increase thermal conductivity, although to the extent that higher filler loading volumes lead to the incorporation of air bubbles into the casting, the increase may not be as significant as anticipated from laboratory air-free test data.

Figure 14-19 and Table 14-24 indicate the effect of various fillers on thermal conductivity. Increasing the loading volume of particulate fillers results in a proportional increase in the thermal conductivity; but greater increases may be obtained by the incorporation of metallic fibers [27], as indicated in Figs. 14-20 and 14-21, or of metallic-plated glass fibers [34]. The metallic fibers should be present

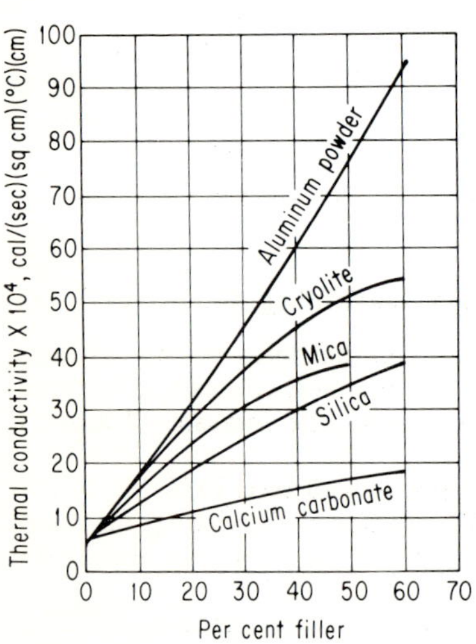

FIG. 14-19. Effect of fillers on thermal conductivity of cured DGEBA [23].

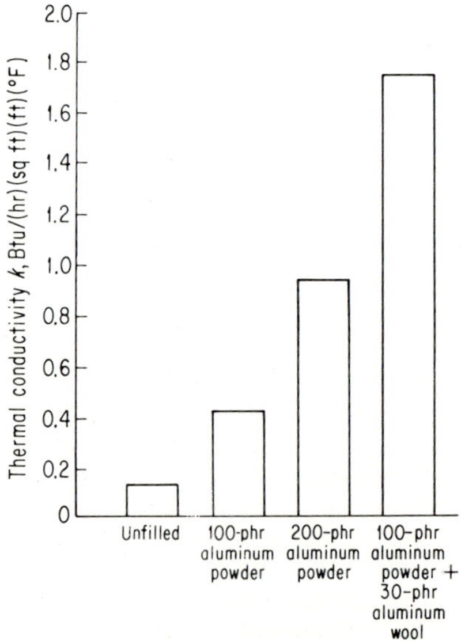

FIG. 14-20. Thermal conductivity of DGEBA cured with MPDA/MDA/PGE vs. filler and loading volume [4–6].

Table 14-24. Thermal Conductivity of DGEBA Cured with MPDA/MDA/PGE vs. Filler [50]

Filler, phr	Volume, % of total	Thermal conductivity, Btu/(hr)(sq ft)(ft)(°F)
None	..	0.13
Aluminum oxide, 325	48	0.82
Atomized aluminum, 200	45	0.94
Iron powder, 300	30	0.52
Copper powder, 1,200	60	0.94
Silica, 150	·39	0.44
Steel	..	26.0
Aluminum	..	116.0
Copper	..	224.0

at least 30 per cent by volume to obtain significant improvements over higher loading volumes of particulate metallic fillers, as shown in Table 14-25.

When the use of metallic powders or conductive fibers is undesirable from an electrical standpoint, it has been suggested that significant increases in thermal conductivity may be obtained with titanium dioxide [10] or with beryllium oxide [61]. Both beryllium oxide particles and fibers have been claimed as epoxy-resin fillers for the production of high-thermal-conductivity electrical insulation [76], but because of their cost and because of toxicity considerations are little used. Magnesium oxide is also quite satisfactory as a high-thermal-conductivity electrical-grade filler. The volume resistivities are equally as good as are experienced with aluminum oxide and only slightly inferior to those of beryllium oxide (10^{14} vs. 10^{16}) [33].

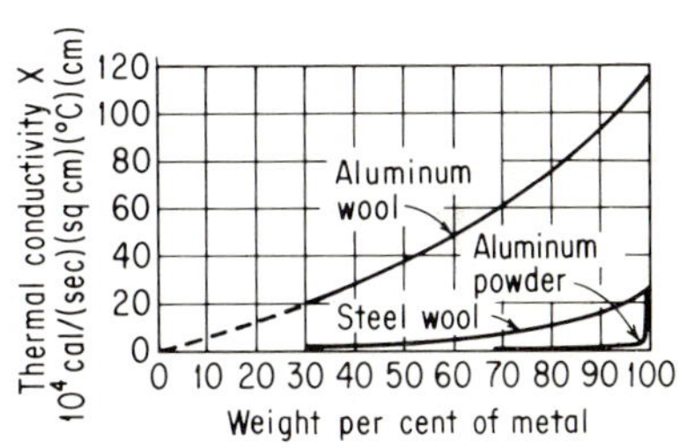

FIG. 14-21. Effect of metallic wools on thermal conductivity of cured DGEBA [40].

Maximum filler loading volumes have been obtained with aluminum oxide by using large-particle-size material (20 to 30 mesh), packing this in a mold with vibration, and subsequently impregnating the filler with the epoxy-resin formulation

Table 14-25. Thermal Conductivity of Epoxy Resins Cured with NMA as a Function of Concentration and Type of Fibrous Metallic Fillers [39]

[In Btu/(hr)(sq ft)(°F)(ft)]

Fiber volume, %	Dicyclopentadiene dioxide					DGEBA	
	Aluminum				Copper	Steel	Aluminum
	⅜ in.			¾ in.	¼ in.	¼ in.	⅜ in.
	× 0.004 in.	× 0.007 in.	× 0.011 in.	× 0.004 in.	× 0.006 in.	× 0.004 in.	× 0.007 in.
5	...	...	...	...	...	0.44	
10	0.53	0.434	0.58	...	0.54		
15	...	...	...	...	...	0.49	
20	0.56	0.59	0.70	0.59	0.57	...	0.48
25	...	...	...	...	...	0.64	
35	0.76	0.73	0.84	...	1.06		

containing 325-mesh aluminum oxide. By this procedure, thermal conductivity has been increased to 60 cal/(sec)(sq cm)(°C)(cm) $\times 10^{-4}$. Volume resistivity was little changed; dissipation factor was somewhat lowered; dielectric constant was increased; and dielectric strength was reduced [70,72].

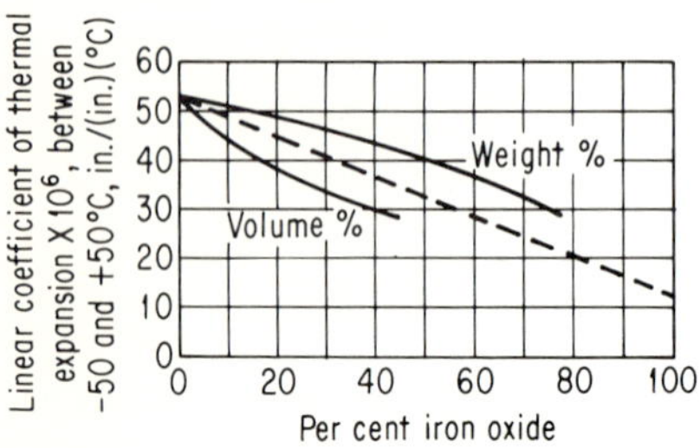

FIG. 14-22. Effect of iron oxide on coefficient of thermal expansion of DGEBA cured with MPDA/MDA/PGE [50].

Thermal conductivity at low temperatures has been determined for DGEBA filled with 100 phr silica and 100 phr alumina and cured with a modified aliphatic primary amine. Silica gave 0.011 g-cal/(sec)(cm²)(°C)(cm) from −148 to −100°C; alumina gave 0.0175 from −142 to −124°C. Heat release was observed during the cooling process at fixed points on the curve, perhaps attributable to a sudden decrease in polymer-chain spacing or to reorientation of interchain bonds [48].

Coefficient of Thermal Expansion

The general effect of fillers is to reduce the coefficient of thermal expansion of the cured epoxy resin in proportion to the loading volume of filler. The reduction is not linear (Figs. 14-22 and 14-23), although in reporting the data curves are usually smoothed out (Fig. 14-24). The nature of the heating media results in variations, as indicated in Figs. 14-25 and 14-26.

The effect of various fillers, loaded to give a specified viscosity at working temperature, is given in Table 14-26. The effect of several curing agents on the thermal

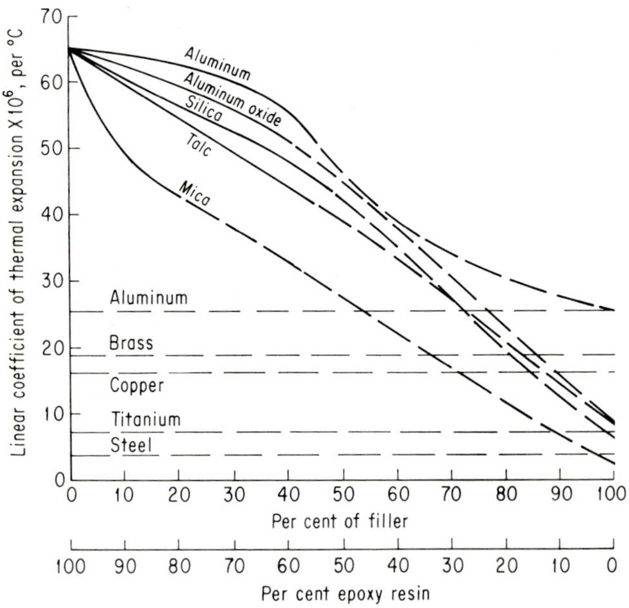

FIG. 14-23. Effect of fillers on coefficient of thermal expansion of cured DGEBA [9].

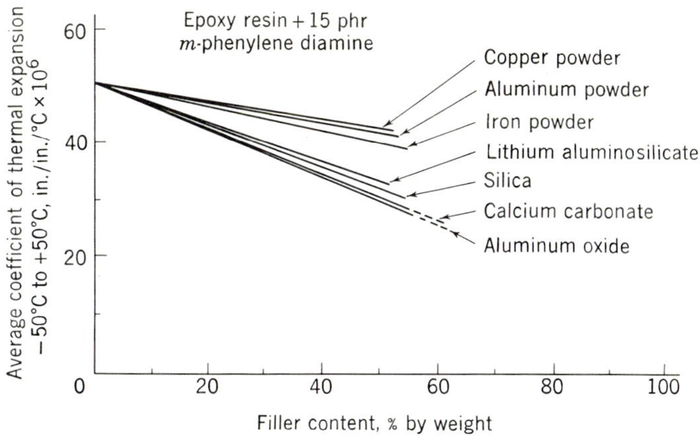

FIG. 14-24. Effect of various fillers on coefficient of thermal expansion of cured DGEBA [4–6].

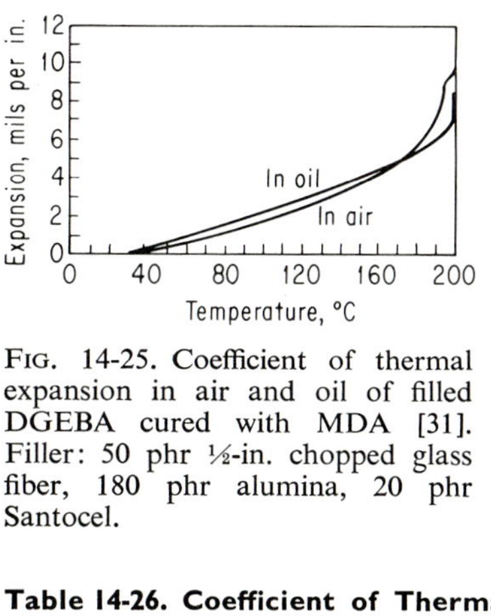

FIG. 14-25. Coefficient of thermal expansion in air and oil of filled DGEBA cured with MDA [31]. Filler: 50 phr ½-in. chopped glass fiber, 180 phr alumina, 20 phr Santocel.

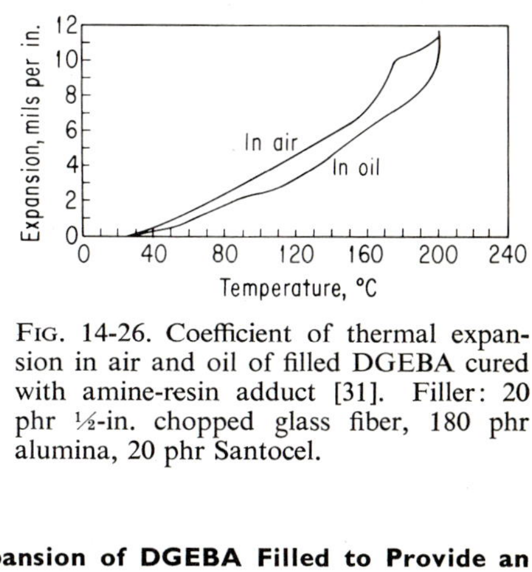

FIG. 14-26. Coefficient of thermal expansion in air and oil of filled DGEBA cured with amine-resin adduct [31]. Filler: 20 phr ½-in. chopped glass fiber, 180 phr alumina, 20 phr Santocel.

Table 14-26. Coefficient of Thermal Expansion of DGEBA Filled to Provide an Initial Viscosity at Working Temperature between 4,000 and 5,000 Centipoises and Cured with Stoichiometric Amounts of Listed Curing Agents [25]

[From +50 to −50°C, in./(in.)(°C) × 10⁶]

Filler	DETA/ acrylonitrile (1/2)	MDA	HHPA + 0.1 phr BDMA	Chlorendic anhydride	Menthane diamine
No filler	57	46	53	53	49
Mica, 1,000 mesh	41	34	..	..	31
Silica, 325 mesh	42	30	22	26	27
Zirconium silicate, 1.5 microns	43	33	19	30	
Hydrated alumina, 0.08–0.12 micron	57	37	35	37	40

Table 14-27. Coefficient of Thermal Expansion of DGEBA Filled with Silica and Cured with Listed Anhydrides [4–6]

Anhydride	Cure schedule, hr/°C	Filler concentration, parts/100 of mix	Coefficient of thermal expansion from +50 to −50°C, in./(in.)(°C) × 10⁶
DDSA + 1 phr BDMA	1/130 + 24/150	0	80
		43	66
		85	55
		128	46
		213	36
NMA + 1 phr BDMA	1/140 + 24/150	0	57
		52	42
		105	36
		157	31
		209	26
Chlorendic anhydride (20) + HHPA (80) + 0.5 phr BDMA	1/120 + 24/150	0	50
		50	39
		100	34
		150	28

expansion of silica-filled systems is shown in Table 14-27. The effect of several fillers on a system cured with MPDA is shown in Table 14-28. Figures 14-27 to 14-29 indicate that the use of diluted resins to increase loading volume of filler does not necessarily result in a corresponding decrease in thermal expansion, insofar as the diluted resins have higher initial expansion rates.

Data have been reported (Fig. 14-30) to indicate that a filler having a negative coefficient of thermal expansion is more effective than silica in lowering the expansion rate of the epoxy when high loading volumes are involved, and a zirconium

Table 14-28. Coefficient of Thermal Expansion of DGEBA Cured with MPDA as a Function of Filler [4–6]

Filler	% filler by weight	Coefficient of thermal expansion, in./(in.)(°C) × 10⁶
Silica	68.6	26.1
	20	45
	17	46
Zirconium silicate	68	28
	40	37.2
	20	42.3
Aluminum powder	63.6	33.5
	40	43.8
	20	47.5
Lithium silicate	72.3	19
	40	33.3
	20	46.9
Calcium carbonate	63.5	23
	22.3	46
Iron powder	72.3	33
	28.1	45
Copper powder	72.3	36
	23.3	47
Hydrated alumina	30.7	41
	16.2	46
Aluminum oxide	53.6	29.5
	31.5	41
Alumina	56.5	30
	31.5	41
Mica	37.0	37.8

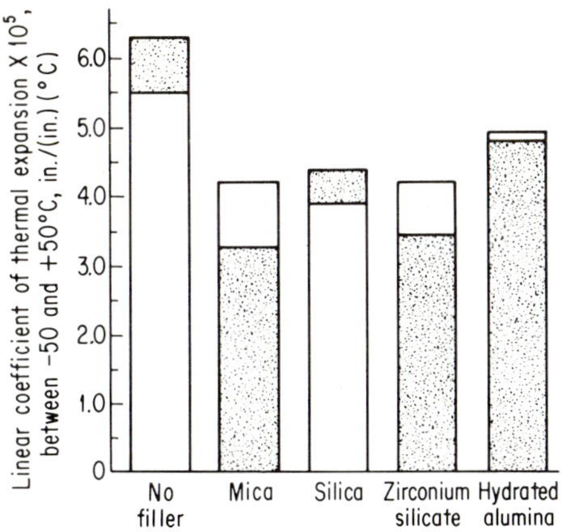

FIG. 14-27. Coefficient of thermal expansion for system shown in Fig. 14-9 [25].

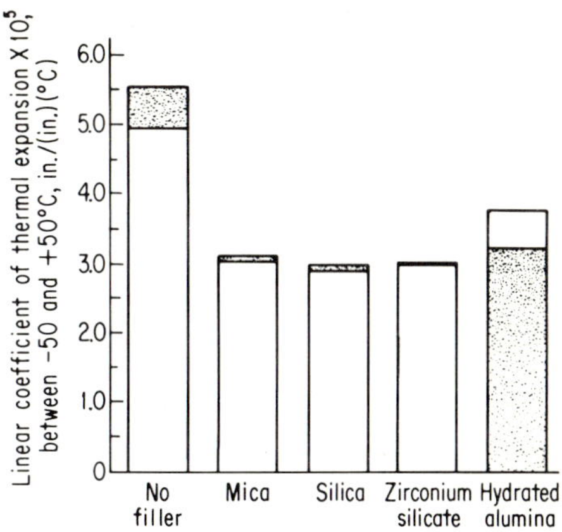

FIG. 14-28. Coefficient of thermal expansion for system shown in Fig. 14-10 [25].

silicate having zero coefficient of thermal expansion is reported to be superior to silica at temperatures above 100°C and to calcium carbonate above 40°C [43]. The effect of lithium aluminum silicate on the coefficient of thermal expansion of DGEBA cured with diethanolamine and MPDA is shown in Table 14-29.

It should be noted, that reduction in thermal expansion rates in the resin does not lead to stress-free castings, even when the expansion coefficient matches that of the insert. Microscopic stresses will be developed at the resin-filler interfaces [37].

Thermal Shock Resistance

Particulate fillers generally improve thermal shock resistance of cured epoxy resins, although insofar as they decrease tensile strength and elongation, the maximum loading volume is not necessarily optimum in terms of thermal shock resistance. Ideally, the filler should be loaded in a volume sufficient to match as closely as

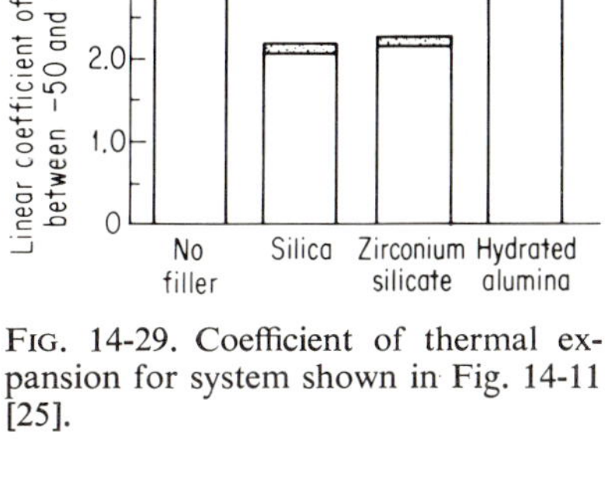

FIG. 14-29. Coefficient of thermal expansion for system shown in Fig. 14-11 [25].

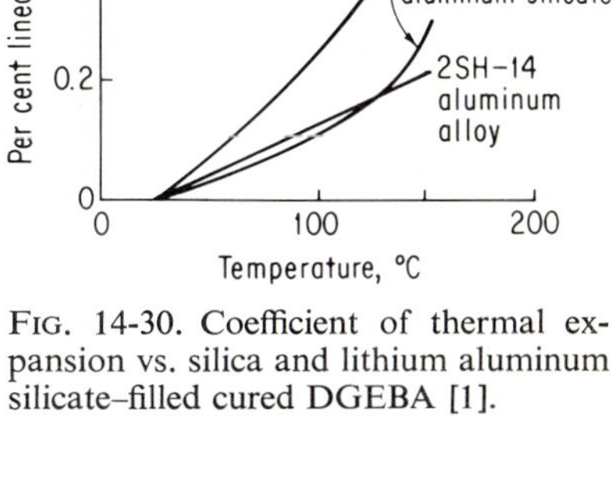

FIG. 14-30. Coefficient of thermal expansion vs. silica and lithium aluminum silicate–filled cured DGEBA [1].

Table 14-29. Effect of Lithium Aluminum Silicate on Coefficient of Thermal Expansion of Two Cured DGEBA Systems [3]

Curing agent, phr	Lithium aluminum silicate, phr	Temperature range, °C	Coefficient of thermal expansion, in./(in.)(°C) × 10^6
Diethanolamine, 12...	100	+100 to −50	36.3
	133	+100 to −50	32.0
MPDA, 14..........	100	+100 to −50	35.0
	133	+26 to −50	25.6
		+64 to −28	33.8
		+100 to −64	44.5

possible the coefficient of thermal expansion of the insert or to achieve an average value for the coefficients of thermal expansion of the various materials involved in the inserts. In the vast majority of the cases this is not practical, and compromises must be arrived at, balancing the reduction of strength against the lower shrinkage and lower thermal expansion obtained from the filler.

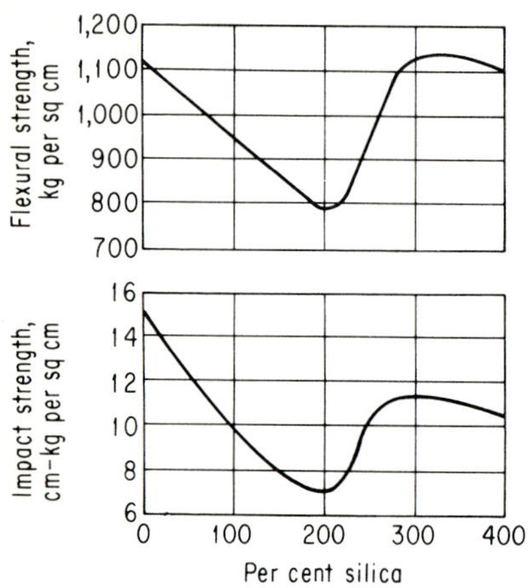

FIG. 14-31. Effect of filler loading volume of impact and flexural strength of DGEBA cured with HHPA [51].

The effect of lithium aluminum silicate in improving thermal shock resistance of several high-temperature (and inherently low-shock-resistant) systems is given in Table 14-30. In this case, the thermal shock tests were performed by curing the epoxy system around the standard Military Specification hex bar and observing whether or not cracks occurred during the high-temperature cure cycle.

Differences will exist between particulate fillers in terms of imparting thermal shock resistance. Silica, for example, has been reported as providing much poorer thermal shock resistance than mica [4], although because of the other variables involved in thermal shock-resistant formulations, meaningful comparisons between fillers are difficult.

Fibrous fillers used at higher loading volumes generally result in dramatic improvement in thermal shock resistance, but unfortunately in many critical applications their use is precluded because of handling difficulties.

Impact Resistance

The effect of particulate fillers on the impact resistance of cured epoxy systems will vary with the nature of the filler, the type of resin, and the type of impact test. Aluminum powder, which appears to impart some degree of ductility to the epoxy, is superior to alumina and gives improvement over the unfilled system. Differences between resins may be observed for a given filler, as indicated in Table 14-31. However, when the Izod (or notched) impact test is employed, the effect of filler on the results will not be particularly great, as indicated when iron oxide is used as a filler in Table 14-32. The relationship between impact and flexural strength vs. filler loading volume is indicated in Fig. 14-31.

Fibrous fillers generally increase impact resistance, particularly at higher loading volumes.

Table 14-30. Properties of Novolac Resin (Functionality 3.3) Filled with Lithium Aluminum Silicate and Cured with Listed Anhydrides [38]

Curing agent, phr	Lithium aluminum silicate, %	Weight loss at 232°C		Shrinkage			DT, °C	Crack on hex bar (MIL-C-16923 A) during cure
		In 900 hr, %	%(hr)	During cure	At 232°C			
					In 900 hr, %	%(hr)		
NMA, 101, +DMP-30, 1.5....	50	...	8.2(692)	0.7	...	2.4(692)	...	Yes
HHPA, 86	60	1.5	1.5(979)	0.85	0.8	0.96(979)	155	No
NMA, 13.3, +HHPA, 75.2....	50	...	7.2(519)*	0.9	...	2.8(519)*	165	Yes
	50	2.7	5.5(1,854)	1.0	1.3	2.6(1,854)	165	Yes
	60	1.8	3.1(1,486)	0.7	0.8	1.5(1,486)	175	No
	70	1.2	1.2(1,068)	0.5	0.7	0.8(1,068)	185	No
NMA, 11.7, +HHPA, 59.8....	60	1.5	1.8(979)	....	0.8	0.9(979)	...	No
NMA, 15.8, +HHPA, 89.7....	60	2.2	2.6(979)	....	1.2	1.3(979)	...	No

* Aged at 260°C.

Table 14-31. Effect of Filler on Impact Resistance of Cured Resin [4–7]

Resin, curing agent, and filler	Impact, in.-lb	Impact as % of unfilled casting
Dicyclopentadiene dioxide, 60%;		
4.4 moles MA + 1 mole trimethylolpropane, 40%:		
Unfilled	14	100
Aluminum, 75%	45	320
Alumina, 75%	20	140
DGEBA + 11 phr DETA:		
Unfilled	29	100
Aluminum, 67%	35	120
Alumina, 67%	16	55

Table 14-32. Effect of Iron Oxide Concentration and Cure Cycle on Impact Strength of DGEBA Cured with MPDA/MDA/PGE [50]

Cure schedule, hr/°C	Filler content, phr	Impact strength, Izod
24/65	0	0.5
24/65 + 2/150	0	0.5
24/65	200	0.4
24/65 + 2/150	200	0.4
24/65	400	0.5
24/65 + 2/150	400	0.4

Adhesive Strength

Fillers, by decreasing shrinkage, are particularly effective in increasing the adhesive strength of epoxy formulations. An effective filler, particularly if ground or suspended into the resin properly, is alumina [44]. Table 14-33 presents a comparison of the shear strength of fluidized DGEBA, unfilled and filled with 100 phr alumina. When the epoxy adhesive is intended for bonding metals, the reduction in thermal expansion obtained by the use of fillers serves to reduce strains on the adhesive during temperature cycling. A further advantage is obtained with fillers, insofar as they can be used in high loading volumes to obtain pastes which eliminate the need for close regulation of the glue-line thickness. Optimum loading volumes, however, may be less than maximum, since incomplete wetting may be experienced at the higher filler concentrations.

Although most fillers serve to increase adhesive strength, zinc stearate, in low concentrations, is frequently used in molding powders where it facilitates mold release.

Table 14-33. Effect of Filler on Shear Strength of Adhesive Formulation [5]
(In pounds per square inch)

Filler	At 25°C	At 105°C	After 1 hr in water at 100°C	After 1 week in water at 25°C
No filler	2,760	450	2,740	2,840
100 phr aluminum oxide	3,750	1,650	3,930	2,890

Specific Gravity

The majority of the fillers, having a higher specific gravity than the resins, increase the specific gravity of the cured systems proportional to the loading volumes. Some fillers give greater bulk at lower volume concentrations than others, and in such systems the specific gravity of the filler will not contribute as heavily as a filler giving lesser bulk. Given the same material, the finer the grind the greater the bulk.

The increase in specific gravity obtained by the use of fillers will, to some extent, offset the cost advantage realized by a lower price per unit of weight.

The effect of several fillers on specific gravity is indicated in Table 14-34 and Figs. 14-32 to 14-34.

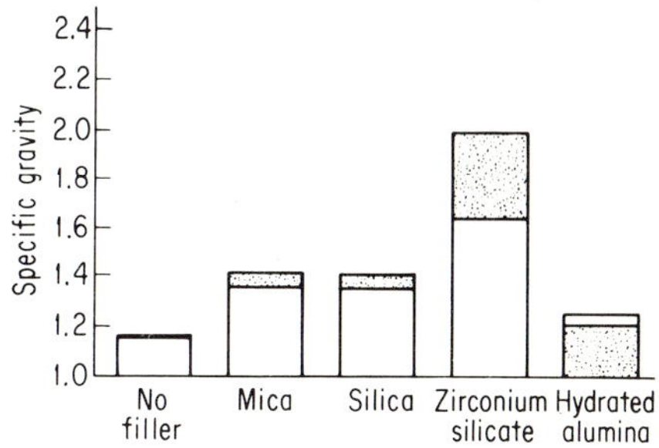

FIG. 14-32. Specific gravity of systems shown in Fig. 14-9 [25].

Fillers with a density lower than that of the resin may be employed to reduce specific gravity. These are usually gas-filled microscopically small balloons and generally bring about a sharp increase in viscosity at relatively low weight concentrations. Properties of one epoxy system filled with glass microballoons are shown in Table 14-35. These balloons have a sieve analysis of 0.5 per cent through +60 mesh, 20 to 50 per cent through +200 mesh, and a particle density of 0.39 to 0.46 gram per cc. This system is compared with a mica-filled system in Table 14-36. A discussion of other low-density fillers is presented in the section on syntactic foams in Chap. 19.

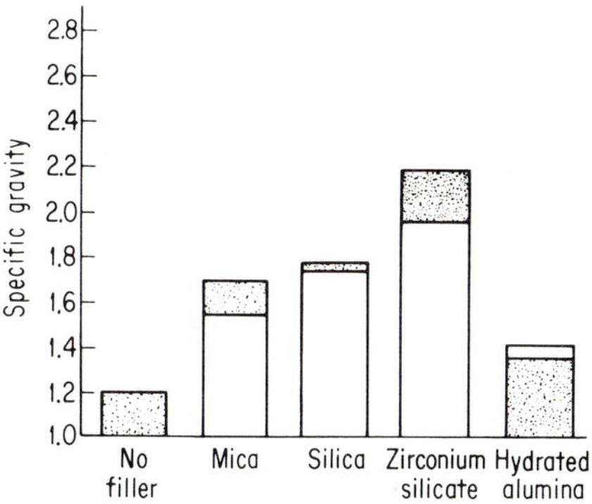

FIG. 14-33. Specific gravity of systems shown in Fig. 14-10 [25].

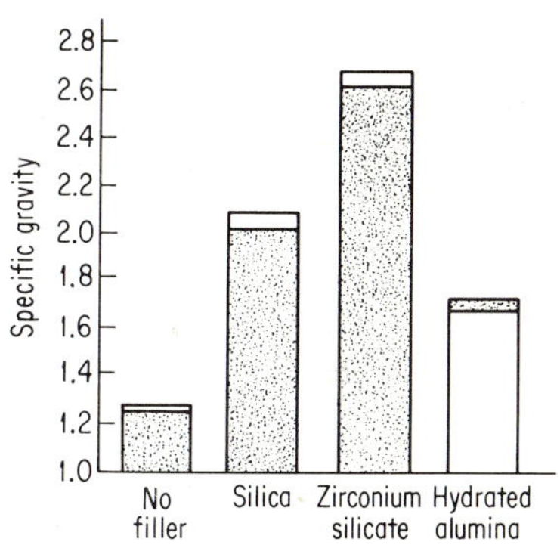

FIG. 14-34. Specific gravity of systems shown in Fig. 14-11 [25].

EFFECT OF FILLERS ON ELECTRICAL PROPERTIES

Nonconductive fillers are employed with electrical-grade epoxy resins and generally result in minor modifications to electrical characteristics.

Metallic fillers degrade electrical values. As little as 2 phr iron oxide will reduce volume resistivity for a typical epoxy system one decade from 3.9×10^{15} ohm-cm [36].

Table 14-34. Specific Gravity of DGEBA as a Function of Filler, Loaded to Provide an Initial Viscosity of 4,000 to 5,000 Centipoises at Working Temperature and Cured with Listed Compounds [25]

Filler	DETA/acrylonitrile (½)	MDA	HHPA + 0.1% BDMA	Chlorendic anhydride	Menthane diamine
No filler	1.18	1.2	1.02	1.52	0.96
Mica, 1,000 mesh	1.24	1.54	....	....	1.32
Silica, 325 mesh	1.56	1.68	1.76	2.0	1.52
Zirconium silicate, <1.5 microns	1.62	1.84	2.48	2.36	
Hydrated alumina, 0.08–0.12 micron	1.28	1.40	1.24	1.60	1.14

Table 14-35. Properties of DGEBA Filled with 43 phr Glass Balloons and Cured with 12 phr Diethanolamine [42]

Property	At −55°C	At 23°C	At 63°C
Tensile strength, psi	5,480	4,660	2,200
Modulus, psi × 10^{-6}	0.53	0.44	0.18
Elongation, %	1.1	1.1	2.6
Flexural strength, psi	7,230	5,850	2,290
Modulus, psi × 10^{-6}	0.47	0.44	0.20
Elongation, %	1.6	1.4	4.6
Compressive strength, psi	14,600	10,600	3,480
Modulus, psi × 10^{-6}	0.44	0.42	0.09
Elongation, %	3.9	3.3	10.6
Coefficient of thermal expansion, in./(in.)(°C) × 10^6:			
At −60 to 0°C		35	
At 0 to 40°C		39	
At 40 to 70°C		42	
Thermal conductivity Btu/(hr)(sq ft)(°F)(in.)		1.1	
Specific gravity		0.85	
DT, °C		77	
Moisture absorption weight, %		0.37	
Fungus resistance		Nonnutrient	

Table 14-36. Properties of DGEBA Filled with Glass Microballoons Compared with Resin Filled with Mica, Both Cured with 12 phr Diethanolamine [42]

Property	Glass microballoon, 43 phr	1,000-mesh mica, 100 phr
Density, pcf	53	102
Tensile strength, psi	4,660	7,700
Modulus, psi × 10^{-6}	0.44	1.2
Flexural strength, psi	5,850	11,300
Modulus, psi × 10^{-6}	0.44	1.2
Compressive strength, psi	10,650	17,970
Modulus, psi × 10^{-6}	0.42	1.0

Metallic fillers, such as powdered copper and silver, may be used in fairly high loading volumes to obtain electrically conductive systems. Flaked silver has been suggested [20], as has flaked silver in combination with silver powder [12] or carbon [21,66] as conductive fillers. A typical silver-filled epoxy film provides a volume resistivity of 158×10^{-3} ohm-cm at $-55°C$, which increases to about 223×10^{-3} ohm-cm at 80°C [49]. Aluminum powder is not so effective as might be expected and may be used at fairly high loading volumes while still preserving electrical resistance (Table 14-37). Electrically conductive epoxies have been obtained by the use of silver-coated copper spheres, these used at about 90 per cent by weight and cured under enough pressure to obtain particle-to-particle contact [69]. Electrically conductive epoxies have also been obtained by the use of metallic wools and conductive fibers and nickel fillers, when the system is cured in a magnetic field [75]. Moldable permanent magnets have been produced by filling an epoxy-resin composition with powdered magnetizable material such as aluminum nickel alloy [78].

Table 14-37. Electrical Properties of the Cured DGEBA vs. Loading Volume of Metallic Fillers [35]*

Filler, phr	Volume resistivity		Dielectric constant
	At 23°C, ohm-cm $\times 10^{-14}$	At 93°C, ohm-cm $\times 10^{-9}$	At 10^3 cps and 23°C
No filler	1.0	6.3	4.4
Aluminum powder, 10	1.1	6.4	4.7
30	0.89	6.1	6.5
50	0.81	5.6	8.7
100	0.45	5.3	10.0
Needlelike aluminum particles, ⅛ in. long, 10	1.2	6.0	4.9
30	0.62	5.4	5.7
50	0.16	3.8	8.8
100	....	...	10.0

Volume Resistivity

The effect of a typical electrical-grade filler, silica, on volume resistivity for several types of cured systems is presented in Figs. 14-35 to 14-37. Although some reduction is brought about, the effect is marginal, and substantially greater changes may be made by varying the curing agent. Under conditions of high humidity, however, fillers may tend to wick in moisture, thus considerably degrading volume resistivity over that obtained with the unfilled system [16].

Arc Resistance

Electrical-grade fillers generally improve the arc resistance of the cured epoxy systems, as indicated in Table 14-38 and Figs. 14-38 to 14-40. When practical, further improvement in arc resistance may be obtained with rough surfaces containing partially exposed filler, and it has been suggested that coarse particles may be sprinkled over the surface of the epoxy before cure to provide even further improvement [22]. Hydrated aluminum oxide and hydrated calcium sulfates [55] will improve arc resistance if cure is sufficiently low to prevent dehydration. DGEBA coatings containing at least 45 per cent by weight of anhydrous magnesium silicate (<0.2 per cent calcium) provide improved tracking resistance [81].

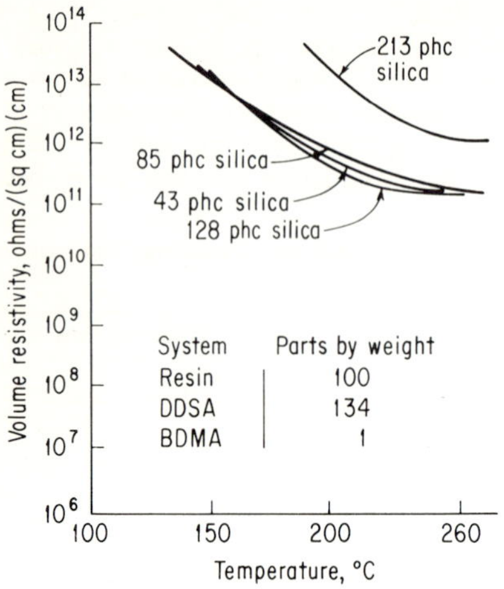

FIG. 14-35. Volume resistivity vs. temperature of DGEBA cured with DDSA as function of filler loading volume [4–6]. Cure cycle: 1 hour at 130°C + 24 hours at 150°C.

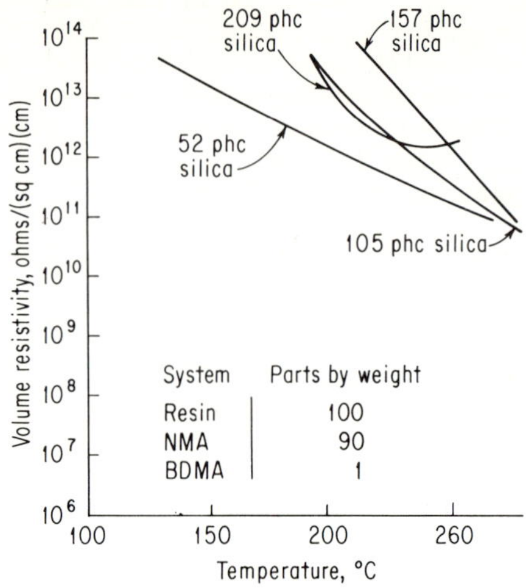

FIG. 14-36. Volume resistivity vs. temperature of DGEBA cured with NMA as function of filler loading volume [4–6]. Cure cycle: 1 hour at 140°C + 24 hours at 150°C.

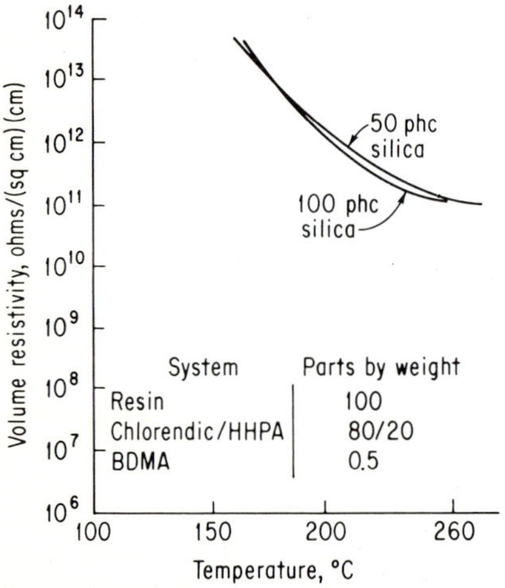

FIG. 14-37. Volume resistivity vs. temperature of DGEBA cured with CA/HHPA as function of filler loading volume [4–6]. Cure cycle: 1 hour at 120°C + 24 hours at 150°C.

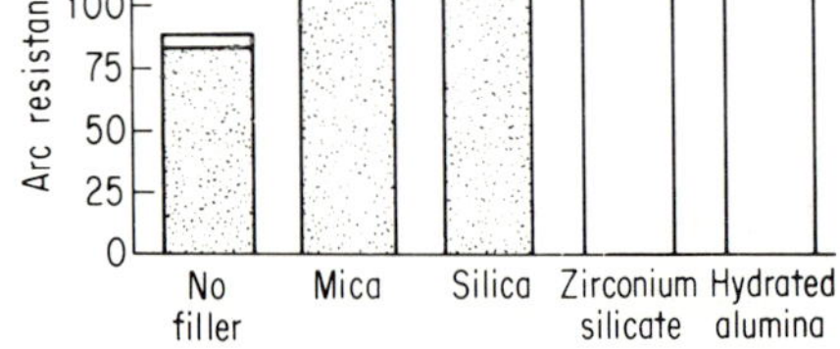

FIG. 14-38. Arc resistance of systems shown in Fig. 14-9 [25].

Table 14-38. Arc Resistance of DGEBA as a Function of Filler, Loaded to Provide an Initial Viscosity of 4,000 to 5,000 Centipoises at Working Temperature and Cured with Stoichiometric Amounts of Listed Curing Agents [25]

Filler	DETA/acrylonitrile (½)	MDA	HHPA + 0.1% BDMA	Chlorendic anhydride	Menthane diamine
No filler	85	115	110	110	95
Mica, 1,000 mesh	125	180	...	...	140
Silica, 325 mesh	140	185	225	190	185
Zirconium silicate, <1.5 microns	125	155	195	140	
Hydrated alumina, 0.08–0.12 micron	135	150	150	130	155

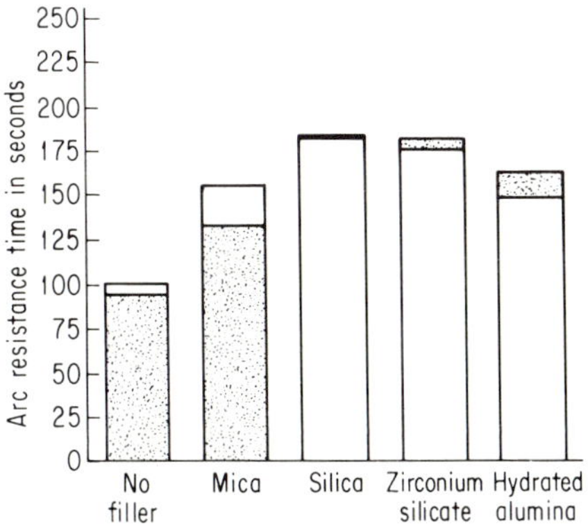

FIG. 14-39. Arc resistance of systems shown in Fig. 14-10 [25].

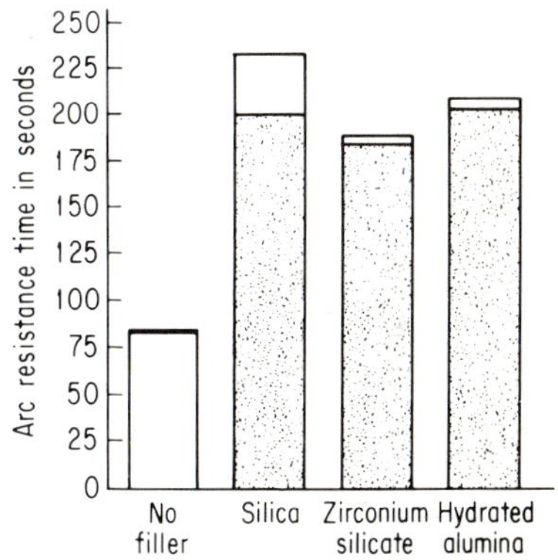

FIG. 14-40. Arc resistance of systems shown in Fig. 14-11 [25].

Dielectric Constant

The general effect of electrical-grade fillers is to increase the dielectric constant of the cured epoxy resin, with minor differences between fillers being observed. This is indicated for several cured systems in Table 14-39 and Figs. 14-41 to 14-49. When the filler is hollow spheres containing air (with a dielectric constant of 1), the

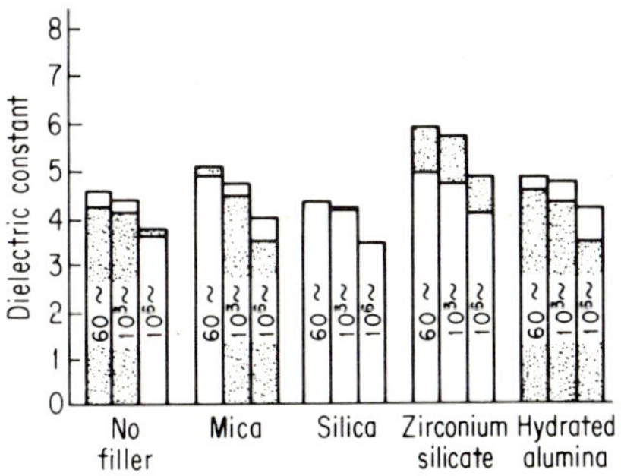

FIG. 14-41. Dielectric constant at 25°C of systems shown in Fig. 14-9 [25].

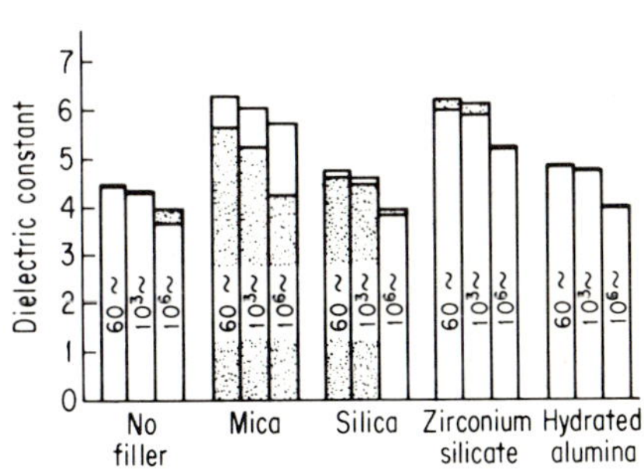

FIG. 14-42. Dielectric constant at 25°C of systems shown in Fig. 14-10 [25].

Table 14-39. Dielectric Constant of DGEBA as a Function of Filler, Loaded to Provide an Initial Viscosity of 4,000 to 5,000 Centipoises at Working Temperature and Cured with Stoichiometric Amounts of Listed Curing Agents [25]

Filler	DETA/acrylonitrile (½)			MDA			HHPA + 0.1% BDMA			Chlorendic anhydride			Menthane diamine		
	25°C	80°C	100°C	25°C	80°C	100°C	25°C	80°C	100°C	25°C	80°C	100°C	25°C	80°C	100°C
No filler:															
At 60 cps	5.4	>10	>10	4.5	4.3	4.4	3.3	3.3	3.3	3.6	3.4	3.4	4.3	4.2	4.2
At 10^3 cps	5.2	>10	>10	4.4	4.3	4.3	3.2	3.2	3.2	3.4	3.4	3.3	4.2	4.1	4.1
At 10^6 cps	4.1	6.2	7.7	3.9	4.0	4.2	3.0	3.0	3.1	3.0	3.1	3.2	3.7	3.8	3.8
Mica, 1,000 mesh:															
At 60 cps	5.5	>10	>10	6.6	7.4	7.5	...	...	...	...	...	...	6.5	7.8	7.5
At 10^3 cps	5.2	>10	>10	6.0	6.4	6.5	...	...	...	...	...	...	5.8	6.9	6.4
At 10^6 cps	4.3	6.3	7.5	4.4	4.5	4.8	...	...	...	...	...	...	3.9	4.4	4.2
Silica, 325 mesh:															
At 60 cps	5.3	>10	>10	4.5	4.5	4.6	4.4	4.4	4.6	4.0	4.0	4.1	4.4	4.4	4.4
At 10^3 cps	5.2	>10	>10	4.4	4.4	4.5	4.2	4.2	4.3	3.8	3.6	3.9	4.3	4.2	4.3
At 10^6 cps	4.4	6.4	7.6	3.8	4.1	4.2	3.6	3.7	3.7	3.4	3.2	3.5	3.7	4.0	3.8
Zirconium silicate <1.5 microns:															
At 60 cps	6.3	>10	>10	5.8	5.6	5.6	6.6	6.8	6.9	5.0	4.5	4.9			
At 10^3 cps	5.9	>10	>10	5.6	5.4	5.5	6.4	6.5	6.5	4.8	4.5	4.8			
At 10^6 cps	5.0	6.9	9.0	5.0	5.1	5.2	5.8	6.3	6.3	4.4	4.3	4.5			
Hydrated alumina, 0.08–0.12 micron:															
At 60 cps	6.5	>10	>10	5.1	5.4	5.8	4.2	4.0	4.0	3.6	3.5	3.7	4.7	5.2	5.8
At 10^3 cps	5.9	>10	>10	5.0	5.1	5.2	4.0	3.9	3.9	3.5	3.4	3.6	4.6	4.6	5.1
At 10^6 cps	4.9	8.4	7.7	4.0	4.4	4.7	3.4	3.5	3.6	3.0	3.3	3.5	4.0	4.1	4.2

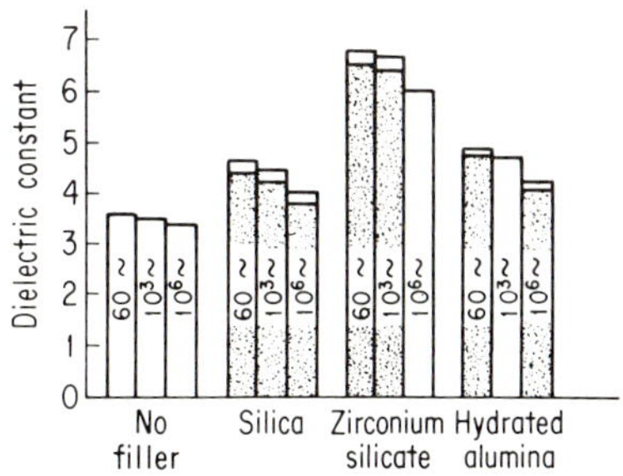

FIG. 14-43. Dielectric constant at 25°C for systems shown in Fig. 14-11 [25].

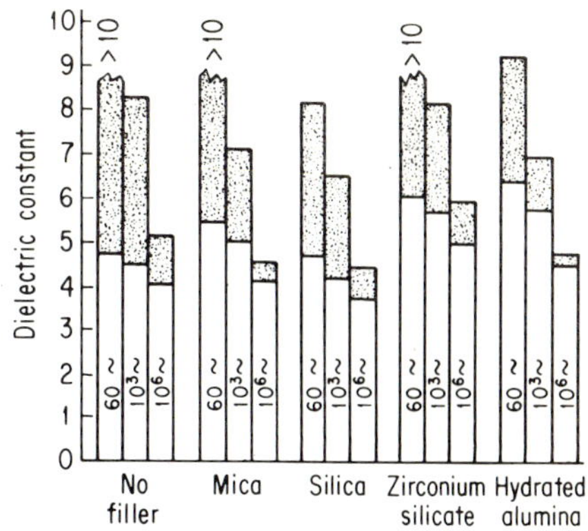

FIG. 14-44. Dielectric constant at 80°C for systems shown in Fig. 14-9 [25].

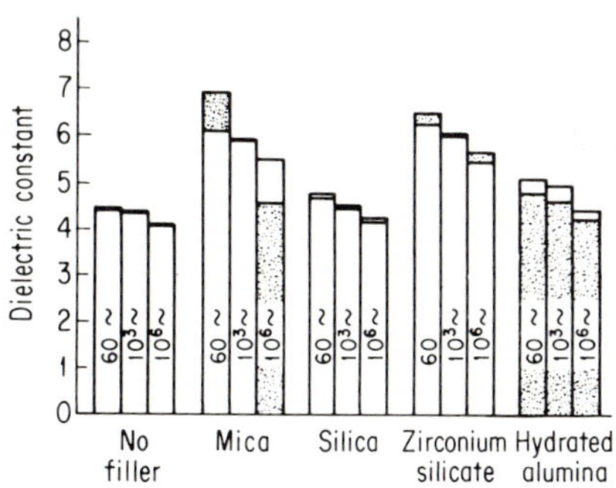

FIG. 14-45. Dielectric constant at 80°C for systems shown in Fig. 14-10 [25].

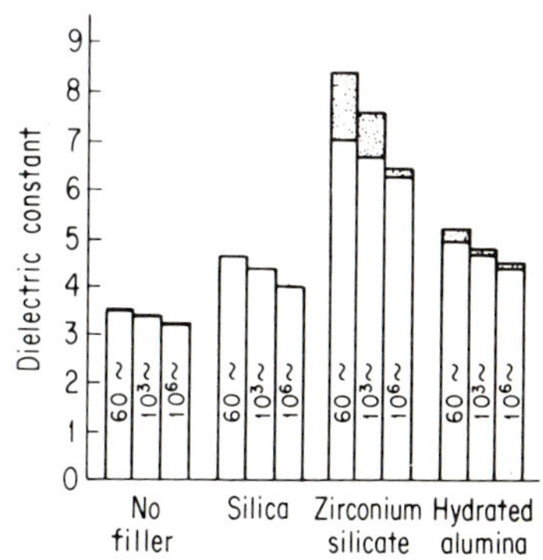

FIG. 14-46. Dielectric constant at 80°C for systems shown in Fig. 14-11 [25].

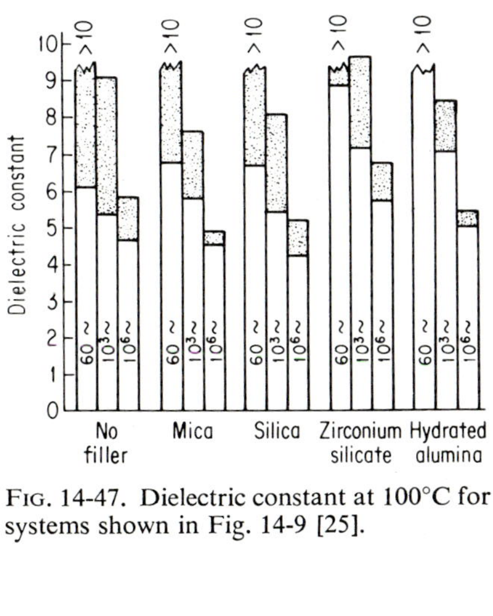

FIG. 14-47. Dielectric constant at 100°C for systems shown in Fig. 14-9 [25].

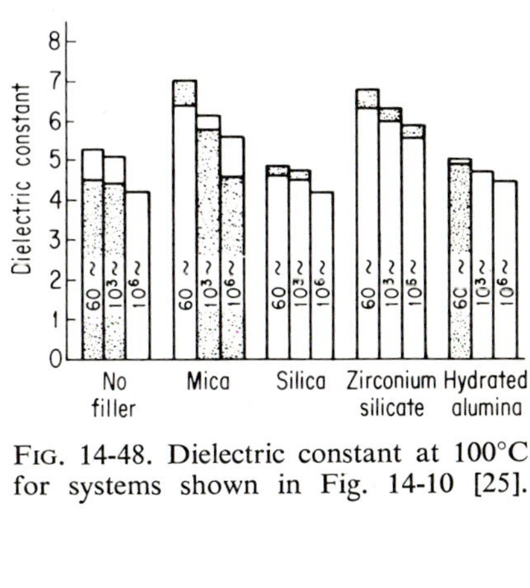

FIG. 14-48. Dielectric constant at 100°C for systems shown in Fig. 14-10 [25].

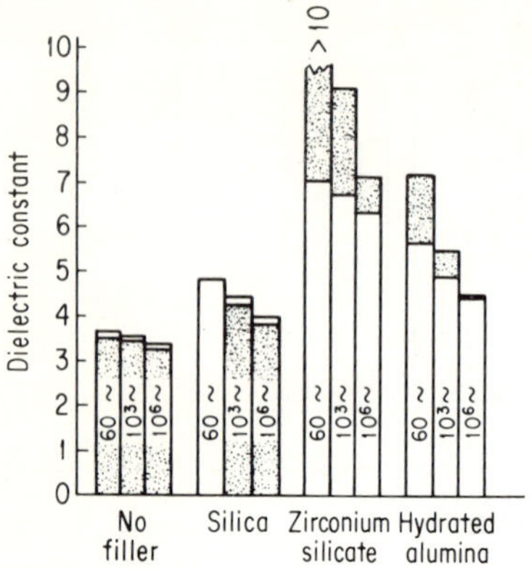

FIG. 14-49. Dielectric constant at 100°C for systems shown in Fig. 14-11 [25].

effect is to reduce the dielectric constant. By use of inorganic microballoons, dielectric constants as low as 1.4 have been reported at a filler loading volume of 20 per cent compared to 4.4 for the unfilled system [14]. Dielectric constant and other electrical properties of a microballoon-filled system are indicated in Table 14-40.

Dielectric Strength

Electrical-grade fillers generally result in little change in the dielectric strength of the epoxy resins, as indicated in Table 14-41 and Figs. 14-50 to 14-52.

In pulse-voltage tests, it was established that filler type and filler particle size influence electrical pulse resistance and that an inverse correlation exists between electrical pulse resistance and thermal shock, leading to the supposition that mechanical failure in the cured compound may be important in stopping partial breakdown [77].

Table 14-40. Electrical Properties of DGEBA Filled with 43 phr Glass Balloons and Cured with 12 phr Diethanolamine [42]

Property	*Value*
Dielectric constant:	
At 10^3 cps	3.02
At 10^6 cps	2.72
Dissipation factor:	
At 10^3 cps	3.96
At 10^6 cps	2.55
Dielectric strength, volts/mil	430
Volume resistivity, ohm-cm:	
At 23°C	2.05×10^{14}
At 66°C	1.97×10^{13}
At 93°C	9.30×10^{10}
At 121°C	2.43×10^{9}
At 149°C	3.68×10^{8}

Table 14-41. Dielectric Strength of DGEBA as a Function of Filler, Loaded to Provide an Initial Viscosity of 4,000 to 5,000 Centipoises at Working Temperature and Cured with Stoichiometric Amounts of Listed Curing Agents [25]

(In volts per mil)

Filler	DETA/acrylonitrile (½)	MDA	HHPA + 0.1% BDMA	Chlorendic anhydride	Menthane diamine
No filler	390	435	390	310	395
Mica, 1,000 mesh	325	425	...	...	355
Silica, 325 mesh	335	440	460	425	460
Zirconium silicate, <1.5 microns	300	395	355	345	
Hydrated alumina, 0.08–0.12 micron	410	400	350	360	410

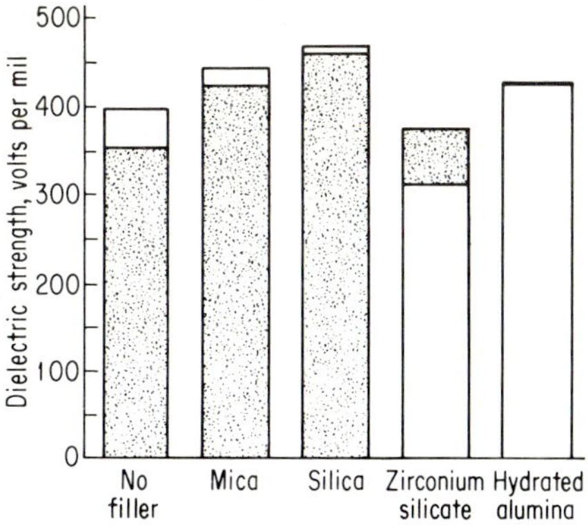

FIG. 14-50. Dielectric strength for systems shown in Fig. 14-9 [25].

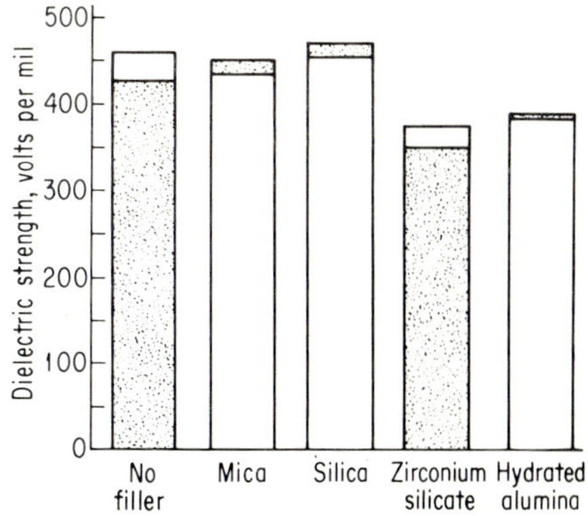

FIG. 14-51. Dielectric strength for systems shown in Fig. 14-10 [25].

Dissipation Factor

Electrical-grade fillers, with exceptions, tend to increase somewhat the dissipation factor of the cured epoxy system, although the extent of the increase is dependent upon both temperature and frequency, as indicated in Table 14-42 and Figs. 14-53 to 14-61. Some oxide fillers, such as chromium, titanium, manganese, and magnesium oxide when added at 1 to 2 per cent to silica-filled epoxies are reported to lower dissipation factor [44].

Surface Resistivity

The effect of electrical-grade fillers on surface resistivity is compared with unfilled castings in Figs. 14-62 to 14-65.

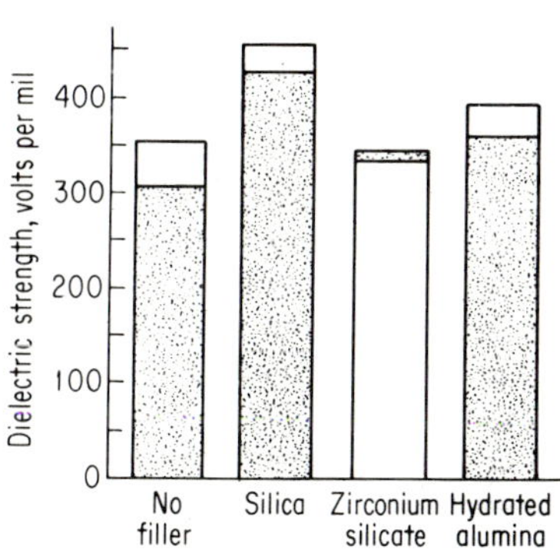

FIG. 14-52. Dielectric strength for systems shown in Fig. 14-11 [25].

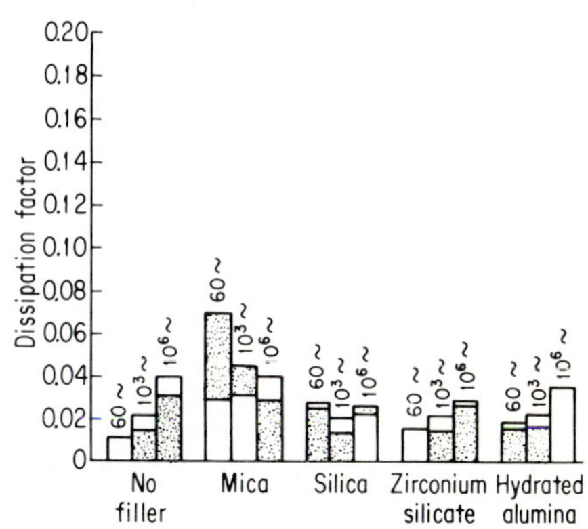

FIG. 14-53. Dissipation factor at 25°C for systems shown in Fig. 14-9 [25].

Table 14-42. Dissipation Factor of DGEBA as a Function of Filler, Loaded to Provide an Initial Viscosity of 4,000 to 5,000 Centipoises at Working Temperature and Cured with Listed Curing Agents [25]

Filler	DETA/acrylonitrile (½)			MDA			HHPA + 0.1% BDMA			Chlorendic anhydride			Menthane diamine		
	25°C	80°C	100°C	25°C	80°C	100°C	25°C	80°C	100°C	25°C	80°C	100°C	25°C	80°C	100°C
No filler:															
At 60 cps	0.016	>0.2	>0.2	0.008	0.003	0.002	0.004	0.004	0.004	0.006	0.002	0.002	0.006	0.004	0.004
At 10^3 cps	0.024	0.159	>0.2	0.014	0.004	0.002	0.011	0.005	0.004	0.009	0.003	0.001	0.012	0.005	0.005
At 10^6 csp	0.052	0.124	0.136	0.032	0.036	0.026	0.012	0.014	0.016	0.030	0.037	0.023	0.032	0.024	0.002
Mica, 1,000 mesh:															
At 60 cps	0.028	>0.2	>0.2	0.060	0.066	0.070	....	....	....	....	....	....	0.064	0.056	0.065
At 10^3 cps	0.028	0.184	>0.2	0.062	0.075	0.074	....	....	....	....	....	....	0.072	0.079	0.072
At 10^6 cps	0.044	0.112	0.115	0.044	0.074	0.073	....	....	....	....	....	....	0.045	0.080	0.073
Silica, 325 mesh:															
At 60 cps	0.024	>0.2	>0.2	0.014	0.024	0.033	0.032	0.054	0.064	0.016	0.032	0.038	0.010	0.019	0.029
At 10^3 cps	0.023	>0.2	>0.2	0.016	0.012	0.016	0.022	0.035	0.044	0.014	0.016	0.021	0.009	0.012	0.013
At 10^6 cps	0.044	0.114	0.104	0.024	0.025	0.020	0.008	0.010	0.011	0.012	0.014	0.014	0.022	0.016	0.012
Zirconium silicate <1.5 microns:															
At 60 cps	0.022	>0.2	>0.2	0.012	0.019	0.014	0.009	0.028	0.044	0.010	0.020	0.021			
At 10^3 cps	0.024	0.192	>0.2	0.016	0.012	0.010	0.006	0.016	0.021	0.012	0.009	0.019			
At 10^6 cps	0.046	0.106	0.18	0.024	0.024	0.019	0.014	0.017	0.013	0.020	0.021	0.014			
Hydrated alumina, 0.08–0.12 micron															
At 60 cps	0.050	>0.2	>0.2	0.010	0.015	0.011	0.008	0.116	0.014	0.012	0.007	0.009	0.017	0.070	0.098
At 10^3 cps	0.046	>0.2	>0.2	0.014	0.007	0.044	0.012	0.012	0.010	0.014	0.007	0.008	0.018	0.043	0.073
At 10^6 cps	0.048	0.112	0.094	0.032	0.028	0.026	0.016	0.016	0.014	0.022	0.025	0.020	0.032	0.032	0.026

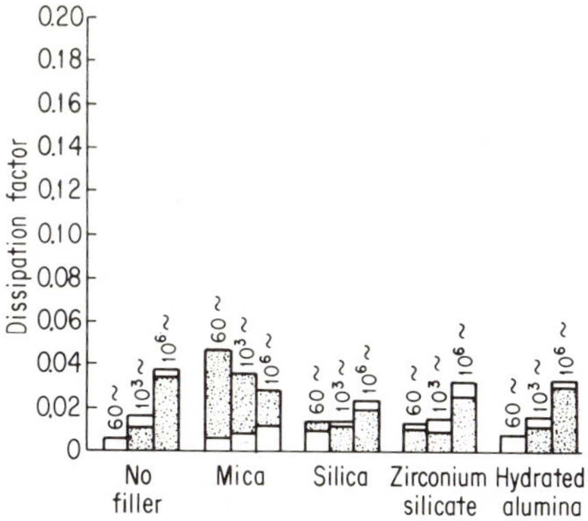

FIG. 14-54. Dissipation factor at 25°C for systems shown in Fig. 14-10 [25].

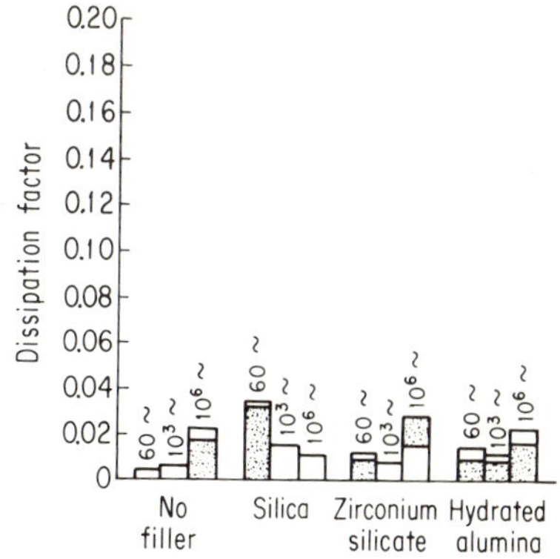

FIG. 14-55. Dissipation factor at 25°C for systems shown in Fig. 14-11 [25].

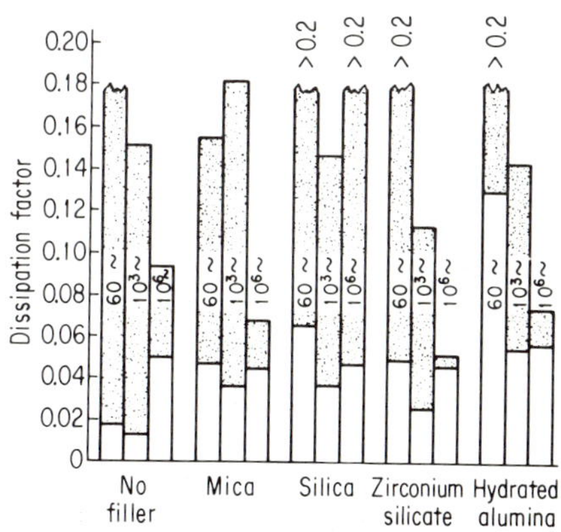

FIG. 14-56. Dissipation factor at 80°C for systems shown in Fig. 14-9 [25].

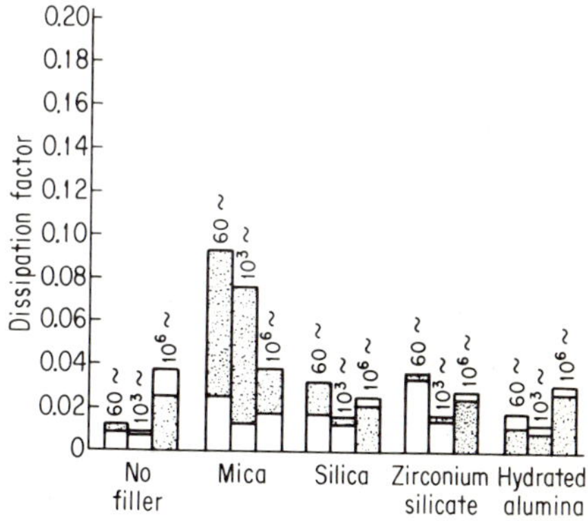

FIG. 14-57. Dissipation factor at 80°C for systems shown in Fig. 14-10 [25].

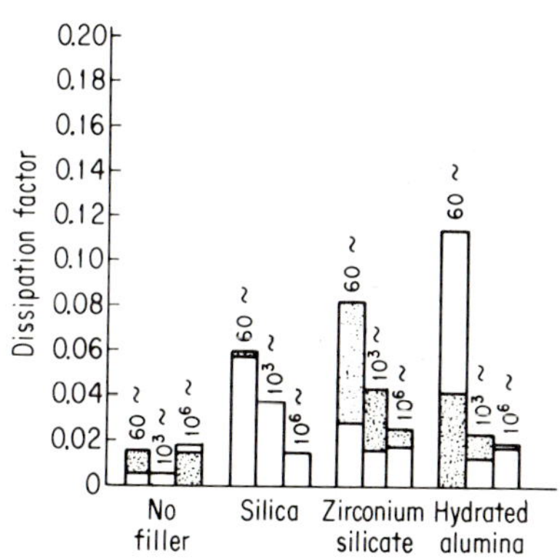

FIG. 14-58. Dissipation factor at 80°C for systems shown in Fig. 14-11 [25].

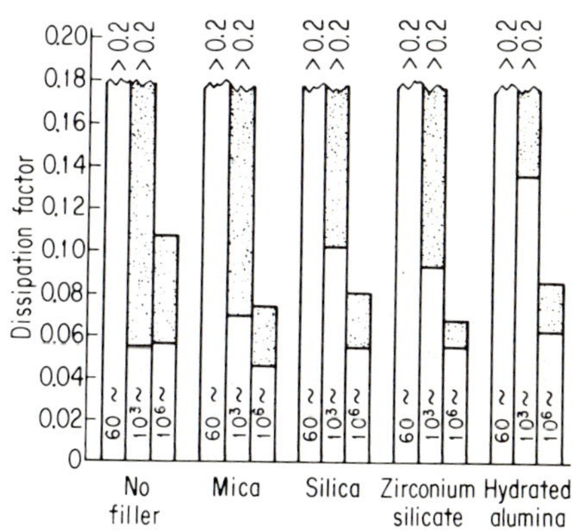

FIG. 14-59. Dissipation factor at 100°C for systems shown in Fig. 14-9 [25].

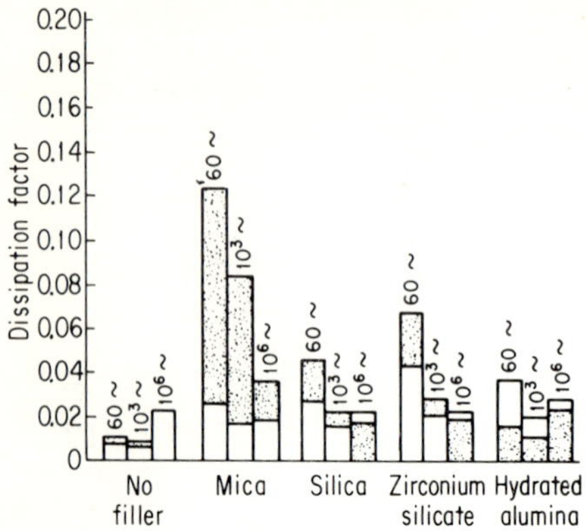

FIG. 14-60. Dissipation factor at 100°C for systems shown in Fig. 14-10 [25].

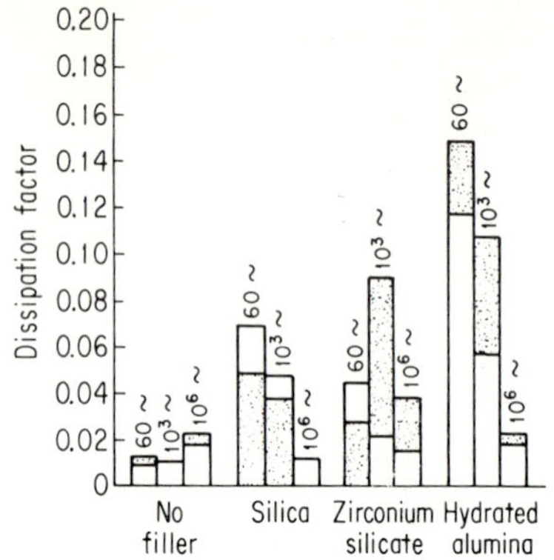

FIG. 14-61. Dissipation factor at 100°C for systems shown in Fig. 14-11 [25].

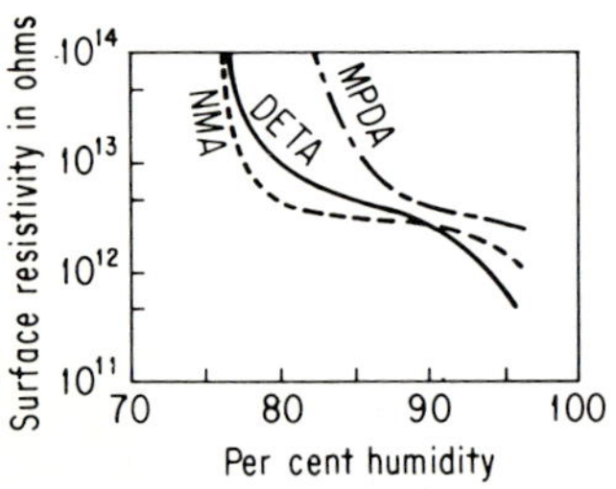

FIG. 14-62. Effect of humidity at 35°C on surface resistivity of cured, unfilled DGEBA [29].

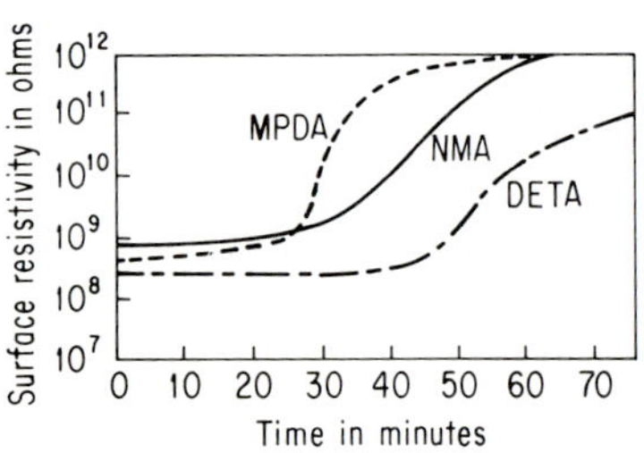

FIG. 14-63. Recovery of surface resistivity at 80 per cent relative humidity and 25°C for cured, unfilled DGEBA [29].

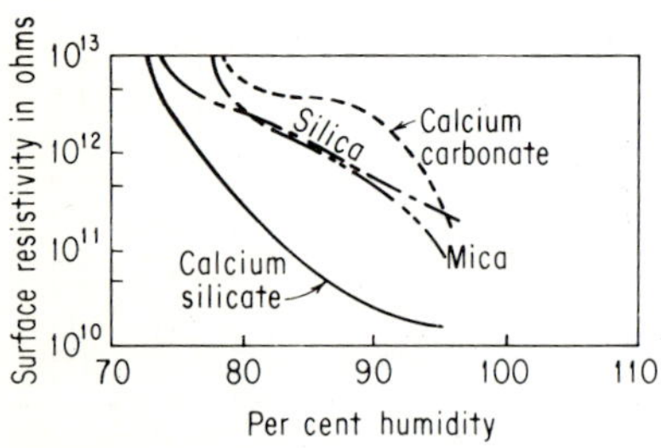

FIG. 14-64. Effect of humidity at 35°C on surface resistivity of filled DGEBA [29].

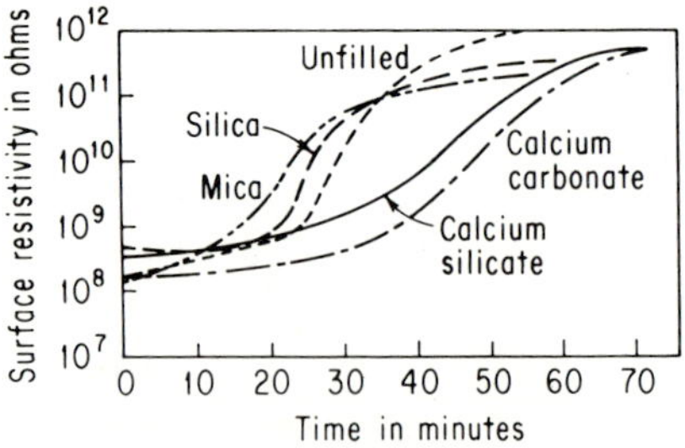

FIG. 14-65. Recovery of surface resistivity at 80 per cent relative humidity and 25°C for cured, filled DGEBA [29].

EFFECT OF FILLERS ON CHEMICAL RESISTANCE AND CORROSIVITY

Fillers may affect the moisture resistance, the moisture-vapor transmission, and the solvent resistance either adversely or advantageously. Some fillers, such as calcium carbonate, tend to lower acid resistance, whereas others, such as silica or aluminum, may tend to lower alkali resistance. Many fibrous fillers exhibit a wicking action for moisture, and this is particularly true of glass fibers upon failure of internal bonds. The effect of several fillers on moisture absorption is indicated

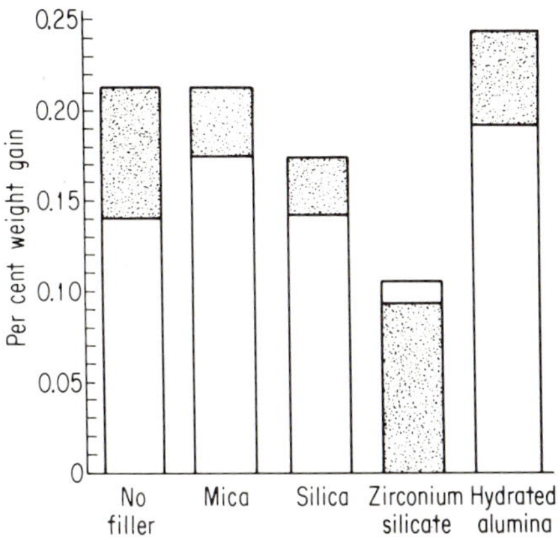

FIG. 14-66. Water absorption after 7 days' immersion for systems shown in Fig. 14-9 [25].

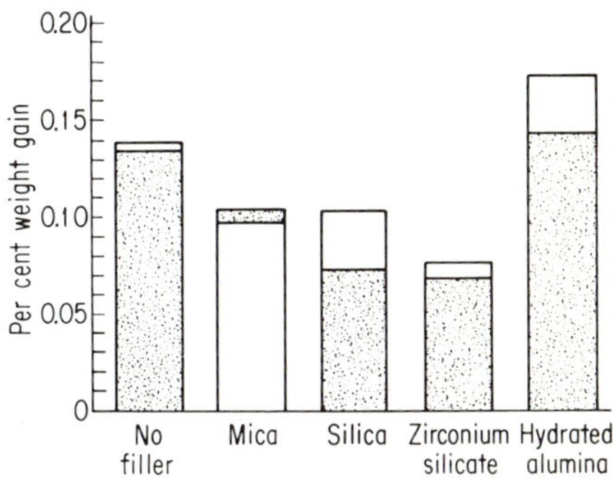

FIG. 14-67. Water absorption after 7 days' immersion for systems shown in Fig. 14-10 [25].

in Figs. 14-66 to 14-69 and in Table 14-43. Figures 14-70 and 14-71 indicate influence of filler on solvent absorption. The effect of fillers on resistance to boiling water and boiling acetone is shown in Table 14-44. Silica flour (1 to 75 microns) has been reported as improving the boiling-water resistance of epoxy films [64]. Permeability constants for filled epoxy films are given in Table 14-45.

Fillers in uncured epoxy-resin films do not appear to improve the permeability to amine vapors [57]. The general effect of fillers on diffusion of water is to reduce the rate. Activation energy U and steric factor K were measured for unfilled and 50 per cent silica-filled DGEBA cured with MDA. Diffusion rate of the former was 1.375×10^{-8} g/(cm)(hr)(torr) and 1.213×10^{-8} for the latter at room temperature. No change in U was noted, and K decreased with filler, leading to the conclusion that filler merely serves to extend the pathway along which the water must diffuse [56]. On the other hand, studies of bonding of cpoxy polymers to titanium dioxide indicate that a restriction of polymer mobility is involved. From studies of absorption vs. partial pressure of solvent vapor, it was concluded that titanium dioxide lowered absorption to a greater extent than would be anticipated from the displacement effect alone [68].

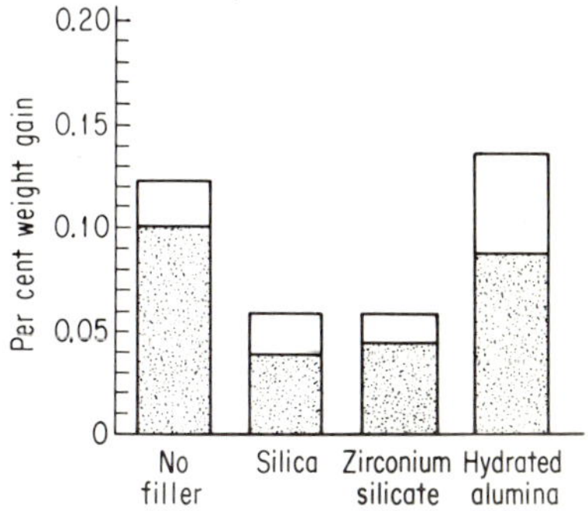

FIG. 14-68. Water absorption after 7 days' immersion for systems shown in Fig. 14-11 [25].

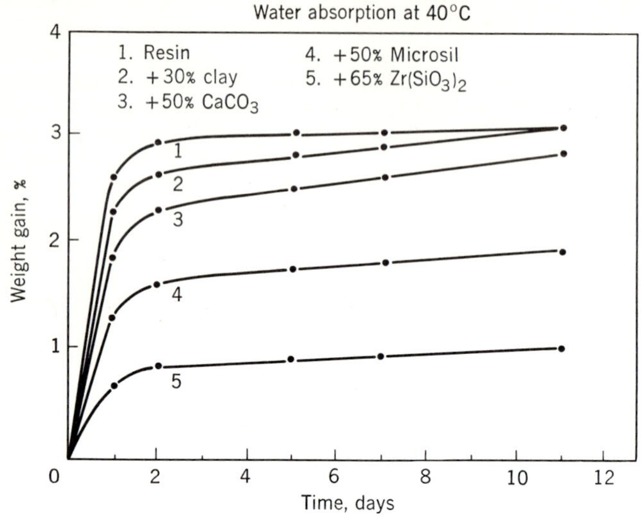

FIG. 14-69. Water absorption vs. filler for cured DGEBA [8].

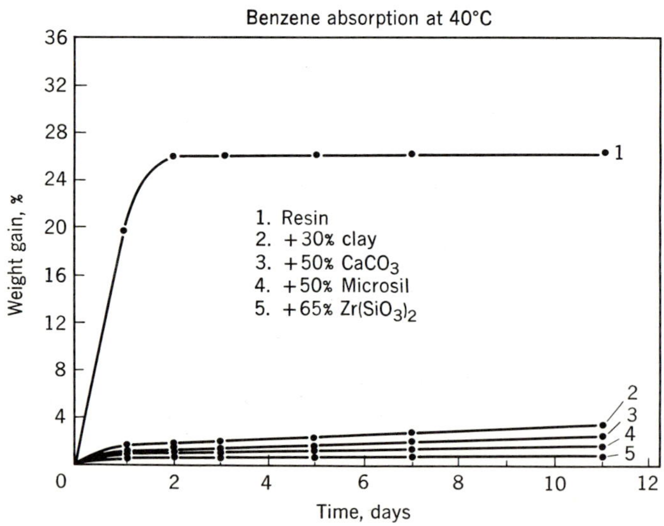

FIG. 14-70. Benzene absorption vs. filler for cured DGEBA [3].

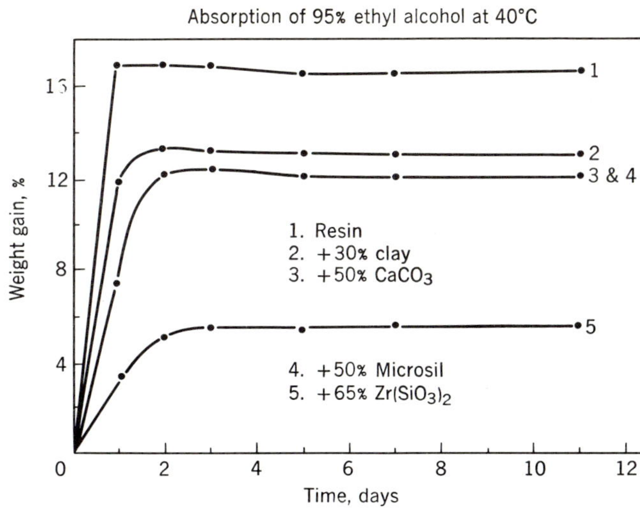

FIG. 14-71. Alcohol absorption vs. filler for cured DGEBA [8].

Table 14-43. Per Cent Moisture Absorption of DGEBA as a Function of Filler, Loaded to Provide an Initial Viscosity of 4,000 to 5,000 Centipoises at Working Temperature and Cured with Listed Curing Agents [25]

(After seven days' immersion)

Filler	DETA/acrylonitrile (½)	MDA	HHPA + 0.1% BDMA	Chlorendic anhydride	Menthane diamine
No filler	0.16	0.165	0.065	0.055	0.165
Mica, 1,000 mesh	0.185	0.09	.....	.....	0.15
Silica, 325 mesh	0.135	0.095	0.05	0.045	0.105
Zirconium silicate <1.5 microns	0.13	0.12	0.10	0.045	
Hydrated alumina, 0.08–0.12 micron	>0.25	0.18	0.175	0.107	0.23

Table 14-44. Effect of Filler on Water and Acetone Resistance of DGEBA Cured with MPDA [47]

Filler, phr	% weight gain after	
	24-hr water boil	3-hr acetone boil
None	1.0	0.8
Mica, 75	1.0	0.5
Calcium carbonate, 175	0.9	0.1
Silica, 5 microns, 100	1.0	0.1
30 microns, 175	0.7	<0.1
44 microns, 175	0.5	<0.1
Hydrated aluminum silicate, 75	1.2	0.3

Table 14-45. Permeability Constants for Epoxy Films [18]

Filler, %	Temperature, °C	Permeability constants		
		Water	Benzene	Ethanol
None	..	0.38	0.60	0.73
Zirconium silicate, 65%	40	0.15		
Microsil, 50%	40	0.16	1.27	
Calcium carbonate, 50%	40	0.25	1.00	0.41
Clay, 30%	40	0.20	0.96	1.00
Silane-treated glass beads, 40%	45	0.52		

REFERENCE: Polyethylene at 40°C 0.35

Fillers may contribute to the corrosivity of plastic compounds in contact with metals. Ions which exist in the cured system or which are created as a result of environmental deterioration are capable of carrying galvanic currents and supporting corrosion. Water-extract-conductance tests have been performed with several cured epoxy compounds. A primary-amine-cured, unfilled epoxy gave negligible conductance at room temperature and about 5.8 ohm-cm^{-1} $\times$ 10^{-4} at 70°C, whereas filled epoxies gave values of 0.5 to 2.8 at room temperature and 7.8 to 9.1 at 70°C. The increase presumably was caused by the ion-releasing capacity of the fillers. Water-extract-conductance tests have been performed with various fillers to establish comparative values (Fig. 14-72).

Considerable work has been expended on the treatment of glass cloth to improve wet-strength retention of laminates, but little work has been done on the treatment

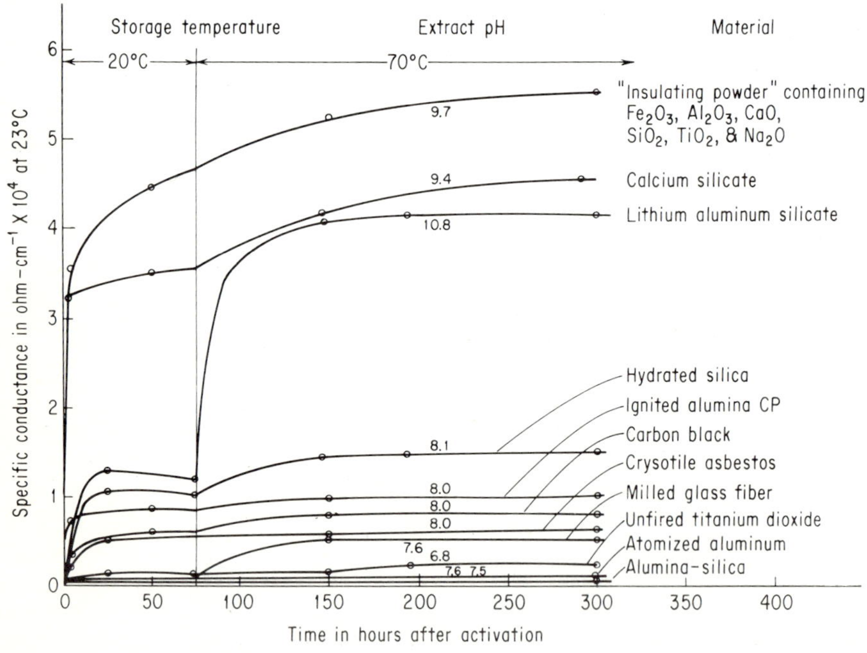

FIG. 14-72. Water-extract conductance for various commonly used epoxy-resin fillers [15].

of particulate fillers to improve chemical resistance; it has been shown, however, that silane compounds improve wet-strength retention of silica-filled epoxy systems, as indicated in Fig. 14-73.

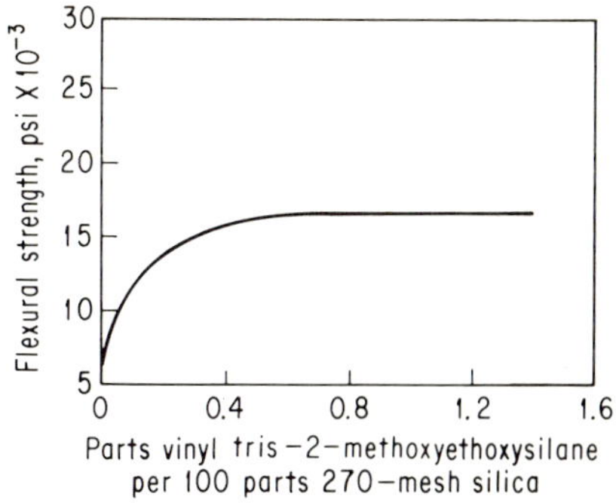

FIG. 14-73. Effect of additive on strength retention of silica-filled DGEBA cured with modified aromatic amine and immersed 7 days in boiling water [53], parts vinyl tris-2-methoxy silane per 100 parts 270-mesh silica.

INCORPORATION OF FILLERS INTO THE RESIN

To obtain best results with fillers, they should be heated to drive off moisture and adsorbed gases before being incorporated into the resin. Once thoroughly dry, they may then be admixed with the liquid resins, preferably by the use of high shearing action. In many instances it will be desirable to employ a grinding device to accomplish the mixing with sufficient thoroughness. A three-roll mill is frequently employed for this purpose. Occasionally it is desirable with larger-particle-size fillers to apply the filler dry and vacuum- and pressure-impregnate through it with the resin.

For optimum results, the filled system should be free of occluded air, and mixing and processing techniques should be designed to minimize air entrapment. The problem becomes increasingly severe when higher loading volumes of fillers are employed, and when the filler provides thixotroping action. Air is usually removed with vacuum or with vibration.

COLORANTS FOR THE EPOXY RESINS

The majority of fillers used in the epoxy technology opacify and give color to the cured systems. When transparent fillers, such as finely divided glasses, are used, clarity may, to some extent, be preserved. The efficiency with which an image is transmitted through the cured epoxy will depend on how well the indexes of refraction and Abbe numbers of the clear filler and the epoxy are matched, but the index of refraction does not, per se, play a vital role in transparency or light transmission [79].

The colorants for the epoxies may be divided into dyes and pigments—dyes being compounds varying from solubility up to the lower colloidal suspension sizes, and pigments being in the higher colloidal range and above 0.1 micron. The materials may be either organic or inorganic, the organic materials being preferred for brightness and strength. The inorganics are generally not strong enough to give darker or brighter shades, particularly in highly filled systems.

The particle size of the commercial pigments may vary quite widely, but control of particle size becomes increasingly important in developing colors which derive their value from a combination of light reflection and light absorption [6], whereas with blacks and whites wider variations in particle sizes may be tolerated.

Since the liquid epoxy resins are fairly good wetting agents, it is sometimes possible to incorporate the coloring agent by mechanical stirring; although, more commonly, it is dispersed into the system with specialized equipment such as three-roll mills. The object generally is not to grind the pigments, that is to say, to reduce their particle size, but rather to disperse them in such a fashion as to approach their primary particle size. Mixing may involve an initial high pigment-to-resin concentration which is blended with high shear action on a rotary mixer. Additional resin is then added, and the compound passed through a paint mill or other suitable device.

Suitably dispersed systems are incorporated in the resin formulation at relatively low percentages (on the order of 1 to 3 per cent for inorganic pigment dispersions or 0.5 to 1 per cent for the organics) to provide the required color. It is generally customary in order to develop hiding power, to use titanium dioxide dispersion along with the colorant. In the case of whites, it is customary to employ a trace of blue tint to help mask the natural slight yellow of the resin.

In general, with the epoxies, the inorganic pigments are preferred. The only organics that work well are the phthalocyanines, Hansa yellows, BON reds, and a variety of specialized vat pigments [32]. The following colorants have been used successfully with the liquid epoxies:

Pigments:

White .... Titanium dioxide
Red .... Cadmium Red Medium (Chemical and Pigment Co.)
Yellow ...... Cadmalith Golden (Chemical and Pigment Co.)
Black .... Hytherm Black (Patent Chemicals)
Gray .... Aluminum powder

Dyes:

Red .... National Fast Red (National Aniline)
Green .... Calco Condensation Green AY (American Cyanamid)
Yellow ...... Calco Condensation Yellow BTC (American Cyanamid)
Blue .... Calco Condensation Blue (American Cyanamid)

The list, of course, is representative, and suitable dyes and pigments are available from a number of other commercial suppliers.

The vehicle for the dispersion may be the liquid epoxy resin, as indicated in Table 14-46, or higher-molecular-weight epoxies may be employed in solvent

Table 14-46. Typical Color Dispersions Used in DGEBA [4-1]

(mol. wt. 420)

Color	Colorant	Pigment concentration, %
White	Titanium dioxide R-110; duPont	60
White	Antimony oxide Red Star; National Lead	64
Black	Carbon black no. 140; Columbian Carbon	25
Red	Toluidine red no. 20-3785; Calco	40
Red	Cadmium red no. 74-5650; Calco	50
Brown	Brown iron oxide no. 1191; Whittaker, Clark & Daniels	60
Yellow	Hansa yellow YT-395; Holland	35
Blue	Phthalocyanine blue BT-284-D; duPont	25
Green	Phthalocyanine green; Fine Colors	25

solutions, typical of which are the following:

	Per cent
Titanium dioxide	60
DGEBA (mol. wt. 700)	15
Butyl alcohol	12.5
Toluol	12.5
or	
Chrome yellow	65
DGEBA (mol. wt. 700)	15
Butyl alcohol	10
Toluol	10

In addition, the colorant may be dispersed in a vehicle other than the epoxy, such as dibutyl phthalate. Suitable dispersed colorants compatible with epoxies are available from a variety of suppliers.

The problems involved in pigmenting epoxy formulations can become quite complex, and the selection, particularly for paint formulations, of the proper pigment or dye to obtain color fastness, thermal stability, tint, gloss, hiding power, etc., is a highly specialized operation.

MOLECULAR SIEVES

Molecular sieves are crystalline-metal aluminosilicates belonging to a class of minerals known as *zeolites*. They may be dehydrated with little or no change of molecular structure, and the network of cavities remaining provides a very high surface area for adsorption of foreign molecules.

Thus, the molecular sieves may be loaded with curing agents for the epoxy resins. When the molecular sieves, so loaded, are dispersed into the epoxy resin, the curing agent is effectively shielded from the system. Release is triggered either by heat or by displacement with another adsorbable compound, such as water or glycols.

A variety of curing agents have been investigated. DETA may be loaded into commercial molecular sieves, and the resultant powder used at about 55 phr with DGEBA. This system will cure in thin films (where moisture from the atmosphere will displace the DETA) overnight at room temperature. If a small amount of water is added, the system is suitable as a tile grout. It will not cure upon application of heat alone, and will remain stable in tightly capped containers for periods in excess of one year [54].

A displacing agent may be employed to obtain heat cures with the DETA-loaded molecular sieves. Hydroxyl-bearing materials have proved satisfactory in this connection, these replacing the DETA at elevated temperatures. However, the addition of the glycols tend to exert a thixotroping action on the system, and pot lives are generally limited to a few days at best.

Special modified systems containing DETA as a curing agent have provided for thermal release, without the use of a displacing agent [2].

Tertiary amine catalysts have been loaded in molecular sieves. Some may suitably be released by heat alone; others will require displacing agents.

SUMMARY

Fillers are employed in the epoxy resin for a variety of reasons. Principally they are used for the control of viscosity, for reducing shrinkage and coefficient of thermal expansion, for effecting cost reduction, and for coloring the resins.

Fillers are normally inert, finely divided inorganic powders in particle size down to about 0.015 micron. They may be employed in loading volumes as low as

0.5 phr to as high as 300 phr or more, depending on the requirements. The addition of progressively more filler results in a corresponding increase in viscosity, the resin being changed from a liquid to a paste to a gunk molding compound. Specialized fillers impart thixotropy. In general, for casting applications, the upper tolerable limit of viscosity is about 5,000 centipoises at working temperature, thus placing a practical limit on filler loading volumes.

Addition of fillers serves to reduce the exotherm and the shrinkage and in most cases to extend the pot life. With particulate fillers, DT is little effected; the tensile and flexural strengths are somewhat reduced; and the compressive strength is marginally improved. With fibrous fillers at higher loading volumes all these values are improved. The surface hardness is increased; the abrasion resistance may be greater or less, and the machining qualities better or poorer, depending upon the specific filler. Thermal conductivity is increased most markedly with fibrous metallic fillers. The coefficient of thermal expansion is lowered, usually in proportion to the loading volume of the filler. The thermal shock resistance and impact resistance may be increased with particulate fillers at some optimum loading volume, and is substantially increased with higher loading volumes of fibrous fillers. Particulate fillers are often used to improve adhesive strengths of epoxy formulations.

The general effect of the addition of fillers is to bring about an increase in the specific gravity, which somewhat offsets the apparent cost savings. Some low-specific-gravity fillers are employed in special applications where weight is critical.

Fillers may be used to convert the normally nonconductive epoxy formulation into an electrically conductive system. For this purpose, particulate metallic fillers, metallic wools, and carbon black are generally used. So-called "electrical grade" (i.e., nonconductive) fillers result in a significant increase in arc resistance, and some increase in dielectric constant; the effect upon volume and surface resistivity, dielectric strength, and dissipation factor is slight.

Fillers may bring about a marginal increase or decrease in chemical resistance, with some fibrous fillers tending to exert a wicking action.

Fillers, as dyes and pigments, may be used as colorants for the epoxy resins. In general, inorganic colorants are preferred for color fastness, whereas the organics are preferred for brighter and stronger colors.

Special fillers may be loaded with epoxy curing agents, the curing agent subsequently being released by heat and/or a suitable displacing agent.

REFERENCES

1. Foote Mineral Co., technical literature.
2. Linde Co., technical literature.
3. Sandia Corp., technical literature.

1950

4. Mueller and Karper, An Improved Epoxy Resin System for Electronic Embedments, *Elec. Mfg.*, February.
5. Wiles, Adhesive Compositions Containing Glycidyl Ethers and Alumina, U.S. 2,528,933.

1954

6. Hoback, Pigment Dispersion: Theory and Practice, *Am. Paint J.*, Sept. 13.
7. Wilcox, Controlling Flow Properties with Fillers, NANTEC of SPI, Chicago.

1955

8. Formo and Bolstad, Where and How to Use Epoxies, *Mod. Plastics*, July.

1956

9. Epstein, Epoxy Based Adhesives, Theoretical Calculations, ACS Symposium, Atlantic City, September.
10. Formo, et al., Epoxy Resin Compositions Containing para, para′ Diaminodiphenylmethane, U.S. 2,773,048.
11. Sokol, Epoxy Castings Replace Metal Forming Tools, *Am. Machinist*, Feb. 27.
12. Wolfson and Elliot, Electrically Conducting Cements Containing Epoxy Resin and Silver, U.S. 2,774,747.
13. Zukas, Hardenable, Fluid Compositions Containing Resinous Polymeric Epoxides and a Bentonite-Amine Reaction Product, U.S. 2,768,992.

1957

14. Dempster, private communication, Standard Oil Co.
15. Perry, et al., The Corrosivities of Plastics and Rubbers, 13th ANTEC of SPE.
16. Parry, et al., Effect of Fillers on High Humidity Insulation Resistance of Epoxy Systems, *SPE J.*, October.

1958

17. Colby, private communication, Union Carbide Plastics Co.
18. Dunbar and Peffer, Permeability and Absorption Studies of Epoxy Resins, ACS Symposium, San Francisco, April.
19. McAdam, New Compositions of Matter Comprising Epoxy Resin, Liquid Polysulfide Polymer, and Buna N, U.S. 2,858,291.
20. Matz, Electrically Conductive Cement and Brush Shunt Connection Containing the Same, U.S. 2,849,631.
21. Peck, Printed Electrical Resistor, U.S. 2,866,057.
22. Radley, Electrical Apparatus with a Thermal and Electrical Insulation Coating, U.S. 2,862,838.
23. Weschler, High Temperature Epoxy Systems for Rotating Equipment, 1st National Conference on Application of Electrical Insulation.

1959

24. Coder, et al., Epoxy Resins in the Tooling Industry, ACS Symposium, Atlantic City, September.
25. Comparative Properties of Filled Epoxy Electrical Potting Compounds, Union Carbide Plastics Co. Report, August.
26. Elarde, Flame Resistant Composition Comprising Chlorinated Polyester Resin, Inorganic Oxide, and an Epoxide Resin, and Method of Preparing, U.S. 2,885,380.
27. Jurras, Molding Compositions Comprising Synthetic Resin and Metallic Filaments, Articles Molded Therefrom and Method of Making Same, U.S. 2,901,455.
28. Naps, Phenolic Resin Compositions Containing a Metal Chelate and Glass Fiber Impregnated Therewith, U.S. 2,876,208.

1960

29. Buchoff and Botjer, Effect of Humidity on Surface Resistance of Filled Epoxy Casting Resins, 15th ANTEC of SPE, Chicago, February.
30. DiSapio, Bonded Coatings Lubricate Metal Parts, *Prod. Eng.*, Sept. 5.
31. Hollinger and Doyle, Stress Response of Glass-filled Resins above Their Glass Transition Temperatures, 15th NANTEC of SPI, Chicago, February.
32. How to Choose the Correct Colorant, *Mod. Plastics*, April.
33. Jansson, Thermally Conductive Potting Compounds for 3-D Welded Electronics, AD 605 679.
34. Novak and McCallum, Method of Making Quick Curing Metal Containing Epoxy Resin Compositions, U.S. 2,956,039.
35. Review of Technical Papers, *Electro-Technol.*, December, p. 187.
36. Ringwood, Epoxy Resins for Encapsulation of Electric Components, *SPE J.*, January.

37. Rutzler, Fillers and Stresses in an Epoxy Resin, ASME Publication 60-RP-11.
38. Warburton and Norman, High Temperature Epoxide Resin Formulations, 3d National Conference on Application of Electrical Insulation, December.

1961

39. Akins, et al., The Effects of Fibre Metal Fillers on the Physical Properties of Epoxy Resin Systems, *Rubber Plastics Age*, September.
40. Aluminum Wool as a Conducting Filler, *Brit. Plastics*, February.
41. Eichenberger, Thermal Aging of Epoxy Resins, *Kunststoffe*, p. 716.
42. Quant, A Low Density Potting Compound, Sandia Corp. Preprint SCR-417.
43. Reynolds, Development of a High Temperature Epoxy Casting Compound, *Insulation*, July.
44. Weigel, Epoxy Resins with Low Dielectric Loss Factor, Ger. 1,120,134.

1962

45. Abelson and Ditzler, Some Properties of Two Epon Resin Systems for Use as Electrical Encapsulants, 4th National Conference on Application of Electrical Insulation.
46. Bowen, Dental Filling Material Comprising Vinyl Silane Treated Fused Silica and a Binder Consisting of the Reaction Product of Bisphenol and Glycidyl Acrylate, U.S. 3,066,112.
47. Ditzler, Fillers for Encapsulation, *Plastics Technol.*, June.
48. Eggert, The Low Temperature Thermal Conductivity of Some Commercial Polymerized Resins and Ceramics, Sandia Corp. Report SCDC-2567.
49. Goodykoontz, Microelectronic Connections, Electronic Circuit Packaging Symposium, Boulder, Colo., August.
50. Parry and Hewitt, The Effect of Fillers on the Physical Properties of Epoxy Resins, Shell Chemical Corp.
51. Rost, Studies in High-temperature-processed Epoxy Resins, *Kunststoffe*, February.
52. Spencer, Cure Shrinkage in Casting Resins, Sandia Corp. Report SCDC 2568.
53. Vanderbilt and Jaruzelski, The Bonding of Fillers to Thermosetting Resins, *Ind. Eng. Chem. Prod. R&D*, September.
54. Waythomas and Mays, Molecular Sieves in Epoxy Resin Tooling, *Mod. Plastics*, November.

1963

55. Arc-Resistant Epoxy Resins (to Siemens-Schuckertwerke A.G.), Belg. 627,330.
56. Bardeleben, The Diffusion of Water through Epoxy Resins, *Kunststoffe*, March.
57. Damusis, et al., Impregnants for Expandable Rigidizable Aerospace Structures, AD 401 066.
58. Davis, Non-fibrous Filler for RP, *Reinforced Plastics*, January–February.
59. Forester, et al., Reinforcement of Epoxy Resins with Metal Fibers, *Mod. Plastics*, April.
60. Houston, Improvements in or Relating to the Production of Hardenable Epoxy Resin Compositions, Brit. 926,888.
61. Kaminskii, Epoxy Resin Compositions, U.S.S.R. 166,064.
62. Kuhlmann, et al., The Effect of Filler Particle-size Distribution and Concentration on Some Properties of Epoxy Casting Formulations, ACS Symposium, New York, September.
63. Larson, Polypropylene in Coating Compositions, Belg. 631,378.
64. Paints Resistant to Boiling Water (to Shell Internationale Research Maatschappij N.V.), Belg. 626,577.
65. Sterman and Marsden, Improved Thermosetting System, *Plastics Technol.*, May.

1964

66. Hartzell and Sakornbut, Conductive Adhesive Flexible at Low Temperatures, U.S. Atomic Energy Commission Report M1M-1207.
67. Kurmaev, Antifriction Compositions from a Phosphoorganic Thiokol FT-1 and an Epoxy Resin, *Plasticheskie Massy*, no. 8, pp. 33–36.

68. Kwei and Kumins, Polymer-filler Interaction: Vapor Sorption Studies, *J. Appl. Polymer Sci.*, pp. 1483–1490.
69. New Plastic Fillers Solve RFI Problem, *Electronics*, Jan. 3.

1965

70. Colletti and Rebori, High Thermal Conductivity Casting Compounds Developed, *Insulation*, January.
71. Delmonte, Radiation Resistant Plastics, RIP Division of SPI, Chicago, February.
72. Development of High Thermal Conductivity Electrical Embedding Compounds, AD 458 127.
73. Fitzgerald, Electrical Charge as a Function of Dynamic Stress in Mica-filled Epoxy, *Mater. Res. & Std.*, pp. 240–242, May.
74. Gilbride and Levinson, Limestone Fillers in Epoxy Systems, *Mod. Plastics*, July.
75. Gul et al., Formation of Electroconducting Structures in a Polymeric Material under the Effect of a Magnetic Field, *Plasticheskie Massy* **1965**(4);46–49.
76. Hessinger and Sharkitt, Beryllium Oxide Coatings on Electrical Elements, Fr. 1,395,618.
77. Milton et al., Pulse Strength of High Polymers, Sandia Corp. Report SC-R-65-851, January.
78. Mouldable Composition for the Production of Permanent Magnets (to Landis & Gyr, A.C.), Brit. 998, 853.
79. Nardone, Investigation of Light Phenomena Associated with Glass Resin Structures, AD 456 669.
80. Silane Coupling Agents Strengthen Metal-filled Epoxies, *Plast. Technol.*, September.
81. Stringfellow and Reighter, Circuit Interrupter Having Track Resistant Coating Thereon, U.S., 3,192,347.

61

Chapter 15

RESINOUS MODIFIERS FOR EPOXY RESINS AND EPOXY RESINS AS MODIFIERS

With compatible resins, the epoxy may interact through active hydrogens to produce a variety of heteropolymers; may react independently to produce a plastic alloy of two thermosetting compounds; or may react independently through the epoxy groups, with heteropolymers being developed through other reactive sites in the epoxy molecule. By these mechanisms, a variety of heteropolymers and plastic alloys may be developed.

A number of resinous modifiers have been mentioned elsewhere in the book. This chapter is concerned primarily with those resins which are not normally considered either curing agents or flexibilizers, with the use of epoxy resins as additives for these materials, and with specialized epoxy-resin molecules. Included also is a discussion of specialized epoxy resins having backbones based on other resin types.

The number of resinous modifiers which have been evaluated in connection with the epoxies is considerable. Cellulose esters, for example, have been used to improve film-curing of epoxidized oils from butadiene [68]; ketone-formaldehyde

[65] and sulfonamide-aldehyde [8] resins have been used as modifiers in coating formulations. Acetone-formaldehyde reaction products have been cooked with polyethylene glycol and used as modifiers for amine- and anhydride-cured DGEBA systems [161]. An alcohol from the reaction of bisphenol A and propylene oxide has been used as an additive for PA-cured epoxidized linseed oil [141]. Addition of 1 per cent by weight of the polymer of methyl-2-hydroxy-4-acryloxybenzoate is

Artists conception of Titan III-C blastoff. DGEBA is one of the resins used in the propellant formulation for the big rocket motors strapped to the sides of the missile. (*United Technology Center, Sunnyvale, Calif., and Dow Chemical Co.*)

reported to stabilize DGEBA-based adhesives against degradation by UV [152], and trace amounts of peroxide inhibit discoloration of epoxy resins produced by the peracetic acid route [160].

Karo syrup and molasses have been suggested as extenders [95]; low-molecular-weight polycarbonates have been added [146]; cellulose acetate butyrate has been used to improve the adhesion of epoxy adhesives to cellulose acetate butyrate [74] and as an anticratering agent [118]; combinations of unsaturated esters of organic acids and esters of cyanuric acid have been employed [101]; polyaldehydes are used together with acid curing agents to improve crease resistance of fabrics and to reduce shrinkage [47]; butadiene-acrylonitrile copolymers are used to improve the peel strength of epoxy/phenolic blends [102]; and various chlorine-, bromine-, and phosphorus-containing molecules have been added to impart fire resistance, typically hexachlorocyclopentadiene as an additive for DGEBA/MA blends [163] or the reaction product of phosphoric acid and epichlorohydrin [192]. Special epoxy resins may be used as special-purpose additives; for example, triglycidyl cyanurate has been used to improve arc resistance, the decomposition products producing a quenching effect [123]. Other additives are epoxy alkyl esters of 3,4-dihydro-1,2-pyran-2-carboxylic acid [183].

Similarly, the use of small amounts of epoxy resins to modify other resins is quite widespread. As plasticizers/stabilizers, they are employed extensively in vinyl formulations, as discussed subsequently, and are used, to a lesser extent, to stabilize chlorinated derivatives of such materials as rubber [13], paraffins [156] and hydrocarbons [11], cellulose plastics [30], cellulose plastics containing combined sulfate groups [42], halogenated fire retardants [41], chloroalkenes [150], etc. Epoxidized novolacs are used as additives for a chloroprene-based contact-adhesive formulation to improve heat resistance [168]. When the synthetic rubber contains suitable reactive groups, the epoxy may be employed in small amounts as a plasticizer [75,133], or in stoichiometric amounts as a curing agent [55] (see also Chap. 16). Polyethylene is stabilized against discoloration by the use of about 0.5 per cent DGEBA to react with the hydrogen chloride evolved from the catalyst used in polyolefin manufacture [178]. Acetals and their polymers are stabilized by the addition of 0.1 to 1 per cent of epoxy resins, with viscosities being unchanged after storage for 5 hours at 40 to 80°C and 3 months at room temperature [179].

Generally, when the epoxy resin is used at low concentrations as an additive, the curing agent is omitted. Suitable reactive sites may be present in the base compound, as for example, when the epoxy is added to sour crude oil to react with mercaptans and kill the odor [73], when the epoxy is used to body thermoplastic polyamide adhesives [93], or when the epoxy is used to treat textile fabrics [48].

The resin may be prereacted, as for example, when it is reacted with an alcohol and this product used at low concentrations as an antistatic additive for plastics [92]; when it is reacted with a polyamine and the product used as a water-insoluble anion-active resin [2,3]; when epoxidized natural oils are reacted with sulfuric acid to produce surface-active agents [143]; when DGEBA (mol. wt. 950) is etherified with trimethylolpropane diallyl ether and the product copolymerized with methyl methacrylate [155]; or when DGEBA is reacted with phosgene to produce a linear polycarbonate [162]. An adduct of DGEBA with excess primary amine has been reported as an ingredient in sustained-release-type medicaments [190]. Natural protein (i.e., casein, peanut protein, etc.) has been reacted with DGEBA and the resultant product used in water-based adhesives [194]. Similarly, aliphatic-based epoxies have been used as additives for water-soluble phenolic or protein-based adhesives [189].

In some cases, the epoxy additives may be used together with a curing agent, as for example, when used in alkaline solution as tanning agents to produce white leather [85], as a treating agent for cellulose fiber [96], as an additive to cellulose acetate to improve the fiber [83], as a dye-receptive modifier for isotactic polypropylene [136], as a treating agent for wool [39], and in combination with a blend of cement and animal glue [184]. Epoxidized polybutadiene is reported to be compatible with solutions of chlorinated atactic polypropylene and a chlorinated copolymer of ethylene-propylene; the solution, containing a curing agent for the epoxy resin, can be cast into a film which provides a contact adhesive capable of being heat cured to provide properties expected of thermosetting compounds [180].

VINYL RESINS

The largest use of vinyl resins (Table 15-1) in epoxy formulations is in solution coatings, where the low cost and high oxidation resistance of the vinyl lends desirable characteristics.

Poly(vinyl formal) is used in low concentrations in DGEBA to improve impact resistance and peel strength of adhesive formulations [89] and as anticratering agents for coating formulations. Likewise polyvinyl acetals, with numerous hydroxyls, have been utilized as anticratering agents [40]. Employed with Lewis-acid-type curing agents the polyvinyl acetals improve impact resistance of DGEBA casting resins [145]; and with phenol-formaldehyde curing agents, improve impact and peel resistance of DGEBA adhesive formulations [132]. These vinyl resins presumably act as high-molecular-weight plasticizers for the epoxy, with some reaction through available hydroxyl groups, although non-hydroxy-bearing resins, such as polyvinyl acetate may also be employed for similar purposes [24]. Reaction

Table 15-1. Structures of Commercial Vinyl Resin

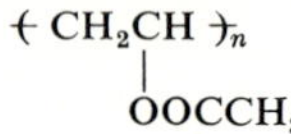

Poly(vinyl acetate)

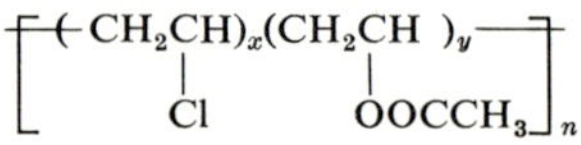

Poly(vinyl chloride–acetate)

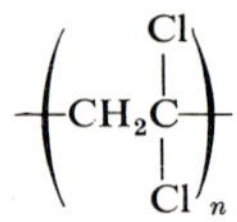

Poly(vinylidene chloride)

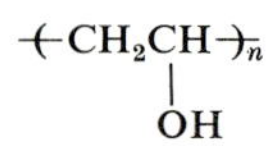

Poly(vinyl alcohol)
[derived by hydrolysis of poly(vinyl acetate)]

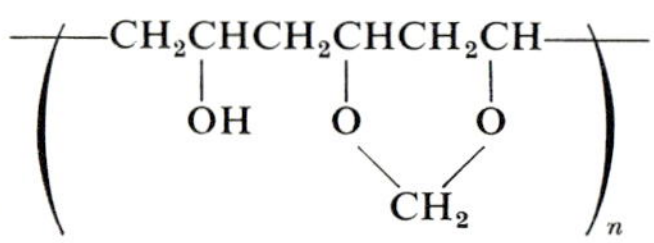

Poly(vinyl formal)
[derived from poly(vinyl alcohol) and formaldehyde]

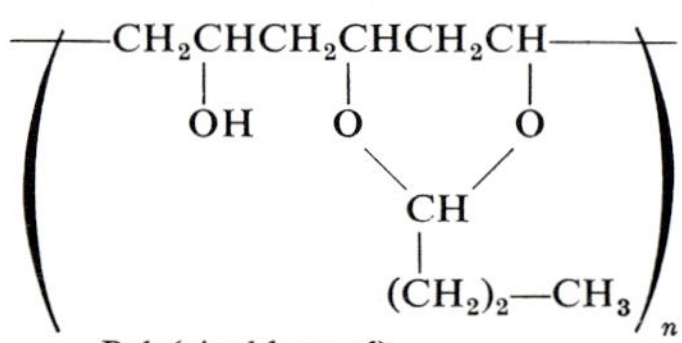

Poly(vinyl butyral)
[derived from poly(vinyl alcohol) and butyraldehyde]

products of vinyl compounds and phenols have been suggested as accelerators for amine curing agents [115]. Solvent-free thermosetting magnet-wire coatings have been claimed from systems using 25 to 135 phr of polyvinylformal, together with polybutadiene and a secondary plasticizer [185].

Polyvinyl chloride has been employed in anhydride-cured DGEBA systems in loading volumes from 200 to 400 phr [105]. Polyvinyl chloride resins, plasticized with primary plasticizers, such as dioctyl phthalate, etc., have been used in DGEBA resins [53] cured with phthalic anhydride or triethanolamine borate [50], with boron trifluoride complexes [97], with aromatic and aliphatic amines [105], and with fatty polyamides [76]. Polyvinyl chloride resins, with and without plasticizers, have been used with DGEBA resins cured with boron trifluoride complexes for calendered films [52] and flame-retardant casting formulations [36].

A typical plastisol-epoxy compound is the following [4]:

	Parts
Dispersion-grade polyvinyl chloride	100
Epoxidized soybean oil (epoxy equivalent 260)	90
Dioctyl phthalate	10
Stabilizer	5
PMDA	See Table 15-2

The effect of the primary/secondary plasticizer ratio on the properties of the vinyl plastisol cured with PMDA at an A/E of 1/1 is indicated in Table 15-3.

Table 15-2. Effect of Curing-agent Concentration on Properties of Vinyl Resin Plasticized with Epoxidized Olefin [12-4]

A/E	Tensile strength, psi	Tensile elongation, %	Hardness, Shore D
0/1	1,300	368	29
0.2/1	600	38	37
0.4/1	1,300	12	58
0.6/1	2,300	3	77
0.8/1	2,800	3	77
1/1	4,000	3	80
1.5/1	4,000	3	80

The use of 3,4-epoxy-6-methylcyclohexylmethyl 3,4-epoxy-6-methylcyclohexane carboxylate resin with vinyl formulations in the development of plastisols has been reported in detail and several formulations are given in [121]. In this case, boron trifluoride–amine complexes were found to provide satisfactory cured properties. Lower-alkyl titanates also have been reported as satisfactory curing agents for plastisols containing alicyclic epoxy resins [187].

Table 15-3. Effect of Epoxidized Olefin/Primary Plasticizer Ratio on Plastisol Cured with PMDA (A/E 1/1) [12-4]

Epoxy/primary plasticizer per 100 parts vinyl resin	Tensile strength, psi	Elongation, %	Hardness Shore D
0/100	1,360	367	17
20/80	1,500	207	29
40/60	1,770	65	49
50/50	1,900	33	59
60/40	2,000	21	67
80/20	2,800	4.8	76
100/0	4,400	2.2	80

Epoxy Resins as Additives to Vinyl Resins

Use of epoxy resins (polyglycidyl ether of glycerol) as vinyl plasticizers was noted as early as 1943 [1]. Shortly after World War II, a variety of epoxies had been suggested as stabilizers/plasticizers, among them epoxies based on diphenols [9] and bisphenol A [12], as well as on polyalcohols epoxidized via epichlorohydrin [10,15] and polyallyl glycidyl ether [23].

When the epoxy plasticizers/stabilizers were introduced commercially into the vinyl technology in 1950, they gained almost immediate acceptance. Such materials are used in very large volumes, but only general remarks are in order, together with the inclusion of a few references indicative of the large body of literature on the subject.

The epoxy plasticizers are employed predominantly with the polyvinyl chloride resins, although they are used also, to some extent, to plasticize polyvinyl acetate. In addition to those mentioned above, among the epoxy compounds which have been suggested as stabilizers/plasticizers are epoxidized fatty acids [104], a variety of monoepoxy [14] and polyepoxy [106] fatty-acid esters and partially epoxidized esters [61], the most useful of the epoxidized esters being the epoxy triglycerides and alkylepoxy stearates. Other stabilizers/plasticizers include epoxidized copolymers of styrene and butadiene [60], epoxy hexahydrophthalates [62], epoxidized olefin resins and monoepoxides derived from cyclohexene [70,79,107], and tin-containing epoxy resins such as dialkyltin epoxy succinates [113].

The aromatic glycidyl ether resins are generally good stabilizers but poor plasticizers, whereas the epoxidized fatty-acid esters are generally good both as stabilizers and plasticizers [63]. Internal epoxy groups on long chains impart greater plasticizing properties than do terminal epoxies on short chains [57]. The epoxy resins with long side chains are usually effective plasticizers because the high molecular weights prevent plasticizer migration, although they may be used in connection with more effective (so-called "primary") nonepoxy plasticizers, such as dioctyl phthalate. Terminal epoxy groups on aliphatic chains are more effective in stabilizing action on vinyls than ring-situated or cyclic epoxies. Internal epoxy groups on aliphatic chains are slightly less effective than terminal epoxy groups on aliphatic chains [126].

The stabilizing action on vinyls arises in part from the ability of the epoxy groups to react with hydrogen chloride liberated during exposure of the vinyl formulation to heat and/or light. The epoxy stabilizers are normally used in conjunction with chlorine-acceptor metallic stabilizers—cadmium, zinc, and strontium salts being most effective [25], with cadmium salts preferred [27]. A synergistic effect is developed, the combination being several times as effective as either stabilizer alone [80,144]. Combination stabilizers have also been used with the epoxies, e.g., zinc stearate plus calcium ethyl acetoacetate [29].

The formation of a chlorohydrin from the epoxy group during the stabilizing process is, in the presence of suitable metallic stabilizers, a transitory phase; tests with barium-cadmium as the co-stabilizer, indicate that although the epoxy improves stability, it may be recovered with an unchanged epoxy content [72]. It is probable that the metallic stabilizer acts as a dehydrohalogenating agent [117].

POLYESTER RESINS

There are two basic types of polyester resins, the saturated and unsaturated.

Unsaturated Polyesters

The unsaturated polyesters are resinous polymers based on the reaction of an unsaturated diacid or anhydride with an alcohol and usually diluted with styrene or similar unsaturated monomer. These resins, referred to as general-purpose polyesters, homopolymerize through the unsaturated double bonds in the presence of a peroxide. Styrene monomer, besides lowering viscosity, provides short-chain cross-linking sites. This type of unsaturated polyester is used in casting and laminating applications where hardenable liquids are required. In handling properties, the unsaturated polyesters somwhat resemble the amine-reactive liquid epoxy resins. They may be used as additives for epoxy resins when the epoxy resin contains reactive unsaturation. Properties of an epoxidized polybutadiene modified with a saturated polyester are given in Table 15-4. An alicyclic epoxy resin, combined with benzoylperoxide, has been reported as satisfactorily reactive with unsaturated polyesters having acid values of about 30 [191].

A second type of unsaturated polyester is the alkyd resins, derived from the esterification of polyols such as glycerol with unsaturated monoacid natural oils. The alkyds are used as coating-vehicle solids and have been suggested as modifiers in DGEBA solution coatings [19]. Natural oils containing conjugated acetylene groups (e.g., boleko oil) have been suggested as additives [172].

Saturated Polyesters

Saturated polyesters are derived from the reaction of saturated aromatic or aliphatic diacids or anhydrides or by combinations of them with polyols.

Table 15-4. Properties of Epoxidized Polybutadiene/General-purpose Polyester Combination [4-3]

(Cure cycle: 1 hr at 80°C + 4 hr at 155°C)

Ingredients and properties	Polyester control	Epoxidized polybutadiene system
FORMULATIONS		
Epoxidized polybutadiene	......	20
General-purpose polyester	100	80
Short-chain acidic polyester accelerator	......	15
Benzoyl peroxide	1	1
Tertiary butyl perbenzoate	......	1
PROPERTIES		
DT, °C	88	118
Tensile strength, psi	6,600	8,200
At 70°C	6,500	5,600
At 120°C	2,400	3,200
Modulus, psi × 10^{-5}	5.71	6.06
At 70°C	3.90	4.55
At 120°C	2.25	2.52
Elongation, %	1.18	1.62
At 70°C	2.78	1.52
At 120°C	6.64	3.15
Flexural strength, psi	15,100	21,500
At 70°C	9,400	11,600
At 120°C	560	1,900
Modulus, psi × 10^{-5}	5.63	5.76
At 70°C	2.32	2.99
At 120°C	0.14	0.98
Hardness, Rockwell M	111	112
Chemical resistance:		
Original values:		
Flexural strength, psi	15,100	21,500
Modulus, psi × 10^{-5}	5.63	5.76
Chemical water:		
Flexural strength, psi	12,500	15,900
Modulus, psi × 10^{-5}	5.68	5.23
Weight change, %	+0.81	+0.099
10% sodium hydroxide:		
Flexural strength, psi	14,000	16,500
Modulus, psi × 10^{-5}	5.52	5.19
Weight change, %	+4.68	+3.66
30% sulfuric acid:		
Flexural strength, psi	13,200	14,800
Modulus, psi × 10^{-5}	5.16	5.22
Weight change, %	+4.59	+5.96
Acetone:		
Flexural strength, psi	4,400	12,400
Modulus, psi × 10^{-5}	1.69	3.30
Weight change, %	−7.75	+6.43

When the diacid or anhydride is aromatic, the polyester is generally a solid, even in low molecular weights, and can only be handled effectively from solutions. Liquid materials based exclusively on the aromatic acids or anhydrides can be developed with some combinations, as for example a combination of iso- and *o*-phthalic anhydride and *p*-hydroxybenzoic acid with a mixture of polyols, but they are less common. Blends of aromatic and aliphatic acids may be used to produce liquid systems [103,171], and many of the aliphatic-based saturated polyesters are liquid [148].

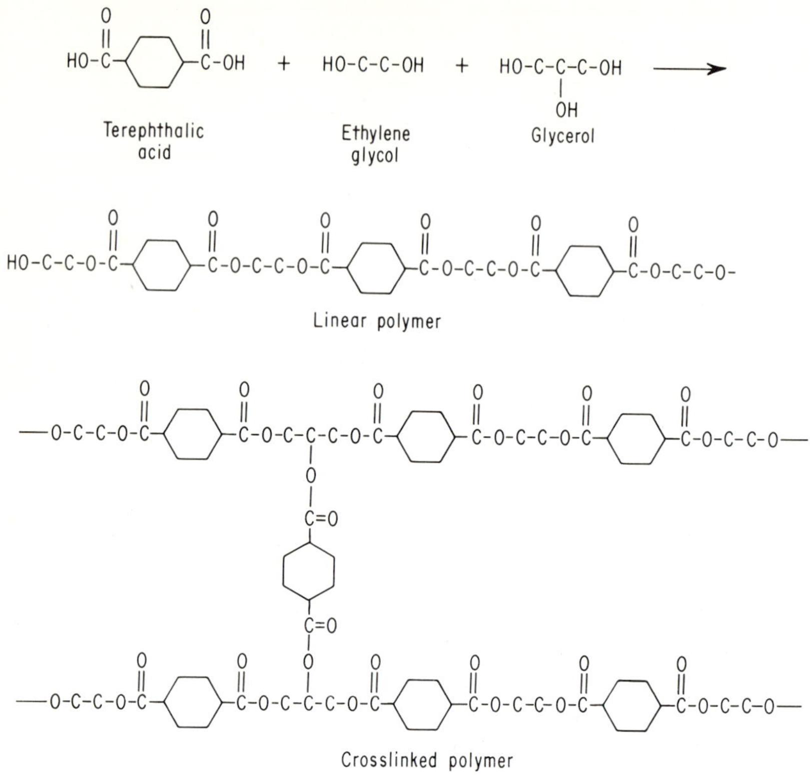

FIG. 15-1. Idealized schematic diagram of saturated polyester resin [116].

The aromatic-based polyesters of the Mylar type, based on terephthalic acid, are high-molecular-weight polymers suitable for casting films, making fibers, and coating magnet wire, but are not suitable for use in the epoxy technology. However, these compounds are well known for their thermal stability to about 150°C and for their excellent toughness and all-round good properties. In this regard, it is of theoretical interest to compare the structure of a cured saturated aromatic-based polyester with that of an aromatic anhydride-cured epoxy resin. The idealized structure of a fully cured polyester of this type is shown in Fig. 15-1. It is very similar to an aromatic anhydride-cured aromatic-based epoxy (Fig. 15-2). The formation of ester linkages when an epoxy reacts with the carboxyl group of an organic acid or anhydride makes the epoxy/anhydride system, in effect, a low-viscosity liquid, heat-curable saturated aromatic polyester, with the exception of the presence of hydroxyl groups and, depending upon the cure conditions, the presence of some amount of ether linkages.

Liquid saturated polyesters may be acid-terminated or alcohol-terminated, or may contain random termination [43]. The hydroxyl-terminated aliphatic polyesters are widely used in the reaction with polyurethane resins, and some of these commercial materials are suitably reactive with the epoxy resins.

In Chap. 16, there is a discussion on the use of hydroxyl-terminated aliphatic polyesters as flexibilizers for the epoxy resins. The use of acid-terminated aliphatic polyesters as curing agents for epoxies is discussed in Chap. 11.

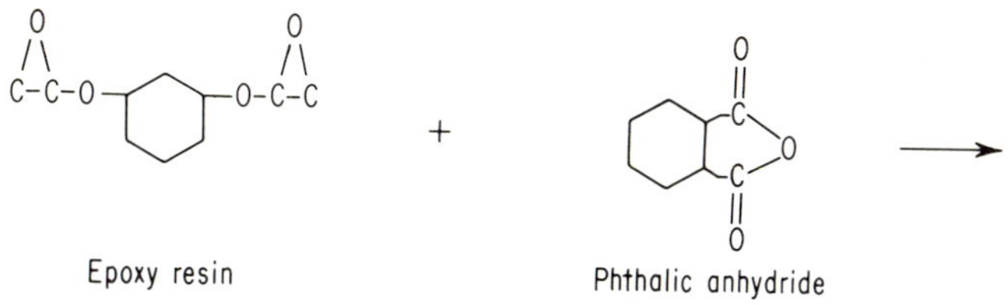

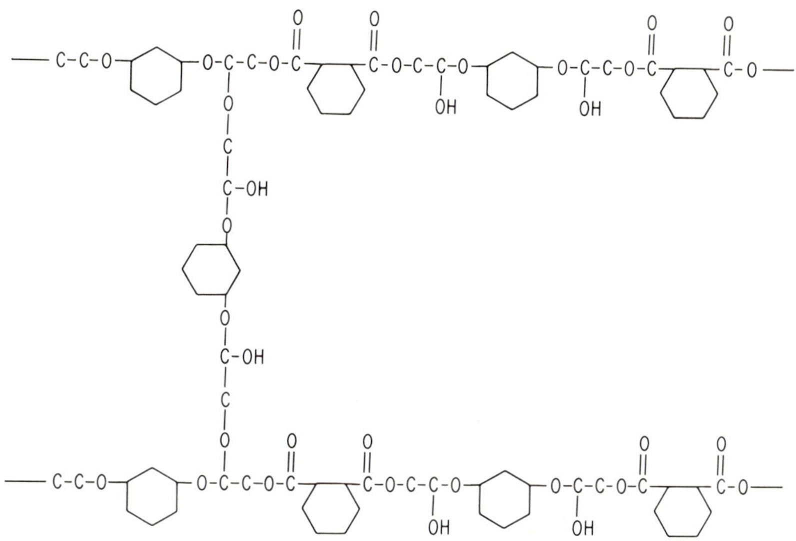

FIG. 15-2. Idealized schematic diagram of epoxy resin cured with a difunctional acid anhydride [116].

ISOCYANATES AND POLYURETHANE RESINS

Polyurethane resins are derived from the reaction of polyisocyanates (i.e., molecules containing —N═C═O groups) with molecules containing active hydrogens. The nomenclature of the polyurethanes was adopted from one of the principle reaction products, the reaction between an alcoholic hydroxyl and an isocyanate.

The chief diisocyanates commercially available are those materials shown in Table 15-5. The chief reactants for isocyanates are polyols, polyether glycols, hydroxyl-terminated polyesters, and primary amines.

The chief reactions of the isocyanate group (illustrated on monofunctional molecules) are shown in Table 15-5.

The urethanes are widely used in the production of foams because of the ease with which reactions (5) and (6), Table 15-5, may proceed, the carbon dioxide being the blowing agent. Urethanes are used in the production of elastomers, because the chain lengths can be controlled exactly as needed by reactions (1) and (2), and the desired amount of cross-linking can be achieved by reaction (3) and hydrogen bonding. Urethanes are also used to an increasing extent in protective coatings, where the very high degree of reactivity provides for good properties at low ambient cure temperatures.

Table 15-5. Chief Commercial Diisocyanates and Chief Isocyanate Reactions [120]

DIISOCYANATES

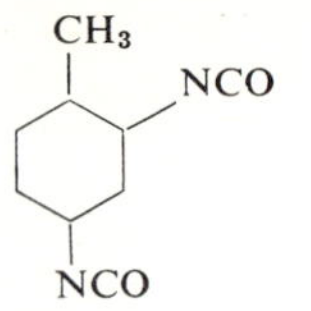

Tolylenediisocyanate

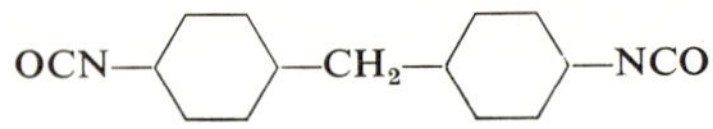

Diphenylmethane-4,4′-diisocyanate

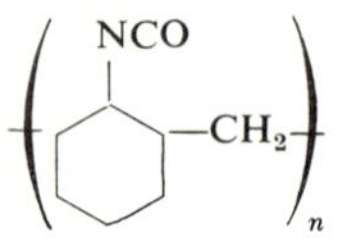

Polymethylenephenylisocyanate

$$OCN{-}(CH_2)_6{-}NCO$$

Hexanediisocyanate

REACTIONS

(1) $R{-}NCO + HO{-}R' \rightarrow R{-}NH{-}C(=O){-}O{-}R'$

Isocyanate, Hydroxyl → Urethane

(2) $R{-}NCO + H_2N{-}R' \rightarrow R{-}NH{-}C(=O){-}NH{-}R'$

Isocyanate, Amine → Substituted ureas

(3) $R{-}NCO + R'{-}NH{-}C(=O){-}NH{-}R'' \rightarrow R{-}NH{-}C(=O){-}N(R''){-}C(=O){-}NH{-}R'$

Isocyanate, Urea → Biurets

(4) $RNCO + R'{-}NH{-}C(=O){-}O{-}R'' \rightarrow R'N(C(=O){-}HNR){-}C(=O){-}O{-}R''$

Isocyanate, Urethane → Allophanates

(5) $3(RNCO) + 3(R'{-}C(=O){-}OH) \rightarrow$

Isocyanates, Carboxylic acid

$R'{-}C(=O){-}O{-}C(=O){-}R' + R{-}NH{-}C(=O){-}R' + CO_2 + R{-}NH{-}C(=O){-}NH{-}R$

Anhydride, Amide, Carbon dioxide, Urea

(6) $RNCO + H_2O \rightarrow [RNH{-}C(=O){-}OH] \rightarrow R{-}NH_2 + CO_2$

Isocyanate, Water, Substituted carbamic acid, Amine, Carbon dioxide

In some cases, urethanes are direct competitors with the epoxy resins. More often, they complement them. Urethanes offer elasticity which, in the main, the epoxy resins do not; and the reactiveness of the systems and the chemical resistance of the cured products are sufficiently different from those obtained with the epoxies to limit the areas of competition.

Considering the large volume production of both urethanes and epoxy resins, it would seem probable that numerous attempts would be made to combine various of the two types to achieve beneficial modifications. This is indeed true. Isocyanates have been reacted with monohydroxyl monoepoxies to produce polyepoxy adducts having the isocyanate "backbone" [71]. When a hydroxyl-containing epoxy resin is reacted with TDI, and a primary amine curing agent added, it is reported that the resultant reaction is sufficiently rapid to provide for cure under water in thin films [81]. Higher-molecular-weight DGEBA resins, considered as polyols, have been reacted with urethanes without curing agent [20,84]. Hydroxyl-bearing epoxies mixed with diphenolic acid [86] and condensation products of diphenolic acid and aliphatic acid aldehydes [87] have been reacted with the polyurethanes. The low-molecular-weight adduct of tolylenediisocyanate and trichlororesorcinol has been reported as a curing agent [153]. Some of the suggested systems are quite complex, as for example, a laminating resin produced from the higher-molecular-weight DGEBA in mixture with an anhydride and a polyol as well as the isocyanate coreactant [127].

Although the isocyanate group has been shown, under some circumstances, to react directly with the epoxy group to produce an oxazolidone [78,149], the reaction is not of consequence in the epoxy-resin technology. When urethanes and epoxy resins are combined in the presence of an active-hydrogen-containing curing agent, both resins compete for the curing agent.

With polyurethane/epoxy blends, the preferred curing agent is MOCA [4,4′-methylene-bis(2-chloroaniline)]. The first reaction is between the urethane and the amine; the subsequent reaction, at elevated temperature, is between the formed urea and the epoxy [130]. Properties of one such epoxyisocyanate (from TDI and 1,4-butylene oxide polyglycol, mol. wt. 2,200) blend is shown in Fig. 15-3. *p*-Nitrosobenzene has also been reported as an excellent catalyst for epoxy-isocyanate adhesives [188].

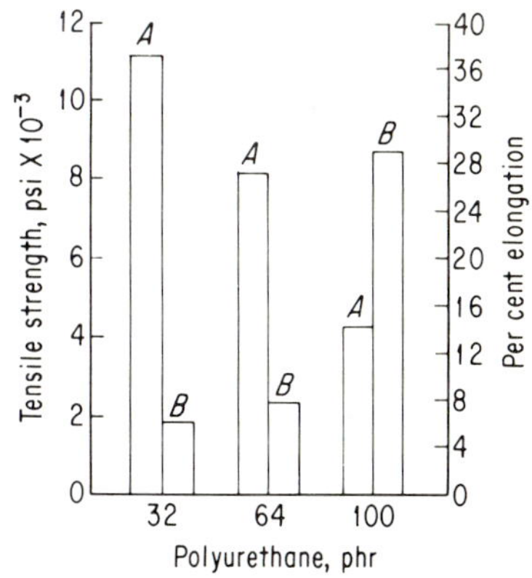

FIG. 15-3. Effect of polyurethane additive on tensile properties of DGEBA cured with modified aromatic diamine [12–4]. *A*, Tensile strength; *B*, per cent elongation.

If higher-molecular-weight epoxy resins are used as sources of hydroxyl groups for the isocyanate, then the epoxy resin may be considered as an expensive polyol, or the isocyanates may be considered as expensive cross-linking agents compared to the lower-cost phenolic resins. Such systems have been investigated but are necessarily suitable only for application in solution. Coatings are satisfactory in baking applications but inferior when air-dried [4-6]. Typical of the film properties are the data given in Table 15-6.

FLUOROCARBONS

The various fluorinated resins commercially available are insoluble in the liquid epoxy resins and do not react with them, but solution coatings have been suggested as offering cured compounds combining the adhesion of the epoxies and the low water transmission and improved chemical resistance and flexibility of the fluorinated materials [88].

Finely divided polytetrafluoroethylene (PTFE) and polytrifluorochloroethylene [54], as well as polyvinylidene fluoride [135], have been added to epoxy solution coatings with baking temperatures being sufficiently high to fuse the fluorinated

Table 15-6. Film Properties of Isocyanate–cross-linked DGEBA Resin Coating Systems Cured 1 Hr at 150°C [4-6]

DGEBA, mol. wt.	Isocyanate, phr			Available hydroxyls cross-linked, %	Film properties		
	Tolylene-diisocyanate	C-5 adduct of polyisocyanate	C-5A adduct of polyisocyanate		Impact, in.-lb	Flexibility, inches	MEK resistance (spot-test method)*
1,600–2,000	42	..	120	20% excess isocyanate	<4	Failed	>24 hr
	35	..	100	...	<4	Failed	>24 hr
	28	..	81	...	>160	1/8	1 min
	20	..	58	...	>160	1/8	<1 min
	15	..	43	...	>160	1/8	<1 min
	..	89	100	...	<4	Failed	>24 hr
	..	72	81	...	20	15/32	>24 hr
	..	54	61	...	60	15/32	>24 hr
	..	37	42	...	>160	1/8	30 min
	..	22	25	...	>160	1/8	5 min
	..	..	100	78	50	15/32	>24 hr
	..	..	79	61	50	15/32	>24 hr
	..	..	52	40	70	15/32	30 min
	..	..	32	25	>160	1/8	5 min
2,500–4,000	35	..	...	100	<4	Failed	>24 hr
	26	..	...	75	30	15/32	>24 hr
	17	..	...	50	140	15/32	2¾ hr
	10	..	...	30	>160	1/8	7 min
Ester from 57.6% resin, mol. wt. 1,600–2,000, 42.4% soybean fatty acids	..	..	93	75	4	Failed	24 hr
	..	..	62	50	40	21/32	24 hr
	..	..	38	30	100	3/16	90 min
	25	..	...	110	>160	1/8	<1 min
	..	..	...	10% excess isocyanate			
	18	..	...	80	>160	1/8	<1 min
	11	..	...	50	>160	1/8	<1 min
	0	..	...	0	>160	1/8	<1 min

* A tuft of cotton was placed on the panel, saturated with solvent, and covered with a watch glass. The time was noted when the film could be scratched by the finger to the bare metal.

compound. PTFE has been added to DGEBA to reduce the coefficient of friction [154], and co- or interpolymers of fluoroölefins with alkyl vinyl ethers and unsaturated monoepoxies have been described [158,175].

Epoxidized fluorinated resins have also been introduced experimentally but have offered no apparent advantage.

SILICONES

Early work on combinations of silicones and epoxies involved the development of materials containing both the silicon atom and the epoxy group for use in treating glass cloth. The goal was to secure a bond to the glass by reacting the silicon with oxygen in the glass, and to achieve reaction between the epoxy group in the treating agent and the epoxy group in the resin, thereby improving adhesion and wet strength in epoxy-glass laminates. Subsequently, silanes containing primary and secondary amine groups were developed. In recent years, specialized silane finishes have been employed for treating glass cloth used with epoxy resins, as discussed in Chap. 22.

Hydroxyl-bearing silicones may be epoxidized with epichlorohydrin to produce materials such as 1,3-bis[3(2,3-epoxypropoxy)propyl]tetramethyldisiloxane, and these may be cured with epoxy-resin curing agents. The net effect of this has not been to obtain a marked improvement in thermal stability, as indicated in Fig. 15-4. The silicone resin may be used in combination with DGEBA resins to give properties as indicated in Table 15-7. It has been reported that adducts of the silicone resin with DADPS provide improved properties in high-temperature adhesive formulations [167], and similar epoxy organosiloxanes have been reported of use in hair dressing preparations to preserve waves [151]. The synthesis of an aromatic-based compound from bis(hydroxyphenyl)diphenylsilane has been reported [173].

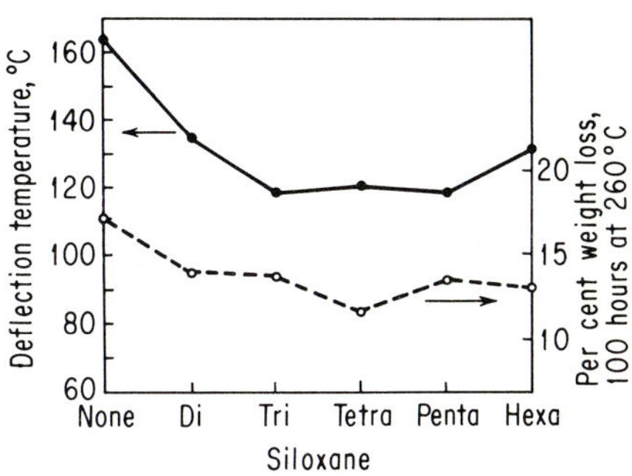

FIG. 15-4. Effect of 33 phr epoxidized silicone on DT and thermal stability of cured DGEBA [108].

The epoxidation of divinyldimethylsilane [16], alkenylacetoxysilane [157], and 1,3,5,7-tetravinyl-1,3,5,7-tetramethylcyclotetrasiloxane [159] has been reported, as has synthesis of siloxane resins having 3,4-epoxycyclohexyl groups attached directly to the silicon atoms [129]. Di- and trichlorosiloxanes have been reacted with higher-molecular-weight DGEBA resins through the hydroxyl groups [177], and dichlorodibutylsilane has been reacted with glycidol to give polyepoxides

Table 15-7. Effect of Concentration of 1,3-Bis[3(2,3-epoxypropoxyl)propyl]tetramethyldisiloxane on Properties of DGEBA Cured with MDA [51]

Epoxidized silicone, %	Hardness, Barcol	DT, °C	Impact strength, Izod	Compressive strength, psi	Flexural strength, psi	Thermal shock at +150 to −55°C	Weight loss after 24 hr at 260°C, %
0	50	140	0.25	36,800	18,600	Fail	19
20	44	121	0.50	44,400	17,800	Pass	10
40	30	97	0.50	38,400	14,200	Pass	8.5
60	16	77	1.24	37,000	10,900	Pass	6.0

according to the following equation [46]:

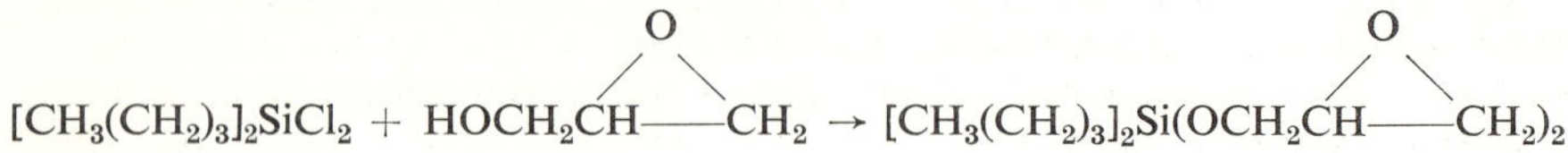

Dichlorodibutylsilane Glycidol DGE of dibutylsilane

Silanes may be reacted with an unsaturated monoepoxy in the presence of platinum to produce polyepoxy resins [108].

A variety of epoxy/silicone combinations have been investigated as potential coating systems [114]. Most of these have contained silanol (Si—OH) [34] or hydroxyl groups [56]. Such products have exhibited improved weathering and chalking characteristics and improved flexibility in films [69], and have been suggested as co-curing agents with other materials such as the fatty polyamides [119].

Silanol groups have been shown not to react directly with epoxy groups [82]. However, when hydroxyl groups are present on the epoxy-resin chain, silanols will condense smoothly with them in the presence of zinc chloride, ferric chloride, etc. [124]:

$$\underset{\text{Silanol}}{-\mathrm{SiOH}} + \underset{\text{Alcohol}}{\mathrm{HOCH_2R}} \xrightarrow{\mathrm{ZnCl}} \underset{\text{Siloxy}}{\mathrm{SiOCH_2R}} + \underset{\text{Water}}{\mathrm{H_2O}}$$

Thus, silanols may be reacted with hydroxyl-bearing epoxy resins to produce a resin adduct which may be further reacted through available epoxy groups. It has also been reported that epoxy groups will catalyze the condensation of silanol groups with each other [111].

In addition to the use of silanol-containing silicone resins, polysiloxanes [169] and silanes [170] containing lower alkoxy groups attached to the silicon atom may be condensed with available hydroxyl groups according to the following equation:

$$\underset{\text{Ethoxysilane}}{\mathrm{RSiOCH_2CH_3}} + \underset{\text{Alcohol}}{\mathrm{HOCH_2R'}} \rightarrow \underset{\text{Alkoxysilane}}{\mathrm{RSiOCH_2R'}} + \underset{\text{Ethyl alcohol}}{\mathrm{CH_3CH_2OH}}$$

The reaction, for example, of phenyltriethoxysilane with DGEBA ($n = 1$) provides the following structure:

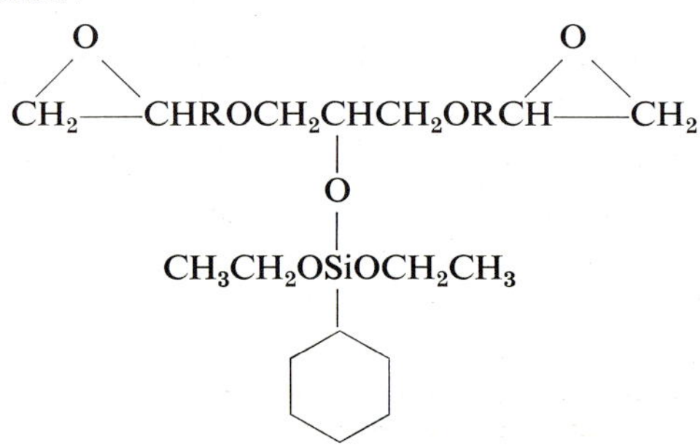

Reaction product of DGEBA ($n = 1$) and phenyltriethoxysilane

The reaction of the lower alkoxy silanes with hydroxyl groups progresses at temperatures from 200 to 280°C, the more bulky chain requiring the higher temperature. With higher temperatures, some of the epoxy groups will be consumed during the reaction. The remaining epoxy groups may be reacted with conventional epoxy-resin curing agents; or a silicone curing agent, such as poly(phenylaluminosiloxane), may also be used when the reaction is conducted at temperatures about 130 to 150°C [111].

In the case of silanol and alkoxy silicone–epoxy reaction products, the siloxyl linkage has been found to be deficient in moisture resistance [77], and to overcome this it is suggested that organosiloxanes containing substituted functional groups (such as amines and carboxyls) at least three atoms removed from the Si atom be employed. Use of aminoalkyl and carboxyalkyl silanes and siloxanes have been described as epoxy curing agents [125].

In addition to these reaction products, high-molecular-weight DGEBA has been esterified with fatty acids and these have been combined with silicones to produce elevated-temperature curing films [31]; silicone alkyd resins (from phenylmethylsiloxane, dimethylterephthalate and isophorone) have been reacted with the higher-molecular-weight DGEBA [33] and DGEBA/phenol-formaldehyde blends [32]; silyl amines (prepared from the reaction of halogenated silicones with ammonia or primary amines) have been shown to react sluggishly with DGEBA resins, cure occurring at about 150°C [147]; and cyclodisilazanes have been reported as curing agents [176].

Modification of silicone adhesives with 10 to 40 per cent of DGEBA ($n = 1$) has resulted in improved shear strengths at 260°C [45], and dicyclopentadiene dioxide has been added to silicone rubber to improve adhesion to metal and prevent cracking of the rubber when exposed to atmospheres containing sulfur gases [134].

Further work appears required to fully exploit the combination of silicone and epoxy.

FURFURAL RESINS

Furfural, itself, is little used, if at all, in the epoxy technology, but furfural resins, particularly from furfural and ketones, have been found suitable as modifiers. Also reported as epoxy modifiers are furfurylidene aminoethanol [140], furfural-MPDA condensation products [174] and furan, itself [142].

Typical of the furfural resins from ketones is the reaction product of furfural and acetone. This may be employed as a modifier with both amine and acid curing agents [67], in the latter case, the resin presumably being incorporated into the epoxy network through the available hydroxyl groups. In addition, the furfural-ketone resins are condensation agents for curing agents based on dicyandiamide [66].

Because of the reaction of the furfural-ketone compound with amines, it is customary to consider the compound as having a pseudo-epoxy equivalent when used in connection with amine curing agents. This dictates the use of greater amounts of the curing agent. Table 15-8 indicates the increased amine requirement when a furfural-ketone modifier is employed with DGEBA. The effect of the modifier on resin viscosity is indicated in Fig. 15-5. The effect of the modifier on various cured properties is given in Figs. 15-6 to 15-10.

Table 15-8. Increased Amine Requirement to Accommodate Reaction with Modifier when Used with DGEBA [37]

DGEBA, %	Furfural-ketone resin, %	Ethylenediamine	DETA	MPDA
100	0	5	10	15
90	10	5.5	11	16.5
80	20	6	12	18
70	30	6.5	13	19.5
60	40	7	14	21
50	50	7.5	15	22.5
40	60	8	16	24

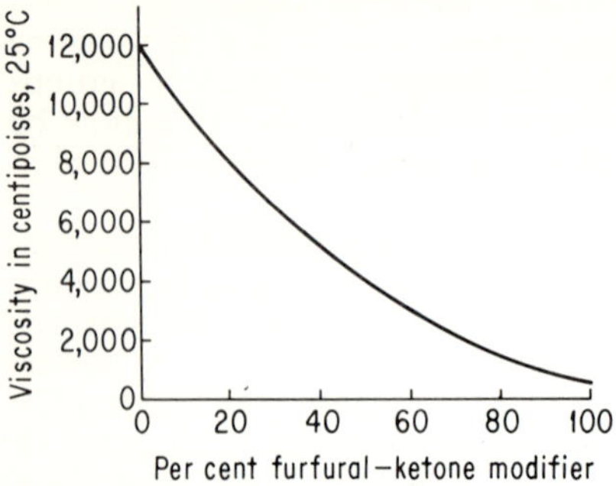

FIG. 15-5. Viscosity of DGEBA vs. concentration of furfural-ketone modifier [90].

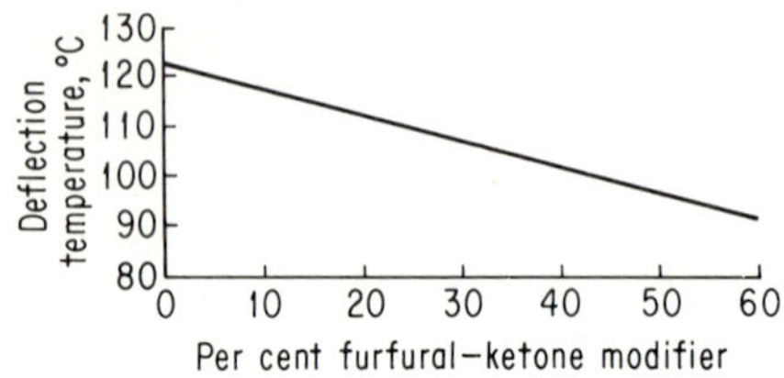

FIG. 15-6. DT of cured DGEBA vs. concentration of furfural-ketone modifier [90].

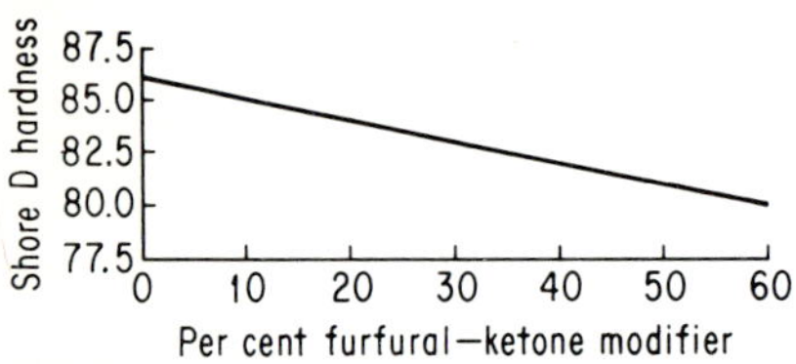

FIG. 15-7. Shore hardness of cured DGEBA vs. concentration of furfural-ketone modifier [90].

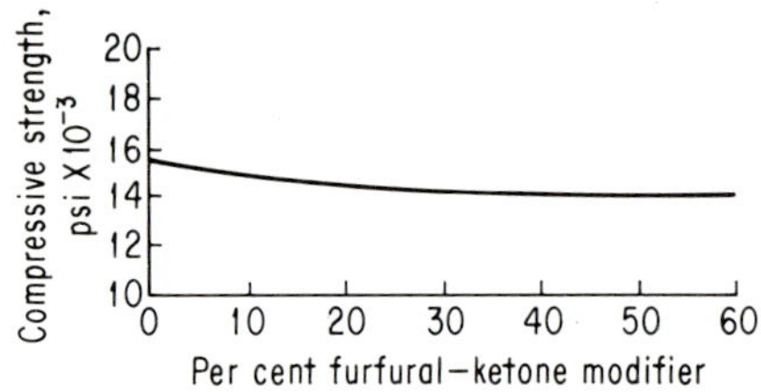

FIG. 15-8. Compressive strength of cured DGEBA vs. concentration of furfural-ketone modifier [90].

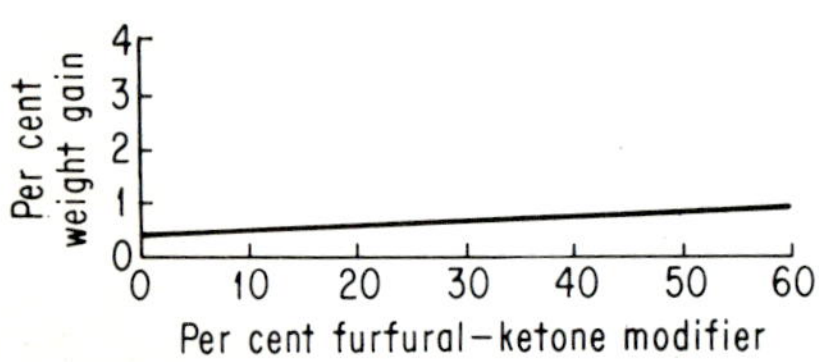

FIG. 15-9. Water absorption of cured DGEBA vs. concentration of furfural-ketone modifier [90].

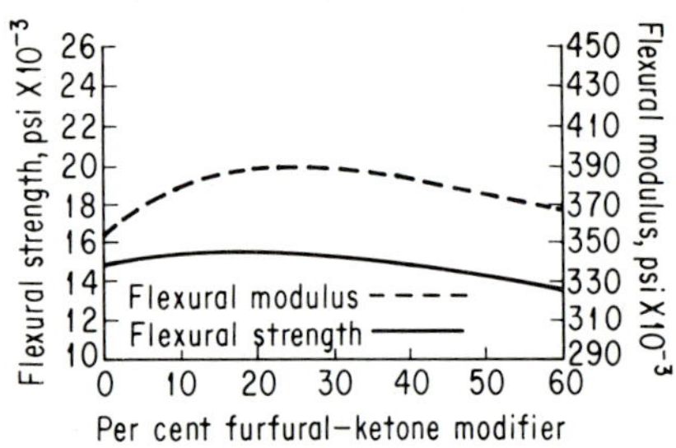

FIG. 15-10. Flexural strength and modulus of cured DGEBA vs. concentration of furfural-ketone modifier [90].

The furfural-ketone reaction products are used principally to reduce costs. An improvement in chemical resistance—particularly to acids, and specifically to hydrofluoric acid—has been reported when the furfural-ketone reaction product is used at about 66 phr, at which concentration there will be some reduction in mechanical strength properties [37].

COAL TAR AND OTHER PHENOLIC MODIFIERS

Phenols and phenolic resins have been previously discussed as acidic-type curing agents; however, a variety of phenolic-based compounds may be considered as modifiers for the resins. With these, it is probable that, at least when amine curing agents are used, the resins are incorporated into the epoxy network merely as nonreactive extenders.

Phenolic resins, of both the resole and the novolac types, can be mixed with polyfunctional amines to obtain systems curing at room temperature but providing significantly reduced exotherm. Typical is a blend of a commercial rosin-modified phenol-formaldehyde resin (25 parts) and DETA (2.5 parts) used at about 100 phr with DGEBA [139].

Typical of other phenolic modifiers are condensation products of polyhydric phenols and fatty-acid esters or drying oils [7]; extracts of pinewood containing rosin acids, phenolic materials, and polyphenols [109]; and alkylated phenolic pitch derived from petroleum [5]. Of chief interest are coal-tar pitches and asphaltic materials.

Natural coal-tar pitch has been described as a modifier for the epoxies [49], as have lower-molecular-weight coal-tar middle-oil fractions (coming over between 200 and 275°C) [91], higher-weight fractions boiling from 250 to 540°C [122], and highly aromatic residues with a precipitation index between 50 and 100 [99]. Solid coal tar, melted together with DGEBA (mol. wt. 1,500), has been pulverized and combined with a powdered curing agent to provide fluidized bed coatings [186], and a mixture of coal tar (melting $\geqq$ 80°C) and anthracene oil (melting 10°C) has been reported as improving the curing of DGEBA under moist environments [182].

The coal-tar products are generally employed in coating formulations, and are applied from solvents, usually at a fairly high solids content. DGEBA is usually the preferred resin in this application, and primary amine curing agents are employed to provide room-temperature curing films. Such formulations are used as chemically resistant coatings for ship bottoms, for pipe and tank coating and general maintenance work, for retopping asphalt drives, etc.

The asphaltic materials are polynuclear unsaturated aromatics derived from petroleum. The liquid species may be used in both epoxy casting and coating formulations. Properties of a typical casting formulation are given in Table 15-9. A formulation involving DGEBA, asphalt, a dimer-trimer fatty acid, and an esterified polyethylene glycol is described in [193].

The epoxy resin may be employed with the higher-melting asphalts to upgrade properties. In these, the epoxy/asphalt combination is usually present as a binder at 8 to 15 per cent with the remainder of the mixture being conventional asphalt aggregate. These systems are processed hot in asphalt handling equipment, and the resultant topping compounds are employed as coatings for jet runways and in other severe-duty applications.

The principal advantage obtained through the use of coal-tar and asphalt additives is cost reduction. In general, the properties of such mixtures are considerably below those obtained with the unmodified epoxy formulations and with the systems employing reactive flexibilizers.

ACRYLIC RESINS

The term acrylic resins covers products derived from acrylic acid, methacrylic acid, or their esters, as shown in Table 15-10.

In addition, the acrylic or methacrylate acid or ester may be reacted with a variety of other unsaturated compounds to produce thermoplastic and thermosetting heteropolymers.

Combinations of epoxy and acrylic offer interesting means of modifying both types of resins. Incorporation of a small number of epoxy groups in acrylic resins provides a convenient route to thermosetting acrylics and makes available the whole

Table 15-9. Properties of DGEBA Modified with Liquid Asphalt [4-1]

(Cure cycle: Gel at 23°C + 8 hr at 100°C)

Ingredients	Formulations 1	2	3	4
Resin	100	100	100	100
Transphalt L-4	50	...	100	
Transphalt L-5	...	50	...	100
TETA	15	15	15	15
Property	**Value**			
Pot life, 400 g at 25°C, minutes	95	69	163	111
Exotherm, °C	140	160	110	120
DT, °C	90	90	81	84
Tensile strength, psi	7,800	8,500	5,300	5,400
Modulus, psi × 10^{-5}	3.2	3.2	2.1	2.1
Elongation, %:				
At failure	3.4	4.8	3.8	3.8
At 0.2% offset	1.9	1.9	1.9	1.7
Yield, psi (at 0.2% offset)	6,100	6,100	3,650	3,600
Flexural strength, psi	12,500	12,700	6,500	6,900
Modulus, psi × 10^{-5}	3.4	3.3	1.9	1.8
Hardness, Barcol	69	71	44	41
Water absorption after 2-hr boil, %	0.75	0.66	0.76	0.65
Dielectric constant:				
At 60 cps	3.23	3.29	3.09	3.21
At 10^5 cps	3.11	3.15	3.02	3.05
Dissipation factor:				
At 60 cps	0.010	0.013	0.012	0.0060
At 10^5 cps	0.017	0.019	0.014	0.0170
Loss factor:				
At 60 cps	0.033	0.042	0.036	0.019
At 10^5 cps	0.052	0.060	0.044	0.053

gamut of epoxy-resin curing agents as modifiers. Acrylic resins may be used as additives [138], as for example when normally liquid acrylic acid and methacrylic acid are mixed with DGEBA and a tertiary amine, and the product is cured [64]. Alternatively, high-molecular-weight polymerized acrylic acid or acrylic acid copolymers may be cross-linked with epoxy resins [164 to 166]. A special epoxy-based system involves esterifying higher-molecular-weight DGEBA with acrylyl chloride or methacrylyl chloride, mixing with an unsaturated diluent such as styrene, and using the mixture as casting resins, etc. [94].

The major interest in the combination of epoxy and acrylics is for vehicle solids in solution coatings, with the primary research being directed to the development of heat-curing appliance enamels and automotive finishes and systems with improved chalk resistance.

Table 15-10. Acrylic Monomers and Resins

$CH_2{=}CHCOOH$
$CH_2{=}CHCOOR$

Acrylic acid and ester

$CH_2{=}C(CH_3)COOH$
$CH_2{=}C(CH_3)COOR$

Methacrylic acid and ester

$CH_2{=}C(CH_3)COO(CH_2)_2OOC(CH_3)C{=}CH_2$

Ethylene glycol dimethacrylate

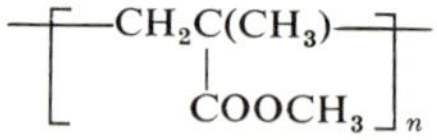

Poly(methyl methacrylate)

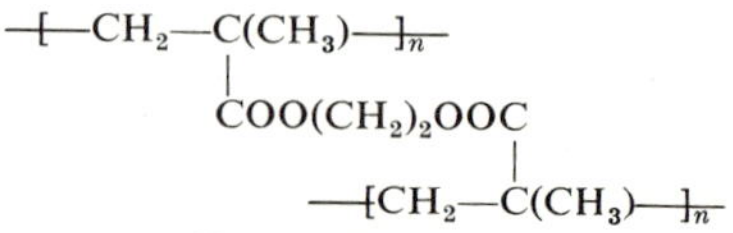

Thermoset acrylic

Epoxy Groups in Thermosetting Acrylics

Acrylic or methacrylic acid may be reacted with epichlorohydrin to produce a molecule containing both an epoxy group and unsaturation [6]. Typical is glycidyl methacrylate:

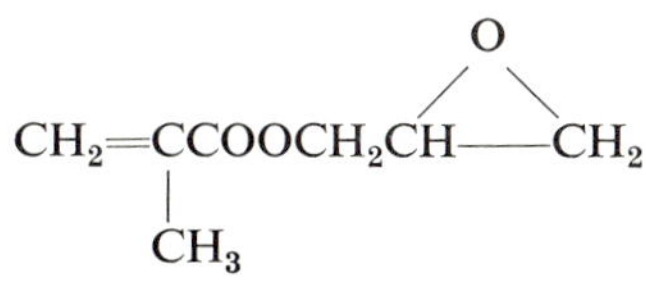

Glycidyl methacrylate

Such compounds may be polymerized through unsaturation to produce polyepoxy resins, in the case of glycidyl methacrylate, of the following structure.

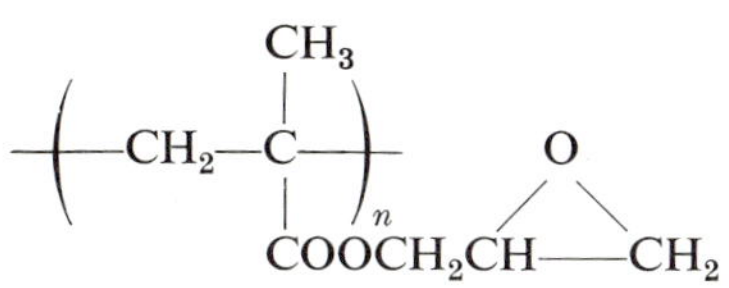

Poly(glycidyl methacrylate)

By copolymerizing glycidyl methacrylate with styrene [18], methyl methacrylate [100], acrylic and methacrylic acid esters [44], acrylonitrile, vinylidene chloride [4], etc., quite complex polyepoxy resins may be developed. Similarly, resins with acrylic backbones may be produced by the polymerization of acrylic acid esters with allyl glycidyl ether [35,131]. Preparation of self-curing epoxy resins from a heteropolymer of glycidyl methacrylate, methylacrylate, styrene, and acrylic acid has been described [17]. These linear polymers cross-link with heating.

The epoxy-containing acrylic resins usually have numerous epoxy groups, approximately 5 to 10 per molecule, with molecular weights from 5,000 to 10,000, and upward. Cure may be obtained with conventional epoxy-resin curing agents, although as shown in Table 15-11, some incompatibility with amines has been

Table 15-11. Amines Tested as Cross-linking Agents for MMA/GMA (70/30) Copolymer [98]

Structural type	Examples	Compatibility
Straight-chain aliphatic	*n*-Hexylamine	Compatible
	n-Decylamine	Slight surface haze
Polyfunctional aliphatic ...	Ethylenediamine	Slight surface haze
	Hexamethylenediamine	Haze
	Bis(4-aminocyclohexyl)-methane	Slight surface haze
	2-Methylhexamethylene-diamine	Oily surface haze
	Diethylenetriamine	Slight surface haze
Hindered aliphatic	Methanediamine	Compatible
	Bis(aminomethyl)mesitylene	Haze
Alkylated aliphatic........	*N*-(2-Phenylethyl)-hexamethylenediamine	Oily surface haze
Hydroxyl-containing	*N*-(2-Phenyl-2-hydroxyethyl)-diethylenetriamine	Slight surface haze
Aromatic	*p′,p′*,-Diaminodiphenylmethane	Compatible

NOTE: All the films were cross-linked after 1 week at room temperature. Five minutes' exposure to 5/1 toluene/butanol did not soften any of the films.

experienced, and even with compatible amines, room-temperature cure is not always too satisfactory. The use of aliphatic dicarboxylic acids such as azelaic seem preferred [58]. Alternatively, the curing agent may be developed by reacting the epoxy groups with polymeric phosphates to produce phosphoric acid esters [28,38], and these compounds subsequently may be reacted with additional resin to produce thermoset films [59].

Properties of a pigmented film based on the copolymer from styrene and glycidyl methacrylate (85/15 by weight) and cured with phosphoric acid are compared with an alkyd baking finish in Table 15-12. Properties of an elastomer from ethyl acrylate and glycidyl methacrylate are given in Table 15-13.

Table 15-12. Resistance Properties of S/GMA (85/15) Phosphate Enamel Evaluated as an Appliance Finish [98]

Property	S/GMA phosphate, 30 min at 150°C, 1.0 mil; Epoxy ester primer, 0.8 mil	Alkyd-UF, 30 min at 150°C, 1.0 mil; Epoxy ester primer, 0.8 mil
Gloss (20°ASTM)	65	50
Overbake discoloration after 30 min at 175°C (low values best)	$\Delta E = 1$	$\Delta E = 2.5$
20 in. impact, duPont tester, convex/concave (high values best)	9/9	7.5/7.5
Mandrel, ASTM D 522–41 (high values best)	8	7.5
Tukon hardness	20	11–13
Salt spray (700 hr) 10—perfect	10	8
Humidity cabinet 10—perfect	7.5*	10
Soap (700 hr, 1.5%) at 75°C	>30 days	Failed in 7 days
Stain resistance:		
Catsup	None	None
Mustard	None	Stain
Vinegar	None	None
Lipstick	None	Stain
Orange dye	None	Stain
Iodine	Slight stain	Stain
Ink	None	
Shoe polish	None	

* Over the best primer available, humidity cabinet = 10.

Epoxy Resins as Curing Agents for Polymerized Acrylic Acids

Acrylic acid or methacrylic acid can be polymerized to high molecular weights, such resins having as many as 25 carboxyl groups per molecule [128]. Acid-containing copolymers [21,22,110] and terpolymers [137] may also be developed. In addition, acrylic copolymers may be developed which contain other groups functional with the epoxy resin [112], and copolymers with vinylpyridine may be employed, these containing built-in amine accelerators [26]. Epoxy resins may then be used to provide cross-linking between the acrylic resin molecules, for example, when DGEBA is used, to increase hardness and heat stability. It has been reported that tertiary-amine-containing acrylics can, in solution with DGEBA, be stabilized by the use of stoichiometric amounts of difluoroacetic acid to give heat-cured coatings [181].

These products bear more resemblance to the thermosetting acrylics than to the epoxy resins.

Table 15-13. Ethyl Acrylate–GMA Copolymers as Elastomers [98]

FORMULATION

EA/GMA copolymer	100
Cross-linking agent	See table
Furnace black	50
Stearic acid (processing aid)	1.0
Catalyst	See table
Processing cycle after blending	5 min at 100°C
Cure	See table

Polymer composition	Cross-linking agent, parts	Catalyst, parts uncompounded gum elastomer	Cure		Ultimate elong. at 1 in./min and 50% relative humidity, %	Tensile strength, psi	Initial modulus, psi
			Time, minutes	Temp., °C			
EA/GMA (94/6)	None	...	..	...	2,200	170	70
			30	150	480	270	400
EA/GMA (94/6)	Azelaic acid, 3.62	Dicyandiamide, 0.11	30	180	250	1,120	590
			30	150	350	610	690
EA/GMA (97/3)	*m*-Phenylenediamine, 1.13	Sulfur, 1.0	30	180	290	940	850
EA/GMA (97/3)*	Trimethyltrimethylenetriamine, 4	Sulfur, 1.0	30	150	440	1,280	480
Commercial acrylic rubber	Trimethyltrimethylenetriamine, 4	Sulfur, 1.0	30	150	270	1,250	650

* This composition had a tendency to cure on the mill.

SUMMARY

Epoxy resins may be reacted with a variety of functional groups and can be modified with resinous materials containing such functional groups. This chapter discusses materials which are generally not regarded as curing agents per se and which are not generally employed primarily because of the flexibility they impart to the epoxy-resin systems. Most are used as low-cost extenders, such as the polyesters, the furfural-ketone reaction products, and the coal-tar products. Others are used to impart specialized properties which are, in a large measure, the contribution of the modifiers—for example, the use of polyurethanes, the fluorocarbons, or silicones in an attempt to upgrade the epoxy resins. Still others, such as the vinyls, are used to provide specialized modifications, such as improved peel resistance in adhesive formulations or anticratering properties in coatings formulations.

Epoxy resins have also been shown to be compatible as additives for a variety of other resins. Generally, the epoxy is present at fairly low concentrations and is employed without the use of a curing agent. The majority of the epoxy resins used as additives are employed as stabilizers/plasticizers for vinyl formulations and for other chlorinated materials, although limited use has been found for more specialized functions in a variety of resins.

REFERENCES

1946

1. Bent and Marple, Polyvinyl Acetal Compositions, U.S. 2,400,333.

1949

2. Dudley, Anion Active Resins and Processes of Producing the Same, U.S. 2,469,683 and U.S. 2,469,684.
3. Lundberg and Dudley, Process of Removing Anions from Liquids, U.S. 2,469,692.
4. Standinger et al., Glycidyl Ester Copolymers, U.S. 2,470,324.

1950

5. Bradley, Epoxy Ether Compositions, U.S. 2,528,417.
6. Dorough, Esters of Epoxy Alcohols with Propenoic Compounds, U.S. 2,524,432.
7. Greenlee, Compositions of Phenol-Oil Condensates and Polyepoxides, U.S. 2,502,145.
8. Greenlee, Compositions of Resinous Epoxides and Aromatic Sulfonamide-Aldehyde Condensates, U.S. 2,494,295.
9. Havens, Thermostable Polymeric Vinylidene Chloride Compositions, U.S. 2,530,353.

1951

10. Abell, Stabilized Compositions Containing High Molecular Weight Halogenated Materials, U.S. 2,559,333.
11. Anderson and Overbaugh, Chlorinated Ethylene Polymers Stabilized with Glycidyl Esters of Fatty Acids and Methyl Glucamine, U.S. 2,541,492.
12. De Nie and Voorthuis, Stabilization of High Molecular Weight Organic Material Containing Inorganic Acid-forming Elements, U.S. 2,564,194 and U.S. 2,564,195.
13. Niederhauser, Chlorinated Rubber and Epoxidized Oils, U.S. 2,556,145.
14. Terry and Wheeler, Plasticized Polyvinyl Resin Compositions, U.S. 2,559,177.
15. Voorthuis, Stabilized Compositions Containing High Molecular Weight Halogenated Materials, U.S. 2,555,169.
16. Wende and Gesierich, Production and Properties of Some Epoxy Alkyl Silanes, *Plaste Kautschunk* (*E. Ger.*), **6**:301–303.

1952

17. Bilton and Segall, Crosslinkable and Crosslinked Vinyl Type Copolymers, U.S. 2,604,463.
18. Erickson and Thomas, Copolymer of Styrene, Glycidyl Acrylate and Glycidyl Methacrylate, U.S. 2,580,901.
19. Greenlee, Resinous Compositions, U.S. 2,591,539.
20. Nelson, Resinous Compositions and Their Use, U.S. 2,594,979.
21. Segall and Dixon, Reaction Product of Diphenylolpropane Diglycidyl Ether with Styrene-Acrylic Acid Copolymers, U.S. 2,604,464.
22. Segall and Dixon, Styrene–Acrylic Acid Copolymers Reacted with Polymeric 4-Vinylcyclohexene Diepoxide, U.S. 2,604,457.
23. Shokal et al., Stabilization of Halogen-containing Organic Materials, U.S. 2,585,506.
24. Wiles and Elam, Compositions Containing Glycidyl Polyethers of Dihydric Phenols and Polyvinyl Acetate, U.S. 2,602,785.
25. Winkler, High Molecular Weight Halogenated Materials Stabilized with a Mixture of an Epoxy Compound and a Carboxylic Acid Salt, U.S. 2,590,059.

1953

26. Allenby, Reaction Products of Polyepoxides with Styrene–Acrylic Acid–Vinyl Pyridine Copolymers, U.S. 2,662,870.

1954

27. Cowell and Darby, Method of Stabilizing Halogen Containing Resins with Cadmium Salts and Epoxy Esters, U.S. 2,671,064.
28. Cupery, Phosphates of Styrene-Epoxy Compound Copolymers, U.S. 2,692,876.
29. Darby, Stabilized Halogen-containing Resins, U.S. 2,669,549.
30. Gearhart, Heat Stable Plastic Composition Containing Lower Fatty Acid Esters of Cellulose, U.S. 2,675,327.
31. Hatcher, Polysiloxane Resin Compositions, U.S. 2,695,276.
32. McLean, Compositions of Silicon-Alkyd Resins, an Epoxy Resin, and a Phenol-Formaldehyde Resin, U.S. 2,687,396.
33. McLean, Compositions of Silicon-Alkyd Resin and an Epoxy Resin, U.S. 2,687,398.
34. Preparation of Dissolved Epoxy Resins Modified with Organosilicone Compounds, Ger. Appl. G13463 (39c, 30).
35. Rothrock and Wilkinson, Allyl–Glycidyl Ether/Acrylate Ester Copolymers, U.S. 2,687,405.

1955

36. Arone, Flame-retardant Insulated Conductors, Method of Making Same, and Compositions Used to Prepare the Same, U.S. 2,717,216.
37. Brenner and Singer, Epoxy-Furane Blends, *Mater. Methods*, June.
38. Cupery, Polymeric Phosphates of Copolymers of Acrylic Ethylenically Unsaturated Epoxy-free Monomers and Ethylenically Unsaturated Epoxy Monomers, U.S. 2,723,971.
39. Mackinney, Treatment of Wool-containing Textile Materials and Products Thereof, U.S. 2,723,924.
40. Scheibli, Composition Containing Glycidyl Polyether and a Polyvinyl Acetal Resin, U.S. 2,713,567.
41. Thompson, Cellulose Derivatives Stabilized with Epoxy Condensation Products, U.S. 2,710,844.
42. Thompson, Cellulose Esters Stabilized with a Resinous Polymeric Epoxide and a Metal Salt, U.S. 2,710,845.

1956

43. Cranker, The Chemistry of Modifiers for Epoxy Resins and Their Effect on Properties, SPE Symposium, Los Angeles, November.

44. Hurwitz, Crosslinked Copolymers of the Esters of Acrylic and Methacrylic Acid, U.S. 2,729,625.
45. Kidwell and McHugh, Investigation of High-temperature Structural Adhesives, U.S. Government Report PB-121657.
46. Martin, Epoxy-substituted Esters of Silicon Acids and Their Preparation, U.S. 2,730,532.
47. Schroeder and Sumers, Treatment of Textile Materials and Product, U.S. 2,774,691.
48. Suen, Shrinkage Control of Cellulosic and Wool Textiles with Diglycidyl Ether Compounds, U.S. 2,730,427.
49. Whittier and Lawn, Coating Compositions Containing Coal Tar Oil and an Epoxy Resin, U.S. 2,765,288.
50. Zukas, Resinous Compositions Comprising Thermosetting Epoxide Resins and Thermoplastic Polyvinyl Resins and Bentonite-Amine Reaction Products, U.S. 2,733,043.

1957

51. Dow Corning Corp. Technical Bulletin Q2-101, December.
52. deGanahl et al., Method of Making Molded Plastic Articles, U.S. 2,815,309.
53. Newey, Method for Preparing Hard, Flexible Vinyl Halide Polymer–Liquid Polyepoxide Compositions and Resulting Products, U.S. 2,795,565.
54. Welch, Method of Coating a Surface with Polyhalocarbon Resin and Article Formed Therewith, U.S. 2,777,783.

1958

55. Bennett, Resins of Epoxy Compounds and Polymers of Alkyl Substituted Vinylpyridines and the Like, U.S. 2,860,120.
56. Bergmann et al., Use of Organosilicone Compounds as Curing Agents for Epoxy Resins, Ger. 1,033,893.
57. Brice and Budde, Epoxy Esters, *Ind. Eng. Chem.*, June.
58. Fang, Process of Making Glycidyl Methacrylate Polymer Composition Containing Same, and Product Coated Therewith, U.S. 2,857,354.
59. Fang, Thermosetting Epoxide Polymer Mixtures, Method of Making Same and Coatings Containing Same, U.S. 2,849,418.
60. FitzGerald and Smith, Resin Composition Containing Epoxidized Hydrocarbon Polymer as Resin Plasticizers, U.S. 2,842,513.
61. Greenspan and Gall, Polyvinyl Resin Compositions Plasticized with Partially Epoxidized Fatty Acid Esters, U.S. 2,857,349.
62. Greenspan and Gall, Vinyl Epoxy Plasticizers, *Ind. Eng. Chem.*, June.
63. Greenspan, Introduction to Symposium on Epoxy Plasticizers-Stabilizers, *Ind. Eng. Chem.*, June.
64. Hall, Acrylic Acid Derivatives of Epoxide Resins and Method of Curing Same, U.S. 2,824,851.
65. Harvey and Rosamilia, Combinations of Ketone Formaldehyde Reaction Products and Glycidyl Ether of Polyhydric Alcohols, U.S. 2,839,481.
66. Harvey and Rosamilia, Compositions Comprising Furfural-Ketone-Amine Reaction Products and an Epoxy Resin, U.S. 2,839,487.
67. Harvey and Rosamilia, Compositions Containing Epoxy Resins and Furfural-Ketone Reaction Products, U.S. 2,839,488.
68. Koenecke and Brady, Cellulose Nitrate–Liquid Polymer Oil Compositions, U.S. 2,842,452.
69. Mika, Cured Silicon-containing Products Prepared from Polyepoxides, U.S. 2,843,560.
70. Mullins, Vinyl Chloride Resin Plasticized with a Bis-(3,4-epoxy-cyclohexane-carboxylate) and Compositions Containing Same, U.S. 2,862,904.
71. Pattison, Poly (Polyalkylene Ether Urethane) Polymers Containing Epoxide Groups, U.S. 2,830,038.
72. Perry, Introduction to Symposium, *Ind. Eng. Chem.*, June.
73. Petty, Process for Sweetening and Stabilizing Hydrocarbons with an Organic Epoxide and an Aqueous Alkaline Phenol, U.S. 2,862,804.

74. Pinder, Lacquer Composition Comprising Epoxide Resin, Polyamide Resin, and Cellulose Acetate Butyrate and Article Coated Therewith, U.S. 2,865,870.
75. Pritchard and Canterino, Epoxy Compounds as Plasticizers for Quaternized Polymers, U.S. 2,860,117.
76. Safford, Compositions Containing Epoxy Resin, Polyvinyl Chloride and Plasticizer, U.S. 2,843,557.
77. Shorr, Siloxane–Epoxide Resin Reaction Products, U.S. 2,819,245.
78. Speranza and Peppel, Preparation of Substituted 2-Oxazolidones from 1,2-Epoxides and Isocyanates, *J. Org. Chem.* **22**:1922–1924.
79. Van Cleve and Mullins, Derivatives of Cyclohexane Oxide as Plasticizers and Stabilizers for Vinyl Chloride Resins, *Ind. Eng. Chem.*, June.
80. Winkler, New and Modified Epoxy Stabilizers, *Ind. Eng. Chem.*, June.
81. Yaeger, Process of Coating a Metal Surface, U.S. 2,829,984.

1959

82. Bond and Brady, Silicone Epoxy Resins in High Temperature Structural Laminates, ACS Symposium, Atlantic City, September.
83. Condo, Compositions Comprising Cellulose Derivative, Epoxy Resin and Epoxy Curing Catalyst and Process for Preparing Same, U.S. 2,879,235.
84. Eitel et al., Process for the Preparation of Coating of Soft Resins from Polyisocyanates and Compounds Containing Hydroxyl Groups, U.S. 2,897,166.
85. Filachione and Harris, Tanning with Epoxy Resins, U.S. 2,869,970.
86. Greenlee, Polyurethane of a Polyisocyanate, an Active Hydrogen Compound, and a Hydroxyaryl Aliphatic Acid, U.S. 2,907,745.
87. Greenlee, Polyurethane of a Polyisocyanate, an Active Hydrogen Compound and a Hydroxyaryl Aliphatic Acid–Aldehyde Condensate and Method of Preparing Same, U.S. 2,907,719.
88. Honn, Compositions Comprising Epoxide Resin and Trifluorochloroolefin Copolymer and Process of Crosslinking Same, U.S. 2,904,528.
89. Hopper and Naps, Epoxy Resin Adhesive Compositions, Their Preparation, and Tape Containing Same, U.S. 2,915,490.
90. Irvington Chemical Co., Technical Bulletin on NX 3157, Extender for Epoxy Resin.
91. Lopata, Corrosion-resistant Coatings Containing Coal Tar Oil and an Epoxy Resin, U.S. 2,889,305.
92. Metzger, Antistatic Resin Compositions, U.S. 2,872,432.
93. Morris and Chaplick, Thermoplastic Polyamide-Epoxy Adhesive, U.S. 2,867,592.
94. Parker, Method of Preparing Acrylate Esters of Epoxy Resins, U.S. 2,890,202.
95. Ramos, Resinous Compositions Comprising Epoxy Resin, Curing Agent and Mixture of Dextrines, Maltose, and Dextrose and Process for Preparing, U.S. 2,894,920.
96. Schroeder, Process for Treating Textile Materials and Resulting Products, U.S. 2,886,473.
97. Shafer, Epoxy Resin–Plastisol Electrical Insulating Materials and Method of Making Same, U.S. 2,892,808.
98. Simms, Epoxide Substituted Vinyl and Acrylate Copolymers, ACS Symposium, Atlantic City, September.
99. Simpson, Compositions Comprising Aromatic Petroleum Residue and Polyepoxide and Process for Treating Surfaces Therewith, U.S. 2,906,720.
100. Staicopoulos, Methacrylate Resin Composition and Method of Making Same, U.S. 2,868,760.
101. Yamada et al., Ethoxyline-Cyanurate-Acrylate Resinous Compositions and Copper Conductor Coated Therewith, U.S. 2,883,308.

1960

102. Been and Grover, Structural Adhesive Composition, U.S. 2,920,990.
103. Berenbaum, Resinous Condensation Product of a Polyepoxy–Polyether Resin and a Hydroxyl Terminated Polyester and Method of Making Same, U.S. 2,944,996.
104. Findley and Ohson, Crosslinked Reaction Product of an Epoxidized Fatty Acid Ester and Polyvinyl Halide or Cellulose Nitrate, U.S. 2,964,484.

105. Fisch et al., Heat-hardenable Compositions Composed of Polyvinyl Chloride, Epoxy Resin and Hardening Agent, Cellular Product, and Method of Preparing Same, U.S. 2,965,586.
106. Mack and Bickford, Maleinated Eleostearic Acids, U.S. 2,865,931.
107. Mullins et al., Vinyl Chloride Resin Compositions, Stabilized and Plasticized with Epoxy Compounds, U.S. 2,924,582.
108. Plueddemann, Epoxysiloxane Casting Resins, *J. Chem. Eng. Data*, January.
109. Reed, Vinsol-Epoxy Compositions, U.S. 2,948,694.
110. Reid and Werner, Epoxide Cure of Rubbery Copolymer Which Contains a Reactive Carboxyl Group, U.S. 2,947,338.

1961

111. Andrianov et al., Graft Copolymers from Polyorganosiloxanes and Epoxy Resins, *Vysokomolekul. Soedin.*, **3** (11): 1692–1697.
112. Applegath, Epoxy Resins in Thermosetting Acrylics, *Ind. Eng. Chem.*, June.
113. Bavley and Knuth, Dialkyl Tin Epoxy Succinate Stabilizers for Polymers of Polyvinyl Chloride, U.S. 2,972,595.
114. Gallagher and Mason, Silicone–Epoxy Resin Compositions, Brit. 864,697.
115. Halewood, Epoxy Resin Compositions, Brit. 878,750.
116. Lee and Neville, New Developments in Epoxy Resins, part II, *Insulation*, January.
117. Lines, Four Epoxy Plasticizers, *Plastics Technol.*, March.
118. Phillips, Epoxy Resins, Brit. 822,889.
119. Rustig, Curing System for Epoxy Resins, Ger. 1,107,400.
120. Toone, Urethane Coatings: Their Chemistry and Composition, *Dyestuffs*, March.
121. "Unox" Epoxy-Vinyl Plastisols, Union Carbide Chemicals Co. Bulletin F-40774, May.

1962

122. Boenau et al., Compositions of Epoxy Resins and Aromatic Hydrocarbon Oils, U.S. 3,062,771.
123. A Circuit Breaker Having Resinous Insulating Members (to Licentia Patent-Verwaltungs, G.m.b.H.), Brit. 903,961.
124. Frye and McLean, Method of Preparing Silicone–Epoxide Resins, U.S. 3,055,858.
125. Kanner, Organofunctional Silicon-Epoxy Compositions, Brit. 906,544.
126. McGuigan, The Effectiveness of Epoxy Stabilizers for Vinyl Coatings as Influenced by Type and Concentration, *Offic. Dig.*, pp. 939–952.
127. Marklow, Epoxy Resins, Brit. 890,745.
128. Murdock and Segall, Thermosetting Compositions Based on Acidic Copolymers Crosslinked with Diepoxides, technical literature, FMC Corp.
129. Organo-silicon Polymers (to Minnesota Mining & Mfg. Co.), Brit. 909,387.
130. Wood, Identification of the Urethane-Epoxy Reaction Product, *SPE Trans.*, July.
131. Woodruff, Mixed Acrylic Polymers Carrying Glycidyl and Carboxyl Groups, U.S. 3,052,659.

1963

132. Been and Grover, Improvements in Structural Adhesives, Brit. 919,549.
133. Bennett, Improvements in Adhesives, Brit. 922,581.
134. Burdick and Polmanteer, Silicone Rubber Compositions, U.S. 3,094,446.
135. Capon and Christofas, Application of Polyvinylidene Fluoride Coatings, U.S. 3,111,426.
136. Cappuccio, Polypropylene Containing a Dye-receptive Modifier Which Comprises Polyalkyleneimine or Mixtures Thereof with an Epoxy Resin, U.S. 3,107,228.
137. Coatings from Acid-containing Terepolymers and Epoxy Resins (to Dow Chemical Co.), Brit. 930,035.
138. Coating Compositions (to De Soto Chemical Coatings, Inc.), Brit. 925,856.
139. Krzesz, Process for Curing Epoxy Resins and Compositions Made Therefrom, Can. 662,682.

140. Mamedov et al., Thermosetting Epoxy-Furan Compositions, U.S.S.R. 156,679.
141. Nishimura and Hata, Copolymerized Inner Epoxy Resin, Jap. 20,989.
142. Oster-Volkov et al., Epoxy-Furan Resins and Some of Their Properties, *Plasticheskie Massy*, no. 9 pp. 52–53.
143. Plapper and Dieckelmann, Novel Reaction Products of a Sulfuric Acid and Epoxides, U.S. 3,096,357.
144. Riser et al., Epoxy Acids and Salts as Stabilizers for Vinyl Plastics, *SPE J.*, August.
145. Rosenberg, Composition Comprising Epoxy Resin, Polyvinyl Acetal, Silica, and BF_3 Amine Complex, U.S. 3,098,054.
146. Schnell and Krimm, Process for the Production of Hardened Polycarbonate-modified Epoxy Resin Bodies, U.S. 3,098,056.
147. Schultz and Zike, Silamine Curing Agents, U.S. 3,072,594.
148. Skora et al., Epoxy Resin for Cold Curing, Pol. 47,694.
149. Smith, Reaction of Phenyl Isocyanate with 2,3-Epoxypropyl Phenylether, *Nature*, Jan. 19.
150. Stabilized Chloroalkenes (to Solvay & Cie.), Belg. 624,361.
151. Steinbach et al., Hair-dressing Agents, Belg. 630,579.
152. Tocker, Polymeric Compositions and Preparations Thereof, U.S. 3,113,907.

1964

153. Aglulin and Gurdzhi, Low-molecular-weight Polyurethane Curing Agent for Epoxide Resins, U.S.S.R. 165,893.
154. Buckley and Johnson, Degradation of Polymeric Compositions in Vacuum to 10^{-9} mm Hg in Evaporation and Sliding Friction Studies, 20th ANTEC of SPE, January.
155. Dalibor, Copolymers from Unsaturated Resinous Epoxy Compounds, Ger. 1,169,135.
156. Ennis and Knabeschuh, Heat Stabilized Chlorosulfonated Polyethylene Containing an Epoxy Compound and a Phenol, U.S. 3,147,315.
157. Epoxyacyloxysilanes (to Dow Corning Corp.), Brit. 954,079.
158. Epoxy Copolymers (to E. I. du Pont de Nemours & Co.), Brit. 948,998.
159. Epoxy Silicones (to General Electric Co.), Brit. 968,204.
160. Goldsmith and Marples, Stabilization of Epoxide Compositions, U.S. 3,142,687.
161. Harvey, Composition Comprising a Glycidyl Polyether and the Reaction Product of an Acetone-Formaldehyde Condensate and a Polyalkylene Glycol, U.S. 3,121,703.
162. Holm, Polycarbonates, U.S. 3,141,863.
163. Humphreys et al., Hardenable Epoxy Resin Compositions, Brit. 966,976.
164. Improvements in Thermoset Compositions Based on Carboxy Polymers and Epoxide Resins (to Devoe & Raynolds Co.), Brit. 948,491.
165. Improvements in Thermoset Compositions Based on Carboxy Polymers (to Devoe & Raynolds Co.), 948,492.
166. Improvements in Thermoset Compositions Based on Carboxy Polymers and Epoxide Resins (to Devoe & Raynolds Co.), Brit. 948,493.
167. Kastrop, High Temperature Adhesive, U.S. 3,160,675.
168. Kost, Contact Adhesion Compositions, U.S. 3,144,428.
169. McWhorter, Siloxane Modified Epoxide Resins, U.S. 3,154,597.
170. Masters, Silane Modified Epoxide Resins, U.S. 3,150,116.
171. Medvedeva et al., Cold-hardening of Epoxy Polyester Resins, *Plasticheskie Massy*, no. 3, pp. 17–18.
172. Modified Epoxide Resins (to Dr. Beck & Co., G.m.b.H.), Fr. 1,359,078.
173. R. G. Neville, private communication.
174. Orobchenko et al., Epoxy-Furan Thermosetting Resins, U.S.S.R. 161,923.
175. Pascal, Epoxy Containing Resins, U.S. 3,132,121.
176. Peake, Cyclodisilazanes, Fr. 1,350,220.
177. Petrashko and Andrianov, Catalytic Copolymerization of Polyorganosiloxanes with an Epoxy Resin in the Presence of Poly(aluminophenylsiloxane), *Plasticheskie Massy*, no. 1, pp. 26–27.
178. Renalli, Method of Preventing Discoloration of Olefin Polymers by Catalyst Residues and the Resulting Polymeric Compositions, U.S. 3,139,414.
179. Thinius and Weichert, Stabilization of Cyclic Acetals of Polyols and Polymers Therefrom, Ger. (East), 27,163.

1965

180. Armour and Diamantopoulos, Epoxy Resin Adhesive Compositions, U.S. 3,169,936.
181. Clemens, Stable Self-containing Epoxy Resin Compositions, U.S. 3,183,281.
182. Coupin and Herzberg, Adhesive for Bonding in Aqueous Medium, Fr. 1,386,437.
183. Curable Polyepoxide Compositions Containing Dihydropyran Derivatives (to Shell Internationale Research), Neth. Appl. 289,831.
184. Donnelly, Inorganic Cement–Epoxy Resin Composition Containing Animal Glue, U.S. 3,198,758.
185. Flowers, Resinous Compositions, U.S. 3,172,921.
186. Goodnight, Coal Tar Coating Composition, U.S. 3,190,845.
187. Hardman and Metten, Vinyl-Epoxide Polymer Mixtures, Brit. 989,375.
188. Improvements in or Relating to Bonding with Epoxy Resin Adhesives (to Henkel & Cie, G.m.b.H.), Brit. 989,390.
189. Lambuth, Process for the Preparation of Epoxide Modified Adhesive Compositions, U.S. 3,192,171.
190. Medicament in Solid Form (to Ciba, Ltd.), Brit. 996,819.
191. Nikki and Ajisada, Polyester Resin Molding, Ger. 1,198,552.
192. Smith, Flame-resistant Epoxy Compositions, Brit. 1,002,185.
193. Street, Asphalt-Epoxy Compositions and Method of Making the Same, U.S. 3,202,612.
194. Weinheimer, Adhesive Composition, U.S. 3,166,523.

Chapter 16

FLEXIBILIZERS AND PLASTICIZERS FOR EPOXY RESINS

Flexibility may be imparted to the cured epoxy-resin structure by two methods: by introducing long-chain molecules into the structure which remain unreacted after cure, or by introducing long-chain elements into the epoxy which react during cure. The former may be considered plasticizers; the latter, flexibilizers. Of the two, the flexibilizers are generally preferred.

The function of the flexibilizers or plasticizers is not to convert the normally rigid-cured epoxy system into a material competitive in flexibility and ultimate elongation with the natural and synthetic rubbers, the silicones, the polyurethanes, or other elastomers. Rather, the purpose is to impart some degree of resiliency and toughness to a normally rather rigid system and thereby to improve thermal shock and impact resistance of castings, improve peel strength of adhesives, and increase flexibility in films. By reducing internal stresses during cure, those flexibilizers not leading to an increase in exotherm permit casting of larger volumes and more complex configurations. Some of the flexibilizers may be used to decrease exotherm and/or shrinkage.

Because of the function of the flexibilizer or the plasticizer, the words themselves are not entirely descriptive. In the main, they bring to mind unrealistic design values and tend to obscure the actual function of the additives. The definition, then, becomes a nice one of degree: at one extreme the words could cover any material which reduces rigidity to any degree; and at the other extreme only those materials which provide x per cent elongation.

In the broader definition, then, it is evident that many compounds may be considered as either flexibilizers or plasticizers, and a variety of these have been presented in other sections of the book.

If the flexibilizer is also an effective curing agent, it has been discussed in the appropriate curing-agent chapter. For example: the polyether diamines as discussed in Chap. 7, the fatty polyamides as discussed in Chap. 10, the polymeric acids and carboxyl-terminated polyesters as described in Chap. 11, and the long-side-chain anhydrides such as DDSA, as discussed in Chap. 12. Many of the reactive modifiers, as discussed in Chap. 15—for example the furfural-ketone reaction products—may also be considered as flexibilizers.

This chapter discusses those additives and epoxy resins whose principal function is to impart a variable degree of flexibility to the cured systems, depending upon the concentration of the additives, rather than to serve as curing agents or modifiers.

PLASTICIZERS

In general, plasticizers have not attained wide usage in the technology. The more common monomeric plasticizers for rubbers and vinyls are generally incompatible with the epoxy resins. They tend to separate out of the system during cure and subsequent aging, and even when retained, result in only a slight reduction in rigidity.

A number of longer-chain molecules, predominantly esters [25] and esterified high-molecular-weight DGEBA resins [1] have been investigated. They provide mechanical plasticizing action, although in some cases they are to some extent coupled into the system through reactive hydroxyls. Reaction products of ethylene glycol ethers with organic acids or anhydrides are reported as effective plasticizers when used on the order of 40 to 60 phr, and these are reported not to exude from the system during aging [23]. Adducts of tetrahydrofurfural alcohol and ethylene oxide have been reported as satisfactory plasticizers for amine-cured casting resins [38], as have polymerized fatty alcohols [46] and diglycerides from glycerol and C_{9-11} alkane monocarboxylic acids [47].

Of the various plasticizers which have been investigated dibutyl phthalate is perhaps the most popular. This material, in addition to providing some plasticizing action together with compatibility, also serves as a diluent for DGEBA, and its use is discussed in Chap. 13.

MONOFUNCTIONAL FLEXIBILIZERS

Monofunctional flexibilizers are somewhat more effective than plasticizers when used with epoxy resins. These materials contain as the reactive point a functional group and consist of open chains containing preferably 16 to 24 atoms [19]. Most often, the functional group is an epoxy. Epoxidized vegetable oils, for instance, and similar compounds containing aliphatic chains over 12 carbons long, are examples. When the curing reaction is predominantly through epoxy groups, such materials serve to decrease system functionality and end-stop polymer chains, creating, in essence, long floating chains in the polymer network, irrespective of the chain length of the monofunctional flexibilizer itself. The long chain forces the molecules of the cured system somewhat further apart than would normally be the case, permitting freer movement of the cross-linked structure under imposed loads. Under these circumstances, most of the reactive diluents may also be considered flexibilizers.

A review of Chap. 13 will suggest suitable monofunctional flexibilizers.

POLYFUNCTIONAL FLEXIBILIZERS

Flexible Epoxy Resins

Most flexibilizers for epoxy resins offer one or more disadvantages in room-temperature curing formulations. In the main, they must be packaged separately from the resin, and often they must be used in stoichiometric amounts (e.g., the amidopolyamines and fatty polyamides). This either results in awkward, three-component systems or else limits the degree of flexibility to a specific value. When stoichiometric considerations are not primary, the choice of curing agent for the epoxy/flexibilizer blend is severely limited, insofar as the reaction should be made to progress in a way so as to couple the flexibilizer into the resin chain; or if a plasticizer is used, problems of compatibility may arise during or after cure, or the plasticizer may dilute the system to the point where cure at room temperature (to some extent dependent upon exotherm) is quite sluggish.

Glycidyl resins are capable of providing flexibilizing properties not conveniently obtainable with lower-cost molecules. Non-glycidyl-type resins would not, in the main, appear of equal value, insofar as such resins are generally unreactive with epoxy curing agents at room temperature.

A number of flexibilizing glycidyl ether resins may potentially be synthesized via the epichlorohydrin route, and other specialized molecules, such as epoxidized thiols of the general structure [3]

Epoxidized thiol

can be similarly synthesized to provide room-temperature stability. In addition, higher-molecular-weight species of DGEBA may, under conditions of reaction not leading to reaction through groups other than the epoxy (i.e., hydroxyls), be employed to impart some degree of flexibility to the cured system.

However, the commercial flexibilizing resins are predominantly glycidyl derivatives of glycols, dimerized acids, or of reaction products of dicarboxylic acids with epoxy resins, with the epoxidation being accomplished by reaction with epichlorohydrin through either the carboxyl or hydroxyl groups. Some, but not all, of these materials provide good reactivity with aliphatic primary amines at room temperature.

Epoxy Resins from Polyols. The commercial flexible epoxy resins from polyols are typified by the following epoxidized glycol:

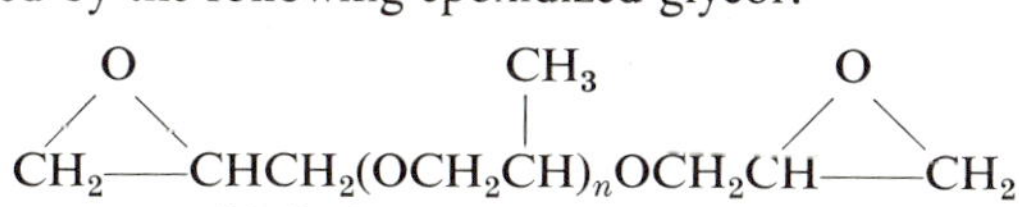

Idealized epoxidized polyether glycol

Such commercial epoxidized glycols vary in molecular weight from about $n = 2$ to $n = 7$. The lower-molecular-weight species do not impart substantial flexibility, but do have some diluting effect on the system. When n is above 7, the epoxidized glycols provide, generally, increased viscosity with little flexibilizing action to justify their use.

Table 16-1. Effect of Concentration of Epoxidized Glycol on Properties of DGEBA Cured with Listed Curing Agent [4-2]

Property	DGEBA/flexible resin			
	100/0	90/10	70/30	50/50
DETA-cured:				
Gel time at 52°C, minutes	9.1	12.5	20.3	34.0
Exotherm, °C	240	213	165	147
DT, °C	104	90	54	
Tensile strength, psi	7,500	4,700	6,200	1,400
Modulus, psi × 10^{-5}	2.1	1.9	2.5	0.20
Flexural strength, psi	12,200	16,000	12,400	1,700 (yield)
Modulus, psi × 10^{-5}	4.5	4.0	3.4	0.29
Compressive strength, psi	26,800	32,000	32,900	16,700
Modulus, psi × 10^{-5}	3.7	3.5	3.0	2.8
Impact, Izod	0.43	0.48	0.76	2.0
Hardness, Shore D	90/87	89/87	88/84	75/64
Dielectric constant:				
At 60 cps	4.19	......	4.14	
At 10^3 cps	4.09	......	3.98	
At 10^5 cps	3.67	......	3.60	
At 10^7 cps	3.30	......	3.13	
Dissipation factor:				
At 60 cps	0.0162	......	0.0208	
At 10^3 cps	0.0239	......	0.0269	
At 10^5 cps	0.0406	......	0.0396	
At 10^7 cps	0.0310	......	0.0319	
Volume resistivity, ohm-cm × 10^{-12}	3.1	1.0	1.1	1.1
MDA-cured:				
DT, °C	150	131	80	68
Tensile strength, psi	5,100	6,700	8,200	6,800
Modulus, psi × 10^{-5}	1.4	1.5	1.3	1.1
Flexural strength, psi	17,700	16,500	17,500	12,200
Modulus, psi × 10^{-5}	4.1	4.0	4.4	3.7
Compressive strength, psi	33,600	33,600	33,700	30,300
Modulus, psi × 10^{-5}	3.2	3.1	3.4	3.3
Impact, Izod	0.076	0.17	0.66	0.42
Dielectric constant:				
At 60 cps	4.09	......	3.86	
At 10^3 cps	4.02	......	3.78	
At 10^5 cps	3.71	......	3.53	
At 10^7 cps	3.42	......	3.24	
Dissipation factor:				
At 60 cps	0.0078	......	0.0093	
At 10^3 cps	0.0153	......	0.0125	
At 10^5 cps	0.0334	......	0.0309	
At 10^7 cps	0.0300	......	0.0315	
Volume resistivity, ohm-cm × 10^{-13}	6.8	6.7	8.7	7.8
PMDA-cured:				
DT, °C	......	170	133	56
Flexural strength, psi	6,700	3,600	8,100	5,800
Modulus, psi × 10^{-5}	4.8	4.4	3.9	0.22
Compressive strength, psi	24,300	24,900	26,600	30,300
Modulus, psi × 10^{-5}	3.7	3.0	2.9	2.3
Impact, Izod	0.18	0.10	0.23	0.34
Hardness, Shore D	90/91	90/88	89/87	86/83

Table 16-2. Effect of Concentration of Epoxidized Glycol on Properties of DGEBA ($n = 0$) Cured with AEP [4-2]

(Blended at 40°C; cured 4 hr at 60°C)

Property	DGEBA/flexible resin				
	100/0	85/15	70/30	50/50	35/65
Tensile strength, psi	7,000	8,200	6,600	2,100	300
Elongation, %	6.6	8.7	8.7	115	66
Color, Gardner	2	1	<1	<1	<1
Condition	Hard	Hard	Flexible	Flexible	Flexible
Weight loss after 4 hr at 205°C	0.48	0.61	0.82	1.23	1.72
Weight loss after 8 hr at 205°C	0.67	0.85	1.20	1.94	3.00

Acetone extraction of 2-g samples after 2 hr at reflux showed no loss below the 50/50 blend. This blend lost 0.83 weight %; at 35/65 loss was 2.0%.

The epoxidized glycols are used in blends with aromatic-based glycidyl ether resins, either from bisphenol A or from the novolacs, and their effect, as is the case with other flexibilizers, is to lower the DT of the systems, to reduce the chemical resistance particularly solvent resistance, to improve impact resistance, to lower surface and volume resistivity, and to increase the coefficient of thermal expansion. The extent of the effect will depend in part upon the aromatic glycidyl ether employed, the curing agent, the concentration of epoxidized glycol, and the service temperature.

The use of epoxidized polyol resins as flexibilizers may be illustrated with an epoxidized glycol having an epoxy equivalent of about 330 [30]. Properties for blends of this compound with DGEBA are shown in Table 16-1; with the resin when

Table 16-3. Effect of Concentration of Epoxidized Glycol on Properties of Epoxidized Novolac (f 3.3) Cured with MDA [4-2]

Property	Novolac/Flexible resin		
	100/0	90/10	70/30
Tensile strength, psi	7,200	6,800	5,400
Modulus, psi $\times 10^{-5}$	1.6	1.3	1.4
Flexural strength, psi	11,000	13,400	16,900
Modulus, psi $\times 10^{-5}$	3.5	3.9	3.2
Compressive strength, psi	33,000	31,200	31,800
Modulus, psi $\times 10^{-5}$	1.6	2.4	1.7
Impact, Izod	0.14	0.23	0.42
Dielectric constant:			
At 60 cps	4.04	......	3.82
At 10^3 cps	3.99	......	3.75
At 10^5 cps	3.71	......	3.49
At 10^7 cps	3.18	......	3.15
Dissipation factor:			
At 60 cps	0.0044	......	0.0085
At 10^3 cps	0.014	......	0.016
At 10^5 cps	0.030	......	0.031
At 10^7 cps	0.023	......	0.029
Volume resistivity, ohm-cm $\times 10^{-13}$	7.0	......	6.3

Table 16-4. Properties of DGEBA Cured with AEP as a function of Concentration of Epoxidized Polyol [4-1]

Property	DGEBA/flexible resin					
	100/0	80/20	60/40	40/60	20/80	0/100
Initial viscosity at 25°C, centipoises	1,530	1,740	2,050	2,320	2,470	2,570
Pot life, 200 g at 25°C, minutes	16	19	22	30	52	120
Gel time, 10 g at 93°C, minutes	......	2	6	11	13	40
Tensile strength, psi:						
Cured 2 hr at 93°C	12,050	9,955	6,285	815	215	35
Cured 8 hr at 93°C	12,560	10,470	8,435	3,975	1,445	110
Elongation, %:						
Cured 2 hr at 93°C	7.1	9.8	13	76	117	160
Cured 8 hr at 93°C	6.6	6.4	11	23	62	68
Volume resistivity, ohm-cm:						
At 25°C	4.4×10^{16}	1.8×10^{16}	9.4×10^{14}	8.2×10^{12}	1.5×10^{11}	2.6×10^{9}
At 100°C	1.6×10^{12}	4.1×10^{10}	2.7×10^{9}	2.9×10^{8}	6.6×10^{7}	3.4×10^{7}
At 121°C	7.6×10^{10}	3.2×10^{9}	7.7×10^{8}	2.5×10^{8}	7.4×10^{7}	2.4×10^{7}
At 150°C	9.8×10^{9}	2.0×10^{9}	7.9×10^{7}	6.6×10^{7}	3.4×10^{7}	2.9×10^{7}
Tensile shear strength,* psi:						
At 0°C	3,080	3,650	4,190	5,350	3,980	1,930
At 25°C	3,690	4,810	6,080	3,560	2,280	710
At 75°C	4,970	3,370	1,770	1,360	970	520
Hardness, Shore D or A:						
At −60°C	90 D	90 D	90 D	87 D	86 D	83 D
At 0°C	89 D	87 D	87 D	86 D	80 D	60 D
At 25°C	89 D	85 D	83 D	79 D	60 D	47 A
At 75°C	83 D	76 D	75 D	46 D	70 A	38 A
Water absorption after 2-hr boil, %	+0.7	+2.3	+2.5	+4.5	+5.8	+6.2

NOTE: Before testing, all specimens were cured 2 hr at 93°C, except for formulations 0/100 and 20/80, which were first gelled at 25°C.
* Aluminum-to-aluminum bonds—3 parts by weight colloidal silica added to control glue-line thickness.

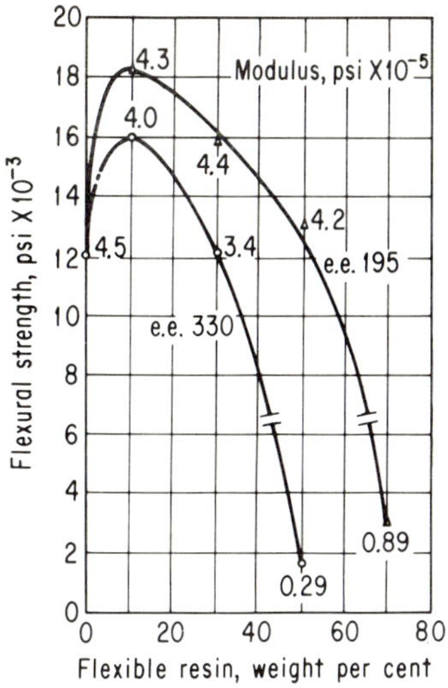

FIG. 16-1. Flexural strength vs. epoxidized polyol concentration for TETA-cured DGEBA [30].

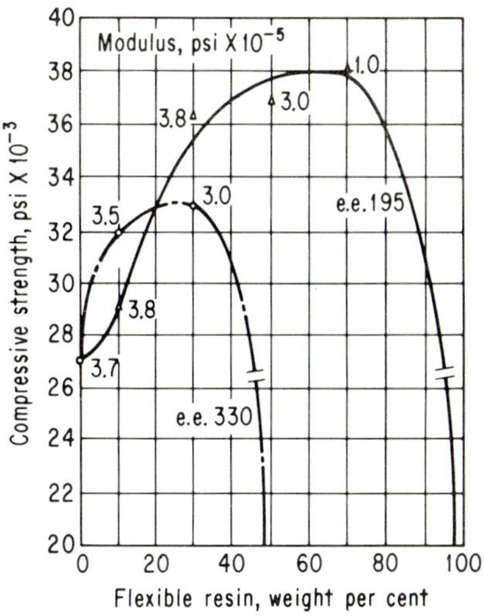

FIG. 16-2. Compressive strength vs. epoxidized polyol concentration for TETA-cured DGEBA [30].

$n = 0$ in Table 16-2; and with an epoxidized novolac resin in Table 16-3. Properties of a system using a slightly higher molecular-weight epoxidized polyol and cured with AEP are shown in Table 16-4. The effect of increasing the molecular weight of the epoxidized glycol on strength properties is indicated in Figs. 16-1 to 16-4. As indicated in Fig. 16-5, use of the epoxidized polyols produces materials having the characteristics of viscoelastic solids rather than true elastomers.

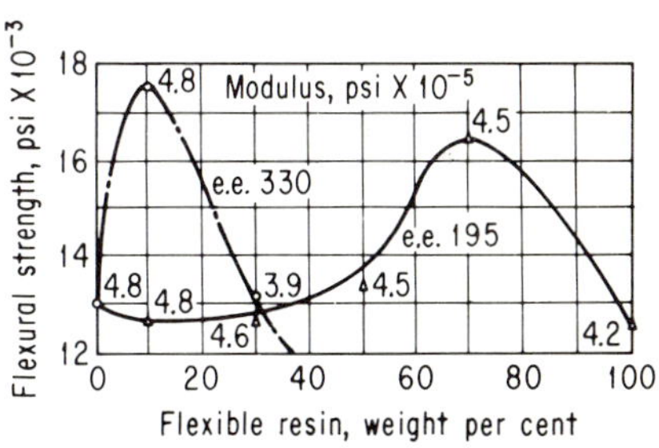

FIG. 16-3. Flexural strength vs. epoxidized polyol concentration for DGEBA cured with NMA and BDMA accelerator [30].

Epoxy Resins from Aliphatic Acids. Epoxy resins may be developed by the epichlorohydrin route from aliphatic acids. Typical is a commercial dimer acid epoxidized to produce an alicyclic diglycidyl ester resin.

Such a material has an epoxy equivalent of 400 to 420 and a viscosity of about 500 cps at room temperature. Tensile properties of the resin in blend with DGEBA and cured with various amines are shown in Table 16-5. The effect of concentration of the two resins is shown in Table 16-6. Tablc 16-7 indicates that a monofunctional amine may be used to reduce functionality and improve elongation. Table 16-8 indicates that with primary amine curing agents the reactivity at room temperature is relatively slow, even in substantial masses. Reactivity may be heightened by the use of phenol accelerators, as shown in Table 16-9. Table 16-10 presents tensile properties when the resin is cured with various anhydrides; and Table 16-11 shows the effect of blending the diglycidyl ester resin with DGEBA when using NMA as a curing agent. The use of a carboxyl-terminated rubber

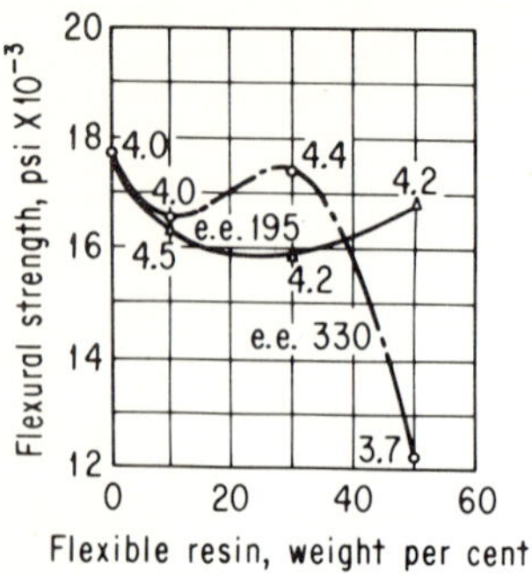

FIG. 16-4. Flexural strength vs. epoxidized polyol concentration for MDA-cured DGEBA [30].

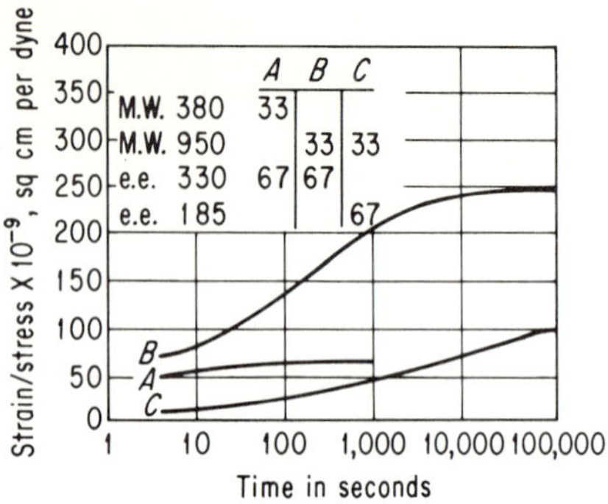

FIG. 16-5. Stress-strain curves vs. epoxidized polyol types for cured DGEBA resins [30]. Temperature 3.5°C.

Table 16-5. Tensile Strength of DGEBA and Glycidyl Ester Resin Blend Cured 4 hr at 125°C with Various Amine Curing Agents [4-6]

Curing agent, phr	Strength, psi	Elongation, %
DETA, 7.0	3,150	28
AEP, 14.5	3,600	48
MXDA, 11.0	1,820	63
N-(2-Hydroxypropyl)ethylenediamine, 13.0	1,670	103
cis-2,5-Dimethylpiperazine, 18.5	1,120	200

Table 16-6. Tensile Properties of Various Blends of DGEBA and Diglycidyl Ester Resin Cured with Typical Epoxy-resin Curing Agents [4-6]

Flexible resin	DGEBA	Phr of curing agent	Tensile properties	
			Strength, psi	Elongation, %
*AEP-cured:**				
100	0	10.8	210	61
80	20	13.2	1,540	99
60	40	15.5	4,140	48
40	60	18.0	8,840	9
*DETA-cured:**				
100	0	5.2	150	39
80	20	6.3	1,180	53
60	40	7.4	4,240	6
40	60	8.4	5,680	2
MDA/MPDA/PGE-cured:†				
100	0	10.0	430	86
80	20	12.3	2,160	47
60	40	14.5	6,700	13
40	60	16.3	9,540	5

* Cure schedule: 4 hr at 125°C.
† Cure schedule: 2 hr at 80°C + 3 hr at 150°C.

Table 16-7. Effect of Monofunctional Amines on the Tensile Properties of DGEBA and Diglycidyl Ester (67/33 blend) Cured with Primary Amines [4-6]

(Cure schedule; 4 hr at 125°C)

Curing agent, phr	Strength, psi	Elongation, %
DETA, 7.0.........	3,150	28
DETA, 4.5 Butylamine, 4.0...	2,150	70
DETA, 3.4 Butylamine, 6.0...	1,240	109
MXDA, 11.0.......	1,820	63
MXDA, 7.5 Butylamine, 4.0...	2,050	114
MXDA, 5.6 Butylamine, 6.0...	1,230	135

Table 16-8. Gel Times for the DGEBA and Diglycidyl Ester Blends [4-6]

Flexible resin	DGEBA	Curing agent, phr	Gel time, minutes
100	0	AEP, 10.5	185
100	0	DETA, 5.2	185
75	25	AEP, 13.8	60
75	25	DETA, 6.6	80
0	100	AEP, 22	15–20
0	100	DETA, 12	15–20

Table 16-9. Effect of Phenol on Room-temperature Gel Time and Tensile Properties of a Blend of DGEBA and Diglycidyl Ester Resin (25/75 Blend) [4-6]

Phenol, phr	AEP, phr	DETA, phr	Gel time for 1-qt mass at 23°C, minutes	7-day cure at 25°C	
				Strength, psi	Elongation, %
0	13.8	...	60	1,900	100
5	13.8	...	15	1,500	90
0	....	6.6	80	1,600	35
5	....	6.6	20–25	750	35

Table 16-10. Tensile Properties of Diglycidyl Ester Resin Cured with Various Anhydrides [4-6]

Curing agent, phr	DMP-30, phr	Cure schedule, hr/C°	Tensile properties	
			Strength, psi	Elongation, %
NMA, 47.............	1.8	3/100	1,250	70
NMA, 23.5...........	2.1	3/100	416	80
DDSA, 69............	4.6	2/110	1,280	120
Adipic anhydride, 33...	1.0	16/110	138	50

Table 16-11. Tensile Properties of Various Blends of DGEBA and Diglycidyl Ester Resin Cured with NMA [4-6]

Flexible resin	DGEBA resin	NMA, phr	Strength, psi	Elongation, %
100	0	44.8	800	70
80	20	54.6	5,090	17
60	40	64.3	9,500	6
40	60	72.5	8,890	3

Table 16-12. Tensile Properties of Diglycidyl Ester Resin Cured 16 hr at 110°C with Carboxyl-containing Synthetic Rubber [4-6]

Rubber, phr	DMP-30, phr	Strength, psi	Elongation, %
32.5	1.0	28	280
32.5	2.0	41	210
32.5	3.0	27	325

Table 16-13. Tensile Properties of DGEBA and Diglycidyl Ester Resin Blends Cured with Boron Trifluoride Monoethylamine [4-6]

Flexible resin	DGEBA	BF_3-MEA phr	Cure schedule, hr/°C	Strength, psi	Elongation, %
100	0	3	2/120 + 1/170 + 3/200	130	35
75	25	3	2/120 + 2/200	600	37
75	25	4	2/120 + 2/200	630	43
75	25	5	2/120 + 2/200	290	57

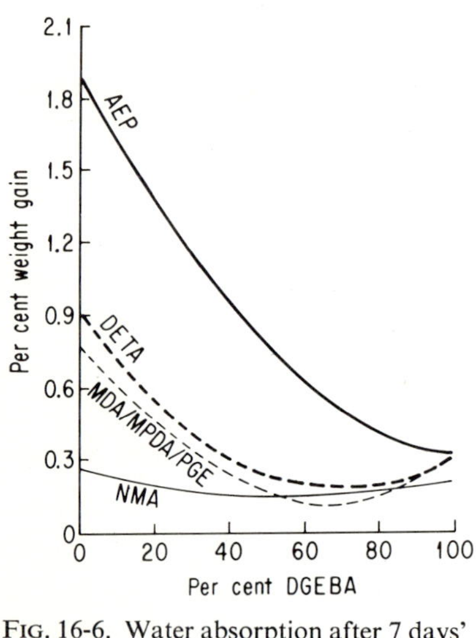

FIG. 16-6. Water absorption after 7 days' immersion vs. diglycidyl ester resin concentration in DGEBA [4-6].

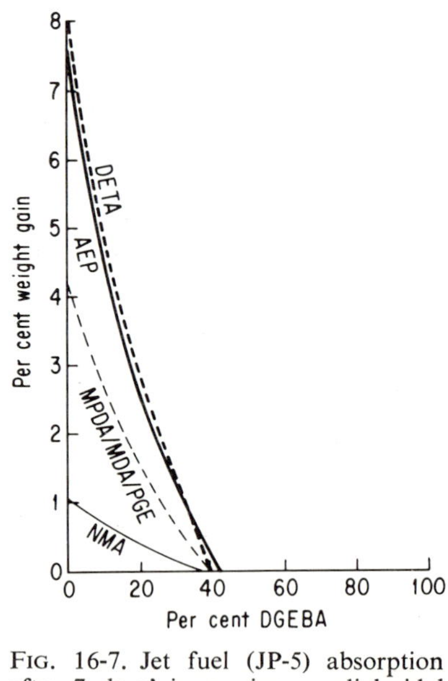

FIG. 16-7. Jet fuel (JP-5) absorption after 7 days' immersion vs. diglycidyl ester resin concentration in DGEBA [4-6].

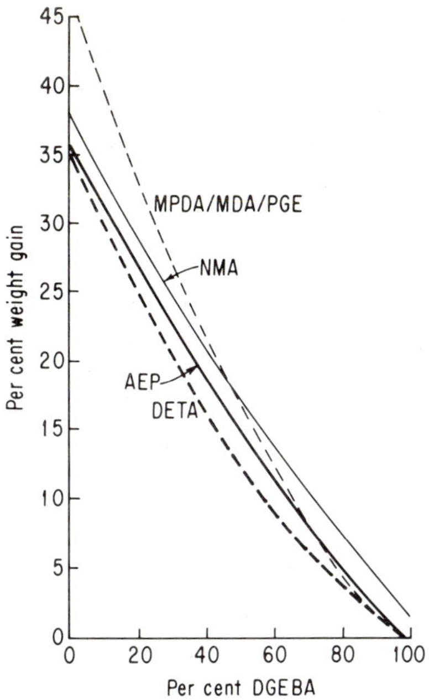

FIG. 16-8. Acetone absorption after 7 days' immersion vs. diglycidyl ester resin concentration in DGEBA [4-6].

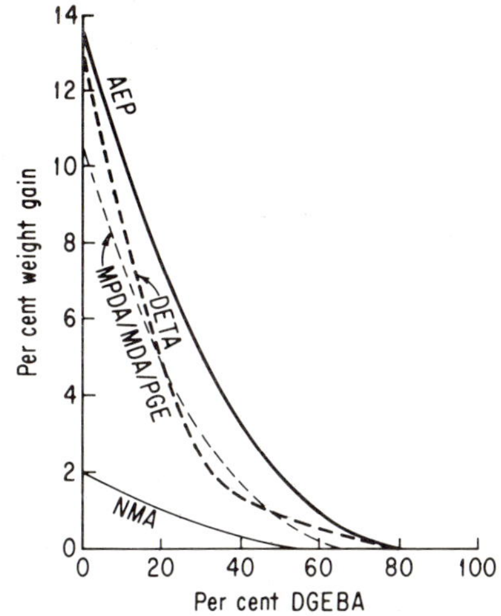

FIG. 16-9. IPA absorption after 7 days' immersion vs. diglycidyl ester resin concentration in DGEBA [4-6].

and a boron trifluoride complex as curing agents is shown in Tables 16-12 and 16-13. Figures 16-6 through 16-9 indicate the effect of the use of the flexibilizer on chemical resistance for various curing agents, and Fig. 16-10 illustrates the effect of varying the amine functionality on a blend of epoxidized aliphatic esters.

Hydroxyl-terminated Flexibilizers

As has been previously indicated, hydroxyl-terminated compounds may be used as coreactants for epoxy resins. Short-chain, polyfunctional polyols are used in quite a number of anhydride-cured formulations, the hydroxyl groups on the polyols serving to open the anhydride rings to provide the reactive carboxyl groups. Use of such short-chain materials is particularly advantageous when employing anhydrides to cure those epoxidized olefin resins which do not, themselves, contain hydroxyl groups along the resin chain. In addition, the short-chain polyols may be used to enhance reactivity of systems catalyzed to favor the epoxy-alcohol reaction and may be used to improve room-temperature stability of some of the boron trifluoride complexes in the epoxy.

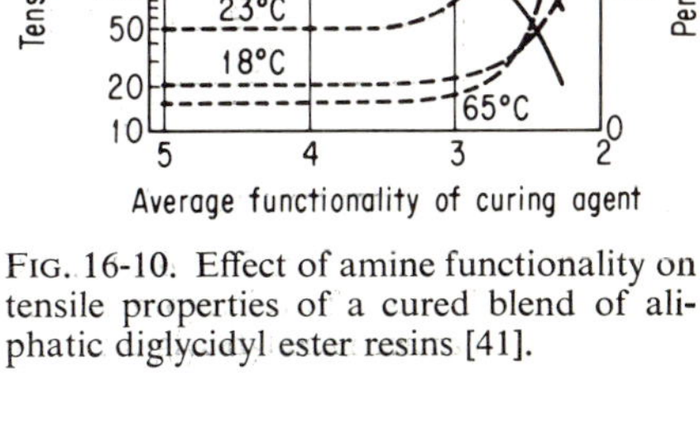

FIG. 16-10. Effect of amine functionality on tensile properties of a cured blend of aliphatic diglycidyl ester resins [41].

Table 16-14. Effect of Polyol on Properties of 2,3-Epoxy-6-methylcyclohexyl, 2,3-Epoxy-6-methylcyclohexane Carboxylate Cured with 4 Per cent BF_3-Piperidine [33]

Polyol, %	Cure, at 175°C, hours	Tensile properties		Hardness, Durometer D	Thermal shock*
		Strength, psi	Elongation, %		
Diol (mol. wt. 400), 30.....	1	2,000	140	72	
Diol (mol. wt. 425), 35.....	3	2,800	170	72	
	1	1,000	185	43	Passed 5
Triol (mol. wt. 700), 40	3	2,000	75	75	
	1	1,600	190	64	Passed
Triol (mol. wt. 1,000 castor oil), 35..................	3	2,300	150	64	Passed
	1	1,400	100	45	Passed
Triol (mol. wt. 1,000), 40...	1	1,000	130	48	Passed
	3	1,300	80	56	
Triol (mol. wt. 3,000),† 30 ..	1	130	70	25	
	3	350	40	30	

* Ten cycles from 130 to −55°C, Olyphant washer.
† Catalyzed with 2% BF_3-piperidine.

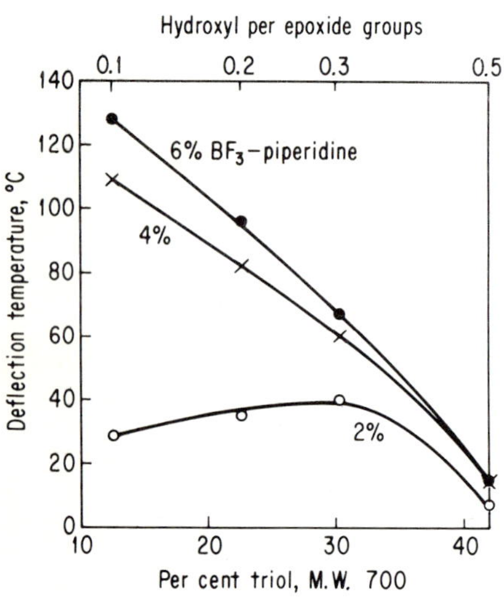

FIG. 16-11. DT vs. polyol and catalyst concentration for 3,4-epoxy-6-methylcyclohexylmethyl-3,4-epoxy-6-methylcyclohexane carboxylate cured with BF_3-piperidine [33]. Cure cycle: 1.5 hours at 80°C plus 1.25 hours at 120°C plus 6 hours at 160°C.

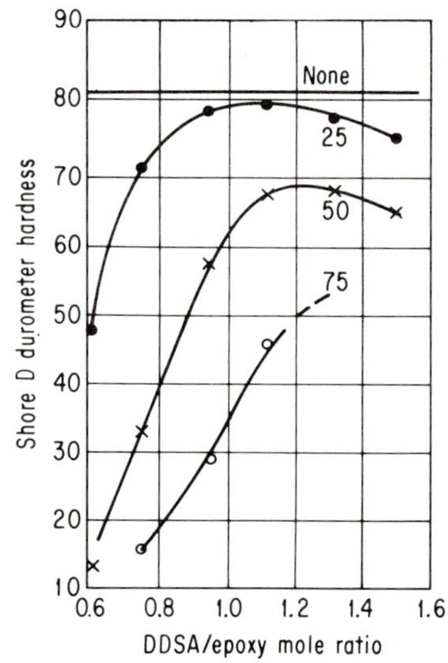

FIG. 16-12. A/E ratio vs. hardness for various concentrations of polyethylene glycol (mol. wt. 1,000) in DDSA-cured DGEBA [13].

Longer-chain polyols suggested as flexibilizers may be divided into two types: polyols without ester linkages, and the hydroxyl-terminated polyesters. Of somewhat lesser utility are the monohydric alcohols, although their use has been reported for imparting flexibility to anhydride cures [21].

Polyols without Ester Linkages. With amine curing agents, the polyols give cured products weak or cheesy rather than flexible, the polyol acting merely as a plasticizer. With Lewis acid curing agents, such as BF_3MEA, performance is improved, since reaction between epoxy groups and hydroxyls is obtained. The use of various polyols with 2,3-epoxy-6-methylcyclohexyl 2,3-epoxy-6-methylcyclohexane carboxylate catalyzed with a Lewis acid complex provides properties as indicated in Table 16-14, and the effect of varying concentrations of the catalyst and one triol on the DT is shown in Fig. 16-11.

The polyol may be used at less than, equal to, or greater than hydroxyl/epoxy ratio of 1/1 in the glycidyl ether resins [5], in epoxidized polybutadiene [20], and in other epoxidized olefin systems.

When the polyols are present with anhydride curing agents, it is probable that optimum results, particularly in terms of chemical resistance, are obtained by varying the curing agent with the varying amounts of the polyol. With the anhydrides, for example, at low concentrations of the polyol, complete reaction will generally be obtained at A/E ratios lower than 1/1; whereas, at high concentrations of the polyol, the system may properly be considered as being cured by an organic acid in a hydroxyl-rich medium, and the optimum concentration of anhydride may exceed an A/E of 1/1. With polyols, anhydride-cured systems employing an A/E as high as 2/1 have been reported [22]. The effect of A/E ratio on hardness of castings modified with a polyol is indicated in Fig. 16-12.

The polyol flexibilizer may be prereacted with the epoxy resin at temperatures of 150 to 200°C for short periods to produce, essentially, a molecule terminated on

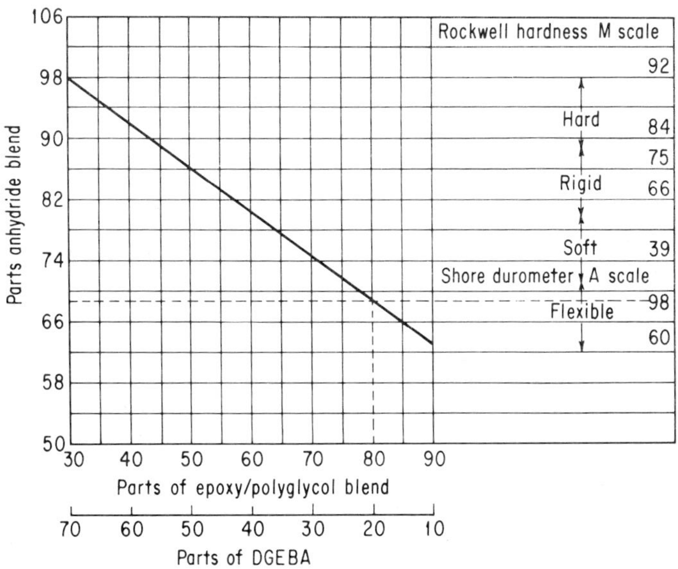

Fig. 16-13. Hardness vs. epoxy/polyglycol blend concentration for DGEBA cured with HHPA/CA (70/30) [4-4].

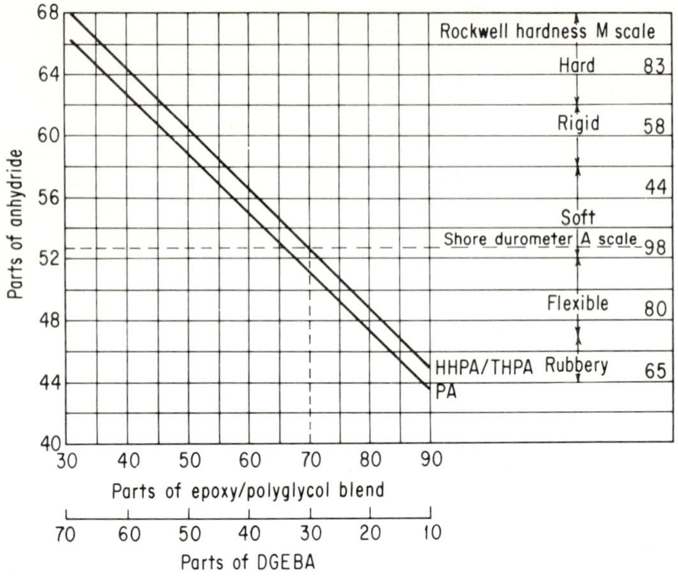

FIG. 16-14. Hardness vs. epoxy/polyglycol blend concentration for DGEBA cured with PA and with HHPA/THPA (90/10) [4-4].

one end with an epoxy group, and by an hydroxyl group at the other end, containing secondary hydroxyls along the chain [24]. Data for a prereacted system of this type are presented in Figs. 16-13 and 16-14 and Table 16-15.

Hydroxyl-terminated Polyesters. Hydroxyl-terminated polyesters [14] may be considered as a special case of the polyols, as discussed above.

A typical hydroxyl-terminated polyester may be synthesized by using 3 moles of adipic acid and 4 moles of glycerol. This reaction product will have 3.4 equivalents of end hydroxyls per 100 grams. By substituting an amine, such as ethyldiethanolamine in the foregoing reaction, a composition containing catalytic tertiary amine groups will be produced.

Table 16-15. Properties of DGEBA Modified with Epoxy/Polyglycol Blend and Cured 3 hr at 120°C with 70/30 Blend of HHPA/CA Catalyzed with 0.25 phr DMP-10 [11]

Property	DGEBA/epoxy-polyglycol/HHPA-CA		
	100/0/116	70/30/98	30/70/75
Impact strength, Izod	0.32	0.39	0.51
Water absorption, %	0.10	0.13	0.58
Weight loss after 24 hr at 150°C, %	0.20	1.04	2.24
Hardness, Shore D	90	87	80
Dissipation factor at 10^6 cps and 23°C	0.015	0.016	0.027
Dielectric constant at 10^6 cps and 23°C	3.83	3.95	4.31
Loss factor at 10^6 cps and 23°C	0.057	0.064	0.115
Surface resistivity after 1 min at 500 volts	2.48×10^{15}	2.07×10^{15}	5.42×10^{14}
Volume resistivity after 1 min at 500 volts:			
At 23°C	Infinite	Infinite	6.9×10^{14}
At 65°C	1.5×10^{16}	4.4×10^{15}	1.2×10^{12}
At 93°C	3.6×10^{15}	2.6×10^{15}	1.1×10^{10}

Table 16-16. Effect of Concentration of Polyester Modifier on Properties of the DGEBA Cured with DETA [10-2]

Property	Polyester/DGEBA/DETA		
	1/1/0.12	1/2/0.24	0/1/0.12
Working time, 200 g, minutes	20–25	15–20	10–20
Set time, minutes	20–25	20–25	15–20
Exotherm, °C	120–135	150–165	150–180
Room temperature cure:			
Tensile strength, psi	1,500–2,200	6,000–8,000	8,000–11,000
Elongation, %	30–60	0–12	0–3
Impact, Izod	50–70	30–50	1–5
Hardness, Shore D	45–55	70–80	85–95
Volume resistivity, ohm-cm	1×10^{11}–1×10^{12}	1×10^{12}–1×10^{13}	1×10^{14}–1×10^{15}
Surface resistivity, ohms	1×10^{12}–5×10^{13}	1×10^{12}–1×10^{13}	1×10^{14}–1×10^{15}
Dielectric constant at 10^3 cps	6.0–6.5	4.5–5.0	3.0–3.5
Dissipation factor at 10^3 cps	0.030–0.040	0.015–0.030	0.005–0.010
Postcure 4 hr at 100°C:			
Tensile strength, psi	2,000	6,500–7,500	
Elongation, %	50	0–10	
Impact, Izod	20–40	10–20	
Hardness, Shore D	55	70–75	
Volume resistivity, ohm-cm	1×10^{11}–1×10^{12}	1×10^{12}–1×10^{13}	
Surface resistivity, ohms	5×10^{11}–1×10^{13}	1×10^{12}–1×10^{13}	
Dielectric constant at 10^3 cps	5.5–5.8	4.5–4.8	
Dissipation factor at 10^3 cps	0.040	0.020–0.030	
Acetone extraction:			
Cured 4 hr at 70°C, %	14.7	6.0	
Postcured 7 days at 120°C	10.3	4.7	
Volume shrinkage on cure (by specific-gravity method) cured 4 hr at 70°C, %	3.9	3.4	
Water absorption after 24 hr at 25°C, %:			
Cured 4 hr at 70°C	2.7	1.3	
Postcured 7 days at 120°C	0.9	0.4	

A typical hydroxyl-terminated long-chain polyester will provide properties as indicated in Table 16-16 for a primary amine cure and as indicated in Table 16-17 for an anhydride cure. Table 16-18 presents a comparison of electrical properties for various curing agents when the polyester modifier is present at 50 phr.

Table 16-17. Effect of Concentration of Polyester Modifier on Properties of DGEBA Cured 22 hr at 150°C with 25 phr PA [10-2]

Property	Polyester/DGEBA		
	1/1	1/1.5	1/2
Tensile strength, psi	1,625	3,400	5,525
Elongation, %	70	20	0
Impact strength, Izod	12.0	7.0	3.0
Hardness, Shore D	63	75	85
Volume resistivity, ohm-cm	3.7×10^{13}	4.3×10^{13}	3.4×10^{12}
Surface resistivity, ohms	4.6×10^{13}	3.2×10^{13}	2×10^{14}
Dielectric constant at 10^3 cps	5.5	4.5	3.69
Dissipation factor at 10^3 cps	0.053	0.039	0.014

Table 16-18. Comparison of Electrical Properties of DGEBA Containing 50 phr Polyester Flexibilizer and Cured with Listed Curing Agents [10-2]

Property	DMP-30 cured 2 hr at 100°C	TETA cured 2 hr at 100°C	DETA cured 5 days at 23°C	PA cured 22 hr at 150°C
Volume resistivity, ohm-cm:				
At 23°C	7.5×10^{13}	6.4×10^{12}	2.0×10^{11}	5.5×10^{13}
At 72°C	4.4×10^{9}	2.1×10^{9}	4.0×10^{8}	1.1×10^{12}
At 100°C	8×10^{7}	1.7×10^{8}	....	2.6×10^{10}
Surface resistivity, ohms:				
At 23°C	1.4×10^{13}	1.2×10^{13}	3.0×10^{11}	7.6×10^{13}
At 72°C	9.1×10^{11}	1.9×10^{11}	5.0×10^{9}	6.4×10^{12}
At 100°C	1.3×10^{10}	1.6×10^{10}	....	1.0×10^{10}
Dielectric constant at 1 kc:				
At 23°C	3.7	4.4	5.6	3.4
At 72°C	6.0	6.4	9.9	5.7
At 100°C	17	9.0	....	5.7
Dissipation factor at 1 kc:				
At 23°C	0.017	0.022	0.051	0.023
At 72°C	0.2310	0.195	0.60	0.049
At 100°C	0.905	0.89	....	0.023

Natural and Synthetic Rubbers

Natural and most synthetic rubbers, such as styrene-butadiene and acrylonitrile-butadiene polymers, contain no groups normally reactive with the epoxy groups directly but they do contain reactive unsaturation. The unsaturation may be epoxidized with peracids [48], or the unsaturated material may be used in blend with an epoxy resin. During the process of vulcanization by sulfur, thiol groups may be developed which are capable of reacting with at least some types of epoxy groups, thus introducing the epoxies into the cured network [6]. Typical is the coreaction of gum nitrile stocks with DGEBA (mol. wt. 900), giving properties as indicated in Tables 16-19 and 16-20. With SBR polymers compatibility toward DGEBA resins is enhanced through the addition of an alkylaniline [50].

It is also possible to prereact the epoxy resin to obtain sulfur-containing reactive sites. For example, glycidyl ether epoxy resins have been reacted with polythiocyanates (the reaction of ϵ-caprolactam, ammonium thiocyanate, and zinc chloride) and these adducts employed as reactants with synthetic rubbers [27]; and DGEBA has been reacted with H_2S and this used as a coreactant giving properties as indicated in Table 16-21.

If an epoxy resin containing unsaturation is employed, then reaction may be brought about by sulfur vulcanization or by peroxide catalysts. Typical formulations and properties are shown in Tables 16-22 and 16-23.

Natural rubber (latex) has been used in mixture with epoxy resin, curing agent, cement, and water to make an expansion-joint mortar [40].

When chlorinated rubbers, such as chloroprene, are mixed with epoxy resins, the curing reaction for the rubber, involving the chlorine, does not appear to promote reaction of the epoxy groups. However, use of 8 to 9 parts DGEBA as a curing agent for chloroprene is reported to provide satisfactory static and improved dynamic properties over conventional curing agents, but with some loss of thermal stability [39]. When chloroprene/epoxy blends are co-cured with a vulcanizing agent reactive with the epoxy, such as polyphenols or aromatic amines, some increase in modulus and chemical resistance is obtained [6]. Within the molecular-weight range of 550 to 3,300, lower-molecular-weight DGEBA species give best properties. With increasing molecular weight a small decrease in strength and a

Table 16-19. Blends of Gum Nitrile Rubber Stocks and DGEBA (mol. wt. 900) [6]

Reactants	Formulations											
Gum nitrile rubber stock	100	100	100	100	100	100	100	100	100	100	100	100
Zinc oxide	5.0	5.0	5.0	5.0	5.0	5.0	5.0	5.0	5.0	5.0	5.0	5.0
Stearic acid	1.0	1.0	1.0	1.0	1.0	1.0	1.0	1.0	1.0			
Sulfur	1.5	....	1.5	1.5	1.5	1.5	1.5	2.5	2.5	2.0	2.0	
Benzothiazole disulfide	1.5	1.5	....	....	....	....	....	....	1.5			
Diphenylguanidine	...	....	....	....	....	....	....	....	....	1.5	1.5	1.5
DMP-30	...	1.5	....	1.5	....	....	....	1.5	1.5			
BF_3-triethanolamine complex	...	....	....	....	1.0							
DGEBA (mol. wt. 900)		25.0	25.0	25.0	25.0	....	25.0	25.0	25.0	....	25.0	25.0
Zinc mercaptobenzothiazole	...	....	....	....	....	....	....	....	....	5.0	5.0	5.0
Hexamethylenediamine	...	....	....	....	....	3.0	3.0					
Optimum cure time, at 145°C, minutes	45	45	30	45	120	60	60	60	45	60	45	
Property	**Value**											
Tensile strength, psi	550	420	1,300	1,405	1,475	600	1,785	1,900	2,000	650	1,400	
300% modulus, psi	350	.....	410	790	530	370	.....	850	1,100	420	1,375	
100% modulus, psi	...	.....	.....	.....	.....	...	1,600					
Elongation, %	500	1,400	100	480	355	400	195	475	525	625	310	820
Permanent set, %	5	83	22	10	14	4	16	8	16	7	10	131
Hardness, Durometer A	20	26	35	49	38	26	45	52	60	27	52	18
Tear strength, psi	60	.....	100	180	.....	.....	.....	.....	210			
Small volume, %:												
Hexane at 25°C after 1 day	29.0	24.0	17.0	10.0	6.5	20.0	16.0	14.0	11.0	21.0	10.0	56.0
Benzene at 25°C after 1 day	112.0	90.0	65.0	37.0	26.0	105.0	27.0	48.0	33.0	97.0	30.0	195.0

Table 16-20. Effect of Epoxy Structure on Blends of Glycidyl Ether Resins and Gum Nitrile Rubber Stock [6]

(Cured 45 min at 145°C)

Reactants	Formulations							
Gum nitrile rubber stock	100	100	100	100	100	100	100	100
Zinc oxide	5.0	5.0	5.0	5.0	5.0	5.0	5.0	5.0
Stearic acid	1.0	1.0	1.0	1.0	1.0	1.0	1.0	1.0
Sulfur	1.5	1.5	1.5	1.5	1.5	2.5	1.5	3.5
Benzothiazole disulfide	1.5	1.5	1.5	1.5	1.5	1.5	1.5	1.5
Glycerol-base glycidyl ether (mol. wt. 300)	25							
DGEBA	...	25						
DGEBA (mol. wt. 900)	...	...	25	...	50	50	100	100
DGEBA (mol. wt. 1,400)	...	...	...	25				
DMP-30	1.5	1.5	1.5	1.5	1.5	1.5	1.5	1.5
Property	**Value**							
Tensile strength, psi	750	985	1,500	1,625	2,100	2,300	2,400	2,600
300% modulus, psi	250	300	790	850	1,050	1,460		
100% modulus, psi	.....	...	.....	.....	.....	.....	1,400	1,600
Elongation, %	1,100	895	480	425	650	315	175	110
Hardness, Durometer A	.....	...	45	49	53	60		72

small decrease in modulus at 300 per cent elongation is observed. The epoxy resin improved fatigue resistance and adhesion [45]. In higher concentrations, the DGEBA resins are incompatible, but the epoxidized polybutadienes are reported to be compatible in all concentrations [44]. Small amounts of chlorinated rubbers have been reported as improving properties of epoxy coal-tar coatings [34] and

Table 16-21. Properties of Gum Nitrile Rubber Stock Cured with H_2S Adduct of DGEBA (mol. wt. 900) [6]

Reactants	Formulations				
Gum nitrile rubber stock	100	100	100	100	100
Zinc oxide	5.0	5.0	5.0	5.0	5.0
Stearic acid	1.0	1.0	1.0	1.0	1.0
Sulfur	1.5	1.5	...	...	0.5
Benzothiazole disulfide	1.5	1.5	...	1.5	1.5
DGEBA (mol. wt. 900)	25	...	20	20	20
DGEBA (mol. wt. 900)–H_2S adduct	...	25	5	5	5
DMP-30	1.5	1.5	1.5	1.5	1.5
Property	**Value**				
Tensile strength, psi	1,400	......	750	1,100	1,800
300% modulus, psi	800	(Cured	400	700	1,100
Elongation, %	480	on	1,200	650	545
Permanent set, %	16	mill)	65	35	10
Hardness, Durometer A	49	......	21	27	60
Swell volume %:					
Water at 150°C after 7 days	3.7	......	.....	.....	3.0
Hexane at 25°C after 1 day	10.0	......	21.0	16.0	9.0
Benzene at 25°C after 1 day	37.0	......	85.0	63.0	31.5

Table 16-22. Properties of SBR Modified with Epoxidized Polybutadiene. Sulfur Cure [4-3]

Component	*Parts by weight*
Styrene-butadiene rubber, 1502	100
High-abrasion furnace black	50
Sulfur	1.8
Benzothiazyl disulfide	1.4
Zinc oxide	5.0
Stearic acid	1.5
Epoxidized polybutadiene	3
MPDA	0.7
Resorcinol	0.1
Tris-β-chloroethyl phosphate	0.3

Property	Control	Epoxidized polybutadiene modified
Tensile strength, psi	3,600	3,630
Modulus, psi	2,630	3,200
Hardness, Shore A	66	74
Tear strength, lb/in.	210	280
Tensile strength after 48 hr at 100°C, psi	2,540	3,030
Tensile strength after 70 hr in D 471–59T fuel B, psi	880	1,690
Tensile strength after 7 days in D 471–59T oil 3, psi	1,200	1,500
Elongation, %	160	220

Table 16-23. Properties of SBR Modified with Epoxidized Polybutadiene. Peroxide Cure [4-3]

Component	*Parts by weight*
Styrene-butadiene rubber, 1502	100
High-abrasion furnace black	50
Dicumyl peroxide	1.4
Octamine	1.0
Epoxidized polybutadiene	15.0
HHPA	12.0
Propylene glycol	1.2

Property	Control	Epoxidized polybutadiene modified
Tensile strength, psi	2,710	1,950
Tear strength, lb/in.	210	280
Hardness, Shore A	67	84
Tensile strength after 48 hr at 100°C, psi	370	1,070
Tear strength, lb/in.	100	200
Tensile strength after 70 hr in D 471–59 fuel B, psi	830	1,580
Tensile strength after 7 days in D 471–59 oil 3, psi	660	890

Table 16-24. Properties of Anhydride Rubbers Cured with DGEBA Resins [6]

Reactants	Formulations							
Butadiene-acrylonitrile rubber	100	100	100	100	100	100	100	100
MA	5	10	5	10	10	5	5	5
DGEBA (mol. wt. 900)	...	...	27.5	55	55	...	27.5	27.5
DMP-30	...	...	1.3	1.3	...	1.3	1.3	
Benzoic acid	...	...	...	...	1.5			
Hexamethylenediamine/DGEBA (mol. wt. 900) adduct	...	...	...	...	...	26		
DGEBA/DDSA precondensate	...	...	...	...	...	...	27.5	
DGEBA/MPDA precondensate	...	...	...	...	...	...	...	27.5
Zinc oxide	20	20						
Sulfur	5	5						
Stearic acid	1	1						
Mercaptobenzothiazole disulfide	1	1						
Optimum cure at 145°C, minutes	60	60	45	45	80	120	45	45
Property	**Value**							
Tensile strength, psi	560	600	1,100	1,300	830	1,250	1,600	1,800
300% modulus, psi	110	85	380	480	270	600	700	820
Elongation %	885	1,100	600	600	800	470	400	330
Hardness, Durometer A	30	33	50	55	42	68	70	74
Swell volume after 1 day, %:								
Water at 100°C	14.0	21.0	6.0	7.0	11.0	4.6	2.4	2.0
Hexane at 25°C	36	33.0	38.0	33.0	47.0	23.0	11.0	7.0
Benzene at 25°C	135	120.0	95.0	85.0	95.0	50.0	36.0	23

small amounts of DGEBA with aliphatic primary amine curing agents have been used to improve performance of chloroprene–phenol-formaldehyde solvent adhesives with respect to bonds to vinyl fabrics [42]. A blend of chlorosulfonated chloroethylene rubber and amine curing agent, fillers, and solvents has been reported as providing satisfactory reactivity with a DGEBA resin containing a curing agent for the rubber such as lead dioxide [49].

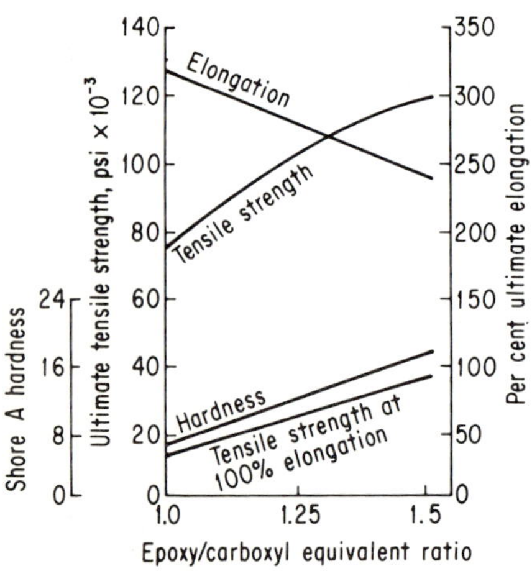

FIG. 16-15. Physical properties of liquid carboxypolybutadiene cured with DGEBA (mol. wt. 700) for 24 hours at 80°C [28].

Synthetic rubbers potentially reactive with the epoxy resins contain methylol, hydroxyl, carboxyl, anhydride, or thiol groups. Of these materials, those containing carboxyl, anhydride, or thiol groups have been investigated. Butadiene-acrylonitrile rubbers, for example, may be reacted with unsaturated anhydrides, such as maleic, and these graft polymers may be cured with epoxy resins. Properties of one such system are given in Table 16-24. Rubbers with carboxyl groups in the chain may be cured with the epoxy resin to give properties as shown in Table 16-25 and Fig. 16-15. The use of one such system as a curing agent for epoxy adhesives is described in [16], and the use of another as curing agent for pressure-sensitive tapes is described in [36].

Although the silicones and the polyurethanes may be considered synthetic rubbers, they are discussed in the previous chapter. Of the remaining rubbers, the polysulfides, being rather widely used in the epoxy technology, warrant extended treatment.

Polysulfides. A variety of mercaptan (SH) containing compounds have been investigated as epoxy resin coreactants. Hydrogen sulfide, itself, for example, has been reported as a reactant to produce high-molecular-weight thermoplastic polymers from glycidyl ether resins [2], and sulfur-containing adducts can be developed as curing agents for nitrile rubber stock, as indicated previously. Thiophenols

Table 16-25. Properties of Carboxylic Rubber Cured with Epoxy Resin [6]

(Cured 80 min at 60°C)

Reactants	Formulations							
Carboxylic rubber	100	100	100	100	100	100	100	100
Philblack 0	40	40	40	40	40	40		
Petrolatum	1.0	1.0	1.0	1.0	1.0	1.0	1.0	
Hexamethylenediamine carboxylate	1.8	.....	.....	.....	.....	.....	1.8	
DGEBA	.....	8.0	8.0	....	8.0	8.0	.....	20
DMP-30	.....	.....	0.5	.....	0.5	.....	.....	0.5
DGEBA (mol. wt. 900)–HMDA adduct	.....	.....	.....	56				
DGEBA/DDSA precondensate	.....	.....	.....	.....	14			
DGEBA/MPDA precondensate	.....	.....	.....	.....	.....	14		
Property	**Value**							
Tensile strength, psi	2,300	2,100	2,950	2,600	2,900	2,400	800	1,240
300% modulus, psi	1,400	1,400	2,475	1,400	1,800	.....	460	1,045
Elongation, %	400	390	345	460	450	200	430	325
Permanent set, %	11	13	7	6	3	2	21	4
Swell volume, %:								
Water at 100°C after 7 days	13.0	7.0	2.0	3.0	2.5	1.8	16.0	3.6
Hexane at 25°C after 1 day	11	7.0	3.0	2.0	2.6	0.5	28.0	3.5
Benzene at 25°C after 1 day	110.0	70.0	26.0	51.0	30.0	20.0	175.0	65.0

have been used in admixture with primary amines as curing agents [26], and compounds containing a single mercaptan group together with dissimilar functional groups on the chain have also been reported as curing agents [15], as have long-chain polymercaptoesters of fatty acids [35].

More commonly, however, the polysulfides employed in the epoxy technology [10] have the following generalized structure [8]:

$$HS—(C_2H_4—O—CH_2—O—C_2H_4—S—S)_{n}C_2H_4—O—CH_2—O—C_2H_4—SH$$

These compounds occasionally contain random mercaptans along the chain, but generally are considered as having a functionality of 2. In practice, then, cross-linking can be made to occur with diepoxy resins only by increasing the functionality of each epoxy group to greater than 1 (i.e., by causing reaction through the formed hydroxyl groups) or by supplying additional functionality separately (i.e., by the use of secondary curing agents).

Special polysulfide curing agents, containing on the average two hydroxyl groups per molecule, are also available. These compounds are somewhat lower in viscosity but give generally similar properties to the non-hydroxyl-containing species [10-1] and will not be further considered here.

The commercial non-hydroxyl-containing polysulfides differ primarily in molecular weight, cross-linking, and viscosity. Originally, they were produced by splitting higher-molecular-weight materials to obtain the liquid species, but the literature indicates that the liquid polymers with controlled molecular weight may be produced without resorting to the splitting operation [17]. Typical of the commercial polysulfides are those shown in Table 16-26. Of these, the material having a molecular weight of 1,000 and a cross-linking of 2 per cent is perhaps the most commonly used, and data presented in this section are based on this compound.

Mercaptans alone react very sluggishly with epoxy groups. The reaction proceeds at a practical rate by the addition of a base which accepts the proton to produce the mercaptide ion. This ion (RS^-) is quite reactive toward the epoxy group and accounts for practically all the mercaptan-epoxy reactions [43].

The polysulfides are most commonly employed in connection with primary or tertiary amines. With these co-curing agents, the curing mechanism is quite specific, even when large amounts of the polysulfide are employed. For example

Table 16-26. Properties of Typical Commercial Polysulfides from Thiokol Corporation [10-2]

Property	Value		
	LP-3	LP-8	LP-33
Physical state	Mobile liquid	Mobile liquid	Mobile liquid
Molecular weight	1,000	500–700	1,000
% cross-linking	2.0	2.0	0.5
Pour point, °C	−20	−32	−12 to −15
Specific gravity after 20 hr at 4°C	1.27	1.27	1.27
Viscosity at 27°C, centipoises	700–1,200	250–350	1,300–1,550
Flash point (open cup), °C	214	182	205
Fire point (open cup), °C	240	205	240
Moisture content, %	0.1	0.2	0.1
pH (water extract)	6.0–8.0	6.0–8.0	6.0–8.0
Color	Clear amber	Clear amber	Clear amber

[4], when using 10 phr of TETA as a room-temperature curing agent for a 2/1 blend of polysulfide/DGEBA, it was found on the basis of toluene extraction data that less than 8 per cent of the mercaptans remained unreacted. When a tertiary amine, DMP-30, was used as a curing agent, the reaction was even more selective. After room-temperature cure, less than 1 per cent of the mercaptans remained unreacted at the 2/1 ratio.

Discounting the random number of mercaptans along the chain, approximately 260 phr of a typical polysulfide is required to provide a mercaptan group for each epoxy group of DGEBA. The usual formulation involves 100 phr or less of polysulfide—most commonly, 25 to 50 phr. A majority of the epoxy groups are therefore left over for reaction with a polyfunctional curing agent or by homopolymerization and cross-linking through formed hydroxyls. In practice, the amount of cross-linking agent employed is usually based on epoxy content and is calculated by ignoring the mercaptan groups.

Differences in properties will be experienced between compounds cured by polyfunctional cross-linking agents and by catalytic-type curing agents. For example, tertiary-amine-cured systems will give better tensile strengths at room temperature; and primary-amine-cured systems, while giving more stable properties initially at elevated temperature, are subject to greater embrittlement with aging.

Polysulfides are usually used with the glycidyl ether resins. The reaction of polysulfides with internal [29] and ring-situated [12] epoxies is sluggish at best, and there are probably better materials for introducing flexibility into these systems.

Table 16-27. Effect of Polysulfide Concentration on Rate of Cure of DGEBA/DMP-30 Systems [10-2]

Property	Polysulfide, %			
	0	33	50	67
Gel time, 50 g, minutes	41	15	21	72
Tack-free, minutes	225	27	38	1,000
Exotherm, °C	45	107	103	61

The polysulfides may be used with anhydride curing agents in glycidyl ether resin systems, but heat cures are required and the free acid may cause subsequent chain scission in the polysulfide. The primary advantage arising from the use of the commercial polysulfides is that controlled degrees of flexibility may be obtained with room- or only slightly elevated-temperature cures.

With the primary aliphatic amines and the more reactive tertiary amines, reactivity is increased, such that lower concentrations of polysulfides will shorten the pot life of the system, as indicated in Table 16-27. Improvements in pot life are obtained only at higher concentrations of the polysulfide, where it exerts a diluting effect on the reaction. Thiuram sulfides and thiazol sulfides have been reported as providing for increased pot lives and still permitting room-temperature cures in the glycidyl ether resin systems [18]. Reaction with at least some of the tertiary amines appears somewhat dependent upon the concentration of the polysulfide, and at higher concentrations a mild heat cure may be required to prevent age hardening.

In general, the tertiary amines and aromatic primary amines are compatible with the commercial polysulfides, thus permitting the convenient attainment of two-component packages with the amine and the polysulfide in one container. The primary aliphatic amines, however, are generally incompatible mixtures, tending to separate and form gel layers at the interface on storage; hence these require three-component packaging [10-2].

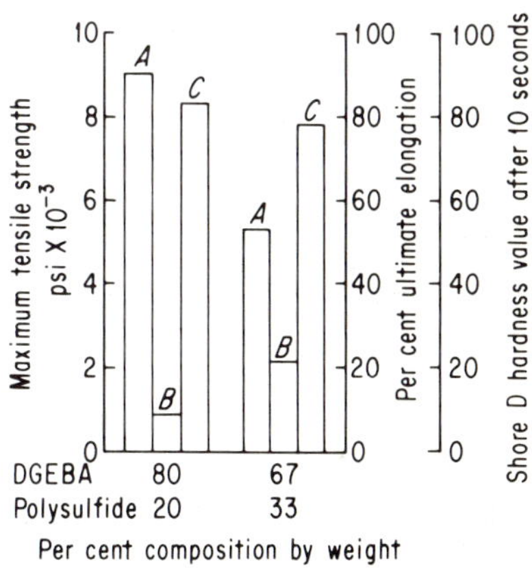

FIG. 16-16. Tensile properties and hardness vs. polysulfide concentration for DGEBA cured with 12 phr DETA [4-6]. *A*, Maximum tensile strength; *B*, ultimate elongation; *C*, Shore D hardness.

Properties for typical polysulfide/epoxy formulations cured with both a primary aliphatic amine and a tertiary amine are given in Table 16-28. The effect on various properties obtained by varying the amount of polysulfide is shown in Table 16-29 and Fig. 16-16 and the effect on clcctrical properties in Fig. 16-17. The effect of different primary amine curing agents on impact strength is shown in Figs. 16-18 and 16-19. The effect of the concentration of the polysulfide on DT is shown in Figs. 16-20 and 16-21. The effect of temperature on various compressive and tensile properties for the polysulfide system cured with MPDA/MDA/PGE is shown in Figs. 16-22 to 16-31.

The effect of various concentrations of the polysulfide on moisture-vapor transmission is shown in Fig. 16-32. The effect of polysulfide on water absorption is given in Fig. 16-33 and in Table 16-30.

Table 16-28. Typical Properties of DGEBA Flexibilized with a Polysulfide [4]

Property	DMP-30, 10 phr		TETA, 10 phr	
Polysulfide, phr	50	100	50	100
Viscosity at 25°C, centipoises	2,650	2,000	2,650	2,000
Pot life, 50 g at 25°C, minutes	20	35	20	40
Exotherm, °C	151	129	154	121
Tensile strength, psi	4,800	2,800	4,580	1,785
After 70 hr at 100°C	4,200	900	5,000	1,800
Elongation, %	10	30	0–5	30
After 70 hr at 100°C	0–5	80	0–5	30
Hardness, Shore D	78	63	60	41
After 70 hr at 100°C	76	40	65	43
Impact strength, ft-lb:				
At 25°C	1.1	78	47	>110
At −34°C	0.5	1.3	2.5	3.5
Dielectric constant at 10^3 cps	4.5	5.5	3.5	6.0
Loss factor at 10^3 cps	0.009	0.012	0.012	0.020
Volume resistivity, ohm-cm	6.1×10^{12}	6×10^{11}	2×10^{12}	1.5×10^{11}
Surface resistivity, ohms	1.5×10^{13}	1.5×10^{13}	4×10^{13}	3.3×10^{12}
Swell volume after 30 days, %:				
Hydrochloric acid, 10%	2	4		
Sulfuric acid, 20%	2	4		
Ammonium hydroxide, conc.	3	7		
Sodium hydroxide, 10%	1	2		
50%	−1	−1		
Sodium chloride, satd. soln.	1	1		
Benzene	38	27		
Carbon tetrachloride	−2	12		
Methyl ethyl ketone	35	20		
Ethyl acetate	26	17		
Water absorption, %:				
After 24 hr	0.2	0.4	0.3	0.4
After 7 days	0.6	1.2	0.8	0.9
After 24 days	1.3	1.5	0.8	1.2

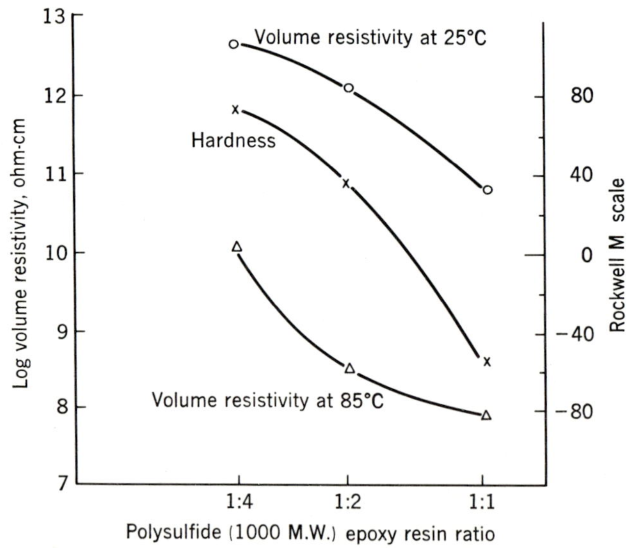

FIG. 16-17. Hardness and volume resistvitiy vs. polysulfide concentration for cured DGEBA [7].

Table 16-29. Effect of Concentration of Polysulfide on Properties of the Cured DGEBA [10-2]

Property	Polysulfide/epoxy					
	100/0	300/100	200/100	100/100	100/200	0/100
Elongation, %	200	100	50	30	10	
Hardness, Shore A	35	50	70	95	98	100
Volume shrinkage, %	5	2.5	2.8	3.5	4.5	6
Dielectric constant at 10^3 cps	7.5	6.5	5.5	4.0	3.8	3.5
Volume resistivity, ohm-cm	10^9	10^{10}	10^{11}	10^{12}	10^{13}	10^{14}

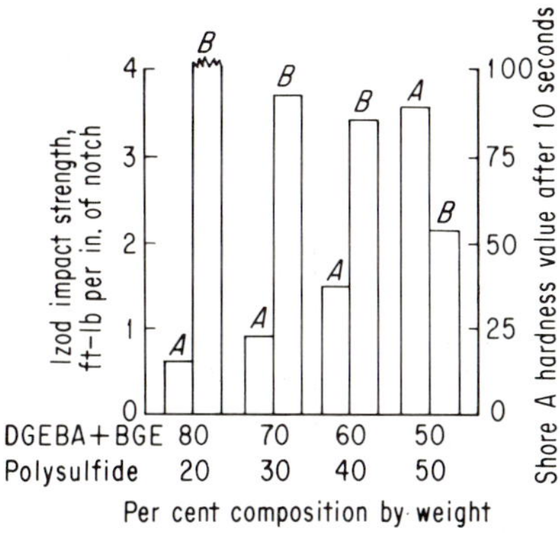

FIG. 16-18. Impact strength and hardness vs. polysulfide concentration for DGEBA diluted with BGE and cured with 10 phr TETA [4-6]. Cure cycle: 20 hours at room temperature plus 60 hours at 80°C. *A*, Izod impact strength; *B*, Shore A hardness.

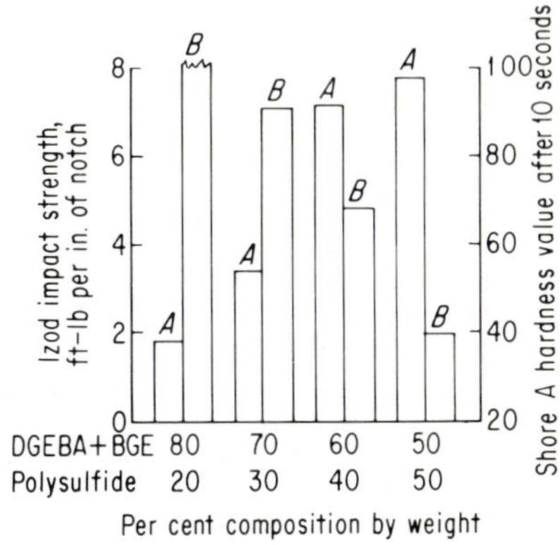

FIG. 16-19. Impact strength and hardness versus polysulfide concentration for DGEBA diluted with BGE and cured with 16 phr DETA. Cure cycle: 20 hours at room temperature plus 60 hours at 80°C. *A*, Izod impact strength; *B*, Shore *A* hardness [4-6].

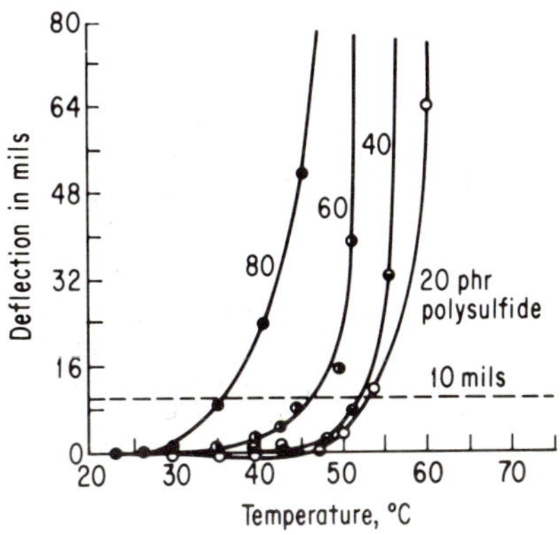

FIG. 16-20. DT vs. polysulfide concentration for DGEBA cured with DMP-30 [9].

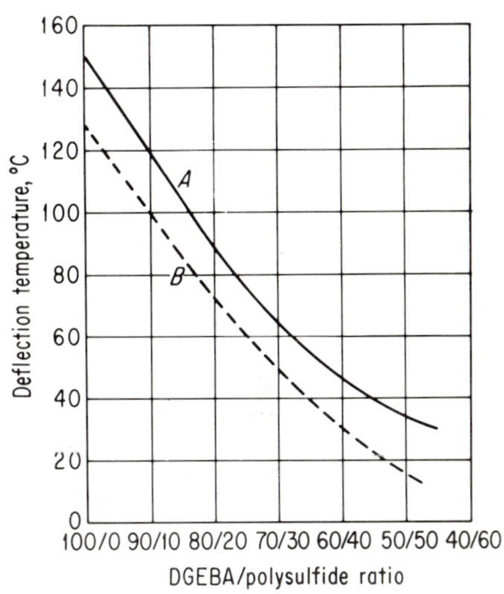

FIG. 16-21. DT vs. polysulfide concentration for DGEBA cured with MPDA/MDA/PGE [4-6]. Cure cycles: *A*, 1 hour at 90°C plus 2 hours at 175°C; *B*, 1 hour at 90°C.

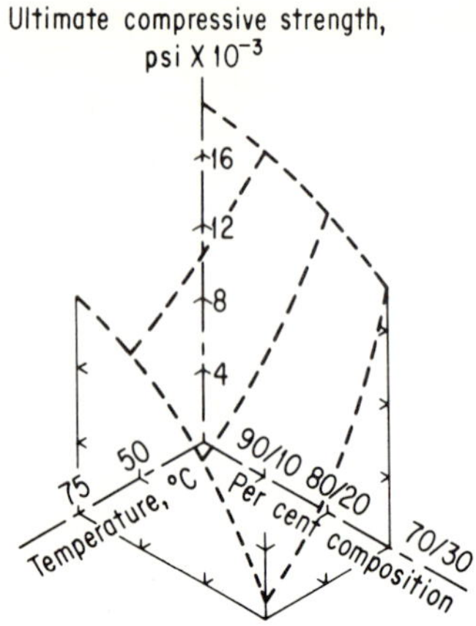

FIG. 16-22. Compressive strength vs. polysulfide concentration and temperature for DGEBA cured with MPDA/MDA/PGE [4-6]. Cure cycle: 1 hour at 90°C plus 2 hours at 175°C.

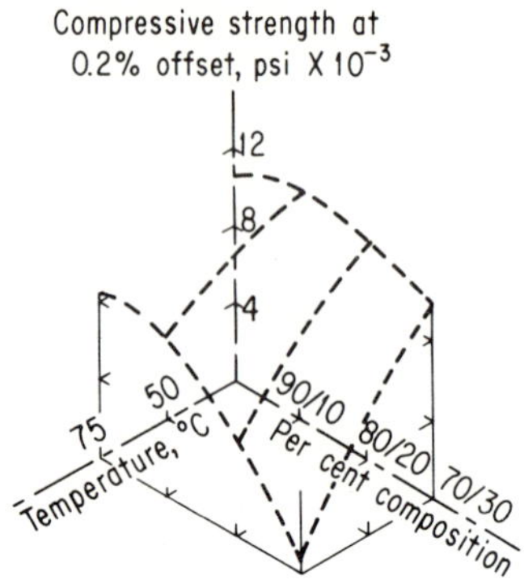

FIG. 16-23. Compressive offset at 0.2 per cent vs. polysulfide concentration and temperature for DGEBA cured with MPDA/MDA/PGE [4-6]. Cure cycle: 1 hour at 90°C plus 2 hours at 175°C.

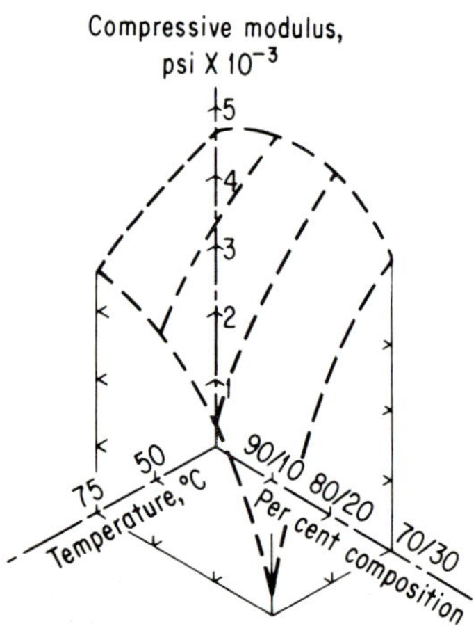

FIG. 16-24. Compressive modulus vs. polysulfide concentration and temperature for DGEBA cured with MPDA/MDA/PGE [4-6]. Cure cycle: 1 hour at 90°C plus 2 hours at 175°C.

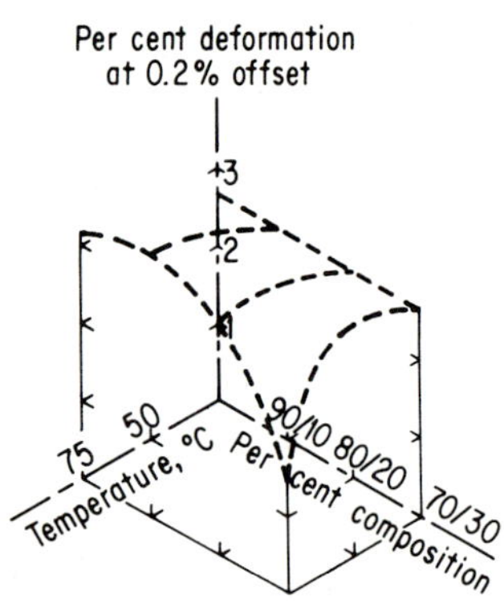

FIG. 16-25. Compressive deformation at 0.2 per cent offset vs. polysulfide concentration and temperature for DGEBA cured with MPDA/MDA/PGE [4-6]. Cure cycle: 1 hour at 90°C plus 2 hours at 175°C.

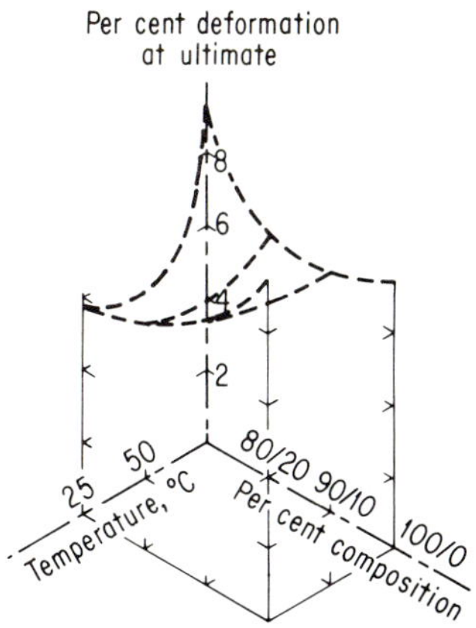

FIG. 16-26. Ultimate compressive deformation vs. polysulfide concentration and temperature for DGEBA cured with MPDA/MDA/PGE [4-6]. Cure cycle: 1 hour at 90°C plus 2 hours at 175°C.

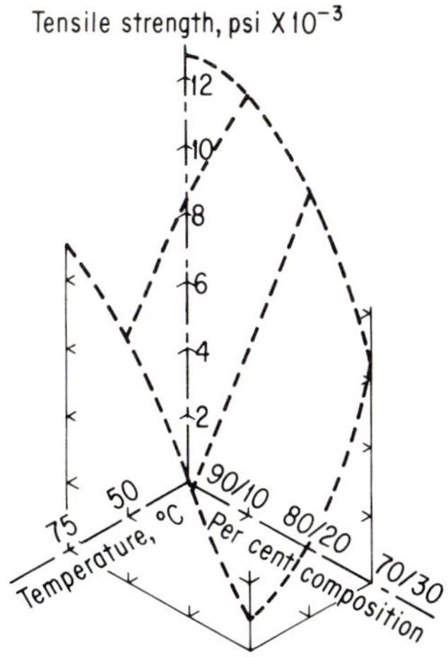

FIG. 16-27. Tensile strength vs. polysulfide concentration and temperature for DGEBA cured with MPDA/MDA/PGE [4-6]. Cure cycle: 1 hour at 90°C plus 2 hours at 175°C.

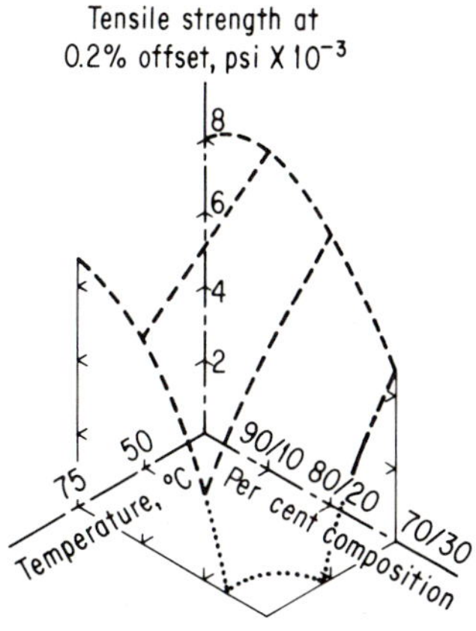

FIG. 16-28. Tensile strength at 0.2 per cent offset vs. polysulfide concentration and temperature for DGEBA cured with MPDA/MDA/PGE [4-6]. Cure cycle: 1 hour at 90°C plus 2 hours at 175°C.

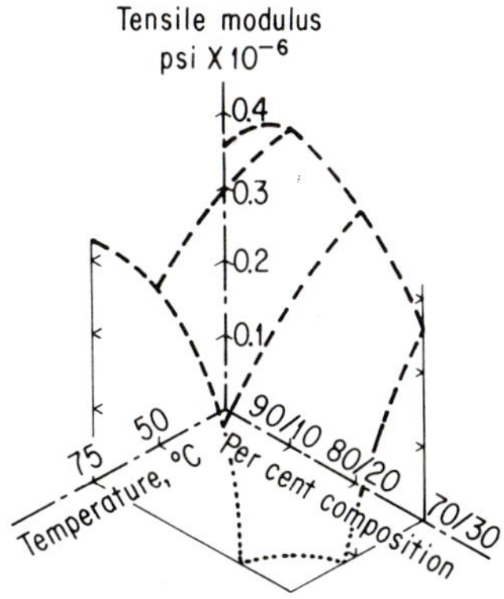

FIG. 16-29. Tensile modulus vs. polysulfide concentration and temperature for DGEBA cured with MPDA/MDA/PGE [4-6]. Cure cycle: 1 hour at 90°C plus 2 hours at 175°C.

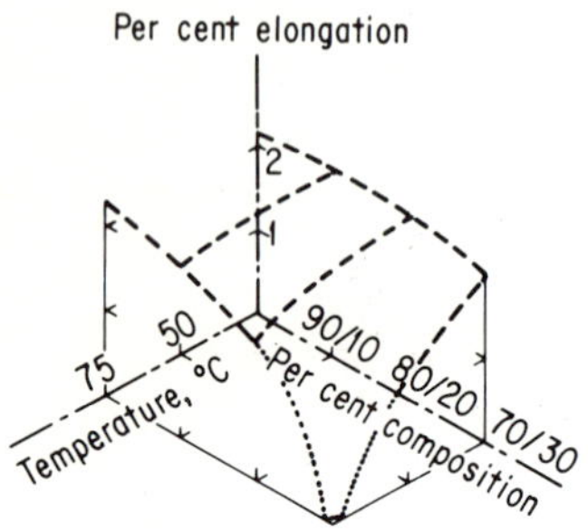

FIG. 16-30. Tensile elongation at 0.2 per cent offset vs. polysulfide concentration and temperature for DGEBA cured with MPDA/MDA/PGE [4-6]. Cure cycle: 1 hour at 90°C plus 2 hours at 175°C.

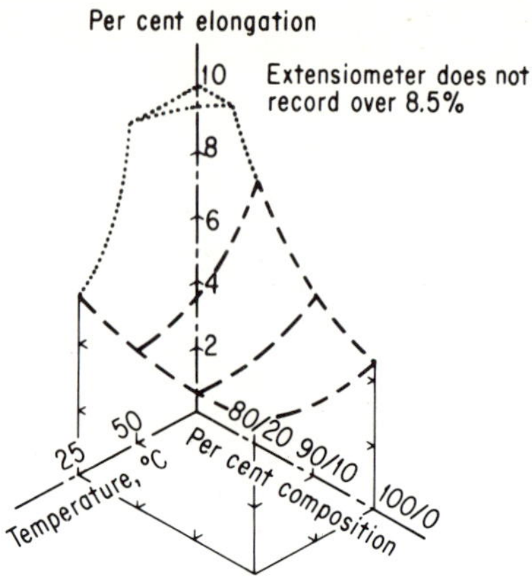

FIG. 16-31. Ultimate tensile elongation vs. polysulfide concentration and temperature for DGEBA cured with MPDA/MDA/PGE [4-6]. Cure cycle: 1 hour at 90°C plus 2 hours at 175°C.

Table 16-30. Per Cent Water Absorption vs. Time for DGEBA Cured with 50 phr Polysulfide and 10 phr DMP-30 [4-4]

Immersion time, days	At 23°C	At 66°C
1	0.37	1.54
2	0.52	2.27
5	0.80	3.00
6	0.87	3.06
7	0.93	3.10
8	1.00	3.13
9	1.04	3.15
12	1.19	3.19
19	1.51	3.20
26	1.75	3.22
33	1.99	3.27
62	2.62	3.52
96	3.13	3.83
124	3.47	4.54
155	3.75	5.36
190	3.99	6.45
365	4.10	11.53

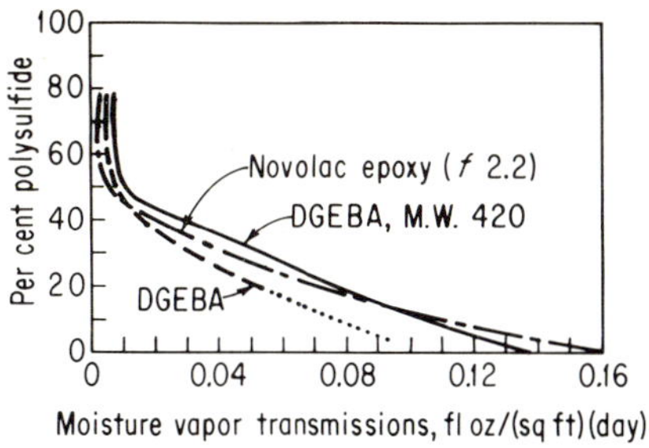

FIG. 16-32. Moisture-vapor transmission vs. polysulfide concentration for various epoxy resins cured with 10 phr DMP-30 [10-2].

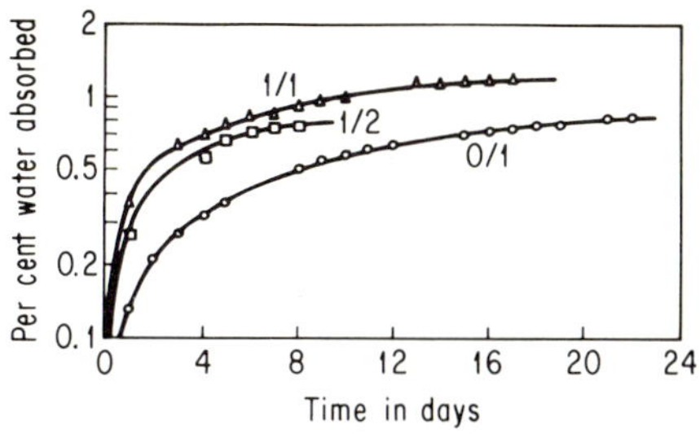

FIG. 16-33. Water absorption vs. polysulfide concentration for DGEBA cured with TETA [10-2] at room temperature. Numbers on curves are polysulfide/epoxy ratios.

In common with other flexibilizers, the polysulfides increase the coefficient of thermal expansion of the cured system in relation to their concentrations. As would be expected from the discussion in Chap. 14, values may be substantially reduced by the addition of fillers, as indicated in Fig. 16-34.

Aniline-formaldehyde resins may be blended with polysulfides at about 50°C to produce upon cooling, a liquid. This blend together with phenol accelerators may then be used to cure DGEBA. This curing agent provides DTs of about 70°C with improved impact resistance. Pot lives of the system range from 90 minutes upward, depending on the accelerator; cure is accomplished in about one hour at 120°C. Blend viscosities are about 20,000 cps [31]. A similar blend of polysulfide and aniline resin has also been used to cure epoxidized novolac resins [32].

Commercial polysulfides have an objectionable odor, and as a partial solution to this problem, styrene oxide has been suggested as an additive to block some of the mercaptans in advance of the use of the polysulfide as a curing agent [37].

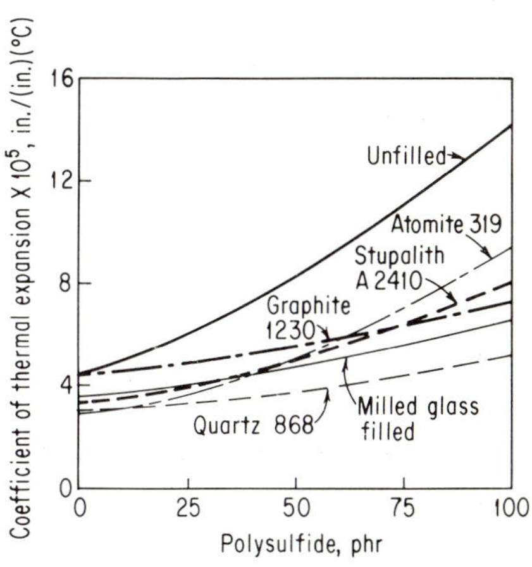

FIG. 16-34. Coefficient of thermal expansion vs. polysulfide concentration for filled and unfilled DGEBA cured with 20 parts by weight of MPDA/MDA/PGE for 24 hours at 66°C [10-2].

SUMMARY

In the epoxy technology, the term *plasticizer* has come to mean an inert, long-chain ingredient which is added to the resin system to provide mechanical plasticizing action. The term *flexibilizer* has come to mean a long-chain ingredient which reacts, at least to some extent, with the resin during cure to provide flexibility by virtue of its increasing the distance between the cross-linking points.

The epoxy resins are not generically flexible materials, and even with the addition of fairly high concentrations of flexibilizer or plasticizer, do not develop cured properties approaching those of the common elastomers. The main reason for employing flexibilizing additives is to reduce the inherent brittleness of the cured epoxy system. Insofar as many of the flexibilizing additives serve to increase elongation rather than to yield true elasticity, a wide variety of reactants may be considered to fall into this classification. Many of these materials are discussed in other parts in the book. This chapter confines itself mainly to a consideration of the use of flexibilizing epoxy resins, of long-chain hydroxyl-terminated molecules, and of natural and synthetic rubbers.

These materials may be used with the aromatic glycidyl ether epoxy resins in varying proportions to obtain varying degrees of flexibility. Some are suitable for use with room-temperature curing agents, whereas others require heat curing or are suitable for use only with the anhydride or Lewis-acid-type curing agents. In general, the flexibilizing epoxy resins and the polysulfide rubbers may be used in formulations curing at or near room temperature, whereas the hydroxyl-terminated molecules require elevated-temperature cures.

The general effect of the use of flexibilizers and plasticizers is to improve the impact resistance and extensibility of the cured system at the expense of tensile strength. Their use permits the casting of larger volumes of resin and of more complex shapes. The cured systems are less strong mechanically than the unflexibilized system, and the chemical resistance, particularly the solvent resistance, and the electrical properties are lowered. In general, too, the temperature resistance of the flexible system is below—often considerably below—what is encountered with unflexibilized systems.

REFERENCES

1953

1. Greenlee, Partial Esters of Polyhydroxyl Polyethers, U.S. 2,653,141.
2. Shokal, Sulfur-containing Resinous Products from Polyepoxides, U.S. 2,622,458.

1956

3. Bender, et al., Diglycidyl Thio Ethers of Dithiols Containing Oxygen, U.S. 2,731,437.
4. Cranker and Breslau, Epoxy Casting Resins Modified with Polysulfide Liquid Polymer, *Ind. Eng. Chem.*, January.
5. Greenlee, Epoxide Resin Cured with a Polyhydric Aliphatic Alcohol, U.S. 2,731,444.
6. Mika, The Use of Epoxy Resins in Synthetic Rubber Compositions, *J. Appl. Chem.*, Sept. 6.
7. Nadler, et al., Potting Aircraft Connectors, *Electron. Equipment*, July.
8. Sorg and McBurney, Polysulfide-Epoxy Compositions, *Mod. Plastics*, October.

1957

9. Breslau, Heat-distortion Properties of Polysulfide Polymer–modified Epoxy Resins, *Plastics Technol.*, April.
10. Fettes and Gannon, Resinous Reaction Products of Polyepoxides and Polysulfide Polymers and Methods of Making the Same, U.S. 2,789,958.

1958

11. Belanger and Klassen, Flexible Epoxy Resins, *Plastics Technol.*, August.
12. Bender, et al., Addition Products of Polyglycidyl Ethers of Phenolic Resins and Polythiols, U.S. 2,849,416.
13. Delmonte, Electrical Properties of Modified Epoxy Resins, *Mod. Plastics*, April.
14. Fisch, Compositions Comprising Epoxy Compounds and Hydroxy Terminated Polyester Resins and Process of Making Same, U.S. 2,830,031.
15. Schroeder, Sulfur-containing Resinous Products from Polyepoxides, U.S. 2,831,830.

1959

16. Been and Grover, Adhesive Composition of Polyepoxide Resins and Butadiene-Acrylonitrile Rubbers and Method of Making Bristle Brushes Using Same, U.S. 2,879,252.
17. Fettes and Bertozzi, Production of Polysulfide Polymers, U.S. 2,875,182.
18. Howard, Method for the Reaction of Polyglycidyl Compounds with Organic Polysulfides, U.S. 2,871,217.
19. Shokal, Process for Preparing Flexible Resinified Products from Polyepoxides and Resulting Products, U.S. 2,915,485.
20. Wheelock, Resins from Epoxidized Polymers and Polyhydric Alcohols, U.S. 2,914,490.

1960

21. Belanger, et al., Epoxide Resin Compositions, U.S. 2,947,712.
22. Belanger and Masters, Epoxide Resin Compositions, U.S. 2,947,717.
23. Greenspan and Light, Epoxy Resins Plasticized with Di(butoxyethyl) Adipate, U.S. 2,923,695.
24. Haberlin, High Temperature Adhesive Containing Polyepoxide Resin Mixture, U.S. 2,951,778.
25. May, Epoxide Resin Compositions, U.S. 2,921,040.
26. Schmitz, Mercaptans as Catalysts for Polymerization of Ethoxylene Resins, U.S. 2,933,473.

1961

27. Antykov, Synthesis of Polythiocyanate-Epoxide Rubber and Polythiocyanate Epoxide Resin, *Jurnal Prikladnoy Khimi* **34**; 227–229.
28. B. F. Goodrich Chemical Corp., Hycar 2000X131, technical bulletin.
29. Greenspan and Light, Resinous Products from Epoxidized Polybutadiene and Polymercaptans, U.S. 2,921,921.
30. Helmreich and Harry, Two Flexible Epoxy Resins, *SPE J.*, June.
31. King and McDowall, Compositions Comprising an Epoxy Resin, an Aniline-Formaldehyde Resin, a Liquid Polysulfide Polymer and a Phenol Curing Agent, U.S. 3,014,007.
32. McDowall and King, Improvements in or Relating to Curing Agents for Epoxy Resins, Brit. 882,016.
33. Patrick, Flexibilized Resin from Unox 201 and Polyols, Union Carbide Chemicals Co.
34. Whittier and Lawn, Bituminous Compositions Containing Epoxy Ether Resin and Chlorinated Rubber, U.S. 2,976,256.

1963

35. Casement, Curing of Epoxy Resins, U.S. 3,090,793.
36. Korpman, Tape Employing an Adhesive Comprising an Epoxy Resin and a Polymer Having Carboxyl Groups, U.S. 3,100,160.
37. Montesano, Polysulfide Polymer Containing Styrene Oxide as an Odor Masker, and Mixture Thereof with an Epoxy Resin, U.S. 3,101,326.
38. Stephens, private communication, Quaker Oats Co.
39. Zakharov and Maiorov, Vulcanization of Chloroprene Rubber by Epoxy Resins, *Kauchuk i Rezina* **22**(8): 8–11.

1964

40. Mortar (to Raybestos-Manhattan, Inc.), Brit. 972,801.
41. Wittenwyler, New Developments in Epoxy Resins, Adhesives and Sealants Council, Cherry Hill, N.J., Mar. 4–6.
42. Yaroch, Neoprene-Epoxy-Phenol-Aldehyde Adhesive Cement, U.S. 3,124,548.

1965

43. H. H. Chen, private correspondence.
44. Improvements in or Relating to Polychloroprene–Epoxidized Polybutadiene Compositions, Brit. 992,992.

45. Makarova et al., Effect of the Molecular Weight of Epoxy Resins on Vulcanization of Chloroprene Rubber, *Kauchuk i Rezina* **24**(5); 6–10.
46. Plasticizers for Epoxy Resins (to Dr. Beck & Co., G.m.b.H.), Brit. 1,001,467.
47. Plasticizers for Epoxy Resins (to Shell Internationale Research), Neth. Appl. 297,207.
48. Roux et al., Epoxidized Elastomers, Fr. 1,383,947.
49. Schenker-Winkler, A Process for the Production of Thermosetting Compositions from Epoxy Resins and Liquid Rubber, Brit. 999,383.
50. Sherman and Toth, Rubber and Epoxy Resin Adhesive Compositions and Method of Manufacture, U.S. 3,171,821.

Chapter 17

CASTING, POTTING, AND ENCAPSULATION WITH EPOXY RESINS

Up to this point in the book, concern has been chiefly with the chemical aspects of the epoxy resins. This and the remaining chapters will consider applications for the epoxy resins in the contemporary technology.

The solid epoxy resins find their chief use in solution coatings, in molding compounds, and in fluidized-bed coatings, and to a lesser extent in adhesive and laminating formulations. The liquid epoxy resins are used in a wider range of applications, including casting, potting, encapsulation, tooling, foams, adhesives, laminates, and coatings.

The technology of the liquid resins involves a number of common considerations which may, for convenience, be considered in detail in this chapter of casting, potting, and encapsulation.

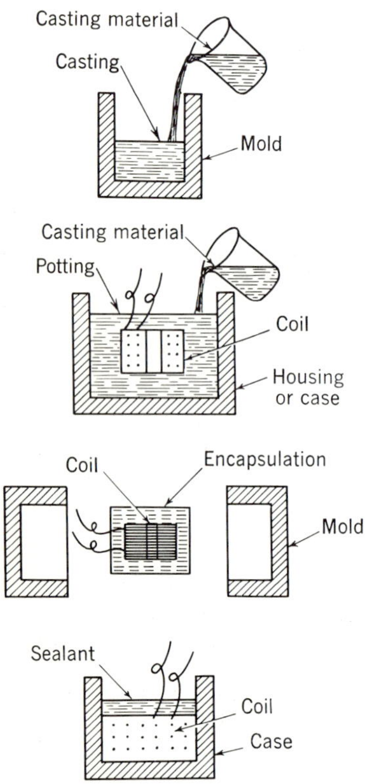

FIG. 17-1. Definition of terms: potting, casting, encapsulation, and sealing.

DEFINITION OF TERMS

The three terms casting, potting, and encapsulation have more or less specific meanings as defined in Fig. 17-1. In actual practice, however, they are used more loosely and often interchangeably. The term *casting* and the term *casting resin* have become generic, embracing most applications and compounds employing the use of pourable liquids. They will be used in the generic sense in most of the following discussion.

APPLICATIONS

There are a variety of other plastic casting resins, among them the polyesters, phenolics, silicones, and acrylics. The liquid epoxy resins possess a number of advantages

(Chap. 1) which make them suitable for a wider range of applications than the other materials. A comparison of the properties of several casting resins is presented in Table 17-1.

The epoxy casting resins are normally, when cured, pale to dark amber, and hence are not suited for the production of clear, transparent castings such as may be made with the acrylics. Special formulations based on DGEBA ($n = 0$) and anhydrides have been produced [35] and special systems have been developed for casting clear

Do-Maker identification plate is cast from resilient epoxy-resin formulation. Casting is made in vinyl mold and cured at room temperature. After cure, raised portion is coated with roller coat to desired color. (Union Carbide Corp.)

films, but in most cases the epoxy casting resins are used in applications where optical clarity is not required. In general, the casting resins do not employ fibrous reinforcement and therefore do not provide the structural strengths which may conveniently be obtained with reinforced molding compounds and with laminates.

Two of the largest-volume uses for epoxy casting resins are for the production of bowling balls and for casting membranes between glass plates of TV tubes to provide unitized construction. The latter application, for example, required 5 million pounds of resin in 1963, the amount declining to about 1 million pounds in 1965 as a result of new production techniques and changed designs.

The remainder of the casting applications, as distinct from potting and encapsulation, are more or less small-volume, involving the production of specialized

Table 17-1. Comparison of Synthetic Casting Compounds*

Property	Class 105 phenolic	Class 105 polyesters	Class 105 epoxy	Class 130 epoxy	Class 180 silicone
Specific gravity	1.30–1.32	1.10–1.46	1.11–1.23	1.15	1.70–1.90
Tensile strength, psi	6,000–9,000	6,000–10,000	12,000	12,000–15,000	3,000–7,000
Elongation, %	1.5–2.0	5	5	1.5	
Modulus of elasticity in tension, psi $\times 10^{-5}$	4.5	3.0–6.4	4.5	6	
Compressive strength, psi	12,500–15,000	13,000–27,000	16,000	15,000	9,000–18,000
Flexural strength, psi	11,000–17,000	8,500–17,000	18,800	15,470	9,800
Impact resistance, Izod	0.25–0.40	0.2–0.4	0.5–1.7	0.192	4.20
Hardness, Rockwell M	120	115	100	112	45
Coefficient of thermal expansion, in./(in.)(°C) $\times 10^{6}$	60–80	80–100	60	60	308
Resistance to heat, continuous, °C	71	121	149	177	249
Deflection temperature, °C	78–82	60–100	120	300	
Volume resistivity, ohm-cm	10^{12}–10^{13}	10^{14}	10^{16}–10^{17}	10^{17}	10^{11}–10^{13}
Dielectric strength, volts/mil	350–400	380–500	400–500	450–500	185
Dielectric constant at 60 cps	6.5–7.5	3.0–4.36	3.8	3.8	4.0–5.0
Power factor at 60 cps	0.10–0.15	0.003	0.0013	0.002	0.0055
Arc resistance, seconds	100–125	125	50–180	100–140	
Water absorption after 24 hr, %	0.12–0.36	0.15–0.60	0.14	1.10	Poor
Effect of weak acids	Slight	Slight	None	None	Slight
Effect of strong acids	Attacked	Attacked	Attacked	Slight	Attacked
Effect of weak alkalies	Slight	Slight	None	None	Slight
Effect of strong alkalies	Decomposed	Decomposed	Very slight	Very slight	Attacked
Effect of organic solvents	Attacked by some	Attacked by ketones and chlorinated solvents	Resistant	Resistant	Attacked by some
Shrinkage, %	8–10	4–6	1–2	1–2	4–8
Adhesion to metals, glass, plastics, ceramics, etc.	Excellent	Fair	Excellent	Excellent	Poor

* From *Mod. Plastics* and trade literature.

Busbar units, complete with cast epoxy switch plug and cable-bar insulators, for 11-kv normal-type switchgear. (*Ciba Products Co.*)

parts with electrical insulating properties [39], small gears and the like [38], and small plastic molds. High-volume production of such items would normally dictate the use of epoxy molding compounds. Epoxy formulations are used for centrifugally cast tubing for epoxy shell molds [75] and cast rods for machining bobbins, etc. Considerable interest has been generated in the production of cast parabolic mirrors by the use of epoxy resins [45,49] and the use of epoxies for backing printing plates [31].

The casting of switchgear from epoxy resins is described in [83]. Vibration damping tiles for shipboard use have been cast from the epoxy resins [89]. Considerable research has been done on the use of epoxy resins as impregnating materials for consolidating loose earth adjacent to well bores [80, 94], for the consolidation

Barrier panel insulation for a new high-speed resistor on-load tap changer, 132 kv (also serves as an oiltight barrier capable of withstanding a vacuum, between the transformer tank and the selector switch). (*Ciba Products Co.*)

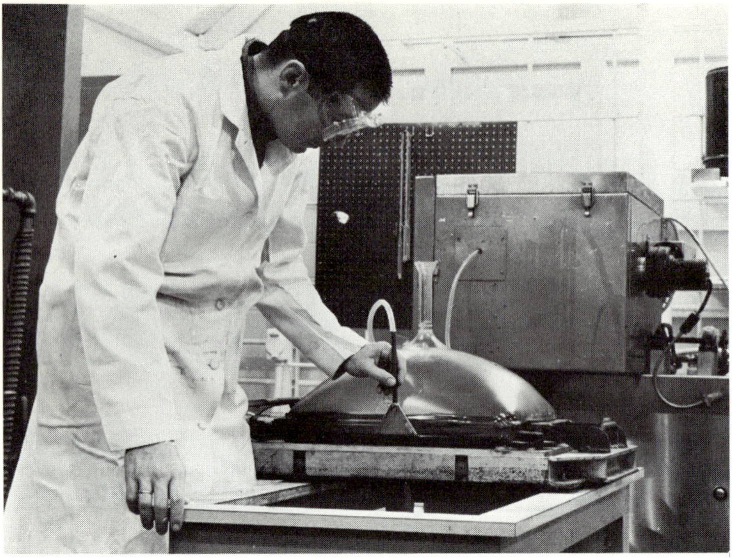

Epoxy-resin formulation is cast from automatic dispenser between safety panel and TV picture tube. (*Dow Chemical Co.*)

of soil samples [68], for the impregnation of graphite to improve strength properties and reduce permeability [77], and as binders for solid propellants [76, 78, 88]. Powdered ceramic particles have been coated with a flexibilized epoxy resin in solvents, and the particles subsequently compacted and machined after the resin is cured. After firing of the machined ware, the process completely burning out the epoxy-resin binder, the finished, machined part is obtained [82]. Epoxy-resin formulations have been filled with a powdered abrasive (alumina) and used for mounting metallographic specimens, which are subsequently ground and polished for microscopic study [87], and special formulations have been developed for embedding tissues for microscopic study [85].

Epoxy castings are particularly suited to photoelastic studies using the stress/freezing technique. The epoxies owe their superiority for this work to the fact they

Railway signal pickup coil encapsulated in epoxy resin as protection against vibration and weather. Coil is mounted under engine near the road bed. (*The Epoxylite Corporation.*)

give high stress-strain optical coefficients of double refraction. This is measured as the optical strain coefficient E/S (E = elastic modulus, S = photoelastic constant). This should be as large as possible, and epoxies give $E/S \simeq 3{,}000$ under load at room temperature and $E/S \simeq 700$ fringes per cm by the stress/freezing method. The values are at the upper limits of E/S values for present-day plastics [59].

As electrical and electronic potting compounds, the epoxies are used throughout the industry in those applications requiring rigid or semirigid plastics. In the electronics field, in particular, they have virtually displaced competition.

Broadly, in the electronics industry, the epoxy resins are used for the encapsulation of welded modules and printed circuits as well as for the protection of individual components: capacitors [44], relay coils [53], loading coils [54], transistor flip-flops [15], and torroid coils [9]. They are widely used in the utility industry for potting cable splices. In the electrical industry, the epoxies are used for the encapsulation of all types of transformers [7,11,17,18,26,30,36,55,58], random-wound motors [6,8,13,14,43,62], and motors using field coils [46] and form coils [21].

COMMON CONSIDERATIONS FOR EPOXY CASTING RESINS

The epoxy casting resins, generically, are liquids at processing temperatures when combined with the curing agents. The mixture is cured at room or elevated temperature to the thermoset state. There are a number of common considerations involved in achieving a good, solid, crack-free, bubble-free casting:

1. Viscosity
2. Reactivity
3. Exotherm
4. Vapor pressure
5. Shrinkage
6. Expansion characteristics
7. Thermal shock
8. Thermal stability

It is useful to examine these common considerations in detail.

Viscosity vs. Temperature

The viscosity of the epoxy-resin compounds decreases rapidly with temperature, as shown in Fig. 17-2. Some curing agents and resinous modifiers produce little effect on the initial mixed viscosity of the resin system, whereas others influence it considerably. Fillers increase viscosity, in general, in direct relationship to the loading volume. Thus, the viscosity of the resin is only one factor determining the viscosity of the formulated compound, and the viscosity ideal for a particular application may yield to other considerations.

In order to assist removal of air bubbles stirred into the compound during mixing, to assist in displacing air or gases contained in any inserts to be embedded in the casting, and to obtain a dense, void-free structure, it is often desirable to heat the compound to some pre-selected temperature (Fig. 17-2).

There are two ways of heating the compound: The resin part (usually marked part *A* in commercially packaged compounds) and the curing agent (marked part *B* or hardener or catalyst) may be heated separately and the two then mixed, or they may be mixed cold (provided that the curing agent does not need to be melted) and then the mixture heated.

In general, it is more desirable to heat the resin and the curing agent separately and to mix them at working temperature. To mix cold and then heat the mixture

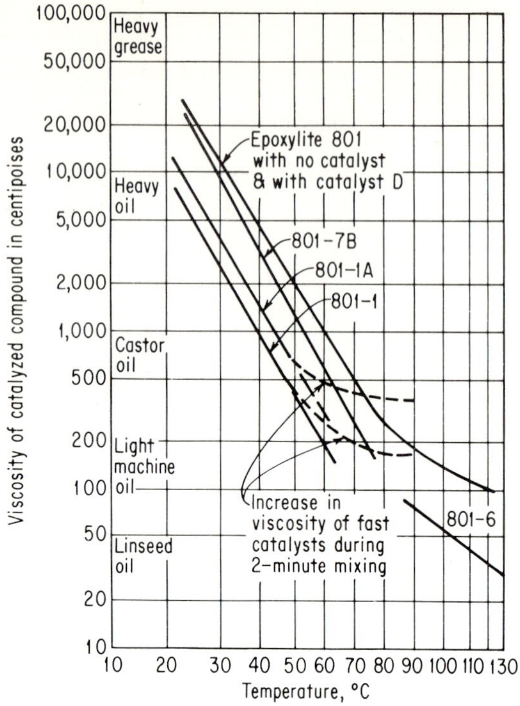

Fig. 17-2. Viscosity of 25%-filled catalyzed DGEBA vs. temperature.

will result in some loss of pot life. Whether or not cold mixing is practical will depend on the reactivity of the ingredients.

Reactivity vs. Temperature

By heating the ingredients of the epoxy-resin compound to a preselected temperature, desirably low working viscosities may be achieved. There are, however, a number of opposing factors which must be considered. The first of these is reactivity.

Epoxy-resin curing reactions ideally obey Arrhenius' law, which in effect states that when the threshold energy of a reaction is exceeded: *For every* 10 *degrees centigrade rise in temperature, the reaction rate doubles.*

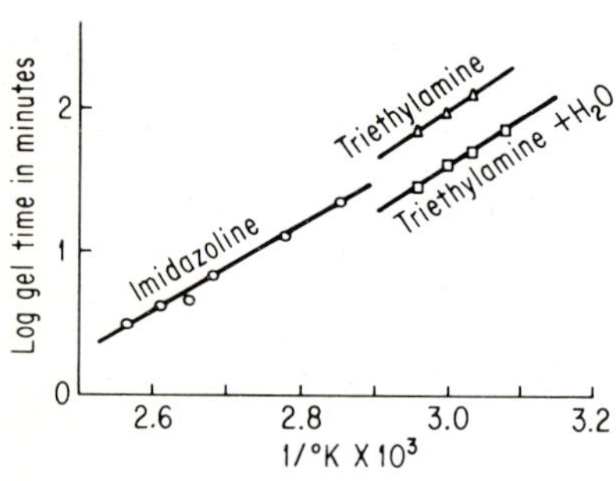

FIG. 17-3. Arrhenius plots of DGEBA reacted with amines [42].

The significance of this is: For every 10°C rise, the pot life is reduced by 50 per cent. Hence, if a compound will gel in 30 minutes at a room temperature of 25°C, then if warmed to 35°C, it will gel in approximately 15 minutes; at 45°C, in 7½ minutes. This is illustrated in Fig. 17-3, where gel time is plotted against the reciprocal of absolute temperature. Thus, when heat is used as a means of achieving low viscosity, reactivity becomes an opposing factor, and low viscosities may not be attainable because of high reactivities.

Reactivity is normally measured by gel time, as in Fig. 17-3. It should be pointed out, however, when using gel time at temperature as a measure of reactivity, two additional factors may affect the values and give rates which deviate from theoretical: molecular weight and mass.

When gel time at elevated temperature is used as the end point, the resin will not gel until a higher degree of polymerization is reached than required for gelation at room temperature. A measure of gel point in closer agreement with theoretical is obtained when a small amount of the reacting mass is removed periodically (on the end of a toothpick conveniently) and examined at room temperature.

The second factor is variation in mass. As mass increases, the temperature increases because of exotherm.

Exotherm

Exotherm is defined as the increase in temperature of the compound above the cure temperature due to the energies released as the epoxy groups react. A typical heat release is 24,000 cal per mole of epoxy oxygen [19,33]. On this basis, one pound of resin (2.39 moles if epoxy assay is 190) and one-tenth of a pound of curing agent would experience a temperature rise expressed by

$$\Delta H = \Delta T C_p m$$

where ΔH is heat evolved from 500 grams reactants = (24,000 × 450)/190 cal per lb. Take C_p as 0.3 cal per gram and m as 500 grams. Therefore

$$\frac{24{,}000 \times 450}{190} = \Delta T 0.3 \times 500$$

$$\Delta T = 374°\text{C}$$

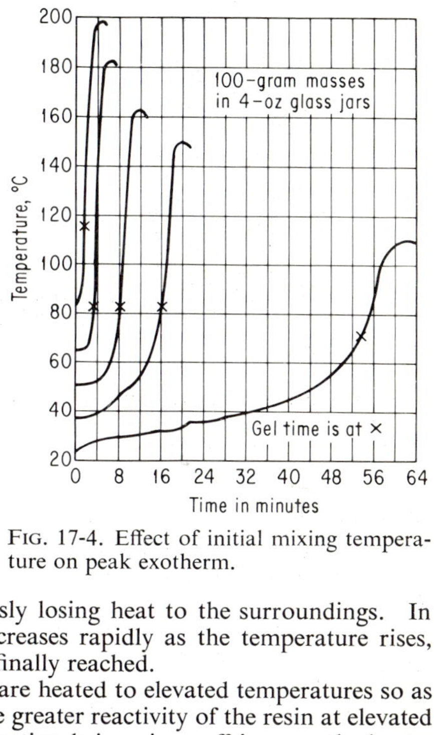

FIG. 17-4. Effect of initial mixing temperature on peak exotherm.

This is a large temperature rise and is indicative of the total heat given off during the cure of the resin. This value, of course, would represent the adiabatic temperature rise, i.e., the temperature rise which would be experienced if the resin were reacted in a perfectly insulated container. In actual practice, however, the resin is continuously losing heat to the surroundings. In addition, the specific heat probably increases rapidly as the temperature rises, further tending to limit the temperature finally reached.

However, if the resin and curing agent are heated to elevated temperatures so as to obtain low working viscosities, then the greater reactivity of the resin at elevated temperatures will result in the heat of reaction being given off in a much shorter period of time, which allows much less time for dissipation to the surroundings. Accordingly, the temperature rise approaches more closely that obtained under adiabatic conditions. This effect is illustrated in Fig. 17-4.

Even though a resin is not worked hot initially, an effect similar to initial heating is obtained when casting large masses. This is due to thermal conductivity considerations—the resin at the outside of the casting is cooled by the environment, but the resin at the center must necessarily rise to a temperature above that of the surrounding resin so as to be able to drive its heat to the outside. When using a fast resin, it is quite possible for the center of a casting to achieve a heat buildup approaching theoretical adiabatic. The casting at this point gases, chars, and explodes.

Even if the reaction does not go to such extremes, numerous detrimental effects can be experienced by high exotherms. Temperature-sensitive transistors can be destroyed; magnet-wire insulation can be thermally abused so that internal short circuits are experienced. Even in plain castings with no inserts, severe gassing and

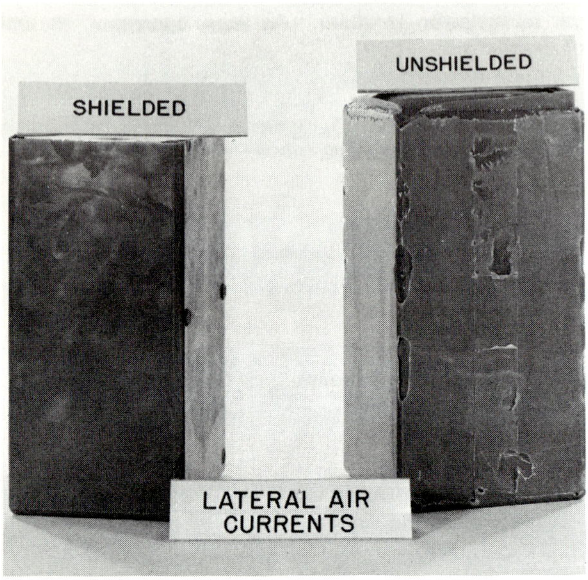

Shielding casting from oven convention currents during cure reduces localized overheating and minimizes convection currents in the casting, leading to markedly improved surface finish. (*Benjamin A. Davis.*)

bubbling can occur because of vapor-pressure considerations. Further, convection currents, set up in the ungelled resin by differences in density, arising from exotherm and oven cooling, can cause uneven gelling and shrinkage voids at the mold surfaces [40].

With DGEBA, amine curing agents, because of high reactivity, liberate heat quickly with correspondingly high exotherms; anhydride curing agents liberate heat slowly and generally give low exotherms even with elevated-temperature cures. With other epoxy resins, exotherms with amines may be reduced and those with anhydrides increased.

Vapor Pressure

The pressure of the gaseous phase of a substance in equilibrium with its liquid phase is defined as the vapor pressure of the substance. At the temperature at which the vapor pressure equals the surrounding or atmospheric pressure, the boundary between the gas and liquid disappears and the substance boils.

If an epoxy-resin and curing-agent combination becomes too hot, this leads to the generation of gas bubbles, which, if they form just before the resin gels, will

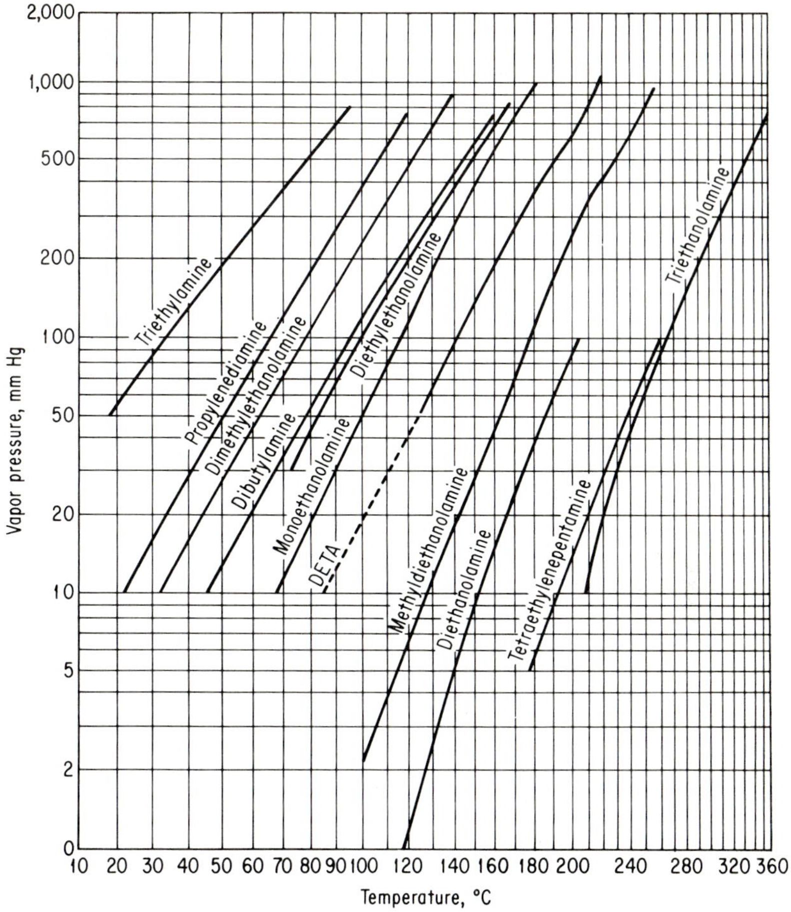

FIG. 17-5. Vapor pressure of various amines [47].

remain trapped in the cured structure. Such bubbles then become sources for mechanical stresses and cracking.

The vapor pressures of a number of curing agents are presented in Fig. 17-5.

Vapor pressure must also be taken into account if vacuum is used to remove air bubbles from the resin/curing-agent mixture. It is quite possible that a significant portion of the curing agent or accelerator may be stripped off inadvertently during vacuum processing.

Shrinkage

Another aspect of working temperature, besides its effect on viscosity, reactivity, exotherm, and vapor pressure, is the effect on shrinkage.

Shrinkage is the reduction in volume or in linear dimensions observed during cure (Figs. 17-6 to 17-8). Shrinkage induces stresses which can damage pressure-sensitive inserts or which can lead to early cracking under thermal cycling [16,20,23, 24,29,41,50,63].

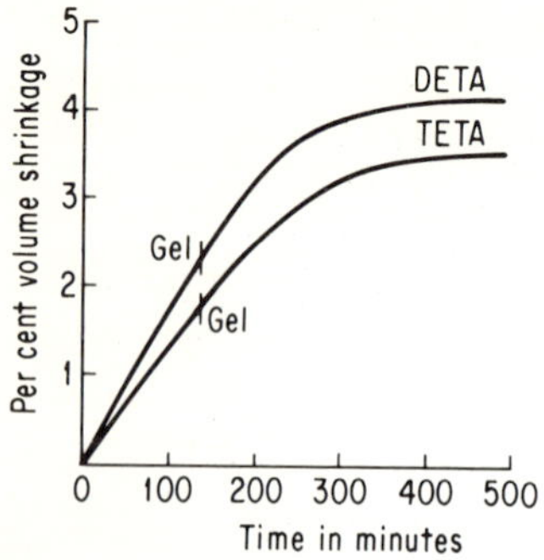

FIG. 17-6. Volume shrinkage at 25°C of DGEBA cured with aliphatic amines [4-6].

Curves of density vs. temperature for a typical casting resin during the curing process are presented in Fig. 17-9. In the figure, the time factor is omitted for the sake of simplification. The right ordinate shows the linear shrinkage expressed in per cent and referred to the cured resin at 20°C. The following explanation provides details on the curves:

We shall first study the course of the curve through points *ABCDEF*. The uncured resin and PA mixture, extrapolated at a room temperature of 20°C, shows a density of 1.229. Assuming that the heating up to the curing temperature is carried out so rapidly that no reaction can take place, the change in density with increasing temperature is linear, according to the following equation:

$$D_T = 1.229[1 - 6.8 \times 10^{-4}(t - 20°\text{C})]$$

If curing is carried out at a constant temperature of 120°C, density will continue to increase until curing is completed (point *D*). When the cured resin is cooled down to room temperature, two different expansion or contraction coefficients have to be taken into consideration. Down to about 105°C (point *D*), the change in density takes place according to the equation:

$$D_T = 1.205[1 - 5.0 \times 10^{-4}(t - 110°\text{C})]$$

Below 105°C, the density change takes place according to the equation:

$$D_T = 1.231[1 - 2.2 \times 10^{-4}(t - 20°\text{C})]$$

The point *E* of the intersection of the two straight lines is, according to definition, the second order transition point. At 20°C, the densities of the uncured and cured mixtures are practically identical. If it were possible to cure at this temperature, the total shrinkage value would be ideal, that is to say, practically zero. If curing could be carried out at 0°C, there would even be a negative shrinkage of expansion of approximately 1% by volume. From the point of view of application, a distinction should be made between density changes: (*a*) in the liquid phase, and (*b*) those in the gelled or solid phase.

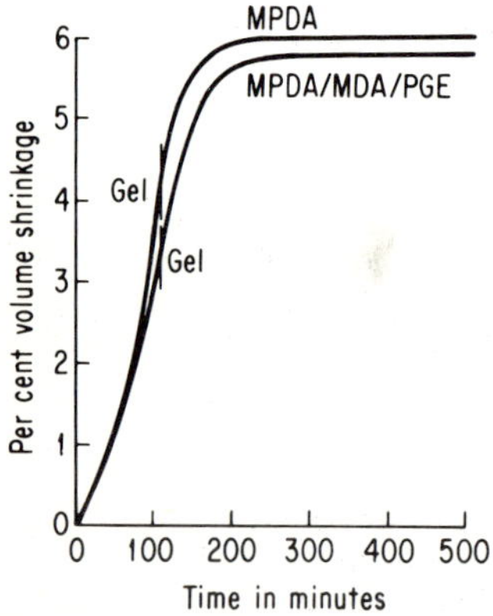

FIG. 17-7. Volume shrinkage at 65°C of DGEBA cured with aromatic amines [4-6].

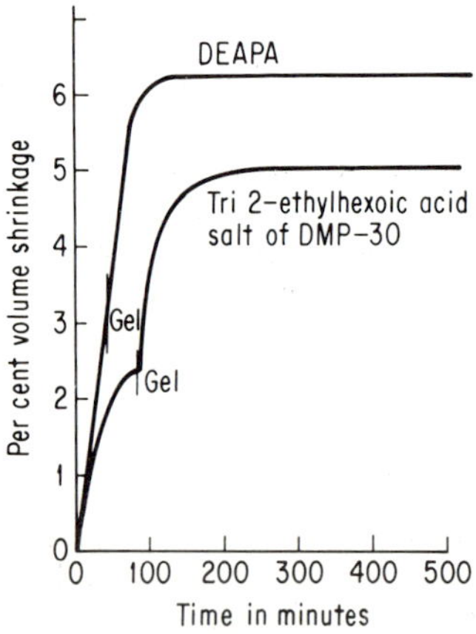

FIG. 17-8. Volume shrinkage at 65°C of DGEBA cured with tertiary amines [4-6].

As long as the density changes take place in the liquid phase, no internal stresses can occur. In the gelled or solid state, however, stresses can always appear and they are inevitable if materials which do not have the same thermal coefficient as the resin, are embedded in it. In our resin mixture the isothermal shrinkage due to curing is distributed about evenly between the liquid phase and the gelled or solid state. A so-called "gel line" is drawn, therefore. The technically important shrinkage, i.e., that which might cause stresses, appears below this gel line. This shrinkage is represented by the line *CG*. Shrinkage at any curing temperature can be read from the figure. For example, if the resin is cured at 180°C, the density curve passes through points *AHIKEF* and the shrinkage in the solid phase is represented by the line *IL*.

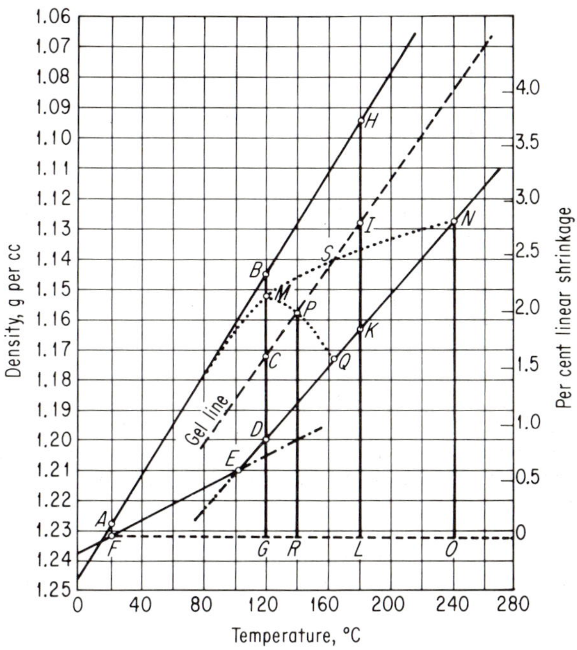

FIG. 17-9. Density changes during cure of DGEBA/PA system [56].

The above observations are only valid if the heating up time is very short and if curing is carried out at a constant temperature. In practice, this is feasible when processing small quantities of resins. In the case of larger castings a partial chemical reaction takes place during the heating up period, and it is impossible to avoid an exothermic reaction. These circumstances are taken care of . . . by the broken lines. Up to 120°C the density curve is not linear to point *B* but goes directly to point *M*. From this point on, the following cases of exothermic reactions are to be considered:

(*a*) The reaction is very fast exothermically so that after having gone through the gel line the decrease in density caused by rise in temperature is greater than the increase in density caused by chemical reaction. The shrinkage in the solid phase is determined by the maximum temperature (line *NO*).

(*b*) The type of reaction whereby after having passed the gel line, chemical shrinkage is greater than expansion caused by rise in temperature. The density shows a continuous increase and a rise in temperature no longer causes a rise in shrinkage.

Shrinkage in the solid phase in (*b*) above is not determined by maximum temperature but by the temperature at the intersection of the gel line (point *P*) and the shrinkage is represented by the line *PR*. In order to achieve low shrinkage, the exothermic reaction

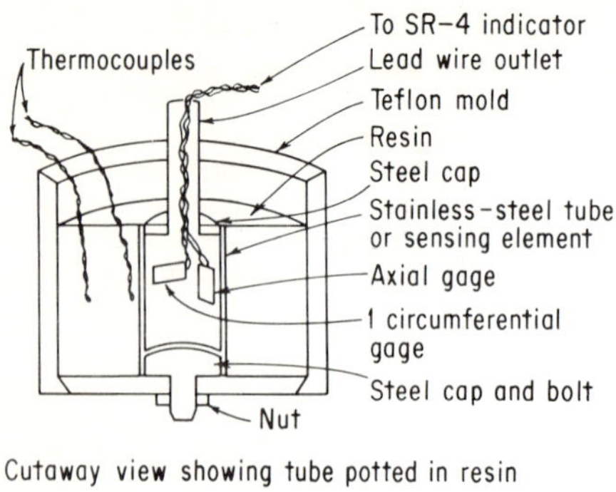

FIG. 17-10. Test apparatus employing strain gage to determine stress in epoxy-resin casting [4-6].

should be contrived, if possible, as follows: (*a*) It should pass the gel line at the lowest possible temperature. (*b*) After having passed the gel line, the reaction should proceed slowly so that chemical shrinkage is always greater than thermal expansion caused by rise temperature [56].

The disadvantages of high shrinkage and the resulting stresses are:

1. Incomplete fill of mold or inaccurate reproduction of mold surfaces
2. Cracking of the casting during cooling
3. Cracking of the casting during thermal cycling
4. Damage to stress-sensitive inserts

Internal stresses have been investigated by the use of strain gages, a typical setup being shown in Fig. 17-10. From strain-gage work, it has been shown that the use of flexibilized epoxy-resin compounds reduces internal stresses as shown in Fig. 17-11, with the reduction being less pronounced as the temperature is lowered and the resin assumes an increasingly rigid character. Cure cycle has been shown to have influence as shown in Fig. 17-12.

Studies of stresses during cure have also been conducted by the use of photoelastic measurements [60] and with solid-state transducers. The latter give data as indicated in Fig. 17-13. In general, stresses exerted on the encapsulated component are more a function of the elastic modulus of the encapsulant than of the difference in expansion coefficients. Lower temperature cures reduce the stresses, since, in effect, the thermal shock from the gel point is less severe [91].

When end requirements dictate use of a rigid system, pressure-sensitive inserts may be embedded in a flexible compound before the final encapsulation, the flexible barrier serving to reduce stresses as indicated in Fig. 17-14. A silicone release agent has been found to provide some stress relief for embedded, stress-sensitive

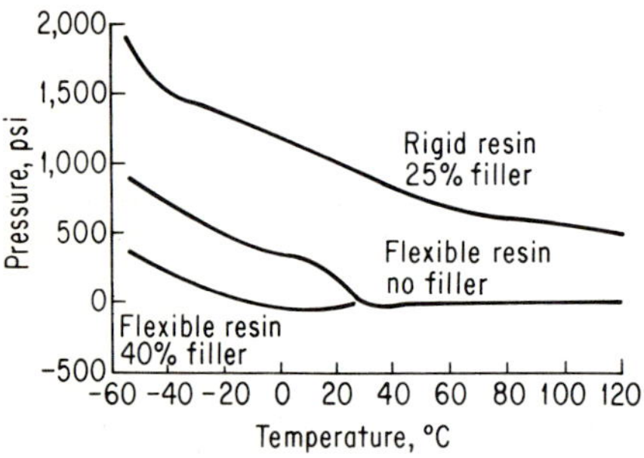

FIG. 17-11. Pressure exerted by rigid and flexible epoxy-resin formulations during thermal cycling [32].

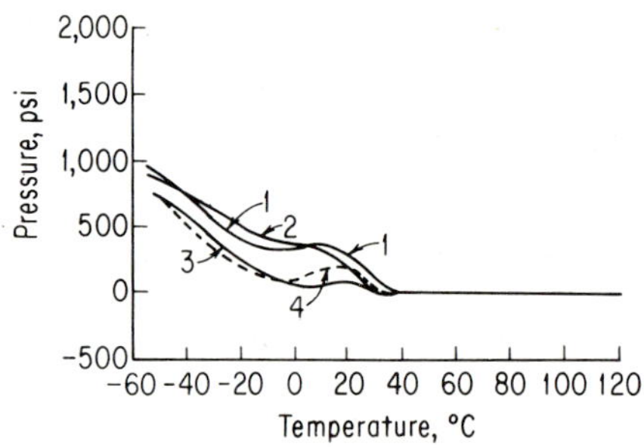

FIG. 17-12. Effect of cure temperature on pressure exerted by flexible epoxy-resin formulation during thermal cycling [32]. 1, Cured 2½ hours at 127°C; 2, cured 2½ hours at 120°C; 3, cured 16 hours at 95°C; 4, gel time 10 hours at 95°C plus cure time 2 hours at 120°C.

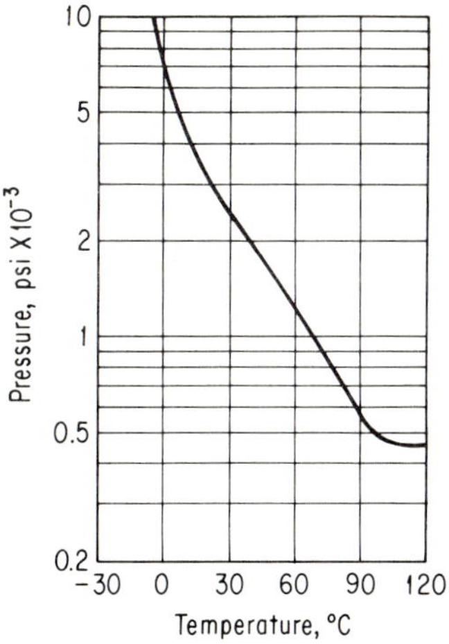

FIG. 17-13. Pressure exerted on embedded transducer by filled DGEBA system during thermal cycling [72].

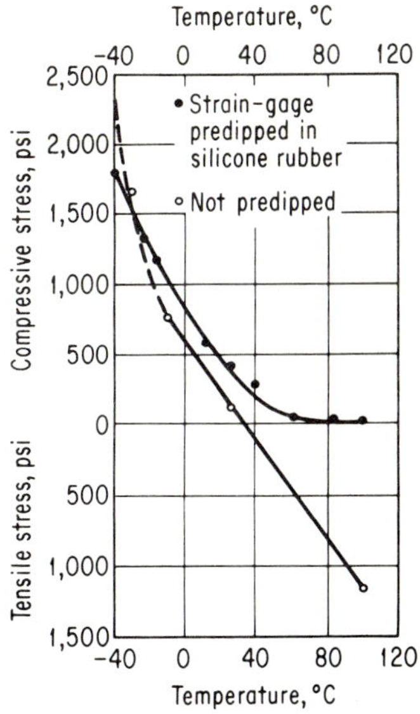

FIG. 17-14. Effect of flexible pre-coat on stress level of DGEBA fatty-polyamide-cured system [17-36.]

capacitors, but RTV silicones are superior. The coating serves to distribute stresses to prevent localized buildups. At lower temperatures, because of higher contraction rates, it will separate from the epoxy interface to prevent stresses from developing. As might be expected, there is a close correlation between modulus of the encapsulating material at lower temperatures and changes in capacitance of embedded capacitors. The lower the temperature required to obtain a stiffness modulus of 5,000 psi, the less capacitance change. Hydraulic tests indicate that the capacitors for this test were stable to readings of 7,000 psi, indicating that the pressure, per se, was not causing the changes in the readings, but rather, that the localized stresses were at fault [86]. Where use of a semirigid or flexible epoxy resin is feasible, the silicone coating may be dispensed with.

In addition to the use of flexibilized resin systems and the control of cure cycles to reduce imposed stresses, fillers may be employed to reduce shrinkage. The effect of fillers on shrinkage has been indicated in Chap. 14.

Expansion Characteristics

Fillers, in addition to reducing shrinkage, lower the coefficients of thermal expansion of the epoxy-resin compound. In a fully cured resin, the variations in pressure with temperature on an embedded insert can be calculated from the differences in thermal expansion rates of the materials involved. In one series of tests, hydraulic pressure was applied to a column of mercury to cause it to provide the same reading as those of an adjacent thermocouple, The pressure on the hydraulic fluid was taken as a measure of pressure exerted on the glass bulb of the thermometer. Typical data are given in Fig. 17-15.

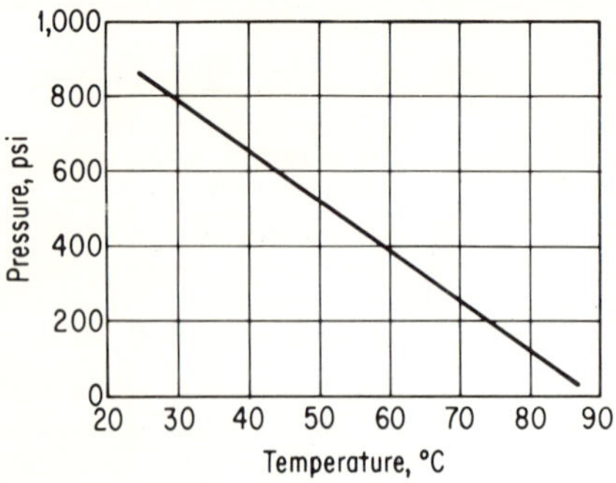

FIG. 17-15. Pressure exerted on glass rod by 20 per cent glass-fiber-filled DGEBA cured with aliphatic amine [41].

Thermal expansion characteristics are a function of the resin, the curing agent, the modifiers, diluents, and fillers present in the formulation. In general, the more highly flexibilized the compound, the higher the expansion rate; the more highly filled, the lower the expansion rate. Ideally, the expansion rate of the resin should be matched to those of the inserts present or should provide an intermediate value in the case of differing expansion rates of the inserts. In practice, when metal inserts are present, this requires extremely high loading volumes of fillers and initially rigid compounds. Expansion rates for various materials are indicated in Fig. 17-16.

Even though it were possible through use of fillers to obtain an exact match of the gross coefficients of thermal expansion between the resin compound and the

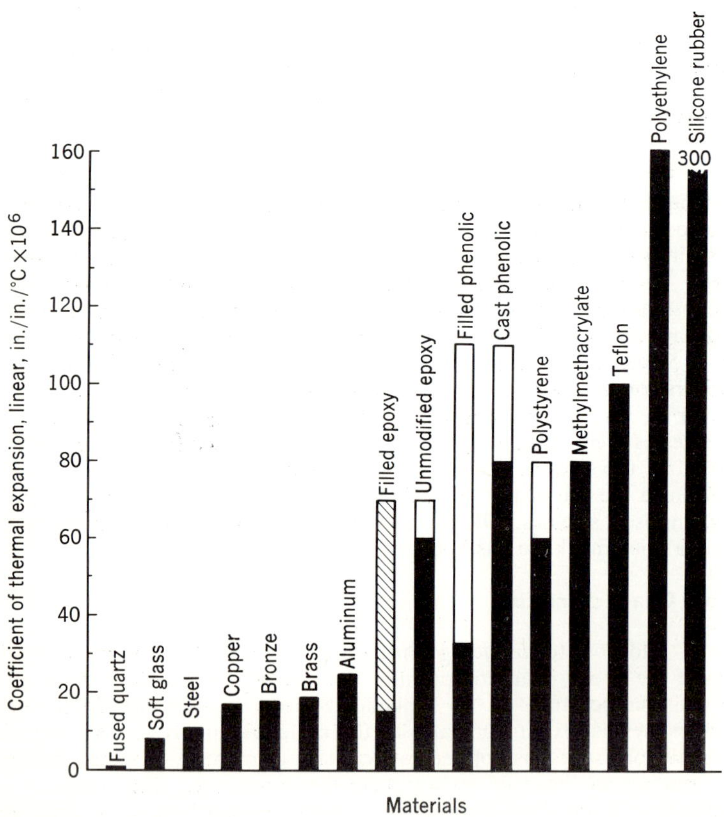

FIG. 17-16. Comparison of coefficient of thermal expansion of various materials.

inserts, a stress-free system would not be obtained. Rather, the stresses would be transferred to the microscopic level, arising between the filler particles and the bulk resin. Thus, while fillers may lead to some improvement in reducing stress levels, their use alone is not necessarily sufficient.

Because the stresses generated by an epoxy compound during temperature cycling are reproducible, the epoxy can be used actively as a method for applying pressure on embedded devices with temperature decrease. In this way, epoxy encapsulation is used for restoring signal capabilities of ferrite cores at low temperatures by means of application of pressure. Such cores are thus capable of operation over a wider temperature range, and the sophisticated electronic compensating circuitry otherwise required is eliminated [92]. In a similar fashion, encapsulated magnetic pressure pickups can be used to monitor rapid temperature changes such as would occur during the tumbling action of a space vehicle with regard to solar radiation.

Thermal Shock

From the above, it can be concluded that when inserts are involved in casting resins, the likelihood of stress cracking during cure or thermal cycling will be a function of the elastic modulus of the resin, the thermal expansion characteristics, and the stresses imposed by the curing temperature.

Thermal shock resistance, indeed, can become a severe problem with the inherently somewhat rigid epoxy-resin systems. The problem is magnified when higher-temperature operation is required, insofar as many flexible systems are inherently unsuited for this service because of relatively poor thermal stability.

To obtain satisfactory thermal shock resistance for severe thermal cycling requires sophistication in the formulation. Initial stresses caused by shrinkage and cure temperatures should be minimized. Reversible stresses caused by the expansion characteristics of the compound should be reduced to the level practical. The basic epoxy polymer should be invested with a combination of flexibility and toughness.

As the stress level on the compound is increased, a point is reached where further relief is not possible through elongation. At this point, only the ultimate tensile strength of the compound is available to resist further imposed stresses, and should this strength be exceeded, failure will result.

On the low-temperature end of the thermal cycle, the elongation is at minimum and the tensile strength is at maximum; on the high-temperature end of the cycle, the elongation is at maximum and the tensile strength is at minimum. Unfortunately, increasing the elongation of the formulation serves to decrease the ultimate tensile strength, and conversely.

In addition to the stresses imposed for reasons discussed above, additional stresses may be applied to the epoxy-resin compound by temperature gradients (e.g., hot internal windings and air-cooled surfaces), and by external factors (e.g., centrifugal forces in rotating electrical equipment).

The types of thermal shock failure which may occur, generally at stress-concentration points imposed by the inserts, are indicated in Fig. 17-17.

Thermal shock testing, therefore, is accomplished through the use of inserts of one type or another which serve to concentrate stresses. In this way, resin formulations may be compared for their relative thermal shock resistance. Data from these tests, however, will not provide absolute design values since stress levels are highly dependent on the size and the configuration of the inserts, the thickness of the casting resin, and the conditions of service.

Thermal Shock Tests. There are numerous testing methods and test fixtures for evaluating comparative thermal shock resistance of epoxy formulations.

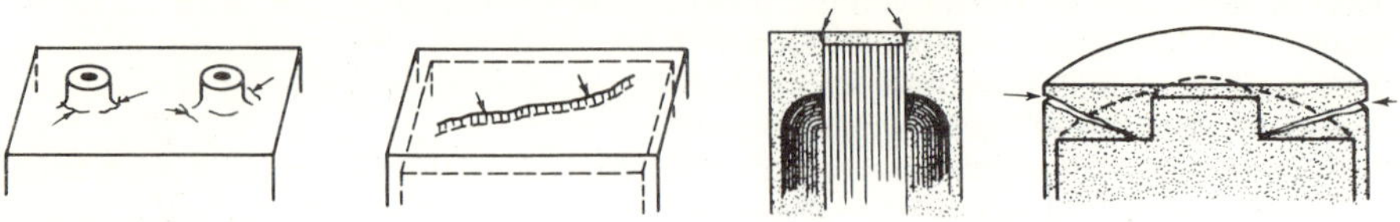

FIG. 17-17. Typical examples of in-service cracking of epoxy formulations [22].

These include:

1. Mil-I-16923 hex bar
2. Pipe with milled feet
3. Hex nut
4. Castellated ½-in. SAE nut
5. Vaned core
6. Nut and bolt in paper coffee cup
7. Olyphant washer

Three of these methods are described below in detail.

Hex-bar Test. One of the earliest thermal shock test devices is the hex-bar test of Mil-I-16923D. In this test, a section of hexagonal ¾-in. steel bar stock 1 in. long is supported centrally in a cylindrical glass mold 1 in. in diameter and 1½ in. high. The bar is cast and cured. Five specimens are heated for a period of time at an upper temperature (90, 105, 130, or 155°C) and then plunged immediately into an alcohol bath at −55°C. If, after 10 such cycles, four of the five specimens are without cracks, the compound is considered to have passed.

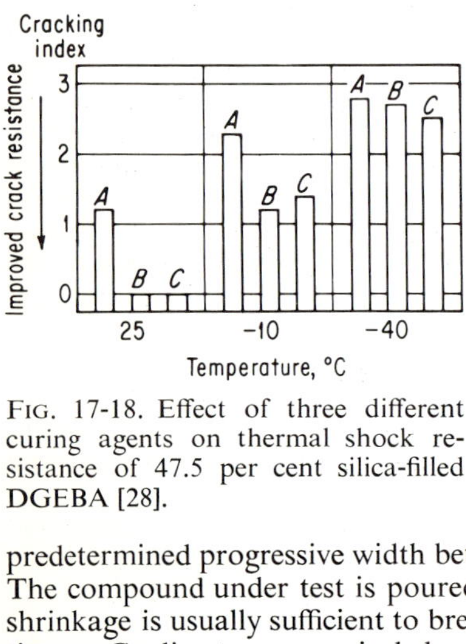

FIG. 17-18. Effect of three different curing agents on thermal shock resistance of 47.5 per cent silica-filled DGEBA [28].

The hex-bar test, although useful for screening out systems with relatively poor thermal shock resistance, is not considered sufficiently severe to discriminate between the better thermal-shock-resistant formulations.

Vaned Core. The vaned core tester is of interest insofar as it provides a numerical rating for thermal shock resistance.

The tester consists of a radially slotted steel cylinder with six hardened and pointed vanes disposed at 60° angles. This cylinder is mounted centrally on a spindle inside a cylindrical shell. The vanes form gaps of predetermined progressive width between the vane edges and the wall of the shell. The compound under test is poured between the core and shell and cured. Cure shrinkage is usually sufficient to break one or more of the thin bridges between sections. Cooling to progressively lower temperatures causes further cracking. Since the thickness of the bridges may be in either arithmetic or geometric progression, a direct and numerical comparison of cracking tendencies is available between successive tests. The tendency of the first breaks to relieve stress on the remaining bridges is lessened by stub inserts with rounded ends which serve as anchors.

Figure 17-18 shows the cracking encountered with three different curing agents at three different temperatures. Figure 17-19 shows the effect of filler concentration on thermal shock resistance.

It is interesting to note that at room temperature and −10°C, the samples with the higher filler content have higher cracking tendencies than the sample with the lower filler content. The reverse is true at −40°C. At lower temperatures, it is probable that thermal expansion rate is more critical than tensile strength. Studies with the vaned core indicate that the more closely the coefficient of thermal expansion approaches steel, the lower the cracking index at −40°C. When aluminum vaned cores are used instead of steel, the cracking index at −10 and −40°C averages one unit lower, which is in agreement with the higher expansion coefficient of aluminum compared to steel.

Olyphant Washer. The Olyphant washers incorporate the various features of the nut, pipe with milled feet, and hex bar into one thermal shock specimen. Two styles are employed, a ⅛- and a ¼-in. thick washer (Fig. 17-20). The latter is the more severe and provides the preferred test for differentiating between resins of high thermal shock resistance. Figure 17-21 shows the outstanding thermal shock resistance of various formulated products in comparison to simple resin/curing-agent blends.

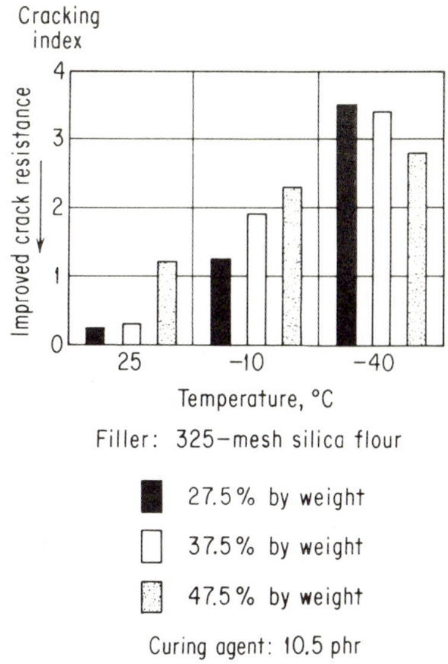

FIG. 17-19. Effect of filler loading volume on thermal shock resistance of DGEBA cured with DMP-30 salt [28].

In addition to thermal shock resistance tests on various inserts, comparative thermal shock resistance may be obtained indirectly from flexural modulus measurements at low temperatures.

Flexural Modulus. Thermal shock tests such as described above have about reached the limit of their usefulness. The better-formulated compounds can take repeated cycling from 180 to −75°C without failure. Thermal shock specimens could be markedly increased in size so as to induce greater stresses, or expensive prototype configuration testing could be employed. Such testing is inherently expensive.

Flexural modulus testing at low temperatures offers promise of providing a simple, low-cost method which gives a scale of values or indexes which can be correlated to thermal shock resistance [73]. Such testing additionally provides for convenient determination of thermal shock resistance after various periods of aging.

Thermal Stability

Thermal stability indicates the ability of the resin to serve at elevated temperature and maintain minimum specified properties.

With elevated-temperature aging, there is generally a progressive loss of strength and an overall reduction in properties. Often there is progressive embrittlement because of loss of volatile fragments and increased cross-linking. As temperature is further increased, surface charring will occur, followed by cracking and decomposition.

Evidently, then, when elevated-temperature service is involved, data from freshly cured samples are inadequate. Meaningful data can be obtained only after longtime aging or as the result of some form of accelerated aging tests at temperatures substantially in excess of the desired service temperature.

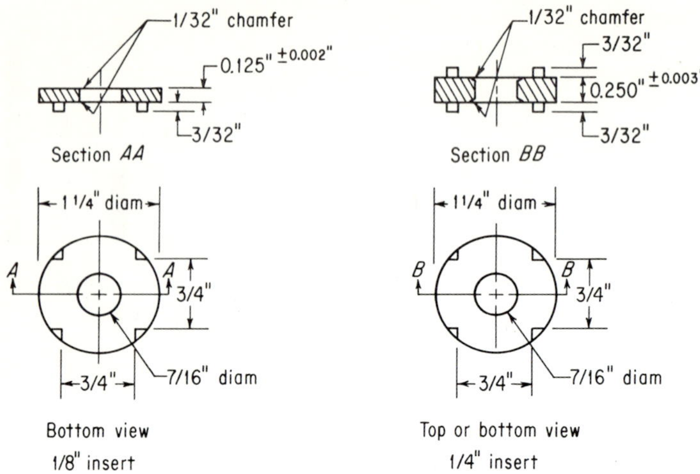

FIG. 17-20. Dimensions of Olyphant washers [22].

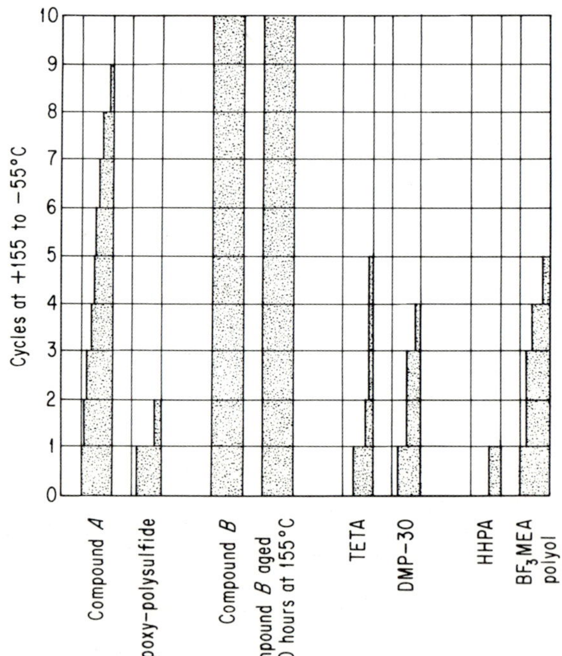

FIG. 17-21. Thermal shock resistance of various DGEBA formulations.

Ideally, the full range of physical properties of importance should be reported after high-temperature aging. This, of course, is expensive and time-consuming. A number of simplified procedures have been employed.

One method for estimating relative thermal stability involves changes of DT with aging; another, changes in flexural strength (Fig. 17-22). The most commonly used method of estimating service life is on the basis of weight loss vs. time. In this case, a weight-loss value is selected more or less arbitrarily as the end point. Typical data are shown in Figs. 17-23 and 17-24. In these figures, weight-loss data are correlated with expected operating life, with 16 per cent loss being selected as the end point.

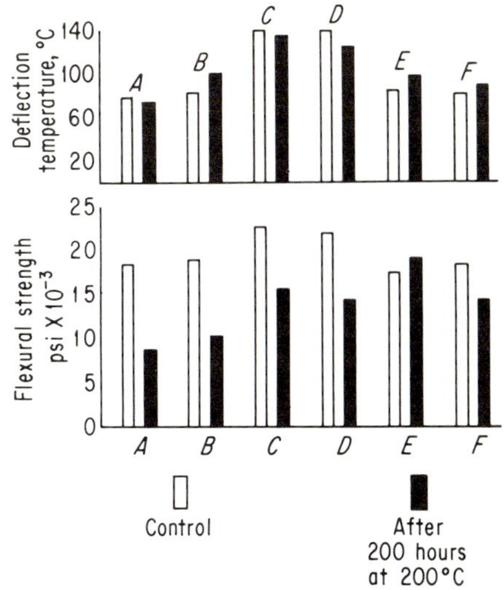

FIG. 17-22. Mechanical properties vs. heat aging of cured DGEBA [4-7]. *A*, DGEBA, glycidyl diluent, and DETA; *B*, DGEBA and cyanoethylate primary amine; *C*, DGEBA and MDA; *D*, DGEBA and MPDA; *E*, DGEBA and bisphenol A; *F*, epoxylated novolac and phenolic resin.

When the epoxy-resin casting compound is used as a primary dielectric barrier, end points in service life may be estimated by considering the degradation mechanisms as being temperature-dependent according to the Arrhenius equation. A convenient elevated temperature is chosen, and testing is continued until minimum acceptable electrical values are reached. The data are then extrapolated backward, assuming that the service life doubles for each 10°C decrease in temperature. A typical life curve for epoxy magnet wire, using breakdown under 1,000 volts alternating current as the end point, is shown in Fig. 17-25.

Obviously, the above methods are open to criticism. Perhaps better thermal stability data can be obtained by use of heat-aged thermal shock samples. Even more promising is the use of flexural modulus testing as a measure of loss of elongation after heat aging. For example [73], one formulation might show 3.5 per cent elongation at −40°C originally and less than half that after 80 hours at 200°C, whereas another would retain flexural elongation greater than 3.5 per cent at −40°C, even after aging 400 hours at 200°C. It is probable that similar data may be developed from tensile test data, where that parameter is more critical than elongation.

It should be noted, however, that thermal stability, per se, may not be the sole consideration in the selection of a casting compound. In service, the combination environment (heat, moisture, mechanical abuse) is often much more deterimental than heat alone. For example, numerous lower-temperature-rated epoxy-resin systems may far outperform the higher-temperature-rated silicones at 180°C since the latter may fail well before their thermal end point because of other environmental factors.

Generally, epoxy-resin systems can be developed to provide good, long-time service life at temperatures up to about 150°C, and special systems have been used for long-time service up to 180°C. At temperatures in excess of this, service life is considerably reduced.

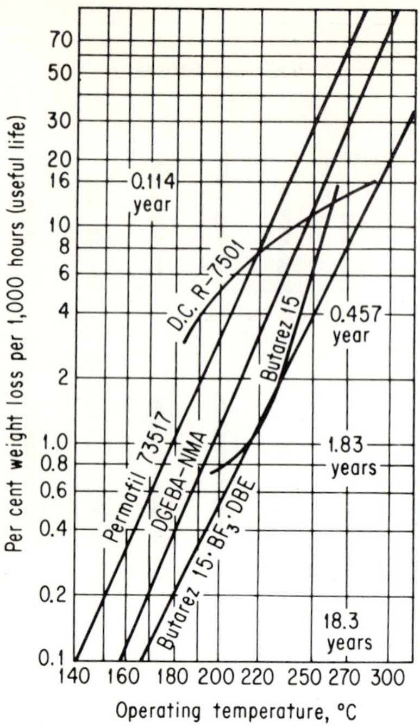

FIG. 17-23. Continuous operating temperatures of five heat-resistant resins at various weight-loss ratios [34].

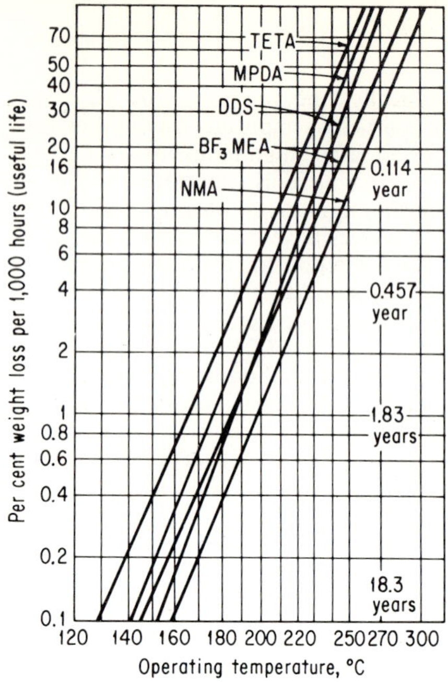

FIG. 17-24. Curing agents vs. continuous operating temperature for cured DGEBA [34].

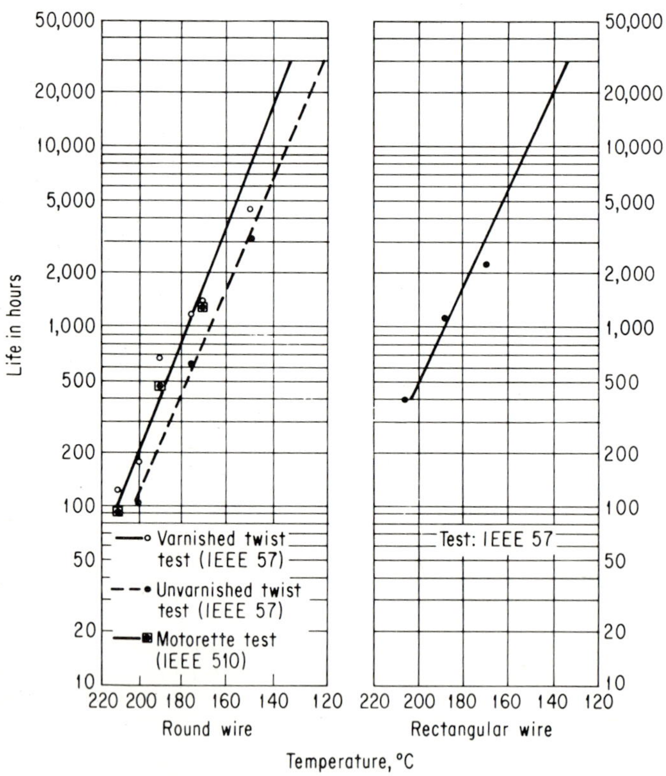

FIG. 17-25. Thermal life curve on epoxy wires [1].

GENERAL CONSIDERATIONS FOR ELECTRICAL AND ELECTRONIC POTTING AND ENCAPSULATING APPLICATIONS

Electrical and electronic applications have in common a number of considerations, as discussed subsequently. Following the general discussion, specific examples will be confined to transformers and motors. In the electronics field, there is a monthly publication covering electronic packaging, with emphasis on the use of modern plastics [2], and at least one excellent book is available [57].

Electrical Properties

In the majority of the electrical and electronic potting and encapsulating applications, the electrical properties of the resin are of secondary importance. Primarily the resin is used to protect the components from the environment and to provide unitized construction and increased strength.

In a limited number of applications, the epoxy-resin compound is used as a primary dielectric barrier. Data applicable to such considerations are given elsewhere in the book.

Of the electrical properties, the one of greatest interest is volume resistivity, since, while electrical properties are usually of secondary importance, the casting resin should possess electrical insulating qualities.

Electrical conductivity in high polymers results from migration of ions through the polymer network [27,37,52]. The ions arise from (1) ionic impurities, (2) ionic products of thermal degradation, and (3) ionic products from degradation caused by background radiation. The migration of these ionic current carriers in an electric field depends upon the mobility of the ion. The mobility is proportional to the reciprocal of the internal viscosity of the cured polymer.

The chief concept in formulating an electrical insulating casting resin then becomes a balance between high thermal shock resistance (which implies a softened or flexibilized system with lessened internal viscosity) and high volume resistivity (which implies a dense, rigid system with high internal viscosity). Careful formulation can result in significant advances over basic resin/curing-agent systems, as can be seen by comparing Fig. 17-21 with Fig. 17-26. As seen from the latter figure, volume resistivity decreases with temperature as internal viscosity decreases. Sharp breaks are usually involved in the region of the DT. Thermal aging, likewise, reduces volume resistivity by introducing ionic products.

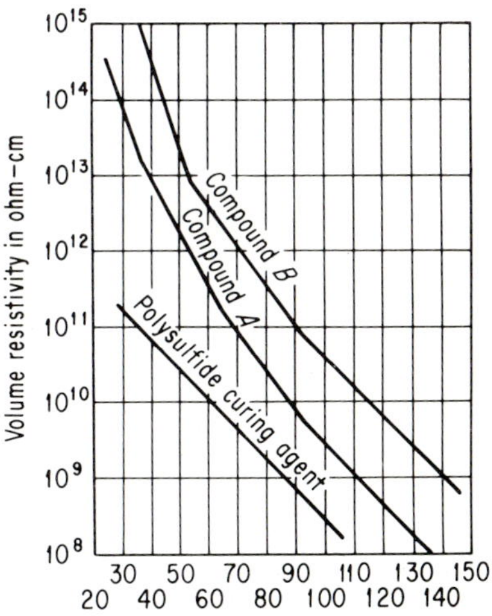

FIG. 17-26. Volume resistivity vs. temperature of DGEBA cured with polysulfide and proprietary formulations.

Chemical Resistance

High humidities constitute one of the more destructive environments normally encountered by electrical-electronic equipment. Epoxy resins, by maintaining good humidity resistance, are particularly well suited for humidity protection. This

Comparison of varnished and encapsulated 5-hp stators. Unit is intended for exposure to fluorinated refrigerants. Highly cross-linked epoxy resins reveal good resistance to degradation by such chemicals. (*The Epoxylite Corporation.*)

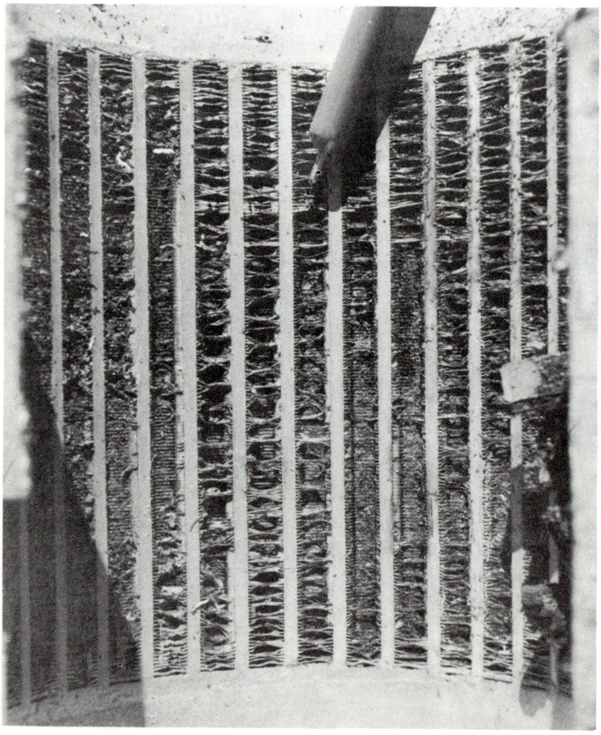

Stator encapsulated in epoxy resin has been exposed to nitric acid bath, which has etched away all iron laminations, leaving only the epoxy resin. Excellent fill achieved is revealed by the thin webbing of epoxy resin which had penetrated between the laminations. (*The Epoxylite Corporation.*)

humidity resistance, in fact, is one of the major reasons for the epoxy resins capturing such a large share of the potential market.

Figures 17-27 to 17-29 compare epoxy and polyester casting resins in terms of humidity resistance. Figure 17-30A presents a comparison of a typical amine and a typical anhydride-cured epoxy-resin system in terms of humidity resistance. Ionic curing agents, such as BF_3 complexes, provide relatively poor properties, as indicated in Fig. 17-30B. The importance of selection and treatment of fillers as regards humidity resistance of the cured resin is illustrated in Fig. 17-31. Determinations of the water-vapor diffusion coefficients of epoxy resins as a function of cross-link density and polarity (as measured by dielectric power factor), give results shown in Table 17-2. The effect of temperature on water-vapor diffusion is shown in Fig. 17-32.

The outstanding resistance of the epoxy resins to chemical attack from solvents, bases, and acids also makes them useful for providing protection in severe industrial

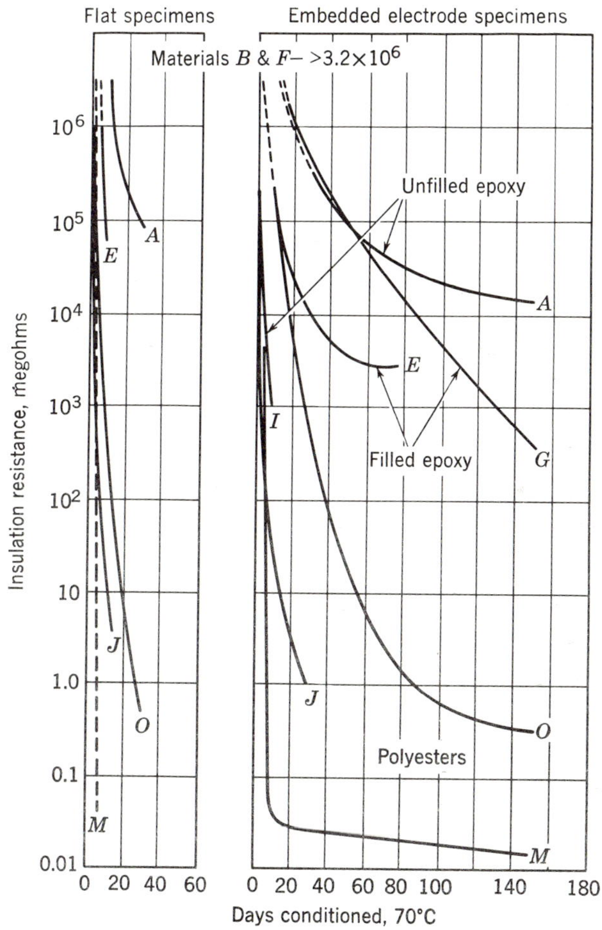

FIG. 17-27. Effect of moisture on insulation resistance of epoxy formulations (*A* to *I*) and polyester formulations (*J* to *M*) [12].

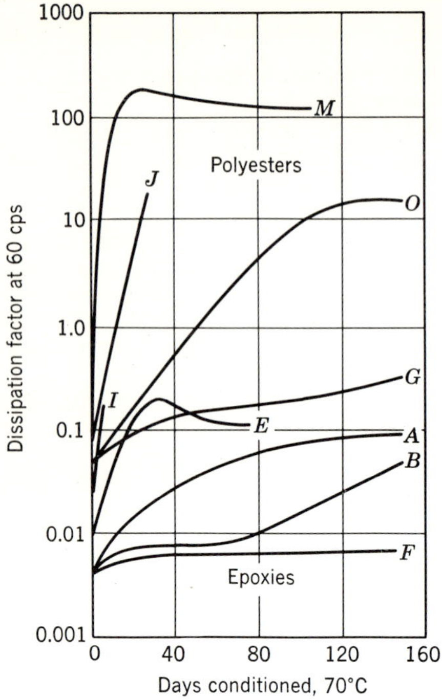

FIG. 17-28. Effect of moisture on dissipation factor of epoxy formulations (*A* to *I*) and polyester formulations (*J* to *M*) [12].

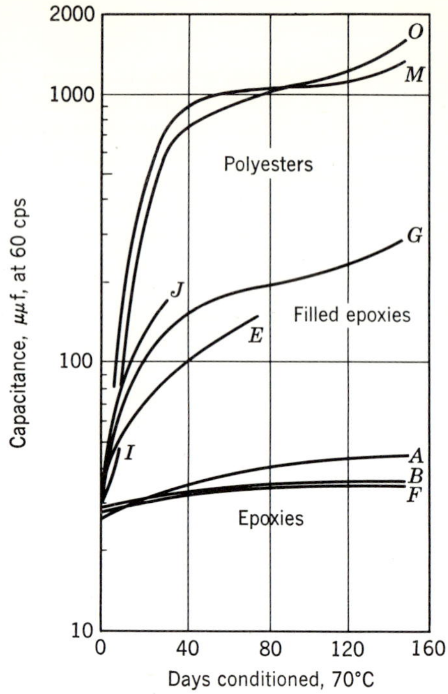

FIG. 17-29. Effect of moisture on capacitance of epoxy formulations (*A* to *I*) and polyester formulations (*J* to *M*) [12].

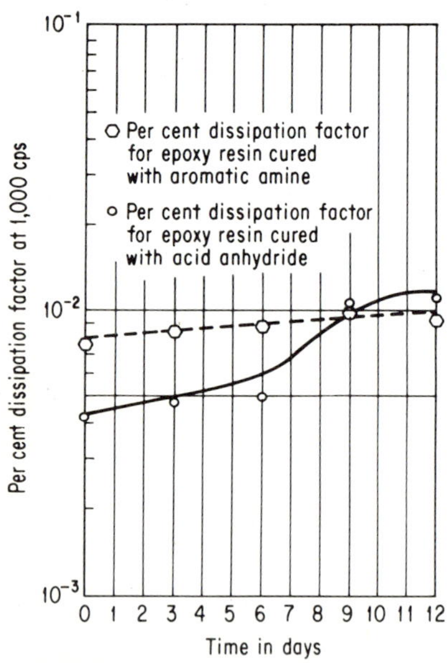

FIG. 17-30A. Dissipation factor at 70°C and 96 per cent relative humidity vs. curing agent for cured DGEBA [65].

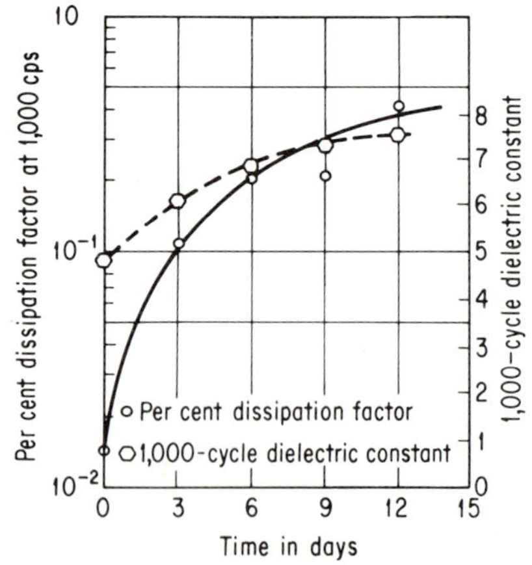

FIG. 17-30B. Dissipation factor and dielectric constant at 96 per cent relative humidity vs. aging time at 70°C for BF_3MEA-cured DGEBA [65].

Table 17-2. Water-vapor Diffusion Coefficient for DGEBA and Six Curing Agents [90]

DGEBA resin cured by	Diffusion coefficient, 10_9 g/(cm)(Torr)(hr)	Power factor
1. Anhydride and flexibilizer	28	0.050
2. Polyamine amine..............	25	0.015
3. Amidopolyamine..............	16	0.009
4. Aromatic amine...............	14	0.011
5. Anhydride	15	0.004
6. Dianhydride..................	13	0.004

environments. The resistance is particularly important for protecting motor windings exposed to fumes and chemical and detergent washdowns.

Thermal Conductivity

In electrical-electronic casting compounds, the thermal conductivity of the applied protection is a critical factor. The epoxy resins generally (unless they contain conductive fillers) are relatively poor thermal conductors. Therefore, when heat-generating units are involved in the inserts, the epoxy may act as thermal insulation. This results in higher internal operating temperatures and the attendant difficulties involved. Depending on the design of the unit, rerating may be required, with the components upgraded to higher-temperature classification, or heat sinks may be used to dissipate the heat. Neither expedient is particularly satisfactory from the standpoint of cost, reliability, and weight.

Alternatively, thermal conductivity may be somewhat improved by the use of fillers, as indicated in Chap. 14. In general, the loading volume of the filler will govern the thermal conductivity obtained, and coarse-particle-size fillers are preferred for best performance.

When using highly filled systems, the processing technique selected should minimize entrapped air, since air is much poorer than the epoxy compound with respect to thermal conductivity. Indeed, by displacing dead air deep in windings

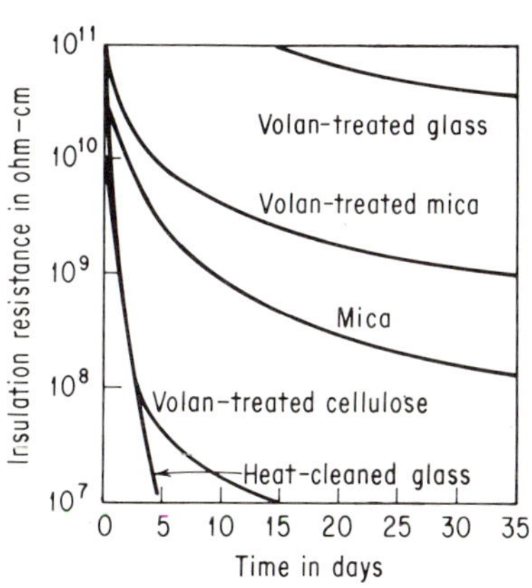

FIG. 17-31. Effect of 50 per cent filler on insulation resistance at 60°C and 96 per cent humidity for DGEBA cured with MPDA [4-6].

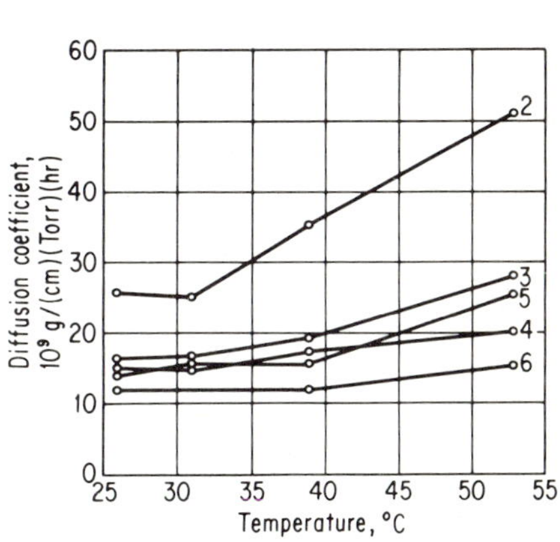

FIG. 17-32. Effect of temperature on diffusion coefficient of epoxy resin (see Table 17-2 for identification) [90].

and interstices with epoxy, some improvement in thermal conductivity may be obtained in a minority of designs over the unencapsulated system.

Compatibility with Magnet Wire

High shrinkage stresses in an epoxy casting resin can lead to other problems besides cracking in thermal shock and crushing of delicate components. The high adhesive strengths of epoxy resins combined with shrinkage stresses can apply very large stresses to magnet-wire insulation, often leading to otherwise mysterious internal short circuits.

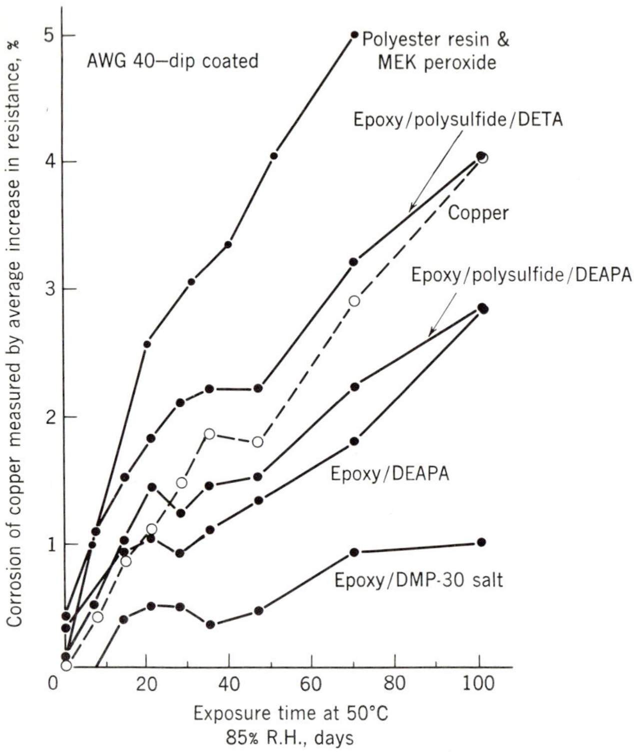

FIG. 17-33. Corrosive effects of various casting compositions on copper wire [10].

Besides the mechanical aspects of shrinkage on the wire and its insulation, there is also the matter of chemical compatibility. Not only will some curing agents corrode bare copper (Fig. 17-33), but also many of the curing agents and diluents used in epoxy-resin formulas are good strippers for the wire insulation, and therefore can seriously attack the wire before cure. Even trace unreacted amounts can attack the wire during service of the cast component, particularly during overload, overtemperature duty cycles.

Modern magnet wire, i.e., the wire used in all magnetoelectric devices such as magnets, coils, transformers, motors, is insulated with a film of organic varnish.

At least eight types of wire varnishes are used today in the United States and Europe. These are:

1. Oleoresinous (referred to often simply as black enamel)
2. Polyvinyl formal (Formex and Formvar)
3. Epoxy
4. Nylon
5. Polyester
6. Urethane
7. Silicones
8. Polyimide and Polyamide-imide

Oleoresinous enamel was the earliest wire enamel introduced. It is characterized by low cost and good service performance at temperatures up to class 105. It is, however, very susceptible to damage during winding operations, and hence is used only in special applications. Polyvinyl formal is one of the most popular all-around wires for class 105. It handles well and gives good overall performance.

Epoxy wires of two types have been introduced, the epoxy–urea-formaldehyde and the epoxy-polyester. The epoxy–urea-formaldehyde wire is still offered for class 130; the epoxy-polyester has been withdrawn in favor of newer, higher-temperature wires. The epoxy–urea-formaldehyde is widely used in oil-filled transformers because of its excellent resistance to hydrolysis.

Nylon wire is preferred for many winding applications because of the lubricity of nylon, which is an aid in positioning wires when winding small coils. Nylon is limited to class 130.

Polyesters, based on ethylene glycol, glycerol, and terephthalic acid, have been widely accepted because of high electrical performance and suitability at temperatures up to class 155. Polyesters are not often used in hermetically sealed units or oil-filled transformers because of tendency of the film to hydrolyze.

Urethane wires have the distinct advantage of ease of stripping for soldering operations. When heated by a gas flame or soldering iron to temperatures of 290°C or above, the film decomposes to gaseous products, leaving a clean wire, rather than a carbonaceous surface as with most films.

Silicone enamels are usually alkyd coatings reinforced with silicone resin. They are usually used in class 155 and class 180 applications.

The aromatic polyimides represent a new class of wires for temperatures up to class 220. The wire is characterized by high chemical resistance and very high cut-through resistance.

The compatibility of these wires with various epoxy-resin systems is measured by preparing twisted-pair specimens and potting them in the resin. The samples are then aged at high temperatures to accelerate any chemical attack according to a schedule recommended by IEEE Method 1-C.

During oven aging, the samples are checked electrically. Usually, a voltage of 1,000 volts is applied to determine whether breakdown has occurred. If no breakdown occurs, the specimens are returned to the oven for further aging and testing. Typical test data are presented in Tables 17-3A, 17-3B, and 17-4.

Experiments in which a fixed percentage of the specimens were tested to determine their breakdown voltage showed that the major percentage of breakdown occurred in the first few hours. After that, the dielectric strength of the wire leveled off at some residual value (Fig. 17-34).

From these data and others [66], it can be seen that the available wires and epoxy-resin/curing-agent systems vary widely in their thermal stability and chemical and mechanical compatibility.

Table 17-3A. Epoxylite® Thermal Aging Program [64]

(Average life in hours)

Film, no. 18 heavy	Temperature, °C	Epoxylite compound no.						
		None	205-1A	205-D2	288	108	293-12	6203
Black enamel	160	744	....	....	>3,870	.....	>3,870	....
	190	172	218	108	>1,747	.....	1,096	....
	210	.....	11	8	>879	.....	326	....
Formvar	160	529	....	1,342	3,027	555	>3,480	1,915
	190	108	1,807	206	448	136	676	283
	210	34	383	107	166	27	247	160
Nylon (Nyform)	160	1,755	....	1,902	......	.....	......	....
	190	395	1,103	264	1,046	.....	754	....
	210	91	188	110	315	.....	283	....
Acrylic (Acrylex)	160	1,550	....	1,903	....	750	4,806	3,197
	190	380	434	236	594	266	438	228
	210	185	279	71	121	53	146	34
Epoxy–urea-formaldehyde	190	350	1,398	523	>1,506	532	918	301
	210	171	212	203	279	90	236	72
	230	84	24	48	59	12	77	40
Polyester (Thermalex-F)	190	.....	263	218	.....	1,431	521	465
	210	1,966	42	62	681	310	120	180
	230	416	18	26	212	154	66	27
Single glass and Thermalex-F	190	.....	....	....	......	.....	3,114	.....
	210	1,574	Epoxy deteriorated	Epoxy deteriorated	3,932	1,973	955	2,802
	230	450	Epoxy deteriorated	Epoxy deteriorated	1,339	807	420	1,966

® is a registered trademark of The Epoxylite Corporation

Table 17-3B. Average Hours of Aging to Fail at 1,000 Volts [61]

Film, no. 18 heavy	Temperature, °C	DGEBA cured with					
		Polyamine	DETA + diluent	DDSA	NMA	DADPS	BF_3MEA
Poly-Thermaleze . . .	165	15,456 8 OK, 2 F	7,728*	5,224	15,456 No F	14,784 9 OK, 1 F	9,524
	200	553	744	1,008	4,000 No F†	1,107	140
Thermaleze B.	165	15,456 8 OK, 2 F	3,528	4,121	8,693	14,784 9 OK, 1 F	13,440*
	200	994	493	203	2,616	951	265
Formvar	165	15,456	4,175	2,675	3,924	5,989	8,722
	200	348	672	331	286	208	259*
Nyleze.	165	5,230	1,322	2,313	4,205	2,826	5,012
	200	409	198	176	234	130	208
Hermeteze	165	15,456 9 OK, 1 F	4,702	4,156	6,725	4,616	6,033
	200	297	294	260	360	154	200
Epoxy	165	15,456 7 OK, 3 F	2,704	4,131	12,550	5,527	6,279
	200	728	133	247	331	196	252

* Average or half-life determined at the point of 5 OK and 5 F (failures).
† This set of samples was lost because of oven failure.
NOTE: DGEBA/polyamine was dip-coated rather than potted.

Compatibility with Insulating Papers

Oftentimes cracking encountered in the field with epoxy-resin-encapsulated components is attributed to thermal shock when close examination will reveal that the cracking was really due to "mechanical incompatibility" of the insulation system with the concept of encapsulation. For example, many insulating papers and cloths are built-up laminates. When used in encapsulated units, the epoxy adheres tenaciously to both sides. Shrinkage and thermal stresses are transferred to the insulation paper. If the insulation paper has an interlaminar strength lower than the forces existing in the resin, the paper will delaminate. If the paper is internal, the delamination becomes a focal point for corona. If the insulating paper projects up near the surface of the casting, then the thin film of resin covering the paper becomes a point of high stress concentration. It will crack and open up to the delaminated paper, thereby providing a passageway for moisture.

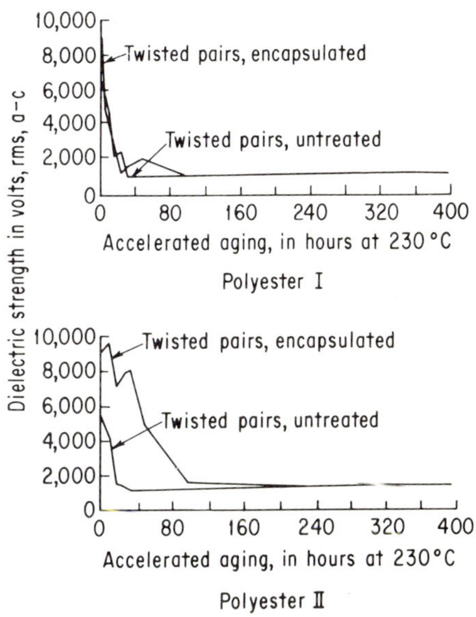

FIG. 17-34. Dielectric strength of heat-aged magnet wires [47].

Reinforcement

Structurally weak epoxy casting compounds may be suitably reinforced to upgrade performance during

Table 17-4. Effects of Various Treatments on Thermal Life of Polyester-coated Magnet Wire [74]

Encapsulant/ impregnant	Test temper-ature, °C	No. of samples tested	No. of samples failed	Time to failure, hours*		Test termi-nated, hours	Average hours per coil
				First	Last		
POLYESTER-COATED MAGNET-WIRE CLASS 155							
Silicone varnish, semirigid	200	5	5	404	1,074	.....	780
	180	5	2	2,663	3,239	3,287	3,152
	160	5	2	2,140	2,308	3,484	2,980
	23	5	0	.....	.....	3,340	3,340
Epoxy resin, rigid	200	5	5	18	334	.....	244
	180	5	5	1,197	2,284	.....	1,874
	160	5	4	1,568	2,980	3,484	2,650
	23	5	0	.....	.....	3,544	3,544
Epoxy resin, flexible	200	5	5	214	786	.....	521
	180	5	2	1,008	1,925	3,149	2,125
	160	5	0	.....	.....	3,484	3,484
	23	5	0	.....	.....	3,339	3,339
Polyester varnish, rigid	200	5	5	256	786	.....	521
	180	5	4	2,101	2,893	2,941	2,394
	160	5	3	1,326	2,896	3,064	2,447
	23	3	0	.....	.....	3,124	3,124
Epoxy resin, rigid	200	5	5	<24	1,399	.....	436
	180	5	5	896	3,464	.....	2,454
	160	5	5	1,310	2,420	.....	2,531
	23	5	1	1,678	1,678	3,694	3,290
EPOXIDE-POLYESTER MAGNET-WIRE CLASS 130							
Epoxy resin, rigid	200	5	5	<24	1,579†	.....	531
	180	5	5	<24	88†	.....	70
	160	5	2	<24	3,083	4,259	3,172
	23	5	1	3,990	3,990	4,494	4,368
Epoxy resin, rigid	200	5	5	<24	194	.....	58
	180	5	5	88	2,035	.....	817
	160	5	5	1,227	3,469	.....	2,395
	23	5	2	2,009	2,446	4,294	3,467
Silicone varnish, semirigid	200	5	5	24	1,017	.....	203
	180	5	5	250	762	.....	482
	160	5	5	243	2,140	.....	1,441
	23	5	2	2,009	2,446	4,294	3,467
Silicone varnish, rigid	200	5	5	89	257	.....	137
	180	5	5	980	1,364	.....	1,124
	160	4	4	1,362	2,142	.....	1,779
	23	5	0	.....	.....	5,954	5,954

EXPECTED SERVICE LIFE OF RESIN/VARNISH MAGNET-WIRE SYSTEM
(Continuous operation in hours)

Operating temperature, °C	Magnet-wire rating			
	Class 130 (B)	Class 155 (F)	Class 180 (H)	At 250°C
200	156	937	5,000	Indefinitely
180	625	3,750	20,000	Indefinitely
160	2,500	15,000	80,000	
130	20,000	20,000		
23	Indefinitely	Indefinitely		

* If the time to first failure is the same as the time to last failure, then there was only one failure.
† NOTE: These are correct hours and not interchanged.

thermal shock and mechanical abuse. Such reinforcement is normally applied by embedding a few layers of glass cloth at points where major stress concentrations are likely to occur. By this device, it is possible to employ the rigid, high-heat-resistant compounds in applications for which they are not otherwise suited. The use of reinforcement is a design consideration that should not be overlooked in the development of cast electrical-electronic units.

Design of Units

From the above considerations, it is evident that successful performance involves a consideration of the properties of the epoxy-resin compound during every phase of design. If the overall design and materials selection are not compatible with the inherent properties of the epoxy resins, unnecessary and time-consuming difficulties will be experienced in the final phases of development. In general, the design should minimize stress concentrations (where thermal shock is a problem), should ensure compatibility of all components, should allow for thermal conductivity considerations, should ensure the protection of pressure-sensitive elements in the package, and should accommodate the handling characteristics of the resin compound.

USE OF EPOXY RESINS IN TRANSFORMERS AND MOTORS

Epoxy resins have seen major use in electrical equipment since the early 1950s. Their use on transformers is probably the largest in volume, closely followed by their use on motors.

Transformers

Epoxy resins are used to impregnate and encapsulate all types of transformers and related magnetic components such as electronic coils and inductances. These may be divided into the following categories, based on design and applications:

1. Electronic transformers and inductors
2. Distribution transformer
3. Current and meter transformers
4. Magnets

Electronic Transformers. Epoxy resins are used to impregnate and encapsulate electronic transformers because of the following advantages:

1. Substantial reduction in transformer size and weight (in many cases up to 50 per cent) may be achieved over older-style units.
2. More dependable service is achieved under high ambient temperatures ranging up to 180°C.
3. Bonding of core coil and terminals into one solid unit provides more rugged construction and better appearance.

The requirements of an epoxy resin for transformer impregnation and encapsulation include the following:

1. Low viscosity and an adequate pot life to permit thorough soaking and wetting of all of the insulation system prior to gelation.
2. Low vapor pressure to permit proper vacuuming and pressure cycling of the compound prior to gelation. Because many of these transformers operate at 12,000 to 20,000 volts, deterioration and degradation of insulation due to corona must be prevented by removing all air from the windings.
3. Good crack and thermal shock resistance. Because transformers contain numerous metallic components—core windings, terminals as well as insulation materials—the resin must absorb high thermal cycling stresses without cracking in

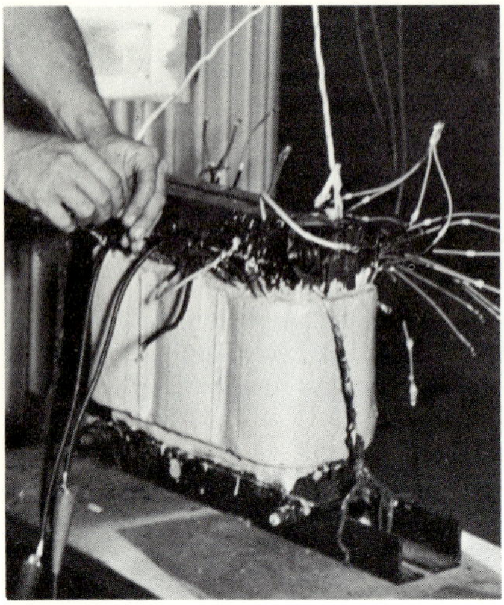

Dry-type transformer is sealed in epoxy resin to protect it against dust. (*The Epoxylite Corporation.*)

Encapsulated open-core transformer: 9-in.-diameter casting, approximately 13 in. high and 11 in. long. (*Components for Research, Inc.*)

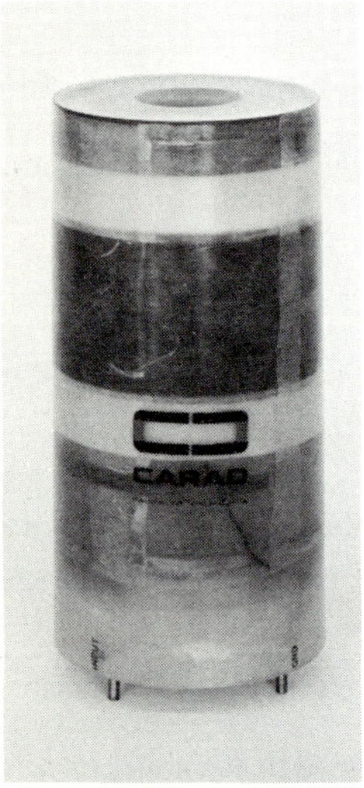

150-kv 2-μsec pulse transformer encapsulated in epoxy-resin formulation. No oil is required. Inside transformer is inserted a neutron generator. Transformer is 3¾ in. O.D. and 7½ in. high and weighs 2½ lb. (*Carad Corp.*)

Pulse-coupling transformer encapsulated in epoxy resin. Specifications: 4 primary, 8 secondary, 200 d-c, secondary working voltage. (*Carad Corp.*)

either an observable manner or at a microscopic level which could cause internal corona.

4. Maximum electrical properties. The volume resistivity of these transformer-impregnating resins must be very high because of the high voltage stresses which they must resist. Similarly the dielectric constant of the insulation is important in transformers because it effects the alternating-voltage stresses on the magnet-wire film, interlayer insulation, epoxy resin, and other insulations. These units may be potted in a one-step operation or alternatively may be first impregnated with a very low viscosity system and subsequently potted or dip-coated with a filled compound.

5. Flame resistance when this is a requirement, as it often is in military and airborne applications. In this connection, formulations may be developed especially to provide the required flame resistance. Typically, tris-β-chloroethyl phosphate may be used as an additive, with hydrated alumina being used as one of the fillers to improve arc resistance [81].

In high-voltage transformers, oil-filled designs are still superior in corona resistance to epoxy-resin units. In order to ensure proper performance, the unit must be designed with the epoxy in mind, and processing must be conducted to ensure thorough impregnation. The epoxy system must be designed to provide low shrinkage, good thermal shock, and maximum retention of electrical properties. Best possible adhesion is required, and cast resin interfaces should be eliminated in areas of high electrical stresses [79].

Distribution Transformers. Distribution transformers are the large transformers used to provide stepdown in voltage from power substations to individual homes or groups of homes. In the past they have been oil-filled and mounted on telephone poles near each group of houses which the unit services.

Use of epoxy resin permits omission of the tank, the tank oil, and the standoff bushings normally used. Experimental 25-kv units weigh 25 per cent less than oil-filled counterparts. Some of the size reduction is attributed to the use of higher operating temperatures, partially the result of the protective effect of the epoxy resins.

The inherent cost of the epoxy resin as well as the cost of precision impregnation required to assure corona-free units has delayed full commercial utilization.

Current and Meter Transformers. Current and meter transformers are transformers designed to measure current or other variables in power networks without disturbing the networks themselves. This type of transformer is one of the first to be encapsulated commercially, and such units have been in service in the United States and Europe for periods ranging up to 15 years. In these transformers the epoxy resin successfully replaced both oil-cooled and dry-type constructions. The advantages of the epoxy resin are elimination of numerous bushings, the ease with which compact designs are achieved, and the high degree of protection afforded to the conductors as well as to personnel who must frequently handle these devices.

Magnets. The term magnets covers a wide variety of sizes and types of electrical devices, ranging from magnets used to erase magnetic movie film to the large heavy-duty magnets used to lift scrap iron in steel manufacturing operations and to handle hot pig-iron billets. The smaller magnets usually consist of magnet wire wrapped on iron pole pieces; the latter type usually employ copper strap conductors. The magnets are sometimes insulated by brushing epoxy-resin formulas layer by layer to provide internal bonding to resist the vibration and mechanical abuse they receive; often they are encapsulated. On some of the larger units impregnation and potting are performed when the strap-wound coils have been mounted in their metal case and connected and are otherwise ready to be returned

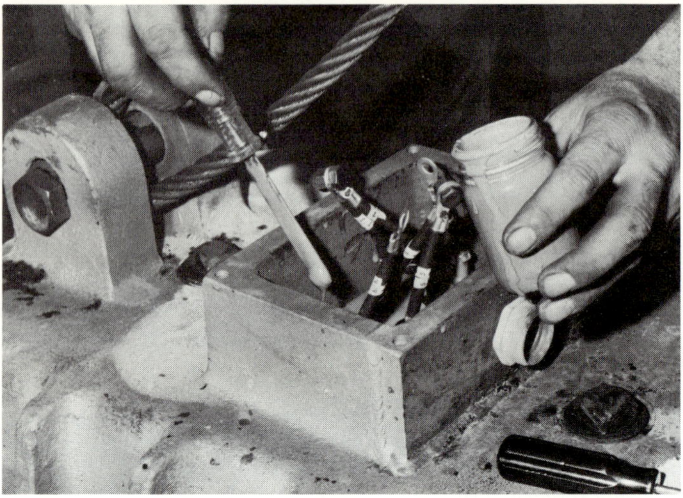

Epoxy resin is used to seal around leads of large lifting magnet. Epoxy resins have found use in potting large magnets exposed to vibration and moisture and high-temperature service. (*The Epoxylite Corporation.*)

to service. The case is then filled with epoxy compound, and the units are thus bonded into a solid, vibration-free structure in one operation.

Resembling the larger lifting magnets are highly specialized coils used in cyclotrons and other nuclear research devices. These coils consist of large copper pipes, water- or gas-cooled, which are bonded into units with epoxy compounds. In these cases, the resin is often brushed in, layer by layer, as the tubes are installed, but in some cases it is applied by pouring.

Electric Motors and Generators

Epoxy resins have produced a virtual revolution in the electric motor field. Their particular virtue is the protection which they afford motor windings from severe environmental contamination.

On a-c equipment a standard varnish will provide less than one-minute service if the equipment is flooded with seawater while operating, whereas standard-production units of high-quality encapsulated motors have served satisfactorily for periods

Three 5-hp a-c motor stators encapsulated in epoxy resin per Mil-M-17060 (Bureau of Ships) as protection against flooding and salt spray. (*The Epoxylite Corporation.*)

A 30-hp stator encapsulated in epoxy resin during rewind of motor to provide protection against moisture. (*The Epoxylite Corporation.*)

exceeding 500 hours when immersed in salt water. Construction of epoxy-encapsulated motors to requirements of Mil-Spec 17060B, for this class of service is described in [84]. Figure 17-35 indicates typical performances of varnished and encapsulated units under conditions of high humidity.

Field coils of d-c motors may be encapsulated and subsequently potted or bonded on the pole pieces. Since the epoxy displaces air between the coil and the pole

Encapsulated motor partially submersed in mud bath. After completion of test, motor was removed, hosed down, and restored to service immediately. (*Allis-Chalmers Mfg. Co.*)

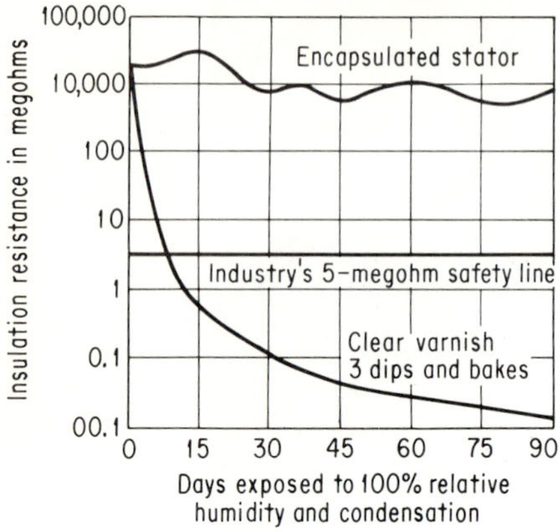

FIG. 17-35. Insulation resistance vs. humidity of encapsulated and varnished stators.

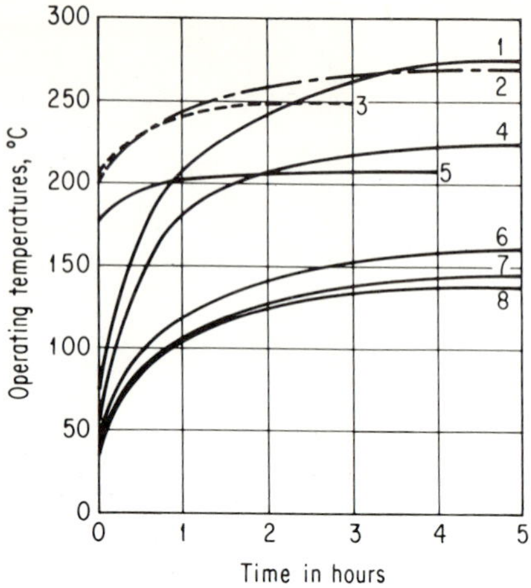

FIG. 17-36. Effect of various resins on motor cooling at full-load operation of traction motors [67]. Field coils molded to pole pieces and treated with synthetic resins. 1, Conventional insulation; 2, polyester; 3, polyester; 4, silicone compound; 5, epoxy; 6, epoxy; 7, epoxy; 8, open-face coil, no resin.

Field coils of rotor of synchronous motor are sealed in epoxy resin as protection against carbon black, which quickly short-circuits unprotected windings. (*The Epoxylite Corporation.*)

piece, not only is there an advantage of high structural integrity, but also there is an improvement in heat transmission, which results in cooler operation (Fig. 17-36).

Synchronous motors normally have d-c field coils and a-c stator coils. The field coils are insulated as above. The stator-coil insulation may be composed of an epoxy-resin–glass-cloth laminate. After the coils have been wound into the unit, the connections are sealed with epoxy and glass tape. Epoxy compounds may also be used for bracing to minimize vibration and eliminate flexing of the coils during full-line starts or other surge conditions. Connections may be sealed with epoxy compounds. The high chemical and humidity resistance of epoxy resins has produced service records for synchronous motors similar to those encountered with encapsulated a-c stators. Figure 17-37 depicts the insulation resistance of a 2,500-hp epoxy-protected synchronous motor as compared to that of a standard type.

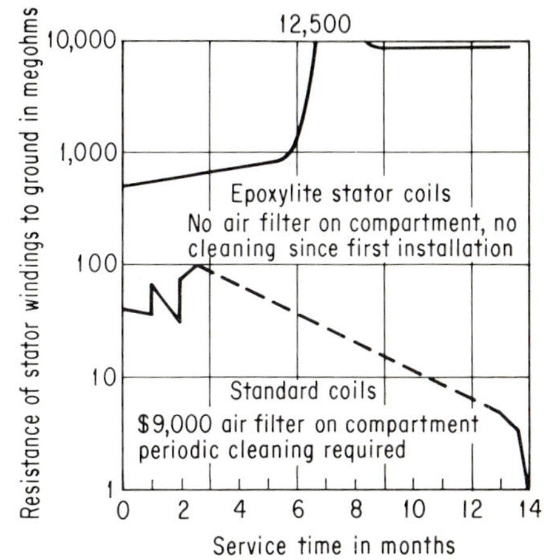

FIG. 17-37. Performance comparison of Epoxylite® molded stator coils vs. motor manufacturer's coils in severe rubber-mill duty. Both motors identical frames, same manufacturer, same application. Rating: 1,800 hp, 40 pole, 2,300 volts. Data taken with TAKK d-c highpot at 2,500 volts.

EPOXY CASTING COMPOUNDS

The general uses for epoxy-resin casting compounds have been indicated previously. In addition to these, considerable interest has been generated in two other areas: their use for making high-voltage bushings and their use in dental applications.

High-voltage Bushings

Epoxy resins are used increasingly for standoff or high-voltage bushings because of their dimensional stability, ease of pouring and curing to the exact size (as contrasted to the high warpage of porcelain), and high bond strengths to copper which permit positive bonding to through studs.

Pioneered in Europe, the application is gaining commercial acceptance in the United States. For example, in the United States cast suspension insulators from epoxy resins have been under test in outdoor environments for some time. At the end of 5 years, the insulators still performed satisfactorily and on the basis of this data, long-time service life was confidently expected [93].

Epoxy Bushings

Although epoxy resins were first synthesized by a dental supply company in Switzerland, they have only recently gained acceptance in dentistry. Early formulations provided disappointing color stability, and later formulations, while not entirely overcoming this disadvantage, revealed inadequate water resistance.

Despite these setbacks and the relatively small market potential, interest has continued in the epoxy resins for dental applications because of their high adhesive strengths, low shrinkage, and good dimensional stability. Interest ranges from crown-and-bridge to direct-filling resins.

Hermetic-seal feed-through bushing is cast from epoxy resin. Sizes and weights are considerably lower than those of bushings fabricated from other materials. (*Components for Research, Inc.*)

Double and single three-phase disconnects cast from epoxy resin. Castings weigh 60 lb. including copper. (*Permali, Ltd., Canada.*)

Tap selector housing for 132-kv on-load tap changer. (*Ciba Products Co.*)

Re-entry-type bushings, up to 700 kv, some designed to withstand mechanical loading of 9½ tons, cast from epoxy resin. (*Ciba Products Co.*)

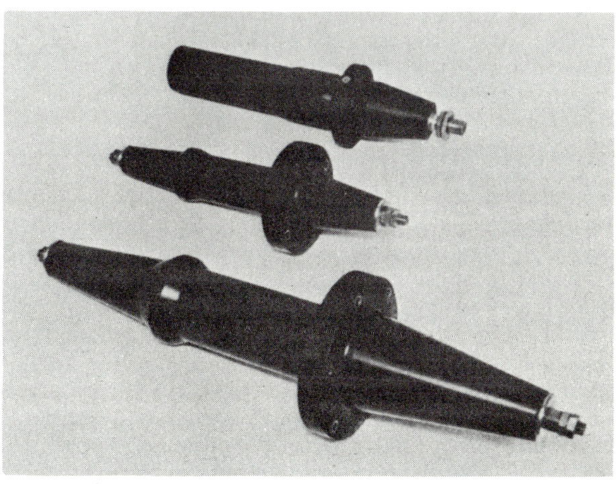

Epoxy bushings, 22, 33, and 66 kv, incorporating stress-control screens. (*Ciba Products Co.*)

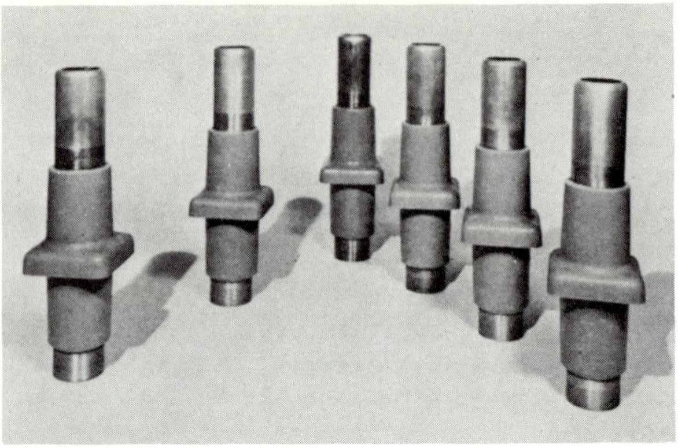

Air circuit breaker bushings with cast epoxy flanges, incorporating fixing inserts, 2,400 amp, 3.3 kv. (*Ciba Products Co.*)

FORMULATION

The proprietary commercial materials for casting, potting, and encapsulation are usually more sophisticated than merely straight resin/curing-agent combinations. Fillers, diluents, and modifiers are used to selectively modify properties. Curing agents may be specially synthesized. The formulations are then optimized by more or less extensive laboratory testing. Data given in the earlier chapters provide useful starting points for the development of suitable compounds.

In general, for the majority of electrical-electronic applications, use is made of anhydride curing agents with flexibilizing modifiers. DGEBA is employed for most general-purpose work, with some of the newer resins being utilized for applications requiring high-temperature service.

The amine curing agents are used for relatively small mass castings and when room-temperature curing is required. One-component systems are based on concepts advanced in Chap. 11.

In the design of suitable compounds, emphasis is divided between cured and handling properties. The latter are of equal importance in cases where system performance depends on proper application, which is only possible with systems optimized to the production method.

PRODUCTION TECHNIQUES

Although the concept of casting or pouring a liquid resin and curing it in a mold is straightforward, the actual doing becomes quite involved when dealing with temperature-sensitive, mass-sensitive, highly reactive resins. Although many production difficulties can be minimized by the proper selection of the resin formulation, frequently an uncompromising decision against optimum handling characteristics must be made. Production techniques must then accommodate the specific compound.

For example, installation of continuous-mixing equipment may be required; efficient ovens or cooling devices may be necessary to reduce exothermic peaks; physical means of eliminating the irritation potential of the curing agent must be devised, etc. In short, consideration must often be given to designing the production process to the compound, rather than the reverse. The factors of general interest in most epoxy formulations are presented in Chap. 25, together with specific handling recommendations. Chapter 22 presents a discussion of laminates, the use of which should be considered alternatively to casting and encapsulation where extremely high structural strengths are required. Chapter 21 discusses adhesives and surface preparations, factors which may require consideration in some potting, encapsulating, and sealing operations.

The following specific considerations will be discussed in this section: the mold, application techniques, the casting of large or complex shapes, and the use of high filler loading volumes.

The Mold

When removable molds are employed for casting and encapsulating operations, the mold should be designed with a slight draft or taper and have a nearly uniform cross section. The taper will facilitate release, and the nearly uniform cross section

Ability of epoxy resins to cure easily at atmospheric pressure permits use of low-cost paper molds in encapsulating stator windings in repair shop, where each motor is a different size. (*The Epoxylite Corporation.*)

will minimize stresses during cure. The interior surface of the mold should be smooth and free of rough edges. If the specific cast configuration is complex and contains undercuts, it will be necessary to employ a two-piece mold, or alternatively, to use one-piece molds made from flexible plastisols or silicone rubber, to use some form of conformal casting technique (using a film applied by vacuum bag) or molds made from special metals or plastics, which may be melted off after the formulation has cured.

Molds may be made from any convenient material possessing the required

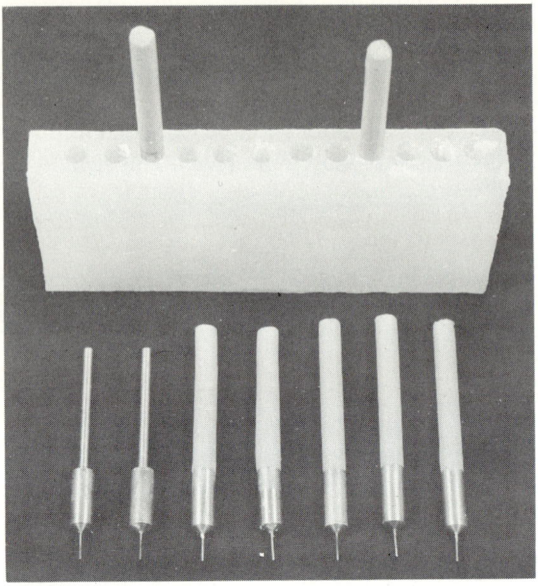

Contact pins with insulating sleeve of epoxy resin applied by using low-temperature cure and low-cost paraffin mold. (*The Epoxylite Corporation.*)

Epoxy casting resin is poured into prepared form and leveled. Casting resin acts as base for huge compressor. Base is required to withstand tremendous vibration. Epoxy compound was selected because of high strength, low shrinkage, resistance to grease and oils, and rapid curing. (*Union Carbide Corp.*)

dimensional stability. Perhaps most widely used are machined metal or cast aluminum molds for operations where production volume warrants. Sprayed metal molds may also be used [3].

Slush-coating molds may be made by dipping a master into molten metal. When the master is withdrawn, a coating of the metal remains, and the master is then removed. The metal is then used as the mold. Following cure, the mold is melted from the encapsulated component by elevating the temperature and is recovered for reuse. Typical of the metals used for slush molding is Cerrotru alloy, having a melting point of about 138°C [71].

Missile transformer potted in semirigid epoxy resin, using diallyl phthalate cup. Epoxy resins afford high protection against vibration, also allow working at high voltages without corona breakdown. (*North American Aviation, Inc.*)

Molds may be constructed from a variety of plastics, including the epoxy resins. Thermoplastic materials, such as polypropylene, may also be employed when curing temperatures permit. Molds from Teflon and silicone may be used, these possessing the advantage of having built-in releasing properties, although they cannot generally be used when it is necessary to hold close dimensional tolerances and are generally unsuited to any but short production runs.

With removable molds it is necessary to employ some form of release agent. Commonly, a silicone oil or grease is employed (see Chap. 25) and must be renewed after every few mold cycles. Alternatively, a thin film of Teflon may be baked on metal molds.

In potting applications, where the mold is to remain permanently in place, the mold-release problem is overcome. However, unless more or less standard potting cases are involved, tooling costs are fairly high, as for example when the potting cases are injection-molded to special requirements. In larger units, the potting cases should not be too rigid; else the formulation will pull away from the case or will crack during cure or thermal cycling. This can, to some extent, be off-set by matching the thermal expansion characteristics of the case with the epoxy compound being employed [48]. For some items, such as resistors and other standard units epoxy shells are commercially available as potting cases [25].

Application Techniques

Excluding the simple application of an epoxy paste by spatula, four distinct application techniques may be employed in forming the systems under discussion:

bench casting, vacuum impregnation, pressure impregnation, and open-mold or dip casting.

Bench Casting. Bench casting is defined as pouring or casting in the open air under workshop conditions. It is relatively difficult for inexperienced personnel to obtain a bubble-free casting by bench techniques, using even a moderately low viscosity compound. Bubbles are entrapped in the compound during mixing; more are incorporated during the pouring operation. The entrapped bubbles can be virtually eliminated by a combination of several techniques. Stirring should be conducted in such a manner as to hold the bubbles to a minimum. The catalyzed system may then be allowed to stand at room temperature, pot life permitting, until the bubbles have escaped. Where practical, warming will expedite bubble release, as will vacuum. Once the bubbles have been allowed to escape to the degree practical, the compound should then be poured slowly and carefully into the molds. Use of heated molds is of considerable help in obtaining good castings. Vibration may be employed if practical. The cast system should immediately be exposed to the initial cure temperature. If this temperature is higher than ambient, there will be a progressive reduction in viscosity, which will facilitate escape of the bubbles, until cross-linking is initiated and the trend reversed.

Vacuum Casting. Vacuum techniques are particularly useful for critical potting applications and may, on occasion, be employed in connection with pressure impregnation.

In determining the degree of vacuum to employ, consideration should be given to the difficulty of the filling operation and to the vapor pressure of the resin system components. Some reactive diluents, for instance, can be pulled off rather easily, and care must always be taken to avoid volatilizing the curing agent. Moisture traps should be provided in the lines to catch any released material; otherwise, it may polymerize in the pump or vacuum chamber. When lower vacuums provide satisfactory fill, they should be used. If high vacuums are necessary, redesign of the resin system may be required to eliminate volatile elements.

It is also sometimes necessary to give consideration to the pretreatment of the foreign components involved. If moisture is critical to system operation, it must be driven off by baking or vacuuming just before the potting operation. If the adhesion to components is critical, these should be pretreated.

Where possible, the compound should be poured under vacuum. This permits the removal of entrapped air in the resin system, and moisture and entrapped air in the component.

If the compound is poured at atmospheric pressure and the vacuum pulled subsequently, foaming, bubbling, and frothing will occur at lower pressures, and the volume will expand considerably. This expanded volume must be accommodated by a reservoir, often ten times the volume of the resin. For optimum results, all foaming, frothing, and bubbling should cease before the component is returned to atmospheric. In practice, the vacuum is usually held until all foaming and frothing ceases, then slowly increased until bubbling ends, then brought rapidly to atmospheric. If foaming and frothing are excessive, too much moisture and entrapped gases are present in the foreign component. If bubbling is excessive, too much material is being volatilized.

Several vacuum cycles may be employed to enhance the fill. It is not advisable, however, to permit the compound to react too far while vacuum is being applied, since higher exotherms may volatilize the curing agent; and cycling should not be continued after viscosity begins to increase.

In the absence of molds, when runoff prior to cure poses a problem, as with small armatures, for example, they may be rotated on a spit during the curing process [70].

Pressure Casting. Pressure impregnation may be employed in connection with vacuum-potting techniques by pressurizing the chamber after the potted system has undergone the vacuum cycle. Pressures of 50 to 100 psi result in thorough impregnation and very small bubbles. The pressures needed to achieve collapse of the bubbles and solvation of the gases in the resin per Henry's law have not been widely investigated.

With electrical coils and similar small components, centrifugal action may be employed to obtain the requisite pressure. A suitable centrifugal apparatus consists essentially of a motor, vertical shaft, chamber, and spinner assembly. The spinner assembly is composed of spinner arms radiating from a central core and attached at the opposite ends to pivoted molds. The molds are poured with an unfilled epoxy compound, and the assembly is spun rapidly. With such equipment, the spinner assembly may be preheated, the resin added, and the assembly spun throughout cure. Spin rates providing pressures on the order of 1,000 psi will form pressure heads that will eliminate foaming and suppress tendency of the curing agent to volatilize. The resultant castings will be dense, void-free, and, in all probability, superior to those obtained by vacuum impregnation [4,5].

Open-mold or Dip Casting. This technique consists in dipping the item to be sealed or encapsulated into a catalyzed resin mix and then removing it. The dip compound is formulated for controlled runoff, and the curing agent should, in most production situations, provide long pot life. This system is further discussed in Chap. 23.

Casting of Large or Complex Shapes

When casting larger volumes, unless higher filler loading volumes are used, there is danger of encountering cracking and runaway exotherms. Temperature variations throughout the mass and localized shrinkage will establish excessive stresses and strains during cure. High exotherms will volatilize the curing agent and/or modifiers to result in entrapped bubbles. Acid anhydride curing agents or epoxy/phenolic blends should be considered for larger castings, because of their lower exotherms. Efficient ovens should be employed during cure. Fillers and flexibilizers should be considered to reduce critical parameters.

In many cases, the most satisfactory solution will be stage casting, by which method the larger casting is formed of layers of compound, each cured, or at least gelled, before the next is added. However, with suitable formulations, large castings can be made. Using sand at high loading volume, Shell Chemical personnel successfully cast a 1,400-lb batch of catalyzed resin in a bathtub. Ciba personnel in Europe have made 2,000-lb castings in transformers.

The stresses and strains established during the cure of complex castings may result in cracking. This can be offset by the incorporation of flexibilizers when cured property requirements permit.

Casting High Loading Volumes of Fillers

For some applications, it is possible to employ low-viscosity systems and work them at elevated temperatures to obtain loading volumes of fillers on the order of 300 phr while still obtaining fairly good fill and penetration using vacuum-pressure techniques.

Alternatively, in some cases, it is possible to employ high loading volumes of filler without increasing the viscosity of the resin formulation. This technique is useful when potting or encapsulating configurations that require high internal impregnation but a highly filled surface sheath. Rather than perform the operation in two steps, it is possible to surround the unit with a coarse dry filler and subsequently vacuum-pressure-impregnate the construction with the unfilled resin formulation. When such operations are performed, vacuums are usually drawn to less than 1 mm mercury.

A third technique is to employ ultrasonic vibration to make highly filled systems pourable. With this technique, it is sometimes possible to eliminate vacuum processing [69].

SUMMARY

Because of the overall balance of properties—physical, chemical, and electrical—the epoxy resins have achieved popularity as casting, potting, and encapsulating compounds.

The epoxies are preferred, in comparison to lower-cost systems, for short production runs, where tooling for injection or transfer molding cannot be justified economically. They are useful in a number of rather specialized applications. Their main volume use, however, is as encapsulating compounds for electrical and electronic equipment.

In electrical and electronic applications, the formulation requirements generally are fairly sophisticated, and considerable engineering thought is required in the design of the component or components to be protected. The design must be such as to minimize stresses to permit satisfactory thermal cycling. All elements contained within the epoxy-resin envelope must be compatible with the epoxy formulation. Special design consideration should be paid to cooling problems, to the extent that this is important, and processing techniques must be established to enable the convenient handling of the optimum formulation.

In general, the epoxies are suitable as encapsulating materials for general-purpose transformers up to about 15 kva and are suitable on designs operating at voltages up to 20,000 volts. They are suitable for most electronic equipment, except possibly some operating at high frequencies. They are useful on large rotating equipment up to class 180, although special techniques may be required for the proper protection of the larger form-wound motors and generators.

REFERENCES

1. Anaconda Wire and Cable Co., sales literature.
2. "Electronic Packaging and Production," Milton S. Kiver Publication, Inc., 222 W. Adams St., Chicago, Ill.

1952

3. Brenner and Hase, Sprayed Metal Molds, *Mod. Plastics*, September.

1954

4. Firth, Centrifugal Method of Impregnating Coils, U.S. 2,695,856.
5. Firth, Electrical Potting and Encapsulation, 10th ANTEC of SPE.
6. Rejda, Epoxy Resins Make Bid for Motor Insulation, *Power*, November.

1955

7. Gallousis and Farneth, Versatility Cast into New CT's, *Allis-Chalmers Elec. Rev.*, 3d quarter.
8. Rejda and Lee, Epoxy Resins, *NISA News*, March.

1956

9. Deimel, Impregnation of Toroids for High Temperature Service, WESCON, Los Angeles, August.

10. Doctor and Franklin, Corrosion of Copper Wire by Resins, *Electron. Equipment*, July.
11. Epoxy Potting Compounds: How and Why They're Used by Jack and Heintz and Freed Transformers, *Insulation*, May.
12. Groves and Pizzino, Effect of High Humidity on Dielectric Properties of Casting Resins, *Elec. Mfg.*, April.
13. Lee, The Use of Epoxy Resins to Encapsulate Electric Motors Operating in Severe Service Environments, AIEE Rubber Group Technical Conference, Akron, Ohio, April.
14. Rejda and Moreland, Epoxy Resins Solve Tough Maintenance Problems, *Plant Eng.*, May.

1957

15. Marshall and Ritt, Epoxy Encapsulation of Transistor Flip-flops, 13th ANTEC of SPE.

1958

16. Bush, Measurement of Stresses in Cast Resins, *Mod. Plastics*, February.
17. Cast Epoxy Resin Transformers, *Allis-Chalmers Trends*, June.
18. Goethe and Henry, Advantages of Filled Resin Embedment Compounds for General Purpose Transformers, NEMA and AIEE, 1st National Conference on Application of Electrical Insulation.
19. Lenchutz et al., Heat of Polymerization of Epon 828, Picatinny Arsenal, Technical Report 2505, June.
20. MCA–MIT Plastics Research Project Report, July 1.
21. New Materials Aid Coil Life, *Mod. Railroads*, April.
22. Olyphant, Thermal Shock Tests for Casting Resins, NEMA and AIEE, 1st National Conference on Application of Electrical Insulation.
23. Parry and Mackay, Cure Shrinkage of Epoxy Systems, *SPE J.*, July.
24. Sampson and Lesnick, Evaluation of Casting Resins Employing Strain Gage Techniques, *Mod. Plastics*, February.
25. Stock, Epoxy Shells Simplify Potting of Resistors, *Electron. Eng.*, Jan. 31.
26. Wahlgren, How Epoxies Are Used to Improve Large and Small Transformer Designs, *Insulation*, June.
27. Warfield, Studying the Electrical Properties of Casting Resins, *SPE J.*, November.

1959

28. Black, private communication, Westinghouse Electric Corp.
29. Chambers and McGarry, Resin Shrinkage Pressures during Cure, 14th NANTEC of SPI.
30. Dixon, Use of One-component Epoxy Systems in High Reliability Transformers, NEMA and AIEE, 2d National Conference on Application of Electrical Insulation, Washington, D.C.
31. Epoxy-backed Printing Plates, *Mod. Plastics*, February.
32. Hanson and Tuzinski, Strain Gage Evaluation of Flexible Epoxy Resins, NEMA and AIEE, 2d National Conference on Application of Electrical Insulation, Washington, D.C.
33. Klute and Viehmann, Heat of Polymerization of Phenyl Glycidyl Ether and of an Epoxy Resin, Diamond Ordnance Fuze Laboratories Technical Report 758, August.
34. Lee, Heat Resistant Encapsulating Resins, 15th ANTEC of SPE, January.
35. Transparent Thermosets, *Plastics World*, April.
36. Wahlgren, Use of Epoxy Resins in High Reliability Transformers, *Electron. Equipment Eng.*, January.
37. Warfield and Petree, The Temperature Dependence and Activation Energy of Electrical Conduction in High Polymers, U.S. Naval Ordnance Laboratory, White Oaks, Md.

1960

38. Buenger, Epoxy Parts Cast by Centrifugal Force, *Prod. Eng.*, May 23.
39. Carroll and Kaufman, Epoxy Casting at Teletype, WE–BTL Casting Resins Conference, North Andover, Mass., May.

40. Davis, Effects of Temperature on Filled Epoxy Encapsulation Materials, *SPE J.*, December.
41. Dewey and Outwater, Pressures on Objects Embedded in Rigid Cross-linked Polymers, *Mod. Plastics*, February.
42. Gough and Smith, A Gel Point Method for the Estimation of Overall Apparent Activation Energies of Polymerization, *J. Appl. Polymer Sci.*, no. 9.
43. Hackney and Walker, Insulation Systems for Naval Shipboard Motors Intermittently Submersed, AIEE Conference Paper 60-338.
44. Iaffaldano, Thixotropic Epoxy Resin Dipping of Capacitors, WE–BTL Casting Resin Conference, North Andover, Mass., May.
45. Kline and Saxton, Moisture and Temperature Effects in Filled-epoxy Mirrors, *J. Opt. Soc. Am.*, November.
46. Kuscharsey, Epoxy Resins as Field Coil Insulation for Heavy Duty Traction Apparatus, Pennsylvania Railroad Co.
47. Lee, Compatibility of Magnet Wire Insulations and Epoxy Encapsulating Resins, *ASTM Bull.*, July.
48. Lunderg, Potting Connectors for Casting Resin Compounds, WE–BTL Casting Resin Conference, North Andover, Mass., May.
49. Saxton and Kline, Optical Characteristics and Physical Properties of Filled Epoxy Mirrors, *J. Opt. Soc. Am.*, November.
50. Scheid, Spherical Strain Gage Development to Evaluate Stresses in Resins, General Electric Co., Technical Memo GEX-255, Milwaukee, Wis., Mar. 28.
51. Segall and Avery, Investigation of Encapsulation Materials and Techniques for High Density Module Program, AIL Report, July.
52. Warfield and Petree, A Study of the Polymerization of Epoxide Polymers by Electrical Resistivity Techniques, *Polymer*, vol. 1.
53. Worley, Progress Report on Epoxy Resin Filled Coils for Wire Spring Relays, WE–BTL Casting Resin Conference, North Andover, Mass., May.
54. Worley, Progress Report on Potting of Loading Coils in Epoxy Resins, WE–BTL Casting Resin Conference, North Andover, Mass., May.

1961

55. Epoxy in Tomorrow's Transformer Design, *Plastics World*, March.
56. Fisch and Hoffman, Reaction Mechanisms, Chemical Structures, and Changes in Properties during the Curing of Epoxy Resins, *Plastics Technol.*, August.
57. Harper, "Electronic Packaging with Resins," McGraw-Hill Book Company.
58. Loring, Modern Materials In High-voltage Pulse Transformers, Joint AIEE, EIA, RIE, and WEMA Conference, San Francisco, May.
59. Moench, Some New Results and Methods in Photoelasticity Using Araldite B, AD 299 877.
60. Outwater and West, Stress Distribution in the Resin of Reinforced Plastics, *Mod. Plastics*, September.
61. Phelps Dodge Copper Products Corp. Wire Engineering Report 666-B.
62. Rejda and Lee, New, Low Cost Method for Encapsulating Motors, *Insulation*, March.
63. Spencer, Cure Shrinkage in Casting Resins, Sandia Corp. Report SC DC-2568, Sept. 6.

1962

64. Thierry, private communication, Essex Wire Corp.
65. Feuchtbaum, Insulation a Critical Factor in High Reliability Design, AIEE, New York, January.
66. Foerster, Project Nancy: Systems Evaluation of Epoxy Encapsulated Coils, 4th National Conference on Application of Electrical Insulation.
67. Kuscharsey and Kelly, Operating Experience with Epoxy Resins in Insulation Systems of Traction Rotating Equipment, AIEE, New York, January.
68. Wells, Resin Impregnation of Soil Samples, *Nature*, Feb. 24.

1963

69. Black and Leo, The Use of Ultrasonic Vibration as an Aid to Resin Compounding and Casting *SPE J.*, August.
70. Fromm and Lauro, Epoxy Resin Impregnation of Electrical Members, U.S. 3,071,496.
71. Wardell, Slush Cast Molds for Epoxy Casting, *Electron. Packaging and Production*, February.

1964

72. Dallimore et al., Measurements of Internal Stress in Electronic Encapsulating Resins with a Small Solid State Transducer, *SPE J.*, June.
73. Gruenwald, Low Temperature Properties of Epoxy Resin Compounds, *SPE J.*, February.
74. Jawitz, Thermal Life of Various Magnet Wire–Encapsulant/Impregnant Combinations Evaluated, *Insulation*, September.
75. McCachern, The Manufacture of Epoxy Shells by Centrifugal Casting, *Western Elec. Eng.*, January.
76. Pruitt et al., Solid Combustible Compositions Containing Epoxy Resin, U.S. 3,130,096.
77. Tashchilov, Use of Epoxy Resins in the Production of Impregnated Graphite, *Khim. Prom.*, no. 4, pp. 303–304.

1965

78. Baker, Solvent Process for Preparing Epoxy-based Propellants, U.S. 3,184,352.
79. Balk, Corona Starting Voltages of Epoxy Encapsulated High Voltage Transformer System, 6th Electrical Insulation Conference (IEEE & NEMA), September.
80. Brandt et al., Sand Consolidation, U.S. 3,176,768.
81. Budovec, Epoxy Insulated Transformer Having Tris-β-chloroethyl Phosphate and Hydrated Alumina in the Insulation, U.S. 3,202,947.
82. Cowan and Ehart, Epoxy Resin-bonded Machinable Ceramics, U.S. Atomic Energy Comm. Report LA-DC-6821.
83. Dorman, Epoxy Resin Insulation Systems for Modern and Future Switch Gear, 6th Electrical Insulation Conference (IEEE & NEMA), September.
84. Foxgrover, New Concepts in Encapsulated Motor Reliability, 6th Electrical Insulation Conference (IEEE & NEMA), September.
85. Grimley et al., Preparation of Large Epoxy Sections for Light Microscopy as an Adjunct to Fine-structure Studies, *Strain Technol.* **40**(6);357–366.
86. Johnson and Ryan, Encapsulated Component Stress Testing, 6th Electrical Insulation Conference (IEEE & NEMA), September.
87. Lehmann and Poynter, Specimen Mounting, U.S. 3,187,444.
88. Parker and Colburn, Solid Propellant, U.S. 3,171,764.
89. Ramos, Vibration Damping Composition and Method of Making Same, U.S. 3,170,887.
90. Steinhaus, Determination of Water Vapour Diffusion Coefficients of Electrical Insulating Materials Based on Unsaturated Polyester Resins and Epoxy Resins as Well as Impregnating Varnishes, *Kunststoffe*, vol. 55, pp. 768–771.
91. Strobel, Evaluation of Resin Materials and Casting Methods for Reducing Low Temperature Stresses on Encapsulated Parts, *Insulation*, April.
92. Stucki, How an Epoxy Can Serve as an "Active Element" in Packaged Equipment, *Insulation*, January.
93. Tetreault, The Effects of Weather on Epoxy Insulators for High Voltage Transmission Lines, *Insulation*, November.
94. Treadway et al., Consolidation of Incompetent Earth Formations, U.S. 3,176,769.

Chapter 18

EPOXY RESINS IN TOOLING

The use of plastics in tooling applications was introduced during World War II to alleviate a critical shortage of tooling metals. The initial tooling plastics were phenolics, and later, polyesters. These materials although fairly satisfactory posed a number of disadvantages which limited their versatility. The epoxies were introduced as tooling resins in about 1950, and during the next 10 years their use grew steadily until, at the end of the decade, they were being employed in about 90 per cent of the plastic tooling applications, at the same time having considerably extended their useful range of applications [23].

The use of epoxy formulations in tooling applications provides several advantages:

1. Speed and ease of manufacture
2. Dimensional stability
3. Excellent strength properties, particularly when suitably reinforced
4. Light weight
5. Low shrinkage during cure which permits close tolerance reproduction

Generally, the epoxy resins have not introduced new concepts in tooling but rather have been used to save time and money over metal tooling or to provide performance superior to that obtained with the older plastic tooling. In a limited number of cases, particularly involving complex designs, the use of the epoxy resins has made possible tooling not economically practical with the older methods and materials.

Tooling is a term which covers a variety of different types of equipment. The tool may be a working model of the part to be produced; it may be a model from which the part is produced, either directly or indirectly; it may be a mold in or against which the part is produced; it may be equipment used to form, under heat and/or pressure, a contour in sheet metal or plastic; or it may be an accessory item used for guiding and checking other production operations.

Epoxy tooling was initially introduced in aircraft and in automobile production, and these areas have continued to account for the majority of the development activity.

In aircraft, where production runs are short and are often complicated by design changes, the epoxies have proved of particular value [12]. Epoxy tooling permits the ready incorporation of design changes, accomplished merely by machining off the excess epoxy and building up, out of new epoxy, the desired modifications, such changes being accomplished, often, within a single working day [9].

In the automobile industry, epoxy metal-forming tools are not suited to long production runs, but such tools are useful as serving for prototypes for metal tooling and can be used to cut the time required for the manufacture of prototype metal parts by as much as 25 to 50 per cent [6]. In addition, they are well suited for the fabrication of drill jigs and other holding and positioning fixtures, and for checking fixtures, where service conditions are less severe than in metal-forming operations.

The next most important area for epoxy tooling has been in foundry work, particularly in the area of producing a number of duplicate patterns in a very short time. The use of epoxy patterns for making molds and cores by means of a smooth facing and a sand-filled epoxy backing has been described [18]. In foundry work epoxies have additionally been used to convert sand cores quickly from the "green" to the cured state. This is accomplished by incorporating a few per cent by weight of an epoxy-resin/polyol blend into the sand and then injecting, under pressure, gaseous BF_3 into the mold. After a few seconds, the gas is released and the core allowed to cure at room temperature for an hour or so [34].

EPOXY TOOLING AS PROTOTYPES AND MODELS

Perhaps the most common application of epoxies for prototype work is their use as splining resins.

The splining resin, a thixotropic paste, is shaped directly on a framework of templates to form a prototype model without the use of molds. The templates are made to size and mounted, usually being held in position by threaded rods. Wire mesh, or other retaining screen, is then placed between the templates, somewhat back from the edges, and the epoxy formulation is combed on, the edges of the templates being used as guides. The final surface coating is slicked up by the use of a solvent-soaked brush, applied when the resin advances to the tacky, semisolid stage. After cure, some hand finishing is involved, the amount depending upon the skill of the applicator.

After design approval, the prototype may be used for making one or more female skins which, in turn, may be used for making male masters for subsequent tooling [17].

Epoxy die-face model cast from aluminum master. Die-face model is used in duplicator machine to reproduce the design in steel. (*Union Carbide Corp.*)

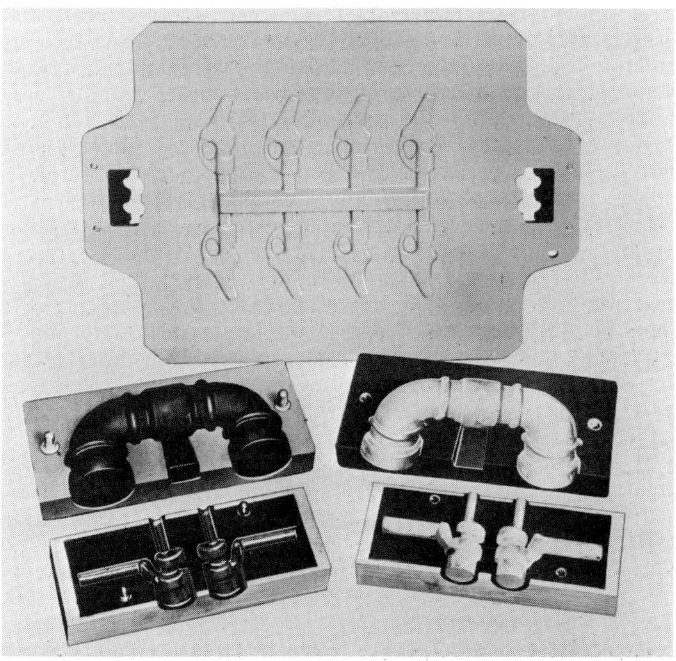

This type ell blow box has been in constant use for approximately 2 years, blowing about 1,000 cores a day and showing no appreciable wear. The match plate (reinforced with glass) and the other corebox were constructed as samples for foundry tryout. (*Ciba Products Co.*)

Epoxy models, usually cast models, are employed in copy milling, using equipment such as the Keller duplicators. The contours of the model are traced by a mechanical arrangement which transfers the pattern to the metal part being machined.

Epoxy patterns, in a sense models, are useful in foundry work. They can be produced at approximately one-third the cost of machined aluminum patterns and are lightweight, noncorrosive, and dimensionally stable [27]. In addition, they may be rapidly duplicated from the same mold to provide for any desired production rate [7]. In duplicating patterns with epoxy casting resins, tolerances of 0.001 to 0.003 in. can be maintained [19]. Suitably formulated materials will provide wear characteristics comparable to those of aluminum. The epoxy patterns give good sand release and easy draws [3], up to 100,000 draws having been made from a single pattern. The epoxy pattern is suitable for use with both ordinary and corrosive sands, although some sticking of green sand has been experienced. This can be overcome by use of parting solutions selected according to the type of sand. The pattern gives good surface finish and nonchilling characteristics. Epoxy formulations are also used for foundry coreboxes, match plates, and cope and drag equipment. Cured epoxy formulations filled with microballoons have been used to replace wood for lightweight, dimensionally stable pattern boards, these being shaped by conventional machining operations [25].

EPOXY TOOLING AS VACUUM-FORMING MOLDS

Epoxy tooling has been used successfully as molds for vacuum forming. In general, the formulation requirements are somewhat more critical than in model work. The relatively poor thermal conductivity of the epoxy has been reported to reduce premature chilling and thereby lead to better vacuum-formed products. On the other hand, surface temperatures during production tend to increase, such that at the end of a day's operation with vinyl sheet, for example, the temperatures may reach as high as 110°C [31]. At higher-temperature operations, aromatic amine curing agents are usually preferred, insofar as the room-temperature-curing amine systems will soften under heat to allow loss of detail [32]. With higher-temperature operation, too, the molding cycle will be increased because of the poor thermal conductivity of the epoxy formulation. Thermal conductivity may be improved by the incorporation of metallic fibers in the formulation, which in effect permits higher-temperature operations [33], but it also is customary to employ a series of cast-in copper cooling tubes. For one of the larger vacuum-forming molds, as much as 150 ft of ¼-in. copper tubing may be employed [21]. On these larger units, stage-casting techniques may be used to prevent excessive exothermic temperatures. Pins coated with release agent may be inserted before cure and withdrawn after cure to provide the necessary vacuum holes [2]. Suitable constructed molds appear to be capable of vacuum forming up to at least 100,000 plastic parts [31].

Molds made of epoxy and metal fibers but not fully impregnated have made inexpensive porous constructions which increase control of air evacuation and give good heat transmission to permit short cycles. A powdered epoxy combined with aluminum powder and cured without pressure also provides a suitable porous mold [35].

EPOXY TOOLING AS MOLDS FOR CASTING AND MOLDING COMPOUNDS

Epoxy resins may be used as tooling for epoxy or other thermosetting plastic encapsulating and potting compounds. Essentially, these are containers to retain the liquid resin during the cure cycle and may be either left in place or removed after cure.

Injection blow molds for polyethylene have been fabricated from the epoxies, but because of low thermal conductivity, a slower rate of production is imposed. Specialized systems have been used for short-run compression and injection tooling for thermosetting plastics. With this tooling, steel backing is normally employed for reinforcement and copper tubing is embedded to cut down the mold cycle. Mold pressure is normally increased more gradually than with metal tooling. Suitably formulated and reinforced systems have been used to mold rubber at temperatures at 170°C and pressures to 3,000 psi, and injection molds for polystyrene have produced up to 3,000 parts, using temperatures of 170°C on a 25-second cycle. Such tooling is suitable for short and experimental runs [14].

EPOXY TOOLING FOR LAMINATED PLASTICS

Epoxy molds for laminated plastics are produced from a male master by a multistep operation. The master is first treated with a release agent, following which a thixotropic surface coating (gel coat) is applied by brush or spatula. The gel coat serves to provide an accurate reproduction of the master. Over the gel coat are applied several layers of glass cloth or of glass mat or fibers and a liquid epoxy resin. After cure, the mold is removed from the master and the cured gel coat treated with several layers of a wax release agent. The plastic-reinforced laminate is then built up to the desired thickness, Mold release may be obtained on larger parts by introducing water between the mold and the part, thus floating the part free.

EPOXY TOOLING FOR JIGS AND FIXTURES

Epoxy tooling for jigs, fixtures, spotting rack, etc., may be cast, but more commonly, rather large items are involved, and it is necessary to employ glass-cloth (or other) reinforcement in the structure.

The procedure is similar to that described above for producing molds for laminated plastics, although a number of important new considerations are introduced.

The procedure generally is to coat the master with release agent and apply a gel coat. Before the application of the laminate, filleting pastes—made by adding glass fibers or cotton flock to the resin system—are used to cover sharp corners and projections. This eliminates the inclusion of air at these points during the laminating operation. The glass cloth preferred is usually open weave, balanced strength, treated with a suitable coupling agent such as Volan A. The application thickness of the laminate depends upon the service conditions. Usually 6 to 25 plies of glass are used, with care being taken to prevent excessive exotherm in the thicker sections. The ratio of glass to resin is about 1/1. Application may involve preimpregnating the glass with the liquid system or applying the glass dry and subsequently working the resin in between plies. In some cases, to obtain laminates of greater density and high glass contents (and consequently of greater strength) the tool may be vacuum-bagged during cure. If the larger tools contain thin sections (⅛ in. or less), metal plates should be molded in at these points. Checking fixtures may embody metal inserts on areas that come in contact with untrimmed edges of metal panels to prevent fraying [16]. Drilling jigs usually employ metal bushings as drill guides. The epoxy tool is drilled oversize, and the bushing bonded with an epoxy adhesive. The bushing may be simply coated with an epoxy

Lightweight epoxy laminates and reinforced epoxy tubing from a huge airplane inspection fixture at about one-fourth the weight and one-half the cost of equivalent steel tooling. (*Union Carbide Corp.*)

paste adhesive and pressed in, although, for greater accuracy, the bushing is usually potted in place with a liquid adhesive system. For larger tools, a reinforced backing structure is attached. This may consist of a highly filled epoxy casting, the cavity being loaded with coarse aggregate and the liquid epoxy poured in over it. Alternatively, when weight considerations are important, use may be made of metal tubing or of glass-reinforced epoxy tubing for the backing structure, these being attached adhesively.

Large tools made by hand lay-up laminating technique have been found to hold to an accuracy of plus or minus 0.002 in., with considerable weight saving over Kirksite—a conventional tooling metal. For example, a 240-lb epoxy-glass tool can replace a 3,100-lb Kirksite tool [1]. Since the Kirksite can be reused, whereas the epoxy cannot, epoxy will not always result in a lower tool cost, but generally

Low-cost highly stable fender welding fixture built up from epoxy resins and glass cloth demonstrates wide usefulness of epoxy tooling. (*Union Carbide Corp.*)

some savings are realized, ranging from 25 to as much as 75 per cent of conventional tooling, depending upon complexity. In general, the epoxy is the least of the cost of fabrication, only about 15 per cent of the tooling cost being material; the remaining, labor [26].

EPOXY TOOLING FOR METAL FORMING

Epoxy systems, although not suitable for shearing tools (unless metal-cutting edges are inserted), may be used in a number of metal-forming operations, in applications where the strength properties (suitably reinforced) are adequate and where surface hardnesses of no greater than about Barcol 65 are required. The

Tough epoxy drop-hammer dies withstand repeated impact and shock of forming sheets of aluminum, annealed steel, or stainless steel. (*Union Carbide Corp.*)

epoxy tool may be cast, or more frequently, for greater strength, be laminated. The epoxy systems may also be used to face existing tools. Epoxy tooling has been used for explosive forming, for stretch-press forms, rubber-press tools, beating blocks, spining chucks, draw dies in single- or double-action presses, drop-hammer punches and dies, etc.

Epoxy tooling is used for forming brass, and other copper alloys, mild steel, stainless steel, titanium, etc. For example, one experimental epoxy-faced die and punch set has been used on 0.064-in. aluminum to produce, over a total surface of about 17 by 20 in., 16 bends over one cross section, including radii of curvature as tight as $\frac{1}{8}$ in. with a maximum depth of draw of 1 in. [10], and it has been reported that epoxy tooling will handle any thickness of aluminum stock press formed to about $\frac{3}{16}$ in. [5]. On stronger sheet metals, thicker than about 0.048 in., and when sharp radii are required (less than $\frac{3}{16}$ in. on 0.036-in. mild steel), it may be necessary to use metal inserts, insofar as the compressive strength of the epoxy may not be great enough to resist the high local stresses imposed by metal wrinkle [29].

Large tools made from special systems, obtained by curing a mixture of steel or aluminum fibers and epoxy resin under heat and pressure, have been used to form 0.090-in. stainless steel sheet at 200 psi and a temperature of 120°C [20]. Use of metal fiber/epoxy (epoxy alloy) has been gaining in acceptance because of the high strength properties that may be obtained, coupled with good thermal conductivity. Typical data on metal fiber/epoxy systems are given in Chap. 14.

The number of pressings that may be obtained with epoxy tooling may be in the thousands, although generalizations are hazardous because of the variations in design; the number of pressings will be less than obtained with good tool-steel tooling. In one application, facing an existing metal punch to restore dimensional accuracy resulted in a tool capable of producing 15,000 outer automobile panels [16].

In formulating for metal-forming applications, the structural strengths of the epoxy are of considerable significance, usually the compressive strengths being most important. Some structural weaknesses may be compensated for by designing proper tool support [28]. In punches, particularly, where impact strength is important, the surface hardness may be the critical factor, rather than the structural strength. In general, for drop-hammer forming operations, some resiliency is imparted to both the punch and the die, the resiliency exerting a smoothing action on the sheet being processed and helping to eliminate wrinkles. Resiliency is usually provided by a facing applied over a rigid core, for the punch, and a facing applied over the laminated construction for the die. Generally, the hardness for the tool is in the range of Shore D 60 to 65 with the punch being less hard than the die. However, resilient punches can, in fact, be used successfully with metal dies, and it has been reported, also, that for short production runs at least, the drop-hammer punch may be hard-surfaced if an all-plastic die is used—there being sufficient yield in the die to make up for slight discrepancies in sheet—provided metal clearance is not a critical factor [22].

Other epoxy tooling, such as explosive forming dies, requires somewhat harder surfaces, in the Shore D range of 70 to 75, and stretch forms and the like are hard-surfaced [24]. When used as surface plates, as for example, granite block coatings, the formulation is designed to provide abrasion resistance [15].

With the epoxy metal-forming tools, creep may cause problems at about 70°C and above, but in general the problems are not so great as those experienced with the phenolics and the polyesters. To eliminate creep, the tooling should be mechanically reinforced, a span-to-depth ratio of 8/1 or less should be maintained, and postcuring (done carefully to prevent warping) can be employed [8].

A variety of molds and techniques may be used for the fabrication of epoxy metal-forming tools. The epoxy system is selected for the required strength and hardness characteristics and is fabricated from some form of master, in a fashion similar to that employed for the manufacture of jigs and fixtures.

When the tools are faced, the procedure involves the construction of a core—or the use of an existing metal tool that has gone out of dimension—which is positioned in a mold, and the facing is subsequently poured in. Fabrication of a typical core might involve the casting of a filled epoxy resin over a mold lined with modeling clay or the like. After the core has cured, the clay is removed, the core repositioned, and the facing cast. Sometimes it is possible to press the core into the facing after it has been applied as a gel coat to the mold. In the case of punch and die sets, the punch can often be used, by covering it with a sheet the thickness of the metal to be handled, as a form for casting the core of the die.

FORMULATION

In tooling applications, three types of formulations may be involved: casting systems, gel-coat systems, and laminating systems.

The majority of tooling formulations are based on DGEBA. The resin is normally cured with room-temperature curing agents, although for special high-temperature applications either aromatic amine or anhydride curing agents may be used. Fillers are employed to reduce cost and to modify cured properties and handling characteristics. Flexibilizers are employed to control surface hardness and to improve impact resistance.

The considerations that enter into the development of suitable formulations have been discussed at some detail in the appropriate earlier chapters. Chapter 22 presents a discussion on laminating applications which will provide additional information on this aspect of tooling formulation.

Casting Systems

The casting resins may be employed to cast the tool directly against a mold, to cast a core which will subsequently be faced, and to cast facings on existing cores.

Cast Tools and Cores. The primary requirements for the resins used to make cast tools and cores are adequate structural properties—if very high structural strengths are necessary, laminated systems are specified—low viscosity, and low exotherm. Low viscosities often cannot conveniently be obtained when fillers are employed, in which case compromises are required.

Table 18-1. Comparison of Taber Abrasion Resistance of Various Tooling Materials [4]

Material	*Loss of weight, mg*
Gypsum	5,000
Amine-cured DGEBA:	
Unfilled	450–500
Silica-filled	200–250
Aluminum alloy	230
Steel, hardened	85
Cast iron (gray)	25

Fillers are used extensively in the cast systems to lower cost and to impart desired properties such as abrasion resistance or improved thermal conductivity. The effect of fillers on abrasion resistance is indicated in Table 18-1. The effect of special fillers, selected for high thermal conductivity, upon thermal conductivity is indicated in Table 18-2. Chapter 14 offers a detailed discussion.

Low exotherm is normally obtained by selection of a curing agent appropriate to the mass being cast and by the incorporation of fillers. Larger masses may require stage-casting techniques or high loading volumes of coarse aggregate to prevent difficulties in this respect.

Table 18-2. Thermal Conductivity of DGEBA vs. Filler and Curing Agent [32]

Specimen	*Thermal conductivity* cal/(sec)(sq cm)(°C)(cm) $\times 10^{-5}$
Cured with aromatic amine eutectic:	
Unfilled	5–9
Capillary casting into precoated and cured blocks of copper and aluminum fiber (34.2% metal)	68
Pressure casting, 65.2% copper and aluminum fiber	384
10% copper and aluminum fiber	24
50% copper and aluminum fiber	87
Cured with NMA:	
10% copper and aluminum fiber	43
50% copper and aluminum fiber	73

In most casting applications shrinkage does not present a severe problem. Most of the shrinkage occurs in the liquid state, showing up as "slump" on the side opposite the tool face, rather than as draw from the master, thus giving good reproduction of detail. However, on pouring risers, if a large mass is in the head just above a thin section of the casting, the mass of the riser will cure first, and draw can be experienced in thin sections [23]. When shrinkage is critical, fillers should be employed, and cure temperature should be as low as practical. If high temperatures

Table 18-3. Volume Shrinkage and Compressive Strength of DGEBA as a Function of Curing Agent, Flexibilizer, and Filler [11]

Property	Cure *A*, TETA at 10 phr				Cure *B*, MPDA/MDA/PGE at 20 phr			
	Polysulfide, phr				Polysulfide, phr			
	0	25	50	100	0	25	50	100
No filler:								
Volume shrinkage, %	......	4.8	3.9	3.6	......	3.9	4.5	2.7
Compressive strength, psi	16,300	8,300	3,400	...	17,100	12,900	5,100	
Modulus, psi × 10^{-6}	0.512	0.23	0.081	...	0.45	0.36	0.11	
Yield, psi	15,900	7,400	2,800	...	15,800	11,600	4,300	
Deformation at yield, % (0.2% offset)	3.4	3.5	4.0	...	4.2	4.3	4.1	
Milled glass fiber (0.031 in.), 10% *with A*, 18% *with B:**								
Volume shrinkage, %	......	3.5	3.4	3.2	......	2.3	2.5	1.5
Compressive strength, psi	19,500	11,300	5,400	...	20,100	16,200	10,000	
Modulus, psi × 10^{-6}	0.77	0.47	0.16	...	1.1	0.67	0.45	
Yield, psi	16,200	8,900	3,600	...	17,300	13,600	8,000	
Deformation at yield, % (0.2% offset)	2.0	2.2	3.0	...	2.9	2.3	2.0	
Lithium aluminum silicate (200 mesh), 23% *with A*, 26% *with B:**								
Volume shrinkage, %	......	2.8	2.8	2.4	......	1.2	2.1	0.6
Compressive strength, psi	16,000	11,000	6,000	...	18,600	16,400	5,550	
Modulus, psi × 10^{-6}	0.78	0.41	0.22	...	0.91	0.70	0.18	
Yield, psi	12,800	9,500	4,800	...	14,800	12,400	4,000	
Deformation at yield, % (0.2% offset)	1.9	2.7	2.4	...	2.0	1.8	2.5	
Graphite (300 *mesh*), 8.5% *with A*, 11% *with B:**								
Volume shrinkage, %	......	3.6	3.1	2.8	......	1.5	1.5	1.2
Compressive strength, psi	15,100	8,500	3,700	...	16,200	13,900	6,000	
Modulus, psi × 10^{-6}	0.63	0.30	0.096	...	0.63	0.53	0.22	
Yield, psi	12,700	6,300	2,100	...	11,000	9,900	4,500	
Deformation at yield, % (0.2% offset)	2.5	2.2	2.4	...	2.0	2.2	2.5	
Quartz (130 *mesh*), 24% *with A*, 31% *with B:**								
Volume shrinkage, %	......	2.9	2.0	1.6	......	1.5	0,6	0.1
Compressive strength, psi	15,600	11,900	5,200	...	20,400	16,200	6,600	
Modulus, psi × 10^{-6}	0.62	0.53	0.23	...	0.98	0.68	0.29	
Yield, psi	13,300	9,800	3,600	...	15,400	12,400	4,100	
Deformation at yield, % (0.2% offset)	2.6	2.0	2.0	...	1.9	1.8	1.6	
Calcium carbonate (325 *mesh*), 24% *with A*, 25% *with B:**								
Volume shrinkage, %	......	2.3	2.4	2.4	......	1.9	2.4	1.8
Compressive strength, psi	16,000	10,600	5,700	...	17,200	13,700	7,400	
Modulus, psi × 10^{-6}	0.65	0.44	0.21	...	0.68	0.52	0.33	
Yield, psi	13,200	9,800	3,400	...	14,300	11,400	5,600	
Deformation at yield, % (0.2% offset)	2.1	2.7	1.8	...	2.5	2.5	2.0	

* Per cent by volume based on total volume of compound.

are used, the liquid resin will gel when it is hot, and thermal contraction on cooling will contribute to total shrinkage.

Cast Facings. Cast facings are normally employed to provide a resilient surface to an otherwise rigid tool. These are normally cast on, and are formulated for controlled flexibility and low viscosity. A variety of flexibilizers may be employed, typical of which are the liquid polysulfides. The effect of a polysulfide flexibilizer on the shrinkage and compressive strength of a system is shown in Table 18-3. Hardness may be modified by varying the amount of flexibilizer. Use of flexibilizers has the general effect of reducing the temperature resistance of the formulation.

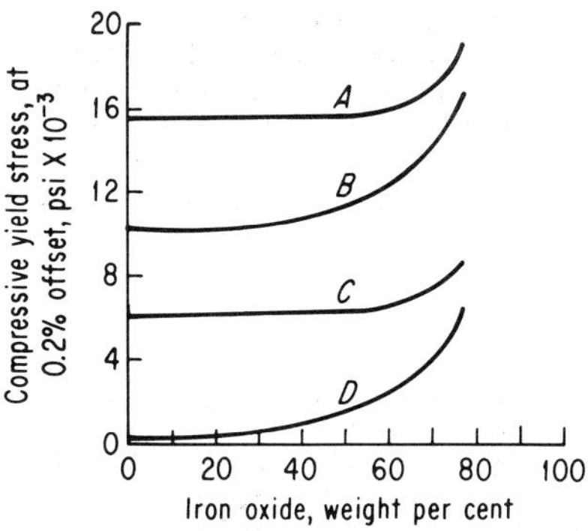

FIG. 18-1. Compressive strength vs. iron oxide concentration and cure temperature of DGEBA cured with MDA/MPDA/PGE [30]. *A*, Tested at 23°C, cured 24 hours at 65°C; *B*, tested at 23°C, cured 24 hours at 65°C plus 2 hours at 150°C; *C*, tested at 120°C, cured 24 hours at 65°C plus 2 hours at 150°C; *D*, tested at 120°C, cured 24 hours at 65°C.

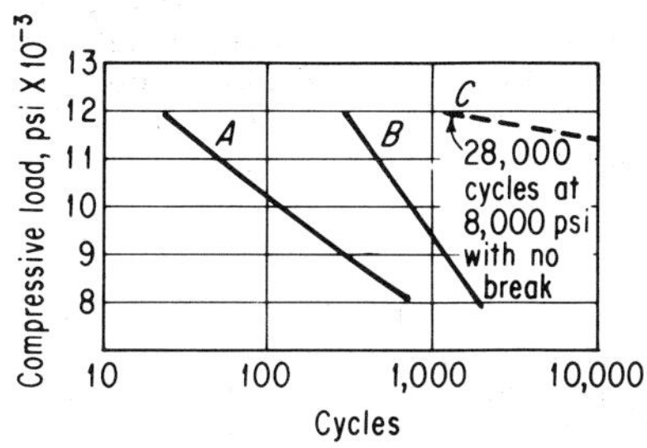

FIG. 18-2. Compressive fatigue of unfilled DGEBA vs. curing agent [30]. *A*, DMP-30 salt cured 24 hours at 65°C; *B*, TETA cured 48 hours at room temperature plus 24 hours at 65°C; *C*, MDA/MPDA/PGE cured 20 hours at room temperature plus 24 hours at 65°C.

Improving Compressive Strengths. When it is necessary to improve compressive strengths, fillers, curing agents, and cure cycles are important. The effect of these variables on typical cast tooling systems is indicated in Figs. 18-1 to 18-9.

Gel-coat Systems

Gel coats may be modified to change surface characteristics from resilient to hard, and by varying filler content and type abrasion-resistant or machinable systems may be developed. In general, the gel coats are formulated with a thixotroping agent to provide nonsag during application and cure. Colloidal silica is frequently used for this purpose in the 5- to 10-phr range. The gel coat is applied usually $\frac{1}{32}$ to $\frac{1}{16}$ in. thick, using a brush or spatula, and is then allowed to gel partially before application of the laminating (or casting resin) reinforcement. If the gel coat is allowed to cure before installation of the reinforcement, it is normally customary to sand the surface lightly to ensure adequate adhesion. Gel coats serve as protection against chemical attack and provide smooth surface layers which prevent the glass fiber from pressing to the surface to cause blemishes in the formed parts.

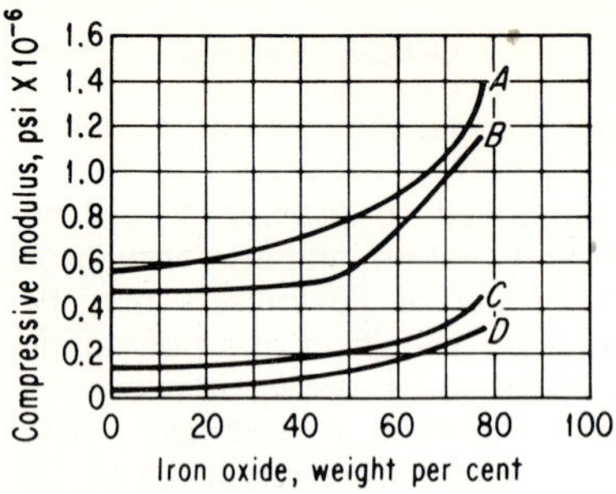

FIG. 18-3. Compressive modulus vs. filler concentration and cure temperature of DGEBA cured with MPDA/MDA/PGE [30]. *A*, Tested at 23°C, cured for 24 hours at 65°C; *B*, tested at 23°C, cured 24 hours at 65°C plus 2 hours at 150°C; *C*, tested at 120°C, cured 24 hours at 65°C plus 2 hours at 150°C; *D*, tested at 120°C, cured 24 hours at 65°C.

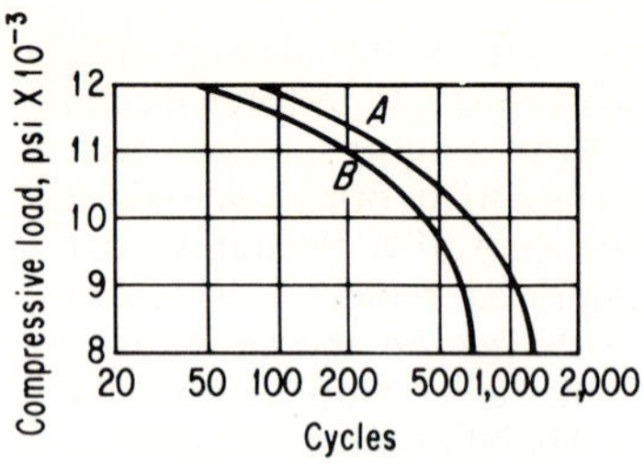

FIG. 18-4. Compressive fatigue vs. filler type for DGEBA cured with MPDA/MDA/PGE [30]. *A*, 10 phr asbestos cured 48 hours at room temperature plus 24 hours at 65°C; *B*, 10 phr mica (water-ground) cured 48 hours at room temperature plus 24 hours at 65°C.

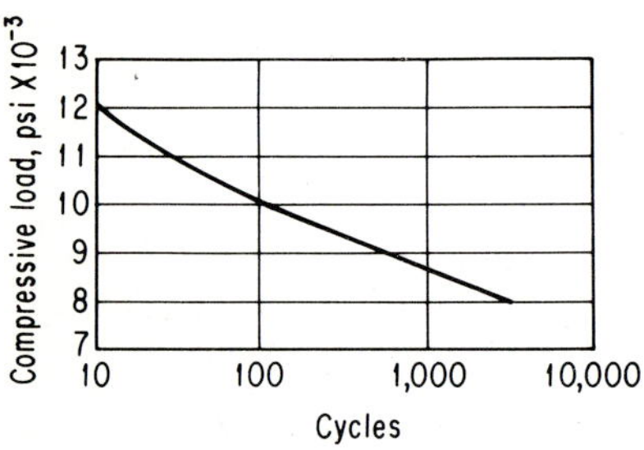

FIG. 18-5. Compressive fatigue of 400 phr sand-filled DGEBA cured with MPDA/MDA/PGE [30]. Cure cycle: 72 hours at 23°C plus 24 hours at 65°C.

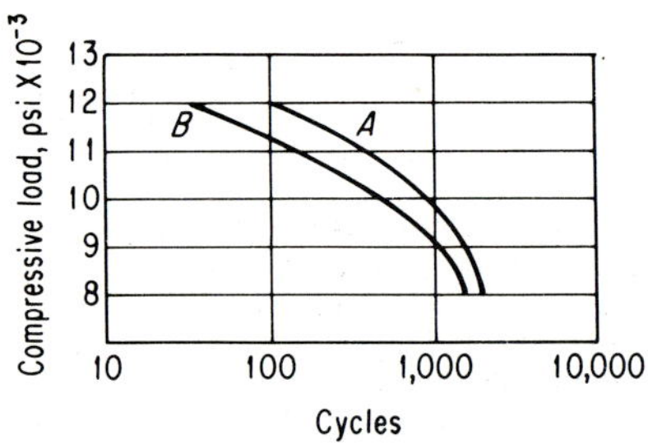

FIG. 18-6. Compressive fatigue vs. loading volume of iron oxide filler of DGEBA cured with MPDA/MDA/PGE [30]. Cure cycle: 20 hours at 23°C plus 24 hours at 65°C. *A*, 400 phr iron oxide; *B*, 200 phr iron oxide.

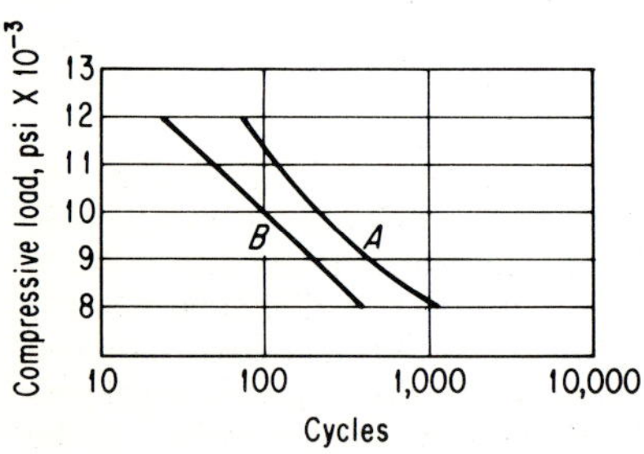

FIG. 18-7. Compressive fatigue vs. filler of DGEBA cured with TETA [30]. *A*, 75 phr calcium carbonate cured 20 hours at room temperature plus 24 hours at 65°C; *B*, 50 phr aluminum silicate cured over 4 days at room temperature.

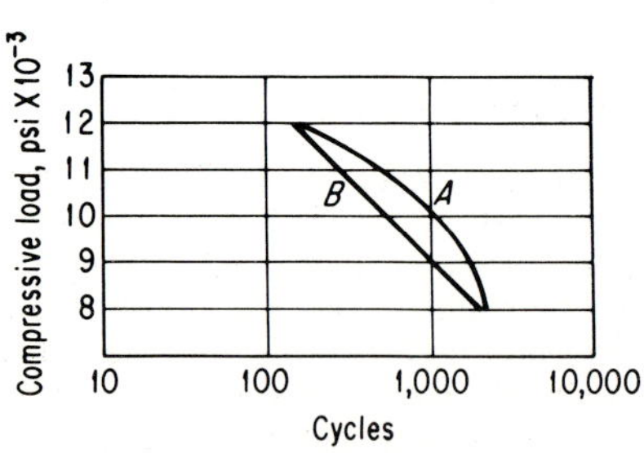

FIG. 18-8. Compressive fatigue vs. cure cycle of DGEBA cured with TETA and filled with 150 phr aluminum oxide [30]. Cure cycles: *A*, 20 hours at room temperature plus 24 hours at 65°C; *B*, over 4 days at room temperature.

Laminating Systems

The usual laminating formulation involves a room-temperature curing agent and DGEBA modified with about 10 phr of a reactive diluent, usually butyl glycidyl ether. The lower viscosity permits better wetting action and facilitates application. If elevated-temperature strengths are important, the reactive diluent is omitted, and aromatic amine or anhydride curing agents are employed.

Laminating formulations, together with glass cloth, glass mat, or glass fibers, are employed where the best possible structural strengths are required, in particular with larger parts where reinforcement is needed and casting techniques are either inconvenient or prohibitively expensive in terms of materials usage. The laminated structures generally possess a weight advantage over comparable cast units.

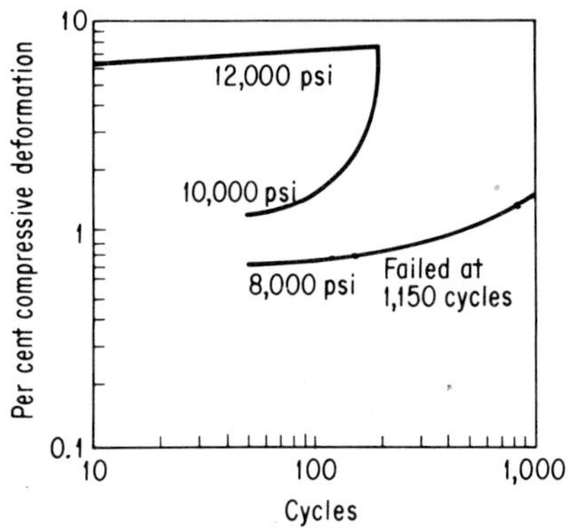

FIG. 18-9. Compressive modulus fatigue of DGEBA cured with TETA for 4 days at 23°C and filled with 100 phr aluminum oxide [30].

FABRICATION OF MOLDS AND MASTERS

Epoxy tooling produced by the casting method requires generally use of an open female mold, although for coreboxes, Keller models, etc., closed molds may be employed. Epoxy tooling produced by the laminating method usually employs a male or female master against which the tool is fabricated.

Molds are made from plaster, epoxy, etc., cast against a male master. The masters are made from wood, gypsum, cement, plaster, metal, or epoxy. One of the advantages of using epoxy tooling is that the mold or master may be made from low-cost conventional materials, unless, of course, elevated temperatures are required to cure the epoxy formulation.

A typical mold may be constructed from a wooden master by taking a plaster splash. The plaster splash may be backed with plaster-saturated hemp fibers and, if necessary, reinforced with tubing. When the plaster-splash mold is removed, it is thoroughly dried at about 50°C, after which any necessary touch-up and sanding is accomplished. The mold then is sealed with quick-drying lacquer or polyvinyl acetate, etc. It is then waxed and polished with a release agent, such as silicone greases or special formulations based on carnauba wax.

Masters may be developed from production parts, from wooden or plastic models, etc. If the model is male and the required master is to be female, the laminate may be built up directly on the model, by treating it first with a sealing agent (if it is porous) and then a release agent. If the model is female and the required master is to be female, a male model is constructed as an intermediate step. When the tooling involves reproducing a single face, the suitable face is developed, a restriction box is installed to contain the resin or the laminate, and the master is then fabricated.

Special molds may be developed for intricate shapes, where draw would prove difficult. Epoxy patterns, for example, may be cast from sand molds, the sand being sealed with polyvinyl acetate or polyvinyl alcohol as release films. In such cases, the patterns are finished by sanding with emery cloth. It has also been reported that the lost-wax process has been used successfully in casting epoxy tools [13].

In constructing the master or mold from which the epoxy tool is to be produced, it is extremely important to maintain dimensional accuracy adequate to the application, insofar as the epoxy resin will faithfully reproduce the master or mold, including existing defects.

After the epoxy is applied and cured, heat may be used to facilitate release. If the master or mold has a lower coefficient of expansion than the resin system, patterns with a concave shape should be heated, and those having a convex shape should be cooled; with plaster or wood patterns having a higher expansion rate, the reverse procedure should be adopted.

SUMMARY

Epoxy tooling, because of its low shrinkage, high dimensional stability, and good overall physical properties which may conveniently be modified to accommodate a rather wide range of requirements, has achieved considerable field success, in large measure replacing other thermosetting plastics and extending the application range of cast plastic tooling.

Epoxy tooling has been used successfully for the manufacture of prototypes and models, and foundry patterns, etc., giving service lives equivalent to those obtained with aluminum. Epoxy tooling, because of the close tolerances which may be held, has found wide usage as drill and welding jigs, spotting racks, and all variety of positioning fixtures. In these applications, where service strength requirements are not particularly severe, epoxy tooling can be expected, barring accident, to provide virtually unlimited service life. Epoxy tooling has also been employed for short-run production of plastic sheets by vacuum-forming techniques, and, suitably reinforced, has been used as tooling for short-run compression and injection molding. Epoxy molds, either open or closed, have been used in connection with a variety of thermoplastic and thermosetting casting resins. Epoxy tooling has been used successfully for short production runs as metal-forming equipment, generally working well with fairly heavy gages of the nonferrous metals and satisfactorily with thinner gages of the ferrous metals. Service life of such tooling is highly dependent upon the specific configuration and the metal being formed. Successful runs of up to 50,000 parts have been reported.

Epoxy tooling may be constructed by the cast method or by the laminating method. The cast method is generally preferred for smaller parts, for items where structural strengths are not too important, and for applying facings to cores or to existing tooling. The laminating method is preferred for larger tooling and in applications where high strength properties are required.

The majority of epoxy tooling is based on DGEBA suitably filled and modified, and cured with a room-temperature curing agent. Room-temperature curing agents permit use of relatively low-cost models.

Use of epoxy tooling often results in a considerable reduction in tooling costs and tooling time as compared to metal tooling; epoxy tooling designed within the inherent limits of the formulations has achieved considerable industrial acceptance since its introduction in the early 1950s.

REFERENCES

1954

1. Shokal, Plastic Fixtures Have Wide Use, ASTE Tool Engineering Conference Paper 22T44.

1955

2. Bertucci, Epoxy Resin Molds for Three Dimensional Effect, *Plastics World*, June.
3. Fieron, Epoxy Patterns, *Foundry*, July.
4. Meyerhans, The Use of Ethoxyline Resins in Modern Tool Manufacture, *Sheet Metal Ind.*, March.

1956

5. AVCO Cuts Costs with Plastic Tooling, *Am. Aviation*, Dec. 18.
6. Esdale, How to Choose Plastic Dies, *Automotive Ind.*, August.
7. Kish, How to Make Duplicate Patterns with Epoxy Resins, *Tool Eng.*, January.
8. Shokal, Epoxy Castings Replace Metal Forming Tools, *Am. Machinist*, Feb. 27.
9. Talbert, How to Make and Repair Epoxy Dies, *Mod. Machine Shop*, January.

1957

10. Price and Peerman, Polyamide Resin Alloys and Tooling Applications, *Tool Eng.*, May.
11. Sorg and Breslau, Polysulfide Liquid Polymer/Epoxy Resin Compounds in Plastic Tooling Applications, *SPE J.*, June.

1958

12. Mele, Cutting Production Costs and Time with Epoxy Resin Press Dies, *Plastics Technol.*, March.
13. Mooring and Lemons, Epoxy Radomes Made by Unique Lost Wax Process, *Plastics World*, April.
14. O'Connor, The Use of Heat Resistant Epoxy Formulations for Producing Low-cost Molds for the Plastics Industry, 14th ANTEC of the SPE.
15. Renaud, Surface Plate, U.S. 2,859,530.
16. Ruddiman, Plastics as Used for Tooling Aids in the Stamping Industry, 14th ANTEC of SPE.
17. Voss, Aircraft Tooling with Plastics, 14th ANTEC of SPE.
18. Wagner, Pattern and Method of Molding, U.S. 2,846,742.
19. Weaver, Plastics Applications in the Foundry, 14th ANTEC of SPE.

1959

20. Bogart, Casting Large Plastic Tools, 15th ANTEC of SPE, January.
21. Giant Epoxy Resin Molds, *Plastics World*, August.
22. Rau, All Epoxy Dies for Short Runs, *Am. Machinist*, June 15.

1960

23. Bogart, Hows and Whys of Plastic Tooling, *Plastic Technol.*, November.
24. Bryan, Epoxy Dies for Explosive Forming, *Tool Eng.*, February.
25. Delmonte, Low-density Epoxy Pattern Boards, RIP Division of SPI, Chicago, February.
26. Hankins, Plastics Materials Most Suitable for Tooling, RIP Division of SPI, Chicago, February.
27. Johnston, The Application of Reinforced Plastics for the Foundry and Pattern Industry, RIP Division of SPI, Chicago, February.
28. Lascoe, Epoxy Plastics for Tooling Investigations, RIP Division of SPI, Chicago, February.
29. Mitchell, Epoxide Tools: an Appraisal, *Brit. Plastics*, September.
30. Parry and Kewitt, The Effect of Fillers on the Physical Properties of Epoxy Resins, Shell Chemical Corp.

1961

31. Lawry, Basic Considerations in Building Epoxy Molds for Vacuum Forming Operations, *Plastics Design and Processing*, September.

32. Schaeffer and Diez, Prototype and Mold Fabrication, *Plastics Technol.*, September.
33. Schaub, Fabrication of High Temperature Molds and Dies Using New Epoxy Resin Systems, 17th ANTEC of SPE, January.

1963

34. Moore, Rapid Curing Resin-Filler System, U.S. 3,107,403.

1965

35. Delmonte, Porous, High-temperature Plastic Tools for Vacuum Forming, *Plastics Design Process.*, November.

Chapter 19

EPOXY-RESIN FOAMS

Epoxy-resin foams are characterized by good adhesive strengths (often of importance when foam in-place formulations are used), low water absorption, good dimensional stability (particularly in the more dense formulations), good heat resistance, and generally good resistance to chemical attack.

The properties of the rigid and semirigid, lower-density epoxy foams are in about the same range as those developed from comparable polyurethane formulations. The epoxy foams appear to hold some advantage in increased water resistance and to provide (when suitably formulated) superior elevated-temperature properties and long-time aging characteristics. But because of availability of lower-cost foams with properties adequate for the majority of commercial applications, and because of the general difficulties involved in achieving an elastomeric epoxy system [11], the epoxy-resin foams have not attained volume usage and are employed only in somewhat specialized applications.

The low-density epoxy foams find use as lightweight potting compounds for electronic equipment and as fillers for reinforcing load-bearing sections of honeycomb members [22].

Higher-density epoxy foams are used as structural supports, in which case they frequently see end service in conjunction with a laminate in a sandwich construction. Preformed blocks of foam have been used in applications such as lightweight rifle stocks, pattern boards (see Chap. 18), checking fixtures, and dielectric aircraft lenses [29]. They may be used as thermally insulating coatings.

Because of the lower water absorption, epoxy foams see some use in flotation equipment for the fishing industry, in other marine applications, and as evaporation retarders on storage tanks. Their use as acoustical insulation is less advantageous than more open foams because of their unicellular structure.

Two basic types of epoxy-foam formulations are available: the chemical foams and the syntactic foams. Chemical foams are produced by the release of a gas

Electronic circuit packages are potted in epoxy syntactic foam as protection against mechanical damage, moisture, and vibration. Molds are shown in the foreground. (*North American Aviation, Inc.*)

through a chemical reaction (or otherwise) during cure. Syntactic foams are obtained by the incorporation of prefoamed—or low-density—fillers into the system.

CHEMICAL FOAMS

The epoxy resin can be made to foam, during the curing reaction, if some element is present which liberates a gas or vapor at the curing temperature prior to gelation.

The curing agent, itself, may react in such a way as to liberate the gas. Examples of this type are the amine boranes, such as dimethylamine borane and pyridine borane, trimethoxyboroxine [17], aconitic acid [1], and substituted phosphitoborohydrides [32]. Compounds may be added which will react with the curing agent to liberate the gas; for example *N,N′*-dinitroso-*N,N′*-dimethyl terephthalamide will react with fatty-polyamide curing agents to liberate nitrogen during the curing reaction [16], and hydrogen peroxide will produce foams with primary amine cures [18]. High-boiling gases (e.g., Freons) may be incorporated in the formulation under superatmospheric pressure and subsequently released [27]. Low-boiling liquids can be added to vaporize on heating. Fillers may be employed which contain sufficient moisture and entrapped air, together with perhaps reactive ingredients, and in this manner foams can be produced. For example, cork and corklike natural materials such as balsa wood may be used to produce foams. Using 80-mesh powdered cork and a room-temperature amine curing agent, products have been obtained that expand to three to four times the original volume within 15 minutes after mixing of the ingredients [23]. *iso*-Dibutylaluminum *o*-acetylphenoxide has been reported as a foaming agent for periodic acid epoxies [33].

The simplest of the foaming agents for the epoxy formulations is water. If the temperature during cure, and prior to gelation, exceeds 100°C, the water, in the vapor phase, will produce a foam. One system of this sort involves use of an aqueous solution of protective colloids (e.g., a copolymer derived from 40 parts vinyl acetate, 60 parts *N*-vinyl-α-pyrrolidone) and a primary aliphatic amine curing agent [21].

However, more sophisticated, so-called blowing agents are usually employed in commercial practice [4]. Typical of these materials are Celogen [*p*,*p*′-oxybis-(benzenesulfonylhydrazide)] and Unicel ND (dinitrosopentamethylenetetramine). A variety of blowing agents are claimed for amine-cured DGEBA foams in [2].

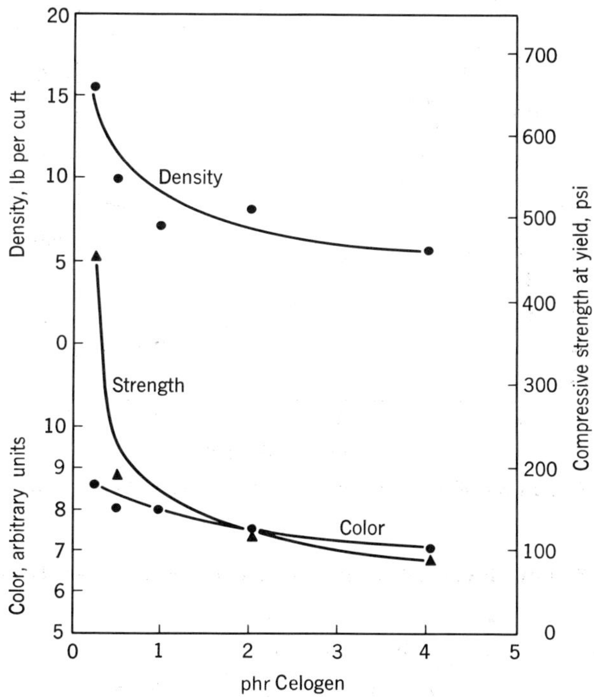

FIG. 19-1. Effect of blowing-agent concentration on properties of DGEBA foam [4-6].

The effect of different blowing agents on a room-temperature-curing epoxy formulation is shown in Table 19-1. The effect of varying the concentration of a typical blowing agent is shown in Fig. 19-1.

In addition to blowing agents, use is frequently made of surface-active agents to control the cell size and to provide a fine and uniform dispersion of the gas bubbles. Several generic types may be employed: (1) partial esters of common fatty acids, (2) polyoxyglycols, (3) hexitane and hexides from sorbitol, and (4) materials derived by adding polyoxyethylene chains to the nonesterified hydroxyls of the hexitans. Also suggested are organopolysiloxanes [25]. Surface-active agents are available under proprietary trade names. For example, polyoxyethylene sorbitanmonolaurate is available as Tween 20. A suitable silicone is Dow Corning 199.

Organic solvents are frequently added to dilute the exothermic reaction and to serve as secondary blowing agents. The effects of varying the concentration of toluene solvent on foam properties is indicated in Fig. 19-2.

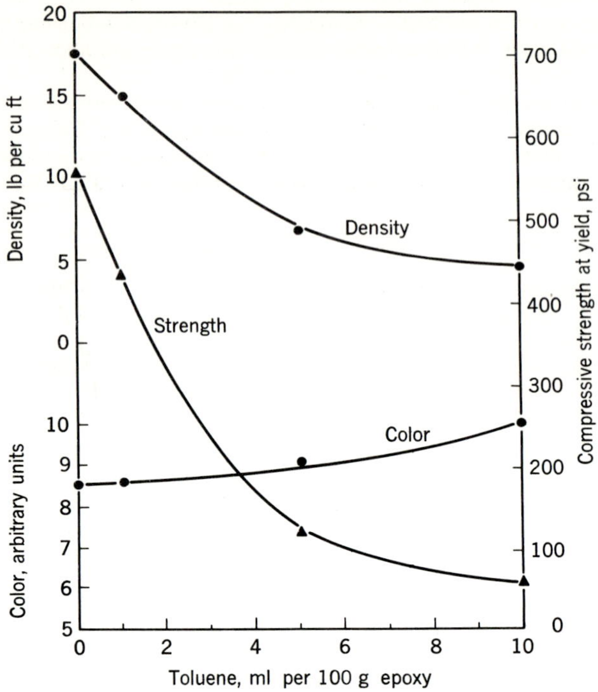

FIG. 19-2. Effect of toluene content of properties of DGEBA foam [4-6].

Table 19-1. Effect of Different Foaming Agents upon Properties of Cured Foam [4-6]

FORMULATION

DGEBA (mol. wt. 450)	100 g
Toluene	5 g
Tween 20	2 drops
DETA	6 g

Foaming agent	Proportion of foamant, phr	Final temperature, °C	Density, pcf	Compressive strength, psi	
				Ultimate	At yield
Ammonium carbonate	1	185	7.5	187	167
Ammonium carbonate	2	...	7.5	159	110
Celogen	1	175	8.5	131	118
Dinitrosopentamethylene-tetramine	1	195	10.7	289	289
Salicylic acid	1	178	9.0	189	127
Diazoaminobenzene	1	182	9.5	180	148
Water	10*	110	4.4	23	22
Water	3	110	14.5	320	252

* 8 phr DETA.

Resinous Modifiers for Epoxy Foams

A variety of the resinous modifiers may be employed with epoxy-resin foams. The phenolic resins, themselves capable of producing rigid foams of good quality, have been used successfully. Foam formulations containing anhydride curing agents and polyvinyl chloride have been developed [15]. A considerable amount of work has been accomplished with isocyanate additives.

The isocyanate/epoxy combinations employ a technique for obtaining the foam that is rather different from that of the other modifiers. Normally, a prepolymer is made from the epoxy and the isocyanate. The reaction occurs between the isocyanate groups and the hydroxyls on the epoxy-resin chain. Other additives, such as an aldehyde may be employed [12]. The prepolymer is then foamed and cured by the addition of a primary amine, or, together with a tertiary amine catalyst, by the addition of water or carboxyl-containing molecules. The reaction is reported to be very heat-sensitive [3].

Fire retardance has been obtained in epoxy foams by addition of 25 phr β,β'-dichlorodiethylvinylphosphonate homopolymer [34].

Fillers for Epoxy-resin Foams

Fillers may be used to obtain some cost reduction with the higher-density epoxy-resin foams. For example, up to about 20 phr of inert, finely divided filler may be used in foam formulations having a density in the 17- to 20-pcf (pounds per cubic foot) range without substantial sacrifice of cured properties. However, overloading with filler will weaken strength because of thin cell walls. With foams having densities as low as 5 pcf, it is probably not advisable to employ filler loading volumes much in excess of 4 to 5 phr [8].

Fillers may be employed as pigments to indicate the thoroughness of the mixing operation, and may be used to impart special properties. Antimony oxide, for example, may be used in connection with chlorinated compounds to improve flame resistance. Metallic fillers are used to control the dielectric constant of the foamed formulation. For example, the dielectric constant of a 20-pcf foam may be as low as 2, whereas by the use of powdered metallic fillers (such as copper, aluminum, or silver) the dielectric constant may be increased to any desired value up to at least 7 [6]. Aluminum powder appears to be the preferred filler in this connection [30,31].

Typical Formulations

As with other epoxy-resin systems, the formulation of chemically foamed epoxies involves a number of considerations. The foams may be supplied in free-flowing powdered form (based on the higher-molecular-weight DGEBA resins) or in the form of "gunk" molding compounds, containing the powdered system wetted with a solvent. The foams also may be supplied as one-component liquids, capable of reaction at elevated temperature; or they may be supplied as two-component systems, which will cure either at room or at elevated temperature, depending upon the curing agent employed.

Typical of the formulations are those listed below [3]:

	Parts
DGEBA (mol. wt. 950)	100
MPDA	10
Celogen	10
Styrene oxide	5
Pluronic L-68	0.1

This product produces very high exotherm at the curing temperatures (120°C) and is most suitable for masses of less than 100 grams. It produces a density of about 6.8 pcf and a compressive strength of 80 psi.

	Parts
Liquid epoxidized novolac resin (*f* 2.1)	100
CA	80
Tolylenediisocyanate	60
Triethanolamine	10
Pluronic L-64	0.5
Water	2.5

The density of this foam may be varied, within limits, by the amount of water employed.

	Parts
DGEBA	100
Tolylenediisocyanate	50
Diethylethanolamine	3
Water	5
Pluronic L-64	0.5

The formulation starts foaming as soon as mixed. Foams having a density as low as 2 pcf have been prepared.

	Parts
DGEBA	80
MPDA	28
Celogen	2
Plyophen 5023	20
Pluronic L-64	0.1

This foam expands on heating to about 100°C and provides a good cell structure.

A room-temperature-curing, two-component epoxy foam, using a chlorofluorocarbon as blowing agent, has been used as a spray-on coating that combines thermal insulation with corrosion protection. Used with dual-nozzle or mixing-head-type spray equipment, the coating foams within 20 seconds after application and is tack-free in 2 minutes at 23°C. Heat liberated during the reaction of the epoxy with the curing agent volatilizes the chlorofluorocarbon. Typical cured properties are given in Table 19-2.

Generally speaking, with room-temperature curing formulations, improved properties may be obtained by applying a postcure to the foamed product. The postcure normally should be initiated before the exothermic heat is completely dissipated, since the low thermal conductivity of the solid makes it slow to reheat, especially at the inner portions of thick sections.

Cure in-place formulations are particularly critical, insofar as the foaming reaction is dependent upon heat. Thus, the amount of material being foamed, the type of container employed, the temperature at which the operation is conducted, the cure temperature used, etc., all, to some extent, require consideration. It is necessary to select the curing agent giving just the required exothermic properties for the mass and the configuration being cast. This becomes increasingly critical as the size of the cast mass is increased, insofar as exothermic heat may cause a problem at the core of the mass because of poor thermal conductivity.

Often, for a given configuration, the stoichiometric amount of a particular curing agent may not provide best properties, and it is therefore necessary to investigate use of the curing agent over a range of concentrations. This is indicated in Fig. 19-3, where DETA is used to cure DGEBA (mol. wt. 450). The effect of processing temperature on the formulation is indicated in Fig. 19-4.

Table 19-2. Properties of Spray-on Foam [4-6]

Property	*Value*
Density, pcf:	
Over-all	1.8–2.0
Core	1.7–1.9
Thermal conductivity at 23°C, Btu/(hr)(sq ft)(°F)(in.)	
Initial *k*-factor	0.11–0.12
Aged *k* factor*	0.16–0.17
Tensile strength, psi	26–31
Elongation, %	34–44
Compressive strength, psi	13–17
Adhesive strength, psi:	
Of cold-rolled steel	27
Of primed steel	29
Of stainless steel	32
Of aluminum	27
Of sandblasted polystyrene	31
Water-vapor transmission at 23°C and 50% relative humidity, permeable inches	0.9–1.2
Water absorption after 48 hr under 10-ft head, psf	0.03–0.05
Dimensional stability, %:	
At 160°C after 72-hr total immersion in liquid nitrogen:	
Volume change	+0.6 to +0.8
Weight change	+0.2 to +0.4
At 60°C and ambient humidity after 72 hr:	
Volume change	+2.9 to +3.7
Weight change	+0.2 to +0.6
Unicellularity of closed cells, %	88–90
Resistance to solvents:	
Aliphatic hydrocarbons	Excellent
Aromatic hydrocarbons	Fair
Low-boiling ketones	Poor
Resistance to chemical agents:	
Acid	Excellent
Alkali	Good

* Equilibrium value reached at ambient temperature.

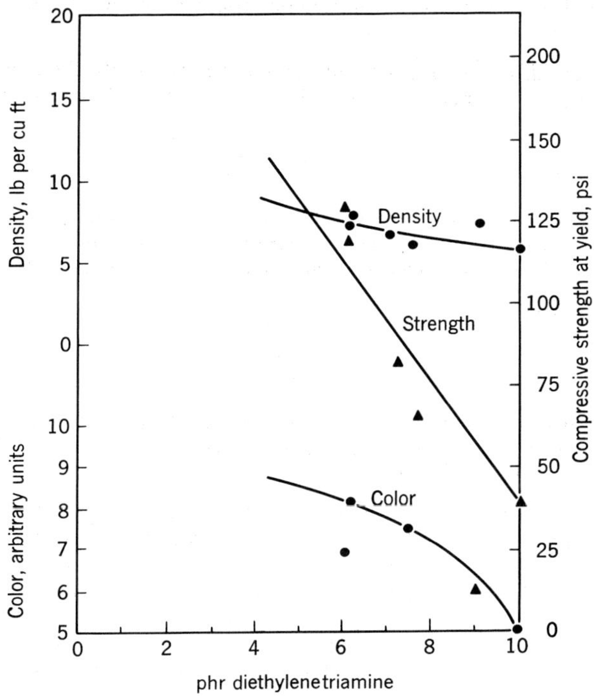

FIG. 19-3. Effect of curing-agent per cent on properties of DGEBA foam [4-6].

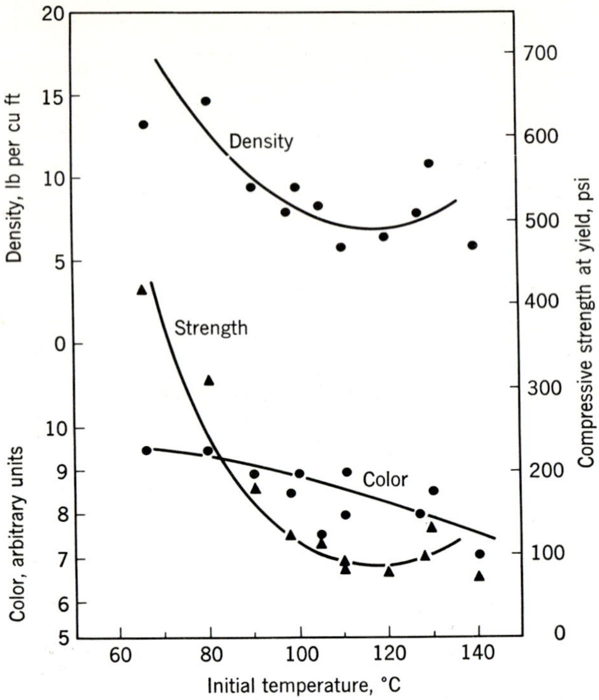

FIG. 19-4. Effect of initial temperature on properties of DGEBA foam [4-6].

During mixing, air or other gas should be incorporated into the blend to provide nucleation. The requirement for nucleation can be understood from the following relationship:

$$\frac{dr}{dt} = \frac{3pr}{8\eta} - \frac{3\gamma}{4\eta} \tag{1}$$

This equation states that the rate at which a bubble grows is a function of its radius, r, the excess pressure, p, inside the bubble, the viscosity, η and the surface tension, γ of the liquid medium. When the bubble neither grows nor contracts, i.e., $dr/dt = 0$,

it can be derived that

$$p = \frac{2\gamma}{r} \tag{2}$$

From Eq. (2),

it can be seen that if no bubbles are present, i.e., $r = 0$, the theoretical pressure becomes infinite. Thus, for smooth foam expansion, preformed nuclei of gas must be introduced into the liquid before the rise of the foam [20].

In view of the variables involved (temperature, mass configuration of the unit, method of mixing, selection of curing agent and additives and concentration of curing agent and additives, selection of processing temperatures and curing temperatures, etc.) the formulation of suitable epoxy-resin chemical foams is a rather sophisticated art. Best results have been obtained when the foams are produced under tightly controlled conditions in molds and ovens designed specifically to accommodate the systems. Thus, the foams which are supplied in prefoamed and cured blocks—rather than those which are supplied as foam in-place formulations—have generally proved most satisfactory in service.

Properties

The physical properties of the epoxy-resin chemical foams may be modified by the selection of the curing agent (in accordance with principles discussed earlier in the book), chiefly to achieve improved heat resistance or increased resiliency.

In addition, the properties of the foamed product will be highly dependent upon the density of the foam itself.

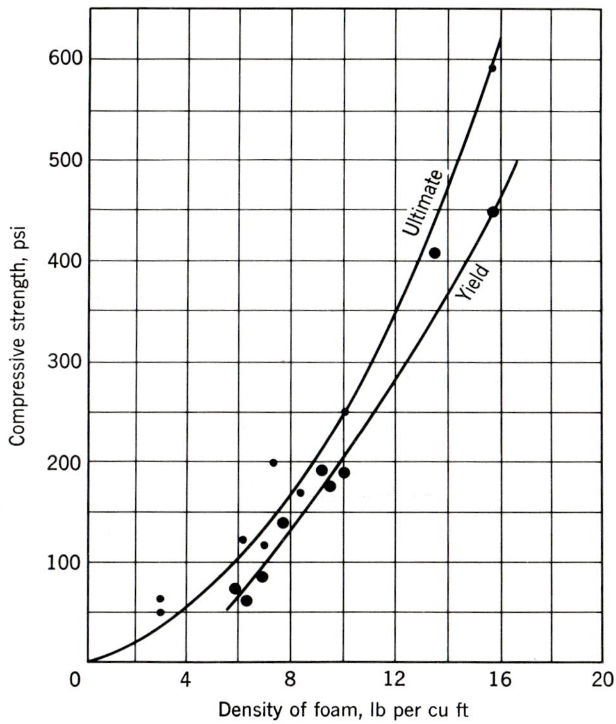

FIG. 19-5. Relation of compressive strength to density of DGEBA foam [4-6].

Compressive strengths at room temperature of the rigid epoxy foams vary from about 60 psi at a density of 5 pcf to about 1,100 psi at a density of 20 pcf. The relationship is very nearly linear between these points. Typical values obtained for compressive strength are those indicated in Fig. 19-5. With a rigid epoxy foam at a density of 5 pcf compressive modulus is about 2,500 psi, whereas at 18 pcf it is about 27,000 psi [8].

Tensile strengths for rigid epoxy foams are from about 50 psi at 5 pcf to about 500 psi at 18 pcf [8].

The temperature resistance of the epoxy foams is largely a function of the curing agent selected, together with the density of the foam. For example, when an aromatic amine curing agent is employed, foams of 5 pcf will compress 40 per cent at 105°C under a 20-psi load; whereas foams of 19 pcf will compress only 8 per cent at 120°C under the same load [8]. At cryogenic temperatures, some increase in tensile strength might be expected over the performance at room temperature; this, however, does not seem correct, with lower strengths generally being obtained at low temperatures—probably because the irregularities in the structure create notch effects where cell walls join each other [7]. Modulus, rigidity, and stress-strain curves for a typical epoxy foam at low temperatures are shown in Figs. 19-6 and 19-7.

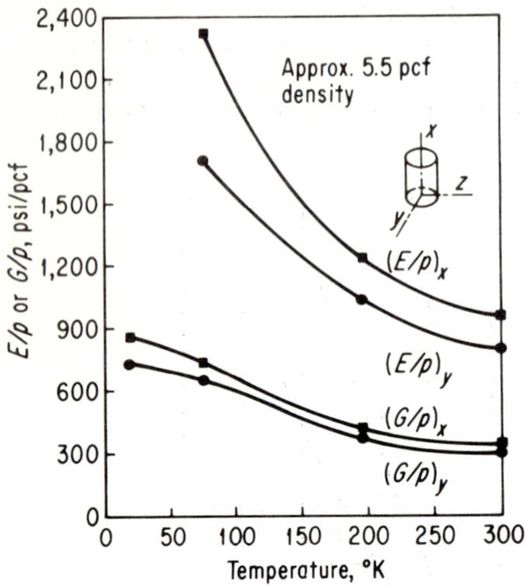

FIG. 19-6. Modulus of elasticity E/p and rigidity G/p vs. temperature for rigid DGEBA foam [7].

Blocks of epoxy foam may be compressed at higher temperatures and cooled under pressure. They will retain their compressed shape until reheated. Thus, they can be worked in the compressed shape and subsequently be expanded in place upon the application of heat.

The electrical properties of the epoxy-resin foams, like the physical properties, are dependent on the density. At lower densities, the dielectric constant is quite low insofar as the majority of the structure is composed of air with a dielectric constant of 1. With the higher-density foams, the dielectric constant increases with the increased amount of the epoxy resin in unit volume.

The thermal insulating properties of the epoxy foams are quite good, increasing from near that of air at lower densities to near that of the unfoamed epoxy resin at the higher densities [9]. Low initial thermal conductivity is dependent on cell size (Fig. 19-8) and density (Fig. 19-9). When heat dissipation is required, it is sometimes possible to incorporate metallic inserts to provide paths for thermal dissipation.

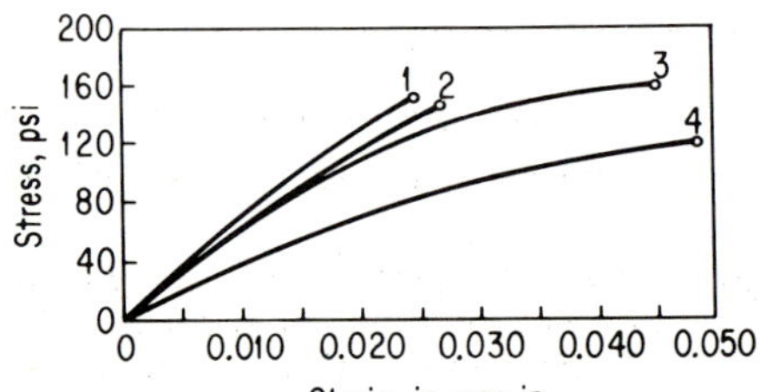

FIG. 19-7. Stress-strain curves for rigid DGEBA foam [7].

Specimen no.	Temperature, °K	Density, pcf	Foam-growth direction
1	195	5.43	Parallel
2	195	5.62	Perpendicular
3	300	6.06	Parallel
4	300	5.06	Perpendicular

In general, the chemical resistance falls in line with that expected with the unfoamed formulations. Specifically, with regard to water resistance, for example, a 5-pcf epoxy chemical foam was immersed in water for 3 weeks and revealed only about a 100 per cent water pickup, thus placing it, after this period, in the buoyancy range equivalent to that obtained initially with a 10-pcf foam [8].

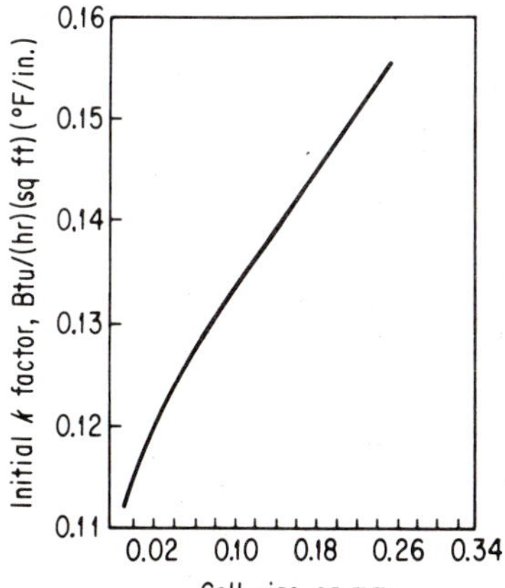

FIG. 19-8. Initial thermal conductivity vs. cell size for DGEBA foam [24].

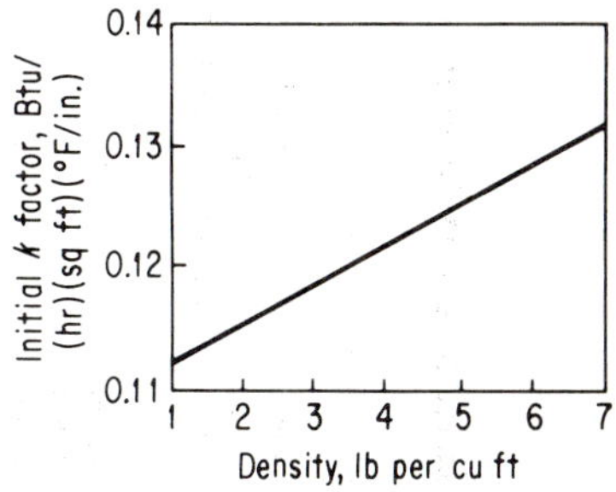

FIG. 19-9. Initial thermal conductivity vs. density for DGEBA foams with cell sizes from 0.20 to 0.04 sq mm [24].

SYNTACTIC FOAMS

The second method for producing foamed epoxy systems involves the utilization of microscopically small hollow spheres made of either organic or inorganic materials. The organic materials are usually made from phenolic, urea-formaldehyde, or polystyrene resins. These are filled with an inert gas, such as nitrogen (in the case of the phenolic spheres) or Freon or pentane (in the case of the polystyrene spheres). The inorganic materials are usually based on aluminum silicate or glass. Typical of the commercial materials are those shown in Table 19-3.

Table 19-3. Low-density Fillers [10]

Name	Description	Particle size	Bulk density, g/ml
Kanamite grade 200 ...	Unicellular hollow clay spheres, primarily aluminum silicate	90% 35–60 mesh 10% 20–100 mesh	0.9
Colfoam microballoons	Urea-formaldehyde spheres	90% 20–100 mesh 10% through 200 mesh	0.2–0.5
Microballoons, BJOA-0840.........	Hollow phenolic spheres	10% 40–100 mesh 74% 40–200 mesh 16% through 200 mesh	0.3–0.4
Glass microballoons, CPR 2077..........	Hollow glass microspheres	89% 60–325 mesh 11% through 325 mesh	0.26

Table 19-4. Physical and Electrical Properties of DGEBA Cured with 2-Ethylhexoate Salt of DMP-30 vs. Various Fillers [10]

PHYSICAL PROPERTIES

Filler	Viscosity at 25°C, centipoises	Shrinkage, linear, %	Hardness, Shore D	Density at 21°C	Tensile strength, psi	Thermal conductivity, Btu/(hr)(sq ft)(°F/in.)	Thermal expansion at 25 to 100°C, in./(in.)(°C) × 10⁵	Weight reduction over silica-filled compounds, %
None	13,500–19,500	0.12	80–85	1.17	8,000	2.68	8.7	26.5
Silica, 325 mesh, 100 phr	43,000–48,000	0.08	80–85	1.59	5,500	6.38	8.6	0
Phenolic spheres, 15 phr	34,000–38,500	0.14	80–84	0.86	3,300	1.91	8.2	46.0
Kanamite, 34 phr	34,000–39,000	0.06	75–80	1.01	2,000	2.47	6.7	36.5
Colfoam, 4 phr	45,000–48,000	0.17	80–85	1.01	4,050	4.15	8.6	36.5
Glass spheres, 14 phr	44,000–47,000	0.25	80–85	0.95	4,200	4.56	8.2	41.0

ELECTRICAL PROPERTIES

Filler	Dielectric constant at 25°C		Power factor at 25°C		Volume resistivity, ohm-cm			Dielectric strength at 25°C, volts/mil
	1 kc	1 Mc	1 kc	1 Mc	25°C	65°C	100°C	
None	3.8	3.7	0.0035	0.015	8.7×10^{14}	....	5×10^{11}	400–500
Silica, 325 mesh, 100 phr	3.4	3.4	0.003	0.012	1.3×10^{14}	6.2×10^{13}	......	>330
Phenolic spheres, 15 phr	3.2	2.7	0.003	0.014	1.0×10^{14}	5.3×10^{13}	......	>330

The organic spheres (or microballoons) are limited in service temperature to the temperature resistance of the organic resin employed in their manufacture. The inorganic materials are not, at least within the limits of the epoxy technology, the limiting factor of the service temperature of the foams. The inorganic materials generally provide more rigid systems with better strength properties, and the organic materials provide lower densities [14]. The handling characteristics of both the organic and inorganic materials are similar, and they can therefore conveniently be considered together. For polystyrene microballoons, however, separate treatment is warranted, insofar as these materials, unlike the others, may be further expanded during the curing reaction.

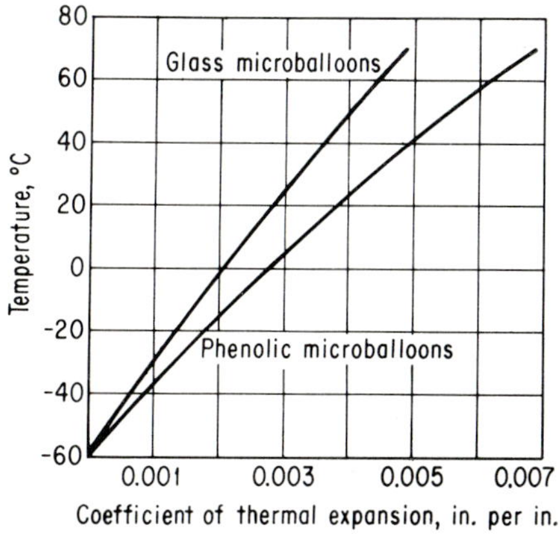

FIG. 19-10. Comparison of effect of organic and inorganic microballoons on thermal expansion of DGEBA cured with 12 phr DEA [14-3].

When the microballoons are employed as filler, the loading volume will determine the density of the cured product. At lower loading volumes, formulations may be developed which are sufficiently liquid to permit pouring, although at these volumes the reduction in density is not dramatic. A comparison of the properties of several microballoon-filled systems with those of a silica-filled system (all loaded to provide a pourable viscosity) is presented in Table 19-4. Further comparison is presented in Table 19-5. The effect of the fillers on thermal conductivity is shown in Table 19-6. The inorganic microballoons generally show a lower thermal expansion rate, as indicated in Fig. 19-10.

The microballoons may be employed in higher loading volumes to provide pack-in-place foams. The loading volume of such foams is rather high, and they are usually tamped in place under considerable pressure. With the phenolic microballoons, for example, the formulation may be as follows:

	Parts
DGEBA	27
Butyl glycidyl ether	3
DETA	3
Phenolic microballoons	120

The curing agent is added and mixed, after which the phenolic spheres are added. The resultant product is a thick putty. Alternatively, high loading volumes of

Table 19-5. Properties of Cured DGEBA at Equivalent Loadings of Filler [19]

Property	Microballoons at 33%	200-mesh silica at 33%
Specific gravity	0.6	1.7
Flexural strength, psi	4,400	4,200
Linear expansion, in./(in.)(°F) $\times 10^6$	17	16
Dielectric constant at 10^{10} cps	1.9	3.6
Dissipation factor at 10^{10} cps	0.015	0.028
Resistivity, ohm-cm $\times 10^{-12}$	1	4

Table 19-6. Comparison of the Hot-spot Temperatures of Unembedded Tubes with Those Embedded in DGEBA Formulation Containing Various Fillers [10]

(In degrees centigrade)

Filler	Operating unembedded tubes at 23°C ambient	Operating embedded tubes at 23°C ambient	Operating embedded tubes at 82°C ambient
Silica	100	75	112
Kanamite	100	85	125
Colfoam	100	100	135
Phenolic microballoons	100	105	140
Glass microballoons	100	95	129

microballoons may be obtained by admixing the balloons with powdered resin systems [26].

Degassing microballoon-filled systems, when required, is somewhat difficult. One suggested method involves vibrating the mixture in a thin film and passing a blast of hot air over the surface to break the bubbles [28].

As with the chemical foam, metallic fillers may be used to provide selected dielectric constants [5]. Flexibilizers and resinous modifiers, likewise, may be used.

Polystyrene Microballoons

Polystyrene microballoons are microscopic, gas-filled spheres which, under heat, will expand to provide substantial volume increases. The microballoons may be added to the epoxy-resin formulation, and the exothermic heat (or the heat during oven cure) can be employed to provide for the expansion. In this manner, foams having densities as low as 5 pcf may be developed.

However, because the expansion of the polystyrene is a function of temperature, close temperature control is required to obtain reproducible results, and, as is the case with the chemical foams, the size and the configuration of the mass being cast exerts a critical influence on the nature of the finished structure. The production of low-density polystyrene foams in sections of less than $\frac{3}{4}$ in. when primary-amine-type room-temperature curing agents are used is generally not practical; conversely, in larger sections, the beads may fuse because of excess heat. Insofar as the exotherm is a function of the concentration of the resin/catalyst mixture, the situation becomes increasingly critical with the higher-density products.

Polystyrene microballoons lend themselves well to the construction of laminate-faced sandwiches in closed-mold production. As the microballoons expand, the surface layer of epoxy resin will saturate the glass-cloth lining of the mold. Sufficient pressure (from 10 to 35 psi) will be exerted (assuming an optimum formulation) to assure a thoroughly saturated structure prior to gelation.

In order to eliminate, to some extent, the requirements for rather critical temperature control during the reaction, the polystyrene microballoons may be preexpanded by steam or radiant heating and these then added to the mold, heated in advance to a convenient temperature, such as 60°C. The mixed epoxy resin is then poured over the preexpanded beads and allowed to cure. In these cases, a limited amount of further expansion will occur as the result of exotherm or cure temperature to permit complete mold fill [13].

A typical (unpreexpanded) formulation will contain 100 to 750 phr of the polystyrene microballoons, the amount of balloons depending upon the desired density and the configuration of the unit. A typical formulation would be [4-4]:

	Parts
DGEBA	100
Polystyrene beads	175
cured with	
60/40 Isopropyl alcohol/water	10
DEAPA	6

The curing agent is selected to obtain the desired exotherm characteristics during cure. The mixed compound is troweled into a 95°C mold, which is closed and heated to 120°C. The resultant cured product will provide a density of approximately 7 pcf, with a compressive strength of 100 psi.

The polystyrene microballoons provide foams with room-temperature physical and electrical properties similar to those obtained with other organic syntactic foam fillers. Temperature resistance, however, is limited to about 70°C, with maximum operating temperatures (contrary to the experience with chemical foams) being obtained with lower-density products, insofar as the higher-density systems distort with increasing temperature because of the increasing gas pressure within the only partially expanded microballoons [4-2].

SUMMARY

Epoxy-resin foams are of two generic types: chemical foams, created by the liberation of gas during cure, and syntactic foams, produced by the incorporation of low-density fillers into the mixed compound. Both types of foam are supplied as liquids or as thick, trowelable, pack-in-place pastes.

The chemical foams are sensitive to a number of processing variables, including temperature, mass, configuration, surface tension, and viscosity. It is customary with each new unit to develop, largely by trial and error, the proper formulation. Factors to be considered are: the type and concentration of curing agent (which may not provide optimum results at stoichiometric concentrations), the type and concentration of the blowing agent (i.e., the gas-releasing element), the type and concentration of the solvent employed (if any), and the type and concentration of the surface-active agent used to provide more uniform foaming action.

Expandable low-density fillers (i.e., polystyrene microballoons) involve considerations common to the chemical foams, insofar as the properties of the expanded product are sensitive to processing temperature, mass size, and configuration. The nonexpandable low-density fillers (both organic and inorganic) involve considerations normally encountered with inert fillers. The inorganic microballoons are generally preferred when rigidity at elevated-temperature operations are important.

The physical properties of the foams are dependent on the density of the foam, with straight-line relationships being involved for most properties. The foams will provide dielectric constants from below 2 to above 7, depending on the density, cell size, and the presence or absence of metallic fillers. Chemical resistance is in line with what would be expected of the unfoamed products.

The epoxy-resin foam formulations may be modified by the use of flexibilizers or resinous modifiers, and fillers may be employed, although the use of fillers, in any substantial concentration, is usually restricted to the higher-density products. The curing agent is selected to provide the desired exothermic temperatures and necessary cured properties.

At the current state of the technology, epoxy-resin foams are special-purpose materials, and have not gained a substantial commercial market.

REFERENCES

1952

1. Koroly, Expanded Cellular Resinous Products Containing Polyepoxides and Aconitic Acid, U.S. 2,623,023.

1956

2. Parry and Blackburn, Foam-forming Compositions Containing Glycidyl Polyether of a Dihydric Phenol, U.S. 2,739,134.

1957

3. Bolstad and Stenerson, Epoxy Foams, 13th ANTEC of SPE.

1958

4. Aase and Bolstad, Foamed Epoxy Resin Compositions and Method of Making, U.S. 2,831,820.
5. Cuming, Radome Sandwich Using Artificial Dielectric Foam, *Electron. Design*, Apr. 16.
6. Cuming and Andress, Plastics and Ceramic Foams for Electronic Applications, *Elec. Mfg.*, May.
7. McClintok, Low Temperature Properties of Plastic Foams, *SPE J.*, November.
8. Nickerson, Epoxy Resin Foams, 14th ANTEC of SPE.
9. Segal, Techniques of Evaluating Epoxy Foams, Epoxy Resin Symposium of SPE, Minneapolis, Minn., October.

1959

10. Parr, Low Density Fillers with Epoxy Resins for Embedding Airborne Electronic Circuits, ACS Symposium, Atlantic City, September.
11. Schnitzer and Richter, Foam: Newest Member of a Versatile Family, *Mod. Plastics*, October.
12. Sekmakas, Aldehyde Resin–Epoxy Resin–Organic Polyisocyanate, Prepolymer, Method of Making Foam Therefrom, and Product, U.S. 2,906,717.
13. Struthers, Epoxy-Styrene Foam: Putting Exotherm to Work, *Mod. Plastics*, October.

1960

14. Cressey, Physical Properties of Filled, Low-density Epoxies, *SPE J.*, May.
15. Fisch, et al., Heat-hardenable Compositions Composed of Polyvinyl Chloride, Epoxy Resin and Hardening Agent, Cellular Product and Method of Preparing Same, U.S. 2,965,586.
16. Kohrn, Preparation of Cellular Polyamide Epoxy Resin Compositions, U.S. 2,936,294.
17. Lee and Neville, New Curing Agents for Epoxy Resins, *SPE J.*, March.
18. Sander and Meyer, Foamed Polyepoxides Resistant to Chemicals, Heat and Mechanical Stress, Ger. 1,110,860.

1961

19. Alford and Veatch, Glass Microballoon Particles, a Low-Density Filler, *Mod. Plastics*, November.
20. Burge et al., Foamable Epoxy Resin, Regional Technical Conference of Buffalo Section of SPE, Oct. 5.
21. Improvements in the Production of Porous Plastics on the Basis of Polyepoxy Compounds (to Badische Anilin-& Soda-Fabrik, Inc.), Brit. 881,257.
22. Pollock and Zahn, Three New Epoxy Resin Foams, *Mater. Design Eng.*, August.
23. Schardt, Epoxy Resin Composition Containing Cork or Balsa Wood and Expanded Products Therefrom, U.S. 2,993,014.
24. Toohy, Low Density Rigid Epoxy Foams, *Chem. Eng. Prog.*, October.

1963

25. Andres et al., Polyepoxide Foams, Belg. 626,937.
26. Feuchtbaum and Dunaetz, Powdered One Component Syntactic Foams for Space Age Packaging, *Insulation*, January.
27. An Improved Process for the Production of Epoxy Resin Foams (to Shell Internationale Research, Maatschappij N.V.), Brit. 919,779.
28. Milewski and Egbert, Method of Making Low Density Epoxy Resin Compositions, U.S. 3,103,406.

1964

29. Goldberg and Schnitzer, Foamed-epoxy Lens Antenna for Doppler System, *Elec. Design News*, February.
30. Nickerson et al., Dielectric Foams, U.S. 3,129,191.
31. Polyepoxide Resin Foams (to De Bell & Richardson, Inc.), Brit. 967,259.
32. Reetz and Dixon, Substituted Phosphitoborohydrides, U.S. 3,119,853.

1965

33. Puchala and Anselm, Epoxy Resin Foams, Ger. 1,199,993.
34. Valgin et al., Foam Plastics, U.S.S.R. 168,881.

Chapter 20

EPOXY-RESIN MOLDING COMPOUNDS AND COATING POWDERS

EPOXY MOLDING COMPOUNDS

Epoxy molding compounds were first made available about 1954. These early compounds had relatively short shelf lives and were of limited usefulness. By 1958, epoxy molding compounds were available which gave shelf lives satisfactory for production operations, and the technology dates from that period.

The epoxy molding compounds are basically of two types: the general-purpose molding compounds and the soft-flow (or encapsulation-grade) molding compounds.

The general-purpose molding compounds may, broadly, be considered to be in competition with other thermosetting molding materials, such as the phenolics and the diallyl phthalates. These epoxy compounds are employed for the production of structural parts, such as cases, housings, etc. The soft-flow molding compounds are used, in the main, for the encapsulation of electronic and electrical components and assemblies or circuits, and thus may be considered to be in competition with epoxy casting resins.

Advantages and Disadvantages Compared to Other Thermosetting Molding Compounds

Epoxy molding compounds offer a number of advantages over the more conventional thermosetting molding powders, among them being [8]:

1. Dimensional stability in the cured product.
2. Good bonds to fillers and inserts when present.
3. Good electrical and mechanical properties which, with suitable formulation, may be retained to about 150°C.
4. Good chemical resistance.
5. Good flow properties such that less than conventional molding pressures may be used. Low pressures permit use of less expensive molds, molding of delicate inserts, and attainment of fine detail [9]. In addition, with transfer molding, through-gating is possible, permitting more compact mold design. Thus, a press

Items manufactured with epoxy-resin molding compounds. (*Hull Corp.*)

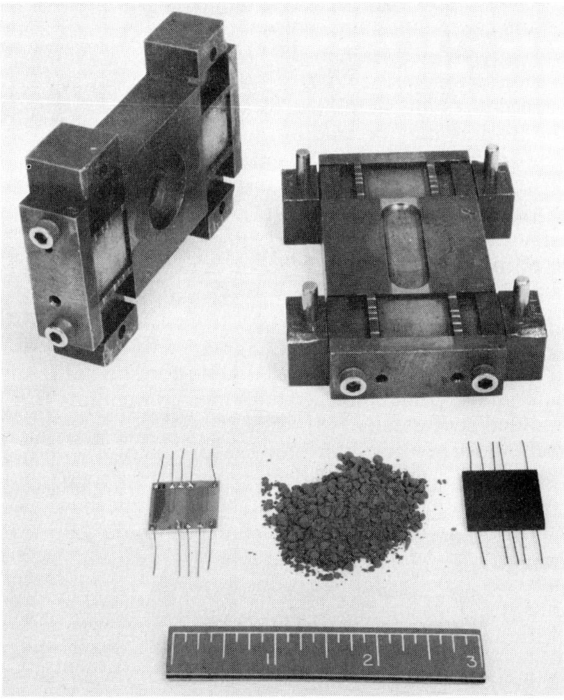

A thin film electronic circuit is encapsulated in epoxy resin by transfer molding. (*North American Aviation, Inc.*)

of 25-ton capacity will have a die space of 18 by 18 in., whereas, for equal die space with phenolics, a 75-ton press would be required [21].

6. Cure with no evolution of by-products and with low shrinkage.

Epoxy molding compounds also possess disadvantages:

1. Epoxy molding compounds are more expensive than conventional thermosetting molding compounds.
2. The inherently high adhesive strengths of the epoxies occasionally cause difficulty with mold release.
3. The materials at processing temperatures are somewhat soft and flexible, particularly on short mold cycles; thus ejector pins may core through the molding [3].
4. Because of flow characteristics, mold design must be tight to prevent excessive flash [14]. Generally, one may expect the epoxy to flow between knockout pinholes and to flow back along any movable side core, thus requiring periodic cleaning [22].
5. Because of the sensitivity of flow rate to temperature, fairly close temperature control must be maintained. Some compounds, for example, provide optimum flow rates in the range of 130 to 150°C and others in the 150 to 165°C range. Temperatures outside these ranges will lead to unsatisfactory performance.

Advantages and Disadvantages of Molding vs. Casting

In some cases, molding with epoxy compounds offers considerable advantage over the use of epoxy resins.

Advantages are as follows:

1. Use of molding compounds eliminates the mixing operation generally required with casting resins.
2. Shot measuring is facilitated.
3. Handling procedures are simplified.
4. Higher filler loading may be employed.
5. Fibrous reinforcing fillers may conveniently be used.
6. The mess associated with the liquid systems is eliminated, but powders from the resin can be a housekeeping as well as a dermatitis problem.

On the other hand:

1. Proper use of molding compounds requires expensive molding equipment.
2. Formulation versatility is reduced.
3. Storage conditions must be closely controlled, often requiring refrigerated rooms.
4. Molding techniques are particularly suitable for smaller items, utilizing 1 to 25 grams of material. With reinforced molding compounds, up to at least 100 grams may be molded successfully [48].
5. Molding techniques are not suited for low production volumes. Presses may cost up to $12,000 or more, and multicavity molds up to $10,000. Production rates on the order of 20,000 units per shift per machine have been obtained on small components with multicavity molds, but the initial expense would not be justified for low-volume production work.

Table 20-1 presents types of applications which are more suitable for casting operations; Table 20-2 presents applications which are suitable for either casting or molding operations, with the selection of one in favor of the other being arrived at from a consideration of economics. Table 20-3 presents a comparison of

With pair of work-holding fixtures, such as the one shown here, an operator can encapsulate approximately 2,400 solenoid coils per shift with epoxy molding compound. (*Hull Corp.*)

Table 20-1. Applications Best Suited to Casting [25]

Small quantities under 10,000 pieces per year
Pressure- and/or temperature-sensitive components, such as transistors, vacuum tubes, wax-impregnated capacitors, components with thermoplastic lead wire or insulation, magnetic cores
Units required to operate continuously at very low or very high temperatures—above 160 or below −30°C
Components containing large metal cores or otherwise subject to severe thermal shock stresses
Components in which it is desirable to obtain intimate impregnation of wire windings in the embedment process
High-voltage applications in which absolutely void-free embedments are necessary to avoid corona
Highly flexibilized epoxy systems
Large embedments (dependent on press capacity, but certainly those requiring more than 10 cu in. of compound)

Table 20-2. Requirements Generally Suitably Met from Either Casting or Molding [25]

Operating voltage below 2,000
Continuous operating temperatures ranging from −30 to +155°C
Most chemical environments in which epoxides are suitable, such as most acids and alkalies, hydraulic fluids, salt water, etc.
Military applications involving MIL-I-16923, types B, C, and/or MIL-E-4970
Underwriters class A, B, temperature ratings
Components such as resistors, capacitors, transformers, switches, solenoids, relays, rectifiers, and combinations of these

Table 20-3. Properties of Epoxy Molding and Casting Resins [32]

Property	Molding compounds		Casting resins	
	Fiber-glass filler	Mineral filler	No filler	Silica filler
Tensile strength, psi	14,000–30,000	5,000–7,000	4,000–13,000	7,000–13,000
Elongation, %	4	....	3–6	1–3
Modulus, psi × 10^{-5}	30.4	....	3.5	
Flexural strength, psi	20,000–26,000	10,000–15,000	13,300–21,000	8,000–14,000
Compressive strength, psi	25,000–30,000	18,000–25,000	15,000–21,000	17,000–28,000
Impact strength, Izod	8–15	0.25–0.45	0.2–1	0.3–0.45
Hardness, Rockwell	M100–M108	M101	M80–M110	M85–M120
Specific gravity	1.8–2	1.6–2.06	1.11–1.4	1.6–2
Thermal conductivity, cal/(sec)(sq cm)(°C)(m) × 10^6	7–10	7–18	4–5	10–20
Specific heat, cal/(°C)(g)	0.19	....	0.25	0.20–0.27
Thermal expansion, in./(in.)/(°C) × 10^5	1.1–3	2.4–5	4.5–6	2–4
Heat resistance, continuous, °C	160–260	150–260	130–280	130–280
DT, °C	200–260	130–230	25–280	60–280
Volume resistivity, ohm-cm	3.8×10^{16}	9×10^{16}	10^{12}–10^{16}	10^{12}–10^{16}
Dielectric strength, ⅛ in. thick, vols/mil	360	330–400	400–500	400–550
Dielectric constant:				
At 60 cps	4–5	4–5	4–5	4–5
At 10^3 cps	4–5	4–5	4–5	4–5
At 10^6 cps	4–5	4–5	4–5	4–5
Arc resistance, seconds	125–140	150–180	45–120	150–300
Water absorption, %	0.05–0.095	0.1	0.08–0.15	0.04–0.10
Burning rate	Self-extinguishing	Self-extinguishing	Slow	Self-extinguishing

properties of epoxy molding and casting compounds. Further comparisons are given in [51].

Molding Equipment

Generally epoxy molding compounds are employed with either compression or transfer molding equipment, although, in the simplest case, when pressure is not employed, they may be used by melt-on techniques.

Melt-on Techniques. When pressure is not used, the molding compound is dusted into a cavity or applied as a preform and then melted to provide the necessary flow. Pellets are most commonly used in these applications [18].

The use of pellets in the melt-on technique has been reported for bonding and sealing capacitors into their cans [26], and an automated production line for the encapsulation of carbon resistors using an epoxy-molded shell sealed with a preform has been described [20].

Compression Molding. Although compression molding may employ compounds containing either fibrous or nonfibrous fillers, the technique is particularly well suited to compounds containing fibrous reinforcement. This discussion will consider the use of fibrous-reinforced molding compounds; the subsequent discussion of transfer molding will confine itself to the use of compounds containing particulate fillers.

In compression molding, the molding compound is placed in a prepared mold, and heat and pressure are applied to cause the compound to flow into the desired configuration. After the compound has hardened, it is then ejected. Conventionally, pressures in the 1,000 to 3,000-psi range are used with most thermosetting molding compounds. With the epoxy molding compounds, pressures as low as 500 psi have been used successfully.

The molding equipment is often modified to optimize handling techniques with the epoxy resins. With compounds having lower bulking factors, the size of the prefill chamber may be reduced. Bleeding sprues should be considerably smaller than those used with phenolic molding compounds. Because of softness of the compound at processing temperatures, the ejection pins should be as large as possible, and the strokes should be gentle. Vents (when used to provide for the release of moisture present in the compound or the mold or to release air in cavities during closing) should be quite small. Contrary to what is experienced with the phenolics, parts tend to remain in the cavity rather than on the force during the opening stroke (presumably because of the low shrinkage), and release mechanisms should accommodate this phenomenon [21].

The strength properties of the molding compounds will be a function of the type of fiber employed, the length of the fiber, and the amount of the fiber present. Effect of the concentration of the fibers is indicated in Figs. 20-1 and 20-2. Compounds containing long fibers are generally bulky and present handling problems. Short fibers are less bulky and provide essentially the same characteristics as the long ones except that there is some sacrifice in physical strength. Long fibers may increase moisture penetration unless densely molded and generally provide fairly rough surfaces [31].

With fibrous molding compounds, preforming is generally undesirable since the operation, at least to some extent, breaks down the fibers [41].

During cure, as the resinous binder becomes liquid, the fibers will orient to produce unidirectional strengths [15]. When molding intricate parts, this orientation may be undesirable insofar as it produces planes of structural weakness in the final product.

Transfer Molding. Transfer molding equipment consists essentially of a chamber, in which the molding powder is placed and liquefied, and a piston which

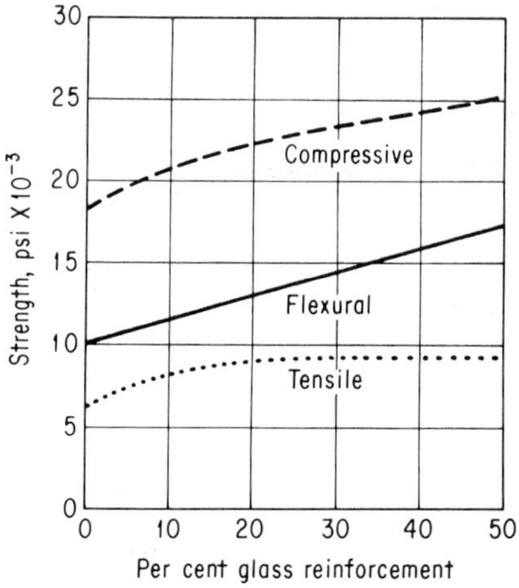

FIG. 20-1. Strength properties vs. glass concentration of DGEBA molding compound [11].

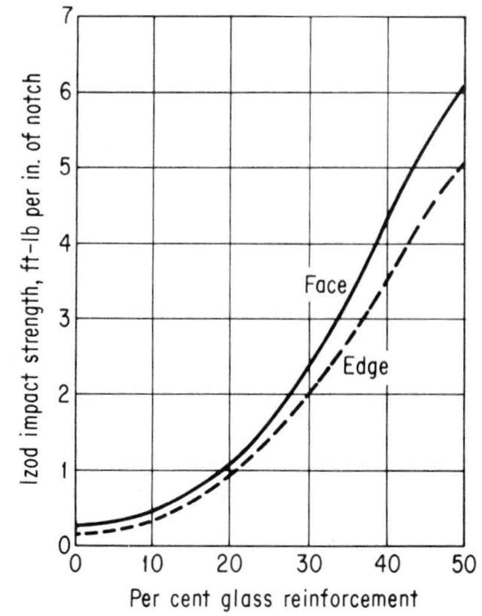

FIG. 20-2. Izod impact strength vs. glass concentration of DGEBA molding compound [11].

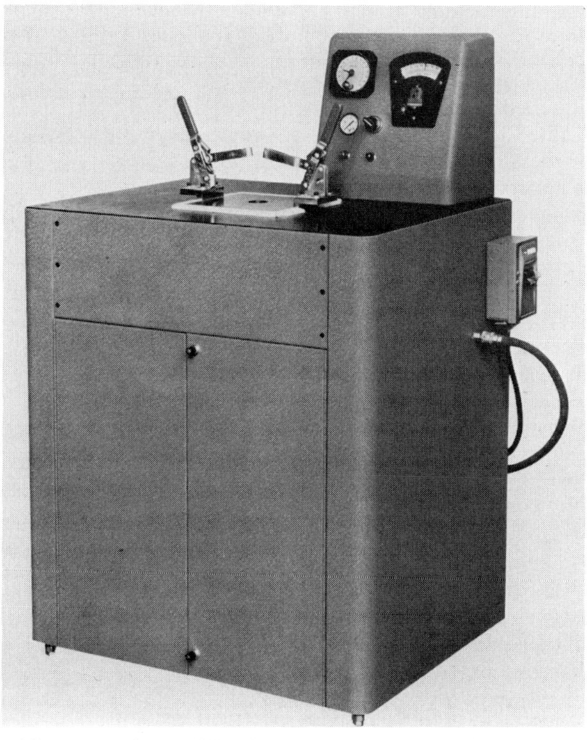

Transfer molding press for molding low-pressure epoxy resins. (*Furane Plastics.*)

Full automatic transfer molding press for automatic production of small electronic and electrical parts from epoxy molding compounds. (*Hull Corp.*)

transfers the liquefied material through runners into mold cavities. A schematic diagram is presented in Fig. 20-3.

As with compression molding equipment, it is necessary to modify transfer molding design approaches for use with epoxies. Runners are customarily smaller than with phenolic compounds. The plunger and pot may be rectangular rather than round to make more cavities possible in a given die space. The plunger and pot should be especially smooth, and the close-fitting part of the plunger in the pot should be lengthened to minimize backflow. The bulk factor of the epoxy compounds may be as low as 1.3, thus dictating use of smaller transfer pots and plungers [21]. Tolerances of knockout pins and pinholes must be kept to less than one-thousandth of an inch.

Conventionally, transfer molding compounds are handled at pressures from 1,000 to 10,000 psi. With the epoxy compounds, because of their soft-flow characteristics, pressures as low as 50 psi may be employed, although normally pressures are about 500 psi.

Epoxy molding compounds designed for soft-flow transfer molding may contain low percentages of fibrous fillers, or may contain rather high loading volumes of particulate fillers. The soft-flow compounds may be preformed for easier handling. With larger charges, say 50 grams or more, preheating to near molding temperature makes possible reduction of molding pressure by as much as two-thirds and reduces curing time by as much as 50 per cent in addition to assisting in the elimination of voids [37].

These so-called encapsulation-grade compounds are used to mold and seal electronic components in order to provide electrical insulation, mechanical strength, and moisture protection. Diodes, wire-wound coils, pulse transformers, and

Typical production installation of encapsulating transfer press. Operator has just removed work-holding fixture from the mold and examines the 40 encapsulated devices. Installations of this type are producing as many as 20,000 encapsulated units per 8-hour shift, using high-speed epoxy molding compounds. (*Hull Corp.*)

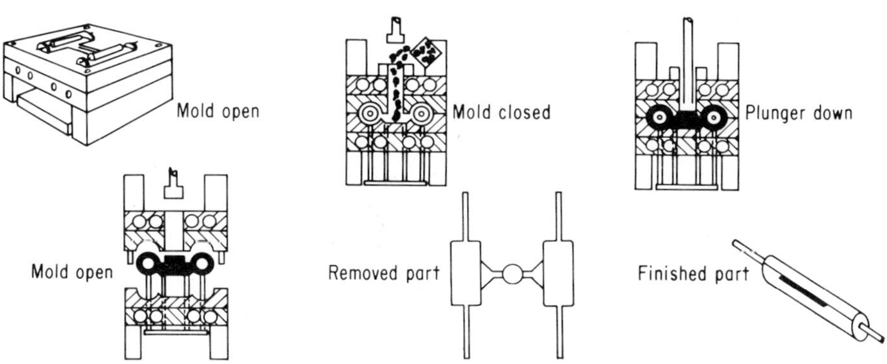

FIG. 20-3. Schematic diagram of transfer molding equipment and molding operation [50].

resistors are some of the parts successfully produced through the use of encapsulation-grade epoxy molding compounds.

Molds

Suggested modifications to molds have been mentioned in the previous section, and as indicated, the molds should be very tight-fitting to avoid excessive flash. In addition, because of the waterlike flow characteristics of the epoxy molding compounds, provisions should be made to facilitate periodic cleanup.

Highly polished chrome-plated molds are recommended. With compression molds, the chrome should cover the cavity and the forces as well as the parting lines; with transfer molds, the surface of the pot and the runner surfaces as well as the cavities should be chrome-plated.

Release agent should be used copiously during the initial cycling of the molds, with carnauba wax being preferred to the synthetic release agents [21], and it is generally recommended that the mold be purged with a general-purpose phenolic compound before the use of the epoxy resins. After initial startup, further use of release agent is unnecessary when the formulation contains an adequate amount.

Formulation

In addition to formulating for cured end properties, it is necessary with epoxy molding compounds to formulate with an aim to producing the required viscosity at processing temperatures, together with a relatively fast gel time. High hot hardness, after gelation, is also desirable to facilitate release. It is also vital to formulate for stable shelf life.

The epoxy molding compound must be supplied to meet the requirements of plasticity needed for each type of molding technique. The plasticity is measured by means of both the ASTM cup closing time and a spiral transfer molding procedure. The ASTM cup method gives information describing the softness of the compounds in compression molding, while the spiral transfer procedure gives an indication of the relative ability of the compound to be transfer-molded. For example, a high-cup closing time combined with a short spiral indicates a stiff material best suited for compression molding, whereas a short-cup closing time with a long spiral indicates a material well suited for transfer molding.

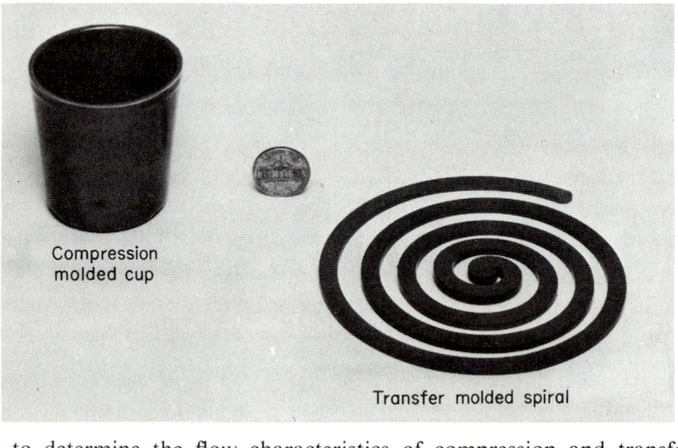

Test items to determine the flow characteristics of compression and transfer molding compounds. (*Mesa Plastics Co.*)

Molding powders may be supplied in flake, free-flowing, or pellet form, depending upon the type of operation and the economics of production. The molding compound may be especially designed for soft-flow transfer molding, or may be highly filled and most suitable for compression molding or pressureless so-called "gunk" molding.

The usual formulation consists of the epoxy resin, fillers, pigments, a release agent, and a curing agent.

Resin. Both liquid and solid epoxy resins are employed in epoxy molding compounds. These may be used alone, or they may be used in blends. In addition, high-molecular-weight adducts may be employed.

Resin blends, usually of epoxidized novolacs and DGEBA are frequently selected in preference to a single type in order to provide the necessary flow characteristics and fast gel times. Where practical, the resin blend should provide high functionality, in order to obtain fast gels, coupled with a sharp melting point to achieve controlled flow.

Typical of the resin adducts which have been used as the resinous portion of the epoxy molding compound is a reaction product of DGEBA and a phenol-formaldehyde novolac. The adduct is prepared by reacting at 140°C 70 phr of a phenol-formaldehyde resin (phenol/formaldehyde ratio 0.85) with DGEBA until a viscosity of 400 cps is obtained [12]. The adduct is then cooled.

Resins other than glycidyl ether resins may be used. Formulation and properties of a molding compound based on epoxidized polybutadiene are given in Table 20-4.

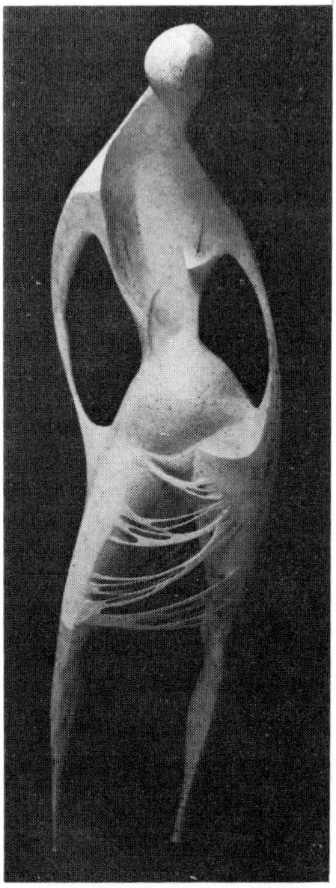

Use of epoxy gunk molding compound as artist medium produces shapes and forms extremely difficult to obtain with other materials. (*Ken Shutt, artist.*)

Curing Agents. Most epoxy molding compounds employ curing agents which permit mold cycles on the order of ½ to 5 minutes or less at temperatures of about 150°C. The curing agents normally preferred for molding powders are the aromatic amines, the solid anhydrides, and the latent curing agents such as dicyandiamide. Chlorine-containing curing agents are used to provide flame resistance [43].

Typical of the aromatic amine curing agents employed is methylenedianiline [1,23]. Properties of molding compounds, based on various molecular weights of DGEBA, cured with MDA are given in Table 20-5. Other aromatic amines which have been suggested as suitable are 2,6-diaminopyridine [5] and *m*-phenylenediamine [24]. In order to reduce gel time, it is customary to employ a phenolic accelerator such as phenol or resorcinol [2].

Typical of the solid anhydrides used in early molding formulations is PMDA, though others have tended to replace it for reasons of economics. Properties of several formulations are given in Table 20-6. Formulation I consists of pulverized DGEBA (mol. wt. 950) containing 18 phr PMDA. Formulations II and III employ DGEBA and PMDA at 56 phr. Formulation II contains 20 phr tetrahydrofurfuryl alcohol (containing 1 per cent dicyandiamide); formulation III contains 60 phr 4,4′-isopropylidenebis(hydroxyethylphenol) and 20 phr polyethylene glycol (mol. wt. 3,500). Formulation I was obtained by milling the pulverized ingredients together. Formulations II and III were obtained by dispersing the

Table 20-4. Properties of Molding Compound Based on Epoxidized Polybutadiene [4-3]

Component	*Parts by weight*
Epoxidized polybutadiene	100
Micronized fumaric acid	17.7
Dicumyl peroxide	1.5
Silica flour	188
Colloidal silica	30
Calcium stearate	1.2

*Property**	*Value*
DT, °C	298
Impact, Izod	0.32
Hardness, Rockwell M	112

Elevated-temperature flexural properties	*Strength, psi*	*Modulus, psi × 10^{-5}*	*Elongation, %*
At 23°C	15,000	13.3	1.3
At 70°C	8,300	7.3	1.5
At 120°C	5,000	3.7	1.8
At 175°C	3,300	3.4	1.1
At 230°C	2,100	3.8	0.6

Elevated-temperature electrical properties at 50 cps	*Dielectric constant*	*Dissipation factor*	*Volume resistivity, ohm-cm*
At 23°C	4.25	0.0288	1.0×10^{15}
At 60°C	4.79	0.0469	3.6×10^{14}
At 93°C	5.18	0.0530	4.6×10^{13}
At 115°C	5.14	0.0679	1.4×10^{13}
At 148°C	5.40	0.0803	4.3×10^{12}
At 198°C	5.31	0.1062	1.0×10^{12}

* Cured 5 minutes at 168°C.

Table 20-5. Effect of Molecular Weight of DGEBA Cured with MDA on Molding Compound [3]

DGEBA, mol. wt.	Shelf life at 27°C	Cure time at 150°C	DT, °C
950	20–30 days	8 min	86
		2 hr	111
850	15–20 days	8 min	100
		20 min	110
650	12–24 hr	8 min	105
		16 min	115
380	½ hr	Too short flow	

Table 20-6. Properties of Molding Compounds Cured with PMDA [12-4]

Formulation (see text)	Shelf life, months	Press conditions, min/°C	Postcure, hr/°C	DT, °C	Impact strength, Izod
I	2	10/100 + 10/185	16/200	120	1.8–2.2
II	>11	10/180	30/200	210	0.9
III	>11	10/190	16/200	185	6.0

PMDA and other ingredients at elevated temperature. The dispersed PMDA systems were then cooled to minimize exotherm, and B-staged at room temperature.

Release Agents. In order to facilitate mold release, zinc stearate and/or other suitable fatty-acid esters or saturated fatty acids [47,52] are included in the formulation.

In addition to these, other compounds may be employed, the type and the concentration being dependent upon a number of factors, such as filler, etc. For example, with wood or quartz flour as a filler, a blend of mineral wax (37.5 parts), zinc stearate (25 parts), and stearic acid (37.5 parts) may be used at 2.5 to 4 phr to obtain satisfactory release. With mica, mineral wax alone is satisfactory at 3 phr. With talc, the mineral wax requirement is increased to 4 phr [12].

In the absence of suitable release agents, considerable difficulty may be experienced because of the low shrinkage and the high adhesion of the epoxy compounds.

Fillers. The effect of particulate fillers on the properties of epoxy molding compounds can be derived, in a general way, from the discussion in Chap. 14. Silica flour is usually preferred since moderate loading volumes can be used without too much increase in viscosity. Talcs and calcium carbonate are used when machinability is a requirement. Aluminum has been most widely used for lowering exotherm of the system when electrical properties are not important [53].

Fibrous reinforcement of the molding compounds is usually achieved with fibers 1⁄16 to 1 in. long. These may be composed of glass, polyamide, polyester, asbestos, etc. The synthetic fibers generally provide good resistance to thermal shock and moisture and good electrical properties; they are somewhat deficient in heat resistance, and they generally give somewhat higher shrinkage during processing than do the inorganics [46]. Asbestos fibers provide good electrical and physical properties but somewhat inferior long-term moisture resistance and some sacrifice in impact strength. Glass fibers provide improved high-temperature performance, but sometimes cause severe gate and mold abrasion.

Production Techniques. *Incorporation of Curing Agent.* When higher-molecular-weight epoxy resins are employed, they may be reduced to powder and then combined with a powdered curing agent. Alternatively, the higher-molecular-weight epoxy may be melted, the curing agent added, and the mixture cooled and pulverized. With liquid epoxy resins, the procedure is to react the resin with the curing agent and advance the mixture to the B-stage. The general procedure for producing the B-staged systems from the aromatic amines, for example, is to dissolve the amine into the epoxy resin at elevated temperature, mixing thoroughly for 2 to 3 minutes. The mixture is then quenched and pulverized.

The pulverizing operation is conducted with standard pulverizing equipment, although with lower-melting-point blends, cooling equipment may be required to prevent softening and blocking of the resin during the pulverizing operation.

Specialized systems have been suggested, such as MPDA encapsulated in spherical particles of higher-molecular-weight DGEBA [49] and blends of filler impregnated with the higher-molecular-weight resin and of filler coated with an aromatic amine curing agent [45].

Incorporation of Fillers. Particulate fillers may be incorporated in the liquid epoxy formulations by the use of two-roll mills or screw-type kneaders. If the formulation is solid, fillers may be blended by the use of a pebble mill or the like, after which if desired they may be rolled under pressure and then ground [42].

Several methods may be used to incorporate fibrous fillers. If low fibrous filler contents are desirable, the fibers and the liquid epoxy system may be blended on sigma-blade mixers to provide gunk molding materials. Alternatively the shorter fibers may be blended into the liquid resin by milling. Dry molding powders may be blended with chopped fibers in a Z-blade mixer at elevated temperature, and the

Physical forms for epoxy-resin molding compounds. (*Mesa Plastics Co.*)

resulting product chopped in a rotating cutter or mascerated to flake form. For higher filler contents, the fibers may be coated with the liquid epoxy resin and then B-staged and chopped [38], or the resin and the chopped fiber may be sprayed alternately into preforms. Similar techniques may be used for the solid resins, the fibers being first coated by the resin in solution and the solvents subsequently evaporated [55].

Storage Life. In general, molding powders (particularly those formed by the B-stage technique) will have somewhat limited shelf lives, and, to assure reproducible flow rates, it is advisable that they be stored at about 5°C or below. With the more reactive systems, even with mild refrigeration, flow will decrease upon aging. With pressure-molding techniques, this can be compensated for by the use of higher pressures; for melt-on techniques, higher temperatures may be used. In critical transfer molding applications, however, changing flow characteristics with aging can cause difficulties.

EPOXY COATING POWDERS

Epoxy coating powders may be applied by a variety of techniques: dusting, spraying, flocking, tumbling, and fluidized bed and fog chamber. For example, spray coating of heated motor stacks with a hydrazide-cured DGEBA resin is described in [54]. Significant variables are: preheat temperature, nozzle position, nozzle air pressure, and speed of rotation. Claimed advantages over fluidized-bed coatings are: thickness and placement control as well as good edge coverage.

Fluidized-bed coating has been the most widely accepted, and this discussion will be confined to that topic.

Fluidized-bed Coating

In fluidized-bed coating [4] a powdered mixture is suspended in a gaseous medium, and into this is dipped an object heated to above the melting point of the mixture. The powdered resin melts over the surface of the object, and the object is then removed. With thermoplastic resins, the object is cooled to room temperature; with thermosetting resins, a postcure is applied.

Fluidized-bed epoxy coatings [44] offer a number of advantages over conventional dip and spray coats:

1. The fluidized powder is a solvent-free, 100 per cent solids coating, such coatings generally having inherently improved film properties over solvent-containing systems.
2. The fluidized-bed epoxy coatings can be formulated to retain good electrical insulation resistance, even at elevated temperatures.
3. Fluidized-bed epoxy coatings provide high moisture and chemical resistance.
4. Fluidized-bed epoxy coatings are economical in terms of material usage, since no solvents are employed and waste is minimized.
5. Fluidized-bed epoxy coatings may be applied at high buildups in a single dip. Film thicknesses up to 40 to 50 mils may be deposited in a few seconds, although more commonly, the film thickness per dip is limited to something on the order of 20 mils. This thickness is 20 to 40 times greater than that normally obtained by a single application of a solvent-containing coating.
6. Fluidized-bed epoxy coatings provide good coverage of complex shapes and will give good coverage of even sharp edges.
7. Fluidized-bed epoxy coatings may be applied at about one-half to two-thirds the labor cost of spray coatings. When a single coat can replace the multiple coats required by a more conventional system, the cost savings become increasingly significant.
8. The dermatitis potential is minimized by use of the epoxy in the powdered form and by the absence of solvents.

Along with the advantages of the fluidized-bed coating technique there are some disadvantages:

1. The fluidized-bed epoxy powders are normally rather expensive. Dry-blend epoxy powders have extremely small particle size, making fluidization difficult. In addition, the high surface area makes moisture pick-up a problem [35].
2. Masking operations (when required) are time-consuming.
3. Low-buildup coatings (below about 5 mils) are difficult to obtain. However, coupled with electrostatic deposition, coatings thinner than this may be deposited.
4. Formulation versatility is limited by the requirements that the epoxy system be supplied in a non-blocking powdered form.
5. With the higher buildup coatings, it is difficult to obtain sufficient flexibility and impact resistance, since these two properties are, in part, a function of coating thickness.
6. The fluidizing-bed process is probably only economically suitable for the processing of relatively small components, but for large units spray techniques have been developed.
7. Accurate control of the processing variables is required if optimum results are to be obtained, and the number of variables is considerable.

Epoxy fluidized-bed techniques have been used for applying insulating layers to motor stacks [39,40], for encapsulating non-temperature-sensitive electronic components [33], and for applying corrosion-resistant finishes to small parts. A description of an automated production line for coating small motor stacks by the fluidized-bed technique is presented in [27]. A comparison of epoxy with other fluidized-bed powders is given in Table 20-7.

Fluidized-bed Equipment

The fluidized-bed equipment, essentially, consists of an arrangement for introducing a gas through a finely divided powder. In simplified form, the equipment is

Table 20-7. Properties of Fluidized-bed Coatings [35]

Property	Vinyl	Cellu-losic	Epoxy	Nylon	Poly-ethylene	Chlo-rinated poly-ether
Chemical resistance:						
Exterior durability	E	E	F	F	F	F
Salt spray	E	E	VG	G	E	E
Water (salt, fresh)	E	VG	G	F	VG	E
Solvents:						
Alcohols	E	E	E	G	E	E
Gasoline	E	G	E	E	VG	VG
Hydrocarbons	G	G	E	E	VG	E
Esters, ketones	P	P	F	G	G	VG
Chlorinated	P	P	E	E	F	P
Salts	E	VG	E	VG	E	E
Ammonia	E	P	P	G	E	E
Alkalies	E	F	VG	G	VG	E
Mineral acids:						
Dilute (10%)	E	G	E	F	E	E
Concentrated (over 30%)	G	P	G	P	VG	E
Oxidizing acids:						
Dilute (10%)	E	P	G	P	VG	E
Concentrated (over 30%)	G	P	P	P	P	G
Organic acids:						
Acetic, formic, etc.	F	P	F	P	VG	E
Oleic, stearic, etc.	E	F	E	VG	VG	E
Mechanical and physical properties:						
Abrasion resistance	G	VG	VG	E	F	VG
Flexibility	E	G	G to P	G	E	F
Impact resistance	E	E	G	VG	F	G
Maximum service temperature, °C	95	80	175	80	70	120
Dielectric strength, volts/mil	VG	VG	E	G	E	VG
Decorative properties:						
Color range	E	E	P	F	G	P
Color retention	VG	E	F	VG	VG	G
Initial gloss	VG	E	G	G	VG	G
Gloss retention	G	E	P			

KEY: E, excellent; VG, very good; G, good; F, fair; P, poor.

NOTE: These data are intended only as a preliminary selection guide. Final selection should be made after consulting with coating formulator and after suitable testing.

illustrated in Fig. 20-4. A discussion of a production setup for fluidized-bed coatings is presented in [36].

In the fluidized state, the bed of powdered epoxy expands about 80 per cent and behaves like a boiling liquid. Vibration is used to keep the powder in a continually agitated state; otherwise, channeling, inhomogenous phases, and irregularly distributed gas pockets will occur, these adversely influencing the quality of the coating [7]. The fluidized powder offers little resistance to immersion, seeks small openings, and coats all exposed surfaces.

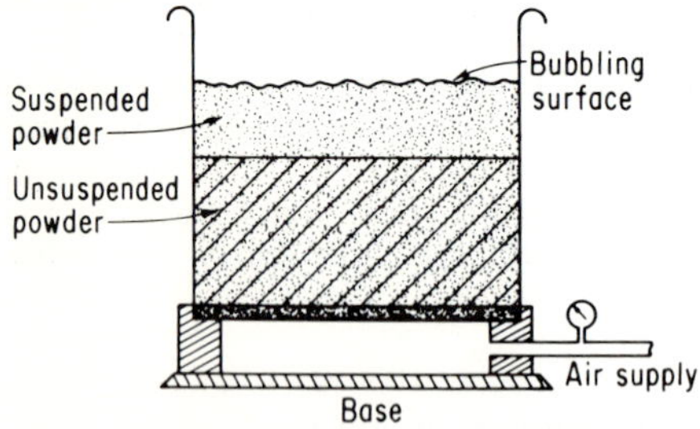

FIG. 20-4. Typical fluidized bed [19].

The gas supply is usually either nitrogen or compressed air. The gas is predried to remove moisture the presence of which would cause formation of undesirable agglomerations of the powder.

Operations Involved in Fluidized-bed Coating

The essential operations involved in fluidized-bed coating are as follows [16]:

1. Cleaning. In order to assure adhesion, the parts must be cleaned by sand-blasting or by some appropriate form of chemical cleaning, as discussed in greater detail in Chap. 21.

2. Masking. Where required, the parts are masked to reduce the difficulties involved in removing the highly adhesive epoxy resin from areas where it is not desired. The masking is generally accomplished with tapes or with thin films of a suitable release agent.

3. Preheating. The object to be dipped in the fluidized bed is preheated. The time of heating is dependent upon the heat capacity of the object and the time elapsing between its removal from the oven and immersion.

4. Dipping. The preheated part is dipped for a selected time in the fluidized bed. On complex parts, uniformity of coating is obtained by rotating the object during the immersion.

5. Blow-off. Where required, the excess powder is blown off with compressed air after the object is removed from the fluidized bed.

6. Postcuring. After the coating has been applied, it is then normally exposed to a postcure cycle, the retained heat in the object seldom being sufficient to assure satisfactory cure. During the postcure cycle, temperatures are selected to provide even flow-out to improve the quality of the coating [10]. Optimized cure cycles have been determined statistically for DGEBA (mol. wt. 1,000 and 1,400) when reacted with phthalic anhydride, tetrahydrophthalic anhydride, chlorendic anhydride, and dicyandiamide [28].

7. Finishing. After the postcure cycle, defects, if any, are patched by sprinkling additional powder over the heated surface. After defects are repaired and the repair cured, the material over the masked areas is removed and final cleanup accomplished.

Variables Involved in Fluidized-bed Coating

The extent of the buildup per coat and the amount of edge coverage obtained are determined by a number of variables: (1) particle size of the epoxy powder, (2) temperature of the object being coated, (3) size of the object being coated, (4) velocity of the airstream, and (5) immersion time. Edge coverage is generally 30 to 70 per cent of flat coverage, compared to about 15 per cent obtained with liquid resin coatings [34]. However, edge coverage remains one of the major problems.

Particle Size of Epoxy Powder. For fluidized-bed coatings, the optimum formulation contains a variety of particle sizes, generally ranging from about 50 to 325 mesh. When thinner coatings are desired, the finer mesh sizes should predominate, since with mesh sizes of 200 or coarser, coating thicknesses of less than 10 mils are difficult to obtain [2-19]. On the other hand, with the thicker coatings, if too many fines are present (200 mesh or finer), undesirable dusting and channeling may occur [17]. Typical of the particle-size distribution of commercial fluidized-bed powders are those indicated in Table 20-8. The effect of particle size on coating thickness is indicated in Fig. 20-5.

Temperature of Object. The temperature of the object being coated will, in part, determine the thickness of the coating and the efficiency with which edge covering is provided. Figure 20-6 illustrates the effect of temperature on edge

Table 20-8. Particle-size Distribution of Commercial Epoxy Powders and Per Cent Retained on Screen [19]

Material	Particle size, mesh (microns)				
	150(85)	180(79)	230(62)	325(44)	Through
A	29.0	23.6	17.1	17.8	12.5
B	6.2	8.1	9.5	26.2	50.0
C	12.7	11.0	8.5	59.3	8.5
D	59.1	10.7	5.4	7.4	17.4
E	0.8	6.2	12.0	22.5	58.5

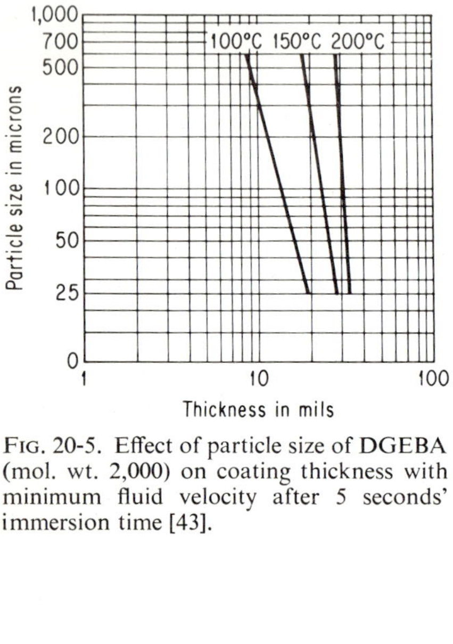

FIG. 20-5. Effect of particle size of DGEBA (mol. wt. 2,000) on coating thickness with minimum fluid velocity after 5 seconds' immersion time [43].

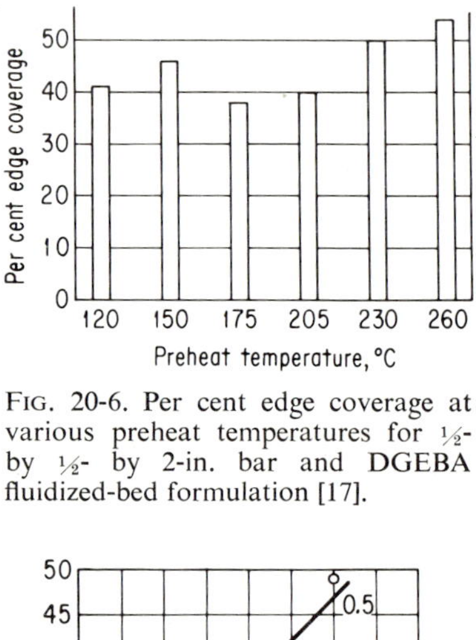

FIG. 20-6. Per cent edge coverage at various preheat temperatures for ½- by ½- by 2-in. bar and DGEBA fluidized-bed formulation [17].

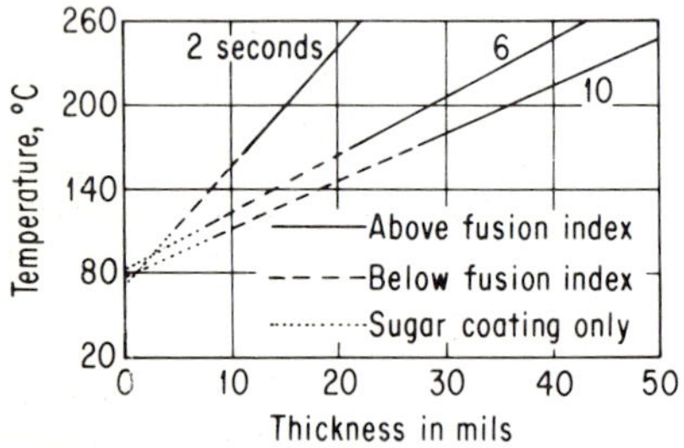

FIG. 20-7. Effect of preheat temperature and immersion time on coating thickness of DGEBA fluidized-bed formulation [17].

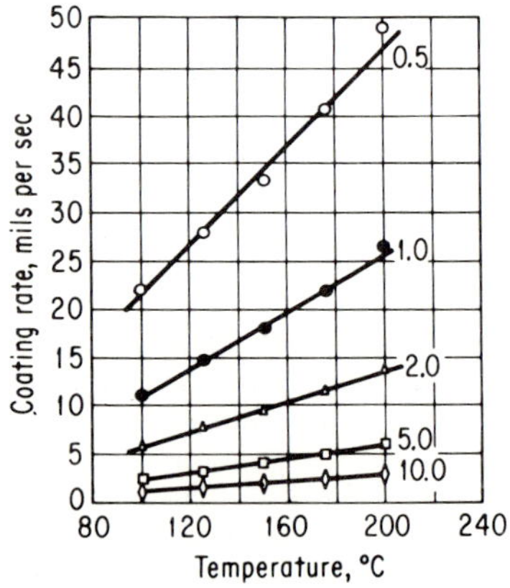

FIG. 20-8. Effect of immersion temperature on coating thickness for DGEBA (mol. wt. 2,000 and 125-micron size) at gas velocity of 1.27 fpm [13].

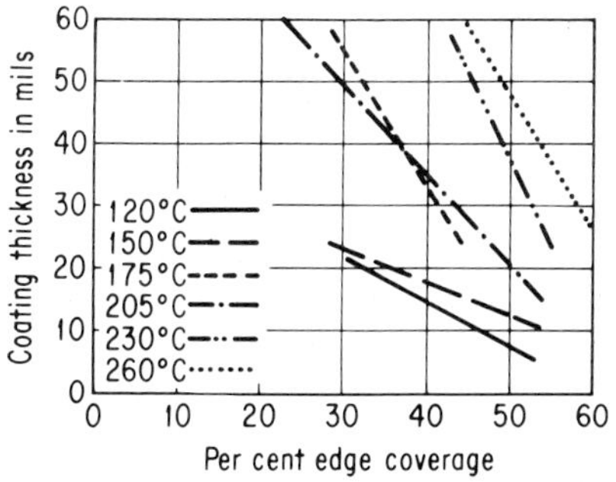

FIG. 20-9. Per cent edge coverage at various coating thicknesses for ½- by ½- by 2-in. bar and DGEBA fluidized-bed formulation [17].

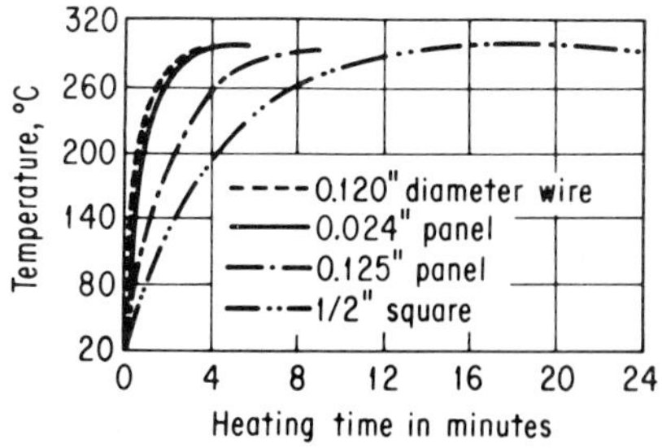

FIG. 20-10. Time to reach presoak heat for articles of different mass shapes [17].

coverage. Figures 20-7 and 20-8 illustrate the effect of temperature on coating thickness. Figure 20-9 illustrates the effect of coating thickness on edge coverage. From these it can be seen that higher temperatures yield thicker coatings and (for a given coating thickness) improved edge coverage. A minimum temperature limit exists with a given system, at temperatures below which, the coating will be powdery (so-called "sugar coatings") and will not properly flow out at postcuring temperatures.

Effect of Size of Object Being Coated. The larger the mass of the object being coated, the longer the time required for preheating to a specific temperature (Fig. 20-10), and, of course, the more of the preheat which is retained during the interval when the unit is being transferred from the preheat oven to the fluidized bed. In general, the smaller objects will, therefore, require a higher preheat temperature, as indicated in Fig. 20-11.

Effect of Velocity of Airstream. Within limits, the higher the gas velocity, the less the coating thickness. Suitable gas velocities fall in the 1- to 3-fpm range. The effect of gas velocity on coating thickness is indicated in Fig. 20-12.

Effect of Immersion Time. The longer the immersion time, again within limits, the greater the coating thickness. However, with the thicker coatings, sufficient heat may not be present at the outer layer to assure adequate flow, and

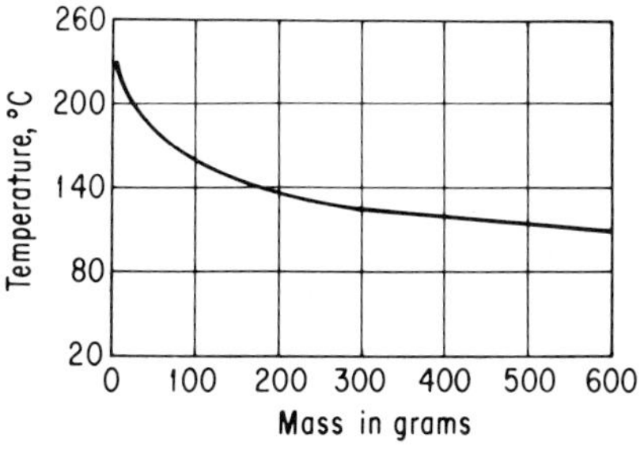

FIG. 20-11. Effect of size of square hex bar on preheat temperature required to produce a continuous film with 3-second dip time and DGEBA fluidized-bed formulation [17].

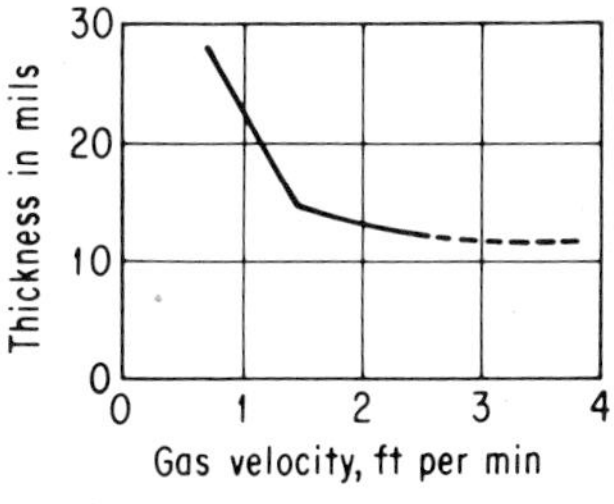

FIG. 20-12. Effect of gas velocity on coating thickness of DGEBA (mol. wt. 2,000 and 25-micron size) after immersion at 150°C for 5 seconds [17].

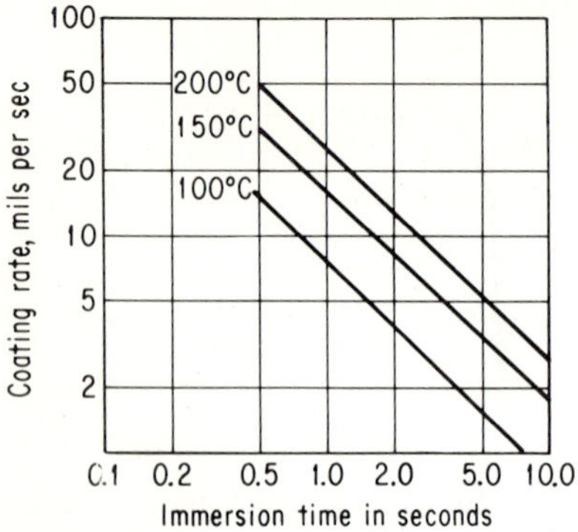

FIG. 20-13. Effect of immersion time on coating thickness of DGEBA (mol. wt. 2,000 and 580-micron size) with gas velocity of 2.8 fpm [13].

the effect at this layer will be as though the preheat temperature were too low. The effect of immersion time on coating thickness is indicated in Fig. 20-13.

Formulation of Fluidized-bed Coatings

The epoxy powders for fluidized-bed coating work may be produced by any of the techniques described for the production of epoxy molding powders. Perhaps the chief difficulty involved is obtaining sufficient flexibility and toughness in the higher-buildup coatings.

Conventionally, a small amount of filler is employed in the fluidized-bed coating to increase the hiding powder and, if desirable, to provide pigmentation. Thixotropic fillers may be used to control flow, and in part, regulate coating thickness [30].

A typical fluidized-bed formulation is the following [4-4]:

	Parts
Micropulverized DGEBA (mol. wt. 2,000)	100
Colloidal silica (for flow control)	3.5
Rutile TiO_2 (for hiding powder and white color)	16
Melamine (curing agent)	4

The materials are ground together in a pebble mill.

Properties of a cured epoxy-silicone coating powder are indicated in Table 20-9.

Table 20-9. Typical Properties of Films Made from Epoxy-Silicone Powder [29]

Property	*Value*
Flame resistance	Self-extinguishing
Weight loss after 100 hr at 200°C, %	1.7
Water absorption, %:	
After 24-hr immersion	0.55
After 2-hr immersion in boiling water	1.2
Chemical resistance (ultrasonic bath at 40°C for 10 min):	
Alcohols	No effect
Aromatics	No effect
Aliphatics	No effect
Chlorinated	No effect except by methylene chloride which causes slight softening—complete recovery occurs after removal from bath
Ketones	Softens—complete recovery after removal from bath
Detergent solutions	No effect
Dilute acid	No effect
Dilute alkali	No effect
Dielectric strength, volts/mil	$>$1,000
Dielectric constant:	
At 60 cps	3.57
At 10^5 cps	3.50
Dissipation factor:	
At 60 cps	0.0071
At 10^5 cps	0.013
Volume resistivity, ohm-cm	4.4×10^{15}

SUMMARY

Powdered epoxy resins, achieving commercial significance in about 1958, have found increasing acceptance as molding powders and as fluidized-bed coatings. The formulation techniques are common to both applications.

In the formulation, the epoxy powder may be produced in one of three ways: (1) by grinding together dry resin and curing agent, (2) by melting together dry resin and curing agent and subsequently grinding the resultant solid, or (3) by B-staging the resin and curing agent to attain the solid state and subsequently grinding it. Primary considerations are given to providing long shelf-life characteristics coupled with the controlled flow properties required of the application. Versatility of formulation is somewhat limited by the requirements.

In the formulation of molding powders, use is frequently made of fillers—both fibrous and particulate—in rather high loading volumes. In fluidized-bed coatings, the fillers are normally employed in lower amounts.

Use of powdered epoxy resins provides a convenient method of processing compounds in preference to conventional casting techniques and spray-on or dip coatings. Both applications are designed for relatively high-volume production rates.

REFERENCES

1956

1. Formo et al., Epoxy Resin Compositions Containing Para, para' Diamino Diphenyl Methane, U.S. 2,773,048.

1958

2. Bolstad, Epoxy Molding Compositions and Method of Molding with Same, U.S. 2,866,768.
3. Dewar, Transfer and Compression Molding of Epoxy Compounds, 14th ANTEC of SPE.
4. Gemmer, Fluidized Bed Coating Process, U.S. 2,844,489.
5. Newey, Process for Curing Polyepoxides and Resulting Products, U.S. 2,824,082.
6. Wagner, Pattern and Method of Molding, U.S. 2,846,742.

1959

7. Delmonte, Fluidized Bed Coatings, 1st National Technical Conference of SPE, Los Angeles, October.
8. Fina, Compression and Transfer Molding, *Plastics Technol.*, May.
9. Fina, The New Epoxy Molding Compounds, 1st National Technical Conference of SPE, Los Angeles, October.
10. Howard, Solventless Coating, *Mod. Plastics*, July.

1960

11. American Marietta Co., EMC Capacitor Molding, technical bulletin.
12. Epoxy Novolac Resins for Application in Molding Powders, Laminates and Castings, Koniklijke/Shell Laboraturium, Amsterdam, September.
13. Gaynor et al., Process Parameters in Fluidized Bed Coating, ACS Symposium, New York, September.
14. Hull, Methods and Equipment for Metering, Dispensing and Molding Epoxies, *Insulation*, July.
15. Karlson, The Use of Prefoam and Premix Molding Powders for Reinforced Plastic Materials, WE–BTL Casting Resin Conference, North Andover, Mass., May.
16. Lee, Application of Electrical Insulation by the Fluidized-bed Process, *Electro-Technol.*, October.

17. Nagel and Richart, Fluidized Bed Coatings: an Analysis of The System, ACS Symposium, New York, September.
18. Sussman, Epoxy Pellets, *Plastics World*, October.
19. Worley, Fluidized Bed Applications Using Epoxy Resins, WE–BTL Casting Resins Conference, North Andover, Mass., May.

1961

20. Automated Production Line, *Plastics World*, September.
21. Hull, Equipment and Tooling for Production with Epoxy Molding Compounds, 17th ANTEC of SPE.
22. Hull, Important Considerations in the Compression and Transfer Molding of Epoxies, *Plastics Design and Processing*, December.
23. Mackinzie et al., New Precondensates of an Epoxy Resin, Its Manufacture and Molding Compositions Containing the Same, U.S. 2,989,498.
24. Mitroshichev, The Composition of Epoxy Based Compounds, *Energet* (*U.S.S.R*), vol. 3, no. 4.
25. Rugg, Molding vs. Casting for Epoxy Encapsulation, *Mod. Plastics*, September.
26. Sussman, Now . . . Preshaped Epoxy Adhesives, *Prod. Eng.*, May 22.

1962

27. Automated Fluidized Bed Process Integrated into Motor Production Line, *Insulation*, August.
28. Bolson, Optimum Curing Conditions for Solid Epoxy Resins, *SPE J.*, July.
29. Dow Corning Corp. Bulletin 07-019.
30. Elbling, Pulverulent Coating Compositions Adapted for Use in Fluidizing Processes Containing Epoxy Resins and Fillers, U.S. 3,039,987.
31. Fleming, Tough Epoxy Shells Encase Pulse Forming System, *Insulation*, July.
32. Reese, Epoxy Resins, *Machine Design*, Sept. 20.
33. Richart, Coating Techniques for Electrical Insulation, Regional Technical Conference of Newark Section of SPE, May.
34. Richart, Insulating Electrical Parts, *Plastics Technol.*, November.
35. Richart, A Report on Fluidized Bed Coating Systems, part II, *Plastic Design and Processing*, July.
36. Rothschild, Production Setups for Fluidized-bed Coatings, *Mod. Plastics*, January.
37. Rugg, Transfer Molding Encapsulated Coils, *Plastics World*, September.
38. Schrub and Coad, High Strength Fiberglas-Epoxy Molding Compound, SPI 17th NANTEC (RIP Division), Chicago.
39. Slater, Fluidized-bed Application of Epoxy Resin Insulation to Rotating Electrical Machinery, 4th National Conference on Application of Electrical Insulation.
40. Slater, Insulating Electric Motors by Fluidized-bed, *Plastics Technol.*, March.
41. Woolf et al., The Effect of Preforming and Electronic Preheating the Glass Fiber Fibrous Diallyl Phthalate, Epoxy, and Other Thermosetting Compounds, 18th ANTEC of SPE.

1963

42. Frichette, Epoxy Resin Molding Compositions, Belg. 629,380.
43. Frichette, Nonflammable Epoxy Resin Molding Compositions, Belg. 629,381.
44. Manasia and Somerville, Fluidized Bed Coating Process Employing an Epoxy Resin and Process for Preparing Said Resin, U.S. 3,102,823.
45. Sussman, Epoxy Resin Compositions, Belg. 633,164.
46. Woolf, private communication, Mesa Plastics.
47. Yost, Molding Compositions, Brit. 920,470.

1964

48. Gaynes, Plastic Compounds for Connector Anti-deflection Discs Evaluated, *Insulation*, May.

49. Guinn, Epoxy Resins for Molding, Brit. 947,050.
50. Hull Corp., technical data.
51. Uglione and Bell, Epoxy Molding Powders: New Method of Encapsulation for Welded Modules, *Mod. Plastics*, August.
52. Yost, Epoxy Molding Compound including a High Molecular Weight Saturated Fatty Acid, U.S. 3,144,421.

1965

53. Hosford, A Guide to Fillers and Reinforcements for Thermosets, *Plastics Technol.* February.
54. Meyer, Factors in the Spray-coating of Stators and Armatures with Powders for Integral Insulation, *Insulation*, October.
55. Willie et al., Epoxy Resin Molding Compositions with Improved Storage Life, Ger. 1,198,999.

1966

56. Improved performance in Redesigned D-C Motors Credited to new Insulation System, *Insulation*, April.

Chapter 21

EPOXY-RESIN ADHESIVES

Epoxy-resin adhesives were patented in international commerce in the late 1940s [1] and gained considerable popularity during the 1950s for use in metal-to-metal joining applications where high structural strengths were required. In these applications, typified by the space and airframe industries, epoxy adhesives offer a number of advantages over the more conventional materials: ease of application, low cure temperature, low cure pressure, and high strength.

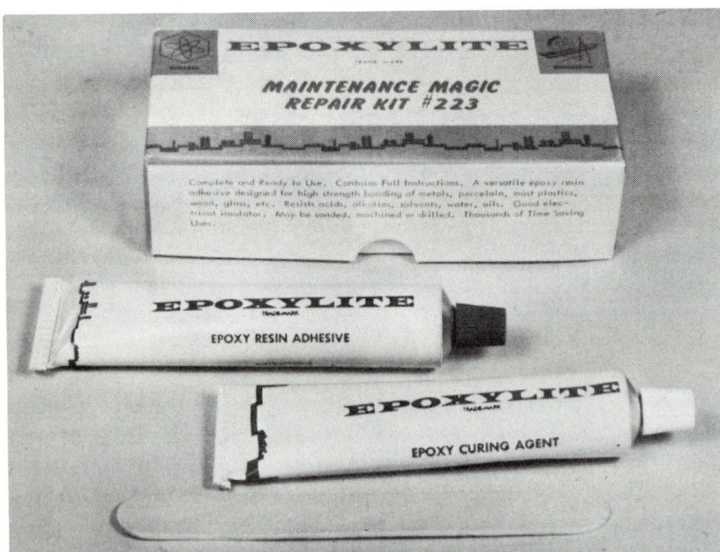

Epoxy-resin adhesives in twin tubes are very popular for both consumer and industrial repair. Such adhesives are usually DGEBA and polyamide or adduct of DGEBA and low-functionality, high-molecular-weight polyamine. (*The Epoxylite Corporation.*)

In the early 1960s, the epoxy-resin adhesives were introduced into an increasing number of fields, notably in the construction industry and the automobile manufacturing industry. In these markets, the chief advantage of the epoxy adhesive is that it permits the joining of dissimilar materials and provides outstanding bonds to nearly all materials, with the exception of a few of the newer, nonpolar thermoplastics. Increased volume use in these markets is expected as research improves processibility with regard to high-volume application equipment. In the consumer market, packaging of epoxy adhesives in squeeze tubes has led to attainment of sizable markets. The total epoxy adhesive market as defined in this chapter consumed 6.5 to 7 million pounds in 1966, the market growing at a rate of approximately 10 per cent per year.

Basically, adhesive strength with the epoxies is obtained by thorough wetting and by low shrinkage during cure. The thorough wetting assures that the adhesive in the liquid form can contact the adherends closely so that adhesive forces may come into play. The close contact is maintained after cure, and stresses which tend to disrupt established adhesive forces are minimized by the low shrinkage of the compound [62,75,92,95].

Epoxy adhesives provide the following advantages:

1. Dissimilar materials may be joined and galvanic corrosion eliminated.
2. Cure can often be accomplished at low temperatures, in many cases at room temperature.
3. Good bonds can be obtained to most relatively hard-to-bond materials, such as glass.
4. Cure can often be accomplished without the application of pressure (because for practical purposes) no volatiles are evolved during the reaction.
5. Leakproof and chemically resistant joints may be fabricated.
6. Good filleting action is obtained in honeycomb bonding.
7. Within broad limits, strengths are independent of glue-line thickness.

Offsetting these, to some degree, are the disadvantages. In addition to the relatively high cost and the processing difficulties are the following:

1. Careful surface preparation is required for optimum results.
2. Good joint design is required for maximum strengths.
3. Peel, cleavage, and impact strengths are generally low, particularly with the materials for elevated-temperature operation.

The epoxy formulations are particularly suited for making high-strength metal joints, for bonding dissimilar materials, and for honeycomb constructions.

A review of the literature indicates that the number and variety of applications for the epoxy adhesives is very large, and the following comments, on generically similar applications, can do no more than indicate certain areas of usefulness.

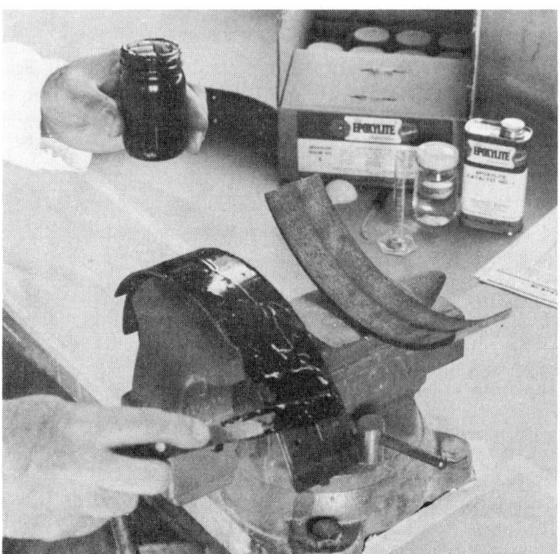

Epoxy-resin adhesive is used to bond brake lining to brake shoes. Method is more reliable than riveting and allows lining to be worn down completely, instead of only to the tops of the countersunk rivet heads as formerly. (*The Epoxylite Corporation.*)

APPLICATIONS FOR EPOXY ADHESIVES

For Bonding Dissimilar Materials

Epoxy adhesives are used in a number of special applications for attaching metals to plastics, generally polyester and phenolic laminates to aluminum or steel. For example, an epoxy adhesive can be used to bond phenolic centrifugal brake

Epoxy adhesive is used to bond light-reflecting buttons onto California freeway. (*Ciba Products Co.*)

linings onto steel shoes with greater convenience than can rivets, and since the rivets are eliminated, the lining can be worn down 100 per cent without scoring the drum on failure. The variety of materials involved may be considerable. For example, one adhesive formulation is used in a playback cartridge to bond polyester film, aluminum, mu metal, sintered iron, alloyed magnets, plastic-insulated copper wire, rubber, and silver-plated brass.

Recently the epoxy adhesives have come into favor in certain construction applications where their use can be economically justified, such as attaching metal studs in concrete, bonding pancake lighting systems on aircraft runways, bonding

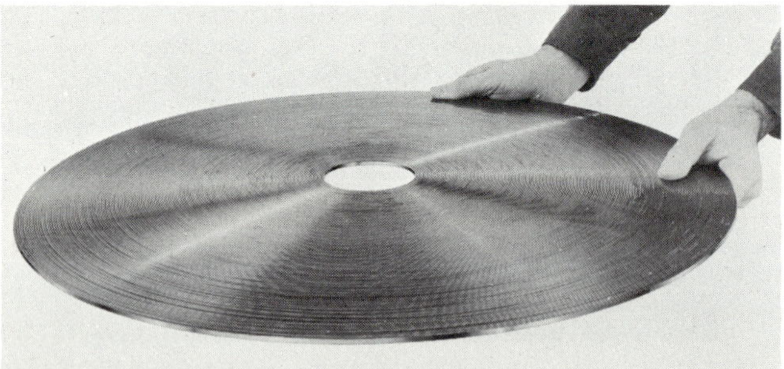

Adhesive-coated epoxy magnet wire cures to provide strong bond between copper conductors. (*G. M. Basford Co.*)

traffic markers on concrete, bonding rubber gaskets on pontoon bridges and in clay and plastic pipes, laying below-grade vinyl tile, and bonding steel tile to concrete. There are a number of aerospace applications where the epoxy adhesives are employed for bonding dissimilar materials, perhaps typical of which is bonding solar cells in satellites and copper to phenolic in printed-circuit boards. Some of these applications are extremely important technologically, but the usage of material is small, and the formulation, for critical jobs, is quite complex. In addition to these applications, the epoxy adhesives are used rather widely, in very small amounts, for mounting gems [64] and for museum restoration work [104]. Of the latter, perhaps the most famous was the restoration work on the Dead Sea Scrolls.

As Binders for Organic or Inorganic Particles

The epoxy adhesives may be used to bond together virtually any type of aggregate, in which case the epoxy is present in relatively small amounts, being used essentially to coat the individual particles and bond them at their points of contact. As binders for sawdust, for example, they have been used in the sculpturing of more or less artistic figures. Decorative, porous structures may be developed by shaking together a large amount of Mexican beach pebbles and a small amount of catalyzed epoxy. The pebbles may then be poured into a form and cured to give a structure of good physical strength. The use of an epoxy binder has been suggested for upgrading the properties of cork compositions [36].

As Primers

The epoxy compositions, usually applied from solution, may be used as primers to promote adhesion of a subsequently applied top coating or to provide improved

Section of curb on highway overpass has cracked and fallen off. After all deteriorated concrete had been removed, a thin coating of an epoxy was applied to effectively bond new concrete to old. The epoxy "tie coat" is actually stronger than the concrete itself. (*Union Carbide Corp.*)

resistance to corrosion. Typical of these applications is the use of an epoxy system as a primer coating over old alkyd paint before application of an alkyd recoat [59]. Such epoxy primers are also used as intercoats to permit the bonding of new concrete to old. Suitable formulations for such work are developed in Chap. 24.

As Chemically Resistant Adhesives and Sealants

Because of the chemical resistance of the epoxies, they are often used in fairly severe chemical environments. Generalizations, however, are to be avoided in this area, since the nature of the chemicals, their concentration, and the temperature of the environment all conspire to influence chemical resistance adversely. Typical examples of the use of epoxy adhesives in chemical environments are for caulking seams and sealing rivets on chemical storage tanks [52], for bonding paintbrush bristles [5], as sealants for pipes handling corrosive fluids—particularly effective with caustics, as shear keys for prestressed concrete bridges [69], and for miscellaneous maintenance and repair applications, such as sealing leaking oil-filled transformers. For severe environments, the formulations can become quite specialized.

For Bonding Plastics

Epoxy resins bond well to thermosetting plastics, except for the silicones, and to most thermoplastic materials except polyolefins, fluorinated compounds, and some

Retrieval of Firebee target drone, a high-performance free-flying jet target used in training by United States Air Force. Epoxy-resin adhesive is used in joining nose-cone sections. (*Ryan Aeronautical Co.*)

plasticized vinyls. They have been used widely for bonding polyester laminates—for example, in the construction of portable military shelters [35], polyester aircraft wing tanks handling aviation fluid [9], and in the repair of polyester boats. Specialized adhesives are used for attaching bondable polytetrafluoroethylene gaskets and for bonding nylon-phenolic ablative coatings to missile nose cones. Boilproof bonds to cellophane have been prepared [94].

For Bonding Glass

Epoxy adhesives provide extremely good bonds to glass, making them particularly suited for the fabrication of glass laminates, as discussed in Chap. 22. In addition, they are used to bond sheet glass in special applications, such as for example in preference to lead for the construction of stained-glass windows [60] and for the

Layers of glass bonded with epoxy adhesive are clearly seen in this close-up of the chancel of the United Church of Squaw Valley, Calif. Use of epoxy adhesives results in cost reduction over leaded stained-glass construction. At night, lights from beneath the floor of the chapel highlight the edges of the stained glass to produce a delicate, prismatic effect. (*Shell Chemical Co.*)

construction of mobile light mosaics [49]. The epoxy adhesives provide a thermal expansion rate substantially higher than that of the glass, and for this reason, unless some provision for expansion and contraction is allowed, cracking of the restrained plate glass may occur.

In Masonry and Stone

Epoxy adhesives have been used in masonry work for bonding bricks, cinder blocks, etc. Here, epoxies should not normally be considered (from the economical standpoint) unless thin, regular glue lines can be employed in the application. Epoxy adhesives have been used successfully to patch and repair polished marble slabs, with powdered marble being used as a filler. An unusual application was the repair and restoration of the Great Stone Face in New Hampshire [56].

For Bonding Wood and Other Structurally Weak Materials

When bonding structurally weak materials such as wood, the epoxy adhesive is likely, in the main, to be no more satisfactory (except from the standpoint of moisture resistance) than more conventional, and lower-cost, glues. When maximum strengths are required, careful joint design coupled with the use of dowels is required, as with conventional glues. The epoxy adhesives, however, possess the

Old bridge at Letts, Iowa, is being replaced by new. Shown is prestressed concrete girder being installed. Girder is bonded together with epoxy compound before delivery to construction site. (*Jones-Dabney Co.*)

advantage of curing without the release of volatiles, hence eliminating the requirement for pressure, and can be used as gap-filling compounds. Epoxy adhesives have been used in attaching aluminum reinforcing pins and lift mountings in wooden shoe heels and to a limited extent in antique furniture restoration. Epoxy adhesives have been utilized in the manufacture of decorative, plywood-faced, foam-filled interior roof panels to provide acoustical and thermal insulation, coupled with high strength and light weight [95].

Epoxy adhesives have also been reported as suitable for the repair of bones in both humans and animals [97].

For Metal Bonding

Epoxy adhesives are employed for a variety of metal-bonding applications. They have been used to bond pipe railings to eliminate visible fastening devices and to avoid the high cost of cleanup of welded joints [96]. They have been used as plastic body solders [23] for automobile repair (although lower-cost adhesive systems are generally favored). They are employed for a variety of miscellaneous metal-bonding applications, such as bonding a mixer support plate on an in-service crude oil tank [63]. They have been used for bonding railroad tracks and for bonding small metals parts in watches [50]. NOTS, China Lake, Calif., has developed a quick-disconnect adhesive system which gives structural strength during axial loading but fails reproducibly under a combination of tensile and peel stresses.

Large aircraft make use of epoxy resins in numerous ways: as adhesives for joining metal parts in place of riveting and brazing; as glass-reinforced composites for wing edges, ducts, drains, etc.; in adhesive-bonded honeycomb; as sealants; and as coatings resistant to high-temperature hydraulic fluids. For aircraft such as the supersonic Valkyrie XB-70, where service temperatures can exceed 450°F when cruising at 2,000 mph, most epoxy-resin types are unsuited, and specialized high-temperature formulations are employed: applications include adhesive bonding of reinforcing doublers and sealing of fuel lines. (*North American Aviation, Inc.*)

Petals on this solar collector are composed of aluminum skin bonded with epoxy adhesive to aluminum honeycomb. After firing into space, collector opens into a 32-ft-diameter parabolic mirror that focuses sun's rays on energy-collection unit. (*Union Carbide Corp.*)

Principally, however, the use of epoxy adhesives for metal-bonding applications has centered in the airframe industry, and it has been toward this industry that the majority of the research work has been directed. For example, 85 per cent of the load-carrying metal exterior of the B-58 is adhesively bonded [27], although not exclusively with the epoxy formulations.

The epoxy adhesives have proved particularly suitable for the bonding of metallic honeycomb structures. One large aircraft company reports having constructed over 40,000 epoxy-bonded sandwich structures during the course of 4 years without the occurrence of a single field failure [18]. Aircraft honeycomb parts, epoxy-bonded, have been used in doors, fuselage side panels, ailerons, elevators, rudders, trim tabs, trailing edges, wing tips, rotor blades, and in wings themselves [45].

MECHANISMS OF ADHESION

A number of theories have some bearing on problems of adhesion of polymers to solid surfaces. Adhesive forces may be expected to arise from van der Waals interactions between absorbant and adsorbate. Assuming satisfactory wetting properties, these forces are more than sufficient to account for observed adhesive strengths.

Uncatalyzed DGEBA has a surface tension of 47 to 49 dynes per cm, whereas catalyzed DGEBA has a surface tension of 30 to 45 dynes per cm, depending on the curing agent [93]. This surface-tension value is below the critical surface energy of most solids except a few plastics and other organic materials, so that the epoxy resins generally wet well-cleaned metallic, inorganic, and organic surfaces. A series of tests conducted with Kel-F illustrates the importance of surface tension [102]. It was found that DGEBA-DEAPA gave a surface tension of 33 dynes per cm at room temperature; DGEBA-DETA, 44 dynes per cm. Under poorest conditions, the best bonds obtained were in the 1,000-psi range. However, as surface tension was reduced, by either temperature or the use of additives, substantial improvements in bonds were realized. When DGEBA-DETA was modified with 0.1 per cent of an anionic fluorochemical surfactant, the surface tension was reduced to 28 dynes per cm and an adhesive maximum strength was obtained at fairly low-temperature cures. With the DEAPA system, up to 4,000 psi was obtained at cures of about 60 and about 200°C—in the first case because the adhesive had sufficient time to wet the Kel-F before cure; and in the second case because with temperatures in the range of the Tg of the Kel-F, the Kel-F was capable of wetting the cured or curing epoxy resin.

When polar adsorbates are involved, specific interactions such as hydrogen bonding may be present, and in some cases even primary valence bonds may be established (e.g., wash primers on metals). Further interactions and the formation of carbon-to-metal bonds may arise by the generation of free radicals by thermal decomposition in the adhesive during cure, and these may then react with the substrate [43]. It has also been suggested that on freshly cleaned metals atoms having unsatisfied binding forces may interact with the adhesive, but since oxidation is quite rapid, the effect of this, in practice, may be slight [57].

In any event, metal oxide surfaces are usually hydroxylated, the hydroxyl groups being acidic or basic, and in addition anions and cations of the lattice surface [82] provide sites on which organic groups may be adsorbed. On aluminum oxide, it has been shown that both hydroxyl and epoxy groups tend to pack closely on the surface [82]. Surface potential measurements made after adsorption of DGEBA onto chrome showed that the molecules at low coverage were adsorbed in different configurations depending on the solvent, but at high coverage were adsorbed vertically irrespective of solvent. Desorption of DGEBA into solvents was very rapid at first, then very slow, suggesting that the vertically oriented layer (one oxygen atom per molecule interacting with the surface) was desorbed quickly, but that when sufficient surface was available for two or more oxygen atoms to interact with the surface, the rate of desorption was much reduced [85].

It has been shown that DGEBA and higher-molecular weight species form compressible films on water with adhesion to the substrate being primarily dependent on the hydroxyl groups [7], and a relationship between hydroxyl content and adhesive strengths of DGEBA resins to aluminum joints has been established [12].

SIGNIFICANT VARIABLES IN ADHESIVE APPLICATIONS

The significant variables to be considered in the application of an epoxy adhesive system are given below.

Joint Design

It is difficult to appreciate the virtually overriding importance of joint design until it is realized that, when using a rigid adhesive, identical bonding areas may yield values ranging from less than 1,000 psi to as high as 12,000 psi or even higher. The difference in the values may be accounted for by the extent to which peel forces are applied to the joint during the test. The four basic types of bonding stresses are shown in Fig. 21-1.

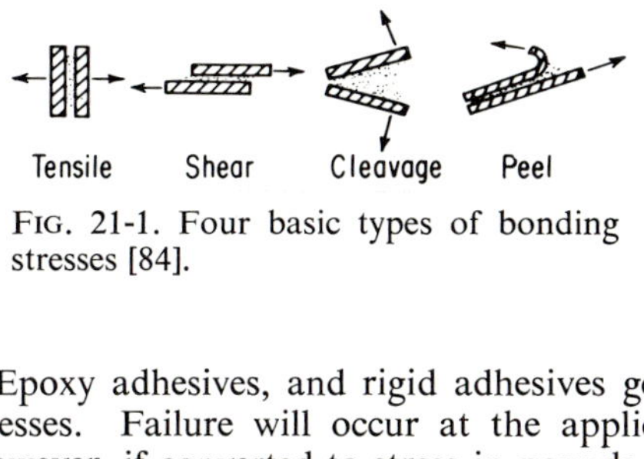

FIG. 21-1. Four basic types of bonding stresses [84].

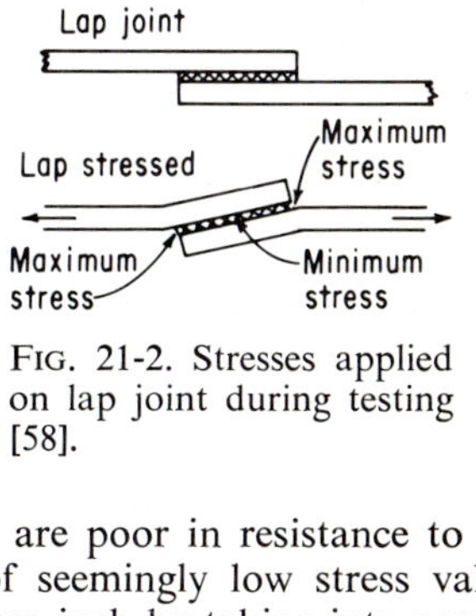

FIG. 21-2. Stresses applied on lap joint during testing [58].

Epoxy adhesives, and rigid adhesives generally, are poor in resistance to peel stresses. Failure will occur at the application of seemingly low stress values. However, if converted to stress in pounds per square inch by taking into account only the actual bonding area to which the stress is applied, the values would be quite reasonable.

The effect of peel can be illustrated by observing the effect of increasing the bonding area in lap shear tests—where peel forces, though minimized, are nonetheless present (Fig. 21-2). If a rigid epoxy adhesive is employed, increasing the length of the overlap results in a decrease in the calculated stress at failure. Thus, a 1-in. overlap provides little more strength than a ¼-in. overlap, as indicated in Fig. 21-3. On the other hand, if a flexible adhesive is employed, the peel stress can be more readily distributed throughout the bonding area, and increasing the overlap will not provide as severe a reduction. This is illustrated in Table 21-1. In this table, also, is indicated the effect of increasing the thickness of the adherends on the concentration of stresses. As the thickness of the adherends increase, the calculated pounds per square inch for the adhesive is reduced. Thinner adherends, obviously, are more capable of distributing the applied stresses than are the thicker adherends. Adhesive strength, therefore, is dependent upon the flexibility of the adhesive layer, the flexibility of the adherends, and the extent to which peel forces are present in the

Table 21-1. Effect of the Length of Overlap on Calculated Stress for Flexible Nylon-Epoxy Adhesive [70]

Metal thickness, inches	Overlap, inches	Lap shear strength at 93°C, psi
0.064	0.25	4,400
0.064	0.50	5,000
0.064	1.50	Metal failure
0.080	1.00	2,980
0.080	1.25	2,660
0.080	1.50	2,340
0.080	1.75	2,220
0.080	2.00	2,540
0.080	2.25	2,260
0.080	2.50	Metal failure
0.125	2.50	2,250
0.125	3.00	2,266
0.125	3.50	1,980
0.125	4.00	1,930
0.125	4.50	1,800

imposed stress. The major problem in joint design, then, is to overcome this concentration of stress and to apply it equally over the total bonding area.

Peel can be avoided in a number of ways. Perhaps the simplest method is to employ a rivet or a mechanical fastener at the end of the overlap, this serving to resist the imposed peel stresses while the full bonding surface resists the other loads. Fuselage body splices and reinforcements for the Boeing 727 jet are joined in this fashion, the combination of epoxy and rivets sharply increasing fatigue strengths over those provided by rivets alone at very little weight penalty [89].

Alternatively, the geometry of the joint may be changed, or the joint may be reinforced, in order to accommodate the peel stresses. Corner joints are normally reinforced. Butt joints may be recessed. When stiffeners are employed, as they

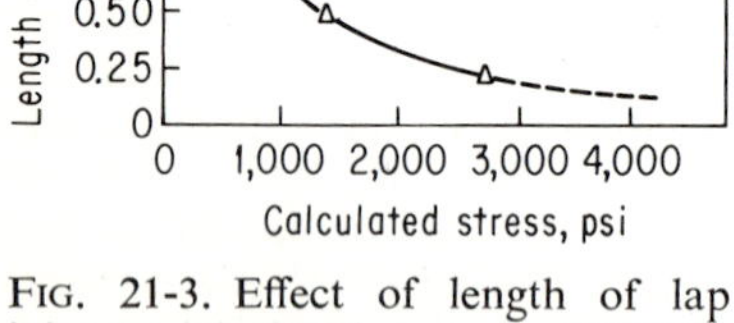

FIG. 21-3. Effect of length of lap joint on calculated stress for DGEBA cured with TETA [33].

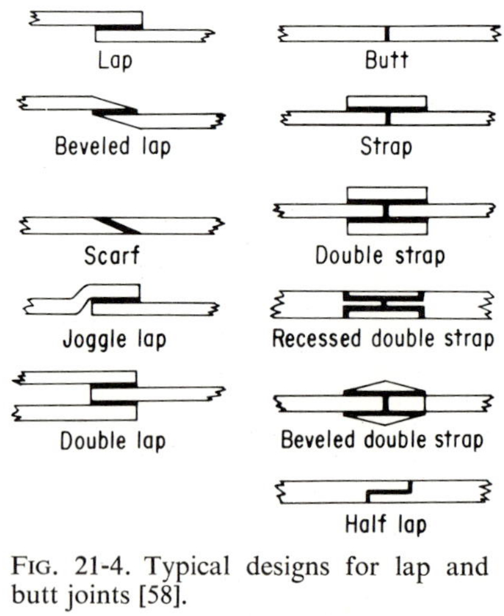

FIG. 21-4. Typical designs for lap and butt joints [58].

frequently are for the support of thin sheets, the stiffeners may have tapered flanges to improve peel resistance. A variety of geometrical changes may be made in simple lap and butt joints to improve performance, as indicated in Fig. 21-4. Of these, when production operations permit, a scarf or beveled lap joint is quite effective in improving peel resistance. The effect of the beveled lap joint in resisting peel stresses imposed during lap-shear loading is indicated in Fig. 21-5. The reinforced joints, such as the beveled double-strap joint, are likewise very effective. As in the case of welding, improved peel resistance can often be obtained by feathering the flash or forming feathered fillets.

Special joint designs may be employed to improve impact resistance, as for example, a circular adhesive joint has been shown to be more impact-resistant than a semicircular joint of the same bonded area [32].

The actual mechanism of failure in an adhesive joint can be attributed to the propagation of cracks from initial flaws and stress cracks in the adhesive. Studies have been conducted, employing built-in flaws, to determine the factors affecting crack propagation, and equations have been developed to indicate the inherent toughness of the adhesive [91]. By extension, it should be possible, assuming surface preparation sufficient to cause cohesive failure, to analyze the strength of any given joint mathematically on the basis of the configuration of the joint and the physical properties of the adhesive and the adherends. In even a simple lap-shear joint, however, the problem is extremely complicated, and simplifying assumptions must be made which reduce the validity of the calculations. The stresses may be examined experimentally, usually with the aid of photoelastic models, but again, the applicability of the results to actual design depends on the validity of unproved assumptions. Therefore, in practice, an empirical approach is taken, in which specimens of the actual configuration to be employed are tested to ensure satisfactory performance. A review of the problems involved in analyzing the mechanics of lap-shear joints is presented in [100].

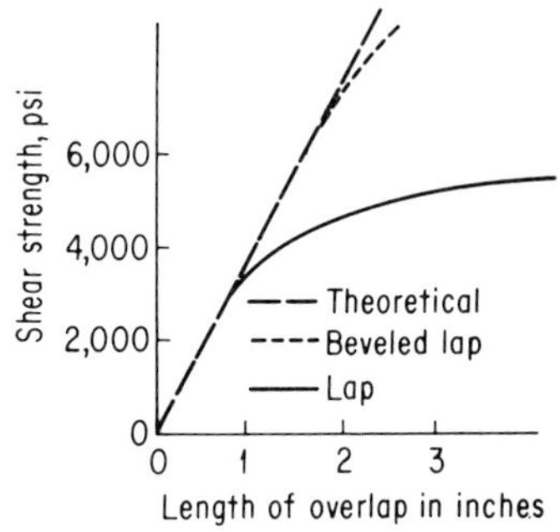

FIG. 21-5. Effect of joint design on shear strength of lap joint [58].

Glue-line Regulation

The epoxy-resin adhesives are somewhat unique in that the thickness of the glue line can be varied from 1 to 12 mils or more without substantially reducing bond strengths. Closer regulation may be required for complex joints, but when pressure is not normally used, or is at best nominal, the independence of strength on glue-line thickness is quite valuable in production operation. The production tooling may, to a very considerable degree, dictate the thickness for a specific application.

When the epoxy resins are used in honeycomb bonding, closer regulation of the glue-line thickness is required. This is often accomplished by the use of either supported or unsupported epoxy tapes of a predetermined thickness. In sandwich constructions, the major part of the adhesive bond consists of the fillet. The amount of adhesive trapped between the edge of the core and the skin is too small to contribute any significant strength to the joint. It is therefore necessary to regulate the thickness of the glue line—and hence the thickness of the fillet—rather closely to obtain reproducible results. The thicker the fillet, the higher the strength of the adhesive—but correspondingly, the more adhesive required and the greater the weight of the adhesive. Optimum values for each specific application can be arrived at usually in terms of core rather than adhesive failure under test. To illustrate the dependence of the strength on the thickness of the film in honeycomb constructions, an unsupported film adhesive having a weight of about 0.05 psf (pounds per square foot) and giving a peel strength on the order of 50 lb per in. width will, at a weight of 0.09 psf give a peel strength on the order of 100 lb. per in. width [71].

Curing Pressure

The liquid, unmodified epoxy adhesives generally do not require more than contact pressure to assure satisfactory joints. With the powdered adhesives, best results are obtained with nominal pressure, on the order of 15 psi. Pressures up to 100 psi may be employed with tape adhesives, although the usual range is from less than 5 to 50 psi. In all cases, pressure should be sufficient to ensure wetting and even distribution of the adhesive over the bonding surface but not high enough to starve the glue line. The selected pressure should be maintained throughout the initial phases of the cure.

With some of the modified epoxy adhesives, particularly those containing phenolic modifiers, volatiles may be released during cure. In these cases, pressures are employed during the cure cycle to suppress the volatiles.

Adhesive Carriers

When the adhesive is supplied in tape form, it may be unsupported or carried on a cloth. The nature of the cloth becomes increasingly important when higher-temperature applications are involved. Conventionally, glass cloth is the preferred carrier. A number of glass cloths and surface treatments have been investigated in an effort to improve elevated-temperature strength performance. As indicated in Table 21-2, the strength appears, at least within limits, independent of glass style and glass finish. Attempts to improve elevated-temperature performance by treating the glass with metallic naphthenates and with metal chelating agents did not prove of value, nor did asbestos cloth provide results superior to those obtained with glass

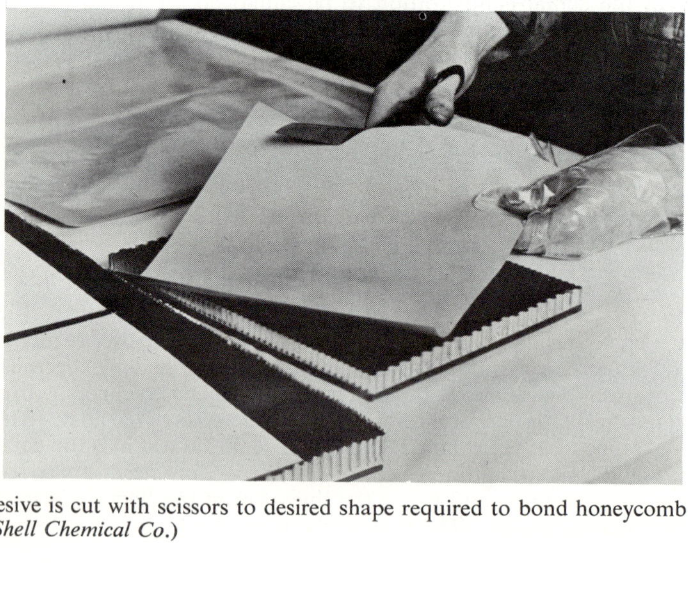

Film adhesive is cut with scissors to desired shape required to bond honeycomb core and skins. (*Shell Chemical Co.*)

[21]. When elevated-temperature service is not involved, lower-cost cloth may be employed as a carrier. Cheesecloth, for example, has been used successfully as a carrier for liquid epoxy systems in honeycomb bonding. When a graphite fabric carrier is used, it can be energized and serves as an internal electrical-resistance heater for curing the adhesive on application of an electric current [105].

Use of cloth carriers generally results in a reduction of peel strength and may introduce a wicking problem in comparison with unsupported adhesives.

Table 21-2. Effect of Heat Aging and Carrier Types on Lap-shear Strength of Epoxy-Phenolic High-temperature Adhesives [21]

(Lap-shear strength at 25°C after aging at 260°C on aluminum coupons, in pounds per square inch)

Aging period at 260°C, hours	Formulation 422-J tape, psi	Formulation FPL-878 liquid, psi	Formulation FPL-878 with		
			112 cloth, heat-cleaned	120 cloth, Volan A	120 cloth, heat-cleaned
0	2,482	1,862	2,080	1,960	1,925
50	2,030	1,466	1,274	1,352	1,068
100	1,764	1,524	965	1,044	1,241
150	1,472	1,576	860	656	920
200	1,186	1,466	638	650	622

Surface Preparation

Surface preparation represents a critical aspect in the use of epoxy-resin adhesives. For most applications, the surface should be free of all grease and loosely held foreign contaminants; in some applications, chemical treatment may be required to deposit a suitable film; and in a few applications, primers may be necessary.

Sufficiently clean surfaces can usually be obtained by thorough surface degreasing with a solvent, or by wire-brushing or sandblasting. This is usually adequate for most metals, for most plastics and for materials of construction.

Care must be taken to prevent the cleaned surface from becoming contaminated before the application of the adhesives. One of the most common sources of contamination, and one of the most serious, is body oils deposited while handling the freshly cleaned parts. Release agents may, however, be intentionally applied to some part of a joint to reduce bond strengths [101]. One ingenious method for applying in situ surface preparation involves using about 20 phr alumina filler in the adhesive. The abrasive filler can be made to scour the surface by rubbing the faying surfaces together (60 strokes in 30 seconds with 3/16 in. displacement) or vibrating the assembly (1/16 in. at 120 cps). This technique provided results equivalent to that obtained by sandblasting [108].

When elevated-temperature service is involved, when chemical environments are encountered, or when adhesion is critical, it is customary to employ special surface-cleaning methods. The treatments listed in Appendix 21-1 have produced generally satisfactory results.

In connection with the preparation of metallic surfaces, it has been noted that difficulty is sometimes experienced at elevated temperature, particularly during long-time aging in air. Oxidative degradation is brought about by certain of the metals. A study of various metals indicates that the presence of copper, iron, and nickel cause most marked degradation. Aluminum had little effect, and zinc and silver tend to bring about some nominal improvement in thermal resistance in air. The effect of various metals on thermal stability in air is indicated in Table 21-3.

Attempts have been made to reduce the oxidative degradation by special surface treatments. For example, the surface of stainless steel has been treated with zinc naphthenate, and the treatment resulted in improvement in long-term bond-strength retention at elevated temperature, but required processing at impractically high temperatures (650°C) to be of use in production [25]. Investigations of antioxidants and adhesive additives perhaps give more promise in obtaining improved resistance to thermal degradation on metals than do improved surface-preparation treatments.

The question of whether or not a smooth surface is preferred for metals is still not answered. Evidence is available on both sides. A rough surface theoretically increases the bonding area and provides for some degree of mechanical adhesion.

Table 21-3. Effect of Metals Deposited on Clad Aluminum on Lap-shear Strength of Bonds before and after Aging [25]

(Shear strength at 75°C, in pounds per square inch)

Metal	Unaged			Aged 100 hr at 288°C		
	Min.	Max.	Avg.	Min.	Max.	Avg.
None	1,610	2,010	1,771	860	1,070	935
Manganese	1,640	1,860	1,712	950	1,080	1,044
Chromium	1,550	1,680	1,622	760	1,040	912
Iron	1,450	1,490	1,465	450	680	552
Nickel	1,570	1,670	1,605	640	790	702
Zinc	1,630	1,990	1,836	1,070	1,200	1,136
Copper	1,280	1,450	1,390	0	0	0
Silver	2,080	2,180	2,120	1,140	1,220	1,162
Cerium	1,850	1,960	1,890	1,150	1,250	1,196

On the other hand, improved wetting may be obtained with a smooth surface, and this will minimize the formation of gas pockets and stress concentrations. Still, the pattern of entrapped bubbles in a randomly roughened surface may not be as likely to set up a line of weakness as the bubbles which may form in a plane between two smooth surfaces [92]. Which is to be preferred in practice is still somewhat a matter of individual choice and convenience, although studies using an ultracentrifuge appear to confirm the superiority of smooth steel surfaces over rough ones [17]. In some cases, however, sufficient roughness to cause improper interfacial contact can create effects that overshadow both wettability and glass-transition temperature effects, as has been shown when bonding to a chlorotrifluoroethylene co- and terpolymer [107], where poorer results were obtained than with an unroughened surface.

Primers

Primers may be used to protect the metal surfaces after cleaning and before the joining operation. Since, in production, the cleaning may precede the bonding operation by many days, the use of primers is quite valuable. They may additionally be used to improve peel, bend strengths, tensile strengths, and humidity and salt-spray resistance, particularly on aluminum. A commonly used primer is a vinyl-phenolic adhesive, and studies on this system indicated that best results were obtained when the primer was thoroughly cured, rather than merely dried, before the application of the epoxy adhesive [38].

In honeycomb bonding, improved peel strengths (over those normally obtained with the epoxy adhesive) may be obtained with phenolic-nitrile compounds. However, these systems have poor flow characteristics, which may lead to reverse filleting

(Fig. 21-6). To utilize the filleting action of the epoxy formulations while preserving the peel strengths of the phenolic-nitrile adhesives, composition film adhesives have been developed in tape form, as illustrated in Fig. 21-7.

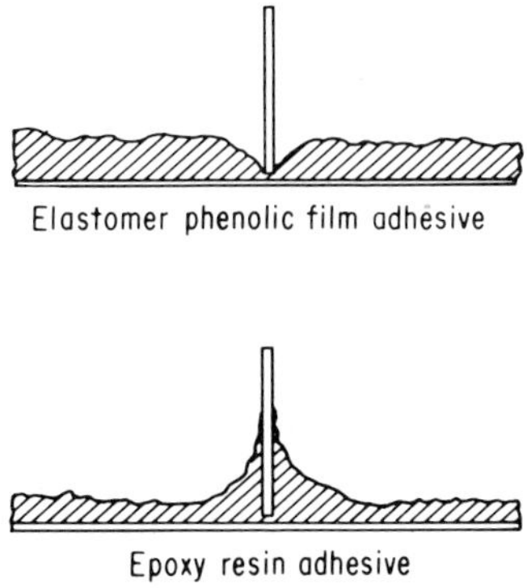

FIG. 21-6. Filleting action of epoxy adhesive compared to that of phenolic adhesive [54].

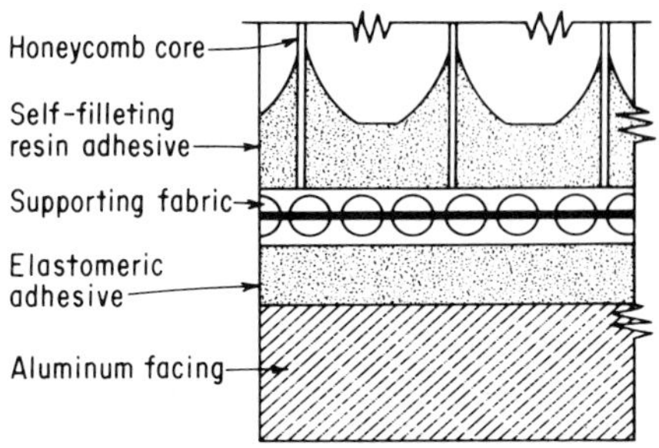

FIG. 21-7. Use of composition film to obtain peel strength with phenolic adhesive [54].

Application Techniques

Depending upon the adhesive system and the production requirements, the adhesive may be applied to the bonding surfaces by tapes or by dipping, brushing, rolling, troweling, or spraying. The last-mentioned technique appears to provide slightly better bonds, but for practical purposes, the application method is dictated by the formulation and production requirements.

When employing the liquid, 100 per cent solids systems, it is customary to coat both faying surfaces, to assure wettability, with the surfaces being immediately joined. If the layers are allowed to stand for an appreciable time before joining, with some systems some of the curing agent may evaporate to weaken the ultimate strength of the joint. On the other hand, with a liquid, solvent-containing system, optimum results are obtained only when time is allowed for the solvent to escape before the surfaces are joined; in all such cases the solvents must be driven off as thoroughly as possible.

Once the bonding surfaces have been joined, they should not be disturbed until cure has progressed well beyond the gel stage. A minimum of handling is recommended, and where possible, clamps or fixtures should be employed to preserve the bond and alignment.

When practical, using the liquid, 100 per cent solids systems, it is advisable to evacuate the adhesive before application, since this can result in as much as 10 per cent overall improvement in bond strengths [38].

Test Procedures

Test procedures, for reasons as discussed above under Joint Design, can properly be considered a variable in adhesive formulations. The majority of the standard test procedures may not provide values applicable to the actual design, although, of course, they are quite useful for evaluating the relative merits of various systems.

The property of adhesives most widely reported is lap-shear strengths, by the test procedure outlined in Specification MIL-A-8331. The procedure consists essentially of making a ½-in. overlap on 1-in.-wide specimens, curing the specimens,

Epoxy adhesives are used extensively in the various spacecraft and boosters of the Apollo project, the American project to put a man on the moon. Shown in the photograph is the Saturn stage booster rockets and propellant tank assembly—approximately 8 stories high and 33 ft in diameter. Bulkhead between the propellant tanks which contain liquid hydrogen (−459°F) and liquid oxygen (−297°F) is bonded with an epoxy adhesive. (*North American Aviation, Inc.*)

Large autoclave is employed to cure epoxy tape adhesive used to bond phenolic honeycomb in the cryogenic bulkhead of the Saturn stage. The bulkhead diameter is 33 ft, making this the longest single adhesive bond ever made. (*North American Aviation, Inc.*)

Command module of which will be main spaceship of American Astronauts on flight to and return from the moon is assembled in dust-free workroom. (*North American Aviation, Inc.*)

Command module is constructed by bonding up chemically milled welded aluminum and expanded aluminum honeycomb. Rivets and other fastening devices were avoided in favor of epoxy adhesives on the basis of weight savings and because of elimination of rivet holes, assures a tighter hermetic seal on the spacecraft. (*North American Aviation, Inc.*)

Typical setup for lap-shear test of epoxy-resin adhesive. (*Ciba Products Co.*)

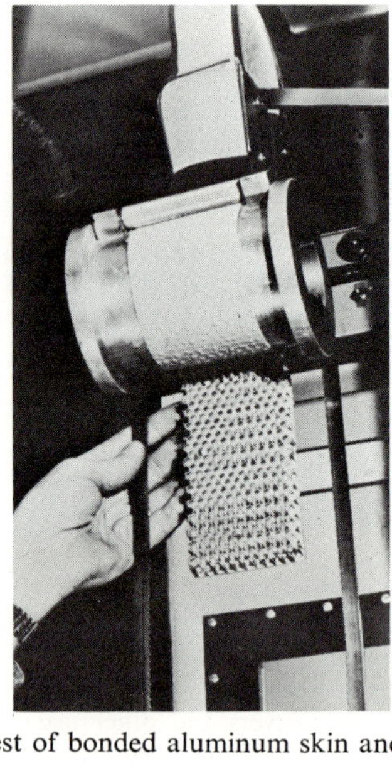

Climbing drum sandwich test of bonded aluminum skin and ¼-in. cell honeycomb core. (*Shell Chemical Co.*)

and pulling them apart in a machine equipped with a pin-loading device or with jaws. A fairly low loading rate is usually specified, although with the rigid epoxy adhesives, the values obtained are relatively independent of loading rate (excepting shock loading); whereas with flexible adhesives, and with adhesives tested in the range of their deflection temperature (DT) or above, the higher loading rates generally favor increased adhesive values [146]. In shock loading (i.e., load to failure varying from 1 to 12 milliseconds), improved performance is usually experienced as against static loading, with, in general, rigid and filled adhesives showing smaller changes between static and dynamic testing than semirigid and unfilled systems [111].

The temperature at which the tests are run will have considerable influence on the values, with values generally declining with increasing temperature. In the present technology, it is possible to obtain lap-shear values on the order of 500 psi at 300°C for fairly long periods of exposure. However, and rather surprisingly, it is

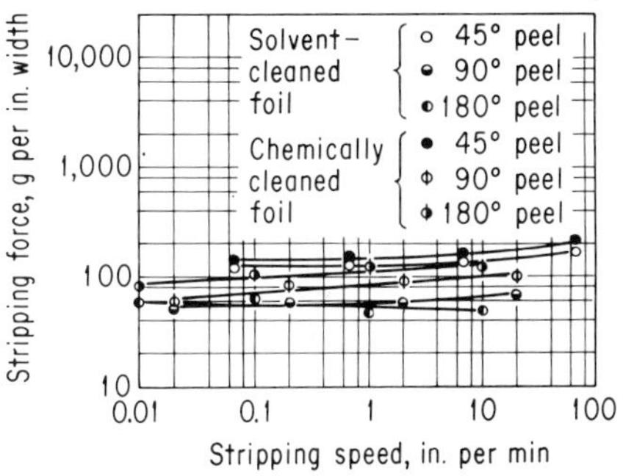

FIG. 21-8. Peel strength of 1.5-mil aluminum foil from DETA-cured DGEBA [24].

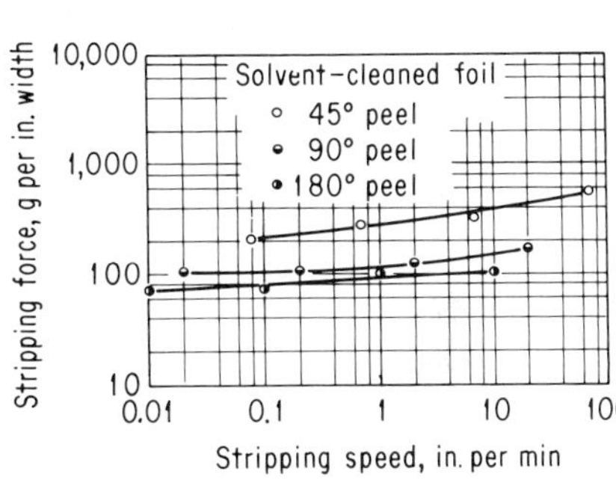

FIG. 21-9. Peel strength of 3.0-mil aluminum foil from DETA-cured DGEBA [24].

often possible to obtain substantially better values at temperatures up to about 800°C, providing the exposure time is only on the order of 1 to 3 minutes. For testing at these elevated temperatures and short-time exposures, radiant heating equipment has been developed [61].

As indicated earlier, the lap-shear tests introduce some degree of peel into the adhesive formulation, and values in the range of 4,000 to 5,000 psi are about maximum. In straight tensile tests employing butt joints, with the test designed to eliminate peel, values up to 12,000 psi have been obtained [79].

A number of other tests have been developed for adhesive formulations, notably peel tests. A description of a standard peel-test procedure can be found in ASTM Method D 903–49 for flexible-to-rigid bonds. A climbing drum peel test, suitable for use on honeycomb and laminates, is described in ASTM D 1781–60T, A stripping test procedure, involving curing the adhesive in contact with metal foil and subsequently stripping the foil from the resin, is discussed in [24]; by this procedure, foil thickness, surface preparation, and stripping angle provide variables (Figs. 21-8 and 21-9). For testing rigid members, 1-in.-wide coupons may be bonded together, with faying surfaces extending 2 in. beyond the glue line. These are then bent at a 90-deg angle, and force is applied.

A variety of tests have been devised for determining creep rupture, fatigue strength, etc. A number of accelerated aging tests and life aging tests are described in [42]. Military specifications for elevated-temperature adhesives are described in MIL-A-5090D. Techniques have been developed to permit the correlation of

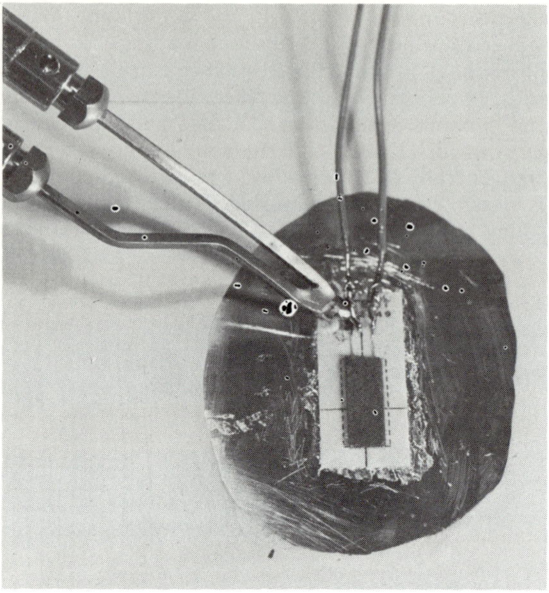

Strain gage is installed to aircraft wing and then insulated with epoxy adhesive to protect gage against moisture during testing. (*The Epoxylite Corporation.*)

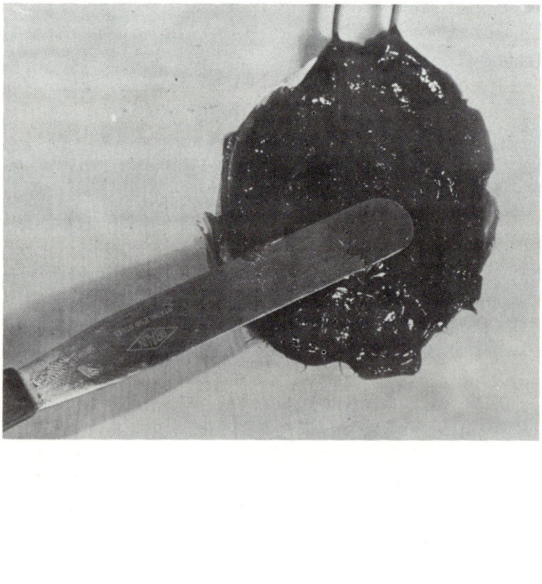

sonic energy absorption to bond quality to permit nondestructive testing [19]. Use of an ultrasonic resonance instrument, Fokker Bond tester, for example, gives a sensitive indication of flaws in adherend or core, voids, porosity, poor coverage, poor wetting, and inadequate gap filling. It is not, however, sensitive to poor adhesion per se, poor surface preparation, contamination, improper mixing, incomplete cure, or weak surface films on the adherends [109].

SELECTIVE IMPROVEMENT IN ADHESIVE CHARACTERISTICS

The commercial epoxy resins, used with an appropriate curing agent, may provide a satisfactory adhesive without modification. However, the handling of such compounds is complicated by the tendency of the resin to flow, particularly during cure, with resultant irregularities in the glue line and starved areas on the bonding surface. The temperature range of such adhesives is limited; the thermal expansion rates are high; the viscosity (depending on the resin or the curing agent) may not be sufficiently low to provide sufficient wetting action; shrinkage rates, while quite low, may be too high for many applications; and peel strengths are poor.

These characteristics may be improved selectively, although often an improvement of one property of the formulation may be obtained only at the expense of another.

Controlling Flow

Flow characteristics may be regulated by the incorporation of fillers, by the use of tapes, or by the careful regulation of the curing cycle. If required, thixotropic pastes may be formulated which will not flow during cure and which are useful for bonding non-close-fitting joints.

Extending Temperature Ranges

Flexibilizers can be employed to improve low-temperature bond strengths to −50°C, and their use is normally recommended for bonding materials having differing rates of thermal expansion. When bonding materials with similar rates of thermal expansion, good bond strengths have been obtained with primary amine curing agents and diethylaminopropylamine (Fig. 21-10) at cryogenic temperatures [80]. High-strength (3,200 psi) aluminum-to-aluminum bonds have been obtained at low temperatures (75°K) with DGEBA (mol. wt. 600) cured with *meta*-phenylenediamine [29]. Increasing the molecular weight of the resin resulted in lower lap-shear strengths at 75°K (perhaps because of an increase in the coefficient of thermal expansion of the cured system), whereas some improvement in impact strength may be obtained with either higher- or lower-molecular-weight resins at 75°K [66]. A series of tests of various commercial adhesives indicates that polyamine-cured filled epoxy systems or glass-cloth-supported epoxy/phenolic blends are suitable for cryogenic temperatures [77]. Fatty-polyamide curing agents give relatively poor performance, particularly when exposed simultaneously to radiation [103]. Best performance has been obtained with epoxy/nylon blends [65], as indicated in Fig. 21-11.

At higher temperatures, curing-agent selection is critical. Generally, the aliphatic amines are suitable for service to temperatures somewhat below 100°C. The tertiary amines are better at elevated temperature, as indicated in the aging studies presented in Tables 21-4 and 21-5. Aromatic amines provide still better elevated-temperature properties, as indicated in Table 21-6. Special anhydrides and amphoteric oxides provide the best properties. The specific epoxy resin is selected to provide high cross-linking density and usually to give a high degree of aromaticity.

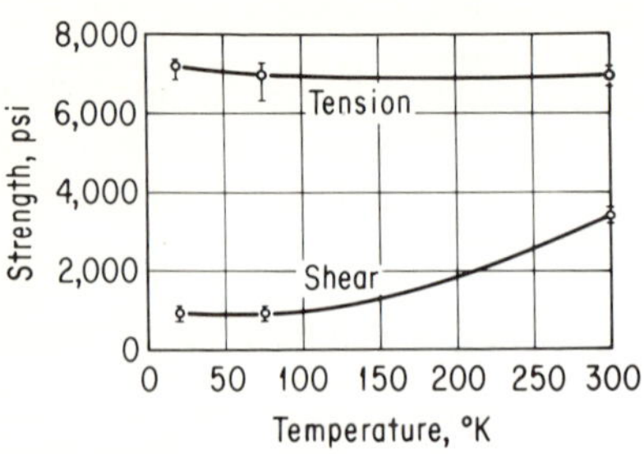

FIG. 21-10. Lap-shear strength vs. temperature for aluminum-filled DGEBA cured with DEAPA [28].

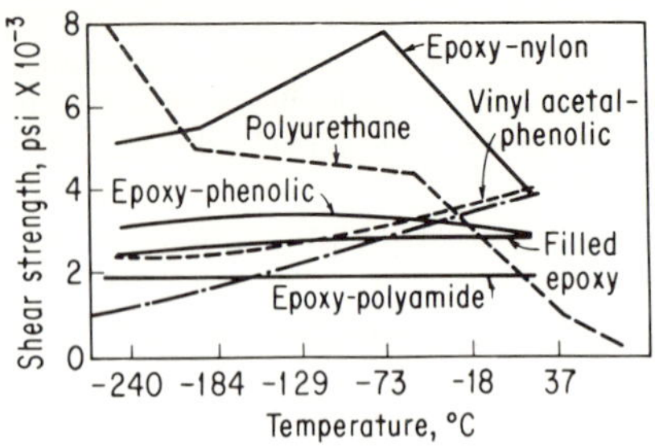

FIG. 21-11. Properties of cryogenic structural adhesive systems [99].

Table 21-4. Effect of Room-temperature Aging on Copper Lap-shear Strength of DGEBA-Triethanolamine at 100°C [30]

Time, in weeks	*Lap-shear strength at 100°C, psi*
0	1,050
10	1,100
20	1,050
30	1,040
40	1,000
50	1,000
60	980

Table 21-5. Effect of Heat Aging on Copper Lap-shear Strength of DGEBA Cured with Tri-2-ethylhexoate Salt of DMP-30 and DEAPA [30]

(Lap-shear strength at 100°C, in pounds per square inch)

Aging time, weeks	DMP-30 salt aged:		DEAPA aged:	
	At 120°C	At 160°C	At 120°C	At 160°C
0	1,300	1,300	900	900
2	.....	900	.....	920
4	.....	800	.....	900
5	1,150	.....	1,250	
6	.....	700	.....	850
8	.....	600	.....	750
10	900	550	1,175	
15	875	.....	1,150	
20	850	.....	1,100	
25	850	.....	1,100	

It has been shown that nonconjugated carbon-to-carbon unsaturation improves thermal stability of high-temperature bonds and the formation of ester cross-links increases thermal stability [11]. But, at temperatures in the range of 260 to 320°C, oxidation of the adhesive becomes a problem (Fig. 21-12). Considerable work has been expended in evaluating the effect of antioxidants, as indicated in Table 21-7. The use of phenolic additives has also been investigated in detail; however, at temperatures of about 290°C the degradation of the epoxy-phenolic adhesives has been attributed to the phenolic component [11]. It is possible that the phenolic component, which evolves gases during cure, thereby reducing the density of the adhesive, provides passageways for the entrance of the oxygen to accelerate degradation.

Table 21-6. Effect of Aromatic Amines on Lap-shear Strengths of DGEBA Adhesive Carried on 112 Volan A Glass Cloth [20]

Curing agent	phr	Bonding conditions			Lap-shear strengths, psi			
		°C	psi	Min-utes	At 23°C	At 127°C	At 177°C	At 204°C
MPDA	15	150	10	45	3,100	3,275	736	507
		150	100	45	2,935	3,510	790	660
		177	100	45	3,330	2,978	830	645
o-Phenylenediamine	14.7	150	10	45	2,780	2,660	710	
		150	100	45	2,901	2,319	797	
		177	100	45	2,568	1,831	695	
4-Chloro-*o*-phenylene-diamine	19.2	150	10	45	2,550	1,510	485	
		150	100	45	2,644	1,470	600	
		177	100	45	1,960	1,583	515	
2,4-Toluenediamine	16.7	150	10	45	2,955	2,750	830	550
		150	100	45	3,168	2,834	885	575
		177	100	45	2,874	2,386	835	570
MDA	24.0	150	10	45	4,200	3,170	930	575
		150	100	45	3,705	3,470	870	590
		177	100	45	3,700	3,150	955	615
	27.0	150	10	45	3,220	4,160	700	550
		150	100	45	3,940	4,630	920	610
		177	100	45	4,050	4,350	960	755
	29.6	150	10	45	4,970	3,800	800	565
		150	100	45	4,630	3,610	740	590
		177	100	45	4,310	3,650	710	530
4-Methoxy-6-methyl-*m*-phenylenediamine	20.7	150	10	45	2,650	2,020	980	405
		150	100	45	2,440	1,970	980	420
		177	100	45	2,465	3,190	1,020	530
	23.0	150	10	45	3,460	3,868	985	560
		150	100	45	3,500	3,480	895	660
		177	100	45	2,270	2,490	555	480
4,4′Bi-*o*-toluidine (toluidine)	25.0	150	10	45	3,080	2,465	695	570
		150	100	45	3,040	2,675	730	565
		177	100	45	3,020	3,010	825	585
	27.0	150	10	45	3,650	2,335	1,050	705
		150	100	45	3,600	2,980	940	605
		177	100	45	3,085	3,435	1,095	670
	29.0	150	10	45	3,140	3,100	920	569
		150	100	45	3,245	3,070	1,040	655
		177	100	45	3,290	3,050	950	535
	31.0	150	10	45	3,120	3,515	1,000	780
		150	100	45	3,340	3,685	1,010	655
		177	100	45	3,330	2,915	1,010	640
	33.0	150	10	45	3,605	3,950	1,235	735
		150	100	45	3,465	4,580	1,390	870
		177	100	45	3,740	3,610	1,265	780
3,3′-Dimethoxyben-zidine (dianisidine)	28.5	150	10	45	3,640	2,300	500	400
		150	100	45	3,480	1,595	510	440
		177	100	45	3,600	1,510	495	415
	30.5	150	10	45	3,625	2,255	765	605
		150	100	45	3,590	2,000	770	575
		177	100	45	3,340	2,545	810	615
	33.1	150	10	45	3,630	2,445	730	525
		150	100	45	3,920	2,940	815	540
		177	100	45	4,135	2,320	790	565

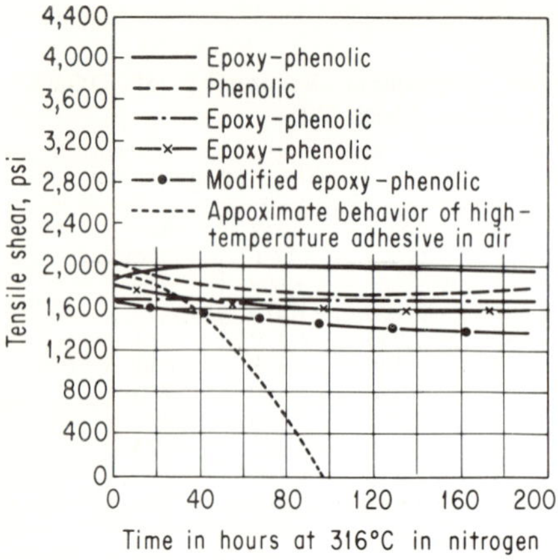

FIG. 21-12. Tensile shear vs. time at 316°C for commercial adhesives [37].

Ideally, elevated-temperature adhesives should be processed in such a way as to minimize oxygen entrapment during application, should be as dense as possible, and should contain antioxidants to improve thermal stability.

Even in the temperature range below 250°C, the upper temperature limit is critical. With systems suitable for operation in the 150 to 200°C range, there is usually a moderate increase in bond strength as temperature rises, but thereafter the bond strength decreases with further temperature increase to a dropoff point beyond which loss of properties is rapid. Since single batches of adhesive will give dropoff strengths that vary over a 5 to 10°C range, a safety margin should be allowed in design work. In addition, if the applications are critical, strength at elevated temperature should be investigated at relatively short time intervals. With some high-temperature adhesive formulations, even after apparently thorough cure, strengths may decline rather sharply during the first 25 to 50 hours of aging, thereafter recovering to provide values at the end of 100 hours of aging about equal to those obtained at the initial exposure temperature.

Table 21-7. Effect of Antioxidants on Lap-shear Strength of Epoxy-Phenolic Adhesive (Formulation FPL-878) [25]

(Lap-shear strength at 25°C after 100 hr, at 288°C, in pounds per square inch)

Compound added	Additive, 1%			Additive, 3%		
	Min.	Max.	Avg.	Min.	Max.	Avg.
None	40	420	130			
Aluminum flake	20	20	20	20	20	20
Aluminum oxide	40	260	162	200	770	474
Manganese dioxide	420	700	566	60	530	326
Zinc dust	0	0	0	40	410	218
Zinc oxide	0	260	52	0	0	0
Stannous oxide	0	220	72	120	740	372
Stannic oxide	80	440	228	200	420	306
Stannous sulfide	140	690	426	80	630	380
Cuprous sulfide	90	430	222	60	460	176
Arsenic trioxide	290	460	320	320	570	440
Arsenic pentoxide	160	190	140	20	140	60
Bismuth trioxide	40	620	304	20	150	66
Antimony trioxide	0	690	250	0	0	0
Antimony trisulfide	140	600	328	30	690	416

Because the high coefficient of thermal expansion of the unmodified epoxy resins does much to disrupt the adhesive bond with changes in temperature, reductions in this value, by the use of fillers, will have the effect of extending the upper temperature limits of the formulations to at least some degree. Antimony trioxide appears to be a particularly satisfactory filler for high-temperature adhesives [72], perhaps because of a catalytic effect [76].

It should be noted that high-temperature epoxy adhesives provide somewhat lower room-temperature strengths than adhesives suitable for lower-temperature service. Other things equal, adhesive formulations with low glass-transition temperatures give higher room-temperature bond strengths than those with higher glass-transition temperatures—while on the other hand, high glass-transition temperatures are necessary to preserve the physical properties and adhesive strengths of high-temperature formulations [88].

Lowering the Coefficient of Thermal Expansion

Thorough cure will result in some decrease in the coefficient of thermal expansion, but if higher temperatures are employed to achieve it, the method may be somewhat self-defeating. Fillers, as discussed in Chap. 14, are the most effective in reducing the value.

Epoxy-resin adhesive used to bond crack in slate switchboard, saving an extremely expensive replacement. (*The Epoxylite Corporation.*)

Ideally, the coefficient of thermal expansion should be lowered to match that of the material being bonded (or, with like-on-unlike bonds, to a value between those of the two materials). In practice, however, it is often not possible to employ a sufficiently large loading volume of filler. High loading volumes increase viscosity to the point where thin, regular glue lines cannot easily be obtained and in other ways tend to degrade, rather than improve, the bond. For some applications and some fillers, loading volumes up to 200 phr may be employed, but optimum shear-strength values are usually obtained with lesser amounts. Suggested loading volumes for some good adhesive fillers used with DGEBA are mica 14 phr, short-fiber asbestos 25 phr, aluminum oxide 50 phr, talc 80 phr, and zinc dust 100 phr. These, of course, are not optimum for all applications or in all formulations.

When temperature limits permit, it is useful to compensate for differences in the coefficient of thermal expansion by the use of flexibilizers to absorb the internal stresses during thermal cycling. For short periods at higher temperatures, supported tapes will somewhat improve performance by more uniform distribution of stresses. When bonding materials with greatly differing rates of thermal expansion, severe warping may be encountered in the absence of holding fixtures; warping will be particularly bad if high-temperature cure cycles are required.

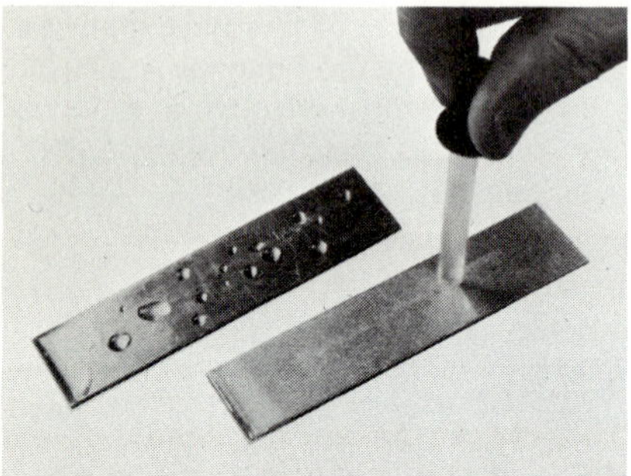

Spreading of water on surface is taken as a measure of the effectiveness of surface preparation. (*Bostik Methods, pub. B. B. Chemical Div. of General Shoe Machinery Corp.*)

Improving Wetting Action

Wetting action can be improved by warming the adhesive and/or by the incorporation of reactive or nonreactive diluents in to the system.

In production situations, with two-component adhesives, there will be a progressive increase in viscosity, which may alter reproducibility of the optimum bond; this can be avoided by mixing small quantities at a time or by using automatic blending equipment. Many of the one-component systems, which tend to advance somewhat under storage, should be checked periodically to assure proper performance. It has been shown that as the epoxy adhesive system nears the end of its pot life, resistance to hot water and salt spray is sharply reduced, although this will not necessarily be reflected in lower initial lap-shear values [106].

When diluents are employed to improve wettability, monofunctional diluents are generally used if the adhesive is required to serve only in the neighborhood of room temperature. At elevated temperatures, difunctional reactive diluents should be employed. The effect of diluents on elevated-temperature strength retention is shown in Figs. 21-13 and 21-14.

Much of the early literature dealing with epoxy-resin adhesives involved the use of resin blends, generally a viscous DGEBA (then not being produced in low-viscosity versions) and the polyglycidyl ether of glycerol. The latter was required to obtain a sufficiently fluid liquid. Greater improvements were obtained by using blends of different molecular-weight species of the aromatic resin in combination with the aliphatic one [8].

Reducing Shrinkage

Not only are fillers useful for controlling flow and for reducing the coefficient of thermal expansion to permit substantial retention of the bond up to the temperature limit of the particular formulation, but also, by bulk displacement, they serve to reduce the rate of shrinkage. Their total effect is impressive. Fillers may improve operational bond strengths by 50 to 100 per cent. The effect of various fillers on the strength properties of epoxy adhesive formulations is indicated in Figs. 21-15 and 21-16. The effect of various fillers loaded at a constant 100 phr is indicated in

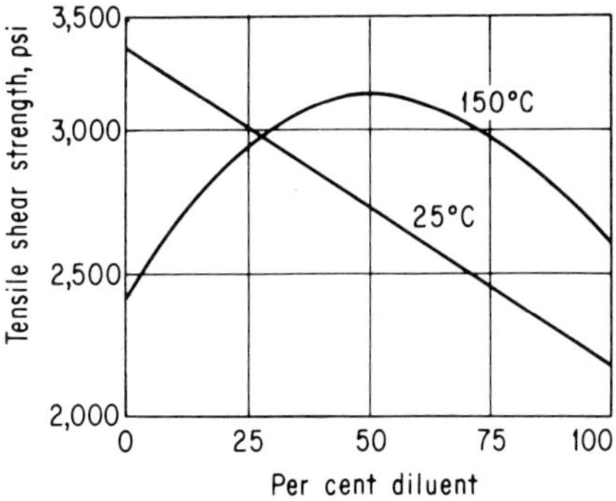

FIG. 21-13. Lap-shear strength vs. concentration of diluent (1/1 blend of polyglycidyl ether of tetraphenylethane and 2-glycidyl phenyl glycidyl ether) in DGEBA containing 30 phr asbestos floats and 6 phr polyvinyl acetate and cured with MPDA/MDA/PGE [67].

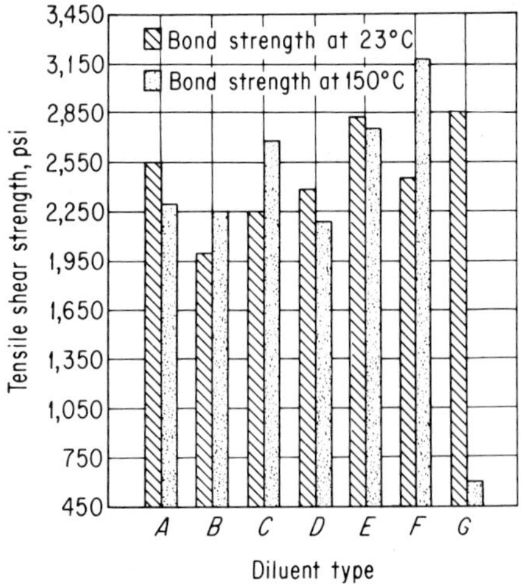

FIG. 21-14. Effect of diluents on lap-shear strengths of DGEBA containing 30 phr asbestos floats and 6 phr polyvinyl acetate and cured with MPDA/MDA/PGE [67]. Diluents: *A*, 2-glycidyl phenyl glycidyl ether; *B*, vinylcyclohexane-3-diepoxide; *C*, 2,6-diglycidyl phenyl glycidyl ether; *D*, 3,4-epoxy-6-methylcyclohexylmethyl-3,4-epoxy-6-methylcyclohexane carboxylate; *E*, diglycidyl ether; *F*, 3,4-epoxycyclohexylmethyl-3,4-epoxycyclohexane carboxylate; *G*, allyl glycidyl ether.

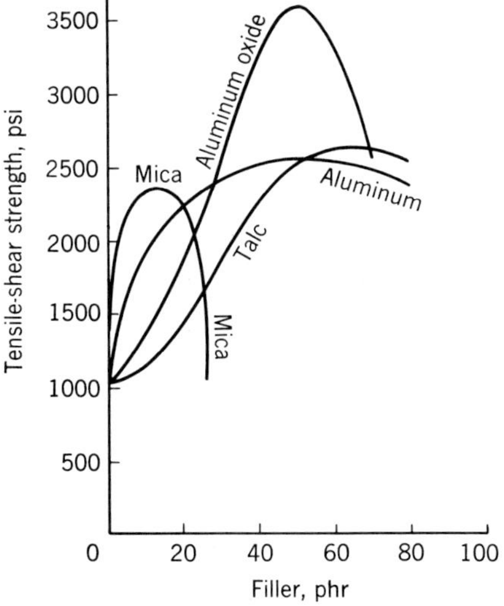

FIG. 21-15. Effect of fillers on lap-shear strength of DGEBA formulation tested at 25°C [13].

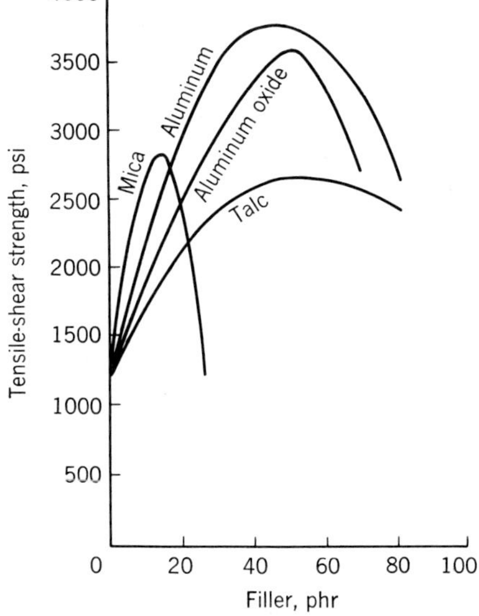

FIG. 21-16. Effect of fillers on lap-shear strength of DGEBA formulation tested at 57°C [13].

Epoxy-resin paste is used to brace knuckles of form-wound coils to prevent vibration and movement. Method eliminates need for costly tying and blocking. (*The Epoxylite Corporation.*)

Table 21-8. The effect of filler blends on adhesive strength is shown in Table 21-9, and the effect of fillers and filler blends on impact resistance is shown in Table 21-10. The effect of fillers on adhesive strength of the heat-cured DGEBA (without curing agent) is presented in Table 21-11.

Table 21-8. Effect of Fillers on Block-shear Strength of Adhesive Formulation [4-6]

(In pounds per square inch)

Fillers at 100 phr	Phenolic linen laminates tested at				Aluminum blocks tested at	
	23°C	75°C	90°C	105°C	23°C	105°C
Aluminum powder....	2,790	.....	1,470	1,390		
Ignited Al_2O_3	4,600	1,360	1,195	530	1,650	1,150
Short-fiber asbestos, 24 phr	1,740	1,270	580	510	2,180	3,910
Carbon black	2,000	555	980	910	2,000	1,170
Silica	2,840	1,600	1,250	830	1,530	600
Zinc dust	2,510	600	300	225	4,080	3,865

Table 21-9. Effect of Filler Blends on Bend Strength of DGEBA Cured with 100 phr Fatty Polyamide (Amine Value 210–230) [31]

(Bend strength in pounds)

Filler blend	30% filler	50% filler	70% filler
Aluminum/clay, 1/3	155	180	150
Aluminum/clay, 1/1	100	95	120
Aluminum/clay, 3/1	200	175	175

Table 21-10. Effect of Filler Blends on Impact Strength of DGEBA Cured with 100 phr Fatty Polyamide (Amine Value 210–230) [31]

(Impact strength in pounds)

Filler blend	10% filler	30% filler	50% filler	70% filler
Aluminum/clay, 1/3	...	4.0	6.3	3.1
Aluminum/clay, 1/1	2.3	2.1	2.0	1.3
Aluminum/clay, 3/2	...	9.4	5.1	1.5
Aluminum	0.7	1.8	2.3	1.1
Iron oxide (black)	4.0	4.0	1.8	1.7
Clay	1.7	1.7	1.2	3.1

Table 21-11. Effect of Fillers on Thermal Stability of Uncatalyzed DGEBA (mol. wt. 950) Cured 16 Hours at 260°C. Adhesive Carried on Glass Cloth [21]

Filler	Amount of filler, %	Shear strength at 24°C psi after aging:		
		At 25°C	100 hr at 260°C	200 hr at 260°C
None	0	2,120	2,090	1,210
Antimony trioxide	50	2,380	2,840	1,982
Zinc powder	50	1,530	670	540
Zinc borate	30	1,600	1,124	1,012
Cuprous sulfide	50	2,050	1,952	1,524
Aluminum flake	25	2,640	1,134	

Increasing Tack

In adhesive formulations, it is sometimes necessary to achieve some degree of initial tack in order to provide self-positioning action.

An obvious method for producing high-tack epoxy adhesives is by precuring the film to the tacky stage. This is patented in the case of a polysulfide-epoxy-amine adduct formulation—the tack adhesive being on one side of the rear-view-mirror mounting bracket, it then being applied to a window glass and held in place for final cure with pressure-sensitive tape [81].

Tack can be achieved by proper selection of formulation ingredients, usually by the use of higher-molecular-weight fatty-polyamide-type curing agents and selected fillers, although with the systems which are capable of supporting, let us say, ceramic tile on a vertical surface, the texture of the formulation makes mixing difficult.

Liquid tackifiers suggested for epoxy adhesives are the following: polyisobutylene resins, polyterpene resins, paracoumarone-indene resins, rosin derivatives such as triethylene glycol ester of hydrogenated resins, hydrogenated methyl ester of rosin, hydroabictyl alcohol, and methyl ester of rosin. Solid tackifiers suggested are rosin, hydrogenated rosin, dehydrogenated rosin, polyterpenes, coumarone-indenc resins, polyalkyl styrenes and oil-soluble phenol-aldehyde resins [87].

Modifying Other Properties

The mechanical and electrical properties and the chemical resistance of the adhesive systems are not specifically cited in this chapter. Such data for a variety of similar compounds are provided elsewhere in the book. One point, however, deserves general comment. The excellent chemical resistance of the epoxies is reflected in the performance of most adhesive systems. For example, with a typical formulation for bonding aluminum, a shear strength of 2,300 psi will be obtained with a room-temperature cure. This value may even increase somewhat after 30 days immersion in tap water, salt water, alcohol, hydraulic oil, etc. Stronger solvents will bring about some decrease in performance with, for example, shear strengths dropping to 1,400 psi after 30 days' immersion in methyl ethyl ketone.

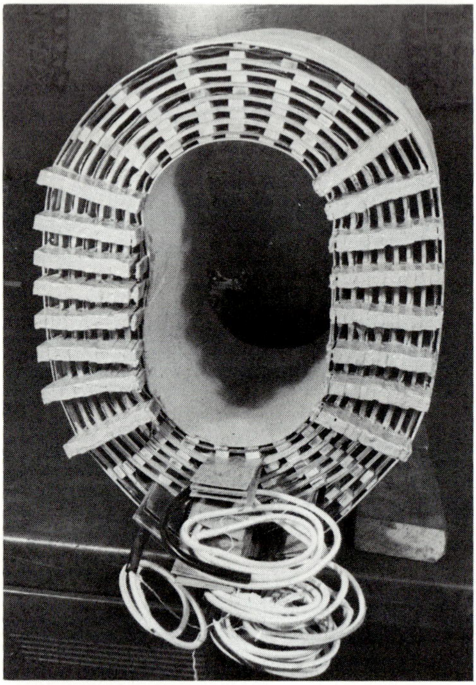

Wooden coil spacers in oil-filled transformer are bonded with epoxy adhesive to eliminate need for tying with string. (*The Epoxylite Corporation.*)

Although chemical resistance of adhesive joints is generally in line with values expected from chemical-resistance data given previously in the text, water resistance and particularly resistance to hot, humid environments may be less than expected from other data. A general discussion of this is given in Chap. 22 as related to adhesion to glass filaments. Humidity resistance of metal-to-metal joints has been found to be a function, in part, of the type of metal. Aluminum joints are particularly bad in this respect, the difficulty apparently arising from the effect of water at the oxide interface. Poor humidity resistance thus appears to be more a function of surface and surface preparation than of the specific epoxy-adhesive formulation. In addition, aluminum metal is subject to attack under conditions of high humidity from highly basic polyamines often used as curing agents.

In addition to the mechanical properties, the chemical resistance, and the electrical properties, it is often desirable to bring about other modifications to the adhesive formulation. The use of modifiers to improve impact strength is indicated in Table 21-12. An example of highly specialized modifiers is the use of 2,5-diphenyloxazole to improve the radiation resistance of an epoxy adhesive: A formulation showing 40 per cent strength loss without the additive showed only 11 per cent loss with the additive [22]. Zinc or lead chromate may be used to improve salt-spray resistance of epoxy adhesive formulations [44]. Examples of other modifications are given in the following section.

Table 21-12. Effect of Flexibilizing Modifier on Strength of Epoxy Resins Cured 2 Weeks at 23°C with TETA [33]

Resin	Temperature, °C	Monoglycidyl ether of 3-(pentadecyl) phenol, %								
		0	5	10	15	20	25	30	40	50
DGEBA:										
Lap-shear strength, psi	−55	433	920	859	1,027	1,025				
	23	408	597	587	601	734				
	70	260	397	393	385	597				
Impact strength, ft-lb/sq in. ..	−55	3.3	9.5	10.8	9.1	14.9	10.5	5.0	5.8	7.1
	23	3.3	4.1	12.8	12.8	18.6	4.5	3.3	2.7	4.9
	70	8.8	8.1	7.5	4.9	5.1	....	4.9	4.7	1.1
DGE of 3-(*pentadecyl*)-*phenol:*										
Lap-shear strength, psi	−55	523	1,006	914	940	931				
	23	845	496	485	568	471				
	70	186	170	157	107	103				
Impact strength, ft-lb/sq in. ..	−55	5.1	7.2	7.4	7.7	6.2	....	6.1	3.2	1.5
	23	10.7	8.8	8.8	11.4	11.8	4.2	5.9	3.6	1.3
	70	3.6	4.5	1.8	3.1	2.4				

FORMULATION

For many general-purpose applications—metal to metal, wood to glass, phenolic to stainless—where service environments are not unusual and temperatures do not vary markedly from normal ambient, the selection of the adhesive formulation can be based on cost and convenience. For many jobs, special compounds, which use commercial recommendations merely as starting points, are necessary; in some cases, their design may require considerable developmental effort.

There are hundreds of epoxy-resin adhesive formulations currently in use by industry. A number of these have been suggested already in earlier sections of the chapter, as illustrations for the various parameters under discussion. It is helpful now to look at some specific classes of formulations in greater detail. For convenience, the epoxy adhesives are considered to fall into either the flexible or the rigid category, with the term flexible including those systems having improved peel resistance.

Flexible Systems

A variety of modifiers has been suggested for improving the flexibility of epoxy adhesives. For example, linear polyamide (nylon type) used at 20 to 30 phr has been reported as providing good peel strengths to temperatures of 100 to 150°C [70], and when used at 300 phr with cures above 160°C to fuse the polyamide, considerably improved room-temperature lap-shear values [12-4]. Epoxy/phenolic blends have been reacted with butadiene-acrylonitrile copolymers to improve peel resistance for printed-circuit-board adhesives, and the copolymer has also been used with the epoxy resins without phenolic modifiers [48]. Long-chain diglycidyl and monoglycidyl ethers have also been used to impart flexibility, as indicated in Table 21-12. More commonly, moderate improvements in peel strength have been obtained by the use of small amounts of polyvinyl formal [68] or polyvinyl acetate [73]. Typical of such an adhesive system would be the following formulation:

	Parts
DGEBA	100
Asbestos, fine fibers	40
Low-molecular-weight polyvinyl acetate	4
Tabular alumina, 325 mesh	40
DEAPA	10

The improvement in peel strength is reflected in higher lap-shear strengths, as indicated in Fig. 21-17. It should be noted, however, that in order to obtain reliable data in this regard, the adhesive system should be thoroughly cured. It has been found, for example, with epoxidized novolac adhesives, modified with

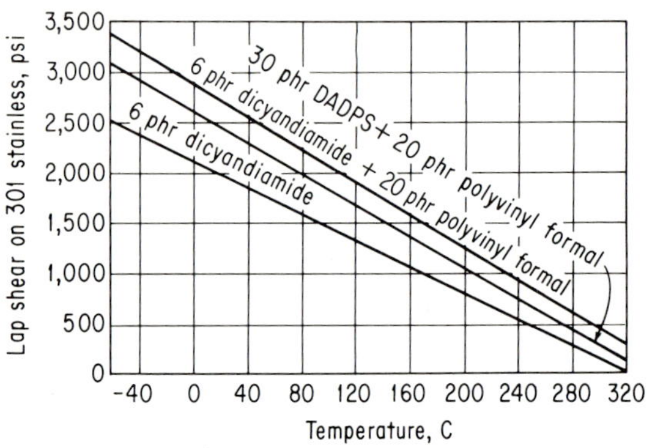

FIG. 21-17. Effect of polyvinyl formal on lap-shear strength of polyglycidyl ether of tetraphenylethane filled with 100 phr aluminum powder [4–6].

polysulfides, that aging at room temperature for 1 year resulted in a 25 per cent strength reduction, the reduction being attributed to increasing cure and the consequent increasing embrittlement [110].

Major improvements in the flexibility of epoxy adhesive systems are usually obtained by the use of flexibilizing curing agents such as the amidopolyamines or fatty polyamides, or of flexibilizing additives such as the polysulfides.

Flexibilizing Amides. Most commonly, the fatty polyamides (Chap. 10) are employed as flexibilizers. The fatty polyamides are excellent thermoplastic adhesives in their own right, since they contain polar amine, carboxyl, and amide groups. As flexibilizers which react directly with the epoxies, they become an integral part of the cured system, thus distributing the stresses and impact loads throughout the adhesive bond. A typical fatty polyamide with fluidized DGEBA and a filler loaded at 40 phr will give 200-lb bend strengths and 16-lb inch-width peel strengths on metal. Peel strengths of 4 to 5 lb are common with rigid adhesives. The shear strength to aluminum after an impact of 80 in-lb will be about four times that of a rigid system and only slightly below that of the preimpact strength.

Higher percentages of fatty polyamide will improve peel strengths; lower percentages will permit retention of bond at somewhat higher temperatures. The more flexible bond required, of course, the higher the amount of fatty polyamide. An optimum formulation might contain as much as 250 phr fatty polyamide.

Thorough cure can be obtained by baking 4 hours at 66°C followed by 30 minutes at 120°C, although for many applications long room-temperature cures will prove adequate. If production situations permit, cure temperatures of 150°C for 20 minutes may be used to give extremely fast cures without significantly reducing the adhesive properties. When room-temperature cures are required, an accelerator, such as a phenol, or a secondary, more reactive curing agent is often employed.

Two typical epoxy–fatty-polyamide formulations are listed below.

For metal-bonding applications

Fatty polyamide (amine value 210–230)	50
DGEBA	45
Butyl glycidyl ether	5
Tabular alumina (325 mesh)	40

This formulation will give such typical shear-strength values at room temperature as:

3,200 psi on aluminum
2,800 psi on cold-roll steel
2,100 psi on Naval brass
1,400 psi on hard copper

The effect of varying the filler type and concentration on adhesive strength and elevated temperature is shown in Table 21-13.

Table 21-13. Effect of Filler on Lap-shear Strength of DGEBA Containing 10 phr Butyl Glycidyl Ether and Cured with 100 phr Polyamide (Amine Value 210–230) [16]

Type of filler	Filler, phr	Lap-shear strength, psi		
		At 23°C	At 120°C*	After mechanical shock of 80 in.-lb
Tabular alumina.......	20	3,200	2,560	2,940
	60	3,080	.....	1,060
	100	2,960	2,070	440
Calcium carbonate.....	20	2,730	1,620	2,240
	60	3,200	.....	300
	100	3,400	1,430	130

* Specimens tested after 5-min conditioning at 120°C. Lower results can be expected after longer exposures, or when bond temperature reaches 120°C.

For bonding rubber to metal. Same as above, with filler omitted. This formulation cured for 48 hours at room temperature will give shear strengths of 1,000 psi on aluminum-rubber bonds, with failures mostly in the rubber.

Because of the relatively high viscosity of the fatty polyamides, it is sometimes preferable to substitute a lower-viscosity amidopolyamine as the curing agent. Typical of such a formulation is the following room-temperature-curing body solder [4-4]:

	Parts
DGEBA............	100
Amidopolyamine	85
Bisphenol A	15
Aluminum flakes	15
Colloidal silica	12.6

The amidopolyamines may also be used to obtain cure under water in filled adhesive systems.

In addition to the use of these commercial amines, specialized amide systems have been developed for limited applications. For example, an adhesive for linear polyester has been developed from a partially amidized acid from a secondary amine, reacted at less than stoichiometric with a DGEBA resin and cured with a dihydrazide [74]. A special system has been developed by coating a plasticized polyamide film with DGEBA (mol. wt. 950) [53].

The amidopolyamines and the fatty polyamides, usually blended with other curing agents or accelerators, are widely used as curing agents for consumer epoxy-resin adhesives, as contained in the two-tube systems.

Polysulfides. Polysulfides provide substantially lower working viscosities than do the fatty polyamides. They may be used over a rather wide range, together with stoichiometric amounts of the selected curing agent, to provide controlled amounts of flexibility in the adhesive joints. The materials are discussed in some detail in Chap. 16.

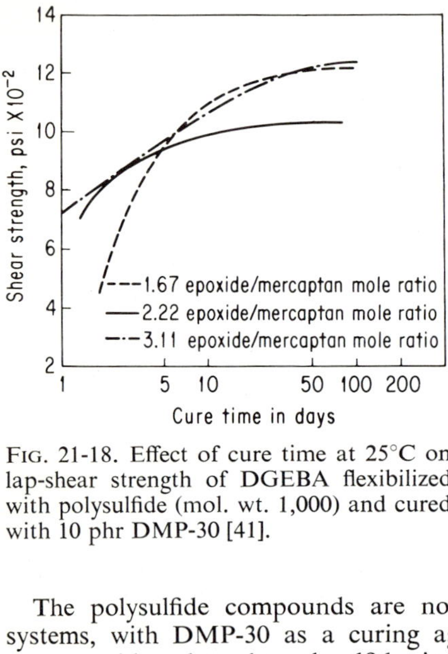

FIG. 21-18. Effect of cure time at 25°C on lap-shear strength of DGEBA flexibilized with polysulfide (mol. wt. 1,000) and cured with 10 phr DMP-30 [41].

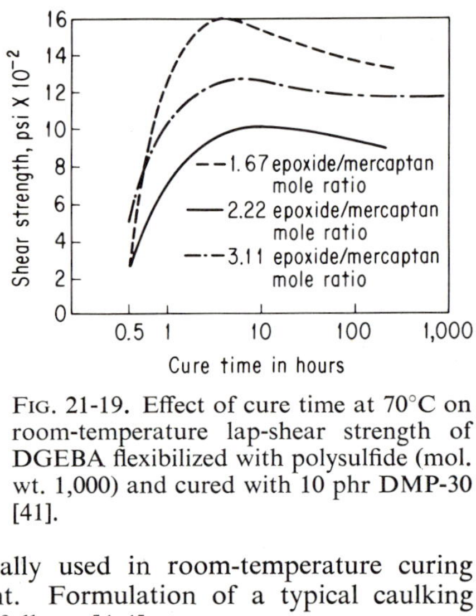

FIG. 21-19. Effect of cure time at 70°C on room-temperature lap-shear strength of DGEBA flexibilized with polysulfide (mol. wt. 1,000) and cured with 10 phr DMP-30 [41].

The polysulfide compounds are normally used in room-temperature curing systems, with DMP-30 as a curing agent. Formulation of a typical caulking compound based on the polysulfides is as follows [4-4]:

Material	*Parts*
Part A:	
DGEBA	85
Butyl glycidyl ether	15
Calcium carbonate	40
Asbestos, fine fiber	19
Part B:	
Polysulfide (mol. wt. 1,000)	200
Calcium carbonate	80
Asbestos, fine fiber	29
DMP-30	6

The above formulation, processed by milling, provides a flexible caulking compound with a heavy paste viscosity.

Although the polysulfide-modified epoxy-resin adhesive formulations are suitable for application where room-temperature curing is required, the optimum cure cycle

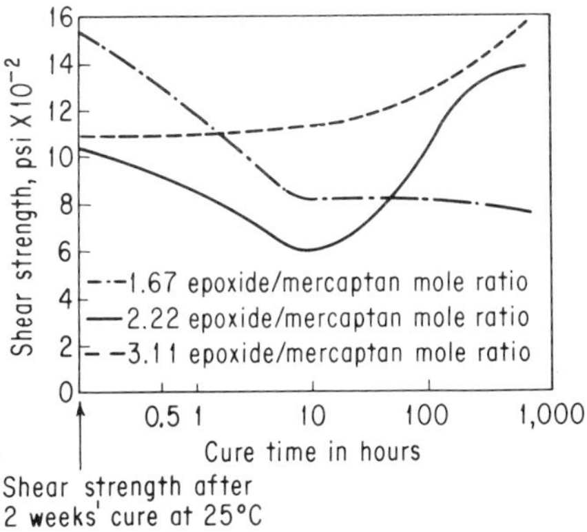

FIG. 21-20. Effect of postcure at 70°C on room-temperature lap-shear strength after 2-week cure at 25°C of DGEBA flexibilized with polysulfide (mol. wt. 1,000) and cured with 10 phr DMP-30 [41].

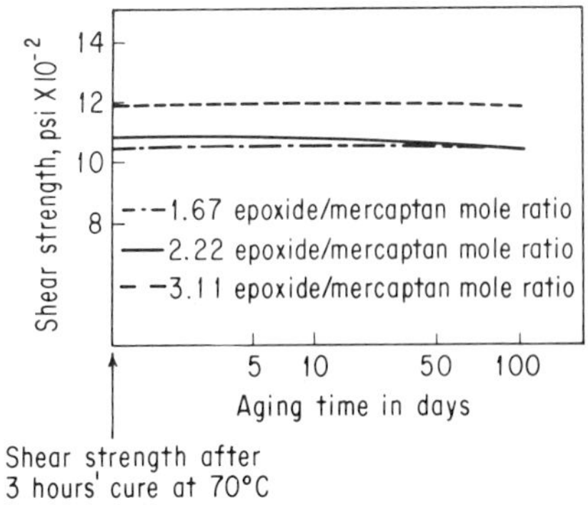

FIG. 21-21. Effect of room-temperature aging on lap-shear strength at 25°C of DGEBA flexibilized with polysulfide (mol. wt. 1,000) and cured 3 hours at 70°C with 10 phr DMP-30 [41].

will be dependent upon the ratio of polysulfide to epoxy. This dependence, for a system using a polysulfide (mol. wt. 1,000), DGEBA, and DMP-30 at 10 phr, is illustrated in Figs. 21-18 to 21-21.

In some applications, the polysulfide and fatty-polyamide flexibilizers may be used in combination. For example, when employing an aromatic diglycidyl ether resin with unsaturated side chains, bond strengths on the order of 6,000 psi and peel strengths of 30 to 40 lb have been obtained with the combination [39]. A room-temperature curing formulation which qualifies to MIL-A-5090-B, having good hot

Epoxy adhesive is used to seal leak in oil-filled transformer case while still in service, eliminating the need for disconnecting and draining the transformer and welding the case. (*The Epoxylite Corporation.*)

strengths to 82°C, consists of the following [26]:

	Parts
DGE of resorcinol	70
Epoxidized blend of 50 mole resorcinol and resorcinol resin	30
Butadiene dioxide	20
Water-ground mica (325 mesh)	60
Fatty polyamide (amine value 210–230)	80
Polysulfide (mol. wt. 1,000)	20.5
2,4-Toluenediamine	20.5

Rigid Systems

Rigid adhesive systems may, arbitrarily, be divided into room-temperature curing type and heat curing type.

Room-temperature Curing Adhesives. Typical of the room-temperature curing agents useful in adhesive formulations are such cross-linking aliphatic amines as DETA, triethylamine, and catalytic amines such as DMP-30.

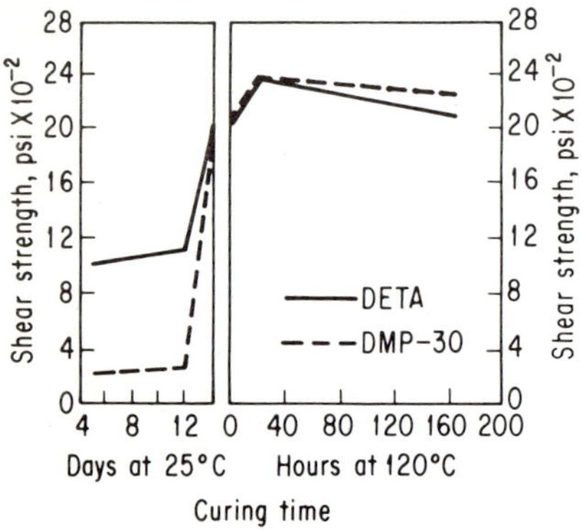

FIG. 21-22. Effect of cure on shear strength of DGEBA formulations [41].

With room-temperature curing agents, the adhesive dropoff point will be between 80 and 120°C, the exact point depending considerably on the filler, the specific resin, and the length of cure. For instance, DETA in a typical adhesive formulation will give a room-temperature shear strength on aluminum-to-aluminum bonds of over 2,000 psi after an 8-hour cure at 25°C. If, however, the temperature is rapidly increased to 85°C, the shear strength will be reduced to about 10 per cent of the room-temperature value. On the other hand, using triethylamine, about 8 days of room-temperature cure will be required to establish comparable room-temperature shear strengths, but at 85°C the value will be somewhat increased over that at room temperature. The effect of cure temperature and time on typical room-temperature-curing rigid adhesives is shown in Fig. 21-22.

Table 21-8 presents a comparison of the effect of selected fillers on shear strengths when used in connection with the following formulation:

	Parts
DGEBA (mol. wt. 450)	75
Polyglycidyl ether of glycerol	25
Allyl glycidyl ether	10
Filler	100
Triethylamine	12.5

In the above formulation, use of aluminum powder and alumina increases substantially the resistance of the adhesive to boiling water. This is also true when DETA is employed as the curing agent [2,3]. In the above system, employing the mixed resin, the addition of the monoepoxy improves the water resistance. Using DETA as the curing agent, a tenfold improvement in bond-strength retention is obtained over values obtained without the monoepoxy [4].

Room-temperature-curing rigid adhesive formulations may be less complex than indicated above. For example, a good general-purpose room-temperature curing

Table 21-14. Effect of Filler Loading Volumes on Lap-shear Strength of Adhesive Formulation [13]

Filler	Filler, phr	Tensile shear strength of aluminum blocks, psi	
		At 23°C	At 60°C
None	..	1,030	1,205
Aluminum	40	2,525	3,725
	60	2,525	3,580
	80	2,390	2,620
Aluminum oxide	30	2,505	2,805
	50	3,615	3,600
	70	2,555	2,665
Mica	6	2,210	1,810
	14	2,360	2,815
	26	1,055	1,210
Silica	40	600	610
	90	1,460	1,225
Talc	30	1,870	2,360
	60	2,650	2,655
	80	2,520	2,400

adhesive may be formulated with DGEBA and 10 parts of tertiary amine. Alumina and thixotroping fillers may be added to obtain the desired consistency. Optimum loading volumes are indicated in Table 21-14.

Heat Curing Adhesives. In many applications, room-temperature curing adhesives are most convenient to production. However, the short pot life of the room-temperature curing formulations often provides a serious drawback, and in these cases heat curing formulations may be employed. In addition, if maximum strengths are necessary to the application, heat curing is generally recommended (insufficient exotherm being generated in thin glue lines to assure complete cure with most room-temperature curing systems), and in these cases, in order to obtain increased pot lives, it is frequently customary to specify heat curing systems. Heat curing systems, too, lend themselves to the formulation of one-component powders, liquids, and tapes, whereas, of course, the room-temperature curing systems do not. And finally, by using heat curing systems, improved strength retention may be obtained at elevated temperatures.

In the latter case, when strength retention is required at temperatures in the range of 150°C or higher, special considerations are involved, and these are treated separately in the section on high-temperature adhesives.

Of the various curing agents which may be employed to obtain heat curing systems, perhaps the most useful in the epoxy technology is DEAPA. This compound is widely used as a curing agent, since it provides properties equivalent to DETA, with better strength retention at increased temperature, and with a pot life on the order of 4 hours at room temperature. Aromatic amines may be employed to obtain increased pot life, as may the anhydrides, although the latter are little used (except for high-temperature work) because of generally poor overall adhesive properties. More commonly, however, latent curing agents are employed, since these permit the formulation of one-container systems.

With heat-curing adhesive formulations employing such curing agents as MDA and MPDA, some increase in the upper operating temperatures will be realized. For instance, MPDA will give shear strengths over four times as great as DETA at temperatures of 105°C, with correspondingly better performance at somewhat higher ranges. Table 21-6 indicates the comparative performance of a number of aromatic diamines at elevated temperature.

However, aromatic diamines possess several disadvantages:

1. They provide increased shear strengths over an intermediate temperature range from 100 to 150°C. For the lower end of this range, DEAPA and room-temperature curing agents are, with the proper selection of fillers, satisfactory; for the upper part of the range, there appear to be a limited number of industrial applications where such constant temperatures are encountered. Aircraft and missile work generally require either adhesives capable of withstanding temperatures of about 100°C maximum or of 260°C and above. It should be noted, however,

Epoxy-resin adhesive seals around heat-exchanger tubes at return end. (*The Epoxylite Corporation.*)

that some work with diaminodiphenylsulfone has indicated elevated-temperature performance above the range normally expected of the aromatic amines. With this curing agent, strengths above 1,000 psi have been obtained after aging 400 hours at 260°C. Resins employed were polyglycidyl of tetraphenolethane [47], and this blends in with DGEBA (mol. wt. 2,000) [40].

2. The aromatic amines, in general, provide poorer bond strengths at room temperature and are more sensitive to temperature cycling than are the aliphatic amines. The shrinkage rate is fairly high and cannot as easily be offset by the addition of fillers. Methods for upgrading the adhesive strength have been suggested, including the addition of polyvinyl formal to overcome the brittleness and the use of additives such as dipentamethylene thiuramtetrasulfide [34].

The largest potential application for the heat curing adhesives does not appear to be necessarily related to the improved elevated-temperature performance; but rather to their use in one-container adhesive formulations.

Typical of the curing agents employed in one-container hot melts and fluid systems is dicyandiamide. This curing agent may be ground, along with the filler, into a solid epoxy resin to provide a powdered adhesive. Likewise it may be milled into a fluid epoxy containing a filler to provide a long-shelf-life one-container fluid. Dimethylcyanamide, a toxic liquid, may be used at about 5 phr to provide stable one-container adhesive systems. Boron trifluoride complexes are also employed.

High-peel-strength epoxy adhesive powder being applied to pretreated rubber by sifting device. (*Shell Chemical Co.*)

A typical proprietary one-container adhesive system, formulated from DGEBA, a latent curing agent, and employing an asbestos filler, requires a cure schedule of 175°C for 90 minutes and provides a shelf life at room temperature in excess of 2 years. The cured adhesive will provide bond strengths on aluminum as indicated in Table 21-15.

Table 21-15. Properties of Typical One-container Adhesive System [4-6]

Property	*Value*
Lap-shear strength, psi:	
At 25°C	3,000
At 82°C	3,600
At 120°C	2,200
At 150°C	1,100
At −58°C	3,100
At 25°C after 200 hr	3,000
At 82°C after 200 hr	2,700
Impact strength at 25°C, ft-lb	21
Bend strength at 25°C, lb	175

A typical stick-solder formulation can be given as follows:

	Parts
DGEBA (mol. wt. 400)	100
MDA	28.5
Powdered aluminum	60
Colloidal silica	10

The ingredients are prepared as a hot melt and cast into small-diameter tubes or are ground to 60-mesh powders and then compressed into void-free sticks [55]. To apply, the area to be bonded is warmed with a torch until it is sufficiently hot to cause the stick solder to flow when drawn across the surface. Postcure will be required for small masses of solder. After cure, the solder may be finished by conventional techniques as required.

In situ heating of film adhesives may be obtained by laminating high-resistance filaments into the film and subsequently applying an electric current [83]. This provides the advantage of a one-component system without the requirement for external heat sources.

High-temperature Adhesives. The high-temperature epoxy-resin adhesives are, in one sense, only distantly related to the rest of the epoxy adhesive technology. A number of approaches have been investigated in the development of adhesive formulations suitable for continuous service from 260 to 315°C. Other work has been accomplished on the development of adhesives which are capable of withstanding short-time exposures to temperatures above 2000°C.

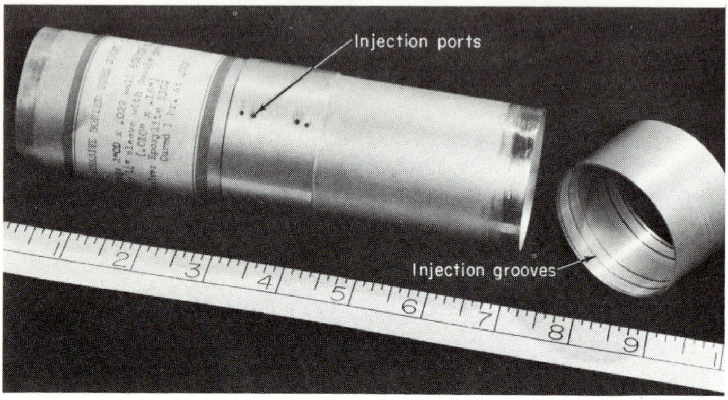

The adhesive bonding of some aluminum fuel lines on the XB-70 was preferred in lieu of conventional fittings because of a favorable weight advantage. The weight advantage of brazed joints as used for steel lines throughout the XB-70 was not applicable for these aluminum lines. The adhesive was added to the joint by injection after mating. (*North American Aviation, Inc.*)

Investigations of high-temperature systems have generally taken one of four courses: (1) development of epoxy coreactants which will increase the temperature resistance of the system; (2) development of antioxidants; (3) development of high-temperature curing agents; and (4) development of special epoxy resins.

Development of Coreactants. A very substantial body of work has been devoted to the investigation of silicone resins of one type or another in an effort to upgrade the temperature resistance of the epoxy-based adhesives [14]. Perhaps the most satisfactory material has been a silicone derived from the condensation of bisphenol A with a pentaethoxytriphenyltrisiloxane [78]. It is probably fair to say, however, that the use of silicone modifiers has not been entirely successful.

Even more work has been expended on the development of epoxy/phenolic combinations. Here, however, as indicated previously, the phenolic component appears to contribute to thermal instability at temperatures in excess of 290°C. The first such epoxy/phenolic blend was Shell Chemical Corp. formulation Epon® 422-J, which consisted of DGEBA (mol. wt. 1,000), about 50 phr of a phenolic resin, and 150 phr of powdered aluminum. The curing agent was 9 phr dicyandiamide, and a stabilizer, copper 8-quinolinolate, was used at 1.5 phr. This system gives bond strengths as indicated in Table 21-16. It is employed as a hot melt, or more conveniently, as supported or unsupported tape.

Further research with epoxy/phenolic blends led to the development of Forest Products Laboratory formulation FPL-878 which is considered one of the best

Table 21-16. Lap-shear Strength of Phenolic-Epoxy Adhesive 422J [4-6]

Test temperature, °C	*Lap-shear strength, psi*
−58	2,300
21	2,300
159	1,900
204	1,900
260	1,600
At 25°C after 100 cycles at 25–250°C.....	1,000
At 25°C after 200 hr at 250°C	700

epoxy-phenolic adhesives [11]:

	Grams
Durez 16227	12.5
Bakelite BV 9700	3.3
Epon 1007	2.0
1-hydroxy-2-napthoic acid	0.2
n-Propyl gallate	0.2
Ethyl acetate	2.0

This compound is brushed on both faying surfaces, precured for 30 minutes at 93°C, and pressed at 160°C for 30 minutes at 50 psi. The 422-J and FPL-878 formulations are compared in Table 21-2. A phenolic/epoxy blend that gives somewhat lower bond strengths throughout the temperature range but better strengths at 315°C on aluminum coupons consists of the DGEBA (mol. wt. 2,000) in combination with 500 phr phenolic resin, 100 phr methyl ethyl ketone, and cured with 20 parts hexamethylenetetramine [6]. Properties of this compound are presented in Table 21-17.

Table 21-17. Lap-shear Strength of Phenolic-Epoxy Adhesive Cured with Hexamethylenetetramine [6]

Test temperature, °C	*Lap-shear strength, psi*
−58	1,366
26	1,497
232	1,328
316	1,074

Development of Antioxidants. As indicated in Table 21-3, a number of metals contribute catalytically to oxidative degradation at elevated temperatures.

Antioxidants and stabilizers have been investigated to ameliorate the situation. The first material to be investigated in detail was 8-quinolinol and its derivatives [15]. Although still not providing completely satisfactory values after long-time aging, the stabilizers increased the shear strengths on the 422-J formulations from the 0- to 350-psi range to the 700- to 800-psi range after 200 hours at 260°C. A number of standardized antioxidants are discussed in [15], and although no satisfactory one was discovered, carbon black appeared the most promising. Antioxidants subsequently investigated are indicated in Table 21-7. The addition of zinc naphthenates appeared to offer the most promise [25].

In connection with the problem of developing antioxidants, the specific composition of the metals being bonded assumes increasing importance. Structural metals such as aluminum are of little utility at higher temperatures, the exotic metals and various steel alloys being generally specified, and these are the ones which appear most readily to catalyze thermal decomposition in the epoxy adhesives.

Development of High-temperature Curing Agents. Another promising avenue of approach in the development of elevated-temperature-resistant adhesives is the selection of high-temperature-resistant curing agents.

Table 21-18. Lap-shear Strength of Epoxy Adhesive Cured with Pentamethyldiethylenetriamine [10]

Test temperature, °C	*Lap-shear strength, psi*
−58	3,240
26	3,110
121	2,950
149	1,040
204	570
260	380

Arsenic pentoxide has been shown to be an effective curing agent for epoxy formulations as well as an antioxidant [78]. Rather surprisingly, pentamethyldiethylenetriamine at 4 phr has been reported to provide good elevated-temperature performance, as indicated in Table 21-18. Diaminodiphenylsulfone also gives good values with highly functional epoxy resins [4]. A number of anhydride curing agents have been investigated for elevated-temperature service, and specialized systems employing PMDA have given credible performances after 1,000 hours aging at 260°C when tested at 260°C [51].

Another class of curing agents suitable for high-temperature service are the substituted phthalocyanines. Copper-PMDA-phthalocyanine, for example, was used as a curing agent for an epoxidized phenol-formaldehyde novolac, to give properties as indicated in Table 21-19.

Table 21-19. Lap-shear Strength of Epoxidized Novolac Cured with Cu-PMDA-Phthalocyanine [90]

(In pounds per square inch)

Cure schedule	No catalyst	Metallic catalyst, 1 phr
At 260°C:		
After 100 hr at 260°C	1,920	2,500
After 200 hr at 260°C	1,720	2,050
At 316°C after 100 hr at 260°C:		
+ 30 min at 316°C	600	960
+ 11 hr at 316°C	1,360	1,650

Development of Special High-temperature Epoxy Resins. A variety of special high-temperature-resistant epoxy resins have been synthesized. These have been based on thermally stable aromatic backbones. The epoxidized novolacs have shown best commercial promise. Similar materials based on the substituted polyphenylene oxides have also been developed [98].

Although a great deal of work has been expended on the development of high-temperature-resistant epoxy adhesive formulations, more work is required for the development of a system suitable for long-time service at temperatures of 300°C and above.

SUMMARY

Epoxy resins are employed in many adhesive applications and are supplied in a variety of forms: room-temperature curing liquids; heat curing liquids; and heat-curing one-component liquids, powders, and tapes. The epoxy adhesives are, perhaps, most widely used for structural metal bonding in the aircraft and missile industry, mostly in honeycomb constructions. In industry generally, they have not significantly contributed to the overall volume of adhesives being produced commercially in the United States.

The epoxy-resin adhesives are high-strength compounds, and because of this, they are very sensitive to factors such as surface preparation which may not be critical with lower-strength materials. In addition, they are relatively weak in peel, and careful joint design—basically to distribute the stress over the total surface area of the adhesive—is required to obtain maximum results.

The room-temperature-curing epoxy adhesives, in general, are limited to service no higher than about 100°C. The heat cured systems generally provide somewhat better properties at elevated temperature and offer considerably greater handling convenience. The high-temperature adhesive compounds—primarily of interest to the airframe and space industries—are highly specialized formulations which may impose special processing requirements.

The epoxy-resin adhesive systems, consisting basically of the curing agent and the resin, may be modified by the incorporation of flexibilizers and by the addition of fillers. The use of flexibilizers serves to improve the peel and impact resistance of the adhesives but at the expense of the temperature resistance. Fillers serve to reduce the thermal expansion of the adhesive systems, lower the cost, and reduce the shrinkage. Virtually all the epoxy adhesive formulations contain fillers and/or flexibilizers of one type or another.

The epoxy adhesives should, in the main, be considered as somewhat specialized systems. They are particularly suitable for bonding dissimilar materials, for bonding otherwise difficult-to-bond materials, for making high-strength joints, and for applications where gap filling, filleting, and other specialized properties are difficult to attain in any but a 100 per cent solids system.

REFERENCES

1949

1. Method of Adhesively Uniting Materials, Especially Metals, and Adhesives Therefor (to Ciba, Ltd.), Australia 133,819.

1950

2. Wiles, Adhesive Compositions Containing a Fluidized Glycidyl Ether and an Amphoteric Oxide, U.S. 2,528,934.
3. Wiles, Adhesive Compositions Containing Glycidyl Ethers and Alumina, U.S. 2,528,933.
4. Wiles, Compositions Containing Glycidyl Ethers, U.S. 2,528,932.
5. Bixler, Epoxide Compositions, U.S. 2,512,996.

1954

6. Black and Blomquist, Metal-bonding Adhesives with Improved Heat Resistance, *Mod. Plastics*, December.
7. Glazer, Monolayer Studies of Some Ethoxyline Resin Adhesives and Related Compounds, *J. Polymer Sci.*, April.
8. Naps, Adhesive Composition Containing Polyepoxide Materials, U.S. 2,682,515.

1955

9. New Look in Wing Tanks, *Aviation Age*, November.

1956

10. Black and Blomquist, Metal-bonding Adhesives for High Temperature Service, *Mod. Plastics*, June.
11. Black and Blomquist, A Study of the Deterioration of Adhesives in Metal Bonds at High Temperature, WADC Technical Report 55-330.
12. de Bruyne, The Adhesive Properties of Epoxy Resin, *J. Appl. Chem.*, July.

13. Epstein, Epoxy-based Adhesives, Theoretical Calculations, ACS Symposium, Atlantic City, September.
14. Kidwell and McHugh, Investigation and Development of High-temperature Structural Adhesives, WADC Technical Report 54-98, part 3.
15. Naps, Elevated Temperature-resistant Modified Epoxide Resin Adhesives for Metals, WADC Technical Report 53-126, parts 1–3.
16. Peerman et al., Polyamide-Epoxy Combinations as Structural Adhesives, *Plastics Technol.*, January.

1957

17. Alter and Soller, Molecular Structure of Epoxy Polymers as a Basis for Adhesion, ACS Symposium, New York, September.
18. Anderson, Non-destructive Testing of Bonded Metal Sandwich Structures, ASTM Special Technical Publication 201.
19. Arnold, An Ultrasonic Technique for Non-destructive Evaluation of Metal-to-Metal Adhesive Bonds, ASTM Special Technical Publication, 201.
20. Bandaruk, Aromatic Amines as Curing Agents, *Plastics World*, November.
21. Black and Blomquist, Development of Metal-bonding Adhesive with Improved Heat Resistance, WADC Technical Report 56-650.
22. Mixer, Nuclear Deterioration of Structural Adhesives, ACS Symposium, New York, September.
23. Plastic Solder for Automobile Sheet Metal Repairs, *Automotive Ind.*, August.
24. Snoddon, Metal-to-Resin Adhesion as Determined by a Stripping Test, ASTM Special Technical Publication 201.

1958

25. Black and Blomquist, Metal Surface Effects on Heat Resistance of Adhesive Bonds, *Ind. Eng. Chem.*, June.
26. Clair and Moult, Room Temperature Cured Resorcinol Epoxide Adhesive for Metals, *Ind. Eng. Chem.*, June.
27. Elam and Elam, The Development of Adhesive Bonding in Aircraft, ACS Symposium, San Francisco, April.
28. Low Temperature Strength of Epoxy Resin Insulation Adhesive, *Insulation*, December.
29. McClintock and Hiza, Epoxy Resins as Cryogenic Structural Adhesives, *Mod. Plastics*, June.
30. McGuiness, Properties of Epoxy Adhesives at Elevated Temperatures, Epoxy Resin Symposium of SPE, Minneapolis, Minn., October.
31. Peerman and Floyd, A Guide to the Formulation of Polyamide-Epoxy Plastic Solders for Automobiles, Epoxy Resin Symposium of SPE, Minneapolis, Minn., October.
32. Stivala, Impact Strength of Adhesives as Affected by Geometry of Joint, Picatinny Arsenal, Technical Notes 18, February.
33. Stivala and Powers, Improved Impact Epoxy Adhesives, *Ind. Eng. Chem.*, June.
34. Tsong, Amine-Epoxide Adhesive Composition Containing Thiuram Sulfide and Process of Making Same, U.S. 2,849,417.

1959

35. Aitken, Portable Military Shelters, *Military Engr.*, July–August.
36. Baumgartner, Cork Compositions Containing Epoxy Resin, U.S. 2,904,524.
37. Black and Blomquist, Relationship of Polymer Structure to Thermal Deterioration of Adhesive Bonds in Metal Joints, NASA Technical Note D-108.
38. Epstein and Litvak, The Influence of Selected Fabrication Variables on Metal-to-Metal Bonding, ACS Symposium, Atlantic City, September.
39. Hart, Interpolymer of Epoxy Resin, Polyamide Resin and Polysulfide Resin, U.S. 2,919,255.
40. Hopper and Naps, Epoxy Resin Adhesive Compositions, Their Preparation, and Tape Containing Same, U.S. 2,915,490.
41. Petronio et al., Glass-to-Metal Bonding, *Adhesives Age*, January.

42. Thomas, Accelerated Aging Tests and Life Aging Properties of Aircraft Metal Adhesives, *ASTM Bull.*, January.

1960

43. Barenholtz et al., Types of Bonds Involved in Adhesion of an Epoxy Resin to Steel and Aluminum, ACS Symposium, Cleveland, April.
44. Blommers et al., Development of Low Pressure Ambient Temperature Curing Adhesive for Field Repair of Metal Airframes, WADC Technical Report 59-15.
45. Bonding Honeycomb Core Laminates, *Adhesives Age*, October.
46. Braune et al., Effects of Rate of Loading on Shear Stress of Resin Adhesives, *SPE J.*, May.
47. Compositions Containing Polyepoxide and Polyaminodiphenyl Sulphones (to N.V. de Bataafsche Petroleum Maatschappij), Brit. 826,534.
48. Epoxide Resin Compositions (to Minnesota Mining & Manufacturing Co.), Brit. 833,649.
49. Epoxies and KLM's Mobile Light Mosaic, *Adhesives Age*, December.
50. Epoxy Bonded Miniature Watch Parts, *Adhesives Age*, March.
51. Epoxylite Corp. Booklet L-800.
52. Finger Painting with Epoxies, *Adhesives Age*, October.
53. Groves, Adhesive Composition, U.S. 2,962,468.
54. Martin, Why Not Weld It Chemically, *Prod. Eng.*, Feb. 29.
55. Paul, Process for Preparing Polyepoxide Solder, U.S. 2,965,930.
56. Reynolds, When the Great Stone Face Began to Rock, *Monsanto Magazine*, February.
57. Rutzler, Adhesion and Surface Preparation, SPI Symposium, Chicago, February.
58. Slosser, Joints, *Adhesives Age*, January.
59. Stuart, Coating Compositions Containing Epoxy-Hydroxy Polyether Resin and the Reoperation of Alkyd Enamels with Same, U.S. 2,949,380.

1961

60. Adhesive-bonded Stained-glass Windows, *Progressive Architecture*, February.
61. Allison et al., Rapid Heating Method for Testing Plastics and Adhesively Bonded Systems, 17th ANTEC of SPE.
62. Bikerman, The Science of Adhesive Joints, Academic Press, Inc., New York.
63. Bonding Replaces Welding, *Pacific Factory*, September.
64. Ciba, Ltd. (Eng.) Technical Notes 218.
65. Hertz, Cryogenic Adhesive Evaluation Study, AD 273 219.
66. Hiza and Barrick, Cryogenic Adhesive Properties of Bisphenol A Epoxy Resins, *SPE Trans.*, April.
67. May and Nixon, Reactive Diluents in Aromatic Amine Cured Epoxy Adhesives, *J. Chem. Eng. Data*, April.
68. Naps, Adhesive Compositions and Their Preparation, U.S. 2,986,546.
69. Prestressed Concrete Bridge Has Epoxy Shear Keys, Waterproofing, *Railroad Track and Structures*, June.
70. Riel, Adhesive Bonding, a New Concept in Structural Adhesives, 17th ANTEC of SPE, January.
71. Riel, New Adhesive Films Give Optimum Glue Line in Honeycomb Structures, *Mater. Design Eng.*, January.
72. Swaffer, Antimony Trioxide as a Filler in Epoxy Adhesives, Brit. 884,578.
73. Trostyanskaya and Uenkova, Curing of Epoxy Resin Adhesives, *Plasticheskie Massy*, no. 2, pp. 16–18.
74. Wear et al., Resinous Compositions, U.S. 2,970,972.

1962

75. Adhesion and Cohesion, ed. Weiss, Elsevier Publishing Company, New York.
76. Barenholtz and Rutzler, Correlation of Some Thermal Properties of Epoxy Adhesives and Shear Strengths on Steel Laminates, ACS Symposium, Washington, D.C., March.

77. Frost, The Strength of Ten Structural Adhesives at Temperatures Down to −424°F., C.E.L. National Bureau of Standards, Boulder, Colo.
78. Levine, Arsenic Pentoxide Cure of an Organic Adhesive, *Ind. Eng. Chem. Prod. R&D*, June.
79. Mathews and Silver, Methods for Testing the Strength Properties of Adhesives and Test Data, NAVORD Report 3923.
80. Miller et al., Properties of Foams, Adhesives, and Plastic Films at Cryogenic Temperatures, ACS Symposium, Washington, D.C., March.

1963

81. Barlet, Method of Bonding Metal to Glass, U.S. 3,075,871.
82. Dear et al., Heat of Wetting of Aluminum Powder, *Trans. Faraday Soc.* **59**:713–720.
83. Epoxy Adhesive Film has Built-in Heat Curing System, *Plastics Design and Processing*, October.
84. Harper, Adhesive Bonding Materials and Applications, *Electron. Packaging and Production*, September.
85. Intorre et al., Sorption Behavior of the Diglycidyl Ether of Bisphenol A, *J. Phys. Chem.*, January.
86. Kern, Investigation of Etching and Cementing of Teflon, AD 425 068.
87. Korpman, Tape Employing an Adhesive Comprising an Epoxy Resin and a Polymer Having Carboxylic Groups, U.S. 3,100,160.
88. May, The Physical Significance of Acid-Anhydride Cures on Epoxy Adhesive Properties, 19th ANTEC of SPE, Los Angeles.
89. New Metalworking Techniques Assist Boeing on Apollo and 727 Projects, *Western Metalworking*, August.
90. Parry, Heat Resistant Adhesives Based on Phthalocyanine Copolymers, T.D. Report ASD-TDR-63-396.
91. Ripling et al., Final Report on Application of Fracture Mechanics to Adhesive Joints, AD 424 289.
92. Zisman, Influence of Constitution on Adhesion, *Ind. Eng. Chem.*, October.

1964

93. Contact Angle, Wettability, and Adhesion, Advances in Chemistry Series 43, American Chemical Society, Washington, D.C.
94. Dean, Chemical Adhesion, *Offic. Dig.* (*Eng.*) **36**(473):664–678.
95. Epoxy Adhesives Invade the Wood Glue Market, *Hitchcock's Woodworking Dig.*, November.
96. Epoxy Bonds Railing Joints, *Maintenance*, October.
97. Epoxy Resin Compositions for Surgical Use (to Unisearch, Ltd.), Brit. 952,842.
98. Hoyt et al., Evaluation of Poly(2,6-dimethyl)-1,4-phenylene Oxides and Related Copolymers for High Temperature Adhesive Application, *J. Appl. Polymer Sci.*, pp. 1633–1641.
99. Kausen, Adhesives for High and Low Temperatures, *Mater. Design Eng.*, September.
100. Kutscha, Mechanics of Adhesive Bonded Lap-type Joints: Survey and Review, AD 607 147.
101. Maly, Joint Adhesives, U.S. 3,146,142.
102. Schonhorn and Sharpe, Surface Energetics, Adhesion and Adhesive Joints, part 2, *Polymer Letters*, pp. 719–722.
103. Space Environmental Effects on Materials and Components, vol. 1, appendix 1, Redstone Scientific Information Center Report RSIC-150.
104. Werner, New Materials in the Conversion of Antiquities, *Museums J.* (*Eng.*), no. 1, pp. 5–16.

1965

105. Bandaruk, Adhesive Bonding with Graphite Fibers as an Internal Electrical Resistance Heat Source, Symposium on Structural Adhesive Bonding, Stevens Institute of Technology, September.

106. Delmonte et al., Influence of Application Variables on Properties of an Epoxy Adhesive, Symposium on Structural Adhesive Bonding, Stevens Institute of Technology, September.
107. Schonhorn and Sharpe, Surface Energetics, Adhesion, and Adhesive Joints, IV: Joints between Epoxy Adhesives and Chlorotrifluoroethylene Copolymer and Terpolymer (Aclar), *J. Polymer Sci*, **A, 3**:3087–3097.
108. Scott, Adhesives Containing Abrasives, U.S. 3,193,424.
109. Smith and Cagle, A Quality Control System for Adhesive Bonding Utilizing Ultrasonic Testing, Symposium on Structural Adhesive Bonding, Stevens Institute of Technology, September.
110. Wegman and Duda, Effect of Surface Preparation and Aging Time on Epoxidized Novolac Bonds to Stainless Steel, AD 460 681.
111. Wegman and Tanner, Strength of Epoxy Adhesives, *Adhesives Age*, September.

Appendix 21-1

RECOMMENDED SURFACE PREPARATIONS

The following surface preparations have been recommended in the literature. Where appropriate, they are followed by a neutralizing rinse, a water rinse, and an oven dry.

METALS

Ferrous

Sandblasting with grease-free sand or wire brushing has been found generally satisfactory for iron. Mild steel may be immersed in a 10 per cent solution of sodium metal silicate for 10 minutes at 60°C or dipped in 10 to 20 per cent phosphoric acid.

One aircraft company, after an extensive test program, concluded that the best method for preparing stainless steel is by vapor honing with trichloroethylene. A chemical cleaning solution which may be used is the following, parts by weight:

Concentrated hydrochloric acid	50
30% hydrogen peroxide	2
Formalin solution	10
Water	45

The stainless steel is immersed in this solution for 10 minutes at 65°C.

Nonferrous

Aluminum and Aluminum Alloys. Degrease and dip for 10 minutes at 70°C in the following solution, parts by weight:

Sodium dichromate	66
96% sulfuric acid	666
Water	1,000

The solution should not be overheated, since at temperatures above 70°C there is a rearrangement of the surface of the aluminum which leads to poor bonds. The sodium dichromate etch gives a controlled and porous layer of aluminum oxide on the surface which provides improved adhesive and cohesive characteristics [1]. The oxide film forms rapidly on exposure to air, and unless bonding or priming is accomplished within a few hours after the surface treatment, the oxide film will become too thick.

Studies conducted to determine the effect of the surface preparation of aluminum on high-temperature bond strengths showed that the sodium dichromate etch was generally satisfactory. However, when clad aluminum bonds employing this treatment are exposed to salt spray, resistance is likely to be deficient. It was concluded that to obtain adequate salt-spray resistance, the cladding should be completely removed, insofar as surface flaws can be shown to exist which permit the entrance of contamination and the failure of the bond. Since the removal of the cladding is generally undesirable, it is suggested that a primer be employed that is capable of diffusing into the metal surface and blocking the contaminant pathways before the application of the adhesive [2].

Beryllium. Degrease and immerse 3 minutes at 80°C in 20 per cent NaOH in distilled water.

Cadmium. Electroplate with silver.

Chromium. Etch 1 to 5 minutes at 90 to 95°C in the following, parts by volume:

Conc. HCl (sp. gr. 1.18)	8.5
Water	10

Copper and Copper Alloys. Etch for 1 to 2 minutes at room temperature in the following solution, parts by volume:

42% ferric chloride	15
Concentrated nitric acid	30
Water	197

In some cases, if the copper or the alloy is heavily scored, improvement in bond strength will be obtained. With copper and copper alloys, degradation at elevated temperatures would be expected. Brass should be degreased and abraded with emery cloth.

Gold. Degrease and roughen surface.

Lead. Degrease and abrade with emery cloth.

Magnesium and Magnesium Alloys. Several treatments for magnesium have been evaluated [3], among them being an Iridite no. 15 dip-and-rinse that gave bond strengths of 2,000 psi with a filled adhesive cured with DEAPA. An anodizing treatment (Dow no. 17), which gives slightly lower bond strengths but better all-around performance, particularly in terms of humidity resistance, is perhaps the preferred treatment. Next best is a deoxidation procedure which involves immersing the degreased magnesium for 3 minutes at room temperature in the following solution:

Chromic acid	26.6 oz
Sodium nitrate	32.0 oz
Glacial acetic acid	1 gal
Distilled water	1 gal

This gives lap shears using the DEAPA-cured adhesive on the order of 2,400 psi with good temperature and salt-spray resistance but fairly poor humidity resistance.

Nickel. Etch 5 seconds in conc. (1.42) HNO_3.

Silver. Degrease, remove tarnish with fine emery cloth.

Solder. Degrease, abrade with emery cloth.

Titanium Alloys. The following solution, parts by weight, has been suggested for preparing titanium alloys:

Sodium fluoride	10
Chromium trioxide	5
Distilled water	250
Conc. H_2SO_4 (sp. gr. 1.86)	50

Research on the bonding of titanium was active from about 1956 to 1958. For the next 7 years, little was reported. Renewed interest in 1965 and 1966 resulted in the development of special proprietary treatments which were made available to the airframe industry.

A chromic acid etch has also been reported as satisfactory.

Tungsten and Tungsten Carbide. Etch 1 to 5 minutes at room temperature in the following, parts by weight:

Conc. HNO_3 (1.41)	30
Distilled water	15
Hydrofluoric acid (1.18)	5
Conc. H_2SO_4 (1.86)	50
H_2O_2	A few drops

Zinc and Zinc Alloys. Degrease and abrade with emery cloth. Apply adhesive immediately. Or etch 2 to 4 minutes at room temperature with 20 parts conc. HCl/80 parts H_2O.

PLASTICS

Thermosetting

In general, an adequate surface preparation consists of a solvent wipe, followed by a light sanding to remove surface contamination. This treatment has been found satisfactory for diallyl phthalate, epoxy, phenolic, polyester, polyurethane and urea-formaldehyde resins.

Satisfactory bonds have not consistently been obtained to silicone resins; but with some silicones, adequate bonds are achieved when cure temperatures are 140°C or above.

Thermoplastic

Cellulose Plastics. After degreasing and abrading, the plastic should be warmed to 100°C for one hour and the adhesive applied before the plastic cools to room temperature.

Chlorinated Polyether. Etch 5 minutes at 70°C in the following, parts by weight:

H_2SO_4 (1.84)	1,500
Potassium dichromate	75
Water	120

Polyacetals. Dip 80 seconds at 70°C in the following solution:

	Per cent
Kieselguhr	0.5
Dioxane	3.0
p-Toluenesulfonic acid	0.3
Tetrachloroethylene	96.2

Dry 5 minutes at 65°C, and rinse immediately for 1 minute in hot water.

Polyamide, Linear. Wet and soften with 15 per cent aqueous solution of phenol in the case of aliphatic-based species. Wholly aromatic polyamides provide good surfaces when lightly sanded.

Polyester, Linear. Linear polyester reorients at temperatures on the order of 135°C, and satisfactory bonds may often be obtained by curing at or above this temperature. A liquid treatment consists of immersing the linear polyester for 10 seconds in a 2 per cent solution of vinyltrichlorosilane in xylene [4]; alternatively, it may be etched for 2 to 10 minutes at 70 to 95°C in the following, parts by weight:

Caustic soda	20
Water	8

Polyethylene. A variety of methods have been suggested for preparing the surface of polyethylene. Generally, the treatments are more effective with the high-density compounds. Among the treatments suggested are the following: Expose the polyethylene to a combination of ultraviolet light and chlorinated solvent [5].

Grit blasting may be used, or etch 2 minutes in the following:

H_2SO_4	100 cc
Potassium dichromate	16 g
Water	50 cc

Strike 15 seconds with trichloroethylene at 80°C to produce a microetch effect (which lasts approximately 48 hours) [6]. Flame-spray by passing a flame once or twice quickly across the surface. NOTE: Toluol or acetone may be used to remove mold release when present. Carbon black pigments in the polyethylene generally make bonding more difficult.

Polyformaldehyde. Dip 10 seconds at room temperature in chromic acid [7].

Polypropylene. The treatments found satisfactory for polyethylene are suitable. A 1-minute dip in chromic acid at 70°C may also be used.

Polystyrene. Expose to an oxidizing flame to render surface hydrophilic. One pass of the flame usually gives optimum results. Alternatively, 2 to 3 phr styrene oxide may be added to the adhesive formulation.

Polytetrafluoroethylene. Polytetrafluoroethylene and other fluorinated hydrocarbons may be treated with colloidal silica [8] and a sodium-etch process [9,10]. A liquid treatment consists of mixing sodium in a solution of napthalene and an ether such as tetrahydrofuran; the sodium reacts with the napthalene to produce a bluish-black complex, and the fluorinated hydrocarbon is then immersed in this. Suitably treated materials are available from suppliers of the fluorinated hydrocarbons.

Vinyl. Good bonds can generally be obtained to vinyl by a light sanding followed by a solvent wipe. With the more highly plasticized vinyls, a polyurethane primer may be used to seal against plasticizer migration on aging.

Other Plastics. Other plastics generally may be treated with a light sanding followed by a solvent wipe. In some cases, the surfaces may be primed with the liquid monomer of the plastic, or small amount of the monomer may be incorporated into the adhesive formulation.

MISCELLANEOUS SURFACES

Carbon. Degrease, abrade with fine emery cloth, and degrease.

Ceramics. Degrease, abrade with carborundum/water slurry.

Concrete. Cast concrete should normally be thoroughly cleaned with a detergent solution and then etched with dilute hydrochloric acid. If silicone curing agents were used in the concrete formulation, the bond generally will be unsatisfactory even with surface preparation. See Chap. 23 for a more detailed discussion. Concrete blocks need no surface preparation if uncontaminated by grease and oils.

Glass. A thorough solvent cleaning with trichloroethylene has been found to provide satisfactory bonds to plate glass. Special finishes are employed with glass fibers, as discussed in Chap. 22.

Jewels. Degrease.

Leather. Degrease and roughen.

Masonry. Masonry and similar porous surfaces do not require special treatment, but sufficient adhesive should be used to allow for absorption.

Rubber. Immerse for 10 (± 5) minutes at room temperature in concentrated sulfuric acid, or apply a sulfuric acid paste of barytes. This treatment is satisfactory for both natural and most synthetic rubbers. The immersion time should be sufficient to provide a brownish surface film. The rubber, when flexed after treatment, should produce tiny surface cracks. For synthetic rubbers which cannot be etched in a reasonable time with sulfuric acid, concentrated nitric acid may be used.

Wood. Plane or sand the surface for maximum bond.

REFERENCES

1. Black and Blomquist, Adhesive Deterioration, in Metal Bonds at High Temperatures, part I, *Adhesives Age*, May, 1959.
2. Riel, et al., A Simplified Theory Explaining Corrosion Resistance of Adhesive Bonded Aluminum Panels, ACS Symposium, San Francisco, April, 1958.
3. Eickner, Effect of Surface Treatment on Adhesive Bonding Properties of Magnesium, Forest Products Laboratory Report 1865 (1958).
4. Sayer, Treatment of Polyethylene Terephthalate, U.S. 2,785,085 (1957).
5. Strzyzewski and Geilenkirchen, Process for Improving the Adhesiveness of Lacquers and Printing Colors on Polyethylene Products, U.S. 2,873,241 (1959).
6. Ziccarelli, Polyolefin Surface Treating Methods, *Mod. Plastics*, November, 1962.
7. Bodner and Powers, Adhesive Bonding of the Newer Plastics, *Plastics Technol.*, August, 1958.
8. Doban, Method for the Preparation of Cementable Fluorocarbon Polymer Surfaces, U.S. 2,906,658 (1959).
9. Kelley, Method for Coating Polyfluoroethylenes, U.S. 2,898,228 (1959).
10. Purvis and Beck, Method of Activating the Surface of Perfluorocarbon Polymers and Resultant Article, U.S. 2,789,063 (1957).

Chapter 22

EPOXY-RESIN LAMINATES AND MONOFILAMENT REINFORCED COMPOSITES

Epoxy resins may be used as binders in the preparation of laminates of glass cloth, paper, synthetic cloth, metal fibers, wood sheet, glass flake or plate, mica, asbestos, and many others. However, with the exception of paper (which is used to make composites for printed-circuit boards), glass cloth or glass filaments are the chief reinforcing materials used in epoxy-resin-based laminates or composites. The majority of this chapter is devoted to a discussion of glass-reinforced laminates and composites. The final section of the chapter provides a survey of other reinforcing materials.

THEORY OF GLASS-FIBER LAMINATES

Although glass has extremely high tensile and compressive strengths, these properties are not readily apparent in a cylindrical specimen. Surface flaws, occurring randomly, cause cracking and failure of the specimen at values far below theoretical. However, if glass is reduced to fiber form, the same number of random flaws will occur in only a few fibers, the remainder of the material being of theoretical strength, and the overall strength of the fiber bundle will reflect more nearly the theoretical value.

To make practical use of the high strength of these fibers, it is necessary to isolate them one from another in order to prevent abrasion and the formation of new surface flaws. Thus by using synthetic resins it is possible to bond the glass fibers—supplied as cloth, mat, roving, or filaments—into a solid laminate having a very high strength-to-weight ratio (i.e., tensile strength divided by specific gravity). Filament windings, for example, can provide tensile strengths of 200,000 psi (with the theoretical possibility of doubling that value), at a specific gravity a little over 2. Steel piano wire gives a tensile strength of 300,000 psi at a specific gravity of 7.80. Further comparisons are given in Table 22-1.

Table 22-1. Comparison of Strengths of Materials

Material	Tensile strength, psi	Specific gravity	Tensile strength/ specific gravity
Steel, structural	60,000	7.80	7,700
Aluminum	15,000–70,000	2.88	5,200–24,000
Epoxy–glass-cloth laminate	100,000	1.80	55,500
Steel, piano wire	300,000	7.80	38,500
Epoxy-glass filament winding	300,000	2.03	148,000

A study of the nature of the bonding forces in laminates [16] indicates that upon initial loading there is a tendency for the adhesive bond between the glass and resin to be broken. The frictional forces between the resin and the fibers account for the high-strength properties of the laminate. This effect is indicated in Fig. 22-1.

With glass laminates, the strengths of both the glass and the resinous binder are of significance, with the glass contributing the major portion. The resinous binders are selected on the basis of their adhesive strengths, their fatigue resistance, their heat resistance, their chemical and moisture resistance, etc. The epoxy resins have proved superior to polyester resins, particularly in relation to moisture resistance [18]; the phenolics or silicones are generally preferred when maximum heat resistance is required. Fatigue resistance of epoxies is superior to all three other types. Epoxy-resin binders provide perhaps the highest interlaminar shear strengths obtainable in the present technology. In filament winding, the epoxy resins appear to have established a clear superiority to the other thermosetting compounds.

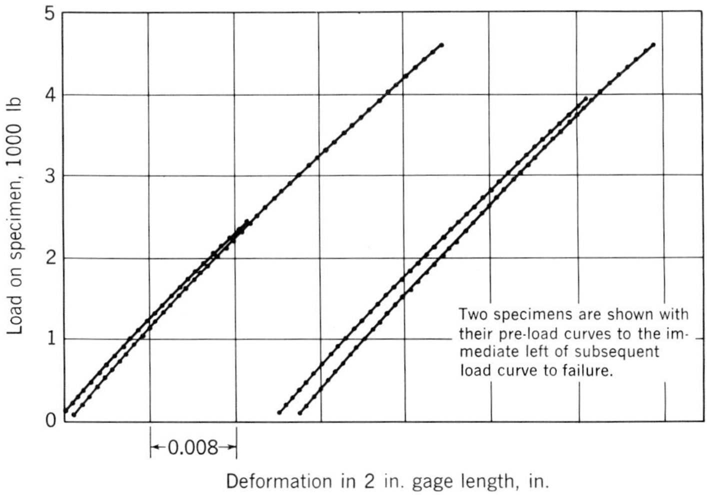

FIG. 22-1. Breaking of resin-glass bond by initial tensile loading [16].

COMMERCIAL APPLICATIONS OF EPOXY–FIBER-GLASS COMPOSITES

The major impetus in the development of epoxy–fiber-glass composites came as a result of the requirements for aircraft and missile applications. Subsequently, the technology was extended into related fields such as printed-circuit boards, and then into the development of corrosive-resistant piping. With the advent of filament winding, the commercial technology has been further extended to the fabrication of all types of spherical and cylindrical tanks and rectangular shapes. In addition, there is a rather large variety of miscellaneous applications.

Use of Epoxy-Glass Laminates in Aircraft

Epoxy–fiber glass composites for aircraft offer simplicity, light weight, low cost, low radar echo, and a minimum of logistic support [112,153,157].

Epoxy-glass laminates have found application in practically all stages of the fabrication of aircraft. They are used for prototype aerodynamic models [67], and in aircraft tooling as discussed in Chap. 18. They are also used as a variety of components, among them being the upper fairing surface of the wings of the Boeing 720 jets [105]; fin tips which house antennas and related equipment on the Boeing 707 jets [33]; fin tips, wing-to-body fairings, tail cone, vertical-stabilizer fairing and leading-wing panels on the Boeing 727—in all, 3,000 lb of epoxy–glass-cloth laminates [121]; double-floor construction on jet liners to provide thermal insulation [45]; aircraft propeller cuff covers [10]; aircraft ducting [61]; radomes [78]; fuel-gage tank probes [17]; lightweight airline tray carriers, and helicopter blades. The listing, of course, is typical rather than definitive.

In general, the epoxy-glass laminates have been used as secondary structures rather than primary ones because of the low modulus of elasticity; however, with the development of the technology, particularly the introduction of high-strength, high-modulus fibers, interest in using epoxy composites has increased markedly, extending to fabrication of components for an all reinforced-plastic aircraft.

Epoxy-glass spinner being used on Grumman "Mohawk" airplanes. Spinners incorporate deicers for the aircraft's propellers. Spinners provide high strength, outstanding vibration resistance, and low production costs. (*Union Carbide Corp.*)

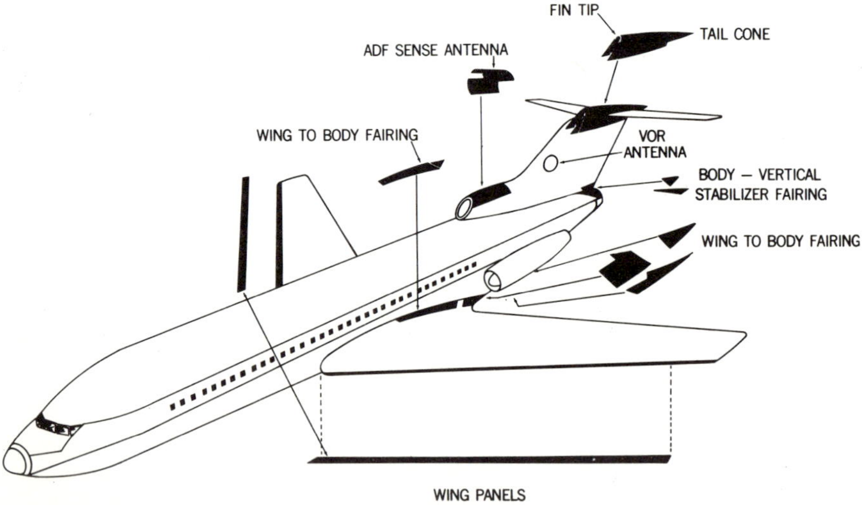

Epoxy–glass-cloth laminates are used in these areas of Boeing 727. Reinforced plastics shown weigh in excess of 1½ tons. (*Union Carbide Corp.*)

Use of Epoxy-Glass Laminates in Missiles

In missile applications, perhaps the most significant area of use is in filament winding, where epoxy-glass structures are employed to reinforce rocket motors and for the construction of rocket cases [24]. The first- and second-stage motor cases of the Polaris are filament-wound, with the second-stage case containing enough roving (268,000 miles) to reach to the moon [102]. Filament-wound rocket cases have been produced for the Nike-Hercules, Bomarc, Little John, Pershing, and Minute Man missiles with a weight saving of 50 per cent over that of comparable steel structures [43]. The thrust chambers for the Atlas missile are reinforced by filament winding at a weight saving of 25 per cent over that of brazed and welded-steel band reinforcement [89]. The Courier satellites are composed of two hemispheres of an epoxy-resin–glass laminate [104], and the Explorers made use of prepreg epoxy-glass honeycomb structures [71]. The epoxy-glass laminates are also used for a variety of components, such as for example supporting brackets on the Redstone missile [62]. The epoxy-glass laminates may also be employed as high-temperature thermal barriers [70]. A comparison of typical space materials, including epoxy-glass laminates, is given in Table 22-2. Filament-wound epoxy-glass structures have been used for air, nitrogen, and helium pressure vessels, although here they require a rubber liner since they are porous under high pressure [37]. Filament-wound tanks have been used to contain the liquid fuel on orbital vehicles [124]. Again, of course, the listing is typical rather than definitive.

Use of Epoxy-Glass Laminates as Corrosion-resistant Piping

A large application for epoxy laminating resins is in corrosion-resistant piping. Epoxy piping has the highest strength and (except for fluorocarbons) the highest operating temperature in the plastic pipe field [48]. The chemical resistance varies with the concentration of chemicals, the specific chemical environment, and the operating temperature. Chemical resistance data for filament-wound pipe are indicated in Table 22-3. These data are presented as indicative only, and should not be considered absolute limits.

As an indication of the present market, it has been estimated that a pipe system (pipe and fittings) capable of handling 150 psi and 120°C can service about 75 per cent of the surface line requirements in oil and gas production; 85 per cent of the chemical process industries' piping needs; and virtually all gas distribution and water mains. With capability of handling 1,000 psi, the reinforced pipe is potentially suitable for an additional 10 to 15 per cent of the surface pipeline requirements in the oil and gas industry [176].

Epoxy piping may be wound from glass cloth or glass tape, may be centrifugally cast, or may be filament-wound. A resin-rich layer or coating is applied to the inner diameter of the pipe to improve the chemical resistance [65].

The use of chemical-resistant epoxy pipe for oil-well tubing and line pipes has been described [91]. A typical case history of a corrosion-resistant epoxy pipe involves its duty in a carbon black slurry. Metal pipe in this application lost approximately 50 per cent of its thickness at the end of 3 months, whereas the epoxy pipe showed no appreciable wear after 10 months' service [73].

Hydrospace Applications

Because of favorable strength-to-weight ratios, epoxy-glass composites are almost uniquely suited to hydrospace applications and offer improved performance, at lower depths, than can be obtained with structural metals. For example, the small submarine Alvin has been constructed for exploring ocean bottoms to the

Table 22-2. Thermal Properties of Some Space Materials [72]

(Measured at about 93°C)

Material	Thermal conductivity, Btu/(°F)(in.)(hr)(sq ft)	Specific heat, Btu/(lb)(°F)	Density, pci	Specific conductivity,* Btu/(°F)(sq ft)(hr)(sq in.)	Heat-sink coefficient,† Btu/(sq in.)(°F)
Aluminum, 245	98	0.22	0.098	0.105	0.022
Beryllium	85	0.52	0.066	0.179	0.034
Copper	211	0.09	0.320	0.015	0.028
Graphite	79	0.17	0.050	0.253	0.085
Magnesium	84	0.25	0.064	0.187	0.016
Steel, 4130	11	0.13	0.290	0.313	0.038
Fiber-glass–epoxy‡	0.89	0.24	0.063	17.9	0.015
Fiber-glass–polyester	0.97	0.28	0.066	15.7	0.018
Fiber-glass–phenolic	0.87	0.25	0.059	19.7	0.015
Fiber-glass–silicone	1.02	0.24	0.059	16.5	0.014
Refrasil modified phenolic	2.3	....	0.060	7.3	
Urethane foam, rigid	0.48	0.41	0.006	347	0.002

* Higher values connote efficient insulators (on a weight basis).
† Higher values connote efficient heat sinks (on a weight basis).
‡ All laminates reinforced with fabric.

Table 22-3. Chemical Resistance of Filament-wound Pipe [56]

Chemical	*Upper temperature limit,* °C
Acetaldehyde	27
Acetic acid, glacial	52
Acetic acid, 3%	38
10%	27
50%	27
Acetone	27
Alcohol, ethyl	66
Alcohol, isopropyl	66
Alcohol, methyl	66
Alcohol, methyl isobutyl	66
Alcohol, secondary butyl	66
Allyl chloride	52
Aluminum chloride	80
Aluminum fluoride	66
Aluminum hydroxide	66
Aluminum nitrate	80
Aluminum sulfate	80
Alums	80
Ammonia, wet	52
Ammonium carbonate	66
Ammonium chloride	80
Ammonium fluoride	52
Ammonium hydroxide, 10%	80
20%	66
28%	52
Ammonium nitrate	80
Ammonium phosphate	66
Ammonium sulfate	80
Amyl acetate	27
Amyl chloride	27
Aniline	27
Antimony trichloride	66
Barium carbonate	80
Barium chloride	80
Barium hydroxide	80
Barium sulfate	80
Barium sulfide	80
Beer	66
Benzene	52
Benzoic acid	80
Black liquor	66
Borax	80
Bromic acid	66
Bromine water	52
Butyl acetate	27
Butyl cellosolve	66
Butyric acid	52
Calcium bisulfite	80
Calcium carbonate	80
Calcium chlorate	80
Calcium chloride	80
Calcium hydroxide	66
Calcium hypochlorite, 5%	52
Calcium nitrate	80
Calcium sulfate	80
Carbon tetrachloride	52
Castor oil	80
Chloroacetic acid, 10%	66
Chlorine water	52
Chlorobenzene	27
Chloroform	52
Chromic acid, 30%	27
10%	52
Citric acid	80
Copper chloride	80
Copper fluoride	80
Copper nitrate	80
Copper sulfate	80
Crude oil, sour	80
Diacetone alcohol	66
Dichlorobenzene	52
Dichloroethylene	27
Ethyl acetate	66
Ethyl cellosolve	52
Ethyl chloride	27
Ethylene glycol	80
Fatty acids	66
Ferric chloride	80
Ferric nitrate	80
Ferric sulfate	80
Ferrous chloride	80
Ferrous sulfate	80
Fluosilicic acid	80
Formic acid	52
Freon	66
Gasoline, sour	80
Glucose	80
Glycerin	80
Glycol, ethylene	80
Glycol, propylene	80
Hexylene glycol alcohol	66
Hydraulic fluid	80
Hydrobromic acid	66
Hydrochloric acid, 25%	80
Hydrofluoric acid	27
Hydrogen peroxide, 30%	NR
10%	66
Hydrogen sulfide, aq.	80
Hypochlorous acid	80
Jet fuel	80
Kerosene	80
Lactic acid	80
Lauric acid	80
Lead acetate	80
Magnesium carbonate	80
Magnesium chloride	80
Magnesium hydroxide	80
Magnesium nitrate	80
Magnesium sulfate	80

Table 22-3. (continued)

Chemical	*Upper temperature limit, °C*	*Chemical*	*Upper temperature limit °C*
Maleic acid, 25%	80	Sodium cyanide	80
Mercury	80	Sodium dichromate, etch. soln.	NR
Methyl ethyl ketone	66	Sodium ferrocyanide	80
Methyl isobutyl carbitol	66	Sodium fluoride	80
Methyl isobutyl ketone	52	Sodium hydroxide, 50%	80
Mineral oil	80	Sodium hypochlorite, 10%	66
		Sodium methoxide, 40%	66
Naphtha	80	Sodium nitrate	80
Naphthalene	52	Sodium nitrite	80
Nickel chloride	80	Sodium peroxide	27
Nickel nitrate	80	Sodium silicate	52
Nitric acid, 10%	52	Sodium sulfate	80
15%	27	Sodium sulfite	80
30%	NR	Sodium thiosulfate	27
		Stannic chloride	80
Oleic acid	80	Stearic acid	66
Oxalic acid, sat	NR	Sulfuric acid, 10%	80
		50%	80
Perchloric acid	80	60%	80
Phenol, sat.	NR	70%	27
Phosphoric acid, 85%	80	95%	NR
20%	80		
Pickling acids	66	Tannic acid	80
Picric acid	27	Tartaric acid	80
Plating solutions	80	Tetraethyllead	52
Potassium bicarbonate	80	Toluol	52
Potassium bromide	80	Trichloroethylene	52
Potassium carbonate	80	Triethylamine	27
Potassium chloride	80	Trisodium phosphate	66
Potassium dichromate	80	Turpentine	27
Potassium hydroxide	66		
Potassium nitrate	80	Urea	66
Potassium permanganate, 10%	66		
25%	52	Vinegar	52
Potassium sulfate	66	Vinyl acetate	57
Silicic acid	80	Water, distilled	80
Silver nitrate	66	Water, fresh	80
Soaps	80	Water, salt	80
Sodium acetate	80	Water, sea	80
Sodium bicarbonate	80		
Sodium bisulfate	80	Xylene	52
Sodium bromide	80		
Sodium carbonate	80	Zinc chloride	80
Sodium chlorate	80	Zinc sulfate	80
Sodium chloride	80		
Sodium chlorite, 25%	52		

NR. Not recommended.

depth of 6,000 ft. The basic construction consists of a 1¾-in. steel pressure sphere coated with a syntactic epoxy foam and surfaced with a ⅛-in.-thick epoxy-glass laminate [146].

Other Applications for Epoxy-Glass Laminates

Epoxy-resin–glass laminates have achieved considerable industrial acceptance in the electronics industry as printed-circuit boards [32,34]. Flat-sheet glass laminates for this application are produced to NEMA grades G-10 and G-11.

Epoxy laminates have also been widely accepted in outdoor sports equipment and in related applications: for manufacturing golf club shafts, arrow shafts,

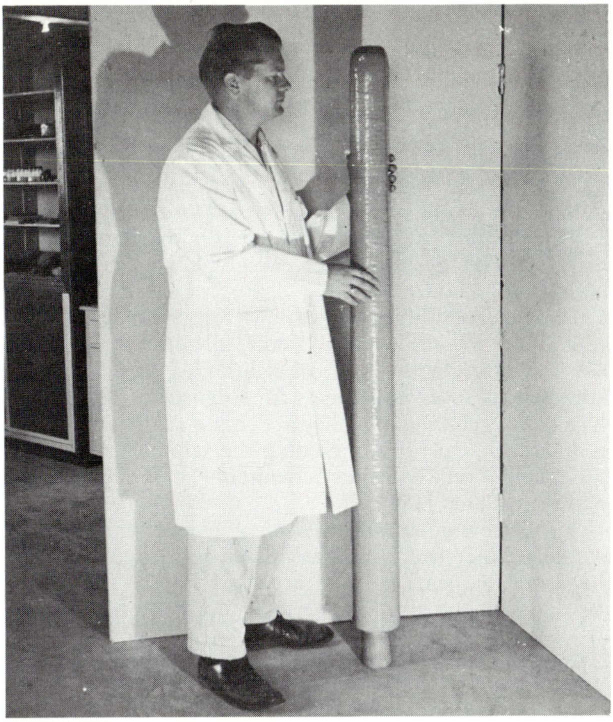

Deep-well oil-well probe insulated in laminate of epoxy resin and glass cloth. (*The Epoxylite Corporation.*)

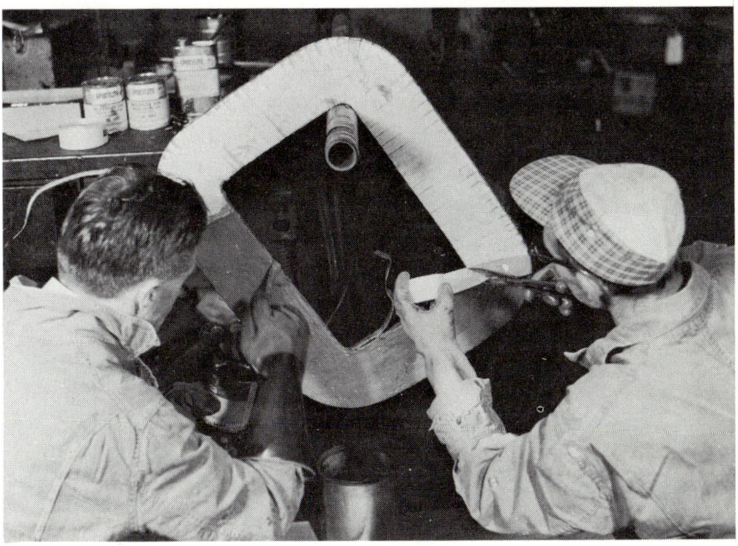

Wet lay-up technique is demonstrated in insulation of 400-kva transformer primary coil. Glass tape and brushable epoxy compound form tough moisture-resistant laminate. (*The Epoxylite Corporation.*)

fishing rods, pole-vaulting poles, etc. They are attached to wooden cores in the fabrication of hunting bows [95]. They have been employed in the fabrication of swimming pools and for the construction of lightweight railing for racetracks. They have been used as polo stick shafts [14]. Epoxy–fiber-glass sports boats have been made [41], although it is possible that the additional cost over polyester would not normally be warranted. Epoxy and glass cloth are the preferred materials for patching polyester-glass boats [40,75] and plastic automobile bodies. Filament-wound rifle barrels are being produced at a lower cost than metal barrels; these have the further advantages of reduced weight, cooler operation, and increased burst strengths [69].

This by no means exhausts the applications for epoxy-glass laminates. They have been used as bogy wheels for army tanks, as lightweight ladders, as railway tank cars, as hot-line poles by the electrical utilities, as ultrahigh-voltage insulators for transmission lines, as roofs and panels for English lorry cabs, as bathtubs for electrotherapy, as battery cases, etc. They are employed in a variety of hand-lay-up repairs to iron and steel structures, including the relining of pump casings [107]; they have gained recognition as an artist's medium [51]; and they serve as inhibitor coatings on solid propellants [137].

An application of growing commercial significance is the use of epoxy-resin electrical tapes made either from preimpregnated fabrics, or for greater unidirectional strengths, from filaments. These are used for the insulation of electrical coils, and particularly for the insulation of form-coil motors [58].

TECHNIQUES OF MANUFACTURE

Essentially, a laminate is prepared by interlayering glass and resin and then curing the resultant structure. The glass may be supplied in the form of unidirectional

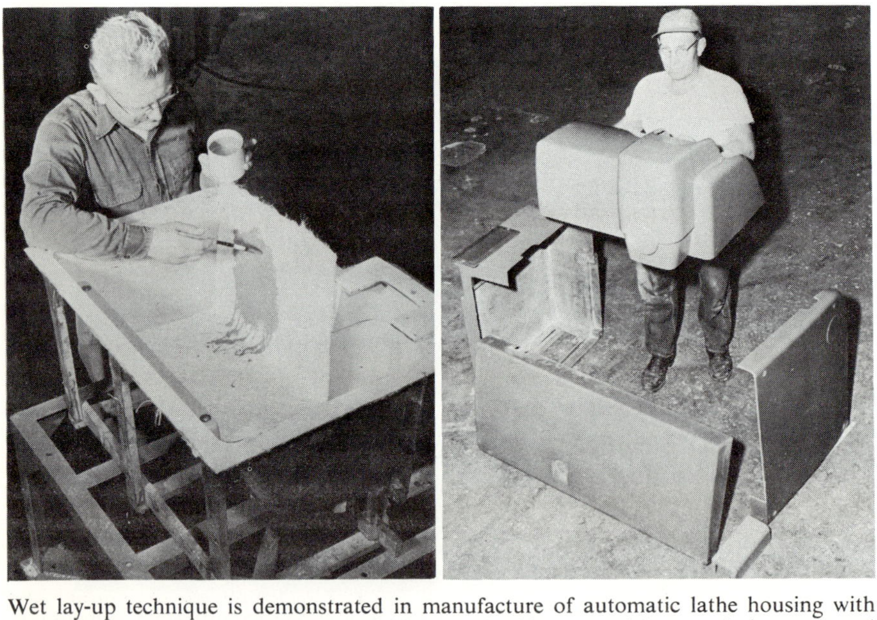

Wet lay-up technique is demonstrated in manufacture of automatic lathe housing with complex curved shape. Concave mold is coated with repeated layers of glass mat and epoxy resin. Finished product, removed from mold, needs only touch-up and painting. (*Union Carbide Corp.*)

strands for filament winding, in random-oriented form as mats, or in woven form as cloth. Two techniques may be used to prepare the laminate: (1) wet lay-ups and (2) dry lay-ups.

Wet lay-ups involve the impregnation of the cloth or mat, or the coating of the strands, with a non-solvent-containing resin and the essentially immediate application of the glass to the laminating operation.

During wet lay-up operations, it is frequently necessary to interrupt production—either to permit exotherm to dissipate or to permit shift changes, etc. If the laminate gels in the interim, it is customary to sand the surface lightly to assure adhesion of the new layers to the old. It has been reported [76] that the necessity for sanding can be eliminated by dusting nylon flock over the wet surface when production is interrupted. After cure, excess flock is blown off and laminating resumed.

Dry lay-ups involve the impregnation of the cloth or filament bundle with either a 100 per cent solids or a solvent-containing resin system in advance of the production operation.

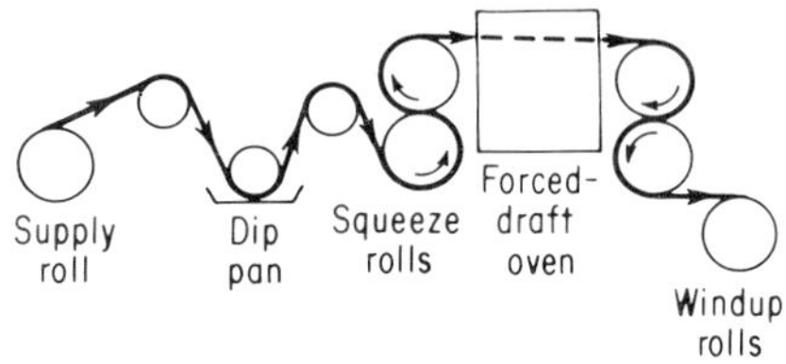

FIG. 22-2. Treating equipment for coating glass cloth with solution epoxy resins [109].

If a 100 per cent solids system is employed, the liquid resin is mixed with the curing agent and applied to the glass, then B-staged to produce a solid resin at room temperature which is capable of flow on the application of heat.

If a solvent system is used, the solvents are removed after application to provide a controlled-tack surface capable of cure by the subsequent application of heat. The mixture is applied to the glass by squeeze rolls or coating bars in such a manner as to control the resin buildup, and all the solvents are driven off to leave a dry but only partially polymerized coating. A typical application system is shown in Fig. 22-2.

If the higher-molecular-weight species of resin are employed, the resultant prepreg cloths will be stiff and possess poor drape characteristics; such cloths are most advantageously used in the preparation of flat-press laminates. To improve drape characteristics, mixtures of high- and low-molecular-weight resins may be employed in connection with solvent-system dry lay-up. Alternatively, liquid resins may be advanced to the solid state by the use of amounts of catalyst too low to effect cure.

When dry lay-ups are employed, pressure is normally required to assure adequate flow and fill.

A so-called packaged prepreg process for filaments has been developed which permits application directly to the spools of filament, thereby eliminating the need for unwinding and rewinding the fragile glass [143].

Formation of the Laminate

The laminate may be formed under contact pressure in which only the weight of the plies is applied to the resin, under low pressure, and under high pressure. In filament windings, the glass strands may be applied under selected tension to provide for prestressing.

The laminate is usually formed over a machined or cast mold coated with a release agent to facilitate removal. A lost-wax process has been developed for making radomes [74], and similar techniques may be employed for complicated structures. A discussion of molds is presented in Chap. 18.

Contact-pressure Laminates. Contact laminates usually possess high resin contents and are chiefly used where the expense of equipment and considerations of production do not justify the increased performance obtained with pressure laminates.

Low-pressure Laminates. Low-pressure laminates may be formed by the "vacuum-bag impregnation" technique. In this process, the laminate is placed over the mold and a thin film of plastic draped over the unit and anchored airtight along its edges, or, alternatively, the mold and laminate are placed into a plastic

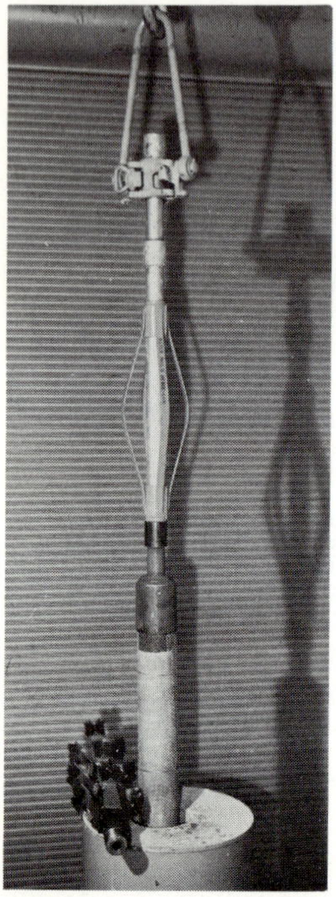

Device for repair of leaking well casings uses epoxy-resin-saturated glass cloth applied over inflatable mandrel. Device is lowered into the casing, mandrel is inflated, and epoxy-glass cloth is then cured in contact with casing wall. After cure, the patch is tough, hard, and leakproof, yet reduces diameter of casings only about 0.6 in. Tubing centralizers keep device centered in hole. Wire scratcher, laying atop simulated well at left of mandrel, is used to prepare surface in advance of application of patch. (*Pan American Petroleum Corp.*)

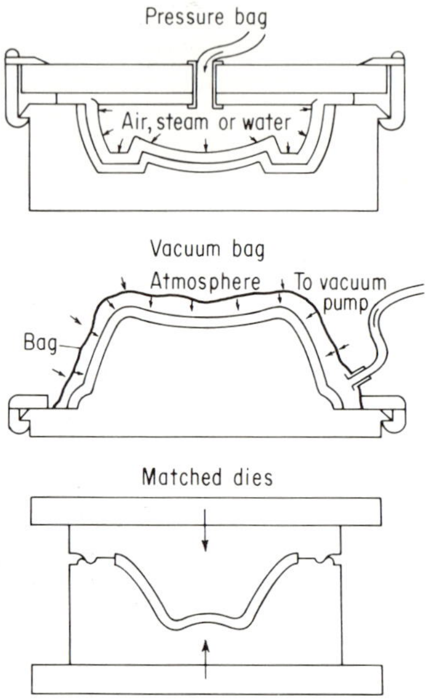

Fig. 22-3. Production equipment for producing low- and high-pressure laminates [6].

bag. Vacuum is then pulled on the laminate, whereupon the plastic sheet is pressed inward to force the laminate to assume the contour of the mold. It may be helpful to apply exterior pressure by means of a spatula. Customary procedure is to cure the laminate under vacuum. In addition to vacuum bag, a pressure-bag technique may be used, both being illustrated in Fig. 22-3.

High-pressure Laminates. High-pressure laminates are formed by the use of matched molds (Fig. 22-3) or platens pressed together mechanically and heated, with the pressure being retained until the laminate has hardened to the desired shape.

Small filament-winding machine in operation. (*Ciba Products Co.*)

Centrifugal Casting. Centrifugal casting techniques may also be employed for obtaining pressure. Such techniques are most frequently used in the manufacture of tubing. Typical of this procedure is the use of glass mat preformed to the general shape and installed in a mandrel. Premeasured resin is then injected and the tubing spun.

Filament Winding. Filament winding is one of the newest and fastest-growing techniques for producing composites and is the subject of highly specialized textbooks [168]. Basically, filament winding consists of impregnating continuous strands of glass filament with a matrix material. The impregnated strands are then wound under tension in a precise and controlled pattern onto a mandrel. The wound structure is subsequently cured. In 1965, about 4 million pounds of epoxy resins were used in filament-winding applications—the most rapidly growing field of epoxy-resin usage.

Since filament-wound structures present their highest strengths when tested in tension, the pattern is designed to assure maximum strength in the direction of maximum in-service stress. Because of the unidirectional strength and the high glass content (80 to 85 per cent), the structures are extremely strong in tension, but are rather brittle and impact-sensitive. They are, additionally, weakest in interlaminar shear, and the winding pattern must be established to accommodate this.

High-strength filament-wound structures are very sensitive to flaw defects built in during manufacturing. For example, on large rocket-motor cases two knots spaced less than 1.2 in. apart will reduce ultimate strength substantially [122].

Because of the high glass content, the design should minimize machine operations after cure, and if machining is required, the structures should be supported as though they were 100 per cent glass.

Filament winding is carried out automatically on special machines with mechanical or electronic control, and winding is conducted under tension. For this reason, the design should avoid, as nearly as possible, flat areas and areas of reverse curvature which will reduce the tension loading during winding [79]. Calculations concerning optimum prestressing loads are presented in [110].

Filament windings may be B-staged, slit from the mandrel, and supplied in flat sheet and tape form [87]. A nonwoven unidirectional fabric, composed of parallel, continuous, longitudinal strands of glass fiber cross-bonded every 3 in. with resin-ladened warp, permits the application of filament-winding design techniques to the production of parts not easily made by the standard filament-winding process [126].

A bibliography of literature on filament winding, covering the years 1962 to 1964, is given in [170].

FORMULATION

The major factors involved in the formulation of epoxy–fiber-glass composites are:

1. Type of glass and its surface treatment
2. Resin, modifier, and curing-agent system
3. Laminating pressure or glass tension, resin content, and curing cycle

Types of Glass and Surface Treatment Employed in Epoxy-Glass Composites

Because the fiber glass employed in the composite, to a large extent, determines the final properties of the laminated structure, the selection of the glass and of the surface finish is a critical factor in the formulation.

Types of Glass. The basic components of fiber glass are called *filaments* and are slender fibers of glass ranging from 0.00023 to 0.00038 in. in diameter [94]. Larger-diameter fibers, however, have been suggested for improved compressive strengths in filament-wound structures. With an epoxy matrix, when testing fibers from 0.001 to 0.010 in. in diameter, best compressive strengths were obtained at fiber diameters of 0.005 in. [183].

Virgin fiber, after formation, does not appear to suffer any appreciable loss of strength during the processing currently used to produce roving; but roving does have about 20 per cent lower strength than predicted from the virgin fibers. This probably indicates that contemporary processing techniques are satisfactory, and that the failure in the fiber bundle is sequential, the weakest fiber first, etc., thus transferring progressively higher loads to the remaining unbroken fibers [185].

The filaments are grouped into strands, each strand usually containing either 102 or 204 filaments [13]. The strands may be used in the filament-winding process. They may be twisted into yarns and the yarns woven into cloths. They may be collected into rovings, usually 20 to 60 strands in a bundle, and supplied in mats of either continuous or chopped ($\frac{1}{4}$ to 2-in.) rovings held together by a binder. The woven cloths are the most expensive, the continuous strands the least expensive.

For epoxy composites, E-glass or S-glass is employed. E-glass is superior to the standard alkali glass used for sheet-glass construction, and an inverse relationship between laminate strength and alkalinity (as Na_2O) has been established [115]. For example, mat laminates employing both types, treated with DGEBA cured with MPDA, give the following values: E-glass, dry flexural strength of 42,000 psi and strength retention after 4 hours boil of 95.2 per cent; alkali glass, dry flexural strength of 30,400 psi and strength retention of 65.1 per cent. The same superiority (36,000 psi and 91 per cent vs. 26,200 psi and 76 per cent) were shown when the curing agent was HHPA [83]. S-glass, a later development, features strengths in the virgin single fiber of 700,000 psi, compared to 500,000 psi for E-glass.

Prototype solar concentrator is backed with aluminum honeycomb between layers of epoxy-impregnated glass fiber for strength coupled with light weight. (*The Boeing Co.*)

In addition to E-glass and S-glass, a high-modulus (or beryllium oxide) glass has been developed [178] which provides superior performance, as illustrated in Tables 22-4 and 22-5. Some 50 per cent increase in modulus over that of E-glass (17×10^6 psi vs. $10\text{-}12 \times 10^6$ psi) coupled with an overall increase in strength properties has been obtained with the high-modulus glass [128]. The high-modulus glass is desirable in critical filament-winding applications, since it provides decreased elongation, thus offering greater freedom in formulation requirements, as discussed subsequently. It does pose special handling problems in view of the toxicity of the beryllium.

High-melting-point 96 per cent silica glass (Refrasil) has been developed for use in high-temperature ablative applications. Low-dielectric glass filaments (having a dielectric constant of 3.5 compared to 5 for E-glass) have been produced to give broad-band characteristics with high microwave transmission [134]. Hollow filaments, containing approximately 40 per cent voids, have been developed to permit fabrication of lightweight structures [169].

Glass Cloths. The yarns are woven into a variety of fabrics, with the strands which run parallel to the length of the fabric designated *warp* and the cross threads

Table 22-4. Effect of E-glass (Borosilicate) on Properties of DGEBA Cured with MPDA, 181-style Glass [101]

Condition of test	Direction with warp, deg	Strength, psi	% of room-temperature strength	Modulus, psi × 10^{-6}	% of room-temperature modulus
Flexural tests:					
Room temperature	0	70,700	.....	3.69	
	45	55,000	.....	2.05	
2-hr boil	0	61,700	87.3	3.68	99.7
30 days in water	0	67,800	95.9	3.69	100.0
½ hr at 150°C	0	40,700	57.6	2.53	68.6
Compressive tests:					
Room temperature	0	53,700	.....	4.19	
2-hr boil	0	46,100	104.1	4.03	93.8
30 days in water	0	47,500	108.2	4.21	96.3
½ hr at 150°C	0	22,100	50.3	....	
Tensile tests:					
Room temperature	0	63,800	.....	3.49	
2-hr boil	0	58,200	91.2	3.45	98.9
30 days in water	0	57,900	90.8	3.30	94.6
½ hr at 150°C	0	40,800	64.0		

Specific gravity 1.90.
Per cent resin content 22.82.

Table 22-5. Effect of High-modulus Beryllium Oxide Glass on Properties of DGEBA Cured with MPDA, 181-style Glass [101]

Condition of test	Direction with warp, deg	Strength, psi	% of room-temperature strength	Modulus, psi × 10^{-6}	% of room-temperature modulus
Flexural tests:					
Room temperature	0	89,800	....	5.00	
	45	62,700	....	2.62	
2-hr boil	0	79,300	88.3	4.79	95.8
	45	59,600	95.1	2.39	91.2
30 days in water	0	75,900	84.5	4.66	93.2
	45	57,200	91.2	2.39	91.2
½ hr at 150°C	0	44,200	49.2	3.69	73.8
Compressive tests:					
Room temperature	0	61,000	....	5.71	
	45	31,500	....	3.05	
2-hr boil	0	46,700	76.6	5.55	97.2
	45	30,200	95.9	3.03	99.3
30 days in water	0	44,200	72.5	5.52	96.7
	45	29,700	94.3	2.77	90.8
½ hr at 150°C	0	32,300	53.0		
Tensile tests:					
Room temperature	0	64,500	....	4.76	
	45	31,300	....	2.90	
2-hr boil	0	62,100	96.3	4.58	96.2
	45	29,300	93.6	2.18	75.2
30 days in water	0	62,600	97.1	4.81	101.1
	45	28,400	90.7	2.01	69.3
½ hr at 150°C	0	51,400	79.7		

Specific gravity 2.02.
Per cent resin content 26.12.

Table 22-6. Effect of Glass Style on High-modulus and E-glass Laminates, DGEBA Cured with MPDA [101]

Properties	128-style fabric		143-style fabric		Parallel strand	
	High modulus	E	High modulus	E	High modulus	E
Flexural:						
Dry strength, psi × 10^{-3}	72.1	57.7	155.2	127.5		
Dry modulus, psi × 10^{-6}	3.83	2.81	7.96	5.83		
Wet strength, psi × 10^{-3}	60.7	50.5	130.7	108.7	252.5	225.8
Wet modulus, psi × 10^{-6}	3.51	2.60	7.85	5.48	10.26	7.62
Compressive:						
Dry strength, psi × 10^{-3}	42.2	39.3	78.3	66.4		
Dry modulus, psi × 10^{-6}	4.78	3.51	8.63	6.02		
Wet strength, psi × 10^{-3}	33.5	35.6	50.5	49.5		
Wet modulus, psi × 10^{-6}	4.58	3.21	8.78	6.25		

designated *filling*. The specific type of weave dictates to some degree the properties of the fabric, and the specific laminating application will in many cases determine the best suitable weave. Weave, crimp, twist, and filament diameter are also important factors in glass fabric design [23]. Strands which are crimped and bundled will, in laminates under stress, transfer the load to the resin, thus resulting in failure at low values. The effects of two fabric styles on strength are shown in Table 22-6 as a function of both E-glass and high-modulus glass.

It is necessary to consider certain inherent properties of the cloth selected for use. Should the warp and filling have unequal breaking strengths, the flexural and tensile strengths of the laminate will depend on how the fabric is cut. Additionally, in building plies, consideration should be given to the proper positioning of each layer. It is generally considered that for optimum strength the warps should be parallel and side by side rather than stacked, although an alternative view holds that fabric nesting (difficult to obtain in production in any event) has little or no significant effect on strength properties [49].

Surface Treatments. During the processing of the glass, a sizing is applied which prevents the fragile filaments from breaking. Conventionally, a textile size of dextrin and an emulsified oil is employed. After the processing operation, the sizing may be removed by chemical treatment but normally is removed by a high-temperature, heat-cleaning process. Heat cleaning usually leaves about 0.2 per cent residue, but residues of less than 0.1 per cent can be obtained. The effect of the presence of sizing on the strength of a laminate using DGEBA and MPDA is indicated in Fig. 22-4. The "caramelized" glass, listed in the figure, is one in which the sizing has been only incompletely removed.

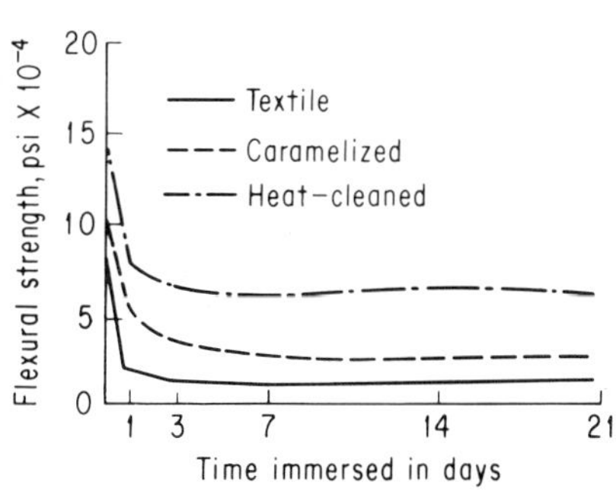

FIG. 22-4. Effect of sizing removal on laminates of MPDA-cured DGEBA [82].

Improved sizings result in the retention of increasingly high percentages of the virgin strength of the glass and thus result in higher-strength laminates. One process involves the application of an epoxy-resin sizing at the point where the glass filaments emerge from the bushing during the drawing operation. The commercial

HTS glass, for example, is E-glass sized with 1 to 2 per cent of DGEBA/A-1100 silicone (1/1). Another process involves the use of silane resins, incorporated in a carrier, and used as a sizing. These have proved particularly effective with rovings [82]. Specially sized glass may be used without further treatment. A review of the literature on glass-cloth sizing, sizing removal, and treating agents, is given in [154]. It should be noted, however, that storage history of the sized glass may be critical in terms of subsequent performance. HTS glass, for example, undergoes polymerization and hydrolysis upon normal storage, with only the latter change being partly reversible and both changes detrimental to laminate properties. Storage under dry refrigeration is recommended to ensure consistent quality [192].

The heat-cleaned glass also may be used directly with the epoxy laminating resin. However, in order to eliminate the film of water which preferentially forms on the glass surface upon exposure to high humidity or moisture, it is customary to apply some form of organic finish.

The first organic finishes investigated were designed especially for use with the polyester resins. These were chromic complexes or silane finishes which condense with the available hydroxyls on the surface of the glass. A discussion of the chemistry of the chromium complexes as coupling agents is given in [111]. A popular glass finish for polyesters, Volan A, is composed of a chromium complex which reacts through the ionization of the chlorine groups in water and ammonia to couple with the surface of the glass [4]. Or again, silane finishes, such as vinyltrichlorosilane (Garan)—applied from a water system containing a small amount of organic acid—react through the silanol group to couple to the glass. The polyester finishes generally contain both unsaturation and hydroxyl groups—the unsaturation permitting reaction with polyester resins, and the hydroxyls, at least to some extent, permitting reaction with the epoxies. It has been indicated [187] that silane finishes must react chemically with thermosetting resins to be effective glass treatments.

Silane finishes have been developed especially for the epoxies, and these contain difunctional reactive sites [84] such as difunctional amines in the case of A-1100 finishes, the hydrolysis product of γ-aminopropyltriethoxysilane [133], and epoxy-containing silanes such as 4-epoxycyclohexylethyltrimethoxysilane, glycidoxypropyltrimethoxysilane, and the reaction products of allyl glycidyl ether or glycidyl methacrylate with methyl-, vinyl-, or allyltrichlorosilane [1]. Also suitable are silanes containing allyl ester and phenolic functional groups, the latter proving particularly effective at least with aromatic-amine-cured aromatic glycidyl ether systems [129].

Although the silane finishes are generally most popular with epoxy resins, a number of other systems have been suggested. Titanium tetrachloride has been used for treating the fibers in advance of application of a thin coating of DGEBA from benzene [162]; and epoxidized vegetable oils, and adducts of these with primary aliphatic amines, have been reported as satisfactory treating agents which serve also as lubricants [171].

Silane glass finishes are normally applied from aqueous solution, but it has been suggested that a superior method is to apply thinner layers which give improved properties in laminates [138].

Although the epoxy resins are not so sensitive to the type of finish as are, for example, the phenolic resins, and good results may be obtained with heat-cleaned glass, there is improvement in physical and electrical properties even when using finishes designed for the polyesters [98]. The effect of various finishes on the strength of epoxy laminates is indicated in Table 22-7. From this it can be seen that the type of epoxy formulation, as well as the finish, influences performance. It has been shown, for example, that when DGEBA is cured with a Lewis acid, such as BF_3MEA, the finish which is optimum may not prove optimum for the resin cured with MPDA, and neither of these finishes may prove optimum for a system cured with DEAPA [27]. With Volan A treated fabric, amine-cured laminates have been shown to provide poor resistance to elevated-temperature storage at high humidities, whereas silane finishes with anhydride curing agents have led to improved performance [64].

Particularly in regard to moisture resistance and to resistance to hot water, improvements have been obtained by the use of surface finishes with the epoxy resins. The reason for this is not entirely obvious. Studies of adhesive failure on glass laminate models has indicated that shrinkage stresses from thermal contraction leads to the formation of passageways by which moisture may enter, stresses of several thousand pounds per square inch being developed during cure, with the stress level being drastically increased by postcure [99]; however, it has been noted that no correlation, in the absence of bond failure, is apparent between cure stresses and moisture resistance, such that cure temperatures on the order of 177°C are practical, when not leading to bond failure, although they do lead to higher stress levels [125] and lower values in adhesion [116]. Tensile loading produces cracks leading to lower moisture resistance whereas compression loading does not, thus leading to the conclusion that lack of ductility in the resin results in adhesive failures. It has been suggested, therefore, that surface finishes impart the required local ductility to the resin at the glass-resin interface [39], since adhesive strength, per se, in the absence of adsorbed moisture, is not particularly dependent on the glass finish [46]. Alternatively, it has also been suggested [50] that specific finishes facilitate better wetting action during the fabrication operation, and conversely suggested that the coupling agent actually provides poorer wetting, thus minimizing localized stresses on the surface of the glass fiber, such as would occur upon failure of the glass-resin bond [127].

Table 22-7. Effect of Glass Finish on Flexural Strength of Laminates Under 25,000-psi Load in 66°C Water [106]

Resin	Type of glass finish	Time to failure, hours
Epoxy I	Chrome	15
	Silane	16
	Improved chrome	15
Epoxy II.....	Chrome	50
	Silane	198
	Improved chrome	461

With regard to wetting, the critical surface energy of uncontaminated glass is sufficiently high that it can be wet by virtually all pure liquids. When improper wetting occurs, it can therefore be attributed to a surface layer of contamination on the glass or to the coupling agent. Thorough wetting should provide adhesive strengths to the glass in excess of the tensile strength of the epoxy matrix, even in the absence of chemical bonding between the two materials. Furthermore, thorough wetting would serve to eliminate voids, bubbles, and other defects which normally occur and reduce bond strength far below theoretical [190].

In studies of both treated and untreated glass, it was found, in fact, that the wettability of silane-treated glass was about as great as that experienced with polyethylene [147]. Critical surface energy for such systems has been reported as being 16 ergs per sq cm for the silane-treated glass as against the 20 ergs per sq cm for glass with an adsorbed water layer [188]. This was further confirmed by measuring contact angles of various epoxy matrix materials to siloxane finishes.

Contact angles were from 60 to 80° on the treated glasses, whereas they were about 28° on the untreated glass [175]. From this it was concluded that to the extent the silane finishes are converted to the siloxanes, an insoluble, nonwettable barrier is presented to the matrix.

The surface tension of MPDA mixed with DGEBA is about 41.9 dynes per cm, and with bis(2,3-epoxycyclopentyl) ether about 41.4—which is the range for most epoxy laminating resins. It would thus appear that even under optimum conditions, when employing conventional coupling agents, inadequate wetting is obtained.

Further, the theory that surface finishes improve performance by chemical bonding is not entirely adequate. The theory that the finish provides a deformable surface layer which absorbs residual stresses, likewise, is not entirely satisfactory, since the layer is too thin to deform the required amount. However, infrared studies have indicated that a deformable layer may be created in the matrix material and extend outward from the surface finish a distance of several hundred angstroms. This deformable layer is created by the surface finish through preferential adsorption, by which mechanism one or more of the matrix constituents, while in the liquid phase, are deactivated or destroyed or adsorbed by the finish. Thus, the properties of the final cured matrix immediately adjacent to the glass finish are quite different from the properties of the bulk matrix [156], and this difference may account for the success obtained through the use of these coupling agents.

The conclusion that a deformable layer in the matrix itself is developed through the action of coupling agents is probably further corroborated by experiments on glass rods. These, spaced more or less equally apart, were employed to study stress distribution with both flexible and rigid matrices. The stress distribution was improved with flexible resins, upon the failure of some of the individual filaments. This, however, does not correlate with experience in actual laminates, where flexible matrices appear to offer no particular improvement over rigid materials. It may be reasoned [141] that in actual laminates the fibers are not equally spaced and in some cases are touching, so that high stress concentrations can be developed irrespective of the matrix; but it also may be that the development of the deformable layer in the matrix, essentially, (for stress-distribution purposes) converts the rigid matrix into a flexible member.

Resin, Curing-agent, and Modifier System

The resin, curing-agent, and modifier system provides a major variable in composite formulations.

The resin is usually selected in terms of viscosity requirements, with the low-viscosity resins being preferred for wet lay-up applications, and the higher-molecular-weight resins, generally, being preferred for dry lay-up work. The majority of the epoxy resins employed in laminates are of the DGEBA type.

When fabricating laminates from the epoxies, the stoichiometric amount of curing agent determined as optimum for casting formulations often does not provide optimum properties in the laminate, at least insofar as flexural and tensile strengths are concerned [50]. A combination of factors may account for this: (1) the higher curing and postcuring temperatures normally employed may shift the curing reaction to favor other than stoichiometric amounts, and (2) the reinforcing effect of the glass cloth may dictate, for optimum performance, a slightly more extensible cured system. However, the specific property of the laminate being tested may also be of consequence. It has been reported for instance, that stoichiometric concentrations are preferred for lowest creep properties [30].

Of all the epoxy laminating formulations developed to date, the one which has unquestionably been most widely tested is DGEBA cured with MPDA. Typical properties of such a laminate are presented in Table 22-8. Such laminates are resistant to the temperature and humidity cycling of MIL-Std-354 [7] and meet the requirements of MIL-R-9300A, class I, type I, [4]. Flexural strength retention of the laminates after 10 million cycles was 40 per cent of original, compared to 22 to 28 per cent for polyester, silicone, and phenolic laminates [21]. Flexural strength values after the fatigue cycle are about 16,000 psi [26]. Extensive work

Table 22-8. Typical Physical Properties of 181-style Glass Cloth with DGEBA and MPDA, Resin Content 33% [4-7]

Property	*Value*
Flexural strength, psi:	
At 23°C	79,000–89,000
At 127°C	52,000–60,000
At 23°C after 2-hr water boil	77,000–82,000
Modulus, psi × 10^{-6}:	
At 23°C	3.6–3.9
At 127°C	3.0–3.5
At 23°C after 2-hr water boil	3.2–3.6
Tensile strength, psi	50,000–58,000
Modulus, psi × 10^{-6}	3.3–3.6
Edgewise compressive strength, psi	49,000–51,000
Modulus, psi × 10^{-6}	3.5–3.6
Hardness, Rockwell M	115–117
Impact strength, Izod	12–15
Water absorption after 24 hr, % weight gain	0.05–0.07

has been done on cure cycles, with a postcure at temperatures of 200°C being recommended for maximum properties at 150°C [5]. The system may be used in wet lay-up operations or may be B-staged for dry lay-up work and may be cured from contact to high pressure.

Another fairly common formulation employed in dry lay-up applications consists of DGEBA (mol. wt. 950) and about 3 phr dicyandiamide. The resin is used as a 50 per cent solution in acetone. The dicyandiamide is dissolved at about 10 per cent solids in methyl Cellosolve. The mixture is applied at a dry contents of about 30 to 35 per cent to give a resin content in the cured laminate from 25 to 30 per cent when cured from 30 minutes to 2 hours at 1 to 30 psi and 160°C [1]. In printed-circuit-board work (i.e., NEMA grade G-10), the formulation is processed to give about a 40 per cent resin content when using pressures of 1,000 psi and temperatures of 160 to 170°C for 30 to 120 minutes.

DADPS is used with DGEBA to give the NEMA grade G-11 laminate.

Typical properties of some commonly used curing agents with DGEBA resins are indicated in Table 22-9 and Figs. 22-5 and 22-6. Porous laminates have been developed through the incorporation of solvents, as indicated in Fig. 22-7. DGEBA has been bodied with a small amount of BF_3MEA to obtain a dry lay-up system which is deposited from solution. Results obtained with several curing agents in this system are indicated in Table 22-10. The chlorinated DGEBA has been cured with a phenolic curing agent to give a flame-resistant system with properties as indicated in Table 22-11.

Epoxidized novolac resins [55] and the epoxidized olefin resins [52] are used to upgrade the temperature resistance of laminates. For example, dicyclopentadiene dioxide, cured with maleic anhydride, has been used to meet the high-temperature requirements of MIL-R-9300A [86], as has epoxidized polybutadiene cured with unsaturated anhydrides and peroxide catalysts [119]. Fatigue properties of an epoxidized novolac laminate, vs. temperature, are shown in Figs. 22-8 and 22-9.

With filament winding, a variety of curing agents and resins has been investigated. The results of one such investigation are indicated in Table 22-12.

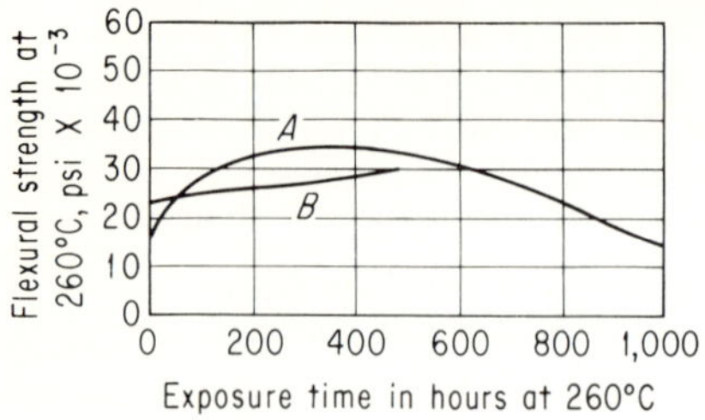

FIG. 22-5. Flexural strength vs. aging for PMDA-cured DGEBA laminates [66]. *A*, 56 phr PMDA, 8 phr diethylene glycol, 0.2 phr DMP-10, low-temperature cure; *B*, 31.2 phr PMDA, high-temperature cure.

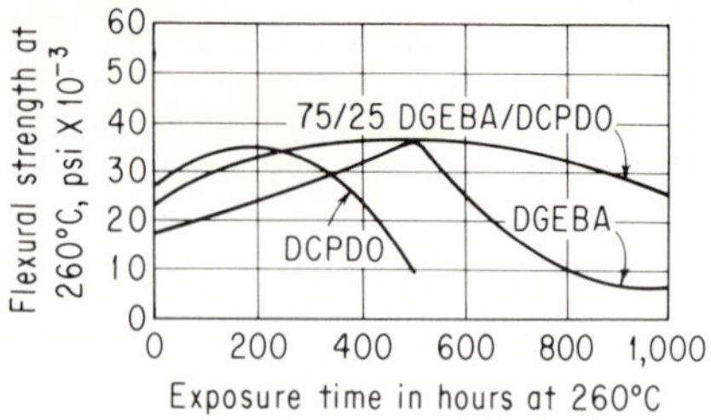

FIG. 22-6. Effect of resin-type on flexural strength vs. aging for PMDA-cured laminates [66]. DCPDO = dicyclopentadiene dioxide.

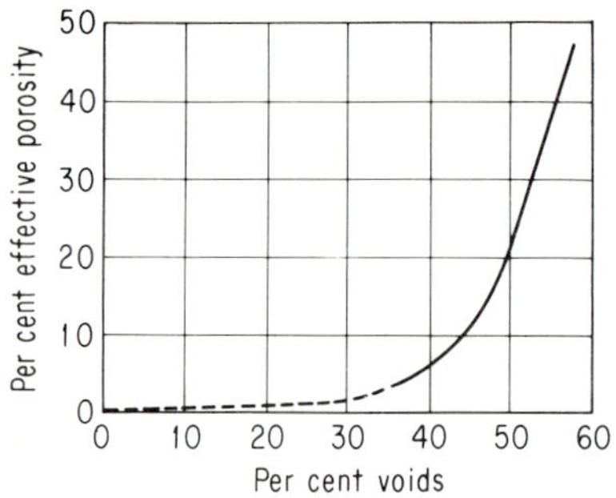

FIG. 22-7. Effective porosity vs. per cent voids of laminates of DGEBA cured with amine-resin adduct containing trichloroethylene [63].

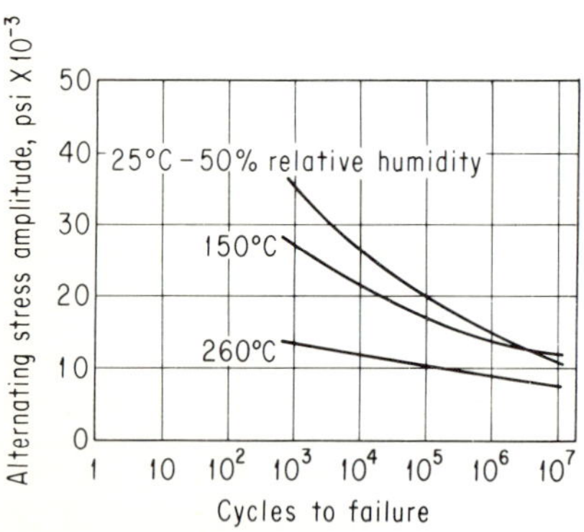

FIG. 22-8. Fatigue resistance vs. temperature for unnotched laminate of cured epoxidized novolac resin [12].

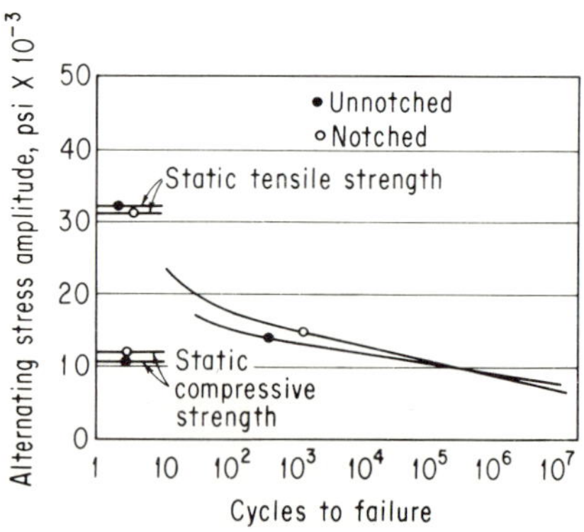

FIG. 22-9. Effect of notch on fatigue resistance at 260°C for laminate of cured epoxidized novolac resin [12].

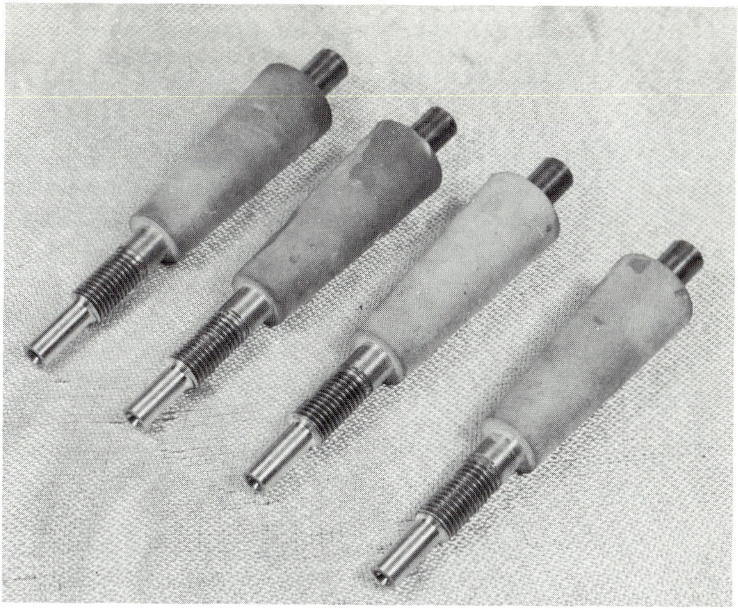

High-temperature insulating sleevings made of epoxy resin and glass cloth are used on large motor powder input connectors. (*The Epoxylite Corporation.*)

Table 22-9. Properties of DGEBA Glass-cloth Laminates as Function of Curing Agent [4-6]

FORMULATIONS

Anhydride	Concentration, phr	Resin content, %	Cure schedule pressure at 165°C, minutes: Contact pressure	Cure schedule pressure at 165°C, minutes: At 200 psi	Oven postcure at 150°C, hours
A. DDSA + 1 phr BDMA...	134	23.0	3	20	24
B. CA.....................	117	27.0	3	15	24
C. HHPA + 1 phr BDMA...	78	25.0	None	20	24
D. NMA	90	25.5	4	15	24

FLEXURAL PROPERTIES

Formulation	At 23°C		At 121°C	
	Strength, psi	Modulus, psi × 10^{-6}	Strength, psi	Modulus, psi × 10^{-6}
A	74,900	3.58	8,000	0.99
B	82,400	3.71	64,100	3.15
C	90,600	3.76	15,450	1.62
D	91,300	3.72	23,400	1.76

Table 22-9. (*continued*)

EDGEWISE COMPRESSIVE STRENGTH PROPERTIES

Formulation	Strength, psi	Modulus, psi × 10^{-6}	Deformation at ultimate, %
A	24,000	1.56	1.1
B	29,100	1.40	1.3
C	34,900	1.20	1.6
D	39,900	2.65	1.4

TENSILE PROPERTIES

Formulation	Strength, psi	Modulus, psi × 10^{-6}	Elongation at ultimate, %
A	52,900	3.55	1.7
B	55,500	3.67	1.8
C	57,700	4.00	1.8
D	60,700	4.20	1.7

Formulation	Supported interlaminar shear strength at 23°C, psi
A	1,700
B	1,600
C	2,000
D	2,300

Table 22-10. Flexural Strength of Bodied DGEBA as Function of Curing Agent [4-6]

Curing agent	Strength, psi		Modulus, psi × 10^{-6}		Resin content, %
	At 23°C	At 150°C	At 23°C	At 150°C	
BF_3MEA	75,500	48,500	2.7	2.3	30
DADPS/BF_3MEA.	74,000	58,000	2.8	2.6	30
Dicyandiamide....	79,000	21,000	3.0	1.7	30

Table 22-11. Properties of H6-28 Glass Cloth with Chlorinated DGEBA and Phenolic Curing Agent, Resin Content 33% [109]

Property	*Value*
Flexural strength, psi:	
Lengthwise	69,800
Crosswise..............................	65,800
Water absorption after 24 hr, %..............	0.049
Dielectric strength, parallel to laminate, kv	
Short-time	57
Step-by-step.............................	53.5
Dissipation factor at 10^6 cps	0.0094
Dielectric constant at 10^6 cps..................	4.45
Copper bond, lb/in..........................	8.6
Blister time at 260°C, solder seconds	60+

Filament winding generally requires a long-pot-life, low-viscosity resin, with the tensile elongation of the cured matrix being critical. The tensile elongation of E-glass is on the order of 3 per cent, and the matrix employed should have a greater elongation than this in order to transmit stress efficiently. If the elongation of the matrix is less than that of the glass, the stresses will be transferred to the matrix. In the intermediate temperature range generally using amine cures, elongations up to 7 per cent have been obtained in the matrix, but in more heat-resistant systems the elongations are still with contemporary matrices rather lower than desired for

Finishing touches are being applied to a large filament-wound storage tank. (*Black, Sivalls and Bryson, Inc.*)

maximum performance [96]. Nonetheless, the matrix material is the limiting factor in the heat resistance of the filament-wound structure, and lesser elongations than ideal must be accepted when maximum heat resistance is required.

For high-temperature applications, the epoxidized novolac resins are preferred, these being cured with anhydride curing agents. Satisfactory heat resistance to 400°C has been obtained in specialized applications for short service times [100], but in the authors' experience, service life of even the best epoxy-resin laminates ranges from 1000 to 2000 hr at 250°C and 25 to 100 hr at 300°C.

Typical mechanical properties of a filament-wound structure are indicated in Table 22-13. Table 22-14 presents typical commercial values for filament-wound pipe. The working properties and mechanical strengths of filament windings, with an epoxidized novolac resin cured with NMA as a matrix, are shown in Table 22-15.

In addition to the selection of the proper resin and curing agent for the application, it may in some cases be desirable to select modifiers.

Often, in wet lay-up laminating operations, reactive diluents are employed, although these are generally felt to degrade the fatigue life of the laminate and to reduce the heat resistance. Typical of systems containing a diluent is the use of dibutyl phthalate in DGEBA cured with TETA. The flexural strength of this system, as a function of curing-agent concentration, is indicated in Fig. 22-10.

Properties of a system modified with an asphaltic additive are given in Table 22-16; properties of a system modified with a polysulfide are given in Table 22-17.

A number of attempts have been made to incorporate silicone resins in epoxy laminates in order to upgrade the temperature resistance of the system. One series of experiments involved bodying DGEBA (mol. wt. 950) with a silanol-containing silicone and curing the resultant product with MPDA. The flexural strength and

Table 22-12. Physical Properties of Filament Windings with Various Matrices [92]

Formulation	Average hoop tensile strength, psi × 10^{-5}					
	Ring stress			Glass stress		
	At 25°C	At 260°C	At 399°C	At 25°C	At 260°C	At 399°C
A	1.57	1.25	0.84	2.32	1.84	1.20
B	1.62	0.69	....	2.34	0.99	
C	1.51	1.11	0.81	2.02	1.49	0.96
D	1.27	....	....	1.87		
E	1.41	1.16	0.97	2.27	1.90	1.53
F	1.36	1.26	0.85	2.24	1.88	1.29
G	1.52	1.40	1.12	2.07	1.92	1.51
H	1.24	1.10	0.63	1.84	1.67	0.97
I	1.66	0.86	0.76	2.43	1.13	0.99
J	1.45	1.25	0.82	2.35	1.89	1.20
K	1.65	1.28	0.85	2·22	1.71	1.11
L	1.81	1.17	0.79	2.42	1.53	1.10
M	2.03	0.81	0.64	2.60	1.07	0.81
N	1.81	1.34	0.53	2.39	1.74	0.66
O	1.62	1.50	0.90	2.21	2.06	1.26
P	1.41	1.32	0.81	1.92	1.79	1.03
Q	1.77	1.28	0.55	2.41	1.81	0.71

electrical properties of such a laminate, as a function of the concentration of silicone, are given in Tables 22-18 and 22-19. Properties of the 25/75 silicone/epoxy blend as a function of curing agent are given in Table 22-20.

The silanol-silicones can also be incorporated in epoxy resins which contain few if any initial hydroxyl groups by the use of a small amount of organic acid to supply the necessary hydroxyls. The effect of the use of such a silicone on the properties of two such resins is presented in Table 22-21. The effect of concentration of NMA on a silanol-silicone/epoxidized novolac blend is indicated in Table 22-22.

Although modifiers are often employed in laminating formulations, the use of fillers is not too common. In general, the glass itself may be considered the filler, the addition of other fillers serving merely to reduce the glass content. In addition, unfilled formulations are preferred in order to obtain best possible wetting and

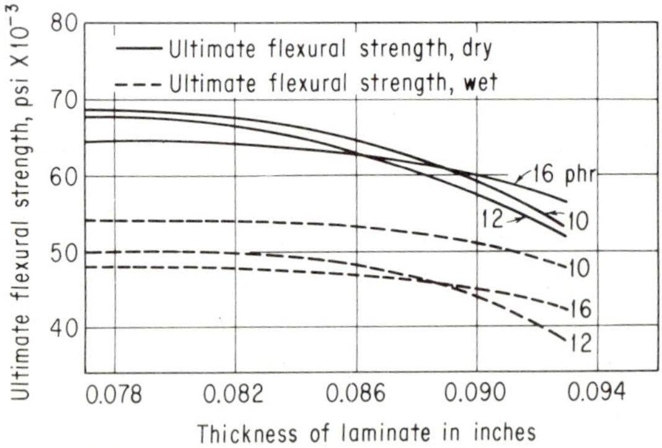

FIG. 22-10. Flexural strength as a function of laminate thickness and concentration of TETA for DGEBA–dibutyl phthalate laminate [4-1].

Table 22-12. (*continued*)

Formulation	Average interlaminar shear strengths, psi					
	Based on cross-sectional area			Based on resin content		
	At 25°C	At 260°C	At 399°C	At 25°C	At 260°C	At 399°C
A	7,710	3,430	680	26,730	11,940	2,350
B	7,450	.....	...	27,800		
C	6,120	2,290	370	30,140	7,620	1,750
D						
E	9,140	3,660	890	23,400	9,450	3,170
F	7,790	2,980	420	26,800	9,990	1,280
G	9,690	4,380	630	33,100	14,900	2,140
H	9,880	2,310	...	27,200	6,680	
I	5,500	1,270	530	30,600	6,920	2,940
J	7,440	3,830	550	26,800	14,400	2,420
K	7,510	2,480	650	34,300	11,250	2,950
L	6,640	1,470	600	35,200	7,570	3,350
M	6,630	730	360	29,700	3,420	1,610
N	4,380	2,310	...	23,850	12,500	
O	5,400	2,620	830	19,850	9,680	3,040
P	6,040	3,010	570	27,180	13,650	2,560
Q	7,740	2,830	610	31,400	11,880	2,540

Compounds

A. DGE of novolac (*f* 3.3)/NMA/DMP-30.
B. DGEBA/MPDA.
C. DGEBA/NMA.
D. Polyglycidyl ether of tetraphenylethane/fatty polyamide (amine value 290–320).
E. Polyglycidyl ether of tetraphenylethane/NMA.
F. Polyglycidyl ether of tetraphenylethane/BF_3MEA.
G. DGE of novolac (*f* 3.3)/BF_3MEA.
H. DGE of novolac (*f* 3.3)/PMDA adduct.
I. DGE of novolac (*f* 3.3)/triallyl cyanurate.
J. Dicyclopentadiene dioxide/MA/glycerol.
K. Epoxidized linear olefin (9% epoxy)/MA.
L. DGE of resorcinol/NMA.
M. DGE of diallyl bisphenol A/NMA.
N. Triglycidyl cyanurate/NMA.
O. Allyl diglycidyl cyanurate/NMA/PMDA.
P. Allyl diglycidyl cyanurate/DGE of novolac (*f* 3.3)/NMA.
Q. Vinyl resorcinol DGE/NMA.

Table 22-13. Typical Mechanical Properties of Filament-wound Structure [97]

Property	Maximum psi	Minimum psi
Hoop stress at 54°C	163,000	135,000
Shear stress:		
Interlaminar	6,000	4,000
Crosslaminar	9,000	7,000
Bonded joint	4,000	1,500
Threaded hole:		
Radial	12,000	10,000
Axial	9,000	7,000
Compressive strength ...	40,000	25,000
Bearing strength	60,000	40,000

Table 22-14. Typical Physical Properties of Filament-wound Pipe [56]

Property	*Value*
Specific gravity	2.03
Density, pci	0.072
Tensile strength, psi	
Unidirectional windings	200,000
Helical-wound cylinder	130,000
Modulus, psi × 10^{-6}	3–6
Compressive strength, psi	70,000
Bearing strength, psi	35,000
Thermal conductivity, Btu/(hr)(°F)(sq ft)(in.)	2–5
Coefficient of thermal expansion, in./(in.)(°F) × 10^6	6
Specific heat	0.25
Hardness, Barcol	40–70
DT, °C	150
Strength-to-weight ratio, composite	1,808,000
Temperature range, °C	−55 to 177

Table 22-15. Working Properties and Mechanical Strengths of Filament Windings with Epoxidized Novolac Resin (*f* 3,3) Cured with 101 phr NMA and 1.5 phr DMP-30 [93]

Temperature, °C	*Gel time*
25	6 days
45	12 hr
54	7 hr
80	82 min
93	55 min
108	40 min
121	25 min

Conditioning, hours soak at 260°C	Hoop tensile strength, psi × 10^{-5}					
	Ring stress			Glass stress		
	At 23°C	At 260°C	At 399°C	At 23°C	At 260°C	At 399°C
0	1.57	1.25*	0.84	2.32	1.84	1.20
6	1.41	1.29	0.88	2.02	1.80	1.17
12	1.56	1.21	0.87	2.09	1.60	1.16
24	1.39	1.23	1.13	1.85	1.66	1.52
48	1.40	1.19	0.97	1.94	1.63	1.26
96	1.37	1.12	0.99	2.05	1.59	1.46
192	1.19	1.04	0.54	1.58	1.36	0.74

Conditioning, hours soak at 260°C	Interlaminar shear strength, psi					
	Based on cross-sectional area			Based on resin content		
	At 23°C	At 260°C	At 399°C	At 23°C	At 260°C	At 399°C
0	7,710	3,430	681	26,730	11,940	2,350
6	4,830	2,420	520	20,500	9,990	2,170
12	4,840	2,520	480	21,400	11,600	2,210
24	4,530	2,690	539	19,700	11,600	2,320
48	5,310	1,860	460	22,470	7,910	1,950
96	4,070	2,460	576	17,800	10,780	2,520
192	1,570	1,390	706	8,440	7,500	3,830

* Soaked 2 hr at 260°C before testing.

Table 22-16. Physical Properties and Chemical Resistance of TETA-cured DGEBA Laminate. Using Transphalt Modifier. 30 Per Cent Resin Content [4-1]

PHYSICAL PROPERTIES

Property	50 phr Transphalt L-4		50 phr Transphalt L-5	
	Value	Retention, %	Value	Retention, %
Tensile strength at 23°C, psi:				
Dry	52,500	...	56,450	
Wet:				
After 2-hr boil	46,400	88	50,200	89
After 30 days' immersion	49,200	94	53,700	95
Modulus at 23°C, psi × 10^{-6}:				
Dry	3.1	...	3.4	
Wet:				
After 2-hr boil	2.9	94	3.1	90
After 30 days' immersion	3.8	100	3.0	87
Flexural strength:				
At 23°C, psi:				
Dry	66,450	...	69,200	
Wet:				
After 2-hr boil	56,100	84	57,850	83
After 30 days' immersion	63,300	95	66,250	96
At 71°C after ½ hr: dry	36,900	89	46,250	67
Modulus; psi × 10^{-6}:				
At 23°C:				
Dry	2.9	...	2.8	
Wet:				
After 2-hr boil	2.4	84	2.8	100
After 30 days' immersion	2.9	100	3.8	100
At 71°C after ½ hr: dry	2.3	79	2.5	88
Compressive strength at 23°C, psi:				
Dry	36,600	...	37,400	
Wet: after 30 days' immersion	37,650	100	38,500	100

CHEMICAL RESISTANCE

Chemical reagent solution	50 phr Transphalt L-4		50 phr Transphalt L-5	
	Weight change, %	Flexural strength retained, %	Weight change, %	Flexural strength retained, %
Sulfuric acid, 3%	+0.7	83	+0.4	84
Sodium chloride, 10%	+0.2	100	+0.1	100
Distilled water	+0.2	95	+0.1	96
Nitric acid, 10%	+4.0	43	+2.7	54
Hydrochloric acid, 10%	+1.4	51	+0.8	58
Acetic acid, 5%	+1.1	87	+0.5	92
Hydrogen peroxide, 3%	+1.5	77	+1.1	78
Sodium hydroxide, 1%	+0.1	96	+0.1	93
20%	+0.1	96	−0.03	93
Petrohol 99 (Mil-F-5566)	−0.02	100	−0.11	100
Hydrocarbon test fluid (Mil-H-3136)	−0.02	100	−0.01	100
Ethylene glycol (Mil-E-5559)	+0.14	100	+0.01	100
Hydraulic oil (Mil-O-5606)	+0.09	100	+0.02	100

Table 22-17. Physical and Electrical Properties of Flexibilized Laminates. Resin Content 24 to 28% [10-2]

Ingredients	Formulations						
DGEBA	100	100	100	100	100	100	100
DMP-10	10	10	10	10	10	10	10
Polysulfide (mol. wt. 1,000)	...	16.7	20	25	33.3	50	100
Property	**Value**						
Flexural strength, psi × 10^{-3}	65	63	61	60	59	35	10
After 2 hr in boiling water	38	37	32	31	30	20	15
Modulus, psi × 10^{-6}	3.1	3.0	2.9	2.8	2.6	2.1	1.3
After 2 hr in boiling water	2.6	2.5	2.5	2.4	2.4	1.7	0.6
Impact, Izod	11	11	12	12	14	15	Greater than 16
Edgewise compressive strength, psi × 10^{-3}	35	35	33	31	30	27	17
Dielectric constant:							
At 10^3 cps	4.4	4.6	4.6	4.7	4.9	5.0	5.2
At 10^9 cps	3.6	3.6	3.8	3.8	4.0	4.1	4.2
Dissipation factor							
At 10^3 cps	0.002	0.0026	0.0028	0.003	0.0033	0.0035	0.0040
At 10^9 cps	0.021	0.011	0.013	0.014	0.016	0.019	0.023

Table 22-18. Flexural Strengths of Epoxy-Silicone Laminate as Function of Concentration of Silicone [38]

(In pounds per square inch)

Conditioning	% silicone			
	0	25	50	75
As afterbaked:				
At 23°C	74,900	80,500	67,600	52,300
At 260°C	6,090	12,400	8,310	10,400
After 200 hr at 260°C:				
At 23°C	Blistered	63,000	47,000	22,400
At 260°C	Blistered	17,900	17,400	13,700
After 500 hr at 260°C:				
At 23°C	Blistered	47,400	12,200	9,740
At 260°C	Blistered	15,200	13,700	9,250

Table 22-19. Electrical Properties of Epoxy-Silicone Laminate as Function of Concentration of Silicone [38]

Property	% silicone			
	0	25	50	75
Arc resistance	15	160	150	110
Dielectric strength, volts/mil				
In air	230	240	230	240
In oil	390	320	370	350
Aged 100 hr at 232°C in air	......	270	370	470
Dielectric constant				
Dry				
At 60 cps	5.70	4.70	4.68	5.18
At 10^3 cps	5.58	4.66	4.64	5.13
At 10^6 cps	5.22	4.60	4.55	5.02
Wet				
At 60 cps	*	5.28	5.88	6.28
At 10^3 cps	6.80	4.98	5.32	5.34
At 10^6 cps	5.29	4.67	4.71	5.08
Dissipation factor:				
Dry				
At 60 cps	0.0241	0.006	0.008	0.007
At 10^3 cps	0.0243	0.004	0.005	0.005
At 10^6 cps	0.0183	0.005	0.008	0.008
Wet				
At 60 cps	*	0.04	0.07	0.291
At 10^3 cps	0.586	0.03	0.04	0.036
At 10^6 cps	0.023	0.01	0.02	0.013

* Too high to measure.

Table 22-20. Flexural Strengths of Epoxy-Silicone Laminates (75/25) as Function of Curing Agent [38]

(In pounds per square inch)

Curing agent	Afterbaked		After 200 hr at 250°C		After 500 hr at 250°C	
	At 23°C	At 260°C	At 23°C	At 260°C	At 23°C	At 260°C
MPDA	80,500	12,400	63,000	17,900	47,400	15,200
PMDA	77,000	12,700	67,700	17,600	64,700	26,200
PMDA + DMP-30	69,700	8,300	63,900	11,400		
NMA + DMP-30	88,000	12,800	73,400	17,400		
Trimellitic anhydride	60,700	13,300	55,900	21,500		

Table 22-21. Comparison of Flexural Strengths of Laminates made with Different Epoxy Resins in Silicone/Epoxy (25/75) Cured with NMA [38]

(In pounds per square inch)

Resin	As afterbaked		After 200 hr at 260°C	
	At 23°C	At 260°C	At 23°C	At 260°C
DGEBA (mol. wt. 950)	88,000	12,800	73,400	17,400
Epoxidized novolac (*f* 3.3)	69,300	26,000	48,000	25,800
Polyglycidyl ether of tetraphenylethane	61,100	27,000	52,300	32,000

Table 22-22. Effect of NMA Concentration on Flexural Strength of Laminate Made with 75% Epoxidized Novolac (*f* 3.3) and 25% Silanol Silicone, DMP-30, at 0.5% [38]

(In pounds per square inch)

Moles NMA/equivalent epoxy	As afterbaked		After 200 hr at 260°C	
	At 23°C	At 260°C	At 23°C	At 260°C
0.75	61,100	24,000	45,900	27,000
0.85	69,300	26,000	48,000	25,800
0.95	58,200	25,000	40,100	25,100

void-free structures, fillers (particularly in higher loading volumes) making this more difficult. The effect of several fillers on the tensile strength of an epoxy–glass-mat laminate is shown in Fig. 22-11. Fillers may provide some improvement in moisture resistance, as shown in Fig. 22-12.

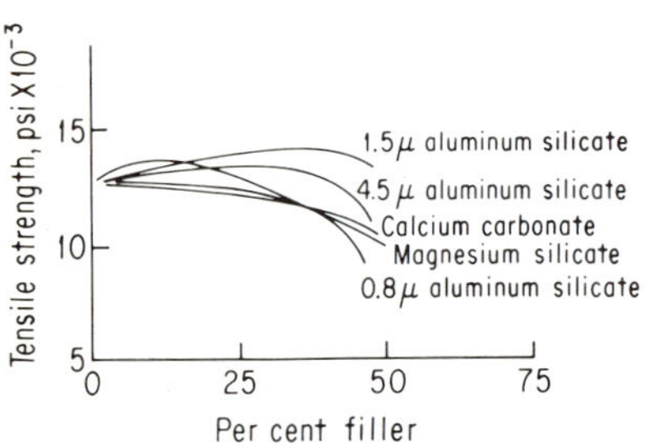

FIG. 22-11. Effect of fillers on tensile strength of DGEBA–glass-mat laminates [130].

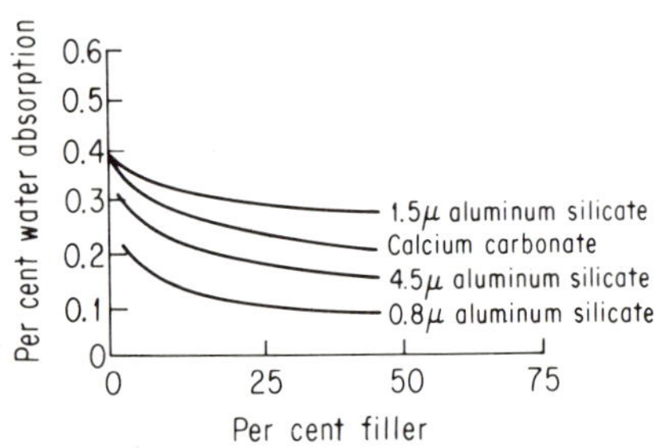

FIG. 22-12. Effect of fillers on moisture resistance of DGEBA–glass-mat laminates [130].

Laminating Pressure, Resin Content, and Curing Cycle

Laminating pressure, resin content, and curing cycle are treated under one heading because they are mutually interdependent with regard to their effect on properties. They are likewise dependent, to some extent, on the application technique.

Laminating Pressure. One of the advantages of the epoxy resins is that they can be laminated with only contact pressure. Seldom, however, is a dense void-free fill achieved, and the laminate will possess resin-rich and resin-poor areas. The resin content will be about 50 to 60 per cent. This results in flexural strengths of 30,000 to 45,000 psi.

Low-pressure laminates, in contrast to contact-pressure laminates, will have correspondingly better strength properties. The following room-temperature strengths may be considered typical of the relationship between laminating pressure and flexural strengths:

Pressure, psi	*Flexural strength, psi*
3	38,000
25	63,000
200	78,000

The effect of laminating pressure and temperature on a laminate of DGEBA cured with MPDA is indicated in Figs. 22-13 and 22-14.

However, when the resin content is held constant, the effect of laminating pressure will be slight:

Pressure, psi	Flexural strength, psi	Flexural modulus, psi × 10^{-6}
30	67,000	2.8
100	67,000	2.8
300	70,000	3.0

With filament-wound systems, the tension of the winding head may be adjusted to provide prestressed structures. This stressing, coupled with internal stress buildups during fabrication because of shrinkage and thermal differentials, might

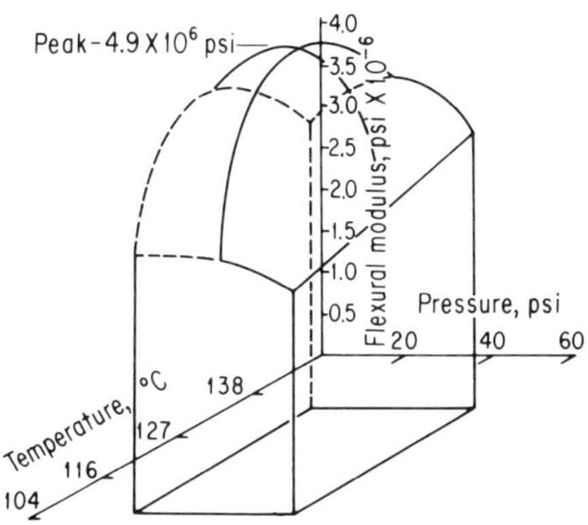

FIG. 22-13. Effect of molding pressure and temperature on flexural modulus of MPDA-cured DGEBA laminate [36].

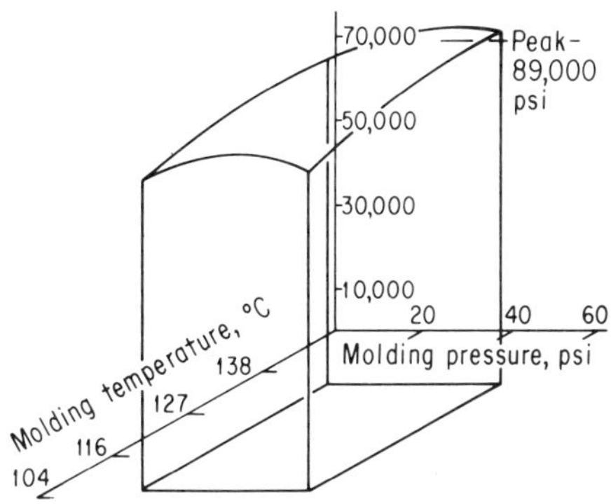

FIG. 22-14. Effect of molding pressure and temperature on flexural strength of MPDA-cured DGEBA laminate [36].

be expected to reduce resistance to externally applied loads, since local stress levels might exceed the cohesive strength of the matrix and lead to local failures which would degrade the laminate and increase permeability to liquids [164]. However, in practice, at least, in terms of interlaminar shear strengths, residual stresses have been shown to be of negligible importance in filament-wound structures [172].

More importantly, the amount of pressure employed will control in part the void content of the composite structure. Voids will severely reduce interlaminar shear strengths. A study of this factor, employing filament-wound structures containing voids controlled by varying the winding pressure during fabrication, indicated that the voids may be of several types; they may arise from entrapped and adsorbed air; from chemically reactive finishes which undergo changes when stored under high humidities; from interstitial voids which occur with stiffer surface finishes and when the surface finish bonds one filament to another so the matrix material cannot penetrate; and from planar voids which occur at discontinuities in the winding pattern and within the structure. Photoelastic studies established that these voids lead to points of high stress concentrations, and when the voids extend along the length of the fiber, provide areas which are initially debonded [186].

Resin Content. The resin content is dependent on the laminating pressure. As an interrelated parameter, it determines the strength of the ultimate laminate.

Increasing the percentage of glass increases the modulus of the laminate. The variations that can be expected with resin content within a working range of 30 to 35 per cent are not overly significant.

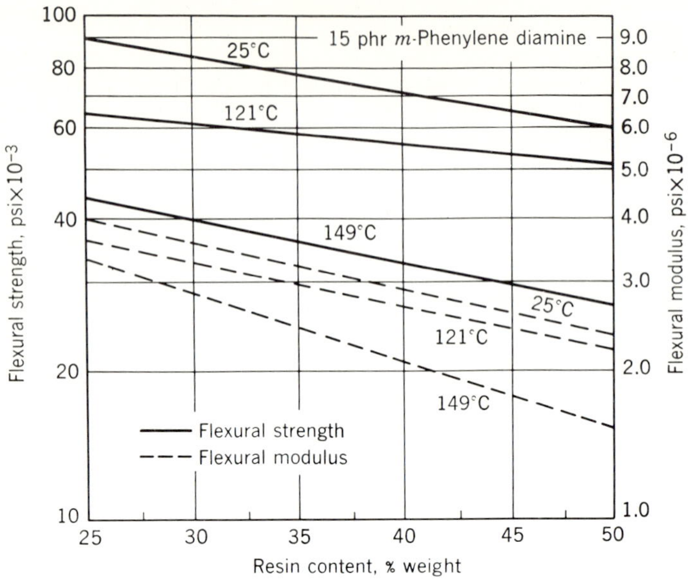

FIG. 22-15. Effect of resin content on flexural strength of MPDA-cured DGEBA laminate [4-6].

The effect of resin content on flexural strengths for various temperatures is indicated in Fig. 22-15 and for various forms of reinforcement in Fig. 22-16. With DGEBA cured with MPDA, room-temperature flexural strengths vary inversely with resin content over the range of 29 to 46 per cent [8]. With DGEBA (mol. wt. 950) cured with dicyandiamide, the best flexural strength is obtained at a glass content of 30 per cent [29].

In general, the optimum glass content for tensile strength is within limits highest (Fig. 22-17), for compression strength lowest. In filament-wound structures, the general effect of resin content on interlaminar shear is indicated in Fig. 22-18.

Cure Cycle. The cure cycle for laminates may be considered to be somewhat more complex than that for castings, coatings, and adhesives. It may be accomplished in one or more of three stages: (1) precure, (2) cure during pressure, and (3) postcure.

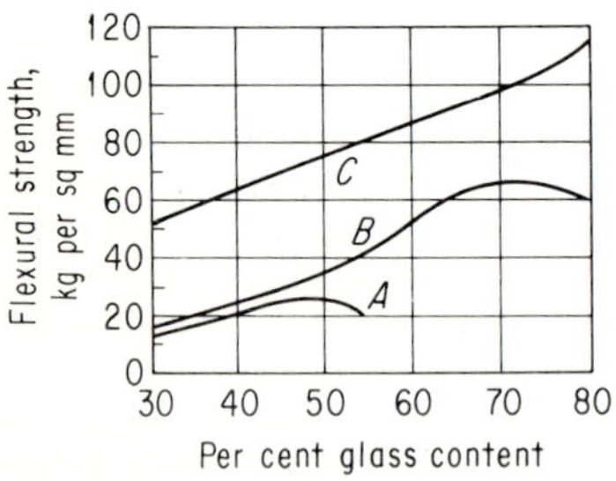

FIG. 22-16. Flexural strength as a function of glass content. *A*, Mat; *B*, cloth; *C*, roving [182].

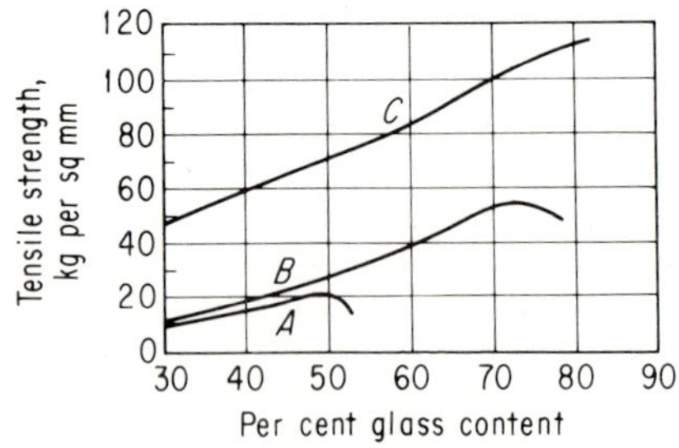

FIG. 22-17. Tensile strength as a function of glass content. *A*, Mat; *B*, cloth; *C*, roving [182].

Specific examples will serve to illustrate typical cure cycles. With a wet lay-up laminate, not involving pressure and employing a polyfunctional primary aliphatic amine as curing agent for DGEBA diluted with butyl glycidyl ether, cure can be accomplished in an oven or under infrared lights at 90 to 120°C in 30 minutes to 1 hour. The infrared lights rather than ovens have been found to be particularly effective with plaster molds, since it is then unnecessary to heat the entire mass to obtain cures. The effect of postcure is indicated in Table 22-23. In one series of

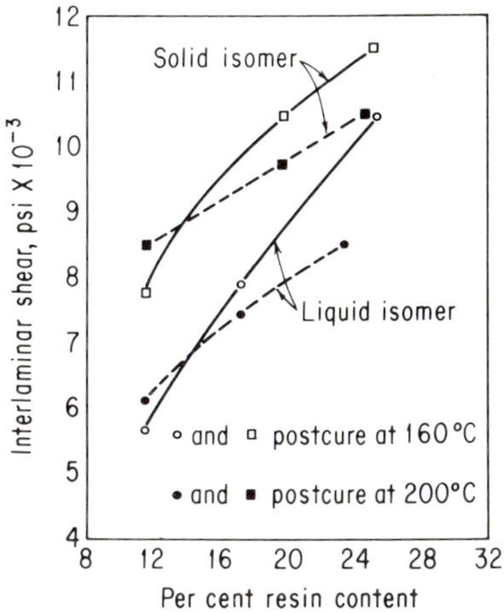

FIG. 22-18. Effect of resin content on interlaminar shear of NOL rings made with aromatic-amine-cured bis(2,3-epoxycyclopentyl) ether [158].

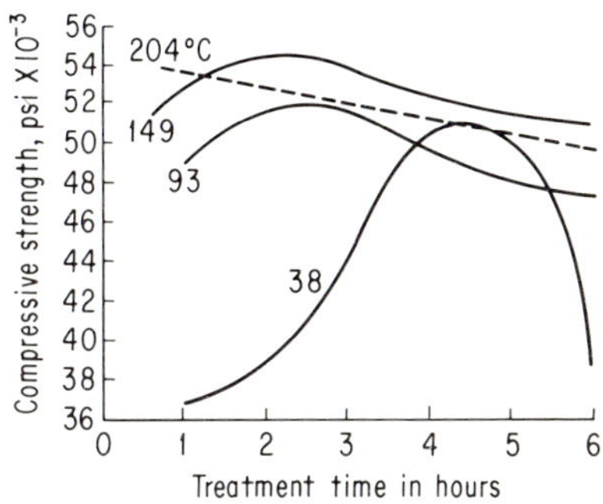

FIG. 22-19. Effect of postcure time on compressive strength of MPDA-cured DGEBA laminate [36].

studies, using DGEBA and 10 phr DETA, it was shown that maximum glass-to-resin bond strengths are obtained with a cure no greater than 55°C, with higher cure temperatures lowering bond strengths [116].

A dry lay-up, employing MPDA with DGEBA in a solvent-impregnated system, cured in matched molds and employing pressure to provide adequate flow, will have physical properties dependent on both cure temperature and cure pressure. Selected data are presented in Table 22-24. The effect of postcure on the same system is presented in Table 22-25. The effect of postcure on a laminate of DGEBA cured with MPDA is shown in Fig. 22-19.

PROPERTIES OF THE CURED COMPOSITE

Data for a number of specific laminates have been given earlier in this chapter, and these, together with data furnished in prior chapters (in particular, Chaps. 7 to 12) will have utility in the formulation of laminate resin systems for specific applications. A detailed guide to the specifications and standards for rigid laminated plastics, with emphasis on the epoxy resins, has been compiled by the Department of Defense [123].

Physical Properties

The physical properties normally reported for laminates are those given in Table 22-26. These test procedures are analyzed and criticized in [103]. In

Table 22-23. Effect of Postcure on Properties of Contact-pressure Laminate, DGEBA with Butyl Glycidyl Ether and Cured with Aliphatic Primary Amine [15]

Mechanical properties	Cured at room temperature	Postcured 8 hr at 160°C
Flexural strength, psi	52,500	61,800
At 70°C	7,600	46,600
After 2 hr in 100°C water	46,200	57,400
Modulus, psi × 10^{-6}	2.51	3.25
At 70°C	0.38	2.37
After 2 hr in 100°C water	2.29	2.79
Edgewise compressive strength, psi	38,100	43,000
Modulus, psi × 10^{-6}	2.94	2.83
Tensile strength, psi	36,100	40,700
Modulus, psi × 10^{-6}	2.03	2.35
Impact strength, Izod	12.4	10.4

Table 22-24. Typical Cure Cycles and Properties of Laminate of DGEBA Cured with MPDA [4-6]

Cure cycles	Cure temperature, °C	Cure time, minutes	Cure pressure, psi
A	107	30	200
B	143	15	200
C	224	15	200

Property	Initial	After 3 hr in boiling acetone	After 3 hr in boiling water
Flexural strength, psi × 10^{-3}:			
A	79.5	59.4	69.1
B	86.5	82.1	72.1
C	80.5	78.7	76.3
Flexural modulus, psi × 10^{-6}:			
A	3.4	1.8	3.8
B	4.0	4.0	3.4
C	3.5	3.1	3.3
Hardness, Barcol:			
A	70	40	70
B	70	70	70
C	70	68	68
Weight increase, %:			
A	....	2.8	0.25
B	....	0.12	0.17
C	....	0.09	0.18

Table 22-25. Effect of Cure Pressure and Postcure on Elevated-temperature Performance of Laminate of DGEBA Cured with MPDA [4-6]

Cure cycle	Cure temperature, °C	Cure time, minutes	Cure pressure, psi	Postcure at 240°C, hr
A	150	15	200	
B	150	15	200	1
C	150	15	25	
D	150	15	25	1

Property	At 25°C	At 150°C	At 260°C
Flexural strength, psi × 10^{-3}:			
A	75.3	18.5	8.8
B	72.3	56.9	11.1
C	74.2	18.1	6.2
D	68.7	26.3	7.0
Flexural modulus, psi × 10^{-6}:			
A	3.2	1.4	1.3
B	3.0	3.2	1.6
C	3.2	1.5	1.3
D	2.8	1.8	1.2

brief, laminate tensile tests are conducted on dumbbell specimens or rectangular panels. Several compression tests are available: The edgewise compression test, for example, involves supporting a columnar specimen to prevent buckling during loading. Flexural tests are conducted by installing the flat specimen across a span and loading the specimen at the center. Interlaminar shear tests normally employ a specimen with a hole drilled or punched in it. In bearing tests, the specimen is pin-loaded, with the size of the pins determining the stress concentration.

For filament-wound structures, somewhat different procedures are employed, with tensile tests being conducted on 5.75-in. I.D., 0.125-in. O.D. ring specimens wound on a special mandrel, the so-called NOL (Naval Ordnance Laboratory) rings [42]. A deficiency of the test is that as the split-disk test fixture is separated, the specimen is forced into a straight section over the separation. To minimize the effect of such bending stresses, the ring diameter may be increased and the specimen thickness reduced [135]. Other effects of fabrication variables on NOL ring testing are discussed in [140]. NOL ring properties do not, in general, correlate very well with properties of castings or cloth laminates except for some correlation between hoop stress and tensile properties determined at 0.2 per cent offset at

Table 22-26. Standard Test Procedures for Glass-cloth Laminates [103]

Test	ASTM	AIA	Federal test method standard
1. Tension	D 638–58T D759–48*	ARTC-11-II	No. 406-1011
2. Compression	D 695–54 D759–48*	ARTC-11-I	No. 406-1021.1
3. Flexure.........	D 790–58T	ARTC-11-VIII	No. 406-1031.1
4. Shear	D 732–46	ARTC-11-V	No. 406-1041
5. Hardness	D 785–51		
6. Impact	D 256–56 D 758–48*	....	No. 406–1071
7. Bearing	D 953–54	ARTC-11-IV	No. 406-1051

* Test for properties at subnormal and supernormal temperatures.

room temperature [150]. There is also some question as to how well NOL ring properties will provide design data for actual cylindrical configurations. Problems have been encountered with end reinforcement and support in the destructive testing of flat filament-wound specimens [167].

Through the use of micrometric eyepieces on compound microscopes, it is possible to measure resin content, glass content, entrapped air, and areas of poor wetting.

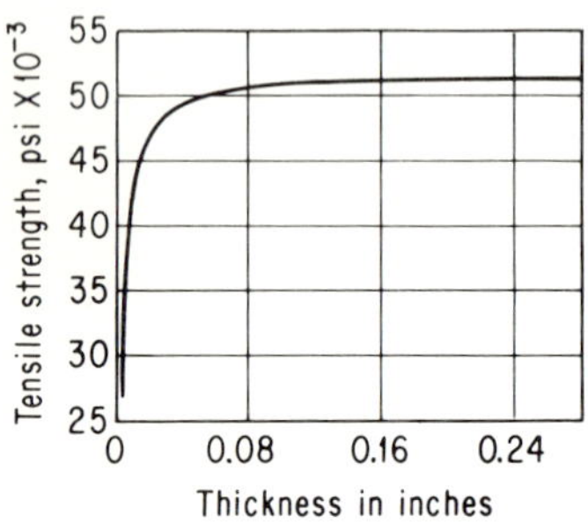

FIG. 22-20. Effect of thickness on tensile strength of MPDA-cured DGEBA laminate [20].

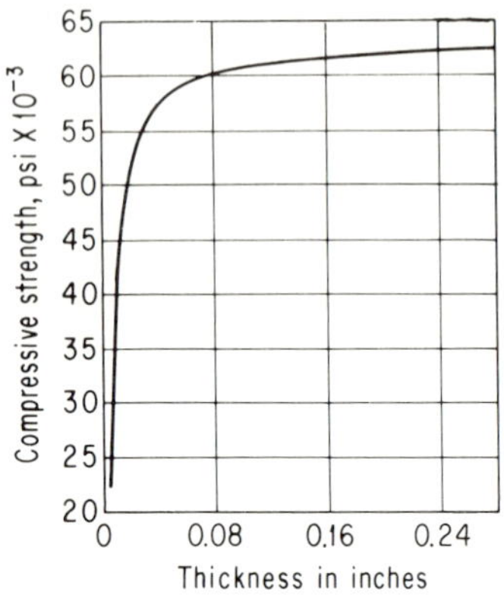

FIG. 22-21. Effect of thickness on compressive strength of MPDA-cured DGEBA laminate [20].

With refinements to the equipment, birefringence phenomena may be measured to indicate the magnitude of inherent strain within the specimen [165].

Nondestructive testing for filament-wound structures employ corona, eddy current, beta rays, and microwaves to measure resin-to-glass ratio, insulator thickness, voids, and delaminations [120,144].

The physical property most widely reported in the literature for glass laminates is the flexural strength. This is determined with relative ease, but the data are essentially worthless for design [103].

Although no single property can be considered as a general criterion for the resin formulation, compressive strength is probably the most sensitive of the tests. In flexural and tensile tests the glass is the controlling variable; whereas in compression the resin is controlling [49]. The thickness of the laminate, itself, of course, will to some extent influence the results, as indicated in Figs. 22-20 to 22-22, as will the type of reinforcement, etc.

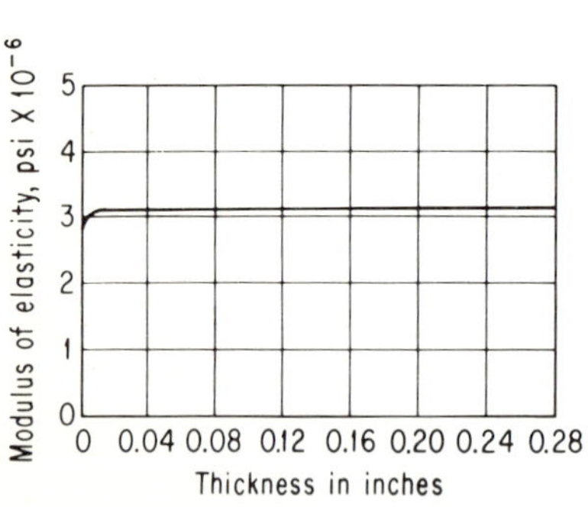

FIG. 22-22. Effect of thickness of modulus of elasticity of MPDA-cured DGEBA laminate [20].

An attempt to correlate physical properties of filament windings with molecular structure of the epoxy matrices by statistical analysis resulted in some correlation between tensile modulus and dynamic shear modulus and between DT and Tg. A high degree of correlation was reported between interlaminar shear strength and tensile strength at break—these latter two properties apparently are a function of the matrix material rather than an interaction between the matrix material and the glass reinforcement [151]. Other tests with filament-wound structures in compression loading indicate that the strength is controlled by the yield point of the matrix [167]. An interesting series of tests with filament-wound bottles indicated that the burst strength in a perfectly designed and fabricated bottle was independent of both the tensile strength and the elongation of the matrix material. Indeed, it was found that bottles containing DGEBA/MPDA matrices performed as well or better when the matrix was only B-staged rather than fully cured [174].

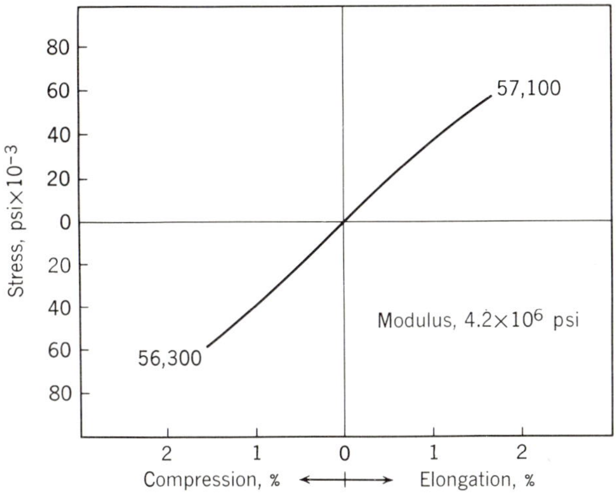

FIG. 22-23. Stress-strain curves for DEAPA-cured DGEBA laminate [123].

Performance of glass composites are, of course, dependent on the temperatures of the testing. Strengths progressively deteriorate as the temperature is increased above room temperature. Below room temperature, tensile, compressive, and flexural strengths increase gradually as the temperature is lowered to −196°C. Flexural strength peaks at this value, while tensile and compressive strengths continue to increase to −253°C [149]. The changes in modulus of elasticity follow the same trend but at a much reduced rate [181]. The high rate of strength increase, combined with the small change in modulus, indicates that the epoxy laminates maintain their room-temperature toughness and ductility even at cryogenic temperatures [149]. At cryogenic temperatures, static fatigue, attributed to a stress-corrosion process resulting from adsorbed water vapor, has not been observed; and its absence has been assumed to arise from water being immobilized at those temperatures [191].

Generally, most cloth-laminate data in the literature are reported on ⅛-in.-thick 12-ply laminates of Volan A 181-style E-glass. Tensile and compressive strengths for typical systems with this glass and this number of plies are indicated in Figs. 22-23 to 22-25. Typical flexural strengths obtained with various aliphatic amine curing agents are indicated in Table 22-27. The effect of curing agent on the flexural-strength retention at elevated temperature is indicated in Figs. 22-26 and 22-27 and in Tables 22-28 and 22-29.

A summary of the physical properties of DGEBA cured with MDA is given in Table 22-30. The effect of resin molecular weight on the system is indicated in

Table 22-27. Typical Properties of Laminates of DGEBA (mol. wt. 450) Cured with Aliphatic Amines [4-6]

Curing agent	phr	Resin, %	Cure		Flexural strength, psi × 10^{-3}	Flexural modulus, psi × 10^{-6}	Flexural yield, psi × 10^{-3}	Water absorption after 30 days' immersion, at 25°C, %
			Time, minutes	Temp., °C				
DMP-30	6	31	60	80	63.4	3.2	63.4	0.9
DETA	8	29	30	115	69.0	2.8	69.0	0.6
DMAPA	6	31	60	80	63.9	3.0	63.9	0.8
DMP-10	16	32	60	80	67.1	3.4	62.5	1.5
DEAPA	6	29	60	115	64.0	3.1	63.8	0.8
BDMA	6	29	60	80	66.7	3.3	66.7	0.9
Piperidine	6	33	180	80	68.3	3.2	64.3	0.2
Diethylamine	12	29	150	115	64.0	3.2	62.0	1.0
Pyridine	15	32	120	80	71.0	3.2	63.3	0.7

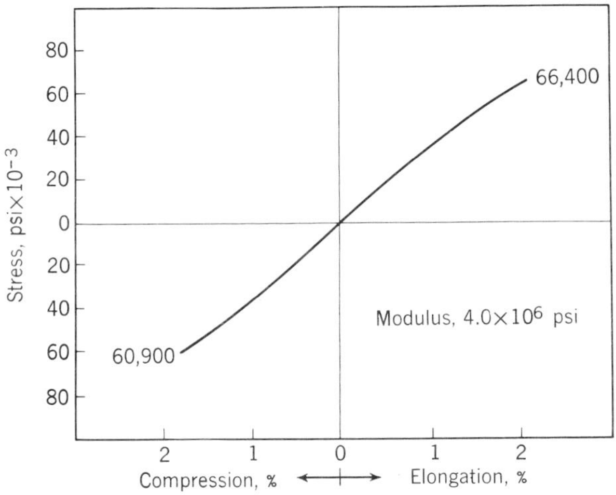

FIG. 22-24. Stress-strain curves for dicyandiamide-cured DGEBA (mol. wt. 950) laminate [3].

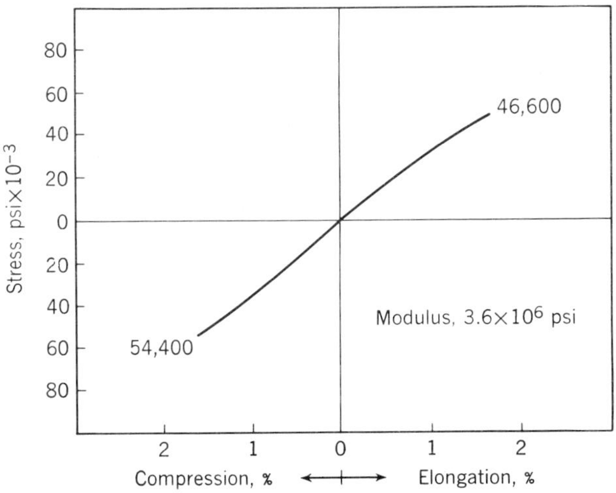

FIG. 22-25. Stress-strain curves for phenolic resin-cured DGEBA laminate [3].

Table 22-28. Typical Properties vs. Temperature of Laminate of DGEBA Cured with 120 phr CA [12-6]

Cure cycle: Cure *A* pressed at 100 psi for 2 hr at 180°C; cure *B* pressed at 100 psi for 24 hr at 180°C

Property	At 23°C	At 138°C	At 177°C	At 204°C	At 260°C
Cure A:					
Flexural strength, psi	87,900	75,500	73,500	29,800	9,700
Modulus, psi $\times 10^{-6}$ ..	4.17	3.28	3.33	2.22	1.20
Cure B:					
Flexural strength, psi	86,700	76,600	74,000	39,300	9,800
Modulus, psi $\times 10^{-6}$ ...	4.01	3.36	3.35	2.71	1.20

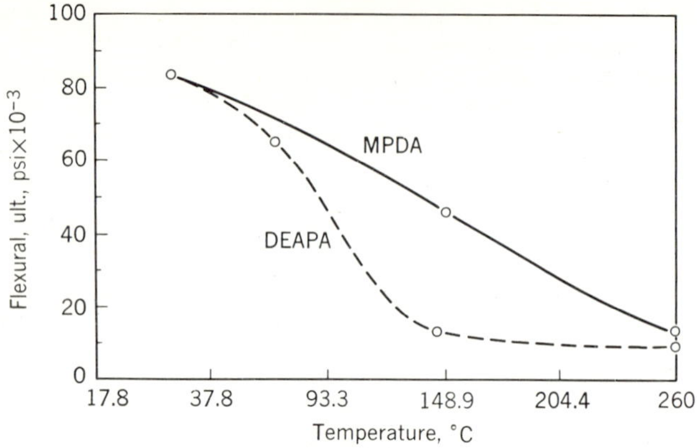

FIG. 22-26. Effect of temperature on flexural strengths of DGEBA laminates [3].

Table 22-31. A comparison of typical properties of cloth laminates and filament-wound structures with those of metals is given in Table 22-32.

The performance of the laminate under imposed short-time load to ultimate failure is of course of less design significance than the performance of the system under dynamic conditions and under continuous loading. Basically, of interest are the physical properties under alternating stresses (fatigue resistance) and under steady-state stresses (creep resistance). Creep and fatigue resistance are generally controlled more by the properties of the resin than by the properties of the glass.

Fortunately, performance under fatigue conditions is generally quite good, as indicated in Figs. 22-8 and 22-9. With cloth laminates, however, performance will, in some measure, be dependent upon the direction of the load, with cloth generally providing best values when stressed at an angle of 0° to the warp dry and when stressed at either 0° or 45° to the warp wet [22]. Resin contents between 26 and 34 per cent are preferred for best performance. With unidirectional filament-wound structures, best fatigue strength is obtained by alternating plies at ±5° about the major axis, with resin content, within limits, being of little significance [113].

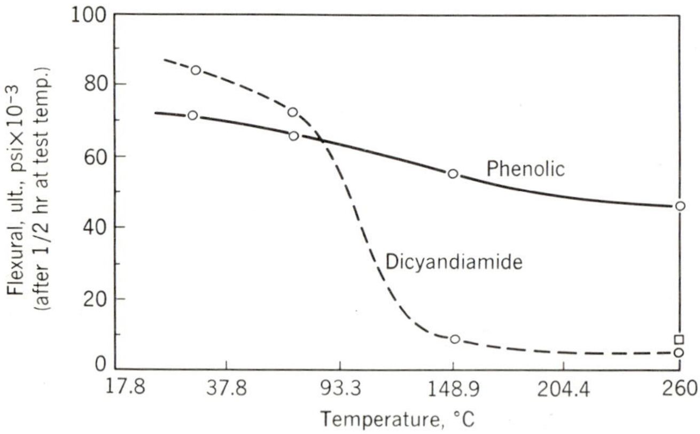

FIG. 22-27. Effect of temperature on flexural strengths of DGEBA (mol. wt. 950) laminates [3].

Table 22-29. Typical Properties vs. Temperature of Laminate of DGEBA Cured with PMDA and PMDA/MA Mixtures [12-4]

Cure cycle: Pressed at 100 psi for 15 min at 150°C with 2 min zero pressure precure; postcured 16 hr at 160°C.

Property	PMDA	PMDA/MA
Flexural strength, psi:		
At 23°C	73,000	74,000
At 150°C	45,000	46,000
At 260°C	15,000	16,000
At 260°C after aging at 200°C	23,000 (500 hr)	24,000 (500 hr)
At 260°C after aging at 260°C	17,000 (192 hr)	26,000 (500 hr)
Modulus, psi × 10^{-6}:		
At 23°C	3.4	3.5
At 150°C	2.6	2.9
At 260°C	2.2	2.3
At 260°C after aging at 200°C	2.4 (500 hr)	2.7 (500 hr)
At 260°C after aging at 260°C	2.4 (192 hr)	2.9 (500 hr)
Weight loss, %:		
After 500 hr at 200°C	1.2	1.2
After 192 hr at 260°C	6.0	
After 500 hr at 260°C	. . . .	7.5

A comparison of the various types of composites in terms of fatigue resistance is given in Fig. 22-28. The woven laminates have about one-half the fatigue resistance of SAE 1060 spring steel; the filament-wound structures give fatigue resistance equal to or slightly better than that of the SAE steel and have nine times greater capacity for stored energy than spring steel [152]. Too, epoxy laminates are superior to polyesters, for example, in maintaining low internal temperatures during dynamic testing, as indicated in Fig. 22-29. And finally, with regard to fatigue resistance, the composites are not notch-sensitive, retaining 80 to 90 per cent of their resistance when notched [152].

The effect of moisture conditioning on fatigue strength is still somewhat debatable. Some data indicate improvement after prestressing and moisture conditioning; others, degradation. The majority of the literature indicates that moisture, with or without prestressing, does little harm to fatigue resistance [160].

Table 22-30. Typical Mechanical Properties of Laminates of DGEBA Cured with MDA [15]

Cure cycle: Pressed 1 hr at 160°C; postcured 2 hr at 150°C + 6 hr at 200°C

Property	*Value*
Tensile strength, psi	50,000–58,000
Modulus, psi × 10^{-6}	3.3–3.6
Edgewise compressive strength, psi	49,000–51,000
Modulus, psi × 10^{-6}	3.5–3.6
Flexural strength, psi:	
At 25°C	79,000 89,000
At 127°C	52,000–60,000
After 2 hr in water at 100°C	77,000–82,000
Modulus, psi × 10^{-6}:	
At 25°C	3.6–3.9
At 127°C	3.0–3.5
After 2 hr in water at 100°C	3.2–3.6
Impact strength, Izod	12–15
Hardness, Rockwell M	115–117
Water absorption after 24 hr at 25°C, weight gain, %	0.05–0.07

Table 22-31. Typical Mechanical Properties of Laminates Cured with MDA vs. Molecular Weight of DGEBA [15]

Cure cycle: Pressed at 400 psi for 1 hr at 160°C; postcured 8 hr at 160°C

Property	DGEBA	
	Mol. wt. 1,000	Mol. wt. 1,500
Tensile strength, psi	52,000–59,000	61,000–66,000
Modulus, psi × 10^{-6}	2.9–3.4	3.5–4.0
Edgewise compressive strength, psi	45,000–50,000	48,000–52,000
Modulus, psi × 10^{-6}	3.5–3.7	3.9–4.5
Flexural strength, psi:		
At 25°C	80,000–85,000	95,000–100,000
At 127°C	69,000–74,000	70,000–75,000
After 2 hr in water at 100°C	77,000–79,000	85,000–90,000
Modulus, psi × 10^{-6}:		
At 25°C	3.6–3.9	4.4–4.6
At 127°C	3.5–3.7	4.0–4.2
After 2 hr in water at 100°C	3.4–3.7	3.8–4.0
Impact strength, Izod	12–15	16–18
Hardness, Rockwell M	110–112	102–107
Water absorption after 24 hr at 25°C, weight gain, %	0.04–0.06	0.04–0.08

It should be pointed out, particularly with filament-wound structures, that much fatigue (and other) testing is conducted on small specimens; and although the scale-up factor has not been established, experience indicates that strength will degrade in per cent about $\frac{1}{20}$ the rate at which the volume increases in per cent. It may also be anticipated that the modes of failure will change as the size and complexity of the specimen increase. This indicates that design detail for stress relief and favorable orientation of the fibers cannot be overemphasized [160].

Thorough cure generally improves creep performance, as indicated in Fig. 22-30, leading to a lower permanent set under constant load. A number of investigations have been conducted to determine creep and rupture properties in laminates [54], a formula for predicting performance has been developed [60], and creep data at low temperatures have been obtained [80].

Table 22-32. Comparative Specific Strengths [159]

Property	Glass-epoxy		Aluminum 7075-T6	Titanium Ti-155A	Tool steel 5GR-MO-V
	Filament-wound	Fabric			
Density, pci	0.070	0.070	0.010	0.163	0.281
Tensile strength, ksi	250	100	82	145	280
Specific tensile strength*	3.6	1.4	0.81	0.89	1.00
Tensile modulus, psi × 10^{-6}	9.0	3.5	10.3	16.0	30.0
Compressive strength, ksi	180	90	72	140	260
Specific compressive strength*	2.6	1.3	0.71	0.86	0.93
Compressive modulus psi × 10^{-6}	9.0	3.0	10.5	18.0	30.0
Flexure strength, ksi	270	100			
Flexure modulus, psi × 10^{-6}	7.0	5.0	3.9	6.2	11.0
Shear strength, ksi	25	30	49	85	170
Specific shear strength*	0.36	0.43	0.49	0.52	0.61
Bearing strength, ksi	65	60	147	273	465
Specific bearing strength*	0.93	0.86	1.46	1.67	1.66

* Strength divided by density, psi × 10^6.

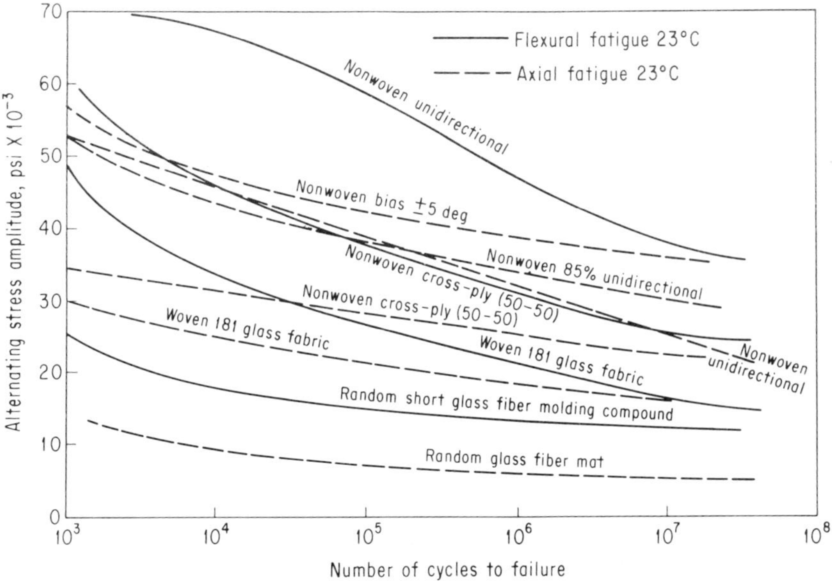

FIG. 22-28. Comparison of the effects of glass reinforcement types on fatigue strengths [152].

FIG. 22-29. Average laminate temperature under dynamic stress. *A*, Epoxy–glass-cloth laminates; *B*, polyester–glass-cloth laminates [182].

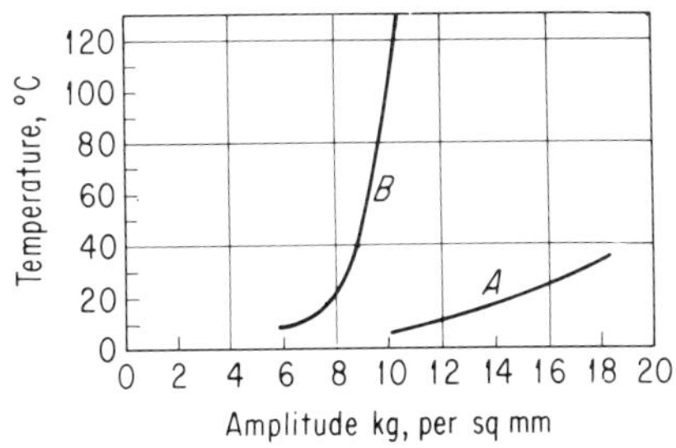

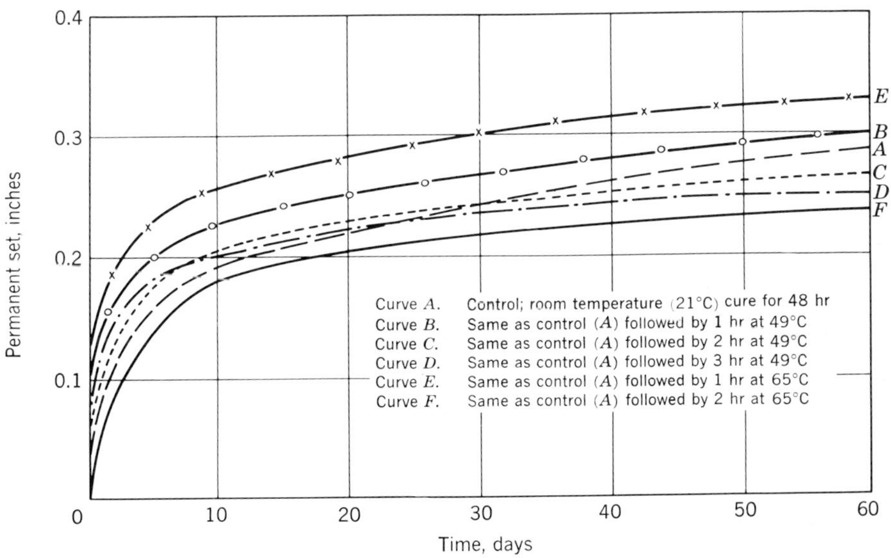

FIG. 22-30. Effect of postcure on creep-filled DGEBA laminates cured with aliphatic primary amine [9].

Air-deflecting countervanes mounted directly on the engines of Sikorsky S-58 helicopters are made with epoxy–glass-cloth laminates. Service life is extended five times over that of aluminum with over 10 per cent weight savings. (*Union Carbide Corp.*)

As the strength properties of the laminate increase through the use of higher glass contents and improved glass, interlaminar shear becomes the weakest link in the chain. Interlaminar shear values generally fall between 5,000 and 7,000 psi [11], although values up to 15,000 psi have been obtained. It appears that the wetting action and the degree of elongation of the resin are controlling factors in interlaminar shear strengths. Correlations between interlaminar shear and compressive strength have been obtained, both types of failure occurring by a debonding process [179], again indicating the influence of the resin on this property. It appears unlikely, however, that dramatic improvements in interlaminar shear will be realized by changes in resin formulation alone. Perhaps even more significant than the matrix is the influence of voids—the interlaminar shear varying inversely with the void content [179]. Additional improvements can be achieved with proper fiber orientation. With filament-wound structures, improvements up to 90 per cent have been obtained with triplanar winding patterns [177].

The coefficient of thermal expansion of laminates is about that of steel, 5 to 6×10^{-6} in./(in.)(C°), but will vary depending upon the direction being measured. It is minimum when determined at an angle of 0° to the major axis. Expansion characteristics may be altered after exposure to a critical temperature range. Typical of the values for thermal expansion are those indicated in Figs. 22-31 and 22-32.

Thermal conductivity of the laminate when determined edgewise is influenced more by the reinforcement than by the resin; when conducted normal to the laminate, the resin is more influential. Glass laminates tested edgewise give considerably higher values for thermal conductivity than do castings [90]. Values of 2.3 (compared to 314 for steel and 1,416 for aluminum) are typical. Specific values are a function, in part, of the glass. S-glass, for example, gives slightly higher values than E-glass.

Of mportance, particularly to the missile and airframe industry, are flame-resistant laminates. Laminates with very short flame-out times can be obtained by the use of chlorinated curing agents, such as CA, chlorinated additives such as chlorinated paraffins, or chlorinated or brominated resins. Conventionally, a few

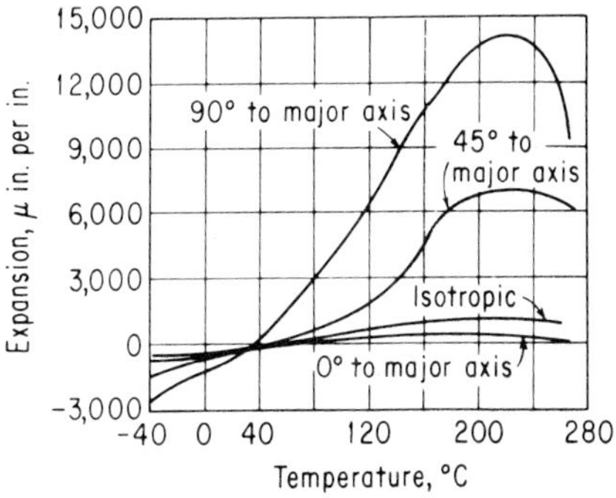

FIG. 22-31. Influence of reinforcement orientation on thermal expansion of DGEBA laminate containing 6ON Roving [90].

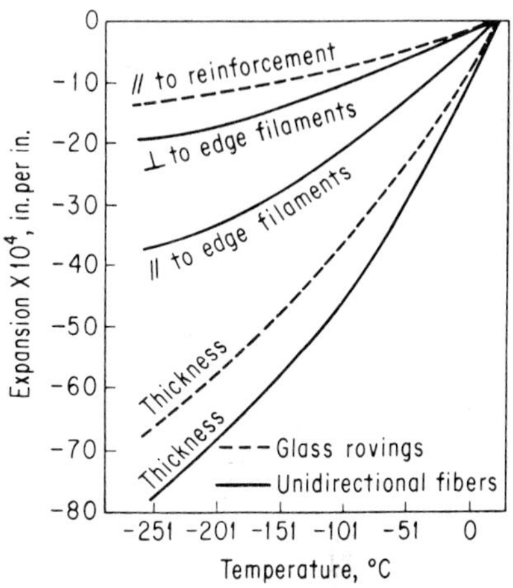

FIG. 22-32. Low-temperature contraction of epoxy laminates [180].

parts of antimony trioxide are added to such formulations to assist in extinguishing the flame after removal of the source. Flame-out times of flame-resistant laminates are from less than 1 to about 3 seconds, depending on the formulation. Typical properties of a laminate based on chlorinated DGEBA and cured with MDA are given in Table 22-33.

Table 22-33. Typical Properties of Chlorinated DGEBA/MDA Laminate [4-7]

Property	*Value*
Flexural properties:	
At 23°C:	
Strength, psi	85,000–97,000
Modulus, psi $\times 10^{-6}$	3.8–4.2
At 127°C:	
Strength, psi	65,000–68,000
Modulus, psi $\times 10^{-6}$	3.6–3.9
At 23°C after 2-hr water boil:	
Strength, psi	92,000
Modulus, psi $\times 10^{-6}$	4.0
Tensile properties at 23°C:	
Strength, psi	66,500
Modulus, psi $\times 10^{-6}$	3.8
Edgewise compressive strength at 23°C, psi	53,300
Modulus, psi $\times 10^{-6}$	4.2
Hardness, Rockwell	M114
Impact strength, Izod	17.7
Water absorption after 24 hr	0.034% gain

Electrical Properties

As with fillers generally, the effect of glass on the electrical properties of the epoxy resins is to increase the dielectric constant in proportion to the amount of glass present, as indicated in Fig. 22-33. Figures 22-34 and 22-35 compare the dielectric constant of DGEBA cured with 3 phr BF_3MEA vs. frequency and temperature for a casting and a laminate. The effect on other electrical properties is, in the main, of less significance. Figures 22-36 and 22-37 compare the dissipation factor of DGEBA cured with 3 phr BF_3MEA vs. frequency and temperature for a casting and a laminate.

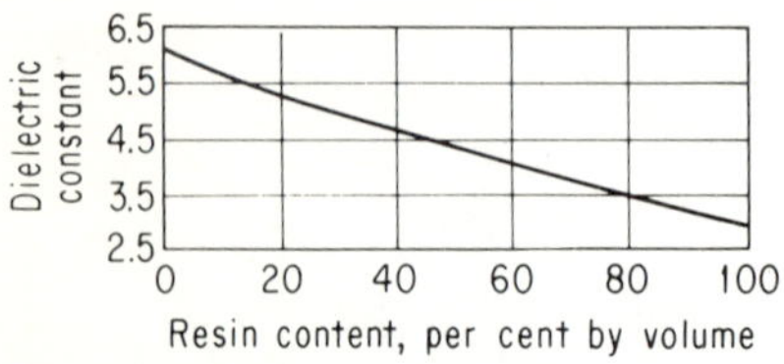

FIG. 22-33. Effect of resin content on dielectric constant of DGEBA laminates at room temperature and 9,375 Mc frequency [68].

The electrical properties for laminates composed of DGEBA cured with DADPS with an accelerator are shown in Figs. 22-38 and 22-39. The effect of resin molecular weight and curing agent on the electrical properties is shown in Figs. 22-40 to 22-43. The electrical properties for a blend of resins cured with DADPS are shown in Figs. 22-44 and 22-45. The electrical properties for several other epoxy laminates are shown in Tables 22-34 to 22-39.

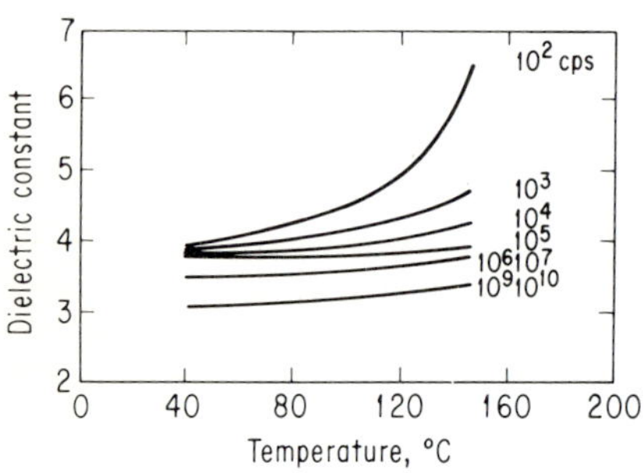

FIG. 22-34. Dielectric constant vs. temperature and frequency for DGEBA casting cured with 3 phr BF_3MEA [118].

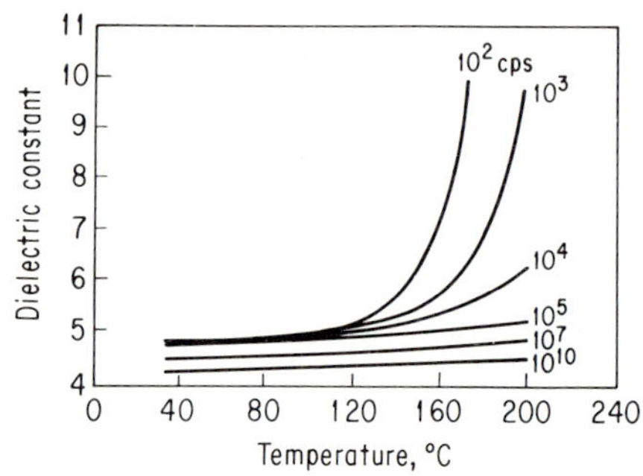

FIG. 22-35. Dielectric constant vs. temperature and frequency for DGEBA laminate cured with BF_3MEA [118].

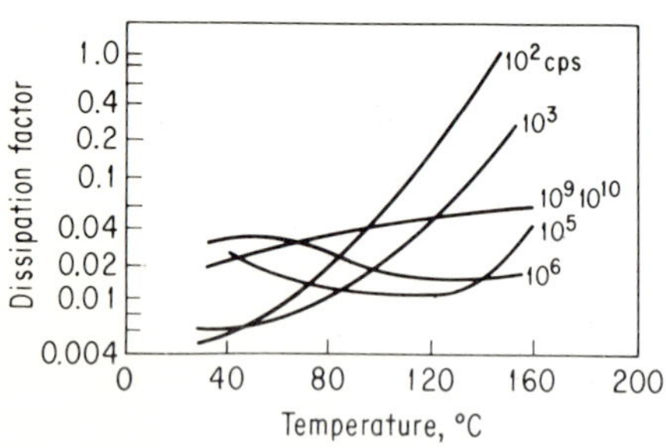

FIG. 22-36. Dissipation factor vs. temperature and frequency for DGEBA casting cured with BF_3MEA [118].

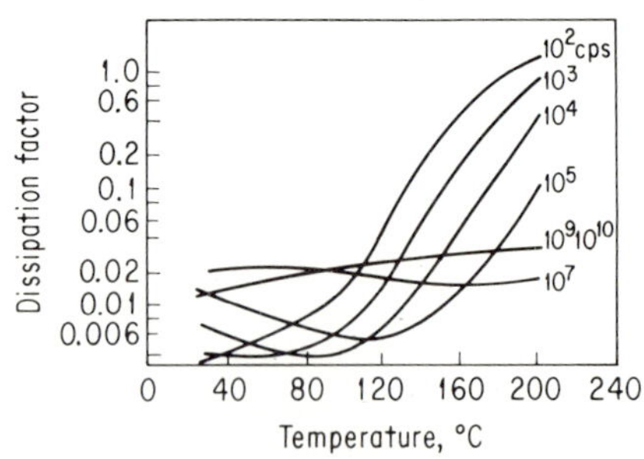

FIG. 22-37. Dissipation factor vs. temperature and frequency for DGEBA laminate cured with BF_3MEA [118].

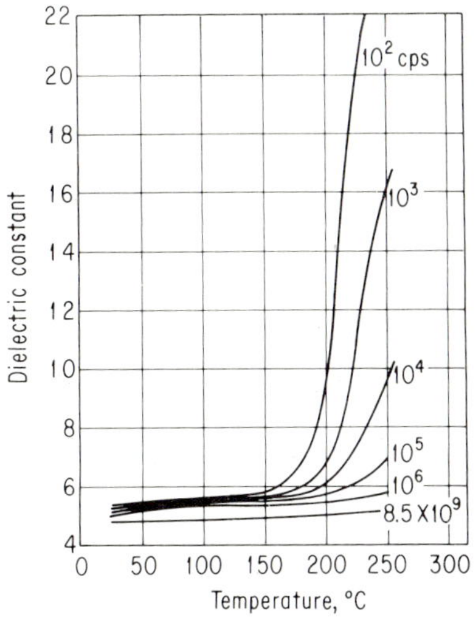

FIG. 22-38. Dielectric constant vs. temperature and frequency for DGEBA laminate cured with DADPS and 1 phr BF_3MEA [4-6].

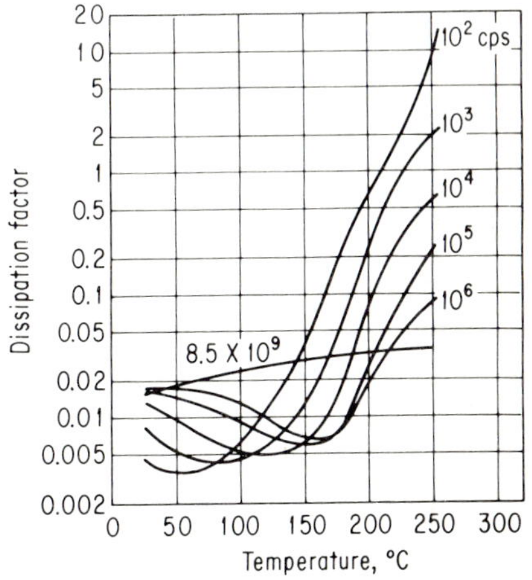

FIG. 22-39. Dissipation factor vs. temperature and frequency for DGEBA laminate cured with DADPS and 1 phr BF_3MEA [4-6].

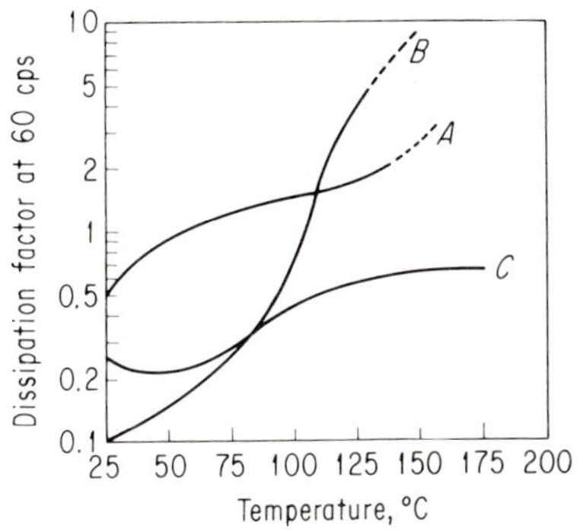

FIG. 22-40. Effect of temperature on dissipation factor of DGEBA laminates [44]. *A*, DGEBA/MPDA; *B*, DGEBA (mol. wt. 950)–dicyandiamide; *C*, DGEBA (mol. wt. 450)–CA.

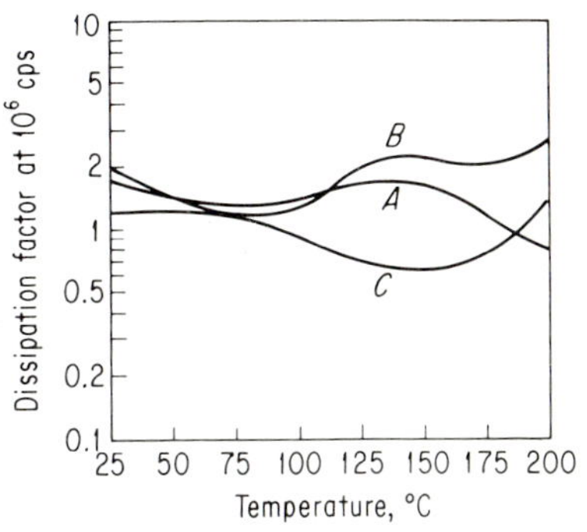

FIG. 22-41. Effect of temperature on dissipation factor of DGEBA laminates [44]. *A*, DGEBA/MPDA; *B*, DGEBA (mol. wt. 950)–dicyandiamide; *C*, DGEBA (mol. wt. 450)–CA.

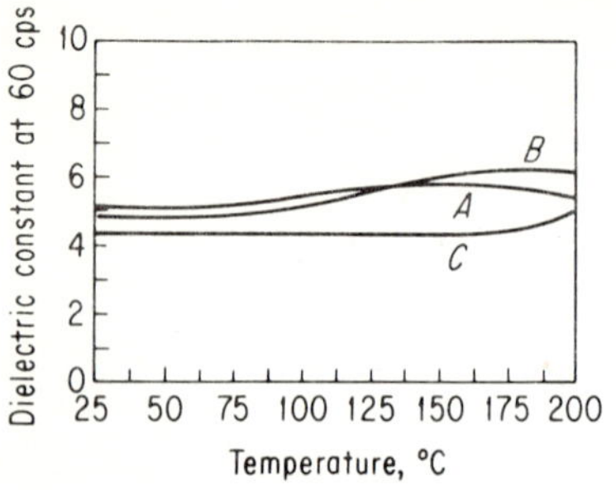

FIG. 22-42. Effect of temperature on dielectric constant of DGEBA laminates [44]. *A*, DGEBA/MPDA; *B*, DGEBA (mol. wt. 950)–dicyandiamide; *C*, DGEBA (mol. wt. 450)–CA.

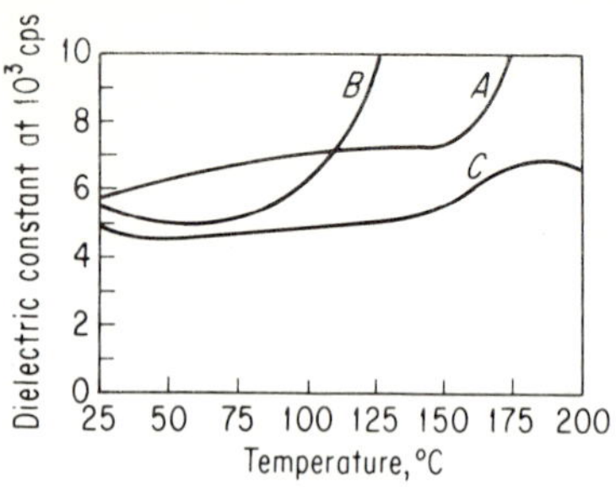

FIG. 22-43. Effect of temperature on dielectric constant of DGEBA laminates [44]. *A*, DGEBA/MPDA; *B*, DGEBA (mol. wt. 950)–dicyandiamide; *C*, DGEBA (mol.wt. 450)–CA.

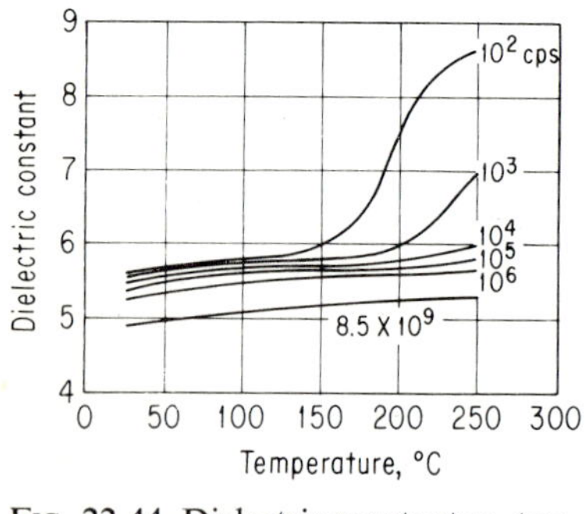

FIG. 22-44. Dielectric constant vs. temperature and frequency for laminates of DGEBA/polyglycidyl ether of tetraphenylethane (1/1) cured with DADPS [4-6].

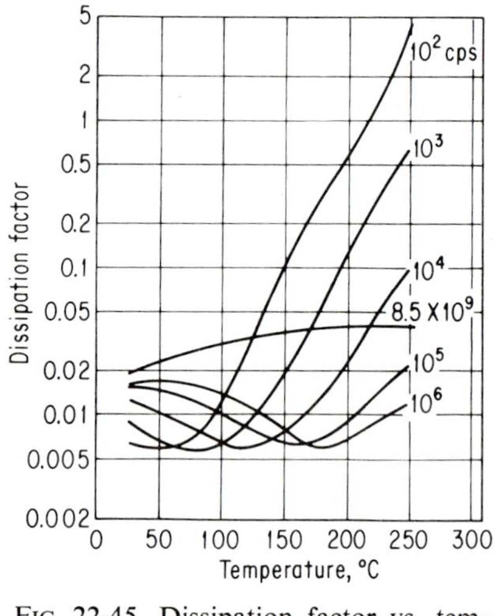

FIG. 22-45. Dissipation factor vs. temperature and frequency for laminates of DGEBA/polyglycidyl ether of tetraphenylethane (1/1) cured with DADPS [4-6].

The electrical properties of the epoxy-resin laminates are of most interest for applications utilizing electrical-grade tapes, for printed-circuit boards, for radome applications [88], and, as hand layups made in the field, involving synthetic-tape reinforcements, in high-voltage cable terminations [194 to 200].

Chemical Resistance

Chemically resistant, rust-free piping and tanks represent one of the largest potential markets for epoxy laminates. Because of this, increasing attention is being given to studies intended to characterize changes in properties caused by exposures to chemical environments.

Water is the most common fluid which must be resisted, and in this connection epoxy laminates are superior to both phenolics and polyesters [15]. At room

Table 22-34. Comparison of Electrical Properties of Four DGEBA Laminates [3]

[DGEBA (mol. wt. 380) with DEAPA and MPDA (mol. wt. 1,000) with other compounds]

Curing agent and property	At 10^2 cps	At 10^3 cps	At 10^4 cps	At 10^6 cps
Diethylaminopropylamine:				
Power factor	0.0192	0.0167	0.0131	0.0222
After 24 hr in water	0.0292	0.0195	0.0141	0.0226
Dielectric constant	5.16	5.01	4.91	4.77
After 24 in water	5.26	5.08	4.96	4.82
Metaphenylenediamine:				
Power factor	0.0231	0.0180	0.0163	0.0202
After 24 hr in water	0.526	0.0347	0.0213	0.0211
Dielectric constant	5.68	5.25	5.38	5.22
After 24 hr in water	6.03	5.66	5.44	5.23
Dicyandiamide:				
Power factor	0.0093	0.0080	0.0102	0.0219
After 24 hr in water	0.0213	0.0106	0.0108	0.0226
Dielectric constant	5.35	5.27	5.22	5.05
After 24 hr in water	5.44	5.35	5.27	5.12
Phenolic resin:				
Power factor	0.0218	0.0197	0.0127	0.0157
After 24 hr in water	0.0866	0.0784	0.0331	0.0213
Dielectric constant	4.65	4.49	4.38	4.33
After 24 hr in water	5.69	5.10	4.76	4.51

Table 22-35. Typical Electrical Properties of Laminate Cured with MDA vs. Molecular Weight of DGEBA [15]

(Cure cycle: Pressed 1 hr at 160°C; postcured 2 hr at 150°C + 6 hr at 200°C)

Property	DGEBA, mol. wt.		
	350–400	1,000	1,500
Power factor at 10^6 cps:			
Dry	0.017–0.019	0.015–0.018	0.010–0.014
After 1 day in water	0.010–0.025	0.019–0.021	0.015–0.018
After 2 days in water at 50°C	0.010–0.035	0.020–0.025	0.015–0.020
Dielectric constant at 10^6 cps:			
Dry	4.2–4.6	4.5–4.7	4.7–4.9
After 1 day in water	4.4–4.8	4.5–4.9	4.8–5.1
After 2 days in water at 50°C	4.6–5.3	4.6–5.2	4.9–5.2
Loss factor at 10^6 cps:			
Dry	0.07–0.09	0.07–0.09	0.05–0.07
After 1 day in water	0.08–0.12	0.09–0.10	0.08–0.09
After 2 days in water at 50°C	0.10–0.20	0.11–0.12	0.08–0.10
Dielectric strength, volts/mil:			
Perpendicular to laminations	500–550	450–480	450–500
After 2 days in water at 50°C	400–500	300–350	350–450
Arc resistance, seconds	130–180	140–180	155–185
Insulation resistance, megohms	$5\text{–}10 \times 10^5$	1×10^6	$1\text{–}3 \times 10^6$

Table 22-36. Typical Electrical Properties of Laminate of DGEBA Cured with MPDA [4-6]

Property	At 23°C	At 100°C
Dielectric constant:		
At 60 cps	5.50	6.10
At 10^2 cps	5.49	
At 10^3 cps	5.42	
At 10^4 cps	5.33	
At 10^5 cps	5.18	
At 10^6 cps	5.05	
At 1.5×10^7 cps	4.89	
At 7.8×10^7 cps	4.81	
Dielectric strength, volts/mil:		
Short time	410	
Step-by-step	310	
Dissipation factor:		
At 60 cps	0.006	0.004
At 10^2 cps	0.0096	0.0139
At 10^3 cps	0.0089	0.0084
At 10^4 cps	0.0135	0.0061
At 10^5 cps	0.0173	0.0072
At 1.5×10^7 cps	0.0204	
At 7.8×10^7 cps	0.0197	
Arc resistance at 50% relative humidity, seconds	127	
Insulation resistance, megohm $\times 10^{-6}$:		
Initial	......	20
After 24 hr at 2°C and 90% relative humidity	......	5
At 60°C and 95% relative humidity	......	1
Volume resistivity, ohm-cm $\times 10^{-12}$:		
Initial	......	1,923
After 24 hr at 2°C and 95% relative humidity	......	2,000
Surface resistivity, ohm-cm $\times 10^{-12}$:		
Initial	......	238
After 24 hr at 2°C and 95% relative humidity	......	240

Table 22-37. Typical Electrical Properties of Laminate of DGEBA/Phenolic Resin Blend, Catalyzed by 0.5 Per Cent α-Methyl-BDMA [15]

(Cure cycle: Pressed at 400 psi for 1 hr at 160°C; postcured 8 hr at 160°C)

Property	*Value*
Power factor at 10^6 cps:	
Dry	0.010–0.018
After 1 day in water	0.010–0.020
After 2 days in water at 50°C	0.014–0.021
Dielectric constant at 10^6 cps:	
Dry	4.2–4.9
After 1 day in water	4.6–5.1
After 2 days in water at 50°C	5.0–5.2
Loss factor at 10^6 cps:	
Dry	0.04–0.09
After 1 day in water	0.05–0.10
After 2 days in water at 50°C	0.06–0.11
Dielectric strength:	
Perpendicular to laminations, volts/mil	450–500
After 2 days in water at 50°C	350–450
Arc resistance, seconds	150–180
Insulation resistance, megohms $\times 10^{-5}$	1–3

Table 22-38. Effect of Temperature on Electrical Properties of Laminate of DGEBA Cured with MA/PMDA and PMDA [12-4]

Property	PMDA		MA/PMDA	
	Cured 24 hr at 180°C	Aged 200 hr at 200°C	Cured 24 hr at 180°C	Aged 200 hr at 260°C
Dielectric constant at 10^3 cps:				
At 23°C	4.06	4.67	4.11	4.20
At 100°C	4.94	4.72	4.43	5.21
At 150°C	4.96	4.69	4.51	5.21
At 200°C	4.97	4.77	5.38	5.21
Power factor at 10^3 cps:				
At 23°C	0.0013	0.0052	0.0041	0.0044
At 100°C	0.0053	0.0031	0.0034	0.0036
At 150°C	0.0071	0.0035	0.0047	0.0038
At 200°C	0.0650	0.0052	0.0136	0.0042

temperature, extrapolated data indicate that complete bond failure would occur between treated glass and epoxy only after 45 years' total immersion [142]. However, even though the epoxies provide superior resistance, the effect of hot water (65 to 85°C), even at low pressures, is very severe, and thus hot water constitutes one of the worst services. Resistance to higher humidities at elevated temperature may be even more critical.

The role of glass surface finishes on water resistance has been investigated in some detail, and the general conclusion is that treating agents do improve performance. On studies with flat plates of E-glass it was found that without coupling agents, the failures appear to be strictly adhesional; with coupling agents, failures were in either the matrix or at the coupling agent, indicating that the beneficial effects were due primarily to the influence of the agent on the resin. Interestingly, under normal conditions an inverse relationship between wetting and bond-strength retention was observed [161]. It has also been noted that alkaline-deficient E-glass provides improved performance over alkaline-rich E-glass [142].

The mechanism of failure appears to be the entrance of water (either through the resin or at the interface). It has not yet been established that the specific failure occurs because of the leaching out of alkaline materials from the glass surface, but hydrochloric acid–etched glass is superior to nonetched. However, although changes in formulation may affect some improvements [189], it is probable that this has only a secondary effect. Water enters at the interface at a rate some 450 times as rapidly as through the resin itself [161].

Table 22-39. Typical Electrical Properties of Laminate of DGEBA Cured with 10 phr DMP-30 and Flexibilized with Polysulfide (mol. wt. 1,000) [10-2]

Property	Polysulfide, phr						
	0	16.7	20	25	33.3	50	100
Dielectric constant:							
At 10^3 cps	4.4	4.6	4.6	4.7	4.9	5.0	5.2
At 10^9 cps	3.6	3.6	3.8	3.8	4·0	4.1	4.2
Dissipation factor:							
At 10^3 cps	0.0020	0.0026	0.0028	0.0030	0.0033	0.0035	0.0040
At 10^9 cps	0.021	0.011	0.013	0.014	0.016	0.019	0.023

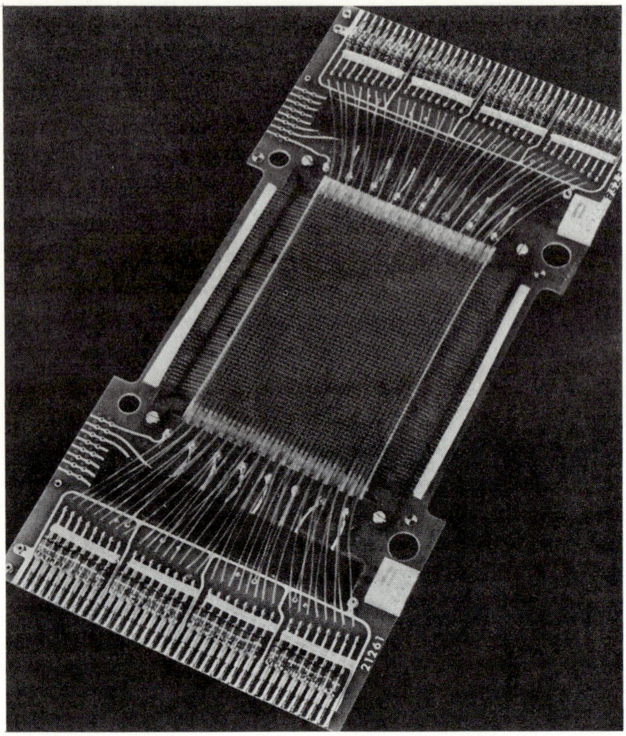

Printed electronic circuit for computer made from epoxy-resin laminate. (*Dow Chemical Co.*)

One man can handle eight 30-ft lengths of Red Thread epoxy-glass pipe produced by Dow Smith, Inc. Pipe is light but strong and withstands corrosion from the ground, petroleum crudes, salt water, etc. (*Dow Chemical Co.*)

In unstressed, filament-wound composites using HTS glass, quite good humidity resistance has been obtained. However, when stressing or temperature cycling is imposed on the structures, microscopic cracks occur at the glass-resin interface. The crazing can occur at as low as 10 per cent of the ultimate load and would appear to occur at levels below the ultimate elongation of the matrix. Crazing occurs routinely in filament-wound vessels and may account for the poor observed moisture resistance [148].

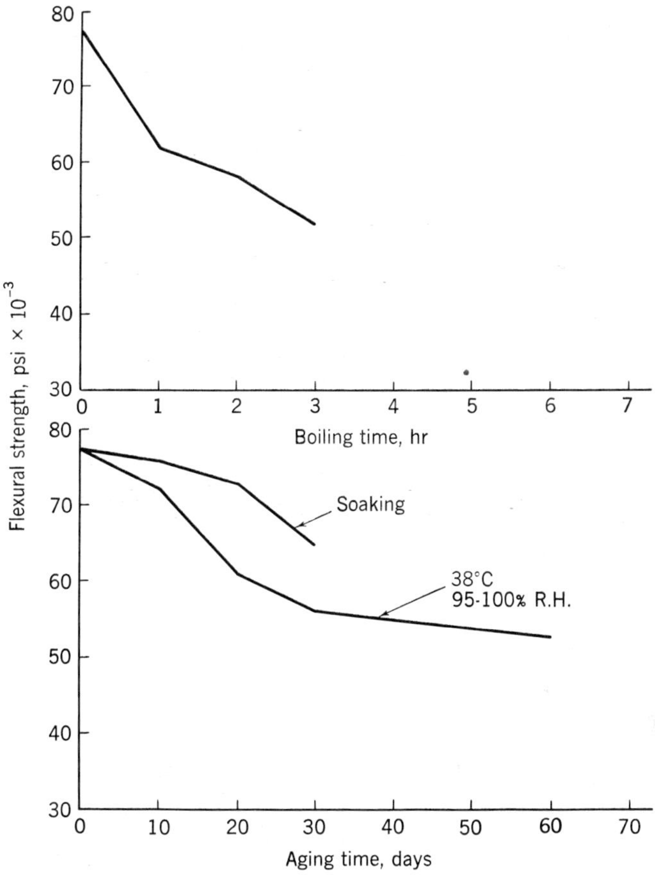

FIG. 22-46. Effect of water on DGEBA laminates cured with 4 phr dicyandiamide [4-6].

The general effect of water on resin strength is augmented by the lubricating effect of water on the coefficient of friction between the resin and glass fiber. As indicated earlier, once the adhesive bond between the glass fiber and resin has been broken by stressing, only the resin shrinkage pressure against the glass fiber, and mechanical locking action, provides for stress transfer. Because this is a friction mechanism and water is a lubricant, the strength of the laminate is appreciably reduced (Fig. 22-46), although moisture pickup may not be too great, as shown in Figs. 22-47 to 22-50. The effect of water and oil on deflection for a laminate under constant load is indicated in Fig. 22-51, and the effect of air and water on deflection is given in Fig. 22-52.

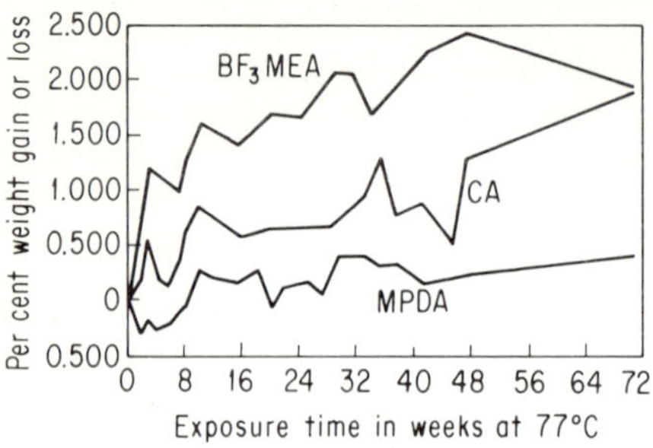

FIG. 22-47. Deionized water resistance vs. curing agent for DGEBA laminates [35].

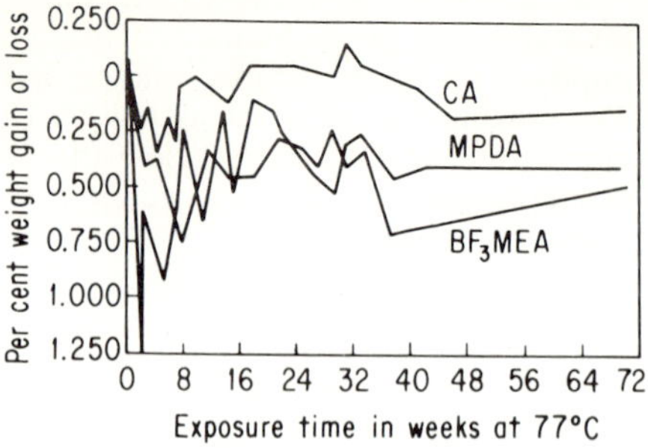

FIG. 22-48. Resistance to water (pH 2) vs. curing agent for DGEBA laminates [35].

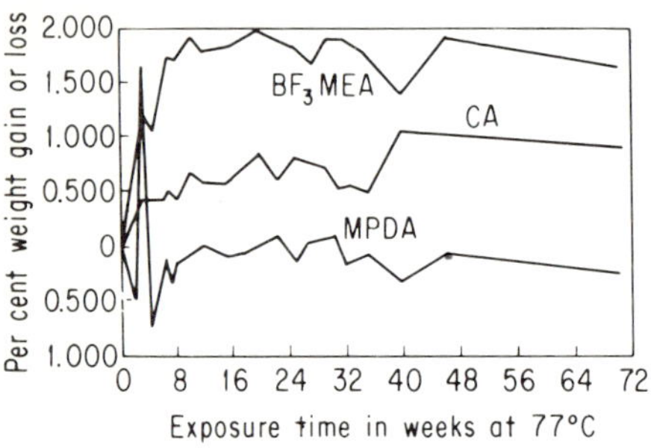

FIG. 22-49. Resistance to water (pH 11) vs. curing agent for DGEBA laminates [35].

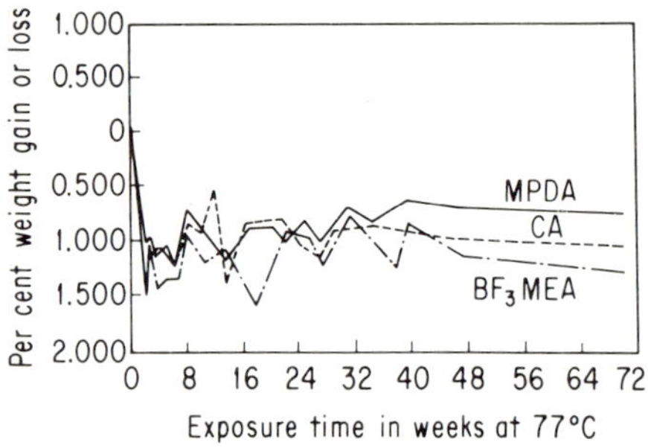

FIG. 22-50. Resistance to moist air vs. curing agent for DGEBA laminates [35].

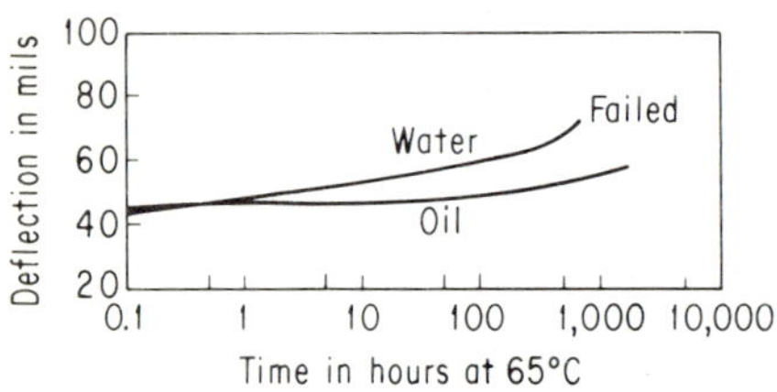

FIG. 22-51. Effect of oil and water on creep resistance of DGEBA laminate loaded at a flexural stress of 25,000 psi [106].

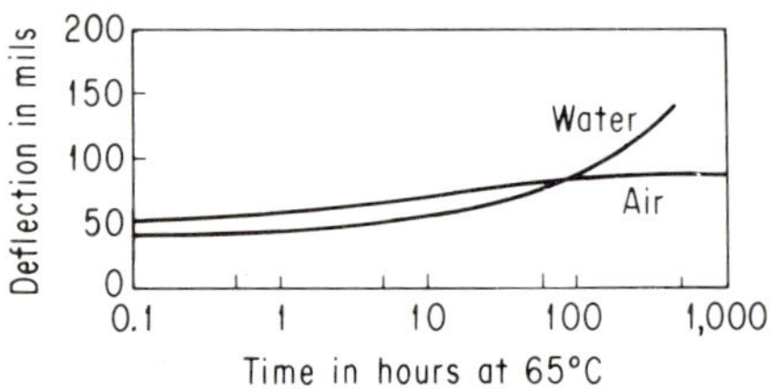

FIG. 22-52. Effect of oil and water on creep resistance of DGEBA casting loaded at a flexural stress of 3,500 psi [106].

Special test procedures, involving elevated temperatures and higher relative humidities, have been developed to establish the service life of filament-wound piping [132]. The surfaces of the laminates are resin-rich to improve moisture and humidity resistance, since this, in effect, shields the glass fibers from the environment and minimizes the effects of internal adhesive failure. The importance of surface protection was demonstrated in a series of hydrostatic tests with cut laminates [163]. Under 7,000 psig, the laminates experienced high water pickups leading to delamination. A brush-on epoxy coating, however, reduced water pickup to only 0.05 per cent after 100 cycles at 1 cpm from 0 to 7,900 psig, and no delamination was experienced.

Other chemicals than water also result in significant reduction in properties. The action of some of these is emphasized by any tendency of the reagent to swell or dissolve the cured resin or attack the glass. Specific data for DGEBA laminates cured with an aromatic amine and an anhydride are shown in Tables 22-40 and 22-41.

Table 22-40. Chemical Resistance of Laminate of DGEBA Cured with MPDA [4-6]

Cure cycle: Pressed at 150°C at 200 psi for 15 min; postcured 1 hr at 204°C

System tested in	% weight gain after immersion for period noted			Flexural strength after 1-year immersion, psi	Flexural modulus after 1-year immersion, psi × 10^{-6}
	30 days	180 days	365 days		
Sulfuric acid, 95%	Delaminated after 7 days				
70%	0.20	0.34	0.42	72,000	3.4
3%	−0.09	−0.45*	0.09	48,000	2.2
Hydrochloric acid, 37%	−0.21	−5.7	.....	40,200†	2.6†
10%	0.00	−0.09	−0.32	49,900	2.6
Nitric acid, 50%	Top layer delaminated				
30%	−0.21	−4.8	.....	13,700†	1.9*
Phosphoric acid, 98%	0.10	0.74	0.90	67,300	2.9
10%	0.27	0.54	0.65	66,600	2.9
Acetic acid, glacial	0.34	0.42	0.25	65,200	3.0
10%	0.29	0.62	0.73	61,300	3.3
Oxalic acid, sat.	0.24	2.5	.....	43,000†	2.4†
Sodium hydroxide, 50%	−0.23	−1.3	.....	66,900†	2.3†
1%	0.17	0.36	0.50	64,500	3.0
Ammonium hydroxide, 28%	0.70	2.3	2.4	24,400	2.8
Sodium chloride, 20%	0.24	0.40	0.67	69,300	3.4
Sodium sulfate, 30%	0.24	0.64	0.74	67,200	3.5
Ammonium nitrate, 50%	0.21	0.60	0.73	56,200	2.7
Acetone	0.33	1.0	.....	53,000†	2.8†
Ethyl acetate	0.45	0.45	0.72	65,500	2.9
Ethylene dichloride	0.35	0.59	0.55	66,000	2.7
Ethylene glycol	−0.03	−0.23	0.09	70,300	3.1
Hydraulic brake fluid	0.18	0.40	0.54	64,800	3.5
Jet fuel	0.08	0.14	0.32	71,800	3.2
108-octane gasoline	0.07	0.13	0.31	66,400	3.1
Methyl ethyl ketone	0.24	0.34	0.35	72,700	2.9
Isopropyl alcohol	−0.02	−0.05	0.01	65,700	3.5
Allyl chloride	0.08	0.72	1.1	69,400	2.9
Hydrogen peroxide	Delaminated				
Water, distilled	0.21	0.67	0.63	70,900	3.7

* 90-day data.
† 180-day data.

Extensive data for filament-wound piping are presented in Table 22-2. Epoxy piping is used in handling sour crude oil, aviation gas, JP-4, diesel fuel, and special Navy fuel oil, even with primary aliphatic amine cures and when modified with about 50 phr of a polysulfide [53].

Outdoor weathering tests on laminates of DGEBA (mol. wt. 950) cured with dicyandiamide have indicated, over a 12-month period, a very nominal reduction in strength, on the order of 5 to 10 per cent [19], and tests with other systems give generally comparable values, although differences between systems are noted and high-temperature systems do not necessarily provide improved weathering performance [131].

Table 22-41. Chemical Resistance of Laminate of DGEBA (mol. wt. 450) Cured with CA [4-1]

(After 30 days' immersion)

System tested in	Weight gain, %	Dimensional change, %	Flexural strength retained, %	Flexural modulus retained, %
Distilled water	0.272	0.000	91	100
Hydrogen peroxide, 3%	0.188	0.000	32	61
Sulfuric acid, 3%	−0.266	0.000	86	100
Hydrochloric acid, 10%	−3.61	0.000	49	70
Nitric acid, 10%	−0.119	0.000	41	82
Acetic acid, 5%	0.173	0.000	100	100
Ammonium hydroxide, 10%	0.236	0.000	91	100
Sodium chloride, 10%	0.015	0.000	100	100
Isopropyl alcohol, 98%	0.017	0.000	100	100
Hydrocarbon fluid, 100%	0.089	.....	98	100
Boiling water after 3 hr	......	.....	83.4	100
Refluxing acetone after 3 hr	......	.....	78.8	70

Laminates generally are somewhat porous to high-pressure gases and may tend to wick in contaminants after the resin-glass bond has been broken by field stresses. Impermeability may be obtained, where practical, by the use of an impermeable outer membrane or by sandwiching metal foil into the laminated structure.

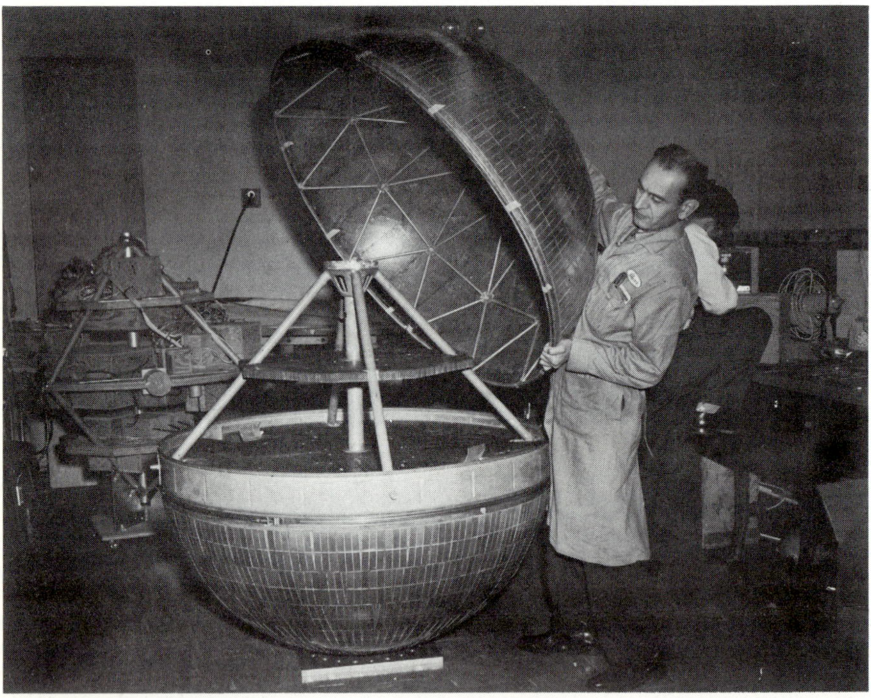

Assembly of Courier simulated mechanical model in preparation for engineering evaluation. Shell is composed of epoxy–glass-cloth laminate. (*Philco Corp. Western Development Labs., Palo Alto, Calif.*)

In some applications, it is of interest to prevent silica from the glass contaminating the working fluid. Tests conducted using epoxy-glass pipe, with synthetic liners, indicate there is a very low transmission rate, as shown in Table 22-42.

Table 22-42. Migration in Parts per Billion per Day of Silica to Demineralized H_2O [145]

Temperature, °C	Weeks			
	1	2	3	4
25	19	5	6	0.5
50	26	10	8	2
65	83	17	16	10
82	69	53	60	27

NOTE: All samples except the one tested at 82°C contained epoxy-impregnated Dynel liners. The 82°C sample contained epoxy-impregnated asbestos liner.

In tests in exotic space environments, epoxy laminates have proved satisfactory. Laminates exposed to ultraviolet radiation from 2,000 to 6,000 angstroms, pressures of 10^{-7} mm mercury, and temperatures up to 160°C have been little affected after 500 hours. One system, employing fluidized DGEBA, for example, provided a weight loss of 0.83 per cent and a reduction in flexural strength from 95,300 to 86,300 psi [108].

EPOXY LAMINATES USING OTHER THAN GLASS-FIBER REINFORCEMENT

Fibrous Reinforcement

Glass fiber, itself, has rather poor chemical resistance, and this will detract from the overall chemical resistance inherent in the epoxy-resin matrix. For this reason, use is sometimes made of a liner of some other synthetic fiber to protect the high-strength glass from the chemical environment. Asbestos fibers are frequently used for this purpose.

A considerable amount of epoxy resin is used in paper-based laminates made to NEMA grade FR-3 [139]. In this case, DGEBA (mol. wt. 950) is cured with CA to provide flame resistance. Epoxy-impregnated paper is also used for the manufacture of paper-resin core condenser bushings in the 15- through 69-kv class.

The properties of various synthetic fibers, suitable for use with the epoxies, are indicated in Tables 22-43 and 22-44.

Specialty applications for synthetic laminates include the use of polyamide fibers to increase impact resistance over that obtained with glass cloth [31], and their use in conjunction with glass mat for the same purpose [57]. Laminates have been made from epoxy resin and acrylic fibers, such laminates providing low moisture absorption and high resistance to boiling water [25]; they are used commercially for the manufacture of potheadless cable terminations in voltage classes up to and including 15 kv [47]. Laminates with polyester fiber have been used to make synthetic reeds for wind instruments [114]. Carbon-based filaments have been produced which are suitable for filament-winding operations. With MPDA-cured systems, acceptable strength properties have been obtained, but with less than optimum wetting; with NMA cured systems, poor wetting was experienced [166].

Table 22-43. Properties of Fibers [136]

Fiber	Specific gravity	Length, inches	Diameter, microns	Tensile strength, psi × 10^{-3}	Modulus of elasticity, psi × 10^{-6}	Heat resistance, °C	Coefficient of linear expansion, in./(in.)(°C) × 10^6
Synthetic inorganic:							
Borosilicate (E-glass)	2.6	*a, b*	5–15	400	10.5	160[c], 435[d]	2.8
Beryllium glass	2.6	*a, b*	5–15	280	12–20	435[d]	6
Quartz (fused silica)	2.2	*a, b*	8–10	100–350	10–25	1040[e]	5–7
Carbon	1.8	*a, b*	1–100	20	1–4	1890[e]	1–3
Aluminum silicate	2.7	Up to 10	2–20	100–600	2–15	985[d]	1–8
Graphite	1.6	Up to 4	2–30	2–20	.....	2060[f]	0.6–4
Rock wool	2.8	Up to 4	1–22	....	.....	835[d], 1425[c]	2–6
Aluminum oxide (Al_2O_3)	3.99	*a, b*	250	100	55	2010[f]	7
Beryllium oxide (BeO)	3.0	Up to 4	100	3.0	40–50	2565[d], 1485[c]	5.8
Titanium oxide (TiO_2)	4.2–4.4	Up to 4	100	....	20–50	1900[d]	7–9
Zirconium oxide (ZrO_2)	5.5	Up to 4	100	....	26	2690[d]	
Magnesium oxide (MgO)	3.65	......	100	70–150	25–32	2800[d]	
Natural inorganic:							
Asbestos	2.5	Up to 3	0.02	100–200	20–25	825[d]	
Metals and refractories:							
Steel	7.8	*a, b*	1–25	200–400	20–30	870[e]	8–10
Aluminum	2.8	*a, b*	4–20	60–90	10	345[e]	17–20
Tungsten	19.3	Up to 1	20	200	58	1865[e]	4.5
Tantalum	16.6	Up to 0.5	5	70–90	28	1635[e]	6.6
Molybdenum	10.2	Up to 0.5	5–20	....	42	1425[e]	5.4
Magnesium	1.8	*a, b*	6.15	40	6	345[e]	8–20
Synthetic organic:							
Fluorocarbon	2.2	*a, b*	20	47	0.4	135[f]	
Polyester	1.4	*a, b*	10–25	100	.....	120[e]	
Acrylic	1.2	*a, b*	10–25	50	.....	110[e]	
Polyamide	1.1	*a, b*	10–40	70–120	.....	120[e]	
Cellulose acetate	1.3	*a, b*	11–44	25	.....	125[e]	
Regenerated cellulose (rayon)	1.5	*a, b*	10–40	30–105	.....	95[d]	
Natural organic:							
Cotton	1.6	Up to 2	17	50–110	.....	58[g]	
Sisal	1.3	Up to 24	19	120	.....	38[g]	
Wool	1.3	Up to 15	28	29	.....	38[g]	
Wood cellulose (paper)	0.7–1.15	Up to 0.5	1–4	....	.....	135[g]	

[a] Filament. [b] Staple. [c] Softens. [d] Decomposes. [e] Melts. [f] Sublimes. [g] Used up to this temperature.

Table 22-44. Comparison of Properties of Synthetic Fibers [155]

Rating	Resistance to abrasion and fatigue	Chemical resistance		Resistance to organic solvents	Heat resistance	Water absorbency	Specific gravity
		To acids	To alkalies				
Excellent	Nylons	Fluorocarbons	Fluorocarbons	Glass	Glass	Wool	Glass
	Polyesters	Olefins	Nylons	Fluorocarbons	Fluorocarbons	Rayons	Fluorocarbons
Good	Saran	Acrylics	Olefins	Acrylics	Acrylics	Cotton	Saran
	Modacrylics	Modacrylics	Saran	Polyesters	Cotton	Acetates	Cotton
	Polyesters	Saran	Polyesters	Nylons	Nylons	Nylons	Rayons
	Acrylics	Polyesters		Cotton	"Nytril"	"Nytril"	Modacrylics
Fair	"Nytril"	Vinyon	Glass	Rayons	Polyesters	Acrylics	Polyesters
	Wool	"Nytril"	Cotton	Wool	Acetates	Polyesters	Acetates
	Rayons	Glass	Rayons	Polyesters	Rayons	Saran	Wool
	Acetates	Wool	"Nytril"	Saran	Wool		"Nytril"
	Cotton	Nylons	Polyesters		Olefins		
Poor	Fluorocarbons	Cotton	Acrylics	"Nytril"	Saran	Olefins	Acrylics
	Glass	Acetates	Acetates	Olefins	Polyesters	Fluorocarbons	Nylons
		Rayon	Wool	Acetates		Glass	Olefins

Experimental composites with boron fibers (having moduli five times that of glass) gave tensile strengths up to 344,000 psi with moduli of 16.6×10^6. High-strength special steel wires (600,000 psi at 0.003 in. diameter with moduli of 26 to 30×10^6 psi) have been prepared [178].

In general, however, applications for fibers other than glass are somewhat specialized.

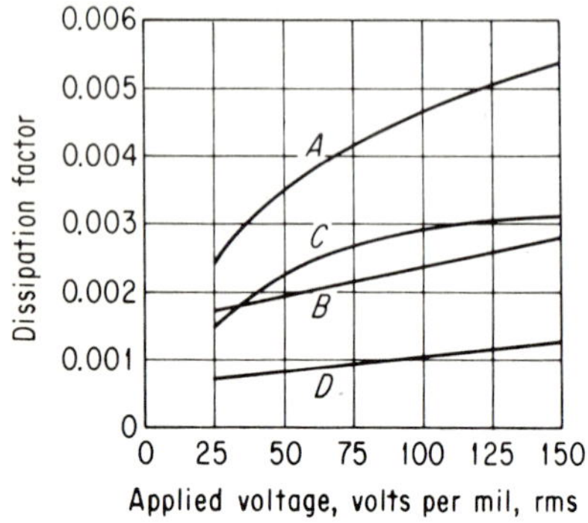

FIG. 22-53. Dissipation factor vs. applied voltage for electrical coil insulations [77]. *A*, Shellac-bonded mica folium; *B*, asphalt-bonded mica tape; *C*, epoxy-bonded mica folium; *D*, epoxy-bonded mica tape.

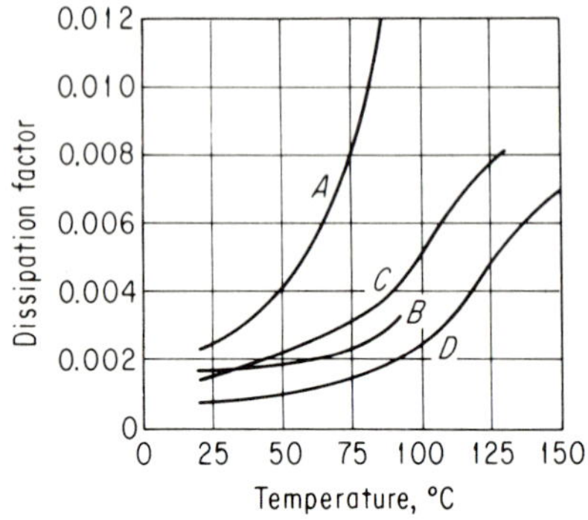

FIG. 22-54. Dissipation factor vs. temperature for electrical coil insulations [77]. *A*, Shellac-bonded mica folium; *B*, asphalt-bonded mica tape; *C*, epoxy-bonded mica folium; *D*, epoxy-bonded mica tape.

Metallic whiskers—small-diameter monocrystals—have occasioned considerable interest in experimental laboratories. Even though the length of the crystals is quite short, their flaw-free nature permits attaining tensile strengths up to at least 74,000 psi as compared to 113,000 psi for a comparable system employing glass fibers [173]. With sapphire (aluminum oxide) whiskers, sharp improvements in strengths are observed at as little as 14 per cent loading volumes, and addition of as little as 1 per cent to glass composites will produce a 10 to 20 per cent increase in interlaminar shear. With loading volumes possible to molding compounds, it is expected that whisker-based composites will give tensile strengths up to 150,000 psi [184].

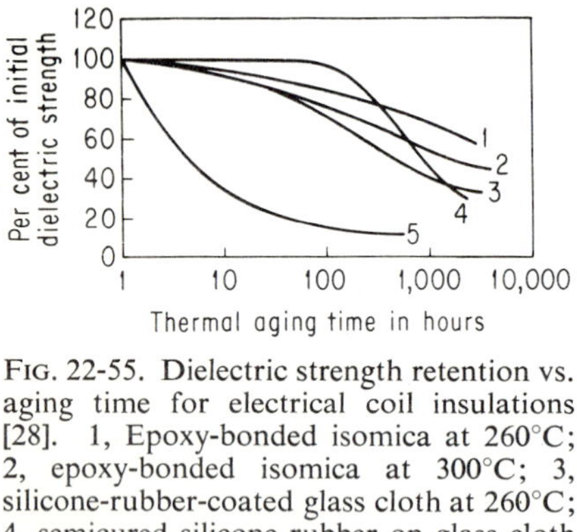

FIG. 22-55. Dielectric strength retention vs. aging time for electrical coil insulations [28]. 1, Epoxy-bonded isomica at 260°C; 2, epoxy-bonded isomica at 300°C; 3, silicone-rubber-coated glass cloth at 260°C; 4, semicured silicone rubber on glass cloth at 260°C; 5, silicone-varnished glass cloth at 260°C.

Flake Reinforcement

In addition to fibers, wires, and whiskers, flakelets have been investigated as reinforcement for the epoxy resins.

Laminates, for example, from glass flakes [59] have been suggested for printed-circuit boards and reinforced coatings [117].

Mica-epoxy laminates have achieved some popularity in the electrical industry for coil insulation. Epoxy-bonded mica insulation is compared with other types in Table 22-45 and Figs. 22-53 to 22-55.

Laminates of mica paper and epoxy provide good properties (Table 22-46) at a potentially lower cost than with glass cloth [85].

Table 22-45. Comparison of Electrical Insulations Tested on Coils [77]

Property	Shellac-bonded mica folium insulation	Asphalt-bonded mica tape insulation	Epoxy-resin-bonded mica folium insulation	Epoxy-resin-bonded mica tape insulation
Voltage breakdown at 50 cps after 60 sec at 90°C, volts/mil (rms)	290	360	490	525
Corona starting voltage, volts/mil (crest)	>35	>53	>45	>92

SUMMARY

Epoxy–glass-fiber composites provide high-strength structures with strength-to-weight ratios comparable to those of metals, and often superior to them. Initially investigated as structural members in the aircraft and missile industry, their use has extended to a variety of commercial applications, perhaps chief of which is chemically resistant piping.

Essentially, the composite is composed of some form of fibrous reinforcement together with a matrix.

Glass is the most commonly used reinforcement and may be in the form of strands (composed of a number of very fine filaments) for filament winding; the strands may be gathered into rovings, and these chopped to provide mats; or the strands may be twisted into yarns and woven into cloth. Filament windings provide the highest (unidirectional) strength properties, giving tensile strengths up to and exceeding 300,000 psi. Glass cloths give laminates having more uniform strength characteristics, which can provide values for tensile strengths up to 120,000 psi. Mat laminates provide lower-cost structures having uniform strength characteristics, but generally give lower strength properties than cloths.

For most laminating, a borosilicate glass is employed, the so-called E-glass which has virgin strengths as high, in filament form, as about 500,000 psi. Special sizing methods have been developed to preserve virgin strength during processing.

After the glass has been processed, it is usually treated with a chemical finish to eliminate the formation of water on the glass surface and to improve performance of the laminate under high humidities and hot-water conditioning.

The matrix is selected to provide low viscosity with convenient processing characteristics. When elevated-temperature performance is desirable, the formulation becomes more critical. For most applications, the DGEBA resins are preferred, but for higher temperature, the epoxidized novolacs and the epoxidized olefin resins are specified. The curing agent is generally selected to provide the desired handling characteristics and the requisite elevated-temperature strength retention.

The glass content of the composite generally determines the tensile and flexural strength properties, whereas the matrix more strongly influences the compressive strength and the heat resistance. With the exception of dielectric constant (which increases with increasing glass content), the electrical properties are a function of the resin formulation employed.

The composite may be constructed by one of several techniques: wet lay-ups may be used, employing a liquid system, which is cured immediately after application; or dry lay-ups may be used, employing either a higher-molecular-weight resin

Table 22-46. Physical and Electrical Properties of Mica Laminates with Amine-cured Epoxidized Novolac Resin [85]

Property	Integrated (phlogopite)	Reconstituted (muscovite)
Flexural strength, psi:		
At 23°C	58,000	60,000
At 260°C	11,500	17,000
Modulus, psi × 10^{-6}:		
At 23°C	12.4	10.5
At 260°C	3.0	5.7
Tensile strength, psi:		
At 23°C	37,000	38,000
At 260°C	10,000	11,000
Modulus, psi × 10^{-6}, at 23°C	15	12
Flatwise compressive strength, psi, at 23°C	20,000	Not run
Edgewise compressive strength, psi:		
At 23°C	8,000	17,000
At 260°C	3,600	5,300
Shear strength, psi, at 23°C	19,000	19,000
Specific gravity	2.4–2.6	2.4–2.6
Flammability, ASTM D 568–56T	Self-extinguishing	Self-extinguishing
Coefficient of expansion, in./(in.)(°C) × 10^6	9–14	9–14
Machinability	Fair–good	Fair–good
Heat resistance, °C	127–260	127–260
Peel strength of clad laminates, 1 oz, lb/in. width	9–12	9–12
Moisture absorption after 28 days' immersion, %	0.04	0.03
Dielectric constant:		
At 60 cps	2.82	3.08
At 10^3 cps	2.76	3.06
At 10^6 cps	2.60	2.94
After 24 hr in water:		
At 60 cps	....	4.25
At 10^3 cps	5.98	3.27
At 10^6 cps	3.04	3.01
Dissipation factor:		
As received, plain board:		
At 60 cps	0.020	0.0039
At 10^3 cps	0.010	0.0042
At 10^6 cps	0.006	0.0056
After 24 hr in water:		
At 60 cps	....	0.15
At 10^3 cps	0.93	0.052
At 10^6 cps	0.034	0.0097
Dielectric strength, 1/32 in. thick, volts/mil	1,000	1,200
Arc resistance after 96 hr at 96% relative humidity, seconds	35	179
Volume resistivity, ohm-cm:		
Plain	1.32×10^{15}	3.71×10^{15}
Clad	1.68×10^{15}	3.40×10^{15}
After 96 hr at 96% relative humidity:		
Plain	9.12×10^{14}	9.28×10^{14}
Clad	1.1×10^{12}	5.23×10^{14}
Surface resistivity, ohms: ...		
Plain	1.32×10^{15}	3.7×10^{15}
Clad	9.12×10^{14}	9.28×10^{14}
After 96 hr at 96% relative humidity:		
Plain	2.37×10^{14}	9.05×10^{14}
Clad	6.5×10^7	6.77×10^{13}

containing a latent curing agent or a B-staged system, these being carried onto the glass in solution, with the solvent being subsequently removed. The dry lay-up prepregs may be stored for subsequent use. With wet lay-ups, either room-temperature or heat cures may be employed, depending upon the nature of the formulation. Prepregs require heat cure.

Production techniques may involve contact-pressure, low-pressure, or high-pressure laminating, the particular technique chosen being based, to an extent, on the ultimate properties but strongly considering the exact production situation.

The epoxy-glass composites provide mechanical properties superior to the polyesters, electrical properties superior to the phenolics, and moisture resistance, fatigue resistance and interlaminar shear strengths superior to both. Better high-temperature strength is obtained with special phenolic and silicone formulations, although research directed to improving the epoxy resins in this respect is continuing.

The properties of the epoxy-glass composite can be varied rather widely, as is the case with other epoxy systems, by the selection of the resin, curing-agent, and modifier complex. Because of this versatility, and the overall satisfactory properties, the epoxy-resin–glass composites are finding increasing use in both military and civilian applications.

REFERENCES

1953

1. Erickson and Silver, Chemical Finishes for Glass Fiber Reinforcement, U.S. Government Report PB-121165.
2. Resin-Glass Tank Is Shatterproof, *Aviation Week*, Oct. 26.

1954

3. Elam and Hopper, Structural Laminates from Epoxy Resin Adhesives, *Mod. Plastics*, October.
4. Erickson et al., Universal Type Chemical Finishes for Glass Fibers Used in Reinforced Plastics, OTS PB-121172.
5. Hopper, High Strength Epon Laminates, WADC Technical Report 52-5, supplement 3, October.
6. Moldable Reinforced Sheet, *Mod. Plastics*, December.

1955

7. Barnet, The Resistance of Glass Reinforced Plastics to Temperature-Humidity Cycling, OTS PB-121173.
8. Hopper, Miscellaneous Aspects of Epon Resin Laminating, Shell Chemical Corp. Report 99-55.
9. Sokol, Epoxies for Tools and Dies, *Tooling and Production*, August.
10. Swackhammer, Reinforced Plastic Replaces Aluminum, *Aero Dig.*, January.
11. Werren, Mechanical Properties of Plastics Laminates, Forest Products Laboratory Report 1820B, September.

1956

12. Boller, Fatigue Properties of Various Glass-fiber-reinforced Plastic Laminates, WADC Technical Report 55-389.
13. Brookfield, Some Aspects of the Use of Glass Fibers for the Reinforcement of Epoxy Resins, British Society of Chemical Industry Symposium, April.
14. Dunn, Epoxy Resins in Glasscloth Laminates, *Reinforced Plastics*, October.
15. Godard et al. (Union Carbide Plastics Co.) Epoxy Resins in Reinforced Plastics, 11th NANTEC (RIP Division) of SPI.

16. Outwater, The Mechanics of Plastics Reinforcement Tension, *Mod. Plastics*, March.
17. Poneman, Thin-wall Epoxy-Glass Tubing, *Mod. Plastics*, November.
18. Sang, et al., High Strength Plastic Laminates for Use on Aircraft H-F Antenna, Stanford Research Institute.
19. Warren and Heebink, Weathering of Glass-fabric-base Plastic Laminates, WADC Technical Report 55-319.

1957

20. Boller, Effect of Thickness on Strength of Epoxy and Phenolic Laminates Reinforced with Glass Fabric, WADC Technical Report 56-522.
21. Boller, Fatigue Properties of Glass Fiber Reinforced Plastic Laminates Subjected to Various Conditions, 12th NANTEC (RIP Division) of SPI.
22. Boller, Tensile Stress-Rupture and Creep Characteristics of Two Glass-fabric-based Plastic Laminates, Forest Products Laboratory, July.
23. Bowditch and Johnson, Investigation of the Effects of Glass Fabric Geometry on the Strength Properties of Low Pressure Glass Fabric Base Structural Laminates, WADC Technical Report 57-270.
24. Epstein, et al., Reinforced Plastics for Rocket Applications, *Western Plastics*, May.
25. Green and Mays, Synthetic Fiber Reinforced Thermosetting Resins, 13th ANTEC of SPE.
26. Pusey, Flexural Fatigue Strengths of Reinforced Thermosetting Laminates, *Plastics Technol.*, September.
27. Salzinger, Recent Advances in the Technology of High Temperature Systems, 13th ANTEC of SPE.
28. Schwabe, Thermal Stability of Flexible Sheet Insulation Using the Wrapped Bar Method, AIEE Winter Meeting, New York.

1958

29. Brown, Physical Properties of Prepreg Laminates, *Plastics Technol.*, July.
30. Delmonte, Creep Characteristics of Laminated Epoxy Plastics, *Plastics Technol.*, October.
31. Gelin, et al., Design of Thermosetting Plastics for Improved Impact Resistance, ACS Symposium, Chicago, September.
32. Green, et al., Epoxy Glass Laminates for Printed Circuits, 1st National Conference on Application of Electrical Insulation.
33. Massie, Fin Tip, *Aircraft and Missiles Mfg.*, October.
34. Molded Printed Circuits Save Missile Space, *Electron. Equipment Eng.*, April.
35. Murray and Stein, Evaluating Epoxy Resin Degradation Systems, *Plastics Techol.*, February.
36. Raech and Harris, Optimum Aromatic Amine–Hardened Epoxy-Glass Laminate, *Plastics Technol.*, May.
37. Zeilberger, Plastics for Liquid Rocket Engines, *Astronautics*, August.

1959

38. Bond and Brady, Silicone Epoxy Resins in High Temperature Structural Laminates, ACS Symposium, Atlantic City, September.
39. Desai and McGarry, Failure Mechanism in Glass Fiber Reinforced Plastics, *ASTM Bull.*, July.
40. duBarry, Epoxy New Miracle Plastic, *Popular Boating*, March.
41. Epoxy Resins Go to Sea, *Plastics World*, February.
42. Erickson, et al., Proposed NOL Ring Test Method for Parallel Glass Roving Reinforced Plastics, Evaluation of Chemical Finishes, NAVORD Report 5680.
43. Filament Winding Rocket-motor-Cases, *Aircraft and Missiles Mfg.*, September.
44. Haroldson and Watrous, Thermal Stability of Glass Based Fabric Epoxy Resin Laminates, ACS Symposium, Atlantic City, September.
45. Ingenious Insulation for Jets, *Mod. Plastics*, September.
46. McGarry, Resin-Glass Bond Characteristics, *ASTM Bull.*, January.

47. Rueckert, et al., Epoxy Resin–Orlon Cable Terminations for 4-KV Service, AIEE Technical Paper 59-660.
48. Skrotzki, Lightweight Plastics Compete with Metal for Piping, Valve and Fitting Market, *Power*, September.
49. Sonneborn, Control of Variables in Heat Resistant Glass Reinforced Plastics, AD 156 069.
50. Sonneborn, et al., Control of Variables in Heat Resistant Glass Reinforced Plastics, vols. I and II, AMC Technical Report 59-7-589.
51. Winfield, Thermosetting Resins: a Dream Material for the Artist and Architect, 15th ANTEC of SPE, January.
52. Zeilberger and Lieb, New Resins Lead Structural Plastics Development, *Space/Aeronautics*, November.

1960

53. Alfers and Graner, Glass Reinforced Plastics Laminates in Contact with Sea Water or Fuels, SPI Symposium, Chicago, February.
54. Antwerp, Presenting Creep Data on Reinforced Plastics, *Plastics Technol.*, June.
55. Banks, et al., The Properties of a New Polyfunctional Epoxide Resin, *Brit. Plastics*, December.
56. Black, Sivalls and Bryson, technical bulletin, July.
57. Bobalek and Evans, The Design of Reinforced Plastic Compositions for Impact Resistance, SPI Symposium, Chicago, February.
58. Charbonneau and Hedrick, New Epoxy Resin Tapes, 3d Annual Conference of Application of Electrical Insulation, December.
59. Flake-Glass Laminates, AD 244 104.
60. Goldfein, General Formula for Creep and Rupture Stresses in Plastics, *Mod. Plastics*, April.
61. Hankins, New Concepts in the Manufacture of Fiber Glass Ducts, SPI Symposium, Chicago, February.
62. Haugan, Why Reinforced Plastics for Redstone Missiles, 15th NANTEC (RIP Division) of SPI, Chicago, February.
63. Hill, et al., Porous Plastic Laminates, *SPE J.* September.
64. Judd, et al., The Deterioration of Epoxy Laminates under Extreme Aging Conditions, SPI Symposium, Chicago, February.
65. Levy and Vanderbilt, Evaluation and Development of Glass Reinforced Plastic Pipe for the Petroleum Industry, SPI Symposium, Chicago, February.
66. Menard and Cooner, Epoxy Resin Laminates with High Thermal Resistance, *SPE J.* March.
67. Model-making with Araldite, Technical Notes 206, Ciba Ltd., (Eng.).
68. Muller and Chase, Epoxy Resins Strong to 500°F. Plus, *Mater. Eng. Design*, April.
69. Now, The Plastic Gun Barrel, *Mod. Plastics*, April.
70. Plant and Goldstein, Plastics for High Temperature Thermal Barriers, SPI Symposium, Chicago, February.
71. Prepregs, *Mod. Plastics*, July.
72. Raech, High Temperature Plastics Current Capabilities and Future Prospects, SPI Symposium, Chicago, February.
73. Robe, et al., Reinforced Epoxy Pipe Tames Abrasive Slurry, *Chem. Processing*, June.
74. Sandwich without Core, *Mod. Plastics*, February.
75. Saunders, Epoxy Rejuvenates a Lightning, *Motor Boating*, January.
76. Sokol, Laminating Process Simplifies Tooling, *Tool Engr.*, July.
77. Tengstrand and Anderson, Foreign Insulation Practices, Insulation of Large Generators in Sweden, 3d National Conference on Application of Electrical Insulation, December.
78. Toellner, Pre-impregnated Fabric as a Raw Material for Reinforced Plastics, SPI Symposium, Chicago, February.
79. Yates and McCarthy, Design of Filament Wound Structures, SPI Symposium, Chicago, February.

1961

80. Antwerp, Equipment for Determining Low-temperature Creep, *Plastics Technol.*, September.
81. Barriner, Cut Costs with Disposable Mandrels, *Plastics Technol.*, October.
82. Brookfield and Pickethall, Finishes for "E" Glass to Improve Resin-Glass Properties, *Brit. Plastics*, February.
83. Brookfield, et al., A New Technique of Testing Resin/Glass Laminates, *Brit. Plastics*, September.
84. Collier, Epoxy–Glass Fiber Systems and Method for Improving the Bonding Relation Therebetween, U.S. 2,974,062.
85. Dingman, Epoxy Mica Paper Laminate Reinforcement, *SPE J.*, September.
86. Epoxide Compound Meets Military Requirements, *Chem. Prog.* (Union Carbide Plastics Co.), May.
87. Filament Winding Goes Commercial, *Mod. Plastics*, October.
88. Fong and Pratt, The Effects of Temperature and Humidity on the Dielectric Properties of Plastics, 17th ANTEC of SPE, January.
89. Hawkinson, Filament Winding as a Fabrication Technique for Rocket Propulsion Components, SAMPE Filament Winding Symposium, Pasadena, March.
90. Howse and Pears, Thermal Properties of Reinforced Plastics, *Mod. Plastics*, September.
91. Kays, Handling Mixed Corrosives with Glass Reinforced Epoxy Well Tubing and Line Pipe, *Corrosion*, March.
92. Kindall and Susman, High Temperature Resin Developments for Filament Winding, 17th ANTEC of SPE, January.
93. Kindall and Susman, High Temperature Resin System for Glass Filament Wound Components, SAMPE Filament Winding Symposium, Pasadena, March.
94. Layton, Review of Various Glass Filaments, SAMPE Filament Winding Symposium, Pasadena, March.
95. Lyon, Thousands of Bows for Modern Archers, *Plastics World*, May.
96. Madden, Bakelite Epoxy Resins for Filament Winding, SAMPE, San Bernardino, Calif., March.
97. Medney, et al., Designing with Filament Winding, SAMPE Filament Winding Symposium, Pasadena, March.
98. Miglarese, The Effect of Finishes on Heat Resistant Phenolic and Epoxy Laminates, *SPE J.*, December.
99. Outwater and West, Stress Distribution in the Resin of Reinforced Plastics, *Mod. Plastics*, September.
100. Peterson, Optimum Filament Wound Plastic Composites, SAMPE Filament Winding Symposium, Pasadena, March.
101. Peterson, Properties of High Modulus Glass Reinforced Plastics, *SPE J.*, January.
102. Plastics Development in American Missiles, *Brit. Plastics*, September.
103. Raech, What's Wrong with Data on Reinforced Plastics, *Mater. Design Eng.*, May.
104. Royal and Matlock, Epoxy Passes Toughest Test: Outer Space, *Mod. Plastics*, October.
105. RP Solves Design Problem for Jet Liner Wing, *Mod. Plastics*, July.
106. Sharp and Weber, Effect of Water on Strength of Structural Plastics, Shell Chemical Corp. Report P-423.
107. Susman and Nusbaum, New Pumps from Old Via Plastics, *Western Plastics*, March.
108. Wahl, The Effects of High Vacuum and Ultraviolet Radiation on Reinforced Plastics, *SPE J.*, August.
109. Welch, Flame-resistant Epoxy Laminates, *Western Plastics*, June.
110. Wills, et al., Prestressed Spherical Pressure Wells, SAMPE Filament Winding Symposium, Pasadena, March.
111. Yates and Trebilcock, The Chemistry of Chromium Complexes Used as Coupling Agents in Fiber-glass Resin Laminates, 16th NANTEC (RIP Division) of SPI, Chicago.

1962

112. Abelin and Wohlberg, Glass Fiber Reinforced Plastics in Highly Stressed Aircraft Parts, 17th NANTEC (RIP Division) of SPI, Chicago.

113. Boller, Fatigue Properties of Plastic Laminates Reinforced with Unwoven Glass Fibers, ASD-TDR-62-464.
114. Brilhart, Notes on Synthetic Reeds, *Plastics World*, May.
115. Brookfield and Pickethall, Some Further Studies on the Effect of Glass-Resin Type and Cure on the Strength of Laminates, 17th NANTEC (RIP Division) of SPI, Chicago.
116. Broutman and McGarry, Glass Resin Joint Strength Studies, *Mod. Plastics*, September.
117. Brown, Glass Flake Reinforced Epoxy Laminates, *SPE J.*, October.
118. Corey, et al., Some Electrical Properties of Epoxy Resin Fiberglas Laminates, Shell Chemical Corp.
119. Cullen and Johnston, Epoxidized Polymers of Polybutadiene in Reinforced Plastics, 17th NANTEC (RIP Division) of SPI, Chicago.
120. Hand, Testing Reinforced Plastics with Ultrasonics, *Plastics Technol.*, February.
121. High Speed Performance and Light Weight, Too, *Bakelite Rev.*, October.
122. Kies and Bernstein, Recent Advances in RP Rocket Motors, *Mod. Plastics*, November.
123. Kobler and McNalley, Guide to Specifications for Rigid Laminated Plastics, OTS PB-171038.
124. Lemons, Large Filament Wound Liquid Propellant Tanks, *Western Plastics*, January.
125. Matta and Outwater, The Nature, Origin, and Effects of Internal Stresses in Reinforced Plastic Laminates, 17th NANTEC (RIP Division) of SPI, Chicago.
126. A New Type of Glass Reinforcement Offers Low Cost and High Strength Design Possibilities, *Plastics Design and Production*, April.
127. Outwater and Ozaltin, The Surface Effect of Various Environments and of Thermosetting Resins on the Strength of Glass, 17th NANTEC (RIP Division) of SPI, Chicago.
128. Peterson, Engineering Properties of High Modulus Reinforced Glass, 17th NANTEC (RIP Division) of SPI, Chicago.
129. Plueddeman, et al., New Silane Coupling Agents for Reinforced Plastics, *Mod. Plastics*, August.
130. Raech, The Role of Mineral Fillers in Reinforced Plastics, *Western Plastics*, December.
131. Ruggar and McNalley, Resistance of Reinforced Plastics to Weathering, 17th NANTEC (RIP Division) of SPI, Chicago.
132. Wolff and Siuta, Some Considerations on Filament Wound Structures, *SPE J.*, January.
133. Young, et al., Interfacial Properties of an Amine-modified Silane and Its Preformed Polymer at E-glass Surfaces, 17th NANTEC (RIP Division) of SPI, Chicago.

1963

134. Carter and Moorfield, The New Glass Fibers, *Plastics Technol.*, September.
135. David and Madig, Epoxy Prepreg Properties, *SPE J.*, November.
136. Davis, Fibrous Reinforcement Run Down, *Plastics World*, November.
137. De Fries, Rocket Motor and Solid Propellant Grain with Woven Polymeric Inhibitor Coating, U.S. 3,108,433.
138. Eakins, Shear Failure Mechanisms in Parallel Filament Glass-Resin Composites, *SPE J.*, April.
139. Jubb, Degradation of Cured Paper-Epoxy Resulting from Lamination, *Mod. Plastics*, December.
140. Kenna, et al., Effect of Fabrication Variables on the NOL Ring, *Western Plastics*, April.
141. Melvin, Effect of Resin Properties on the Fracture of FRP Laminate Models, AD 412 339.
142. Nelson, Glass Surface Chemistry for Glass Fiber Reinforced Plastics, AD 403 795.
143. New Epoxy-Glass Prepreg, *Plastics Technol.*, April.
144. System Uses Four Nondestructive Tests to Measure Quality of Filament Wound Structures, *Plastics Design and Processing*, September.
145. Young, Filament Wound Fiberglass Tanks and Allied Structures, 18th Annual Petroleum Mechanical Engineering Conference, Tulsa, Okla., September.

1964

146. Alvin, RP Vital in Creating Submarine Research, *Reinforced Plastics*, November–December.
147. Bascom, Some Surface Chemical Aspects of Glass-Resin Composites, AD 604 682.
148. Brelant, et al., Combined Effects of Pre-stress and Humidity Cycling upon Filament-wound Internal Pressure Vessels, *SPE J.*, September.
149. Brink, Mechanical Behavior of Reinforced Plastics at Cryogenic Temperatures, *SPE J.*, October.
150. Christie, et al., A Study of the Factors Affecting the Performance of Epoxy Resins Used in Filament Winding, 20th ANTEC of SPE, January.
151. Cooney, Development of Improved Resin Systems for Filament-wound Structures, AD 605 179.
152. Davis, et al., The Fatigue Resistance of Reinforced Plastics, *Mater. Design Eng.*, December.
153. Development of a Structural Fiberglass Aircraft Component, North American Aviation Report NA 64H-366, A.I.A.A. Meeting, Wichita, Kans., May 25.
154. Eakins, Glass/Resin Interface: Patent Survey, Patent List, and General Bibliography, Plastic Report 18, September.
155. Ehlers, Woven Fabrics: Their Properties and Uses, *Mater. Design Eng.*, May.
156. Erickson, et al., Effects of Glass Surfaces on Laminating Resins, *Mod. Plastics*, August.
157. Hackman, et al., Structural Fiberglass Aircraft Component-Program Results, AIAA. Paper 64-442, 1st AIAA Annual Meeting, Washington, D.C., June–July.
158. Harvey, Research on Improved Epoxy Resins, AD 603 748.
159. Howard, Filament Winding, *Prod. Eng.*, Oct. 26.
160. Kies, A Navy Analysis of Glass Reinforced Plastics for Hydrospace Applications, AD 609 708.
161. Laird and Nelson, The Effect of Glass Surface Chemistry on Glass-Epoxy Systems, *SPE Trans.*, April.
162. Lipatova and Skorynina, The Grafting of Three-dimensional Polymers onto Glass Surfaces, *Vysokomolekul. Soedin.*, no. 8, p. 1536.
163. MacDonald and Glowacki, Protective Coatings for Deep Submergence Vessels, AD 608 873.
164. McGarry, Resin Cracking in Composites, *Chem. Eng. (London)* **182**:236–242.
165. McNiel and Bennett, Light Microscopy and Reinforced Plastics, *Mod. Plastics*, December.
166. Prosen, Some Properties of NOL Rings of Carbon-based Fibers and Epoxy Resins, AD 454 635.
167. Romstad, Methods of Evaluating Tensile and Compressive Properties of Plastic Laminates Reinforced with Unwoven Glass Fibers, AD 603 486.
168. Rosato and Grove, Filament Winding: Its Development, Manufacture, Applications, and Design, Interscience Publishers, Inc., New York, 360 pp.
169. Rosen, Hollow Glass Fiber Reinforced Laminates, AD 451 684.
170. Shibley, Filament Winding Bibliography, Plastec Report 19.
171. Starmann and Findley, Pretreatment of Glass Fibers with Epoxidized Compounds Having an Oxiran Content above about 8.5 Percent, U.S. 3,119,711.
172. Stone and Greszczuk, Study of Residual Stresses in Thick Glass Filament Reinforced Laminates, AD 609 610.
173. Sutton, et al., Whisker-reinforced Plastics for Space Applications, *SPE J.*, November.
174. Whisenhunt, Filament-winding Plastics, AD 609 770.

1965

175. Bascom, Some Surface Chemical Aspects of Glass-Resin Composites, RIP Division of SPI, Chicago, February.
176. Corey, Stressed Applications for Fiber Glass Reinforced Plastics, *Plastics Design and Processing*, January.
177. Elkin, Interlaminar Shear of Filament-wound Reinforced Plastics, RIP Division of SPI, Chicago, February.

178. Filament Winding High Strength Products, *Plastics World*, February.
179. Fried, The Response of Orthogonal Filament Wound Materials to Compressive Stress, RIP Division of SPI, Chicago, February.
180. Hauck, Epoxy Plastics, *Mater. Design Eng.*, January.
181. Hertz, The Effect of Cryogenic Temperatures on the Mechanical Properties of Reinforced Plastic Laminates, *SPE J.*, February.
182. Jellinek, Glass Reinforced Epoxy Resins, *Kunststoffe*, February.
183. Levenetz, Large Diameter Fiber Reinforced Plastic Composite Structures, RIP Division of SPI, Chicago, February.
184. Milewski and Shyne, Whiskers Make Reinforced Plastics Better than Metals, RIP Division of SPI, Chicago, February.
185. Patrick and Hood, Glass Reinforcements for Filament Wound Composites, RIP Division of SPI, Chicago, February.
186. Paul and Thompson, The Importance of Voids in the Filament-wound Structure, *Mod. Plastics*, December.
187. Plueddemann, Silane Coupling Agents for Thermoplastic Resins, RIP Division of SPI, Chicago, February.
188. Throckmorton and Browne, The Effect of Surface Tension on Strength of Glass Fiber Reinforced Plastics, RIP Division of SPI, Chicago, February.
189. Welch and Pitt, A New Approach to Epoxy Laminating Systems, RIP Division of SPI, Chicago, February.
190. Zisman, Improving the Performance of Reinforced Plastics, *Ind. Eng. Chem.*, January.

1966

191. Keys et al., Cryogenic Properties of Glass Reinforced Plastics, in "Advances in Cryogenic Engineering," vol. 11, Plenum Press.
192. Paul and Thompson, Influence of Environment on Glass Roving and Laminate Properties, *Mod. Plastics*, January.
193. Rugger and Titus, Weathering of Glass Reinforced Plastics, Plastec Report 24, Plastics Technical Evaluation Center, Picatinny Arsenal.

Bibliography on Epoxy Resin–Orlon Tape Cable Terminations

194. Betz et al., Epoxy Resin-Orlon Cable Terminations for 33-kv Service, IEEE paper TP-59-659, May 19, 1959.
195. Rueckert et al., Epoxy Resin–Orlon Cable Terminations for 4-kv Service, IEEE paper TP-59-660, May 19, 1959.
196. Ziegenfuss et al., Miscellaneous Joints and Terminations Using Epoxy Resin-Orlon Insulation, IEEE paper DP-59-661, May 19, 1959.
197. Claggett et al., Cable Sheath and Accessory Repairs Using Epoxy Resin-Orlon Tape, IEEE paper DP-59-662, May 19, 1959.
198. Betz et al., Epoxy Resin–Orlon Applications in Cable Systems, IEEE paper 63–51, Jan. 27, 1963.
199. Betz and Lusby, Epoxy Tape Used in Cable Terminals, *Elect. World*, Sept. 28, 1964.
200. Neville and Herreras, Field Experience with High Voltage Power Cable Joints and Terminations Insulated with Epoxy Resins, SPE Conference, Montreal, Canada, Mar. 7, 1966.

Chapter 23

SOLVENTLESS EPOXY-RESIN COATINGS

Because many epoxy resins and numerous of their curing agents are low-viscosity liquids, they may be applied as high-buildup 100 per cent solids coatings.

With high-buildup coatings, the addition of minor amounts of solvent to control viscosity and to permit more convenient application is sometimes resorted to. The solvent largely remains entrapped in the coating and may thus be considered as a nonreactive diluent.

In solution coatings, as discussed in Chap. 24, the solvent is present in substantial amounts and must be removed before the cure, thus imposing limitations on coating thicknesses. In addition, the solution coatings in the main employ the higher-molecular-weight resins which are solid at room temperature and which must be dissolved to permit their use as coating vehicles.

The non-solvent-containing or 100 per cent solids coatings based on epoxy resins are used for dip coats, gel coats, reinforced coatings, "varnishes" for electrical equipment, corrosion-resistant brush and spray coats, decorative coats, monolithic concrete topping and repair compounds, and chemically resistant trowel-on coatings.

The 100 per cent solids coatings provide for thick buildups in a single application, and since cure is accomplished without the release of solvents, there is less chance of pinholes and passageways forming in the film to admit contaminants to the protected components. Thorough coverage, together with the high electrical properties and chemical resistance of the liquid epoxy resins, may be achieved without baking cycles and within a very few hours at ambient temperature. The coatings can be applied to heat- or solvent-sensitive surfaces, and they are particularly convenient for coating small areas and for the protection of in-place structural members. One unusual application involved the selection of a fatty-polyamide-cured system for application to protect solvent-sensitive overpainting from detection in an attempt to upgrade a minor seventeenth century painting [39].

These coatings do have disadvantages. They can be summarized as (1) brittleness, (2) poor resistance to higher temperatures and thermal cycling, and (3) comparatively high costs. If room-temperature cure is required, the short pot lives also work to their disadvantage.

Obviously the disadvantages may be overcome, to some extent, by proper formulation, and progress is made yearly in this regard. Brittleness, for example, can be reduced—at the expense of chemical and temperature resistance—by the incorporation of flexibilizers, but if high buildups and high orders of flexibility are required, the formulation becomes increasingly critical.

Physical properties of the 100 per cent solids and high-buildup coating systems are similar to those obtained with filled systems, as discussed in Chap. 14, and are largely a function of curing agent and modifiers. Considerable curing-agent and modifier data are reported elsewhere in the book and are not repeated here.

The largest area of use for these coatings is in pipe-lining applications where the epoxy resin provides improved chemical resistance. A typical example is use for lining cement-asbestos sewer pipe. The city of San Buenaventura, Calif., for example, has installed almost 30,000 linear feet of this type of piping after a considerable laboratory test program [37].

DIP COATS

Dip coats for electronic equipment are based on DGEBA, the lower-molecular-weight epoxidized novolacs, and these resins in combination with the liquid epoxidized olefin resins.

The liquid resin or resin blend is modified with a thixotroping filler or filler combination, the aim being to obtain a system which will provide a regular coating even over sharp projections, and which will not run off or drip excessively during the cure cycle. Flexibilizing modifiers are used when thermal shock is to be encountered in service.

Room-temperature-curing dip coatings may employ the primary aliphatic polyamines as curing agents, but more commonly the fatty polyamides are used. A simple formulation consists of:

	Parts
DGEBA	100
Chemically treated bentonite	15
Black iron oxide	5
DETA	10

For heat curing systems, anhydrides such as DDSA and more complicated curing-agent blends are used. For single-container systems, BF_3MEA or dicyandiamide is preferred. A simple one-component formulation is:

	Parts
DGEBA	90
Butanediol diglycidyl ether	10
Colloidal silica	10–20
BF_3-MEA	3

Ideally, the dip-coating material should be sufficiently fluid to permit degassing under vacuum and yet sufficiently thixotropic to permit the depositing of a regular buildup of up to 60 mils per dip. Accomplishing this goal requires considerable specialization in the formulation.

In addition to the dip coating of electronic components such as capacitors, etc., the dip coating of printed-circuit boards, to preserve electrical properties of the mounted components during humidity and salt-spray cycling, is an application of interest. Dip-coating the printed-circuit boards presents rather unique problems. The board contains a number of mounted components and raised connections which cannot, with ease, be covered uniformly in a single-dip application. If the coating provides a sufficiently high buildup to give satisfactory coverage, weight and compound usage will be increased proportionately, and the problems of deaerating the compound will be magnified. Conversely if the coating at the high spots is not sufficiently thick, inadequate moisture resistance will be obtained [19]. The problem is substantially magnified when it is necessary to provide flameproof coatings which have sufficient clarity to permit readout of numbers on the components on the boards. Although epoxy formulations are used as printed-circuit-board coatings, polyurethanes appear to offer a number of advantages and may be expected to replace the epoxies in this application [23], with the epoxy being used in a paste form to protect the high spots.

In dip-coat applications, the component being dipped is withdrawn from the coating tank at an optimum speed in order to assure uniformity of buildup.

In spite of the number of variables involved, dip coatings offer an economic solution to some high-volume production applications for the epoxy resins.

GEL COATS

Gel coats are 100 per cent solids coatings which are brushed on or troweled onto molds before the laminating or casting application. These provide the required aesthetic surface finish and reproduce the contour of the mold. They are pigmented to the desired color and are formulated from the liquid resins with a thixotroping filler. Additional fillers are selected on the basis of the desired surface characteristics. Modifiers are used to impart flexibility when that is a requirement.

Most commonly, primary polyamine curing agents are employed to effect cure at room temperature.

REINFORCED COATINGS

For severe chemical environments and for applications involving temperature extremes, it is occasionally necessary to employ relatively high buildup coatings without the addition of flexibilizing additives. In these cases, the coating system may be composed of a rigid epoxy resin and suitable curing agent and reinforced with a woven synthetic fabric.

The surface to be coated is thoroughly cleaned—metal surfaces are preferably sandblasted. The epoxy formulation is then brushed on or applied with a roller

Monolithic topping compound for dairy. Flooring employs tempered Masonite, over which is placed glass cloth. Glass cloth flashed up the sides. The epoxy topping compound is then applied to give a smooth, skid-resistant surface. (*Union Carbide Corp.*)

coat. The fabric is pressed in, and additional resin is applied to the surface. One or more layers of cloth may be used, depending upon the desired thickness of the buildup.

Generally, such reinforced coatings are applied as tank linings, etc., in environments where it is not convenient to employ elevated-temperature cures. Frequent mixing operations must be allowed for. The application is somewhat messy, but with proper surface preparation and careful application, such coatings provide maximum obtainable chemical resistance with the room-temperature-curing epoxy-resin formulations.

Clay pipe coated with epoxy resin and glass cloth is used in drainage system of large chemical plant, where acid leakage would destroy unprotected clay pipe. (*The Epoxylite Corp.*)

NON-SOLVENT-CONTAINING VARNISHES

For many electrical applications, it is desirable to employ a 100 per cent solids impregnant rather than one containing solvents. The compound must have the lowest possible viscosity at processing temperatures to provide thorough impregnation, and some method must be developed to prevent it from draining out of the unit after application.

In extremely tight electrical windings, the impregnation may be accomplished with vacuum and pressure, and the compound will be retained in the interstices. In units with more open windings, the coils may be wrapped in tight-weave glass tape, and this structure is then impregnated. The tape will prevent excessive runoff during cure. In other cases, a filler may be added which filters out of the compound to form a highly filled layer on the surface of the coil; or, alternatively, the coil may be surrounded by a dry filler and the impregnating compound forced through it by vacuum and pressure. The filler retains the impregnant during cure.

Because of the rather lengthy vacuum-pressure cycle involved, the 100 per cent solids impregnants are long-pot-life systems, or employ latent curing agents. They are used for the impregnation of transformers, especially high-voltage and high-temperature transformers [4] and for the impregnation of form-coil motors.

Typical of the use of the materials for the impregnation of a high-voltage transformer is the use of a special anhydride-cured resin formulation applied by the following processing techniques [18]:

1. Warm transformer, and mold to 80°C.
2. Warm epoxy to 80°C. Mix thoroughly for 5 minutes.
3. Transfer epoxy compound to mold in vacuum tank, making sure tank has adequate reservoir.
4. Pull a vacuum of 1 to 3 mm mercury for 5 minutes. Release to atmosphere. (Use nitrogen if available.)
5. Pull a vacuum of 1 to 3 mm mercury for 10 minutes. Release to atmosphere. (Use nitrogen if available.)
6. Repeat step 5. (Subsequent pressurization to 100 psi with nitrogen is desired if available.)
7. Remove transformer and mold.
8. Gel overnight (15 to 16 hours) at 60°C. (Avoid the stresses of higher gel temperatures.)
9. Cure 24 hours at 95°C. (Avoid curing stresses of higher cure temperatures.)

Using this specification, together with properly formulated epoxies, high-voltage transformers have been produced which withstand proof tests at 1,200 to 1,500 pulses at 150,000 volts.

High-temperature form-coil motors have been satisfactorily insulated by the following technique: The coils are formed conventionally and wrapped. They are then dropped easily into the slots and the connections made. The connected motor is immersed in a vat of long-pot-life, high-temperature-resistant epoxy, and a vacuum and pressure cycle is used to ensure complete impregnation. The motor is removed, drained and baked, and delivered for service. The process is quite similar to that employed with conventional varnishes, with the exception that the coils are not laminated under heat and pressure in advance of winding, since the non-solvent-containing epoxy formulation provides the desired fill without the necessity for curing under pressure.

CORROSION-RESISTANT BRUSH AND SPRAY COATS

Brush and spray coatings are usually supplied containing small amounts of nonreactive diluent to control viscosity, and to permit the use of higher loading volumes of pigment. The liquid resin is filled with a combination of fillers, one at least providing the required thixotroping action to permit application to vertical and overhead surfaces. Small amounts of flow-control agents are used to prevent cratering in thin films.

Spray equipment has been developed (see Chap. 25) that permits the application of fast-curing liquid systems.

When working with higher-molecular-weight resins, satisfactory 100 per cent solids coatings have been obtained by a flame-spray technique [29], the powdered resin/curing-agent/filler blend being liquefied with heat before deposition on the substrate.

Cracked concrete in feed trough is repaired by nonsolvent epoxy-resin coating being applied to feed trough. (*The Epoxylite Corporation.*)

The 100 per cent solids brush or spray-on coatings provide enhanced chemical resistance over that obtained with solution coatings, by virtue of the higher buildups involved. They are generally somewhat brittle and are, because of the amount of compound used, rather expensive.

In addition to 100 per cent solids high-buildup coatings, there are a number of so-called 100 per cent solids medium-buildup coatings employing ketimine curing agents. These curing agents are complexes of primary aliphatic polyamines with ketone solvents (Chap. 7). When admixed with the liquid DGEBA/epoxy resins, they provide pot life of 4 to 5 hours at room temperature and yet cure rapidly in films 4 to 8 mils thick. The curing reaction is initiated when the complex is dissociated by atmospheric moisture. Thus, the material in a large mass does not obtain the required moisture, whereas, in a film, the moisture is obtained rapidly and dissociation occurs. A typical resin used with the ketimine curing agents is DGEBA containing about 18 phr of dibutyl phthalate. An unpigmented coating with this resin when cured in a 7-mil film gives properties as indicated in Table 23-1. To the unpigmented coating may be added pigment dispersions as described in Chap. 14.

Table 23-1. Properties of Typical Ketimine-cured Epoxy Coating [4-1]

Property	*Value*
Impact resistance, in.-lb	Pass 48
Flexibility: conical mandrel	Pass
10% NaOH resistance:	
After 24-hr immersion	Slight softening
After 24-hr recovery	Complete
MEK resistance after 1-hr immersion	Fail
Boiling water resistance:	
After 1-hr immersion	Softening, whitening
After 24-hr recovery	Complete
Salt-spray resistance after 200 hr	Good: some undercutting at score mark

Thickness of the coating is, however, restricted by the requirement for atmospheric moisture pickup, and spraying is the preferred application method.

DECORATIVE COATINGS

High-buildup epoxy-resin coatings have been used for glazing building blocks to provide a finished appearance similar to that obtained with ceramic tile. The coating is applied in two coats, the first being worked into the porous face of the block to provide a smooth surface, the second coating being sprayed to provide the desired surface finish [16]. A typical formulation for this work has been reported as [1]:

	Parts
DGEBA	61.5
Xylene	18.7
Paraffin wax (5% soln. in xylene)	3.6
Titanium dioxide	25
Cadmium red	0.4
Bentonite	4.6

The resultant mixture, cured with an aliphatic polyamine, is used for both first and second coats by the above-described method.

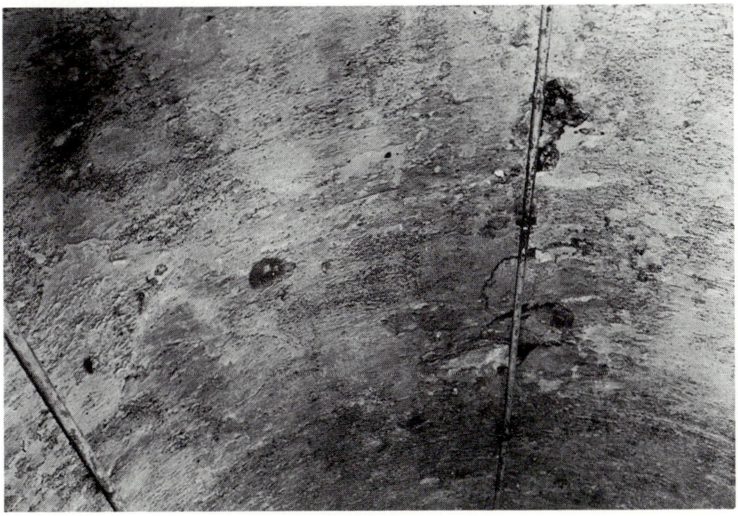

Underpass coating I. Fifty-year-old concrete of underpass at 191st Street and Broadway, New York City, showing condition before application of epoxy coating. (*Ciba Products Co.*)

Underpass II. Application of epoxy coating to sandblasted concrete of pedestrian underpass. (*Ciba Products Co.*)

Use of trowel-on coatings for glazing cinder blocks has also been reported. These formulations, employing aliphatic primary polyamines and fatty polyamides as curing agents, and pigmented white, gray, and maroon, have shown good performance after 2 years of outdoor exposure. Moderate chalking occurred during the first few months of exposure, but thereafter the coatings were unaffected [7].

Another decorative application for the epoxies—which like the glazing operation offers a potential cost savings—is the use of a 100 per cent solids–epoxy formulation as a binder for terrazzo aggregate [11].

The topping is applied in thickness of $\frac{3}{16}$ to $\frac{1}{4}$ in., compared to the several inches employed with conventional terrazzo mixes. The weight reduction permits use of

Underpass III. 860-ft underpass at 191st Street and Broadway, New York City, refinished with epoxy-resin coating at one-fourth cost of ceramic tile. (*Ciba Products Co.*)

Multicoat system with speckle finish is used to give tilelike surface to concrete-block walls. (*Ciba Products Co.*)

above-grade terrazzo flooring without heavy structural penalties, and the relatively thin coating permits installation in existing areas where higher buildups would present difficulties. The coating provides toughness, chemical resistance, and more varied color possibilities than are available with regular terrazzo mixes.

If the topping is to be applied over existing concrete, the discussion of surface preparation, given in the following section of this chapter, should be consulted.

The formulation of terrazzo mixes involves the filled DGEBA. A suitable filler is 100 phr calcium carbonate. To this is added approximately 700 phr of marble chips, a blend of no. 0, 1, and 5 at 25/25/50 by weight being satisfactory. Dry, oil-free pigments are incorporated to meet matrix color requirements. The blend

Epoxy Terrazzo I. Epoxy terrazzo mix is troweled between stripping in flooring section at La Guardia Air Terminal Complex. A total of 175,000 sq ft of monolithic epoxy terrazzo was applied in this installation. (*Ciba Products Co.*)

Epoxy Terrazzo II. Finishing of epoxy terrazzo flooring after cure is accomplished with standard terrazzo grinding equipment. Epoxy matrix withstands grinding and flooding with cooling water. (*Ciba Products Co.*)

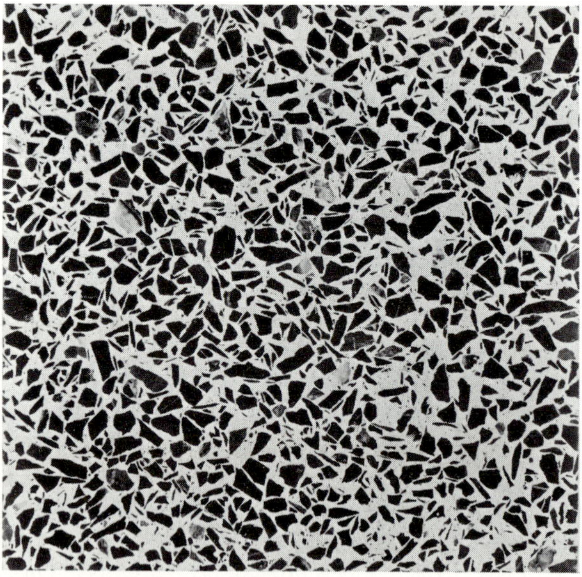

Epoxy Terrazzo III. Appearance of finished block of epoxy terrazzo. (*Union Carbide Corp.*)

Epoxy Terrazzo IV. Typical finished installation of epoxy terrazzo. (*Union Carbide Corp.*)

is mixed thoroughly with a fatty-polyamide curing agent, the marble chips being added at the end of the mixing operation. The mixture is then spread by screeding or troweling. After a cure of 24 to 48 hours, the topping is ground. It is generally advisable that grinding not be delayed excessively, since after a cure of 7 days or so at room temperature grinding becomes progressively difficult.

Other decorative topping compounds may be formulated and used with good effect. The basic formulation, as given above, less the terrazzo chips, may be

Rough-aggregate-faced epoxy-coated panels provide aesthetic surfaces and give excellent thermal insulation by use of polystyrene foam core. (*Ciba Products Co.*)

bodied with finely divided white sand and troweled over an existing surface, and then, before cure, a decorative aggregate may be broadcast over the surface. Pleasing effects have been obtained with colored rocks and gravel and with small Mexican beach pebbles.

TROWEL COATINGS AND CONCRETE PATCHING AND REPAIR COMPOUNDS

Epoxy-based, 100 per cent solids trowel coatings cured with room-temperature curing agents have been used for chemical and moisture-resistant silo linings, coatings for underwater erosion and cavitation protection, moisture-resistant plasters [6], protection of splash-zone areas of existing marine structures [36], and for other related jobs. Typical of these troweling compounds is a formulation consisting of the following [2]:

	Parts
DGEBA	125
Chemically treated bentonite	32.9
cured with	
Polysulfide (mol. wt. 1,000)	100
Zinc sulfide	52.8
Chemically treated bentonite	34.7
DMP-30	12.5

When epoxy coatings of this type are applied to underwater structures, the formulation conventionally contains excess amine curing agent. Probably the fatty polyamides are the preferred curing agent, but the fatty monoamides have also been suggested as have the aliphatic primary amines. The excess amine migrates

Section of offshore drilling rig protected with epoxy-coal-tar coating. (*Shell Chemical Co.*)

to the surface of the metal being treated to displace the water, and the remaining amine acts as the curing agent [38].

With regard to cavitation protection through the use of the epoxy resins, the literature is somewhat conflicting. Good results have been reported, but laboratory and field tests have also indicated that performance is inferior to that obtained with polysulfide and neoprene rubbers [40].

The largest use for epoxy troweling compounds is in concrete patching and repair compounds.

Concrete Patching and Repair Compounds

Epoxy-resin formulations, used as monolithic coatings over concrete surfaces, have found application in two general areas: as floor-topping compounds where the flooring is exposed to severe chemical environments, and as coating materials for highways. Additionally, they have been used for the bonding of new to old concrete, for bonding cast-in-place concrete curbing, and for bonding traffic markers to concrete slabs. The applications all have many considerations in common.

The epoxy-resin formulations are particularly suited as concrete topping and repair materials because of their physical properties and their high adhesive strengths. The adhesive strengths to properly prepared concrete will considerably exceed the cohesive strength of the concrete itself. Test cores and laboratory specimens have repeatedly demonstrated that under tensile, compressive, and impact loading, failure will occur in the concrete rather than at the epoxy-concrete interface, and in some cases improved strengths will be obtained over those of unbonded samples because of the reinforcing effect of the epoxy [28].

The tensile strengths of the epoxy formulations, and compressive strengths, even of epoxy systems containing as much as 700 phr sand filler, will exceed that of most concrete at room temperature. Highly filled systems for example, give compressive strengths on the order of 3,500 psi.

Epoxy-resin systems have coefficients of thermal expansion up to 10 times that of concrete. Although incorporation of fillers will help reduce problems resulting from this difference, steps must also be taken to provide an epoxy formulation capable of stress relief without cracking either itself or the concrete to which it is bonded. It should be noted that many failures of epoxy patches and surface breakdown attributed to poor cleaning of the concrete and weak concrete around an epoxy patch may actually be caused by high stresses applied to the epoxy-concrete interface by a poorly designed epoxy system. Broken concrete adjacent to an epoxy cement patch or in a sample which has been freeze/thaw–cycled is not necessarily a sign of a superior epoxy. It may be a sign of an improper one.

Floor Topping and Repair. Use of the epoxy resins for floor topping and repair has been limited by the relatively high cost of such systems, both in terms of application labor and material usage and, perhaps even more importantly, by the all-too-frequent field failures because of improper formulation and/or application.

Epoxy floor-topping formulations, consisting of the liquid epoxy resin filled more or less highly with a coarse filler (usually fine sand), should be applied to the concrete in thicknesses of at least $\frac{1}{4}$ in., and for heavy-duty service in thicknesses up to $\frac{1}{2}$ in. Thinner coatings are subject to impact damage and subsequent chemical attack [14]. The thicker coatings correctly applied provide outstanding serviceability.

For example, a $\frac{1}{4}$-in. epoxy topping was applied over an 8,400-sq-ft area in a plating room, where it was exposed to solutions of hydrogen sulfide, hydrochloric and nitric acids, various other chemicals, and water. After 2 years of exposure of

16 to 20 hours daily, the coating was unaffected by the environment [21]. In another application, epoxy floor toppings have been used to withstand fruit acids and sugar concentrates, live steam, and hydrochloric acid [10]. Under these severe conditions, concrete would normally erode very rapidly.

It is customary, in floor-topping operations, to minimize epoxy usage. This can be done by several methods.

If the existing flooring is badly damaged, it is cleaned and coated with an epoxy adhesive membrane. Such membrane consists of a lightly filled liquid epoxy system diluted with a minor amount of solvent, and cured with a room-temperature curing agent, either a fatty polyamide accelerated by a tertiary amine or a blend of a tertiary amine and a liquid polysulfide. The membrane is applied by brush or squeegee at an application rate of 150 to 200 sq ft per gal. After application, sufficient open time is allowed to permit escape of solvents (if present) and new concrete is then cast on the membrane while the membrane is still in the tacky and uncured state. When the membrane is cured, the new and old concrete are firmly bonded. The new concrete should then be cleaned and an epoxy topping compound applies.

When it is necessary to fill only one or two badly eroded areas before the application of the topping, this is done by priming the area with the epoxy membrane and tamping in a sand-cement mortar as fill [2].

The membrane, of course, may be used to bond new to old concrete when it is necessary merely to resurface the existing flooring. The use of such a membrane in resurfacing the flooring of the fire-gutted South Bronx Marine Transfer Terminal has been reported [8].

In addition to the resurfacing technique prior to topping, it is possible to reduce

Epoxy primer coat being applied in advance of epoxy topping compound. Finished epoxy topping shown in background. Topping withstands pressures in excess of 10,000 psi at ⅜ in. thickness as aircraft engines are trundled about. Topping cures for service in only 2 days. (*Union Carbide Corp.*)

material cost by using a very heavily filled epoxy topping compound. The concrete flooring is primed with an epoxy membrane, and then, before cure of the membrane, is topped with a tamp-in-place mixture of sand and epoxy that is virtually dry because of the high filler loading volume.

With epoxy floor-topping compounds a compromise between ease of maintenance and nonskid characteristics is required. If the topping is applied so that the surface finish is smooth and slick, maintenance will be quite easy; but when wet, the surface will be very slippery. In order to obtain nonskid, the surface should be roughened by selecting the proper mixture of aggregate or by broadcasting a relatively coarse

Strips of nonskid epoxy are used to provide traction for service vehicles at John F. Kennedy International Airport. (*Union Carbide Corp.*)

aggregate over the surface before the cure of the topping. The rough-surfaced topping compounds are difficult to maintain. A compromise which has proved satisfactory in some cases, with thinner toppings, has been to apply a relatively fluid topping compound with a short-nap roller coat to leave an orange-peel surface which provides ease of maintenance with skid resistance.

Highway Maintenance and Repair. Epoxy systems are used in highway maintenance and repair to fill chuck holes in the concrete slab, to repair cracks, to seal spalled areas, to provide waterproofing membranes, to apply skid proofing in hazardous areas, and to bond safety markers. Comprehensive descriptions have been provided by the Corps of Engineers covering both procedures and materials [5], and the Department of Interior has issued a detailed report on the use of epoxy resins in concrete construction and repair applications [20].

Chuck holes are prepared and filled by first cleaning out the existing hole, cutting back to good concrete if the exposed concrete is spalled and powdery, and then applying a thin epoxy primer coat. Before the primer coating is cured, a highly filled epoxy compound is tamped in.

Cracked slabs are repaired first by widening the crack by sawing and cutting. The crack should be opened to two or more times its original width to prevent subsequent excessive stress concentrations during thermal cycling in service. The crack is then blown clean with oil-free compressed air, and a fluid epoxy system with a room-temperature curing agent is poured in, or in the case of very thin cracks, when veeing out is impractical, forced in under pressure. A reapplication is required after the compound has gelled if settling is pronounced. In the case of large cracks, the bottom portion of the crack is filled with an inexpensive filler. Rubber pipe has been reported as a good filler for this purpose, the pipe being stretched during installation so that it provides a close fit when relaxed. The epoxy compound is then applied over the filler.

Spalled area in concrete is repaired by brushing on a fluid epoxy system containing a room-temperature curing agent. The liquid system settles through the spalled surface and hardens, incorporating the concrete dust into the coating as filler. Such repairs have been made by the New Jersey Highway Authority Maintenance Department with good initial success [9], but it was later reported that adjacent portions of concrete had failed, indicating the need for sealing the entire area [17].

Epoxy systems are applied at relatively thin buildups as protection over existing slabs. The wearing surface gives weather protection and skid resistance, inhibits dusting, and provides resistance to deicing salts, fuels, and lubricants [3]. Application of such a membrane, employing a spray-on coating of an epoxy system containing a coal-tar additive, for resurfacing the entire 28-lane miles of the Triborough Bridge in New York has been described in [31]. The 25-lane miles of the upper deck of the Bay Bridge in San Francisco has been resurfaced, as has the Dalles Bridge over the Columbia River at the Oregon-Washington state line [24]; the 40-year-old 23d Street Viaduct Bridge [22] and the new Intercity Viaduct [33] in Kansas City, Mo.; the Passaic River Bridge in New Jersey [34]; the Canadian River Bridge in Texas [32]; and others throughout the United States.

The thickness of the epoxy topping most suited has been a subject of some discussion in the literature. Initial installations tended to minimize epoxy usage, and for light traffic and in areas such as California where there are not severe freeze/thaw cycles these have been satisfactory. When heavier traffic is encountered and weather conditions are more severe, the epoxy topping should be spread at a rate of 2.5 to 3 ($1/16$ in. thick) lb per yd [30]. The skid-proofing aggregate originally employed was fairly coarse, but this led to excessive wear, and it was found that use of materials from 12 to 30 mesh gave comparable results with improvement in wearability [30].

Epoxy-resin coatings have been used as waterproofing membranes for concrete before the application of asphalt topping. DGEBA has been used in mixture with hot asphalt, heated aggregate, and long-chain unsaturated fatty acids [15] and fatty polyamides [35] as road-topping compounds.

Since the use of coal tar in combination with epoxy resins reduces the cost of the polymer by about 50 per cent, it is an attractive combination [13,29]. These compositions were first used [12] in 1956 and have been improved for construction usage since then. A widely used combination [4-6] employs two preblended components mixed in a 1/1 weight ratio just before use. A number of mobile metering-mixing tank wagons have been built for the purpose of coating roadways which can apply the resin at rates up to 30 gal per minute.

Railroad siding ramp is coated with skidproof epoxy topping ½ in. thick. After over one-year service to truck traffic, there are no signs of wear or deterioration. (*Union Carbide Corp.*)

Preparation of Concrete. In concrete topping and repair applications, one of the more critical factors is the attainment of a properly prepared concrete surface. In the absence of good surface preparation, adhesion will be inadequate and performance of the epoxy compound will be poor.

Not all concrete mixes react the same toward the epoxy. Those containing silicone additives, for faster curing, are particularly bad insofar as they tend to act as release agents. Water contents may vary widely, as will surface finishes and ultimate strength properties in the concrete itself. The concrete may have been subjected to a variety of abuses and may be contaminated to various depths by chemicals and greases.

Application of epoxy-polysulfide compound as a sealer for cracks and spalled areas on concrete bridge deck. (*Thiokol Chemical Corp.*)

In view of the variables involved, it is not possible to recommend a single surface-preparation technique which is satisfactory for all applications. In addition, optimum surface-preparation techniques are not always practical under field conditions, and compromises are frequently required.

Surface preparation consists of several phases:

Phase One: Removal of so-called dead concrete and/or sealing of existing spalled areas. If dead concrete is not removed and the spalled areas are not sealed, these will separate from the good concrete in service, leading to lifting of the topping.

Phase Two: Filling of cracks and patching existing holes. If existing defects are not patched and filled in advance, the applied coating will not be of a monolithic nature and the surface will be irregular.

Phase Three: Application of suitable cleaning procedures to obtain a surface to which adhesion will be adequate. If surface contamination exists, this will lead to bond failure and subsequent lifting in the presence of moisture and contaminants.

Dead concrete is normally removed by chipping out and brushing or vacuuming off. Spalled areas are sealed with a fluid epoxy compound, although if severe spalling is observed, the surface should be removed mechanically to the good concrete beneath. Cracks are filled by procedures described earlier.

Final surface preparation is accomplished by one or more of the following techniques:

1. Sandblasting.
2. Grinding or mechanical scarification.
3. Acid etch and high-pressure washdown, the etch consisting of dilute hydrochloric acid, spread by a broom and allowed to remain on the concrete some 10 minutes or so.
4. Detergent or solvent cleaning. Grease-cutting detergents of the nonionic type can be used, or in particularly stubborn instances solvents such as perchloroethylene or high-flash naphtha may be required. They must be used with suitable safety precautions.

On new concrete, a high water content is initially present, and the concrete should be allowed to age before the application of the epoxy topping. Under ideal conditions, a minimum of 10 days is required, but at least 30 days, and even longer, is preferred. The surface of new concrete is always weak, no matter how strong the remainder of the slab may be. This is because the method of laying the slab by vibrating and/or troweling will encourage compaction and consequent rise to the surface of the lighter components such as surplus cement, water, and fine sand. Known as *laitance*, this may be as much as $\frac{1}{16}$ in. thick. Unless laitance is removed, it will limit the performance of anything applied to it [26].

In view of the above discussion, it is apparent that it is advisable to make test patches and cores before application of the final topping. A test procedure is described in detail in [39].

Methods of Application. The application method is determined by the consistency of the coating compound, the desired surface finish, and the desired thickness of the coating. For membranes applied on the order of $\frac{1}{16}$ to $\frac{1}{8}$ in. or less, application is usually by roller coat, brush on, or by spray equipment.

For the higher-buildup coatings self-leveling compounds are used, these being poured on. They are low in viscosity, and the surface finish will be quite glossy. Alternatively, the more highly filled systems may be applied with a trowel or by the use of screed bars. The very highly filled systems must be tamped in place.

Particularly with trowel coatings, the best consistency in the formulation is obtained by the use of various particle sizes of fillers. Since the handling characteristics will depend, in part, on the ambient temperature, it is often useful to

Application of sand-filled epoxy-resin floor-topping compound by screed bar. (*Dow Chemical Co.*)

provide the fillers separately, allowing the workman to select the best blend and filler concentration (within limits) for the effect desired.

Temperature is critical during application and during the initial cure cycle. If the temperature falls much below 15°C, handling difficulties are increased and cure will be considerably protracted. It is preferred to apply the coating when the ambient temperatures are such as to provide for satisfactory cure. In some cases, it is necessary to heat the slab in advance of application by the use of torches—a time-consuming, relatively expensive, and inefficient method. For small-area repairs, heat can conveniently be applied by covering the area with a slightly bowed piece of sheet steel and heating the steel with a torch. When the surface is being cooled rapidly by a breeze, the patch may be covered by a plastic film to reduce the cooling action.

Conversely, of course, in very thick buildups it is possible to obtain excessive exothermic reactions. These may be encountered, for example, when pouring rather large and deep concrete patches, particularly on warm days. In this event,

Trowel-on application of sand-filled epoxy-resin floor-topping compound. (*General Mills, Inc.*)

low-exotherm systems should be specified or a multistage casting should be made, the first pour being allowed to harden and cool before addition of the second pour. It is advisable on multistage pours, particularly with highly filled systems, to apply a primer coating between the pours.

Cleanup of equipment after application of the epoxy coating can be accomplished with solvents, but with the usual formulations, soap and warm water will prove satisfactory if accomplished before the epoxy resin has started to set up.

In the application of epoxy topping compounds, it is most helpful to have prepackaged formulations, with adequate room being allowed in one of the containers for all the mixing operations. This eliminates errors in mixing ratios which, otherwise, are virtually sure to occur on larger jobs.

The mixing procedure is to combine the curing agent and the resin, stirring with electrically driven stirring apparatus. After these have been thoroughly mixed, the fillers are added volumetrically, and stirring is continued until the fillers are completely wetted. Particular care must be taken to scrape down the sides and bottoms of the mixing container several times during the operation. Since rather large batches are mixed at one time, on the order of 5 gal or even more, mixing is relatively time-consuming.

Formulation. Formulation of concrete-topping and repair compounds and concrete membranes involves two separate selections, to some degree interdependent (1) selection of the epoxy–curing-agent–modifier complex and (2) selection of the proper fillers and optimum filler ratio.

Selection of Epoxy–Curing-agent–Modifier Complex. The epoxy resin employed in the concrete-topping repair applications is DGEBA. Often it is reduced in viscosity with a reactive diluent, such as about 10 phr butyl glycidyl ether, or with somewhat larger concentrations of a flexibilizing diluent. The additional use of a small amount of solvent and/or pine oil as a nonreactive diluent (usually in the range of 10 phr) does not appear to reduce cured strength of the topping compound [11].

Customarily, some form of flexibilizer is employed. Current practice prefers amidopolyamines, fatty polyamides, or polysulfides (mol. wt. 1,000). When using the epoxy as a concrete sealer for highway applications, coal tar is employed as the flexibilizing additive, the system being cured with a primary aliphatic polyamine. The coal tar offers a cost advantage over most other flexibilizers, but possesses the disadvantages of color and loss of flexibility at low temperatures.

No single flexibilizer is satisfactory for all concrete-topping applications. It is probable that the polysulfide flexibilizer (the most expensive selection) is preferred from the standpoint of overall chemical resistance vs. flexibility. The polysulfide (mol. wt. 1,000) is used in the 30- to 50-phr range. The concentration of tertiary amine curing agent (DMP-30) may be varied to vary the working life and the cure time, but optimum chemical resistance is obtained with concentrations of 5 to 6 phr. Pot lives can be extended, at the expense of cure time, by the selection of less active curing agents.

The amidopolyamines and fatty polyamides provide longer pot lives than do the polysulfides; this is at the expense of increased cure times. Accelerators may be used to speed the cure—usually tertiary amines—but these degrade the chemical resistance of the resultant coating. Stoichiometric amounts of the fatty-polyamide curing agents provide optimum chemical resistance, thus limiting the variation in cured flexibility which may be realized with a given system. The polyamide curing agents are not skin sensitizers, thus restricting the dermatitis potential in an essentially messy operation.

Typical properties of two concrete topping formulations, compared with concrete, are given in Table 23-2. Chemical resistance of systems containing various

Table 23-2. Typical Physical Properties of Epoxy Flooring Compounds in Comparison to Concrete [4-1]

FORMULAS USED

Concrete	*Parts by volume*
Type-2 cement	1
Sand	2
⅜-in. gravel	3

Ingredients	*Formula* 1, *PBW*	*Formula* 2, *PBW*
DGEBA (mol. wt. 400)	210	220
Butyl glycidyl ether	49	
Epoxidized 3-(pentadecyl) phenol		65
Fatty polyamide (amine value 290–320)	38	41
DMP-30	13	15
Sand aggregate	1,440	1,580

TESTS PERFORMED AFTER 28 DAYS' AGING

Property	*Value*
Rockwell hardness, M:	
Concrete	64
Formula 1	71
Formula 2	59
Shrinkage, in./in.:	
Concrete	0.0012
Formula 1	0.0010
Formula 2	0.0015
Compressive strength, psi:	
Concrete	
After 3 days	3,110
After 7 days	3,590
After 28 days	3,730
Formula 1:	
After 3 days	6,610
After 7 days	8,260
After 28 days	8,270
Formula 2:	
After 3 days	2,500
After 7 days	6,570
After 28 days	8,320
Vicat softening, maximum penetration from 23 to 204°C, mm	
1,000-g load:	
Formula 1	0.08
Formula 2	0.063
3,000-g load:	
Formula 1:	
At 23°C	Taken as zero point
At 52°C	0.01
At 66°C	0.025
At 93°C	0.122
At 121°C	0.340 and 0.5820 (2 specimens)
At 149°C	0.835 and 1.015 (2 specimens)
At 177°C	1.020 and 1.595 (2 specimens)
At 204°C	1.845 and 1.755 (2 specimens)
Formula 2:	
At 23°C	Taken as zero point
At 52°C	0.018
At 66°C	0.055
At 93°C	0.142
At 121°C	0.155
At 149°C	0.22 and 0.36 (2 specimens)
At 177°C	0.31 and 0.505 (2 specimens)
At 204°C	0.370 and 0.590 (2 specimens)
Flexural strength, psi:	
Formula 1	20,200
Formula 2	22,800
Flexural modulus, psi × 10^{-6}:	
Formula 1	3.97
Formula 2	4.55
Maximum deflection, inches:	
Formula 1	0.437
Formula 2	0.465
Tensile strength, psi:	
Concrete	250–450
Formula 1	1,480
Formula 2	1,430
Adhesion to steel, psi	
Concrete	Specimen breaks
Formula 1	1,260
Formula 2	990
Impact-falling ball, ft-lb:	
Concrete:	
At 24°C	10.1
At 0°C	10.0
Formula 1:	
At 24°C	9.5
At 0°C	9.5
Formula 2:	
At 24°C	12.3
At 0°C	12.3

Abrasion*, loss in thickness per 1,500 cycles, inches:	
Concrete	0.0255
Formula 1	0.0085
Formula 2	0.0040

* Using no. 60-grit carborundum cloth, load of 88 oz. Test run on ¼-in.-thick specimens.

Application of thin epoxy-resin membrane to bridge deck. Potholes in deck were filled with sand-filled epoxy compound before application of membrane. (*Union Carbide Corp.*)

flexibilizers suitable for concrete floor-topping applications is discussed in Chaps. 10 and 16.

Fillers and Filler Loading Volume. As indicated previously, coarse aggregates are used to provide skid proofing on epoxy surfacing. In addition, when large voids are to be filled with the epoxy compound, extremely coarse fillers such as pea gravel and even rocks may be used to good advantage. Thixotroping fillers are added when nonsag requirements exist. Powdered perlite, in this connection, has been reported as being superior to colloidal silica in that it does not interfere as much with the wetting action. But these are specialized cases.

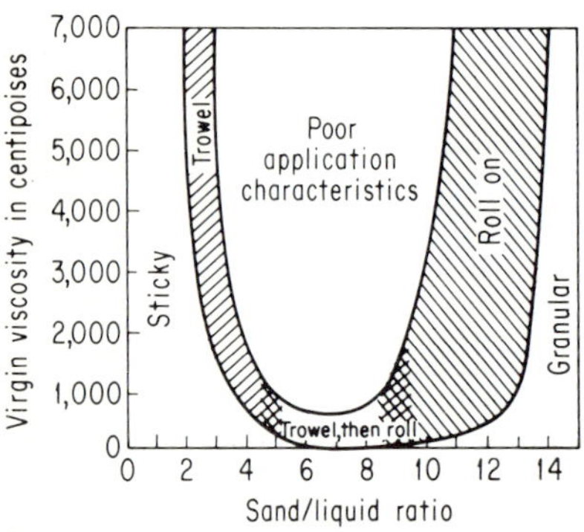

FIG. 23-1. Guide to loading volume of sand for various application techniques 4-4].

In general, the filler employed in the epoxy monolithic topping compounds is sand.

Clean, dry sand is used in a blend of mesh sizes, the blend ranging from 140 or finer up to about 80 mesh or coarser. Because the void volume of tightly packaged sand is about 34 per cent, the loading volume of filler is limited to about 66 volume per cent of the formulation if, theoretically, a dense, void-free structure is to be obtained. However, some air is trapped and this effectively reduces the void volume, allowing the use of additional filler. Selection of the proper types of sand and the proper ratios to use is critical to successful performance.

Figure 23-1 indicates the effect of loading volume of sand on the handling characteristics of the epoxy formulations.

Sand has the following effects on the cured physical properties of the floor-topping compounds [25]:

1. Decreases the tensile, flexure, and compressive strength.

2. Increases the moduli.
3. Decreases the shrinkage.
4. Improves abrasion resistance.
5. Improves thermal shock resistance.
6. Lowers the coefficient of thermal expansion.

The effect of filler loading volumes on the shear strengths of an epoxy mortar formulation is indicated in Fig. 23-2.

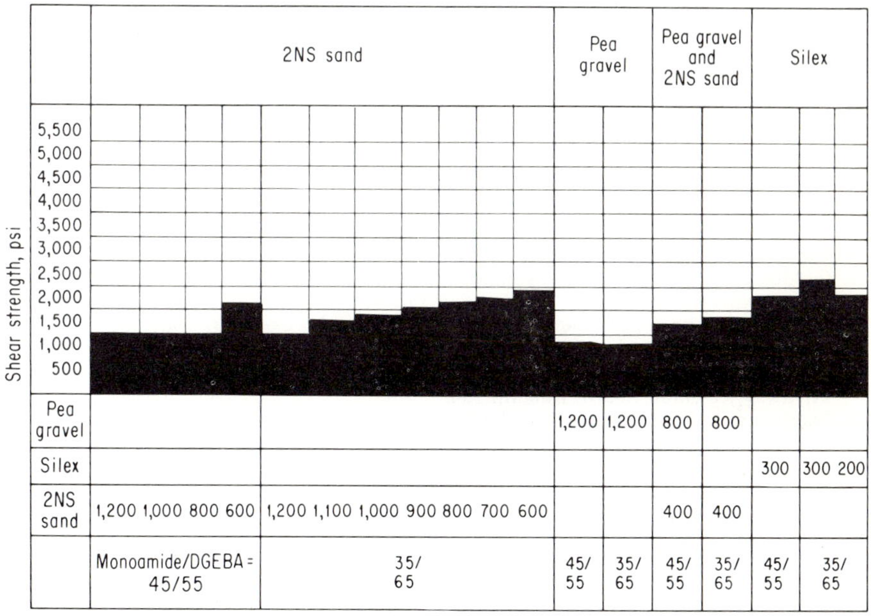

FIG. 23-2. Effect of filler and loading volume on shear strength of DGEBA mortars [4-2].

SUMMARY

Epoxy formulations based on the liquid epoxy resins and substantially free of solvents are employed for a number of industrial applications, including dip coats, spray coats, impregnating compounds, and reinforced and unreinforced trowelable coatings.

The 100 per cent solids coatings (except for those cured with ketimines) are applied in relatively high buildups ($\frac{1}{16}$ in. or more) and offer the ultimate in chemical resistance obtainable with epoxy coating formulations.

The principal use for such coating formulations is as concrete topping and repair compounds. Such formulations are based on the diluted DGEBA and contain a flexibilizing additive or a flexibilizing curing agent. They are loaded with fine sand to the maximum volume consistent with the application and service requirements. These formulations must be carefully designed to accommodate the stresses encountered in service.

Of particular importance in the application of the concrete-topping and repair compounds is proper preparation of the surface. Surface preparation requires mechanical removal of the weak surface layer, followed by a chemical etch.

The 100 per cent solids epoxy compounds, when used in higher-buildup coatings, offer many advantages in return for their relatively high cost.

REFERENCES

1958

1. Asbeck and Jordan, Coated Building Blocks and Methods of Producing Same, U.S. 2,861,011.
2. De Franco, Ends Concrete-Floor Pitting, *Food Eng*., December.
3. Dorman and Gruber, Epoxies Big Potential, *Mod. Plastics*, November.

1959

4. Hirsch, Evaluation of Class H Impregnation Systems, *Insulation*, May.
5. Instructions for Repair Procedures for Uncontrolled Cracks and Spalls in Rigid Pavements Using Epoxy Resin, Grouts, Mortars, and/or Concretes, Technical Memorandum, No. 2-11. Corps of Engineers, Rigid Pavement Laboratory, May.
6. Le Fave, Those Amazing Epoxies, *Western Architect and Engr*., August.
7. Richardson, The Use of Liquid Epoxy Resins in Coatings, Union Carbide Plastics Co.

1960

8. Fire Damage to Concrete Repaired by New Techniques, *Civil Eng*., June.
9. Goldberger, The Use of Epoxy Resins in Repair of Concrete Pavements, *Roads and Eng. Construct*., July.
10. Hodges, Epoxies, Their Use in Floor Maintenance, *Western Ind*., May.
11. Milligan, Floor Compositions, Brit. 855,007.
12. Nagin et al., The Development of Resinous Skid Resistant Surfaces for Highway, Highway Research Board, Bulletin 184.
13. Nagin and Russell, Pavement and Method of Producing the Same, U.S. 2,948,201.
14. Severance, Tops in Floor Topping, *Mod. Plastics*, February.
15. Simpson, Bituminous Compositions, U.S. 2,956,034.

1961

16. At Lloyd's Shopping Centre Epoxy Wall Coatings Clean Easily with Durability, *Building Maintenance and Modernization*, November.
17. Goldberger, Corrective Measures Employing Epoxies on Concrete Bridge Decks of New Jersey's Golden State Parkway, N.J. Highway Authority, Highway Research Board, January.
18. Lee and Neville, New Developments in Epoxy Resins, *Insulation*, May.
19. Ringwood, Moisture Resistant Coatings for Class B Printed Circuitry, 17th ANTEC of SPE.
20. Selander and Mitchell, Epoxy Resins for Concrete Construction and Repair, General Report No. 28, U.S. Dept. of the Interior Bureau of Reclamation, Apr. 17.
21. Smith, How Plastic Shielding Protects Plating Room Floor, *Plant Eng*., June.

1962

22. Epoxy Coating Seals New Deck, *Construct. Methods and Equipment*, February.
23. Farris and Mansur, Protective Coatings for Printed Circuit Assemblies, *Insulation*, May.
24. Mackenzie, Thin Epoxy Layer Restores Concrete Bridge Deck, *Civil Eng*., March.
25. Nordby and Kriegh, Guide for Use of Epoxy Compounds with Concrete, *J. Am. Concrete Inst*., September.
26. Scales, M.C., Private Correspondence, Ciba Products Co.
27. Spraying the Unsprayable Plastics, *Plastics World*, August.
28. Wakeman et al., Glue! For Concrete Repair, *Mater. Res. Std*., February.
29. Wittenwyler, Compositions Comprising A Coal Tar Product, A Polyepoxide and Abrasive Particles, and Process for Treating Surfaces Therewith, U.S. 3,303,088.
30. Wittenwyler, A Progress Report on Epoxy Road Surfacing, *Highway Res. Abstr*., July.

1963

31. Big Resurfacing Job Handled with Epoxy Cement, *Contractors and Engrs. Mag.*, September.
32. Ferrari, Memo to a Pock-Marked Bridge: Epoxy on You, *Texas Highways*, April.
33. Minarik, Epoxy Resin Paving Cements, *Public Works*, July.
34. Riley, Bridge Deck Repair Technique, *Public Works*, May.
35. Simpson, et al., Surfacing Compositions Comprising a Mixture of a Polyepoxide, a Polyamide, and a Petroleum Derived Bituminous Material, U.S. 3,105,771.
36. Splash Zone Coating Protects Corroded Steel Piling, *Mater. Protection*, October.

1964

37. Hotz, Epoxy-Lined Pipe Reduces Cost of Trunk Sewer, *Civil Eng.*, September.
38. Jorda, Process for Treating and Repairing Surfaces Immersed in Water, U.S. 3,160,518.
39. Keck and Feller, Detection of Epoxy Resin Coatings on a 17th Century Painting, *Studies Conserv.* (1), 1–8.
40. Selander, Resistance of Cavitation Damage—Laboratory and Field Tests on Four Selected Coatings, AD 450 699.

1966

41. Crowley et al., A Three-dimensional Approach to Solubility, Eastman Chemical Products, Inc., *J. Paint Technol.*, vol. 38, no. 496, May.
42. Suggitt and Graft, Ten Years Experience in the Evaluation of Coating for Underwater Steel Structures, Hydro-Electric Power Commission of Ontario, *J. Paint Technol.*, vol. 38, no. 494, March.

Chapter 24

EPOXY-RESIN SOLUTION COATINGS

Solution coatings consist, essentially, of three separate components: the film-forming binder (or vehicle solids), which together with the solvent is known as the vehicle—and the pigments.

Epoxy solution coatings, of various kinds, have year after year, continued to account for at least one-half of the total epoxy-resin production. The following listing is typical of the uses: for coating various types of metal furniture, including wrought iron [11] and aluminum [7] lawn furniture and laboratory furniture [18]; for appliance finishes, such as on evaporative coolers [9] and washing machines [8]; for copper and brass hardware [10]; for specialized wood coatings proof against alcohols and for heat-resistant aluminum-backed veneers [46]; for can and drum linings [5] and for coating collapsible aluminum tubes [32]; for pipe coatings [20]; for general-purpose industrial coatings [12] used for plant maintenance [55] and offshore drilling rigs [19]; for specialized applications such as coating air-rifle barrels to produce traditional rifle-barrel blue [13]; for magnet-wire coatings [50]; for humidity-proofing printed-circuit boards [59]; for marine coatings, floor coatings (for both masonry and wood), swimming pool coatings, metal primers, etc.

As of 1965, the automobile industry was the largest single user of epoxy resins—about 30 per cent of the total epoxy solution coatings market. The use is mainly in primers for car bodies and chassis. About 40 per cent of the total solutions coatings went to general-purpose industrial maintenance and about 20 per cent in pipeline coating and marine applications. The remaining 10 per cent encompasses all other maintenance coating activities [91].

Marine applications have received considerable attention in the literature. Case histories of shipboard tank service are given in [94], where the epoxies have proved quite successful. Use of epoxy coal-tar compounds, in conjunction with an antifouling topcoat, for coating ship bottoms is described in [84]. For offshore metal structures, poor bonds have been reported with epoxy coatings, with special vinyls providing better service in at least one author's opinion [90].

In spite of this overall interest in epoxy solution coatings, only one booklength treatment was available as of the end of 1965—and that on varnishes [96].

The number of miscellaneous applications is quite large. Some measure of the versatility of the epoxy-resin solution systems can be gained from the following brief remarks.

A very considerable amount of attention has been devoted to developing epoxy-resin systems for treating fabrics. They are used to provide bleach resistance [36], waterproofing [29], and crease and shrink resistance [34,35], and to treat wool fabrics to obtain "easy-care" cloths that can compete with the synthetics [51]. In the latter case, for example, a blend of DGEBA and the polyglycidyl ether of glycerol is mixed, in water, with a fatty polyamide and applied to the wool at about 0.1 per cent. The effect on the fabric appears to be attained by spot welding at the fiber interstices while the remaining portions of the fibers are substantially free of the resin. The conventional epoxy resins may be used in this application, but it is preferred, at least in part, to use aliphatic-based resins. Special epoxy resins have been developed, a number based on aliphatic amines [48]. A general review of this field is given in [85].

A number of epoxy coatings have been investigated for equalizing emissivity of electronic components. With suitable coatings, hot-spot temperatures may be detected by infrared devices, thus simplifying troubleshooting [86].

Epoxy resins have been used to seal pores of permeable materials such as soils and dams [76] and to loosely bind together fillers to give structures which have been suggested, among other things, as cigarette filters [78]. Solution systems have been used to seal loose sand in the well bores of oil wells. In the latter case, the sand is flushed in advance with alcohol to remove the contaminants. The epoxy is then applied. When the solvent flashes off and the epoxy cures, the sand is still porous but tightly bonded, and the problem of sand influx into the oil is eliminated. Some wells have been in operation over 2 years, thus treated, and have produced over 2 million barrels of oil without notable sand influx [68].

Solution coatings of DGEBA cured with amine curing agents have been used for the impregnation of paper card stock [81]. Solution coatings have been used to print watermarks on paper [82]. They have been used as golf ball paints [47]. Mixtures with polypropylene, solvents, pigments, and curing agents have been used for printing inks for untreated polyethylene films [53]. And uncured DGEBA (mol. wt. 1,500–4,000) has been used as a thermoplastic layer over recording tape. Upon heating, the DGEBA flows in accordance with the electronically recorded pattern, and on cooling is available in a form suitable for optical readout [58].

Epoxy varnishes have been used for treating aluminum offset printing plates [88]. Although quite far down the list as being easy to clean after radioactive contamination, they are easier to clean than stainless steel surfaces and have been used with good success by the AEC in contaminated applications [87]. The list could be extended.

However, in spite of the interesting secondary applications, the primary function of the epoxy solution coatings is to protect against corrosion, and it is useful to consider the nature of corrosion by way of introduction to the solution coatings.

CORROSION

All corrosion of metals, with the exception of direct chemical attack by strong acids or bases, is caused by the formation of local galvanic cells [52]. To prevent corrosion, therefore, it is necessary to prevent the formation of these cells. This

Epoxy-resin coating being applied to exterior of railroad freight car. (*Ciba Products Co.*)

can be accomplished by (1) avoiding differences in potential in the metal, (2) avoiding the formation of electrolytes, and (3) preventing reaction on the anode or cathode.

In the first case, surface contaminants, such as mill scale and foreign matter, must be removed, and dissimilar metals must be isolated from one another. In the second case, the protective film must be tough, durable, and chemically resistant, and provide the highest electrolytic resistance and the lowest permeability consistent with the other requirements. In the third case, rust-inhibiting compounds must be incorporated into the film to prevent reaction on the anode or cathode and to sacrifice themselves in preference to the metal being protected.

As is evident, fulfilling these three requirements requires great care and attention to formulation and application.

To consider some of these variables specifically:

The cured film should be designed to be pinhole-free, to adhere tightly to the surface, and to possess the best possible resistance to environmental attacks.

Pinhole-free solvent-containing coatings can best be obtained by the use of multiple coats.

Adhesion is obtained either through the use of primers or by the use of surface-preparation techniques compatible with the film being used. In addition, adhesion may be influenced by the number and thickness of the coats. Some debate exists regarding optimum film thicknesses for maximum adhesion, and it is possible that this may vary with the formulation. With at least some epoxy films, it has been shown that adhesion of the film increases approximately 60 per cent per mil with increasing thickness of the first coat, up to, of course, a limit; whereas increased total film thickness without increasing the thickness of the first coat results in only a moderate (5 per cent) improvement per mil [45]. With regard specifically to adhesion, epoxy paints, like all other paints, lose adhesion on exposure to high humidities and total immersion. However, loss of adhesion, per se, does not automatically cause corrosion of metallic members. A fatty-polyamide-cured DGEBA retained perhaps no more than 10 per cent of initial adhesion after total immersion; yet the sample was uncorroded.

From these data, it would follow that high initial values for adhesion of paints is a worthless criterion to judge performance by, and that adhesion may really be of only secondary importance [95]. It has, in fact, been shown that long-term adhesion, under total immersion, is relatively independent of filler and of the osmotic pressure of the medium in which the film is immersed, both influencing merely the rate of loss of adhesion, but not the ultimate value. Higher water absorption and water permeability usually give more rapid adhesion losses, but it seems fair to say that adhesion loss occurs before any gross water pickup has occurred. On the basis of data showing that water penetrates the paint film very quickly and probably spreads along the boundary layer, it was theorized [95] that adhesion loss is caused by the action of water on the boundary layer, which is probably different in composition from that of the bulk polymer. Data were generated showing that this boundary layer was more water-sensitive than the bulk polymer and contained water-soluble constituents. The exact mechanism by which water operates on the boundary layer was not established. It might displace polar groups and reduce van der Waals forces; the water might swell the boundary layer to set up high stress concentrations; or the water might physically alter the boundary layer in a way that weakens it mechanically. In any event, the observations in this series of tests [95] are consistent with work accomplished regarding adhesion to glass fibers as reported in Chap. 22.

Best possible resistance to environmental attack requires that the coating be designed specifically for the service environment involved. Coatings providing satisfactory solvent resistance may be deficient in alkali or acid resistance; coatings suitable for room-temperature service may fail rapidly at only slightly elevated temperatures.

A particularly severe environment for epoxy-resin coatings is total immersion in water. Failure in service may be attributed to improper formulating, but it is also caused by a variety of other factors: inadequate film thickness, insufficient cure, low temperature of application, inadequate surface preparation, insufficient induction period, entrapped solvents, and contamination of the substrate. As the purity of the water increases, the problem becomes even more difficult, particularly when water-soluble compounds are present in the substrate. As these go into solution and cross the membrane to the more dilute side, blistering will almost inevitably result [72].

Perhaps, from the formulator's standpoint, the most difficult requirement in corrosion resistance is the attainment of the lowest possible moisture and oxygen permeability [60].

In addition to what might be called the permeability constant of the cured film, the effect of environment on this parameter must be considered. Infrared studies of oxygen permeability in films indicate that simple oxidation of alcoholic hydroxyls to keto groups account, in part, for increased oxygen absorption and permeability, although the total reaction is more complex [17]. Unreacted epoxy groups will act as focal points for oxidation, so that thorough cure is required for optimum performance. Although with phenolic- or amine-cured epoxy films temperatures of 180 to 200°C are required to get substantial oxidation in a short time, at temperatures above this some readily oxidizable group is introduced into the resin structure, indicating that these higher temperatures should be avoided during cure. Chemical changes in epoxy films also can be observed as a result of exposure to ultraviolet [89], and these changes might well be expected to adversely affect performance.

A small amount of work has been reported on diffusion of liquids, in addition to water. For example, diffusion rates of acids through epoxy films may be

Epoxy-resin coating provides protection to aboveground pipeline. (*Ciba Products Co.*)

determined by applying the pigmented film of interest over a clear film containing an acid indicator, such as methylene red. Diffusion rate is then determined with a microscope by observing color changes. With the epoxy film under investigation, the observed rates of 10 per cent sulfuric, nitric, and hydrochloric acids were 200, 140 and 100 microns, respectively, after 5 days [80]. In another study, involving penetration of nitroglycerin through a 15-mil epoxy coating (used to prevent nitroglycerin from contacting the insulator of a rocket case and thereby increasing the char rate of the insulator), it was found that penetration was only about 17 mils at 55°C after 2 weeks and that the effect on the char rate was negligible [54].

VEHICLE SOLIDS

Vehicle solids for coating applications are, in general, higher-molecular-weight resins melting above room temperature; these, upon the evaporation or release of the volatiles, are tack-free, with further polymerization being accomplished with or without heat through available reactive sites. Most solution-coating work in the epoxy technology has involved the use of DGEBA resins, the preference being for molecular weights from 900 upward.

DGEBA resins with molecular weights of 900 to 1,000 are used primarily with amine curing agents, the curing reaction occurring predominantly through the epoxy groups. DGEBA resins with molecular weights of 1,500 to 2,000 are normally used to esterify fatty acids. The resultant epoxy esters are then cured in films through reaction of the unsaturation of the fatty acids under the influence of conventional driers. DGEBA resins with molecular weights of 3,000 to 4,000 are normally employed in conjunction with urea-melamine-, or phenol-formaldehyde resins, the curing reaction occurring by condensation of the methylol groups with the hydroxyl groups on the epoxy chain, the epoxy groups themselves contributing little if anything to the reaction. Even higher-molecular-weight DGEBA species, ranging upward to 80,000 or so, are used as vehicle solids without the addition of any curing agents or coreactant, the molecular weight of the film being sufficiently high, upon application, to provide good protection. A comparison of some of these coating systems is given in Table 24-1.

Table 24-1. Comparison of Solvent-borne DGEBA Resin-hardener Coating Systems [33]

Property	Amine room temperature, low or bake (90–120°C)	Fatty polyamide room temperature, low or bake (90–120°C)	Urea- and melamine-formaldehyde medium bake (150–175°C)	Phenolic high bake (175–230°C)
Film color	2	3	2	5
Flexibility	2	1	2	1
Hardness	2	3	1	1
Abrasion	2	2	1	1
Water resistance	2	1	1	1
Solvent resistance	2	3	1	1
Corrosion resistance	1	2	1	1
Chemical resistance	3	4	2	1

LEGEND: 1, best; 5, poorest. Based on laboratory and field development work.

In addition to DGEBA resins, a variety of other epoxies has been investigated for use in solution coatings. Epoxidized novolacs and silicones have been used for the production of more heat-resistant films. Epoxidized aliphatic polyols have been used in emulsion coatings. Cyclohexene dioxide has been adducted with a fatty acid, the final polymerization being accomplished in solution through the unreacted epoxy groups [64]. Partially epoxidized linseed oil has been used together with anhydride curing agents [65]. A variety of low-molecular-weight diepoxies has been reacted with polyphenols to produce higher molecular-weight resins [3]. Epoxidized polybutadiene has been reported as providing satisfactory solution coatings when reacted with phenol- and urea-formaldehyde resins and to give good anhydride-cured baking films. They may also be esterified with fatty acids to produce ester coatings [4–3]. Epoxidized dimer acids and adducts of dimer acids and DGEBA are offered commercially to provide coating vehicle solids of improved flexibility. Bulk-polymerized epoxidized drying oils have been reported as being used to improve low-temperature flexibility of epoxy enamels [70].

In addition to the resin, a number of coreactants and modifiers may be present in the vehicle solids. Some of these are discussed in other sections of the book. A wide variety of specialized coreactants (such as the zein fraction of corn gluten [62]) will not be considered, such details being beyond the intention of the present work. The compatibility of DGEBA resins with various other coating materials is indicated in Table 24-2.

Table 24-2. Compatibility of DGEBA Resins with Various Coating Materials [4-6]

Coating material	DGEBA resin, mol. wt. and % present											
	1,000			1,500			3,000			4,000		
	90%	50%	10%	90%	50%	10%	90%	50%	10%	90%	50%	10%
Typical commercial alkyd resins:												
Alkyd, short, soy	I	I	I	I	I	I	I	I	I	I	I	I
Alkyd, medium, soy	SI	I	I	I	I	I	I	I	I	I	I	I
Alkyd, long, soy	C	I	I	I	I	I	I	I	I	I	I	I
Alkyd, styrenated	C	I	I	I	I	I	I	I	I	I	I	I
Alkyd, nondrying	C	C	C	C	C	C	C	C	C	C	C	C
Alkyd, nondrying	C	I	C	I	I	I	I	I	I	I	I	I
Typical commercial phenolic resins pure and modified:												
Phenolic	C	C	C	C	C	C	C	C	C	C	C	C
Phenolic, rosin-modified	I	I	I	I	I	I	I	I	I	SI	I	I
Typical commercial cellulose derivatives:												
Cellulose acetate	I	I	I	I	I	I	I	I	I	I	I	I
Cellulose nitrate,	I	I	I	I	I	I	I	I	I	I	I	I
Ethylcellulose	SI	I	I	I	I	I	I	I	I	I	I	I
Typical commercial amino resins:												
Melamine-formaldehyde	SI	I	I	I	I	I	I	I	I	I	I	I
Sulfonamide	C	C	C	C	C	C	C	C	C	C	C	C
Urea-formaldehyde	C	C	C	C	C	C	C	C	C	C	C	C
Typical commercial chlorinated derivatives:												
Chlorinated biphenyl	C	C	I	C	C	I	C	C	I	C	I	I
Chlorinated rubber	SI	SI	SI	I	I	I	I	I	I	I	I	I
Typical commercial vinyl derivatives:												
Polyvinyl acetate	C	C	C	C	C	C	C	I	I	I	I	I
Polyvinyl chloride acetate	I	I	I	I	I	I	I	I	I	I	I	I
Polyvinyl formal	C	C	C	C	C	C	C	C	C	C	C	C
Typical miscellaneous commercial compounds:												
Blown linseed oil	C	I	I	I	I	I	I	I	I	I	I	I
Butyl acetyl ricinoleate	C	I	C	C	I	I	I	I	I	I	I	I
Dammar	C	SI	C	SI	I	C	SI	I	SI	SI	I	SI
Ester gum	I	I	SI	I	I	I	I	I	I	I	I	I
Ester gum, maleic	I	C	C	I	I	C	I	I	I	I	I	I
Terpene resin	I	I	I	I	I	I	I	I	I	I	I	I
Coumarone resin	I	I	C	I	I	C	I	I	I	I	I	I
Styrenated polyester	SI	I	I	I	I	I	I	I	I	I	I	I

C. Compatible, bright clear film   I. Incompatible, cloudy film.   SI. Slightly incompatible, slight haze or smoke appearance.

When the epoxy vehicle and modifiers have been selected, it may be necessary to employ small amounts of resinous flow-control agents. These additional vehicle solids are used on the order of 1 to 2 per cent, typical materials being polyvinyl butyral and special silicone resins; the latter being preferred for amino-resin–epoxy vehicles, both being suitable for the phenolic-resin–epoxy and epoxy-reactive vehicles. Urea-formaldehyde resins at 3 to 4 phr are used to provide flow control for the lower-molecular-weight resins. Motor oil, paraffin, lanolin, castor oil, etc., are other additives suggested as flow-control agents [26].

As might be expected from the discussion in Chap. 15, epoxy resins may be used as modifiers for a variety of other coating resin systems, principally the acrylics and alkyds, although modifications may be required. For example, the higher-molecular-weight DGEBA resins are somewhat too functional for use in the manufacture of epoxy-modified alkyds. When these are used with glycerol–phthalic alkyds, for example, they may be defunctionalized by partial esterification to reduce the number of hydroxyls per molecule to about three [38]. Coumarone-indene resins, themselves suitable [93], may be improved by phenolic modifications [92].

Solution Coatings Based on DGEBA (mol. wt. 900–1,000)

DGEBA (mol. wt. 900–1,000) contains sufficient epoxy groups to permit thorough cross-linking with conventional epoxy-resin curing agents.

Most of the curing agents discussed earlier in the book are, at least theoretically, suitable for use in solution coatings based on this molecular-weight range of the resin. In practice, however, the selection is more limited. The primary advantage of the use of epoxy-reactive solution coatings is that high (if not maximum) chemical resistance can be obtained with room-temperature curing systems. If heat curing is permissible, it is probable that a solution coating based on one of the higher-molecular-weight species would prove more satisfactory from the standpoint of both cost and cured properties. Therefore, the selection of curing agents is, in practice, narrowed to the aliphatic primary polyamines, the fatty polyamides, and room-temperature curing catalysts, such as tertiary amines and strong acids, and various blends thereof.

Epoxy-resin paint used to coat 200,000 ft of concrete jet-hanger floor at John F. Kennedy International Airport. Paint protects surface from jet fuels, lubricants, and cleaning agents. (*Ciba Products Co.*)

Table 24-3. Immersion Tests on Polyamine-Epoxy Films [15]

DGEBA, mol. wt.	Curing agent	phr	Cure schedule (1)	Chemical resistance							
				10% HCl at 52°C		10% NaOH at 52°C		Distilled water at 52°C		Methyl ethyl ketone at 24°C	
				24 hr	48 hr	24 hr	7 days	24 hr	7 days	24 hr	7 days
380	DETA	8	*a*	4B	4B	OK	4R	4B, 4R, 4Bl	4B, 4R, 4Bl	3S	3S
380	DETA	14	*a*	4L	4L	OK	4B, 4R	4B, 4R, 2Bl	4B, 4R, 3Bl	OK	OK
950	DETA	6	*a*	4B	4B	3R	3R	4B, 4R	4B, 4R	3S	4S
380	DETA	8	*b*	OK	OK	OK	2B	OK	2B, 2R	3S	3S
380	DETA	14	*b*	4L	4L	OK	3R	2B	4B, 4R	OK	OK
950	DETA	6	*b*	OK	OK	OK	4B, 4R	4B, 4R	4B, 4R	3S	3S
380	MDA	27	*b*	4C	4C	OK	2B, 2R	2B, 2R	3B, 3R	OK	OK
950	MDA	10	*b*	OK	2C	OK	2B, 2R	3B, 2R	3B, 3R	3S	3S
380	MDA/DETA(1/1)	20.8	*a*	4L	4L	3R	2B, 4R	4B, 4R, 3Bl	4B, 4R, 3Bl	2S	OK
950	MDA/DETA(1/1)	7.5	*a*	4B	4B	OK	4R	4B, 4R	4B, 4R	3S	3S
380	MDA/DETA(1/1)	20.8	*b*	4B, 4C	4L, 4C	OK	2B	3B	3B, 4R	Cracking	OK
950	MDA/DETA(1/1)	7.5	*b*	OK	OK	OK	2B	3B, 3R	4B, 4R	2S	3S

1. Cure schedule: *a*, 7 days at 24°C.
   b, 30 min at 150°C.
2. Slight.
3. Moderate.
4. Severe.

B. Blistering.
Bl. Blushing.
C. Discoloration.
L. Lifting.

OK. Unchanged.
R. Rusting.
S. Softening.

Table 24-4. Physical Properties of Polyamine-Epoxy Films [15]

DGEBA, mol. wt.	Curing agent	phr	Cure schedule (1)	Flexibility	Reverse impact, in.-lb	Adhesion	Pencil hardness
380	DETA	8	*a*	Pass ⅛ in.	Pass 30	Poor	2H
380	DETA	14	*a*	Pass ⅛ in.	Pass 30	Poor	2H
950	DETA	6	*a*	Pass ⅛ in.	Pass 30	Good	2H
380	DETA	8	*b*	Pass ⅛ in.	Pass 30	Excellent	4H
380	DETA	14	*b*	Pass ⅛ in.	Pass 30	Poor	6H
950	DETA	6	*b*	Pass ⅛ in.	Pass 30	Excellent	3H
380	MDA/DETA (1/1)	20.8	*a*	Pass ⅛ in.	Pass 30	Poor	3H
950	MDA/DETA (1/1)	7.5	*a*	Pass ⅛ in.	Pass 30	Good	2H
380	MDA/DETA (1/1)	20.8	*b*	Pass ⅛ in.	Pass 30	Fair	6H
950	MDA/DETA (1/1)	7.5	*b*	Pass ⅛ in.	Pass 30	Fair	2H

1. Cure schedule: *a*, 7 days at 24°C.
b, 30 min at 150°C.

Typical properties of solution coatings cured by amines and fatty polyamides are given in Tables 24-3 to 24-8. Properties for films flexibilized with a polysulfide are given in Tables 24-9 and 24-10. In these tables, also, DGEBA (mol. wt. 340) is compared with DGEBA (mol. wt. 950).

Aliphatic Primary Polyamines. Of the aliphatic primary polyamines, the most frequently used are DETA, TETA, and tetraethylenepentamine. Amines such as DEAPA are also used, these combining advantages of the primary and tertiary amine in one molecule.

The primary aliphatic polyamines are used at about 6 phr with DGEBA (mol. wt. 900–1,000). Since the systems are reactive at room temperature, there will be a progressive increase in viscosity after the solutions are mixed. The amines are therefore packaged separately usually, for convenience, in a solvent system. For

Epoxy-resin coating being applied to railroad flatcar deck. (*Ciba Products Co.*)

Table 24-5. Immersion Tests on DMP-30/DGEBA Resins Films [15]

DGEBA, mol. wt.	phr	Cure schedule (1)	Chemical resistance							
			10% HCl at 52°C		10% NaOH at 52°C		Distilled water at 52°C		Methyl ethyl ketone at 24°C	
			24 hr	48 hr	24 hr	7 days	24 hr	7 days	24 hr	7 days
380	10	*a*	3L	4L	4Bl	4Bl	4R, 3Bl	4R, 3Bl	4S, 2Bl	4B, 4S
950	2	*a*	2R	3R	2D	4D	4Bl	4B, 4R, 3Bl	4S	4S, 4D
380	10	*b*	4B	4L	OK	2R	2B, 2R	2B, 4R	OK	OK
950	2	*b*	2B	2B	OK	2B	3B, 3R	4B, 4R	2S	2S

1. Cure schedule: *a*, 7 days at 24°C.
   b, 30 min at 150°C.
2. Slight.
3. Moderate.
4. Severe.

B. Blistering.
Bl. Blushing.
C. Discoloration.
D. Dissolving.

L. Lifting.
OK. Unchanged.
R. Rusting.
S. Softening.

Table 24-6. Physical Properties of DMP-30 Epoxy Films [15]

DGEBA, mol. wt.	phr	Cure schedule (1)	Flexibility	Reverse impact, in.-lb	Adhesion	Pencil hardness
380	10	*a*	Fail ½ in.	Fail 2	Poor	2H
950	2	*a*	Fail ½ in.	Fail 2	Excellent	2H
380	10	*b*	Pass ⅛ in.	Pass 30	Excellent	4H
950	2	*b*	Pass ⅛ in.	Pass 30	Excellent	2H

1. Cure schedule: *a*, 7 days at 24°C.
b, 30 min at 150°C.

DETA, for example, the solvent might be *n*-butanol and toluene at 1/1 as a 50 per cent solution.

Pot life of the amine/resin blend will depend upon the specific amine and the solvent concentration and will range from 8 hours to several days at room temperature. However, the progressive increase in viscosity may require periodic thinning with the consequent reduction in solids content. Viscosity data for solution coatings containing two typical amines are presented in Fig. 24-1.

If humid conditions prevail, or if the amine-catalyzed system is applied too quickly after mixing (it must normally be allowed an induction period of about an hour or so before application), blushing is likely to occur. In pigmented coatings, gloss will not be too satisfactory, and the amines in general tend to provide relatively poor color characteristics in clear films.

In comparison with the fatty-polyamide curing agents, the aliphatic polyamines are preferred for chemical and solvent resistance; however, in general, unless proper

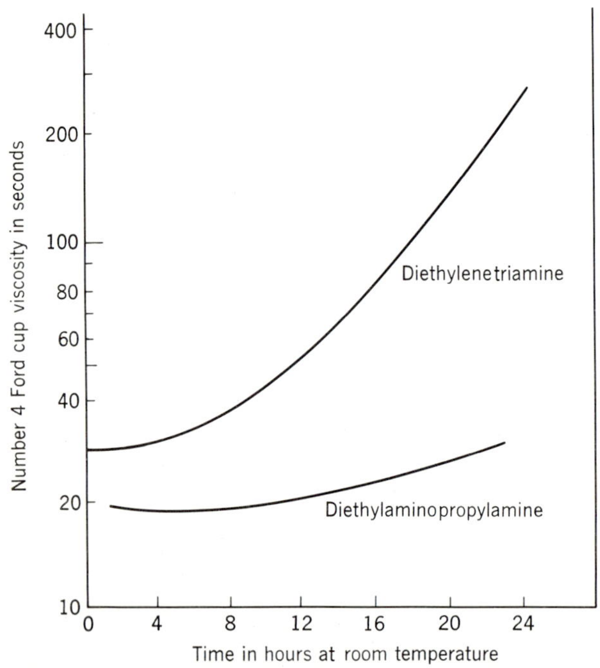

FIG. 24-1. Effect of amine curing agents on room-temperature viscosities [4-7].

Table 24-7. Immersion Tests on Fatty-polyamide–DGEBA Resin Films [15]

DGEBA, mol. wt.	Fatty polyamide, amine no.	E/P	Cure schedule (1)	Chemical resistance							
				10% HCl at 52°C		10% NaOH at 52°C		Distilled water at 52°C		Methyl ethyl ketone at 24°C	
				24 hr	48 hr	24 hr	7 days	24 hr	7 days	24 hr	7 days
380	83–93	50/50	*b*	3B	4B	OK	2R	2B	3B, 2R	OK	2S
950	83–93	50/50	*b*	3B	4B	OK	2B, 2R	2B	4B, 2R	OK	2S
380	210–230	70/30	*a*	2B	3B	OK	Lost gloss	3B, 2Bl	3B, 3R	3S, 3W	4S, 4W
950	210–230	70/30	*a*	2B	4B	OK	2R	4B, 4R	4B, 4R	2S	3S
380	210–230	70/30	*b*	2B	4B	OK	2B	2B	3B, 2R	OK	2S
950	210–230	70/30	*b*	2B	4B	2B	2B	2B	4B, 2R	OK	OK
380	210–230	50/50	*a*	4B	4L	2R	2R, 2B	3B	3B, 3R	OK	2S
950	210–230	50/50	*a*	4L	4L	OK	2B, 2C	4B	4B, 4R	OK	2S, 3C
380	210–230	50/50	*b*	4B, 3L	4L	OK	OK	2B	4B, 2R	OK	OK
950	210–230	50/50	*b*	4L	4L	2B, 2R	2B, 2R	3B	4B, 3R	2C	4S, 3C
380	290–320	80/20	*a*	3B, 2Bl	3B, 2Bl	OK	OK	2B, 2R, 2Bl	2B, 2R	2S	4S
950	290–320	80/20	*a*	2B	3B	OK	OK	3B, 3R	4B, 3R	2S	3S
380	290–320	80/20	*b*	2B, 2D	2B, 2D	OK	OK	2Bl	OK	OK	OK
950	290–320	80/20	*b*	OK	OK	OK	OK	2B	2B, 2R	3S	2S
380	290–320	60/40	*a*	3B	4L	OK	OK	3B, 3R	3B, 3R	OK	2S
950	290–320	60/40	*a*	4L	4L	OK	OK	4B, 3R, 3Bl	4B, 4R	2S	4S, 4C
380	290–320	60/40	*b*	2B	2B, 2L	OK	OK	OK	3B, 2R	OK	OK
950	290–320	60/40	*b*	3B, 3L	4L, 3C	OK	OK	2B, 2R	4B, 4R	OK	3S

1. Cure schedule: *a*, 7 days at 24°C.
   b, 30 min at 150°C.
2. Slight.
3. Moderate.
4. Severe.

B. Blistering.
Bl. Blushing.
C. Discoloration.
D. Dissolved.

L. Lifting.
OK. Unchanged.
R. Rusting.
W. Wrinkling.

Table 24-8. Physical Properties of Fatty-polyamide–DGEBA Resin Films [15]

DGEBA, mol. wt.	Fatty polyamide, amine no.	E/P	Cure (1)	Flexibility	Reverse impact, in.-lb	Adhesion	Pencil hardness
380	83–93	50/50	*b*	Pass ⅛ in.	Pass 30	Excellent	H
950	83–93	50/50	*b*	Pass ⅛ in.	Pass 30	Excellent	2H
380	210–230	70/30	*a*	Pass ⅛ in.	Pass 30	Excellent	H
950	210–230	70/30	*a*	Pass ⅛ in.	Pass 30	Excellent	2H
380	210–230	70/30	*b*	Pass ⅛ in.	Pass 30	Excellent	H
950	210–230	70/30	*b*	Pass ⅛ in.	Pass 30	Excellent	2H
380	210–230	50/50	*a*	Pass ⅛ in.	Pass 30	Excellent	2H
950	210–230	50/50	*a*	Pass ⅛ in.	Pass 30	Excellent	2H
380	210–230	50/50	*b*	Pass ⅛ in.	Pass 30	Excellent	2H
950	210–230	50/50	*b*	Pass ⅛ in.	Pass 30	Excellent	2H
380	290–320	80/20	*a*	Pass ⅛ in.	Pass 22	Excellent	H
950	290–320	80/20	*a*	Pass ⅛ in.	Pass 30	Excellent	2H
380	290–320	80/20	*b*	Pass ⅛ in.	Pass 30	Excellent	H
950	290–320	80/20	*b*	Pass ⅛ in.	Pass 30	Excellent	2H
380	290–320	60/40	*a*	Pass ⅛ in.	Pass 30	Fair	2H
950	290–320	60/40	*a*	Pass ⅛ in.	Pass 30	Excellent	H
380	290–320	60/40	*b*	Pass ⅛ in.	Pass 30	Excellent	4H
950	290–320	60/40	*b*	Pass ⅛ in.	Pass 30	Excellent	2H

1. Cure schedule: *a*, 7 days at 24°C. *b*, 30 min at 150°C.

stoichiometry is observed, they provide poor resistance to water, and they tend to be sensitive to high humidities during cure and to produce rather brittle and inadequately cured films at lower ambient temperatures [16]. Amine adducts containing phenol accelerators have been used to produce some improvement in curing rates.

Ideally (if optimum properties are to be obtained) elevated temperature should be employed when curing these systems. Cure cycles are quite short, the following being typical:

Cure time, minutes	*Temperature, °C*
120	38
90	45
20	93
10	120
4	177
2	204
1	232

At room temperature, the coating will dry tack-free within a few hours (depending on solvents) and will cure sufficiently to permit recoating within 24 to 48 hours, depending on the severity of the environment to be encountered. Full room-temperature cure will be obtained in about 7 days at temperatures of 23°C or above.

A typical polyamine-cured formulation consists of the following:

Material	Pounds	Gallons
Part A:		
DGEBA (mol. wt. 950)	456	46.01
Toluol	153	21.15
Methyl isobutyl ketone	137	20.45
Butyl Cellosolve	16	2.13
Flow-control agent, 60% nonvolatile	23	2.77
Part B:		
DETA	28	3.50
Xylol	14	1.94
Butanol	14	2.06
Total	841	100.01

Table 24-9. Immersion Tests on DGEBA Resin–Polysulfide (mol. wt. 1,000) Film [15]

DGEBA, mol. wt.	E/P	DMP-30, phr	Cure schedule (1)	Chemical resistance							
				10% HCl at 52° C		10% NaOH at 52°C		Distilled water at 52°C		Methyl ethyl ketone at 24°C	
				24 hr	48 hr	24 hr	7 days	24 hr	7 days	24 hr	7 days
380	80/20	10	*a*	3B, 3L, 3S	4L, 3D	4Bl	4Bl	3B, 4R	4B, 4R	2S	2S
950	80/20	2	*a*	OK	4B, 3Bl	OK	2Bl	3B, 4Bl	3B, 4Bl	4D	4D
380	80/20	10	*b*	4B	4B	OK	2R	3B	4B, 4R	OK	2S
950	80/20	2	*b*	3Bl	2Bl	OK	2R	3B, 3Bl	4B, 4Bl, 2R	2S	3S
380	50/50	10	*a*	3S	3D	4Bl	4Bl, 4S	4R, 4B	4R, 4B	4S, 2D	4S
380	50/50	10	*b*	4D	4D	OK	3R, 2C	4B, 2Bl	4B, 4R, 3Bl	3S	4S
950	50/50	2	*b*	4D	4D	OK	3R, 4Bl	4Bl	4Bl, 4R	4S	4S

1. Cure schedule: *a*, 7 days at 24°C.
   b, 30 min at 150°C.
2. Slight.
3. Moderate.
4. Severe.

B. Blistering.
Bl. Blushing.
C. Discoloration.
D. Dissolving.

L. Lifting.
OK. Unchanged.
R. Rusting
S. Softening.

Table 24-10. Physical Properties of Epoxy-Polysulfide Films [15]

DGEBA, mol. wt.	E/P	DMP-30, phr	Cure schedule (1)	Flexibility	Reverse impact, in.-lb	Adhesion	Pencil hardness
380	80/20	10	*a*	Fail ½ in.	Fail 2	Poor	2H
950	80/20	2	*a*	Pass ⅛ in.	Fail 2	Excellent	2H
380	80/20	10	*b*	Pass ⅛ in.	Pass 30	Excellent	2H
950	80/20	2	*b*	Pass ⅛ in.	Pass 30	Excellent	2H
380	50/50	10	*a*	Pass ⅛ in.	Pass 30	Excellent	H
950	50/50	2	*a*	Pass ⅛ in.	Pass 30	Good	H
380	50/50	10	*b*	Pass ⅛ in.	Pass 30	Excellent	2H
950	50/50	2	*b*	Pass ⅛ in.	Pass 30	Excellent	H

1. Cure schedule: *a*, 7 days at 24°C.
b, 30 min at 140°C.

This formulation gives the following physical characteristics:

Nonvolatile, %	59.6
Viscosity, full body, no. 4 Ford cup at 25°C, seconds	104
Viscosity, spraying, no. 4 Ford cup at 25°C, seconds	18
Nonvolatile at spraying viscosity, %	40.0
Curing time at 25°C, days	7
Film thickness, mils	1.2
Impact resistance, in.-lb	160+
Flexibility, conical mandrel	Passes
Abrasion resistance, falling sand, liters/mil	63.1

Typical properties of a clear amine-cured coating of this type are given in Table 24-11.

With the primary aliphatic polyamines, the concentration of the amine is critical in terms of chemical resistance. Best moisture resistance is obtained by the use of slightly less than stoichiometric, thereby reducing the number of free amine groups in the system. Best solvent resistance, on the other hand, is obtained by the use of something in excess of stoichiometric to achieve the highest per cent reaction of the epoxy groups. For exterior films, excess amine will bleed out to cause initial chalking of pigmented systems, and when chalking is critical, something less than stoichiometric is normally preferred. In any event, the exact concentration for optimum performance will be dependent on the cure cycle [22].

Fatty Polyamides. The fatty polyamides are generally considered to offer several advantages over the primary aliphatic polyamines for solution coatings based on DGEBA (mol. wt. 950). The fatty polyamides provide higher orders of flexibility and impact resistance. Additionally use of fatty polyamides makes it possible to deposit firmly adhering films even on wet and tightly rusted surfaces. Humidity and surface preparation are thus somewhat less critical during application.

The fatty polyamides also are considered to provide improved moisture resistance over the polyamine-cured films and have been claimed as more suitable for total immersion. They provide superior gloss and somewhat better blush resistance, although initial color is inferior.

The fatty polyamides provide for somewhat extended pot lives, but in most other handling characteristics they are similar to the primary amines.

Like the polyamines, the fatty polyamides must be packaged separately. The higher-molecular-weight materials may be dissolved in a solvent such as isopropyl alcohol and toluene at 1/1. The liquid fatty polyamides may be packaged with the solvent-containing flow-control agent (when used) or may be cut with a thinner such as xylene and butanol at 9/1.

Table 24-11. Physical Properties and Chemical Resistance of Typical Amine-cured DGEBA Coating Formulation [4-6]

(Test specimen: Multicoat clear coatings on rounded-end probes, average film thickness 6 to 8 mils)

PHYSICAL PROPERTIES

Flexibility	Good
Impact resistance	Good
Hardness	Excellent
Outdoor durability	Excellent
Color	Very good
Abrasion resistance	Good

CHEMICAL RESISTANCE

Solvent or reagent	Exposure time			
	1 week	2 weeks	1 month	2 months
Organic:				
Acetone	ss	ss	ss	
Methyl ethyl ketone	ss	ss	ss	
Methyl isobutyl ketone	ss	ss	ss	
Ethyl alcohol	u	u	u	
Secondary butyl alcohol	u	u	u	
Methyl isobutyl carbinol	u	u	u	
Toluene	u	u	u	
Xylene	u	u	u	
Mineral spirits	u	u	u	
Carbon tetrachloride	u	u	u	
Allyl chloride	u	s	s	
Formaldehyde, 37% w	Failed in 1 week			
Ethylenediamine	Failed in 1 week			
Diethylenetriamine	Failed in 1 week			
Linseed fatty acid	u	u	u	
Acetic acid, 5% w	u	u	u	
50%w	u	u	u	
glacial	Failed in 1 day			
Sodium methoxide, 40% w	u	u	u	
Detergent solution	u	u	u	
Citric acid 10%	u	u	u	
Inorganic:				
Sodium hydroxide, 20% w	u	u	u	u
Ammonium hydroxide, 27% w	Failed in 1 week			
Ammonia vapor	u	u	u	u
Phosphoric acid, 85% w	Failed in 1 day			
Hydrochloric acid, 10% w	u	u	u	u
20% w	u	u	u	u
36% w	u	u	u	u
Hydrochloric acid vapor	u	u	u	u
Sulfuric acid, 25% w	u	u	u	Failed
50% w	u	u	u	Failed
70% w	u	Discolored	Discolored	Discolored
80% w	Failed in 1 week			
90% w	Failed in 1 day			
Nitric acid, 10% w	u	u	u	Failed
20% w	u	u	Failed	
30% w	u	Failed		
Sodium chlorite, 25% w	u	u	u	Failed
Sodium hypochlorite, 5% w	u	u	u	Failed
Calcium hypochlorite, 5% w	u	u	u	u
Water	u	u	u	u

s. Softened. ss. Slightly softened. u. Unaffected.

Like the polyamines, the fatty polyamides require an induction period after mixing and prior to application. This should be a minimum of 30 minutes, and preferably one hour. Induction period is particularly critical when total immersion in water is required [56].

A typical clear fatty-polyamide-cured formulation consists of the following:

Material	Pounds	Gallons
Part A:		
DGEBA (mol. wt. 950)	315.0	31.80
Toluol	108.0	15.00
Methyl isobutyl ketone	93.7	14.00
Butyl Cellosolve	13.1	1.75
Flow-control agent	16.7	2.00
Part B:		
Fatty polyamide (amine value 210–230)	176.00	21.40
Toluol	104.2	14.50
Cellosolve	11.5	1.50
Total	838.2	101.95

Properties of this formulation are as follows:

Total nonvolatile, %	59.8
Weight per gallon, pounds	8.2
Resin/polyamide ratio	64/35.7
Pot life at spray viscosity, (part A + part B), days	2½–3

Typical performance data for a clear fatty-polyamide-cured film are presented in Table 24-12.

A typical fatty-polyamide-cured formulation for gray epoxy floor enamel consists of the following:

Material	Pounds	Gallons
Part A:		
Titanium dioxide	100.0	2.86
Calcium carbonate	200.0	8.90
Lampblack	1.5	0.11
Thixotropic agent	3.5	0.23
Fatty polyamide (amine value 210–230)	129.0	15.55
Xylol	88.0	12.00
Butyl Cellosolve	41.5	5.50
Part B:		
DGEBA (mol. wt. 950)	240.0	24.80
Xylol	118.0	16.30
Methyl isobutyl ketone	52.0	7.75
Butyl Cellosolve	72.1	9.50
Total	1,045.6	103.50

This formulation provides the following properties:

Pigment/binder (by weight) ratio	45/55
Pigment volume concentration, %	23
Total nonvolatile, %	65
Viscosity part A + part B (Stormer), KU	76
Weight per gallon, pounds	10.1
Pot life (part A + part B), hours	8

Typical performance data are given in Table 24-13.

Table 24-12. Performance Data for Clear DGEBA Resin–Polyamide Film [4-1]

Property	*Value*
Impact resistance, in.-lb	Pass 160
Flexibility, conical mandrel	Pass
Abrasion resistance, liters of sand/mil	77.7
Chemical resistance:	
10% NaOH, after 24 hr	Unaffected
10% HCl after 7 days	Soft, no. 6 blisters
Water after 1-hr boil	Unaffected
Salt-spray resistance after 200 hr:	
Of scribed panel	Good
Of unscribed panel	Unaffected
Film properties, rate of dry (hr:min):	
Set to touch	0:05
Cotton-free	0:15
Paper-free	6:00

Table 24-13. Performance Data for DGEBA Resin–Polyamide Film Pigmented with Titanium Dioxide, Calcium Carbonate, and Lampblack [4-1]

Property	*Value*
Abrasion resistance, liters of sand/mil:	
After 2 days' cure	47.2
After 9 days' cure	69.5
After 30 days' cure	66.3
Alkali resistance:	
10% NaOH after 24 hr	Unaffected
Water immersion after 1-hr boiling*	Unaffected
Gasoline resistance after 7 days†	Unaffected
Oil resistance after 7 days†	Unaffected
Gloss, Gardner, 60°	84

* Films air-dried at room temperature for 13 days. Dry film thickness 1.75 mils.

† Films air-dried at room temperature for 7 days. Dry film thickness 1.75 mils.

Catalytic Curing Agents. Catalytic curing agents may be used with DGEBA (mol. wt. 950), but this is not so common as the use of the polyamines and the fatty polyamides. Typical of catalytic curing agents is the use of DMP-30 as indicated in Tables 24-5, 24-6, 24-9, and 24-10. Tertiary amines frequently serve as accelerators for the fatty-polyamide curing agents and are sometimes utilized in blends with the polyamines.

Other catalysts may be employed with the resin, and some rather sophisticated systems may be developed. When 2-ethylhexoic acid is the catalyst, the DGEBA resin (or the resin condensed through the hydroxyl groups with a silanol-bearing silicone resin) cures at temperatures on the order of 150°C and above [23]. Such systems and the use of such catalysts are, however, reserved for specialized applications.

Coatings Based on Esters of DGEBA

A very substantial amount of DGEBA resin production is devoted to the synthesis of higher-molecular-weight DGEBA for application in epoxy-ester coatings. When used in this connection, the resins can be considered polyhydric alcohols. These are then reacted with suitable unsaturated fatty acids.

The epoxy-ester coatings are employed as metal primers, floor coatings, air-dry maintenance finishes, and appliance enamels, and to upgrade the properties of emulsion coatings made from polyvinyl acetate, etc.

These oil-modified epoxies combine adhesion, toughness, abrasion resistance, flexibility, high gloss, and durability. Although the epoxy esters do not provide as good chemical resistance as the other thermosetting DGEBA solution coatings, they are equal in this respect to unplasticized short-oil phenolic varnishes and superior to the more flexible long-oil type. The initial color and color retention of the epoxy esters is better than that of the phenolics, although slightly inferior to

Epoxy-resin-coated panels undergoing exterior weathering. (*Ciba Products Co.*)

that of the soybean alkyds. The epoxy esters under constant submersion are inferior in alkali resistance to the chlorinated rubber and vinyl paints, and they tend to chalk more rapidly than the alkyds.

The esterification process consists of reacting the hydroxyl-bearing resins with fatty acids, such as linseed, dehydrated castor oil, or soybean to produce the ester-linked system, with water being given off as a by-product.

The reaction is accomplished by cooking the two intermediates at 200 to 300°C until the reaction has progressed to the desired stage. A description from the literature illustrates the process:

> To 1 equivalent of a polymeric polyhydric alcohol (e.g., that prepared from bisphenol, epichlorohydrin and phenol) in an apparatus equipped with a mechanical stirrer is added 0.9 to 0.95 mole of an unsaturated acid mixture, composed of the chain acids derived from drying oils, from semidrying oils, or from nondrying oils, of the glyceride type, or mixtures of such acids. The apparatus is equipped with a condenser attached thereto through a water trap, with a thermometer and with means for bubbling inert gas such as CO_2 through the reaction mixture during esterification. Heat is applied, and after the polymeric polyhydric alcohol has melted, stirring is started and continued during esterification. A slow stream of an inert gas such as CO_2 is bubbled through the reaction mixture to aid in removing water formed during esterification. The reaction mixture is heated for, e.g.,

4 to 5 hours at, e.g., 250 to 260°C, water of esterification being removed through the water trap. The resulting product is cut to the desired solid content by adding hydrocarbon solvents, and the resulting solutions form excellent compositions for making protective films adapted to harden on air oxidation or heat polymerization or both [1].

A number of refinements may be introduced into the procedure, as, for example, the addition of an organic chelating agent to permit cooking to low acid numbers without gelation [42]. Alternative methods are also available: for example, steam may be bubbled through the combination of epoxy and fatty acid with metallic lead or lead oxide as a catalyst [6]. Esters of closely controlled molecular weight may be developed in situ by reacting DGEBA and a charge of bisphenol A with the selected fatty acid. The amount of bisphenol A is varied to control the molecular weight of the final ester. DGEBA esters may be modified during the cook with styrene to produce styrenated esters, the styrene being present on about an equal weight basis with the DGEBA resin. Such styrenated esters combine very fast drying time with hardness, mar resistance, and flexibility.

Preferred DGEBA resins for esterification are those in the 1,500- to 2,000-molecular-weight range. Higher-molecular-weight resins, containing more hydroxyls, will body during cook to yield high-viscosity products unless used with nonbodying acids.

DGEBA resin esters may be dissolved in hydrocarbon solvents if sufficient fatty-acid modification is present. For short-oil esters, having less than half the available hydroxyls esterified, an aromatic hydrocarbon is required. However, for long-oil esters, for example an ester having 0.8 of the available hydroxyls esterified with linseed fatty acids, a completely aliphatic hydrocarbon such as mineral spirits can be used as a solvent.

DGEBA esters may be used in air-dry formulations (for which a small amount of conventional drier is added), or they may be used with urea resins and melamine resins to provide heat-convertible films [2]. In addition, partial esters may be developed which can be co-cured at room temperature with metal driers and amines [37].

The properties of the specific DGEBA ester film will be determined by a variety of factors, among them being the acid and/or oil acid selected, the ratio of reactants, the cooking cycle, the type of equipment used (i.e., open or closed kettle), the type of cook (i.e., azeotropic or fusion), and the molecular weight of the resin.

The effects of oil length and acid selection on the film properties are similar to the effects experienced with alkyd resins. Table 24-14 indicates the general effects.

As indicated previously, epoxy esters are normally based on DGEBA (mol. wt. 1,500–2,000) although both higher- and lower-molecular-weight species may be employed, as indicated in the formulations below [4-6]:

Air-dry Enamel Vehicle

Formulation:

DGEBA (mol. wt. 1,000)	43%
Linseed fatty acid	47%
W. G. gum rosin	10%

Typical data:

Acid number	7.0 mg KOH/g
Viscosity, Gardner-Holdt	Y–Z
Color, Gardner	8–9
Weight per gallon, pounds	7.9
Solvents	Mineral spirits
Nonvolatile, %	60%

Table 24-14. Comparative Properties of DGEBA Esters [4-1]

Acid	Equivalent weight	Ester type	Processing temperature, °C	Color and retention	Flexibility	Chemical resistance	Characteristics
Coconut	215	B	249	EX	EX	G	Pale, baking
Soybean	280	AD	260	VG	VG	G	Pale, air-dry
Linseed	280	AD	260	F	VG	G	Durable
Castor (dehydrated)	285	B & AD	249	VG	EX	VG	Fast baking or pale, air-dry chemical resistance
Tung	285	AD	249	F	VG	EX	Fast, air-dry chemical resistance
Tall oil	300 (AV)	AD & B	274	G	F	F	Low cost
Rosin	350	Resin	274	F	P	P	Low cost, fast, air-dry hardness caustic resistance
Hydrogenated rosin	345	Resin	274	G	P	F	As above but better color
Maleic anhydride	49	Modifier	...	..	..	..	Improves viscosity and color retention
Dimer fatty acids	305	Modifier	...	..	..	..	Improves viscosity and drying

AD. Air-dry. AV. Average. B. Baking. EX. Excellent. F. Fair. G. Good. P. Poor. VG. Very good.

A cobalt drier at about 0.04 per cent metal is used with the above.

NONDRYING BAKING VEHICLE

Formulation:

DGEBA (mol. wt. 3,000)	75%
Coconut fatty acid	25%

Typical data:

Cooking temperature, °C	232
Acid number	1 mg KOH/g
Viscosity, Gardner-Holdt	W–Y
Color, Gardner	7–9
Weight per gallon, pounds	9.0–8.1
Solvent	Xylene
Nonvolatile, %	50

Although the epoxy esters are similar to more conventional alkyd resins, they offer three advantages: higher adhesion, flexibility, and chemical resistance. Since ester linkages are vulnerable to caustic attack, alkali-resistance tests are a fair measure of chemical resistance. Table 24-15 presents alkali resistance for several cured films. The films were immersed at 70°C in a 2 per cent soap solution.

Table 24-15. Comparison of Alkali Resistance of Finishes Based on DGEBA Esters and Finishes Based on Alkyd Resins [4-6]

Vehicle	Age of film, days	Condition of film after immersion for period noted		
		48 hr	186 hr	336 hr
85% epoxy ester 15% melamine	1	OK	Small blisters	Some loss of adhesion
85% alkyd 15% melamine	1	Some loss of adhesion	Off panel	
85% epoxy ester 15% melamine	30	Slight blistering	Off panel	
85% alkyd 15% melamine	30	...	Completely off panel	
60% epoxy ester 40% melamine	1	OK	Slight blistering	Slight blistering
60% alkyd 40% melamine	1	Some loss of adhesion	Off panel	
60% epoxy ester 40% melamine	30	Slight blistering	Off panel	
60% alkyd 40% melamine	30	...	Completely off panel	

The DGEBA ester coatings provide indefinite shelf life without requiring two-container systems. The heat-convertible coatings require only short moderately high bakes. For a melamine-converted film, for example, 30 minutes at 150°C is adequate.

DGEBA esters may be applied by all techniques employed with coatings based on more conventional resins and may be handled as conveniently.

Although the overall properties of the DGEBA ester films are inferior to the properties of the other thermosetting epoxy films, they possess the advantage of lower cost, arising from the reduced epoxy content and the feasible use of inexpensive solvent vehicles.

Solution Coatings Based on DGEBA (mol. wt. 3,000–4,000)

In addition to the reaction with the carboxyl groups of fatty acids, the hydroxyl groups of the higher-molecular-weight DGEBA resins may also be reacted, by condensation, with molecules containing methylol groups [30].

Of the methylol-bearing resins, two types are used most frequently in the epoxy solution coatings technology: urea-formaldehyde and melamine-formaldehyde resins (considered together as amino resins), and resole-type phenol-formaldehyde resins. These resins are reacted with DGEBA (mol. wt. 3,000–4,000), the DGEBA resin being present at 40 to 75 per cent.

Steel launch structure at Cape Kennedy is protected with epoxy-based paint. (*Ciba Products Co.*)

Amino Resins. The amino resin epoxy coatings are used in appliance finishes, drum linings, decorative finishes, etc., with the urea-formaldehyde/epoxy systems providing best color and the melamine-formaldehyde/epoxy systems providing greatest flexibility and impact resistance.

The melamine-formaldehyde resins, more expensive than the urea-formaldehyde resins, are available commercially in solution at solid contents in the neighborhood of 50 per cent. Xylol is a common solvent for further reducing viscosity. The melamine-formaldehyde resins provide pot lives of 3 to 4 months in the epoxy resin.

Hexamethoxymethylmelamine, from methanol and hexamethylolmelamine, can be used as a curing agent for DGEBA resin coatings—the reaction being through the hydroxyls on the resin and the methoxy groups—methanol being eliminated [69].

Several species of urea-formaldehyde resins, varying in reactivity, are available commercially in solutions at solid contents in the neighborhood of 60 per cent. Common solvents are xylene and butanol at 4/6. The products of higher reactivity give pot lives in the resin mix on the order of 1 to 2 months; of lower reactivity, pot lives in excess of 2 years. A typical formulation using the more reactive curing agent will achieve satisfactory cure at the following schedules:

Cure time, minutes	*Temperature, °C*
60	120
20	150
6	177
3	204

For the less reactive urea-formaldehyde resins, cure temperatures will be somewhat higher for a given time period.

The amino resins may be used without catalyst, or acid catalysts can be employed to provide accelerated cure [4]. For example, when *p*-toluenesulfonic acid dissolved in *n*-butyl acetate is used with a urea-formaldehyde/DGEBA resin blend, coatings may be developed that are dry to the touch in 10 minutes and hard enough for sanding at the end of one hour. Such systems, being highly reactive, are best handled with dual-nozzle spray equipment.

Phenol-Formaldehyde Resins. Phenol-formaldehyde curing agents provide the maximum in chemical and abrasion resistance obtainable with DGEBA coatings and superior elevated-temperature performance, but because of the poor color retention of phenolic resins they are unsuited to decorative applications. They are most used in corrosion-resistant piping, wire coatings, electrical varnishes, collapsible tube coatings, etc.

The phenol-formaldehyde resins require somewhat higher or longer bakes than do the amino resins. They are sluggishly reactive, if at all, in the uncatalyzed epoxy system, requiring use of acid catalysts to obtain convenient reaction rates. Use of such catalysts will reduce the pot life of the compounds to the 1- to 3-month range, depending upon the catalyst selection.

The phenol-formaldehyde resins are available commercially in solution. A typical clear formulation consists of the following [4-6]:

PHENOL-FORMALDEHYDE COATING

	% by *Weight*
Vehicle nonvolatiles, 35%:	
DGEBA (mol. wt. 3,000)	73.1
Phenol-formaldehyde resin	24.4
Flow-control agent	1
Phosphoric acid	1.5
Vehicle volatiles, 65%:	
Methyl isobutyl carbinol	25
Methyl isobutyl ketone	25
Toluene	25
Xylene	25

A typical formulation containing 1.5 per cent phosphoric acid cures according to the following schedule:

Cure time, minutes	*Temperature, °C*
60	150
30	177
20	190
10	204

Typical chemical resistance for a phenol-formaldehyde/DGEBA coating is given in Table 24-16.

Solution Coatings Based on DGEBA (mol. wt. 80,000)

Bisphenol A and epichlorohydrin may be reacted in solution to produce linear polymers with molecular weights on the order of 30,000 to 80,000. Such materials are commercially available in solutions at 30 to 40 per cent nonvolatiles. They may be further thinned by solvent blends, such as

	Per cent
Cellosolve acetate	60
Methyl ethyl ketone	30
Toluene	10

Table 24-16. Resistance to Chemicals and Solvents of DGEBA Resin Solution Coating Heat-converted by Phenol-Formaldehyde Resin [4-6]

Film unaffected by a 3 months' immersion in the following reagents at room temperature (all solutions are in water):

Alcohol:	Bases:
Ethyl	Sodium hydroxide (all conc.)
Isopropyl	Ammonium hydroxide, 10%
Secondary butyl	
n-Butyl	Acids:
Methyl isobutyl	Acetic, 1%
Neosol	Linseed fatty
Neosol A	Sulfuric, up to 75%
Diacetone	Hydrochloric, up to 20%
Hexylene glycol	Nitric, up to 10%
Glycerine	Phosphoric, up to 85%
Organic chlorides:	Miscellaneous reagents:
Carbon tetrachloride	Liquid detergent, 100%
Allyl chloride	Liquid detergent, 50%
Ketones and aldehydes:	Solid detergent, 1%
Methyl isobutyl ketone	Resitoxaphene
Formaldehyde, 30%	Sodium methoxide, 40% in methanol
Cyclic hydrocarbons:	Sodium chlorite, 25%
Toluene	Sodium hypochlorite, 5%
Xylene	Calcium hypochlorite, 5%
Ethers:	Ferric chloride, 5%
Diethyl ether	Water
Bis(β-chloroethyl) ether	Salt spray at 38°C for 500 hr

Film unaffected by the following materials, all exposed for 3 weeks at 66°C, except as noted:

Isopropyl alcohol	Methyl isobutyl ketone
Secondary butyl alcohol	Allyl chloride
Methyl isobutyl carbinol	20% sodium hydroxide boiling for 24 hr
Neosol	73% sodium hydroxide at 138°C for 2 weeks
Neosol A	Glycerin
Diacetone alcohol	Glycerin at 77°C for 6 weeks
Hexylene glycol	Water

Films slightly soften in 1 month's exposure at room temperature to acetone, methyl ethyl ketone, ethylene dichloride, hydrochloric acid (36%), sulfuric acid (78%), and hydrogen peroxide (15%). The coating has limited resistance to sulfuric acid (at concentrations greater than 85%) and hydrogen peroxide (at concentrations greater than 15%).

Because of the high molecular weight of the vehicle solid, no curing agent is required. The coatings dry upon the evaporation of the solvent to provide a continuous film with good toughness and abrasion resistance. Chemical resistance, in particular solvent resistance, is inferior to that obtained with thermosetting films.

Properties of a clear coating based on one of these high-molecular-weight resins are compared with those other coating systems in Table 24-17. A typical pigmented formulation is given in Table 24-18.

DGEBA COATINGS MODIFIED WITH THERMOSETTING ACRYLICS

Baking-type solution coatings based on combinations of DGEBA resins and thermosetting acrylic resins [57] offer gloss retention and color comparable to alkyd-melamine finishes with chemical resistance similar to the DGEBA resin/urea-formaldehyde coatings. The acrylic-modified finishes cure in about 30 minutes at 150°C.

The preferred DGEBA for use with acrylic resins has a molecular weight of about 950 to 1,000, and the system provides a coating of extreme hardness. Lower-cost

Table 24-17. Comparative Performance of Various Solution Coatings [4-6]

Test conditions	DGEBA (mol. wt. 80,000)	Lacquer resins (all systems force-dried)			Converted DGEBA resin systems			
		Vinyl resin	Acrylic resin	Nylon resin	DGEBA resin amine-cured at 23°C	DGEBA resin phenolic-cured at 204°C	DGEBA ester melamine-cured at 204°C	DGEBA resin urea-cured at 190°C
Chemical resistance:								
20% NaOH after 11 days at 93°C	8	6	1	1	8	9	1	5
30% H_2SO_4 after 14 days at 23°C	10	10	7	1	10	10	8	3
10% HCl after 14 days at 23°C	9	9	6	1	8	5	5	2
Distilled H_2O after 20 days at 23°C	10	7	5	1	8	10	10	10
Distilled H_2O after 11 days at 93°C	6	7	1	1	4	10	8	4
Environmental resistance:								
Humidity after 1,000 hr at 38°C and 100% relative humidity	8	7.5	8.25	1				
Humidity after 2,500 hr at 38°C and 100% relative humidity	8	7	6	1				
Heat resistance	7	1	9	5	7	6	4	3
Tabor abrasion resistance, weight loss/100 rev, mg	0.4	0.7	3.3	..	1.9	..	2.1	0.6

Ratings used: 10. Unaffected.
0. Complete failure.

Table 24-18. Typical Air-dry White Enamel Formulation Based on DGEBA (mol. wt. 80,000) [4-6]

Ingredients	*Gallons*
Titanium dioxide	3.31
Phthalocyanine blue	Trace
Magnesium silicate	0.50
Bentonite	0.20
DGEBA, 40% nonvolatile	42.32
Cellosolve acetate	31.23
Solvesso 100	21.53
Neosol proprietary solvent	0.91
Total	100.00
Constant	*Value*
Pigment volume concentration	25%
Solids, % weight	29.96
Weight per gallon, pounds	8.74
Consistency, Krebs units	60–75
Thinner composition	*% wt.*
Cellosolve acetate	20
MEK	30
Toluene	30
Methyl isobutyl carbinol	20
Total	100

systems with improved flexibility, although somewhat less hard, can be obtained by the use of polyglycol or DGEBA ester modifiers.

The gloss retention of various DGEBA resin/thermosetting acrylic resin coatings is shown in Fig. 24-2.

Other acrylic/epoxy systems which have been developed include a thermoplastic reaction product of an epoxy resin and a carboxyl-containing acrylic which is

Luxury hotel at Fort Lauderdale, Fla., is protected from moist environment with exterior-grade epoxy-resin paint. (*Ciba Products Co.*)

suitable for reaction with urea-formaldehyde resins [75]; the addition of the reaction product of DGEBA (mol. wt. 1,400) and an unsaturated alcohol to the acrylic resin [77]; and the use of special tertiary amines for developing acrylic/epoxy blends with shelf stability in excess of one month [74].

EPOXY EMULSIONS

Epoxy esters are used to upgrade the properties of emulsion coatings from polyvinyl acetate, acrylic, and butadiene-styrene copolymers [43]. Water-soluble aliphatic epoxies, may be used to upgrade the properties of emulsion coatings from polyvinyl acetate, acrylic, and butadiene-styrene copolymers [25] and other water-insoluble acrylic coatings [49].

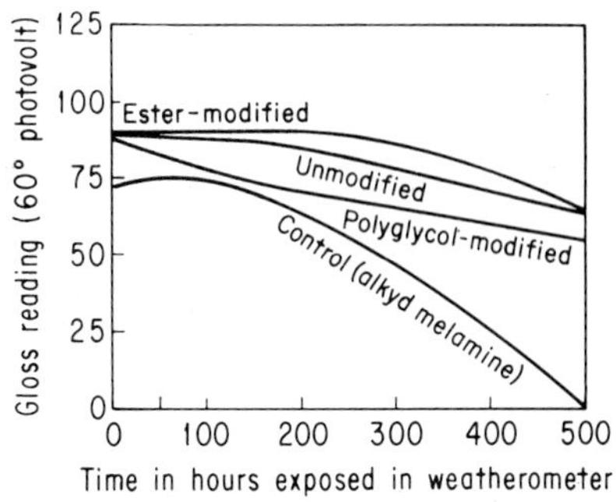

FIG. 24-2. Effect of accelerated weathering on gloss of white DGEBA resin/acrylic enamels [4-6].

Emulsions with DGEBA resins are prepared with acid- and alkali-stable nonionic emulsifiers (i.e., partial esters of fatty acids) and a stabilizing colloid such as polyvinyl acetate. Emulsions based on DGEBA (mol. wt. 1,000) have used fatty polyamides, organic solvents, and organic or inorganic acid stabilizers [27,31]. Bake coatings can be developed from epoxy emulsions and water-soluble phenolic resins [4-7]. Aromatic-based polyepoxy resins have been used with amine [41] and acid [40] curing agents for treating fabrics. The preparation of DGEBA ester emulsions has been described [4-1]. The use of epoxy emulsions as developers for lithograph printing plates has been reported [14]. Epoxy/coal-tar emulsions have been developed for use as roofing materials and for repairing highway cracks [73].

Generally, the epoxy emulsions suffer from poor stability and have not gained widespread industrial acceptance, though the use of epoxy esters to upgrade other emulsion coating systems is an application of commercial significance.

SOLVENTS

The uncured DGEBA resins are soluble in oxygenated solvents, such as ketones, esters, and ethers, and in highly halogenated hydrocarbons, although the efficiency of the solvents declines with increasing molecular weight. Fast-evaporating solutions are based on low-boiling ketones, such as acetone or methyl ethyl ketone; slow-evaporating solutions are based on the higher-boiling materials. Aromatic solvents, such as toluene and xylene, as well as simple alcohols, such as isopropyl and *n*-butyl alcohol, are not active solvents for the DGEBA resins; however, in combination, they develop latent solvency. Table 24-19 presents data on various solvents vs. viscosity for DGEBA (mol. wt. 1,000–4,000).

Coating formulations involve a combination of true solvents and thinners, the thinners employed to reduce cost, and to regulate the viscosity. When used to improve leveling properties, the thinner may contain up to 25 per cent of a leveling agent such as cyclohexanol. Figure 24-3 presents viscosity vs. solids content for two typical solvent solutions and various molecular-weight DGEBA resins.

Determination of proper solvent balance is somewhat of an art, and improper selection will cause difficulties. For example, it has been reported that solvent balance plays a considerable role in chalk resistance. A proper solvent blend will hold in the curing agent while providing a continuous film. A solvent blend of

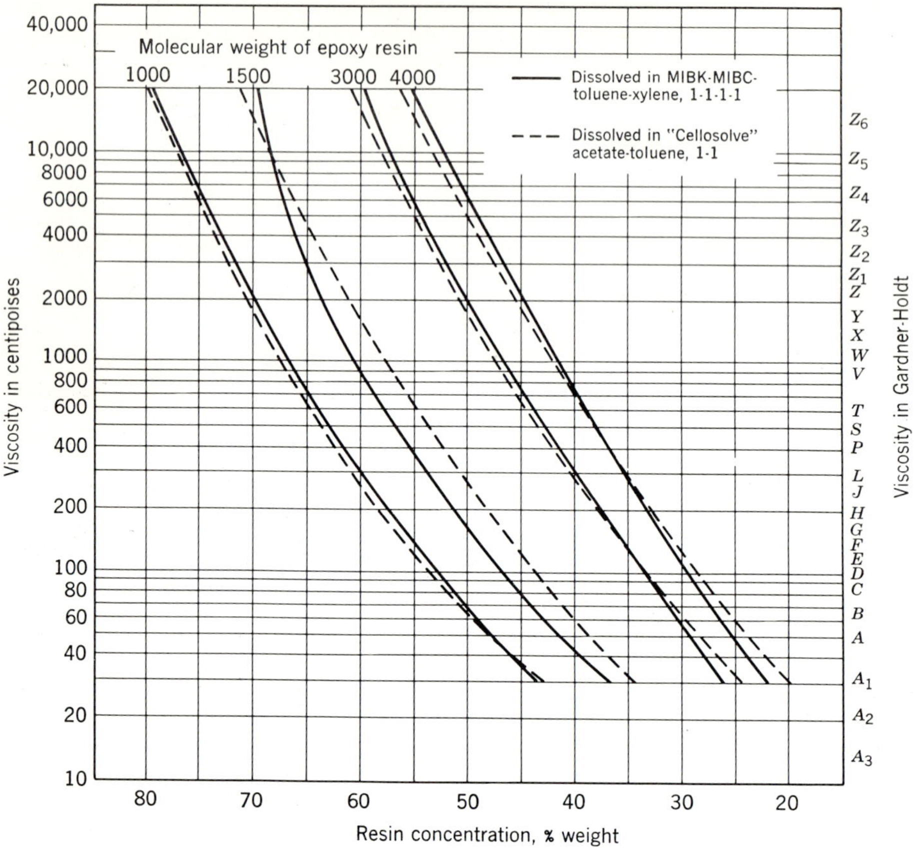

FIG. 24-3. Viscosity of DGEBA resins in solution at 25°C [4-6].

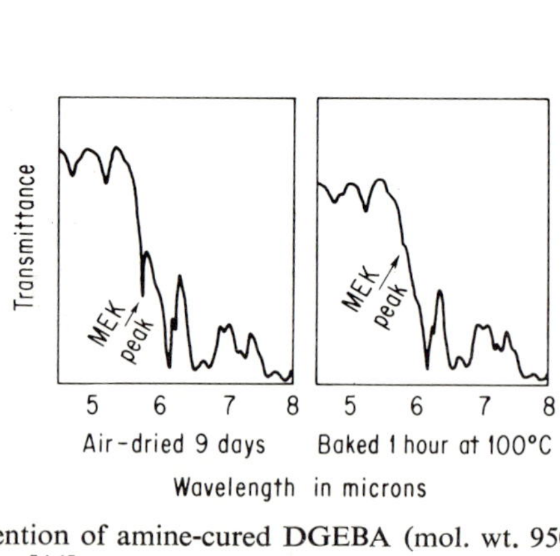

FIG. 24-4. Solvent retention of amine-cured DGEBA (mol. wt. 950) films as determined by infrared spectroscopy [44].

Table 24-19. Solubility Characteristics of DGEBA Resins [4-6]

Solvent	Viscosity of 50% w resin solution (Gardner-Holdt at 25°C) DGEBA resins, mol. wt.			
	1,000	1,500	3,000	4,000
Ketones:				
Acetone	...	...	G–H	
Methyl ethyl ketone	A_3	A	I–J	U–V
Methyl isobutyl ketone	A	F	W	Z_{1-3}
Diacetone alcohol	...	...	Z_{5-6}	
Isophorone	...	...	Z_6	
Esters:				
Ethyl acetate	...	...	Y	
n-Butyl acetate	...	...	X	
Cellosolve acetate	F	Q	Y	Z_5
Ether alcohols:				
Methyl Cellosolve	...	...	Y	
Ethyl Cellosolve	F	Q	Y–Z	Z_{4-5}
Butyl Cellosolve	...	...	Z_3	
Ethyl Carbitol	...	...	Z	
Butyl Carbitol	...	...	Z_5	
Chlorinated solvents:				
Trichloropropane	...	...	Z_6	
Chloroform	H–I	W	Z_{5-6}	Z_6
Mixed solvents:				
Toluene/acetone, 1/1	...	...	K	
Toluene/methyl ethyl ketone, 1/1	...	...	N	
Toluene/methyl isobutyl ketone, 1/1	A_1	G–H	U–V	Z_{1-2}
Toluene/diacetone alcohol, 1/1	...	...	Y	
Toluene/isophorone, 1/1	...	...	Z_{1-2}	
Toluene/isopropyl alcohol, 1/1	...	...	X	
Toluene/*sec*-butyl alcohol, 1/1	...	...	Z–Z_1	
Toluene/methyl isobutyl carbinol, 1/1	...	...	Z_1	
Toluene/Cellosolve acetate, 1/1	A	K	Z	Z_2

xylene and MIBK gives poor chalk resistance, whereas the blend containing cyclohexanone at 7 per cent gives substantial improvement [24].

Obviously, the variety in solvent blends is virtually without limit; one typical blend, for DGEBA (mol. wt. 950), consists of [21]:

	Parts
Xylene	32
Methyl isobutyl ketone	32
Cellosolve	32
Cyclohexanol	4

Polymer solutions made up in this solvent system were found to possess good leveling and flow properties and an evaporation rate such as to avoid "orange peel," bubbles, and other blemishes.

Even with proper solvent balance, however, it is probable that some solvent will be entrapped after cure. Heat cures are normally required to eliminate the solvent, as indicated by the infrared spectrum shown in Fig. 24-4 where the MEK peak is in evidence at 5.7 microns [44].

It is necessary, of course, not only to select the proper solvents for the given application, but to determine, as well, the solids content most suitable for the job.

1½ million sq ft of gantry installation at Port of Long Beach, Calif., is protected with epoxy-resin-based paint. (*Ciba Products Co.*)

The solids content of commercial formulations will vary from about 20 to 70 per cent for most applications. The concentration will determine the thickness of a single buildup, but should not be considered as a qualitative guide to performance. Several coats of a low-solids system will provide increased resistance over single coats of high-solids systems, primarily because the multiple coats promote solvent evaporation and reduce the possibility of continuous passageways (formed during the evaporation of the solvent) from the surface of the coating to the surface of the component being protected.

Coating thicknesses are generally dictated by service requirements. Mild industrial atmospheres require about 3.5-mil films, severe industrial atmospheres about 5-mil films, and submersed duty 5 to 15 mils. Specialized applications, such as metal primers, etc., may require coating thicknesses of substantially less than 3.5 mils—sometimes as little as 0.5 mil.

By proper combination of solvents and thinners, it is possible to reduce viscosity of a given resin system while preserving the solids content at a given per cent. For instance, in a typical formulation involving 70 per cent solids DGEBA resin may be combined with:

	Parts
Methyl isobutyl ketone	33
Cellosolve	33
Xylene	34

at a 70/30 ratio to give a brushing viscosity of 5,000 centipoises at room temperature. At the same ratio, combined with

	Parts
Methyl isobutyl ketone	45
Butyl Cellosolve	5
Toluene	50

a varnish having a spraying consistency of 2,400 centipoises at room temperature will be achieved.

If reduced solids content is permissible, the solution may be cut until the desired consistency is reached. A thinner for the above spraying composition would consist of *n*-butanol and toluene at 1/1. A solvent for the above spraying composition might consist of methyl isobutyl ketone, cyclohexanol, and toluene at 1/1/2. If the solids content is reduced to 40 per cent with such thinners, or solvents, the room-temperature viscosity of the coating will be reduced to about 30 centipoises.

For some applications where sensitive surfaces are involved, the solvent system will be selected with that parameter in mind; likewise, the method of pigmentation employed may dictate use of the less volatile systems.

PIGMENTS AND EXTENDERS

Clear vehicles are seldom used for corrosion protection because their outdoor durability is limited; therefore, most of the epoxy solution coatings, with the exception of magnet-wire coatings, electrical varnishes, and other specialized cases, are pigmented. In addition to improving the stability against ultraviolet, the pigments provide hiding power, color, and anticorrosive properties.

Most conventional paint pigments can be used with epoxy coating formulations. Typical materials are phthalocyanine blues and greens, titanium dioxide, carbon blacks, iron oxides, toluidine red, etc. Extenders consist of calcium carbonate, silicates, diatomaceous earth, etc. Bentones and colloidal silica are frequently used for flow control. For optimum exterior service, aluminum pigments may be employed. Leafing pigments may be used when overcoating primers to provide additional protection for sacrificial metal. Metal primers may be based on red lead, red iron oxide, zinc dust, or zinc oxide.

The effect of a typical pigment at various loading volumes on stress-strain properties is indicated in Fig. 24-5.

As is the case of other coating formulations, gloss may be adjusted by varying the binder/pigment ratio as well as by the selection of curing agents.

The general effect of chemically inert pigments is to increase the cure time. With room-temperature curing agents, for instance, cure times are sometimes about doubled.

Some typical pigment dispersions for use in DGEBA resin solution coatings are given below:

Red Lead Primer Dispersion

Material	*% by weight*
Red lead	69
Diatomaceous earth	6.5
Magnesium silicate	5.4
Aluminum stearate	0.3
Resin Vehicle	18.8
DGEBA, 55%	
Toluene, 22.5%	
Methyl isobutyl ketone, 20.3%	
Butyl Cellosolve, 2.2%	

Used at 79.5 per cent in letdown.

Dark-green Enamel Dispersion

Material	*% by weight*
Phthalocyanine green	25
Resin Vehicle	75
Resin, 33⅓%	
Butanol, 33⅓%	
Toluene, 33⅓%	

Used at 44 per cent in letdown.

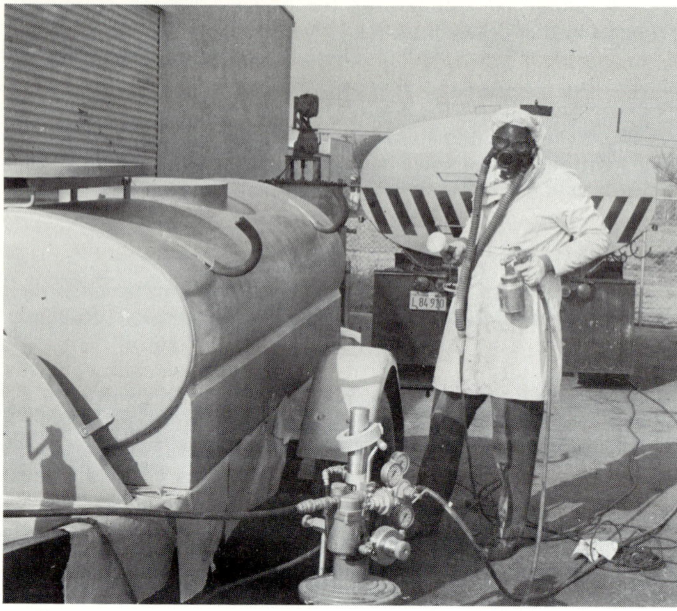

Epoxy resin being applied to insides of distilled water tanks. (*The Epoxylite Corp.*)

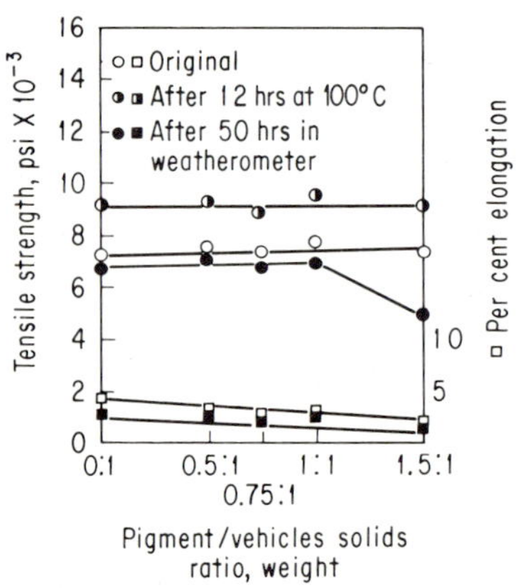

FIG. 24-5. Effect of pigmentation (rutile TiO_3) on stress-strain properties of DGEBA (mol. wt. 3,000) cured with urea-formaldehyde resin [39].

For metal-primer applications, it is frequently desirable, in view of the settling encountered with the high-specific-gravity fillers, to supply the sacrificial metal separately to be mixed in just before application. The following formulation, of this type, is suitable for spray application at ½ mil as a shop primer for metals:

Material	*Pounds*
Part A:	
Pigment:	
Colloidal silica	25.4
Resin solution (60% nonvolatile):	
DGEBA (mol. wt. 950)	112.4
MIBK	25.0
Cellosolve	25.0
Xylene	25.0
Solvents and flow-control agent:	
MIBK	42.3
Cellosolve	42.3
Xylene	42.3
Beetle 216-8	9.4
Total base vehicle component	349.1
Part B:	
Fatty polyamide (amine value 210–230)	50.9
MIBK	28.2
Cellosolve	28.2
Xylene	28.2
Total curing-agent component	135.5
Part C:	
Zinc dust	2372.0
Total zinc dust component	2372.0
Total formulation	2856.6
Properties of mixed formulation:	
Nonvolatile content, % by weight	89.8
Pigment/binder, weight ratio	14.2/1.0
Pigment volume concentration, % by volume	70.0
Viscosity, Krebs units	100
Weight per gallon, pounds	28.6

SURFACE PREPARATION AND PRIMERS

As has been indicated elsewhere, particularly in Chaps. 21 and 23, proper surface preparation is vital to obtain maximum adhesion with the epoxy resins.

For most metals, the best surface preparation consists of sandblasting to clean, white metal, with the coating being applied shortly after the sandblasting. Preparation techniques for plastics and other surfaces are given in the above-cited chapters.

Because of the outstanding adhesion of the epoxy-resin solution coatings, they are often used over metals, plastics, and construction materials as primers for other coating systems [28]. For example, used with acrylic topcoats, they are expected to double the life of aluminum siding over that obtained by the acrylic alone [66]. When used over metals in conjunction with polyester topcoats, good intercoat bonds may be obtained when the primer is only partially cured; uncured primer gives incompatibility, and cured primer offers poor adhesion [97].

If the epoxy possesses such outstanding adhesion, it would seem that it would be unnecessary to employ a primer with it. However, particularly over metals, the primer may serve a triple function: it may be applied at a time when the metal surface can conveniently be cleaned; it may contain a sacrificial metal, which, upon the failure of the film or upon exposure to electrolytes, will perferentially react to inhibit corrosion of the base metal; and it may be formulated to improve long-term film flexibility.

Red lead, zinc dust, zinc yellow, and zinc-chromate-filled epoxy primers have been used with success in connection with epoxy topcoats [67], although whether the increased cost is warranted is still a matter of debate. Such primers appear to provide little additional protection in totally submersed environments, and their

Swimming pool filter tanks lined with heavy epoxy coating as protection against corrosion and chlorine. (*The Epoxylite Corporation.*)

effectiveness otherwise will depend upon the selection of the curing agent. Zinc-rich primers have been reported to provide poor mechanical film properties [83].

Tests have indicated that, in terms of corrosion resistance to salt spray for periods up to 672 hours, both a single epoxy primer coat and a single epoxy primer coat over one coat of zinc chromate primer provide equal performance, no corrosion being detected with either [61]. The military generally recommends a wash primer for ferrous metals, if the surfaces are not to be subjected to high heat or hot diester oil. Aluminum surfaces are treated with the chromate conversion process. Adhesion to fiber glass was found to be satisfactory without the wash primer [71]. However, with epoxy/fatty-polyamide coatings, the wash primer will provide poor adhesion if it is applied under conditions of low humidity. The situation can be corrected by the addition of distilled water to the diluent portion of the primer [63].

The primary advantage of the use of epoxy primers under epoxy top coatings would appear to be that they permit the coating of surfaces conveniently in advance of field assembly at a time when surface preparation can be accomplished with the desired thoroughness.

It has been reported, however, that possible difficulties may be experienced with amine-cured epoxy primers with regard to intercoat adhesion. Laboratory data indicated that migration of the amine to the primer surface might cause difficulties, although field results over a 2½-year period did not indicate any actual trouble. To eliminate the potential problem it is recommended that the topcoat be applied within 4 hours after application of the primer coat, that the primer be baked for one hour at 175°C, or that the primer be lightly sanded [79].

SUMMARY

The epoxy resins most widely used in solution coatings are the higher-molecular-weight species of DGEBA, ranging upward from a molecular weight of 950. Species in the molecular-weight range of 950 to 1,000 are normally employed with primary aliphatic amines and their adducts and the fatty polyamides in air-drying solution coatings. Species in the molecular-weight range of 1,500 to 2,000 are most frequently esterified. Species in the molecular range of 3,000 to 4,000 are used with the phenol-formaldehyde and amino resins in baking coatings. Very high molecular-weight species may be used without curing agent or coreactant.

In general, the air-dry coatings provide the chemical resistance and toughness normally associated with the more conventional baking coatings. The phenol-formaldehyde/epoxy systems provide the maximum abrasion resistance and chemical resistance; and the amino/epoxy systems provide the maximum in flexibility and color retention. The epoxy esters may be considered as upgraded alkyds.

REFERENCES

1948

1. Greenlee, Synthetic Drying Compositions, U.S. 2,456,408.

1950

2. Greenlee, Polyepoxide-containing Compositions and Reaction Products, U.S. 2,528,359.
3. Greenlee, Polymeric Polyhydric Alcohols, U.S. 2,503,726.

1952

4. Scheibli and Dannenberg, Room Temperature Curing "Epon" Resin Coatings, ACS Symposium, Los Angeles, April.

1953

5. Hartman, New Resin Linings Give Better Container Protection, *Iron Age*, June 11.
6. Rinse, Epoxide Resins Heated with Vegetable Oils in the Presence of Steam, U.S. 2,652,376.

1954

7. Deisenroth, Engineering Metal Finishes, *Steel*, July 19.
8. Farber, New Epoxy Primer Has High Resistance to Alkali, *Mater. & Methods*, February.
9. Rodick, New Type Resin Finish as Applied to Evaporative Coolers, *Finish*, October.

1955

10. Christ, Epoxy Coatings Protect Copper and Brass, *Org. Finishing*, August.
11. Keefer, Tough Plastic Coating, *Furniture Mfr.*, July.

12. Knight, All Purpose Industrial Coatings, *Maintenance*, August.
13. New Gun Metal Blue, *Production*, April.

1956

14. Hall, Aqueous Composition of an Unmodified Hydroxyl-containing Resinous Glycidyl Polyether of a Dihydric Phenol, U.S. 2,754,279.
15. Henson et al., Liquid Epoxy Resins in Coatings, ACS Symposium, Atlantic City, September.
16. Ott and Zumstein, Epoxide Resins and Their Use in Coldsetting Coating System, *J. Oil Colour Chemists' Assoc.*, May.
17. Park and Blount, Oxidative Degradation of Epoxy Resin Coatings, ACS Symposium, Atlantic City, September.
18. Poe, Epoxy Coating Safeguards Laboratory Furniture, *Metal Prog.*, February.
19. Sternberg, Epoxy Corrosion Preventers, *Plant Eng.*, October.
20. Theis, How New Internal Coatings Are Proving Their Worth, *Oil Gas J.*, Feb. 13.

1957

21. Alter and Soller, Molecular Structure of Epoxy Resin Polymers as a Basis for Adhesion, ACS Symposium, New York, September.
22. Bruin, Chemical Reactions Involved in the Curing of Epoxide Amine Systems and Their Significance of Various Applications, FATIPEC Congress.
23. Dow Corning Corp. Bulletin U-7-704.
24. Lebbink, Chalking Characteristics of Pigmented Amine-cured Ethoxyline Resins, FATIPEC Congress.
25. Schroeder, Laminating Compositions and Laminate Made Therefrom, U.S. 2,784,128.
26. Taylor, Beneficiated Chemically Curing Polyepoxy Ether Resinous Compositions, U.S. 2,794,007.
27. Wittcoff and Renfrew, Mixed Suspensoids of Epoxy Resins and Polyamide Resins, U.S. 2,811,495.

1958

28. Evans and Pappalardo, Coated Metal Articles, U.S. 2,847,323.
29. Schroeder, Treatment of Wool Containing Textiles, U.S. 2,829,071.
30. Widmer and Zuppinger, Reaction Product of a Compound Capable of Reacting with Hydroxyl Groups with an Etherification Product of a Polyglycidyl Ether, U.S. 2,824,850.

1959

31. Aelony and Wittcoff, Polyamide Resin–Epoxy Resin and Emulsion, U.S. 2,899,397.
32. The Aluminum Collapsible Tube, Its Development and Use, Ciba Ltd., Basel.
33. Buege and Drubel, Solvent Solution Coatings: Versatile Workhorse, *Mod. Plastics*, July.
34. Coe, Shrinkproofing Textiles with Polyepoxides and Ammonia, U.S. 2,886,402.
35. Coe, Shrinkproofing Textile with Volatile Polyamines and Epoxides, U.S. 2,869,971.
36. Condo, Treatment of Textile Materials, U.S. 2,886,472.
37. Lederman, Air Drying Wrinkle Coating Compositions Containing Epoxy Resin Esters, U.S. 2,867,591.
38. Mika, Epoxy Alkyd Modifications, *Paint Varnish Production*, March.
39. Mika, Free Film Properties of Epoxide Resin Coating Systems, *Offic. Dig.*, April.
40. Schroeder, Polyepoxide Emulsions and Method of Treating Textiles Therewith, U.S. 2,872,427.
41. Schroeder, Polyepoxide Emulsions and Method of Treating Textiles Therewith, U.S. 2,872,428.
42. Stoesser et al., Method of Esterifying Epoxy Resins with Unsaturated Acids, U.S. 2,875,165.
43. Thorstad, Emulsions: Why and How They Are Used, *Mod. Plastics*, July.

1960

44. Dannenberg et al., Infrared Spectroscopy of Surface Coatings in Reflected Light, *Anal. Chem.*, March.
45. Dannenberg, Measurement of Adhesion by a Blister Method, ACS Symposium, Cleveland Ohio, April.
46. DeBat, Finshield Construction, *Furniture Design and Mfg.*, November.
47. Dosmann and Streed, Painting of Golf Balls, U.S. 2,939,710.
48. Floyd, Process for Preventing Shrinkage and Felting of Wool, U.S. 2,961,347.
49. Hurwitz, Aqueous Dispersion of Acrylic Acid Copolymer and Aliphatic Polyepoxide, U.S. 2,945,358.
50. Lee and Neville, The Chemistry of Modern Magnet Wire Insulation, *Insulation*, April.
51. Pardox, Improvements in Epoxy–Amino Polyamide Finishes for Easy-care Wool Fabrics, *J. Textile Inst.*, no. 12.
52. Pistor, Principal Types of Protective Coatings, AIA Report 25-C.
53. Schulde et al., Process of Applying a Coating of an Atactic Polymer of Polypropylene to a Polyethylene or Polypropylene Surface, U.S. 2,927,047.

1961

54. Boynton, Study to Determine the Penetration of Casting Solvent into Epoxy Barrier Coating, AD 453 856.
55. David, Epoxies Fight Acids, *Plant Eng.*, March.
56. Fortener, Epoxies vs. Water, Pittsburgh Society for Paint Technology, Apr. 3.
57. Fry, Composition Comprising an Acrylic Acid Polymer and a Polyglycidyl Ether and Products Thereof, U.S. 3,008,914.
58. Katchman, Recording Medium Having an Image Receiving Coating of a Thermoplastic Epoxy Resin, U.S. 3,142,585.
59. McGuiness, Improved Reliability for Printed Circuits with Protective Coatings, 17th ANTEC of SPE, January.
60. Pittsburgh Society, A Study of Factors Affecting Rusting of Steel and Blistering of Organic Metal Coatings, *Am. Paint J.*, Nov. 2.

1962

61. Comparison of Corrosion Resistance of Zinc Chromate Primer to Epoxy Primer, AD 425 105.
62. Howland, and Reiners, Preparation and Properties of Epoxy-cured Zein Coatings, *Paint and Varnish Production*, February.
63. Lyon, Effect of Low Humidity on the Adhesion of Epoxy-Polyamide Coating Systems to Wash Primer, AD 283 948.
64. McGary et al., Coatings from Cyclohexene Oxide Derivatives, *Ind. Eng. Chem.*, April.
65. Pheinneck, Studies of Partially Epoxidized Linseed Oil, ACS Symposium, Washington, D.C., March.

1963

66. Epoxy Coatings Extend Aluminum Siding Life, *Prod. Finishing*, April.
67. Floyd and Glaser, Some Observations Concerning Zinc Dust Primers for Metals, *Offic. Dig.*, January.
68. Havenaar and Meijs, Sand Consolidation with Epoxy Resin Solutions, *J. Inst. Petrol.*, no. 480, pp. 392–397.
69. Hensley, Hexamethoxymethylmelamine with Alkyd and Epoxy Resins for Surface Coatings, *Offic. Dig.*, January.
70. Hood and Jefferies, Catalyzed Epoxide Resin/Epoxidized Hydrocarbon Drying Oil Gloss Enamels and Metal Article Coated Therewith, U.S. 3,077,461.
71. Lyon, Evaluation of the Adhesion of the Specification Epoxy System, with and without Wash Primer, to a Variety of Ferrous and Non-ferrous Substrates, AD 411 125.
72. Wray and Tator, Epoxies, *Mater. Protection*, April.

1964

73. Bituminous Epoxide Emulsions (to Shell Internationale Research Maatschappij N.V.), Brit. 960,911.
74. Brockman and Gesall, Thermosetting Polymeric Coatings, Brit. 971,209.
75. Bussell, Thermosetting Mixture of a Styrene–Acrylic Acid–Epoxide-containing Compound and an Amine-Aldehyde Resin, U.S. 3,156,740.
76. Cured Epoxy Compositions for Sealing (to Shell Internationale Research Maatschappij N.V.), Fr. 1,367,851.
77. Dalibor, Copolymers of Unsaturated Resinous Epoxy Compounds, Ger. 1,163,550.
78. Maly, Curing Epoxide Resin Compositions, U.S. 3,130,657.
79. Rakaczky, Adhesion of Specification Finishes to Epoxy Resin Primer, MIL-P-52192 (ORD), AD 608 133.
80. Svoboda et al., Investigating the Stability of Cured Epoxy Coatings in Liquid Media, *Lakokraschnye Materialy i ikh Primenenie*, no. 5, pp. 57–58.
81. Tang and Sheng, Method of Impregnating Card Stock, U.S. 3,131,083.
82. Vaurio, Paper Product with Chemical Watermark and Means for Making Same, U.S. 3,140,959.

1965

83. Adomenas, Epoxy Anticorrosive Priming System, *J. Oil Colour Chemists' Assoc.* 48: 423–439.
84. Banfield, Heavy Duty Coatings on Ships' Bottoms, *Corrosion Prevention Control* 12(7): 23–25.
85. Berni et al., Novel Finishes Imparted to Cotton through Cellulose-Epoxide Reactions, *Am. Dyestuff Reporter*, June 7.
86. Burrows, Emissivity Equalization by Thermosetting Coatings, Research Report, U.S. Naval Applied Science Laboratory, Naval Base, Brooklyn, N.Y.
87. Crilly and Bassett, Evaluation of Paint Systems for Use in Radioisotope Laboratories, AD 457,917.
88. Giudicelli, Wear-resistant Offset Printing Plates, Fr. 1,408,854.
89. Hearst, Degradation of Organic Coatings by Irradiation with Light: II, Attenuated Total Reflectance Spectra of Coatings Exposed to Ultraviolet Light, AD 460,142.
90. Mackdanz, Protection of Offshore Structures, *Mater. Protection*, vol. 4, no. 10.
91. Polyesters and Epoxies in Corrosion Control in the West, *Western Plastics*, November.
92. Shackelford, Epoxy Resin Coating Compositions, Belg. 660,982.
93. Shackelford, Epoxy Resin Compositions, Belg. 660,983.
94. Todd, Two-step Selection of Coatings for Shipboard Tanks, *Mater. Protection*, vol. 4, no. 10.
95. Walker, Effect of Water on the Adhesion of Surface Coatings, *Offic. Dig.*, December.
96. Weigel, "Epoxy Resin Varnishes," Wissenschaftliche Verlagsgesellschaft M.B.H., Stuttgart.
97. Workman, Process for Bonding Polyester Coatings to Metal Substrates Using a Cured Epoxy Primer and Product Obtained Thereby, U.S. 3,164,488.

Chapter 25

HANDLING METHODS AND SAFETY PRECAUTIONS FOR EPOXY RESINS

The solid epoxy resins without curing agents are almost exclusively the concern of paint and varnish formulators familiar with synthetic resins generally. The handling characteristics of the solid epoxy resins are very similar to those of the other solid coating materials, and their application in production poses no unusual problems.

The liquid resins, on the other hand, are extremely versatile materials and have invaded a number of industries previously unfamiliar with plastics in general; and because many of the users are unfamiliar with epoxies in particular, their use in production situations may involve a number of initial inconveniences. Most of these can be precluded by foresight. The subject is of sufficient technological importance to have warranted at least one full book-length treatment [28].

It is generally advisable for engineering and production employees to work with laboratory samples in order to become familiar with the general handling properties of the resin systems; to experience the exotherm phenomena, the pot-life characteristics, the effect of temperature on viscosities; and to assess the other variables influencing the handling characteristics of the systems. Cured laboratory samples can be conveniently employed in the anticipated service environment or evaluated by suitable test procedures, to establish the suitability of the particular formulation and the specific effect of handling variables on ultimate properties. The total of this activity will indicate the handling characteristics that must be accommodated in subsequent production.

STORAGE

The liquid epoxy resins, in general, pose no particular storage problem, provided moisture is excluded. Formulated compounds may require special storage; solution coatings, involving volatile solvents, present definite fire hazards.

The liquid epoxy resins generally have rather high flash points—for example for DGEBA about 240°C. The lower-molecular-weight materials have somewhat lower flash points, DGEBA diluted with butyl glycidyl ether having a flash point of about 90°C. These, therefore, do not normally pose fire hazards, and may be transferred with heating to improve flow—the lower-viscosity materials being pumped at around 40 to 50°C and the higher-viscosity materials at temperatures up to 70°C, or in some cases even higher.

Carbon steel is satisfactory for pipe and tank construction and will not rust beneath the epoxy. Such rust as occurs above the level of the epoxy may be removed by filters, with epoxy-coated iron pipe, aluminum or stainless steel pipe, or epoxy-glass tubing used downstream of the filter to prevent further rusting [4-6].

Formulated products containing fillers may require periodic agitation to prevent hard settle; and those systems containing latent curing agents should normally be stored in a cool place to provide maximum shelf life. Curing agents, usually supplied as part *B* of formulated products, may require special handling: some are quite sensitive to atmospheric moisture and should be stored tightly closed; others may crystallize at lower temperatures and require reheating prior to use; others may require storage in glass or other special containers. In these cases, the recommendations of the formulator, or of the curing-agent supplier, should be adhered to.

Storage conditions for solution coatings will, in large measure, be dictated by the nature of the solvents employed, and again manufacturer's recommendations should be followed.

Epoxy resins and their curing agents do not pose the same industrial hazards as are encountered with polyester formulations (i.e., explosive contamination of organic peroxides with promoters) and the literature has reported no serious accidents resulting from improper storage. However, many of the ingredients going into epoxy formulations are highly reactive, and these constitute a potential hazard if inadvertently mixed with other reactive chemicals.

INCORPORATION OF FILLERS

It is normally inconvenient to incorporate the fillers during the initial stages of the production process. Optimum results require heating the fillers to drive off any moisture present and, after their incorporation by thorough mixing, allowing the resin to stand for a suitable period of time until the bubbles have escaped. In practice, it is most convenient to work the resin warm during the filling operation, and if the production situation requires room-temperature pot lives, the filled system must be allowed to return to ambient before addition of the curing agent. For these reasons, the fillers are usually added well in advance of production by a separate compounding operation. Fillers may be added by conventional paint-mixing equipment, although three-roll mills are frequently employed to obtain thorough wetting.

Fillers may be incorporated more conveniently as a third component at the time of the blending operation with some of the more sophisticated mixing and metering equipment. In special cases, particularly when handling highly filled concrete coatings or sacrificial metal primers, the filler is added at the time of mixing. In such cases, the severe settling tendencies of the filler at the loading volume required prohibit earlier addition.

OPTIMUM BATCH SIZE

The optimum batch size will be dependent on the curing agent employed in the system and the temperature at which the system is worked. Generally, the epoxy resins maintain a workable viscosity for the majority of their pot life; once appreciable exotherm is generated, however, gelation occurs rapidly thereafter. For the more highly exothermic compounds the size of the batch will establish the pot life. For given mixing equipment, the larger the mass, the longer the mixing time, thus further reducing the available time for the application operations. Alternatively, a system of batch replenishment may be employed, involving the periodic addition of freshly catalyzed mix to the material being worked. Figures 25-1 and 25-2 indicate

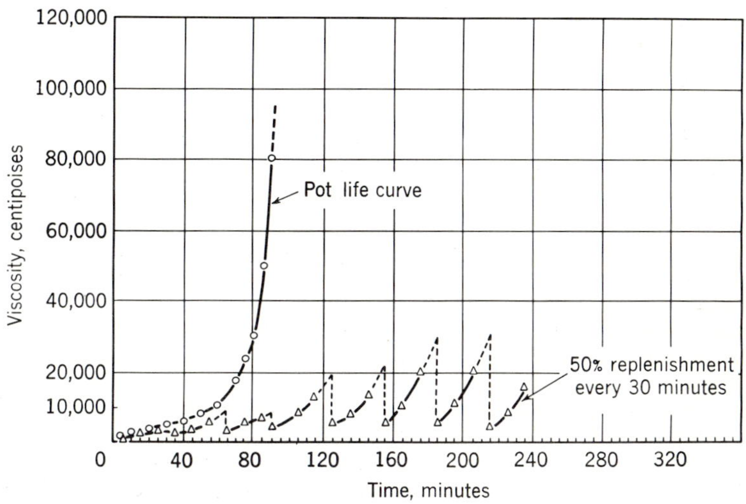

FIG. 25-1. Effect of batch replenishment on viscosity of ethylenediamine-catalyzed DGEBA [1].

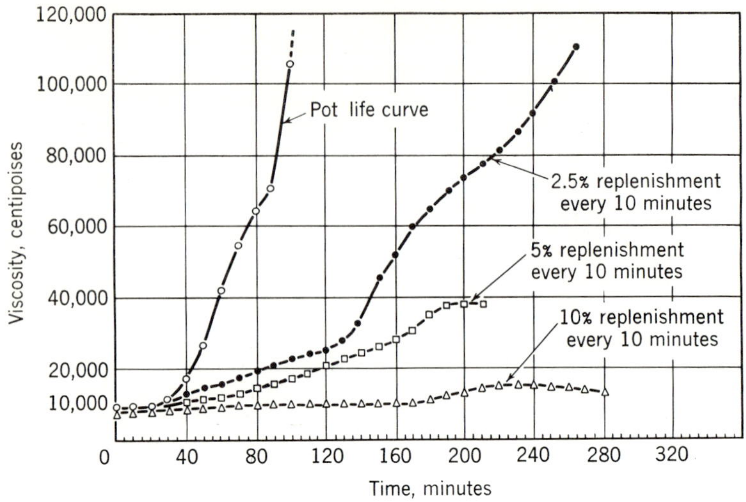

FIG. 25-2. Effect of batch replenishment on viscosity of amine-resin adduct-catalyzed DGEBA [1].

the effect of various amounts of batch replenishment on two typical aliphatic-amine-catalyzed systems. These figures, likewise, illustrate the nature of the transition period as reflected in a more or less steep and sudden increase in viscosity. With the more reactive amines and DGEBA, batches are usually 1 to 2 lb, with larger amounts (up to 5 lb) sometimes being used.

Anhydride-cured systems, in general, permit the use of larger batches; and with latent curing agents, the batch size is "unlimited." Often, highly filled systems, or those diluted with solvents, will provide a pot-life extension sufficiently great to permit the use of fairly large batches of even room-temperature curing compounds. The batch size, of course, should be selected to permit the convenient use of all the material prior to excessive increase in viscosity.

METERING AND MIXING

The least complicated method of blending the epoxy-resin formulation is to weigh the ingredients separately and then mix very thoroughly by hand. This method, requiring no more than a scale and container, represents little equipment investment, but in production the labor involved is often excessive, and the mixing operation, in particular, may not provide reproducible results. Mixing must be very thorough or results will not be satisfactory. Hand mixing is generally accomplished by stirring the ingredients with a spatula. Deformable plastic containers, which may be kneaded, are also used [8].

In order to assure thorough mixing, it is often possible to pigment the two components in different colors, so that improper mixing may be detected visually. Commonly, one component is blue and the other yellow, the combination producing green. It has been suggested [25] that triphenylmethane dyes may be incorporated into the epoxy resin, these causing a color change as an amine curing agent is blended into the system.

In most cases, for even low-volume production operations, some form of electric-operated mixing equipment is employed. At its simplest, this can be an electric hand drill to which is attached a mixing blade. More complicated equipment may

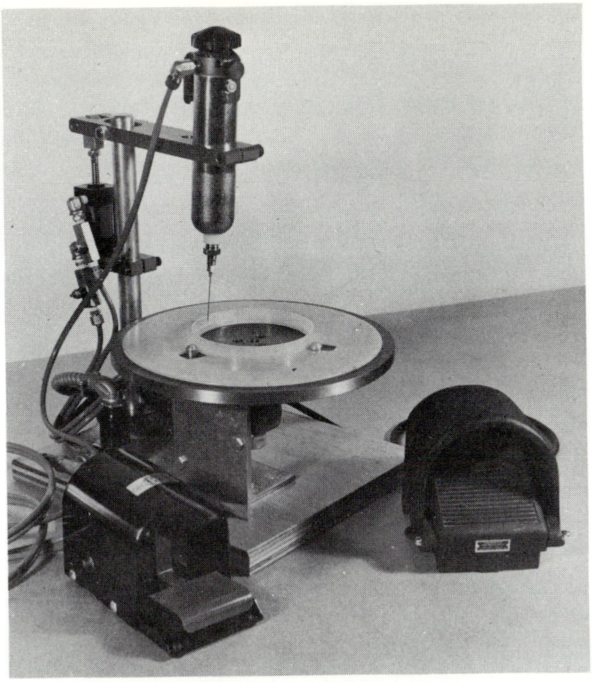

Hypodermic needle is used to disperse tiny quantities of mixed epoxy-resin formulation. (*Semco, Inc.*)

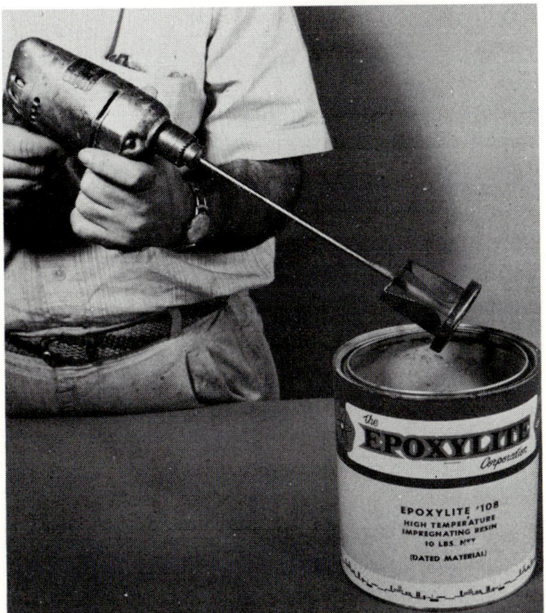

Epoxy resin and curing agent are often mixed in original shipping containers using simple stirrer driven by hand drill. (*The Epoxylite Corporation.*)

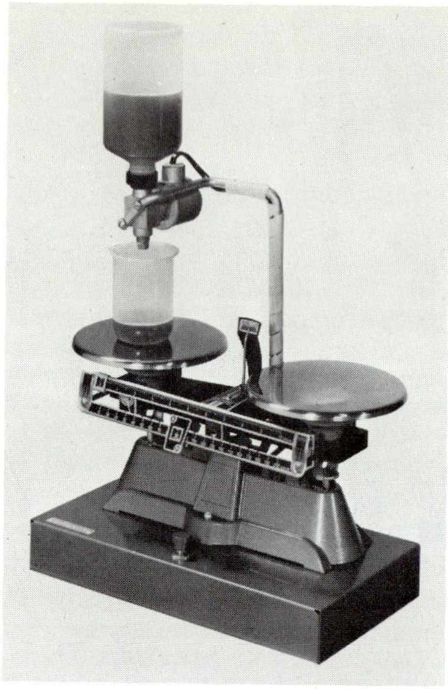

Relatively uncomplicated equipment facilitates batch weighing. Liquid compound is dispensed under gravity feed with automatic cutoff provided electrically. (*Delson Corp.*)

Typical desk-model metering and mixing device for epoxy resins. (*Hull Corp.*)

be developed. There are several commercial suppliers, for example, of special mixing blades designed to minimize air entrapment and to scour the sides and bottoms of containers, and of mixing equipment which will handle standard container sizes. One of the latter, for example, involves a fixed blade and a rotating platform: the container being placed on the platform, the blade inserted, and the platform rotated.

It is also possible to employ continuous dispensing equipment in order to reduce labor and to improve accuracy. Such equipment may range from the elementary to the elaborate. The dispensing machine may, for example, consist of separate gear pumps set to deliver fixed ratios which may be changed by changing the gears; it may be connected to a valve-operated scale which cuts off flow at a designated weight; or it may consist of a micrometer valve in the flow line, the valve being adjusted to regulate the flow under a constant pressure and viscosity [24].

From this it is only a step to connecting the metered compound to a mixing head to develop a metering and mixing machine.

Metering and Mixing Machines

Metering and mixing machines have met with considerable success in the polyester technology, but such equipment, without adjustment, is unsatisfactory for dispensing the epoxy resins.

Polyesters may be formulated as two components, each of low and equal viscosity and specific gravity and combined at equal ratios. Although thorough mixing and accurate proportion is desirable, considerable leeway is tolerable.

With the epoxies, such is not the case. The viscosity and the specific gravity of the resin and curing agent are seldom equal; often it is necessary to handle systems having mixing ratios as low as 5/100; and mixing must be quite thorough to provide adequate cured properties. Thus, throughout the dispensing and mixing system much greater accuracy is required than with the polyesters.

In addition, many thousands of epoxy formulations, filled and unfilled, are employed in the technology, and obviously a system designed to handle one will not

Metering and mixing device in operation. (*Semco, Inc.*)

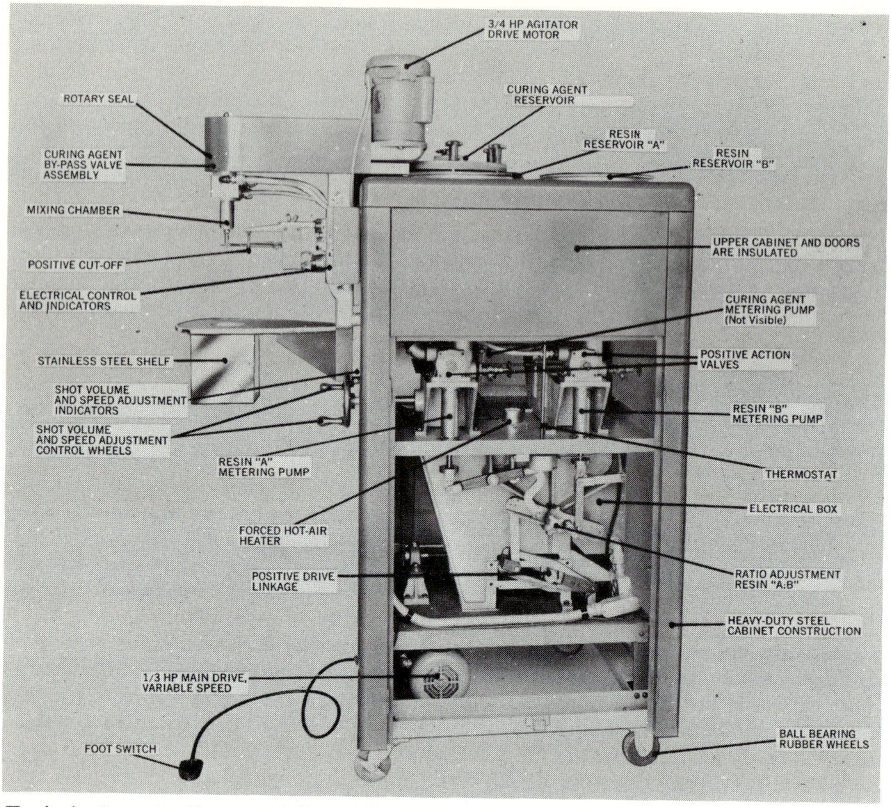

Typical schematic diagram of dispensing equipment. This is the model 600, introduced in 1960 by H. & V. Hardman Co. (*H. & V. Hardman Co.*)

necessarily handle all the rest. This, coupled with the need for great accuracy, has delayed the development of universal epoxy metering and mixing equipment.

At the present state of the art, best results appear to be obtained by adjusting equipment to the specific formulation. The adjustments may be minor or may be rather complex. Proper procedure, however, dictates the selection of the formulation in advance of the metering and mixing equipment, rather than the reverse. Under these circumstances, satisfactory results have been obtained with quite a number of commercial metering and mixing machines.

A few of the suppliers of commercial equipment designed for metering and mixing formulations for casting and coating applications are indicated in Table 25-1.

Dispensing Equipment. With continuous dispensing equipment, the resin and curing agents are loaded into separate ratio-control devices and are then transferred at a preselected ratio to a mixing chamber and thence to the ejection nozzle.

Ideally, the mixing should be conducted quite rapidly with a minimum batch in the mixing head, and the ejector should be as simple as possible in order to facilitate cleanup.

The equipment should be constructed of chemically resistant piping and gasketing; and if filled and abrasive systems are to be handled, they should be abrasion-resistant as well. Some form of pressure is normally required to deliver the material

Table 25-1. Partial List of Suppliers of Mixing and Metering Equipment for Epoxy Systems

Dispensing equipment	*Spraying equipment*
Auto Air Industries, Inc.	Auto Air Industries, Inc.
Automatic Process Control	Binks Manufacturing Co.
Bell and Gossett Co.	DeVilbiss Co.
Castor Engineering Co., Inc.	A. Gusmer, Inc.
C.P.M. Specialty Machinery Corp.	Gray Co., Inc.
Delsen Corp.	Hodges Chemical Co.
G. Diehl Mateer Co.	Hupp Engineering Associates
Gabriel Williams Co., Inc.	Plaster Supply House
H. & V. Hardman Co.	Pyles Industries, Inc.
Hull Corp.	Rand Development Co.
Martin Sweets Co., Inc.	SealZit of America
Meter Mix Corp., Div. Federal Machine Tool Co.	Stuart Marine Service
Mid-States Industrial Plastics Co.	
Mitchell Specialty Co., Div. Industrial Enterprises, Inc.	
Murphy and Muller, Inc.	
National Instruments Laboratories	
Pyles Industries, Inc.	
Semco Sales and Service, Inc.	

to the mixing chamber, and it may be necessary to control the temperature of the compounds and to provide agitation to prevent filler settling. Additionally, degassing apparatus may be required. Equipment of considerable complexity may therefore be developed.

Much of the commercial equipment employs metering pumps of the positive-displacement type—gear, piston, diaphragm, or worm pumps—and has variable

Metering, mixing, and dispensing equipment. (*Semco, Inc.*)

Mixing equipment for sand-filled compound meters three components separately and performs mixing in screw mixer before dispensing. (*Union Carbide Corp.*)

Equipment for vacuum processing epoxy formulations. Equipment provides prebake under vacuum together with vacuum casting. Lower casting chamber has planetary turntable for positioning of molds. (*Hull Corp.*)

ratio adjustments and variable speed drives. The positive-displacement piston pump appears to be preferred [21]. The mixing designs are varied: baffle chambers, rotary shafts with a helix blade or pinholes, wobble-type impellers, and shear plates [14].

The commercial equipment is capable of handling, in some designs at least, viscosities up to 100,000 centipoises on the part *A* (or resin) side and from 0.5 to 1,000 centipoises on the part *B* (or curing-agent) side [12]. Commercial equipment has been reported to be capable of handling diluted resins containing up to 55 per cent silica dust with amine-adduct curing agents [13].

When handling filled systems, it is customary to employ agitation to prevent filler settling, although in some cases it may be practical to deliver the filler separately into the mixing chamber. One such system, for example, recommended for use in the construction industry for handling concrete topping and repair compounds, consists of resin and curing agent metered through positive-displacement pumps into one side of a screw-type mixer; from the other side, the sand filler is metered by a vibrating feeder. The three components are then mixed as they travel down the rotating screw to be delivered at the bottom of the screw barrel [18].

The compound may be dispensed from vacuum pots when complete deaerating is necessary; or vacuum chambers may be attached to the injection nozzle to vacuum the component and compound during the encapsulating operation [4]; or an on-the-fly degasser may be installed in the line [20]. This latter is a device through which moving films of liquid are allowed to pass. During passage they are exposed to vacuum. Equipment consists essentially of a cylindrical column mounted in a glass cylinder. In the space between the two cylinders are mounted metal-ribbon heating elements to reduce the viscosity of the fluid. The fluid flows in the bottom between the two cylinders and is forced upward over the heating coils. It then flows over the top of the inner cylinder and down the inner walls where final degassing is accomplished.

Spraying Equipment. Spraying equipment for two-component epoxies consists of metering equipment attached to a mixing-spraying nozzle. The compounds are fed to the mixing head, atomized, and ejected under pressure.

Atomization can be accomplished by blowing jets of air either around or against the issuing fluid stream to shape it into a cone or a fan-shaped spray pattern. Such atomization is suitable for low-viscosity liquids, but does cause considerable overspray. Alternatively atomization can be accomplished hydraulically by causing the fluid to issue through a correctly designed nozzle at a high velocity. In this case, overspray is minimized and higher-viscosity systems may be delivered, although, of course, with the use of higher pressures.

Mixing may be accomplished hydraulically, the liquid velocity providing the turbulence and fluid shear necessary to develop an adequate admixture; alternatively, mixing may be accomplished mechanically just before atomization.

Of the available equipment, those spray guns which provide for mixing within a mixing chamber prior to ejection, rather than the dual-nozzle type which depend upon mixing to occur after ejection by the impingement of atomized individual sprays, appear to be the preferred equipment for handling the epoxy-resin formulations. An analysis of six commercial items of spray equipment is presented in [11].

As with the dispensing equipment, fillers may in some cases conveniently be supplied separately. They are ejected into the spray from a separate orifice and are wetted during the air passage. Most commonly, this technique is used to spray with fiber-glass roving, the roving being chopped and ejected with a special item of equipment attached to the spray gun [7].

One-component epoxies and the longer-pot-life two-component systems may be handled with conventional spray equipment, but thorough cleanup at intervals is

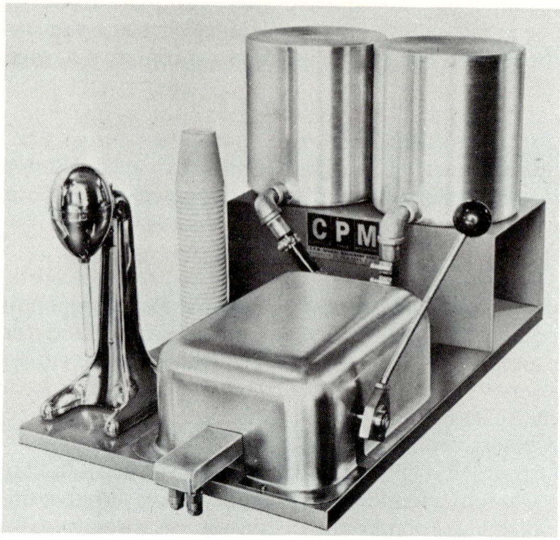

Dual-nozzle epoxy-dispensing equipment. After epoxy is dispensed, it is mixed separately by use of blender on left. (*CPM Special Machinery Corp.*)

required to prevent the more reactive formulations from polymerizing in the lines or spray head.

PACKAGING

Quite frequently many of the handling difficulties involved with epoxy-resin systems may be eliminated by special packaging techniques. Basically, two different packaging arrangements may be used:

1. Two-container kits
2. One-container cartridges

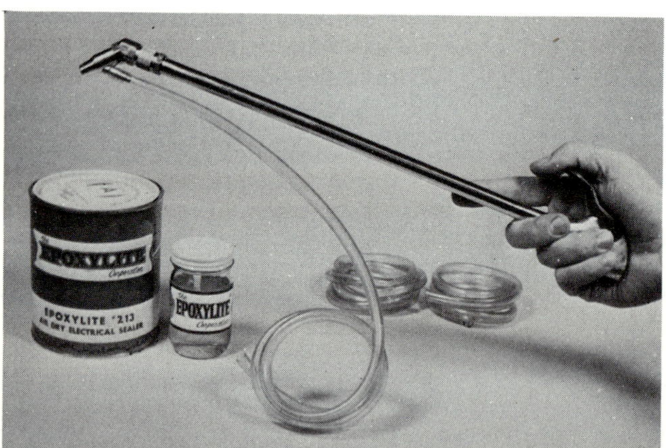

Low-cost air-operated hand gun sprays heavy epoxy coatings, employs disposable feed hose and nozzle tip feeding, needs minimum of cleanup. (*The Epoxylite Corporation.*)

The two-container kits provide properly proportioned ingredients separated physically one from the other. The most widely used system is to package the resinous component (part *A*) in a can or jar, attaching thereto a can or a vial of the correct amount of the curing agent (part *B*). More sophisticated arrangements, of course, are possible: Two components may be packaged together in a single plastic bag, separated by a removable clip or by a heat seal which may be ruptured under pressure; they may be packaged in separate tubes, equal lengths of ingredients being extruded; they may be packaged in tube-stirrer arrangements, the curing agent being carried in the hollow stirrer rod, the resin in the tube itself; or the two ingredients may be packaged in a single container separated by a removable or collapsible divider, or in separate bubble packs on a flat sheet.

One-container systems, most frequently, consist of prepackaged formulations containing latent curing agents. These come ready for installation in application equipment, such as caulking guns. Refrigerated one-component mixes are also available. These usually are small, custom packages of room-temperature curing systems, mixed and then held under refrigeration until ready for use [15].

Powdered epoxy systems, generally, are supplied in bulk or pelletized form and pose no particular handling problems.

CURING

If temperatures higher than ambient are employed to cure the resin system, four devices are feasible: ovens, infrared heat lamps, high-frequency heaters, and electron beams. From the standpoint of ultimate properties, forced-air recirculating ovens provide most satisfactory control of exothermic heat and result in castings relatively free of residual stresses. Infrared lamps are convenient cure accelerators for room-temperature-type curing agents and provide a degree of cure adequate for most industrial applications. High-frequency heaters are in limited use, the power factor of the epoxy resin being too low to provide for efficient operation. Temperature rises for some DGEBA-based systems are shown in Table 25-2. Electron beams have been reported as accelerating cure of paint films [34].

Production-line tunnel ovens of various constructions may be used in connection with conveyor equipment to provide rapid and continuous processing of smaller potted, sealed, and encapsulated components.

In any event, the curing equipment should be adequate to accommodate the cure cycle best suited to the material being worked. In some cases, temperature variations from batch to batch may be reflected in critical parameters of the curved system.

TROUBLESHOOTING

During the application of the liquid epoxy resins in production, not infrequently trouble develops. Sometimes it is difficult to isolate the cause of the difficulty, and careful detective work may be required. A list of common troubles encountered with encapsulating compounds together with corrective actions is presented in Table 25-3.

MACHINING

It may be necessary to perform a variety of finishing operations on epoxy castings [19] or laminates [10,17].

Epoxy castings may be machined with considerably greater ease than epoxy laminates, insofar as the glass in the latter is extremely abrasive.

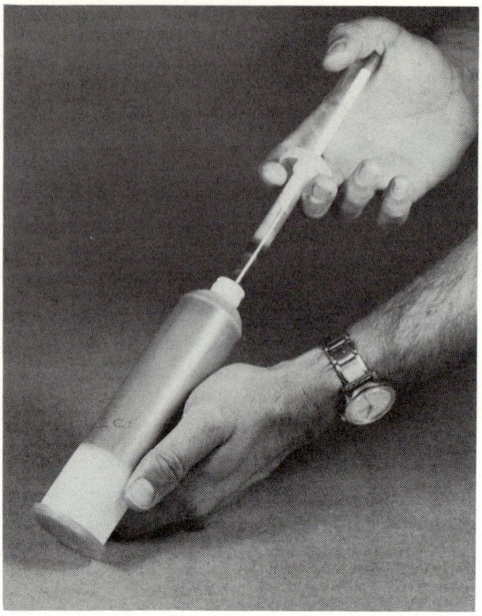

Self-contained two-component package permits convenient use of room-temperature-curing epoxy resin. Curing agent is packaged in hollow mixing rods. After curing agent is forced out, compound is mixed in cartridge. (*Semco, Inc.*)

Table 25-2. Temperature Rise of DGEBA Formulations in Dielectric Heating [33]

Curing agent	phr	ΔT, °C	Gel time, minutes
DMAPA	6	65	3–5
Aromatic amine eutectic	20	72	60
NMA	90	76	120
Fatty polyamide	50	109	0.5
DMP-30 salt	13	83	15

NOTE: 30-g mixtures; dielectric heating cycle 30 sec; distance between plates approximately 2 in.

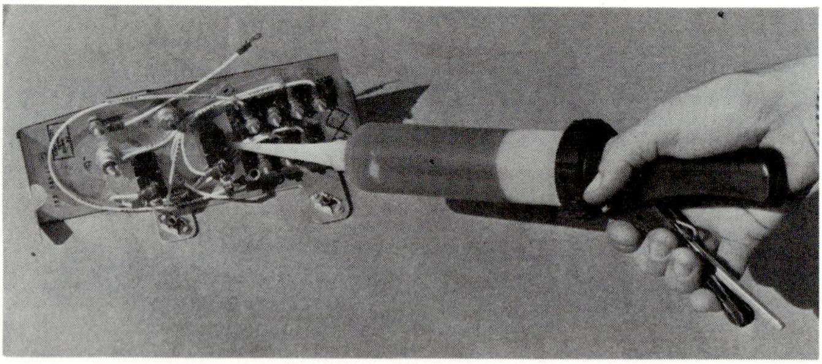

After mixing, cartridge is installed in hand gun and applied to component. (*Semco, Inc.*)

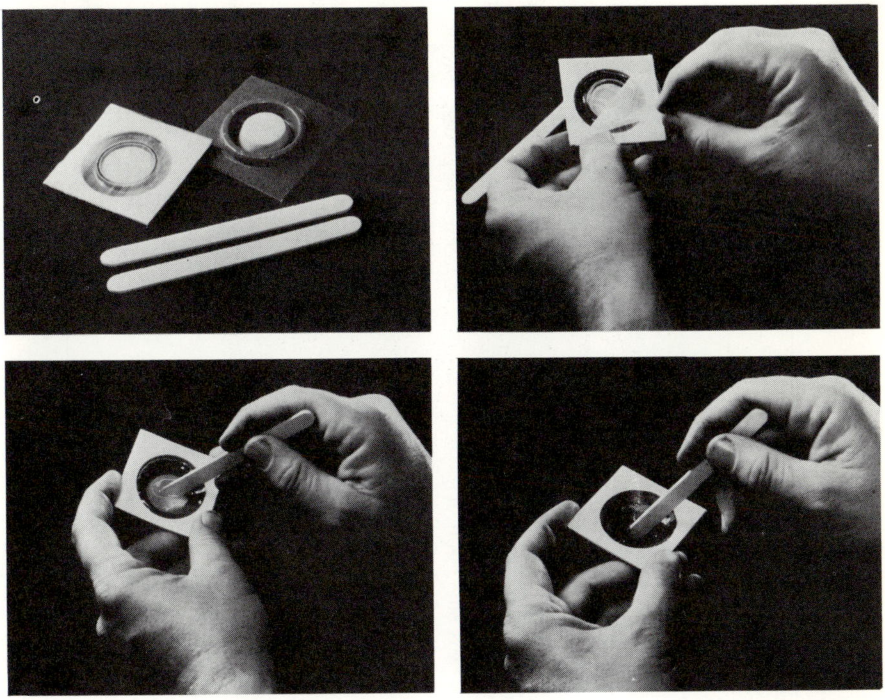

Small unit package of epoxy adhesive contains separate compartments which pop open to larger mixing container. These and similar packages permit convenient packaging of room-temperature curing systems for one-shot applications. (*Plastics Associates.*)

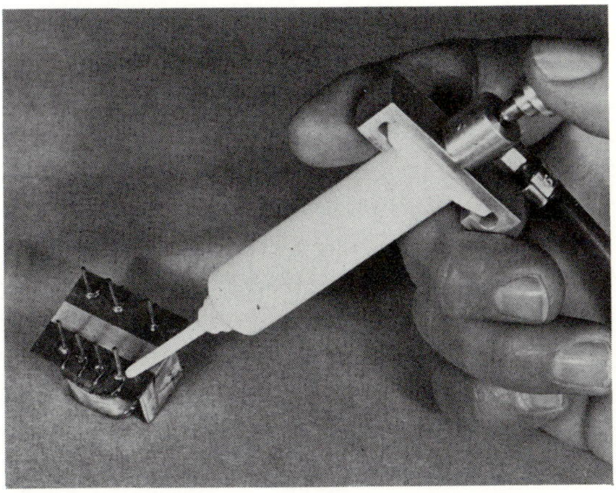

Polyethylene cartridge contains frozen adhesive. When adhesive is brought to room temperature, compound is dispensed by air pressure. (*Semco, Inc.*)

Table 25-3. Common Causes for Defects in Encapsulating Formulations [23]

Defect	Appearance	Probable causes (listed in order to be checked)	Corrective actions
Uncured	Resin remains liquid or very soft and sticky	Cure temperature too low	Check oven temperature. Increase temperature
		Cure time too short	Extend cure time
		Mix ratio incorrect, dispensing equipment out of adjustment	Check mixing process. Cure a small sample with a fresh control sample. Adjust equipment
		Insufficient mixing	Mix thoroughly. Color, if pigmented, should be uniform
		Separate parts not mixed	Stir separate system constituents before blending
		Contamination	Keep parts, mold, resin clean. Practice good housekeeping
Thermoplastic	Casting appears normal at room temperature but becomes liquid or tacky at elevated temperatures. Resin not completely cross-linked	Mix ratio incorrect	Check mixing process. Adjust equipment
		Excess mold release	Use release agents sparingly
		Moisture	Thoroughly dry component just before casting
		Contamination	Check resin area for possible contaminants: oils, dirt, greases, waxes, etc.
		Incompatible insulations	Check component for plasticized films, such as vinyl acetate, thermoplastics that liquefy at operating temperatures of component, tar, waxes. Dissection of unit while hot worthy. Replace questionable insulation with known compatible materials
			Check magnet-wire compatibility with wire and resin manufacturer

Oozing	Casting opens. Liquid oozes out fissure or around terminals, lugs, etc.	Incompatible insulation	Same corrective action as previous entry
		Moisture	Inner insulation laden with moisture. Dry component thoroughly. Paper or fiber parts prime suspects
Blemishes	Surface rough, spotted	Rough mold surface	Clean or polish mold
		Excess mold release	Use thin release, such as silicone oil diluted with solvent
Surface soft spots	Exterior of casting has small or large soft or sticky areas, with or without some voids	Excess mold release	Use release agents sparingly. Dilute with solvent
		Dirty mold	Clean mold after use
Voids (surface)	Bubbles or holes in resin surface	Rough mold surface	Polish mold surface
		Leaky mold. Air enters during vacuum cycle	Seal molds, polish joints, replace gaskets. Release vacuum slowly
		Poor mold design. Horizontal "shelf" areas trap air	Redesign mold to eliminate air traps. Taper "shelves" for air exit
Voids (throughout interior)	Bubbles, voids, or "dry" areas in casting. May not be visible on surface. Low corona starting voltage	Insufficient resin used	Provide for a "head" of resin over component to allow for air escaping
		Insufficient vacuum	Evacuate resin and part before casting, and pour under vacuum. Apply lower vacuum or extend time. Try pressure after vacuum
		Resin cured before air escaped	Use slower-curing resin system or lower processing temperatures
		Resin too thick	Heat component, mold, and resin if needed to reduce viscosity, or use thinner resin for impregnation
		Component design	Areas blocked from impregnant: for example, cuffed interlayer coil insulation gives tight coil ends, prevents impregnation

Tabel 25-3. (*continued*)

Defect	Appearance	Probable causes (listed in order to be checked)	Corrective actions
Sticking	All or part of mold difficult to remove	Undercuts in mold	Remove undercuts and repolish mold
		Insufficient mold release	Apply release agent carefully and completely
		Mold not "broken in"	Reapply mold release agent before each casting. Use mold several times
		Mold surface rough	Polish mold surface
		Permanent-type release agent worn or abraded	Reapply release agent. Regrind TFE-type coatings
Cracking	Fissures develop in casting, during cure, cooldown, or subsequent thermal shock	Resin not cured	Increase cure time to develop maximum strength
		Sticking	Resin damaged during mold removal
		Wrong choice of resin	Flexible and/or filled resin preferable for severe mechanical and thermal shock exposure
		Oven temperature too high	Check oven temperature
		Gel temperature too high	Gel resin at lowest possible temperature to minimize stresses
		Poor component or mold design	Design component or mold so that resin thickness is uniform around component, 1/16 in. minimum. Fillet in sharp internal corners with heavily filled resin before casting. Reinforce crack areas with glass cloth or glass-pressure-sensitive adhesive tape
Crazing	Surface exhibits fine cracks, ripple, roughness	Cure too hot	Reduce oven temperature in 10°C stages
		Too large a mass (primarily high-exotherm systems)	Pour casting in stages. Allow first pouring to gel before pouring second
Charring or discoloration	Resin appears burned, especially in center	Cure too hot-mass of high-exotherm resin too large	Reduce cure temperature. Reduce mass of resin. Use external cooling with room cure systems. Refrigerate, air-cool, heat-sink

Adhesion	Resin releases from components at edges, corners, terminals, etc.	Contamination of surfaces with oil, grease, mold release, skin oil, etc.	Degrease component before casting. Handle with gloves. Silicones prime suspect
		Nonstick surfaces	Replace or prime surfaces of non-bondable materials
		Oxidized surfaces (metal) resin bonds to oxide which releases from metal	Abrade or chemically clean surface just before casting, or use primers
Distortion	Casting warps or distorts from mold shape	Cure temperature too hot	Reduce cure temperature. Keep mold temperature uniform
		Insufficient resin	Allow larger sprue volume or recap casting
		Poor design	Keep resin wall thickness uniform. Add ribs
		Excessive shrinkage	Use more filler. In extreme cases, pack mold with porous filler, impregnate with resin

Cast Epoxies

Turning. Cast epoxies may be turned with conventional steel tools, although best results are obtained with diamond or tungsten carbide tips. It is important that the tool be correctly ground, with the feather edge being removed. The speed should be medium to high and the feed should be low. A speed of 800 rpm, a feed of 0.0025 in. per rev, and a cut of 0.004 in. have been found to be satisfactory for use with unfilled systems and those containing nonabrasive fillers, although a better surface finish can be obtained if the feed is reduced to 0.001 in. per rev. Surface-finish values (rms) of 20 to 90 are attainable easily [22]. Filled epoxies require more gentle handling, and a speed of 200 to 300 rpm is recommended. The working piece should be supported, and care should be taken to prevent overheating.

Drilling. Normal high-speed and high-spiral drills are preferred. The feed should be slow. Excessive heat should be avoided to prevent cracking.

Cutting Threads. With a tool bit on a lathe, the tool should be ground with just enough angle to clear the lead of thread. A radius of $3/16$ in. or more is needed to produce the necessary cutting or shaving action on both cutting edges. A cut of between 0.005 and 0.010 in. is recommended.

When using a tap, it may be necessary to grind the hook angle to obtain the correct cutting or shaving angle. When using a die, it may be helpful to warm the workpiece to 70°C or so in an oven if difficulty is encountered at ambient.

Sawing. The thickness of the workpiece will determine the selection of the saw. For general purposes, 12 to 14 teeth per in. will be found satisfactory.

Grinding. Diamond-faced wheels are the most satisfactory, and it is important to use a coolant and a slow feed. If this is not done, there may be sufficient heating to produce undesirable stresses or dimensional changes.

Milling, shaping, routing, and cutting may be undertaken, but, bearing in mind the equipment available, the operator is the best person to determine the type of tool to be used.

Laminates

Epoxy-glass laminates are preferably machined with diamond tools; conventional tool steel or carbide is satisfactory for working epoxy laminates of paper and other soft fibers.

Turning. The epoxy laminates can be turned at speeds from 400 to 800 fpm surface speed. The laminates should be clamped to prevent delamination. The dust is very abrasive and irritating and should be collected. Wet machining is preferred when possible.

Drilling and Punching. High-spiral drills have been found to be best, with point angles greater than 90° being recommended when inspection standards are high.

For punching, tool-steel dies are satisfactory for short run, with hard-chrome plating on both punch and die being used for production. Only very large production runs would justify the expense of carbide-tipped punches. The epoxy laminates are punched at ambient conditions, and there is, therefore, no need to design dies to allow for coefficient of thermal expansion. In general, punches and dies should be oversize by about 3 per cent of the material thickness. The tendency of the laminate to yield, causing pierced holes to be smaller than the punch and blanked parts to be larger than the die, is markedly less with epoxy-glass laminates than with, for example, paper- or cellulosic-based laminates.

Sawing. Bandsaws are most economical for production runs. The life of steel bandsaws can be prolonged five- or tenfold by hard-chrome-plating the blades, these running at about 800 surface fpm. Diamond-tipped bandsaws are most economical for long production runs, these operating at about 3 surface fpm.

Diamond-tipped circular saws used for cutting epoxy-glass laminates generally operate at about 13,000 rpm, and for cutting sheet thicker than about ½ in. should be operated with a coolant.

RELEASE AGENTS

Because the epoxy resins are extremely adhesive materials, release agents should be used to protect molds and other nonintegral components which contact the resin during cure. Release agents may be liquid, paste, or films. In general, the release agent should be insoluble in the resin mixture, should not melt at the curing temperature employed, and should be applied in a continuous film.

Most liquid and paste release agents are based on the silicone resins, although in some cases waxes and polyvinyl alcohol solutions may be used. Paste release agents based on special clays have also proved effective.

Release agents can be made from polyethylene and Teflon, and these provide some improvement in service life, but repeated reuse will lead to adhesion problems. Cellophane or similar films can be used conveniently to separate layers of catalyzed resin or to protect uncured surfaces during production. Uncoated cellophane is water-soluble and strips off quite easily. Cardboard trays, laminated with special release films, have been developed as low-cost molds [30].

STRIPPERS

The epoxy resins, because of their extreme resistance to solvents, are very difficult to remove from surfaces, once cured. Thoroughly cured and filled systems must be soaked for relatively long periods (i.e., from several days to a week) in special solvents before they can be removed, and even then removal may be a time-consuming job.

Commercial strippers are based on methylene dichloride and other chlorinated solvents, these usually containing strong caustics or acids [32] to attack the chemical bonds. Such systems work fairly well with aliphatic-amine-cured epoxies; aromatic-amine-cured systems and some anhydride-cured systems are exceedingly difficult, if not impossible, to strip. In general, the greater the heat resistance of the cured epoxy, the more difficult the stripping operation.

From a practical standpoint, therefore, it is often less expensive to scrap small low-cost components than to remove the plastic. On larger components, the epoxy resins can be sandblasted off about as easily as aluminum. Most efficiently, resins containing little or no chlorine are burned off, using temperatures in excess of 400°C. Adequate ventilation should be provided, since, in burning, the resin produces a black, acrid smoke.

When the cured material has contaminated a surface, such as a spatula blade, it may be ground off without difficulty. The uncured resin may be removed with a convenient solvent, such as xylene, or with hot, soapy water, or, if polymerization has not progressed too far, with an active solvent, such as methyl ethyl ketone.

Film strippers for solution coatings are available in proprietary grades, and these are more effective. A typical compound will soften or lift films within an hour to the degree that the coating can then be removed by slight scraping.

HANDLING PRECAUTIONS

The health hazards normally associated with the handling of chemicals may be divided into two general types:

1. Those producing systemic damage and serious physiological injury
2. Those producing local irritation and eliciting allergenic responses

Toxicity

The epoxy resins of the DGEBA class are relatively free of the first hazard. In fact these resins, both liquids and solids, have been used commercially for nearly 20 years, with, as far as the literature reveals at least, no serious physiological injury being encountered by any worker.

Not all the epoxy resins, however, are as relatively innocuous as DGEBA resins. Some of the newer resins, particularly the more reactive liquid species, may pose some degree of physiological hazard.

In discussing the physiological hazard, it is necessary to consider two aspects: (1) the toxicity of the compound, and (2) the carcinogenic properties. DGEBA resins have been found to be relatively nontoxic by a series of animal experiments, the results of which are indicated in Table 25-4. As can be seen, the probable lethal dose for man would be approximately one quart, an amount which could not accidentally be ingested. Even with the more reactive epoxies, such as the polyglycidyl ether of glycerol, the lethal dose would be in the range of one pint. In tests to determine chronic toxicity of the uncured resins, feeding studies were conducted on rats for up to 26 weeks. From this it was estimated that a diet containing at least 10 grams per day would be required to produce toxic manifestations in an average man [4-6]. For all practical purposes, then, these epoxies may be considered practically nontoxic when taken internally. Similar results were obtained when test animals were exposed to air saturated with vapors of the DGEBA resins for a total of fifty 7-hour-day exposures [4-6].

Table 25-4. Acute Intragastric Toxicity of Uncured DGEBA Resins [4-6]

DGEBA, mol. wt.	Experimental species	Intragastric dose	AIHA* mortality	Classification	Probable lethal dose for man†
950	Rat	30 g/kg	None	Relatively harmless	1 qt
950	Mouse	20 g/kg	None	Relatively harmless	1 qt
3,000	Rat	30 g/kg	None	Relatively harmless	1 qt
3,000	Mouse	20 g/kg	None		
700	Mouse	20 ml/kg	None		
450	Mouse	30 ml/kg	None		
380	Rat	11.4 g/kg	LD_{50}	Practically nontoxic	1 qt

* Toxicity classifications by Hodge and Sterner, *Am. Ind. Hyg. Assoc. Quart.*, vol. 10, no. 4, December, 1949.

† National Safety Council's Proposed Procedures for Evaluating the Acute Toxicity of New Chemical Products, National Safety Council, 425 North Michigan Avenue, Chicago 11, Ill.

Some of the newer epoxy resins have somewhat lower LD_{50} figures and can therefore be said to be more toxic on ingestion. The LD_{50} of DGEBA on rats is 11.4 grams per kg, whereas for 3,4-epoxy-6-methylcyclohexylmethyl 3,4-epoxy-6-methylcyclohexane carboxylate it is 4.92, for vinylcyclohexene dioxide 2.83, for

DGE of resorcinol 2 to 3, and for butadiene dioxide 0.088. A daily dose of 5 mg of epoxidized soybean oil (corresponding to 50 mg of diepoxystearic acid) produced growth inhibition in rats, and a daily dose of 45 mg was lethal within 8 days for 80 per cent of the specimens. The following subsidiary effects were observed: lowering of protein efficiency, increased water consumption, storage of epoxy acids in the body fats, reduction of blood cholesterol level, and increase in the total blood lipids and phosphatides [26].

Some of the newer resins may present definite hazards from skin penetration (vinylcyclohexene dioxide), or from exposure to the vapors, as for example butadiene dioxide which gave a 4-hour inhalation LD_{50} on rats of 56 ppm. Because of potential hazards with lower-molecular-weight epoxies, principally diluents with molecular weights below 200 such as butyl glycidyl ether, styrene oxide, and phenyl glycidyl ether, special labeling is required under interpretations of the Federal Hazardous Substances Act of 1962, and workers should avoid breathing hot fumes from systems containing these compounds.

To some extent, too, some of the newer resins may present carcinogenic properties. This danger does not appear to be present with DGEBA resins, but it may be present, to some extent, with DGE of resorcinol, and with liquid epoxies in which the epoxy groups are separated by four or fewer carbon atoms and perhaps with some others.

It is apparent, therefore, that before beginning evaluation of the newer epoxy resins, careful attention should be given to the possible adverse systemic effects which may occur from prolonged skin contact or from exposure to vapors.

In addition to the possibility of the epoxy resins producing serious physiological injury, some of the other ingredients employed in the formulations may be capable of producing serious effects.

Generally, the lower-viscosity curing agents and modifiers should be regarded with respect, in the absence of toxicological data. Fortunately those materials most commonly used in the epoxy technology are not severely toxic, as indicated in Table 25-5. and from the standpoint of accidental ingestion do not pose a significant industrial health hazard in this respect, although primary amines such as DETA and TETA require special labels under interpretations of the Federal Hazardous Substances Act of 1962. Many of these materials most commonly employed in the epoxy technology have been used industrially for a number of years and have not been shown, under normal usage, to produce long-term systemic effects, although if accidentally ingested in quantity, some are capable of producing liver damage and severe gastrointestinal disturbances.

The newer chemicals should be handled with care until toxicological data are available.

Cured epoxy-resin formulations, in general, have been shown to be inert and nontoxic, as indicated in Table 25-6. Rats fed diets of up to 10 per cent of the cured resins for period of 6 weeks showed no significant changes over controls.

Cured epoxy-resin systems have been evaluated in a number of applications for contact with various food substances and potable water. DGEBA cured with an aliphatic polyamine has been found unsuitable for contact with wines, since alcoholic 1 per cent solutions of tartaric acid or 2 per cent acetic acid extracted the curing agent [29]. The following curing agents have been found suitable for DGEBA systems that come in contact with foods: diphenylamine, DETA, ethylenediamine, isophthalyl dihydrazide, TEPA, TETA, DMP-30, DMP-30 fatty-acid salts, and TEPA reacted with fatty acids [31]. Xylene-formaldehyde/epoxy-resin blends may be used on food-contact coating applications provided the total extractables from water and ethanol (50 per cent) at 83°C for 24 hours do not exceed 0.05 mg per sq in. of food-contact surface [31].

Table 25-5. Acute Oral Toxicity of Common Epoxy-Resin Curing Agents [4-6]

Material studied	LD_{50}* g/kg	AIHA classification
DETA	2.3†	Slightly toxic
TETA	4.34†	Slightly toxic to practically nontoxic
Tetraethylenepentamine	3.90†	Slightly toxic
DEAPA	1.70‡	
Ethylene oxide adduct of DETA	4.80†	Slightly toxic to practically nontoxic
DGEBA adduct of DETA	2.5†	Slightly toxic
MPDA	0.3§	
MDA	0.126–0.252†	Moderately toxic
Aromatic amine eutectic	0.15–0.25¶	Moderately toxic
DDSA	3.2†	Slightly toxic
PA	1–6 g produced no fatalities in dogs	

* The dose in grams per kilograms of body weight which was fatal to 50% of the total number of animals in the experimental group.
† LD_{50} value for rats.
‡ LD_{50} value for mice.
§ Value reported for cats and rabbits by Hanzlik, *J. Ind. Hyg.*, 4, 1922.
¶ LD_{50} value of most toxic component.

Table 25-6. Chronic Feeding Studies on Rats with Cured DGEBA Resin [4-6]

Diet	Feeding period, weeks	Resin in feed, %	Feed consumed, grams	Mean weight gain, %	Organ/body weight ratios, %	
					Liver	Kidney
Controls for 6-*week feeding study*	..	....	900	189	3.89	0.696
DGEBA cured with BF_3MEA	6	1	978	168	3.71	0.695
		5	901	204	3.59	0.685
		10	1,010	173	3.53	0.724
DGEBA cured with MPDA	6	1	966	168	3.71	0.726
		5	925	175	3.79	0.745
		10	1,053	181	3.65	0.703
DGEBA cured with aromatic amine eutectic	6	1	980	177	3.71	0.710
		5	917	202	3.48	0.705
		10	1,019	203	3.48	0.729
DGEBA (mol. wt. 950) cured with DETA	6	1	903	173	3.93	0.715
		5	1,062	160	4.04	0.712
		10	1,100	189	3.89	0.721
DGEBA (mol. wt. 950) cured with fatty polyamide (amine value 210–230)	6	1	1,000	178	3.97	0.727
		5	987	185	3.86	0.732
		10	1,062	177	3.82	0.720
Controls for 12-*week feeding study*	..	....	.....	183		
DGEBA (mol. wt. 3,000) cured with DETA	12	2.0	.....	203		
DGEBA (mol. wt. 3,000) cured with urea formaldehyde resin	12	2.0	.....	213		

Studies on the effect of the pyrolysis and combustion products of cured DGEBA on rats indicated a LD_{50} value of 3.2×10^5 mg min per cu m for the pyrolysis products, although no deaths were observed when rats were exposed to slightly higher concentrations of the combustion products. It was concluded that the pyrolysis products may constitute a hazard to human personnel in circumstances realizable in practice [27].

Irritation and Sensitization

Although the liquid epoxy resins are not particularly severe primary skin irritants, many of the curing agents employed in the epoxy-resin technology are.

Considering the curing agents as divided between amines and acids, both classes are capable of causing chemical burns. Some of the amines are strong caustics, and they will produce local injury on short exposures and can inflict severe damage to the eyes. The organic acids, and to a lesser extent, the anhydrides, are primary irritants with the degree of hazard more or less related to the volatility. The liquid anhydrides are particularly damaging to the eyes. Vapors from the liquids and heated solids are also very irritating to the eyes, mucous membrane, and lungs.

Toxic effects and primary skin irritation, however, are minor concerns in the epoxy technology, and the risks (in general) in handling commercial formulations are no greater than those encountered in handling many household chemicals. The major health hazard connected with the epoxy resins and their curing agents is dermatitis.

Dermatitis from the epoxies and their curing agents closely resembles poison ivy, and usually first affects the exposed areas of the face and hands, particularly the eyelids, around the lips, and on the wrists and forearms. In very severe cases, it can affect other parts of the body.

The situation is of sufficient concern to elicit the comment, in 1959 that "epoxy resins are causing more occupational dermatitis than any other new chemical material within the past 10 years" [6].

No process utilizing reactive resin systems can be made completely free of hazard since chemical and biological activity are inseparable [9]. The epoxy groups may react with primary amine groups associated with amino acid lysine incorporated into a protein; the amine curing agents may react with proteins where amino acid residues carry terminal carboxyl groups, and acid anhydrides likewise have the property of chemically combining with protein. Such reaction with body protein may in part account for sensitization, which may cause both dermatitis and respiratory distress [16] and it follows that the risk of sensitizing individuals to these materials is directly proportional to the exposure time, concentration, volatility, and solubility of the materials.

In addition, the epoxy resins are often handled in connection with solvents, the solvents frequently—and unnecessarily—being employed for cleanup purposes. *Solvents, themselves, account for a substantial part of the total cases of industrial dermatitis* [3].

It is apparent, then, when considering dermatitis (and the less frequent allergenic reaction, respiratory distress) that a number of ingredients are responsible and that the total effect may be synergistic.

Once an individual has been "sensitized" to the epoxy resins and to their coreactants, reexposure, even in very low concentrations, will produce a reaction. The period of time required for sensitization will depend upon the individual and may vary from a few days to months. It is the authors' conclusion, however, that continuous and repeated skin contact with the chemicals employed in epoxy-resin formulations will, ultimately, produce sensitization in virtually all individuals.

Once sensitized, the individual must avoid further skin contact. It has been reported that some individuals may become desensitized after initial dermatitis, and for this reason it is recommended that operators are better not removed from the workbench at the first symptoms so that they have the opportunity of adapting or desensitizing themselves [4-1]. This desensitization, in the authors' opinion, based on personal experiences extending over 10 years, is more probably accounted for by the individual being more careful of skin contact after initial sensitization. We believe that sensitization will disappear only over a substantial period of time during which contact is completely eliminated.

The sensitization potential of the ingredients in epoxy formulations may be roughly listed in the following order, from severest to least severe [5].

1. Volatile primary polyamines
2. Nonvolatile primary polyamines
3. Reactive diluents such as styrene oxide, butyl glycidyl ether, allyl glycidyl ether, and phenyl glycidyl ether
4. Liquid epoxy resins, with, in general, the lower-molecular-weight species being more severe than the higher-molecular-weight species
5. Amine adducts
6. Acid anhydrides and tertiary amine salts
7. Solvents
8. Cured epoxies—nuisance hazards in grinding and finishing only
9. Solid epoxy resins

In this list, it should be observed that solvents are placed quite far down, and this is probably a fair representation of their sensitizing potential. However, solvents in combination with the ingredients of epoxy formulations magnify the difficulty out of proportion: The solvents defat the skin, rendering it vulnerable to attack, and also serve as liquid carriers to spread the chemicals and bring them into intimate contact with the skin. Also, once a person is sensitized, the solvents, due to their volatility, are a frequent source of marked skin irritation.

From the above it is evident that a problem does exist. Two approaches may be taken to lessen the hazard:

1. The epoxy formulation may be selected to minimize the irritation potential of the ingredients.
2. The epoxy formulations may be handled in such a fashion as to minimize skin contact and thus delay indefinitely the onset of sensitization.

Where practical, the ingredients with the lowest sensitization potential should be selected. Systems of very low sensitization potential may be developed from ethylene oxide adducts of primary amines and DGEBA; likewise, low irritation-potential systems may be developed from anhydrides and partially reacted anhydrides and liquid, undiluted resins.

In those cases where the handling or cured properties dictate the use of systems with higher sensitization potential, skin contact should be avoided in order to prevent the development of sensitization.

It is most important to avoid sensitization—for once this takes place, subsequent resistance will be lowered.

The majority of individuals appear to have satisfactory initial resistance, which diminishes with continued contact. Hence, the full irritation potential will not immediately appear in a production situation. This, of course, encourages careless handling. Precautions must be exercised from the beginning of use rather than after the problem appears. If there is no skin contact, there will be no dermatitis [2]. The following conditions, if observed, will prevent the development of sensitization.

AVOID SKIN CONTACT. The epoxy resins and resin formulations should not be allowed to come in contact with the skin. This can be accomplished by the use of plastic gloves and/or suitable barrier creams. Should contact occur, the contaminated area should be washed immediately and thoroughly with soap and water or alcohol—never with a solvent! If repeated washings are necessary, a lanolin-based skin cream should be provided to restore natural oils.

Paper towels should be readily available and should be used for wiping up spilled resin and for cleaning equipment. These should not be reused. Care should be taken to avoid spreading the resin throughout the working area. This will occur if employees handle foreign items (e.g., doorknobs, desk drawers, spatulas, light fixtures, scales, etc.) while wearing contaminated gloves.

PRACTICE PERSONAL HYGIENE. Work clothes should be changed and laundered every few days. Resin that contaminates the clothes will almost inevitably contact the skin. The undersides of shirt cuffs accumulate resin and particularly the fumes from the curing agents. The cuffs abrade the wrist to assist the sensitizing process. Long-sleeved shirts should be changed daily. Frequent washings, uncontaminated clothing, and workmanlike habits will control inadvertent contact with the resin and eliminate its effect.

During warmer weather, even increased attention should be given to the practice of personal hygiene.

PROVIDE ADEQUATE VENTILATION. Displaced-air ventilation should be provided, and when resin mixes are worked hot, it should be considered indispensable. Hoods should be used to cover the working benches, and the fumes should be exhausted to the atmosphere through connecting air ducts.

Of course, it may be expected that even if the foregoing procedures are scrupulously adhered to, a fraction of a per cent of individuals will still develop allergenic reactions. These individuals must be transferred from further contact.

If the foregoing procedures are *not* observed, a high incidence of dermatitis will inevitably occur.

Dilute Burow's solution has been found to be of value in controlling the itching associated with the poison-ivy-like rash. The rash will disappear, except in severe cases, within a few days without further medication, provided contact is not resumed. In severe cases, medical workers have found water-based cortisone ointments to provide effective control.

In the event of eye contact with ingredients in epoxy formulations, the eye should be flushed with water for approximately 15 minutes to assure complete removal of the chemical, and medical attention should be obtained immediately.

The thoroughly cured epoxy resins are not skin irritants or sensitizing agents; however, partially cured compounds breathed in or in contact with moist skin may trouble sensitized individuals. For this reason, grinding and similar operations should be conducted in a way so as to minimize atmospheric contamination with dust, and molding powders and fluidized-bed coatings should be handled with care.

CONCLUSION

The excellent properties and the extremely versatile nature of the epoxy resins in casting, potting, sealing, adhesive, laminating, and coating formulations have resulted in their widespread use in industries hitherto unfamiliar with thermosetting plastics.

It has, therefore, been necessary for these industries to reorient their materials-handling processes to accommodate unfamiliar materials. This is accomplished by an experimental period, followed by the determination of the important variables in the system and the integration of the system into a feasible production process employing specific equipment. Special consideration is given to eliminating the possibility of dermatitis by enforcing proper hygiene and installing adequate ventilation.

When such a sound and well-thought-out approach is employed, the epoxy resins can be used with complete satisfaction and without difficulty to produce all varieties of products wherein their superior properties are utilized to the maximum extent.

REFERENCES

1955

1. Susman, Processing of Epoxy Resins, *Mod. Plastics*, April.

1956

2. Dorman, Dermatoses and Epoxy Resins, 12th ANTEC of SPE.

1957

3. Morris, Epoxy Resins, *AMA Arch. of Dermatol.*, December.

1958

4. Gammel, Material Handling of Epoxy and Polyester Resins, 14th ANTEC of SPE.
5. Stifter, A Control Program for Epoxy-caused Dermatitis, *Plastics Technol.*, February.

1959

6. Birmingham, New Causes of Occupational Dermatoses, *AMA Arch. of Ind. Health*, December.

1960

7. Butler and Hawkins, Large Epoxy Reinforced Plastic Structures, 15th NANTEC (RIP Division) of SPI, Chicago, February.
8. Daniel, Flooring Composition and Method of Making the Same, U.S. 2,943,953.
9. Grillo, The Biochemical and Toxicological Problems Associated with Potting Resins, WE–BTL Casting Resins Conference, North Andover, Mass., May.
10. Martin, Do's and Don'ts for Fabrication of Laminates, *Mod. Plastics*, April.
11. Pitt and Loesser, Recent Advances in Epoxy Handling Equipment, University of Wisconsin Seminar on Epoxy Resins, May 29.
12. Reed, Mixing, Metering and Dispensing Equipment for Styrene-Polyester and Epoxy Resin Compounds, WE–BTL Casting Resins Conference, North Andover, Mass., May.
13. Rott, The Use of the Automatic Process Control Machine for Proportioning and Dispensing Epoxy Resins, WE–BTL Casting Resins Conference, North Andover, Mass., May.
14. Worley, Metering, Mixing and Dispensing Equipment, WE–BTL Casting Resins Conference, North Andover, Mass., May.

1961

15. Carr et al., Frozen Epoxy Adhesive Solves Instrument Bonding, *Research/Development*, September.
16. *Dyestuffs*, September.
17. Laminate Cutting, *Plastics World*, December.
18. Loeser, Epoxy Troweling Compounds: New Application Techniques, *SPE J.*, March.
19. Thomson, Machining Cast Epoxide Products, *Brit. Plastics*, September.
20. Zukor, Exploring Fabrication Techniques, *Plastics Technol.*, September.

1962

21. Breslau and Schneider, The Mechanic Chemical Aspects of Reactive Resins, 18th ANTEC of SPE.
22. Conrath, A Feasibility Study of Epoxy Machining Characteristics, AD 408 739.
23. Dosser, Trouble Shooting Defects in Epoxy Resin Castings, *Insulation*, November.
24. Gammel, Proportioning Multiphase Adhesives, *Plastics Design and Processing*, February.
25. Warnsdorfer, Method for Indicating Complete Mixing of Epoxy Resins and Amine Hardener and Compositions Resulting Therefrom, U.S. 3,030,329.

1963

26. Kieckebusch et al., *Fette, Seifen, Anstrichmittel*, pp. 919–924.
27. Lelong and MacFarland, Hazards from Thermodecomposition of Epoxy Resins, *Arch. of Environmental Health*, December.
28. Rost, "Processing Techniques for Epoxy Resins," Carl Hanser, Verlag, Munich, 132 pp.

1964

29. Aleksina and Chubarova, Epoxy Resins Used in the Food Industry, *Gigiena i Sanit.*, no. 6, pp. 100–103.
30. Epoxy Tiles, *Plastics World*, November.
31. Food Additives, Resinous and Polymeric Coatings, *Federal Register*, pp. 2540, 8144, 9329.
32. Goldsmith and Roessler, Coating Remover Composition, U.S. 3,138,556.
33. Goodemote et al., How to Increase Your Output of Small Epoxy Castings, *Plastics Technol*, September.
34. Milly, Drying of Polymerizable Coating Compositions, Fr. 1,370,173.

GLOSSARY OF ABBREVIATIONS

A/E	Anhydride-to-epoxy ratio
AEP	*N*-Aminoethylpiperazine.
A/F	Aniline-formaldehyde resins (see Chap. 8)
A/H/E	Anhydride-to-hydroxyl-to-epoxy ratio
BDMA	Benzyldimethylamine
BF_3MEA	Boron trifluoride monoethylamine
BGE	Butyl glycidyl ether
BTMAC	Benzyltrimethylammonium chloride
CA	Chlorendic anhydride (see Chap. 12)
cps	cycles per second
DADPS	Diaminodiphenylsulfone
DBTA	Dimethyl-substituted butenyltetrahydrophthalic anhydride
DEA	Diethanolamine
DEAPA	Diethylaminopropylamine
DETA	Diethylenetriamine
DDSA	Dodecenylsuccinic anhydride
DGE	Diglycidyl ether
DGEBA	Diglycidyl ether of bisphenol A (mol. wt. approx. 380), the standard commercial epoxy resin. Higher- and lower-molecular-weight species are identified by molecular weight or by degree of polymerization (*n* number) (see Chap. 2)
DMAPA	Dimethylaminopropylamine
DMP-10	Dimethylaminomethylphenol
DMP-30	Tris(dimethylaminomethyl)phenol
DT	Deflection temperature, formerly heat distortion temperature (HDT) (see Chap. 6)
f	functionality of resin or curing agent
HHPA	Hexahydrophthalic anhydride
MA	Maleic anhydride
MDA	4,4'-Methylenedianiline
MPDA	*meta*-Phenylenediamine
NMA	Maleic anhydride adduct of methylcyclopentadiene (see Chap. 12)
PA	Phthalic anhydride
PAPA	Polyazelaic polyanhydride
PGE	Phenyl glycidyl ether
phr	parts per hundred of resin
PMDA	Pyromellitic dianhydride
PSPA	Polysebacic polyanhydride
TEPA	Tetraethylenepentamine
TETA	Triethylenetetramine
TMA	Trimellitic anhydride
TMOBX	Trimethoxyboroxine

INDEX